PATTY'S TOXICOLOGY

Fifth Edition

Volume 7

Glycols and Glycol Ethers
Synthetic Polymers
Organic Sulfur Compounds
Organic Phosphates

PATTY'S TOXICOLOGY

Fifth Edition
Volume 7

EULA BINGHAM
BARBARA COHRSSEN
CHARLES H. POWELL
Editors

CONTRIBUTORS
Rodney J. Boatman
Laureen Burton
Finis L. Cavender

Steven T. Cragg
James B. Knaak
Howard G. Shertzer

Edward J. Sowinski
Jan E. Storm
Bailus Walker, Jr.

A Wiley-Interscience Publication
JOHN WILEY & SONS, INC.
New York / Chichester / Weinheim / Brisbane / Singapore / Toronto

DISCLAIMER: Extreme care has been taken in preparation of this work. However, neither the publisher nor the authors shall be held responsible or liable for any damages resulting in connection with or arising from the use of any of the information in this book.

This book is printed on acid-free paper. ∞

Copyright © 2001 by John Wiley & Sons, Inc. All rights reserved.

Published simultaneously in Canada.

No part of this publication may be reproduced, stored in a retrieval system or transmitted in any form or by any means, electronic, mechanical, photocopying, recording, scanning or otherwise, except as permitted under Sections 107 or 108 of the 1976 United States Copyright Act, without either the prior written permission of the Publisher, or authorization through payment of the appropriate per-copy fee to the Copyright Clearance Center, 222 Rosewood Drive, Danvers, MA 01923, (978) 750-8400, fax (978) 750-4744. Requests to the Publisher for permission should be addressed to the Permissions Department, John Wiley & Sons, Inc., 605 Third Avenue, New York, NY 10158-0012, (212) 850-6011, fax (212) 850-6008, E-Mail: PERMREQ @ WILEY.COM.

For ordering and customer service, call 1-800-CALL-WILEY.

Library of Congress Cataloging in Publication Data:

Patty's toxicology / [edited by] Eula Bingham, Barbara Cohrssen, Charles H. Powell.— 5th ed.
    p. ; cm.
  "A Wiley-Interscience publication."
  Includes bibliographical references and index.
  ISBN 0-471-31940-6 (cloth: v. 7 : alk.paper); 0-471-31943-0 (set)
    1. Industrial toxicology—Encyclopedias. I. Bingham, Eula. II. Cohrssen, Barbara.
III. Powell, Charles. H. IV. Patty's industrial hygiene and toxicology
    [DNLM: 1. Occupational Medicine. 2. Occupational Diseases. 3. Poisons. 4. Toxicology. WA 400 P3222 2000]
RA1229 .P38 2000
613.6′2—dc21   99-053898

Printed in the United States of America.

10 9 8 7 6 5 4 3 2 1

Contributors

Rodney J. Boatman, Ph.D., DABT, Health and Environmental Laboratories, Eastman Kodak Company, Rochester, New York

Laureen Burton, MPH, Health Scientist, United States Environmental Protection Agency, Washington, D.C.

Finis L. Cavender, Ph.D., DABT, CIH, Raleigh, North Carolina

Steven T. Cragg, Ph.D., DABT, Tox Temps, Alexandria, Virginia

James B. Knaak, Ph.D., Health and Environmental Laboratories, Eastman Kodak Company, Rochester, New York

Howard G. Shertzer, Ph.D., University of Cincinnati, Cincinnati, Ohio

Edward J. Sowinski, Ph.D., DABT, CIH, Hudson, Ohio

Jan E. Storm, Ph.D., New York State Department of Health, Bureau of Toxic Substances, Troy, New York

Bailus Walker, Jr. Ph.D., MPH, Howard University College of Medicine, Washington, DC

Preface

In this Preface to the Fifth Edition, we acknowledge and note that it has been built on the work of previous editors. We especially need to note that Frank Patty's words in the preface of the second edition are cogent:

> This book was planned as a ready, practical reference for persons interested in or responsible for safeguarding the health of others working with the chemical elements and compounds used in industry today. Although guidelines for selecting those chemical compounds of sufficient industrial importance for inclusion are not clearly drawn, those chemicals found in carload price lists seem to warrant first consideration.
>
> Where available information is bountiful, an attempt has been made to limit the material presented to that of a practical nature, useful in recognizing, evaluating, and controlling possible harmful exposures. Where the information is scanty, every fragment of significance, whether negative or positive, is offered the reader. The manufacturing chemist, who assumes responsibility for the safe use of his product in industry and who employs a competent staff to this end, as well as the large industry having competent industrial hygiene and medical staffs, are in strategic positions to recognize early and possibly harmful exposures in time to avoid any harmful effects by appropriate and timely action. Plant studies of individuals and their exposures regardless of whether or not the conditions caused recognized ill effects offer valuable experience. Information gleaned in this manner, though it may be fragmentary, is highly important when interpreted in terms of the practical health problem.

While we have not insisted that chemical selection be based on carload quantities we have been most concerned about agents (chemical and physical) in the workplace that are of toxicological concern for workers. We have attempted to follow the guide as expressed by Frank Patty in 1962 regarding practical information.

The expansion of this edition to include biological agents, e.g., wood dust, Histoplasma, not previously covered, reflects our concern with their toxicology and potential for adverse health effects in workers. In the workplace of the new century, physical agents and human factors appear to be of more concern. Traditionally, these agents or factors, ergonomics, biorhythms, vibration, and heat and cold stress were centered on how one measures them. Today, understanding the toxicology of these agents (factors) is of great importance because it can assist in the anticipation, recognition, evaluation, and control of the physical agent. Their mechanisms of actions and the assessment of adverse health effects are as much a part of toxicology as dusts and the heavy metals.

Chapters on certain topics such as reproduction and development, and neurotoxicology reflect the importance of having at hand for practical use such information to help those persons who are responsible for helping to safeguard health to better understand toxicological information and tests reported for the various chemicals. As noted in Chapter One, the trend in toxicology is increasingly focused on molecular biology and, for this "decade of the genome," molecular genetics. Therefore, it seemed crucial to have a chapter that would help to explain the dogma of our teachers in industrial toxicology that, frequently, there are two workers side by side, and one develops an occupational disease and the other does not. Hence the chapter on genetics was authored by an expert in environmental genetics.

The thinking and planning of this edition was a team effort by us: Charlie, Barbara, and Eula. Over many months we worked on the new framework and selected the contributors. When Charlie died in September, 1998, we (Barbara and Eula) knew that we had a road map and, with the help of our expert contributors, many of whom the three of us have known for 10, 20, or even 30 years, would complete this edition. The team effort was fostered among the current editors by many of the first contributors to Patty's such as Robert A. Kehoe, Francis F. Heyroth, William B. Deichmann, and Joseph Treon, all of whom were at Kettering Laboratory at the University of Cincinnati sometime during their professional lives. The three of us have a long professional association with the Kettering Laboratory: Charles H. Powell received a ScD., Barbara Cohrssen received a MS, and Eula Bingham has been a lifetime faculty member. Many of the authors were introduced to us through this relationship and association.

The authors have performed a difficult task in a short period of time for a publication that is as comprehensive as this one is. We want to express our deep appreciation and thanks to all of them.

Kettering Laboratory, Cincinnati, Ohio   EULA BINGHAM, Ph.D.

San Francisco, California   BARBARA COHRSSEN, MS

                                                                                                                                              CHARLES H. POWELL, ScD.

Contents

85 Glycols   1

Finis L. Cavender, Ph.D., DABT, CIH and
Edward J. Sowinski, Ph.D., DABT, CIH

86 Ethers of Ethylene Glycol and Derivatives   73

Rodney J. Boatman, Ph.D., DABT and James B. Knaak, Ph.D

87 Glycol Ethers: Ethers of Propylene, Butylene Glycols, and Other Glycol Derivatives   271

Rodney J. Boatman, Ph.D., DABT

88 Synthetic Polymers   397

Bailus Walker, Jr. Ph.D., MPH, and Laureen Burton, MPH

89 Synthetic Polymers—Olefin, Diene Elastomers, and Vinyl Halides   425

Bailus Walker, Jr. Ph.D., MPH, and Laureen Burton, MPH

90 Polyvinyl Acetate, Alcohol, and Derivatives, Polystyrene, and Acrylics   487

Bailus Walker, Jr. Ph.D., MPH, and Laureen Burton, MPH

91 Synthetic Polymers—Cellulosics, Other Polysaccharides, Polyamines, and Polyimides   521

Finis L. Cavender, Ph.D., DABT, CIH

92 Synthetic Polymers, Polyesters, Polyethers, and Related Polymers	573
Steven T. Cragg, Ph.D., DABT	
93 Polyurethanes, Miscellaneous Organic Polymers, and Silicones	619
Steven T. Cragg, Ph.D., DABT	
94 Organic Sulfur Compounds	681
Howard G. Shertzer, Ph.D.	
95 Organophosphorus Compounds	767
Jan E. Storm, Ph.D.	
Subject Index	967
Chemical Index	999

USEFUL EQUIVALENTS AND CONVERSION FACTORS

1 kilometer = 0.6214 mile
1 meter = 3.281 feet
1 centimeter = 0.3937 inch
1 micrometer = 1/25,4000 inch = 40 microinches
   = 10,000 Angstrom units
1 foot = 30.48 centimeters
1 inch = 25.40 millimeters
1 square kilometer = 0.3861 square mile (U.S.)
1 square foot = 0.0929 square meter
1 square inch = 6.452 square centimeters
1 square mile (U.S.) = 2,589,998 square meters
   = 640 acres
1 acre = 43,560 square feet = 4047 square meters
1 cubic meter = 35.315 cubic feet
1 cubic centimeter = 0.0610 cubic inch
1 cubic foot = 28.32 liters = 0.0283 cubic meter
   = 7.481 gallons (U.S.)
1 cubic inch = 16.39 cubic centimeters
1 U.S. gallon = 3,7853 liters = 231 cubic inches
   = 0.13368 cubic foot
1 liter = 0.9081 quart (dry), 1.057 quarts
   (U.S., liquid)
1 cubic foot of water = 62.43 pounds (4°C)
1 U.S. gallon of water = 8.345 pounds (4°C)
1 kilogram = 2.205 pounds

1 gram = 15.43 grains
1 pound = 453.59 grams
1 ounce (avoir.) = 28.35 grams
1 gram mole of a perfect gas ≎ 24.45 liters
   (at 25°C and 760 mm Hg barometric pressure)
1 atmosphere = 14.7 pounds per square inch
1 foot of water pressure = 0.4335 pound per
   square inch
1 inch of mercury pressure = 0.4912 pound per
   square inch
1 dyne per square centimeter = 0.0021 pound per
   square foot
1 gram-calorie = 0.00397 Btu
1 Btu = 778 foot-pounds
1 Btu per minute = 12.96 foot-pounds per second
1 hp = 0.707 Btu per second = 550 foot-pounds
   per second
1 centimeter per second = 1.97 feet per minute
   = 0.0224 mile per hour
1 footcandle = 1 lumen incident per square foot
   = 10.764 lumens incident per square meter
1 grain per cubic foot = 2.29 grams per cubic meter
1 milligram per cubic meter = 0.000437 grain per
   cubic foot

To convert degrees Celsius to degrees Fahrenheit: °C (9/5) + 32 = °F
To convert degrees Fahrenheit to degrees Celsius: (5/9) (°F − 32) = °C
For solutes in water: 1 mg/liter ≎ 1 ppm (by weight)
Atmospheric contamination: 1 mg/liter ≎ 1 oz/1000 cu ft (approx)
For gases or vapors in air at 25°C and 760 mm Hg pressure:
   To convert mg/liter to ppm (by volume): mg/liter (24,450/mol. wt.) = ppm
   To convert ppm to mg/liter: ppm (mol. wt./24,450) = mg/liter

CONVERSION TABLE FOR GASES AND VAPORS[a]
(Milligrams per liter to parts per million, and vice versa; 25°C and 760 mm Hg barometric pressure)

Molecular Weight	1 mg/liter ppm	1 ppm mg/liter	Molecular Weight	1 mg/liter ppm	1 ppm mg/liter	Molecular Weight	1 mg/liter ppm	1 ppm mg/liter
1	24,450	0.0000409	39	627	0.001595	77	318	0.00315
2	12,230	0.0000818	40	611	0.001636	78	313	0.00319
3	8,150	0.0001227	41	596	0.001677	79	309	0.00323
4	6,113	0.0001636	42	582	0.001718	80	306	0.00327
5	4,890	0.0002045	43	569	0.001759	81	302	0.00331
6	4,075	0.0002454	44	556	0.001800	82	298	0.00335
7	3,493	0.0002863	45	543	0.001840	83	295	0.00339
8	3,056	0.000327	46	532	0.001881	84	291	0.00344
9	2,717	0.000368	47	520	0.001922	85	288	0.00348
10	2,445	0.000409	48	509	0.001963	86	284	0.00352
11	2,223	0.000450	49	499	0.002004	87	281	0.00356
12	2,038	0.000491	50	489	0.002045	88	278	0.00360
13	1,881	0.000532	51	479	0.002086	89	275	0.00364
14	1,746	0.000573	52	470	0.002127	90	272	0.00368
15	1,630	0.000614	53	461	0.002168	91	269	0.00372
16	1,528	0.000654	54	453	0.002209	92	266	0.00376
17	1,438	0.000695	55	445	0.002250	93	263	0.00380
18	1,358	0.000736	56	437	0.002290	94	260	0.00384
19	1,287	0.000777	57	429	0.002331	95	257	0.00389
20	1,223	0.000818	58	422	0.002372	96	255	0.00393
21	1,164	0.000859	59	414	0.002413	97	252	0.00397
22	1,111	0.000900	60	408	0.002554	98	249.5	0.00401
23	1,063	0.000941	61	401	0.002495	99	247.0	0.00405
24	1,019	0.000982	62	394	0.00254	100	244.5	0.00409
25	978	0.001022	63	388	0.00258	101	242.1	0.00413
26	940	0.001063	64	382	0.00262	102	239.7	0.00417
27	906	0.001104	65	376	0.00266	103	237.4	0.00421
28	873	0.001145	66	370	0.00270	104	235.1	0.00425
29	843	0.001186	67	365	0.00274	105	232.9	0.00429
30	815	0.001227	68	360	0.00278	106	230.7	0.00434
31	789	0.001268	69	354	0.00282	107	228.5	0.00438
32	764	0.001309	70	349	0.00286	108	226.4	0.00442
33	741	0.001350	71	344	0.00290	109	224.3	0.00446
34	719	0.001391	72	340	0.00294	110	222.3	0.00450
35	699	0.001432	73	335	0.00299	111	220.3	0.00454
36	679	0.001472	74	330	0.00303	112	218.3	0.00458
37	661	0.001513	75	326	0.00307	113	216.4	0.00462
38	643	0.001554	76	322	0.00311	114	214.5	0.00466

CONVERSION TABLE FOR GASES AND VAPORS (*Continued*)
(*Milligrams per liter to parts per million, and vice versa;*
25°C and 760 mm Hg barometric pressure)

Molecular Weight	1 mg/liter ppm	1 ppm mg/liter	Molecular Weight	1 mg/liter ppm	1 ppm mg/liter	Molecular Weight	1 mg/liter ppm	1 ppm mg/liter
115	212.6	0.00470	153	159.8	0.00626	191	128.0	0.00781
116	210.8	0.00474	154	158.8	0.00630	192	127.3	0.00785
117	209.0	0.00479	155	157.7	0.00634	193	126.7	0.00789
118	207.2	0.00483	156	156.7	0.00638	194	126.0	0.00793
119	205.5	0.00487	157	155.7	0.00642	195	125.4	0.00798
120	203.8	0.00491	158	154.7	0.00646	196	124.7	0.00802
121	202.1	0.00495	159	153.7	0.00650	197	124.1	0.00806
122	200.4	0.00499	160	152.8	0.00654	198	123.5	0.00810
123	198.8	0.00503	161	151.9	0.00658	199	122.9	0.00814
124	197.2	0.00507	162	150.9	0.00663	120	122.3	0.00818
125	195.6	0.00511	163	150.0	0.00667	201	121.6	0.00822
126	194.0	0.00515	164	149.1	0.00671	202	121.0	0.00826
127	192.5	0.00519	165	148.2	0.00675	203	120.4	0.00830
128	191.0	0.00524	166	147.3	0.00679	204	119.9	0.00834
129	189.5	0.00528	167	146.4	0.00683	205	119.3	0.00838
130	188.1	0.00532	168	145.5	0.00687	206	118.7	0.00843
131	186.6	0.00536	169	144.7	0.00691	207	118.1	0.00847
132	185.2	0.00540	170	143.8	0.00695	208	117.5	0.00851
133	183.8	0.00544	171	143.0	0.00699	209	117.0	0.00855
134	182.5	0.00548	172	142.2	0.00703	210	116.4	0.00859
135	181.1	0.00552	173	141.3	0.00708	211	115.9	0.00863
136	179.8	0.00556	174	140.5	0.00712	212	115.3	0.00867
137	178.5	0.00560	175	139.7	0.00716	213	114.8	0.00871
138	177.2	0.00564	176	138.9	0.00720	214	114.3	0.00875
139	175.9	0.00569	177	138.1	0.00724	215	113.7	0.00879
140	174.6	0.00573	178	137.4	0.00728	216	113.2	0.00883
141	173.4	0.00577	179	136.6	0.00732	217	112.7	0.00888
142	172.2	0.00581	180	135.8	0.00736	218	112.2	0.00892
143	171.0	0.00585	181	135.1	0.00740	219	111.6	0.00896
144	169.8	0.00589	182	134.3	0.00744	220	111.1	0.00900
145	168.6	0.00593	183	133.6	0.00748	221	110.6	0.00904
146	167.5	0.00597	184	132.9	0.00753	222	110.1	0.00908
147	166.3	0.00601	185	132.2	0.00757	223	109.6	0.00912
148	165.2	0.00605	186	131.5	0.00761	224	109.2	0.00916
149	164.1	0.00609	187	130.7	0.00765	225	108.7	0.00920
150	163.0	0.00613	188	130.1	0.00769	226	108.2	0.00924
151	161.9	0.00618	189	129.4	0.00773	227	107.7	0.00928
152	160.9	0.00622	190	128.7	0.00777	228	107.2	0.00933

CONVERSION TABLE FOR GASES AND VAPORS (Continued)
(*Milligrams per liter to parts per million, and vice versa; 25°C and 760 mm Hg barometric pressure*)

Molecular Weight	1 mg/liter ppm	1 ppm mg/liter	Molecular Weight	1 mg/liter ppm	1 ppm mg/liter	Molecular Weight	1 mg/liter ppm	1 ppm mg/liter
229	106.8	0.00937	253	96.6	0.01035	227	88.3	0.01133
230	106.3	0.00941	254	96.3	0.01039	278	87.9	0.01137
231	105.8	0.00945	255	95.9	0.01043	279	87.6	0.01141
232	105.4	0.00949	256	95.5	0.01047	280	87.3	0.01145
233	104.9	0.00953	257	95.1	0.01051	281	87.0	0.01149
234	104.5	0.00957	258	94.8	0.01055	282	86.7	0.01153
235	104.0	0.00961	259	94.4	0.01059	283	86.4	0.01157
236	103.6	0.00965	260	94.0	0.01063	284	86.1	0.01162
237	103.2	0.00969	261	93.7	0.01067	285	85.8	0.01166
238	102.7	0.00973	262	93.3	0.01072	286	85.5	0.01170
239	102.3	0.00978	263	93.0	0.01076	287	85.2	0.01174
240	101.9	0.00982	264	92.6	0.01080	288	84.9	0.01178
241	101.5	0.00986	265	92.3	0.01084	289	84.6	0.01182
242	101.0	0.00990	266	91.9	0.01088	290	84.3	0.01186
243	100.6	0.00994	267	91.6	0.01092	291	84.0	0.01190
244	100.2	0.00998	268	91.2	0.01096	292	83.7	0.01194
245	99.8	0.01002	269	90.9	0.01100	293	83.4	0.01198
246	99.4	0.01006	270	90.6	0.01104	294	83.2	0.01202
247	99.0	0.01010	271	90.2	0.01108	295	82.9	0.01207
248	98.6	0.01014	272	89.9	0.01112	296	82.6	0.01211
249	98.2	0.01018	273	89.6	0.01117	297	82.3	0.01215
250	97.8	0.01022	274	89.2	0.01121	298	82.0	0.01219
251	97.4	0.01027	275	88.9	0.01125	299	81.8	0.01223
252	97.0	0.01031	276	88.6	0.01129	300	81.5	0.01227

[a] A. C. Fieldner, S. H. Katz, and S. P. Kinney, "Gas Masks for Gases Met in Fighting Fires," *U.S. Bureau of Mines, Technical Paper No. 248*, 1921.

PATTY'S TOXICOLOGY

Fifth Edition

Volume 7

Glycols and Glycol Ethers,
Synthetic Polymers,
Organil Sulfur Compounds,
Organil Phosphates

CHAPTER EIGHTY-FIVE

Glycols

Finis L. Cavender, Ph.D., DABT, CIH, and
Edward J. Sowinski, Ph.D., DABT, CIH

1 INTRODUCTION TO CLASS OF CHEMICALS

The glycols are hydrocarbons that have two hydroxyl groups attached to separate carbons in an aliphatic chain or alicyclic ring. They are used as antifreeze agents, cryoprotectants, solvent carriers, chemical intermediates, and vehicles in a number of chemical formulations. They vary from slightly viscous liquids to waxy solids, are soluble in water, alcohols, and ketones, and are insoluble in hydrocarbons (1–10).

They have low vapor pressures; therefore inhalation of vapors and aerosols is of little concern unless they are heated, agitated, or sprayed. They have little or no odor; thus only their irritancy at high concentrations is a warning property (11–15).

Because of the irritant properties and the potential for central nervous system (CNS) depression at high concentrations, the threshold limit value (TLV) for glycols is generally set as a ceiling limit. The chemical and physical properties of several of the more common glycols are given in Table 85.1 (1–15).

1.0 Ethylene Glycol

1.0.1 CAS Number: [107-21-1]

1.0.2 Synonyms: 1,2-Ethanediol; 1,2-dihydroxyethane; ethane-1,2-diol; ethylene alcohol; ethylene dihydrate; 2-hydroxyethanol; monoethylene glycol

1.0.3 Trade Names: Dowtherm, Fridex, Lutrol-9, Norkool, Tescol, Ucar 17

1.0.4 Molecular Weight: 62.068

Patty's Toxicology, Fifth Edition, Volume 7, Edited by Eula Bingham, Barbara Cohrssen, and Charles H. Powell.
ISBN 0-471-31940-6 © 2001 John Wiley & Sons, Inc.

1.0.5 Molecular Formula: $C_2H_6O_2$

1.0.6 Molecular Structure: HO~~~OH

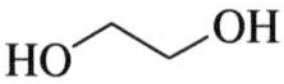

1.1 Chemical and Physical Properties
1.1.1 General

Ethylene glycol is a colorless, slightly viscous, hygroscopic liquid that has a bittersweet taste. It is miscible with water, lower aliphatic alcohols, aldehydes, and ketones and is practically insoluble in hydrocarbons and similar compounds (1–5, 7–10, 15). Additional properties are given in Table 85.1.

1.1.2 Odor and Warning Properties

Ethylene glycol is an odorless liquid that is irritating to the eyes and skin. The aerosol is nonirritating to the eyes and throat. Irritancy of the liquid at high concentrations is a warning property.

1.2 Production and Use

Historically, ethylene glycol has been manufactured by hydrolyzing ethylene oxide. Presently, it is also produced commercially by oxidizing ethylene in the presence of acetic acid to form ethylene diacetate, which is hydrolyzed to the glycol, and acetic acid is recycled in the process (10). Production in 1993 was 5.23 billion lb (16).

Table 85.1. Physical and Chemical Properties of Common Glycols (Diols)

Property	Ethylene Glycol	Diethylene Glycol	Triethylene Glycol	Propylene Glycol
CAS Number	[107-21-1]	[111-46-6]	[112-27-6]	[57-55-6]
Molecular formula	$C_2H_6O_2$	$C_4H_{10}O_3$	$C_6H_{14}O_4$	$C_3H_8O_2$
Molecular weight	62.07	106.12	150.1	76.1
Specific gravity (25/4°C)	1.11	1.12	1.125 (20/20°C)	1.033
Boiling point, °C (760 mmHg)	197.4	245	287.4	187.9
Freezing point, °C	−13.4	−8.0	−4.3	−31.0
Vapor pressure, mmHg (25°C)	0.06 (20°C)	<0.01	0.001	0.13
Refractive index (25°C)	1.432		1.456	1.431
Flash point, °F (o.c.)	240	290	330	215–225
Percent in saturated air (25°C)	0.017	0.0013 (20°C)	0.00013 (20°C)	0.038
1 ppm ≎ mg/m³ at 25°C and 760 mmHg	2.54	4.35	6.14	3.11
1 mg/l ≎ ppm at 25°C and 760 mmHg	365.0	230.7	162.8	321.6

More than 25% of the ethylene glycol produced is used in antifreeze and coolant mixtures for motor vehicles. This use may decrease in the coming years because of environmental concerns for ethylene glycol released or spilled from motor vehicles. It is used in hydraulic fluids and heat exchangers and as a solvent. Large amounts are used as a chemical intermediate in the production of ethylene glycol esters, ethers, and resinous products, particularly polyester fibers and resins (10, 15).

Contact with the skin and eyes is most likely to occur in industrial handling. Inhalation may be a problem if the material is handled hot or if a mist is generated by heat or by violent agitation. Swallowing is not likely to be an industrial problem unless the material is stored in unmarked or mislabeled containers (10, 15).

1.3 Exposure Assessment

1.3.1 Air

Although numerous methods are available for determining ethylene glycol when it is present in substantial amounts, few of the older methods are applicable to determining small amounts significant for industrial hygiene (15, 17–22). The National Institute for Occupational Safety and Health (NIOSH) method 5500 (23) uses gas chromatography with a flame ionization detector.

1.3.2 Background Levels

Davis et al. (24) and Spitz and Weinberger (25) developed gas chromatographic methods for determining ethylene glycol in an aqueous solution. These methods permit the detection of ethylene glycol in the nanogram range and should be useful in determining atmospheric concentrations because sampling by scrubbing air with water is effective.

1.3.3 Workplace Methods

In addition to the NIOSH method 5500 (23), Adams and Collins (26) developed an automated gas chromatographic method for quantifying ethylene glycol in the workplace.

1.3.4 Community Methods

Chairova and Dimov (27) and Esposito and Jamison (28) also developed gas or gas–liquid chromatographic methods for detecting glycols and their derivatives in environmental mixtures. The U.S. Environmental Protection Agency (USEPA) method 8015B (29) is also suitable for environmental analysis.

1.3.5 Biomonitoring/Biomarkers

1.3.5.1 Blood. Peterson and Rodgerson (30) and Cummings and Jatlow (31) recommend gas chromatographic methods for serum or plasma analysis (sensitivity 0.02 mg/mL). A capillary method using gas chromatography was reported by Johnson (32). Because mannitol has been used to ethylene glycol poisoning, it is important to know that, when present, mannitol gives a false positive response for ethylene glycol (33).

1.3.5.2 Urine. Flanagan (34) developed a gas chromatographic method for quantifying ethylene glycol in urine or dialysis fluid.

1.4 Toxic Effect

Ethylene glycol presents negligible hazards to health in industrial handling, except possibly where it is being used at elevated temperatures. It is low in acute oral toxicity, is not significantly irritating to the eyes or skin, is not readily absorbed through the skin in acutely toxic amounts, and its vapor pressure is sufficiently low so that toxic concentrations cannot occur in the air at room temperatures. Mists or aerosols generally are considered low in toxicity also, but exposure to high concentrations may result in more or less serious effects. Ethylene glycol is not considered carcinogenic. Available studies suggest that it is not mutagenic. The principal health hazard of ethylene glycol is from ingesting large quantities in single doses. Lesser quantities ingested, inhaled, or absorbed through the skin repeatedly over a prolonged period of time can also present a significant health hazard (3–15). In addition to some older reviews (35–39), a more recent review has been published by the Agency for Toxic Substances and Disease Registry (15).

1.4.1 Experimental Studies

1.4.1.1 Acute Toxicity. Acute Oral. The oral LD$_{50}$ ranges between 1.65 g/kg for the cat to 15.28 g/kg for some mice (14). Oral toxicity data are summarized in Table 85.2.

Eye Irritation. Carpenter and Smyth (45) reported that ethylene glycol failed to cause appreciable irritation when instilled into the eyes of rabbits. McDonald et al. (46), using the rabbit, placed one drop (0.05 mL) of eythlene glycol onto the eye every 10 min for 6 h (36 applications). They found that a 4% solution in a balanced salt medium caused mild conjunctival redness, mild chemosis, minor flare, and iritis. A 0.4% solution caused no detectable effects when compared to controls. Later work by McDonald et al. (47) showed that concentrations higher than 4% caused more serious injury. Grant (48) found that levels of 265 mg/m^3 caused no ocular damage to the eyes of chimpanzees; however, rabbits and rats developed severe eye irritation, edema of the lids, and some corneal opacity when exposed to 12 mg/m^3 for several days.

Table 85.2. Acute Oral Toxicity of Ethylene Glycol to Laboratory Animals

Reference	\multicolumn{5}{c}{LD$_{50}$ Values (g/kg of Body Weight)}
	Mouse	Rat	Guinea Pig	Rabbit	Dog
Laug et al. (36)	14.6	6.14	8.20		
Smyth et al. (40)		8.54	6.61		
Bornmann (41)	15.28				
Page (10)					>8.81
Pochebyt		5.89			
Plugin (42)	8.0	13.0	11.0	5.0	
Bove (10)		10.0–13.4			
Antonyuk (43)		10.88			
Kersting and Nielsen (10)					Minimum lethal dose 7.36; some dogs survived 14.7
Dow (44)		11.3			

GLYCOLS 5

Skin Irritation. Ethylene glycol produces no significant irritant action upon the skin. A slight macerating action on the skin may result from very severe, prolonged exposures, which is comparable to that caused by glycerin under similar conditions (15).

Dermal Absorption. The dermal LD_{50} in the rabbit is 9.53 g/kg (14). Hanzlik et al. (49) have shown in animal studies that toxic amounts of ethylene glycol can be absorbed through the skin.

Acute Inhalation. The 1-h LC_{50} in rats is 10.9 mg/L (14). Flury and Wirth exposed rats for 28 h during 5 days to an atmosphere essentially saturated (0.5 mg/L) with ethylene glycol (39). No deaths occurred, but the animals exhibited slight CNS depression.

1.4.1.2 Chronic and Subchronic Toxicity. *Oral Toxicity.* Antonyuk (43) found that the no-effect level in rats fed for 3 months was 1.08 g/kg/day. This was the highest level fed and represents a dose of one-tenth the acute oral LD_{50}. Gaunt et al. (50) also conducted subchronic dietary feeding studies on rats. The no-effect level when fed for 16 weeks was 0.1% (71 mg/kg/day for male rats and 85 mg/kg/day for female rats). Based on this study, the acceptable daily intake for a 70-kg person was suggested as 50 mg ethylene glycol per day. The males given a diet containing 0.25% (178 mg/kg/day) and higher levels developed oxaluria, oxalate crystals in their urine, renal damage, and at the 1.0% level, increased kidney weight and altered renal function. These effects were seen in females, but to a lesser degree, and only in those dosed at the 1.0% level (850 mg/kg/day). Nagano et al. (51) fed ethylene glycol to mice in doses ranging from 62.5 to 4000 mg/kg, 5 days/week for 5 weeks and observed no effects upon the testes or blood.

Morris et al. (38) maintained rats for 2 years on diets containing 1 or 2% ethylene glycol. The findings were of shortened life span, calcium oxalate bladder stones, severe renal injury, particularly of the tubules, and centrilobular degeneration of the liver at both levels. Blood (52) in a follow-up study found that the no-effect level was no more than 0.2% in the diet (100 mg/kg/day for 2 years). He found calcification of the kidneys and oxalate crystal stones at the 0.5% level in male rats; however, females showed only calcification of the kidneys at the dietary level of 1% and higher, but oxalate stones only at the 4% level. There was increased water consumption and protein appeared in the urine of males that received the 1 and 4% diets and in females on the 4% diet. Blood et al. (53), using three rhesus monkeys, conducted 3-year feeding studies. The two males were fed a diet containing 0.2% ethylene glycol, which equates to an average daily dose between 0.02 and 0.07 g/kg. The one female was fed a diet containing 0.5%, which is an average daily dose between 0.14 and 0.17 g/kg. No adverse effects were observed, and no oxalate crystals were found in the urine. Roberts and Seibold (54) found that a macaque monkey, given ethylene glycol at a concentration of 0.25% (about 0.24 g/kg/day) in its drinking water for 157 days, developed oxalate deposition in the kidneys. Two other monkeys did not show renal oxalate crystals when they were given 0.25% (about 0.4 g/kg/day) for up to 60 days. Higher doses, given even for short periods, caused morphological tubular changes of the kidney, even though oxalate crystals were not detected. A limited number of the monkeys given 10% ethylene glycol in their drinking water for a few days, followed by lower concentrations, developed calcium oxalate crystals in the brain. Those monkeys given an average dose of less than 0.4 g/kg/day developed mild glomerular damage and azotemia, even though no calcium oxalate crystals were detected.

Inhalation Toxicity. Felts (55) reported on studies in which four chimpanzees were exposed for 28 days to saturated vapors of ethylene glycol (256 mg/m^3) at 5 psi in an atmosphere consisting of 68% oxygen and 32% nitrogen, and observed only minor depression of the white blood cell count 2 weeks after terminating exposure. Two other chimpanzees in a similar but separate experiment experienced some signs of minor renal effects but no eye, behavioral, or clinical effects. Wiley et al. (56) exposed rats and mice to concentrations of 0.35–0.40 mg/L (140–160 ppm), 8 h/day for 16 weeks without producing injury. Antonyuk (43) conducted studies in which animals (probably rats) were exposed to ethylene glycol vapor for 24 h/day for 3 months. The "no-effect" level was 0.3 mg/m^3. Growth depression and blood effects were seen at a level of 8.4 mg/m^3. Coon et al. (57) report observing no adverse effects when rats, rabbits, guinea pigs, squirrel monkeys, and dogs were exposed to 57 mg/m^3 of ethylene glycol vapor, 8 h/day, 5 days/week for 6 weeks. When a separate set of these animals was exposed continuously 24 h/day for 90 days, the rats and rabbits experienced moderate to severe eye irritation. Corneal opacity was seen in the rabbits after 3 days of exposure and in the rats in 8 days. Otherwise, there were no significant effects seen in any of the animals. Wills et al. (58) found that monkeys survived exposures of 500 to 600 mg/m^3 for 5 to 7 months. Harris (59) exposed rats, rabbits, mice, guinea pigs, dogs, and monkeys to 100 ppm ethylene glycol continuously for 3 months in a chamber at 5 psi in an atmosphere of 100% oxygen to which ethylene glycol was added. Careful examination of the animals revealed no adverse effects other than some pulmonary irritation in the rats and mice; no behavioral changes were detected in the monkeys.

1.4.1.3 Pharmacokinetics, Metabolism, and Mechanisms.

Metabolism. Many workers have explored the metabolism of ethylene glycol to understand its toxicological effects better and to discover effective methods of treatment. As a result, a number of schemes have been developed (15, 59, 60–69). Basically, all agree that the main steps in degradation of ethylene glycol are as follows:

$$\text{Ethylene glycol} \rightarrow \text{glycolaldehyde} \rightarrow \text{glycolic and glyoxylic acid}$$

Glyoxylic acid is then metabolized into a number of chemicals that have been identified in expired air, urine, or blood.

Several reviews (15, 66, 68) report that the metabolism is as shown in Figure 85.1. The Roman numerals in Figure 85.1 indicate the following: (I) The metabolism of ethylene glycol to glycolaldehyde is mediated by alcohol dehydrogenase. (II) Glycolaldehyde is metabolized to glycolic acid by aldehyde oxidase (IIa), or to a lesser extent to glyoxal (IIb). (III) Glyoxal is changed both to glycolic acid in the presence of lactic dehydrogenase, aldehyde oxidase, or possibly both enzymes (IIIa), and to glyoxylic acid via some oxidative mechanism (IIIb).

The main path of the degradation of glycolic acid is to glyoxylic acid. This reaction is mediated by lactic dehydrogenase or glycolic acid oxidase.

V–VIII. Once glyoxylic acid is formed, it is apparently degraded very rapidly to a variety of products, a few of which have been observed. Its breakdown to 2-hydroxy-3-oxoadipate (V) it is thought, is mediated by thiamine pyrophosphate in the presence of

GLYCOLS

Ethylene glycol (HOCH$_2$CH$_2$OH)
↓ I
Glycolaldehyde (HOCH$_2$CHO)
↙ IIa   ↘ IIb
(HOOCCH$_2$OH) Glycolic acid ⟵ IIIa ⟶ Glyoxal (OHCCHO)
↓ IV   ↙ IIIb
(HOOCCOOH) Oxalic acid ⟵ VIII ⟶ Glyoxylic acid (HOOCCHO)
↙ VI   ↓ VII   ↘ V ⟶ Oxalomalate
(HCOOH) Formic acid   2-Hydroxy-3-oxoadipate
↓ O$_2$ Glycine (H$_2$NCH$_2$COOH)
CO$_2$ + water   +   2-Oxo-4-hydroxygluconate
            Benzoic acid   ↓ +CO$_2$
            ↓   Malate
            Hippuric acid
(C$_6$H$_5$CONHCH$_2$COOH)

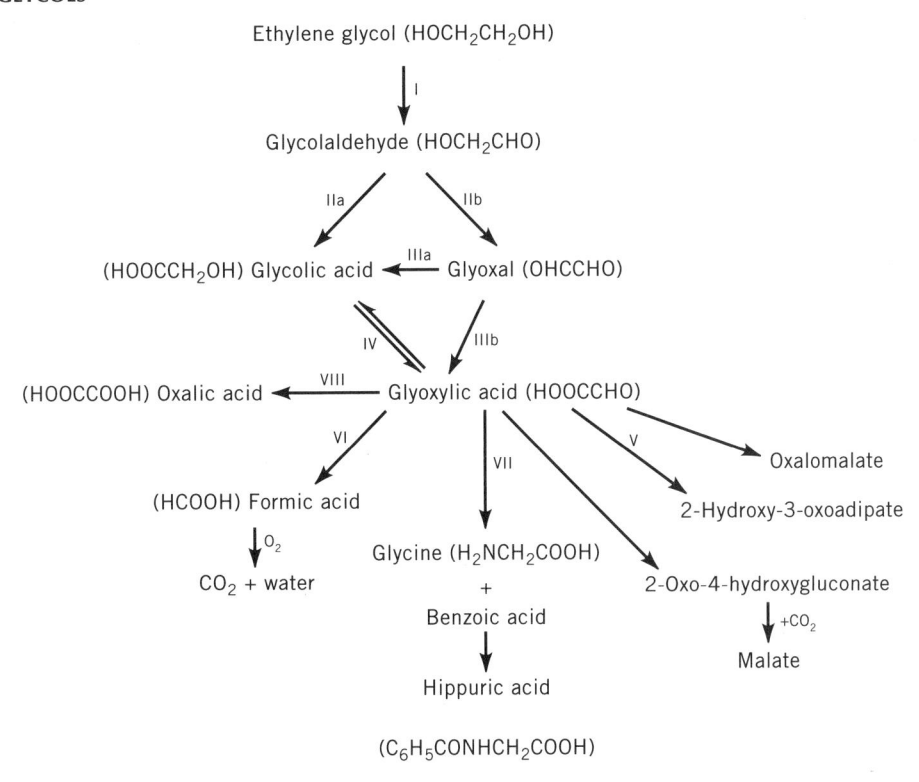

Figure 85.1. Combined scheme of ethylene glycol metabolism. Based on Refs. 59–69. See text for explanation of the mechanism of reaction indicated by the Roman numerals.

magnesium ions (40). The formation of glycine involves pyridoxal phosphate and glyoxylate transaminase (VII), whereas the formation of carbon dioxide and water via formic acid (VI) apparently involves coenzyme A (CoA) and flavin mononucleotides. Oxalic acid (VIII) formation from glyoxylic acid, has been considered to be the results from the action of lactic dehydrogenase or glycolic acid oxidase.

When ethylene glycol is introduced into the body, it is rapidly distributed into the water in the body. Initially, ethylene glycol is excreted unchanged in the urine for a few hours, accounting for as much as 22% of the amount given. Carbon dioxide is also promptly excreted via the lungs and may account for a large amount of the ethylene glycol administered (70). Glycolic acid in the form of salts is also excreted via the urine, in an amount equivalent to as much as 34 to 44% of the dose given. Neither glycolaldehyde nor glyoxylic acid has been found in the urine. Oxalic acid, if formed, it also excreted via the urine. The amounts formed depend on the species and on the dose and account for 0.27 to about 3% of the dose given to laboratory animals (67), and possibly up to 2.3% of the dose in humans (67). Rhesus monkeys and chimpanzees excrete larger amounts of oxalic acid salts, as much as 18 to 20% of the administered dose. This may have been due to the large doses given (59). The liver and the kidney are the main sites of metabolism (66).

Mechanism of Action. To evaluate the mode of action of ethylene glycol, it is necessary to understand its clinical effects. Parry and Wallach (68), POISINDEX (71), Ellenhorn and Barceloux (72), and Jacobson and McMartin (73) summarized the ethylene glycol as follows:

The effects of ethylene glycol poisoning appear in three fairly distinct stages. The severity of these stages and the advance from one stage to the next depend greatly on the amount of ethylene glycol that enters the body. The initial stage consists of "central nervous system (CNS) effects." This stage usually occurs shortly after exposure, within 30 min or so, and lasts up to 12 h. The CNS effects are characterized by signs of drunkenness, although there is no alcoholic odor on the breath; nausea; vomiting; and if large doses are experienced, coma followed by convulsions and death in some cases. Mild hypotension, tachycardia, low-grade fever, depressed reflexes, generalized or focal seizures, myoclonic jerks, and tetanic contractions can occur. Ocular signs such as nystagmus, ophthalmoplegia, papilledema, and subsequent optic atrophy have all been reported. This first stage is usually attributed to the presence of unchanged ethylene glycol which, it is thought, contributes to CNS effects. The other signs seen are considered due to metabolic products.

The second stage is usually labeled the "cardiopulmonary effects" stage and ordinarily lasts from 12 to 36 h after exposure. Studies of the blood usually reveal a moderate leukocytosis (10 to 40,000/mm^3), predominance of polymorphonuclear cells, and a normal hematocrit. Serum sodium bicarbonate levels are often depressed, an anion gap greater than 10 is usually present, and hypercalcemia along with acidosis is often found. Hypocalcemia is often observed also. The stage is initiated by the onset of coma and is characterized by symptoms such as tachypnea, tachycardia, mild hypotension, and cyanosis. In severe cases, pulmonary edema, bronchopneumonia, cardiac enlargement, and congestive failure are present. Death in this stage, if it occurs, usually comes between 24 and 72 h after exposure. This stage is usually attributed to acidosis and hypocalcemia caused by the metabolic products.

The final stage is known as the "renal failure" stage and occurs if the patient survives the first two stages. Urinalysis usually shows a low specific gravity with proteinuria, microscopic hematuria, pyuria, and cylindruria. In some cases, calcium oxalate and hippuric acid crystals are seen. Although oxalate is a minor metabolic product in humans, urinary oxalate crystals are common, but not invariable. There are two forms of calcium oxalate crystals, the tent shape of the dihydrate and the prism shape of the monohydrate. The dihydrate form occurs only during high urinary calcium and oxalate concentrations (72).

Costovertebral tenderness may be observed if the patient is alert. The renal damage may vary from a mild increase in blood urea nitrogen and creatinine, followed by recovery, to complete anuria with acute tubular necrosis that can lead to death. Originally it was thought that renal failure could be attributed solely to the presence of calcium oxalate. However, much work has been done to show that renal damage can occur at levels of exposure where no or few crystals are detected. Therefore, it is now felt that the various metabolic products, other than calcium oxalate, contribute largely to the renal damage that occurs. The renal necrosis is characterized by dilated proximal tubules, degeneration of tubular epithelium, and intratubular crystals. Distal tubular degeneration is present usually but is less prominent. Peterson et al. (74) reported that an extract of human liver containing alcohol dehydrogenase actively catalyzed the oxidation of ethylene glycol, that the rate of oxidation of the glycol was slower than that of ethanol, that the purified enzyme oxidized ethylene glycol, and that this could be inhibited by ethanol. They then demonstrated in *in vivo* experiments in rats and squirrel monkeys that ethanol was effective in reducing the toxicity of ingested ethylene glycol and markedly increased its excretion. The mechanism for this protection is twofold; the ethanol competitively inhibits the oxidation of the glycol to more toxic metabolites and enhances the excretion of the glycol *per se*. In 1992, Cox et al. (75) have shown 1,3-butylene glycol to be more effective than ethanol in the treatment of ethylene glycol toxicity. Butylene glycol binds more efficiently to alcohol dehydrogenase and is less inherently toxic than ethanol. Parry and Wallach (68) urge that the rapid recognition and immediate treatment of ethylene glycol poisoning are necessary to reduce its toxicity. Ethylene glycol poisoning should be suspected if there is nondiabetic ketoacidosis, an alcohol-like intoxication without the presence of alcohol odor on the breath, coma, and metabolic acidosis coupled with an anionic gap and/or oxalate found in the urine.

1.4.1.3.1 Absorption. Marshall and Cheng (76) exposed rats (nose only) to ^{14}C-ethylene glycol vapor at a concentration of 32 mg/L for 30 min. Approximately 61% of the amount inhaled was retained, largely in the nasopharyngeal region. The total amount retained was calculated to be equivalent to 0.9 + 0.3 g/kg for the males and 0.6 + 0.5 g/kg for the females.

1.4.1.3.2 Distribution. Because of its water solubility, ethylene glycol distributes to all tissues. Marshall and Cheng (76) found that once distributed, blood levels remain fairly constant as ethylene glycol in the tissue stores equilibrates with the blood. After six hours, blood levels declined by apparent first-order kinetics with a half-life of 53 h.

Frantz et al. (77) found that oral and interperitoneal doses are rapidly absorbed in rats and mice and that tissue levels are essentially the same following each route of administration.

1.4.1.3.3 Elimination. Hewlett et al. (78) reported that the elimination half-life for ethylene glycol in rats and dogs was 1.7 hr and 3.4 h, respectively. Ethylene glycol concentration in the plasma peaked at 2 h and glycolate concentration peaked at about 5 h.

Marshall and Cheng (76) found that the predominant routes of elimination of the radioactivity were 55 to 70% as ^{14}CO$_2$ in the expired air and 14 to 26% in the urine.

1.4.1.4 Reproduction and Developmental. DePass et al. (79) found no abnormalities in the reproductive performance of Fischer 344 rats that received 40, 200, or 1000 mg/kg/day for three generations. When administered to mice in drinking water for two generations, Lamb et al. (80)) found that ethylene oxide was fetotoxic at 1% but not at 0.25 or 0.5%. Maronpot et al. (81) reported a no-observed-effect, level (NOEL) of 1000 mg/kg/day in female rats exposed on days 6 to 15 of gestation. Price et al. (15) reported increased skeletal malformations in rat fetuses from dams exposed to 1250 mg/kg/day on days 6 to 15 of gestation. Dams exposed to 5000 mg/kg/day showed reduced litter size and increased implantation loss. In mice, the fetuses of dams exposed to 750 mg/kg/day on days 6 to 15 of gestation exhibited increased skeletal malformations. In rabbits, the developmental NOAEL was 2000 mg/kg/day, and the maternal toxicity NOAEL was 1000 mg/kg/day. Tyl et al. (82) found that nose only exposure to ethylene glycol resulted in fetotoxicity and maternal toxicity. The NOAEL for maternal toxicity was 1000 mg/m^3 for rats and 150 mg/m^3 for mice and the NOAEL for developmental effects was 150 mg/m^3 for both. Apparently, teratogenicity via inhalation is not observed because an exposure to an aerosol for 6 h results in blood levels much lower than those following oral gavage (83). Teratogenicity is associated with the accumulation of the metabolite, glycolic acid, in the blood (31) and the development of metabolic acidosis (84).

1.4.1.5 Carcinogenesis. NTP (77) found no evidence of carcinogenicity in rats and mice fed diets of up to 5% for 2 years. DePass et al. (85) found no evidence of carcinogenicity in rats and mice fed diets resulting in ethylene glycol dosages of up to 1 mg/kg/day for 2 years. Mason et al. (86) found no chemically related tumors when ethylene glycol was injected into Fischer rats subcutaneously twice a week for 52 weeks, then held for 18 months for examination. Blood (52) also found no tumors associated with ethylene glycol when it was fed to rats at 1% in the diet for 2 years. Derse (87) found no chemically related tumors in mice given ethylene glycol subcutaneously and observed for 15 months. Homburger (88) also found no tumors related to ethylene glycol when it was injected subcutaneously into mice for 8 to 12 weeks; then the injection sites were excised, homogenized, and injected into 25 mice. He also found that ethylene glycol did not increase lung tumors when injected once intravenously into mice.

1.4.1.6 Genetic and Related Cellular Effects. McCann et al. (89) reported that ethylene glycol was nonmutagenic in the *Salmonella*/microsome mutagenicity test. Pfeiffer and Dunkelberg (90) found that ethylene glycol was inactive when tested against *S. typhimurium* strains TA98, TA100, TA1535, and TA1537. There were no dominant lethal effects in Fischer 344 rats that received 40, 200, or 1000 mg/kg/day for three generations (79).

1.4.2 Human Experience

1.4.2.1 General Information. When one considers the huge volumes of ethylene glycol that are handled and used industrially, there have indeed been few instances of adverse effects. Only one episode is reported that can be considered of industrial origin (91), and this involves inhalation of vapors from heated material. One case of alleged absorption through the skin involved a medicinal product (92).

1.4.2.2 Clinical Cases

1.4.2.2.1 Acute Toxicity. Pons and Custer (93) in their review of 18 human cases stressed effects on the brain. Grant (94) also noted cerebral injury from acute poisoning and believes it may have caused permanent injury. Ross (95) described a fatal case in which 1/4 to 1/2 pint of an antifreeze solution was ingested; acute meningoencephalitis followed by anuria and death from renal failure resulted after 12 days. Tadokoro et al. (96) report that acute renal failure occurred after a 18-yr old male inadvertently ingested antifreeze. When admitted, the patient exhibited nausea, vomiting, and convulsions, but recovered in 8 days. The single oral dose lethal for humans has been estimated at 1.56 g/kg

(36). It is apparent that ethylene glycol is more acutely toxic for humans than for the laboratory animals studied. The American Association of Poison Control Centers reported six and five deaths due to ethylene glycol ingestion in 1989 and 1990, respectively (97, 98).

Grant (48) reports that exposure of humans to the vapor or spray of ethylene glycol for 4 weeks at a concentration of 17 mg/m^3 produced no ill effects. Volkmann (92) reports a case of what was believed to be ethylene glycol poisoning resulting from the massive application of an eczema remedy containing ethylene glycol. A comatose condition accompanied by miosis and slowed pulse occurred 4 h after application but no oxalate was found in the urine.

Wills et al. (58) exposed 22 volunteers to aerosols of ethylene glycol (particle size 1 to 5 mm) at concentrations varying from 3 to 67 mg/m^3. The exposures were essentially continuous (20 to 22 h/day) for 1 month. The volunteers reported some irritation of the nose and throat and occasionally slight headache and low backache, but there were no other significant adverse effects. Further tests found that aerosol levels of 200 mg/m^3 were intolerable and were very noticeable at levels of 140 mg/m^3. Wills et al. (58)) exposed humans under ambient conditions 22 h/day, for 28 days, to 68.5 mg/m^3 of ethylene glycol as a mixture of vapor and aerosol dispersed in air. The subjects were unable to detect the ethylene glycol and experienced no adverse effects. When Harris (48) increased the concentration to 137 mg/m^3, the volunteers reported throat and eye irritation and were able to detect the ethylene glycol by taste. A concentration of 205.5 mg/m^3 was considered intolerable because of irritation of the eyes and throat. Human volunteers, exposed to about 28 mg/m^3 of ethylene glycol as an aerosol for about 4 weeks, reported some complaints of irritation of the throat, mild headache, and low backache. At concentrations of about 140 mg/m^3 for part of a day, the complaints became marked; levels of above 140 up to 200 mg/m^3 caused reluctance to be exposed for more than a few minutes, and levels of 250 to 300 mg/m^3 were intolerable even for one breathing cycle (58).

1.4.2.2.2 Chronic and Subchronic Toxicity. Troisi (91) reports the results of an investigation of complaints among women workers in an electrolytic condenser factory and attributes the trouble to the inhalation of ethylene glycol vapor. The work involved coating aluminum and paper with a mixture containing 40% ethylene glycol, 55% boric acid, and 5% ammonia at 105°C. Nine of 38 women exposed to the vapors suffered frequent attacks of loss of consciousness and nystagmus, and five of them showed an absolute lymphocytosis. Further examination revealed five additional cases of nystagmus among the other 29 workers. Proper enclosure of the system to prevent inhalation of vapors resulted in complete recovery of all affected individuals, although the two most severely affected were removed to other work. The work of Gaunt et al. (50) and others suggest that the daily intake of ethylene glycol in the diet of a 70-kg person may be 50 mg/day.

1.4.2.2.3 Pharmacokinetics, Metabolisms, and Mechanisms. The treatment of ethylene glycol poisoning has been greatly enhanced by knowledge of its metabolic and pharmacokinetic parameters. These were discussed earlier. Several authors (71–73, 95 99–102) have described their experience in treating cases of human poisoning. Basically, the suggested treatment involves infusing ethanol to inhibit competitively the metabolism of ethylene glycol to oxalic acid, infusing alkali to control the profound acidosis typical of such poisoning, infusing a diuretic such as mannitol to help prevent or control brain edema, and hemodialysis to remove the ethylene glycol from the circulation and to help manage renal insufficiency. The importance of early diagnosis and treatment cannot be overemphasized.

Recently, Harry et al. (103) reported that the alcohol dehydrogenase inhibitor, 4-methylpyrazole is an effective antidote for ethylene glycol ingestion.

1.5 Standards, Regulations, or Guidelines of Exposure

The American Conference of Governmental Industrial Hygienists (ACGIH) TLV is a ceiling limit of 50 ppm (100 mg/m^3) (104). In 1989 the Occupational Safety and Health Administration (OSHA) proposed a permissible exposure limit (PEL) of 50 mg/m^3 but this was struck down in court thus, there is no OSHA PEL for ethylene glycol (105). NIOSH questions whether a ceiling of 50 ppm is adequate to protect workers. The Environmental Protection Agency's (EPA) oral reference dose (RfD) for ethylene glycol is 2.0 mg/kg/day (106) and the World Health Organization's (WHO) acceptable daily intake is 25 mg/kg (107). Ethylene glycol is on the EPA Toxic Substances Control Act (TSCA) Chemical Inventory (108) and Test Submission Data Base (98).

1.6 Studies on Environmental Impact

Evans and David (109) exposed ethylene glycol to typical river water and 20°C and found that it was completely degraded in 3 days. At temperatures of 8°C, complete or nearly complete degradation occurred in 7 days, depending on the river water used. In similar tests Pitter (110) also showed that ethylene glycol is rapidly degraded.

2.0 Diethylene Glycol

2.0.1 CAS Number: [111-46-6]

2.0.2 Synonyms:
2,2'-Oxybis-ethanol; 2,2'-oxydiethanol; bis(2-hydroxyethyl) ether; carbitol; glycol ether; ethylene diglycol; 2,2'-dihydroxydiethyl ether; beta; beta'-dihydroxydiethyl ether; 2,2'-dihydroxyethyl ether; 2-hydroxyethyl ether; 3-oxapentane-1,5-diol; 3-oxa-1,5-pentanediol; brecolane ndg; deactivator e; deactivator h; DEG; Dicol; dissolvant apv; TL4N; dihydroxyethylether

2.0.3 Trade Names: NA

2.0.4 Molecular Weight: 106.12

2.0.5 Molecular Formula: $C_4H_{10}O_3$

2.0.6 Molecular Structure: HO$\frown$O$\frown$OH

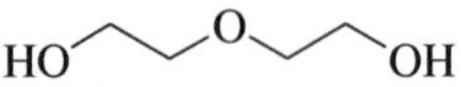

2.1 Chemical and Physical Properties

2.1.1 General

Diethylene glycol is a colorless, essentially odorless, viscous, hygroscopic liquid. Initially, it has a sweetish taste, but its aftertaste is bitter. It is miscible with water, lower aliphatic alcohols, and ketones, and is practically insoluble in aliphatic hydrocarbons and fatty oils. Additional properties are given in Table 85.1.

2.1.2 Odor and Warning Properties

Diethylene glycol is odorless and does not irritate the eyes.

2.2 Production and Use

Diethylene glycol is produced commercially as a by-product of ethylene glycol production. It can also be produced directly by reaction between ethylene glycol and ethylene oxide (111). Diethylene glycol is used in gas conditioning and in permanent antifreeze formulations; as a constituent of brake fluids, lubricants, mold-release agents, and inks; as a softening agent for textiles; as a plasticizer for cork, adhesives, paper, packaging materials, and coatings; as an intermediate in the production of the explosive diethylene glycol dinitrate; and as an intermediate in the production of certain resins, morpholine, and diethylene glycol esters and ethers.

Diethylene glycol presents practically no hazard from the standpoint of industrial handling. It is quite stable chemically and does not present a hazard due to flammability, except at high temperatures or where mist may be involved. It is not appreciably irritating to the eyes and skin and is not absorbed through the skin in appreciable amounts except possibly under adverse conditions where extensive and prolonged skin contact occurs. Its vapor pressure at room temperatures is so low that toxic concentrations of vapor are impossible. It should be noted that a hazard from repeated prolonged inhalation may exist in operations involving heated material or where mists or fogs are generated. However, any reasonable industrial hygiene control would eliminate this possibility. Although the principal hazard to health presented by diethylene glycol is ingesting a substantial amount, this should not occur in industrial handling unless the material is put in unlabled or mislabeled containers.

2.3 Exposure Assessment

Methods for determining small amounts of diethylene glycol, of industrial hygiene significance are few. The chromatographic methods described by Davis et al. (24) and Spitz and Weinberger (25) should be useful for determining workplace exposures.

2.4 Toxic Effects

Diethylene glycol presents negligible hazards to health in industrial handling except possibly where it is being used at elevated temperatures. It has low acute oral toxicity, is not irritating to the eyes or skin, is not readily absorbed through the skin, and its vapor pressure is sufficiently low so that toxic concentrations of vapor cannot occur in air at room temperatures. Aerosols generated by violent agitation and/or heat may result in excessive amounts in air. The principal health hazard of diethylene glycol is from ingesting large quantities in single doses.

Before 1937, human toxicological information on diethylene glycol and other glycols was rather incomplete. In 1937, more than 100 deaths were caused by the ingestion of an elixir consisting of sulfanilamide and diethylene glycol as one of the major solvents. As a result of this tragedy, a large number of investigations were conducted in an effort to clarify the toxicological picture of the various glycols involved.

2.4.1 Experimental Studies

2.4.1.1 Acute Toxicity. *Acute Oral.* Most of the older published articles that deal with the toxicity of diethylene glycol and the other common glycols have been ably reviewed (53). The findings of Laug et al. (36) and Smyth et al. (40) on small animals are representative. Laug (36) reports that the acute oral LD_{50} values for rats, guinea pigs, and mice is 16.6, 8.7, and 26.5 g/kg, respectively. Smyth et al. (40) report similar values of 20.8 g/kg for rats and 13.2 g/kg for guinea pigs. Plugin (112) has also studied diethylene glycol and reports finding that the LD_{50} for mice, rats, rabbits, and guinea pigs are 13.3, 15.6, 26.9, and 14.0 g/kg, respectively. Laug (40) reports that the symptomatology for rabbits, dogs, mice, and guinea pigs is quite similar; first noted were thirst, diuresis, roughened coat, and refusal of food, followed days later by suppression of urine, proteinuria, prostration, dyspnea, a bloated appearance, coma, decrease in body temperature, and death. Plugin (112) reports that rats given 3.1 g/kg/day for 20 days orally were not affected. He concluded that there were no cumulative effects, suggesting that diethylene glycol may be readily metabolized under these conditions.

Eye Irritation. Carpenter and Smyth (45) reported that diethylene glycol failed to cause appreciable irritation when introduced into the eyes of rabbits. No cases of injury to human eyes have been reported nor would any be expected.

Skin Irritation. Diethylene glycol produces no significant skin irritation; however, prolonged contact over an extended period of time may produce a macerating action comparable to that caused by glycerol (4).

Dermal Absorption. Hanzlik et al. (49) have shown that commercial diethylene glycol of unknown purity can be absorbed in toxic amounts through the skin of rabbits. The data are erratic and therefore difficult to quantify. A more definitive dermal LD_{50} value for rabbits is 13.3 g/kg (113).

Inhalation Toxicity. An essentially saturated atmosphere generated at approximately 170°C and a fog generated at about 70°C caused no deaths of rats exposed for 8 h (113).

2.4.1.2 Chronic and Subchronic Toxicity. *Subchronic Oral.* Wegener (114) gave rats 1 mL of a 20% aqueous solution of diethylene glycol per 100 g of body weight daily for a period of 12 weeks and concluded that it had no influence on the reproductive ability of the animals or on their offspring. Bornmann (41) administered the material to rats in concentrations of 1, 2, 5, 10, and 20% in their drinking water. He found that the material had a CNS depressant effect and caused central paralysis of the respiratory and cardiac centers. Loesser et al. (115) found that concentrations of 5 to 20% in the drinking water of rats caused weight loss and death, but that 1 and 2% had no such effect. They also report finding 2-naphthol in the urine and bile of animals treated with diethylene glycol.

Subchronic Dermal Toxicity. Marchenko (116, 117) applied 2.8 g/kg/day for 2 months to the skin of rats and found that they developed edema of the brain, plethora, and minute brain hemorrhages.

Subchronic Inhalation Toxicity. Sanina (118) reported that 10 of 16 mice exposed to levels of 4 to 5 mg/m^3 (0.92 ppm) 2 h/day for 6 to 7 months developed malignant mammary tumors, whereas no tumors developed in the 20 controls. The concentrations stated were generated by heating diethylene glycol at 30 to 35°C in a dish inside the 100-L chamber that housed the mice. The significance of these findings is minimal because the report gives no indication that the control animals were subjected to the same extraneous stresses, particularly heat, as the experimental animals. Further, there is no indication of the purity of the diethylene glycol tested.

Marchenko (116, 117) exposed mice and rats to about the same concentration as Sanina (118) for 3 to 7 months and found structural changes in the central nervous system and endocrine and parenchymatous organs. The purity of the diethylene glycol was not given, and the results appear to conflict with the other available data.

Chronic Oral. A comprehensive study of the chronic oral toxicity of diethylene glycol was reported by Fitzhugh and Nelson (119), who maintained rats for 2 years on diets containing 1, 2, and 4% diethylene glycol. The purity of the diethylene glycol was not described and hence some of the effects noted may not, in fact, be attributable to diethylene glycol. At the dietary level of 1% of diethylene glycol, they observed slight growth depression, a few bladder stones identified as calcium oxalate, slight kidney damage, and infrequent liver damage. At the 2% dietary level, slight growth depression, a number of bladder stones and bladder tumors, moderate kidney damage, and slight liver damage were noted. At the 4% dietary level, there was some mortality, marked growth depression, frequent bladder stones and bladder tumors, marked kidney damage, and moderate liver damage.

Weil et al. (120, 121) fed diethylene glycol containing only 0.031% of ethylene glycol to weanlings and to 2-month-old and year-old rats for up to 2 years at levels of 4.0 and 2.0% in a laboratory chow. Because of the difference in chemical intake, the year-old rats were given a diet containing 6% diethylene glycol after 1 month on the 4% diet. Although the weanling rats developed more bladder stones than the other groups, the difference was insignificant. The yearling rats developed their bladder stones somewhat earlier. The highest stone formation was eight in 20 rats at the 4% dosage level. None was found in the rats fed the 2% level. Weil et al. (120, 121) conclude that diethylene glycol substantially free of ethylene glycol does not cause bladder stones, suggesting that it is not metabolized to any great degree to ethylene glycol.

2.4.1.3 Pharmacokinetics, Metabolism, and Mechanisms. *Metabolism.* Repeated administration to dogs for a week did not lead to consistent increases in urinary oxalate. However, the urinary oxalate increased in rats maintained on water containing diethylene glycol. Wiley et al. (122) were unable to demonstrate the presence of oxalic acid in the urine following large doses of diethylene glycol to rabbits and dogs. These apparent discrepancies may well be attributable to the purity of the material studied.

Mode of Action. Diethylene glycol in large doses is a depressant to the central nervous system. Deaths from large single doses that occur within 24 h it is believed, result from this action. Acutely toxic doses, not immediately fatal, may exert their effect primarily on the kidney and, to a lesser extent, on the liver. Deaths or serious injuries are associated primarily with renal insufficiency caused by swelling of the convoluted tubules and plugging of the tubules with debris. Chronic effects from prolonged and repeated exposure to the commercial product are most likely to be centered in the kidney and to a lesser degree in the liver. In metabolism studies with the dog, Haag and Ambrose (123) found that a large portion of the diethylene glycol administered was excreted unchanged in the urine.

2.4.1.5 Carcinogenesis. Weil et al. (10, 120), in their long-term studies on rats of three different age levels, found only one bladder tumor in those fed diets that contained 4% diethylene glycol. This tumor was in a rat that also had bladder stones (119). To clarify the question of the cause of the tumor, Weil et al. (120, 121) implanted calcium oxalate stones or glass beads into the bladders of rats. They found that bladder tumors never developed without the presence of a foreign body in the bladder. This led to the conclusion that diethylene glycol essentially free of ethylene glycol is not a primary carcinogen.

2.4.1.6 Genetic and Related Cellular Effects Studies. Mutagenicity. Pfeiffer and Dunkelberg (190) found that diethylene glycol is inactive when tested against *S. typhimurium* strains TA98, TA100, TA1535, and TA1537.

2.4.2 Human Experience

2.4.2.1 General Information. Human experience in industrial handling and use of diethylene glycol has been excellent except for the "elixir of sulfanilamide tragedy," in which more than 100 deaths were attributed to its ingestion. Since then, a great number of articles have been published dealing with the clinical and experimental aspects of diethylene glycol poisoning. These have been well summarized by Geiling and Cannon (37), and none reflect the neurological observations in rats and mice reported by Marchenko (116, 117). A few cases have also been reported from other uses of diethylene glycol in medicinals (124) and from accidents (125). In general, pathology observed in human victims resembles closely that which has been described previously for laboratory animals and consists primarily of degeneration of the kidney and fewer lesions in the liver. Death in practically all cases was due to renal insufficiency.

2.4.2.2.1 Acute Toxicity. As a result of the "elixir of sulfanilamide" episode, the acute lethal dose for humans has been estimated by Calvery and Klumpp (126) at about 1 mL/kg.

2.4.2.3 Epidemiology Studies. Telegina et al. (127) studied the effect of exposure of diethylene glycol on 90 workers who produced aromatic hydrocarbons from crude oil and found that the workers exposed for 1 to 9 years did not experience any differences in the incidence of skin tumors, nervous system, or internal organs. There is, however, no mention of how long after exposure the workers were studied.

2.5 Standards, Regulations, or Guidelines of Exposure

Because of the low vapor pressure of diethylene glycol, its low toxicity when studied by most investigators, and the lack of adverse human experience in industrial settings, there

GLYCOLS

has been no industrial hygiene standard established. Because significant atmospheric contamination with vapor can occur only under extreme conditions, an industrial hygiene guide of 10 mg/m^3 of particulate material is suggested. Amounts greater than this generally would be considered a nuisance to be avoided.

3.0 Triethylene Glycol

3.0.1 CAS Number: [112-27-6]

3.0.2 Synonyms:
2,2'-[1,2-Ethanediylbis(oxy)]bis-ethanol; 2,2'-(ethylenedioxy)diethanol; triglycol; bis(2-hydroxyethoxy)ethane; glycol bis(hydroxyethyl) ether; di-beta-hydroxyethoxyethane; 1,2-bis(2-hydroxyethoxy)ethane; 3,6-dioxaoctane-1,8-diol; ethylene glycol dihydroxydiethyl ether; TEG; trigen; triglycol; trigol; ethylene glycol-bis-(2-hydroxyethyl)ether; 3,6-dioxa-1,8-octanediol

3.0.4 Molecular Weight: 150.1

3.0.5 Molecular Formula: C$_6$H$_{14}$O$_4$

3.0.6 Molecular Structure: HO~~O~~O~~OH

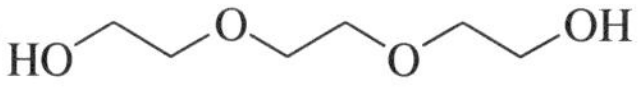

3.1 Chemical and Physical Properties

3.1.1 General

Triethylene glycol is a colorless to pale straw-colored, essentially odorless, viscous, hygroscopic liquid. It is miscible with water and many common solvents. It is practically insoluble in aliphatic hydrocarbons and fats. Additional properties are given in Table 85.1.

3.1.2 Odor and Warning Properties

Triethylene glycol has low vapor pressure and is not irritating to the eyes.

3.2 Production and Use

Triethylene glycol, like diethylene glycol, is produced commercially as a by-product of ethylene glycol production. Its formation is favored by a high ethylene oxide to water ratio (111).

Triethylene glycol is used for many of the same applications as diethylene glycol, but it has two distinct properties of importance; it is less volatile and less toxic. It is used as a humectant in tobacco, as a plasticizer, as a dehydrating agent for natural gas, and as a selective solvent. It is a valuable intermediate for the manufacture of plasticizers, resins, emulsifiers, demulsifiers, lubricants, explosives, and many others.

The industrial handling and use of triethylene glycol present no significant problem from ingestion, skin contact, or vapor inhalation. Its low oral toxicity suggests that it may be considered safe for many applications where intake is limited. Similarly, its negligible skin irritation and absorption properties make it suitable for use to some extent in preparations intended to be applied over appreciable areas of the body. Furthermore, it is stable chemically and does not present a hazard due to flammability, except possibly at high temperatures or where fogs or mists are involved.

3.3 Exposure Assessment

There is no need for determining atmospheric concentration of triethylene glycol for industrial hygiene purposes. If analysis of the atmosphere were to be made, however, the methods noted in Section 2.3 probably would be useful.

3.4 Toxic Effects

Triethylene glycol has very low acute and chronic oral toxicity, it is not irritating to the eyes or skin, and the inhalation of amounts that conceivably could cause injury does not seem likely.

3.4.1 Experimental Studies

3.4.1.1 Acute Toxicity. Oral Toxicity. Latven and Molitor (128), Smyth et al. (40), Laug et al. (36), and Stenger et al. (129) studied the acute oral toxicity of triethylene glycol and found that it is less toxic than diethylene glycol. Smyth et al. (40) reported the oral LD_{50} for rats and guinea pigs as 22.06 and 14.66 g/kg, respectively. Laug et al. (36) found that the LD_{50} values for rats, guinea pigs, mice, and rabbits are 16.8, 7.9, 18.7, and 8.4 mL/kg, respectively. Stenger et al. (129) state that Woodard found that the LD_{50} values for mice, rats, guinea pigs, and rabbits are 21.0, 18.9, 8.9, and 9.5 g/kg, respectively.

Eye Irritation. Latven and Molitor (128) tested triethylene glycol for its effect on the rabbit eye and found that it is similar to glycerin and diethylene glycol and less irritating than propylene glycol. Carpenter and Smyth (45) reported that triethylene glycol failed to cause appreciable irritation when introduced into the eyes of rabbits.

Skin Irritation. Triethylene glycol produces no significant irritation of the skin. However, prolonged contact over an extended period of time may result in a macerating action similar to that caused by glycerin.

Dermal Absorption. No studies have been reported that deal with the skin absorption of triethylene glycol. Although it is possible that, under conditions of severe prolonged exposure, some of the material may be absorbed through the skin, it is extremely doubtful that a quantity sufficient to produce an appreciable systemic injury would be absorbed.

Inhalation Toxicity. Interest in the toxicity of inhaled triethylene glycol was initiated by the observation by Robertson in 1943 (130) and later in 1947 (131) that triethylene glycol was an effective air sterilizer. In 1947, Robertson et al. (132) reported extensive experiments with monkeys and rats showing that prolonged inhalation of saturated vapors, as in air disinfection (about 1 ppm), had no toxic effect. Antonyuk (43) exposed test animals (species not given) continuously for 3 months to triethylene glycol vapors at concentrations of 5 and 1 mg/m^3 (approximately 0.814 and 0.163 ppm). He found that the 5-mg/m^3 level caused minor effects at most and that the 1-mg/m^3 level caused no effects.

3.4.1.2 Chronic and Subchronic Toxicity. Subchronic Oral Toxicity. Lauter and Vrla (133) reported that rats could tolerate 3% in their drinking water for 30 days without effect,

GLYCOLS

but 5% caused ill effects. These dosages are equivalent to about 5 and 8 g/kg/day. They described the material they used as "commercial grade." Because commercial-grade triethylene glycol may contain several percent of diethylene glycol, the possibility that the toxic effect may have been caused by diethylene glycol rather than by triethylene glycol cannot be overlooked.

Chronic Oral Toxicity. The most comprehensive chronic study of the oral toxicity of triethylene glycol is that reported by Fitzhugh and Nelson (119). These investigators fed the material at concentrations of 1, 2, and 4% in the diet of rats for 2 years without producing adverse effects. These dosage levels are equivalent to as much as 3 to 4 g/kg/day without effect.

3.4.1.3 Pharmacokinetics, Metabolism, and Mechanisms. McKennis and co-workers (134) studied the fate of ^{14}C-labeled triethylene glycol in rats and of unlabeled material in rabbits. They found that in both species a high percent of small doses was eliminated in the urine unchanged and possibly as the mono- and dicarboxylic acid derivatives of triethylene glycol. In the studies with rats, little if any, ^{14}C-oxalate or ^{14}C-triethylene glycol in conjugated form was found in the urine. Small portions of the administered ^{14}C activity were found in the feces (2–5%) and in expired air (1%). Recoveries of the administered ^{14}C activity ranged from 91 to 98%.

3.4.1.4 Reproductive and Developmental. *Teratogenicity.* Stenger et al. (129) treated female mice, rats, and rabbits during selected periods of pregnancy with triethylene glycol at levels of 2.25 g/kg/day by subcutaneous injection and rats at a level of 4.5 g/kg/day by oral administration and found that there was no evidence of teratogenic effects.

3.4.2 Human Experience

3.4.2.1 General Information. During the studies on the effectiveness of triethylene glycol as an air sterilizer, numerous persons were exposed, and according to Jennings et al. (135) and Harris and Stokes (136), none were adversely affected. The developments in the field of air sterilization have been well reviewed by Polderman (137).

3.4.2.3 Epidemiology Studies. Grant (48) reports that splash contamination in the eyes of humans causes acute pain. This may be followed by transitory disturbance of corneal epithelium and gradual diminishing signs of irritation, but no persistent injury is expected.

3.5 Standards, Regulations, or Guidelines of Exposure: NA

4.0 Tetraethylene Glycol

4.0.1 CAS Number: [112-60-7]

4.0.2 Synonyms: Tetraglycol; ethanol, 2,2′-[Oxybis(2,1-ethanediyloxy)]bis-; 3,6,9-trioxaundecan-1,11-diol

4.0.4 Molecular Weight: 194.11

4.0.5 Molecular Formula: $C_8H_{18}O_5$

4.0.6 Molecular Structure: HO~O~O~O~OH

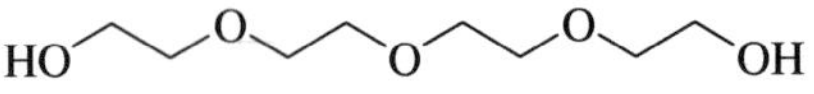

4.1 Chemical and Physical Properties

4.1.1 General

Tetraethylene glycol is a colorless liquid that is miscible with water and many common solvents. It is practically insoluble in aliphatic hydrocarbons and oils.

4.1.2 Odor and Warning Properties

Tetraethylene glycol has no vapor pressure and is not irritating to the eye.

4.2 Production and Use

Tetraethylene glycol is prepared commercially by adding ethylene oxide to ethylene glycol, diethylene glycol, or water in the presence of a suitable catalyst. Tetraethylene glycol is used as a plasticizer and solvent where a high boiling point and low volatility are important. It can be an effective coupling agent in formulating water-soluble and water-insoluble materials. It is used as a chemical intermediate for manufacturing glycol esters and ethers and certain resins.

Industrial exposure to tetraethylene glycol is mainly topical.

4.3 Exposure Assessment

Methods described for ethylene glycol probably can be adapted. The method described by Ramstad et al. (138) for polypropylene glycols may also be applicable.

4.4 Toxic Effects

Tetraethylene glycol has very low acute oral toxicity; the LD_{50} value for rats is 30.8 mL/kg (14). It is nonirritating to the eyes and skin and is not likely to be absorbed through the skin in toxic amounts. The dermal LD_{50} for rabbits is 20 mL/kg (14). Essentially saturated vapors of this material caused no significant adverse effects when rats were exposed to them for 8 h (113).

Yoshida et al. (139) reported that chicks fed a diet containing 5% of tetraethylene glycol for 27 days as an energy source were unaffected. They postulate that chicks apparently were unable to metabolize it. Weifenbach (140), in his study of the suitability of the use of tetraethylene glycol as a solvent for drugs, found that it caused a slow drop in blood pressure, late production of arrhythmias just before death, and kidney and blood effects when it was infused intravenously into anesthetized rats at a rate of 22 mL/h.

4.5 Standards, Regulations, or Guidelines of Exposure: NA

GLYCOLS

5.0 Polyethylene Glycols

5.0.1 CAS Number: *[25322-68-3]*

5.0.2 Synonyms:
alpha-Hydro-omega-hydroxy poly (oxy-1,2-ethanediyl) PEG 1000; Polyoxyethylene 1000; Polyglycol 1000; Polyethylene glycol 400; poly ethylene oxide; Polyethylene Glycol 8000; Polyethylene glycols; Carbowax PEG 8000; ethoxylated 1,2-ethanediol; ployoxyethylene ether; Carbowax PEG 400; Poly(ethylene glycol) 100; Poly(ethylene glycol) 1000; Poly(ethylene glycol) 1500; Poly(ethylene glycol) 200; Poly(ethylene glycol) 300; Poly(ethylene glycol) 6000; Poly(ethylene glycol) 900; Polyoxyethylene; carbowax 200; emkapol 200; gafanol e 200; pluriol e 200; polydiol 200; polyethylene gylcol; PEG; polyyox WSR-301; PEG 200; Macrogol; Polyethylene glycol 20,000; Poly(ethylene glycol) 4000; Poly(ethylene glycol) 600; Poly(ethylene glycol) 2000; Poly(ethylene glycol)3400; Ethylene glycol 8000 polymer; poly(ethylene glycol) 10000; polyoxyethlene

5.0.4 Molecular Weight: 62.068

5.0.5 Molecular Formula: $[C_2H_6O_2]_n$

5.0.6 Molecular Structure: HO~~O~~O···~OH

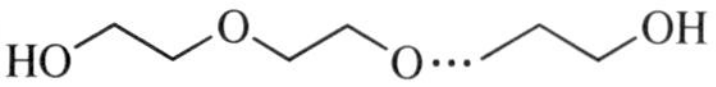

5.1 Chemical and Physical Properties

Polyethylene glycols may be represented by the formula $H(OCH_2CH_2)_nOH$. Those that have average molecular weights of 200, 300, 400, and 600 are viscous, nearly colorless, odorless, water-soluble liquids that have very low vapor pressures. Polyethylene glycols that have average molecular weights of 1000 and more are nearly colorless, water-soluble, waxy solids at room temperature. All of the unstabilized polyglycols are inherently susceptible to oxidative degradation, which occurs with increased rapidity as temperature increases and as the availability of oxygen increases. Their physical and chemical properties are given in Table 85.3.

5.2 Production and Use

Polyethylene glycols are prepared commercially by adding ethylene oxide to ethylene glycol, diethylene glycol, or water in the presence of caustic or other catalysts. The molecular weights of the product can be controlled by the proportions of the reactants used (111).

Polyethylene glycols of average molecular weight 600 or less exist as liquids at room temperature. They are used primarily as reactive intermediates for manufacturing fatty acid ester surfactants and as solvents for gas processing.

Polyethylene glycols of average molecular weight 1000 to 2000 exist at room temperature as soft to firm solids that have low melting points. They are primarily bases for cosmetic creams and lotions, as well as pharmaceutical ointments and toothpaste formulations.

Polyethylene glycols of average molecular weight 3500 to 20,000 exist at room temperature as firm to hard, brittle, waxlike solids. They are used as binders, plasticizers,

Table 85.3 Physical and Chemical Properties of Polyethylene Glycols

Material Designation[a]	Physical State	Mol. Wt. Range	Specific Gravity (25/25°C)	Freezing Range (°C)	Refractive Index (25°C)	Flash Point[b] (°F)	Refs.
200	Liquid	190–210	1.125	Supercools	1.459	340–360	44, 141
300	Liquid	285–315	1.125	−15 to −6	1.463	385–415	44, 141
400	Liquid	380–420	1.125	4–8	1.465	435–460	44, 141
600	Liquid	570–630	1.125	20–25	1.466	475–480	44, 141
1000	Solid	956–1050	1.117	36–40		490–510	44, 141
1450	Solid	1300–1600	1.210	43–46		490	44
1500	Solid	1300–1600	1.21	43–46		490	141
1540	Solid	1300–1600	1.21	43–46		510	141
2000	Solid	1900–2300	1.211	47–50		510	44
4000	Solid	3000–3700	1.204	53–56		520	141
4000	Solid	4200–4800	1.212	54–57		515	44
6000	Solid	6000–7500		60–63		520	141
6000	Solid	7000–8000	1.212	56–59		515	44
9000	Solid	9000–10,000	1.212	60–64		520	44
10,000	Solid						
14,000	Solid						
20,000	Solid						
4,000,000	Solid						

[a]Generally designates average molecular weights.
[b]Cleveland open cup.

molding compounds, stiffening agents, and paper adhesives. Some use of the esters of higher molecular weight polyethylene glycols as thickeners for cosmetic formulations has also been noted.

Industrial exposure to polyethylene glycols is limited almost entirely to topical contact.

5.3 Exposure Assessment

Polyglycols can be quantified in aqueous solution by the method of Duke and Smith (22) based upon the reaction of alcoholic hydroxyl groups with ammonium hexanitratocerate to form a red product. Infrared spectrophotometry may also be useful. Shaffer and Critchfield (142) have described a method for quantitatively determining solid polyethylene glycols in biological materials. Ramstad et al. (138) described a method for determining polypropylene glycol that could be adaptable.

5.4 Toxic Effects

Polyethylene glycols present practically no health hazards in industrial handling and use. They are not significantly irritating to the eyes, skin, or mucous membranes; they have exceptionally low oral toxicity; and their vapor pressures are so low that there is no hazard from inhalation.

GLYCOLS

5.4.1 Experimental Studies

5.4.1.1 Acute Toxicity. Oral Toxicity. All polyethylene glycols have very low acute oral toxicity. It is noteworthy that toxicity decreases as molecular weight increases. Representative data are given in Table 85.4.

Eye Irritation. Carpenter and Smyth (45) reported that polyethylene glycols do not appreciably irritate to the eyes of rabbits. This has been confirmed (48, 144).

Skin Irritation and Sensitization. Although early reports by Smyth et al. (145) reported that skin sensitization was observed in guinea pigs tested with certain polyethylene glycols, later studies (144, 146, 147) show that currently produced materials have no irritating or sensitizing properties.

Dermal Absorption. As concluded by Smyth et al. (146), the lethal dose via dermal application of the polyethylene glycols is so large as to defy establishing of LD_{50} values. Unpublished studies (144) that employed essentially the technique of Draize et al. (148), showed that single doses of 20 g/kg of the various polyethylene glycols ranging from molecular weights of 200 through 9000 had no toxic effects.

Table 85.4. Acute Oral Toxicity of Polyethylene Glycols

(Aproximate LD_{50} Values) (g/kg)[a]

Material	Rats M	Rats B	Rats F	Guinea Pigs M	Guinea Pigs B	Guinea Pigs F	Rabbits M	Rabbits B	Rabbits F	Mice	Ref.
200		34						20			142
200	34		28			17	14			34	44
300		39			20			21			142
300	30		29	21					21	31	44
400		44			16			27			142
400		44				21	22			36	44
400		33								29	143
600	33		30			28	19			36	44
1000		42			22						142
1000	45		32			41			>50	>50	44
1500		44			29			30			142
1540		51			37					50	142
2000	45		>50	>50			>50			>50	44
4000		59			51			76			142
4000	>50		>50		50		>50			>50	44
6000		>50			>50						142
6000	>50		>50			>50	>50			>50	44
9000	>50		>50		>50				>50	>50	44
10,000		>50									142
14,000		>32									44
20,000		>16									44
4,000,000		>4.0									143

[a] M = males; B = both sexes; F = females.

5.4.1.2 Chronic and Subchronic Toxicity. *Subchronic Dermal Toxicity.* The studies reported by Luduena et al. (149) in 1947 indicated that toxic amounts of polyethylene glycols 200, 400, 1500, and 4000 were quite readily absorbed through the skin of rabbits when applied by inunction 6 days a week for 5 weeks. However, because part of this study was on animals on a deficient diet and because there were no controls, its significance cannot be evaluated. This is emphasized by the fact that other well-controlled studies by Smyth et al. (150) that used, similar materials, and by Tusing et al. (151) that used, a wider spectrum of polyethylene glycols found no adverse effects from larger doses for a longer period of time.

Schmidt (152) also tested E-300 and E-600 on mice 24 times for an 8-week period. He too found that doses of 0.05 mL to an area of 4 to 6 cm^2 had no effect.

Chronic Oral Toxicity. Smyth and co-workers (146,150) summarized the extensive feeding studies they conducted with polyethylene glycols. Polyethylene glycols that have average molecular weights of 400, 1540, and 4000 caused no adverse effect in dogs when fed 2% in their diet for 1 year (150). When fed to rats for 2 years as a part of their diet, polyethylene glycols 1540 and 4000 had no effect at a level of 4%, and polyethylene glycol 400 had no effect at a level of 2% (146). The other polyethylene glycols have been studied by dietary feeding techniques using rats, but for shorter periods of 3 to 4 months (44, 150). It appears that several percent of these materials can be tolerated in the diet of rats without appreciable adverse effects, indicating that they are exceptionally low in chronic oral toxicity (See Table 85.5).

5.4.1.3 Pharmacokinetics, Metabolism, and Mechanisms. Polyethylene glycols in general are slow-acting parasympathomimetic-like compounds, according to Smyth et al. (146), When they are given intravenously, they tend to increase the tendency of the blood to clot and if given rapidly, cause clumping of the cells, and death occurs from embolism.

5.4.1.3.1 Absorption. Shaffer and Critchfield (153) reported that polyethylene glycols that have average molecular weights of 4000 and 6000 were not absorbed from the rat intestine within 5 h, whereas small amounts of lower molecular weight materials (1000 and 1540) were absorbed. Shaffer et al. (154) reported on studies with human subjects using polyethylene glycol that had an average molecular weight of 400. They were able to recover 77% in the urine in 12 h following the administration of 1 g intravenously and to recover 40 to 50% in the urine when a 5- to 10-g dose of the material was given orally. Ethylene glycol was not a metabolite of polyethylene glycol 400.

In 1968, Principe (143) applied polyethylene glycol 400 and 4000 to the skin of horses and assumed that they were not absorbed through the skin because they were not found in the urine. A dose of 4.26 g of polyethylene glycol given orally gave rise to only trace amounts in the urine during the 72-h test period. If injected intramuscularly, polyethylene glycol 4000 was eliminated in the urine essentially completely in about 30 h. Chemical studies of the urine indicate that it was excreted essentially unchanged. Carbon-14-labeled polyethylene glycol 4,000,000 was not absorbed from the digestive tract of the rat and dog when it was administered orally (155).

5.4.1.3.1 Absorption. Schutz (156) observed that polyethylene glycol 400 markedly reduces the percutaneous absorption of certain materials that readily penetrate the intact

GLYCOLS

Table 85.5. Summary of Repeated Oral Dose Toxicity of Polyethylene Glycols

Mean Molecular Weight	Species	Sex[a]	Duration of Study (months)	Dosage Level (% in Diet) Without Effect	Dosage Level (% in Diet) With Effect	First Sign of Adverse Effect Noted[b]	Ref.
200	Rat	B	3	8	16	LW	146
300	Rat	B	3	4	8	W	146
400	Rat	B	3	8	16	W	146
	Rat	M	24	2	4	W, LW, LP	146
	Rat	F	24	4			146
	Dog	B	12	2			146
600	Rat	B	3	8	16	KW, W	146
1000	Rat	B	3	8	16	W	146
	Rat	B	3	10	15	W	44
1500	Rat	B	3	4	8	W	146
1540	Rat	B	3	4	8	W	146
	Rat	B	24	4	8	LP	146
	Dog	B	12 (possibly 8)	2			146
2000	Rat	B	4	15			44
4000	Rat	B	3	4	8	W	146
	Rat	B	4	5	10	W, LW	44
	Rat	B	24	4	8	W	146
	Dog	B	12(possibly 8)	2			146
6000	Rat	B	3	16	24	KW, W	146
	Rat	M	3	10	15	W	44
	Rat	F	3	15			44
9000	Rat	B	3	15			44
10,000	Rat	B	3	>2.5			44
4,000,000	Rat	B	24	5 (or 2.76)[c]		None	143
4,000,000	Rat	B	3		10	LP, KP	143
4,000,000	Dog	B	24	2 (or 0.56)[c]		None	143

[a] B = both sexes; M = male; F = female.
[b] W = decrese in body weight; LW = increase in liver weight/100 g of body weight; KW = increase in kidney weight/100 g of body weight; LP = slight histological changes in the liver; KP = slight histological effects in the kidney.
[c] g/kg/day.

skin in toxic amounts. He found that phenol, dimethylaniline, phenol red, barbital, salicylic acid, and γ-hexachlorocyclohexane were very poorly absorbed from solutions of the glycol.

Polyethylene glycol E-300 (PEG-300) and PEG-300/methylated spirits(PEGIMS) [6:1 v/v ratio of PEG-300 to methylated spirits (95% ethanol, 5% methanol)] are reportedly superior to water as decontaminants of the skin of rats exposed to phenol (157,158). Other similar studies using the skin of swine, believed to be more similar to the skin of humans than the skin of rats, indicates that a water shower for 15 min is practically equivalent to decontamination with PEG-300 or PEG-IMS. Both treatments were effective in reducing mortality, skin injury, plasma concentration, and retention time of absorbed phenol compared to animals exposed to phenol but not decontaminated. These data plus the universal availability of water indicate that water is the decontaminant of choice (44).

Carpenter et al. (147) also tested for absorption and excretion of polyethylene glycol 4000 in rats, and found that when given orally, it was excreted via the feces. However, when it was given by intravenous injection, it was excreted via the urine.

5.4.2 Human Experience

Oral Toxicity. When 10-g doses of materials that had average molecular weights of 6000 and 1000 were given orally to five human subjects, none of the 6000 molecular weight material was found in the urine in the following 24 h, whereas about 8% of the 1000 molecular weight material was found (153)

Eye Irritation. No cases of injury to human eyes have been reported nor would any be expected.

Skin Irritation and Sensitization. Although early reports by Smyth et al. (145) reported that skin sensitization was observed among a few human subjects tested with certain polyethylene glycols, later studies (144, 146, 147) show that currently produced materials have no irritating or sensitizing properties. This has been borne out by their very wide application in cosmetics without difficulty.

5.5 Standards, Regulations, or Guidelines of Exposure: NA

6.0 Propylene Glycol

6.0.1 CAS Number: [57-55-6]

6.0.2 Synonyms: 1,2-Propanediol; methyl ethylene glycol; 1,2-dihydroxypropane; trimethyl glycol; 1,2-propylene glycol; monopropylene glycol; propane-1,2-diol; alpha-propyleneglycol; dowfrost; PG 12; sirlene; solar winter ban; propanediol

6.0.4 Molecular Weight: 76.1

6.0.5 Molecular Formula: $C_3H_8O_2$

6.0.6 Molecular Structure:

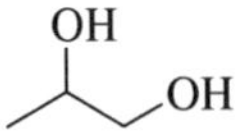

6.1 Chemical and Physical Properties

6.1.1 General

Propylene glycol is a colorless, almost odorless, slightly viscous liquid that has a slightly acrid taste. It imparts no odor or taste of its own when used in food colors and flavors. Additional physical and chemical properties are given in Table 85.1.

GLYCOLS

6.1.2 Odor and Warning Properties

Propylene glycol has no odor and is not irritating to the eyes.

6.2 Production and Use

Propylene glycol generally is synthesized commercially by starting with propylene, converting to the chlorohydrin, and hydrolyzing to propylene oxide, which is then hydrolyzed to propylene glycol. It can also be prepared by other methods (111).

Propylene glycol is used in antifreeze formulations, heat exchangers, and brake and hydraulic fluids; in the manufacture of resins, which accounts for a large portion of its use, polypropylene glycols, and propylene glycol ethers and esters; as a solvent in pharmaceuticals, foods, cosmetics, and inks; as a plasticizer for resins and paper; and as a humectant in textiles, tobacco, and pet foods. It is also used in the vapor form as an air sterilizer for hospitals and public buildings.

Industrial exposures are from direct contact or inhalation vapors and mists where the material is heated or violently agitated. Other exposure is by ingestion resulting from its use in foods and drugs.

6.3 Exposure Assessment

Propylene glycol in air can be quantified by several of the methods noted under ethylene glycol, although it would seem unnecessary for industrial hygiene purposes.

A method for detecting propylene glycol in body fluids is described by Lehman and Newman (159).

6.4 Toxic Effects

The health hazards in industrial handling and use of propylene glycol are negligible. Its systemic toxicity is especially low and, since 1942, it has been considered an acceptable ingredient for pharmaceutical products (161). The Food and Drug Administration approves its use in food products and in cosmetics (162). Inhalation of atmospheres that contain propylene glycol presents no health hazard. Exposures created by operations that producing hot vapors or by high-speed mechanical action in which a fog of propylene glycol is produced have not been studied. However, it is difficult to visualize how this condition could create a hazard, because the material has extremely low systemic toxicity. The toxicology of propylene glycol has been reviewed extensively (9, 15, 162, 163).

6.4.1 Experimental Studies

6.4.1.1 Acute Toxicity. Oral Toxicity. The acute oral toxicity of propylene glycol has been studied by a number of investigators (36, 113, 159, 164–166). The acute oral LD_{50} values for rats range from 21.0 to 33.7 g/kg; for mice, from 23.9 to 31.8 g/kg; for guinea pigs, from 18.4 to 19.6 g/kg; for rabbits, from 15.7 to 19.2 g/kg; and for dogs, from 10 to 20 g/kg.

Laug et al. (36) report observing minimal kidney changes from large doses.

6.4.1.2 Chronic and Subchronic Toxicity. *Subchronic Oral Toxicity.* Seidenfeld and Hanzlik (167) gave groups of rats drinking water containing 1, 2, 5, 10, 25, and 50% propylene glycol for a period of 140 days. Animals that received water containing either 25 or 50% propylene glycol died in 69 days, whereas those that received 1, 2, 5, or 10% appeared normal throughout the observation period. The calculated average daily intakes for the latter four groups were about 1.6, 3.7, 7.7, and 13.2 g/kg/day of propylene glycol, respectively. Histopathological examination of the tissues from these animals revealed no renal or other pathological disturbances. Weatherby and Haag (164) confirmed that rats tolerate 10% propylene glycol in drinking water without toxic effects.

Hanzlik and associates (165) found that rats could tolerate up to 30 mL/kg daily of propylene glycol when fed in the diet for a 6-month period. This is equivalent to 1.8 lb daily for a 70-kg person. Morris et al. (38) fed rats 2.45 and 4.9% propylene glycol in the diet, allowing, respectively, average daily intakes of 0.9 to 1.77 mL/kg for a 24-month period without significant effect on growth rate. Microscopic examination of the tissues revealed very slight liver damage but no renal pathology.

Whitlock et al. (168) found that a diet containing 30% of propylene glycol was not well tolerated by young rats and that producing females were unable to bring their young to weaning. Diets containing 40, 50, or 60% of propylene glycol were lethal after a few days.

Van Winkle and Newman (169) showed that propylene glycol, given in concentrations of 5 or 10% in the drinking water of dogs for 5 to 9 months, caused no adverse effects. Criteria employed were liver function, kidney function, and histopathological examination of the visceral organs. Further, they found no alterations in the serum calcium levels of cats and dogs fed large doses of propylene glycol.

Eye Irritation. Propylene glycol is not injurious to the eyes of rabbits (156, 170).

Skin Irritation and Absorption. Propylene glycol generally produces no significant irritant action upon the skin. One of the many medicinal uses of propylene glycol is as a solvent in eardrops. Morizono and Johnstone (171) found that a concentration of 10% or more of propylene glycol in Ringer's solution instilled into the inner ear cavity of guinea pigs caused apparent irreversible deafness. They recommended that the concentration of propylene glycol used in eardrops should be less than 10%.

Chronic Oral Toxicity. Gaunt et al. (172) fed both male and female rats propylene glycol in their diet at levels of 6,250, 12,500, and 50,000 ppm for a period of 2 years. They found no significant ill effects based on mortality, body weight, food consumption, hematology, urinary cell excretion, urine-concentrating ability of the kidneys, organ weights, or histopathology, including tumor incidence.

Weil et al. (173) found that male and female dogs fed diets providing propylene glycol at a dose level of 2.0 g/kg for 2 years were unaffected as judged by mortality, body weight change, diet utilization and water consumption; histopathology; organ weights of liver, kidney, and spleen; and measurement of blood, urine, and biochemical parameters. At a daily dose of 5 g/kg, the dogs gained more weight than the controls, especially during the early part of the experiment, because of the higher caloric intake. An increase in the rate of erythrocyte hemolysis and a slight increase in total bilirubin were also noted.

Hemoglobin, packed cell volume, and total erythrocyte count were lowered slightly, whereas the incidence of anisocytes, poikilocytes, and reticulocytes was increased, suggesting that erythrocytes were being destroyed, accompanied by accelerated replacement from the bone marrow. This effect was not sufficient, however, even at the 20% dietary level to result in any irreversible changes, and there was no evidence of damage to the bone marrow or spleen. They state that dogs fed propylene glycol at approximately 8% in their diet (equivalent to 2 g/kg/day) can use it as a carbohydrate energy source with no adverse effects.

The use of propylene glycol as a source of carbohydrate energy in animal feed has led to other dietary feeding studies involving cattle, sheep, chickens, and cats. It is used in pet foods as a humectant as well as a source of energy. In lactating cattle, cows tend to develop hyperketosis, which may be due to low carbohydrate stores in the body that are used to produce lactose excreted in the milk. Fisher and co-workers (174, 175) summarized much of the work done to determine the usefulness of propylene glycol as a dietary supplement to alleviate this condition.

Fisher et al. (174) reported on studies in lactating cows in which propylene glycol was added to food concentrates at levels of 3, 6, and 9%. The concentrate was fed to cows for 8 weeks during the early stages of lactation, and it was found that there was no consistent effect on feed intake, body weight change, or efficiency of ration use. Propylene glycol at the 3 and 6% levels increased the yield of milk but caused a slight decrease in milk fat and an increase in milk lactose content. There was also a significant decrease in hyperketosis in the propylene glycol-treated animals compared to control animals. Sauer et al. (175), in a study of 120 cows for a 2-year period in which propylene glycol was added to the diet at 3 and 6% levels for 8 weeks postpartum, substantiated the suitability of propylene glycol as a food additive for such cattle. It depressed the blood ketone and free fatty acids slightly below those of the control cattle level when the cows were not stressed by either high lactation yield or low concentrate food intake. However, if the cows were stressed by adverse environmental factors and slightly reduced food concentrate intake, the addition of propylene glycol at 3 and 6% to the concentrate ration caused a significant reduction of blood ketones and plasma fatty acids and increased the blood glucose concentration. They suggest that propylene glycol used as a feed additive at 3 and 6% of the concentrate should be desirable because of its ability to decrease significantly the incidence of clinical and subclinical ketosis in cows during early lactation, when they are most susceptible to such a metabolic disorder.

Shiga et al. (176) fed male lambs propylene glycol at levels of 0.5 and 1.0 g/kg in feed every other day for 104 days. Neither level produced any adverse effects. They did increase the weight gain, especially in the early part of the experiment (about 30 days). Evaluation of the rumen liquor revealed an increase in propionate and total fatty acid concentrations. The weight percentage of wool, dressed carcass, and red meat-fat ratios were also greater than those of the controls.

Propylene glycol, by short-term feeding tests with chicks and chickens, is a suitable source of energy; the suggested level is being in the range of 2.5 to as much as 8% in their diet. However, most authors feel that the level should be no more than 2 to 3%. Bailey et al. (177) fed chicks from 1 to 26 days of age with diets containing 8 and 16% of propylene glycol. The 8% level did not cause significant ill effects as judged by live weight gain, food consumption, and carcass analysis. The 16% level caused growth depression and lower fat and higher protein content in the carcass. Persons et al. (178) also fed chicks from day 1 for 3 weeks with levels of 5 and 10% of propylene glycol in the diet. Both levels showed adverse effects such as depressed body weight gains at the 5% level and depressed bodyweight gains and decreased food efficiency, but no mortality at the 10% level. They also fed hens for two periods of 28 days each with diets containing propylene glycol at levels of 2.5 and 5%. The 2.5% level decreased food consumption because the hens compensated for the change in energy intake, but there was no effect on egg production or general health. The 5% level caused decreased food and energy intake and reduced egg production. The authors suggest that a level of 2.5% propylene glycol is tolerated well by both chicks and hens. Based upon 21- to 28-day feeding studies of chicks on diets that contained 2.5, 5, and 10% propylene glycol, Waldroup and Bowen (179) found that chicks can utilize 5% propylene glycol in their diet without ill effects. The 10% level caused depressed body weight gains, reduced efficiency of feed utilization, diarrhea, and the development of deformed toes.

Yoshida et al. (139) also found that chicks fed 5% propylene glycol in their diet for 27 days had no ill effects. Higher levels caused inferior well-being and diarrhea.

Harnisch (180) feels that the no-effect level is in the range of 2 to 3% propylene glycol. This resulted from his 8-week feeding studies on 4- to 5-week-old broilers in which he found that a diet containing 5% propylene glycol depressed food intake and reduced feed conversion efficiency.

Because of the use of propylene glycol as a humectant, as well as a source of energy in prepared cat food, a study (44) was conducted in which groups of two male cats each were maintained for 94 days on diets containing various amounts of propylene glycol. Two groups of two male cats each served as controls. The average calculated doses of propylene glycol consumed, based on food intake and body weights, were 0, 80, 443, 675, 1763, or 4239 mg/kg/day. The primary treatment-related effect was noted in the red blood cells (RBC), which exhibited Heinz body formation. This effect in the RBC was accompanied by increased amounts of hemosiderin pigment in the Kupffer cells of the liver and reticuloendothelial cells of the spleen. The formation of Heinz bodies and increased hemosiderin occurred in a dose-related manner at doses of 675 mg/kg/day and higher. A daily dose level of 443 mg/kg/day caused a very slight increase in Heinz body formation without detectable increased hemosiderin in the liver or spleen compared to the incidence in the controls. No treatment-related effects of any type were observed in the group that ingested 80 mg/kg/day. Other hematologic parameters that were evaluated and unaffected by any level of treatment included packed cell volume, RBC count, hemoglobin, RBC morphology to evaluate polychromasia, RBC reticulocyte count, RBC osmotic fragility, methemoglobin, total and differential white blood cell counts, and light microscopic examination of tissue bone marrow. In addition, serum clinical chemistry values of blood urea nitrogen, glutamic pyruvic transaminase activity, alkaline phosphatase activity, glutamic oxaloacetic transaminase activity, glucose concentration, and total bilirubin, as well as routine urinalysis, were unaffected by treatment. The clinical appearance and demeanor, body weights, organ weights, gross pathology, and histopathology of tissues other than liver and spleen were unaffected by treatment at any of these levels of propylene glycol.

The cat is much more sensitive than other species to the formation and/or retention of Heinz bodies. Even so, their presence in significant numbers does not seem to affect cats adversely.

GLYCOLS

Inhalation Toxicity. Robertson et al. (137) exposed sizable groups of rats and monkeys for periods of 12 to 18 months to atmospheres saturated with propylene glycol vapor and produced no ill effects.

6.4.1.3 Pharmacokinetics, Metabolism, and Mechanisms.
Ruddick (162) in his review of the toxicity of propylene glycol, summarized the work done to establish its metabolism in the body. It is oxidized to lactic acid or pyruvic acid by two pathways. These two metabolites are then used by the body as sources of energy either by oxidation through the tricarboxylic acid cycle or by generation of glycogen through the glycolytic pathway. The metabolic pathways are, it is thought, as shown in Figure 85.2. From one-fourth to one-half of an oral dose given to rats, dogs, or human beings appears unchanged in the urine within 24 h (159, 165, 181, 182).

In ruminants, such as sheep and cattle, research has shown that the metabolism of propylene glycol is carried out to a large extent by the microbial flora in the rumen (174, 176, 183, 184). The metabolite is primarily propionate. In the chick, excessive intake of propylene glycol results in its passage to the cecum, where it is metabolized by bacteria to propionaldehyde (185).

Browning (39) states that the studies on propylene glycol indicate that about one-third is excreted via the kidneys as a conjugate with glucuronic acid and the rest is metabolized or excreted in the urine unchanged. This suggests that the organic injury and the CNS depressing action are probably due to the excessive presence of the propylene glycol and not to its metabolites or its glucuronide.

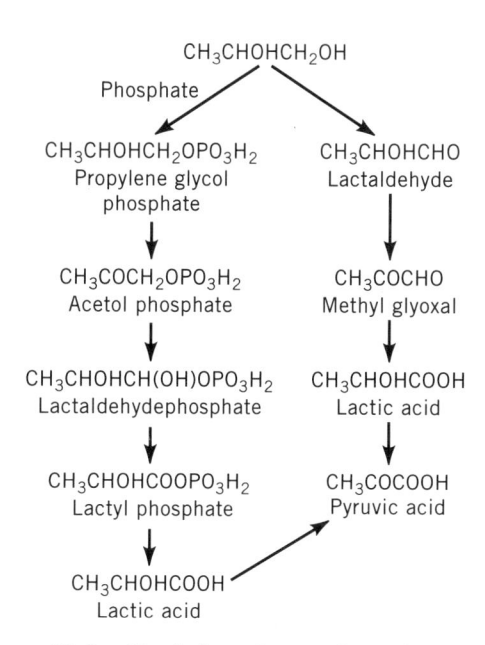

$$
\begin{array}{c}
CH_3CHOHCH_2OH \\
\swarrow \text{Phosphate} \searrow \\
CH_3CHOHCH_2OPO_3H_2 \qquad CH_3CHOHCHO \\
\text{Propylene glycol phosphate} \qquad \text{Lactaldehyde} \\
\downarrow \qquad\qquad \downarrow \\
CH_3COCH_2OPO_3H_2 \qquad CH_3COCHO \\
\text{Acetol phosphate} \qquad \text{Methyl glyoxal} \\
\downarrow \qquad\qquad \downarrow \\
CH_3CHOHCH(OH)OPO_3H_2 \qquad CH_3CHOHCOOH \\
\text{Lactaldehydephosphate} \qquad \text{Lactic acid} \\
\downarrow \qquad\qquad \downarrow \\
CH_3CHOHCOOPO_3H_2 \qquad CH_3COCOOH \\
\text{Lactyl phosphate} \qquad \text{Pyruvic acid} \\
\downarrow \qquad \nearrow \\
CH_3CHOHCOOH \\
\text{Lactic acid}
\end{array}
$$

Figure 85.2. Metabolic pathways of propylene glycol.

6.4.1.4 Reproductive and Developmental.

Reproductive Effects. Guerrant et al. (186) checked the reproductive capacity of rats fed up to 30% propylene glycol in their diets through six generations. No adverse effects on reproduction were found when concentrations of propylene glycol in the diet were less than 7.5%. At higher levels, the rats that received the propylene glycol consumed less food, grew slower, had young at an older age, produced smaller litters, on average, and weaned fewer young than the control animals. At the 30% level, the females did not breed normally and when they had young, they did not feed them properly. They failed to wean third-generation young. Emmens (187) fed mice 0.1 mL of a 50% water solution of propylene glycol for several days before mating and found that it reduced the mating to as little as 30% of normal and litters to 15%. The mice swelled visibly with intestinal gases and then recovered.

Teratogenicity. Gebhardt (188) found that 0.05 mL of propylene glycol was not teratogenic when injected into the yolk sac of chick embryos. However, it caused a high mortality of the embryos when injected into the air sac on the fourth day of development and unilateral micromelia in about 20% of the survivors. Waldroup and Bowen (179) reported that chicks fed high levels of propylene glycol in their diets developed a high incidence of toe deformities, 57 out of 168 chicks compared to the control group, 9 out of 168.

6.4.1.5 Carcinogenesis.

Dewhurst et al. (189) and Baldwin et al. (190) in studies on the carcinogenicity of other chemicals used propylene glycol as the solvent. As a result they tested propylene glycol alone for carcinogenic activity in rats and mice. Dewhurst et al. (189) used a single injection of 0.2 mL, whereas Baldwin et al. (190) gave rats and mice three to five subcutaneous injections, amount not specified. In neither case were tumors observed during a period of about a year (191) or 2 years (189).

Wallenious and Lecholm (191) applied propylene glycol to the skin of rats three times a week for 14 months but found no tumor formation. Stenback and Shubik (192) confirmed these findings when they applied propylene glycol at undiluted strength and as a 50 and 10% solution in acetone to the skin of mice during their lifetimes.

No tumors have been reported in the lifetime dietary feeding studies (38, 172, 173). In fact, Gaunt et al. (172) specifically state that no tumors were found in the rats. Thus, propylene glycol is without carcinogenic properties.

6.4.1.6 Genetic and Related Cellular Effects Studies.

Kennedy et al. (193) studied dominant lethal effects in mice injected intraperitoneally with propylene glycol at a level of 10 mg/kg. The results suggest that it is nonmutagenic at this level. Pfeiffer and Dunkelberg (90) found that propylene glycol is inactive when tested against *S. typhimurium* strains TA98, TA100, TA1535, and TA1537.

6.4.2 Human Experience

6.4.2.1 General Information.

There has been no reported injury to humans from industrial use. However, when used in large doses as a vehicle for repeated medication of a 15-month-old youngster, it caused adverse signs characterized by hypoglycemia and CNS depression. Recovery was prompt upon cessation of treatment (144). This observation has no significance as far as the hazard from industrial exposure is concerned.

GLYCOLS   33

6.4.2.2 Clinical Cases

6.4.2.2.1 Acute Toxicity. Eye Irritation. Propylene glycol has not caused any eye injury in human beings, nor would such be expected, but it may cause transitory stinging, blepharospasm, and lacrimation (111, 195).

Dermal Effects. Based on the results of extensive studies by Warshaw and Herrmann (196) on some 866 human subjects who had various dermatologic backgrounds, propylene glycol may cause primary skin irritation in some people, possibly due to dehydration, but the material is not a sensitizer. Because of the very low systemic toxicity of propylene glycol, no problem from percutaneous absorption can be anticipated. Propylene glycol has been used widely in preparations for topical application, and no evidence of systemic injury to humans has been reported. These findings are substantiated by other workers (113, 170, 181–183, 197–199).

Inhalation Studies. Human beings also have been exposed to saturated and supersaturated atmospheres for prolonged periods in an air-sterilization program without adverse effect (134).

6.4.2.2.3 Pharmacokinetics, Metabolism, and Mechanisms. The uptake of propylene glycol mist by humans was studied using a 10% solution in labeled deionized water nebulized into a mist tent (199). Less than 5% of the mist entered the body, and of this 90% lodged in the nasopharynx and rapidly disappeared into the stomach. Very little was found in the lungs.

6.5 Standards, Regulations, or Guidelines of Exposure: NA

7.0 1,3-Propanediol

7.0.1 CAS Number: [504-63-2]

7.0.2 Synonyms: Trimethylene glycol; 1,3-dihydroxypropane; propane-1,3-diol; 1,3-propandiol

7.0.4 Molecular Weight: 76.1

7.0.5 Molecular Formula: $C_3H_8O_2$

7.0.6 Molecular Structure: HO⁀⁀OH

7.1 Chemical and Physical Properties

1,3-Propanediol, an isomer of propylene glycol, is a viscous, colorless, odorless, hygroscopic liquid that has a brackish irritating taste (113). Additional properties are given in Table 85.1.

7.2 Production and Use

1,3-Propanediol is prepared as a by-product in the manufacture of glycerin by the saponification of fat (111). It is used to lower the freezing point of water and as a chemical intermediate. Industrial exposure is limited.

7.3 Exposure Assessment

Methods applicable to the other glycols should be useful.

7.4 Toxic Effects

Van Winkle (194) found that 1,3-propanediol is about twice as toxic as the 1,2 isomer, is nonglycogenic, and causes marked CNS depression in nearly fatal doses.

7.4.1 Experimental Studies

7.4.1.1 Acute Toxicity. Oral. When 1,3-propanediol was given in single oral doses, the LD$_{50}$ for rats was estimated at about 14.7 to 15.8 g/kg, and the lethal range was 9.5 to 19.0 mL/kg. For cats the LD$_{50}$ was not determined, but 3 mL/kg was fatal to cats. For reasons unknown, cats are particularly sensitive to single oral doses of the material.

1,3-Propanediol was tested as an energy source for chicks as a replacement for corn. It was added to the diet at a level of 10% and was fed for 16 days. At this level, the chicks experienced extremely poor growth, one-half of that seen in the controls, but no other specific signs (200).

7.4.1.2 Chronic and Subchronic Toxicity. Subchronic Oral. When the material was fed to rats as a part of their diet for 15 weeks, a concentration of 5% was without grossly apparent toxic effects, but no autopsies were performed. Twelve percent in the diet was not well tolerated, as evidenced by poor growth. A daily dose of 5 mL/kg given by intubation also caused poor growth; a daily dose of 10 mL/kg was fatal to all rats within 5 weeks (194).

7.4.1.3 Pharmacokinetics, Metabolism, and Mechanisms. Metabolism. Gessner et al. (201) gave 1,3-propanediol to rabbits and analyzed the urine for it and its metabolites; none was found. They suggested that it might have been oxidized completely to carbon dioxide in the body.

Mode of Action. Ulrich and Mestitzova (202) state that 1,3-propanediol is different in its action from propylene glycol because 1,3-propanediol harms the kidneys and the liver at much lower doses.

7.4.1.4 Reproductive and Developmental. Teratogenicity. Gebhardt (188) found that 1,3-propanediol caused micromelia when injected at a dose of 0.05 mL (0.053 g) into the air sac or yolk sac of fertile eggs on the fourth day or at the beginning of incubation. He suggested that 1,3-propanediol, unlike propylene glycol, may be teratogenic.

7.5 Standards, Regulations, or Guidelines of Exposure: NA

8.0 Dipropylene Glycol

8.0.1 CAS Number: *[25265-71-8]*

8.0.2 Synonyms: Oxybis propanol

8.0.4 Molecular Weight: 134.2

8.0.5 Molecular Formula: $C_6H_{14}O_3$

8.0.6 Molecular Structure: :

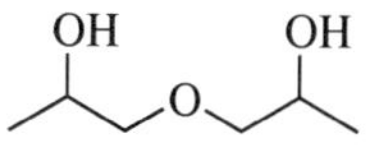

8.1 Chemical and Physical Properties

Dipropylene glycol is a colorless, odorless, slightly viscous liquid. See Table 85.1 for additional properties.

8.2 Production and Use

Dipropylene glycol is prepared commercially as a by-product of propylene glycol production. Three linear isomers are possible, but these have not been separated and studied, and the exact composition of the commercial product is not known. It is also possible to prepare cyclic isomers such as 2,6-dimethyl-1,4-dioxane and 2,5-dimethyl-1,4-dioxane, but these are not likely to form under conditions employed commercially.

Dipropylene glycol is used for many of the same purposes as the other glycols, but mostly in particular applications where its solubility characteristics (greater hydrocarbon solubility) and lower volatility are useful. It is not used in drugs, pharmaceuticals, or food applications because its toxicological characteristics have not been clearly defined. Industrial exposure is most likely to be from direct contact and possible inhalation of mist from heated or violently agitated material.

8.3 Exposure Assessment

Dipropylene glycol undergoes the same general reactions as other glycols. The chemical methods for determination are the same nonspecific methods described for other glycols. Fair specificity can be obtained by spectrographic methods. Chromatographic procedures such as described by Ramstad et al. (138) are adaptable.

8.4 Toxic Effects

Although dipropylene glycol is more active physiologically than propylene glycol, it still has very low toxicity. Industrial handling and use of dipropylene glycol should present no significant problems from ingestion, skin contact, or vapor inhalation. The information available, however, is not considered adequate to allow evaluating whether this material is suitable for use in foods, drugs, or cosmetics.

8.4.1 Experimental Studies

8.4.1.1 Acute Toxicity. *Acute Oral.* Dipropylene glycol has low acute oral toxicity. The oral LD$_{50}$ for rats is reportedly 14.8 g/kg (165) and 15.0 g/kg (203). Yoshida et al. (139) were able to feed chicks a diet containing 5% dipropylene glycol for 27 days without adverse effects. The chicks were unable to use it as an energy source.

Eye Irritation. Dipropylene glycol has not caused significant eye irritation or injury when tested in the eyes of rabbits (139).

Skin Irritation. When dipropylene glycol was applied repeatedly to the skin of rabbits for prolonged periods (10 applications in 12 days), it had a negligible irritating action, and there was no indication that toxic quantities were absorbed through the skin (48).

Inhalation. Experimental data are not available for the inhalation toxicity of dipropylene glycol. However, it is not likely to produce injury because of its low vapor pressure and low systemic toxicity.

8.4.1.2 Chronic and Subchronic Toxicity. *Subchronic Oral.* Rats were not affected by 5% dipropylene glycol in their drinking water for 77 days (141). In those animals administered a level of 10%, some died from hydropic degeneration of kidney tubular epithelium and liver parenchyma. These effects were similar to those of diethylene glycol but less severe and less uniformly produced.

In study by NTP (NTP, 1992), New Zealand white rabbits given 1200 mg/kg/day of dipropylene glycol via oral gavage on gestation days 6 through 19 did not show maternally toxicity even though preliminary study data indicated that this dosage was in the toxic range for this species. No fetal toxicity was noted in any of the offspring from dams exposed to dipropylene glycol (204).

8.4.2 Human Experience

No untoward effects have been reported from the use of dipropylene glycol nor would any be expected.

8.5 Standards, Regulations, or Guidelines of Exposure: NA

9.0 Tripropylene Glycol

9.0.1 CAS Number: [1638-16-0]

9.0.2 Synonyms: (1-Methyl-1,2-ethanediyl) bisoxybispropanol

9.0.3 Trade Names: NA

GLYCOLS

9.0.4 Molecular Weight: 192.3

9.0.5 Molecular Formula: C$_9$H$_{20}$O$_4$

9.0.6 Molecular Structure:

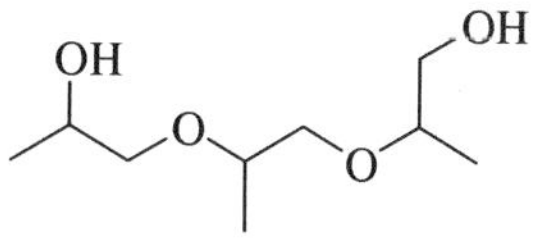

9.1 Chemical and Physical Properties

Tripropylene glycol is a colorless, slightly viscous liquid.

9.2 Production and Use

Tripropylene glycol is made commercially as a by-product of propylene glycol production. It is used much as the other glycols as an intermediate in the production of ethers, esters, and resins; as a nonvolatile solvent; and as a polyester humectant and a plasticizer. It is not used in drugs, pharmaceuticals, or food applications because its toxicological properties have not been clearly defined. Industrial exposure is limited almost entirely to direct contact with the liquid. Vapors or mists may be encountered under unusual conditions.

9.3 Exposure Assessment

Tripropylene glycol undergoes the typical reactions of polyols that do not have vicinal hydroxyl groups. Chromatographic procedures such as described by Ramstad et al. (138) are adaptable.

9.4 Toxic Effects

9.4.1 Experimental Studies

9.4.1.1 Acute Toxicity. Oral. Tripropylene glycol has low acute oral toxicity; the oral LD$_{50}$ for rats is 12.0 g/kg (113). Yoshida et al. (139) found that chicks were unaffected when they were fed a diet containing 5% tripropylene glycol for 27 days.

Eye Irritation. Tripropylene glycol is nonirritating to the eyes or skin or rabbits (44, 113).

Dermal Absorption. It has low toxicity by absorption through the skin; the dermal LD$_{50}$ for rabbits is greater than 16.3 g/kg (113). Essentially continuous contact for 2 weeks did not indicate absorption of toxic amounts through the skin of rabbits (44).

Inhalation. Rats exposed to essentially saturated vapors generated at room temperature for 8 h were unaffected (113); hence it is doubtful that hazardous conditions would occur when it is handled under industrial conditions.

9.4.2 Human Experience

There have been no reports of human experience.

9.5 Standards, Regulations, or Guidelines of Exposure

No exposure standard is believed necessary.

10.0 Polypropylene Glycols

10.0.1 CAS Number: [25322-69-4]

10.0.2 Synonyms: Poly(propanediols); poly[oxy(methyl-1,2-ethanediyl)], alpha-hydro-omega-hydroxy-; propylene oxide, propylene glycol polymer; Poly(propylene glycol) 1000; Poly(propylene glycol) 3000; Poly(propylene glycol) 400; Poly(propylene glycol) 2000; Poly(propylene glycol) 4000

10.0.4 Molecular Weight: 76.095

10.0.5 Molecular Formula: $[C_3H_8O_2]_n$

10.0.6 Molecular Structure:

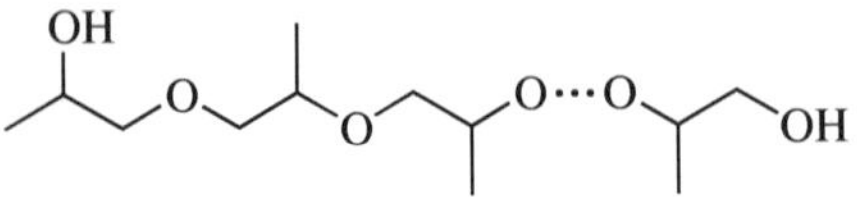

10.1 Chemical and Physical Properties

Polypropylene glycols are clear, lightly colored, slightly oily, viscous liquids that have very low vapor pressures. All of these materials are quite stable chemically and do not present hazards of flammability except at elevated temperatures. Additional physical and chemical properties are given in Table 85.6. Hereafter in this section, polypropylene glycols are referred to as "P," followed by the average molecular weight.

Table 85.6 Physical and Chemical Properties of Polypropylene Glycols

Material Designation[a]	Specific Gravity[b]	Pour Point (°C)	Refractive Index (25°C)	Flash Point (°F)[c]	Fire Point (°F)[c]	Ref.
400	1.007	−49	1.445	330		138
				390	405	205
750	1.004	−44	1.447	495	525	203
1200	1.007	−40	1.448	224[d]		141
				460	505	205
2000	1.002	−30	1.449	390[d]		141
				445	510	205
3000	1.001	−29	1.449	440	505	203
4000	1.005	−26	1.450	365[d]		141
				445	515	205

[a] Average molecular weight.
[b] 25/25°C
[c] Cleveland open cup.
[d] Pensky Mertens closed cup.

10.2 Production and Use

Polypropylene glycols are prepared commercially by reacting propylene glycol or water with propylene oxide. They are used as lubricants, solvents, plasticizers, softening agents, antifoaming agents, mold-release agents, and intermediates in the production of resins, surface-active agents, and a large series of ethers and esters. They are widely used in hydraulic fluid compositions.

Industrial exposure is most likely to be by direct contact with the skin and eyes. Ingestion should not be a problem except from an accident. The very low volatility of these materials makes inhalation improbable except perhaps where mists are formed from violent agitation or high temperatures.

10.3 Exposure Assessment

The method described by Ramstad et al. (138) should be useful.

10.4 Toxic Effects

Low molecular weight polypropylene glycols (200 to 1200) have appreciable acute oral toxicity, are mildly irritating to the eyes, are not irritating to the skin, and although they are absorbed through the skin to some extent, skin penetration does not present a serious industrial hazard. The inhalation of mists or vapors from heated material, particularly low molecular weight material, could be hazardous. These materials are not like the low molecular weight polyethylene glycols in their physiological activity; they are rapidly absorbed from the gastrointestinal tract, are potent CNS stimulants, and readily cause cardiac arrhythmias. The higher molecular weight materials whose average molecular weights are 2000 or more have very low toxicity by all routes and do not have the stimulant effect upon the CNS typical of the lower molecular weight materials.

10.4.1 Experimental Studies

10.4.1.1 Acute Toxicity. *Oral.* The acute oral toxicities of polypropylene glycols are given in later on. The low molecular weight materials (400 to 1200) are rapidly absorbed; excitement and convulsions appear within minutes after administration. No excitement or convulsions were observed with the higher molecular weight material. Necropsy of animals treated with the largest doses 1 to 8 days after exposure revealed nothing remarkable. P2025 was fed to rats for 7 days in a diet at levels of 1.58 and 5g/kg/day. There were no effects seen at the 1.58-g/kg/day level; however, there was definite body weight depression at the 5 g/kg/day level (113).

Skin and Eye Irritation. Tests conducted on rabbits have indicated that these materials are not significantly irritating to the skin, even for prolonged and repeated exposures (205, 206). Direct contact with the eyes may cause slight transient pain and conjunctival irritation but no corneal damage. The response is similar to that caused by a mild soap (113).

Dermal Absorption. Acute skin absorption tests conducted by a "sleeve" technique similar to that developed by Draize et al. (148) indicated that the materials are all poorly absorbed through the skin. When single doses of 30 g/kg were applied for 24 h, four of five animals treated with P400, P750, or P1200 survived, whereas all six animals so treated with polypropylene glycol 2000 survived (205). The dermal LD_{50} values for rabbits are reportedly 20 g/kg for P425 and more than 20 g/kg for P1025, P1950, and P2025 (113).

10.4.1.2 Chronic and Subchronic Toxicity. *Subchronic Oral.* Small groups of male rats were maintained for 100 days on diets containing 0.1 and 1.0% of P750 and 0.1, 0.3, 1, and 3% of P2000. Those animals that received the diet containing 0.1% of P750 were unaffected, as judged by studies of mortality, growth, organ weights, and gross and microscopic examination of the principal internal organs. The animals that received the diet containing 1% of P750, when judged by the same criteria, exhibited only a slight increase in the weights of the livers and kidneys but without histological changes. Hematologic studies of this latter group of animals did not reveal any abnormalities. One percent of P750 in the diet was well accepted by the rats, and it is especially worthy of note that there was no evidence of any of the pharmacological signs (excitement, tremors, convulsions) seen in the acutely poisoned animals. It is postulated that the material is readily metabolized or eliminated when absorbed in small doses; this probably accounts for its lack of physiological effect (207).

The rats that received 0.1, 0.3, and 1% of P2000 suffered no ill effects, as judged by the criteria listed earlier. Hematologic studies were conducted only at the 1% level, and all values fell in the normal range. Although the growth of those animals maintained on the diet containing 3% of P2000 was slightly below normal during most of the test period, there were no other changes attributable to the experimental diets (205).

Polypropylene glycol P1200 was fed to dogs and rats in their diets for 90 days at concentrations of 0.0, 0.1, 0.3, and 1%. Groups of two dogs of each sex received the medicated diets, and a group of three of each sex received unmedicated diet. Groups of 25 rats of each sex constituted all four of the groups. Male dogs on the 1% diet received from 317 to 380 mg/kg/day of the P1200, whereas the females received from 275 to 501 mg/kg/day. The average daily dose for the male rats on the 1% diet was 526 mg/kg and for the female rats, 810 mg/kg. On the 0.3% diet, the male dogs received daily doses that averaged from 77 to 99 mg/kg/day; the female dogs, 90 to 123 mg/kg; the male rats, 157 mg/kg; and the female rats, 189 mg/kg (205).

This study showed no evidence of adverse histopathological, hematologic, or clinical chemical or other effects from consumption of the chemical, except for body weight gains in dogs and rats at the high level (1%). When the experiment was complete, the high level group of dogs showed net losses in body weight. The high-level rats, compared with the nonmedicated controls, showed less gain in body weight (44, 205).

Subchronic Dermal. P2000 was bandaged onto the shaved abdomens of groups of five rabbits each, 24 h/day, five times a week for 3 months. At a dosage level of 1.0 g/kg, there were no adverse effects, as judged by studies of growth, hematology, weights of organs, and gross and microscopic examination of the lungs, heart, liver, kidney, adrenal, testes, stomach, intestine, and skin taken from the site of the prolonged and repeated exposure. Judged by the same criteria, dosage levels of 5 and 10 g/kg caused slight depression of growth. At the 10-g/kg level, mortality was increased but because the cause was respiratory infection, the significance of the observation is questionable (205).

GLYCOLS

10.4.1.3 Pharmacokinetics, Metabolism, and Mechanisms. Metabolism. In the studies by Yoshida et al. (139) in which various glycols including P2000 and P3000 and propylene glycol itself were evaluated as energy sources for chicks, it was found that propylene glycol was a good energy source. However, neither P2000 nor P3000 was used. This led the authors to suggest that chicks were unable to metabolize the polymers. Extensive investigation of the pharmacological activity of these polypropylene glycols has indicated that they are all CNS stimulants. P400, P750, and P1200 are quite potent in this respect, whereas P2000 has little such activity (205, 208).

10.4.2 Human Experience

Skin Sensitization. Not all of the polypropylene glycols have been tested for human skin sensitization properties. However, P2000 caused neither skin irritation nor signs of skin sensitization when applied both continuously and by repeated application to 300 human volunteers (205). P425, P1025, and P2025 caused no responses when applied to 50 human volunteers (113). Thus all polypropylene glycols may be considered without skin sensitization properties.

No cases of toxicity have resulted from the manufacturing, handling, and use of polypropylene glycols.

10.5 Standards, Regulations, or Guidelines of Exposure: NA

11.0 Butanediols

11.0.1 CAS Number: [25265-75-2]

11.0.2 Synonyms: Butylene glycols

11.0.4 Molecular Weight: 90.12

11.1 Chemical and Physical Properties

Commercial 1,2-, 1,3-, 1,4-, and 2,3-butanediols are clear, viscous liquids that are miscible with water and alcohol. They all have a molecular weight of 90.12. The most important chemical and physical properties are given in Table 85.7.

11.2 Production and Use

1,2-Butanediol is a relatively new product that is produced commercially by hydrating the corresponding 1,2-butylene oxide in a manner similar to that described for other simple glycols (144).

According to Curme and Johnston (111), 1,3-butanediol is prepared commercially by the catalytic reduction of acetaldol but may be produced by other routes as well. 1,4-Butanediol is produced in Germany by hydrogenating 2-butyne-1,4-diol, but other methods can also be used. 2,3-Butanediol is produced by fermentation, and the distribution of optical isomers depends on the species of bacteria used.

Table 85.7. Physical and Chemical Properties of Butylene Glycols (Butanediols)

Butanediol Isomer	CAS Number[a]	Boiling Point (°C) (760 mmHg)	Flash Point (°F) (o.c.)[b]	Freezing Point (°C)	Refractive Index, 20°C	Vapor Pressure (mmHg) (20°C)	Specific Gravity (20/20°C)	Ref.
1,2	[584-03-2]	193.5–195			1.4369 (25°C)		1.0017	44
1,3	[107-88-0]	207.5	250	< −50	1.4401	0.06	1.0059	111
1,4	[110-63-4]	230	247	16(m.p.)			1.020	111
2,3	[513-85-9]	182	185	19	1.4377	0.17	1.0093	111

[a]General CAS Number is [25265-75-2].
[b]Open cup.

Butanediols are not used extensively commercially, but there is considerable interest in them as intermediates for polyester resins. Because of its low toxicity, the 1,3-isomer, has been proposed for cosmetic and pharmaceutical applications.

Exposure is from direct contact in handling and use.

11.4 Toxic Effects

The toxicity of 1,2- and 1,3-butanediol is very low when administered in both single and repeated doses. In single oral doses, 1,4-butanediol is about 10-fold more toxic than either the 1,2- or the 1,3-isomer, and the 2,3-isomer is intermediate in toxicity. Data on the 1,3- and 1,4-isomers indicate that they are not significantly irritating to the eyes, skin, or mucous membranes, nor are they likely to be absorbed through the skin in hazardous amounts. The undiluted 1,2-isomer is not significantly irritating to the skin but is irritating to the eyes. Dilution to 10% with water eliminates the effect on the eyes.

Butanediols do not present any appreciable handling hazards other than possible eye irritation from contact with the 1,2-isomer.

11.0a 1,2-Butanediol [584-03-2]

11.0.2a Synonyms: 1,2-Butylene glycol; 1,2-dihydroxybutane

11.0.4a Molecular Weight: 90.122

11.0.5a Molecular Formula: $C_4H_{10}O_2$

11.0.6a Molecular Structure:

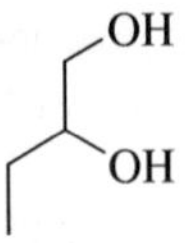

11.4.1a Experimental Studies

11.4.1.1a Acute Toxicity. Acute Oral. Single-dose oral toxicity studies have shown that this material has very low acute oral toxicity for rats; the LD_{50} is about 16 g/kg. In large doses, the material causes CNS depression, irritation of the gastrointestinal tract, profound vasodilation of the visceral vessels, and a marked congestion of the kidneys. No hemorrhage was apparent. Deaths that occurred within a few hours are believed due to CNS depression, and those that were delayed are believed due to kidney injury (144).

Schlussel (209), along with his work on the 1,3 isomer, found that young female rats could tolerate a basic diet in which up to 30% of the calories were replaced by 1,2-butanediol, but that 40% replacement caused death in 11 to 29 days.

Eye Irritation. When applied to the eyes of rabbits, the undiluted liquid was painful, irritating, and injurious, whereas a 10% aqueous solution caused no response (144).

Skin Irritation. The material was not irritating to the skin of rabbits even when exposures were prolonged or repeated (44).

Inhalation. Rats were unaffected by a single 7-h exposure to an atmosphere essentially saturated at 100°C and then cooled to room temperature (44).

11.4.1.3a Pharmacokinetics, Metabolism, and Mechanisms. Gessner et al. (201) were unable to find any glucuronides or other metabolites in the urine of rabbits given 1,2-butanediol orally. Strack et al. (221) gave 1.0 g/kg of 1,2-butanediol intravenously during a 2-min period to rabbits, and found that it was metabolized slowly, as shown by the fact that the blood level fell slowly and that there was only slow excretion via the urine. That which was excreted in the urine was found either as the glucuronide or unchanged. Tissue analysis did not show accumulation of 1,2-butanediol.

11.4.1.6a Genetic and Related Cellular Effects Studies. Mutagenic studies revealed no chromosomal aberrations (224).

11.0b   1,3-Butanediol *[107-88-0]*

11.0.2b   *Synonyms:* 1,3-Dihydorxybutane; 1,3-butylene glycol

11.0.4b   *Molecular Weight:* 90.122

11.0.5b   *Molecular Formula:* $C_4H_{10}O_2$

11.0.6b   *Molecular Structure:*   HO$\frown$OH

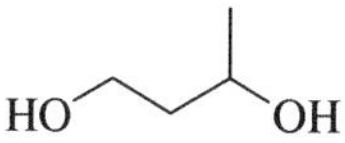

11.4.1.1b Acute Toxicity. *Acute Oral.* This material has very low oral toxicity. When given in single oral doses, Loeser (210), Fischer et al. (211), and Bornmann (41) all report that the LD_{50} is 23.44 g/kg for mice and 29.59 g/kg for rats. Smyth et al. (212) report that the oral LD_{50} for rats is 22.8 g/kg. Smyth et al. (40) found that the oral LD_{50} for guinea pigs is 11.0 g/kg.

Harnisch (180) and Yoshida et al. (140) both found that young chicks grew well and were unaffected when fed a diet containing 5% of 1,3-butanediol for 10 (171) or 27 days (37).

Skin Irritation. Smyth et al. (212) report that the material is not irritating to the rabbit skin.

Inhalation. Smyth et al. (212) exposed rats to saturated vapors of 1,3-butanediol for 8 h without any adverse effects.

11.4.1.2b Chronic and Subchronic Toxicity.

Subchronic Oral. 1,3-Butanediol also has very low toxicity when given in repeated oral doses. Loeser (210) reports feeding an accumulated dose equal to two to three times the median lethal dose to rats during a 6-week period without organic damage or growth depression. Fischer et al. (211) report that 20% in the drinking water of rats for 44 days was without any effect when judged by studies of growth, hematology, liver, kidney, and bladder. However, Bornmann (41) states that 20% in drinking water caused slight depression of growth, but no effect at 10% or less. Kopf et al. (213) fed rats orally 0.5 or 1.0 mL of 1,3-butanediol twice a week for 45 to 185 days without any effect, and dogs were fed 2.0 mL of a 50% aqueous solution twice a week for 5 to 6 months without effect. Schlussel (209) reports that young rats tolerated a basic diet in which 1,3-butanediol accounted for up to 40% of the total calories. Smyth et al. (212) fed groups of 10 rats for 90 days on diets containing sufficient 1,3-butanediol to cause a daily ingestion of up to 5.6 g/kg without any adverse effect as judged by growth, mortality, food consumption, liver and kidney weight changes, and histopathological examination of the liver, kidney, spleen, and testes. Chagovets et al. (214) fed 1,3-butanediol at a 10% level for 4 weeks without adverse effects. Harris et al. (215) found also that rats fed 47% of the calories in their diet for 62 days as 1,3-butanediol experienced no visible adverse effects. However, they found that the levels of glutamate, lactate, pyruvate, and glucose in the brain were less than those found in the controls.

Miller and Dymsza (216) fed rats a diet for 30 weeks containing 30% fat, to which 20 to 30% 1,3-butanediol was added. The 20% level was without significant adverse effects, but the 30% caused an impairment in the use of the diet. The addition of 1,3-butanediol caused an increase in liver phosphohexase and no significant change in serum glucose but an increase in liver glycogen. These findings led the authors to suggest that 1,3-butanediol may not be oxidized via the β-hydroxybutyric acid route.

Yoshida et al. (217) fed 8-week-old calves a diet containing 5.88% 1,3-butanediol for 6 weeks as a substitute for the usual energy source and found that the calves were unaffected as judged by growth and digestibility of the diet. However, Harnisch (180) found that the 5% level in chicks resulted in depressed food intake when fed for an 8-week period. He suggests that the suitable level of 1,3-butanediol for chicks is 2 to 3%. At levels higher than 5%, diarrhea occurred.

Chronic Oral. Scala and Paynter (218) fed rats 1,3-butanediol in their diets at levels of 1, 3, and 10%, and dogs at 0.5, 1, and 3% for 2 years. They found that the no-effect levels were 10% for the rats and 3% for the dogs. The criteria evaluated were body weight, food consumption, pharmacological effects, blood studies, urinalysis, and gross and histopathological evaluation of 16 tissues in the rat and 18 in the dog.

11.4.1.3b Pharmacokinetics, Metabolism, and Mechanisms.

Mehlman and co-workers (222) studied the metabolism of 1,3-butanediol extensively. They conclude that 1,3-

butanediol is metabolized by the liver in the following steps (223):

$$CH_3CHOHCH_2CH_2OH \text{ 1,3-Butanediol} <290> \text{ Liver alcohol dehydrogenase}$$

$$CH_3CHOHCH_2CHO <290> \text{ Aldol } CH_3CHOHCH_2COOH \text{ } \beta\text{-Hydroxybutyric acid}$$

β-Hydroxybutyric acid is further metabolized in the tricarboxylic acid cycle to carbon dioxide, which accounts for about 90% of the dose administered. In other studies (222) in which rats were fed 1,3-butanediol for 3 to 7 weeks, it was found that the blood level of β-hydroxybutyrate increased significantly. The blood level of acetoacetate, a possible metabolite of β-hydroxybutyrate, was also higher than normal.

Mode of Action. Many studies have shown that animals fed 1,3-butanediol develop a significant decrease in adipose tissue (224). This suggests that either 1,3-butanediol or its metabolites interfere with lipogenesis. This reduction of adipose tissue decreases the resistance of the animals to the stress of cold temperatures.

1,3-Butanediol, like other glycols, has CNS depressant effects at high doses (224). Ayres and Isgrig (225) concluded from their work that 1,3-butanediol at high doses depresses motor activity. They believe that this effect may be due to CNS depression and/or muscle relaxation.

11.4.1.4b Reproductive and Developmental.

Teratogenicity. Dymsza (224) reports that teratogenic studies of 1,3-butanediol in rats and rabbits showed no teratologic effects. Gebhardt (118) also found no teratogenic effects in chick embryo tests.

Reproduction. Dymsza (224) reported that rats fed a semipurified diet containing 24% of 1,3-butanediol for three generations were not adversely affected in their ability to reproduce. Later studies showed that there was no effect on the reproduction of rats fed a diet containing 24% of 1,3-butanediol for four generations, nor in dogs fed 20% for one generation.

11.4.1.5b Carcinogenesis.

There were no tumors found in the 2-year feeding studies on dogs and rats (218). Thus it appears that 1,3-butanediol is not carcinogenic.

11.4.2b Human Experience

Human Oral. Tobin et al. (229) found that human volunteers tolerated a diet containing up to 10% of 1,3-butanediol for 5 to 7 days with no adverse effects as judged by blood tests; however, there was a slight decrease in blood glucose levels.

Human Eye Irritation. Grant (48) reports that 1,3-butanediol can cause severe stinging of the eye in humans which is cleared rapidly by irrigation. There was no irritation or injury.

Human Skin Irritation. Loeser (210), Husing et al. (230), and Fischer et al. (211), have reported that the 1,3-isomer is not irritating to human skin or mucous membranes. This is substantiated by Shelanski (231), who used 200 humans ages 6 to 65. These latter studies also indicate that 1,3-butanediol is not a skin sensitizer.

11.0.c 1,4-Butanediol *[110-63-4]*

11.0.2c Synonyms: 1,4-Butylene glycol; 1,4-dihydroxybutane; 1,4-tertramethylene; tetramethylene glycol; butanediol; butane-1,4diol; diol 14b; sucol b; 1,4-tetramethylene glycol; butylene glycol; tetramethylene-1,4-diol

11.0.4c Molecular Weight: 90.122

11.0.5c Molecular Formula: $C_4H_{10}O_2$

11.0.6c Molecular Structure: HO~~~~OH

11.4.1.1c Acute Toxicity. *Acute Oral.* This isomer is about ten times more toxic when administered to animals than the 1,2- or the 1,3- isomer. The oral LD_{50} is 2.18 g/kg for mice (211) and 1.78 g/kg for rats (44). Kryshova (219) gives the following LD_{50} values: mice 2.06; rat 1.52; guinea pig 1.20; and rabbit 2.53 g/kg.

Eye Irritation. Application to the eyes of rabbits showed that the material is slightly irritating; it caused a very slight conjunctival irritation but no corneal injury (48).

Skin Irritation. Repeated application to the rabbits' skin, both intact and abraded, resulted in no appreciable irritation and no evidence of absorption of acutely toxic amounts (144). Judging from these observations, there is no appreciable hazard from skin contact associated with ordinary industrial handling. Schneider (220), however, reports finding that the material highly toxic on the skin. Perhaps this apparent discrepancy can be attributed to the quality of the test material (210).

11.4.1.2c Chronic and Subchronic Toxicity. *Subchronic Oral.* Kryshova (219) fed guinea pigs and rats for 6 months on diets containing 1,4-butanediol at levels of 25 and 30 mg/kg/day. The no-effect level was 25 mg/kg/day. At the 30-mg/kg/day level there were intitial effects, such as blood cholinesterase depression, changes in the protein fraction of the serum, and decrease in the sulfhydryl groups of the blood. There were no changes in the weight, behavior, or general condition of the rats.

11.4.1.3c Pharmacokinetics, Metabolism, and Mechanisms. Gessner et al. (201) report that when butane-1,4-diol was fed to rabbits, most of it was degraded; small amounts of succinic acid were found in the urine. Sprince et al. (226) found that its action was similar to that of γ-hydroxybutyrate and γ-butyrolactone. Maxwell and Roth (227) found that slices of rat brain, liver, kidney, and heart could metabolize 1,4-butanediol to γ-hydroxybutyrate, and the liver was the most active. They postulated that the CNS depressing action of 1,4-butanediol is due primarily to the presence of γ-hydroxybutyrate.

Hinrichs et al. (228) found that the material caused deep CNS depression, constriction of the pupils, total loss of reflexes, and kidney injury; they attributed death to paralysis of the vital centers. This has been confirmed (144). Ulrich and Mestitzova (202) also report that animals experienced accelerated breathing, loss of muscle tonus, drowsiness, and areflexia. In humans given 15 g rectally, miosis, unconsciousness, and coma were seen.

11.4.1.4c Reproductive and Developmental. Gebhardt (188) reports that 1,4-butanediol caused no signs of teratogenic action when injected into the yolk sac of fertile chicken eggs.

11.0d 2,3-Butanediol *[513-85-9]*

11.4.1.1d Acute Toxicity. Fischer et al. (211) report that the oral LD_{50} for mice to be 9.0 mL/kg. Gessner et al. (201), in their study of the metabolism of glycols, state that they found levels of glucuronides of 2,3-butanediol in the urine of rabbits equivalent to about 20% of the dose given. Gebhardt (188) found no evidence of teratogenic action in chick embryo tests.

11.5 Standards, Regulations, or Guidelines of Exposure: NA

12.0 Polybutylene Glycols

12.0.1 CAS Number: [25190-06-1]

12.0.2 Synonyms: alpha-Hydro-omega-hydroxypoly (oxy-1,4-butanediyl)

12.1 Chemical and Physical Properties

Polybutylene glycols are clear, viscous, oily, slightly yellow liquids that have sweetish tastes. They are less than 1.0% soluble in water and more than 25% soluble in methanol and ether. They are quite stable chemically and do not present hazards of flammability except at elevated temperatures (132).

12.2 Production and Use

The polybutylene glycols considered here are prepared commercially by reacting 1,2-butylene glycol or water with 1,2-butylene oxide. Their numerical designations indicate their average molecular weights. Other polybutylene glycols can, of course, be made using other butylene oxides (32).

Uses for these materials are being developed. Industrial exposure is most likely to be from direct contact with the skin and eyes.

12.4 Toxic Effects

Limited toxicological information is available on the two polybutylene glycols known as polyglycol B-1000 and polyglycol B-2000 that have average molecular weights of 1000 and 2000, respectively. Both have low single-dose oral toxicity; their oral LD_{50} values for rats are higher than 4.0 g/kg. In such oral doses, however, they do produce marked injury to the kidneys. They are slightly irritating but not damaging to the eyes of rabbits. Prolonged and repeated contact with the skin of rabbits failed to cause any significant topical effect, and there was no evidence of absorption of toxic amounts through the skin. These materials do not present any appreciable hazards in industrial handling and use. However, until more data become available, care should be taken to avoid ingestion, particularly repeated ingestion (144).

12.5 Standards, Regulations, or Guidelines of Exposure: NA

13.0 1,5-Pentanediol

13.0.1 CAS Number: [111-29-5]

13.0.2 Synonyms: 1,5-Pentylene glycol; pentamethylene glycol

13.0.3 Trade Names: NA

13.0.4 Molecular Weight: 104.2

13.0.5 Molecular Formula: $C_5H_{12}O_2$

13.0.6 Molecular Structure: HO~~~~~~OH

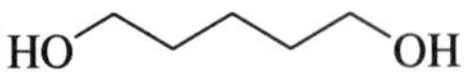

13.1 Chemical and Physical Properties

1,5-Pentanediol is a clear liquid. Following are some of its chemical and physical properties:

Melting point: −16°C
Boiling point: 242°C
Flash point: 129°C
Specific Gravity: 0.994

13.2 Production and Use

1,5-Pentanediol is used largely as a chemical intermediate. Industrial exposure is likely to be from direct contact.

13.4 Toxic Effects

1,5-Pentanediol has low acute oral toxicity; the oral LD_{50} value for rats is 5.89 g/kg. It is essentially nonirritating to the skin and only very mildly irritating to the eyes. It is not likely to be absorbed through the skin in toxic amounts because the dermal LD_{50} value for rabbits is higher than 20 mL/kg. Rats exposed to essentially saturated vapors generated at room temperature for 8 h survived. Exercising reasonable care and caution in handling 1,5-pentanediol should be adequate to avoid serious toxic effects (232).

13.5 Standards, Regulations, or Guidelines of Exposure: NA

14.0 2-Methyl-2,4-pentanediol

14.0.1 CAS Number: [107-41-5]

14.0.2 Synonyms: 2-Methypentane-2,4-diol; hexylene glycol

14.0.3 Trade Names: Diolane; Isol; Pinakon

14.0.4 Molecular Weight: 118.18

GLYCOLS 49

14.0.5 Molecular Formula: C₆H₁₄O₂

14.0.6 Molecular Structure:

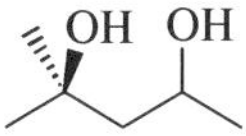

14.1 Chemical and Physical Properties

2-Methyl-2,4-pentanediol is a mild-odored liquid. Following are some of its chemical and physical properties 111, 233:

Melting point: −40°C
Boiling point: 198°C
Flash point: 101°C
Specific Gravity: 0.925
It is soluble in water.

14.2 Production and Use

2-Methyl-2,4-pentanediol is prepared commercially by the catalytic hydrogenation of diacetone alcohol (111). It is used as a chemical intermediate, a selective solvent in petroleum refining, a component of hydraulic fluids, a solvent for inks, and as an additive for cement (111).

Industrial exposure is likely to be from direct contact or from inhalation, particularly if the material is heated.

14.3 Exposure Assessment

Determination can be accomplished by the usual methods applicable to polyols. Infrared spectrophotometry and gas chromatography may be applicable if specificity is necessary.

14.4 Toxic Effects

2-Methyl-2,4-pentanediol has low acute oral toxicity; it is appreciably injurious to the eyes, slightly irritating to the skin, not readily absorbed through the skin, and has a sufficiently low vapor pressure at ordinary temperatures so as not to present an appreciable hazard from inhalation. Atmospheres essentially saturated at room temperature (about 66 ppm) are detectable by odor and may be slightly irritating to the eyes. Atmospheric contamination resulting from handling at elevated temperature causes marked irritation of the eyes and hence warning of excessive concentrations. Pharmacologically, the material is a hypnotic. It is slowly excreted, largely as the glucuronic acid conjugate.

14.4.1 Experimental Studies

14.4.1.1 Acute Toxicity. Oral. When the material is given orally, the LD_{50} for mice is 3.5 g/kg and for rats approximately 4.79 g/kg. Spector, according to Larsen (234) reported that the oral LD_{50} value for guinea pigs and rabbits is from 2.6 to 3.0 g/kg, and for mice and rats, 3.7 to 4.2 g/kg. Hypnosis occurred in mice following single doses of 1.85 g/kg;

the effect of larger doses was profound. The material caused irritation of the lungs and large intestine, but no gross effects were apparent in the kidney, brain, or heart (233, 235).

Eye Irritation. When introduced into the eyes of rabbits, the undiluted material caused appreciable irritation and corneal injury that was slow to heal (113, 233,235).

Skin Irritation. When the undiluted material was applied to uncovered rabbit skin, it caused minor skin irritation (113).

Dermal Absorption. A single 24-h application of 1.84 g/kg to rabbits caused transitory mild edema and erythema, but no deaths (233). The dermal LD$_{50}$ in rabbits was 12.3 g/kg (233, 235) and 7.90 g/kg (113). The inunction of 0.92 g/kg/day for 90 days to the skin of rabbits caused no effects; the application of 1.85 g/kg/day caused reversible cloudy swelling of the liver.

Inhalation. Rats exposed for 8 h to saturated vapor survived (113,233,235). The calculated vapor concentration was approximately 160 ppm (113). Exposure of rats to an atmosphere generated by heating the material to 170°C for 8 h also caused no deaths (113). Rats and rabbits that were exposed to an aerosol at a level of 0.7 mg/L for 7 h/day for 9 days survived and showed no adverse effects pathologically (113).

14.4.1.2 Chronic and Subchronic Toxicity. According to Larsen (234), Brown et al. administered hexylene glycol to rats for 8 months at a level of 590 mg/kg/day with no detectable adverse effects. Larsen (234) fed mice orally 20 mg/day in 2 mL of whole milk for up to 81 days and found only minor effects in the kidneys of a few animals. Rats were also fed the material in milk for 4 months at the average rate of 98 and 150 mg/day. The acceptance of the milk containing the hexylene glycol decreased in proportion to the hexylene glycol content. None of the rats showed adverse effects in growth nor were there histopathological changes in the liver and testes, but there were minor changes in the kidneys. It was also shown that approximately 40% of the hexylene glycol was accounted for in the urine, but only 4% of the amount excreted was free glycol; the other 36% was conjugated with glucuronic acid. The fertility of male rats given an average oral dose of 148 to 190 mg/day of hexylene glycol for 130 days was unchanged when compared to that of the control group.

14.4.1.3 Pharmacokinetics, Metabolism, and Mechanisms. Deichmann and Dierker (236) found that the oral administration of hexylene glycol to rats and rabbits resulted in a substantial increase in the amount of hexuronates in the plasma and in the urine. This was substantiated by Larsen (234).

14.4.2 Human Experience

Jacobsen (237), in studies on five human subjects, found both free and conjugated hexylene glycol in the urine after single or repeated oral doses. When the daily dose was 600 mg or less, none was detected in the urine. At daily doses up to 5 g/day, substantial amounts of the free hexylene glycol and the conjugated form were found in the urine. Excretion was slow and persisted for up to 10 days after dosing stopped.

Most human beings exposed 15 min. to 50 ppm in the air were able to detect the odor and a few noted eye irritation. At a concentration of 100 ppm, the odor was plain, and some noted nasal irritation and respiratory discomfort; at 1000 ppm (4840 mg/m^3), there was irritation of the eyes, nose, and throat and respiratory discomfort (233).

GLYCOLS 51

14.5 Standards, Regulations, or Guidelines of Exposure

In the absence of adequate data upon which to establish a hygienic standard, maintaining atmospheric concentrations below those that cause discomfort (233).

15.0 2,2,4-Trimethyl-1,3-pentanediol

15.0.1 CAS Number: [144-19-4]

15.0.2–15.0.3 Synonyms and Trade Names: 2,2,4-Trimethylpentane-1,3-diol; TMPD glycol

15.0.4 Molecular Weight: 146.229

15.0.5 Molecular Formula: $C_8H_{18}O_2$

15.0.6 Molecular Structure:

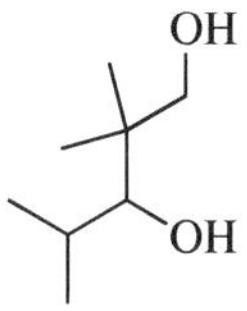

15.1 Chemical and Physical Properties

2,2,4-Trimethyl-1,3-pentanediol is a somewhat volatile liquid that has the following properties: melting point, 49°C; boiling point, 228.9°C; flash point, 125°C, and density, 0.94.

15.2 Production and Use

2,2,4-Trimethyl-1,3-pentanediol is made by reacting isobutyryl chloride with ethyl isobutyrate. Its main uses are as a component of polyester resins used in food packaging and as an insect repellent. Industrial exposure is likely to be by direct contact with the liquid. Vapors or mists may be encountered under some conditions.

15.3 Exposure Assessment

This material can be determined by the usual nonspecific methods used for polyols. Infrared spectrophotometry and gas chromatography may be applicable if specificity is desired.

15.4 Toxic Effects

2,2,4-Trimethyl-1,3-pentanediol has low acute and chronic oral toxicity, is essentially nonirritating to the eye and only slightly to moderately irritating to the skin, is low in toxicity by skin absorption, and is readily metabolized. It should pose no significant health hazard in industrial handling. Because of lack of information on inhalation, reasonable care and caution to avoid prolonged or repeated exposure to its vapors should be taken.

15.4.1 Experimental Studies

15.4.1.1 Acute Toxicity. Terhaar et al. (238) report that the oral LD_{50} for rats, mice, and guinea pigs is nearly 20 g/kg.

Eye Irritation. A mixture of 25% solution of 2,2,4-trimethyl-1,3-pentanediol 8% glycerin and 67% ethanol caused moderate but transient irritation to the eyes of rabbits; the response was similar to that caused by the solvent mixture alone (238).

Skin Irritation. Repeated applications to the skin of rabbits and guinea pigs caused only slight to moderate reddening of the skin. No evidence was mentioned that this material was absorbed through the skin in acutely toxic amounts (238).

15.4.1.2 Chronic and Subchronic Toxicity. Rats of both sexes were fed for 60 days on diets containing 0.5, 1, or 2% 2,2,4-trimethyl-1,3-pentanediol (238). The females at the 2% level ate less and therefore experienced growth depression. This was accompanied by a slight increase in the average weights of the liver, adrenal, kidney, heart, and brain. The males at this level also increased liver, adrenal, and kidney weights. Those given the 0.5% diet showed no adverse effects by the usual criteria. The group fed the diet containing 1% of the test material were used for the teratology study discussed later.

15.4.1.3 Pharmacokinetics, Metabolism, and Mechanisms. Astill and Fassett (239) found that when given orally to rats, 94 to 99% of the 2,2,4-trimethyl-1,3-pentanediol was eliminated in the urine in 3 to 4 days, 2% in the feces, and less than 0.1% as carbon dioxide in expired air. The urinary metabolites were 0.8 to 1.7% as unchanged 2,2,4-trimethyl-1,3-pentanediol, 72 to 73% as its *O*-glucuronide, 2.9 to 3% as 2,2,4-trimethyl-3-hydroxy-valeric acid, and 4.3 to 4.4% as the valeric acid glucuronide.

15.4.1.3.1 Absorption. Studies using rabbits and guinea pigs were undertaken to establish the fate of this material (239). In rabbits given a single dose of 2.0 g/kg on the skin, there was increased urinary glucuronide output accounting for about 30% of the dose given. At a dose of 0.18 g/kg, most of the material was absorbed and eliminated via the urine (75–82% of dose), less than 0.5% as carbon dioxide, and 2% in the feces. Volatilization at the application site accounted for about 14% of the dose given. The guinea pig tests (dose 0.09 mg/kg) showed that the material was absorbed and excreted similarly but to a lesser extent.

15.4.1.4 Reproductive and Developmental. Rats were fed 2,2,4-trimethyl-1,3-pentanediol at a dose of 1% in their diet through three generations. During each generation, the rats were bred twice. The only consistent finding was a decreased pup weight. In the six breedings, pup mortality was greater than in the controls three times, similar twice, and less once. Thus this material at 1% in the diet has some adverse effect on the reproductive capability of rats. More studies are needed to resolve the qualitative and quantitative aspects. The data available suggest that a hazard from likely industrial exposure would be very small (238).

15.4.2 Human Experience

Human tests using 3.7 to 4.9 g/person indicated that the material was not absorbed through the skin in appreciable amounts. There was no detectable increase in urinary glucuronides. Of the dose applied to the skin, 82 to 93% was recovered from the application site.

GLYCOLS 53

15.5 Standards, Regulations, or Guidelines of Exposure

No inhalation data are reported.

16.0 2-Ethyl-1,3-hexanediol

16.0.1 CAS Number: [94-96-2]

16.0.2 Synonyms: 2-Ethyl-1,3-hexanediol; ethyl hexylene glycol; ethohexadiol; 2-ethyl-1,3-hexylene glycol; octylene glycol; 6-12 insect repellent; EHD; ethyl hexanediol; Rutgers 6-12; carbide 6-12; eh diol; 2-ethylhexane-diol-1,3; 2-ethyl-3-propyl-1,3-propanediol; 3-hydroxymethyl-*n*-heptan-4-ol; Repellent 6-12; 2-ethylhexane-1,3-diol

16.0.3 Trade Names: NA

16.0.4 Molecular Weight: 146.2

16.0.5 Molecular Formula: $C_8H_{18}O_2$

16.0.6 Molecular Structure:

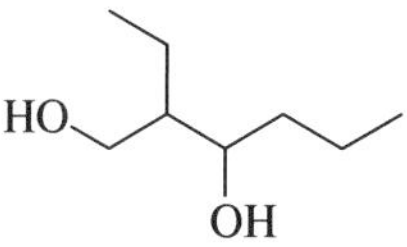

16.1 Chemical and Physical Properties

2-Ethyl-1,3-hexanediol is an oily, colorless, slightly viscous liquid, that is soluble in alcohol and ether but poorly soluble in water (4.2% at 20°C).

16.2 Production and Use

2-Ethyl-1,3-hexanediol is produced commercially by hydrogenating butyraldol (2-ethyl-3-hydroxycaproaldehyde). It is used largely as an insect repellent but also as a solvent for resins and inks, a plasticizer, and a chemical intermediate in the production of polyurethane resins (111).

Industrial exposure is largely by direct contact. Extensive experience with human beings has been acquired through its extensive use as an insect repellent (111).

16.3 Exposure Assessment

Methods applicable to other polyols that have terminal hydroxyl groups should be useful.

16.4 Toxic Effects

2-Ethyl-1,3-hexanediol has low acute and chronic oral toxicity, is not appreciably irritating to the human skin, is somewhat irritating to mucous membranes, is and slowly absorbed through the skin. Once absorbed, it causes CNS depression but little systemic injury. It is considered safe for use as an insect repellent.

16.4.1 Experimental Studies

16.4.1.1 Acute Toxicity. *Oral.* Lehman (240) reports that when 2-ethyl-1,3-hexanediol was fed in single doses to various species, the oral LD_{50} values obtained were as follows: rats, 2.5; mice, 4.2; guinea pigs, 1.9; and chicks, 1.4 g/kg. Smyth et al. (212) report the oral LD_{50} for rats as 2.71 g/kg. In large doses, the material causes deep CNS depression, and this is believed to be the cause of death.

Eye Irritation. Contact with undiluted 2-ethyl-1,3-hexanediol caused moderate irritation and possibly some corneal injury to rabbits (113).

Skin Irritation. 2-Ethyl-1,3-hexanediol is somewhat irritating to the skin of rabbits, but human skin is quite resistant (240). Mucous membranes, however, are quite sensitive to the material (240). This is confirmed by Carpenter and Smyth (45) in their studies with the rabbit eye.

Dermal Absorption. The dermal LD_{50} by skin absorption for rabbits is reportedly higher than 9.42 g/kg. However, when rabbits were inuncted daily for 90 days, 1.88 g/kg caused about 50% mortality, and somewhat greater mortality resulted at the dosage level of 3.76 g/kg. Continued contact caused appreciable irritation to the skin of rabbits, and animals that died exhibited moderate liver and kidney injury. Leukocytosis was observed in only one animal treated at the 3.76 g/kg level.

Skin Sensitization. Guinea pigs were not sensitized by this material (113).

Inhalation. Rats exposed to essentially saturated vapors generated at room temperature for 2 h survived; those exposed for 8 h died. Rats exposed to a mist generated at 170°C for 2 h also survived, whereas those exposed for 8 h died. Those exposed to a fog generated at 70°C with a nebulizer at a concentration estimated at 4800 ppm for 8 h all survived (113).

16.4.1.2 Chronic and Subchronic Toxicity. Smyth and co-workers (212) report that rats fed for 90 days a diet that provided a daily intake of 0.70 g/kg of the glycol did not grow as well as the controls but apparently suffered no systemic injury. When rats were maintained on a diet that supplied a daily intake of 0.48 g/kg, growth was normal and no detectable adverse effects were noted.

Lehman (240) reports that rats were fed for up to 2 years on diets containing 2, 4, and 8% 2-ethyl-1,3-hexanediol. Growth was depressed at all levels. At the 8% level, all animals were dead within 18 weeks; death was due to inanition. Those at the 4 and 2% levels survived, and autopsy revealed no adverse effects attributable to the glycol.

16.4.2 Human Experience

Two hundred twenty-three human volunteers were patch tested for sensitization with the undiluted product. Three of the volunteers developed mild skin reddening; otherwise no significant signs of sensitization occurred (113). When tested as a repellent in stick formulations containing sodium stearate, none showed signs of irritation, sensitization, or fatiguing of the skin (113).

GLYCOLS 55

Lehman (240) concludes, "The fact that the compound is poorly absorbed through the skin of humans and is nonirritating warrants the conclusion that it may safely be used as a component of an insect repellent product."

16.4.3 Pharmacokinetics, Metabolism, and Mechanisms

Gessner et al. (201) report that they could not detect any metabolites from 2-ethyl-1,3-hexanediol when it was fed to rabbits. In other studies, there was no increase in excretion of ascorbic acid by rats. Rabbit urine was thought to have a mercaptan odor, and there was a definite increase in blood urea and blood and urine glucuronates (113).

16.5 Standards, Regulations, or Guidelines of Exposure

Given the limited data available, precautions to avoid prolonged or repeated exposures to high concentrations of vapor, mists, or fog should be taken.

17.0 Styrene Glycol

17.0.1 CAS Number: [93-56-1]

17.0.2 Synonyms: 1-Phenyl-1,2-ethanediol; phenyl glycol

17.0.3 Trade Names: NA

17.0.4 Molecular Weight: 138.2

17.0.5 Molecular Formula: $C_8H_{10}O_2$

17.0.6 Molecular Structure:

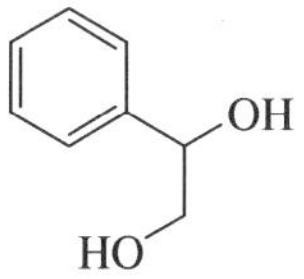

17.1 Chemical and Physical Properties

Styrene glycol is a white, practically odorless solid.

17.2 Production and Use

Styrene glycol is made commercially from styrene oxide by hydrolysis. Styrene glycol is used largely as a chemical intermediate. Industrial exposure is likely to be by direct contact with the solid or solutions of the material. Vapors and mists may be encountered under certain conditions.

17.3 Exposure Assessment

Methods for determining small amounts of styrene glycol have not been developed. It is believed, however, that chemical methods applicable to other glycols could easily be modified.

17.4 Toxic Effects

Unpublished range-finding toxicological studies (44) indicate that styrene glycol has low oral toxicity and is not significantly irritating to the skin. It is not expected to present an appreciable hazard in ordinary industrial handling and use.

17.4.1 Experimental Studies

17.4.1.1 Acute Toxicity. The oral LD_{50} for guinea pigs is between 2.0 and 2.6 g/kg. When given by intubation as a solution in olive oil five times a week for 1 month, dosage levels of 0.5 and 1 g/kg were tolerated by rabbits and rats. A dosage level of 1 g/kg caused minor liver injury in the rabbit.

Prolonged and repeated contact with a 20% solution of styrene glycol in propylene glycol caused no injury to the skin of rabbits, nor did it penetrate the skin in toxic amounts.

17.4.1.3 Pharmacokinetics, Metabolism, and Mechanisms. According to Ohtsuji and Ikeda, and Ikeda and Imamura, as quoted by Milvy and Garro (241), styrene glycol could be metabolized as shown in Figure 85.3.

17.4.1.6 Genetic and Related Cellular Effects Studies. Milvy and Garro (241), using a spot test on *Salmonella* strains developed by Ames and his co-workers for screening potential carcinogens, found that styrene glycol was not mutagenic. In addition, it did not inhibit the growth of bacteria. This was confirmed by Vainio et al. (242).

17.5 Standards, Regulations, or Guidelines of Exposure: NA

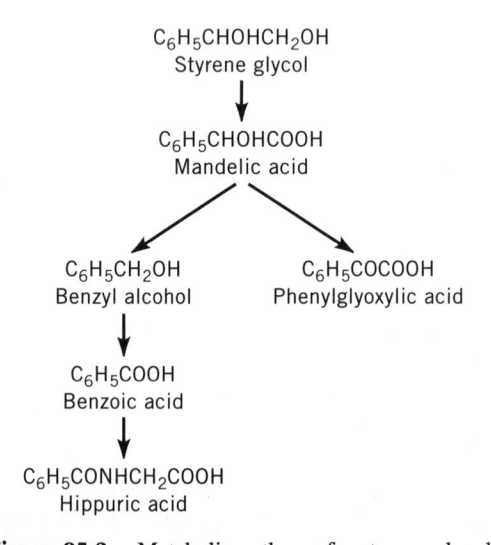

$C_6H_5CHOHCH_2OH$
Styrene glycol
↓
$C_6H_5CHOHCOOH$
Mandelic acid

$C_6H_5CH_2OH$
Benzyl alcohol

$C_6H_5COCOOH$
Phenylglyoxylic acid

↓
C_6H_5COOH
Benzoic acid
↓
$C_6H_5CONHCH_2COOH$
Hippuric acid

Figure 85.3. Metabolic pathway for styrene glycol.

GLYCOLS

18.0 Mixed Polyglycols

Mixed polyglycols have been designed to supply products that have properties which are desirable for special needs. Basically they consist of reaction mixtures of ethylene oxide and propylene oxide, and sometimes butylene oxide with various polyols. They have a wide variety of uses, such as antifoam agents, synthetic oils, industrial lubricants, surfactants, mold-release agents, cosmetic oil substitutes, pharmaceuticals, chemical intermediates, and others.

Industrial exposure is expected to be from direct contact. Because of the great variation in these compounds, they are discussed by product or groups of related products.

18.0a Polyglycol 11 and 15 Series

18.1a Chemical and Physical Properties

Polyglycols are clear viscous liquids that have the properties shown in Table 85.8.

18.2a Production and Use

The polyglycol 11 series is prepared by reacting glycerol with different amounts of propylene oxide. The polyglycol 15 series is prepared by reacting glycerol with various mixtures of propylene oxide and ethylene oxide (44).

18.4a Toxic Effects

None of the mixed polyglycols in the 11 or 15 series described presents any handling hazards of significance. The toxicological information is summarized in Table 85.9.

18.5a Standards, Regulations, or Guidelines of Exposure

The low volatility and low toxicity of these materials makes a hygienic standard unnecessary.

Table 85.8. Physical and Chemical Properties of Polyglycols of the 11 and 15 Series

Commercial Designation, Polyglycol	Average Molecular Weight	Specific Gravity (25/25°C)	Refractive Index (25°C)	Flash Point (°F)	Solubility (g/100 g) (25°C) Water	Solubility (g/100 g) (25°C) Methanol or Ether
11–100	1030	1.026	1.452	435	<0.1	100
11–200	2700	1.018	1.452	435	<0.1	100
11–300	4000	1.017	1.450	440	<0.1	100
11–400	4900	1.017	1.450	445	<0.1	100
15–100	1100	1.070	1.460	510	Misc.	Misc.
15–200	2600	1.063	1.460	470	Misc.	Misc.
15–500	5000	1.051	1.459	475	Misc.	Misc.
15–1000	9000	1.053	1.458	480	Misc.	Misc.

Table 85.9. Summary of Toxicological Information on Certain Mixed Polyglycols

Polyglycol	Rat LD$_{50}$ (g/Kg)	Eye Irritation (Rabbits)[a]	Effect on Skin Irritation (Rabbits)[b]	Sesitization (Humans)[c]	Percutaneous Absorption (Rabbits)[d]
11–100	2.0	None	Slight		None
11–200	>4.0[e]	Trace	None		None
11–300	>4.0[e]	Trace	Slight	None	None
11–400	>4.0[e]	None	None	None	None
15–100	31.6	Trace	Very slight		None
15–200	15–20	Trace	Very slight	None[f]	None[g]
15–500	16	Trace	None		None
15–1000	>4.0[e]	None	None	None	None

[a] One to two drops directly in eye; "trace" = conjunctival irritation but no corneal injury.
[b] Contact 24 h/day for 12 days.
[c] Repeated insult test in 50 human beings.
[d] None apparent from skin irritation test.
[e] No deaths at this, the largest dose fed.
[f] Repeated insult test on 50 human beings plus "Swartz" test on 200 human beings.
[g] LD$_{50}$ by "Draize" test = >30 g/kg.

18.0b Ucon Fluids

Ucon fluids are Union Carbide Corporation's trade name for polyalkylene glycols and diesters (243).

18.1b Chemical and Physical Properties

Fluids 50-HB-260 and 50-HB-5 100 are water-soluble monobutyl ethers of polymers of ethylene oxide and propylene oxide whose approximate mean molecular weights are 940 and 4000, respectively. Fluids 25-H-2005 and 75-H1400 are polymers that have two terminal hydroxyl groups and approximate molecular weights of 4100 and 2200, respectively. The number 25 or 75 in the fluids' names indicates the percentage of units of ethylene oxide in the polymers.

18.4b Toxic Effects

18.4.1.1b Acute Toxicity. These products have low acute oral toxicity. The results of oral toxicity tests are summarized in Table 85.10. The only sign of nonlethal effect was that of depression in varying degrees, depending on the product.

18.4.1.2b Chronic and Subchronic Toxicity. Male and female rats and dogs were fed diets containing these materials for a 2-year-period. In the rat studies, the materials were added to the diet so that each product was consumed at a constant level throughout the experiment. The highest levels in all experiments was 0.5 g product/kg/day. The only

Table 85.10. The Acute Oral Toxicity of Four Ucon Fluids to Various Laboratory Animals

Product	Molecular Weight	LD$_{50}$ Values (mL/kg Body Weight)			
		Rat Male	Rat Female	Mouse Female	Rabbit Male
50-HB-260	940	5.95	4.49	7.46	1.77
50-HB-5100	4000	>64	45.2	49.4	15.8
25-H-2005	4100	14.1	35.9	22.6	35.6
75-H-1400	2200	>64	>16	45.2	35.4

adverse effect seen in the studies on these four products was slight growth depression in the female rats that consumed 0.5 g/kg/day of 25-H2005. The other products caused no adverse effects at this level. Criteria observed were behavior, diet consumption, mortality, life span, incidence of infections, terminal liver and kidney to body weight ratios, body weight gain, hematocrit, total red cell count, incidence of neoplasms, and gross and histopathology in 20 tissues. Dogs were fed diets containing 0.15 and 1.67% of these products. This percentage was kept constant throughout the experiments. For the high dosage, the levels were an average of 0.62, 0.61, 0.62, and 0.50 g/kg/day for 50-HB-260, 50-HB-5100, 25-H-2005, and 75-H-1400, respectively.

The observations made on the dogs were of appetite, body weight change, mortality, terminal liver and kidney to body weight ratios, hematocrit, hemoglobin, red and white cell total counts, differential white cell counts, serum urea nitrogen, serum alkaline phosphatase, 15 min BSP retention, and gross and histopathology of 18 tissues. The only dose-related response seen was granular degeneration of the cytoplasm of the smooth muscle in the intestinal wall in the dogs fed the 1.67% level of 25-H-2005. The significance of this finding is unknown. The other products were without adverse effects at the 1.67% level.

18.4.1.3b Pharmacokinetics, Metabolism, and Mechanisms. Rats were fed these products, labeled primarily on the ethylene oxide moiety with ^{14}C, at a dosage of 67 mg/kg of body weight. The urine, feces, carbon dioxide, and carcasses were analyzed for ^{14}C during a 7-day period. The total ^{14}C recovery from all areas ranged from 80 to 95% of the dose given.

These data show that only 50-HB-260 was metabolized to ^{14}CO$_2$ in significant amounts. Fluid 50-HB-260 was absorbed from the gut in large amounts, fluid 25-H-2005 in a lesser amount, and fluids 50-HB-5100 and 75-H-1400 in insignificant amounts. The identities of the ^{14}C compounds found in the urine and feces were not determined.

18.4.1.5b Carcinogenesis. In addition to the negative carcinogenic results seen in the dietary feeding in rats, none of these products produced papillomas or carcinomas in mice that were painted on the clipped skin of the back three times a week until death.

18.5b Standards, Regulations, or Guidelines of Exposure: NA

18.0c Pluronic Polyols

Pluronic polyols are BASF Wyandotte's trade name for polyoxyalkylene derivatives of propylene glycol (244).

18.1c Chemical and Physical Properties

These materials are relatively nonhygroscopic, vary from water-insoluble to water-soluble products, and are soluble in common aromatic hydrocarbons, chlorinated hydrocarbons, acetone, and low molecular weight alcohols. They are good solvents for a number of diversified materials such as iodine and perfume oils. Usually those whose molecular weights are up to 3400 to 3800 are liquids, those in the range of 4000 to 6500 are pastes, and those whose molecular weights of are higher than 6600 are solids. Some of the physical properties of these products are given in Table 85.11.

18.2c Production and Use

These products are block polymers made by the condensation of propylene oxide onto a propylene glycol nucleus followed by the addition of ethylene oxide to the ends of the base at levels ranging from 10 to 80% of the final molecule. The products have molecular weights ranging from 1000 to more than 15,000.

18.4c Toxic Effects

Pluronic polyols can be considered low in health hazard in industrial operations. Generally, low molecular weight polyols are possibly very slightly more toxic than those of a higher molecular weight. In the feeding studies, the lower molecular weight products are less tolerated than those of higher molecular weight. They are not eye irritants and those tested for skin contact properties were neither irritants nor sensitizers. They are not readily absorbed through the skin.

Table 85.11. Some Properties of Pluronic Polycols

Pluronic Designation	Approximate Molecular Weight	Taste and Odor	Chemical Stability to Acids	Physical Form
L44	2,200	Mild bitter taste	Good	Liquid
L61	2,000	Mild bitter taste	Good	Liquid
L62	2,500	Mild bitter taste	Good	Liquid
L64	2,900	Mild bitter taste	Good	Liquid
L101	3,800	Mild bitter taste	Good	Liquid
F68	8,350	Practically odorless and tasteless	Good	Solid
F108	14,000	Mild bitter taste	Good	Solid
P85	4,600	Mild bitter taste	Good	Paste
P127	11,500	Mild bitter taste	Good	Solid

GLYCOLS

Table 85.12. Acute Oral Toxicity of Pluronic Polyols

Product	Average Molecular Weight	Species	LD$_{50}$ (g/kg)	Signs
L44	2,200	Rat	5	Depression and prostration; death and respiratory paralysis probable
L61	2,000	Rat	2.1	Salivation, hypoactivity; ruffed fur
L62	2,500	Rat	5	Depression and prostration; death and respiratory paralysis probable
L64	2,900	Rat	5	Depression and prostration; death and respiratory paralysis probable
F68	8,350	Mouse	>15	At 15 g/kg, about 20% died
		Rat	>15	No effects seen
		Dog	>15	No effects seen other than diarrhea
		Rabbit	>15	No effects seen
		Guinea pig	>15	Mild sedation
P85	4,600	Rat	>34	
F127	11,500	Rat	15.4	

18.4.1.1c Acute Toxicity. All of these products have low in acute oral toxicity. The oral LD$_{50}$ values are given in Table 85.12.

18.4.1.2c Chronic and Subchronic Toxicity. A number of pluronic polyol products have been fed to rats and/or dogs and were found moderate to low in chronic oral toxicity. The results of the tests are summarized in Table 85.13. In these tests, the criteria evaluated were growth, mortality, behavior, hematologic studies and urinalysis, gross pathology, and histopathology, and in the dogs, EKG evaluations.

Eye Irritation. Pluronic polyol F68 tested as 5 and 10% aqueous solution was not irritating to the eyes of rabbits. pluronic polyols L44, L62, and L64 were tested as 25, 50, 75, and 100% in the rabbit eye and caused only transient mild to moderate irritation but no corneal injury. The degree of irritation increased with concentration. Pluronic polyols L33, F68, L101, and F127 did not cause local anesthetic effects when tested in the eyes of rabbits.

Skin Irritation. Pluronic polyols F68, L44, L62, and L64 were tested for skin irritation and skin sensitization using human volunteers. None was a skin irritant or a skin sensitizer. Pluronic polyol F127 at a level of 25% in a formulation caused a trace of irritation on rabbit skin but was not considered a primary skin irritant.

Dermal Absorption. The dermal LD$_{50}$ for pluronic polyol F127 was higher than 2.0 g/kg for rabbits.

Inhalation. No inhalation studies have been reported. There should be no significant health hazard from inhalation in ordinary industrial operations primarily because of their low volatility.

18.4.1.4c Reproduction and Developmental. Pluronic polyols L101 and F108 were fed to rats through three generations at levels of 100, 250, and 500 mg/kg of L101 and 300,

Table 85.13. Results of Dietary Feeding of Pluronic Polyol Products to Rats/or Dogs

Product	Species	Dose (g/kg/day)	Duration	Effect
L33	Dog	0.04	90 days	No effect
	Dog	0.2	90 days	Some emesis; no deaths
	Dog	1.0	90 days	Emesis; growth depression; no deaths
	Rat	0.2	90 days	No effects
	Rat	1.0	90 days	Not palatable, reduced food intake resulting in severe growth depression
L62	Dog	0.04	2 years	No significant effects
	Dog	0.2	2 years	Emesis, some periods of salivation; no significant pathological changes
	Dog	0.5	2 years	Deaths due to poor nutritional state; four of six dogs died; no pathology seen
	Rat	0.04	2 years	No effects
	Rat	0.2	2 years	Only growth depression
	Rat	0.5	2 years	Severe growth depression; some deaths due to inanition; no pathology seen
F68	Dog	0.1	6 months	No effect
	Rat	5% in diet	2 years	No effect
	Rat	7.5% in diet	2 years	Decrease in growth rate; no pathology seen
P85	Dog	1.0	90 days	No effect
	Rat	1.0	90 days	No effect
L101	Dog	0.5	90 days	No effect
	Rat	0.5	90 days	No effect
L108	Dog	5.0	90 days	No effect
	Rat	5.0	90 days	No effect

1000, and 2500 mg/kg of F108. No adverse effects were seen as judged by growth and the ability to mate and produce viable normal offspring or by gross examination and histopathology of the parents at the end of the test period.

18.4.1.5c Carcinogenesis. Only pluronic polyol L62 was fed to rats for 2 years. In this study, there was no mention of tumors during gross or histopathological examination.

Metabolism. Pluronic polyol F68 was the only material studied. When given to rats intravenously, essentially all the material given was excreted unchanged via the urine in 24 h. When given orally, essentially all of it was excreted unchanged in the feces.

18.4.2c Human Experience

No adverse human experience has been reported.

18.5c Standards, Regulations, and Guidelines of Exposure

None seems necessary for Pluronic polyols because of their very low volatility and their low toxicity.

GLYCOLS

Table 85.14. Physical and Chemical Properties of Pluracol Polyols V-10, TP-440, and TP-740

Property	V-10	TP-440	TP-740
Form	Pale yellow viscous liquid	Liquid	Liquid
Average molecular weight	> 20,000	418	732
Specific gravity	1.089 (60/60°F)		
Viscosity	45,000 cgs at 100°C	625 cgs at 25°C	325 cgs at 25°C
Flash point	510°F		
Solubility in water	up to 75%	0.2%	0.1%

18.0d Pluracol V-10, TP-410, and TP-740 Polyols

Pluracol is BASF Wyandotte's trade name for polyalkalene derivatives of trimethylol propane (245).

18.1d Physical and Chemical Properties

The physical and chemical properties are summarized in Table 85.14.

18.2d Production and Use

Pluracol V-10 polyol is a polyoxyalkalene derivative of trimethylol propane whose average molecular weight is more than 20,000. It is used as a thickening agent in hydraulic fluids.

Pluracol TP-440 and TP-740 polyols are polyoxypropylene derivatives of trimethylol propane. They are used in the urethane industry. Industrial exposure to these products is likely to be from direct contact.

18.4d Toxic Effects

Pluracol V-10, TP-440, and TP-740 polyol should present no significant health hazards in anticipated industrial handling. The reported toxicological information on them is summarized in Table 85.15.

Table 85.15. Summary of Toxicological Information on Three Pluracol Products

Product	Rat LD$_{50}$ (g/kg)	Eye Irritation in Rabbits	Skin Effects — Irritation	Skin Effects — Sensitization	Skin Absorption in Rabbits
Pluracol V-10	> 10.0[a]		Humans, none	None	
Pluracol TP-440	3.72[b]	Moderate, no corneal injury	Rabbits, trace		> 3/kg
Pluracol TP-740	2.5[b]				

[a] No adverse effects seen at this level, the highest dose fed.
[b] High doses that caused death resulted in sedation, then convulsions, salivation, lacrimation, tremors, amd muscular weakness.

18.5d Standards, Regulations, and Guidelines of Exposure

No toxicological testing has been reported that would permit a suggested hygienic standard. It is believed that none is likely to be needed because of the low toxicity and the low volatility of Pluracol products.

BIBLIOGRAPHY

1. D. R. Lide, ed., *CRC Handbook of Chemistry and Physics*, CRC Press, Cleveland, OH, 1992–1993.
2. S. Budavari, ed., *The Merck Index*, 11th ed., Merck and Company, Rahway, NJ, 1989.
3. *Registry of Toxic Effects of Clinical Substances*, U.S. Dept. Health, Education, and Welfare, Cincinnati, OH, Sept. 1991.
4. N. Sax, *Dangerous Properties of Industrial Materials*, 7th ed., Van Nostrand Reinhold, New York, 1987.
5. *Toxicology and Hazardous Industrial Chemicals Safety Manual for Handling and Disposal with Toxicity Data*, International Technical Information Institute, publ. Japan, 1976.
6. *Fire Protection Guide on Hazardous Chemicals*, National Fire Protection Association, Quincy, MA, 1991.
7. N. I. Sax and R. J. Lewis, Sr., eds., *Hawley's Condensed Chemical Dictionary*, 11th ed., Von Nostrand Reinhold, New York, 1987.
8. E. W. Flick, *Industrial Solvents Handbook*, 3rd ed., Noyes Publications, Park Ridge, NJ, 1985.
9. J. H. Kuney, ed., *Chemcyclopedia 90*, American Chemical Society, Washington, DC, 1990.
10. *Hazardous Substances Data Bank*, TOXNET System, National Library of Medicine, Bethesda, MD, 1993.
11. T. J. Haley and W. O. Brandt, *Toxicology*, Hemisphere Publishing Corporation, New York, 1987.
12. J. Rosenberg, Solvents. In J. LaDou, ed., *Occupational Medicine*, Appelton & Lange, Norwalk, CT, 1990, pp. 359–286.
13. P. D. Bryson, *Comprehensive Review of Toxicology*, 2nd ed., Aspen Publishers, Rockville, MD, 1989, pp. 283–293.
14. *Registry of Toxic Effects of Chemical Substances*, On-Line, TOXNET System, National Library of Medicine, Bethesda, MD, 1993.
15. ATSDR, *Toxicological Profile for Ethylene Glycol and Propylene Glycol*, U.S. Dept. of Health and Human Services, Atlanta, GA, 1993.
16. *Chem. Eng. News* **71**(15), 11 (April 12, 1993).
17. K. G. Bergner and H. Sperlich, *Z. Lebensm. Unters. Forsch* **97**, 253 (1953).
18. B. Worshowsky and P. J. Elving, *Ind. Eng. Chem.* **18**, 253 (1946).
19. R. C. Reinke and E. N. Luce, *Ind. Eng. Chem.* **18**, 244 (1946).
20. W. H. Evans and A. Dennis, *Analyst* **98**, 782 (1973).
21. P. Lepsi, *Prac. Lek.* **25**, 330 (1973).
22. F. R. Duke and G. F. Smith, *J. Ind. Eng. Chem. Anal. Ed.* **12**, 201 (1940).
23. *NIOSH Manual of Analytical Methods*, 4th ed., U.S. Dept. Health, Education, and Welfare, U.S. Printing Office, Washington, DC, 1994.

24. A. Davis, A. Roaldi, and L. E. Tufts, *J. Gas Chromatogr.* 306 (Sept 1964).
25. H. D. Spitz and J. Weinberger, *J. Pharm. Sci.* **60**, 271 (1971).
26. M. Adams and M. Collins, Anal. Proc. (*London*) **25**, 190–191 (1988).
27. N. Chairova and N. Dimov, *Heft. Khim.* **361**, 361 (1973).
28. G. G. Esposito and R. G. Jamison, *Soc. Automot. Eng. J.* **79**, 40 (1971).
29. *US EPA Method 8015B, Physical Chemical Methods SW846 Methods*, 1986.
30. R. L. Peterson and D. O. Rodgerson, *Clin. Chem.* **20**, 820 (1974).
31. K. C. Cummins and P. I. Jatlow, *J. Anal. Toxicol.* 6, 324–326 (1982).
32. J. A. Johnson, *J. Anal. Toxicol.* **13(1)**, 25–26 (1989).
33. I. J. Gilmour, R. J. W. Blanchard, and W. F. Perry, *N. Eng. J. Med.* **291**, 51 (1974).
34. B. J. Kersting and S. W. Nielsen, *Am. J. Vet. Res.* **27**, 574 (1966).
35. W. F. von Oettingen, *U.S. Public Health Bull. No. 281*, 1943.
36. E. P. Laug, H. O. Calvery, H. J. Morris, and G. Woodard, *J. Ind. Hyg. Toxicol.* **21**, 173 (1939).
37. E. M. K. Geiling and P. R. Cannon, *J. Am. Med. Assoc.* **111**, 919 (1938).
38. H. J. Morris, A. A. Nelson, and H. O. Calvery, *J. Pharmacol. Exp. Therap.* **74**, 266 (1942).
39. E. Browning, *Toxicity and Metabolism of Industrial Solvents*, Elsevier, Amsterdam, 1965.
40. H. F. Smyth, Jr., J. Seaton, and L. F. Fischer, *J. Ind. Hyg. Toxicol.* **23**, 259 (1941).
41. C. Bornmann, *Arzneim. Forsch.* **4**, 643, 710 (1954), **5**, 38 (1955); *Chem. Abstr.* **49**, 7131 (1955).
42. V. P. Plugin, *Gig. Sanit.* **33**, 16 (1968).
43. O. K. Antonyuk, *Gig. Sanit.* **39**, 106 (1974).
44. The Dow Chemical Company, unpublished data.
45. C. P. Carpenter and H. F. Smyth, Jr., *Am. J. Ophthalmol.* **29** 1363 (1946).
46. T. O. McDonald, M. D. Roberts, and A. R. Borgmann, *Toxicol. Appl. Pharmacol.* **21**, 143 (1972).
47. T. O. McDonald et al., *Bull. Parenter. Drug Assoc.* **27**, 153 (1973).
48. W. M. Grant, *Toxicology of the Eye*, Charles C. Thomas, 3rd ed., Springfield, IL, 1986.
49. P. J. Hanzlik et al., *J. Ind. Hyg. Toxicol.* **29**, 325 (1947).
50. J. F. Gaunt et al., *BIBRA* **14**, 109 (1975).
51. K. Nagano et al., *Jap. J. Ind. Health* **21**, 29 (1979) (in Japanese, English summary).
52. F. R. Blood, *Food Cosmet. Toxicol.* **3**, 229 (1965).
53. F. R. Blood, G. A. Elliott, and M. S. Wright, *Toxicol. Appl. Pharmacol.* **4**, 489 (1962).
54. J. A. Roberts and H. R. Seibold, *Toxicol. Appl. Pharmacol.* **15**, 624 (1969).
55. M. Felts, *Proceedings of the 5th Annual Conference on Atmospheric Contamination in Confined Spaces*, *AMRL TR-69-130*, paper No. 9, Aerospace Med. Res. Lab., Wright-Patterson Air Force Base, OH, 1969, p. 105.
56. F. J. Wiley, W. C. Hueper, and W. F. von Oettingen, *J. Ind. Hyg. Toxicol.* **18**, 123 (1936).
57. R. A. Coon et al., *Toxicol. Appl. Pharmacol.* **16**, 646 (1970).
58. J. H. Wills et al., *Clin. Toxicol.* **7**, 463 (1974).
59. E. Harris, *Proceedings of the 5th Annual Conference on Atmospheric Contamination in Confined Spaces*, *AMRL TR-69-130*, paper No. 8, Aerospace Med. Res. Lab., Wright-Patterson Air Force Base, OH, 1969, p. 9.

60. P. K. Gessner, D. V. Parke, and R. T. Williams, *Biochem. J.* **79**, 482 (1961).
61. L. H. Smyth, Jr., T. D. R. Hochaday, M. L. Efron, and J. E. Clayton, *Trans. Assoc. Am. Physicians* **77**, 317 (1964).
62. L. Hansson, R. Lindfors, and K. Laiko, *Deut. Z. Genchtl. Med.* **59**, 11 (1967).
63. J. S. King and A. Wainer, *Proc. Soc. Exp. Biol. Med.* **128**, 1162 (1968).
64. J. M. Melon and J. Thomas, *Therapie* **26**, 985 (1971).
65. C. C. Liang and L. C. Ou, *Biochem. J.* **121**, 447 (1971).
66. E. W. McChesney, L. Goldberg, and E. S. Harris, *Food Cosmet. Toxicol.* **10**, 655 (1972).
67. F. Underwood and W. M. Bennetl, *J. Am. Med. Assoc.* **226**, 1453 (1973).
68. M. F. Parry and R. Wallach, *Am. J. Med.* **57**, 143 (1974).
69. K. L. Clay and R. C. Murphy, *Toxicol. Appl. Pharmacol.* **39**, 39 (1977).
70. K. Hagstam et al., *Acta. Med. Scand.* **178**, 599 (1965).
71. B. H. Rumack and D. G. Spoerke, *POISINDEX Information System*, Micromedex Inc., Denver, CO, 1993.
72. M. J. Ellenhorn and D. G. Barceloux, *Medical Toxicology-Diagnosis and Treatment of Human Poisoning*, Elsevier, New York, 1988.
73. D. Jacobson and K. E. McMartin, *Med. Toxicol.* **1**(5), 309–334 (1986).
74. D. I. Peterson et al., *J. Am. Med. Assoc.* **186**, 955 (1963).
75. S. K. Cox, K. E. Fenslew, and L. J. Bolen, *Vet. Hum. Toxicol.* **34**(1), 36–42 (1992).
76. T. C. Marshall and Y. S. Cheng, Inhalation *Toxicology Research Institute Annual Report, LMF-84*, Dec. 1980, P.493.
77. S. W. Frantz, J. L. Beskitt, C. M. Grosse et al., *Xenobiotica* **26**, 1195–1220 (1996).
78. T. P. Hewlett, D. Jacobson, T. D. Collins et al., *Vet. Human Toxicol.* **31**, 116–120 (1989).
79. L. R. DePass, M. D. Woodside, and R. R. Maronpot, *Fundam. Appl. Toxicol.* **7**, 566–572 (1986).
80. J. C. Lamb IV, R. R. Maronpot, D. K. Gulati et al., *Toxicol. Appl. Pharmacol.* **81**, 100–112 (1985).
81. R. R. Maronpot, J. P. Zelnak, and E. V. Weaver, *Drug Chem. Toxicol.* **6**, 579–594 (1983).
82. R. W. Tyl, B. Ballantyne, L. C. Fisher, et al., *Fundam. Appl. Toxicol.* **27**, 49–62 (1995).
83. R. W. Tyl, B. Ballantyne, L. C. Fisher et al. *Fundam. Appl. Toxicol.* **20**, 402–412 (1993).
84. W. Breslin, E. Carney, N. Freshour et al., *Toxicologist* **36**, 100 (1997).
85. L. R. Depass, R. H. Garman, M. D. Woodside et al., *Fundam. Appl. Toxicol.* **7**, 547–565 (1986).
86. M. M. Mason, C. C. Cate, and J. Baker, *Clin. Toxicol.* **4**, 185 (1971).
87. P. H. Derse, *U. S. Clearinghouse Fed. Sci. Tech. Inform.*, *P. B. Rep. No. 195153* 1969.
88. F. Homburger, *U. S. Clearinghouse Fed. Sci. Tech. Inform.*, *P. B. Rep. No. 183027*, 1968.
89. J. McCann, E. Choi, E. Yamasaka, and B. N. Ames, *Proc. Natl. Acad. Sci. US* **72**, 5135 (1975).
90. E. H. Pfeiffer and H. Dunkelberg, *Food Cosmet. Toxicol.* **18**, 115 (1980).
91. F. M. Troisi, *Br. J. Ind. Med.* **7**, 65 (1950).
92. E. Volkmann, *Hippokrates* **21**, 549 (1950); *Chem. Abstr.* **47**, 9567 (1953).
93. C. A. Pons and R. P. Custer, *Am. J. Med. Sci.* **211**, 544 (1946).
94. A. P. Grant, *Lancet* **263**, 1252 (1952).

95. I. P. Ross, *Br. Med. J.* **1**, 1340 (1956).
96. M. Tadokoro, et al., *Nippon Jinzo Gakkai Shi* **37**, 353–356 (1995).
97. T. L. Litovitz, B. F. Schmitz, and K. M. Bailey, *Toxicology* **8**(5), 394–431 (1990).
98. T. L. Litovitz, B. F. Schmitz, and K. M. Bailey, *Toxicology* **9**(5), 461–500 (1991).
99. M. F. Michelis, B. Mitchell, and B. B. Davis, *Clin. Toxicol.* **9**, 53 (1976).
100. W. E. C. Wacker, H. Haynes, R. Druyan, W. Fisher, and J. E. Coleman, *J. Am. Med. Assoc.* **194**, 173 (1965).
101. J. Pinter, J. Csaszar, and E. Wolfer, *Zeit. Urologie* **59**, 885 (1966) (translation by Sunlone Inc., Arlington Heights, IL).
102. H. C. Aquino and C. D. Leonard, *J. Kentucky Med. Assoc.* **70**, 463 (1972).
103. P. Harry, E. Jobard, M. Briand et al., *Pediatrics* **102**, E31 (1998).
104. *Threshold Limit Values for Chemical Substances and Physical Agents and Biological Exposure Indices*, American Conference of Governmental Industrial Hygienists, Cincinnati, OH, 2000.
105. OSHA, *Permissible Exposure Limits-Final Listing, 54 FR 2920*, Jan. 1, 1989.
106. EPA, *IRIS Data Base*, 1993.
107. WHO, *Tech. Info. for Prob. Spills*, Geneva, 1984.
108. EPA, *TSCA Test Submission Data Base*, Jan. 1993.
109. W. H. Evans and E. J. David, *Water Res.* **8**, 97 (1974).
110. P. Pitter, *Collect Czech. Chem. Commun.* **38**, 2665 (1973).
111. G. O. Curme, Jr. and F. Johnston, eds., *Glycols*, American Chemical Society Monograph Series 114, Reinhold, New York, 1952.
112. V. P. Plugin,*Gig. Sanit.* **33**, 16 (1968).
113. Union Carbide Corporation, unpublished data.
114. H. Wegener, *Arch. Exp. Pathol. Pharmakol.* **220**, 414 (1953); *Chem. Abstr.* **48**, 2919 (1954).
115. A. Loesser et al., *Arch. Exp. Pathol. Pharmakol.* **221**, 14 (1954); *Chem. Abstr.* **48**, 4698 (1954).
116. S. A. Marchenko, *Vrach. Delo* **2**, 138 (1973).
117. S. A. Marchenko, *Farmakol. Toksikol. (Kiev)* **1973**(8) (1975).
118. Y. P. Sanina, *Gig. Sanit.* **33**(2), 36 (1968).
119. O. G. Fitzhugh and A. A. Nelson, *J. Ind. Hyg. Toxicol.* **28**, 40 (1946).
120. C. S. Weil, C. P. Carpenter, and H. F. Smyth, Jr., *Arch. Environ. Health* **11**, 569 (1965).
121. C. S. Weil, C. P. Carpenter, and H. F. Smyth, Jr., *Ind. Med. Surg.* **36**, 55 (1967).
122. F. H. Wiley, W. C. Hueper, D. S. Bergen, and F. R. Blood, *J. Ind. Hyg. Toxicol.* **20**, 269 (1938).
123. H. B. Haag and A. M. Ambrose, *J. Pharmacol. Exp. Therap.* **59**, 93 (1937).
124. M. D. Bowie and D. McKenzie, *S. Afr. Med. J.* **46**, 931 (1972).
125. P. Auzepy, H. Taktak, P. L. Toubas, and M. Deparis, *Sem. Hop. Paris* **49**, 1371 (1973).
126. H. O. Calvery and T. G. Klumpp, *Southern Med. J.* **32**, 1105 (1939).
127. K. A. Telegina et al., *Gig. Tr. Prof. Zabol* **15**, 40 (1971).
128. A. R. Latven and H. Molitor, *J. Pharmacol. Exp. Therap.* **65**, 89 (1939).
129. E. G. Stenger, L. Aeppli, E. Peheim, and F. Roulet, *Arzneim. Forsch.* **18**, 1536 (1968).
130. O. H. Robertson, *Harvey Lect. Ser.* **38**, 227 (1943).
131. O. H. Robertson, *Wisconsin Med. J.* **46**, 311 (1947).

132. O. H. Robertson et al., *J. Pharmacol. Exp. Therap.* **91**, 52 (1947).
133. W. M. Lauter and V. L. Vrla, *J. Am. Pharm. Assoc.* **29**, 5 (1940)
134. H. McKennis, Jr. et al., *Toxicol. Appl. Pharmacol.* **4**, 411 (1962).
135. B. H. Jennings, E. Biggs, and F. C. W. Olson, *Heating, Piping, Air Conditioning* **16**, 538 (1944).
136. T. N. Harris and J. Stokes, Jr., *Am. J. Med. Sci.* **209**, 152 (1945).
137. L. D. Polderman, *Soap Sanit. Chem.*, 133 (July 1947).
138. T. Ramstad, T. J. Nestrick, and R. H. Stehl, *Anal. Chem.* **50**, 1325 (1978).
139. M. Yoshida, H. Hoshii, and H. Morimoto, *Nippon Kakim Gakkaishi* **6**, 73 (1969).
140. H. Weifenbach, *Arzneim. Forsch.* **23**, 1087 (1973).
141. H. D. Keston, N. C. Mulinos, and L. Pomerantz, *Arch. Pathol.* **27**, 447 (1939).
142. C. B. Shaffer and F. H. Critchfield, *Ind. Eng. Chem. Anal. Ed.* **19**, 32 (1947).
143. A. H. Principe, *J. Forensic Sci.* **13**, 90 (1968).
144. The Dow Chemical Company, *Dow Polyethylene Glycols*, 1959.
145. H. F. Smyth, Jr., et al., *Ind. Hyg. Toxicol.* **24**, 281 (1942).
146. H. F. Smyth, Jr., C. P. Carpenter, and C. S. Weil, *J. Am. Pharm. Assoc. Sci. Ed.* **39**, 349 (1950).
147. C. P. Carpenter et al., *Toxicol. Appl. Pharmacol.* **18**, 35 (1971).
148. J. H. Draize, G. Woodard, and H. O. Calvery, *J. Pharmacol. Exp. Therap.* **82**, 377 (1944).
149. F. P. Luduena, J. K. Fellows, G. L. Laqueur, and R. L. Driver, *J. Ind. Hyg. Toxicol.* **29**, 390 (1947).
150. H. F. Smyth, Jr., C. P. Carpenter, and C. S. Weil, *J. Am. Pharm. Assoc. Sci. Ed.* **44**, 27 (1955).
151. T. W. Tusing, J. R. Elsea, and A. B. Sauveur, *J. Am. Pharm. Assoc. Sci. Ed.* **43**, 489 (1954).
152. O. Schmidt, *J. Soc. Cosmet. Chem.* **21**, 835 (1970).
153. C. B. Shaffer and F. H. Critchfield, *J. Am. Pharm. Assoc. Sci. Ed.* **36**, 152 (1947).
154. C. B. Shaffer, F. H. Critchfield, and J. H. Nair, *J. Am. Pharm. Assoc. Sci. Ed.* **39**, 340 (1950).
155. H. F. Smyth, Jr., C. S. Weil, M. D. Woodside, J. B. Knaak, L. J. Sullivan, and C. P. Carpenter, *Toxicol. Appl. Pharmacol.* **16**, 442 (1970).
156. E. Schutz, *Arch. Exp. Pathol. Pharmakol.* **232**, 237 (1957); *Ind. Hyg. Dig. Abstr.* **22**, 720 (1958).
157. D. M. Conning and M. J. Hayes, *Br. J. Ind. Med.* **27**, 155 (1970).
158. V. K. H. Brown, L. B. Valeris, and D. J. Simpson, *Arch. Environ. Health* **30**, 1 (1975).
159. A. J. Lehman and H. W. Newman, *J. Pharmacol. Exp. Therap.* **60**, 312 (1937).
160. Council on Pharmacy and Chemistry of the American Medical Association, *New and Nonofficial Remedies*, Lippincott, Philadelphia, 1949.
161. Food and Drug Administration, *Food, Drug, Cosmet. Law J.* **13**, 856 (1958).
162. J. A. Ruddick, *Toxicol. Appl. Pharmacol.* **21**, 102 (1972).
163. *GRAS (Generally Recognized as Safe) Food Ingredients-Propylene Glycol and Derivatives*, National Technical Information Service, U.S. Dept. of Commerce, Rep. No. FDABFGRAS-033, 1973.
164. J. H. Weatherby and H. B. Haag, *J. Am. Pharm. Assoc.* **27**, 466 (1938).
165. P. J. Hanzlik et al., *J. Pharmacol. Exp. Therap.* **67**, 101 (1939).
166. C. Martin and L. Finberg, *J. Pediatr.* **77**, 877 (1970).

167. M. A. Seidenfeld and P. J. Hanzlik, *J. Pharmacol. Exp. Therap.* **44**, 109 (1932).
168. C. P. Whitlock, N. B. Guerrant, and R. A. Dutcher, *Proc. Soc. Exp. Biol. Med.* **57**, 124 (1944).
169. W. Van Winkle, Jr., and H. W. Newman, *Food Res.* **6**, 509 (1941).
170. C. S. Weil and R. A. Scala, *Toxicol. Appl. Pharmacol.* **19**, 276 (1971).
171. T. Morizono and B. M. Johnstone, *Med. J. Aust.* **18**, 634 (1975).
172. I. F. Gaunt et al., *Food Cosmet. Toxicol.* **10**, 151 (1972).
173. C. S. Weil et al., *Food Cosmet. Toxicol.* **9**, 479 (1971).
174. L. J. Fisher et al., *Can. J. Anim. Sci.* **53**, 289 (1973).
175. F. D. Sauer, J. D. Erfle, and L. J. Fisher, *Can. J. Anim. Sci.* **53**, 265 (1973).
176. A. Shiga et al., *Nippon Chikusan Gakkai-Ho* **46**, 341 (1975).
177. H. S. Bailey, S. J. Slinger, and J. D. Summers, *Poultry Sci.* **46**, 19 (1967).
178. J. N. Persons et al., *Poultry Sci.* **47**, 351 (1968).
179. P. W. Waldroup and T. E. Bowen, *Poultry Sci.* **47**, 1911 (1968).
180. S. Harnisch, *Arch. Gefluegelkd.* **37**, 187 (1973).
181. W. Van Winkle, Jr., *J. Pharmacol. Exp. Therap.* **72**, 344 (1941).
182. H. W. Newman and A. J. Lehman, *Proc. Soc. Exp. Biol. Med.* **35**, 601 (1936-1937).
183. J. L. Clapperton and J. W. Czerkawski, *Br. J. Nutrit.* **27**, 553 (1972).
184. D. Giesecke, *Arch. Int. Physiol. Biochim.* **82**, 645 (1974).
185. M. Yoshida and H. Ikumo, *Agric. Biol. Chem.* 35, 1628 (1971).
186. N. B. Guerrant et al., *Bull. Natl. Formulary Comm.* **15**, 205 (1947).
187. C. W. Emmens, *J. Reprod. Fert.* **26**, 175 (1971).
188. D. O. E. Gebhardt, *Teratology* **1**, 153 (1968).
189. F. Dewhurst, D. A. Kitchen, and G. Calcutt, *Br. J. Cancer* **26**, 506 (1972).
190. R. W. Baldwin et al., *Br. J. Cancer* **22**, 133 (1968).
191. K. Wallenious and U. Lecholm, *Odontol. Revy* **24**, 39, 115 (1973).
192. F. Stenback and P. Shubik, *Toxicol. Appl. Pharmacol.* 30, 7 (1974).
193. C. L. Kennedy, Jr. et al., *Toxicology* **5**, 159 (1975).
194. W. Van Winkle, Jr., *J. Pharmacol. Exp. Therap.* **72**, 227 (1941).
195. W. Bartsch et al., *Arzneim. Forsch.* **26**, 1581 (1976).
196. T. C. Warshaw and F. Hermann, *J. Invest. Dermatol.* **19**, 423 (1952).
197. F. N. Marzulli and H. I. Maibach, *Food Cosmet. Toxicol.* **12**, 219 (1974).
198. J. P. Nater, A. J. M. Baar, and P. J. Hoedemaeker, *Contact Derm.* **3**, 181 (1977).
199. S. K. Bau, N. Aspin, D. E. Wood, and H. Levison, *Pediatrics* **48**, 605 (1971).
200. R. D. Creek, *Poultry Sci.* **49**, 1686 (1970).
201. P. K. Gessner, D. V. Parke, and R. T. Williams, *Biochem. J.* **74**, 1 (1960).
202. L. Ulrich and M. Mestitzova, *Prac. Lek.* **7**, 299 (1963).
203. H. E. Christensen, ed., *Registry of Toxic Effects of Chemical Substances*, 1976 ed., U.S. Dept. of Health, Education, and Welfare, National Institute for Occupational Safety and Health, Rockville, MD, 1976.
204. Developmental Toxicity Evaluation of Dipropylene Glycol (CAS No.25265-71-8) Administered by Gavage to New Zealand White (NZW) Rabbits from Gestation day 6 through Gestation Day 19, Final report and appendix, NTIS #PB92-238294/AS, Springfield, VA, 1992.

205. The Dow Chemical Company, *Polypropylene Glycols from Dow*, Form No. 118-1014-80, 1980.
206. C. B. Shaffer et al., *Arch. Ind. Hyg. Occup. Med.* **3**, 448 (1951).
207. The Dow Chemical Company, *Polypropylene Glycols*, Form No. 125-129-57, 1959.
208. F. E. Shideman and L. Procita, *J. Pharmacol. Exp. Therap.* **103**, 293 (1951).
209. H. Schlussel, *Arch. Exp. Pathol. Pharmakol.* **221**, 67 (1954); *Chem. Abstr.* **48**, 5315 (1954).
210. A. Loeser, *Pharmazie* **4**, 263 (1949); *Chem. Abstr.* **43**, 8558 (1949).
211. L. Fischer et al., *Z. Ges. Exp. Med.* **115**, 22 (1949); *Chem. Abstr.* **44**, 9070 (1950).
212. H. F. Smyth, Jr., C. P. Carpenter, and C. S. Weil, *Arch. Ind. Hyg. Occup. Med.* **4**, 119 (1951).
213. R. Kopf, A. Loeser, G. Meyer, and W. Franke, *Arch. Exp. Pathol. Pharmakol.* **210**, 346 (1950); *Chem. Abstr.* **45**, 5308 (1951).
214. R. V. Chagovets et al., *Urk. Biokhim. Zh.* **46**, 275 (1974).
215. R. L. Harris, M. A. Mehlman, and R. L. Veech, *Fed. Proc.* **31**, 670 (1972).
216. S. A. Miller and H. Dymsza, *J. Nutr.* **91**, 79 (1967).
217. M. Yoshida, K. Osada, S. Fujishiro, and R. Oda, *Agric. Biol. Chem.* **35**, 393 (1971).
218. R. A. Scala and O. E. Paynter, *Toxicol. Appl. Pharmacol.*, 10, 160 (1967).
219. S. P. Kryshova, *Gig. Sanit.* **33**, 37 (1968).
220. W. Schneider, *Pharm. Ind.* **12**, 226 (1950); *Chem. Abstr.* **45**, 3998 (1951).
221. E. Strack, D. Biesold, and H. Thiele, *Z. Ces. Exp. Med.* **132**, 522 (1960).
222. M. A. Mehlman et al., *J. Nutr.* **101**, 1711 (1971).
223. R. L. Tate, M. A. Mehlman, and R. B. Tobin, *J. Nutr.* **101**, 1719 (1971).
224. H. A. Dymsza, *Fed. Proc.* **34**, 2167 (1975).
225. J. J. B. Ayres and F. A. Isgrig, *Psychopharmacologia* **16**, 290 (1970).
226. H. Sprince, J. A. Josephs, Jr., and C. R. Wilpizeski, *Life Sci.* **5**, 2041 (1966).
227. R. Maxwell and R. H. Roth, *Biochem. Pharmacol.* **21**, 1621 (1972).
228. A. Hinrichs, R. Kopf, and A. Loeser, *Pharmazie*, **3**, 110 (1948); *Chem. Abstr.* **42**, 5567 (1948).
229. R. B. Tobin et al., *Fed. Proc.* **34**, 2171 (1975).
230. E. Husing, R. Kopf, and A. Loeser, *Fette Seifen*, **52**, 45 (1950); *Chem. Abstr.* **44**, 7999 (1950).
231. M. V. Shelanski, *Cosmet. Perfum.* **89**, 96 (1974).
232. H. F. Smyth, Jr. et al., *Am. Ind. Hyg. Assoc. J.* **23**, 95 (1962).
233. Shell Chemical Corporation, *Ind. Hyg. Bull.*, *Hexylene Glycol*, SC:57–101 and SC:57–102, 1958.
234. V. Larsen, *Acta Pharmacol. Toxicol.* **14**, 341 (1958).
235. H. F. Smyth, Jr. and C. P. Carpenter, *J. Ind. Hyg. Toxicol.* **30**, 63 (1948).
236. W. B. Deichmann and M. Dierker, *J. Biol. Chem.* **163**, 753 (1946).
237. E. Jacobsen, *Acta Pharmacol. Toxicol.* **14**, 207 (1958).
238. C. J. Terhaar, W. J. Krasavage, and R. L. Roudabush, *Toxicol. Appl. Pharmacol.* **29**, 87 (1974).
239. B. D. Astill and D. W. Fassett, *Toxicol. Appl. Pharmacol.* **29**, 151 (1974).

240. A. J. Lehman, *Assoc. Food Drug Offic. U.S. Quart. Bull.* **19**, 87 (1955).
241. P. Milvy and A. J. Garro, *Mutat. Res.* **40**, 15 (1976).
242. C. Vainio et al., *Scand. J. Work Environ. Health* **2**, 147 (1976).
243. H. F. Smyth, Jr., et al., *Toxicol. Appl. Pharmacol.* **16**, 675 (1970).
244. BASF Wyandotte Corporation, Bulletin No. 05-3012, 765.
245. C. W. Leaf, *Soap and Chemical Specialties*, 48 (Aug. 1967).

CHAPTER EIGHTY-SIX

Ethers of Ethylene Glycol and Derivatives

Rodney J. Boatman, Ph.D., DABT, and James B. Knaak, Ph.D.

INTRODUCTION

A Background

There are currently seven U.S. manufacturers of ethers and other derivatives of ethylene glycol (EG), diethylene glycol (DEG), and higher glycols. Five of them are members of the American Chemistry Council (ACC) Glycol Ethers' Panel (Table 86.1).

B Production and Use

The glycol ethers most commonly encountered industrially are colorless liquids that have mild ethereal odors. Alkyl glycol ethers are manufactured in a closed, continuous process by reacting ethylene oxide with an anhydrous alcohol in the presence of a suitable catalyst. Depending on the molar ratios of the reactants and other process parameters, the product mixtures obtained contain varying amounts of the monoethylene-, diethylene-, triethylene-, and higher glycol ethers. Typically, the products in these mixtures are separated and purified by fractional distillation. Ethylene-based glycol ether production capacity in 1996 was about 1500 million lb. In 1995, output of the butyl ether of ethylene glycol, the most widely produced glycol ether was about 413 million lb (1).

The miscibility of most of these ethers with water and with a large number of organic solvents makes them especially useful as solvents in oil–water compositions. Their relatively slow rate of evaporation also makes them useful as solvents and coalescing agents in paints (2). Other uses include inks, cleaners, chemical intermediates, process

---

Patty's Toxicology, Fifth Edition, Volume 7, Edited by Eula Bingham, Barbara Cohrssen, and Charles H. Powell.
ISBN 0-471-31940-6 © 2001 John Wiley & Sons, Inc.

Table 86.1. U.S. Manufacturers of Ethylene Glycol Ethers and Derivatives

Manufacturing Company	Trade Name
The Dow Chemical Company[a]	Dowanol® solvent
Eastman Chemical Company[a]	Eastman® solvent
Ferro Corporation	
Olin Chemical Corporation	Poly-solv® solvent
Equistar Chemical Company[a]	
Shell Chemical Company[a]	Oxitol solvent® (ethers of EG)
	Dioxitol® solvent (ethers of DG)
Union Carbide Corporation[a]	Cellosolve® solvent (ethers of EG)
	Carbitol® solvent (ethers of DEG)

[a]Member of the American Chemistry Council, Ethylene Glycol Ethers Panel.

solvents, brake fluids, and deicers (2). The ethers of the higher glycols are used as hydraulic fluids. An estimate of the U.S. production and use of representative ethylene glycol ethers is presented in Table 86.2. Production of ethylene glycol ethers (total) in Western Europe amounted to 245 thousand metric tons in 1995 (1). Japanese production of ethylene glycol ethers (total) amounted to 59.5 thousand metric tons in 1995 (1). Lesser amounts of glycol ethers are also produced in Brazil, Canada, and Eastern Europe (1).

C Occupational Exposure

Occupational exposure to glycol ethers occurs dermally and by inhalation. Ingestion is not a concern in industrial exposure, although some cases of intentional ingestion of consumer products containing ethylene glycol ethers have been reported. Several review articles were recently published dealing with house painters' exposure to glycol ethers in water-based paints (3) and exposure of workers in the semiconductor industry to glycol ethers during manufacturing (4–12).

Analytical Methods

A number of analytical methods have been published that are suitable for detecting glycol ethers in environmental air samples (13). These methods rely principally on adsorption onto activated charcoal or other suitable material followed by desorption and gas chromatographic analysis. Lower limits of detection generally range from 0.1 to 2 ppm in air depending on the specific analytical method employed and on the gas sampling procedure used. Typical methods include NIOSH method 1403 (14) and OSHA method 83 (15). A typical example is an industrial hygiene monitoring method developed by Dow Chemical Company (16) that can analyze for nine airborne glycol ethers: ethylene glycol methyl ether (EGME), ethylene glycol ethyl ether (EGEE), ethylene glycol phenyl ether (EGPhE), ethylene glycol butyl ether (EGBE), diethylene glycol methyl ether (DGME), diethylene glycol ethyl ether (DGEE), diethylene glycol phenyl ether (DGPhE), diethylene glycol butyl ether (DGBE), triethylene glycol methyl ether (TGME), and triethylene

Table 86.2. Ethylene Glycol Ethers and Ether Acetates: Production Volumes and Uses in the United States[a]

Chemical	Company Trade Name	CAS Number	Production Volume (10⁶ lbs)	Type of Consumer End Products	Consumer Products Volume (10⁶ lbs)	Consumer Products (% of Production)	Consumer Products (Approx. Weight %)	Percent Industrial/ Consumer Use
Ethylene glycol monomethyl ether (EGME)	Methyl Cellosolve® Solvent Glycol Ether EM	[109-86-4]	15	None				100/0
Ethylene glycol monoethyl ether (EGEE)	Cellosolve® Solvent Glycol Ether EE	[110-80-5]	32	None				100/0
Ethylene glycol monoethyl ether acetate (EGEEA)	Glycol Ether EE Acetate	[111-15-9]	< 30	None				100/0
Ethylene glycol monobutyl ether (EGBE)	Eastman® EB Solvent Butyl Cellosolve® Solvent Dowanol® EB Butyl Oxitol® EB Glycol Ether EB	[111-76-2]	408.5	Paints/coatings Cleaners Solvent Polishes	51 74 34.3 0.1	12.5 18.1 8.4 0.2	2–25 0.1–25 — 5–10	60.8/39.2
Ethylene glycol monobutyl ether acetate (EGBEA)	Eastman EB Acetate Butyl Cellosolve® Acetate Glycol Ether EB Acetate	[112-07-2]	15.6	Paints/coatings Solvent	0.1 0.3	0.6 1.9	5–25 —	97.5/2.5
Ethylene glycol mono-propyl ether (EGPE)	Eastman® EP Solvent Propyl Cellosolve® Solvent n-Propyl Oxitol	[2807-30-9]	<15	Paints/coatings Solvents Cleaners	< 1 < 1 < 1	0.7 4.0 2.0	10–20 — 1–10	93.3/6.7
Ethylene glycol mono-phenyl ether (EGPhE)	Dowanol® EPH	[122-99-6]	<20	Paints/coatings Cleaners Dyes	< 10 < 5 < 1	37.5 19.0 6.5	5–15 5–15 5–15	37/63
Ethylene glycol monohexyl ether (EGHE)	Hexyl Cellosolve® Solvent	[112-25-4]	<10	Cleaners	< 5	50.0	0.1–3	50/50
Diethylene glycol mono-methyl ether (DGME)	Eastman® DM Solvent Methyl Carbitol® Solvent Dowanol® DM Glycol Ether DM	[111-77-3]	69.5	Paints/coatings Solvents	5.6 0.1	8.0 0.1	0.1–20 —	91.8/8.2

Table 86.2. (*Continued*)

Chemical	Company Trade Name	CAS Number	Production Volume (10⁶ lbs)	Type of Consumer end Products	Consumer Products Volume (10⁶ lbs)	Consumer Products (% of Production)	Consumer Products (Approx. Weight %)	Percent Industrial/ Consumer Use
Diethylene glycol monoethyl ether (DGEE)	Eastman® DE Solvent Carbitol® Solvent-low gravity Glycol Ether DE	[111-90-0]	33	Hair colorant Floor polish Paints/coatings Cleaners Solvents	1.3 4.7 0.1 0.2 0.2	3.9 14.2 0.3 0.6 0.6	1–10 2–10 10–20 1–10 —	80.3/19.7
Diethylene glycol monoethyl ether acetate (DGEEA)	Eastman® DE Acetate	[112-15-2]	<5	Solvents	<1	7.0	—	93/7
Diethylene glycol monobutyl ether (DGBE)	Eastman® DB Solvent Butyl Carbitol® Solvent Dowanol® DB Butyl Dioxitol® Glycol Ether DB	[112-34-5]	101.5	Paints/coatings Cleaners Solvents Brake fluids	26 18.5 1.8 5.5	25.6 18.2 1.8 5.4	2–20 1–20 — 10–65	49/51
Diethylene glycol monobutyl ether acetate (DGBEA)	Eastman® DB Acetate Butyl Carbitol® Acetate Glycol Ether DB Acetate	[124-17-4]	9.9	Paints/coatings Solvents	0.9 0.6	9.1 6.1	1–75 —	85/15
Diethylene glycol monopropyl ether (DGPE)	Eastman® DP Solvent	[6881-94-3]	<5	Floor polish Brake fluids	<1 <5	10.0 60.0	2–10 10–65	30/70
Diethylene glycol monohexyl ether (DGHE)	n-Hexyl Carbitol® Solvent	[112-59-4]	<5	Cleaners	<5	30.0	0.1–3	70/30
Triethylene glycol methyl ether (TGME)	Methoxytriglycol Glycol Ether TM	[112-35-6]	27	Brake fluids	19.8	73.3	20–70	26.7/73.3
Triethylene glycol ethyl ether (TGEE)	Ethoxytriglycol Glycol Ether TE	[112-50-5]	7.2	Brake fluids	2.2	30.6	10–65	69.4/30.6
Triethylene glycol butyl ether (TGBE)	Butoxytriglycol Glycol Ether TB	[143-22-6]	38.9	Brake fluids	31.4	80.7	10–65	19.3/80.7
Triethylene glycol monopropyl ether (TGPE)		[23305-64-8]	<5	Brake fluids	<5	100.0	10–65	0/100

[a]This information was compiled by the Chemical Manufacturers Association, Ethylene Glycol Ethers Panel.

glycol phenyl ether (TGPhE), collected on charcoal or silica gel using an air pump. The procedure involves pulling air that contains the volatile glycol ethers through charcoal or silica gel adsorbent tubes for periods of time up to 8 h. The adsorbed glycol ethers are desorbed from charcoal using water/CS_2 and from silica gel using a 25/75% mixture of methanol/water. The aqueous layer is analyzed for glycol ethers by gas chromatography using a flame-ionization detector (GC-FID). A 3.3 ft × 2 mm i.d., glass column, packed with 5% SP-1000 on 60/80 mesh Carbopack B may be used for the analysis or the packed column may be substituted with a SPB-1000 (Supelco, Inc., Bellefonte, PA) capillary column (30 m × 0.53 mm i.d., 0.5 μm film thickness). Supelco (17) recommends using a Nukol® fused silica capillary column for glycols (15 m × 0.53 mm i.d., 0.5 μm film). Water can be directly injected into this column. The column may be used to separate free acids, phenols, glycol ethers, and glycols dissolved in water. Several other similar capillary columns would also be suitable for this analysis.

Eckel et al. (18) reported the presence of glycol ethers in groundwater samples from hazardous waste sites. These authors point out that the general-purpose USEPA methods (GC/MS) used for either volatile or semivolatile pollutants may underestimate the true amounts of glycol ethers in environmental samples. This is due to the inefficiency of either the common purge-and-trap or solvent extraction methods for isolating glycol ethers. These authors suggest that more appropriate techniques would involve direct aqueous injection, extractive alkylation or salting-out extraction with derivatization.

D Pharmacokinetics, Metabolism, and Mechanism of Action

Pharmacokinetics — Absorption

Inhalation. Glycol ethers as a class are polar solvents that have generally low volatility and higher boiling points than most comparable solvents. Respiratory retention of glycol ethers is high and accounts for 50 to 80% of inhaled amounts due to high blood/air partition coefficients (i.e., 34913 for EGME) in lungs. Inhalation of 5.13 ppm EGME at the Occupational Exposure Limit (OEL) concentration of 16 mg/m^3 for 8 h resulted in an uptake of 57 mg in human volunteers at rest (19). Work (50 W, 100 W, 150 W) results in an increase in respiratory volume and the inhalation and retention of substantially larger amounts of volatile glycol ethers (EGME, EGEE, and EGBE).

Dermal — Liquid. Glycol ethers (neat) are absorbed by the dermal route, and 0.2 to 3% of topically applied EGME is absorbed by humans under occlusive conditions (area (27 cm^2) × concentration   (1.0 g/cm^3) × K_p (28.9 × 10^{-4} cm/h) × time   (10 h) = 780 mg absorbed; 780 mg/27000 mg applied × 100 = 2.9%) (19). The amount absorbed is determined by the exposed area, applied concentration, permeation constant (K_p), and total contact time. Microgram quantities are absorbed per cm^2 of skin per h from concentrations (mg/cm^3) normally encountered in the work place. For those glycol ethers that have been adequately tested, dermal administrations generally result in toxicities that are similar to those following oral administrations, but require higher equivalent dermal doses and longer periods of exposure. Skin penetration is determined primarily by the stratum corneum. The *in vitro* rates of permeability of a homologous series of glycol ethers

Table 86.3. Permeation of Glycol Ethers Through Human Epidermis *In Vitro*[a]

Glycol Ether	Mol. Wt.	Vapor Pressure (mmHg)	Boiling Pt. (°C)	Permeation Constant (cm/h × 10^4)	Relative Rate
Ethylene glycol series:					
Methyl ether	76	9.7	124	28.9	131
Ethyl ether	90	5.75	135	8.42	38
Butyl ether	118	0.88	171	2.14	9.7
Diethylene glycol series:					
Methyl ether	120	0.18	194	2.06	9.4
Ethyl ether	134	0.14	202	1.32	6.0
Butyl ether	162	0.043	230	0.357	1.6
Triethylene glycol series:					
Methyl ether	164	<0.01	249	0.34	1.5
Ethyl ether	178	<0.01	256	0.24	1.1
Butyl ether	206	<0.01	279	0.22	1.0

[a] Taken from Refs. 20 and 21.

through human stratum corneum have been summarized in Table 86.3. The methyl, ethyl, and butyl ethers of diethylene and triethylene glycol are absorbed to a lesser extent than the corresponding ethers in the ethylene glycol series (20, 21). An increase in either the chain length of the alkyl substituent or in the number of ethylene glycol moieties (ethylene oxide adducts) results in a decreased rate of percutaneous absorption. It can be seen that EGME is absorbed 131 times more rapidly than the triethylene glycol butyl derivative. Increased absorption of the lower molecular weight homologues is offset by an increased rate of evaporation from the skin, as indicated by the lower boiling points for these derivatives.

Dermal-Vapor. Glycol ethers are absorbed through skin in accordance with the following rate equation:

$$\text{Flux} = Dk_m C/l = K_p C$$

where C is the concentration gradient of the chemical in skin (g/cm^3), l is the thickness of the skin (cm), k_m is the partition coefficient of the chemical in skin (unitless), D is the diffusion coefficient (cm^2/h), and K_p is the permeability constant (cm/h). In the case of glycol ether vapors, skin/air partition coefficients need to be determined and used in the absorption equation to account for higher concentrations in skin due to partitioning. The equation is as follows for whole body exposure at the OEL for EGME of 16 mg/m^3 (5.13 ppm): area (18,000 cm^2) × concentration (16.0 × 10^{-6} mg/cm^3) × PC (11,734) × K_p (28.9 × 10^{-4} cm/h) × time (8 h) = 78 mg absorbed (19). The combined amount (135 mg) results in a dose of 1.9 mg/kg for a 70-kg person during an 8-h workday, assuming total body contact with vapors.

Table 86.4. Tissue Partition Coefficients for EGME, EGEE, and EGBE and Alkoxyacetic Acid Metabolites

	Tissue Partition Coefficients					
	EGME[a]	MAA[a]	EGEE[b]	EAA[b]	EGBE[c]	BAA[c]
Blood/air	34,913	0.99	22,093	NA	7,965	NA
Liver/blood	1.02	1.26	1.0	1.1	1.46	1.30
Fat/blood	0.05	—	0.04	0.32	2.03	0.77
Poorly perfused/blood	0.93	—	0.94	0.05	0.64	1.31
Richly perfused/blood			1.1	1.06	1.46	1.30
Placenta/blood	1.20	1.05				
Extraembryonic fluid/blood	1.48	1.33				
Embryo/blood	1.52	0.94				

[a] Clarke et al. (41).
[b] Hayes et al. (42).
[c] Corley et al. (23).

Pharmacokinetics — Distribution to Tissues

Regardless of the route of administration, glycol ethers are absorbed and distributed to tissues when administered to animals whose blood/air and skin/air partition coefficients for EGME are of the order of 34,900 and 11,700. Except for adipose tissue, measured tissue to blood partition coefficients for glycol ethers and alkoxyacetic acid metabolites are generally near unity (11700/34900 = 0.335 for EGME skin/blood). See Table 86.4. Notable exceptions include EGME and methoxyacetic acid (MAA), which attain greater concentrations in developing embryos and embryonic fluids than in plasma, and EGBE, which has an increased affinity for lung and gastrointestinal tissue (22, 23).

Metabolism and Elimination

A general metabolic scheme for ethylene glycol monoalkyl ethers is summarized in Figure 86.1. Glycol ether acetates are rapidly hydrolyzed to their corresponding glycol ethers by carboxylesterases. This occurs in a variety of tissues, including blood and olfactory mucosa and as a result, both the toxicity as well as patterns of metabolic elimination are nearly identical for glycol ethers and their acetates. The parent glycol ethers are substrates for alcohol dehydrogenase (ADH) which catalyzes the conversion of their terminal alcohols to aldehydes. Further conversion by aldehyde dehydrogenase produces alkoxyacetic acids. The conversion by ADH and the resulting toxicity of the glycol ether can be blocked by pyrazole and other ADH inhibitors. Alcohol also inhibits this conversion. The higher molecular weight glycol ethers, (di- and triethylene glycol ethers), are believed to be poor substrates for ADH and are partially cleaved by the action of P-450 isozymes. Alkoxyacetic acids are significant if not major urinary metabolites of glycol ethers in both animals and humans and are the only toxicologically significant metabolites that have been detected *in vivo*. Although indirect evidence suggests that the aldehydes form under *in vivo*

```
       RCO₂H ◄─── RCHO              +            HOCH₂CH₂OH* ──► CO₂*

    ROCH₂CH₂OSO₃H*                                ROCH₂CH₂OGluc*
                         ↖                      ↗
            Conjungation with            Conjungation with
                sulfate                    glucuronic acid

       ROCH₂CH₂OC(O)CH₃  ──►  ROCH₂CH₂OH
     Ethylene glycol ether acetate   Ethylene glycol ether
                                          ↓
                                    ROCH₂CHO
                                 2-Alkoxyacetaldehyde
                                          ↓             ?
                                    ROCH₂CO₂H*  ──► ──► CO₂*
                                   2-Alkoxyacetic acid
                                          ↓
                         2-Alkoxyacetylglycine* (rodents)
                                          or
                         2-Alkoxyacetylglutamine* (humans)
```

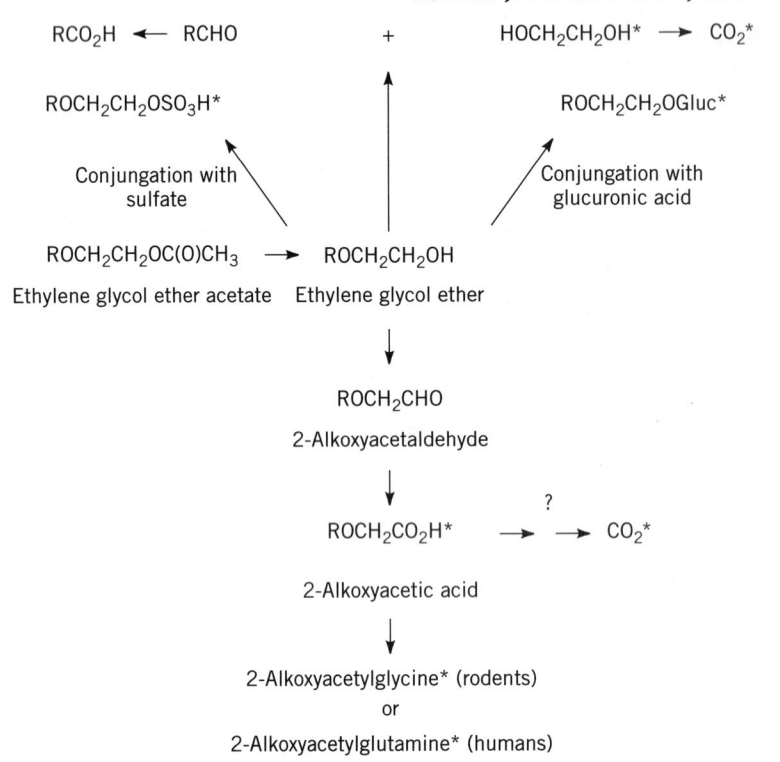

Figure 86.1. Metabolic pathways for ethylene glycol ethers in rodents and humans. Metabolites marked by an asterisk (*) have been identified for one or more glycol ethers.

conditions, these have not yet been detected in whole animal studies. In animal studies, ethylene glycol has been identified as a minor metabolite of glycol ethers. Sulfate and glucuronide conjugation of the parent glycol ethers has been reported as have the glycine (rodents) and glutamine (humans) conjugates of the alkoxyacetic acid metabolites. The formation of ethylene glycol under normal conditions of exposure is unlikely to contribute to the toxicity of glycol ethers. None of the other conjugates that have been identified has been associated with a toxicological response, and the formation of these is presumed to represent detoxification (24–26).

Mechanism of Action

The alkoxyacids, methoxyacetic and ethoxyacetic acid (MAA and EAA), formed during the metabolism of EGME and EGEE, are presumed responsible for the toxic effects observed in reproductive and developmental studies (27, 28). Butoxyacetic acid formed from EGBE is responsible for red cell hemolysis in sensitive species (28). The effects of MAA and EAA on development and reproduction may involve the introduction of these acids into the Kreb's cycle by the formation of methoxy- or ethoxyacetyl-CoA and by the formation of methoxy- or ethoxycitrate by mitochondrial enzymes. The effects are partially attenuated by administering acetate, serine, or other oxidizable carboxylic acids (29).

E Toxic Effects of Glycol Ethers

Structure–Activity

Glycol ethers generally have low acute, single-dose toxicity, and LD$_{50}$ values generally range from 1.0 to 4.0 g/kg of body weight. In animals and humans, high-dose administrations (>350 mg/kg) result in central nervous system depression, although the results from many studies show no specific damage to neural tissues. Other toxicological effects attributable to glycol ethers are associated with metabolism to the corresponding alkoxyacetic acids. In the case of EGME, EGEE, and certain other glycol ether derivatives, significant reproductive, developmental, hematologic, and immunologic effects have been associated with the formation of either methoxyacetic (MAA) or ethoxyacetic acids (EAA). For other glycol ether derivatives substituted with propyl, butyl, or higher homologues, developmental effects secondary to maternal toxicity (without teratogenic effects), as well as hematologic effects, are observed.

Reproductive/Developmental

Testicular toxicity is consistently observed in animal studies with EGME, EGEE, and their acetates as well as with certain higher molecular weight homologues such as glyme and DGME. This effect is related to the formation of either MAA or EAA. EGME is the most potent in the series, and it produces testicular degeneration described histopathologically as moderate to severe degeneration of the germinal epithelium of the seminiferous tubules. These effects are seen in mice, rats, and rabbits following oral, dermal, or inhalation exposures. Acute doses larger than 100 mg/kg of EGME are required to produce effects in the mouse. No consistent testicular toxicity has been observed from glycol ether derivatives substituted with alkyl groups that have more than two carbons, (30–32).

As in the case of testicular toxicity, ethylene glycol ethers substituted with alkyl groups that have more than two carbon atoms are not considered selectively toxic to the fetus or developing embryo but, in certain cases, display toxicities that are secondary to maternal toxicity. In contrast, EGME has produced significant reproductive and developmental effects in animals at maternally nontoxic levels higher than 175 mg/kg and from all routes of administration. Effects include reduced sizes and numbers of litters, resorptions, and malformations. With EGME, exencephaly, paw anomalies and other skeletal effects are observed in mice. With EGEE, cardiovascular defects and skeletal anomalies were seen in rats and rabbits after inhalation exposures; however, effects are observed at considerably higher dose levels than for EGME. All glycol ethers that produce frank teratogenicity can also metabolize to either MAA or EAA. The notable exception is DGEE, which showed no significant developmental toxicity in mice, rats, or rabbits. This latter result may be due to the low amounts of the EAA metabolite formed, as well as to the lower relative activity of this acid versus that of MAA (33, 34).

Hematologic Effects

In vivo administration of glycol ethers to certain animal species causes acute hemolytic anemia. Ethylene glycol monobutyl ether (EGBE) is the most potent of the glycol ethers, and it produces hemolytic toxicity in sensitive species by all routes of administration.

Hemolysis is preceded by swelling of red blood cells as evident from increased mean cell volume (MCV) and hematocrit (HCT), increased osmotic fragility and morphological alterations followed by a decrease in the number of circulating red blood cells. Hemolysis results in elevating of plasma hemoglobin, hemoglobinuria, focal necrosis of the liver, and deposition of intracytoplasmic hemoglobin in kidneys. Butoxyacetic acid (BAA), the major metabolite of EGBE, is largely responsible for its hemolytic activity. *In vivo* administration of ethylene glycol ethers demonstrates that hemolytic activity increases as the length of the alkyl carbon chain increases and EGBE is the most potent. Human red blood cells are more resistant to the hemolytic effects of BAA than the rat's (28). *In vitro* studies indicate that approximately 0.5 mM BAA is required to produce slight hemolysis of rat red blood cells, and nearly complete hemolysis occurs at 2 mM following 4-h incubations. In contrast, concentrations of 8 mM BAA cause only slight hemolysis of human red cells.

The primary effects of EGBE and EGPE are hemolysis of red blood cells, but the shorter alkyl-chain glycol ethers cause myelotoxicity. EGME and EGEE administration caused bone marrow hypocellularity, suppression of granulocyte-macrophage progenitor colony formation, and inhibited erythropoiesis in both sexes of B6C3F1 mice. The relative potency ranking of the three chemicals was EGME > EGEE > EGBE, and EGBE was relatively inactive.

Immunologic Effects

Recent studies have revealed that EGME and its metabolite MAA are immunosuppressive in rats, and dosages of 50–200 mg/kg/day for 10 days resulted in thymic involution in the absence of reductions in body or spleen weight; reduced lymphoproliferative (LP) responses to the T cell mitogens concanavaline A, phytohemaglutinin, and several other mitogens; depressed production of interleukin-2; and alterations in primary antibody plaque-forming cell (PFC) response to sheep red blood cells and trinitrophenyl lipopolysaccharide. The PFC response was one of the most sensitive immunologic end points for EGME and MAA-induced immunosuppression in the rat. Mice are not affected by exposure to EGME. The underlying reason(s) for the difference in sensitivity is unknown (35).

Genetic Toxicity/Carcinogenicity

A 2-year chronic EGBE inhalation study was conducted recently in F344/N rats and B6C3F1 mice (36, 37). Whole body exposure to air concentrations of 0, 31.2, 62.5, or 125 ppm (rats) and 0, 62.5, 125, and 250 ppm (mice) were carried out 6 h/day, 5 days/week for up to 18 months. No treatment-related effects were observed in male rats. In female rats, combined or malignant pheochromocytomas of the adrenal medulla increased at the highest exposure level; however, this increase was not statistically significant compared with concurrent (unexposed) controls. In female mice and to a lesser extent in males, increased incidences of forestomach squamous cell papillomas or carcinomas were observed. In male mice, hemangiosarcomas of the liver also significantly increased. All effects seen in rats or mice were only in the highest exposure groups. There were no indications of carcinogenicity in a 2-year rat dietary feeding study of EGEE at a dose level equivalent to 0.9 g/kg/day (38). The ethylene glycol ethers and their alkoxyacid metabolites are generally inactive in tests involving *Salmonella typhimurium* his-locus mutation, chromosomal aberrations in CHO cells, sister chromatid exchange in CHO cells, and mouse lymphoma L5178Y cell tk locus mutation. The alkyoxyacetaldehydes induced sister-chromatid exchanges, chromosomal aberrations, aneuploidy, and micronuclei in mammalian cells *in vitro* (39).

F Physiologically Based Pharmacokinetic/Pharmacodynamic Models

A number of oral, i.v., inhalation, and percutaneous physiologically based pharmacokinetic (PBPK) models were developed for EGME, EGEE, EGEEA, and EGBE to describe and predict the time course of glycol ethers and their metabolites in the body and/or the developing conceptus. The models were used to extrapolate data from one route of administration (i.e., oral, dermal, inhalation) to another and from one animal species to another (i.e., rat to humans) (40). A typical inhalation PBPK model consisted of five tissue compartments and lung space. Flows (liters) are alveolar, cardiac, and those to fat, liver, muscle, and poorly and richly perfused tissues. Compartment volumes are in liters for blood, liver, fat, muscle, poorly and richly perfused tissues. Venous blood concentrations depend on the nature of the glycol ether and glycol ether metabolite concentrations in tissues and their tissue/blood partition coefficients. Metabolic removal of glycol ethers and metabolites are described kinetically by maximum enzymatic rates. V_{max} values and Michaelis constants (K_m) for metabolic reactions lead to the formation and removal of MAA, EAA, and BAA (23, 41, 42). Diagrams of the structure of these PBPK models are given in Figures 86.2 and 86.3 for EGME and EGBE.

Pharmacodynamic Considerations, Organogenesis, and Hemolysis

The pharmacodynamic properties of the glycol ethers (i.e., effects on reproduction and development, red blood cells, etc.) in part depend on the maximum concentration of MAA, EAA, and BAA (i.e., C_{max}) in target tissues (i.e., testes, embryo, fetus, and blood) and/or the area under the time–concentration curve (AUC) for target tissues. Other factors such as the stage of development (organogenesis) play even more important roles (22, 23, 43).

G Environmental Risk Assessment

Ethylene glycol ethers and acetates may enter the environment from manufacturing effluents and emissions and as a result of their use in commercial products. In a recently published review by Staples et al. (44), environmental fate and toxicity data were reviewed and summarized for a series of ethylene glycol ethers and acetate derivatives. Based primarily on physical and chemical properties, these materials, if released to the environment, would tend to remain in the water column and would not have a significant potential to bioconcentrate within aquatic organisms or to be adsorbed onto soils or sediments. These materials will not be persistent and would generally be classified as "practically nontoxic" according to generally accepted criteria. In the aquatic environment, aerobic biodegradation constitutes the primary removal mechanism; volatilization, hydrolysis, and photolysis are negligible. In the atmosphere, reaction with photochemically produced hydroxyl radicals results in rapid removal rates.

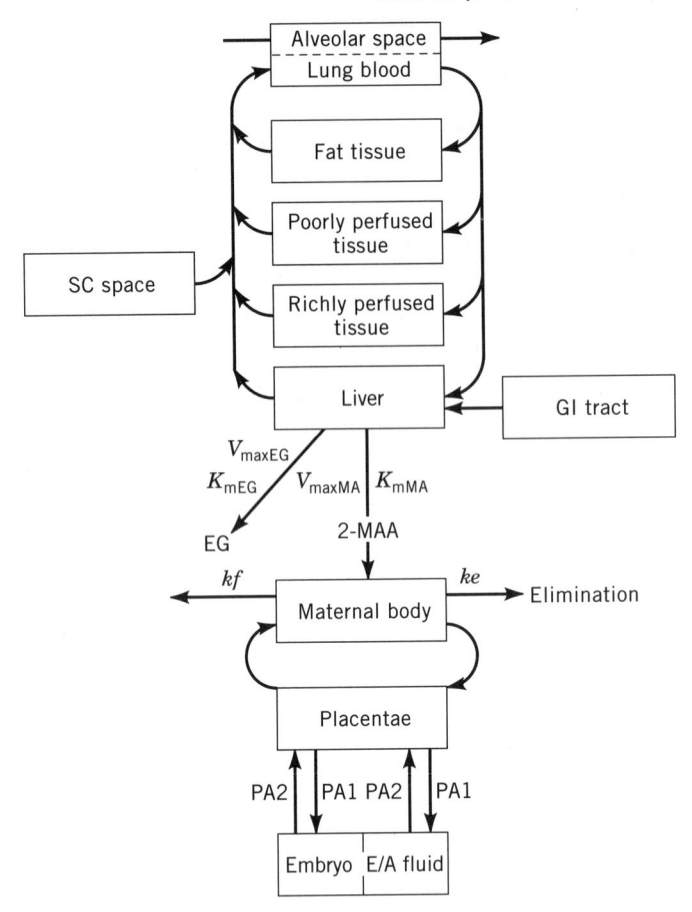

Figure 86.2. Physiologically based pharmacokinetic model used to describe the disposition of 2-methoxyethanol and its major metabolite, methoxyacetic acid, in the pregnant mouse. E/A fluid is the combined exocoelomic and amniotic fluids that surround the embryo (from Clarke et al. (41), reproduced with permission).

1.0 Ethylene Glycol Monomethyl Ether

1.0.1 CAS Number: [109-86-4]

1.0.2 Synonyms: 2-Methoxy ethanol; monomethyl ether of ethylene glycol; methyl glycol

1.0.3 Trade Names: Methyl Cellosolve®; Glycol Ether EM

1.0.4 Molecular Weight: 76.09

1.0.5 Molecular Formula: $C_3H_8O_2$

1.0.6 Molecular Structure: ⁄O⁀OH

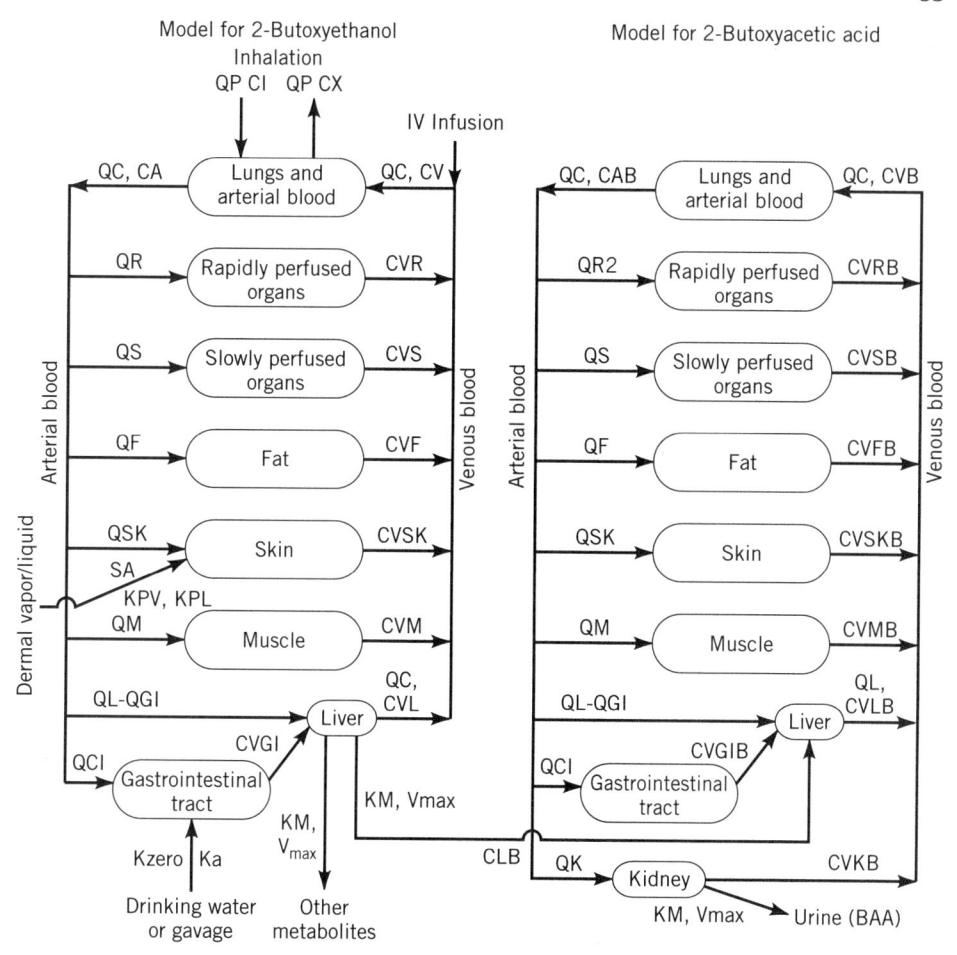

Figure 86.3. Physiologically based pharmacokinetic model for 2-butoxyethanol and its major metabolite, butoxyacetic acid, in rats and humans (from Corley et al. (23), reproduced with permission).

1.1 Chemical and Physical Properties

1.1.1 General

See Table 86.5.

1.1.2 Odor and Warning Properties

Under controlled conditions and using human volunteers, May (45), found that the odor threshold for EGME was approximately 60 ppm and the level of strong odor was 90 ppm. In another study, human volunteers were exposed to levels of 25 and 115 ppm (46) for a full working day. The 25-ppm level was considered the maximum odor level tolerated by three of five people, whereas the intolerable level was more than 115 ppm.

Table 86.5. Physical and Chemical Properties of Ethers of Ethylene Glycol

Property	Methyl	Ethyl	n-Propyl	Isopropyl	n-Butyl	Isobutyl	n-Hexyl	2-Methyl-pentyl	2-Ethylhexyl	2,6,8-Trimethyl-4-nonyl	Dimethyl	Diethyl	Phenyl
CAS Number	[109-86-4]	[110-80-5]	[2807-30-9]	[109-59-1]	[111-76-2]	[4439-24-1]	[112-25-4]	[29290-45-7]	[1559-35-9]	[10137-98-1]	[110-71-4]	[629-14-1]	[122-99-6]
Molecular formula	$C_3H_8O_2$	$C_4H_{10}O_2$	$C_5H_{12}O_2$	$C_5H_{12}O_2$	$C_6H_{14}O_2$	$C_6H_{14}O_2$	$C_8H_{18}O_2$	$C_8H_{18}O_2$	$C_{10}H_{22}O_2$	$C_{14}H_{30}O_2$	$C_4H_{10}O_2$	$C_6H_{14}O_2$	$C_8H_{10}O_2$
Molecular weight	76.09	90.12	104.15	104.15	118.17	118.18	146.23	146.2	174.3	230.4	90.122	118.18	138.17
Specific gravity 25°C	0.962 0.960 @ 20°C	0.926 0.9311 @ 20°C	0.913	0.91	0.898 0.9012@ 20°C	0.887	0.889 (20/20°C)	0.888 (20/20°C)	0.883	0.886 (20/20°C)	0.8683 (20°C)	0.8484	1.102
Boiling point (°C)	124.2	135.6	149	139.5–144.5	171.2	160.3	208.3	197.1	215	227 (300 mmHg)	85.0	121.4	245
Freezing point (°C)	−85	−170			−70		−45	< −80 (sets to glass)		< −50 (sets to glass)	−58	−74	14
Vapor pressure (mmHg at 25°C)	9.7	5.75 3.8 torr@ 20°C	2.9	5.2 2.6 torr@ 20°C	0.88 0.76 torr@ 20°C	1.6	0.1 (20°C)	0.9 (20°C)	0.12	< 0.01 (20°C)	61.2 (20°C)	12.5	0.0073
Refractive index (25°C)	1.400	1.406	1.412	1.407	1.417	1.413	1.4290	1.428 (20°C)		1.439 (20°C)		1.3922 (20°C)	1.536
Flash point (°F), open cup (closed cup)	41.7°C (46.1°C)	49°C (44°C)	51°C	69°C (60°C)	165	142	190	88	208		−6°C	35°C	130°C
Autoignition temperature (°C)	285	235		244.44	244				227				
Flammability limits (vol. % in air)	2.5–19.8 (125–140°C)	1.82–14.0 (140–150°C)			1.13–10.6 (160–180°C)				0.61 (121°C)				
Vapor density (air = 1)	2.6	3.0			4.1	4.1	5.04	5.06	6.0	7.97	3.11	4.1	4.8
Percent in saturated air (25°C)	1.28	0.76	0.38	0.68	0.093	0.21		0.01				1.64	0.00096
1 ppm ≈ mg/m³ at 25°C, 760 mmHg	3.11	3.68	4.25	4.25	4.83	4.83	5.98	5.98	7.13	9.42	3.68	4.83	5.65
1 mg/l ≈ ppm at 25°C, 760 mmHg	322	272	235	235	207	207	168	168	140	106	272	207	177

1.2 Production and Use

See Table 86.2 for production and use data. The manufacturers of EGME recommend that it not be used in consumer products.

1.3 Exposure Assessment

1.3.1 Air

See the Introduction for a general description of air sampling and analytical procedures.

1.3.3 Workplace Methods

See the Introduction for a general description of sampling and analytical procedures for the workplace.

1.3.5 Biomonitoring/Biomarkers

1.3.5.2 Urine. Analysis of Methoxyacetic Acid in Urine. A selective and sensitive method for determining methoxyacetic acid (MAA) in urine was developed by Groeseneken et al. (47) as a biological monitoring method. The method involved freeze-drying urine adjusted to pH 8 to 8.5 with KOH, acidifying the dried urine with a small volume of HCL in methylene chloride, and methylating the acid (MAA) using a solution of diazomethane. The methylated acid was analyzed by gas chromatography on a CP WAX 57 CB, WCOT fused silica column (25 m × 0.33 mm i.d., 0.22 mm film thickness). More recently, Shih et al. (48) reported an improved method for analyzing MAA, ethoxy acetic acid (EAA), and butoxyacetic acid (BAA) in urine that involves acidfication (HCl) and extraction with a mixture of methylene chloride and isopropanol followed by gas chromatography/mass spectrometry without preparing of the methylated derivative.

1.4 Toxic Effects

1.4.1 Experimental Studies

Ethylene glycol monomethyl ether (EGME) has low single-dose oral toxicity and moderate repeated-dose oral toxicity. It is mildly irritating to the skin and eyes. EGME is absorbed through the skin, but it has low acute toxicity by this route. It is appreciably toxic when inhaled. Its vapors are irritating in acutely toxic concentrations. Concentrations that may cause serious systemic toxicity upon prolonged and repeated inhalation have negligible warning properties. The material exerts its principal toxicological action on the brain, blood, kidneys, thymus, and testes. EGME is embryo toxic and teratogenic in experimental animals by all routes of exposure.

1.4.1.1 Acute Toxicity. Oral. Gross (49) fed rabbits repeated daily doses of EGME and found that seven doses of 0.1 mL/kg caused temporary hematuria. Larger doses caused exhaustion, tremors, albuminuria, hematuria, and death. Autopsy revealed severe kidney injury. According to Carpenter and co-workers (50), the single-dose oral toxicity of EGME is 3.4 g/kg for rats, 0.89 g/kg for rabbits, and 0.95 g/kg for guinea pigs. Union Carbide

(51) lists the LD_{50} for rats as 2.46 g/kg, whereas Saparmamedov (52) found that the LD_{50} for mice is 2.8 g/kg when administered via an oil solution. If the ether is given in water, the LD_{50} is slightly higher. In massive doses, the material is narcotic, but at lower dosage levels, deaths are delayed and are accompanied by lung edema, slight liver injury, and marked kidney injury. Hemoglobinuria may occur from single doses.

Nagano et al. (32) fed mice doses ranging from 62.5 to 4000 mg/kg, 5 days/week for 5 weeks, and observed testicular atrophy and leukopenia; the intensity was dose-related. Testicular injury from higher doses was reported earlier by Wiley et al. (30). In another unpublished report (53), the LD_{50}s in fasting and fed rats were 2.3 and 3.9 g/kg, respectively, and the LD_{50} values for fasting and fed mice were 3.9 and 4.5 g/kg, respectively.

EGME administered orally to rats for four consecutive days at dose levels of 100 or 500 mg/kg produced thymic atrophy, lymphocytopenia, and neutropenia (54). Treatment with EGME also abolished splenic extramedullary hematopoiesis. Toxicity is related to a direct effect on bone marrow. Recovery occurred after treatment terminated. Testicular atrophy was evident in animals treated with 500 mg/kg and persisted after treatment terminated.

Injection. Gross (49) gave repeated subcutaneous injections of EGME to guinea pigs and rabbits. His results show that, whereas seven daily injections of 0.25 mL/kg caused no adverse signs in the guinea pig, five injections of either 0.5 or 1.0 mL/kg caused prostration, labored breathing, and death. The response of rabbits was quite similar, except that the rabbit is slightly more resistant than the guinea pig to the materials effects; deaths occurred after seven injections of 1.0 mL/kg.

Carpenter et al. (50) found that EGME in concentrations of more than 25% in 0.75% sodium chloride solution was hemolytic to rat erythrocytes. When this 25% solution was given intravenously to rats, the LD_{50} was 2.7 g/kg; for the undiluted material, the LD_{50} was 2.2 g/kg. When the undiluted material was given intraperitoneally to the rat, the LD_{50} was 2.5 g/kg. Wiley et al. (30) injected (site unspecified but believed to be intramuscular) two dogs with 6 mL and two rabbits with 2 mL daily. One rabbit died after two injections, and one dog was anuric after the third (last) injection. Clinical examination and autopsy of the animals treated with EGME revealed anuria, calcified casts in the urine, irritation of the bladder mucosa, hemorrhage in the gastrointestinal tract, lung edema, and liver and testicular injury.

Inhalation. The toxicity of EGME by inhalation was determined for several animal species. Perhaps the most pertinent studies in this area have been conducted by Werner and co-workers (55–57). This work has been well summarized by Smyth et al. (58) and Browning (59). Werner et al. (55) exposed mice for 7 h to various concentrations and found that the LC_{50} is 1480 ppm based on deaths 3 weeks after exposure. They concluded that lung and kidney injury was generally the cause of death.

Gross (49) found that a few repeated exposures to 800 or 1600 ppm produced serious systemic intoxication that was characterized for the most part by irritation of the respiratory tract and lungs, hematuria, albuminuria, cylindrical casts in the urine, and severe glomerulitis.

Werner et al. (56) exposed groups of rats 7 h/day, 5 days a week for 5 weeks to a concentration that averaged 310 ppm. After 1 week of exposure, they noted that the percentage of immature granulocytes in the circulating blood increased. They observed no changes in the kidneys or lungs.

Goldberg et al. (60) found that a single exposure of mice to 125 ppm for 4 h potentiated the hypnotic effects of barbiturates; at 500 ppm for 4 h, spontaneous motor activity decreased. This concentration had no effect on rolling roller performance but did produce a 100% increase in barbiturate sleeping time. Goldberg et al. (60) studied the effects of repeated (2, 5, 8, 11, and 14 days) 4-h inhalation exposures at 0, 125, 250, and 500 ppm on the inhibition of conditioned avoidance-escape behavior. Eight days of repeated exposure at 500 ppm was required to inhibit the conditioned response in 50% of the animals. Exposures longer than 8 days failed to inhibit the conditioned response further. No impairment of motor capacity was observed. Dajani (61) also studied the effect of this material on the behavior of mice and rats. Inhibition of conditioned avoidance response in rats was seen only when lethal or near-lethal exposure occurred at 3317 ppm by inhalation.

Dermal. EGME is absorbed readily through the skin in toxic amounts. Quantification by the "sleeve" technique, essentially as described by Draize et al. (62), indicated that the LD_{50} was approximately 2 g/kg for rabbits (63). In similar tests, LD_{50} values for rabbits were 1.29 g/kg (51). In a later study, the dermal LD_{50} in rabbits was reportedly 3.9 g/kg (64). Saparmamedov (52), however, found that when the compound was applied to mouse skin, there were no signs of toxicity. The signs of intoxication from absorption through the skin are essentially the same as those from other routes of administration.

Eye Irritation. When EGME was introduced into the eyes of rabbits, it produced immediate pain, conjunctival irritation, and slight transitory cloudiness of the cornea, which cleared within 24 h (65). Carpenter and Smyth (66) classify the material as similar to ethanol for its effect on the rabbit eye. Grant (67) reported one recorded human eye exposure incident in which complete recovery occurred within 48 h after exposure.

Skin Irritation. EGME in repeated and prolonged contact with the skin of rabbits failed to cause any appreciable irritation (65). Krasavage and Terhaar (64) reported slight skin irritation according to the method of Draize et al. (62).

1.4.1.2 Chronic and Subchronic Toxicity.
Oral. In studies in which rats were fed diets containing 0.01 to 1.25% EGME for 90 days, the lowest level depressed appetite. Initial pathological effects were seen at the 0.05% level. The 1.25% level caused death (51).

Drinking Water. Dieter (68) administered EGME in drinking water to F344/N rats and B6C3F1 mice in 2-week and 13-week studies. In the 2-week studies, groups of rats and mice (5 males and females) received EGME at dosages that ranged from 100 to 400 mg/kg and 200 to 1300 mg/kg, respectively. No chemically-related effects on survival were observed for either the rats or mice. Decreased body weight gains were noted for both male and female rats along with a dose-related decrease in water consumption. Thymic atrophy was noted in male and female rats and testicular atrophy in male rats and mice. In the longer 13-week EGME drinking water studies (750 to 6000 ppm) in rats, groups of 10 males and females consumed 70 to 800 mg/kg/day. Mice consumed amounts ranging from 300 to 1800 mg/kg. Mortality occurred in male and female rats administered 4500 to 6000 ppm EGME, but no mortality in mice occurred at all levels. In rats, treatment-related histopathological changes were observed in the testes, thymus, and hematopoietic tissues (spleen, bone marrow, and liver). Dose-related degeneration of the germinal epithelium in the seminiferous tubules of the testes was present in rats administered EGME for 60 days at 3000 and 1500 ppm, and degeneration at 3000 ppm was marked. Recovery periods of 30 and 56 days resulted in only partial recovery from testicular degeneration. Administration of EGME to rats in drinking water caused a progressive anemia associated with cellular depletion of bone marrow and fibrosis of the splenic capsule. EGME had similar effects on the testes, spleen, and adrenal glands (females only) of mice. A no-observed-effect level (NOAEL) was not reached in the 13-week rat study because testicular degeneration in males and decreased thymus weights in male and females occurred at the lowest administered concentration (750 ppm). In male mice, the 13-week NOAEL for testicular degeneration and increased hematopoiesis in the spleen was 2000 ppm. A NOAEL was not reached for female mice because adrenal gland hypertrophy and increased hematopoiesis in the spleen occurred at the lowest administered concentration (2000 ppm).

Inhalation. Werner et al. (57) exposed two dogs to a vapor concentration of 800 ppm of EGME for 7 h/day, 5 days a week for 12 weeks. Again, the most significant changes were in the blood. The hemoglobin concentration, cell volume, and the number of erythrocytes decreased. The red cells showed increased hypochromia, polychromatophilia, and microcytosis. The blood picture, in regard to white cells, was characterized by a greater than normal number of immature forms. These authors point out that the methyl ether, which is significantly less hemolytic *in vitro* than the propyl and butyl ethers of ethylene glycol, altered the red cells in the dogs the most. This is quite a different finding from that in mice (55) and rats (56), where there is a definite correlation between hematologic effects and hemolytic potency. The site at which these blood changes occur is obscure. The lack of significant amounts of hemosiderin in the spleen suggests that it is not hemolytic. Although no serious damage to the bone marrow was observed in these studies, it is doubtful that the studies were sufficiently prolonged to demonstrate that the effect was not centered in the marrow.

Subchronic vapor inhalation studies were conducted by Miller et al. (69) using male and female Sprague–Dawley rats and New Zealand white rabbits. The animals were exposed for 6 h/day, 5 days/week, for a total of 13 weeks at concentrations of 0, 30, 100, and 300 ppm. Hematologic analyses revealed decreases in hemoglobin concentration, packed cell volume, white blood cell counts, and platelet counts, as well as a tendency toward decreased red blood cell counts in both male and female rats and rabbits exposed to 300 ppm EGME. There were no significant hematologic changes in either male or female rats or rabbits in the 100- or 30-ppm groups. Body weights of male and female rats in the 300-ppm group and females in the 100-ppm group were significantly lower than those of controls. Relative thymus weights of male and female rats and relative testes weights of males in the 300-ppm groups were significantly lower than controls. Liver weights were not altered at concentrations up to 300 ppm. Relative thymus and testes weights of rabbits exposed to 300 ppm EGME were significantly lower than controls. Body weight and relative liver weights of rabbits were not significantly altered by EGME. There were no treatment-related histopathological observations in rats exposed to 100 or 30 ppm EGME. Histopathological observations of rats in the 300-ppm group revealed bilateral, severe degeneration of testicular germinal epithelium, as well as lymphoid tissue atrophy and decreased liver glycogen in rabbits in the 300-ppm group. Similar but less severe testicular changes were seen in three of five rabbits exposed to 100 ppm EGME, and slight degenerative changes were found in one of five rabbits exposed to 30 ppm EGME. A larger follow-up study sponsored by the CMA (70) resulted in no testicular effects in rabbits exposed to 3, 10, or 30 ppm EGME vapors for 13 weeks. A concentration of 30 ppm is considered the no-adverse-effect level for testicular effects in rabbits.

1.4.1.3 Metabolism, Pharmacokinetics, and Mechanisms. Metabolism. According to Zavon (71), EGME is metabolized to methoxyacetic acid. This was confirmed in the male rat by Miller et al. (72) by orally administering ^{14}C-labeled EGME at dose levels of 76.1 and 660 mg/kg and analyzing urine, feces, and respiratory air for metabolites. Urine was the major route of elimination; more than 50% of the administered dose was excreted in urine, 12% as respiratory CO_2, 0.4% as volatiles, 2.7% in the feces, and 18% remained in tissues after 48 h. The urinary metabolites were separated by an ion-exclusion chromatographic system using a cation exchange resin. Gas chromatography and (or with) mass spectrometry were used to identify the separated products. One major peak that contained 90 to 95% of the urinary metabolites was identified as methoxyacetic acid. A study by Moss et al. (73) revealed the presence of methoxyacetic acid and methoxyacetylglycine in the urine of rats that were administered EGME, and methoxyacetylglycine was the smaller of the two metabolites.

Mebus et al. (74) studied the metabolism of EGME (3.3 mmole/kg) in the presence of tracer doses (2 µCi) of [1-^{14}C]-methoxyacetic acid, [2-^{14}C]-2-methoxyethanol and [methoxy-^{14}C]-2-methoxyethanol using glass metabolism cages to collect respiratory $^{14}CO_2$. Five to six percent of the administered radiolabeled dose (all three labels) was eliminated as respiratory CO_2. The majority of the administered radiolabeled dose was eliminated in urine (87.8 to 93.7%), and 70 to 80% was present in 24-h urine samples. Two metabolites were separated by HPLC and identified after derivatization and analysis by GC/MS as 2-methoxyacetic acid (2-MAA) and 2-methoxy-*N*-acetylglycine (2-MNA). A third metabolite that eluted on the HPLC column just before 2-MAA and comprised 12 to 18% of the dose was not identified. Mouse embryos incubated *in vitro* with 2-MAA converted [1-^{14}C]-2-MAA and [2-^{14}C]-2-MAA to CO_2, but not [methoxy-^{14}C]-2-MAA (74).

Sumner et al. (25, 26) fully characterized the urinary metabolites of [1,2-methoxy-^{13}C]-2-methoxyethanol administered to male F-344 rats and pregnant CD-1 mice using NMR spectroscopy. The metabolic pathway for EGME is given in Figure 86.4 and the identified metabolites (attenuated in some cases by addition of L- and D-serine or acetate) as a percentage of total excreted metabolites in 24-h urine samples are shown in Table 86.6. Most of the urinary metabolites (82.4–92.4%) from the rat and mouse were derived by metabolism of EGME to 2-MAA.

Moslen et al. (75) studied the biotransformation of EGME by alcohol dehydrogenase in the testes in a number of animal species including humans. EGME biotransformation was detected in the testes of the Wistar rat and one strain of mice but not in testes from hamster, guinea pigs, rabbits, dogs, cats, or humans. Hepatic capacities for EGME biotransformation by alcohol dehydrogenase varied from 22 to 2.5 nmol/mg of protein/minute in a species rank of hamsters ≫ rats = mice > guinea pigs = rabbits. The results do not explain species and strain differences in susceptibility to EGME toxicity.

```
H₃COCH₂CH₂OH  ──────────→  H₃COCH₂CO₂H
Methoxyethanol              Methoxyacetic acid
```

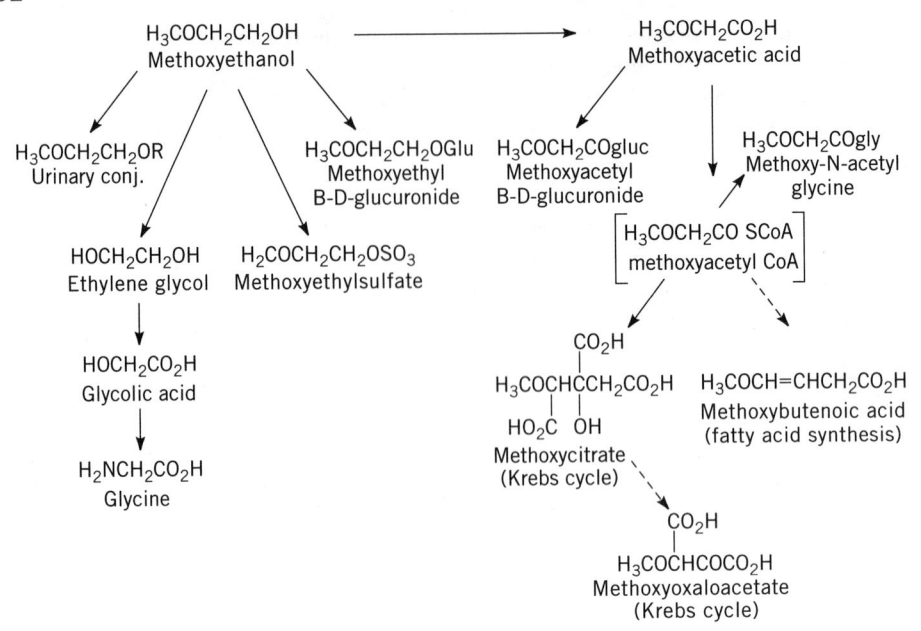

Figure 86.4. Proposed metabolic pathway for 2-methoxyethanol (from S. Jenkins-Summer et al. (25), reproduced with permission).

Pharmacokinetics. Dajani (61) reported that EGME was found equally distributed in brain, plasma, lung, and liver 1 h after administration and that the half-life in the body was approximately 1 to 2 h, unless dose levels were near the lethal level. In inhalation studies, the plasma levels of EGME increase nearly linearly from exposures of 1, 2, 4, and 6 h to 3317 ppm. When the exposure was extended to 8 h, the concentration in the plasma was more than double that found after the 6-h exposure, suggesting that metabolic and/or excretory mechanisms were saturated. The material was detected in rat urine 30 min after an intraperitoneal administration of 1.0 mg/kg and remained present for a total of 7 h after the injection, 3 h after it was last seen in the blood. Aasmoe and Aarbakke (76) found gender differences in the elimination of EGME and 2-MAA by male and female rats. The investigators conclude that the rate of EGME elimination is significantly faster in female than in male rats. No gender difference was seen in the elimination of 2-MAA, and the elimination of 2-MAA was much slower than EGME.

Physiologically Based Pharmacokinetics. In the last few years, interest has increased in developing physiologically based pharmacokinetic (PBPK) models for describing the pharmacokinetics of chemicals and their metabolites and for extrapolating the results from one route of administration to another and between animal species including humans. PBPK models require metabolic pathway data, enzymatic and metabolic rate constants (V_{max} and K_m), and tissue partition coefficient data for parent compounds and their metabolites. PBPK models applicable to organogenesis were developed by Clarke et al. (41) and Welsch et al. (77) to extrapolate the risk of adverse prenatal effects across doses or species following exposure to EGME.

Table 86.6. Metabolites in 24-Hour Urine (% of Total Excreted Metabolites)[a]

Metabolite[b]	Mice 250 mg/kg	Mice 25 mg/kg	Rat 250 mg/kg	Rat 25 mg/kg	Mice 250+ L-serine	Mice 250+ D-serine	Mice 250+ acetate	Mice 25+ acetate
EG	2.1	6.4	6.1	13.1	1.7	5.3	2.4	5.2
2-MAA	59.0	50.1	60.4	55.1	50.1	56.4	59.1	42.2
2-MNG	25.5	29.5	12.1	15.9	31.9	20.9	20.9	15.4
KC	6.4	9.1	10.0	14.1	5.8	7.0	3.3	3.6
FAS	1.5	N.D.[c]	0.2	N.D.	1.6	1.0	1.0	N.D.
2-ME-derived	7.6	11.3	17.3	15.9	10.6	14.7	15.6	38.9
2-MAA-derived	92.4	88.7	82.7	84.1	89.4	85.3	84.4	61.1

[a] Jenkins-Sumner et al. (24).
[b] EG, ethylene glycol; 2-ME-derived (glycine, glucuronide and sulfate conj. and glycolic acid); KC, Kreb cyclic intermediates (methoxyacetyl-CoA, methoxycitrate, etc.); FAS (fatty acids); 2-MNG, 2-methoxy-N-acetyl glycine; 2-MAA-derived (metabolites requiring activation of 2-ME to 2-MAA).
[c] N.D., nondetectable.

Physiological Organogenesis Models. The original flow-limited placental transfer organogenesis model by Clarke et al. (41) was developed to describe the deposition of EGME and 2-MAA following orally or subcutaneously administered EGME in pregnant CD-1 mice. Pharmacokinetic data on the disposition of MAA (2-methoxyacetic acid) were used to develop the model. A description of the PBPK model is given in Figure 86.2. This model that describes the dosimetry of 2-MAA for a single day (gd 11) in mouse development was further developed by Terry et al. (78–80) (see Fig. 86.5). Data were

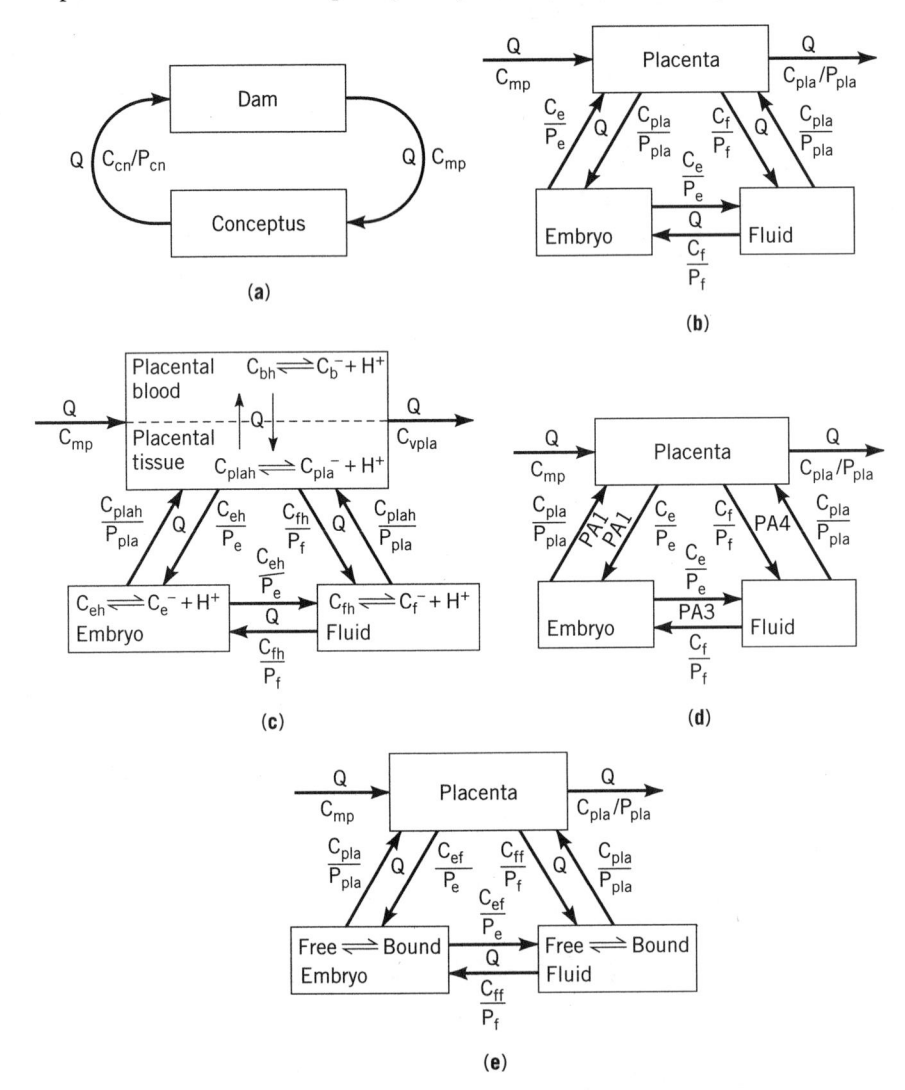

Figure 86.5. Model variants for the mouse gestation PBPK model. (A) Flow-limited. 9d 8 model for 2-MAA. (B) Flow-limited gd 11 and 13 model for 2-MAA. (C) pH trapping model for 2-MAA on gd 11 and 13. (D) Active transport model for 2-MAA on gd 11 and 13. (E) Reversible binding model for 2-MAA on gd 11 and 13 (from Terry et al. (79), reproduced with permission).

collected on gestation days (gd) 8, 11, and 13. Pharmacokinetics and tissue partition coefficients for 2-MAA were determined in maternal plasma and conceptus on gd 8, and in plasma, embryo, and extraembryonic/amniotic fluid (EAF) on gd 11 and 13. For simulation of the gd 8 data, the conceptus was described as a single compartment and analyzed as a "decidual swelling." On gd 11 and 13, the placenta, embryo, and EAF were described. Several hypotheses were modeled and tested to predict 2-MAA dosimetry. The model variants shown in Figure 86.5 depict (A) a flow limited gd 8 model for 2-MAA, (B) a flow limited gd 11 and 13 model for 2-MAA, (C) a pH trapping model for 2-MAA on gd 11 and 13, (D) an active transport model for 2-MAA and gd 11 and 13, and (E) a reversible binding model for 2-MAA on gd 11 and 13. The flow-limited, single-compartment model adequately predicted 2-MAA dosimetry on gd 8. On gd 11 and 13, the flow-limited, muticompartment model underpredicted all embryo data for 2-MAA by at least 40%.

When the original Clarke et al. (41) model for gd 11 was evaluated by using an expanded database, it failed to simulate 2-MAA kinetics accurately on gd 13. The pH trapping model showed clearly that pH gradients across embryonic membranes could not account for 2-MAA accumulation in the embryo. Models that describe active transport and reversible binding provided the best simulations of chemical deposition on both gd 11 and 13. The mouse gestation PBPK model was extrapolated to pregnant rats by incorporating data for the physiology of rat pregnancy (77). The model simulations were validated with biological data of maternal and embryo pharmacokinetics of 2-MAA collected on gd 13 and 15. Welsch et al. (77) extended their organogenesis PBPK modeling work in mice and rats to humans. This work is discussed in Section 1.4.2.3.3.

Physiological Parameters, Partition Coefficients, and Pharmacokinetic Constants. In the Clark et al. (41) PBPK model, tissue volumes were obtained by weighing individual tissues from five gd 11 mice. Approximately 90% of the total body weight was accounted for, including the weights of the individual conspectus and their component tissues, i.e., embryo proper, yolk sac membranes, and extraembyronic fluid. *In vitro* metabolic rate constants from Sleet et al. (27) and Green et al. (81) were initially used by Clarke et al. (41) in their PBPK flow-limited organogenesis model. Model simulations revealed that V_{max} and K_m values failed to describe the faster *in vivo* metabolism of EGME. The PBPK model was refined using rate constants obtained by optimization. The partition coefficients, (Table 86.4), were determined by *in vitro* methods, rather than by steady-state tissue/blood concentrations *in vivo*. The high EGME blood/air PC (34,913) in the pregnant mouse is similar to that obtained in humans (32,836) by Johanson and Dynesius (82) and in the pregnant rat (31, 342) by Clark et al. (40). The very low 2-ME fat/blood PC in the mouse (0.05) compared well with that in the pregnant rat (0.04)(40) and with an oil/human blood partition coefficient (0.016)(82).

New Pharmacokinetic Constants for PBPK Models. Green et al. (83) recently developed pathways and relative rates of metabolism of EGME, EGEE, and EGBE using rat and human hepatocytes. The concentrations used for each glycol ether in the metabolic rate studies were 0.02, 0.2, 2.0, and 10.0 mM. Metabolites were analyzed by HPLC. The percentage of EGME converted to MAA in rat hepatocytes was similar at all concentrations, whereas in the human hepatocytes, the percentage converted to MAA decreased between 0.2 and 2.0 mM. The V_{max} value for converting EGME to methoxyacetic acid by rat hepatocytes was similar to those obtained for converting EGEE to EAA and EGBE to

Table 86.7. Kinetic Parameters for Alkoxyacetic Acid Formation by Rat and Human Hepatocytes[a]

	Rat		Human	
Glycol Ether	V_{max}[b]	K_m[b]	V_{max}[b]	K_m[b]
EGME	1511[c] (1081, 1941)	6.3[c] (6.1, 6.5)	61.3[c] (61.3, 61.3)	1.7[c] (2.3, 1.1)
EGEE	1519[c] (1078, 1941)	6.6[c] (7.0, 6.1)	70.8 ± 76.0[d]	1.2 ± 1.0[d]
EGBE	741 ± 265[d]	0.9 ± 0.5[d]	113 ± 58[e]	0.9 ± 0.3[e]

[a]Green et al. (83).
[b]The units for V_{max} are nmol/h × 10^6 hepatocytes. The units for K_m are mM.
[c]Data represent the mean of two experiments. Individual values are presented in ().
[d]Data are the mean and SD of three experiments.
[e]Data are the mean and SD of four experiments.

BAA (Table 86.7). In human hepatocytes, V_{max} followed the order of EGBE > EGEE > EGME. V_{max} was 15- to 20-fold higher in rat than in human hepatocytes. Aasmoe et al. (84) studied the role of male and female rat liver alcohol dehydrogenase isozymes (ADH) in oxidizing of EGME, EGEE, EGBE, 2-pentyloxyethanol (2-PE), and 2-hexyloxyethanol (2-HE). The ADH activity (V_{max}, K_m) of a 100,000-g supernatant (cytosol) fraction (0.7–1.0 mg of protein) was determined in a reaction mixture composed of 0.1 M glycine–NaOH buffer, pH 10.4, 1.3 mM NAD, and glycol ethers as substrates (EGME, 0.50–20 mM). The rate of NADH formation was followed over time. Starch gel electrophoresis was used to separate the proteins and determine their activity One enzymatically active zone (ADH-3) was found. V_{max} and K_m values are given in Table 86.8. Differences in activity were found between males and females and between substrates, and isozymes from females possessed greater activity (V_{max}) against substrates and equivalent affinities (K_m).

Table 86.8. Rat Liver Alcohol Dehydrogenase Activities (V_{max}) and Affinity Constants (K_m) for a Series of Ethylene Glycol Ethers in Liver from Male and Female Rats[a]

	V_{max}[c] (nmol NADH/min/mg protein)		K_m[c] (mM)	
Substrate[b]	Male	Female	Male	Female
Ethanol	11.76 ± 3.75	20.41 ± 4.65	1.26 ± 0.29	1.23 ± 0.40
EGME	7.76 ± 2.81	14.27 ± 3.04	3.61 ± 1.13	2.73 ± 0.51
EGEE	7.82 ± 3.18	14.89 ± 2.84	2.23 ± 0.56	1.83 ± 0.51
EGBE	5.78 ± 1.79	8.65 ± 0.46	0.27 ± 0.11	0.18 ± 0.05
EGPhE	3.04 ± 0.78		0.23 ± 0.16	
EGHE	1.66 ± 0.27		0.15 ± 0.07	

[a]Aasmore et al. (84)
[b]EGME, ethylene glycol monomethyl ether; EGEE, ethylene glycol monoethyl ether; EGBE, ethylene glycol monobutyl ether; EGPhE, ethylene glycol monophenyl ether; and EGHE, ethylene glycol monohexyl ether.
[c]Values are means ±SD for five animals of both sexes.

Mechanisms. Ethylene glycol monomethyl ether and methoxyacetic acid were equipotent in causing teratogenicity. Alcohol dehydrogenase (ADH) catalyzes the initial rate-limiting oxidation of EGME to methoxyacetic acid that leads to embryo toxicity. According to Sleet et al. (27), the oxidation of a dose of 3.3 to 4.6 mmol/kg was nearly complete in the pregnant mouse in 1 h, and the concentration of methoxyacetic acid in conceptus' tissues was 1.2 times the level in maternal plasma. When ethanol, the preferred substrate of ADH, was administered with EGME, the conversion of EGME to methoxyacetic acid was delayed, and the level in the embryo was reduced by 50% of the peak in embryos from dams administered only EGME. This resulted in incorporating less ^{14}C-labeled products into natural products (as a percentage of dose) synthesized by the embryo and a reduction in the teratogenic outcome. On the basis of the findings of this study, it appears that further metabolism of MAA determines the embryonic outcome of gestational exposure to EGME. The mechanism involved is currently unknown, but simple physiological compounds such as formate, acetate, glycine, and glucose ameliorate the embryo toxicity of ethylene glycol monomethyl ether, according to Mebus and Welsch (85). The common link for these materials may be oxidation pathways involving tetrahydrofolic acid as a catalyst for one carbon transfer into purine and pyrimidine bases. Beattie et al. (86) showed that the inhibition of Sertoli cell lactate production by EGME could account for its effect on spermatogenesis. Lactate is preferentially metabolized by spermatocytes. Beattie and Brabec (87) studied the effects of methoxyacetic and ethoxyacetic acids on mitochondrial function. At concentrations beginning at 3.85 mM, 2-MAA and 2-EAA inhibited state 3 respiration (+ADP) and the respiratory control ratio (RCR) [ratio of state 3 (+ADP) to state 4 ($-$ADP)] in hepatic mitochondria using either succinate or citrate/malate as substrates.

Kirsten et al. (88) demonstrated that fluoroacetate, a structural analog of 2-MAA, is metabolized to fluorocitrate, which binds to mitochondrial membranes (50 pmoles of ($-$) erythro-fluorocitrate per mg of mitochondrial protein) and prevents the transfer of citrate in the mitoplast, citrate metabolism, and citrate-dependent ATP synthesis. Simultaneous incubation of 5 mM citrate and 100 nM ($-$) erythro-fluorocitrate delays the onset of inhibition. In several later related studies, compounds that contained ($-$SH) groups were effective in removing bound fluorocitrate, thereby reducing the toxicity of fluoroacetate. Methoxycitrate formed by the reaction of methoxyacetyl-CoA with oxalacetate may inhibit the mitochondrial transfer of citrate, citrate metabolism, and ATP synthesis. The introduction of these closely related, but nonbiological acids (i.e., fluoroacetate and methoxyacetate), into biological systems most likely disrupts the maintenance of steady-state levels of important metabolites (hemostasis) and gives rise to the adverse effects observed in toxicity studies. A number of studies were conducted that deal with the attenuation of 2-MAA toxicity by simple physiological compounds (29, 89–92). Ambrose et al. (92) examined the protective effects of the amino acid L-serine against EGME-induced neural tube defects (NTD). Concurrent administration of 16.5 mmol L-serine/kg (2-amino-3-hydroxypropionic acid) provided nearly complete protection against EGME-induced NTD, and approximately 48 h was required for the protective effect to become fully evident. This concentration is large when one considers the concentration of free serine ($\sim$0.1 mM) in tissues. Serine is readily metabolized to acetic acid (i.e., by transamination and decarboxylation reactions, to form citric acid via acetyl-CoA and oxalacetate. Methoxyacetic acid is not readily metabolized to CO_2 in the liver and results in high levels of circulating metabolite.

1.4.1.4 Reproductive and Developmental.

The adverse reproductive effects described by Nagano et al. (32) in the adult male mouse and the developmental toxicity seen in the mouse embryo (34) triggered numerous fertility and teratogenicity studies by various routes of administration. These experiments showed that EGME and the monoethyl ether of ethylene glycol caused embryo toxicity in several test species by several routes of exposure (93–95).

Reproductive Effects. Wiley et al. (30) first reported that glycol ethers cause testicular damage to rabbits, and Stenger et al. reported the same on rats and dogs (31). Nagano et al. (32) investigated the effects of a number of ethylene glycol monoalkyl ethers and found that EGME, EGME acetate, and ethylene glycol monoethyl ether and its acetate were the most potent in their effects on the testes of mice. The dose–response relationship between testicular damage and oral ingestion of EGME was studied in 4-week-old male rats using single and multiple (2, 4, 7, and 11 days) doses of 0, 50, 100, 250, and 500 mg/kg (96, 97). Dosing with 50 mg/kg/day for 11 days produced no testicular abnormalities, whereas dose levels of 100, 250, and 500 mg/kg/day resulted in degeneration of pachytene spermatocytes 24 h after a single dose. Continued dosing progressively depleted the early spermatid population. The degenerative changes consisted of cellular shrinkage, increased cytoplasmic eosinophilia, and nuclear pyknosis. The changes were restricted to secondary spermatocytes and to the pachytene, diplotene, diakinetic, and dividing stages of primary spermatocyte development. Recovery was followed after the 4-day treatment at the 500 mg/kg/day dose level. Maturation-phase spermatids were seen after 4 weeks, and by 8 weeks full spermatogenesis was present in the majority of the tubules from the exposed animals. Animals treated for 4 days with methoxyacetic acid exhibited changes similar to those caused by EGME given at an equimolar dosage for the same period of time. The results showed that EGME affected primarily the meiotic cells of the testis, whereas the cells in mitotic division were spared.

Testicular Toxicity—Target Cells. Chapin and Lamb (98) conducted studies in the F344 rat designed to measure the target cell for EGME action in the rodent testes. Adult F344 rats were treated with 150 mg/kg/day, 5 days/week, and sacrificed 1, 2, 4, or 10 days after the first dose. The most sensitive cells were the premeiotic and meiotic spermatocytes. As the length of treatment continued, the younger transitional spermatocytes and more pachytene spermatocytes were affected. Early- and late-stage spermatids were not affected, nor was the visual appearance of spermatogonia changed by exposure. The Sertoli cells were relatively unaffected.

Reproductive Toxicity—In Vivo Dermal. Feuston et al. (99) studied the reproductive toxicity of EGME applied dermally for seven consecutive days to occluded and nonoccluded sites in male rats at dose levels of 0, 625, 1250, or 2500 mg/kg/day and 0, 1250, 2500, or 5000 mg/kg/day, respectively. Sperm were examined on weeks 4, 7, 10, and 15; fertility was assessed on weeks 1, 4, 7, 10, and 14. A dose-related decrease in epididymal sperm count, testicular spermatid count, and testes and epididymal weights was observed. An increase in abnormal sperm morphology and a reduction in fertility accompanied these changes. The study did not report the area of exposed skin or the number of milligrams of EGME applied per unit area of skin.

Developmental — Oral. Developmental phase-specific and dose-related teratogenic effects in CD-1 mice were studied by Horton et al. (33) using EGME. Multiple doses of 250 mg/kg or single doses of 500 mg of EGME/kg induced significant embryo toxicity, as measured by reduced gestation day 18 fetal weights. Embryo lethality, measured by an increased percentage of resorbed implantations, occurred in all groups. Phase-specific teratogenic effects were evident among the groups. Administration of 250 mg/kg EGME from gestation days 7 through 14 resulted in gross malformations (exencephaly and paw lesions). Exposure to three doses of 250 mg EGME/kg between gestation days 7 and 10 resulted in an increase in exencephaly. The groups treated on gestation days 7, 8, and 9 or 8, 9, and 10 had significantly more exencephalic fetuses than controls. The incidence of digit malformations rose markedly as developmental age advanced at the time of exposure.

The gross malformations in the Horton study were similar to those reported by Nagano et al. (34), who administered EGME by gavage at dose levels of 31.25, 62.5, 125, 250, 500, or 1000 mg/kg to pregnant mice on days 7 to 14 of gestation. Nagano observed complete fetal mortality at 1000 mg/kg. All fetuses except one were dead at 500 mg/kg, and fetal mortality increased at 250 mg/kg. Exencepahaly, umbilical hernia, and abnormal digits were present at 250 mg/kg. Fetal body weight was reduced at dose levels ≥ 125 mg/kg, and skeletal malformations of the ribs and vertebrae were seen at dose levels ≥ 62.5 mg/kg. Skeletal variations were noted in all treatment groups.

Scott et al. (100) studied the teratological potential of EGME in pregnant monkeys (*Macaca fascicularis*) at dose levels of 12, 24, and 36 mg/kg to assess the risk of exposure to this solvent to women of childbearing age. The female monkeys were treated daily throughout gestation days 20 to 45 by gavage, and the fetuses were removed at day 100 by Caesarean section. At the high-dose level, all eight pregnancies ended in the death of the embryo, whereas at the middle- and low-dose levels, embryonic death occurred in 3 of 10 pregnancies and 3 of 13 pregnancies, respectively. Maternal toxicity (loss of body weight) occurred at all dose levels. Daily doses of EGME during gestation days 20 to 45 resulted in an increase in the concentration of EGME and methoxyacetic acid in serum during the first 5 to 8 days of administration. The slow elimination of methoxyacetic acid ($t_{1/2}$, 20 h) by the monkey was believed largely responsible for the embryonic levels of methoxyacetic acid in blood.

In Vivo Vapor. Hanley et al. (94) studied the effects of exposure to low vapor concentrations of EGME on embryonal/fetal development in F344 rats, CF-1 mice, and New Zealand white rabbits. Rabbits and rats were exposed to vapor concentrations of 0, 3, 10, or 50 ppm for 6 h/day on days 6 through 18 or days 6 through 15 of gestation, respectively. Mice were exposed to 0, 10, or 50 ppm on days 6 through 15 of gestation. Exposure concentrations up to 50 ppm EGME for 6 h/day during gestation produced minimal maternal effects in the rat. Litter sizes, resorption rates, and fetal body weight measurements of exposed animals were comparable to controls. There was slight fetotoxicity at 50 ppm in the form of skeletal variations in the rat. Exposure of rabbit embryos/fetuses to 50 ppm EGME by inhalation for 6 h/day on gestation days 6 to 18 induced malformations that involved numerous organ systems, increased the incidence of developmental variation and resorptions, and decreased body weights. In mice, increases in the frequency of minor developmental variations were interpreted as evidence of slight fetotoxicity at 50 ppm. Hanley et al. (94) established 10 ppm as a no-observed-effect level (NOEL) for the rat, mouse, and rabbit. The rabbit was the most sensitive of the species tested. Exposure concentrations of 200 ppm resulted in complete embryo lethality in rats and rabbits.

Nelson et al. (101) exposed pregnant rats to three concentrations of EGME (50, 100, and 200 ppm) for 7 h/day on gestation days 7 to 15. Data were analyzed on a litter basis. EGME was highly embryo toxic and produced increased resorptions at 50 and 100 ppm and complete resorptions at 200 ppm. Reduced fetal body weights and skeletal and cardiovascular defects were seen at 100 and 50 ppm.

In Vivo Dermal. Wickramaratne (102) used a modified Chernoff–Kavlock assay to determine the teratogenic potential of ethylene glycol monomethyl ether via the dermal route. No malformed offspring were observed, but litter size was reduced. Feuston et al. (103) determined the effects of a single application of EGME applied to the backs of pregnant rats on gestation days 10, 11, 12, 13, or 14 at dose levels of 0 and 2000 mg/kg and at dose levels of 0, 250, 500, and 1000 mg/kg on gestation day 12. The only maternal effect observed was a transient loss in body weight. Resorption was significantly increased in dams exposed to EGME on gestation day 10. Fetal weights were reduced at dose levels of 1000 and 2000 mg/kg when EGME was applied on gestation days 10 and 12. EGME was teratogenic regardless of the gestation day of administration. The NOEL for developmental toxicity of a single dermal dose applied on gestation day 12 was 250 mg/kg.

Ethylene glycol monomethyl ether was used as a positive control in a recent dermal developmental toxicity study that involved 2-ethylhexanol (104). Undiluted EGME was applied topically to pregnant F344 rats for 6 h/day on gestation days 6 through 15 at doses of 420 and 1260 mg/kg/day in the range-finding study and 840 mg/kg/day in the main study. The treatment area was approximately 10 cm^2. Application of EGME resulted in reduced weight gain, increased postimplantation loss, reduced fetal weights, and increased fetal malformations and variations. Results were similar to those of Feuston et al. (103) after dermal exposure of pregnant Sprague–Dawley rats to single doses of 250, 500, 1000, and 2000 of EGME/kg on day 12 and to 2000 mg/kg/day on gestation days 10 through 14.

Developmental Neurotoxicity. Nelson and Brightwell (105) conducted behavioral testing and neurochemical evaluations on pups prenatally exposed to EGME. After exposure at 25 ppm EGME, maternal animals showed no treatment-related effects. The only significant effects that involved pups exposed on days 7 to 13 of gestation were in the avoidance conditioning tests. However, numerous neurochemical alterations were noted in the brainstem and cerebrum. The authors indicated that, whereas additional studies are needed to elucidate the significance of the neurochemical changes, functional alterations may be a more sensitive measure of effects than observing changes in tissues.

Embryo Dosimetry and Teratogenic Responses. Welsch et al. (22) examined the relationship between EGME/2-MAA dosimetry in maternal plasma and in embryos and its association with teratogenic outcome at two different stages (gd 11, digital malformations; gd 7-9, exencephaly) of prenatal development in CD-1 mice. According to Clarke et al. (43), the total dose given by different routes and rates of entry is not well correlated with teratogenic outcomes. Pharmacokinetic profiles indicate that single bolus s.c. injections compared to constant-rate s.c. infusions produce C_{max} values that are indistinguishable, whereas the concentration–time curves for 2-MAA and the resultant AUC values can be quite different. The 2-MAA tissue dosimetry in maternal plasma, embryo, and EAF indicates that, AUC is better correlated on gd 11 with the incidence and severity of dysmorphogenesis than C_{max} (43). On gd 8, the toxicokinetics of 2-MAA associated with exencephaly induction were determined for a variety of dosing regimens. In both maternal plasma and conceptus, the correlation coefficients between C_{max} and exencephaly rates or other end points of developmental toxicity induced on gd 8 were higher than for AUC. The toxicokinetic differences between gd 8 and gd 11 indicate that dosimetry–teratogenicity determinants are quite distinct for the developmental phase during which a particular organ differentiates and a specific chemical teratogen acts on the embryo (22). Sleet et al. (106–110) conducted a series of studies in rats and mice to establish the relationships among dose, time of exposure, and critical period of development, so that treatments could be developed for investigating the mechanism of EGME teratogenicity in species other than the mouse. The findings of these studies indicate that the developmental phase specificity of EGME was not influenced by the route of administration, whereas the toxicity of EGME to the developing concepti was influenced by the route of administered. The similar effect of administration route on EGME toxicity in the rat and 2-MAA toxicity in the mouse suggests that the mechanism of toxicity of EGME is comparable between the rat and mouse.

Whole Embryo Culture. Pitt and Carney (111) evaluated MAA, the teratogenic metabolite of EGME, using rabbit whole embryo cultures *in vitro*. At a concentration in culture of 5.0 mM, MAA markedly inhibited morphological development and produced an increased percentage of malformed embryos (90% for MAA versus 3% for the negative control). Altered morphology included swollen brain regions, lack of or delayed facial development, and delayed and/or swollen caudal neural tube. These researchers indicate that the effects are similar to those seen following *in vivo* treatment. Thus work from this same laboratory (Dow, unpublished) revealed neural tube defects in rabbit embryos following treatments on gd 7–20. Other published work (discussed earlier) indicates the presence of a wide variety of limb defects. However, the degree of developmental delay induced by MAA in these *in vitro* studies precluded any evaluation of limb development.

1.4.1.5 Carcinogenesis. There are no experimental carcinogenicity or cancer epidemiology data relating to this chemical (112), but some short-term test data are available and are summarized in the section on genetic and related cellular effects.

1.4.1.6 Genetic and Related Cellular Effects Studies. EGME was assayed for genetic activity using the Ames test, unscheduled DNA synthesis (UDS) assay in human embryo fibroblasts, the sex-linked recessive lethal (SLRL) test in *Drosophilia*, dominant lethal tests in male rats, and the sperm abnormality test in mice (113–115). Mutagenicity studies that involving EGME and *S. typhimurium*, with or without S9 activation, were essentially negative (112, 116–119).

In a dominant lethal study (115), EGME-exposed male rats (30, 100, or 300 ppm EGME, 6 h/day, 5 days/week for 13 weeks) were mated to unexposed females. The fertility of the 30- and 100-ppm groups was similar to that of controls, but the fertility of the 300-ppm group was suppressed. In the mouse sperm abnormality test (114), the test animals were exposed to air concentrations of 25 or 500 ppm EGME, 7 h/day for 5 days and sampled 35 days later. Sperm abnormalities increased from 5.2% in the air control to 9.4% in the 500-ppm group. The predominant abnormalities were banana-shaped heads and amorphous heads. Point mutations in the Ames test and unscheduled DNA synthesis (UDS) in human embryo fibroblasts were not increased by methoxyethanol. The sex-linked recessive lethal (SLRL) test gave ambiguous results. Chromosomal aberration frequencies did not increase in rat bone marrow. It was suggested that the evidence available is sufficient to conclude that glycol ethers are not strongly genotoxic agents (13).

Au et al. (120) conducted studies to determine the clastogenic effects of EGME and its metabolite methoxyacetaldehyde (MALD) in bone-marrow cells of B6C3F1 mice after acute and subchronic exposure to the chemicals by the oral route. Mice treated with cyclophosphamide under similar conditions were used as controls. Mice treated acutely with EGME and MALD were implanted with bromodeoxyuridine to label cells, so that only the cells at their first posttreatment mitoses could be selected for chromosomal examination. Neither EGME (35–2500 mg/kg) or MALD (20–1000 mg/kg) caused induction of chromosomal aberrations *in vivo*. In a later but separate study, Chiewchanwit and Au (121) demonstrated that MALD produces chromosomal aberrations *in vitro* in human lymphocytes (10–40 mM treated for 1 h or 0.05–2.5 mM for 24 h) and CHO cells (1.25–20 mM for 3 h).

A summary of current *in vitro* genotoxicity studies was presented by Elias et al. (122). The data indicate that the alkoxyacetaldehydes are the active metabolites of the ethylene glycol ethers in *in vitro* cell systems. They express a genotoxic activity and an aneugenic potential in a dose-dependent manner at noncytotoxic low concentrations (below 1 mM). More recently, Dhalluin et al. (123) reported that sequential treatment of Syrian hamster embryo (SHE) cells by methoxyacetaldehyde (MALD) followed by TPA increased Bcl-2 oncoprotein expression and led to superinduction of ornithine decarboxylase (ODC) activity. The parent alcohol EGME was inactive in this assay.

1.4.1.7 Other: Neurological, Pulmonary, Skin Sensitization. Immunotoxicity. Studies with EGME indicate that animals exposed to this glycol ether develop leukopenia, reduced thymus weight, and atrophy of the thymus (32). Henningsen et al. (124) reported significant depression in thymus weight in F344 rats that were exposed to EGME in amounts greater than 200 mg/kg and in B6C3F1 mice exposed to 1000 mg/kg (males) or 1200 mg/kg (females) EGME in drinking water, whereas immune function studies in B6C3F1 mice failed to demonstrate suppressive effects on humoral or cell-mediated immunity or host resistance to bacterial challenge (125). According to Houchens et al. (126) and Dieter et al. (127), cell-mediated immunity assessed by the mixed lymphocyte reaction was enhanced 100-fold by EGME in mice. Dose-related increases in natural killer cells were reported by Exon et al. (128). The mechanism by which EGME enhances NK-cell activity is unknown.

Rats but not mice are immunosuppressed by exposure to EGME. Metabolism of EGME to 2-MALD (methoxyacetaldehyde) and 2-MAA via alcohol dehydrogenase and aldehyde dehydrogenase is required for immunosuppression in the rat. Oral dosing of male and female rats with EGME or 2-MAA, at dosages of 50–200 mg/kg/d for 10 days, resulted in the following: thymic involution in the absence of reductions in body or spleen weight; reduced lymphoproliferative (LP) responses to the T cell mitogens concanavaline A (Con A) and phytohemaglutinin (PHA), the B cell mitogen *S. typhimurium* mitogen (STM), and the T and B cell mitogen pokeweed mitogen (PWM); depressed production of interleukin

2 (IL-2) by splenocytes; decreased numbers of splenic CD4+ helper/inducer T lymphocytes; delayed onset of expulsion of *Trichinella spiralies* adult worms; and alteration in the primary thymocyte-dependent antigen response to sheep red blood cells (SRBCs) and the thymocyte-independent antigen response to trinitrophenyl lipopolysaccharide (TNP-LPS). In contrast, no alterations were observed in NK cell activity, the MLR response or the *in-vitro*-generated cytotoxic T lymphocyte (CTL) response. These results suggest that humoral immunity in rats is more sensitive to ME and MAA exposure than cell-mediated immunity. The primary antibody plaque-forming cell (PFC) response was one of the most sensitive immunologic end points for EGME and 2-MAA-induced immunosuppression (35, 129).

1.4.2 Human Experience

1.4.2.2 Clinical Cases. The only fatal case of poisoning in a human from ingesting EGME is recorded by Young and Woolner (130). The amount of material consumed is speculative, but it is believed that the man consumed about 200 mL of the material mixed with rum. He was admitted to the hospital in a comatose condition and died 5 h later without regaining consciousness. The urine contained ethanol but no methanol, thus supporting the conclusion of Wiley et al. (30) that the ether is not hydrolyzed. Autopsy revealed hemorrhagic gastritis, marked degeneration of the kidney tubules, and fatty degeneration of the liver. In 1936, Donley (131) described a case of "toxic encephalopathy" suffered by a female who was employed in "fusing" shirt collars by dipping them in a solution composed of ethylene glycol monomethyl ether, isopropanol, and cellulose acetate. She suffered from headache, drowsiness, lethargy, generalized weakness, irregular and unequal pupils, disorientation, and psychopathic symptoms. Two years later Parsons and Parsons (132) described two cases of poisoning that resulted from inhaling ethylene glycol monomethyl ether vapors again encountered in the manufacture of "permanently starched" collars. The symptoms experienced by these two men were weakness, sleepiness, headache, gastrointestinal upset, nocturia, loss of weight, burning of the eyes, and a complete change of personality from one of sharp intelligence to one of stupidity and lethargy. Clinical examination revealed macrocytic anemia. Both patients recovered completely. As a result of the two cases reported (132), Greenburg and co-workers (133) examined these two, and 17 other workers employed in the same factory who were using ethylene glycol monomethyl ether and 67% denatured ethanol; no benzene was present, and the denaturants are not suspect. All had blood abnormalities, suggesting macrocytic anemia, and all suffered some degree of excessive fatigue, abnormal reflexes, and tremors. Examination revealed general immaturity of the leukocytes in every case.

Unfortunately, from the industrial hygiene researcher's viewpoint, ventilation of the operation causing these effects was improved before Greenburg et al. (133) could measure the concentrations to which the men were actually exposed. After changes in the ventilation system had been completed, concentrations ranged from 25 to 76 ppm. Apparently these findings led the investigators to suggest a threshold limit of 25 ppm, even though the concentrations to which the affected persons were exposed were, in all probability, much higher.

Zavon (71) reported five cases of illness in workers. A simulated exposure suggests that these workers were exposed to levels that ranged from 61 to 3960 ppm. These men practiced poor hygienic controls and worked 9 to 10 h/day, 6 days a week. Four of the men, those best studied, showed symptoms of central nervous system depression, and in one case cerebral atrophy, as shown by ataxia, a positive Romberg test, slurred speech, and tremors. The changes in personality seen were similar to those described by Donley (131) and by Parsons and Parsons (132). All showed anemia and in one case a hypocellular bone marrow and a decrease of the erythroid elements.

Nitter-Hauge (134) reported on two cases of men who drank an estimated dose of 100 mL EGME. The first clinical symptoms occurred after at least 8 to 18 h and were similar to those reported by Parsons and Parsons (132). Marked acidosis was seen, and one patient developed marked oxaluria.

Ohi and Wegman (135) reported on two cases of poisoning that followed the substitution of EGME for acetone in a mandrel cleaning operation. Because vapor concentrations were well below the accepted safe level and there was appreciable skin contact, the authors believe that cutaneous absorption may have been a significant factor. The signs and symptoms in these cases were typical, and it is noteworthy that recovery occurred when exposure stopped. The high frequency of mental retardation, neurological symptoms, drowsiness, fatigue, macrocytic anemia, and the abnormal leukocyte picture presented by persons excessively exposed to EGME leaves little doubt that the effects are centered primarily in the brain, blood, and kidneys.

1.4.2.2.3 Pharmacokinetics, Metabolism, and Mechanisms. Physiologically Based Pharmacokinetics. Welsch et al. (77) extended their organogenesis PBPK modeling work in mice and rats to humans. A physiological model of human pregnancy (provided by Drs. M. Andriot and E. O'Flaherty, University of Cincinnati) was coupled to the 2-MAA disposition model developed and validated in the mouse and rat. The human model included tissue growth characteristics during pregnancy, blood flow to various tissues, including the embryo during pregnancy; rates of oxidation of 2-EGME to 2-MAA from *in vitro* experiments; and the 2-MAA elimination rate constant of $\sim$77 h determined in humans. Inhalation exposures of 5 ppm EGME (ACGIH-recommended threshold limit value) were conducted in pregnant mice and rats to determine the effect of repeated 8 h/day, 5 days/week, 3-week exposures. The $t_{1/2}$ values for elimination of 2-MAA ($\sim$6 h in mice, $\sim$24 h in rats and $\sim$77 h in humans) had a profound effect on the accumulation of 2-MAA in plasma. This regimen in pregnant mice and rats caused end-of-exposure concentrations (peak concentrations) of 2-MAA in maternal plasma, embryo, and EAF that approached 50 µM. At the end of the fifth day of exposure, human plasma concentrations that reached 50 µM were comparable to the concentrations seen in mice and rats after 8 h of exposure. According to Welsch et al. (77), teratogenic levels in mice are reached when maternal plasma concentrations exceed 1.0 mM. Simulations that involved the current OSHA permissible exposure level of 0.1 ppm, reduced from the previous standard of 25 ppm, caused accumulations of 2-MAA in plasma of the order of 1.0 µM. This concentration is well below the level that produces developmental toxicity in laboratory animals.

In Vitro and In Vivo Skin Absorption — Liquid. The percutaneous absorption of EGME was evaluated *in vitro* using 1.8 cm^2 human epidermal membrane disks, in glass diffusion cells (21). The permeability constant was 28.9×10^{-4} cm/h using infinite quantities of

EGME. In the work environment, finite quantities of solvent are usually present, resulting in nonlinear rates of absorption. The amount per unit area of skin, size of the exposure area (cm^2), evaporation, and the length of the exposure period are important parameters that are needed to determine the amount of solvent absorbed by skin. *In vivo* percutaneous absorption studies that involved finite amounts of EGME have not been conducted with humans. Kezic et al. (19) measured the rate of absorption of liquid EGME (1.0 g/cm^3) on the skin of two men and three women volunteers. The liquid EGME was applied in a glass chamber to a forearm area of 27 cm^2 for a period of 15 minutes. The rate of percutaneous absorption was 2.8 mg/cm^2/h on the basis of cumulative urinary excretion of 2-MAA (27 cm^2 × 1.0 g/cm^3 × 0.25 h × 28.9 × 10^{-4} cm/h = 19.5 mg). Kezic et al. (19) found 20 mg equivalents in urine. Percutaneous absorption is calculated by multiplying the concentration (g/cm^3) applied to skin (i.e., concentration within skin equivalent to surface concentration) by the permeability constant in cm/h (136). The rate of absorption depends directly on the concentration of liquid (mg/cm^3) absorbed by skin; high concentrations produce high rates of absorption. Human percutaneous absorption studies are generally carried out by applying microgram to milligram quantities of chemical per cm^2 of skin (137). Evaporative losses further reduce the amount (amount/cm^3 × exposed skin area) of chemical available for percutaneous absorption when finite doses are applied. No K_p values are available for EGME dissolved in water. K_p values for dilute water solutions of EGME are expected to be more than the value obtained with neat material (28.9 × 10^{-4} cm/h) because EGME is more soluble in hydrated skin.

In Vivo Skin Absorption — Vapors. Kezic et al. (19) measured the dermal uptake of EGME vapors using five human volunteers (two men and three women). The forearms and hands of these volunteers (1037 cm^2) were exposed for 45 minutes to vapor concentrations of 4854 mg/m^3 (48.54 × 10^{-4} mg/cm^3 at the skin surface). On the basis of the K_p (28.9 × 10^{-4} cm/h) value determined by Dugard et al. (21), the amount of EGME absorbed would be 10.9 µg (48.54 × 10^{-4} mg/cm^3 × 28.9 × 10^{-4} cm/h × 1037 cm^2 × 45/60 h = 10.9 µg). Kezic et al. (19) found 128 mg equivalents of EGME (2-MAA) in urine. This finding suggests that the effective concentration in the skin (skin/air partition coefficient ≅ 11,734) was approximately 11,734 times greater (~56.96 mg/cm^3) than the measured air concentrations at the surface of the skin. This value multiplied by the K_p value of 28.9 × 10^{-4} cm/h × 1037 cm^2 of exposed skin × 45/60 h gives an absorbed dose of 128 mg of EGME (equal to the 128 mg found by Kezic et al. (19)). Whole body exposure (8 h exposure; 18,000 cm^2 of skin) to 16 mg/m^3 should result in a dermal uptake of 78 mg of EGME (77 mg calculated by Kezic et al. (19)) or approximately 1 mg/kg of body weight. This calculation assumes a steady-state concentration of 16 mg/m^3 of EGME in air, an air:skin partition coefficient of 11,734, and total skin contact with no evaporative losses from the skin to reduce the effective concentration in skin.

In Vivo Inhalation. Kezic et al. (19) measured the respiratory uptake of 42 mg/m^3 of EGME (0.042 mg/L or 13.5 ppm) by five volunteers (2 men and 3 women) during the course of 4–15 minute periods. The high blood/air partition coefficient of 34,913 resulted in the retention of 80% of the inhaled EGME and a retained dose of 19.0 mg of EGME (2-MAA urinary equivalents). Extrapolation of an 8-h inhalation exposure at the OEL concentration of 0.016 mg/L (5.13 ppm) resulted in an uptake of 57 mg. Interestingly, these same values were obtained using the Corley et al. (23) PBPK model for butoxyethanol by setting model values for MW = 76, vapor conc. = 5.13 or 13.5 ppm, QPC = 30 (respiratory volume), and running the model for 1-h (4 × 15 minutes) and 8-h exposures (James Knaak, unpublished results).

1.4.2.3 Epidemiology Studies. Concern by the semiconductor industry for the use of known reproductive toxicants (such as certain ethylene glycol ethers) in photolithography and for cleaning, degreasing, and stripping operations prompted NIOSH (138) to reexamine recommended exposure limits for EGEs. The photoresist components and solvents include but are not limited to ethylene and propylene glycol ethers, *n*-butyl acetate, and xylene. Other chemicals include hexamethyl disilizane (HMDS), isopropyl alcohol, 1,1,1-trichloromethane, tetrachloromethane, and methyl ethyl ketone. The ethylene glycol ethers (EGEs) used in the work sites and evaluated for their adverse effects on reproduction included Cellosolve® acetate (ethyleneglycol monoethyl acetate), diglyme (diethylene glycol dimethyl ether), EGME, EGEE (ethyleneglycol monoethyl ether), and the acetate of EGME (4, 5, 7, 9, 10, 139–141). Because of their low volatility, dermal absorption was considered the most likely route of entry. A Massachusetts study conducted by Pastides et al. (4) reported an excess of spontaneous abortions among photolithography (31.3%) and diffusion (38.9%) fabrication (fab) workers. Spontaneous abortion was characterized as fetal loss prior to 29 weeks. Retrospective studies of reproductive health conducted among female and male employees and the wives of the male employees, who worked in two semiconductor manufacturing plants, were carried out by Gray et al. (5). The spontaneous abortion rate in female employees (589 pregnancies), was 33.3% for pregnancies conceived during work on processes with the highest EGE exposure potential, 18.8% for intermediate exposure potential processes, and 14.8% for processes requiring no exposure. This trend was highly significant. In contrast to the female employees, there was no positive trend in spontaneous abortion among the wives of male employees. Subfertility (> 1 yr delay in pregnancy) also increased among female employees exposed to EGEs; 27.3% of women who had the highest exposure potential experienced conception delays. A small but nonsignificant increase was noted among couples in which the man worked on processes that resulted in exposure to EGEs. Schenker et al. (9, 140) investigated pregnancy outcomes among women aged 18–44 who became pregnant while working in the industry between 1986 and 1989. Women were selected from 14 participating semiconductor companies that ranged in employee size from 100 to over 6000. Retrospective and prospective studies were carried out. Analysis of fab versus nonfab female workers in the historical cohort found a small but independent increase in spontaneous abortions after adjustment for other risk factors. Women who were exposed to EGE at any level had a slightly higher relative risk of spontaneous abortion. Results from the prospective study were similar to those from the historical cohort. Schenker (10) concluded that a positive dose–response relationship exists between exposure to EGEs and spontaneous abortion and that the highest exposure group has an adjusted relative risk of spontaneous abortion nearly threefold that of women not exposed to these solvents. Lamm et al. (7) summarized the results of three studies (4, 5, 9) by saying that "an increased spontaneous abortion risk appears to be associated with exposure to the photoresist process (photolithography), even though the spontaneous abortion risk cannot at present be independently attributed to any specific chemical." In this regard, a later study conducted in the United Kingdom by Elliot et al. (142) found no evidence of increased risk of spontaneous abortion among fabrication room employees in the U.K. semiconductor industry.

1.5 Standards, Regulations, or Guidelines of Exposure

The American Conference of Governmental Industrial Hygienists (ACGIH) (143) recommends a threshold limit value (TLV) of 5 ppm (16 mg/m^3) with a skin notation for ethylene glycol monomethyl ether. The Occupational Safety and Health Administration (OSHA) has adopted 25 ppm (80 mg/m^3) as a permissible exposure level (PEL); rule-making is currently underway to adopt new workplace exposure standards. In 1991, the National Institute for Occupational Safety and Health (NIOSH) (144) recommended limiting exposure to EGME in the workplace to 0.1 ppm of air (0.3 mg/m^3) as a time-weighted average (TWA) up to 10 h/day during a 40-h workweek (10-h TWA). OSHA proposed the same limits. In March 1993, the Occupational Safety and Health Administration (OSHA) published a proposed standard for regulating occupational exposures to four glycol ethers: 2-methoxyethanol, 2-ethoxyethanol, and their acetates (145). The primary element of the proposal was the Permissible Exposure Limit or PELs. There are three basic limits: 8-hour time-weighted averages, 15-minute excursion limits, and dermal exposure limitations. OSHA reduced the PELs for EGEE and EGEE acetate to 0.5 ppm. OSHA used a traditional noncancer risk assessment approach (NOAELs and uncertainty factors) to derive the proposed PELs. The male reproductive and developmental NOAELs for EGEE/EGEE acetate were 100 and 50 ppm, respectively. An uncertainty factor of 100 was applied to these values, and developmental data represented the most sensitive end point. The benchmark dose approach was suggested as an alternative to the NOAEL approach. Fifteen-minute excursion limits of five times the 8-hour averages (2.5 ppm) were proposed to prevent peak exposures as high as 32 times the 8-hour limit. OSHA also proposed that "no employee be exposed to glycol ethers through dermal contact," though no numerical limits were set for dermal exposure. The proposal also included requirements that employers provide workers with appropriate personal protective equipment to prevent dermal exposure. Requirements for periodic medical examinations and counseling were also added to the proposal along with requirements for hazard communication programs to educate and inform workers about the hazards of glycol ethers. Biological monitoring of urine for metabolites was suggested as way of enforcing the dermal exposure requirements. Refer to Table 86.9 for a complete listing of international exposure standards for EGME.

2.0 Ethylene Glycol Monovinyl Ether

2.0.1 CAS Number: [764-48-7]

2.0.2 Synonyms: 2-(Ethenyloxy)ethanol; 2-(vinyloxy)ethanol; EGVE

2.0.4 Molecular Weight: 88.1

2.0.5 Molecular Formula: C$_4$H$_8$O$_2$

2.0.6 Molecular Structure:

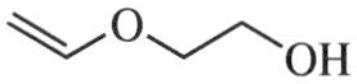

Table 86.9. International Standards and Regulations Regarding Inhalation Exposure to Certain Glycol Ethers and Dioxane[a]

Ether, ppm

	EGME	EGEE	EGBE	EGiPE	EGMEA	EGEEA	EGBEA	Dioxane
ACGIH (TLV/TWA)[e]	5 (s)	5 (s)	20 (s)	25 (s)	5 (s)	5 (s)		20 (s)
NIOSH	0.1 (s)	0.5 (s)	5 (s)		0.1 (s)	0.5 (s)	5 (s)	1 (ceil., 30 min)
OSHA	25 (s)	200 (s)	50 (s)		25 (s)	100 (s)		100 (s)
Australia (OEL/TWA)	5 (s)	5 (s)	25 (s)	25	5 (s)	5 (s)		25 (s)
Austria (OEL/TWA)			20 (s)				20	50 (s)
Belgium (OEL/TWA)	5 (s)	5 (s)	25 (s)	25	5 (s)	5 (s)		25 (s)
Denmark (OEL/TWA)	5 (s)	5 (s)	25 (s)	25	5 (s)	5 (s)		10 (s)
Finland (OEL)	25 (s)	50 (s)	25 (s)		25 (s)	50 (s)		25 (s)
(STEL)	40	100	75		40	100		40
France (OEL/TWA)	5 (s)	5 (s)	25 (s)	25	5 (s)	5 (s)		10 (carc.)
(STEL)								40
Germany (OEL/TWA)[b]	5 (s)	5 (s)	20 (s)	5 (s)	5 (s)	5 (s)	20 (s)	20 (s; carc.)
Hungary (OEL/TWA)	15 mg/m^3 (s)	70 mg/m^3 (s)	100 mg/m^3 (s)		25 mg/m^3 (s)	25 mg/m^3 (s)		10 mg/m^3 (s; carc.)
(STEL)	30 mg/m^3	140 mg/m^3	200 mg/m^3		50 mg/m^3	50 mg/m^3		
Ireland (OEL/TWA)[d]	5 (s)	5 (s)	25 (s)	25 (s)	5 (s)	5 (s)		25 (s)
Japan[b]	5 (s)	5 (s)			5 (s)			100
(STEL)					5			
Netherlands (OEL/TWA)[b]	0.3 (s)	5 (s)	20 (s)	10 (s)	0.3 (s)	5 (s)	20 (s)	11 (s)
(STEL)			40				40	22

108

Philippines (OEL/TWA)	25 (s)	200 (s)	50 (s)	25 (s)	100 (s)	100 (s)
Poland (OEL/TSA)[c]	15 mg/m³ (s)	20 mg/m³ (s)	100 mg/m³ (s)	25 mg/m³ (s)	20 mg/m³ (s)	10 mg/m³ (s)
(STEL)	60 mg/m³	80 mg/m³	360 mg/m³	100 mg/m³		80 mg/m³
Russia (OEL/TWA)	5	5	5 mg/m³	5	5	10 stet
(STEL)		5 mg/m³				10 mg/m³
Sweden (OEL)	5 (s)	5 (s)	20 (s)	5 (s)	5 (s)	25 (s; carc.)
(STEL)	10	10	50	10	10	50
Switzerland (OEL)	5 (s)	5 (s)	20 (s)	5 (s)	5 (s)	25 (s)
(STEL)	10	10	40	10	10	50
Turkey (OEL)	25	200	50	25 (s)	100	100 (s)
United Kingdom (MEL/OEL)[b]	5 (s)	10 (s)	25 (s)	5 (s)	10, (s)	25 (s)
(STEL)						100

[a]1993 standards unless noted. Refer to the text for additional information. s = skin; carc. = carcinogen.
[b]1999 standard.
[c]1998 standard.
[d]1997 standard.
[e]2000 standard.

NOTE: In the case of OEL values for Bulgaria, Columbia, Jordan, Korea, New Zealand, Singapore, and Vietnam, refer to the ACGIH TLV value.

2.1 Chemical and Physical Properties

2.4 Toxic Effects

2.4.1 Experimental Studies

2.4.1.1 Acute Toxicity. The single-dose oral LD$_{50}$ value in mice for ethylene glycol monovinyl ether (EGVE) is 2.9 g/kg. The LC$_{50}$ value for inhalation (length of exposure not given) was 29 mg/L or 8150 ppm. Adverse changes were seen in the liver, kidneys, and lungs (146).

Irritation resulted from instillation of small quantities of EGVE into the eyes of rabbits without apparent corneal injury. The material was irritating to the skin of rats, but it did not cause skin sensitization in guinea pigs (146).

2.4.1.3 Pharmacokinetics, Metabolism, and Mechanisms

2.4.1.3.3 Excretion. Gadaskina and Rudi (147) administered 400 mg/kg of EGVE by gavage (species unspecified) and found that ethylene glycol was excreted in the urine during the next 24 h. The amount of ethylene glycol represented approximately 24% of that which could be formed from complete hydrolysis of the dose given. When both EGVE and ethyl alcohol were administered simultaneously, the amount of ethylene glycol excreted in the urine increased. The results suggest that the toxic effects of EGVE may be due mainly to its metabolites such as ethylene glycol and its other breakdown products.

2.5 Standards, Regulations, or Guidelines of Exposure

EGVE has been classified as dangerous on the basis of its toxicity in rats, mice, guinea pigs, and rabbits (148). Based on organoleptic properties, the maximum suggested permissible concentration of EGVE in water bodies is 2 mg/mL (149).

3.0 Ethylene Glycol Monoethyl Ether

3.0.1 CAS Number: [110-80-5]

3.0.2 Synonyms: 2-Ethoxyethanol

3.0.3 Trade Names: Cellosolve® Solvent

3.0.4 Molecular Weight: 90.12

3.0.5 Molecular Formula: C$_4$H$_{10}$O$_2$

3.0.6 Molecular Structure: ╲╱O╲╱OH

3.1 Chemical and Physical Properties

3.1.1 General

See Table 86.5.

3.1.2 Odor and Warning Properties

According to May (45), EGEE has an odor threshold of about 25 ppm and a strong odor at about 50 ppm. Human volunteers who had some work experience in industrial environments reported that levels of 125 ppm were noticeable and that the intolerable odor level was higher than 255 ppm (46). Thus the warning properties should not be relied upon to prevent prolonged daily exposures to a concentration of vapors that could cause adverse affects.

3.2 Production and Use

EGEE is used in natural and synthetic resins, as a mutual solvent for formulating of soluble oils, lacquers and lacquer thinners, dyeing and printing textiles, varnish removers, and anti-icing additive for aviation fuels. See Table 86.2 for other uses and production volumes. Manufacturers recommend that EGEE not be used in consumer products.

3.3 Exposure Assessment

3.3.1 Air

See the Introduction for a general description of air sampling and analytical procedures.

3.3.2 Workplace Methods

See the Introduction for a general description of sampling and analytical procedures for the workplace.

3.3.5 Biomonitoring/Biomarkers

Clapp et al. (150) monitored the urine and blood of workers in a metal casting shop who were exposed to airborne concentrations of EGEE that ranged up to 20 ppm in the workplace throughout the week. Hand dippers were exposed to the highest continuous levels that ranged from 12 to 19 ppm. No detectable levels of EGEE were found in blood samples. Urine monitoring of exposed workers did reveal positive evidence of EGEE absorption. Samples collected from unexposed workers were consistently negative. In another monitoring study that involved workers in a varnish production plant, Sohnlein et al. (151) found 2.9 ppm as the average concentration of EGEE in air and mean urinary EAA (ethoxyacetic acid) concentrations of 53.2 mg/L in exposed workers. This value was slightly above the current German biological tolerance value (BAT value) of 50 mg/L.

Human exposure to chemicals in the workplace has traditionally been assessed by determining the concentration of airborne residues. More recently, biological monitoring has been used to assess worker uptake of chemicals by all routes of exposure. Both approaches are complementary. The relationship was recently examined and the advantages and limitations of using biomonitoring to assess chemical exposure in the workplace was discussed by Lowry (152) and Lowry et al. (153). The concept of the biological exposure index (BEI) was developed by the American Conference of Governmental Industrial Hygienists (ACGIH).

3.3.5.2 Urine. Recently, Shih et al. (48) reported a method for analyzing ethoxyacetic acid (EAA) in urine (refer to Section 1.3.5.2).

3.4 Toxic Effects

3.4.1 Experimental Studies

Ethylene glycol monoethyl ether (EGEE) has low acute oral toxicity, is not significantly irritating to the skin, is slightly irritating to the eyes and mucous membranes, is readily absorbed through the skin but has low acute toxicity by this route, and is somewhat toxic when inhaled. Its vapors are irritating and disagreeable at acutely toxic concentrations. The material affects the blood, liver, kidneys, and testes. EGEE is considered a developmental toxicant and causes teratogenicity in experimental animals. In addition, EGEE has testicular effects in male mice and rats from oral and inhalation routes of administration. The results of mutagenicity testing of EGEE are equivocal, although it was not carcinogenic in a 2-year dietary study in rats.

3.4.1.1 Acute Toxicity. Oral. Carpenter and co-workers (50) found that the single-dose oral toxicity of EGEE was 5.5 g/kg for rats, 3.1 g/kg for rabbits, and 1.4 g/kg for guinea pigs. The value given for rats, 5.5 g/kg, is somewhat higher than the 3.0 g/kg previously reported by Smyth et al. (154). Laug and co-workers (155) reported the following oral LD_{50} values: 3.46 g/kg for rats, 4.31 g/kg for mice, and 2.79 g/kg for guinea pigs. Others have reported the following LD_{50} values: 2.8 g/kg for rats (63); 4.8 g/kg for mice; 4.45 g/kg for rats; 2.13 g/kg for guinea pigs; 1.48 g/kg for rabbits; and 3.5 g/kg for mice when administered as an oil solution, and slightly higher as a water solution (31); 3.53 g/kg for fasting rats; 2.45 g/kg for fasting mice; 8.10 g/kg for fed rats; and 5.35 g/kg for fed mice (53). Laug et al. (155) observed that the animals displayed no immediate signs of distress. However, they did observe hemorrhage of the stomach and intestine, mild liver injury, severe kidney injury, and hematuria in animals that were seriously affected or died. They concluded from their study that EGEE should not be used in applications where ingestion by humans is expected. Others found reversible liver and kidney injury at levels as low as 0.25 g/kg (156); however, the lesions were not considered compound-related. Stenger et al. (31) reported that the animals tested showed signs of dyspnea, somnolence, and ataxia. In addition, both the rats and rabbits exhibited signs of writhing.

Gross (49) fed rabbits repeated doses of EGEE and found that seven doses of 0.1 mL/kg (0.093 g/kg) caused temporary albuminuria, whereas seven doses of 0.25 mL/kg (0.23 g/kg) caused both albuminuria and hematuria after the seventh feeding. When the dosage was increased to 1 mL/kg (0.93 g/kg), albuminuria and hematuria were observed after the seventh day, followed by death on the eighth day from kidney injury. Two doses of 2 mL/kg (1.86 g/kg) caused exhaustion, refusal to eat, and albuminuria; casts were observed in the urine; and death was believed to be due to kidney injury.

Nagano et al. (32) fed mice doses ranging from 62.5 to 4000 mg/kg, 5 days/week for 5 weeks, and observed testicular atrophy and leukopenia, whose intensities are dose-related. They concluded that the ethyl ether of ethylene glycol was less potent than the methyl ether.

Injection. Von Oettingen and Jirouch (157) concluded that EGEE given subcutaneously to mice was less toxic than ethylene glycol. They found that the minimum lethal dose was about 5.0 mL/kg (4.66 g/kg) for the monoethyl ether and 2.5 mL/kg (2.78 g/kg) for ethylene glycol. They observed, however, that large doses could cause severe kidney injury.

Gross (49) gave repeated subcutaneous injections of EGEE to rabbits. The results showed that seven doses of 0.25 mL/kg (0.23 g/kg) produced no apparent effect. However, higher dose levels produced essentially the same response as observed from oral administration, although the intensity of response was somewhat greater.

Carpenter et al. (50) found that EGEE in concentrations of more than 18% in 0.75% sodium chloride solution were hemolytic to rat erythrocytes *in vitro*. When such an 18% solution was given intravenously to rats, the LD_{50} was 3.3 g/kg, and when the undiluted material was given, the LD_{50} was 2.4 g/kg. When given intraperitoneally to rats, the LD_{50} of the undiluted material was 2.14 g/kg.

Stenger et al. (31) administered EGEE subcutaneously for 4 weeks and found that dose levels up to 0.38 g/kg/day caused no deaths in rats. However, dose levels of 0.185 and 0.38 g/kg caused dyspnea, somnolence, mild ataxia, some growth depression in females and some reduction of hemoglobin levels and hematocrit values. At the 0.38-g/kg level, interstitial testicular edema, dissociation of liver parenchyma, and tubular lesions of the kidney were observed. Dogs given EGEE intravenously at levels of 0.093 and 0.46 g/kg/day for 22 days developed local irritation at the injection site and ataxia, but no change in hemoglobin levels or hematocrit values. Stenger et al. (31) also established the following LD_{50} values: for subcutaneous injection in the rat, 3.16 mg/kg and in the rabbit, 1.85 g/kg. The LD_{50} by intravenous injection was 4.8 g/kg in the mouse, 4.45 g/kg in the rat, 2.13 g/kg in the guinea pig, and 1.48 g/kg in the rabbit.

Inhalation. The acute response of guinea pigs to EGEE in air was studied by Waite et al. (158). They found that guinea pigs could survive exposure to 6000 ppm for 1 h, 3000 ppm for 4 h, and 500 ppm for 24 h without apparent harm. More intense exposures caused injury of the lungs, hemorrhage in the stomach and intestines, and congestion of the kidneys. They concluded that air essentially saturated with EGEE vapor at room temperature was sufficiently disagreeable and irritating to the eyes to provide adequate warning to prevent acute poisoning. Gross (49) reported that the majority of animals repeatedly exposed to 1400 ppm EGEE for 8 h/day died after 4 to 12 exposures. Cats were most susceptible and died 2 days after cessation of exposure to 1400 ppm for 4 or 5 days. One of two mice died after nine exposures, but the other survived 12 exposures without evident effects. Two rabbits survived 12 exposures, one died 7 days later, but two guinea pigs survived 12 exposures without evidence of injury.

Werner and co-workers (55) exposed mice for 7 h to various concentrations and found that the LC_{50} was 1820 ppm. They attributed death to lung and kidney injury.

Werner et al. (56) exposed groups of rats for 7 h/day, 5 days/week for 5 weeks to a concentration that averaged 370 ppm EGEE vapor and noted only a slight effect on the cellular elements of the blood.

Goldberg et al. (159) found that rats exposed to EGEE vapors for 4 h/day, 5 days/week for 10 exposures experienced an early transient weight gain at 2000-ppm exposure level. However, after 3 days of exposure, the rats returned to a normal growth pattern. Those exposed to 4000 ppm showed no inhibition of conditioned avoidance-escape behavior but did exhibit consistent reduced weight throughout the exposure period.

Kasparov et al. (160) reported an LC_{50} for mice of 77 mg/L (20,900 ppm); the length of exposure was not given.

Dermal. EGEE is readily absorbed through the skin of rabbits. Quantification by the "sleeve" technique, essentially as described by Draize et al. (62), indicated that the LD_{50} was 3.6 mL/kg (3.35 g/kg) (50). Others confirm these earlier results and reported the dermal LD_{50} as 3.53 g/kg in rabbits (64).

Eye Irritation. Carpenter and Smith (66) evaluated EGEE for eye irritation. These results, which agree with those of others (158), indicate that when the material is introduced directly into the eye, it produces immediate pain, some conjunctival irritation, and a slight transitory irritation of the cornea, which clears within 24 h. Weil and Scala (161) summarized the work of a number of laboratories, in which eye tests showed that only slight transient eye irritation was found. These observations indicate that the material does not present a serious hazard to the eyes, although it may be painful and uncomfortable.

Skin Irritation. EGEE, even in prolonged and repeated contact with the skin of rabbits, failed to cause more than a very mild irritation (50). These findings have been confirmed by others (52, 64, 161, 162).

3.4.1.2 Chronic and Subchronic Toxicity. *Chronic Toxicity.* Morris et al. (38) reported studies in which EGEE was fed in the diet of rats for a 2-year period. At the dosage level of 1.45%, equivalent to about 0.9 g/kg/day, they observed only slight kidney damage, but did see appreciable tubular atrophy in the testes, accompanied by marked interstitial edema in about two-thirds of the animals. They did not find any oxalate crystals in the kidneys or bladders, as reported for ethylene glycol.

Oral. Smyth et al. (154) reported maintaining rats for 90 days on drinking water that contained EGEE. They found that the maximum dose that had no effect was 0.21 g/kg/day. They also reported that a dose level of 0.74 g/kg reduced growth and appetite, altered liver and kidney weights, and produced microscopic lesions in these organs. Mortality was observed when the dosage was increased to 1.89 g/kg/day. In 90-day dietary feeding studies in rats, a level of 0.25% did not cause significant ill effects, whereas a level of 1.25% in the diet depressed body weight (163).

Drinking Water. The NTP (68) studied the effects of EGEE on groups of five male and five female F344/N rats and B6C3F1 mice by administering the solvent in drinking water for 2 and 13 weeks. Estimates of EGEE consumption based on water ingested by male and female rats ranged from 200 to 1600 mg/kg/day; mice consumed 400 to 2800 mg/kg/day. There were no chemically-related effects on survival for rats or mice in the 2-week studies. Decreased body weight gains and decreased water consumption were noted in both male and female rats administered EGEE. Thymic atrophy was observed in male and female rats and testicular atrophy in male rats that ingested EGEE.

In the 13-week EGEE drinking water studies, groups of 10 male and female rats consumed 100 to 2200 mg/kg/day (1250 to 20,000 ppm), and groups of 10 male and female mice consumed 600 to 11,000 mg/kg/day (2500 to 40,000 ppm) of EGEE. Chemically-related mortality occurred in male and female rats administered 20,000 ppm in water, but no deaths occurred in mice at all levels. Treatment-related histopathological changes were observed in the testes, thymus, and hematopoietic tissues (spleen, bone marrow, and liver). In special 60-day stop-exposure studies, marked to moderate degeneration of the seminiferous tubules was observed in rats treated with 10,000 or 20,000 ppm EGEE but no effects were observed at 5000 ppm. Only partial recovery from testicular degeneration was observed after 30 and 56 days of recovery. Anemia was seen with EGEE. However, evidence for an adaptive response was indicated by increased hematopoiesis in the bone marrow, spleen, and liver. EGEE had similar effects on the testes, spleen, and adrenal gland (females only) of mice. A dose-related degeneration of the germinal epithelium in seminiferous tubules of the testes was observed. EGEE caused a prominent lipid vacuolization of the X-zone of the adrenal gland in female mice and a dose-related increase in splenic hematopoiesis in both male and female mice.

In the 13-week EGEE drinking water study in rats, the NOAEL for decrease thymus weights in males was 1250 ppm; for female rats the NOAEL for all histopathological and hematologic effects was 5000 ppm. For male mice administered EGEE for 13-weeks, the NOAEL was 20,000 ppm for testicular degeneration and increased hematopoiesis in the spleen. The NOAEL for adrenal gland hypertrophy and increased hematopoiesis in the spleen was 5000 ppm in female mice.

Stenger et al. (31) fed rats EGEE for 13 weeks at doses of 0.093 to 0.73 g/kg/day. The no-effect level established was 0.093 g/kg/day, whereas a level of 0.185 g/kg/day resulted in the beginning of signs of adverse effects that consisted of growth depression, reduced food intake, reduced hemoglobin content, and low hematocrit values. Histological changes were observed in the liver, kidneys, and testes at the 0.73-g/kg/day level. Stenger et al. (31) also fed dogs for 13 weeks at levels of 0.046 to 0.185 g/kg/day and found reduced hemoglobin levels and hematocrit values after 5 weeks. The dogs also developed pathological changes in the kidneys and testes similar to those seen in the rats.

Inhalation. Werner et al. (57) exposed two dogs to an EGEE vapor concentration of 840 ppm for 7 h/day, 5 days a week for 12 weeks and observed a slight decrease in hemoglobin and red cells. The blood picture was characterized by a greater than normal number of immature white cells. There was no evidence of kidney or bone marrow injury. There was, however, an increase in the number of calcium oxalate crystals in the urine.

In tests using mice, repeated exposures to 3000 mg/m^3 (8100 ppm) depressed growth, lowered cholinesterase levels, and increased protein excretion in the urine (161).

Barbee et al. (164) reported on rats and rabbits exposed to EGEE vapor at concentrations of 25, 100, or 400 ppm for 6 h/day, 5 days/week for 13 weeks. Animals (both species) had increased incidence of lacrimation and nasal discharge at all doses but intensity of response was not dose-related. The only effect in rats at the high dose (400 ppm) was a significant decrease in pituitary weight in males. The results from rabbits indicated that it was the species more sensitive to subchronic exposure to EGEE. Rabbits had decreased body weight, hemoglobin, hematocrit, and erythrocytes in both sexes. Male rabbits had a significant decrease in the weight of the testes at the 400-ppm level. Histopathological examination of the testes revealed testicular changes that consisted of degenerated tubular epithelium.

3.4.1.3 *Pharmacokinetics, Metabolism, and Mechanisms.* *Metabolism. Oral Gavage.*
Cheever et al. (165) reported on the various routes of excretion of radiolabeled EGEE

following a single oral dose of 230 mg/kg body weight to male rats. Animals were administered EGEE labeled as either [^{14}C]-ethanol or [^{14}C]-ethoxy. Urinary excretion accounted for 76 to 80% of the [^{14}C]-EGEE dose. The major metabolites in the urine were ethoxyacetic acid and N-ethoxyacetylglycine; both metabolites accounted for 73 to 76% of the dose. The amount of the radiolabeled compound recovered in expired air varied, depending on the location of the label. The ethoxy-labeled compound was recovered at 11.7%, and the ethanol-labeled compound was 4.6% of the administered ^{14}C-labeled compound. The biological half-life for the ethoxy-labeled compound was 9.9 ± 1.5 h and 12.5 ± 1.9 h for the ethanol-labeled compound. Upon administration of the ethanol-labeled compound, a radiolabeled component was detected in the rat testes. This component was later identified as ethoxyacetic acid, which suggests that the testicular effects observed in the rat may be due to ethoxyacetic acid.

Inhalation. An inhalation study designed to determine the effect of dose on the absorption, metabolism, and excretion of 2-ethoxy [U-^{14}C]-ethanol by F344/N rats was conducted by Kennedy et al. (166). Rats were exposed to either 5 ppm EGEE for 5 h and 40 min or 46 ppm for 6 h. The uptake and metabolism of EGEE were linear in the concentration range studied. Of the inhaled dose, 28 to 29% was retained (not exhaled as parent compound) by the rat. Significant percentages of the retained dose were exhaled during exposure (22%) and after exposure (16%) as ^{14}CO$_2$. Forty-six percent of the retained dose was excreted in urine, and ~10% remained in tissues 66 h after exposure. EG (ethylene glycol), EAA, and EAG (glycine conjugate of EAA) were identified in urine along with a minor but unknown metabolite. The major metabolite was EAA (~30%) and lesser percentages of EG (~10%) and EAG (~5%).

Oral—Drinking Water. Medinsky et al. (167) evaluated the disposition of 2-ethoxy[U-^{14}C]-ethanol (EGEE) by administering the compound via drinking water to male rats. Animals were allowed access for 24 h to water that contained ^{14}C-radiolabeled EGEE at 220, 650, or 1940 ppm, which was replaced with water devoid of EGEE at the time of removal. Elimination of radioactivity was monitored for 72 h. The majority of the ^{14}C was excreted in urine or exhaled. As much as 25 to 40% of the EGEE was eliminated as ethoxyacetic acid and 20% as CO$_2$; 18% of the EGEE was eliminated in the urine as ethylene glycol. Medinsky et al. (167) concluded that finding ethylene glycol in the urine suggests that dealkylation of EGEE occurs before oxidation of ethoxyacetic acid.

Moslen et al. (75) examined the capacity of liver and testes from hamsters, rats, mice, guinea pig, dogs, cats, and humans to metabolize EGME and EGEE. Species differences between rats and hamsters were found for testicular and hepatic biotransformation of EGEE and EGME.

Dermal. Sabourin et al. (168, 169) studied the uptake, metabolism, and excretion of dermally administered EGEE by F344/N rats as a function of externally applied dose. Three different amounts of ^{14}C-labeled EGEE (400–4000 mμmole/kg) were applied to uniform areas of clipped skin. The applied dose was unoccluded. Approximately 20–25% of the applied dose was eliminated in urine and smaller amounts were excreted in feces and as respiratory ^{14}CO$_2$. Ethoxyacetic acid was the major metabolite found in urine. A proposed metabolic pathway for EGEE through the Kreb's cycle is similar to that by Sumner and Fennell (170) and Sumner et al. (25) for EGME in which ^{14}CO$_2$ is a product of the formation and metabolism of 2-methoxycitrate (Figure 86.4).

Pharmacokinetics. During the past few years, interest has increased in developing physiologically based pharmacokinetic (PBPK) models for extrapolating data from one route of administration to another and between animal species including humans. Gargas et al. (42, 542) developed inhalation PBPK model for EGME, EGEE and EGEEA in pregnant rats based on the models of Clarke et al. (41) and Terry et al. (79) that involved intravenously and orally administered EGME to pregnant mice. The models developed for EGME kinetics in rats and humans (542, 543) were used as the basis for the EGEE/EGEE/ EAA model. In the inhalation studies, pregnant rats were exposed to either 10 or 50 ppm EGME or 50 or 100 ppm EGEEA for 6 h/day for 5 days beginning on gestation days 11–15 (42). The concentrations correspond to a known no observed exposure level (NOEL) and a lowest observed effect level (LOEL). Maternal blood was collected at 1, 3, and 5 h during the exposure on day 5 and for up to 42 or 90 h postexposure for animals exposed to EGEEA and EGME, respectively. Fetuses were collected at the time of blood collection and urine was collected 0–18 h postexposure from each rat scheduled for the 18 h post exposure blood collection. The PBPK models predicted observations from the inhalation studies. Internal dose measurements (peak C_{max} and AUC for MAA/EAA maternal venous blood and fetal tissue curves) for the NOELs and LOELs for both EGME and EGEEA were calculated using the PBPK model. The rat PBPK model was scaled to humans using human metabolic and pregnancy parameters. The human model was used to determine concentrations in air (EGME: 12 ppm; EGEEA: 25 ppm) that would give internal values equivalent to those obtained with NOELs in critical rat studies.

Green et al. (83) developed pathways and relative rates of metabolism of EGEE using rat and human hepatocytes (see Table 86.7). The concentrations of EGEE used were 0.02, 0.2, 2.0, and 10.0 mM. Metabolites were analyzed by HPLC. Ethylene glycol (EG) was the major metabolite of EGEE (30%). The percentage of EGEE converted to EAA in rat hepatocytes was similar at all concentrations, whereas in human hepatocytes, the percentage converted to MAA decreased between 0.2 and 2.0 mM. The V_{max} value for the conversion of EGEE to ethoxyacetic acid with rat hepatocytes was similar to those obtained for the conversion of EGME to MAA and EGBE to BAA. In human hepatocytes, V_{max} followed the order of EGBE > EGEE > EGME. V_{max} was 15- to 20-fold higher in rat than in human hepatocytes.

In a separate study conducted by scientists at the Dow Chemical Company, the metabolism of EGEE was evaluated *in vitro* with an equine liver alcohol dehydrogenase assay obtained from Sigma Chemical Company. The V_{max}, K_m, and V_{max}/K_m were 7.97 µM h^{-1}, 7.12 × 10^{-3} µM, and 1.1, respectively. The activity of other sources of alcohol dehydrogenase toward EGEE were not evaluated. Aasmoe et al. (84) compared the activities of liver alcohol dehydrogenase isozymes for male and female rats with a series of glycol ethers. This work is summarized in Table 86.8. Female rats displayed greater activity than males.

Mechanisms. The developmental and reproductive toxicity of EGEE is believed to be due to its metabolism to ethoxyacetic acid. Beattie and Brabec (87) studied the effects of methoxyacetic and ethoxyacetic acids on mitochondrial function. At concentrations beginning at 3.85 mM, 2-MAA and 2-EAA inhibited state 3 respiration (+ADP) and the respiratory control ratio (RCR) [ratio of state 3 (+ADP) to state 4 (−ADP)] in hepatic mitochondria using either succinate or citrate/malate as substrates. Kirsten et al. (88)

clearly demonstrated that fluorocitrate, formed during the metabolism of fluoroacetate, binds to mitochondrial membranes and prevents the formation of ATP and the transfer and metabolism of citrate. Compounds containing (–SH) groups were effective in removing bound fluorocitrate, thereby reducing the toxicity of fluoroacetate. Ethoxycitrate formed from ethoxyacetic acid may also inhibit the mitochondrial transfer of citrate. A more complete discussion may be found in Section 1 (EGME mechanisms of activity) which deals with the toxicity of methoxyacetate and methoxycitrate and the attenuation of toxicity by simple physiological chemicals such as formate, acetate, and D- and L-serine.

3.4.1.4 Reproductive and Developmental. *Reproductive.* Foster et al. (96) exposed male rats by gavage to 0, 250, 500, or 1000 mg/kg/day for 11 days. Six animals from each treatment group were sacrificed at 6 and 24 h after a single dose. Additional groups of animals received repeated-daily doses and were then sacrificed after 2, 4, 7, or 11 days of treatment. Testicular effects were observed at 500 and 1000 mg/kg/day after 11 days of treatment. The effects are believed to be reversible on the basis of the results from the treatment of male rats with 2-methoxyethanol. Methoxyethanol in other studies caused testicular changes 24 h after a single dose of 100 mg/kg. Male rats administered 500 mg/kg/day 2-methoxyethanol for 4 days developed testicular changes. These changes were reversed by 8 weeks after treatment stopped.

Barbee et al. (164) reported that male rabbits exposed to 25, 100, or 400 ppm EGEE via inhalation for 13 weeks had significantly reduced testes weights at the 400-ppm level, but no effects were seen at the lower levels. Histopathological examination indicated degeneration of the tubular epithelium in 3 of 10 rabbits. Exposures of rats to 25, 100, or 400 ppm had no observable effects on the testes.

Oudiz et al. (171) treated rats with 936, 1872, or 2802 mg/kg EGEE for 5 days and analyzed samples of semen periodically over the following 2 weeks. They found a rapid decline in sperm count, and most males became azospermic by the seventh week at the two highest dose levels. At the lowest dose level, significantly decreased sperm counts and abnormal sperm morphology were seen by week 7. Partial or complete recovery of the sperm parameters was evident by week 14.

Lamb et al. (172) reported that administering EGEE in drinking water at 1 and 2% in a continuous breeding study of mice resulted in testicular effects in the male that included testicular atrophy, a decrease in sperm motility, and an increase in the number of abnormal sperm morphologies. These effects were not observed at 0.5% in the males, and no effects were observed in female mice at 0.5, 1, or 2% EGEE in drinking water.

Zenick et al. (173) administered EGEE by gavage at 0, 936, 1872, or 2808 mg/kg for five consecutive days to male rats. Males were then mated weekly during the next 14 weeks to ovarectomized females. Semen samples recovered from the female reproductive tract and copulatory behavior were monitored on weeks 1, 4, 7, 10, and 14 post-treatment. Males were then sacrificed at week 16. Results indicate that EGEE caused a rapid decrease in sperm count, and a majority of the animals were azospermic by week 7. At week 7 in the low-dose group, there was also a significant increase in the number of amorphic sperm. By week 14, partial or complete recovery was observed in all males that was later confirmed by histological examination of the epididymis and testes.

Ninomiya et al. (174) conducted a study to evaluate the validity of the Hamilton–Thorne sperm analyzer (HTM-IVOS) that comprised a combined optical and computer system. EGEE was administered to Sprague–Dawley male rats (seven per group) at 500 mg/kg/day for 7 days. At 0, 1, 2, and 4 weeks after the end of treatment, preparations of caudal and epididymal tissues were examined. At the end of treatment, there were no changes in either the percentage of motile sperm or the sperm count. At four weeks after treatment, the percentage of motile and progressively motile sperm and the total sperm count decreased significantly. Testicular weight decreased at 1, 2 and 4 weeks after treatment.

Developmental. Stenger et al. (31) evaluated EGEE for possible developmental effects in mice injected subcutaneously from day 1 to 18 of gestation with a dose level of 0.093 g/kg/day. He also injected rats with the same dose from day 1 through day 21 of gestation and rabbits at a level of 0.023 g/kg/day on days 7 through 16 of gestation. None of the animals developed any serious malformations indicative of teratogenic effects. However, only in the rats, were there skeletal aberrations at 0.093 g/kg/day, which was considered an indication of fetotoxicity.

In a study that initially implicated EGEE as a teratogen, New Zealand white rabbits were exposed to 160 or 617 ppm EGEE via inhalation for 7 h/day on days 1 to 18 of gestation, and Wistar rats were exposed to 150 or 649 ppm EGEE for 3 weeks before mating. The sperm-positive females were subsequently exposed to 202 or 767 ppm EGEE for 7 h/day on days 1 to 19 of gestation. Five of 25 rabbits died at the 617-ppm exposure level, and all litters of the surviving dose were resorbed. At the 160-ppm level, there was mild maternal toxicity. Rabbit fetuses at this level had a significant increase in cardiovascular, ventral body wall, and renal defects, and minor skeletal changes increased. Premating exposure to EGEE had no effect on mating success or the rate of pregnancy of the rats. Maternal toxicity was mild at 767 ppm, but all litters were totally resorbed. There was some evidence of slight maternal toxicity at 200 ppm, and the rat fetuses at this dose level had a significant increase in the incidence of cardiovascular defects and wavy ribs (93, 175, 176).

Wier et al. (177) exposed mated female mice to EGEE by oral gavage at concentrations that ranged from 0 through 4.2 g/kg body weight from day 8 through day 14 of gestation. At dose levels of EGEE that did not produce maternal toxicity (1.8 and 2.6 g/kg), EGEE reportedly produced malformations (i.e., fused or missing digits and kinked tail). Fetotoxicity, present as a decrease in fetal body weights, was marginally apparent at the 1.0-g/kg dose level and at all other dose levels tested. Embryo lethality occurred at 1.8 g/kg body weight. The NOEL was 1.0 g/kg body weight for mice by oral administration.

EGEE was applied dermally at 0.25 or 0.50 mL four times a day to gravid rats on days 7 through 16 of gestation. At 0.50 mL of EGEE, there was 100% intrauterine resorptions of litters in all exposed dams. At 0.25 mL four times a day (178, 179), maternal toxicity, fetotoxicity, embryo lethality, and visceral (predominantly cardiovascular) and skeletal (vertebrae) malformations were observed.

Rats and rabbits were exposed to EGEE vapor for 6 h/day on days 6 through 15 of gestation for rats and on days 6 through 18 of gestation for rabbits (180). Exposure levels of EGEE for rats were 10, 50, or 250 ppm and for rabbits were 10, 50, or 175 ppm. Rats had increased pre- and postimplantation loss, fetotoxicity (i.e., reduced fetal weight, delayed ossification, and renal-pelvic dilation), and maternal toxicity at 250 ppm. At all dose levels, there was a significant increase in preimplantation losses; however, this was statistically significant only at 50 ppm. In rabbits, 175 ppm produced minor fetotoxicity (i.e., delayed ossification and extra rudimentary ribs) and an increase in extranormal-length ribs, which may be interpreted as a possible teratogenic effect. No overt toxicity was observed in the does at any dose tested that could be attributed to exposure to EGEE. No teratogenic effects were attributed to treatment with EGEE in rats at concentrations of 250 ppm or in rabbits at 50 ppm. The NOEL for teratogenicity was 250 ppm for rats and 50 ppm for rabbits. The NOEL for developmental toxicity was 10 ppm (marginal fetotoxicity observed at 50 ppm) for rats and 50 ppm (marginal teratogenic effects at 175 ppm) for rabbits.

Goad and Cranmer (181) reported that oral doses of 200 mg/kg EGEE for short periods of time (gestation days 7 to 9 or 10 to 12) caused teratogenic effects (cardiovascular and skeletal) in rats similar to those seen after longer periods of exposure with EGEE.

Whole Embryo culture. During the last few years, whole embryo culture has been used to screen chemicals for their teratogenic hazard (182). Adverse embryonic outcomes (malformation or embryotoxicity) are directly related to the serum concentration of the compound tested. Brown-Woodman et al. (183) conducted a study to determine whether plasma solvent levels during industrial exposure pose a risk to the embryo. A no-effect and effect concentration were determined and compared with known blood levels in industrially exposed or otherwise exposed individuals. The effect level for EGEE was 75.5 mmol/L. No human values were available for comparison.

Behavioral Teratology. In a behavioral teratology study, Nelson et al. (106, 184) exposed pregnant rats to 100 ppm EGEE for 7 h/day on days 7 to 13 or 14 to 20 of gestation. Behavioral testing of the offspring of the dams exposed on days 7 to 13 showed impaired performance on the rotorod and in the open field tests and altered performance in avoidance conditioning. Offspring of the dams exposed on days 14 to 20 of gestation had impaired performance on the activity wheel and in avoidance conditioning. Neurochemicals were elevated in the brains of neonates and 21-day-old rats.

Nelson et al. (185, 186) investigated possible interactions between ethanol and EGEE. Using similar exposure regimen and testing protocols for both studies, they exposed pregnant rats via drinking water to EGEE alone, to 10% ethanol alone, or to a mixture of ethanol and EGEE. The investigators concluded that ethanol given early in gestation (days 7 to 13) reduced the effects of EGEE and when given later in gestation (days 14 to 20) ethanol enhanced the effects of EGEE.

3.4.1.5 Carcinogenesis.
In 2-year dietary feeding studies in rats at a dose level equivalent to about 0.9 g/kg/day, as reported by Morris et al. (38), there were no indications of carcinogenicity.

3.4.1.6 Genetic and Related Cellular Effects Studies.
Genetic Toxicity. Shimizu et al. (117), Zeiger et al. (118), Hoflack et al. (187), Elias et al. (122), and McGregor (113, 112) reported the results of mutagenicity testing with EGEE. EGEE was evaluated in the Ames assay and the *Escherichia coli* reverse mutation assay with and without metabolic activation and was not mutagenic under the conditions of the studies. EGEE was also

evaluated in the Chinese hamster ovary *in vitro* system with and without metabolic activation to determine its potential to induce sister chromatid exchange and chromosomal aberrations. EGEE tested positive for inducing sister chromatid exchanges and chromosomal aberrations; however, the response was weaker with metabolic activation (112). Hoflack et al. (187) tested EGEE and its corresponding aldehyde and acid derivatives up to 10^{-4} mole/plate or up to cytotoxic concentrations. All tested substances gave negative results with TA 98, TA 100, and TA 102 tester strains with or without S9 mix. Elias et al. (122) used Chinese hamster lung (V79) cells to study the genotoxic effects of EGEE (cell growth and cell cycle kinetics, gene mutations, cytogenetic and aneugenic effects, and inhibition of metabolic cooperation) and Syrian hamster embryo cells to study morphological transformations. The genotoxic results were negative with some exceptions. Positive results were obtained with high concentrations of EGEE in both chromosomal aberrations (166 mM) and sister chromatid exchange assays (10 to 100 mM) in Chinese hamster lung (V79) cells. The rat liver S9-mix commonly used to activate chemicals used in the *in vitro* tests did not activate EGEE because dehydrogenase enzymes were lacking for alcohol oxidation. The genotoxicity of EGEE at high concentrations is due to the presence of ethyoxyacetaldehyde (EALD) or acidic metabolites. EALD was genotoxic at concentrations that ranged from 0.1 to 1.0 mM.

3.4.2 Human Experience

3.4.2.2 Clinical Cases. According to Browning (59), examinations of workers employed in manufacturing lacquers and paint revealed very little evidence of any injury to health from using this material. Browning also reported that operators who applied EGEE with a spray gun worked all day without discomfort or ill effects.

Nitter-Hauge (134) reported on two men who ingested approximately 100 mL EGEE. After an initial 8- to 18-h lag time, they developed symptoms that included mental confusion, general weakness, and nausea. Other signs and symptoms were deep and frequent respiration and profound metabolic acidosis. Both men recovered.

Fucik (188) cited a case of a 44-year-old female who drank about 40 mL of a product that contained EGEE by mistake. She became vertiginous and became unconscious shortly after exposure. Upon examination, she was cyanotic and had tachypneumonary edema, repeated tonic-clonic spasms, and acetone on her breath. She was given oxygen and other supportive treatment which resulted in essentially complete recovery after some 44 days.

A group of 17 persons exposed to glycol ethers in a varnish production plants were studied for workplace exposures and urinary metabolites (189). The workers in the plant were exposed to average concentrations of EGEE, EGEE acetate, EGBE, 1-methoxypropanol, 2-methoxypropyl-1-acetate, and xylene of 2.8, 2.7, 1.1, 7.0, 2.8 and 1.7 ppm, respectively. EGBE was measured in blood, and EAA and BAA in urine. Postshift samples contained concentrations of 121.3 µg/L of EGBE in blood and 167.8 and 10.5 mg/L of EAA and BAA, respectively, in urine.

Two groups of workers occupationally exposed to glycol ethers in a varnish production plant or the ceramic industry were examined (151). In the varnish production area, the average concentrations of EGEE, EGEE-acetate, and EGBE in air were 2.9, 0.5, and 0.5 ppm, respectively, on Monday, and 2.1, 0.1, and 0.6 ppm, respectively, on Tuesday. At the same workplaces, the mean urinary EAA and BAA concentrations were 53.2 and 0.2 mg/L on Monday preshift and 53.8 and 16.4 mg/L on Tuesday postshift. The authors concluded that urinary values should not exceed 50 mg/L EAA. This value is the current German biological tolerance value (BAT value) for EAA in urine.

3.4.2.2.1 Acute Toxicity. Acute toxic effects include depression of the central nervous system (CNS) and metabolic acidosis (190). Ethanol has been used as a competitive inhibitor to alcohol dehydrogenase in treating poisoning cases.

3.4.2.2.2 Chronic and Subchronic Toxicity. Chronic effects include CNS dysfunction, bone marrow suppression, anemia, and granulocytopenia (190).

3.4.2.2.3 Pharmacokinetics, Metabolism, and Mechanisms. In Vitro Dermal Testing. Dugard et al. (21) evaluated glycol ethers *in vitro* for dermal absorption across isolated human cadaver abdominal epidermis. They found that 2-methoxyethanol was absorbed across the epidermis most readily at a rate of 2.82 ± 2.63 mg/cm^2/h and the lag time was approximately 1 to 3 h. EGEE and its acetate followed in absorption rates at 0.796 ± 0.46 and 0.800 ± 0.43 mg/cm^2/h, respectively, and the approximate lag time was less than 1 h.

In Vitro and In Vivo Skin Absorption—Liquid. Kezic et al. (19) measured the rate of absorption of liquid EGEE (1.0 g/cm^3) on the skin of two men and three women volunteers. The liquid EGEE was applied in a glass chamber to a forearm area of 27 cm^2 for a period of 15 minutes. The rate of percutaneous absorption was determined by Kezic et al. (19) as 0.7 mg/cm^2/h on the basis of cumulative urinary excretion of 2-EAA. Percutaneous absorption may also be calculated by multiplying the concentration (1.0 g/cm^3) applied to skin by the permeability constant (K_p) of 8.42×10^{-4} cm/h derived by Dugard et al. (21) to give 0.842 mg/cm^2/h and a total absorbed dose of 5.0 mg (27 cm^2 × 1.0 g/cm^3 × 8.42×10^{-4} c/h × 15/60 h = 5.7 mg). Kezic et al. (19) found 5.2 mg of EGEE equivalents (2-EAA) in urine. The rate of percutaneous absorption directly depends on the concentration of liquid (mg/cm^3) applied to the surface of the skin and within skin, and high concentrations produce high rates of absorption (136). Evaporative losses often reduce the amount of chemical available for percutaneous absorption. No K_p values are available for EGEE dissolved in water. K_p values for dilute aqueous solutions of EGEE are expected to be greater than the value obtained with neat material (8.42×10^{-4} cm/h) due to skin hydration. In both guinea pigs and rats, the percutaneous absorption rate of EGBE was higher and dermal breakthrough occurred earlier when the solvent was applied in water than when applied neat (191, 192). Polar substances such as glycol ethers diffuse through aqueous regions of the skin.

In Vivo Skin Absorption—Vapors. Kezic et al. (19) measured the dermal uptake of EGEE vapors using five human volunteers (two men and three women). The forearms and hands of these volunteers (1040 cm^2) were exposed for 45 minutes to vapor concentrations of 3648 mg/m^3 (36.48×10^{-4} mg/cm^3 at skin surface). On the basis of the K_p (8.42×10^{-4} cm/h) value determined by Dugard et al. (21), the amount absorbed would be 2.40 µg (36.48×10^{-4} mg/cm^3 × 8.42×10^{-4} cm/h × 1040 cm^2 × 45/60 h = 2.40 µg). Kezic et al. (19) found 58 mg of EGEE equivalents (2-MAA) in urine. This finding suggests that the effective concentration within the skin (skin/air partition coefficient = 24,217) was approximately 24,217 times greater (~8.834 mg/cm^3) than the measured air concentrations. This value multiplied by the K_p value × 1040 cm² of skin × 45/60 h of exposure gives an absorbed dose of 58 mg of EGME (equal to the 58 mg found by Kezic et al. (19)). Whole body exposure (18,000 cm²; 8 h exposure) to 19 mg/m³ (OEL value) results in dermal uptake of 55.7 mg of EGME (55 mg calculated by Kezic et al., (19)). This calculation assumes a steady-state concentration of 19 mg/m³ of EGME in air, total skin contact (18,000 cm²), a skin partition coefficient of 24,217, 8 h of exposure, a K_p value of 8.42×10^{-4} cm/h, and no evaporative losses from the skin to reduce the effective concentration in the skin.

In Vivo Inhalation. Kezic et al. (19) measured the respiratory uptake of 53 mg/m³ of EGEE (0.053 mg/L or 14.7 ppm) by five volunteers (two men and three women) during the course of 4–15 minute periods. The high blood/air partition coefficient of 34,913 resulted in the retention of 81% of the inhaled EGEE and a retained dose of 24.0 mg of EGEE (2-EAA urinary equivalents). Extrapolation of an 8-h inhalation exposure at the OEL concentration of 0.019 mg/L (5.2 ppm) resulted in uptake of 70 mg. These values were slightly larger (24 vs. 22.6 mg and 70 vs. 64 mg) than those obtained using the PBPK model of Corley et al. (23) for butoxyethanol by setting MW = 90, exposure conc. = 5.2 or 14.7 ppm, QPC = 30 (respiratory volume), and running the model for 1-h (4×15 minutes) and 8-h exposures (Knaak, unpublished).

Pharmacokinetics — Inhalation. Groeseneken et al. (193) exposed human male volunteers via inhalation at various concentrations and under various physical loads. Steady-state levels of retention, atmospheric clearance, and rate of clearance were reached immediately after the start of exposure under all experimental conditions. Retention was high (64% resting) and increased as physical exercise was performed during exposure. Atmospheric clearance increased as pulmonary ventilation rate increased, and the rate of uptake was higher as either or both concentration and pulmonary rate increased. Postexposure breath concentrations declined rapidly during the first few minutes after exposure ceased, then much more slowly thereafter, suggesting at least a two-compartment clearance.

Metabolism — Inhalation. The experimental group of Groeseneken et al. (194) that consisted of 10 healthy male subjects (aged 19–28) were divided into two groups of five. The first group (n = 5) was exposed 4 h at rest to EGEE air concentrations of 10 mg/m³, 20 mg/m³, and 40 mg/m³, and the second group (n = 5) to 20 mg/m³ at rest and during standard exercises at 30 W and 60 W. The time course of EAA was followed for 42 h. Maximal excretion of EAA was reached 3 to 4 h after the end of the 4-h exposure period. Afterward, EAA excretion declined slowly, and the half-life was 21–24 h. On average, 23.1 ± 6.3% of absorbed EGEE was excreted as EAA within 42 h. No relationship was found between EAA excretion and body fat.

Laitinen et al. (195) examined the relationship between urinary biochemistry (metabolites) and occupational exposure to glycol ethers. Urinary oxalic acid and alkoxyacetic acid excretion together related to the decrease of succinate dehydrogenase activity as an indicator of renal mitochondrial effects. The excretion of ammonia by exposed workers doubled compared to controls.

3.4.2.2.4 Reproductive and Developmental. A case control study was conducted by Veulemans et al. (196) among first-time patients at a clinic for reproductive disorders. The study group consisted of 1019 cases, defined as patients diagnosed infertile or subfertile on the basis of a spermiogram and 475 control individuals who were diagnosed as normal by the same procedure. Exposure to glycol ethers was assessed by the presence of the urinary metabolites of EGME and EGEE (i.e., MAA and EAA). EAA was detected in 39 cases and six controls, and the highly significant odds ratio was 3.11 ($p = .004$). MAA was found only in one case and in two controls. The presence of EAA in urine was associated with exposure to paint products. The absence of a significant correlation between urinary values and sperm quality was explained on the basis of the latent period between exposure and the observed effects.

Welch et al. (197) examined the semen of 73 painters and 40 controls who worked in a large shipyard to determine whether EGEE and EGME affected reproduction. The industrial hygiene survey indicated that painters were exposed to EGEE at a time-weighted average of 0–80.5 mg/m^3 and a mean of 9.9 mg/m^3 and to EGME at a time-weighted average of 0–17.7 mg/m^3 and a mean of 2.6 mg/m^3. Painters had an increased prevalence of oligospermia and azoospermia and an increased odds ratio for lower sperm count per ejaculate. Ratcliff et al. (198) evaluated the effects of long-term exposure to EGEE on semen quality among men who worked in a metal casting process. Full-shift breathing zone exposures to EGEE ranged from nondetectable to 24 ppm (geometric mean 6.6 ppm). Concerns for skin absorption prompted the investigators to analyze urine for 2-EAA. Urine measurements showed levels of 2-EAA that ranged from nondetectable to 163 mg EAA/g creatinine. The average sperm count per ejaculate among exposed workers was significantly lower than that of the unexposed group (113 vs. 154 million sperm per ejaculate, respectively). When modifying factors were taken into consideration (age, use of tobacco, alcohol and caffeine, urogenital disorders, etc.), the mean sperm concentrations of the exposed and unexposed did not significantly differ from each other. No effects of exposure to EGEE were noted on semen volume, sperm viability, motility, velocity, and normal morphology, or testicular volume.

3.4.2.3 Epidemiology Studies. Refer to Section 1.4.2.3 for a discussion of this topic.

3.5 Standards, Regulations, or Guidelines of Exposure

The ACGIH (143) recommends a TLV of 5 ppm with a skin notation for ethylene glycol monoethyl ether. OSHA adopted 5 ppm (90 mg/m^3) as a PEL; rule-making is in progress to adopt a new workplace exposure standard. In 1991, NIOSH (144) recommended limiting exposure to EGEE in the workplace to 0.5 ppm of air (1.8 mg/m^3) as a TWA up to 10 h per day during a 40-h workweek (10-h TWA). Refer to Section 1.5 (EGME) for additional discussion. Also, refer to Table 86.9 for a listing of international exposure standards for EGEE.

4.0 Ethylene Glycol Mono-*n*-propyl Ether

4.0.1 CAS Number: [2807-30-9]

4.0.2 Synonyms: EGPE; monopropyl ether of ethylene glycol; propoxyethanol; 2-propoxyethanol; propyl glycol

ETHERS OF ETHYLENE GLYCOL AND DERIVATIVES

4.0.3 Trade Names: Eastman® EP solvent; *n*-Propyl Oxitol® Glycol; Propyl Cellosolve®

4.0.4 Molecular Weight: 104.15

4.0.5 Molecular Formula: $C_5H_{12}O_2$

4.0.6 Molecular Structure: ~~O~~OH

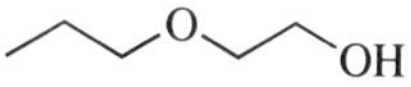

4.1 Chemical and Physical Properties

4.1.1 General

EGPE is miscible with water and soluble in alcohol and ether. Refer to Table 86.5 for a listing of chemical and physical properties of EGPE.

4.1.2 Odor and Warning Properties

EGPE is a volatile liquid that has a mild ethereal odor.

4.2 Production and Use

EGPE is used as a solvent in the coatings industry (2). The U. S. production of this material in 1995 amounted to 23 million pounds (1). Refer to Table 86.2 for more complete production and use information.

4.3 Exposure Assessment

4.3.1 Air

See the Introduction for a general description of air sampling and analytical procedures.

4.3.2 Workplace Methods

See the Introduction for a general description of sampling and analytical procedures for the workplace. There are no published standard methods for assessing exposures to this glycol ether; however, NIOSH Method 1403 can be adapted to measure EGPE (14).

4.4 Toxic Effects

4.4.1 Experimental Studies

Ethylene glycol mono-*n*-propyl ether (EGPE) causes red blood cell hemolysis in guinea pigs, mice, rabbits, and rats, and this effect was first identified by Gross (49) in studies of single animals. Effects secondary to hemolysis are seen in the kidney, liver, and spleen in repeated-dose oral and inhalation studies in rats. The hemolytic activity of EGPE is due to the acid metabolite, propoxyacetic acid (PAA), and is similar to the effects seen with the butyl derivative (butoxyacetic acid, BAA). The human red blood cell is less sensitive to these hemolytic effects, so that adverse effects observed in experimental animals may be less relevant to human exposure. EGPE is rapidly eliminated regardless of the route of administration. It is not neurotoxic in rats and is not a reproductive toxicant in rats, mice, or rabbits, although fetotoxicity was observed in rats at maternally toxic concentrations. EGPE is irritating to the rabbit eye, but single exposures are only slightly irritating to the skin of guinea pigs, mice, and rabbits. There is evidence of some increase in irritation from repeated exposures of guinea pigs. It was not a skin sensitizer in guinea pigs.

4.4.1.1 *Acute Toxicity.* According to Smyth (58), the LD_{50} for rats of EGPE is 4.45 g/kg. Other unpublished data (199) indicate that the oral LD_{50} for rats is between 0.5 and 1.0 g/kg. The acute oral LD_{50}s for this material when given undiluted by gavage in fed rats and mice were 6.18 g/kg and 3.09 g/kg, respectively, whereas these values for fasting animals were 3.09 g/kg and 1.77 g/kg, respectively (53, 200). Clinical signs of toxicity in this latter study included weakness, anorexia, hemoglobinuria, tremors, prostration, and death. Saparmamedov (52) reported that the LD_{50} for mice is 2.4 g/kg.

Administration of EGPE by gavage (water as the vehicle) to female mice, once daily, for eight consecutive days at five dose levels produced significant mortality only at dose levels in excess of 2.0 mg/kg (201). Clinical signs of toxicity included lethargy and prostration; surviving animals generally appeared normal throughout the study.

Inhalation. Rats exposed to a saturated atmosphere of EGPE (single 7-h exposure) survived, but all showed bloody urine within 2 h after the exposures terminated, together with lung, liver, and kidney injuries (199). Hematologic studies revealed a reduction in packed-cell volume but no evidence of hemolysis. Animals exposed for only 4 h were normal 7 days later, whereas those exposed for 7 h exhibited severe kidney injury 2 weeks later. Werner et al. (55) reported that the LC_{50} value for mice for a 7-h exposure is 1530 ppm. They reported that the toxic action was that of dyspnea and hemoglobinuria. Carpenter et al. (202) found that some of the rats exposed to 2000 ppm for 4 h died. Katz et al. (200) reported that the inhalation LC_{50} value for this material is in excess of 2132 ppm, the highest practical vapor concentration that could be achieved without aerosol formation. No mortality was observed in this latter study.

Katz et al. (200) found slight hematologic changes in male and female rats that were exposed to a series of vapor concentrations of EGPE up to a maximum of 800 ppm (6 h/day, 5 days/week, for 2 weeks). Decreased red blood cell counts were observed in both male and female rats at 800 ppm. Male rats displayed significant increases in mean corpuscular volume at both 400 and 800 ppm. Changes in red blood cell morphology were observed in both sexes of rat; the numbers of animals that showed changes, and the severity of the changes were concentration-dependent. In these studies, absolute and relative liver, kidney, brain, heart, and gonad weights of both sexes were normal at all exposure levels. Hematologic parameters of male rats exposed to 100 ppm of EGPE (total of twelve 6-h exposures) were unaffected (203). The NOEL for these studies was reportedly 200 ppm (200).

A pulmonary function battery of tests was performed on male Sprague–Dawley rats exposed to 784 ppm EGPE for 6 h on 10 consecutive days (204). No significant differences in spontaneous mechanics, flow characteristics, or nitrogen washout were seen. A slight decrease in functional residual capacity and marginally lower values for residual volume and total lung capacity may have indicated the presence of slight pulmonary edema.

Dermal. EGPE may be absorbed through the skin of animals in lethal amounts. Smyth et al. (205) stated that the LD$_{50}$ by skin absorption in rabbits is 0.87 g/kg. Katz et al. (200) estimated that the dermal LD$_{50}$ in guinea pigs for this glycol ether is between 1 and 5 mL/kg. The dermal LD$_{50}$ in rabbits determined under an occluded wrap for 6 h for this material is 1.34 g/kg (64).

Eye Irritation. EGPE is irritating to the eyes of rabbits and causes injury to the conjunctival membranes and the cornea and also some iritis; healing appeared complete within a week. When EGPE was introduced into the eyes of rabbits, Gross (49) found that it caused reddening and swelling of the conjunctiva and the lids and corneal damage. In other studies, these same effects were noted but, in addition, iritis was also observed (205). Katz et al. (200) reported severe erythema and moderate edema of the conjunctivas in the rabbit following treatment with 0.1 mL EGPE. Iritis and staining of the adnexa and cornea and some corneal opacity were also observed. All effects, however, were resolved by 14 days.

Skin Irritation. Gross (49) and Smyth et al. (205) reported that EGPE is not irritating to the skin. Other work confirms this for ordinary exposure but indicates also that if the material is confined to the skin for prolonged periods of time, it may produce irritation and possibly a burn. Saparmamedov (52), using mice, reported that neither skin irritation nor signs of skin absorption of toxic amounts occurred. Katz et al. (200) reported that this glycol ether produces slight skin irritation when applied to the depilated abdomen of guinea pigs (occlusive wrap, 24 h). This effect is exacerbated following 10 daily applications without occlusion to the clipped backs of guinea pigs.

4.4.1.2 Chronic and Subchronic Toxicity. Katz et al. (200) administered EGPE orally to groups of male rats at dose levels of 195, 390, 780, and 1560 mg/kg/day (5 days/week for 6 weeks). Clinical signs of toxicity at all dose levels included red discolored urine diagnosed as hemoglobinuria, whose frequency decreased as the study progressed. Other signs of toxicity seen only at the middle- and high-dose levels included weakness, labored breathing, prostration, and rales. Dose-dependent decreases in hemoglobin concentration and red blood cell counts were seen in surviving rats at all dose levels. Dose-dependent increases in spleen weights were also recorded. Histological examination revealed congestion and extramedullary hematopoiesis of the spleen in animals that received the three highest dose levels of EGPE. Histological changes in the kidney included proteinaceous casts (high-dose level only) and hemosiderin (all dose levels) in the convoluted tubules. Focal hemosiderin was seen in the liver at the highest dose level. No testicular injury was noted at any of the dose levels tested. These workers concluded that the primary site of toxicity of EGPE in the rat is the red blood cell and that secondary effects are elicited in the liver, spleen, and kidney.

Nagano et al. (206) reported that oral administration of EGPE to male mice at a maximum dose level of 2.0 g/kg/day (5 days/week for 5 weeks) had no effects on testicular weights. These authors report a strong correlation between chemical structure and the ability of a glycol ether to induce testicular atrophy. The methyl and ethyl ethers of ethylene glycol (either as the free alcohol or esterified) were testicular toxicants. Higher homologues in the series had no effect.

Inhalation. The red blood cell was the site of toxicity in rats exposed subchronically to EGPE by inhalation. Katz (207) exposed groups of male and female rats to 0, 100, 200, or 400 ppm EGPE for 6 h/day, 5 days/week for 14 weeks. Body weight gains for males at the highest concentration were slightly but significantly reduced. A transient and concentration-dependent hemoglobinuria was observed for both sexes at 200 and 400 ppm. In general, more females than males displayed these signs, and the signs were most frequently observed on the first morning of exposure following a weekend of nonexposure. Indicative of hemolysis, red blood cell counts, hemoglobin concentration, and hematocrit values decreased, and reticulocyte counts increased at the two highest concentrations. Absolute and relative spleen weights significantly increased in males exposed to 400 ppm and females exposed to 200 or 400 ppm. Other than pigment deposits seen in the kidneys, liver, and spleen (presumably resulting from hemolysis), no gross or histopathological effects were observed for the test material. The NOEL for toxicity in this study was reported as 100 ppm.

4.4.1.3 Pharmacokinetics, Metabolism, and Mechanisms.

In Vitro Skin Absorption. The material is absorbed readily through the skin of animals. The *in vitro* dermal absorption rates for this glycol ether determined by using Franz-type glass-diffusion cells with both human stratum corneum (from abdominal skin) and full-thickness rat skin are 0.58 ± 0.39 and 2.30 ± 0.79 mg/cm^2/h, respectively (208). The corresponding permeability constants, derived from these values, are 6.43 (± 4.84) $\times 10^{-4}$ and 2.52 (± 0.86) $\times 10^{-3}$ cm/h. The estimated *in vivo* dermal absorption rate for this glycol ether in male Sprague–Dawley rats following occluded (and contained) application of ^{14}C-labeled material was 0.73 mg/cm^2/h, based on the recovery of total radioactivity following a 6-h exposure (209).

Metabolism. The metabolism and pharmacokinetics of EGPE were determined following either intravenous, oral, dermal, or inhalation exposures of male Sprague–Dawley rats to 2-[ethylene-1,2-^{14}C]-EGPE ([^{14}C]-EGPE) (209). Regardless of the route of administration, absorbed radioactivity was eliminated rapidly and the majority was present in 12-h sample collections. Following oral administration of either 15 or 150 mg/kg [^{14}C]-EGPE, 97% and 96%, respectively, of the administered doses were recovered by 72 h. In this case, urinary elimination accounted for 75 to 81% of the activity recovered.

Inhalation exposures at either 25 ppm or 175 ppm [^{14}C]-EGPE resulted in similar patterns of elimination. Of the estimated administered dose, 80% was eliminated in the urine at 25 ppm, and 79% was eliminated in the urine at 175 ppm. Less than 27% of dermally administered radioactivity was absorbed during a 6-h exposure period. The majority (74%) of the administered radioactivity was recovered either as unabsorbed liquid or in washings of the application sites.

The acid metabolite, 2-propoxyacetic acid (PAA), and its glycine conjugate, *N*-(2-propoxyacetyl)glycine, were the principal urinary metabolites identified regardless of the route of administration. In the case of 12-h urinary metabolites (oral, dermal, or inhalation exposures), PAA accounted for 42 to 61% of the total urinary radioactivity, and the glycine conjugate for an additional 24 to 38%. Ethylene glycol accounted for up to 14% of the radioactivity in urine regardless of the route of administration. Glucuronidase treatment of urine revealed the presence of the glucuronide of EGPE at levels of 2 to 6% of the recovered radioactivity.

Pharmacokinetics. The half-lives for the first-order elimination of EGPE and PAA from rat blood (209) were 0.12 and 0.75 h, respectively (i.v. administration). The apparent volume of distribution and clearance rate for EGPE were 98.7% and 1.3 kg/h, respectively, following i.v. administration. Oral administration of EGPE at 15 mg/kg yielded similar pharmacokinetic parameters. At an oral dose of 150 mg/kg, the parent alcohol had an elimination half-life of 0.20 h, suggesting saturation of metabolic or excretory processes. Elimination of the acid, PAA, was less rapid than that of the parent alcohol but displayed first-order elimination kinetics in all cases except after the high oral dose. In this latter case, elimination of the acid appeared to be saturated. Feedback inhibition of metabolism by the acid is a possible explanation for the effects at the high oral dose. Given the similarity of pharmacokinetic parameters for EGPE compared with EGBE (210), differences in the rodent toxicities of these two materials may reflect inherent differences in the toxicity of the primary metabolites, PAA and BAA. However, increased elimination of PAA as the glycine conjugate may provide an alternative explanation for the observed differences. The measured pharmacokinetic parameters for EGPE are listed in Table 86.10 and are compared with similar results for EGBE.

Table 86.10. Pharmacokinetic Parameters Derived from the Analysis of EGPE and PAA Concentrations in Rat Blood Following Intravenous and Oral Exposures of EGPE: Comparison With Results for EGBE and BAA in Blood Following Intravenous Exposure to EGBE[a]

Dose mg/kg	C_{max} µg/mL	AUC mg-h/kg	$t_{1/2}$ h	Clearance, Cl_b kg/h	Apparent Volume of Distribution, V_d L (or %)[e]
EGPE					
15 (i.v.)	10.8 ± 0.7	3.1 ± 0.2	0.12 ± 0.01	1.3 ± 0.1	0.23 ± 0.03 (98.7)
15 (oral)	6.9 ± 3.4	2.0 ± 1.4	0.13 ± 0.01	1.3 ± 1.2	0.24 ± 0.24 (120)
150 (oral)[b]	34.5 ± 7.0	N.C.[c]	0.20 ± 0.03	N.C.	N.C.
EGBE[d]					
31.25 (i.v.)	22.1 ± 4.4	4.1 ± 0.8	0.11 ± 0.002	3.2 ± 0.6	0.49 ± 0.09
62.5 (i.v.)	41.6 ± 4.8	9.2 ± 1.2	0.14 ± 0.03	2.6 ± 0.44	0.52 ± 0.07
125.0 (i.v.)	120.6 ± 5.3	32.0 ± 0.5	0.17 ± 0.01	1.4 ± 0.02	0.34 ± 0.02
PAA					
15 (EGPE, i.v.)	21.5 ± 3.0	36.95 ± 7.55	0.75 ± 0.06	N.C.	N.C.
15 (EGPE, oral)	24.3 ± 1.4	62.0 ± 8.7	1.32 ± 0.12	N.C.	N.C.
150 (EGPE, oral)	149.3 ± 17.7	N.C.[c]	2.37 ± 0.23	N.C.	N.C.
BAA[d]					
31.25 (EGBE, i.v.)	38.3 ± 3.2	94.4 ± 16.3	1.5 ± 0.25	—	—
62.5 (EGBE, i.v.)	74.4 ± 4.6	270.0 ± 25.1	3.2 ± 0.54	—	—
125.0 (EGBE, i.v.)	147.6 ± 6.0	653.4 ± 57.8	2.7 ± 0.3	—	—

[a]Ref. 209.
[b]It was not possible to obtain a reasonable fit of either EGPE or PAA using multicompartmental pharmacokinetic modeling when administered orally at 150 mg/kg. Only terminal elimination half-lives were calculated in these cases.
[c]N.C. = Not calculated. It was not possible to fit EGPE or PAA blood concentration data to multicomponent exponential decay equations.
[d]Taken from the results of Ghanayem et al., 1990 (210).
[e]Values expressed in units of liters assuming 1 kg ≅ 1 L.

4.4.1.4 Reproductive and Developmental. *Reproductive.* EGPE administered by gavage (water vehicle) at 2000 mg/kg to timed-pregnant CD-1 mice during days 7 through 14 of gestation resulted in no compound-related effect on reproductive outcome, as measured by the delivery index (number of live litters produced/number of mice determined to be pregnant) (201). However, this treatment regime resulted in significant increases in the number and percent of dead pups at birth. No effects were seen on the testes of male mice given dose levels as high as 2 g/kg EGPE for 5 days/week for 5 weeks (32) or on the testes of male rats given dose levels as high as 1560 mg/kg/day, 5 days/week for 6 weeks (200) or rats exposed by inhalation at 400 ppm for 14 weeks (207). This glycol ether was also negative when tested in the Chernoff–Kavlok teratogenicity screening test (211) and showed no adverse effects on maternal reproductive index (mortality, weight change, and viable litters) or neonatal indexes, including live-born per litter, percent survival, birth weight per pup, and weight gain per pup.

Developmental. Inhalation exposure of pregnant rabbits to nominal concentrations of 125, 250, or 500 ppm of EGPE (6 h/day on gestation days 6 to 18) resulted in no observed teratogenicity or developmental toxicity (212). The dose levels chosen for this study were based on a probe study in which five of six pregnant rabbits were killed by an exposure to 800 ppm. Slight maternal toxicity was observed at the high-dose level in the definitive study; however, hematologic determinations, absolute and relative organ weights, and observations at necropsy revealed no dose-related maternal effects.

Inhalation exposure of pregnant rats to nominal concentrations of 100, 200, 300, and 400 ppm (6 h/day on gestation days 6 to 15) resulted in no observed teratogenic effects (213). The ossification of certain skeletal elements were delayed and the incidence of supernumerary ribs increased significantly at or above 200 ppm. These latter effects may have been associated with maternal toxicity.

4.4.1.6 Genetic and Related Cellular Effects Studies. Two different groups of researchers demonstrated the *in vitro* toxic effect of EGPE on cultured Chinese hamster V79 cells. Welsch and Stedman (214) reported that this glycol ether inhibits the metabolic cooperation between V79 cells sensitive and insensitive to 6-thioguanine in a concentration-dependent manner when grown in culture. Loch-Caruso et al. (215) reported similar findings. In addition, these workers found that both the degree of inhibition of cell-to-cell communication and cytotoxicity increased with increasing chain length of the alkyl ether. These investigators speculate that for EGPE, interrupted cellular communication may occur with cytotoxicity in the corresponding *in vivo* situation, thus making this effect less important as a mechanism of teratogenesis.

The corresponding alkoxyacetic acid metabolites of ethylene glycol monomethyl and monoethyl ethers, methoxyacetic and ethoxyacetic acids, respectively, have been implicated as the proximal teratogenic metabolites (216). In studies with a series of alkoxyacetic acids, *n*-propoxyacetic acid was markedly less embryo toxic than the corresponding methyl or ethyl homologues when tested in postimplantation rat embryo cultures (217). In contrast to the results for methoxyacetic and ethoxyacetic acids, *n*-propoxyacetic acid did not cause degeneration of pachytene and dividing spermatocytes either in the testes of adult rats or in mixed cultures of Sertoli and germ cells from immature rats (218, 219), in agreement with results from *in vivo* studies.

The *in vitro* hemolytic activity of the propyl derivative is less than that of the corresponding butyl derivative, and human red blood cells are relatively insensitive to the effects of these acids. Ghanayem and co-workers (220) studied the hemolytic effects of several alkoxyacetic acids on rat erythrocytes and found the following ranking of activities: butoxyacetic acid > ethoxyacetic acid > methoxyacetic acid. The common mechanism of toxicity is early swelling followed by hemolysis. Neither propoxy nor butoxyacetic acid causes significant hemolysis of human red blood cells *in vitro* at concentrations up to 5.0 mM (221).

Dieter et al. (222) studied the chemotherapeutic potential of several glycol ethers in the Fischer-rat leukemia transplant model. In this model, leukemic cells are injected subcutaneously into susceptible rats, and the expression of leukemia is quantified by measuring relative spleen weights, WBC counts, RBC indexes, and platelet counts. EGPE was ineffective in this assay; however, both the methyl and ethyl homologues provided protection.

4.4.1.7 Other: Neurological, Pulmonary, Skin Sensitization. *Skin Sensitization.* This material was not a skin sensitizer in the guinea pig when tested using the Buehler method (223) or the alternative footpad method (224). Only slight skin irritation was observed in rabbits and guinea pigs when this glycol ether was administered at 2.2 and 0.9 gm/kg, respectively (64).

Neurotoxicity. A guideline (40 CFR Part 798.6050, published on September 27, 1985 and amended May 20, 1987) subchronic inhalation neurotoxicity study was performed to determine the neurotoxic potential of EGPE (225). Groups of male and female Fischer 344 rats were exposed for 6 h/day, 5 days/week, for 14 weeks to nominal concentrations of 0, 100, 200, and 400 ppm of EGPE. A functional observational battery (FOB) was employed to detect changes in activity, coordination, behavior, and sensory function. Forelimb and hindlimb grip strength were also quantified. At the end of the exposure period, selected animals were perfused, and tissues from both the central and peripheral nervous systems were histopathologically examined after staining with hematoxylin and eosin or Luxol Fast Blue and Bodian silver impregnation. All animals from each exposure group were necropsied and were examined for macroscopic lesions. Brain, liver, kidney, and spleen weights were also measured. Clinical signs of irritation due to EGPE exposure included red discoloration of facial hair, porphyrin nasal discharges and tears, sialorrhea at 200 and 400 ppm in all rats and discolored facial hair at 100 ppm in female rats. At exposure concentrations of 200 and 400 ppm, both males and females demonstrated signs of hematotoxicity, including red discolored urine, pigment deposition in the kidneys and Kupffer cells of the liver, and increased pigmentation of the spleens. Spleen weights were increased at the highest exposure concentration. No neurological deficits in motor or sensory function were detected at exposure concentrations that were systemically toxic to male and female rats. Neurological lesions were not observed in any rats exposed up to 400 ppm for 14 weeks. The NOEL for neurological effects was greater than 400 ppm.

4.4.2 Human Experience

4.4.2.1 General Information. There are no reports of adverse effects in humans.

4.5 Standards, Regulations, or Guidelines of Exposure

The occupational exposure limit for this material in the Netherlands (OEL/TWA) is 10 ppm (44 mg/m^3, skin notation) (226), and in Germany (MAK) it is 20 ppm (86 mg/m^3, skin notation) (227).

5.0 Ethylene Glycol Monoisopropyl Ether

5.0.1 CAS Number: [109-59-1]

5.0.2 Synonyms: 2-(1-Methylethoxy)ethanol; 2-isopropoxyethanol; isopropylethanediol; EGiPE

5.0.3 Trade Names: Isopropyl Oxitol® Glycol; Isopropyl Cellosolve® Dowanol® eipat

5.0.4 Molecular Weight: 104.15

5.0.5 Molecular Formula: $C_5H_{12}O_2$

5.0.6 Molecular Structure:

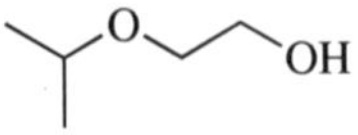

5.1 Chemical and Physical Properties

5.1.1 General

See Table 86.5 for chemical and physical properties. EGiPE is soluble in all proportions in water, alcohol, ether, and acetone.

5.1.2 Odor and Warning Properties

EGiPE is a colorless liquid that has a mild ethereal odor.

5.2 Production and Use

EGiPE is a component of lacquers and other coatings and is used as a solvent for resins.

5.4 Toxic Effects

5.4.1 Experimental Studies

5.4.1.1 Acute Toxicity. The single-dose oral toxicity of EGiPE in the rat is low, the LD$_{50}$ is between 0.5 and 5.6 g/kg. Wolf (228) reports that the single-dose oral LD$_{50}$ of EGiPE in the rat is between 0.5 and 1.0 g/kg, severe kidney and liver injury are observed, and large amounts of blood pass in the urine. Smyth et al. (205) reported an LD$_{50}$ for rats of 5.66 g/kg. The LD$_{50}$ for mice fed EGiPE in oil was 2.3 g/kg, and in water, 2.18 g/kg (52). Kennedy and Graepel (229) classify EGiPE as having low acute oral toxicity in the rat.

Inhalation. Smyth et al. (205) reported that rats exposed for 2 h to 4000 ppm survived, whereas a 4-h exposure caused the death of four of six rats. Table 86.11 summarizes

Table 86.11. Summary of Results of Single Exposures of Rats to EGiPE Vapors[a]

Concentration	Duration of Exposures (h)	No. Dying/No. Exposed	Comments and Observations
Essentially saturated at 100°C and cooled to room temperature	7	4/4	After 3 h, blood in the urine; one animal autopsied after exposure had black and severely enlarged kidneys
Essentially saturated at 100°C and cooled to room temperature	4	2/3	Passed bloody urine immediately after exposure; excess urine; 3 days after exposure, the kidneys were severely necrotic and dark in color
Essentially saturated at 100°C and cooled to room temperature	1	0/3	Bloody urine, slight weight loss, severely injured kidneys
Essentially saturated at 100°C and cooled to room temperature	0.5	0/3	Bloody urine, slight weight loss, questionable kidney injury
160 ppm	6.7, 4	0/5	Bloody urine during exposure; slight weight loss; one rat was autopsied 2 days after exposure, and the liver and kidneys were pale in color; urine in the bladder was clear; the four survivors autopsied 15 days after exposure exhibited evidence of slight to moderate kidney injury
80 ppm	7.0	0/5	Slight weight loss; much blood passed in urine; one rat sacrificed immediately after exposure had no gross pathological changes; another sacrificed the day after exposure had severely affected kidneys, a pale colored liver, and bloody urine in the bladder; a rat sacrificed 9 days after exposure showed no evidence of gross changes of the kidney
80 ppm	4.0	0/5	Questionable evidence of blood in urine, slight weight loss; one rat sacrificed immediately after exposure had severely affected kidneys; one sacrificed the day after exposure had slight pathology of the kidneys, but no blood in urine; another sacrificed 9 days after exposure appeared grossly to have slightly injured kidneys
Saturated vapor	4.0	N.A.[b]	No testicular effects noted

[a] Refs 228 and 230.
[b] N.A. = Information not available.

further single-dose inhalation data on rats. Samuels et al. (230) reported that exposure for 4 h to saturated EGiPE vapors caused no reductions in rat testicular weights 14 days following exposure. Kennedy and Graepel (229) classify EGiPE as having low acute toxicity in the rat following inhalation exposure.

Werner et al. (55) reported that the 7-h LC_{50} for mice is 1930 ppm. Gage (231) reported that rats exposed for 6 h/day for 15 days to 100, 300, or 1000 ppm experienced nasal irritation, lethargy, hemoglobinuria, porphyrinuria, initially at the highest concentration low hemoglobin (fourth day), and reduced mean corpuscular hemoglobin concentration, reticulosis, and congestion of the lungs. At autopsy, organs appeared normal, and no histopathological injury was present. The adverse effects noted for both the blood and urine returned to normal later in the exposure. Those animals exposed to 300 ppm were less affected, whereas those exposed to 100 ppm under the same conditions were unaffected. In studies reported by Doe (232), male rats exposed to 1000 ppm of EGiPE (6 h/day for 9 days) displayed marked hematuria on the first day of exposure that disappeared on subsequent days. In this same study, hemoglobin concentrations, red blood cell counts, and mean corpuscular hemoglobin concentrations were reduced on day 10 following exposure of rats to 1000 ppm but not 300 ppm. Body weight gains of the 1000-ppm-treated animals were lower than controls; however, neither testes weights nor appearance differed from those of the controls.

Dermal. This material may be absorbed through the skin in lethal amounts. The dermal LD_{50} was in the range of 1.08 to 2.37 gm/kg in rabbits (205). Saparmamedov (52) reported that mice, when treated on the skin, experienced no significant irritation or signs of absorbing toxic amounts.

Eye Irritation. When the material was instilled into the eyes of rabbits, it caused marked conjunctival irritation, marked corneal injury, and some iritis. Healing was essentially complete in about 7 days (205).

Skin Irritation. EGiPE is not appreciably irritating to the skin under ordinary conditions of exposure, but if it is confined for prolonged periods, it may cause appreciable irritation, even a burn (205). Zissu (233) reports that EGiPE is a skin irritant in rabbits under the EEC testing protocol and a moderate irritant under the Draize method. In addition, this material produced no skin sensitization in guinea pigs (233).

5.4.1.2 Chronic and Subchronic Toxicity. Although details are limited, Moffet et al. (234) reported that hemolytic effects occurred marginally at 25 ppm and clearly at 50 ppm and 200 ppm when rats were exposed by inhalation to EGiPE for 6-h/day for 26 consecutive weeks. Only minimal clinical chemistry changes were noted in rats. In this same study, guinea pigs, rabbits, and dogs were generally unaffected by 200 ppm EGiPE. Although poorly documented, Lykova et al. (235) report a 3 mg/m^3 threshold of toxic action for EGiPE in a 3-month chronic study in rats in which rats were exposed for 24 h/day.

5.4.1.3 Pharmacokinetics, Metabolism, and Mechanisms. *Metabolism.* Hutson and Pickering (236) studied the metabolism of [1,2-^{14}C] EGiPE in both the dog and the rat. The metabolism was similar in both animals. In the rat, this material was rapidly metabolized and 88% was excreted from the body in 24 h (average for male and female), 73% via the urine, and 14% via the lungs as carbon dioxide. The major urinary metabolites

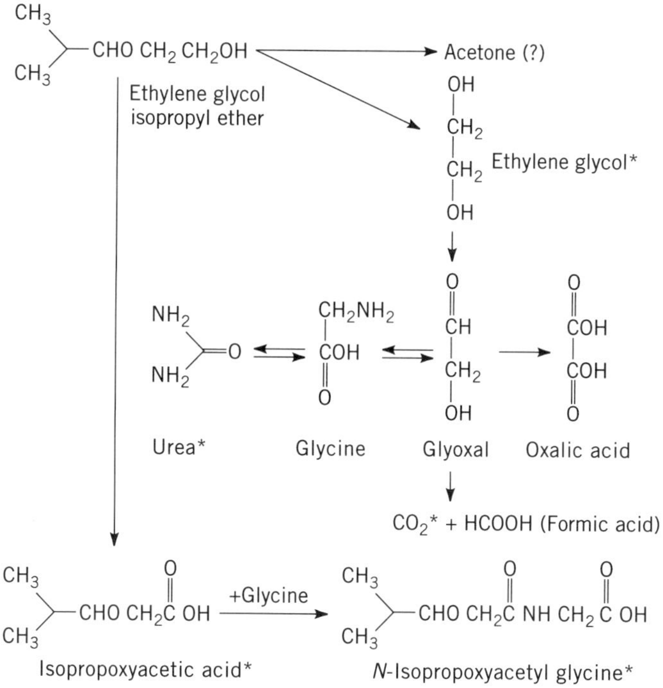

Figure 86.6. Metabolic scheme for EGiPE. (The asterisks indicate those compounds that are excreted by rats.)

were isopropoxyacetic acid (30% of urinary radioactivity), *N*-isopropoxyacetylglycine (46% of urinary radioactivity), and ethylene glycol (13% of urinary radioactivity). Only the first two of these metabolites were identified in the urine from a beagle hound treated with this glycol ether. The occurrence of the glycine derivative of isopropoxyacetic acid was unexpected and suggests conjugation of the acid within the mitochondrion by glycine *N*-acylase. Based on these results, the authors suggest the metabolic scheme shown in Figure 86.6 for the rat.

5.4.1.4 Reproductive and Developmental. EGiPE vapor was administered for 6 h/day on 10 consecutive days, to timed-pregnant Sprague–Dawley rats on gestational days 6 to 15 at nominal concentrations of 0, 100, 300, or 600 ppm (237). Pregnancy rates were high and approximately equivalent across exposure groups. Treatment-related signs included blood in the urine at 300 and 600 ppm and piloerection at 600 ppm. Maternal body weights were significantly reduced at 600 ppm on gestational days 9, 12, and 15 and at 300 ppm on gestational day 9. Both absolute and relative maternal spleen weights increased significantly at 600 ppm. Maternal feed consumption was also significantly reduced at 300 and 600 ppm. There were no treatment-related effects on gestational parameters. The percentage of adversely affected implants/litter increased at 600 ppm. There was no evidence of teratogenicity at any concentration. The NOAEL for maternal toxicity was at or above 100 ppm. The NOAEL for developmental toxicity was 300 ppm.

5.4.1.6 Genetic and Related Cellular Effects Studies. *In Vitro Toxicity.* Welsch and Stedman (214) reported that EGiPE was the most potent of five glycol ethers tested in inhibiting metabolic cooperation between Chinese hamster V79 cells. These same authors reported that the inhibition of intercellular communication between normal human embryonal palatal mesenchyme cells was reduced as determined by decreased [^{3}H]uridine transfer (238). In contrast to the results for ethylene glycol monomethyl ether, recipient-cell to donor-cell ratios remained constant in this study over a series of concentrations, suggesting that the effect of this glycol ether may be to reduce cell viability rather than cell-to-cell communication. EGiPE was negative (repeat assays) in the *S. typhimurium* reverse mutation assay using tester strains TA98, TA100, TA1535, and TA1537 both with and without metabolic (S9) activation and in the *E. coli* WP2 *uvr*A tester strain with and without metabolic activation (239).

5.4.2 Human Experience

5.4.2.1 General Information. No reports of adverse effects in humans from handling or using EGiPE have been published.

5.5 Standards, Regulations, or Guidelines of Exposure

ACGIH (143) recommends a threshold limit value (TWA) of 25 ppm (106 mg/m^3) for this glycol ether. Occupational Exposure Limit (OEL) values in a number of countries including Australia, Belgium, Denmark, France, and Switzerland are also 25 ppm, in accord with the ACGIH TLV. Refer to Table 86.9.

6.0 Ethylene Glycol Mono-*n*-butyl Ether

6.0.1 CAS Number: *[111-76-2]*

6.0.2 Synonyms: EGBE; 2-butoxyethanol; 2-butoxy-1-ethanol; butyl glycol ether; BGE

6.0.3 Trade Names: Butyl Cellosolve®; Butyl Oxitol® Glycol; Dowanol® EB; Eastman® EB Solvent; Glycol Ether EB

6.0.4 Molecular Weight: 118.17

6.0.5 Molecular Formula: C$_6$H$_{14}$O$_2$

6.0.6 Molecular Structure: ~~~O~~~OH

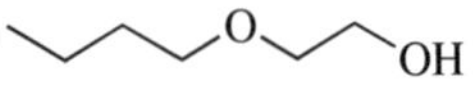

6.1 Chemical and Physical Properties

6.1.1 General

See Table 86.5.

6.1.2 Odor and Warning Properties

EGBE is a colorless liquid that has a mild, ether-like odor. Controlled human exposure studies have indicated that air concentrations of 100 ppm or higher are irritating.

6.2 Production and Use

EGBE remains the single most widely produced glycol ether. U.S. production amounted to 413 million pounds in 1995 (1). EGBE is used in hydraulic fluids; as a coalescing agent for many water-based coatings; as a component of acetate esters as well as phthalate and stearate plasticizers; as a coupling agent to stabilize immiscible ingredients in metal cleaners, textile lubricants, cutting oils, and liquid household products; as a solvent for nitrocellulose resins, spray lacquers, quick-drying lacquers, varnishes, enamels, dry-cleaning compounds, and varnish removers; as a mutual solvent for "soluble" mineral oils to hold soap in solution and to improve the emulsifying properties; as a solvent for vinyl and acrylic paints; in aqueous cleaners to solubilize organic surfactants; and as a solvent in cosmetics (2, 240). Refer to Table 86.2 for a summary of the production and use of EGBE.

6.3 Exposure Assessment

6.3.1 Air

See the Introduction for a general description of air sampling and analytical procedures.

6.3.3 Workplace Methods

A number of analytical methods have been published that are suitable for detecting EGBE in environmental air samples (13). These methods rely principally on adsorption onto charcoal or Tenax® followed by gas chromatographic analysis. Lower limits of detection generally range from 0.1 to 2 ppm depending on the specific analytical method employed and on the gas sampling procedure used. Typical methods include NIOSH method 1403 (14) or OSHA method 83 (15). See the Introduction for a general description of sampling and analytical procedures for the workplace.

6.3.5 Biomonitoring/Biomarkers

6.3.5.1 Blood. Bormett et al. (241) published a method for detecting EGBE and 2-butoxyacetic acid (BAA) in rat and human blood using gas chromatography/mass spectrometry (GC/MS). Johanson and Johnsson (242) describe a similar method for BAA in human blood. Johanson et al. (243) also describe a method analyzing EGBE in human blood. Angerer et al. (244) reported postshift blood levels of EGBE in workers that ranged from a mean of <5 µg/L for laboratory workers to 121 µg/L in the blood of varnish production workers. Although workplace monitoring indicated airborne concentrations well below the German MAK value of 20 ppm, significant urinary excretion of BAA led these authors to suggest that there is a significant dermal component in EGBE exposure.

6.3.5.2 Urine. NIOSH has suggested that the determination of BAA in urine could be used as a method for biological monitoring of EGBE exposure, although a sound relationship between urinary BAA and inhalation exposure to EGBE could not be established (245). A number of published methods exist for analyzing BAA in urine (13, 246). All of them rely on derivatizing the acid followed by gas chromatographic analysis. Angerer et al. (244) reported both pre- and postshift urinary levels of BAA in workers.

Internal versus external exposures were poorly correlated on the basis of BAA excretion. Significant excretion of BAA was measured, even when air concentrations were low or absent. A dermal component of exposure was suggested as an explanation of these discrepancies. Urinary levels of BAA were greatest in varnish production workers and had mean pre- and postshift values of 3.3 mg/L and 10.5 mg/L, respectively. Sohnlein et al. (151) report similar pre- and postshift values of 0.2 mg/L and 16.4 mg/L, respectively, for varnish production workers whose excretion of BAA correlated poorly with ambient air monitoring results. These investigators also suggest that there is a significant dermal component of exposure. Vincent et al. (247) report urinary levels of BAA in workers who employ window cleaning solutions containing EGBE. Urinary levels reached 111 mg BAA/g creatinine in end-shift samples, although the highest mean air concentrations of EGBE were only 2.44 ppm, again suggesting significant dermal uptake of EGBE. Levels of EGBE exposure based on end-shift urinary BAA depended mainly on the use time and the quantity of cleaning agent used. Haufroid et al. (248) reported a good correlation of external (air levels) versus internal (BAA excretion) exposures to EGBE in workers who were exposed at low levels. In this last case, the mean airborne EGBE concentration was 0.59 ppm and urinary BAA levels (post-shift) were 10.4 mg/g creatinine. Rettenmeier et al. (249) and Sakai et al. (250) reported that as much as 92% of BAA in human urine exists as the conjugate with glutamine. The determination of total BAA excreted requires acid hydrolysis before analysis. Thus, earlier published values for BAA excretion in humans may have underestimated the amounts of this metabolite present.

Collinot et al. (251) reported that the increased urinary excretion of D-glucaric acid (DGA) in foundry workers correlated with exposures to low concentrations of EGBE from paints. This may represent an adaptive rather than toxic response. Due to the small number of subjects used in this study, it is difficult to prove unequivocally a connection between EGBE exposure and DGA excretion. Laitinen et al. (252) report minor but significant decreases in urinary glucosaminoglycan (GAG) levels and increases in N-acetylglucosaminidase (NAG) levels in silk-screen printers exposed to glycol ethers and acetates. It was reported that 60% of these printers were exposed above the Finnish OEL for EGEE, but all were below the OEL for EGBE exposure. Urinary BAA excretion of 60 mmole/mole creatinine corresponded to an inhalation exposure level of 5 mL/m^3 of EGBE and its acetate.

Recently, Shih et al. (48) reported an improved method for analyzing butoxyacetic acid (BAA) in urine (refer to Section 1.3.5.2). This method does not involve derivatization.

6.4 Toxic Effects

6.4.1 Experimental Studies

Early studies by Werner (55, 56, 57) and by Carpenter (50) identified the red blood cell as the primary target tissue of EGBE toxicity. Hemolysis that leads to hemoglobinuria as the commonly observed clinical sign of toxicity is produced in rodents and rabbits. It has been adequately demonstrated *in vitro* that the acid metabolite, BAA, is responsible for the hemolytic effects of EGBE. In blood, red blood cell counts decrease with corresponding increases in mean corpuscular hemoglobin concentrations (MCHC) and mean corpuscular

ETHERS OF ETHYLENE GLYCOL AND DERIVATIVES 139

volumes (MCV). Tolerance to these hematotoxic effects has been reported in rats following multiple exposures to EGBE. Effects in other tissues, including the kidney, liver, and spleen, are considered secondary to hemolysis. Effects seen in female rats are often more pronounced than in males. A number of species, including humans and guinea pigs, are relatively insensitive to the hemolytic effects of EGBE.

6.4.1.1 Acute Toxicity. The acute toxicity of EGBE has been extensively reviewed (13, 245, 246, 253, 254). Doses at or higher than the LD_{50} in rats cause sluggishness, ruffled fur, prostration, and narcosis (50). The major clinical effects reported in acute, as well as in subchronic animal studies, resulted from intravascular erythrocyte hemolysis. These effects are observed regardless of the route of administration. Hemolysis is associated most often with secondary effects such as spleen enlargement and attendant hemosiderin deposition, nephropathy, and liver enlargement (hepatocytomegaly) with hemosiderin deposition. Species vary greatly in their susceptibility to EGBE toxicity, primarily due to differences in sensitivity to erythrocyte lysis by BAA, the toxic metabolite of EGBE (255). The rabbit is the most sensitive species to EGBE-induced toxicity followed (in decreasing order of sensitivity) by the rat, mouse, and guinea pig (253). Refer to Table 86.12 for a summary of acute toxicity values for EGBE.

Oral. EGBE was moderately toxic by the oral route in a variety of sensitive species (Table 86.12). Deaths shortly after administration resulted from narcosis; necropsies of animals that died later showed hemorrhagic lungs, mottled livers, congested kidneys, and hemoglobinuria. In rats, much of the variability in reported LD_{50} values is in part due to the greater susceptibility of older animals compared to younger animals (50, 253, 256). Thus, Carpenter et al. (50) measured an LD_{50} in weanling rats of 3000 mg/kg, whereas the corresponding value for yearling rats was 560 mg/kg. Female rats are considered more

Table 86.12. Acute Toxicity of EGBE in Various Laboratory Animals

Species	Gender	Route of Administration	LD_{50} or LC_{50} (mg/kg or ppm)	Reference
Rat	M	Oral	560–3000	50, 354
Rat	M	Oral	1746 (fasting or fed)	53
Mouse	M	Oral	1230	50
Mouse	M	Oral	1519 (fasting) 2005 (fed)	53
Guinea pig	M/F	Oral	1200–1414	50, 257
Rabbit	M	Oral	370; 320	50
Rat	M/F	Inhalation (4 hr)	486 (male) 450 (female)	258
Mouse		Inhalation (7 hr)	700	55
Rabbit	M/F	Dermal	680	261
Rat	F	i.v.	340; 380	50
Rat	F	i.p.	550	50
Mouse	M/F	i.v.	1130	50
Rabbit	M	i.v.	280; 500	50

sensitive to the hemolytic effects of EGBE and display symptoms of toxicity at lower dose levels (50, 253). Mice are less sensitive than rats and display acute oral LD_{50} values that range from 1.2 to 2 gm/kg, and clinical signs of toxicity identical to those observed in rats. Measured oral LD_{50} values for the guinea pig indicate a sensitivity similar to that of the mouse; however, hemolysis is not observed in this species following acute oral dosing with EGBE at levels that cause significant hemolysis in the rat (255). In a recently reported study, mortality in guinea pigs administered aqueous solutions of EGBE was due to necrosis and hemorrhage of the gastric mucosa without evident signs of hemolysis (257). In a single gavage study in the rabbit, LD_{50} values of 0.32 and 0.37 gm/kg were reported (50).

Inhalation. Acute exposure of mice (55) or rats (50, 258) caused respiratory distress, hemoglobinuria, and splenic congestion. Significant mortality was seen in rats at inhalation exposures of 500 ppm and greater. However, increased osmotic fragility of rat erythrocytes was observed following a single 4-h exposure to 62 ppm of female rats (50), or following repeated exposures (total of 30) at 54 ppm and higher of both male and female rats (females more sensitive). Necropsies of animals that died following acute inhalation exposures (4-h) revealed the presence of red-stained fluid in urinary bladders and enlarged and congested spleens (258). These effects (as well as other blood parameters) return to normal after a 14-day recovery period. A single, 1-h exposure to a saturated EGBE vapor produced no mortality or clinical signs of toxicity in guinea pigs (257).

Dermal. Effects seen following dermal administrations of EGBE are similar to those that occur after oral administration and are caused primarily as a secondary effect of hemolysis. Bartnik et al. (191) reported increased MCV and decreased erythrocyte counts and Hb concentrations in female rats exposed for 6 h dermally (semioccluded) to EGBE. No clinical signs of toxicity were noted in rats exposed for 24 h (semioccluded) to 2000 mg/kg, but some deaths were reported when exposures were occluded (259, 260). An LD_{50} of 0.68 gm/kg is reported for the rabbit (261). Hemoglobinuria was reported following repeated (6-h, occluded) dermal exposures in rabbits at doses of 180 and 360 mg/kg. A single 24-h application to rabbits (semi-occluded) also produced some mortality and signs of hemolytic toxicity (262). These effects were more severe following occluded applications of EGBE (263). In rabbits, nine applications (6-h, occluded) produced hemoglobinuria in both genders at 180 and 360 mg/kg (253). This was accompanied by other clinical signs of hemolysis, as well as reduced body weight gains. A report by Roudabush et al. (261) indicates LD_{50} values in the guinea pig of 0.23 mL/kg after application (occluded) to intact skin versus a value of 0.30 mL/kg following application to abraded skin. A more recent study, however, indicates no effect in this species following a single, 24-h occluded application at 2000 mg/kg (257).

Intravenous/Intraperitoneal. Administration of EGBE either intravenously or intraperitoneally caused toxicity similar in magnitude to that following oral administration (Table 86.12).

Skin and Eye Irritation. EGBE was mildly irritating when applied unoccluded to rabbit skin for 4 h but was a moderate irritant when applied for 24 h under occlusive conditions (253). Zissu (233) reports that EGBE is a skin irritant in rabbits under the EEC testing protocol and a severe irritant under the Draize method. EGBE was severely irritating to the eyes of rabbits (253, 264).

6.4.1.2 Chronic and Subchronic Toxicity. Oral. Adult male rats were given EGBE at dose levels of 222, 443, and 885 mg/kg/day for 5 days/week for 6 weeks (265). At the highest dose level, there were two deaths and decreased body weight gains associated with decreased food consumption. The most significant effect reported at all dose levels was on the red blood cells, which is consistent with intravascular hemolysis. Secondary effects included increased spleen weight and hemosiderin accumulation in liver and kidney. Relative liver weight increased with an increase in some serum enzymes, consistent with mild liver damage. There were no effects on the testes, thymus, white blood cells, or bone marrow, which are target organs for the analogous methyl and ethyl ethers of ethylene glycol.

In a study conducted by the U.S. Navy (266) in which rats were orally given EGBE for 13 weeks at a daily dose of 25% of the single oral LD_{50} (about 90 mg/kg/day), hemolysis was the major effect observed in association with secondary effects in the kidneys. The lowest NOEL in an oral subchronic study was 80 mg/kg (253). In this study, EGBE was administered to rats in their feed during a period of 90 days.

Drinking Water. In studies conducted by the NTP (267), groups of male and female F344/N rats or B6C3F$_1$ mice (five males or females/species/dose group) were given EGBE in drinking water for 14 days at targeted dose levels of 0 (control), 100, 150, 250, 400, or 650 mg/kg/day. Due to reduced palatability in rats, measured doses were below targeted levels. Absolute and relative thymus weights slightly decreased in high-dose female rats. Absolute and relative thymus and testis weights of male rats were unaffected by the treatment. There were no chemically related gross lesions in rats in these 2-week studies, nor were there microscopic lesions of the testis or epididymis (only organs examined). In mice, there were no clear treatment-related changes in water consumption of male mice; in female mice, water consumption decreased at all but the highest concentration, where consumption slightly increased. Absolute and relative thymus weights of male mice treated at 400 or 650 mg/kg/day EGBE were significantly lower than controls. Thymus weights of females and testis weights of males were not different from controls.

In stop-exposure drinking water studies conducted by the NTP (267) in male F344/N rats, 30 animals/dose group were treated with EGBE at 0 (control), 1500, 3000, or 6000 ppm for 60 days. Rats were evaluated at the end of the treatment period and at 30 and 56 days following the last treatment. There was a dose-related decrease in water consumption in EGBE-treated rats. Mean body weights of rats in the 6000-ppm group at day 60 and at the study end were lower than controls but remained within 10% of control values. Organ and body weights were unaffected after the 60-day exposures (10 animals evaluated). No lesions of the testis or epididymis of rats were recorded in these stop-exposure studies.

In 13-week drinking water studies conducted by the NTP (267), EGBE was administered to groups of male and female F344/N rats or B6C3F$_1$ mice (10 males or females/species/dose group) at concentrations of 0 (control), 750, 1500, 3000, 4500, or 6000 ppm. Due to reduced palatability, average daily water consumption by rats decreased with increasing concentrations of EGBE, and a dose-related decrease occurred for female rats. All rats survived to the study's termination, but mean body weights and body weight gains were less than those of controls for male and female rats at the two highest exposure concentrations. EGBE induced anemias which were regenerative, macrocytic, and generally hypochromic, indicating erythrocyte swelling. Absolute thymus weights decreased in males in the 4500-ppm group and in males and females in the 6000-ppm groups. A reduction in the size of the uterus of female rats in the 4500- and 6000-ppm groups included a minimal to mild uterine atrophy considered secondary to body weight reductions. Histopathological lesions were present in the liver, spleen, and bone marrow of male and female rats. Liver lesions included cytoplasmic alterations, hepatocellular degeneration, and pigmentation. These lesions were more pronounced in the 3000-ppm and higher groups and were more severe in female rats. Hyperplasia of the bone marrow was present in dosed animals corresponding to increased hematopoiesis. Pigmentation present in Kupffer cells of the liver stained positive for iron. Epididymal weights for males in the 4500- and 6000-ppm groups decreased, but weights were appropriate for corresponding reductions in body weights. Sperm concentration decreased slightly in all treated males, but this parameter did not show a dose-related trend. Estrous cycle lengths for treated females did not differ from those of controls, but animals in the 4500- and 6000-ppm groups differed significantly from controls in the amount of time spent in estrous stages: more time in diestrus and less in other stages. In the opinion of the NTP, EGBE was considered relatively nontoxic at the doses tested, affecting primarily erythrocytes.

In mice, 13-week drinking water studies conducted by the NTP indicated similar but less severe hematologic effects than those recorded for rats (267). In particular, water consumption was variable, and no treatment-related patterns were identified. Male and female mice that received 3000 ppm EGBE and more had slightly lower mean body weights than controls. All other recorded organ weight changes were considered secondary to reduced body weights. There were no other reported, chemically related gross or microscopic lesions in either gender of mouse in these 13-week studies. No biologically significant changes were observed in sperm morphology or vaginal cytology evaluations of mice.

Inhalation. Only one of ten guinea pigs died when exposed repeatedly to 376 ppm EGBE (30 exposures, 7 h/day, 5 days/week) and only two of ten died when exposed repeatedly at 494 ppm (50). In this latter study, erythrocyte fragility was not increased nor was hemoglobinuria observed.

In studies conducted by Dodd et al. (258), rats were exposed whole body to EGBE vapors at 20, 86, or 245 ppm for 9 days or to 5, 25, or 77 ppm for 13 weeks, each for 6 h/day, 5 days a week. No deaths occurred in either study. Similar toxic effects were seen in each study and were consistent with older studies (50, 56). These effects included increased liver weight and hematologic changes consistent with intravascular hemolysis (decreased red cell counts and blood hemoglobin concentration and increased number of immature red cells). The NOEL for the subchronic study was 25 ppm. There was no histopathological evidence of damage to any other organ system, including the liver, kidneys, bone marrow, testes, or other reproductive organs.

In 14-week inhalation toxicity studies conducted by the NTP (268), male and female F344/N rats or B6C3F$_1$ mice (10 male or female/species/dose group) were exposed to 0 (control), 31, 62.5, 125, 250, or 500 ppm EGBE for 6 h/day, 5 days/week for 14 weeks. A total of five female rats at the 500-ppm exposure concentration and one at the 250-ppm level died or were sacrificed moribund before study termination. Clinical observations in rats indicated either hemolytic toxicity or irritation and included red-stained urine, nasal and eye discharge, and increased salivation and/or lacrimation. In female rats at 500 ppm, tails became necrotic and were either removed by chewing or from sloughing. Kidney and liver weights increased at 125 ppm and higher for both male and female rats, and thymus weights in females were significantly less at 500 ppm. Anemia was present in male rats at 125 ppm or higher and in female rats of all exposed groups. The anemia was characterized as macrocytic, normochromic, and regenerative. In addition, there were exposure-related increases in Kupffer cell pigmentation, forestomach inflammation accompanied by epithelial hyperplasia, bone marrow hyperplasia, splenic hematopoietic cell proliferation, and renal tubule pigmentation. Based on the results of these studies, a maximum exposure level of 125 ppm was chosen for rats for subsequent 2-year inhalation toxicity and carcinogenicity studies.

In 14-week inhalation toxicity studies in mice (268), a total of four males and four females died or were sacrificed moribund at the highest exposure level before study termination. Final mean body weights for males at 125 ppm and higher were less than those of controls. Clinical findings in only those animals at 500 ppm that died or were sacrificed moribund included abnormal breathing, red urine stains, and lethargy. Anemia was characterized as normocytic, normochromic, and regenerative in mice exposed at 62.5 ppm or higher and effects were more pronounced in female mice. Liver weights of males at 500 ppm were significantly greater than those of controls. In mice that died or were sacrificed moribund, there was inflammation, necrosis, and ulceration of the forestomach which extended into the body cavity; liver necrosis; renal tubule degeneration; atrophy of the spleen, thymus, and mandibular and mesenteric lymph nodes; and testicular degeneration. Changes secondary to the hemolytic toxicity of EGBE included Kupffer cell hemosiderin pigmentation, hemosiderin pigmentation of the liver and renal tubules, and inflammation and epithelial hyperplasia of the forestomach. Based on the results of these studies, a maximum exposure level of 250 ppm was chosen for mice for subsequent 2-year inhalation toxicity and carcinogenicity studies.

Chronic Inhalation Toxicity. The National Toxicology Program (268) has completed 2-year chronic toxicity and carcinogenicity studies of EGBE in which it was administered 6 h/day, 5 days/week, by inhalation to groups of 50 male and 50 female F344/N rats at concentrations of 0 (chamber control), 31.2, 62.5, or 125 ppm and to groups of 50 male and 50 female B6C3F$_1$ mice at 0, 62.5, 125, or 250 ppm. As in other acute and subchronic studies, exposure to EGBE produced regenerative anemia in both species and sexes of test animals. Survival rates for exposed rats did not decrease versus unexposed chamber controls, but mean body weights of female rats exposed at the highest concentration were less than those of controls. Survival rates for exposed male mice at both the 125- and 250-ppm concentrations were significantly less than those of controls. Mean body weights of both male and female mice were less than those of the unexposed, and decreases appeared earlier in females than in males.

The most consistent exposure-related effect noted in both rats and mice was on the hematopoietic system. This effect was manifested in rats as a concentration-related macrocytic, normochromic, regenerative anemia and was present at 3, 6, and 12 months. Female rats were affected to a greater extent than males. Similarly, this effect on the hematopoietic system in mice was described as a concentration-related normocytic, normochromic, regenerative anemia present at 3, 6, and 12 months. Again, female mice displayed this effect more prominently than males.

Other nonneoplastic lesions of significance in rats were observed in the olfactory epithelium, liver, and spleen. Hyaline degeneration of the olfactory epithelium increased significantly in males at all exposure concentrations and in females at 62.5 or 125 ppm. The change is presumably an adaptive response to the irritant nature of EGBE. The changes noted were primarily of minimal severity and were generally confined to the dorsal meatus or, in more extreme cases, the ethmoid turbinates. Kupffer-cell pigmentation in the liver significantly increased over chamber control values for both male and female rats exposed at 62.5 or 125 ppm. In males, splenic fibrosis significantly increased over that of controls at exposure concentrations of 62.5 or 125 ppm. In mice, significant nonneoplastic lesions were reported in bone marrow, olfactory and respiratory epithelia, and the urogenital system. Incidences of hyperplasia in the bone marrow of male mice exposed to 125 or 250 ppm increased significantly relative to those of the controls. The incidence (but not severity) of hyaline degeneration of olfactory and respiratory epithelia increased in all groups of exposed females. In male mice, a mouse-specific urological syndrome was identified that consisted of glomerulosclerosis and hydronephritis of the kidney, inflammation of preputial and prostate glands, chronic inflammation and ulceration of the prepuce skin, and ulceration of the transitional epithelium of the urinary bladder. The syndrome could represent a bacterial infection (269) often associated with mice housed individually in wire-mesh cages. It was suggested that this syndrome was exacerbated by the irritative effects of EGBE or a metabolite in urine.

Dermal. In a 90-day dermal application study of rabbits, EGBE was applied as a 43% aqueous solution for 6 h/day, 5 days/week under occlusion at dose levels up to 150 mg/kg (253, 270); there was no evidence of systemic toxicity or of skin irritation at the application site. Shorter term applications of 180 mg/kg as a 50% aqueous solution did result in systemic toxicity including hemolysis (253, 270).

6.4.1.3 Pharmacokinetics, Metabolism, and Mechanisms. Pharmacokinetics. The effects of age, dose, and metabolic inhibitors on the toxicokinetics of EGBE were studied in male F344 rats (210). Rats of either 3 to 4 months of age or 12 to 13 months of age were dosed by gavage at 31.2, 62.5, or 125 mg/kg. Pretreatments included pyrazole, cyanamide, or probenicid (an inhibitor of renal anion transport). Toxicokinetic parameters for EGBE, including the area under the curve (AUC), maximum plasma concentration (C_{max}), and clearance rate (Cl_s) were dose-dependent. AUC and C_{max} increased, and Cl_s decreased at increasing dose levels. Other measured parameters were unaltered by dose. Age had no effect on the half-life ($t_{1/2}$), volume of distribution (V_d), or systemic clearance (Cl_s) of EGBE but C_{max} and AUC increased with increasing age. As expected from previous studies, inhibition of EGBE metabolism with either pyrazole or cyanamide resulted in significantly increased $t_{1/2}$ and AUC and decreased Cl_s. BAA toxicokinetics were also altered by dose and age and by administering of metabolic inhibitors. Slight but statistically significant increases in C_{max}, AUC, and $t_{1/2}$ were seen at higher doses and were more pronounced in older rats. Probenecid pretreatment at EGBE dose levels of 31.2 and 62.5 mg/kg produced no changes in the measured toxicokinetic parameters of EGBE but produced two- to three-fold increases in AUC and two- to sixfold increases in $t_{1/2}$ for BAA. The results of these studies further confirm BAA as the proximate hemolytic agent in

EGBE-induced hematotoxicity and indicate that renal organic acid transport is vital to the detoxification mechanism. In addition to the metabolic considerations discussed previously, compromised renal clearance in older rats may also significantly contribute to their increased sensitivity to EGBE-induced hematotoxicity.

Inhalation Pharmacokinetics. Johanson (271) exposed male SD rats continuously to either 20 or 100 ppm EGBE for up to 12 days. At various times, groups of animals were sacrificed by decapitation, and EGBE and BAA concentrations were determined in samples of blood, muscle, liver, and testis. Urine was collected at 24-h intervals. The concentrations of EGBE in blood were slightly higher, and concentrations of BAA in the blood were markedly higher than for either of these materials in the other tissues analyzed, suggesting weak (EGBE) and pronounced (BAA) binding to blood protein. EGBE was rapidly metabolized and cleared from blood. Clearance was estimated at 2.6 L/h, and there was no difference between the 20-ppm and 100-ppm groups. Concentrations of the metabolite BAA reached 30 to 40 µmole/L in blood, 15 to 20 µmole/L in the liver, and 10 µmole/L in the testis. The urinary excretion of BAA accounted for 64% of the estimated respiratory uptake and there was no difference between exposed groups. Renal clearance of BAA was estimated at 0.53 L/h. EGBE disposition kinetics were linear at the exposure concentrations used.

Dill et al. (272) recently reported on the toxicokinetics of inhaled EGBE in male and female F344 rats and B6C3F1 mice exposed at concentrations of 31.2, 62.5, or 125 ppm (rats) or 62.5, 125, or 250 ppm (mice) for 1 day, 2 weeks and 3, 6, 12, and 18 months. Mice eliminated both EGBE and BAA from blood more rapidly than rats. Half-lives of elimination in mice and rats following 1-day of exposure to 62.5 ppm were 3 min (mice) versus 9 min (rats) for EGBE, and 31 min (mice) versus 40 min (rats) for BAA. Female rats eliminated BAA more slowly than males, and this effect was related to a reduced rate of renal elimination. Gender differences were less pronounced in mice. Proportional increases in measured AUCs for the parent compound EGBE indicated linear kinetics. Elimination of BAA decreased with increasing exposure concentrations. After prolonged exposures, the elimination rates of both EGBE and BAA decreased in both species, resulting in longer blood residence times. After a single day of exposure, 19-month-old naive mice eliminated BAA approximately 10 times more slowly than young mice. After 3 weeks of exposure, this difference was less obvious, suggesting that factors other than age alone determine the elimination kinetics of EGBE.

Metabolism. In rats, EGBE is rapidly absorbed following either gavage, percutaneous administration, or inhalation exposure and is eliminated primarily in urine as 2-butoxyacetic acid (BAA) with lesser amounts of the glucuronide and sulfate conjugates of the parent alcohol (167–169, 191, 273–274). Ethylene glycol, a previously unreported metabolite, was also excreted at about 10% of the dose following administration in drinking water. At low-dose levels, oxidation of the alcohol moiety to BAA via the aldehyde, mediated by alcohol dehydrogenase and aldehyde oxidase, predominated (see also Table 86.8); this was saturated at a gavage dose between 125 mg/kg and 500 mg/kg (273, 275). At the higher dose level, EGBE conjugation with glucuronic acid and oxidative dealkylation to form ethylene glycol became quantitatively more important (167). The acid, BAA, is also the primary metabolite in rats following administration of EGBE in drinking water (167). No significant differences in the urinary levels of BAA were found

Butyric acid ← Butyraldehyde + HOCH$_2$CH$_2$OH → CO$_2$*
　　　　　　　　　　　　　　　　　　Ethylene glycol*

CH$_3$CH$_2$CH$_2$CH$_2$OCH$_2$CH$_2$O Gluc　　　　CH$_3$CH$_2$CH$_2$CH$_2$OCH$_2$CH$_2$O SO$_3$H

EGBE Glucuronide*　　　　　　　　　　　　　　EGBE Sulfate*

CH$_3$CH$_2$CH$_2$CH$_2$OCH$_2$CH$_2$OH
EGBE

↓

CH$_3$CH$_2$CH$_2$CH$_2$OCH$_2$CHO
2-Butoxyacetaldehyde

↓

CH$_3$CH$_2$CH$_2$CH$_2$OCH$_2$COH →$^?$ → CO$_2$*
2-Butoxyacetic acid* (BAA)

↓

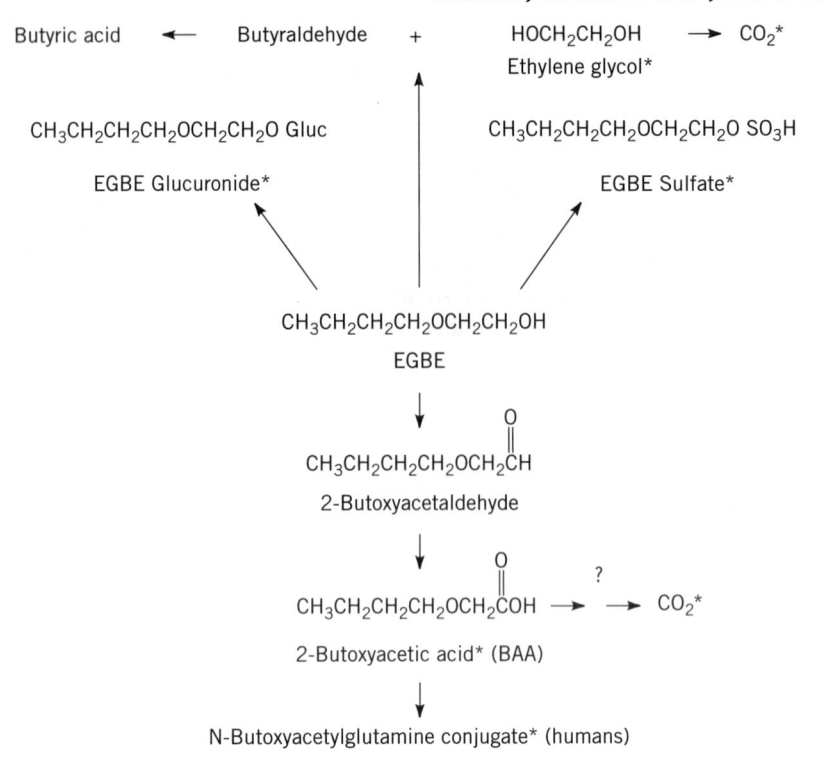

N-Butoxyacetylglutamine conjugate* (humans)

Figure 86.7. Proposed metabolic pathway for EGBE. Materials identified with an asterisk (*) have been positively identified in either rodents or humans.

following administration of equivalent doses of EGBE either dermally or in drinking water (167, 168). Consequently, exposure to an equivalent absorbed dose of EGBE by either the oral or dermal routes is expected to result in similar toxicity (168). In addition, the uptake and metabolism of EGBE following 6-h inhalation exposures was essentially linear up to 438 ppm, a concentration that caused mortality (274). In humans, the glutamine conjugate of BAA has been reported as a urinary metabolite (249). No similar glycine or glutamine conjugate has been identified in rodents. A summary of the metabolic pathway for EGBE is given in Figure 86.7.

Physiologically Based Pharmacokinetic Models. Several physiologically based pharmacokinetic (PBPK) models have been published for EGBE which describe the kinetics of either EGBE or EGBE and BAA in rats and humans. The first of these, published by Johanson (276), was developed from parameters measured in male rats (277), as well as in humans (243). This model allowed determining EGBE kinetics following inhalation exposures and was successfully applied to humans by appropriate scaling of rat-derived parameters. Shyr et al. (278) incorporated distribution and metabolism data for EGBE derived from drinking water, dermal, and inhalation exposures of rats (167–168, 274) into the Johanson (276) model. The amounts of individual metabolites, including BAA, ethylene glycol, and the glucuronide conjugate of EGBE, were incorporated into the model, which successfully explained changes in the urinary profiles of these metabolites following different routes of exposure. Corley et al. (279) further modified the Johanson (276) rat model by incorporating additional routes of exposure and measured partition coefficients for BAA (see Fig. 86.3). Unlike previous models, the Corley model was the first able to simulate blood levels of the hemolytically active metabolite, BAA. This latter model adequately simulated published data in either rats following intravenous, oral, or inhalation exposures or in humans following dermal or inhalation exposures. Higher peak blood concentrations, as well as higher areas under the blood concentration–time curves (AUC), were predicted for rats than for humans.

Lee et al. (280) recently published a PBPK model designed to describe EGBE and BAA pharmacokinetics after long-term exposures in rats and mice. The model was based initially on that of Corley et al. (279) but incorporated data from rats and mice (272) exposed by inhalation for various times up to 18 months. Adjustments were made to account for differences in the metabolism of EGBE by female rats (281), as well as adjustments to parameters used for mice. The model simulated the data of Dill et al. (272) reasonably well for EGBE in blood and BAA in urine but underestimated BAA in blood. Saturation of elimination of BAA is suggested as the reason for these results; alternatively, plasma binding in females could account for the higher levels of BAA in this gender of rat. The model for mice used parameters scaled from rats; V_{max} for the EGBE to BAA conversion, however, was doubled (281). Several age-related parameters were investigated with this model. Parameters such as cardiac output, body composition, metabolic capacity, protein binding, and renal excretion were adjusted on the basis of age. Neither reducing cardiac output nor increasing the volume of the fat compartment (both age-related changes) had any significant effect on the fit of the model to the chronic rodent data (272). However, keeping plasma protein binding capacity constant in rats with age and increasing this capacity with age in mice gave improved fits.

Mechanisms. Early *in vitro* studies conducted by Carpenter et al. (50) confirmed BAA as the active hemolytic agent. Subsequent investigations showed that hemolytic blood concentrations of the acid may be produced following either oral or dermal administration of EGBE. Percutaneous absorption of EGBE in rats is rapid and produces measured blood levels of the acid sufficient to produce hemolysis (191). Metabolism, disposition, and pharmacokinetic studies in male F344 rats conducted by Corley et al. (279) produced hemolytic blood concentrations of the acid following a single oral dose of 126 mg/kg or following simulated 6-h inhalation exposures in excess of 200 ppm.

The hematotoxicity of EGBE is due to its oxidative metabolite BAA (50, 282). Although high concentrations of EGBE can hemolyze rat and human red blood cells *in vitro*, presumably due to a solvent effect on the membrane, BAA lysed rat (but not human) red blood cells at much lower concentrations than EGBE (191). BAA at 2 mM concentrations and below (*in vitro*) readily lysed rat erythrocytes, but hemolysis of human red blood cells by EGBE required a concentration of 20 mM (191, 283). Lysis by BAA and by EGBE *in vivo* was preceded by cell swelling (spherocytosis), but not by EGBE *in vitro*, further suggesting that hemolysis *in vivo* is mediated by the metabolite BAA (283).

Older rats are more sensitive to the hemolytic effects of EGBE *in vivo* than younger rats (50). This may be attributed in part to a metabolic difference: young rats exhaled more unchanged EGBE than old rats, and excreted significantly less of the hemolytic metabolite

BAA in the urine (256, 210). Urinary excretion of BAA appeared to be impaired in older rats, resulting in a larger area under the BAA time–concentration blood curve (AUC) than in younger rats and thus a greater susceptibility to hemolysis (210). Newly formed erythrocytes were also less sensitive to BAA-induced hemolysis than older erythrocytes; when rats were bled to stimulate erythropoiesis, the resulting blood removed 7 days later was rich in young erythrocytes and was much more resistant to hemolysis by BAA than blood from control animals (284). Similarly, treatment of rats that had hemolytic doses of EGBE for 3 days caused stimulation of erythropoiesis, and the treated/recovered rats were much less sensitive to hemolysis *in vivo* or *in vitro*. In this regard, Sivarao and Mehendale (285) reported that administering a single, near-lethal dose of EGBE to rats (500 mg/kg) protected against toxicity from a lethal dose administered 7 days later. These findings confirm earlier animal studies, in which hemolysis was detected on the first day or so of treatment, often as hemoglobinuria, but afterwards hemolytic effects were not observed (54, 258, 253).

In vitro studies indicate that several other alkoxyacetic acids, but not alkoxypropionic acids, also cause hemolysis of rat red blood cells (283). The butyl derivative was the most potent, followed by the propyl, pentyl, and ethyl derivatives. Swelling of red cells and a concomitant decrease in ATP was reported. It was speculated that BAA may interact with the rat red cell membrane causing disruption of the erythrocyte osmotic balance, leading to swelling, secondary ATP depletion, and hemolysis (283).

The red blood cells of species other than the rat, including mice, rabbits, and baboons, were susceptible to *in vitro* hemolysis by BAA at 1 and 2 mM, whereas blood from pigs, dogs, cats, guinea pigs, and humans was resistant (286). Other studies showed that red cells from a large number of humans examined (191), including those who had hereditary red cell disease (sickle cell, spherocytosis), and older individuals (287, 288), were not susceptible to hemolysis under conditions that caused extensive hemolysis of rat cells. These results indicate that hemolysis and secondary effects reported in rats are unlikely to occur in humans under anticipated exposure conditions (279).

Dermal Absorption. Acute dermal application to rats of 200 mg/kg of undiluted EGBE over 12 cm^2 under a perforated capsule for 24 h produced no evidence of hemolysis; higher dose levels resulted in hemoglobinuria in at least some rats. Approximately 25 to 29% of the applied dose was estimated to have been absorbed dermally under these conditions (191). The extent of dilution of EGBE with water modified the rate of dermal penetration; undiluted and 5, 10, and 20% dilutions were absorbed to the same extent through guinea pig skin *in vitro*, whereas 40 and 80% solutions were absorbed twice as fast (289). When various EGBE formulations were applied semioccluded to rat skin, estimated absorption was about 20% of that when it was applied occluded, indicating that, when applied semioccluded, much of the EGBE evaporated before absorption could occur (191). When the rate of dermal absorption was determined *in vitro*, it was two to three times higher for rat skin than for either pig or human skin (191). The dermal permeation constant for undiluted EGBE through occluded human skin *in vitro* is reportedly 2.14×10^{-4} cm/h (21).

Tissue Distribution. Ghanayem et al. (273) administered [^{14}C]-EGBE orally to male F344 rats at either 125 or 500 mg/kg EGBE to determine metabolism and tissue distribution. After 48 h, tissues were collected, and EGBE-derived radioactivity was determined. The highest residual tissue concentrations were found in the forestomach (7.6 µmol eq.

EGBE/gm tissue at the 500 mg/kg dose), and lesser but significant amounts were in liver, kidney, lung, spleen, and blood.

6.4.1.4 Reproductive and Developmental. *Summary of Reproductive Effects Studies.* A number of experimental animal studies demonstrated that EGBE does not cause the adverse effects on the male reproductive system produced by the analogous methyl and ethyl ethers of ethylene glycol. Also, unlike the metabolites 2-methoxyacetic and 2-ethoxyacetic acid, which are responsible for the reproductive effects of the corresponding glycol ethers, the metabolite 2-butyoxyacetic acid (BAA) is not a testicular toxicant. In a continuous breeding assay in mice, the only adverse reproductive effects occurred at maternally lethal doses, indicating that EGBE is also not a reproductive toxicant for the female.

Oral/Dermal. Mice were orally dosed with EGBE for 5 days a week for 5 weeks; at the top dose level (2000 mg/kg/day), all mice died, but at the lower dose levels (1000 and 500 mg/kg/day), no effects on testicular weight or histopathology were reported (32). When rats were administered oral doses up to 1000 mg/kg/day for 4 days (54), or up to 855 mg/kg/day for 6 weeks (265), or in any of the other subchronic studies reviewed before, no effects on the testes were seen. The rabbit, which is the species most sensitive to the testicular toxicity of some other glycol ethers, also showed no evidence of toxicity to any organ, including the testes, after repeated dermal application of EGBE at doses up to 150 mg/kg that were applied occluded as a 43% aqueous solution (270). Rats administered single oral doses of BAA up to 868 mg/kg showed no evidence of testicular toxicity, although there was evidence of hemoglobinuria at the top dose (290).

Drinking Water. The reproductive toxicity of EGBE has been determined in a continuous breeding assay in the mouse in which EGBE was incorporated in drinking water at concentrations of 0.5, 1, and 2% corresponding to estimated daily dose levels of 700, 1300, and 2100 mg/kg/day (291). Both male and female mice were administered EGBE-containing water for 7 days before and during a 98-day cohabitation period. The only affected reproductive parameter at the low-dose level (700 mg/kg/day) was slightly decreased pup weight; however, the two higher dose levels (1300 and 2100 mg/kg/day) were maternally toxic and caused decreased weight gain and water consumption and lethality in 30 and 65%, respectively, of the females. Fertility was also adversely affected at these two lethal concentrations. Crossover studies with survivors showed that this was due to effects on the female. Dosing continued with the offspring of the low-dose level group through weaning until they were mated at 74 days; there were no effects on any reproductive parameter. The only reported effect was an increase in liver weight, which may have been an adaptive response to the increased metabolic load at this dose level. It is possible that the toxic effects were secondary to hemolysis, which is likely to have occurred at all the dose levels administered, but which was not reported in this study (291).

Developmental. The apparent effect on the embryo/fetus discussed before has previously been addressed by definitive developmental toxicity studies. The results imply that EGBE is not a primary developmental toxicant; effects on the embryo/fetus were reported only in the presence of maternal toxicity, including maternal lethality.

Inhalation. Pregnant rats and rabbits were exposed by inhalation to EGBE concentrations of 25, 50, 100, or 200 ppm for 6 h/day on gestational days 6 to 15 (rats) or 6 to 18 (rabbits) (292). At the two higher concentrations, there were signs of maternal toxicity (irritation, decreased food and water intake, hemoglobinuria) in rats, and embryo/fetal toxicity (decreased pup viability and delayed ossification). At the 200-ppm concentration, there was some maternal toxicity and embryo toxicity (decreased number of live fetuses) in rabbits. At concentrations that did not cause maternal toxicity (50 ppm rat, 100 ppm rabbit), there was no evidence of developmental toxicity.

In another inhalation study (101), pregnant rats were exposed for 7 h/day on gestation days 7 to 15 to 150 or 200 ppm EGBE. The only maternal toxic effect was transient hemoglobinuria; there was no evidence of embryo or fetal toxicity. The corresponding methyl ether (EGME) was teratogenic at 50 ppm under the conditions of this study.

Dermal. Undiluted EGBE was repeatedly applied to the shaved skin of pregnant rats (179). At a dose rate of approximately 7000 mg/kg/day (four applications of 0.35 mL each day), there was lethality (secondary to hemolysis), but at a single application of 2400 mg/kg/day there was no hemolysis, no maternal lethality, and no evidence of embryo/fetal toxicity.

Oral. The effect of EGBE on mice was compared to the known teratogenic ethyl ether of ethylene glycol (EGEE) in two short-term probe studies of teratogenic and postnatal effects (177). In the teratology probe, there was embryo lethality only at maternally toxic doses; in the Chernoff–Kavlock assay for postnatal effects, when administered by gavage at 1180 mg/kg/day, EGBE resulted in fewer litters, but there were no other effects on the offspring.

In a study to determine the potential developmental effects on the cardiovascular system, EGBE was administered to pregnant rats at dose levels of 30, 100, or 200 mg/kg/day on gestational days 9 to 11 or days 11 to 13 (specific periods of cardiac development) (293). The two top dose levels were maternally toxic (causing decreased body weight and severe hematoxicity), but prenatal viability was reduced only at the top dose level and only during the day 9 to 11 dosing period. Neither heart nor great vessel defects, the potential developmental effects investigated in this study, were observed at term. Thus the NOEL for maternal toxicity was 30 mg/kg/day, and it was 100 mg/kg/day for developmental toxicity (293).

6.4.1.5 *Carcinogenesis.* In 2-year inhalation studies conducted by the NTP (268), there were no treatment-related increases in any tumor types in male F344/N rats. In female rats, combined or malignant pheochromocytomas of the adrenal medulla increased at the highest exposure concentration (125 ppm). A single malignant pheochromocytoma was present at this level. Although pheochromocytomas in females were not statistically significantly increased in this study over concurrent chamber controls, the observed rate for this tumor type exceeded historical control values. This increased incidence of combined benign and malignant pheochromocytomas resulted in a rating by the NTP of "*equivocal* evidence of carcinogenic activity" in the female F344/N rat.

Hemangiosarcomas of the liver in male mice increased over chamber control rates in all treatment groups, and the observed rate (8%) at the highest exposure concentration of 250 ppm reached minimal statistical significance ($p = .046$). Although the hemangiosarcomas observed in this study were morphologically similar to spontaneously occurring hemangiosarcomas in this species and occurred at a relatively low rate, a rating of "*some* evidence of carcinogenic activity" of EGBE in the male B6C3F1 mouse was assigned by the NTP on the basis of this tumor type. In a retrospective evaluation, sections of liver from male mice from this study were histopathologically analyzed for the presence of *Helicobacter hepaticus* infection by using a silver staining technique (Steiner's modification of the Warthin–Starry stain). This bacterial infection has been associated with increased rates of liver neoplasms, including hemangiosarcomas (294). No bacterial infection was identified in the selected tissue samples which were analyzed in the EGBE study. It is important to note that no fresh or frozen liver samples were available from this study, thus precluding the use of more sensitive assays for this bacterial infection (295).

In this same gender of mouse (male), a dose-related increased incidence of hepatocellular carcinomas combined with a decreased incidence of hepatocellular adenomas (benign) suggested of a progression toward malignancy in this organ. However, no similar effect was observed in female mice. In addition, the incidence of hepatocellular carcinoma in male mice was within the historical control range for this tumor type. Thus, the effects seen in male mice were interpreted as normal variations for this tumor type in this gender and strain of mouse.

Incidences of forestomach squamous cell papilloma and papilloma or carcinoma (combined) significantly increased in female mice in the NTP studies at the highest exposure concentration (250 ppm) relative to the chamber controls. In addition, these incidences exceeded historical control rates for chamber controls. Increased incidences of ulceration and epithelial hyperplasia of the forestomach accompanied the neoplasms in females (and to a lesser extent in males). Based on the increased incidences of forestomach squamous cell papillomas or carcinomas (mainly papillomas), a rating by the NTP of "*some* evidence of carcinogenic activity" of EGBE in the female B6C3F1 mouse was assigned.

6.4.1.6 Genetic and Related Cellular Effects Studies. Glycol ethers as a class do not contain structural alerts for genotoxicity (296) and are generally regarded as nongenotoxic (113). The genotoxicity of EGBE and two of its metabolites, 2-butoxyacetaldehyde and BAA, has recently been reviewed by Elliot and Ashby (297). EGBE was found inactive or only weakly active in a variety of *in vitro* and *in vivo* assays. A single positive result in *S. typhimurium* TA97a (a variant of TA97) was reported by Hoflack et al. (298) at concentrations of 2.2 mg/plate or higher. However, Gollapudi et al. (299) were unable to repeat this result and found that EGBE was negative in the TA97a strain at plate incorporation levels up to 10 mg/plate. EGBE was also negative in other *in vitro* mammalian cell mutagenicity assays and *in vivo* mouse bone marrow micronucleus assays. In a paper by Elias et al. (300), EGBE is reportedly either weakly positive or positive in several nonstandard genotoxicity assays, among which are included DNA repair, sister chromatid exchange, and cell transformation assays. These assays are of limited value for a number of reasons, including unusually shaped dose–response curves, lack of reproducibility, or lack of primary data. All of the reported results are seen at high (8 to 34 mM) EGBE concentrations where artifactual responses due to osmolarity changes in the medium could have influenced the results. In a recent paper, Hoflack et al. (301) reported that a concentration of 5 mM EGBE may increase the genotoxicity of other DNA damaging agents by an unspecified mechanism. The authors of this paper suggest that under normal occupational exposure conditions, such a high tissue concentration is unlikely. Based on the weight of all available data, Elliot and Ashby (297) and the NTP (268) rated EGBE as nongenotoxic.

The EGBE metabolite, 2-butoxyacetaldehyde, is genotoxic in several *in vitro* assays under conditions that produced no similar effects for the parent EGBE (298). It is possible that some of the weakly positive results seen at high concentrations for EGBE are due to this metabolite; however, further work is needed to confirm this. The acid metabolite BAA was nongenotoxic except in two *in vitro* cytogenetic assays reported by Elias et al. (300). The limited data presented by these workers precludes an accurate assessment of these findings.

Immunologic Effects. EGBE or ethylene glycol monomethyl ether (EGME) was administered for 21 days to rats in drinking water at concentrations up to 6000 ppm (EGBE) or 2000 ppm (EGME) to determine potential effects on the immune system (128). EGME caused a decrease in thymus weights and in specific antibody and interferon production (as well as decreased testicular weight), suggesting an immunotoxic effect. However, other than decreased body weight gain at the top concentration, the only reported effect of EGBE was to stimulate natural killer cell activity, implying that EGBE may have antitumor activity *in vivo* (128).

6.4.1.7 Other: Neurological, Pulmonary, Skin Sensitization. Skin Sensitization. This material did not produce skin sensitization in guinea pigs (233).

6.4.2 Human Experience

6.4.2.1 General Information. The toxicity of EGBE has been extensively examined in animal studies. There is corroborating evidence from studies in human volunteers that if occupational exposure is limited to the ACGIH TLV of 25 ppm TWA and consumer products are used normally, no adverse effects should occur. Undiluted EGBE was a mild skin irritant, and dermal exposure should be minimized. As a vapor at 100 ppm and higher, EGBE was a sensory irritant, limiting the potential for extensive human exposure. Studies have shown that hemolysis is not a likely effect for humans under most exposure conditions. EGBE has been investigated extensively in experimental animals for potential male reproductive toxicity and for developmental toxicity. Unlike the related methyl and ethyl ethers of ethylene glycol, the butyl ether is without these effects. EGBE is metabolized primarily in animals and humans to the corresponding acid metabolite 2-butoxyacetic acid (BAA) or to its glutamine conjugate (humans) and has no tendency to bioaccumulate. BAA is the hemolytic agent in rodents, but under simulated human occupational inhalation exposure conditions at the TLV, blood concentrations of BAA were at least two orders of magnitude less than those that might cause hemolysis. When hard-surface cleaners containing EGBE are used, the systemic dose from dermal absorption is limited because a significant amount of EGBE is likely to evaporate.

6.4.2.2 Clinical Cases

6.4.2.2.1 Acute Toxicity. Ingestion. Several reports in the literature indicate that intentional ingestion of EGBE-containing products can be toxic to humans. In two earlier cases that involved women (302, 303), cleaning formulations were ingested. In the first of these, 250 to 500 mL of a window cleaning product that contained 12% EGBE was ingested. When admitted to the hospital, the individual was comatose. Symptoms included metabolic acidosis, hemoglobinuria, and oxaluria. These clinical signs improved gradually, and the patient was discharged after 10 days. EGBE and BAA were assayed in urine. Oxalate levels remained high in urine even by day 8 following exposure. The appearance of unchanged EGBE in urine also suggested that metabolism to BAA had been saturated, and oxalate formation due to cleavage of the ether linkage of EGBE was proposed. In the second case (303), a female ingested about 500 mL of a household product that contained EGBE and ethanol. The dose was estimated at about 500 mg/kg. It has been shown in animals that ethanol interferes with the metabolism and elimination of EGBE (304). Forced diuresis and hemodialysis were performed. Symptoms were similar to the first case, and metabolic acidosis was attributed to lactate production. BAA was excreted in the urine at levels significantly lower than in the previous case, and oxalate levels were normal. Subsequent recovery was uneventful, and the patient was discharged after 8 days. It was suggested that forced diuresis and early hemodialysis in this latter case avoided the metabolite saturation and oxalate formation seen in the earlier poisoning case. In both of these cases, there was transient hemoglobinuria, and in one case a 25% decrease in blood hemoglobin, indicating that hemolysis occurred under these conditions. It is not known whether this was accompanied by erythrocyte swelling, as is the case with rat cells exposed to BAA, or was due to a direct solvent effect of EGBE and possibly other components of the ingested cleaning formulations on the red cell membrane. It has been suggested by Udden (305) that hemodilution from intravenous fluid therapy may contribute to a nonhemolytic anemia in these clinical cases.

Burkhart and Donovan (306) reported a case of massive EGBE ingestion by a 19-year-old, mentally retarded patient. This individual consumed 20 to 30 ounces of a product that contained 25 to 35% EGBE. Other components of the product were propylene glycol, monoethanolamine, and potassium hydroxide. The patient was deeply comatose upon admittance to the hospital. Ethanol treatment was not employed due to the initial physical condition of the patient, but hemodialysis was started after 24 h due to persistent metabolic acidosis. Hematuria developed on the second day and hematocrits fell to 23% on the fourth day. Mild, diffuse cerebral edema was present on the second day of hospitalization. The patient had a significant recovery, although neurological sequalae were reported. Based on blood analyses, blood concentrations of BAA as high as 170 mg/dL (13 mM) were recorded in this patient. It was reported that hemodialysis did not influence BAA concentrations, although acidosis did improve during treatment. Metabolic acidosis correlated with BAA levels and not lactate production, which was not significantly affected. The authors believe that the initial hypotension and hypoxia experienced by this patient may account for the neurological injury; however, a contribution from BAA is also possible. In a second recent case of poisoning by EGBE (307), an 18-year-old male, ingested glass cleaner that contained 22% EGBE on two occasions within a 12-day period. The doses received were estimated at 1.0 to 1.34 gm/kg body weight. Following the first ingestion, hepatic clinical abnormalities were present including elevated SGOT, SGPT, and total bilirubin; these returned to normal within 3 days. These clinical effects were not observed following the second incident. The initial acid–base imbalance seen in this patient resolved rapidly following hemodialysis and ethanol treatment. Most notably, hematologic and renal abnormalities were absent in both ingestions. Peak serum levels of BAA and EGBE were 4.86 mM and 0.00038 mM, respectively. Ethylene glycol was not detected in serum samples.

Bauer et al. (308) reported a case of adult respiratory distress syndrome in a 53-year-old man that resulted from EGBE intoxication. Symptoms included metabolic acidosis, shock, and noncardiogenic pulmonary edema. Following supportive treatment including diuresis and hemodialysis, the patient recovered and was discharged 15 days later. It is not clear from this report whether the pulmonary effects (not seen in other poisonings) were due to EGBE or to the severely unstable condition of the patient.

Recent reports (309, 310) found no clinical effects in 24 pediatric patients who accidentally ingested small amounts of commercial cleaning formulations that contained 10% or less EGBE. In two cases, larger amounts (> 15 mL) were ingested. In these cases, patients did well with gastric lavage or emptying followed by observation. It was suggested that poisonings with amounts of 10 mL or less can be managed in the home by fluid administration and observation.

Evidence is limited and inconclusive regarding the potential toxicity of EGBE due to workplace exposures. Sohnlein et al. (151) reported that no cytogenetic effects (sister chromatid exchange, micronuclei) could be found in varnish production workers exposed to EGBE, as well as to EGEE and EGEE acetate. Haufroid et al. (248) reported slight but statistically significant decreases in hematocrits and increases in MCHC in workers exposed to low levels of EGBE. Urinary BAA excretion was low under these conditions. The results suggest membrane damage to the erythrocyte, even though such effects in rodents are manifested as increased hematocrit and decreased MCHC. The authors recommended that these results be confirmed by further studies. These workers also reported a genetic polymorphism for cytochrome P450 2E1 (CYP2E1) in a single individual for which pre- versus postshift urinary BAA levels were equivalent even in the presence of significant exposure to EGBE. The importance of this genetic variation requires further research.

Inhalation. Male and female human volunteers were exposed to either 100 or 195 ppm of EGBE for 8 h (50); there was no reported evidence of hemolysis, osmotic fragility (a precursor to hemolysis), or other systemic toxic effects, even though hemoglobinuria was reported in rats exposed under the same conditions. The volunteers did complain of immediate and continued discomfort from irritancy to the nose and eyes at both concentrations, and some individuals reported nausea and headache. This suggested that irritancy would warn against significant inhalation exposures in the workplace.

Johanson et al. (243) exposed male volunteers who were exercising lightly to EGBE for 2 h at the Swedish TLV of 20 ppm. Respiratory uptake was estimated at 57% of the inspired amount, and BAA that accounted for 17 to 55% of the estimated dose was detected in the urine. No adverse health effects were reported in these exposed volunteers. In a separate study by Johanson and Boman (311), the dose received by either the inhalation or dermal routes by workers exposed to EGBE vapors at 50 ppm were compared. Male volunteers were exposed via a mouth valve for 2 h and allowed a 1-h rest. A second 2-h exposure period followed during which the men were seated in an exposure chamber wearing only shorts to maximize dermal exposure but wearing respiratory protection. Under these exaggerated conditions, it was concluded that dermal exposure to vapors is a significant route of exposure, and 75% of total EGBE uptake was attributable to the skin. However, there was no reported evidence of hemolysis or any other systemic toxicity in any of these volunteers (311). More recently, Corley et al. (312) applied PBPK modeling to the dermal uptake of EGBE from vapors. These workers concluded that the 75% uptake estimated by Johanson and Boman was most likely due to an artifact of the blood sampling procedure, which involved finger-prick collections from an exposed arm. When venous blood (from an unexposed arm) was sampled from human volunteers exposed dermally to EGBE vapors, it was estimated that no more than 15–27% of total EGBE uptake could be attributable to dermal absorption in humans.

No effects were observed when human blood was incubated *in vitro* with BAA at a concentration of 2 mM; there was some erythrocyte swelling at 4 and 8 mM, and this was followed by minimal hemolysis at 8 mM (283).

An analytic method for detecting BAA in blood was reported by Johanson and Johnsson (313). BAA was determined in the blood of human volunteers exposed by inhalation to 20 ppm EGBE for 2 h. The level of BAA in blood peaked at 4 to 6 h and had an elimination half-life of about 4 h. The peak level detected averaged 45 µM; this is more than two orders of magnitude below the concentration that reportedly causes swelling and minimal hemolysis in human erythrocytes *in vitro* (8 mM) (283).

Dermal–Liquid/Vapor. Dermal absorption was determined in five human volunteers (314) who immersed two or four fingers in undiluted EGBE for 2 h. Blood levels of EGBE and urine levels of BAA were determined, and the percutaneous penetration rate was estimated at 0.4 to 5.7 µmol/h/cm^2. No adverse effects were reported in these individuals as a result of exposure. In an *in vitro* permeation study using human skin, Dugard et al. (21) reported a permeation constant K_p of 2.14×10^{-4} cm/h for liquid EGBE. The K_p value equates to a maximum permeation rate of 1.8 µmol/h/cm^2 (0.214 mg/cm^2/h) of skin for an infinite dose of liquid EGBE. Corley et al. (279, 312) used a K_p of 2.0 cm/h to describe the percutaneous absorption of EGBE vapors (50 ppm ≈ 241.5 ng/cm^3). The K_p value of 2.0 takes into consideration the partitioning of EGBE vapor between air and skin (10,000) leading to a concentration of ~40.1 mg/cm^3 in the skin and a percutaneous absorption rate of 0.52 µg/h/cm^2 (4.3 nmol/h/cm^2). The PBPK model of Corley et al. (312) predicted that no more than 15 to 27% of EGBE uptake can be attributable to absorption through skin. Exercise (50–100 W) results in a three- to sixfold increase in the amount of EGBE absorbed by inhalation without a corresponding increase in dermal absorption. Under these conditions, only 5 to 9% of whole body exposure could be attributable to skin absorption (312).

6.4.2.2.3 Pharmacokinetics, Metabolism, and Mechanisms. Several PBPK models were developed that describe the pharmacokinetics of EGBE or EGBE and BAA following human exposures. All of these models were initially developed for rodents (see 6.4.1.3 above) and have rodent-derived parameters appropriately scaled to humans. Measured plasma concentrations of EGBE in human volunteers exposed by inhalation to 20 ppm EGBE were modeled by Johanson (276). The model accurately predicted blood EGBE concentrations in humans who were performing light exercise. The study concluded that increased physical activity and respiratory rate increased the inhaled dose of EGBE and that ethanol inhibited metabolism and clearance from the blood, but the risk of accumulation of unmetabolized EGBE in the body was low. The PBPK model of Shyr et al. (278) also reasonably predicted urinary BAA excretion in humans following inhalation exposures but over-predicted that following dermal exposures. In this latter case, glutamine conjugation could account for the discrepancy. Corley et al. (279) extended the model of Johanson (276) and successfully applied it to humans. This model was the first to incorporate plasma BAA levels as an internal dose surrogate and predicted that maximum human BAA plasma concentrations would not result in even slight hemolysis following dermal or inhalation exposures. More recently, Dill et al. (272) extended a model derived from chronic rodent data to humans and obtained reasonable fits with published exposure data.

6.4.2.2.7 Other: Neurological, Pulmonary, Skin Sensitization. Skin Sensitization. In a clinical study, 201 adults were exposed dermally (occluded) to a 10% aqueous solution of EGBE, the highest concentration used in cosmetic products (315). The solutions (0.2 mL) were applied to the infrascapular area of the back, to the right or left of the midline, and the area was covered with a plastic film adhesive bandage. Patches were removed at 24 h following application. Each test subject received nine applications during a 3-week period. Following a 2-week rest period, the challenge phase of the study was commenced and consisted of identical patches applied to the previously exposed areas. The patches were again removed after 24 h, and signs of erythema were graded at 48 and 72 h after application. Applications of EGBE (induction phase) initially caused few signs of irritation but by the ninth applications, approximately 25% of the subjects showed slight to definite erythema. The challenge phase caused only slight erythema in a few subjects, and only one showed signs of definite erythema. There was no evidence of skin sensitization in this study.

6.4.2.3 Epidemiology Studies

6.4.2.3.4 Reproductive and Developmental. Cordier et al. (316) published the results of a large, case-controlled study conducted in Europe to examine the relationships between maternal occupational exposure to glycol ethers and congenital malformations. Case (and control) mothers were selected from one of six registries who were participants in the European Registration of Congenital Anomalies (EUROCAT). Exposure or potential exposure to glycol ethers was estimated from the results of questionnaires. Women were classified as exposed primarily to two groups of glycol ethers: the first consisted of EGBE and EGPE and their acetates; the second included methoxypropanol and its acetate, as well as polyethylenic and polypropylenic compounds (317). Glycol ether exposure correlated with a number of congenital malformations. The risk of cleft lip increased with increased level of exposure. The authors stated that this represents evidence of the human teratogenicity of EGBE, as well as of compounds from the propylene glycol group, but it was recognized that occupational exposures to a variety of other solvents were possible, and the possibility of this as a confounder could not be dismissed. In particular, it was suggested that EGBE exposure, rather than being the causative agent, could represent a marker of exposure to a wider range of occupational exposures. In addition, the studies of Cordier et al. (316) lack convincing biological plausibility because the agents implicated have been tested in animals without eliciting teratogenic responses. Selection and recall bias are other confounding influences that make interpretations of these results questionable (Maldinado et al., in press).

6.5 Standards, Regulations, or Guidelines of Exposure

The current ACGIH TLV TWA is 20 ppm, and OSHA PEL is 50 ppm with a skin notation. ACGIH based this value on the inhalation NOEL in a rat study (258), recognizing that the rat was much more sensitive than humans to the hemolytic effects of EGBE (318). NIOSH (245) recommended an occupational standard of 5 ppm, again based on the potential for hemolytic effects reported in rats. Table 86.9 summarizes recommended or regulatory guidelines of exposure for EGBE and also provides a complete listing of international exposure standards.

6.6 Studies on Environmental Impact

Staples et al. (319) reported the results of an environmental risk assessment for a series of ethylene glycol ethers. In all cases, this review indicated that these materials are not persistent and will not bioaccumulate. They are classified as "practically nontoxic" to aquatic organisms. Conservatively calculated exposures to EGBE were less than concentrations of concern for aquatic life. In addition, modeling of environmental distribution indicated that EGBE will concentrate in water (96%), and only slight amounts will distribute to soil, sediment, biota, and suspended solids.

7.0 Ethylene Glycol Monoisobutyl Ether

7.0.1 CAS Number: [4439-24-1]

7.0.2 Synonyms: 2-(2-Methylpropoxy)ethanol; isobutoxylethanol; 2-isobutoxyethanol

7.0.3 Trade Names: Ektasolve — EIB Solvent; Isobutyl Cellosolve® Solvent;

7.0.4 Molecular Weight: 118.18

7.0.5 Molecular Formula: $C_6H_{14}O_2$

7.0.6 Molecular Structure:

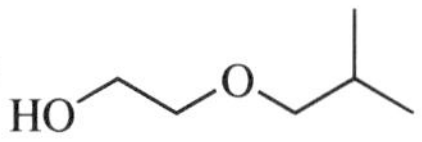

7.1 Chemical and Physical Properties

7.1.1 General

See Table 86.5 for chemical and physical properties.

7.2 Production and Use

7.3 Exposure Assessment

7.3.1 Air

See the Introduction for a general description of air sampling and analytical procedures.

7.3.3 Workplace Methods

See the Introduction for a general description of sampling and analytical procedures for the workplace.

7.4 Toxic Effects

Ethylene glycol monoisobutyl ether (EGIBE) has moderate acute oral toxicity, is irritating and injurious to the eyes, and is essentially nonirritating to mildly irritating to the skin. It may be absorbed through the skin in toxic amounts. It is moderately toxic by inhalation. Like ethylene glycol *n*-butyl ether, it can cause intravascular hemolysis and secondary effects in the liver and kidney.

7.4.1 Experimental Studies

7.4.1.1 Acute Toxicity. Oral. EGIBE reportedly has an oral LD_{50} for rats of 400 mg/kg (320) to 1000 mg/kg (321). The oral LD_{50} for mice was greater than 1600 mg/kg (321). When ingested, it resulted in hemolysis, as evidenced by hemoglobinuria and associated liver and kidney injury.

Injection. When administered intraperitoneally, the LD_{50} was 200 to 400 mg/kg (rats) or 400 mg/kg (mice).

Inhalation. Rats exposed for 0.5 h to air essentially saturated with EGIBE vapor, estimated in the range of 2000 ppm, developed liver and kidney injury accompanied by passage of blood in the urine. One-third of the rats died. All rats exposed for 4.9 h died, and two-thirds of those exposed for 1 h died (321). A 6-h exposure to 200 ppm caused signs of irritation, whereas a similar exposure to 1600 ppm caused death of one-third of the rats tested (320).

Dermal. The dermal LD_{50} for rabbits was in the range of 200 to 400 mg/kg (321). Guinea pigs were more resistant to toxicity by skin absorption; the LD_{50} value was reportedly 10 mL/kg (8870 mg/kg).

Eye Irritation. EGIBE caused considerable eye injury; instillation into the eyes of rabbits resulted in slight iritis and moderate irritation and corneal injury (321). Eye injury may be slow in healing and may even result in some permanent damage (320).

Skin Irritation. EGIBE caused moderate skin irritation when held in contact with guinea pig skin for 24 h (320). However, other data indicate that this material was not significantly irritating to the skin of rabbits even when in continuous contact for 72 h (321). In these dermal studies, even though there was no significant irritation, the rabbits became prostrate, had difficulty in breathing, and passed blood in their urine.

7.4.2 Human Experience

7.4.2.1 General Information. There have been no reported cases of adverse effects in humans from exposure to EGIBE.

7.5 Standards, Regulations, or Guidelines of Exposure

No occupational exposure standards have been established for EGIBE. However, because of its similarity in chemical structure and toxicity to ethylene glycol mono-*n*-butyl ether, it

ETHERS OF ETHYLENE GLYCOL AND DERIVATIVES

is prudent to consider using the current occupational exposure standard (25 ppm PEL TWA, skin notation) for EGBE.

8.0 Ethylene Glycol Mono-*tert*-butyl Ether

8.0.1 CAS Number: *[7580-85-0]*

8.0.2 Synonyms: 2-*tert*-Butoxyethanol; 2-(1,1-dimethylethoxy)ethanol; *tert*-butyl 2-hydroxyethyl ether, ethylene glycol; ethylene glycol *t*-butyl ether; *tert*-butyl cellosolve; *t*-butoxyethanol

8.0.3 Trade Names: Swasolve® ETB; *tert*-Butyl Cellosolve®

8.0.4 Molecular Weight: 118.18

8.0.5 Molecular Formula: $C_6H_{14}O_2$

8.0.6 Molecular Structure:

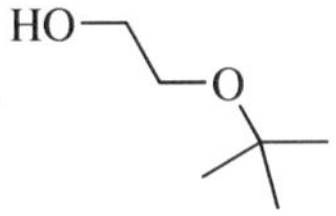

8.1 Chemical and Physical Properties

8.3 Exposure Assessment

8.3.1 Air

See the Introduction for a general description of air sampling and analytical procedures.

8.3.3 Workplace Methods

See the Introduction for a general description of sampling and analytical procedures for the workplace.

8.4 Toxic Effects

8.4.1 Experimental Studies

8.4.1.1 Acute Toxicity. Rats exposed to an essentially saturated atmosphere (2400 ppm) for 5 h became comatose, developed hemoglobinuria, and had reduced blood hemoglobin, followed by death (231). Rats exposed four times to 250 ppm for 6 h showed similar effects, but survived. Rats exposed 15 times for 6 h to 100 or 50 ppm developed increased osmotic fragility of the red blood cells but were otherwise normal; at 20 ppm, no adverse effects were seen. These effects were reportedly similar to those caused by the related ethylene glycol *n*-butyl ether.

8.4.2 Human Experience

8.4.2.1 General Information. No adverse effects have been reported in humans from exposure to EGTBE.

8.5 Standards, Regulations, or Guidelines of Exposure

No occupational exposure standards have been established for EGTBE.

9.0 Ethylene Glycol Mono-2-methylpentyl Ether

9.0.1 CAS Number: [29290-45-7]

9.0.2 Synonyms: 2-(1-Methylpentyl)oxyethanol

9.0.3 Trade Names: 2-Methylpentyl Cellosolve® Solvent

9.0.4 Molecular Weight: 146.2

9.0.5 Molecular Formula: $C_8H_{18}O_2$

9.0.6 Molecular Structure:

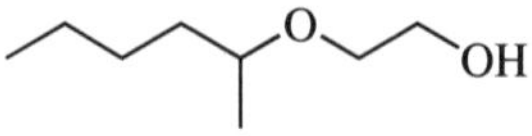

9.1 Chemical and Physical Properties

9.1.1 General

Ethylene glycol mono-2-methylpentyl ether is a clear liquid that is infinitely soluble in alcohol and many other solvents. It is slightly soluble in water, 0.63% at 20°C. More chemical and physical properties are given in Table 86.5.

9.3 Exposure Assessment

9.3.1 Air

See the Introduction for a general description of air sampling and analytical procedures.

9.3.3 Workplace Methods

See the Introduction for a general description of sampling and analytical procedures for the workplace.

9.4 Toxic Effects

Ethylene glycol mono-2-methylpentyl ether has low single-dose oral toxicity; the LD_{50} for rats is 3.73 mL/kg (205). It is markedly irritating and injurious to the eyes; 0.005 mL caused severe eye injury in the rabbit. However, it is only mildly irritating to the skin; it is rated 2 on a scale of 10 in rabbit skin tests but moderately toxic by skin absorption; the 24-h covered skin contact LD_{50} for rabbits is 0.44 mL/kg. Under the conditions of this test, necrosis of the skin was observed. One death was noted in six rats exposed for 8 h to an essentially saturated atmosphere (205). Precautions should be taken to prevent eye exposure and to avoid skin contact. It is prudent to avoid prolonged or repeated inhalation of vapor. This material is no longer commercially produced.

ETHERS OF ETHYLENE GLYCOL AND DERIVATIVES

9.5 Standards, Regulations, or Guidelines of Exposure

No regulatory hygiene standards exist for this glycol ether.

10.0 Ethylene Glycol Mono-*n*-hexyl Ether

10.0.1 CAS Number: [112-25-4]

10.0.2 Synonyms: 2-(Hexyloxy)ethanol; ethylene glycol *n*-hexyl ether

10.0.3 Trade Names: Hexyl Cellosolve®

10.0.4 Molecular Weight: 146.23

10.0.5 Molecular Formula: $C_8H_{18}O_2$

10.0.6 Molecular Structure: HO~~O~~

10.1 Chemical and Physical Properties

10.1.1 General

See Table 86.5.

10.1.2 Odor and Warning Properties

The same as other glycol ethers.

10.2 Production and Use

EGHE is a speciality glycol ether and is produced in relatively small quantities compared with others in the series. The combined U.S. production of EGHE and diethylene glycol hexyl ether was reportedly 12 million pounds in 1995 (1). Refer to Table 86.2 for additional production and use information.

10.3 Exposure Assessment

10.3.1 Air

See the Introduction for a general description of air sampling and analytical procedures.

10.3.3 Workplace Methods

See the Introduction for a general description of sampling and analytical procedures for the workplace.

10.4 Toxic Effects

Toxicological studies of ethylene glycol mono-*n*-hexyl ether (EGHE) indicate that it has moderate single-dose oral toxicity and is severely injurious to the eye. It is essentially nonirritating to the skin by single contact. However, occluded or sustained contact may produce more severe irritation. EGHE is absorbed through the skin, and animals display signs of potential central nervous system involvement following dermal applications. Because of its low volatility, exposure by inhalation is anticipated to be relatively minor. This material did not cause adverse effects on the developing fetuses of rats or rabbits nor is it genotoxic. Care should be taken when handling EGHE to prevent contamination of the eyes and prolonged or occluded contact with skin.

10.4.1 Experimental Studies

10.4.1.1 Acute Toxicity. Oral. The single-dose oral toxicity (LD$_{50}$) of EGHE in the rat has been reported by Smyth et al.(322) as 1.48 g/kg and by Ballantyne and Myers (323) as 1.67 mL/kg (for male rats) and 0.83 mL/kg (for female rats). Additionally, signs of mucosal irritation are present at doses of 1 mL/kg and higher, indicating the potential for lung damage by aspiration.

Inhalation. Smyth et al. (322) reported that an 8-h exposure to essentially saturated vapor generated at room temperature did not kill any of six rats. Ballantyne and Myers (323) reported that male and female rats exposed for 6 h to a statically generated saturated vapor at room temperature (approximate concentration 85 ppm) exhibited neither signs of toxicity nor irritancy. In addition, no signs of toxicity were observed during the 14-day postexposure observation period, nor were any pathological features seen at necropsy following the observation period. In a short-term, repeated-dose study, Fischer 344 rats exposed for 6 h/day during an 11-day period to 0 (control), 20, 40, or 80 ppm exhibited depressed body weight gain and increased liver weights at the highest exposure concentration. However, no alterations in hematologic parameters or in the morphology of the testes or liver were observed.

Dermal. A number of dermal LD$_{50}$ values were determined for EGHE. Smyth et al. (322) reported an LD$_{50}$ of 0.89 mL/kg in rats. Ballantyne and Myers (323) reported an LD$_{50}$ of 0.81 mL/kg in male rabbits and 0.93 mL/kg in female rabbits. Signs of toxicity included salivation, sluggishness, unsteady gait, and comatose appearance. The onset of signs occurred within 20 to 30 min of application of EGHE.

Undiluted EGHE was applied topically to New Zealand white rabbits (five of each sex/ group) at dose levels of 0, 44, 222, or 444 mg/kg body weight/day (324). Nine applications (6 h/day) were given during an 11-day period. A clear dose–response relationship was observed for local skin irritation at dose levels of 44 mg/kg/day and higher. Systemic toxicity (decreased food consumption and loss of body weight) was observed in both sexes of the high-dose group. In addition, decreases in erythrocyte counts and hemoglobin and hematocrit values and increased mean corpuscular volume were reported in both sexes in the high-dose group. A similar trend in hematologic parameters was also noted in males at all dosage levels and in females at the two highest dosage levels (324).

Eye Irritation. EGHE is severely injurious to the eye. Smyth et al. (322) found that a 5% solution instilled into the rabbit eye caused severe injury, and a 1% solution caused minor injury. Ballantyne and Myers (323) reported that instillation of 0.005 mL undiluted EGHE into the rabbit eye caused severe conjunctivitis, iritis, and diffuse corneal injury that healed in 3 to 7 days.

Skin Irritation. Smyth et al.(322) reported that EGHE was essentially nonirritating to the skin of rabbits from short-term contact. However, Ballantyne and Myers (323) reported that 0.5 mL of EGHE applied to the shaven dorsal skin of each of six rabbits and covered with an occlusive dressing for 4 h produced erythema and edema that disappeared within 2 to 7 days. In addition, necrosis was observed in three rabbits between 1 and 7 days postapplication that persisted through 7 days.

10.4.1.2 Chronic and Subchronic Toxicity. *Inhalation.* Rats were exposed to 0 (control), 20, 40, or 70 ppm for 6 h/day, 5 day/week for 13 weeks (325). Sensory irritation was observed in the males and females, as well as decreased body weight gain and increased liver weights in the highest dosage group. Increased liver weight was observed in both sexes of the 40-ppm group, and a slight decrease in body weight gain was observed in the females of this group. However, hematologic abnormalities and testicular atrophy were not observed in any exposure group (325). Thus, the 40-ppm exposure concentration was considered the NOAEL.

10.4.1.3 Pharmacokinetics, Metabolism, and Mechanisms. Aasmoe et al. (281) report that a single isozyme of alcohol dehydrogenase (ADH-3) in rat liver is responsible for oxidizing glycol ethers. A V_{max} value of 1.66 nmol NADH/min/mg protein and a K_m value of 0.15 mM were reported for EGHE. V_{max} and K_m values decreased with increasing chain lengths in the series of glycol ethers tested, suggesting that for equivalent glycol ether concentrations, metabolism of EGHE will be less rapid than that of other shorter chain homologues and will be saturated at lower substrate concentrations (see Table 86.8).

10.4.1.4 Reproductive and Developmental. Tyl et al. (326) exposed timed-pregnant Fischer 344 rats and New Zealand white rabbits to EGHE vapor for 6 h/day on gestational days 6 through 15 or on gd 6 through 18 to concentrations of 0 (control, air), 20, 40, or 80 ppm, then sacrificed on gestational days 21 or 29, respectively. Maternal toxicity, as shown by transient weight gain reduction, was observed in the rabbit at 80 ppm and at 40 ppm in the rat only during exposure. However, there were no treatment-related effects with respect to hematology, necropsy, gestational parameters, malformations, or variations. It was concluded that exposure to EGHE vapor at near-saturated levels, during the period of organogenesis, produced maternal toxicity in the rat and rabbit but no evidence of developmental toxicity or teratogenicity. A NOEL for maternal toxicity was established at 40 ppm for rabbits and 20 ppm for rats.

10.4.1.6 Genetic and Related Cellular Effects Studies. EGHE has been tested in a number of *in vivo* and *in vitro* genotoxicity tests. In a *Salmonella*/microsome bacterial mutagenicity assay (Ames test), EGHE was tested with or without metabolic activation at concentrations that ranged from 0.3 to 15 mg/plate. Mutagenic activity was not observed in any of the five bacterial strains with or without metabolic activation (327). In an *in vitro* cytogenetic assay, EGHE was tested with Chinese hamster ovary (CHO) cells at concentrations that ranged from 0.1 to 0.8 mg/mL without metabolic activation and from 0.08 to 0.4 mg/mL with rat liver S9 added. In this assay, EGHE failed to produce increases in chromosomal aberrations with or without the S9 metabolic activation system. EGHE was not considered a clastogenic agent in these assays (328). EGHE failed to produce a dose-related or reproducibly mutagenic response in CHO cells when tested for mutagenicity or for evidence of sister chromatid exchange. Hence in these assays, EGHE was considered negative for mutagenic activity and was considered inactive in producing DNA damage (329).

10.4.2 Human Experience

10.4.2.1 General Information. There has been no reported instance in which ethylene glycol monohexyl ether caused any adverse effect in humans.

10.5 Standards, Regulations, or Guidelines of Exposure

No occupational standard has been established for permissible vapor exposure to EGHE.

11.0 Ethylene Glycol Mono-2,4-hexadiene Ether

11.0.1 CAS Number: [27310-21-0]

11.0.2 Synonyms: 2-(2,4-Hexadienyloxy)ethanol

11.0.3 Trade Names: 2,4-Hexadienyl Cellosolve®

11.0.4 Molecular Weight: 142.2

11.0.5 Molecular Formula: $C_8H_{14}O_2$

11.0.6 Molecular Structure:

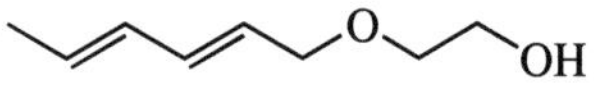

11.1 Chemical and Physical Properties

11.1.1 General

11.1.2 Odor and Warning Properties

The same as other glycol ethers.

11.2 Production and Use

This material is no longer commercially produced.

11.3 Exposure Assessment

11.3.1 Air

See the Introduction for a general description of air sampling and analytical procedures.

11.3.3 Workplace Methods

See the Introduction for a general description of sampling and analytical procedures for the workplace.

11.4 Toxic Effects

11.4.1 Experimental Studies

11.4.1.1 Acute Toxicity. The single-dose oral toxicity of 2-(2,4-hexadienyloxy)ethanol is low; the LD_{50} value for rats is 3.36 mL/kg. Toxic amounts may be absorbed through the skin if exposure is prolonged and excessive; the LD_{50} for 24-h covered skin contact in rabbits is 1.01 mL/kg. Because of its low volatility, no mortality occurred in rats exposed for 8 h to essentially saturated vapor generated at 22.4°C (330).

The material is moderately to severely irritating to the eyes of rabbits; 0.02 mL caused moderate corneal injury (rated 5 on a scale of 10), whereas 0.005 mL caused minor corneal injury. It is only slightly irritating to rabbit skin.

11.5 Standards, Regulations, or Guidelines of Exposure

No occupational exposure standard has been established for this material. The major potential hazards from 2-(2,4-hexadienyloxy) ethanol exposure are from eye and dermal contact. Suitable precautions should be taken to avoid such contact. It is also prudent to avoid repeated or prolonged exposure to its vapor, especially if handled hot or if a fog or mist is generated.

12.0 Ethylene Glycol Mono-2-ethylhexyl Ether

12.0.1 CAS Number: [1559-35-9]

12.0.2 Synonyms: 2-(2-Ethylhexyloxy)ethanol

12.0.3 Trade Names: Ethyl Hexyl Cellosolve® Solvent

12.0.4 Molecular Weight: 174.32

12.0.5 Molecular Formula: $C_{10}H_{22}O_2$

12.0.6 Molecular Structure:

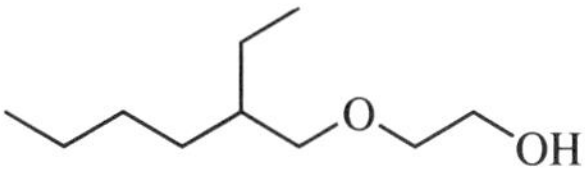

12.1 Chemical and Physical Properties

12.1.1 General

See Table 86.5 for chemical and physical properties.

12.1.2 Odor and Warning Properties

12.2 Production and Use

Ethylene glycol mono 2-ethylhexyl ether (EGEHE) is used as antifungal agent in silicone rubber.

12.3 Exposure Assessment

12.3.1 Air

See the Introduction for a general description of air sampling and analytical procedures.

12.3.3 Workplace Methods

See the Introduction for a general description of sampling and analytical procedures for the workplace.

12.4 Toxic Effects

Ethylene glycol mono-2-ethylhexyl ether is a colorless liquid that has an etheral odor. It has low single-dose oral toxicity and moderate repeated-dose oral toxicity. Blood cells may be the primary site of toxicity for this material, and secondary effects are elicited in the kidney. Eye contact may result in severe necrosis. This glycol ether is a moderate-to-severe skin irritant, but it is not a skin sensitizer. This material is no longer commercially produced.

12.4.1 Experimental Studies

12.4.1.1 Acute Toxicity. *Oral.* Smyth et al. (322) reported an LD_{50} of 3.08 g/kg in rats in a range-finding study. In studies conducted by Krasavage and Terhaar (53), this glycol ether had a single-dose oral LD_{50} for fasting male rats and male mice of 7.83 and 7.31 g/kg, respectively. Corresponding values in fed animals were 5.15 and 3.90 g/kg, respectively.

Oral administration of undiluted ethylene glycol mono-2-ethylhexyl ether for 5 days/week for 6 weeks to groups of 10 male rats at dose levels of 3828, 1914, and 957 mg/kg/day resulted in complete mortality by day 33 at the highest dose level (331). Signs that suggested intravascular hemolysis were recorded in this study. Necropsy revealed blood in the urinary bladders of three of the animals from the high-dose group. Body weights were significantly reduced at all dose levels. Thymic atrophy was observed in animals (5/10) that received the high dose. Reduced hemoglobin concentrations were observed for rats in the intermediate- and low-dose groups, whereas animals in the low-dose group displayed a slight but significant increase in leukocytes. Three of the 10 animals in the intermediate group had enlarged livers, and one of these also had enlarged kidneys. Splenic congestion was also observed at the mid-dose level, but not at the low-dose level in this 6-week study.

Inhalation. Exposure of rats to concentrated vapors of this glycol ether for periods up to 8 h caused no deaths during a 14-day recovery period (322).

Dermal. Following a 24-h dermal exposure and an additional 14-day observation period, Smyth et al. (322) found that the dermal LD_{50} for this material in rabbits is 2.12 g/kg. A similar study conducted in male New Zealand white rabbits found a dermal LD_{50} value of 2.58 g/kg (64).

Eye Irritation. Application of 0.5 mL of an aqueous solution of this glycol ether (containing 250 mg of material) to the eyes of rabbits resulted in severe necrosis (322). A single drop of the undiluted material applied to the eyes of rabbits produced moderate eye irritation and some corneal staining (64).

Skin Irritation and Sensitization. Smyth et al. (322) found that application of undiluted (0.01 mL) ethylene glycol mono-2-ethylhexyl ether to the shaved skin of rabbits produced severe irritation within 24 h of application. More recent studies indicate that this material is moderately irritating to the skin of both male rabbits and guinea pigs following a single 24-h occluded exposure. Repeated dermal application (uncovered) to the shaved skin of guinea pigs resulted in severe irritation. However, the material was not a skin sensitizer in guinea pigs (64).

12.5 Standards, Regulations, or Guidelines of Exposure

No occupational standard for repeated or prolonged inhalation exposure has been established for EGEHE. Given the severe effects produced by this material in animals, skin and eye contact should be avoided.

13.0 Ethylene Glycol Mono-2,6,8-trimethyl-4-nonyl Ether

13.0.1 CAS Number: [10137-98-1]

13.0.2 Synonyms: 2-((1-Isobutyl-3,5-dimethylhexyl)oxy)ethanol; 2-[3,5-dimethyl-1-(2-methylpropyl)hexyloxy]ethanol; 2-(2,6,8-trimethyl-4-nonyloxy)ethanol

13.0.4 Molecular Weight: 230.4

13.0.5 Molecular Formula: $C_{14}H_{30}O_2$

13.0.6 Molecular Structure:

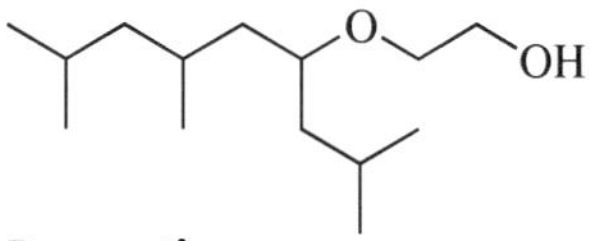

13.1 Chemical and Physical Properties

13.1.1 General

Ethylene glycol mono-2,6,8-trimethyl-4-nonyl ether is a clear liquid that is infinitely soluble in alcohol and many other solvents. It has very slight solubility in water, $< 0.01\%$ at 20°C. More chemical and physical properties are given in Table 86.5.

13.2 Production and Use

This material is no longer commercially produced.

13.3 Exposure Assessment

13.3.1 Air

See the Introduction for a general description of air sampling and analytical procedures.

13.3.3 Workplace Methods

See the Introduction for a general description of sampling and analytical procedures for the workplace.

13.4 Toxic Effects

Ethylene glycol mono-2,6,8-trimethyl-4-nonyl ether has a low order of single-dose oral toxicity; the LD$_{50}$ for rats is 5.36 mL/kg. It is moderate to markedly irritating to the eye and may cause corneal injury. It is rated 5 on a scale of 10 in rabbit eye tests. It is only very slightly irritating to the skin (rating of 2 on scale of 10) and only slightly toxic by skin absorption; the LD$_{50}$ value for rabbits is 3.15 mL/kg. No mortality resulted from exposing rats for 8 h to essentially saturated atmospheres (205). Thus the main hazard from handling this material is eye exposure. Precautions should be observed to avoid eye contact; reasonable caution, cleanliness, and good industrial hygiene practices should be adequate to avoid injury from skin contact and inhalation exposure.

13.5 Standards, Regulations, or Guidelines of Exposure

No occupational exposure standard has been established for this compound.

14.0 Ethylene Glycol Monophenyl Ether

14.0.1 CAS Number: [122-99-6]

14.0.2 Synonyms: 2-Phenoxyethanol; phenoxyethyl alcohol; ethyl glycol phenyl ether; phenoxetol; phenoxyethyl alcohol; Arosol; 1-hydroxy-2-phenoxyethane; β-hydroxyethyl phenyl ether; Euxyl K 400; Phenyl cellosolve; Phenoxethol; phenoxyl ethanol; glycol monophenyl ether; phenoxytol; phenylmonoglycol ether; 2-hydroxyethyl phenyl ether; beta-phenoxyethyl alcohol; dowanol ep; dowanol eph; emeressence 1160; emery 6705; rose ether

14.0.3 Trade Names: Dowanol® EPH Glycol Ether; Phenyl Cellosolve® Solvent

14.0.4 Molecular Weight: 138.17

14.0.5 Molecular Formula: C$_8$H$_{10}$O$_2$

14.0.6 Molecular Structure:

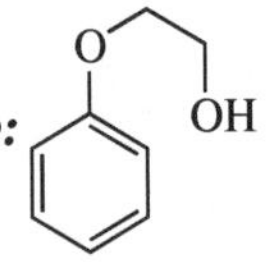

14.1 Chemical and Physical Properties

14.1.1 General

See Table 86.5 for chemical and physical properties.

14.1.2 Odor and Warning Properties

This material has a faint aromatic odor.

14.2 Production and Use

2-Phenoxyethanol (EGPhE) is used as a solvent for inks, resins, and cellulose acetate and as a perfume fixative. Phenoxyethanol has antibacterial properties and is effective against strains of *Pseudomonas aeruginosa*, even in the presence of 20% serum. It has been used as a preservative at 1% concentration (332, 333). EGPhE is an indirect food additive for use only as a component of adhesives (21 CFR 175.105). Refer to Table 86.2 for additional production and use information.

14.3 Exposure Assessment

14.3.1 Air

See the Introduction for a general description of air sampling and analytical procedures.

14.3.3 Workplace Methods

See the Introduction for a general description of sampling and analytical procedures for the workplace.

14.4 Toxic Effects

Studies of the technical or commercially available product showed that it has low single-dose toxicity; the LD_{50} for rats was between 1.0 and 4.0 g/kg. It is not appreciably irritating to the intact skin and is not readily absorbed through the skin in acutely toxic amounts. However, repeated dermal exposure to high doses (1000 mg/kg/day) produced death and red blood cell hemolysis in rabbits. Only slight eye irritation has been reported. Rats tolerated one 7-h exposure to vapors saturated at 100°C and cooled to room temperature without apparent adverse effects. No developmental, reproductive, or mutagenic effects have been observed. Reasonable handling precautions and care to prevent contact with the eyes should prevent any serious toxic effects.

14.4.1 Experimental Studies

14.4.1.1 Acute Toxicity. *Oral.* The oral LD_{50} in rats is likely to be in the range of 1300 to 4000 mg/kg. An early study from Dow (334) reported complete mortality at 3000 mg/kg but no deaths at 1400 mg/kg. Later studies (335) found two deaths out of five rats exposed at 4000 mg/kg and one death out of five at 1000 mg/kg. In a study conducted by Nipa Laboratories (336), the LD_{50} was estimated at 1300 mg/kg.

Oral administration to rabbits of 100, 300, 600, or 1000 mg/kg/day up to 10 days resulted in a dose-dependent (severity and time to onset) intravascular hemolytic anemia characterized by decreased red blood cell count, packed cell volume, and hemoglobin, as well as hemoglobinuria, splenic congestion, renal tubule damage, and a regenerative erythroid response in bone marrow and spleen (337). Female rats given doses up to 2500 mg/kg/day for a maximum of 14 days exhibited lethality associated with lethargy and ataxia; no signs of overt hemolysis were reported (337). Thus rabbits are more sensitive to treatment-induced hemolytic effects than rats.

Inhalation. Because of its low vapor pressure, exposure to significant concentrations at room temperature is not likely. No effects were seen in rats exposed to saturated vapors at room temperature for 7 h (338). Exposure of rats for 7 h to vapors generated at elevated temperature produced lethality in one out of four rats, along with drowsiness and eye and nose irritation (338).

Dermal. The potential systemic toxicity from acute dermal absorption is low; the dermal $LD_{50}s$ in rabbits were estimated at more than 2000 mg/kg (337) and 13 mL/kg (336). Repeated doses of 1000 mg/kg/day to the skin of female rabbits produced hemolysis of erythrocytes and deaths of most animals within 2 weeks (339). A diminished response was noted at 600 mg/kg/day; no deaths or hemolytic effects were seen at 300 mg/kg/day.

Eye Irritation. Instillation of material into the rabbit eye resulted in slight irritation, iritis, and slight to moderate corneal injury that healed in about 1 week (335).

Skin Irritation. Older test reports indicated slight to moderate irritation on repeated confined contact to rabbit skin; however, these effects were not noted in later studies that involved repeated skin contact (338).

14.4.1.2 Chronic and Subchronic Toxicity. *Oral.* In a 13-week subchronic study, rats were treated with doses of 80, 400, and 2000 mg/kg/day as a suspension in gum tragacanth mucilage (340). At 2000 mg/kg, body weights decreased and relative liver, kidney, and thyroid weights increased in both males and females. Minor kidney toxicity was noted at 400 mg/kg.

Dermal. In a 90-day study, rabbits were exposed to dermal applications of 50, 150, or 500 mg/kg/day of undiluted test material (337). All animals survived with no signs of systemic toxicity. Minor skin effects at the application site were not considered toxicologically significant. The NOEL for systemic toxicity following subchronic dermal treatment was 500 mg/kg/day.

14.4.1.3 Pharmacokinetics, Metabolism, and Mechanisms. *Metabolism.* Phenoxyacetic acid has been identified as a major metabolite in the blood of rabbits given a single oral dose of 800 mg/kg of ethylene glycol monophenyl ether (337). In an associated *in vitro* study (337), the parent compound was more hemolytic to rabbit erythrocytes than the major phenoxyacetic acid metabolite.

Skin (post mitochondrial fraction) metabolized EGPhE to phenoxyacetic acid at 5% of the rate for liver. Metabolism was inhibited by 1 mM pyrazole suggesting involvement of alcohol dehydrogenase (341).

14.4.1.3.1 Absorption. *Dermal — In vitro.* The dermal penetration of EGPhE was studied *in vitro* using unoccluded rat skin mounted in static and flow-through diffusion cells during the course of 24 h (341, 342). With both unoccluded cells, EGPhE was lost by evaporation, but occlusion of the static and flow-through cells reduced evaporation and increased total absorption (94 and 85%, respectively). The choice of receptor fluid did not greatly influence penetration (ethanol/water or tissue culture medium).

14.4.1.4 Reproductive and Developmental. *Reproductive.* No significant reproductive effects were reported in a study where male mice were given oral doses of 500, 1000, or

2000 mg/kg/day for 5 weeks (32). Effects on fertility were evaluated in a continuous breeding study using CD-1 mice administered 0.25, 1.25, or 2.5% (0.4, 2.0, or 4.0 mg/kg/day) ethylene glycol monophenyl ether in the feed (291). Both sexes were treated for 7 days before and during a subsequent 98-day cohabitation period. Fertility was slightly decreased (10 to 15% reduction in the number of pups/litter) only at the high-dose level. Systemic toxicity was not generally apparent in the adult parental mice. However, severe neonatal toxicity, reflected by increased mortality and decreased weight gain, was observed in both the mid- and high-dose group F1 male and female pups. Second-generation reproductive performance was not affected in the mid-dose group.

Developmental. Dermal application (0, 300, 600, or 1000 mg/kg/day) to rabbits on days 6 to 18 of gestation resulted in intravascular hemolysis and death at the highest dose (343). No adverse effects were seen in the fetuses of the surviving high-dose animals (5/25). Maternal toxicity was also reported at 600 mg/kg/day in the absence of fetal effects. No adverse maternal or fetal effects were seen at 300 mg/kg/day. EGPhE was not embryo toxic, fetotoxic, or teratogenic at the concentrations tested.

14.4.1.6 Genetic and Related Cellular Effects Studies. No evidence of mutagenicity was observed in an Ames bacterial assay (concentrations up to 5000 mg/plate) (344) and in a mammalian cell CHO/HGPRT gene mutation assay (concentrations up to 3500 mg/ml) (345) with and without metabolic activation. A micronucleus test in mice did not produce any treatment-related responses at dose levels of 300 to 1200 mg/kg (344). Negative results were also obtained in a rat bone marrow chromosomal aberration assay after oral doses of 280, 933, or 2800 mg/kg (346).

14.4.1.7 Other: Neurological, Pulmonary, Skin Sensitization. Guinea pig sensitization tests have not indicated any potential for skin sensitization (336).

14.4.2 Human Experience

14.4.2.1 General Information. Ethylene glycol monophenyl ether did not cause skin irritation, sensitization, or phototoxicity in human clinical trials. This material is regarded as safe when used as a cosmetic ingredient at concentrations less than 1% (347).

14.5 Standards, Regulations, or Guidelines of Exposure

No occupational exposure standards have been proposed or established for ethylene glycol monophenyl ether in the United States. The DFG MAK (Germany) TWA is 20 ppm (110 mg/m^3, skin notation).

15.0 Ethylene Glycol Monomethylphenyl Ether

15.0.1 CAS Number: None found

15.0.2 Synonyms: 1-(2-Hydroxyethoxy)methylphenol

15.0.3 Trade Names: 2-Methylphenyl Cellosolve® Solvent

15.0.4 Molecular Weight: 152.19

15.0.5 Molecular Formula: C₉H₁₂O₂

15.0.6 Molecular Structure:

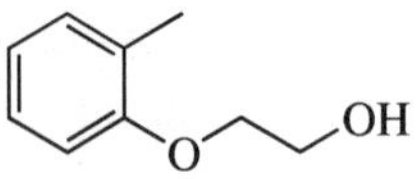

15.2 Production and Use

This material is no longer commercially produced.

15.3 Exposure Assessment

15.3.1 Air

See the Introduction for a general description of air sampling and analytical procedures.

15.3.3 Workplace Methods

See the Introduction for a general description of sampling and analytical procedures for the workplace.

15.4 Toxic Effects

The single-dose oral LD_{50} in rats of ethylene glycol monomethylphenyl ether is 3.73 g/kg, indicating a low order of acute toxicity. The material is severely irritating to the eyes but only mildly irritating to the skin. It is moderately toxic by absorption through the skin; the LD_{50} for rabbits is 0.44 mL/kg. One of six rats exposed for 8 h to essentially saturated vapors generated at room temperature died (348). These data indicate that ethylene glycol monomethylphenyl ether poses a potential health hazard from eye contact and possibly from inhalation. Precautions should be taken to prevent eye contact and to avoid inhalation, especially prolonged or repeated exposure.

16.0 Ethylene Glycol Dimethyl Ether

16.0.1 CAS Number: [110-71-4]

16.0.2 Synonyms: Glyme; Monoglyme; dimethoxyethane; 1,2-dimethoxyethane; 2,5-Dioxahexane; EGdiME; dimethoxyethane; glycol dimethyl ether; dimethyl Cellosolve; DME; alpha, beta-dimethoxyethane; ethylene dimethyl ether; monoethylene glycol dimethyl ether; 1,2-ethanediol, dimethyl ether; ansul ether 121; EGDME; GDME

16.0.3 Trade Names: Dimethyl Cellosolve® Solvent

16.0.4 Molecular Weight: 90.122

16.0.5 Molecular Formula: C₄H₁₀O₂

16.0.6 Molecular Structure:

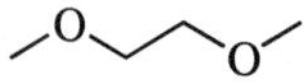

ETHERS OF ETHYLENE GLYCOL AND DERIVATIVES

16.1 Chemical and Physical Properties

16.1.1 General

See Table 86.5.

16.1.2 Odor and Warning Properties

EGdiME is a colorless liquid that has a mild odor.

16.2 Production and use

16.3 Exposure Assessment

16.3.1 Air

See the Introduction for a general description of air sampling and analytical procedures.

16.3.3 Workplace Methods

See the Introduction for a general description of sampling and analytical procedures for the workplace.

16.4 Toxic Effects

16.4.1 Experimental Studies

Ethylene glycol dimethyl ether (EGdiME) is moderately toxic following single-dose oral administration. There is little data on skin and eye contact or on acute inhalation exposure. The limited evidence from repeated-inhalation exposure indicates that the material may cause narcosis. Experimental evidence from *in vivo* tests suggest that EGdiME may be clastogenic. The existing data indicate that EGdiME is developmentally toxic, is embryo toxic, and has caused testicular atrophy in laboratory animals.

16.4.1.1 Acute Toxicity. Oral. EGdiME has moderate to low toxicity by single-dose oral administration in laboratory animals. The LD_{50} value for female mice is reportedly 2525 mg/kg (349). Nagano et al. (206) treated male JCL-ICR mice by gastric intubation 5 days/week for 5 weeks with dose levels of 0 (control), 250, 500, or 1000 mg/kg EGdiME. The animals were then sacrificed the day after the final administration. A dose-dependent decrease in testicular weight and a slight decrease in the combined weight of the seminal vesicles and the coagulating gland were observed in the highest dosage group. Additionally, histopathological examination of the testes revealed dose-related atrophy of the seminiferous epithelium. Hematologic examination revealed a significant decrease in white blood cell count and in the two highest dosage groups, a decrease in red blood cell count.

16.4.1.4 Reproductive and Developmental. Developmental Toxicity. Timed-pregnant mice were dosed orally from day 7 to 10 of gestation with 0 (control), 250, 350, or 490 mg/

kg EGdiME. A dose-related increase in the incidence of intrauterine deaths was observed at the highest dose level and increases in external malformations, primary exencephaly, and skeletal defects were observed in surviving fetuses. The fetuses were not examined for internal defects (350). Timed-pregnant CD-1 mice were dosed orally from day 7 through 14 of gestation with 2000 mg/kg. Maternal mortality was nearly 25%, and none of the surviving mice delivered any viable offspring following administration (100% fetal lethality) (349). Time-mated CD-1 mice were dosed orally on gestation day 11 (plug = 0) with distilled water (control) or 4 mmol/kg (361 mg/kg) EGdiME. The animals were sacrificed on gestation day 18. No signs of maternal toxicity or decreased intrauterine survival were observed. Fetal body weights were significantly reduced. The only gross external malformations were paw defects in 86% of the treated litters (33% of the fetuses). Hind paw defects, were observed more frequently than forepaw defects and syndactyly was the most common malformation. The incidence of oligodactyly and short digits also increased significantly. The authors postulated that *in vivo* conversion of EGdiME to methoxyacetic acid, a known teratogen, is the probable metabolic pathway responsible for the effects observed (351). Timed-pregnant Sprague–Dawley rats were administered EGdiME by gavage on gestation days 8 through 18 at dose levels of 30, 60, 120, 250, 500, or 1000 mg/kg/day. Maternal deaths occurred at the highest dosage level, and fetolethality occurred at dose levels from 120 through 1000 mg/kg/day. The 60-mg/kg/day dose level produced a 7% decrease in body weights and severe edema in the pups that survived to birth. In the dams allowed to go to term, the 60-mg/kg/day group produced litters that had only one-third the number of live births compared to the controls. Dosages of 30 mg/kg/day were not fetolethal on gestation day 19. However, in the dams allowed to go to term, litters were reduced by an average of two pups. A close correlation was observed between fetotoxicity at the various concentrations and maternal body weight gain (352).

16.4.1.6 Genetic and Related Cellular Effects Studies. Genetic Toxicity. The genotoxicity of EGdiME was tested in a number of *in vitro* tests. EGdiME did not produce dose-related or repeatable mutagenic effects in either a Chinese hamster ovary (CHO) cell gene mutation assay or in an assay for unscheduled DNA synthesis. However, in an *in vitro* sister chromatid exchange (SCE) test in CHO cells, EGdiME produced statistically significant increases in the frequency of SCE with and without an S9 metabolic activation system. In addition, microscopic studies indicated that the material also produced increased amounts of chromosomal damage with S9 metabolic activation. The results suggest a possible clastogenic (chromosome-breaking) effect in mammalian cells tested *in vitro* (353).

16.4.1.7 Other: Neurological, Pulmonary, Skin Sensitization. Inhalation. Goldberg et al. (159) exposed trained female rats for 4 h/day, 5 days/week for 2 weeks to concentrations of 1000, 2000, 4000, and 8000 ppm of EGdiME vapor. A single exposure to 8000 ppm caused a significant decrease in the avoidance response; however, it had no effect on the escape response. Further exposures decreased both responses, and half the rats died after five daily 4-h exposures. Autopsy revealed massive hemorrhage of the lungs and the gastrointestinal tract. Several of the rats exposed to 4000 ppm for 10 exposures died. They also showed similar gross pathology. Rats exposed to 1000 and 2000 ppm all

ETHERS OF ETHYLENE GLYCOL AND DERIVATIVES 175

survived, but both groups showed significant, though slight, avoidance response, but no escape response. Complete recovery from the behavioral effects occurred within a few days after exposure stopped. In addition, there was a definite reduction in growth, depending on the concentration.

16.4.2 Human Experience

16.4.2.1 General Information. There have been no reported instances of adverse effects in humans caused by ethylene glycol dimethyl ether. The reproductive and developmental effects noted in animal studies have not been demonstrated in humans.

16.5 Standards, Regulations, or Guidelines of Exposure

The Russian occupational exposure limit (OEL) for EGdiME is 10 mg/m^3 (STEL).

17.0 Ethylene Glycol Diethyl Ether

17.0.1 CAS Number: [629-14-1]

17.0.2 Synonyms: 1,2-Diethoxyethane; ethyl glyme; glyme-1; EGdiEE; diethyl cellosolve; diethoxyethane; 1,2-diethoxyethane; 3,6-dioxaoctane

17.0.3 Trade Names: Diethyl Cellosolve® Solvent

17.0.4 Molecular Weight: 118.18

17.0.5 Molecular Formula: C$_6$H$_{14}$O$_2$

17.0.6 Molecular Structure: ∧_O∧_O∧

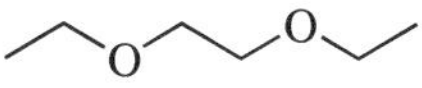

17.1 Chemical and Physical Properties

17.1.1 General

See Table 86.5.

17.1.2 Odor and Warning Properties

EGdiEE is a colorless liquid that has a sweetish odor.

17.2 Production and Use

EGdiEE is used as a solvent for ester gums, shellacs, and for some resins and oils (59). It is also used as a solvent for organic synthesis and as a diluent for detergents. U.S. production of this specialty chemical is low.

17.3 Exposure Assessment

17.3.1 Air

See the Introduction for a general description of air sampling and analytical procedures.

17.3.3 Workplace Methods

See the Introduction for a general description of sampling and analytical procedures for the workplace.

17.4 Toxic Effects

Ethylene glycol diethyl ether (EGdiEE) vapor is irritating to the eyes and mucous membranes. It has a low order of toxicity by oral administration and is appreciably irritating to the eyes, but not to the skin. This material is moderately toxic when inhaled. EGdiEE can cause narcosis. Administration of the material to the dog did not increase the amount of oxalic acid excreted. The kidney, however, was the principal target organ. EGdiEE produced evidence of adverse effects on the developing fetuses of mice and rabbits. The irritating nature of the vapor is sufficient to warn of toxic effects from single exposure but does not provide an adequate warning for exposure to vapor concentrations that could produce adverse effects from prolonged and repeated exposure.

17.4.1 Experimental Studies

17.4.1.1 Acute Toxicity. Oral. Single-oral doses of EGdiEE as a 10% aqueous solution were administered to rats, guinea pigs, and rabbits. The LD_{50} values were 4.39, 2.44, and 2.52 g/kg for the three species, respectively (354, 355). Gross (49) fed a dog and a rabbit 1 mL/kg EGdiEE six times within a period of a week without observing signs of toxicity. A cat that received four 1-mL/kg doses exhibited signs of toxicity after each dose and died after the fourth dose. Necropsy revealed no abnormal findings.

Injection. Gross (49) reported that guinea pigs survived seven subcutaneous injections of 0.5 mL/kg even though they suffered serious weight loss. Death resulted after seven 1-mL/kg injections. After four injections, the animals showed signs of transient narcosis; prostration was apparent just before death. Necropsy revealed kidney injury characterized by parenchymatous and interstitial nephritis. Wiley et al. (30) gave two dogs 9.5 mL/day EGdiEE subcutaneously for 7 days and did not observe an increase in the oxalic acid content of the urine. This treatment reportedly caused no noticeable effects in the animals. However, necropsy revealed injury to the vasculature, liver, brain, testes, and particularly the kidneys.

Inhalation. Gross (49) reported that inhalation of 10,000 ppm for 1 h caused irritation of the mucous membranes and a suggestion of narcosis. Cats were more sensitive than rabbits, guinea pigs, or dogs, but all survived the 1-h exposure. Twelve daily 8-h exposures of mice, guinea pigs, rabbits, and cats to 500 ppm resulted in the deaths of one of two rabbits and the two cats, but no overt injury was seen in the mice and guinea pigs. Microscopic examination of the tissues of both cats showed evidence of kidney injury and, in one, a serious purulent inflammation of the trachea, which may have been of infectious etiology. Exposure to 4000 ppm, as a metered concentration, did not kill any of six rats at 4 h and five of six at 8 h (356).

ETHERS OF ETHYLENE GLYCOL AND DERIVATIVES

Dermal. When the undiluted material was applied for 24 h to the rabbit skin under an impervious covering, the LD$_{50}$ was 8.0 mL/kg (357).

Eye Irritation. Severe injury resulted when 0.1 mL EGdiEE was instilled into the eyes of rabbits; 0.02 mL caused minor injury (358). Hence if this material should be splashed into the eyes, it would cause conjunctival irritation and slight transitory injury of the cornea.

Skin Irritation. Gross (49) reported that the material did not injure the skin of guinea pigs, rabbits, or dogs. A second study similarly found that the material did not irritate the skin of the rabbit (359).

17.4.1.4 Reproductive and Developmental. *Developmental Toxicity* EGdiEE was administered orally to pregnant CD-1 mice at dose levels of 50, 150, 500, or 1000 mg/kg/day and to New Zealand white rabbits at dose levels of 25, 50, or 100 mg/kg/day on gestational days 6 to 15 or 6 to 19, respectively. The mice were sacrificed on day 17 of gestation and the rabbits on day 30. Maternal body weight in the mice was reduced at the 1000-mg/kg/day dose level. At 150 mg/kg/day, the number of litters that had malformed fetuses marginally increased compared to the controls. At ≥500 mg/kg/day, fetal body weight was reduced, and the incidence of malformations increased, particularly the incidence of exencephaly and fused ribs. Maternal body weight in rabbits was unaffected at all dose levels. At dosages of ≥50 mg/kg/day, increased fetal malformations included small spleens, short tails, and fused rib cartilage (360). The developmental NOAEL was 25 mg/kg/day. In a short-term *in vivo* screening assay for reproductive toxicity, EGdiEE reduced the number of live pups per litter, reduced pup survival, and reduced pup weights and pup weight gains in time-mated pregnant CD-1 mice dosed on gestation days 7 to 14 (361). However, due to the small sample size in this study, it was not possible to determine statistical significance.

17.5 Standards, Regulations, or Guidelines of Exposure

An occupational exposure standard has not been established for EGdiEE.

18.0 Dioxane

18.0.1 CAS Number: *[123-91-1]*

18.0.2 Synonyms: 1,4-Dioxane; *p*-dioxane; diethylene-1,4-dioxide; tetrahydro-1,4-dioxin

18.0.3 Trade Names: NA

18.0.4 Molecular Weight: 88.1

18.0.5 Molecular Formula: C$_4$H$_8$O$_2$

18.0.6 Molecular Structure:

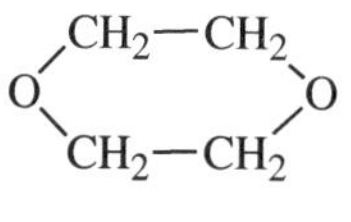

$$\text{O} \underset{CH_2-CH_2}{\overset{CH_2-CH_2}{\diagup\hspace{-0.5em}\diagdown}} \text{O}$$

18.1 Chemical and Physical Properties

18.1.1 General

Dioxane is a colorless liquid miscible with water and most organic solvents; it forms a constant boiling mixture with water that contains 81.6% dioxane and boils at 87.8°C at 760 mmHg. Additional properties are given in Table 86.13.

18.1.2 Odor and Warning Properties

The odor of dioxane in low concentrations is faint and generally inoffensive and has been described as being somewhat alcoholic. According to Silverman et al. (362), it was concluded from studies on 12 subjects exposed 15 min to various concentrations of dioxane that 200 ppm was the highest that they considered acceptable; at 300 ppm, dioxane caused irritation of the eyes, nose, and throat; and at 500 ppm, it was objectionable. Even at higher concentrations, the initial irritation to eyes and respiratory passages is transitory, and it is certain that the warning properties of dioxane are completely inadequate to prevent short-term exposure to toxic amounts. Yant et al. (363) reported immediate slight burning of the eyes accompanied by lacrimation and slight irritation of the nose and throat from an exposure of 1600 ppm for 10 min; at 5500 ppm, eye irritation and a burning sensation in the nose and throat were noted; and at 10,000 ppm or more, pulmonary irritation occurred.

May (45) reported that the odor threshold for dioxane was 170 ppm and 270 ppm was considered a strong objectionable odor. Laing (364), however, reports that the odor threshold is about 7 ppm for rats and that the odor detection is about the same in rats and humans. This level was substantiated by Thiess et al. (365), who reported that the odor recognition level for humans was about 5.6 ppm.

18.2 Production and Use

Dioxane can be made by several routes. Probably the most common are by dimerizing ethylene oxide and by dehydrating ethylene glycol. These and other methods are discussed by Curme and Johnston (366). The material is available in large amounts and is used largely in industry as a solvent for lacquers, plastics, varnishes, paints, dyes, fats, greases, waxes, and resins. When perfectly dry, it is stable indefinitely. However, it is hygroscopic and because of its ether linkages, it produces peroxides and other degradation products upon standing in the presence of oxygen. The U.S. production of dioxane in 1982 was reportedly 6.75×10^6 pounds (U.S. International Trade Commission). More recent production volumes are not available.

18.3 Exposure Assessment

Satisfactory chemical methods for determining low concentrations of dioxane vapor in the air have been developed. It can be determined by interferometer, adsorption, or a combustible gas indicator. It is also reported (367) that dioxane reacts with tetranitromethane to form a bright yellow color. It is possible that this reaction can be adapted for determinations in air. Spectrographic techniques and gas chromatography (368) also offer promising possibilities.

Table 86.13. Physical and Chemical Properties of Some Common Ethers of Di- and Triethylene Glycol

Property	Ethers of Diethylene Glycol									Ethers of Triethylene Glycol		
	Methyl	Ethyl	Propyl	Ethyl Vinyl	n-Butyl	Iso-Butyl	n-Hexyl	1,4-Dioxane		Methyl	Ethyl	Butyl
CAS Number	[111-77-3]	[111-96-6]	[6881-94-3]	[10143-53-0]	[112-34-5]	[18912-80-6]	[112-59-4]	[123-91-1]		[112-35-6]	[112-50-5]	[143-22-6]
Molecular formula	C$_5$H$_{12}$O$_3$	C$_6$H$_{14}$O$_3$	C$_7$H$_{16}$O$_3$	C$_8$H$_{16}$O$_3$	C$_8$H$_{18}$O$_3$	C$_8$H$_{18}$O$_3$	C$_{10}$H$_{22}$O$_3$	C$_4$H$_8$O$_2$		C$_7$H$_{16}$O$_4$	C$_8$H$_{18}$O$_4$	C$_{10}$H$_{22}$O$_4$
Molecular weight	120.15	134.17	148.2	160.21	162.23	162.23	190.28	88.1		164.2	178.23	206.3
Specific Gravity 25°C	1.025	0.999	0.963 (24°C)	0.941 (20/20°C)	0.967			1.035		1.052	1.018	0.983
								1.0329 at 20°C				
Boiling point (°C at 760 mmHg)	193	197	215.0	191.2	230.4			101.1		249.2	256.3	279.4
Freezing point (°C)	−70	−76.0		−80 (sets to glass)	−68.1			11.8		−44.0	−18.8	−35.2
Vapor pressure (mmHg at 25°C)	0.18	0.14	0.02 (20°C)	0.2 (20°C)	0.043		< 0.01	29 torr at 20°C		< 0.01	< 0.01	0.0025
Refractive index (25°C)				1.429 (20°C)			1.437	1.422				
Flash point (°F) (open cup)	83°C	93°C	93 (closed cup)	91	200	200	271	12–22°C (18–33°C)		245	275	
Autoignition temperature (°C)	250	250	204	201	228	228		180				
Flammability limits (vol. % in air)	1.5–9.5	1.2–8.5 (lower limit)	0.85 (lower limit)	0.4 (lower limit)				1.97–22.25				
Vapor density (air = 1)	4.14	4.64	5.1	5.54	5.6	5.58	6.56	3				
Percent in saturated air (25°C)	0.048	0.018		0.03	0.0057	0.0057		4.75		0.021	0.0026	0.00032
1 ppm ≈ mg/m³ at 25°C, 760 mmHg	4.91	5.49	6.06	6.55	6.64	6.64	7.78	3.60		6.72	7.29	8.44
1 mg/L ≈ ppm at 25°C, 760 mmHg	204	188.2	165	152.6	150.8	150.8	128.5	278		148.9	137.2	118.5

18.3.3 Workplace Methods

NIOSH Method 1602 is recommended for determining exposures to dioxane (14).

18.4 Toxic Effects

Dioxane has low single-dose oral toxicity. The liquid is painful and irritating to the eyes, is irritating to the skin upon prolonged or repeated contact, and can be absorbed through the skin in toxic amounts. Dioxane vapor has poor warning properties and can be inhaled in amounts that may cause serious systemic injury, principally in the liver and kidney. This latter effect is largely responsible for the hazardous nature of this solvent. Serious and fatal exposures can be experienced without forewarning; illness sometimes becomes apparent hours after exposure. It may also cause cancer in experimental animals, especially when ingested repeatedly in large doses.

18.4.1 Experimental Studies

18.4.1.1 Acute Toxicity. Oral. Gross (49) reported that rabbits and guinea pigs fed 0.1 mL/kg, 10 times by gavage, exhibited dropsical changes in the liver and that some animals repeatedly fed 0.52 g/kg died after 5, 16, or 20 feedings. Laug and co-workers (155) determined that the single-dose oral LD_{50} of dioxane is 5.66, 5.17, and 3.90 g/kg for mice, rats, and guinea pigs, respectively. LD_{50}s of 2.0 g/kg were also estimated for cats and rabbits. Signs of toxicity in dosed animals progressed from weakness, depression, incoordination, and coma to death. Autopsy revealed hemorrhagic areas in the pyloric region of the stomach, bladders distended with urine, enlarged kidneys, and slight proteinuria, but no hematuria.

Injections. De Navasquez (369) reported that intravenous injection of dioxane in guinea pigs, rabbits, and cats caused selective action on the convoluted tubules of the kidneys that was characterized by acute hydropic degeneration and liver cell degeneration. Deaths were due to uremia caused by intrarenal obstruction and anuria. LD_{50} values reported for dioxane when injected are listed in Table 86.14.

Inhalation. Yant et al. (363) exposed guinea pigs for 3 h to concentrations of 1000 to 30,000 ppm. Gross (49) exposed rats, mice, guinea pigs, and rabbits for 8 h to concentrations that ranged from 4000 to 11,000 ppm. Marked irritation of the mucous membranes was apparent at the higher concentrations. Deaths occurring during exposure or shortly afterward were usually due to respiratory failure because of lung edema, but the animals also exhibited congestion of the brain. Delayed deaths were attributed to

Table 86.14. Acute Toxicity of Dioxane Following Injection

Species	Route	LD_{50} (g/kg)	Reference
Mouse	i.p.	0.75	368
Rat	i.p.	5.5–6.0	368
Rabbit	i.v.	1.55	368

pneumonia. Histological evidence of liver and kidney toxicity was observed in animals that died after exposure, as well as in surviving animals that were evaluated several days after exposure.

Wirth and Klimmer (370), as cited by the U.S. Dept. HEW (368), exposed cats to dioxane at concentrations of 12,000 ppm for 7 h, 18,000 ppm for 4.3 h, 24,000 ppm for 4 h, and 31,000 ppm for 3 h and observed loss of equilibrium, increased salivation, and lacrimation. Narcotic effects were also seen, and the rapidity of development of such effects depended on the dioxane concentration. Activity of all of the cats decreased gradually after exposure and was followed by death. Fatty livers and inflamed respiratory tracts were reported on a necropsy. Similar narcotic effects were also reported in rabbits (371). An LC_{50} of 14,260 ppm was reported for a 4-h exposure of rats (372).

Gross (49) exposed cats, rabbits, and guinea pigs (two of each species) to 45 daily 8-h exposures of 1350 ppm dioxane. Only one cat became ill and had to be sacrificed; it exhibited typical liver and kidney toxicity. The other animals killed after exposures ceased showed slight evidence of liver and kidney injury. This investigator also exposed a similar group of animals plus two mice to 2700 ppm for 8 h/day. Seven of the 10 animals died after 4 to 26 exposures, but the rest survived 34 exposures. Irritation of mucous membranes, emaciation, narcosis, albuminuria, and severe liver and kidney injury were reported.

Dermal. Kidney and liver injury were observed in guinea pigs and rabbits from repeated topical application of dioxane (373). A dermal LD_{50} of 7.6 g/kg was reported for rabbits (374).

Eye Contact. Dioxane, when tested in the eyes of rabbits, caused irritation and transient corneal injury (66). It there were no serious effects from eye contact in humans (67).

Skin Irritation. Dioxane is not considered a skin irritant. However, it is a fat solvent, and prolonged and repeated contact can cause eczema.

18.4.1.2 Chronic and Subchronic Toxicity. *Inhalation.* Fairley et al. (373) exposed rats, mice, guinea pigs, and rabbits for 1 1/2 h/day to concentrations of 10,000, 5000, 2000, or 1000 ppm of dioxane. Mortality was high at the higher levels, and deaths were usually due to lung injury. Animals that survived repeated exposures at all levels suffered marked liver and kidney injury. Refer to Section 18.4.1.5 for a summary of clinical and nonneoplastic effects observed in animal carcinogenicity studies with dioxane.

18.4.1.3 Pharmacokinetics, Metabolism, and Mechanisms. *Metabolism.* Using a combination of analytical techniques (thin layer, ion exchange, and gas chromatography; mass spectrometry; and NMR spectrometry), Braun and Young (375) reported that approximately 85% of a 1000-mg/kg dose of dioxane to rats was excreted urine as a single metabolite, β-hydroxyethoxyacetic acid (HEAA). Using similar techniques, Woo and coworkers (376) identified *p*-dioxan-2-one as the major urinary metabolite in rats. This metabolite, however, is likely to be formed as an artifact from on-column heating of HEAA during gas chromatography (375). The identification of HEAA as the primary metabolite in animals was consistent with an earlier report that HEAA was also the primary urinary metabolite in humans exposed to dioxane via inhalation (1.6 ppm time-weighted average for 7.5 h) (377). In this study, dioxane was extensively metabolized to

HEAA at this low exposure level, and the concentrations of dioxane and HEAA were 3.5 and 414 mmol/L urine, respectively.

Pharmacokinetics. A subsequent pharmacokinetic study in rats demonstrated that the metabolism of dioxane to HEAA was dose-dependent (378). These investigators characterized the plasma half-lives of dioxane following either intravenous dosing (3 to 1000 mg/kg) or a single inhalation exposure to 50 ppm vapor for 6 h. The plasma half-life of dioxane was about 1 h at intravenous doses less than 10 mg/kg and following inhalation exposure. However, as the intravenous doses were increased above 10 mg/kg, the plasma clearance of dioxane was extended, the ratio of dioxane to HEAA recovered in the urine increased, and additional amounts of dioxane were detected in the expired air. These observations suggested that the metabolism of dioxane to HEAA was saturated at high exposure doses. Assuming that the metabolism of dioxane to HEAA represents a detoxification route, these authors concluded that exposures to low doses of dioxane were unlikely to present a significant health risk.

The detoxification of dioxane by metabolism to HEAA is likely to be mediated by hepatic mixed-function oxidase enzymes. Powar and Mungikar (379) showed that two daily doses of 2 g dioxane/kg to mice increased liver weight, liver microsomal protein, and the activities of liver aminopyrine-*N*-demethylase and acetanilide hydroxylase and decreased both NADPH-linked and ascorbate-induced lipid peroxidation. Woo et al. (376) studied the effects of inducers and inhibitors of hepatic mixed-function oxidases on the excretion of *p*-dioxan-2-one in rats and concluded that these oxidases were involved in the metabolism of dioxane. Evidence that multiple high doses of dioxane can result in inducing its own metabolism was provided by Young et al. (378), who showed that 1000 mg/kg/day, but not 10 mg/kg/day, caused more rapid excretion of metabolites.

18.4.1.4 Reproductive and Developmental. Reproductive. The reproductive toxicity of dioxane has not been directly evaluated. However, Lane and co-workers (380) reported a two-generation study of reproduction in which 1,1,1-trichloroethane that contained 3% dioxane in drinking water was administered to mice. The dioxane concentrations in this study corresponded to daily oral doses of 3, 10, or 30 mg/kg dioxane. No treatment-related effects were reported for fertility, gestation, or development in the parental and first- and second-generation test animals.

Developmental. The teratogenicity of dioxane has been evaluated in rats orally given 0.25, 0.5, or 1.0 mL/kg/day on days 6 to 15 of gestation (381). Slight maternal toxicity that was characterized by reduced body weight gain was reported only at the high dose and was accompanied by a slight reduction in fetal body weights. However, no treatment-related effects were observed in implantation numbers, postimplantation loss, live births, or fetal malformations. Schwetz et al. (382) conducted inhalation teratogenicity studies on both rats and mice given 1,1,1-trichloroethane that contained 3.5 percent dioxane. The calculated concentration of dioxane was 32 ppm. At this level, there was no evidence of maternal, embryonal, or fetal toxicity, nor were there any signs of teratogenic response.

18.4.1.5 Carcinogenesis. Drinking Water. Argus and co-workers (383) gave rats dioxane at 1% in their drinking water for 63 weeks (average daily dose of 821 mg/kg; range 517 to 2000 mg/kg). Six of the 26 experimental animals developed hepatomas; histological studies also revealed kidney changes that resembled glomerulonephritis. Groups of 30 rats were given drinking water that contained 0.75, 1.0, 1.4, or 1.8% of dioxane for a period of 13 months and then were examined histologically at 16 months (384, 385). One rat each at the 0.75 and 1.0% level and two rats each at the 1.4 and 1.8% level developed nasal squamous cell carcinomas. A dose-dependent increase of hepatocellular carcinomas in the liver was also observed (4/30 at 0.75% to 23/30 at 1.8%). Both precancerous liver lesions and marked kidney alterations that were characterized by epithelial proliferation of Bowman's capsule, periglomerular fibrosis, and distension of tubules were observed at all dose levels.

Kociba and co-workers (386) gave male and female rats dioxane in drinking water at levels of 0.01, 0.1, and 1.0% for 2 years. These exposure concentrations produced estimated dose levels of 9.6 mg/kg/day for male rats and 19.0 mg/kg/day for female rats at the 0.01% level, 94 and 148 mg/kg/day at the 0.1% level, and 1015 and 1599 mg/kg/day at the 1.0% level. At the 1% level, the rats exhibited decreased body weight gains, decreased survival (only 1 of 60 male rats survived), and decreased water consumption. In addition, hepatocellular and renal tubular epithelial degeneration and necrosis, as well as hepatic regeneration, were also seen. Nasal carcinomas were observed in three of the top-dose group, and 10 high-dose rats developed hepatic tumors compared to 1 of 106 of the control group. Degenerative liver and kidney lesions were seen at the 0.1% dose level, but there was no evidence of treatment-induced carcinogenicity. The 0.01% level was the no-effect level (NOEL). Thus the authors concluded that the toxicity of dioxane was dose-dependent and that liver injury preceded the development of tumors in this organ.

The National Cancer Institute (387) also evaluated the carcinogenicity of dioxane in Osborne–Mendel rats and B6C3F1 mice given 0.5 and 1.0% dioxane in drinking water. Rats received dioxane for 110 weeks and mice 90 weeks. A significant increase in nasal tumors was seen in both sexes of rats at both treatment concentrations; the incidence of liver adenomas also increased in both sexes at the two dose levels. The incidence of hepatocellular carcinomas in mice was significantly elevated in both sexes at both exposure concentrations.

Inhalation. Dioxane did not produce a carcinogenic response in rats exposed by inhalation. Torkelson et al. (388) exposed rats to 111 ppm dioxane for 7 h/day, 5 days/week for 2 years and observed no changes in general appearance, demeanor, growth, mortality, hematologic and clinical chemistry parameters, organ weights, and gross or histopathological examination of the tissues. No hepatic or nasal carcinomas were found.

Dermal. Several studies were conducted to evaluate the long-term effects of dioxane contact on the skin. Application of four grades of dioxane in an ethanol solvent to the skin of mice for 78 weeks produced no evidence of skin tumor formation; hence the authors concluded that dioxane was not likely to cause skin tumors by skin contact (389, 368). A study reported by King et al. (390) suggests that dioxane can be a tumor promoter in a skin bioassay model. Application of dioxane in acetone to mouse skin for 59 weeks after a single initiating dose of 7,12-dimethylbenzanthracene significantly increased the incidence of skin tumors compared to dioxane alone. The relevance of the tumor-promoting property of dioxane to human risk, however, is not understood.

18.4.1.6 Genetic and Related Cellular Effects Studies. Genetic Toxicity. Dioxane had minimal to no genetic toxicity when evaluated in a series of *in vitro* test systems. No genotoxic activity of dioxane was found in *S. typhimurium* tester strains TA1535, TA1537, TA1538, TA98, and TA100, with and without metabolic activation (391) and in the *Saccharomyces cerevisiae* yeast assay (392, 393). Dioxane also tested negative in the *in vitro* L5178Y mouse lymphoma cell forward mutation assay (394) and in *in vitro* hepatocyte unscheduled DNA synthesis assays (391, 395). No chromosomal aberrations were observed in Chinese hamster ovary cells, although a slight increase in sister chromatid exchanges was noted in these cells when they were exposed only to an extremely high (10 mg/mL) concentration of dioxane (396). Sina and co-workers (397) reported that dioxane increased DNA damage rat hepatocytes, as measured by an alkaline elution assay. However, these authors also noted that the positive response was observed only in the presence of significant cytotoxicity, and thus the implications for a direct genotoxic effect of dioxane were uncertain. Dioxane was also reported to increase the incidence of type III-transformed foci slightly in cultured BALB/3T3 cells (398). Dioxane inhibits DNA repair synthesis (DI_{50}, 50% inhibition) in cultured HeLa 83 cells, but at a relatively high concentration of 400 mM (399). Dioxane was negative in an *E. coli* (K-12 *uvrB/recA*) DNA repair assay (400).

Dioxane also has no direct-acting genotoxic activity in *in vivo* test systems. Negative findings were obtained in a *Drosophila* recessive lethal assay and in an intraperitoneal single-dose dominant lethal test (401). Dioxane did not alkylate DNA or induce DNA repair in livers of rats given a single oral dose of 1 g/kg dioxane (391). Goldsworthy and co-workers (395) found that dioxane was inactive in an *in vivo* rat hepatocyte DNA repair assay (single oral dose of 1000 mg/kg or 2% dioxane in drinking water for 2 weeks). These investigators also reported that dioxane did not induce DNA repair in turbinate and maxilloturbinate nasal epithelial cells of rats treated with 1% in drinking water for 8 days followed by a single oral dose of 1000 mg/kg dioxane. Dioxane reportedly caused DNA damage, as determined by a DNA alkaline elution assay, in the livers of rats treated twice in 24 h with extremely high (2550 and 4200 mg/kg) oral doses of dioxane (402). The positive response, however, was associated with evidence of treatment-induced hepatotoxicity. Dioxane was judged negative in an *in vivo–in vitro* rat hepatocyte replicative DNA synthesis (RDS) test in which rats were dosed up to the MTD and DNA synthesis was measured subsequently in isolated hepatocytes (403). In an assay that reportedly provides a high degree of concordance (73%) with rodent cancer bioassays, dioxane produced damage to hepatic DNA (alkaline elution) and elevated hepatic ornithine decarboxylase activity and hepatic cytochrome P-450 content in rats dosed up to 2550 mg/kg (404).

In a series of articles (405), the genotoxicity of dioxane was reviewed and independent results were presented from three different groups on the activity of dioxane in the mouse bone marrow micronucleus assay. Based on a weak positive response in one study, it was not possible to rate dioxane as either positive or negative in this genotoxicity assay.

18.4.1.7 Other: Neurological, Pulmonary, Skin Sensitization. Behavioral Effects. Female rats exposed to dioxane at levels of 1500, 3000, or 6000 ppm for 4 h/day, 5 days a week for 2 weeks were evaluated for avoidance and escape responses before, during, and 2 h after exposure each day (159). No weight changes in the rats were noted until the last day of exposure to 6000 ppm. Only one rat showed variable avoidance responses at the

1500-ppm level. At the 3000-ppm level, two, and on occasion three, rats showed inhibition of avoidance response. At the 6000-ppm level, maximum inhibition of avoidance and escape responses was seen on the second day. Subsequent exposures no longer produced escape responses, and there was a reduced but variable effect on avoidance response. This suggests that all inhibitory effects on behavior were temporary and reversible.

18.4.2 Human Experience

18.4.2.1 General Information

18.4.2.2.1 Acute Toxicity. In 1933, five cases of fatal industrial poisoning by dioxane were reported in England (406, 407). These reports are summarized and discussed by Browning (59). In general, men who worked in a synthetic textile factory apparently inhaled excessive amounts of dioxane. The signs and symptoms were irritation of the upper respiratory passages, coughing, irritation of the eyes, drowsiness, vertigo, headache, anorexia, stomach pains, nausea, vomiting, uremia, coma, and death. Autopsy revealed congestion and edema of the lungs and brain and marked injury of the liver and kidney. Death was attributed to kidney injury. Blood counts of three of the men who died showed no abnormalities other than considerable leukocytosis. The exposure that these men received is still unknown, and it is debatable whether the deaths were caused by chronic exposures or whether they resulted from relatively few intense exposures. On the basis of previous experience in this same factory during several months during which no trouble was encountered and because others in the factory at the same time did not show any serious symptoms, De Navasquez (369) believes, that the persons who died were exposed to high concentrations during a relatively short period of time. It is obvious from this incident that dioxane does not have warning properties adequate to prevent short-term exposures that are dangerous to life.

Johnstone (408) reported the death of a worker who was exposed for 1 week to about 500 ppm of dioxane in the air. In addition, the worker used dioxane to wash glue from his hands, which probably resulted in absorption of some dioxane through the skin. Autopsy revealed damage to kidneys, liver, and brain.

18.4.2.2.3 Pharmacokinetics, Metabolism, and Mechanisms. In conjunction with rodent studies, Young and co-workers (409) also determined the pharmacokinetics of dioxane in human volunteers exposed to 50 ppm dioxane for 6 h. The plasma half-life was approximately 59 min after exposure ended, which closely agreed with the previous observations in rats. In urine, HEAA accounted for 99% of the recovered material, and the remainder was represented by dioxane. The total absorbed dose was estimated at 5.4 mg/kg dioxane, although because of its rapid metabolism to HEAA, it was estimated that no more than 1.2 mg/kg dioxane was present in the body at steady state. Plasma dioxane concentration, it was estimated, achieved 99% of steady-state concentration by the end of the 6-h exposure period. Using a pharmacokinetic model developed from the single inhalation exposure data, these investigators predicted that repeated 8 h/day exposures to 50 ppm dioxane would not lead to an accumulation of dioxane in the body.

A physiologically based pharmacokinetic model was developed to provide insight into the implications for human risk of the dose-, species-, and route-dependent animal tumorigenicity of dioxane (410). Based on an analysis of the nonlinear pharmacokinetics of dioxane observed in animals and correcting for metabolic and physiological differences between rodents and humans, this model predicted that humans continuously exposed to 740 to 3700 ppm dioxane in air or to 20 to 120 ppm in water are unlikely to be at risk of cancer from exposure to this agent. Leung and Paustenbach (411) also developed a physiologically based pharmacokinetic model that further supports the hypothesis that conventional risk assessment approaches used in estimating human risk from animal cancer bioassays (e.g., linearized multistage model) overestimate the human cancer potential from environmental exposures to dioxane by as much as 80-fold.

18.4.2.2.5 Carcinogenesis. Human epidemiological studies to date have produced no indication that exposures of humans to dioxane in industrial operations have caused an increased incidence of tumors. Based on the results from adequate animal studies, dioxane was rated by the U.S. EPA as a Class B2, probable human carcinogen (1994). This classification was based on the induction of nasal cavity and liver carcinomas in multiple strains of rats, liver carcinomas in mice, and gall bladder carcinomas in guinea pigs. Human carcinogenicity data were considered inadequate.

18.4.2.2.6 Genetic and Related Cellular Effects Studies. Thiess et al. (365) found no evidence of chromosomal changes in six workers who were exposed to dioxane in a manufacturing plant.

18.4.2.3 Epidemiology Studies. Thiess et al. (365) conducted an epidemiological study of 74 workers who had been working in a dioxane manufacturing factory from 3 to 41 years. Analysis of the workroom air in the factory in 1974 indicated that the level of exposure was up to 14.24 ppm of dioxane. Of the 74 workers studied, 24 were still working in the plant, 26 were working elsewhere, 15 had retired, and 12 were dead. The study of these workers showed no evidence of adverse effects because of their exposure to dioxane. The studies included extensive medical and physical examinations, chromosomal analysis in six of the actively working employees, and a careful analysis of the causes of death of the 12 dead workers.

Buffler and co-workers (412) conducted a study in 1975 of 165 employees who had worked in a dioxane manufacturing plant in Texas between April 1954 and June 1958. The levels of exposure ranged up to 32 ppm dioxane. Again, there were no indications of adverse effects due to exposure to dioxane (and in some workers exposure to other solvents), as revealed by careful medical and physical examinations and by evaluation of the causes of death of those who had died.

Dernehl and Peele (368), in a similar epidemiological study of workers in a dioxane manufacturing plant that involved 80 workers, again found no evidence of adverse effects due to their possible exposure to dioxane vapor, as shown by complete physical examinations, chest X rays, electrocardiograms, a series of liver profile tests, and, in the case of those who had died, a careful review of the causes of their deaths. Air analysis studies revealed that the exposure levels ranged from 0.05 to 51 ppm.

18.5 Standards, Regulations, or Guidelines of Exposure

A number of recognized authorities have set the following exposure limits for dioxane: ACGIH TLV TWA 20 ppm, 72 mg/m^3; OSHA PEL TWA 100 ppm, 360 mg/m^3; NIOSH REL STEL/CEIL (c) 1 ppm, 30 minute; IDLH Value 500 ppm; DFG MAK (Germany) TWA 20 ppm, 73 mg/m^3; HSE OES (United Kingdom) TWA 25 ppm, 91 mg/m^3 STEL/CEIL (c) 100 ppm, 336 mg/m^3. These are summarized in Table 86.9. Most of these exposure limit are accompanied by a skin notation and (in some cases) a carcinogen notation.

19.0 Diethylene Glycol Monomethyl Ether

19.0.1 CAS Number: [111-77-3]

19.0.2 Synonyms: 2-(2-Methoxyethoxy)ethanol; methoxydiglycol; 2,2′-oxybis-ethanol monomethyl ether; DGME

19.0.3 Trade Names: Methyl Carbitol® Solvent; Dowanol® DM Glycol Ether; Eastman® DM Solvent;

19.0.4 Molecular Weight 120.15

19.0.5 Molecular Formula: C$_5$H$_{12}$O$_3$

19.0.6 Molecular Structure: \O/\/O\/\OH

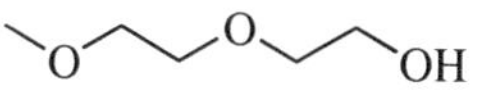

19.1 Chemical and Physical Properties

19.1.1 General

See Table 86.13.

19.2 Production and Use

Diethylene glycol monomethyl ether is used as an industrial solvent, when a high boiling point is needed. It is also used in jet fuels as an anti-icing additive, as a brake fluid diluent, and as coupling agent for making miscible organic–aqueous systems. Production figures and additional use information are given in Table 86.2.

19.3 Exposure Assessment

19.3.1 Air

See the Introduction for a general description of air sampling and analytical procedures.

19.3.3 Workplace Methods

See the Introduction for a general description of sampling and analytical procedures for the workplace.

19.4 Toxic Effects

Diethylene glycol monomethyl ether (DGME) has low oral toxicity, is painful but not seriously injurious to the eyes, and is not irritating to the skin. Although it can be absorbed in toxic amounts through the skin, excessive exposure would be required before serious effects would be expected. Hazardous amounts are not likely to be inhaled under ordinary conditions, but where heated material is encountered, care is warranted. The material exerts its principal toxicological action upon the brain, kidney, liver, thymus, and testes. DGME is embryo toxic and teratogenic by the oral route and fetotoxic and embryo toxic by the dermal route in experimental animals. No adverse human experience has been reported.

19.4.1 Experimental Studies

19.4.1.1 Acute Toxicity. Oral. Smyth et al. (354) orally administered DGME as a 50% aqueous solution to rats and guinea pigs and found that the LD_{50} values are 9.21 g/kg for rats and 4.16 for guinea pigs. The LD_{50} value for rabbits is 7.19 g/kg (413). Others fed the material undiluted to groups of 10 male and 10 female rats and found that the LD_{50} value is between 6.5 and 7.0 g/kg for males and between 5.5 and 6.0 g/kg for females (63). Deaths usually occurred within 48 h or not at all and were believed due to either profound narcosis or kidney injury. In a more recent study, the LD_{50}s in fasting and fed rats were 7.1 and 12.4 g/kg, respectively, and the LD_{50} values for fasting and fed mice were 7.1 and 8.2 g/kg, respectively (53).

Inhalation. The only information available (413) states that rats exposed to an essentially saturated atmosphere of DGME generated at room temperature suffered no deaths or significant ill effects. Because of the low volatility of this material at room temperatures and because of its low oral toxicity and its apparent low toxicity from single exposures to vapors, the material it is believed, presents no unusual hazards from inhalation when handled at room temperature. However, vapors of DGME generated at elevated temperatures or breathed repeatedly over a prolonged period may present a hazard from inhalation.

Dermal. The acute dermal toxicity of DGME was evaluated in three groups of 10 (five females and five males) New Zealand white rabbits at dose levels of 4.0, 8.0, and 16.0 mL/kg. Three males and three females died at each of the two highest dose levels. The LD_{50} value for males and females was 8.98 g/kg (414).

Eye Irritation. DGME is somewhat painful to the eyes but causes only transitory injury (63, 66). Thus the material presents no serious hazard from eye contact under ordinary industrial handling conditions.

Dermal Toxicity/Skin Irritation. DGME is not appreciably irritating to the skin. Although the material can be absorbed in toxic amounts through the skin of rabbits, extensive and prolonged contact is required to cause serious effects. The LD_{50} was reportedly about 20 g/kg (63). In a more recent unpublished report (64), the dermal LD_{50} in the rabbit was 9.4 g/kg, and skin irritation was reportedly slight for both the rabbit and the guinea pig.

19.4.1.2 Chronic and Subchronic Toxicity. Oral. Smyth and Carpenter (415) reported administering commercial DGME to rats for 30 days in their drinking water. They found that the maximum dose that had no effect was less than 0.19 g/kg based upon a microscopic study of the tissues. Animals survived the highest dosage level, 1.83 g/kg. In a 6-week gavage study reported by Krasavage and Vlaovic (331), male rats were given dose levels of 900, 1800, and 3600 mg/kg, 5 days/week for 6 weeks. The highest level was half the oral LD$_{50}$. No significant toxicological effects were seen at 900 or 1800 mg/kg, but 3600 mg/kg killed two of the ten rats. The testes weight was reduced, and 6 of 10 rats had testicular atrophy. Kawamoto et al.(416) orally administered DGME at dose levels of 500, 1000, and 2000 mg/kg/day for 20 days. Daily administration of 2000 mg/kg DGME produced significant decreases in the liver, spleen, thymus, and testis weights by day 5. The authors believe that DGME causes the same adverse effects as EGME, but at a higher dose. However, the effects of EGME on kidney and blood were not observed with DGME. The NOEL for DGME was 500 mg/kg/day.

Nagano et al. (206) administered several glycol ethers in drinking water to male mice. DGME was given to male mice at 2.0% in drinking water for 25 days. No toxicity was seen in the treated animals.

Inhalation. According to Miller et al. (417), exposure of male and female F344 rats to 0, 30, 100, or 216 ppm DGME vapors for 6 h/day, 5 days/week, for 13 weeks resulted in no adverse treatment-related effects. The highest concentration used (216 ppm) was 60% of the theoretical maximum vapor concentration for DGME at 25°C and 1 atm pressure and was the "maximum practically attainable." DGME did not adversely affect hematology, bone marrow, lymphoid tissues, or testes.

Dermal. Hobson et al. (418, 419) exposed male guinea pigs dermally to DGME at levels up to 1.0 mL/kg/day for 13 weeks. Reduced growth was the only toxic sign detected in the DGME-treated animals.

19.4.1.3 Pharmacokinetics, Metabolism, and Mechanisms. Metabolism. The metabolism of DGME has not been investigated, but it is believed to be a substrate for alcohol dehydrogenase and results in forming 2-methoxy-ethoxyacetaldehyde and 2-methoxy-ethoxyacetic acid. The 2-methoxyethoxyacetic acid may be a direct-acting metabolite that is less potent than methoxyacetic acid. However, DGME may be metabolized to give EGME and ethylene glycol. Knaak et al. (420, 421) demonstrated that various higher molecular weight glycol ethers that contain ethylene oxide and also various ethylene-propylene oxide polymers terminated with nonylphenol or butanol are metabolized to lower molecular weight polymers and are oxidized to carboxylic acids by the rat. Cheever et al. (422) reported on the formation of 2-methyoxyethanol (EGME), 2-methoxyacetic acid, 2-(2-methoxyethoxy) ethanol, and 2-(2-methoxyethoxy) acetic acid from bis (2-methoxyethoxy) ether (diglyme). Toxicity studies that involved 2-(2-methoxyethoxy) ethanol and 2-(2-methoxyethoxy) acetic acid per se in the rat led Cheever et al. (422) to conclude that 2-methoxyethanol (EGME) and 2-methoxyacetic acid were entirely responsible for the reported adverse reproductive effects of bis (2-methoxyethoxyethyl) ether. Consequently, the reproductive effects of DGME are most likely to be due to the metabolic formation of 2-methoxyacetic acid.

According to Kawamoto (423), the enzyme that takes part in the metabolism of DGME is different from that of EGME. Administration of DGME increased hepatic microsomal protein and induced cytochrome P-450, but not cytochrome b5 or the reduced form of nicotinamide adenine dinucleotide phosphate cytochrome c reductase. Scientists at The Dow Chemical Company (424) evaluated the metabolism of DGME *in vitro* with an equine liver alcohol dehydrogenase assay obtained from Sigma Chemical Company. The V_{max}

(μmoles/h), K_m (μM), and V_{max}/K_m were 3.92, 7.00 × 10^{-2}, and 0.06, respectively. The scientists concluded that alcohol dehydrogenase has a very low affinity for DGME and is probably not metabolized to any extent by the enzyme. Jelnes and Soderlund (425) reported that mice metabolized and eliminated DGME as (2-methoxyethoxy) acetic acid (66%) and as methoxyacetic acid (33%). In rats 86–90% of an oral dose of DGME was excreted in urine after 96 h.

Pharmacokinetics. The pharmacokinetics of DGME have not been studied in laboratory animals despite the large amount of DGME currently being used worldwide. Consequently, little information is available on the fate (kinetics) of an administered dose of DGME in animals relative to its absorption, distribution, and excretion.

Mechanism. The metabolic conversion of DGME to 2-methoxyacetic acid (2-MAA) is believed responsible for the developmental toxicity observed in laboratory animals (426).

19.4.1.4 Reproductive and Developmental. *Reproductive Toxicity.* A 25-day drinking water study in mice was conducted by Nagano et al. (206) to evaluate the testicular toxicity of DGME. Treatment with 25% DGME in drinking water (approximately 190 mg/kg/day) resulted in no statistically significant changes in testicular weight, combined weight of seminal vesicles and coagulating gland, or white blood cell count. In a study by Kawamoto et al. (416), the daily oral administration of 2000 mg/kg of DGME to rats for 20 days produced a significant decrease in testicular weights. Krasavage and Vlaovic (331) reported testicular atrophy in the rat when 3600 mg/kg/day DGME was administered by gavage for 6 weeks. No testicular atrophy was reported when DGME was administered during this time period at 1800 mg/kg/day.

Developmental Oral. In a developmental toxicity study reported by Hardin et al. (427), DGME was teratogenic in rats after oral doses of 720 and 2165 mg/kg/day given on days 7 to 16 of gestation. No maternal toxicity was observed. The spectrum of malformations reported by Hardin et al. (427) is similar to that previously reported as induced in the rat by EGME (101). The marked similarity of EGME and DGME fetal effects supports a common metabolite, methoxyacetic acid, as the proximate teratogen. A NOEL for developmental toxicity was not established in the Hardin et al. (427) study. However, 720 mg/kg/day was considered a LOEL for developmental toxicity, and 2165 and 5175 mg/kg/day were considered the lower and upper limits for the appearance of maternal toxicity. The A/D ratio (adult toxic dose to developmentally toxic dose) for DGME was 3.0 to 7.2. The A/D ratio for EGME was reportedly 16.0 in mice. DGME was embryo toxic, fetotoxic, and teratogenic to rats.

Yamano et al. (428) administered 0, 200, 600, or 1800 mg DGME/kg/day orally to Wistar rats by gavage on gd 7 to 17. Fourteen dams in each group were killed on gd 20. Maternal animals were unaffected at doses of 200 or 600 mg/kg/day. At 1800 mg/kg/day, maternal animals showed decreased body weight and weight gain and decreased thymus weight. Parturition was delayed approximately 2 days in dams given 1800 mg/kg/day. Developmental toxicity was observed at 600 and 1800 mg/kg/day. At 1800 mg/kg/day, there was an increase in resorptions, edema, and other anomalies. At 600 mg/kg/day, decreased fetal body weight was seen along with an increase in thymic remnants in the neck and decreased ossification of sternebrae and thoracic, sacral and caudal vertebrae. The

Table 86.15. Comparison of Reproductive and Developmental Toxicity for Diethylene Glycol Methyl and Ethyl Ethers[a]

DGME		DGEE	
Rabbit developmental toxicity — dermal (257)		Rat developmental toxicity — dermal	
MNOAEL[b]	250 mg/kg/day (2.1 mmol/kg/day) death, weight, hematology[c]	MLOAEL[b]	6615 mg/kg/day (41.9 mmol/kg/day) weight gain
DNOAEL[d]	50 mg/kg/day (0.4 mmol/kg/day) fetal weight, skeletal changes	DLOAEL/ NOAEL[d]	6615 mg/kg/day (41.9 mmol/kg/day) slight skeletal defects
Rat developmental toxicity — oral		Rat developmental toxicity — inhalation	
MNOAEL	600 mg/kg/day (5.0 mmol/kg/day) weight, thymus weight, delayed parturition	MNOAEL	> 100 ppm
DNOAEL	200 mg/kg/day (1.7 mmol/kg/day) body weight, malformations, viability, delayed landmarks	DNOAEL	> 100 ppm
		Mouse reproductive toxicity — oral	
		LOAEL	4400 mg/kg/day (32.8 mmol/kg/day) fertility, pup weight, caudal epididymal sperm motility
		NOAEL	2200 mg/kg/day (16.4 mmol/kg/day) (sperm motility not examined at this dose or lower dose)

[a]Kimmel (426).
[b]MNOAEL, MLOAEL-NOAEL, or LOAEL for maternal toxicity.
[c]End points are those seen at the LOAEL.
[d]DNOAEL, DLOAEL-NOAEL, or LOAEL for developmental toxicity.

maternal and developmental NOAELs in the rat are summarized by Kimmel (426) in Table 86.15.

In a developmental toxicity screening study, 4000 mg/kg/day administered to pregnant mice produced 10% maternal mortality, a very high incidence of intrauterine death, and reduced neonatal survival of the few pups born alive (361).

Injection. In a developmental toxicity screening assay, pregnant rats were injected subcutaneously on gestation days 6 to 20 with 0.25, 0.5, or 1.0 mL DGME/kg/day (429). No maternal toxicity was detected, and there were no statistically significant fetal effects at 0.25 or 0.5 mL/kg. However, neonatal survival declined at 1.0 g/kg, which may indicate a slight fetotoxic effect at this high dose. Because skeletal and visceral examination were not done, this study was not considered adequate to establish a NOAEL.

Dermal. DGME was evaluated for its potential to produce a teratological response when topically administered to rabbits (430). Groups of 25 inseminated rabbits were treated topically with 50, 250, or 750 mg/kg/day of DGME on days 6 through 18 of gestation. Slight embryo toxicity, fetotoxicity, and maternal toxicity were produced at 750 mg/kg/day. Maternal effects were decreased body weight gain and a decrease in red blood cells and packed cell volume. A slight increase in embryonic resorptions was observed. The fetal alterations observed, mild forelimb flexure, slight to moderate dilation of the renal pelvis, retrocaval ureter, cervical spurs, and delayed ossification of the skull and sternebral bones were considered indicative of fetotoxicity but not teratogenicity. Slight fetotoxicity was observed in the 250-mg/kg/day group. No adverse maternal, embryonic, or fetal effects were observed at 50 mg/kg/day. Results in the rabbit (MNOAEL and DNOAEL) are summarized in Table 86.15.

19.4.1.6 Genetic and Related Cellular Effects Studies. The mutagenicity of DGME was evaluated in *Salmonella* tester strains TA98, TA100, TA1535, and TA1537 (Ames Test) with and without added metabolic activation by AROCLOR-induced S9 liver fractions at concentrations of 0, 20, 100, 500, 2500, and 5000 µg per plate. DGME did not cause a reproducible positive response in any of the bacterial tester strains with or without added activation (431).

19.4.2 Human Experience

19.4.2.1 General Information. Kligman, as reported by Opdyke (432), found that a concentration of 20% DGME in petrolatum applied as a closed-patch test for 48 h caused no irritation or sensitization in 25 human subjects. The most probable routes of humans exposure to DGME are inhalation, dermal contact, and ingestion. Nonoccupational exposures may occur among populations that ingest contaminated water or have contact with formulated products that contain this solvent (433).

19.4.2.2.3 Pharmacokinetics, Metabolism, and Mechanisms. Dugard et al. (21) studied the percutaneous absorption of DGME through isolated human epidermal membranes using diffusion cells (see Table 86.3). The receptor fluid was distilled water. The permeability constant was 0.206 cm/h. The linear absorption rate for DGME during an 8 h period was 0.206 mg/cm^2/h (for an applied concentration of 1.0 mg/mL on 1.0 cm^2 of skin). The absorption rate of DGME is approximately one-tenth that of EGME.

19.4.2.2.4 Reproductive and Developmental. Margin of Exposure. Kimmel (426) assessed the risk of dermal exposure to DGME using the results of the percutaneous absorption studies conducted by Dugard et al. (21). Total dermal exposure to both hands (1300 cm^2) of a 70 kg man for 1, 2, and 6 h of exposure, and the dermal NOAEL for rabbit developmental toxicity (50 mg/kg/day) are shown in Table 86.16. The margin of exposure for DGME is low and suggests a high level of risk from dermal exposure to this glycol ether.

19.5 Standards, Regulations, or Guidelines of Exposure

No occupational exposure standards have been developed for DGME. The MOE (margin of exposure) developed by Kimmel (426) suggests that a standard is needed for skin exposure.

ETHERS OF ETHYLENE GLYCOL AND DERIVATIVES

Table 86.16. Margins of Exposure Based on Comparison of Dermal Absorption Rates and NOAELs for Reproductive and Developmental Toxicity

Glycol Ether	Dermal Absorption Rate[a] (mg/cm^2/h)	Total Absorption[b] (mg/kg) 1 h	2 h	6 h	Margins of Exposure[c] 1 h	2 h	6 h
DGME	0.206	3.82	7.64	22.9	13.1	6.5	2.2
DGEE	0.125	2.32	4.64	13.9	948.2	474.1	158.3
TGME	0.034	0.63	1.26	3.78	476.2	238.1	79.4

[a]Based on data from Dugard et al. (21) and Leber et al. (20).
[b]Based on exposure of both hands (1300 cm^2) and 70 kg average body weight.
[c]Based on lowest NOAEL for each glycol ether as shown in Tables 86.15 and 86.17.

20.0 Diethylene Glycol Monoethyl Ether

20.0.1 CAS Number: [111-90-0]

20.0.2 Synonyms: 2-(2-Ethoxyethoxy)ethanol; diglycol monoethyl ether

20.0.3 Trade Names: Carbitol® Solvent; Dowanol® DE Glycol Ether; Eastman® DE Solvent

20.0.4 Molecular Weight: 134.17

20.0.5 Molecular Formula: C$_6$H$_{14}$O$_3$

20.0.6 Molecular Structure:

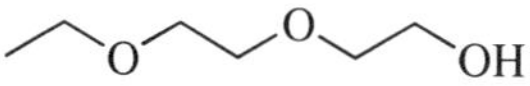

20.1 Chemical and Physical Properties

20.1.1 General

Refer to Table 86.13.

20.2 Production and Use

Diethylene glycol monoethyl ether is used as a solvent for nitrocellulose, resins, mineral oils, dyes, soaps, wood stains, textile printing, lacquers, organic synthesis and as a diluent for brake fluid. Production figures are given in Table 86.2.

20.3 Exposure Assessment

20.3.1 Air

See the Introduction for a general description of air sampling and analytical procedures.

20.3.2 Workplace Methods

See the Introduction for a general description of sampling and analytical procedures for the workplace.

20.4 Toxic Effects

Diethylene glycol monoethyl ether (DGEE) has low oral toxicity and is not appreciably irritating to the eyes or skin, but it may be absorbed in toxic amounts through the skin. Its volatility is sufficiently low that acutely hazardous vapor concentrations do not occur at ordinary temperatures. DGEE does not affect fertility or reproductive performance and is not a developmental toxicant by oral or inhalation administration. DGEE has been evaluated in a 2-year dietary feeding study in rats, where no indication of tumor production was reported. Although this study was not specifically designed for carcinogenicity, it is an indication that DGEE has little or no carcinogenic potential. The material is readily oxidized in the body. Experience with human subjects has been uneventful. It is generally agreed that the relatively pure ether does not present any serious industrial hazards. Reasonable precautions are adequate to ensure safe handling.

20.4.1 Experimental Studies

20.4.1.1 Acute Toxicity. Oral. Smyth et al. (434) fed specially purified DGEE as a 50% aqueous solution and found that the LD_{50} values are 8.69 g/kg for the rat and 3.67 g/kg for the guinea pig. The LD_{50} value for rabbits was 3.62 g/kg (435). When a 40% aqueous solution of a commercial product was fed, the LD_{50}s were 9.74 g/kg for the rat and 4.97 g/kg for the guinea pig. Gross (49), who worked with an industrial product, found that 1 mL/kg was lethal to cats. It caused disturbance of equilibrium, gastrointestinal inflammation, pneumonia, and kidney injury, but no albuminuria. A dog that received a single dose of 2 mL/kg as a 20% aqueous solution was unaffected, except for slight vomiting 3 h after feeding. Laug and co-workers (155) found that the oral LD_{50} values are 5.54 g/kg for rats, 6.58 g/kg for mice, and 3.87 g/kg for guinea pigs. They likened the response of animals fed this ether to that caused by ethanol, but also noted pneumonia and kidney injury in treated animals. They concluded that this material is not suitable for use in food or drugs. Rowe (63) found that the LD_{50} for rats is 5.4 g/kg and observed that the effects of acutely toxic doses were characterized by ataxia, depression, and little apparent injury to visceral organs. Smyth and Carpenter (415) reported maintaining groups of five rats for 30 days on drinking water that contained various amounts of purified DGEE. Dosage levels ranged from 0.21 to 3.88 g/kg/day. They found that 0.49 g/kg was tolerated without any adverse effect, that 0.87 g/kg caused reduction in appetite, and that 1.77 g/kg caused some organic injury. In more recent studies, Krasavage and Terhaar (53) administered DGEE undiluted by gavage and reported oral LD_{50}s of 10.5 and 15.9 g/kg for fasting and fed rats, respectively, and 6.0 g/kg for both fasting and fed mice.

Injection. Various workers have reported LD_{50} values following parenteral administration of DGEE (374), as shown in Table 86.18.

Inhalation. Gross (49) reported that rabbits, cats, guinea pigs, and mice were not injured by 12 daily exposures to an essentially saturated atmosphere of DGEE.

Dermal. Hanzlik et al. (436) reported the results of extensive skin absorption studies with DGEE. They reported that the LD_{50} is 8.5 mL/kg for rabbits when applied by inunction for 2 h to an area of approximately 100 cm^2. When such applications were repeated daily for 30 days, the LD_{50} was estimated at 0.32 mL/kg, and the no-effect level was between 0.04 and 0.08 mL/kg. They reported transient dermatitis and both

Table 86.17. Comparison of Reproductive and Developmental Toxicity for Triethylene Glycol Methyl and Ethyl Ethers[a]

TGME	TGEE
Rabbit developmental toxicity — oral (Hoberman et al., 495)	Rat developmental toxicity — oral (Leber et al., 20)
MNOAEL[b] 500 mg/kg/day (3.1 mmol/kg/day) death[c]	No derived MNOAEL[b] 1000 mg/kg/day, inadequate study, highest dose
DNOAEL[d] 1000 mg/kg/day (6.1 mmol/kg/day) implants and resorptions	No derived DNOAEL[d] 1000 mg/kg/day, inadequate study, highest dose
Rat developmental toxicity — oral (1996)	
MNOAEL 1650 mg/kg/day (10.0 mmol/kg/day) kidney weight and delayed parturition	
DNOAEL 300 mg/kg/day (1.8 mmol/kg/day) pup weight changes, weight gain, and age at testes descent	

[a]Kimmel (426).
[b]MNOAEL, MLOAEL-NOAEL, or LOAEL for maternal toxicity.
[c]End points are those seen at the LOAEL.
[d]DNOAEL, DLOAEL-NOAEL, or LOAEL for developmental toxicity.

Table 86.18. Toxicity of Parenterally Administered DGEE

Species	Route	LD$_{50}$ (g/kg)
Rat	i.v.	2.9
Mouse	i.v.	3.9
Dog	i.v.	3.0
Rabbit	i.v.	0.9
Mouse	s.c.	5.5
Rabbit	s.c.	2.0

microscopic injury and impairment of kidney function. In more recent studies, Krasavage and Terhaar (64) reported a dermal LD$_{50}$ of 9.1 g/kg for rabbits when DGEE was applied to the shaved back under occlusive wrap for 6 h.

Irritation. DGEE is slightly painful but causes no more than a trace of irritation of the conjunctival membranes (63, 66, 157). Others have indicated that 500 mg of DGEE instilled into the eye of a rabbit produced moderate irritation (437).

Skin Irritation and Sensitization. DGEE is not irritating to the skin of rabbits even after prolonged and repeated contact (63). When tested in a dermal LD$_{50}$ study, DGEE under an occlusive wrap was only slightly irritating to rabbit skin (64).

20.4.1.2 Chronic and Subchronic Toxicity. *Subchronic Toxicity/Oral.* Hall et al. (438) fed rats DGEE that contained 0.4% ethylene glycol as an impurity in their diet at concentrations of 0.25, 1.0, and 5.0% for 90 days. They stated that the no-effect level was 1%. At the 5% level, they found retarded growth in both sexes, reduced food intake, no hematologic changes, slightly impaired renal function, especially in males, increased kidney weights in both sexes and of the testes in the males, degeneration of kidney tubules, fatty changes in the liver, and testicular edema.

In studies of male rats (331), DGEE whose purity was in excess of 99.5% was administered by gavage at doses of 0 (control), 1340, 2680, or 5360 mg/kg for 6 weeks (5 days/week). DGEE produced no hematologic changes and only a slight but significant increase in serum urea nitrogen at the highest dose. Slight increases (dose-related) in relative liver and kidney weights were seen; however, terminal body weights were unaffected. Testes weights were unaffected by these treatments. Gross pathological changes noted only at the high dose of DGEE included blood in the urinary bladder and dark and congested spleen (at intermediate- and high-dose levels). Hyperkeratosis of the stomach and hepatocytomegaly and anisokaryosis of the liver, as well as proteinaceous casts and hemosiderin in the kidneys, were seen at the high dose. The 1340-mg/kg/day level may be considered a NOEL for this study.

Gaunt et al.(439), using DGEE that contained less than 0.4% ethylene glycol, fed the material for 90 days to rats at dietary levels of 0.5 and 5.0%; to mice at dietary levels of 0.25, 0.6, 1.8, and 5.4%; and by intubation to guinea pigs at doses of 167, 500, and 1500 mg/kg. Three guinea pigs given 1500 mg/kg/day for 14 to 21 days died showing signs of uremia, and 6 of the 20 male mice fed at the 5.4% level died with signs of severe renal damage. Oxaluria developed in both the rats and mice at the highest level. A reduction of hemoglobin developed in all three species at the highest levels. They stated that the no-effect levels seen in their studies were about 250 mg/kg/day for rats, 850 to 1000 mg/kg/day for mice, and 167 mg/kg/day for guinea pigs.

Dermal. Purified DGEE was applied to the clipped abdomens of rabbits five times per week for 3 months at dosages of 0.1, 0.3, 1.0, and 3.0 mL/kg/day (440). The criteria used in judging effects were growth, mortality, hematologic studies, organ weight studies, blood urea nitrogen determinations, and gross and microscopic examination of the treated areas of skin and of the principal organs. The only effect at the top dosage level of 3.0 mL/kg was an increase in blood urea nitrogen and severe kidney injury in one of the four surviving animals. Minor kidney changes were seen in two of the animals, and one was unaffected. At the 1-mL/kg dosage level, moderate kidney changes were seen in three of the four surviving animals, and at the lower dosage levels, no adverse effects were seen. Thus the no-effect dosage level for repeated exposures during a 90-day period approximates 0.3 mL/kg/day (0.3 g/kg/day). This figure is about 10 times larger than that found by Hanzlik (436) and might well reflect a difference in the purity of the material tested.

Chronic Toxicity/Oral. Morris et al. (38) maintained rats for 2 years on a diet that contained 2.16% purified DGEE. This is estimated to be equivalent to slightly more than 1.0 g/kg/day. The only adverse effects noted were a few oxalate crystals in the kidney of one animal, slight liver damage, and some interstitial edema in the testes. Because the quality of the material tested was not established, the possibility that the oxalate crystals were caused by the presence of small amounts of ethylene glycol in the test sample cannot be overlooked.

Hanzlik et al. (436) administered the pure ether to rats at 1.0% in their drinking water and to mice at 5.0% in their food for a 2-year period without causing significant adverse effects. They also found that when this same ether containing ethylene glycol was fed, kidney injury typical of the glycol occurred. They concluded that the ether was relatively noninjurious.

Smyth et al. (441) fed two grades of DGEE to rats through three generations during a 2-year period; one contained less than 0.2% ethylene glycol, the other 29.5%. The drinking water levels were 0.01, 0.04, 0.2, and 1% (calculated at approximately 0.01, 0.04, 0.20, and 0.95 g/kg). The adverse effects agreed with those seen in earlier studies. The sample that contained the high level of ethylene glycol (29.2%) was considerably more toxic than the purer grade. The maximum safe dosage for the impure material was 0.01 g/kg/day, whereas it was about 0.20 g/kg/day for the purer sample (less than 0.2% ethylene glycol). Thus they consider that the cumulative toxicity of DGEE is largely due to the ethylene glycol content. On the basis of their work, the authors suggest that the permissible intake for humans may be 1.4 g/day (using a 10-fold margin of safety).

Butterworth et al. (442) studied the effect of DGEE on ferrets, a nonrodent species. The material was fed in the diet for 9 months at levels of 0.5, 1.0, 2.0, and 3.0 mL/kg/day (approximately 0.5, 1.0, 2.0, and 3.0 g/kg/day). No adverse effects such as body and organ weight changes, hematologic effects, or histopathological injury were seen. Based on these and other data and using the 100-fold margin of safety, they stated that the acceptable daily intake for a 70-kg person is 2.1 mL/day (approximately 2.1 g/day).

Inhalation. Sprague–Dawley rats were exposed to DGEE vapor concentrations of 0.09, 0.27, and 1.1 mg/L for 6 h/day, 5 days/wk for 28 days by inhalation in a nose-only exposure chamber (443). Respirable droplets were present at the highest concentration. A series of toxicological evaluations revealed no evidence for a systemic effect related to exposure. Mild nonspecific irritation was observed histopathologically in the upper respiratory tract at the two upper levels. The changes consisted of foci of necrosis in the ventral cartilage of the larynx and an increase in eosinophilic inclusions in the olfactory epithelium of the nasal mucosa. The NOAELs for systemic effects and irritation of the upper respiratory tract were 1.1 and 0.09 mg/L, respectively.

20.4.1.3 Pharmacokinetics, Metabolism, and Mechanisms. Metabolism. Fellows et al. (444) found that DGEE is largely destroyed by the body or is excreted as the glucuronate. They noted that when given to rabbits orally or by injection, the urinary content of glucuronic acid increased as it did when propylene glycol was given. Why this should occur with these two materials and not with ethylene and diethylene glycol and glycerol is unknown.

The oral administration of 1503 mg DGEE to a normal adult man resulted in the excretion of 1140 mg of 2-ethoxyethoxyacetic acid, 69% of the total dose, within 12 h (445).

Miller (424) evaluated the metabolism of DGEE *in vitro* by using an equine liver alcohol dehydrogenase preparation obtained from Sigma Chemical Company (1982). The V_{max} (µmoles/h), K_m (µM), and V_{max}/K_m were 6.94, 6.31×10^{-2}, and 0.11, respectively.

It was concluded that alcohol dehydrogenase has a low affinity for DGEE and the compound is probably not metabolized to a significant extent by this enzyme.

20.4.1.4 Reproductive and Developmental Studies. Reproductive. In a continuous breeding study in mice, Williams et al. (446) administered DGEE in drinking water at concentrations of 0, 0.25% (0.44 g/kg), 1.25% (2.2 g/kg), or 2.5% (4.4 g/kg). At the highest concentration (2.5%), the results indicated that the F_1 males had a significant decrease in sperm motility. However, there were no significant effects on fertility or reproductive performance. Other effects noted were significant decreases in absolute brain weights and an increase in absolute liver weights, although there were no histopathological changes noted in either organ. In the same study, Williams tested diethylene glycol (DG) and found that, unlike DGEE, DG was a reproductive toxicant (i.e., it affected fertility and reproductive performance) to mice when administered at high doses (6.1 g/kg/day). Kimmel (426) considered that the 2.5% exposure level (4.4 g/kg/day; 32.8 mmol/kg/day) was a LOAEL based on sperm motility, whereas the 1.25% exposure level (2.2 g/kg/day; 16.4 mmol/kg/day), it was thought, represented a NOAEL. This assignment is questionable because sperm count was not examined at this level or at 0.25% in the F_1 males or the P males. See Table 86.16, Section 19.

Continuous breeding mouse studies (447, 448) were used to determine the relationships between the functional indicators of reproduction (pup measures) and various necropsied end points collected for males and females. Effects on females were noted, but no effects on males.

Developmental. Hardin et al. (179) dermally administered 0.35 mL (2.6 mmol) of DGEE four times a day to rats on days 7 through 16 of gestation (a total of 1.4 mL/day, 6615 mg/kg/day, or 41.9 mmol/kg/day). No signs of developmental toxicity, including teratogenic effects, were observed. The 41.9 mmol DGEE/kg/day was considered near the NOAEL (minimal effects) for the dermal exposure of rats for both maternal and developmental toxicity (426). See Table 86.15 in Section 19.

Nelson et al. (101) reported on gravid rats exposed to 100 ppm DGEE by inhalation on days 7 through 15 of gestation. Because of the low vapor pressure of DGEE, concentrations greater than 100 ppm could not be achieved. No toxicity to the dams or to the offspring was observed. The investigators concluded that DGEE is not likely to be a teratogenic hazard through inhalation exposure. Kimmel (426) considered that the maternal and developmental toxicity of DGEE is > 100 ppm. See Table 86.15, in Section 19.

Whole Embryo Culture. DGEE was tested in rat whole embryo cultures at concentrations that ranged from 0.3 to 1.0 mg/mL. Embryo toxicity increased with the length of the alkyl chain. This contrasts with *in vivo* data where metabolism results in forming ethoxy acids (449).

20.4.1.5 Carcinogenesis. In the 2-year dietary feeding study of Morris et al. (38), in which purified DGEE was administered to rats at a level of slightly more than 1.0 g/kg/day, there were some adverse effects. However, the authors did not mention the development of tumors. Although this study was not designed specifically to evaluate carcinogenicity, the studies suggest that DGEE has little or no carcinogenic potential.

20.4.1.6 Genetic and Related Cellular Effects Studies.
Genotoxicity. The mutagenicity of DGEE was evaluated in *Salmonella* tester strains TA98, TA100, TA1535, and TA1538 (Ames Test) with and without added metabolic activation by AROCLOR-induced rat liver S9 fraction. The material was tested for mutagenicity at concentrations of 0, 3,000, 6,000, 9,000, 12,000, 15,000, and 20,000 µg/plate using the direct plate incorporation method. DGEE did not produce a reproducible positive response in any of the bacterial tester strains with or without added metabolic activation (450).

20.4.2 Human Experience

20.4.2.1 General Information.
No toxic effects have been reported in humans from the industrial use of DGEE.

Browning (59) reported that an alcoholic man who drank a liquid that contained 47% DGEE (about 300 mL) and less than 0.2% methanol developed severe symptoms of central nervous and respiratory injury, thirst, acidosis, and albumin in the urine but no oliguria. He recovered upon symptomatic treatment.

20.4.2.2.3 Pharmacokinetics, Metabolism, and Mechanisms.
The absorption of DGEE through human abdominal skin was measured *in vitro* using Franz-type glass diffusion cells, Epidermal layers from human skin were exposed for 8 h to a solution that contained radiolabeled DGEE in the donor chamber. The appearance of radioactivity was followed in the receptor chamber. Skin damage was determined by using water absorption rates before and after exposure to DGEE. DGEE was evaluated in a separate *in vitro* procedure using human cadaver abdominal epidermis (21). Under the conditions of the study, the mean rate of absorption through the skin was 0.125 mg/cm^2/h (21). The rate of absorption reached steady state at just under 1 h.

20.4.2.2.7 Other: Neurological, Pulmonary, Skin Sensitization.
Cranch et al. (451) found that the material is neither a primary irritant nor a sensitizer and is no different from wool, fat, or glycerine when applied to the skin of 98 human subjects. Furthermore, they reported that a 70% aqueous material did not retard wound healing. Meininger (452) confirmed the lack of skin irritation in humans and was unable to find evidence that DGEE was absorbed through the skin of human subjects. Kligman, according to Opdyke (432), also found that human volunteers showed neither irritation nor signs of sensitization when the material was tested at a 20% level in petroleum in a 48-h closed-patch test.

20.5 Standards, Regulations, or Guidelines of Exposure

The following occupational exposure standards have been established for DGEE: AIHA WEEL TWA 25 ppm; The Netherlands TWA 32 ppm (180 mg/m^3). However, in view of the low vapor pressure and the low toxicity of the material, reasonable and ordinary precautions to avoid inhalation of vapors or mists are adequate to prevent excessive vapor exposure.

21.0 Diethylene Glycol Mono-*n*-propyl Ether

21.0.1 CAS Number: *[6881-94-3]*

21.0.2 Synonyms: 2-(2-(Propoxyethoxy)ethanol; DGPE

21.0.3 Trade Names: Eastman® DP Solvent; Propyl Carbitol® Solvent

21.0.4 Molecular Weight: 148.2

21.0.5 Molecular Formula: $C_7H_{16}O_3$

21.0.6 Molecular Structure: ~~O~~O~~OH

21.1 Chemical and Physical Properties

21.1.1 General

See Table 86.13.

21.2 Production and Use

21.3 Exposure Assessment

21.3.1 Air

See the Introduction for a general description of air sampling and analytical procedures.

21.3.2 Workplace Methods

See the Introduction for a general description of sampling and analytical procedures for the workplace.

21.4 Toxic Effects

The single-dose oral toxicity of diethylene glycol mono-*n*-propyl ether (DGPE) is low. The acute dermal toxicity of this glycol ether to rabbits is low. Repeated oral administration in rats resulted in decreased body weights and red blood cell abnormalities. DGPE causes slight skin irritation in both the rabbit and the guinea pig, and repeated skin application causes exacerbation of the effect in the guinea pig. This material displayed weak activity in a guinea pig skin sensitization study and is a moderate to strong eye irritant in the rabbit.

21.4.1 Experimental Studies

21.4.1.1 Acute Toxicity. Oral. The LD_{50} values determined for this glycol ether when given undiluted by gavage to fed rats and mice were 9.59 and 5.70 g/kg, respectively, whereas these values for fasting animals were 6.66 and 3.81 g/kg, respectively (53). Oral administration of an aqueous solution of this material to rabbits at 1.95 mL/kg resulted in hemoglobinuria, nephrosis, bloody tears, reduced lymphopoiesis, and lowered hematocrits (453). In this same study, these effects were significantly reduced at 0.97 mL/kg, and incipient nephrosis and fatty liver were reported.

Rats fed this compound in the diet at 1.0 and 0.1% during an 11-day period displayed normal weight gains (207). Hemoglobin concentration and hematocrit were slightly reduced in two animals in the high-dose group, but these changes were not statistically significant. Oral administration of undiluted DGPE to male rats (5 days/week, for 6 weeks) at dose levels of 814, 1628, and 3256 mg/kg resulted in decreased body weights at the two higher dose levels and decreased hemoglobin, hematocrit, and red blood cell counts at the highest dose level (331). Slight narcosis was observed in all animals after administration of the first three dose levels, but subsequent treatments produced no compound-related effects. Also in this study, the intermediate- and low-dose levels of this glycol ether produced no clinical signs of toxicity. The no-observed effect level was 814 mg/kg. By direct comparison, DGPE was less toxic to rats than the corresponding ethylene glycol derivative.

Dermal. DGPE is practically nontoxic when applied undiluted under an occlusive wrap on the skin of rabbits (LD_{50} = 5.06 g/kg) (64).

Eye Irritation. DGPE elicited moderate to strong eye irritation when administered undiluted to the conjunctival sacs of rabbits. Prompt washing of the eye reduced the irritant effect (207).

Skin Irritation and Sensitization. Only slight skin irritation was observed during a 2-week period following the application (occluded) of undiluted material to the depilated skin of guinea pigs for a 24-h period (207). Repeated application of this material to guinea pig skin (0.5 mL/day, 10 days) resulted in slight to moderate erythema in 5/5 animals and slight edema in 3/5 animals.

21.4.1.7 Other: Neurological, Pulmonary, Skin Sensitization. This material did not cause skin sensitization when tested in guinea pigs (Eastman Kodak Co., unpublished data, 1993).

21.5 Standards, Regulations, or Guidelines of Exposure

No occupational exposure standard for repeated or prolonged inhalation exposure to DGPE has been established. In view of the effects produced by this material in animals, eye contact should be prevented.

22.0 Diethylene Glycol Mono-*n*-butyl Ether

22.0.1 CAS Number: [112-34-5]

22.0.2 Synonyms: 2-(2-Butoxyethoxy)ethanol; butoxydiglycol; diglycol monobutyl ether; DGBE; butyl diglycol ether; BDGE

22.0.3 Trade Names: Butyl Carbitol® Solvent; Butyl Dioxitol® Solvent; Dowanol® DB Solvent; Eastman® DB Solvent

22.0.4 Molecular Weight: 162.23

22.0.5 Molecular Formula: $C_8H_{18}O_3$

22.0.6 Molecular Structure:

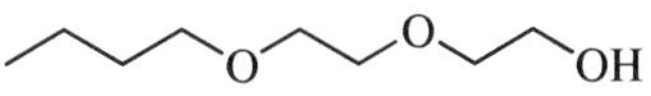

22.1 Chemical and Physical Properties

22.1.1 General

DGBE is miscible with water and is very soluble in a variety of other organic solvents including alcohol, acetone, and benzene. For additional properties, refer to Table 86.13.

22.1.2 Odor and Warning Properties

DGBE is a colorless liquid that has a faint odor.

22.2 Production and Use

Diethylene glycol mono-*n*-butyl ether (DGBE) is used primarily as a solvent in hard-surface cleaners and inks and as a solvent and coalescing agent in paints and other coatings (2). DGBE is also a solvent for nitrocellulose, oils, dyes, gums, soaps, polymers, and plasticizer intermediates. Refer to Table 86.2 for additional production and use information.

22.3 Exposure Assessment

22.3.1 Air

Concentrations of DGBE in room air have been determined under simulated conditions when hard-surface cleaners that contained up to 9% DGBE were used (454). The analytical method involved charcoal tube adsorption, desorption with 5% methanol/95% dichloromethane, and gas–liquid chromatographic separation using dodecanol as an internal standard. Under these simulated use conditions, the peak air concentration was 1.6 ppm, and the breathing zone concentrations did not exceed the 0.8 ppm level of quantitative detection. Refer to the Introduction for other applicable procedures.

22.4 Toxic Effects

DGBE has a low order of acute toxicity by the oral and dermal routes and was not toxic by the inhalation route, possibly due in part to the low vapor concentration that could be generated. The dermal route is the most likely route for human exposure because of the extensive use of DGBE in cleaning products and coatings; DGBE has been extensively tested by this route and caused no target organ, fertility, developmental, or nervous system toxicity in animal studies. Hematotoxicity, associated with the butyl ether of ethylene glycol (EGBE) in rats, was reported in only one study at very high doses of DGBE. Like other glycol ethers, DGBE was not genotoxic. There are no reports of adverse effects in humans from using DGBE-containing products (455).

Because of the large volume of production and widespread use of DGBE in consumer products where significant dermal contact could occur, the Environmental Protection Agency (EPA) (456) issued a test rule requiring dermal subchronic and fertility studies and a dermal neurotoxicity study. Oral and dermal kinetic studies and additional genotoxicity studies were also required. The results of these studies confirmed that DGBE presents little

Table 86.19. Oral Toxicity of DGBE in Laboratory Animals

Species	Gender	LD$_{50}$ (mg/kg)	Reference
Rat		6560	434
Rat (fed)	M	9623	53
Rat (fasting)	M	7292	53
Rat	M	6530	266
Rat	F	5080	266
Mouse (fed)	M	5526	53
Mouse (fasting)	M	2406	53
Rabbit		2200	458
Guinea pig		2000	434

risk of systemic toxicity from occupational exposure to undiluted material or from consumer exposure to formulated products that contain DGBE.

22.4.1 Experimental Studies

22.4.1.1 Acute Toxicity. DGBE has low acute toxicity by the oral (Table 86.19), dermal, and inhalation routes. Undiluted DGBE, 99.5% pure, was administered orally to rats 5 days a week for 6 weeks at doses up to 3564 mg/kg/day, one-half the acute LD$_{50}$ (331). No deaths occurred, but some rats were moribund and/or had decreased body weight gain and food consumption at the highest dose. Symptoms of hemolysis, including hemoglobinuria, enlarged congested spleens, proteinaceous casts, and hemosiderin in the kidney tubules were reported only at the top dose. These effects are the same as those caused by the butyl ether of ethylene glycol (EGBE). All doses produced hyperkeratosis and acanthosis of the stomach in a few animals from repeated irritation at the site of application, but no effects on testis weight or histopathological effects were observed (331).

In earlier studies (415), DGBE was administered to rats in drinking water for 30 days. No toxic effects were reported at a daily dose level of about 50 mg/kg; however, higher concentrations were unpalatable, and liver, kidney, spleen or testis damage was reported at an estimated dose of 650 mg/kg. However, the test material used in this study was made by an older process, involving BF$_3$, which is currently not used. It is likely that the observed effects were due to toxic impurities, such as crown ethers (cyclic oligomers of ethylene oxide) and fluoroethanol (253, 457).

Dermal. The LD$_{50}$ in rabbits by the dermal route was 4000 mg/kg, implying that a significant proportion of the dermally applied dose can be absorbed under the conditions used (458). The dermal LD$_{50}$ in rabbits was reportedly 2764 mg/kg when applied undiluted under occlusive conditions for 24 h (64).

Eye Irritation. Instillation of 0.1 mL of undiluted DGBE into the rabbit eye caused moderate irritation and some tissue damage, but the eye returned to normal within 14 days (459). Aqueous dilutions were less irritating, and a 5% solution produced no detectable irritation.

Skin Irritation. Under the conditions of the dermal acute toxicity study (64), undiluted DGBE was a slight skin irritant in both the rabbit and the guinea pig.

22.4.1.2 Chronic and Subchronic Toxicity. *Oral.* In a U.S. Navy study (266), rats were dosed 5 days/week for 13 weeks with 1, 5, or 25% of the acute oral LD_{50} dose of DGBE as aqueous solutions; there was a 6-week interim sacrifice group. The authors conclude that the 1% LD_{50} dose level (65 and 51 mg/kg/day for males and females, respectively) administered for 13 weeks produced no or minimal toxicity, whereas the mid-dose produced mild effects and the top dose (1625 and 1275 mg/kg/day, respectively) produced excessive mortality. Mortality occurred regularly throughout the study and was probably due to "gavage trauma or dosing misadventure." It did not result from systemic toxicity. The reported pulmonary congestion and edema suggests that DGBE inadvertently entered the lung, possibly by aspiration. Systemic toxicity was suggested by increased blood urea nitrogen and urine N-acetyl-β-glucosaminidase only in males and decreased white blood cell and lymphocyte counts only in females at the mid dose. These clinical signs of kidney and lymphoreticular damage in male and female rats, respectively, were not confirmed by histological examination of the tissues (266). The reported deficiencies in this study make further interpretation difficult.

Dermal. In a combined dermal subchronic/fertility study (460), male and female rats were dosed dermally with up to 2000 mg/kg/day DGBE under occlusion, 5 days/week for 13 weeks. DGBE was irritating to the skin; the severity depended on the concentration and was more irritating in females. Other than slight hemoglobinuria in some females at the end of the study, there was no evidence of systemic toxicity.

Inhalation. No rats died when exposed for 7 h to the maximum attainable vapor concentration of DGBE, estimated at 18 ppm (461). Rats were exposed by inhalation to DGBE for 6 h/day, 5 days/week for 5 weeks at concentrations of 2, 6, and 18 ppm, the highest vapor concentration that could be reliably maintained (461). No significant toxic effects were reported.

22.4.1.3 Pharmacokinetics, Metabolism, and Mechanisms. *Metabolism.* No metabolism studies after oral administration of DGBE have been reported. However, the metabolism of DGBE and its acetate ester, diethylene glycol butyl ether acetate (DGBEA), has been examined after dermal administration to rats (462), and DGBEA has been examined after oral administration (463). When incubated with whole rat blood *in vitro*, DGBEA was hydrolyzed to DGBE, and the half-life was less than 3 min. This suggested that the subsequent metabolism and toxicity of DGBEA would be identical to that of DGBE (463). DGBEA was, and presumably DGBE would be, rapidly absorbed from the gastrointestinal tract, metabolized, and excreted primarily (85% of the dose) in the urine. The major urinary metabolite was 2-butoxyethoxyacetic acid, together with diethylene glycol, hydroxybutoxyethoxyacetic acid, two unidentified metabolites, and exhaled CO_2 (5% of the dose). No DGBE, DGBEA, or 2-butoxyacetic acid was detected (463).

When DGBE and DGBEA were applied to the skin for 5 min and then washed, most (90%) of the material was recovered unabsorbed (462). When applied under occlusion for 24 h, the calculated absorption rates for DGBEA (1.58, 1.28 mg/cm^2/h for males and females, respectively; mean = 1.43) were similar to those for DGBE (0.73, 1.46;

mean = 1.10 mg/cm^2/h). The major metabolite of both materials was again 2-butoxyethoxyacetic acid, but a previously unidentified metabolite — the glucuronide conjugate of DGBE — was detected at 5 to 10% of the applied dose after dermal administration of DGBE (462).

22.4.1.4 Reproductive and Developmental. *Reproductive.* In only one of the several subchronic studies reviewed earlier was there any evidence of damage to reproductive organs (415), and that may have been due to a toxic impurity. This lack of reproductive toxicity was confirmed with pure DGBE in a functional test in rats. DGBE to males was administered at oral doses up to 1000 mg/kg/day for 60 days before mating; females were dosed 14 days before mating and through sacrifice at day 13 of gestation, and a separate group was dosed through weaning. Untreated males were bred with treated females, and vice versa. DGBE had no effect on fertility in either sex and, except for a slight decrease in pup weight toward the end of lactation at the 1000 mg/kg/day dose, there was no adverse effect on embryos, fetuses, or neonates (464).

In the combined dermal subchronic/fertility study mentioned before (460), male and female rats were dosed dermally with up to 2000 mg/kg/day of DGBE under occlusion, 5 days/week for 13 weeks. DGBE was irritating to the skin, the severity depended on the concentration, and females were more sensitive than males. Other than slight hemoglobinuria in some females at the end of the study, there was no evidence of systemic or reproductive toxicity (460).

Developmental. Pregnant rats were fed a diet containing up to 1% DGBE, providing an estimated dose of 633 mg/kg/day on gestational days 0 to 20 (465). Most rats were terminated on day 20 for fetal evaluation, but several were allowed to deliver at term and suckle their young through weaning on postnatal day 21. The offspring were reared until 10 weeks after birth. There was an apparent decrease in maternal body weight gain at all three doses, but it was not dose-related, and there was no decrease in food consumption. There were no statistically significant differences between the control group and any treatment group indicative of prenatal or postnatal developmental toxicity (465).

Pregnant rabbits were dosed dermally with DGBE at doses up to 1000 mg/kg/day on gestational days 7 to 18 in a standard segment II study (464). Maternal toxicity consisted only of skin irritation, especially at the top dose. No adverse effects were seen on intrauterine survival or the incidence of fetal malformation.

22.4.1.6 Genetic and Related Cellular Effects Studies. *Genotoxicity.* Ethylene glycol ethers as a class are not considered genotoxic. DGBE has been specifically tested in many *in vitro* and *in vivo* systems and, apart from one equivocal result, was clearly not genotoxic (466). The test systems that produced negative results included the Ames bacterial reversion assay with *S. typhimurium*, the unscheduled DNA repair assay in rat hepatocytes, and cytogenetics in mammalian (Chinese hamster ovary — CHO) cells. Equivocal results occurred in the mouse lymphoma mammalian cell mutagenicity assay, but in a secondary *in vivo* mutagenicity assay (*Drosophila*), DGBE did not induce sex-linked recessive lethal mutations (466). Recently, in response to the EPA test rule, DGBE was tested for forward gene mutation at the HGPRT locus of CHO cells in culture and in an *in vivo* mouse bone marrow micronucleus assay for cytogenetic damage after a single intraperitoneal dose up to 3300 mg/kg, the maximum tolerated dose. There was no evidence of genotoxicity in either assay (346).

22.4.1.6 Other: Neurotoxic, Pulmonary, Skin Sensitization. Neurotoxicity. No evidence of neurotoxicity or other systemic toxicity was reported when rats were dosed dermally with a 10 or 30% aqueous solution of DGBE or with undiluted DGBE at doses up to 2 g/kg/day, 5 day/week for 13 weeks (467). The study that followed EPA Toxic Substances Control Act neurotoxicity guidelines included a functional observational battery, motor activity, and special neuropathology. The only treatment-related effect was eschar formation at the treatment site in females who received the top dose. These results agree with the results of the subchronic studies reviewed earlier, where no clinical signs were reported that would indicate neurotoxicity, and with the subchronic oral study (266), where neuropathological examination of brain, spinal cord, and peripheral nerve revealed no lesions.

22.4.2 Human Experience

22.4.2.1 General Information. Even though DGBE has been used for many years in consumer products such as paints and hard-surface cleaners, there is only a single report of an adverse health effect associated with a possible exposure in the workplace to DGBE-containing products. There is little likelihood of systemic toxicity to DGBE in normal use in consumer products (455). There has been one report of a suicide when a large amount of a DGBE-containing cleaning product was intentionally ingested, but the death was not definitely attributed to this product. Most cases of accidental ingestion of DGBE-containing cleaning products were asymptomatic (468).

22.4.2.2 Clinical Cases. A woman who worked as an office clerk in a school reported irritation of the upper airways, erythema of the face, and swollen eyelids (469). These symptoms developed shortly after renovation of the workplace that involved painting, wallpapering, and replacement of mats. Symptoms would develop during the workweek and disappear on weekends. She was diagnosed as having general hyperreactivity with atopy. She tested positive by prick test for cat, horse, and dog allergens. Patch testing indicated a positive response to a house paint additive containing DGBE, as well as to DGBE itself. Workplace monitoring is not available, and thus it is difficult to prove definitively that DGBE was responsible for this particular case of hypersensitivity.

22.5 Standards, Regulations, or Guidelines of Exposure

The German MAK value for DGBE is 100 mg/m^3 (1996), and the Netherlands TWA is 9 ppm (50 mg/m^3, skin notation). An ACGIH TLV for this material has not been established.

22.6 Studies on Environmental Impact

In the case of DGBE, it is estimated that 96% of the material released to the environment is distributed to water and only small additional amounts to air, soil, biota, suspended solids, or sediment (319). Conservatively calculated environmental exposures to DGBE are less than concentrations of concern for aquatic life.

23.0 Diethylene Glycol Mono-isobutyl Ether

23.0.1 CAS Number: [18912-80-6]

23.0.2 Synonyms: 2-[2-(2-Methylpropoxy)ethoxy]ethanol; 2-(2-Isobutoxyethoxy)ethanol; DGIBE

23.0.3 Trade Names: Dowanol® DiB

23.0.4 Molecular Weight: 162.23

23.0.5 Molecular Formula: $C_8H_{18}O_3$

23.0.6 Molecular Structure:

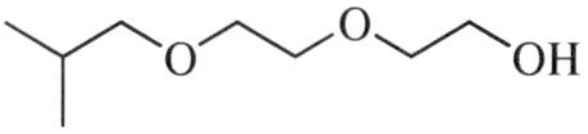

23.1 Chemical and Physical Properties

23.1.1 General

See Table 86.13.

23.2 Production and Use

Diethylene glycol monoisobutyl ether (DGIBE) is not commercially available on an industrial scale at this time.

23.3 Exposure Assessment

23.3.1 Air

See the Introduction for a general description of air sampling and analytical procedures.

23.3.2 Workplace Methods

See the Introduction for a general description of sampling and analytical procedures for the workplace.

23.4 Toxic Effects

DGIBE has low acute oral toxicity; the LD_{50} is in the range of 1600 (321) to 4000 mg/kg (330). DGIBE caused severe eye irritation and corneal damage when instilled into the eye, but it was only mildly irritating to the skin (321). The dermal LD_{50} for rabbits was higher than 4000 mg/kg (321).

No effects were reported after exposure of rats for 7 h to a saturated vapor generated at room temperature, but when generated at 100°C, rats survived but had hemoglobinuria during and for a day following exposure (321). All rats exposed for 6 h to approximately 6700 ppm also survived (374).

23.5 Standards, Regulations, or Guidelines of Exposure

No occupational standard has been established for inhalation exposure to DGIBE.

24.0 Diethylene Glycol Mono-*n*-hexyl Ether

24.0.1 CAS Number: *[112-59-4]*

24.0.2 Synonyms: 2-(2-(2-Hexyloxy)ethoxy)ethanol; *n*-hexyloxyethoxyethanol; hexyl carbitol; DGHE

24.0.3 Trade Names: *n*-Hexyl Carbitol® Solvent

24.0.4 Molecular Weight: 190.28

24.0.5 Molecular Formula: $C_{10}H_{22}O_3$

24.0.6 Molecular Structure:

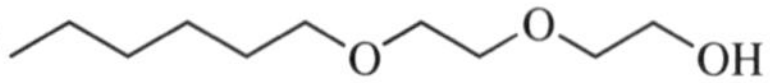

24.1 Chemical and Physical Properties

24.1.1 General

See Table 86.13.

24.2 Production and Use

Refer to Table 86.2 for production and use information.

24.3 Exposure Assessment

24.3.1 Air

See the Introduction for a general description of air sampling and analytical procedures.

24.3.2 Workplace Methods

See the Introduction for a general description of sampling and analytical procedures for the workplace.

24.4 Toxic Effects

Diethylene glycol-mono-*n*-hexyl ether (DGHE) has low acute oral toxicity and low toxicity by skin absorption. This material is severely irritating to the eyes and can cause marked corneal injury that may be slow to heal. It is mildly irritating to the skin by acute contact; however, prolonged or occluded contact can cause severe irritation. Because of its relatively low vapor pressure, vapor inhalation is not anticipated to present a significant hazard at ambient temperatures. DGHE did not produce evidence of genotoxic activity.

24.4.1 Experimental Studies

24.4.1.1 Acute Toxicity. Oral. DGHE has low toxicity by single-dose oral administration to laboratory animals. The LD$_{50}$ value for rats is reportedly 4.92 g/kg (470), 4.92 mL/kg in the male, and 3.73 mL/kg in the female (323).

Inhalation. Exposure to essentially saturated vapor concentrations generated at room temperature for 8 h killed 0/5 rats. No significant ill effects were reported (470). Ballantyne et al. (323) reported that, when groups of five male and five female rats were exposed to essentially saturated vapor at 26°C for 6 h, no signs of toxicity or irritancy were noted during exposure or during the subsequent 14-day postexposure observation period. In addition, no gross pathological features were observed at necropsy.

Dermal. DGHE is slightly toxic when applied to the clipped skin of rabbits under an occlusive dressing for 24 h. Ballantyne et al. (323) reported LD$_{50}$ values in the rabbit of 2.14 mL/kg (males) and 2.37 mL/kg (females). Union Carbide (470) reported an LD$_{50}$ value of 1.5 mL/kg. In an unpublished Union Carbide report (471), New Zealand white rabbits were exposed to DGHE by occluded cutaneous exposure at dose levels of 0, 100, 300, or 1000 mg/kg body weight/day for 9 days (6 h/day) during an 11-day period. Severe irritation was noted in the 1000-mg/kg/day group and mild irritation in the 100-mg/kg/day group. However, there were no treatment-related clinical signs of systemic toxicity.

Eye Irritation. DGHE was severely irritating to rabbit eyes. Ballantyne et al. (323) reported that instillation of 0.1 or 0.005 mL caused severe conjunctivitis and corneal injury to six of six rabbits and required 7 to 21 days for complete corneal healing. Similar findings have also been reported by Union Carbide (470).

Skin Irritation. Mild erythema and edema of about 24 h duration was observed when 0.5 mL of undiluted material was applied under an occlusive dressing to the clipped skin of rabbits for 4 h (470). Thus, DGHE was not considered corrosive under the conditions of this test. However, under more rigorous conditions, 4 mL/kg applied under an occlusive dressing for 24 h led to more severe irritation and persistent erythema, edema, and necrosis.

24.4.1.6 Genetic and Related Cellular Effects Studies. DGHE has been tested in a number of *in vivo* and *in vitro* tests. In an Ames assay, mutagenic activity was not observed in any of the five strains of *S. typhimurium* tested with or without a metabolic activation system (472). In a Chinese hamster ovary cell (CHO) gene mutation test and an *in vitro* sister chromatid exchange (SCE) test in CHO cells, DGHE produced slight increases in gene mutations with and without a rat-liver S9 activation system. However, this effect was not reproducible in duplicate cultures. A linear regression analysis of the CHO dose-response data with S9 indicated that the results represented an essentially significant, although very weak, positive response. In the SCE assay, DGHE did not reproducibly increase the incidence of SCEs with or without an S9 metabolic activation system (473).

DGHE was tested in an *in vivo* mouse micronucleus assay in both male and female Swiss–Webster mice at dosages of 640, 400, and 200 mg/kg. Under conditions of the test, no positive or dose-related increases were produced in the incidence of micronuclei in peripheral blood polychromatic erythrocytes. Hence DGHE was considered inactive as a clastogenic agent (474). An *in vivo* bone marrow chromosomal aberration assay in rats found no reproducible statistically significant or dose-related increases in the incidence of chromosomal aberrations among male and female Sprague–Dawley rats (475).

25.0 Diethylene Glycol Dimethyl Ether

25.0.1 CAS Number: [111-96-6]

25.0.2 Synonyms: Bis(2-methoxyethyl) ether; 1,1'-oxybis(2-methoxyethane); diglyme; 2,5,8-trioxanonane; DYME; DGdiME

25.0.3 Trade Names: Dimethyl Carbitol® Solvent

25.0.4 Molecular Weight: 134.2

25.0.5 Molecular Formula: $C_6H_{14}O_3$

25.0.6 Molecular Structure:

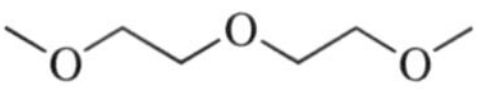

25.1 Chemical and Physical Properties

25.1.2 Odor and Warning Properties

DGdiME is a colorless liquid that has a mild odor.

25.2 Production and Use

DGdiME is produced by the reacting of diethylene glycol with dimethyl sulfate.

25.3 Exposure Assessment

25.3.1 Air

See the Introduction for a general description of air sampling and analytical procedures.

25.3.2 Workplace Methods

See the Introduction for a general description of sampling and analytical procedures for the workplace.

25.4 Toxic Effects

Toxicological studies of diethylene glycol dimethyl ether (DGdiME) indicate that it has moderate single-dose oral toxicity. There are currently no published reports of the material's potential to produce eye or skin irritation. Repeated-dose inhalation studies with laboratory animals demonstrated that the material can cause sperm head abnormalities and testicular atrophy. In addition, the material has caused teratogenicity in laboratory mice and rabbits. This material did not produce evidence of genotoxicity in a series of *in vitro* and *in vivo* assays.

25.4.1 Experimental Studies

25.4.1.1 Acute Toxicity. DGdiME is moderately toxic by single-dose oral toxicity. The LD$_{50}$ for mice is reportedly 4.1 g/kg (349).

Gage (231) reported an inhalation study in which rats exposed to 600 ppm DGdiME vapor experienced irregular weight gain and thymic atrophy. The reported no-observable effect level was 200 ppm. In an *in vivo* mouse sperm abnormality test, groups of 10 male B6C3F1 mice were exposed for 7 h/day to vapor concentrations of 250 ppm for 5 days or 1000 ppm for 4 days. Because of toxicity in the high-exposure group, exposure was reduced by 1 day. The mice were maintained for 35 days without further exposure, then sacrificed and examined. Significant increases in sperm head abnormalities were observed in the highest exposure group, but no significant exposure-related effects were observed in the low-exposure group (113, 114).

25.4.1.2 Chronic and Subchronic Toxicity. Cheever et al. (476) dosed male Sprague–Dawley rats orally with 5.1 mmol DGdiME/kg (684 mg/kg) or 5 mL/kg of distilled water for up to 20 consecutive days. Groups of five animals were sacrificed at 2-day intervals during dosing and at weekly intervals for 8 weeks after exposure stopped. Significant reduction in body weight gain was observed after 18 doses, and testes weights were reduced after 10 daily doses. The relative weights of the testes, epididymides, and thymus glands were reduced after 20 daily doses. In addition, LDH-X (lactate dehydrogenase-X), a pachytene spermatocyte marker enzyme, was significantly reduced by the eighteenth daily exposure. The relative weights of the testes, epididymides, and thymus glands remained significantly reduced in the group dosed for 20 days followed by an 8-week recovery period. In a second study, these same investigators showed that DGdiME is metabolized to a major urinary metabolite, (2-methoxyethoxy) acetic acid (70% of the dose), and a minor metabolite, methoxyacetic acid. Other studies demonstrated no testicular damage when the major metabolite was administered by gavage to rats at a dosage of 5.1 mmole/kg for 20 days (422). Therefore, the authors concluded that metabolic conversion of DGdiME to methoxyacetic acid, the minor metabolite and a known testicular toxicant, accounted for the adverse effects observed.

25.4.1.3 Pharmacokinetics, Metabolism, and Mechanisms. Richards et al. (477) reported on the comparative metabolism of DGdiME using either cultured rat hepatocytes (*in vitro*) or whole animals (*in vivo*). Male Sprague–Dawley rats were given oral doses of 5.1 mmole [^{14}C]-DGdiME/kg, and urine was collected for 96 h and analyzed for metabolites by high-performance liquid chromatography. A spectrum of metabolites was identified in which diglycolic acid (3.9%), methoxyacetic acid (MAA, 6.2%), and 2-(methoxyethoxy)acetic acid (MEAA, 68%) predominated. A similar proportion of these same metabolites formed in hepatocyte cultures that led these researchers to conclude that the *in vitro* system is a good model for predicting the urinary metabolites of DGdiME. In addition, it was found that cultured hepatocytes from rats pretreated with ethanol caused a marked increase overall in the *in vitro* metabolism. The amounts of the teratogenic MAA formed in cultures of hepatocytes from ethanol pretreated rats increased two to four times compared to untreated rats.

The comparative metabolism of DGdiME by rat and human hepatic microsomes has been reported by Tirmenstein (478). Microsomes from phenobarbital-pretreated or ethanol-pretreated rats exhibited an increased capacity to cleave the central ether bond of DGdiME to yield 2-methoxyethanol (EGME). This increase was not observed if incubations contained the P450IIE1 inhibitor isoniazid. Pretreatment of rats with DGdiME caused significantly increased P450 levels and an almost 30-fold increase in P450IIB1/2 activity (pentoxyresorufin dealkylase activity). Human hepatic microsomes also catalyzed the ether cleavage to EGME. Formation of EGME in human microsomes correlated with P450IIE1 (aniline hydroxylase) activity. Thus, hepatic P450IIE1 is implicated in both rats and humans in the central ether cleavage of DGdiME to EGME.

25.4.1.4 Reproductive and Developmental Studies. *Reproductive Toxicity.* Lee et al. (479) exposed male rats to vapor concentrations of 0, 110, 370, or 1100 ppm for 6 h/day, 5 days/week for 2 weeks and then sacrificed at 10, 14, 42, or 84 days postexposure. They found adverse effects on spermatogenesis at all exposure concentrations. Complete resolution of the injury occurred over the 84-day recovery period for concentrations of 110 and 370 ppm. Only partial recovery was noted after 84 days for the high-exposure group. Gross lesions and adverse effects on organ weights were observed. However, hematology, urinalysis, and clinical chemistry results were not reported (480).

Developmental Toxicity. Pregnant CD-1 mice were treated by gavage at a dosage of 3000 mg/kg/day from day 7 to 14 of gestation. Maternal mortality was nearly 25% in the treated mice, and none of the surviving mice delivered any viable offspring (349). Similar results were also reported by Schuler et al. (361). Price et al. (481) treated timed-pregnant CD-1 mice by gavage at dosages of 0 (control), 62.5, 125, 250, or 500 mg/kg/day (dose volume of 10 mL/kg) from days 6 to 15 of gestation. The mice were sacrificed and examined on day 17 of gestation. Reduced maternal weight gain occurred at doses > 250 mg/kg/day. Average fetal body weight/litter was reduced at doses $\geq$125 mg/kg/day. The percentage of postimplantation loss and the percentage of malformed live fetuses/litter also increased at doses $\geq$250 mg/kg/day. The primary malformations observed included defects of the neural tube, limbs, digits, cardiovascular system, urogenital organs, and skeleton. No adverse developmental effects were observed at the 62.5-mg/kg dose level. Hardin and Eisenmann (351) treated time-mated CD-1 mice with an oral dose of 4 mmol/kg (537 mg/kg) on day 11 (plug = 0) of gestation. The mice were sacrificed and examined on day 18 of gestation. Paw defects were observed in 77% of the litters (39.7% of fetuses), and hind-paw defects were more common than forepaw defects. Syndactyly, oligodactyly, and short digits were the most common malformations observed. No other treatment-related gross external malformations were observed. Schwetz et al. (482) treated timed-pregnant New Zealand white rabbits with oral dosages of 0 (control), 25, 50, 100, or 175 mg/kg/day from days 6 to 19 of gestation. At doses $\geq$ 50 mg/kg/day, there was clear evidence of maternal (reduced body weights) and embryo-fetal toxicity. Malformations most frequently observed were in the axial skeleton, kidney, spleen, and cardiovascular system. No adverse effects were observed at 25 mg/kg/day.

Groups of pregnant CD rats were exposed by inhalation to DGdiME at 0 (room air), 25, 100, or 400 ppm for 6 h/day on days 7 through 16 of gestation (483). Female rats were sacrificed on gestation day 21, and fetuses were examined. Maternal toxicity was evident as depressed food consumption at 400 ppm and increased liver weights at 100 ppm. There were no fetuses delivered from dams treated at 400 ppm. At lower concentrations, embryo-fetal toxicity was expressed as decreased fetal weights at 100 ppm. There were low incidences of structural malformations (primarily delayed skeletal ossification and rudimentary ribs) at 25 and 100 ppm. At 25 ppm, fetal defects detected were not significantly different from controls. The NOEL for maternal toxicity was 25 ppm. Although it was felt that 25 ppm produced a minimal embryo-fetal response, it was not possible to establish this concentration as the embryo-fetal NOEL.

Drosophila were exposed in culture vials throughout development to a range of concentrations of DGdiME, MEAA (the principal urinary metabolite of DGdiME), or EGME, and developmental toxicity was assayed by examining for bent humeral bristles in emerging offspring (484). MEAA had no effect. DGdiME produced effects at the two highest concentrations of 0.66 and 1.31 mM. Effects of EGME were observed at 0.33 mM. These findings parallel *in vivo* findings and suggest the usefulness of this assay as a screen for developmental toxicants.

25.4.1.6 Genetic and Related Cellular Effects Studies. *Genetic Toxicity.* DGdiME has been tested in a number of *in vivo* and *in vitro* genetic toxicity tests. In a *Salmonella/* microsome bacterial mutagenicity assay (Ames test), DGdiME was tested with or without Aroclor 1254-induced rat and hamster metabolic activation in triplicate at concentrations up to 10 mg/plate. Mutagenic activity was not observed in any of the six bacterial strains tested at any concentration with or without metabolic activation (485). McGregor et al. (114) reported similar findings. DGdiME was also inactive in an unscheduled DNA synthesis test in human embryo fibroblasts with or without Aroclor-induced metabolic activation at doses up to 10 mg/mL (113, 114). In a sex-linked recessive lethal test, diglyme was tested at a single concentration of 250 ppm for 2.75 h. Results were equivocal (113, 114). In a cytogenetic analysis of rat bone marrow cells, groups of 10 male and 10 female rats were exposed to vapor concentrations of 250 or 1000 ppm DGdiME for 7 h/day for either 1 or 5 days. There was no significant evidence of chromosomal damage. Thus the material was not considered a clastogenic agent. When DGdiME was tested for genotoxic activity in a dominant lethal study at concentrations of 250 or 1000 ppm, no dominant lethality was seen at the 250-ppm exposure level. At 1000 ppm, pregnancy frequency was not consistently affected up to week 3 of mating, but pregnancy frequency fell to 10% in weeks 5, 6, and 7 of mating and there was full recovery to normal by week 10. Implantations were notably reduced only in weeks 6 and 7 of mating. The investigators considered that these results indicated a weak dominant lethal effect (113).

25.5 Standards, Regulations, or Guidelines of Exposure

The following occupational exposure standards have been established for DGdiME: DFG MAK (Germany) TWA 5 ppm (28 mg/m^3, skin notation); The Netherlands TWA 5 ppm (27 mg/m^3).

26.0 Diethylene Glycol Divinyl Ether

26.0.1 CAS Number: [764-99-8]

26.0.2 Synonyms: 1,1-[Oxybis(2,1-ethanediyloxy)bis]ethene; 3,6,9-trioxaundeca-1,10-diene; bis[2-(vinyloxy)ethyl] ether; divinylcarbital

26.0.3 Trade Names: Divinyl Carbitol® Solvent

26.0.4 Molecular Weight: 158.1

26.0.5 Molecular Formula: $C_8H_{14}O_3$

26.0.6 Molecular Structure:

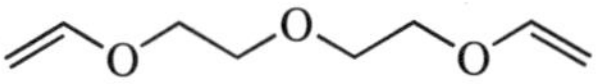

26.1 Chemical and Physical Properties

26.4 Toxic Effects

The single-dose oral LD_{50} of diethylene glycol divinyl ether in rats is 3.73 mL/kg, and the LD_{50} for the mouse is 2.57 mL/kg (486). This material is nonirritating to the eyes and only slightly irritating to the skin. It is not likely to be absorbed through the skin in toxic amounts; the percutaneous LD_{50} value for rabbits is 14.1 mL/kg. Exposure of rats for 8 h to an essentially saturated atmosphere generated at room temperature did not result in deaths (348). Reasonable care and caution in handling should prevent any potential adverse effects from this material. In addition it is prudent to avoid repeated prolonged exposure to vapor, mist, or fog, especially if the compound is being handled under heated or agitated conditions. This material is no longer commercially produced.

27.0 Diethylene Glycol Ethyl Vinyl Ether

27.0.1 CAS Number: [10143-53-0]

27.0.2 Synonyms: 2-(2-(2-Ethoxyethoxy)ethoxyl)ethene; 1-(2-ethoxyethoxy)-2-vinyl-oxyethane

27.0.3 Trade Names: Vinyl Ethyl Carbitol® Solvent

27.0.4 Molecular Weight: 160.2

27.0.5 Molecular Formula: $C_8H_{16}O_3$

27.0.6 Molecular Structure:

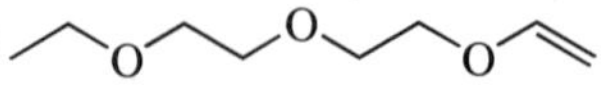

27.1 Chemical and Physical Properties

27.1.1 General

Diethylene glycol ethyl vinyl ether is miscible with alcohol and is soluble in many solvents. Its maximum solubility in water is about 9.4% at 20°C. Additional chemical and physical properties are given in Table 86.13.

ETHERS OF ETHYLENE GLYCOL AND DERIVATIVES

27.2 Production and Use

This material is no longer commercially produced.

27.4 Toxic Effects

Diethylene glycol ethyl vinyl ether has low single-dose oral toxicity; the LD$_{50}$ for rats is 11.3 mL/kg. Narcosis was observed in treated animals. This material is essentially nonirritating to the eyes. It is rated 1 on a scale of 10 in rabbit eye tests. Unoccluded contact for 4 h with 0.5 mL was only mildly irritating to the skin (rated 2 on a scale of 10) in a rabbit irritation test. Occluded 24-h skin contact demonstrated low toxicity; the LD$_{50}$ value for rabbits is 8.41 mL/kg. Necrosis of the skin at the site of application was observed in the 24-h occluded challenge. No mortality resulted from an 8-h exposure of rats to essentially saturated vapors of diethylene glycol ethyl vinyl ether (348, 487).

27.5 Standards, Regulations, or Guidelines of Exposure

Good hygienic practices and avoiding prolonged or repeated skin contact should be adequate to avoid health problems when the material is handled under anticipated industrial conditions. No occupational exposure standard has been established for this compound.

28.0 Diethylene Glycol Monomethylpentyl Ether

28.0.1 CAS Number: [10143-56-3]

28.0.2 Synonyms: 2-[2-(2-Methylpentyl)oxy]ethoxy ethanol; diethylene glycol mono-2-methylpentyl ether

28.0.3 Trade Names: 2-Methylpentyl Carbitol® Solvent

28.0.4 Molecular Weight: 190.32

28.0.5 Molecular Formula: C$_{10}$H$_{22}$O$_3$

28.0.6 Molecular Structure: HO~~~O~~~O~~~

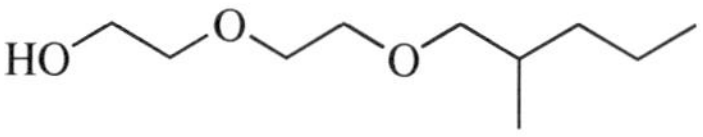

28.1 Chemical and Physical Properties

28.1.1 General

28.2 Production and Use

This material is no longer commercially produced.

28.4 Toxic Effects

The single-dose oral toxicity of diethylene glycol monomethylpentyl ether is considered low; the LD$_{50}$ value for rats is 5.66 g/kg. This material is severely irritating and injurious to the eyes of rabbits. It is rated 6 on a scale of 10 but is only very slightly irritating to the skin. This material has a moderate to low order of toxicity by absorption through the skin; the LD$_{50}$ value for a 24-h covered skin application in rabbits is 1.58 mL/kg. Thus it may be absorbed through the skin, but it is unlikely to pose a serious health hazard from absorption unless skin exposure is widespread and prolonged. In a test for acute lethal toxicity, no mortality occurred when six rats were exposed for 8 h to essentially saturated vapor (348). The primary health hazard in handling this material results from contamination of the eyes. Therefore, precautions should be taken to prevent eye contact. It is also prudent to avoid prolonged or repeated exposure to vapor, mist, or fog.

29.0 Triethylene Glycol Monomethyl Ether

29.0.1 CAS Number: *[112-35-6]*

29.0.2 Synonyms: 2-[2-(2-Methoxyethoxy)ethoxy]ethanol; methyltriglycol; triglycol monomethyl ether; Methoxy triglycol

29.0.3 Trade Names: Dowanol® TM Glycol Ether; Poly-Solv® TM

29.0.4 Molecular Weight: 164.20

29.0.5 Molecular Formula: $C_7H_{16}O_4$

29.0.6 Molecular Structure: HO~~~O~~~O~~~O~

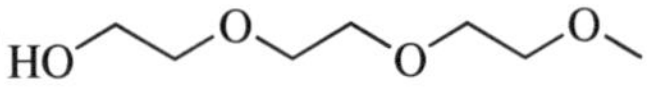

29.1 Chemical and Physical Properties

29.1.1 General

See Table 86.13 for chemical and physical properties.

29.2 Production and Use

Triethylene glycol monomethyl ether (TGME) is used as a coupling solvent (water-based paints), brake/hydraulic fluid blending component, plasticizer/wetting agent (resins, paints), and solvent (488). See Table 86.2 for additional production and use information.

29.4 Toxic Effects

Triethylene glycol monomethyl ether (TGME) has low oral toxicity and is not injurious to the eyes and skin. It is slowly absorbed through skin. Excessive exposure involving large areas of skin for extended periods of time would be required before serious effects would be expected. Hazardous amounts are not likely to be inhaled under ordinary conditions because of its low vapor pressure. In animals, chronic oral exposure involving 4 g/kg/day for 90 days resulted in changes in liver and testes. At maternally toxic oral doses, TGME did not produce malformations and was not selectively toxic to the developing conceptus. TGME is neither mutagenic nor neurotoxic.

29.4.1 Experimental Studies

29.4.1.1 Acute Toxicity. Oral. Smyth et al. (348) found that TGME has low single-dose oral toxicity; the LD$_{50}$ for rats is 11.8 g/kg.

Male Sprague–Dawley CD rats were given TGME for 14 days via drinking water at measured doses of 0, 0.75, 1.6, 3.9, and 8.0 g TGME/kg/day by Gill and Hurley (489). Based on the results of the study, TGME produces severe toxicity at doses of approximately 8 g/kg/day and more and is mildly to moderately toxic at doses of approximately 4 g/kg/day. Under the conditions of the study, the NOEL for subacute toxicity of TGME was 1.6 g/kg/day.

Injection. The LD$_{50}$ value for rats injected intravenously is 8.1 g/kg and 7.4 g/kg when injected intraperitoneally (348).

Inhalation. Exposure of rats to essentially saturated vapors for 8 h caused no significant effects. Repeated inhalation studies have not been reported (348).

Dermal. TGME has low toxicity by absorption through the skin; the LD$_{50}$ value for rabbits was 7.4 g/kg (348). Leber et al. (20) reported that 1000 mg/kg TGME produced no testicular, kidney, or hematologic effects in rabbits when administered topically to the shaved backs (15-cm strip) of female and male rabbits daily for 5 days/week for 3 weeks. Repeated skin applications of undiluted material produced signs of slight irritation.

Skin and Eye Irritation. TGME is not injurious to the eyes or skin (348).

29.4.1.2 Chronic and Subchronic Toxicity. Subchronic Toxicity/Oral. In an adult rat study (490–492), TGME was administered continuously for 90 days in drinking water to male and female rats at target doses of 0, 0.4, 1.2, and 4.0 g/kg/day. Treatment did not result in overt clinical signs of toxicity or changes in behavior or motor activity. The results of this neurotoxicity study are discussed more thoroughly in Section 29.4.1.7. Histological examination of nervous tissue revealed no gross or microscopic lesions attributable to TGME, whereas the testes of 12 of 15 males in the high-dose group were affected. The testicular results are presented in Section 29.4.1.4. Increased liver weight and hepatocellular vacuolation and/or hypertrophy were observed in the mid- and high-dose groups. Thus TGME produces moderate toxicity at 4.0 g/kg/day and minimal to mild toxicity at 1.2 g/kg/day.

Dermal. In a 13-week dermal toxicity study (492–494), TGME was applied undiluted under an occluded wrap for 6 h/day, 5 days/week to the shaved backs of adult male and female Sprague–Dawley rats at dose levels of 0, 400, 1200, or 4000 mg/kg/day. There were no clearly defined indications of hematologic or reproductive toxicity following 13 weeks of treatment. The highest dose level tested, 4000 mg/kg body weight/day, was considered the NOEL for systemic toxicity.

29.4.1.3 Pharmacokinetics, Metabolism, and Mechanisms. Metabolism. No information is available on the metabolism of TGME. However, metabolism studies that involved higher molecular weight polyethylene glycol ethers indicate that the ether linkages are metabolized and the terminal alcohols oxidized to carboxylic acids by the rat (420, 421). In a more recent study by Cheever et al. (422) that involved bis (2-methoxyethyl) ether in the adult male rat, two metabolites, 2-methoxyethanol and 2-methoxyacetic acid, were the agents responsible for the male reproductive effects. These two metabolites are believed to be formed to some extent from diethylene glycol monomethyl ether (DGME) and to a lesser extent from TGME. The principal metabolite of DGME is 2-(2-methoxyethoxy) acetic acid, whereas the principal metabolite of TGME is believed to be 2-[2-(2-methoxyethoxy) ethoxy] acetic acid.

29.4.1.4 Reproductive and Developmental. *Reproductive.* TGME produced testicular effects in a 90-day drinking water study at a dose of 4 g/kg/day (490, 491). Degeneration and/or atrophy of the seminiferous tubules (spermatocytes or developing spermatids) were observed for most males in the high-dose group. Treatment related lesions were not found in the low- or mid-dose group. These testicular effects were not seen when 4 g/kg/day TGME was applied dermally to the backs of rats under an occlusive wrap for 6 h/day, 5 days/week during a 13-week period (492–494).

Developmental. The results of a developmental toxicity screening test in rats was reported by Leber et al. (20). Daily doses of 250 and 1000 mg/kg of TGME were administered by gavage to pregnant rats on days 7 to 16 of gestation. No significant changes in maternal body weights were observed. No adverse effects were noted in the *in utero* development of the conceptus or on the viability or postpartum development of offspring.

In a definitive developmental toxicity study, pregnant CD rats (495) were dosed with TGME via gavage at doses of 0, 625, 1250, 2500, or 5000 mg/kg/day on days 6 to 15 of gestation. No treatment-related effects were seen at 625 mg/kg/day. Doses as high as 5000 mg/kg/day did not significantly affect pregnancy rate, implantations, corpora lutea, live fetuses, or fetal sex ratios, produce malformations, or increase the incidence of external or internal soft tissue variations. The resorption rate increased slightly, but significantly, at the 5000-mg/kg/day dose level. Fetal body weights were significantly reduced at the 2500- and 5000-mg/kg/day dose levels and were slightly reduced at 1250 mg/kg. Doses of 1250 mg/kg/day and higher increased the incidence of skeletal variation. The NOAEL for maternal and developmental toxicity was 625 mg/kg/day; 1250 mg/kg was a NOAEL for maternal toxicity and may be very near the NOAEL for developmental toxicity. TGME was not selectively toxic to the developing rat conceptus.

An additional study was conducted using pregnant New Zealand white rabbits and was reported by Hoberman et al. (495). TGME was administered by gavage at doses of 0, 250, 500, 1000, or 1500 mg/kg/day on days 6 to 18 of gestation to determine potential effects on development. The high dose, 1500 mg/kg/day, did not significantly affect pregnancy rate, implantations, corpora lutea, live fetuses or resorptions, fetal sex ratios, fetal body weights, produce malformations, or increase the incidence of external or internal soft tissue variations.

However, the high dose caused maternal death, abortions, clinical signs of treatment, gross gastrointestinal tract lesions, and reduced gravid uterine weight. One death occurred in the 1000-mg/kg/day group on day 18 of gestation. No treatment-related clinical signs were seen in the other doses in this group. Doses of 500 mg/kg/day and higher increased maternal body weight gain and feed consumption during the postdose period. The NOEL and NOAEL for maternal toxicity were 250 mg/kg/day and 500 mg/kg/day, respectively, and for developmental toxicity, the NOEL and NOAEL were 1000 mg/kg/day and 1500 mg/kg/day, respectively. TGME was not selectively toxic to the developing rabbit conceptus.

Developmental Neurotoxicity. In a developmental neurotoxicity study conducted pursuant to an EPA Test Rule and sponsored by the CMA Glycol Ethers Panel, Bates (496) administered TGME by gavage to pregnant and lactating Sprague–Dawley rats (gestational day 6 through postnatal day 21) at dose levels of 0, 300, 1650, or 3000 mg/kg/day. No overt signs of maternal toxicity were observed, and no discernible effects were noted in the offspring on motor activity, active avoidance, or neuropathology at exposures up to 3000 mg/kg/day. Exposure to TGME at 3000 mg/kg/day resulted in increases in auditory startle amplitude and decreases in latency to maximum startle, but no changes in the habituation process evaluated in the auditory startle test. According to the author, the significance of these auditory startle observations for the health of the animal is unclear. Developmental toxicity was produced in the offspring at TGME levels of 1650 and 3000 mg/kg/day, as evidenced by a decrease in postnatal weight gains. The NOEL for the study was 300 mg/kg/day or more.

A summary of the critical studies, species, NOAELs, and BMDs for TGME is shown in Table 86.17 (426).

29.4.1.6 Genetic and Related Cellular Effects Studies. TGME was evaluated in the Ames *Salmonella* mutagenicity assay (497) using tester strains TA98, TA100, TA1535, and TA1537. The test was conducted with and without an S-9 activation system. TGME did not induce a change in the frequency of revertants to histidine independence. In a mouse bone marrow micronucleus test (498), single oral doses of 0, 500, 1667, and 5000 mg/kg of body weight of TGME were orally administered. The ratio of polychromatic erythrocytes (PCE) to normochromatic erythrocytes (NCE) observed in the treated groups were not significantly different from those of the controls. TGME was judged negative in the test. The Chinese hamster ovary test (499) was used to determine the potential of TGME to induce gene mutations at the HGPRT locus in cultured cells. The mutational frequencies in cultures treated with TGME with and without S-9 were not significantly different from negative control values. TGME was judged nonmutagenic in the CHO/HGPRT assay.

29.4.1.7 Other: Neurological, Pulmonary, Skin Sensitization. The results of the 90-day subchronic neurotoxicity drinking water study (490, 491) are reported in this section. Evaluations included functional observational battery (FOB), motor activity (MA), body weight, food and water consumption, clinical signs of toxicity, mortality, gross lesions, organ weights, and microscopic diagnoses of liver, testes, and the nervous system. TGME did not cause clinical signs of toxicity, FOB alterations, or pathological lesions in the nervous system. Decreases in body weight and food consumption were observed in the mid- and high-dose groups, and minor decreases in MA were observed in the high-dose group during the latter half of the study. MA changes were not considered neurotoxicologically significant. The authors concluded that TGME does not produce neurotoxicity at doses as high as 4.0 g/kg/day.

29.4.2 Human Experience

29.4.2.2 Clinical Cases. Sprague (500) reported an incident that involved two soldiers who drank brake fluid containing TGME in place of alcoholic beverages. The soldiers became ill and were admitted to a hospital for gastric lavage. Blood dialysis was performed on one soldier who was still symptomatic after gastric lavage. Recovery from symptoms occurred rapidly during treatment.

29.4.2.2.3 Pharmacokinetics, Metabolism, and Mechanisms. Dermal Absorption. The penetration of triethylene glycol monomethyl ether through human skin was studied *in vitro* using the epidermis from human abdominal skin (20). An area of 2.54 cm^2 was exposed in a glass diffusion apparatus for treatment. The diffusion rate for TGME is 34 µg/cm^2/hr. The penetration rate was 65 times less than that of EGME. The skin damage ratio for TGME was comparable to that for EGME.

29.5 Standards, Regulations, or Guidelines of Exposure

No occupational exposure standards have been proposed or established for TGME.

30.0 Triethylene Glycol Monoethyl Ether

30.0.1 **CAS Number:** *[112-50-5]*

30.0.2 **Synonyms:** 2-[2-(2-Ethoxyethoxy)ethoxy] ethanol, ethoxy triglycol; TGME

30.0.3 **Trade Names:** Ethoxytriglycol; Dowanol® TE; Poly-Solv® TE

30.0.4 **Molecular Weight:** 178.23

30.0.5 **Molecular Formula:** $C_8H_{18}O_4$

30.0.6 **Molecular Structure:**

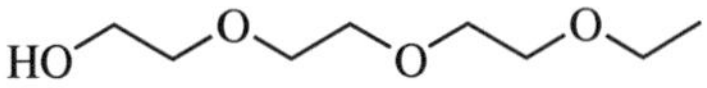

30.1 Chemical and Physical Properties

30.1.1 General

See Table 86.13 for chemical and physical properties.

30.2 Production and Use

TGME is used as a solvent for lacquers, paints, hydraulic brake fluids, printing inks, and chemical specialties (488). Refer to Table 86.2 for additional production and use information.

ETHERS OF ETHYLENE GLYCOL AND DERIVATIVES

30.4 Toxic Effects

Triethylene glycol monoethyl ether (TGEE) has a low order of toxicity by the oral, dermal, and inhalation routes of exposure. TGEE is slightly irritating to the eyes but is not considered a skin irritant. However, a reversible mild irritation may occur from repeated exposure of the skin. TGEE was not a potential developmental toxicant when evaluated by gavage in rats in a developmental toxicity screening assay.

30.4.1 Experimental Studies

30.4.1.1 Acute Toxicity. Oral. Smyth et al. (154, 415) reported that the oral LD_{50} is 10.6 g/kg for rats. An oral LD_{50} of 8.5 g/kg for the rat has been reported (501). Smyth and co-workers (154, 415) also administered TGEE to rats in drinking water (estimated daily intakes were 0.18 to 3.30 g/kg) for 30 days. The NOEL for this study was reportedly 0.75 g/kg/day. However, the nature of effects at the higher dose level was not disclosed.

Inhalation. Rats were exposed via inhalation to a nominal concentration of 200 mg/L of TGEE for 1 h (501). All animals survived the exposure, and the necropsy of the animals appeared normal. In another inhalation study, six male rats were exposed at a nominal concentration of 10.17 mg/L for 7 h. Air was metered through TGEE at 100°C and into the exposure chamber at a rate of 1.0 L/min. Mortality was not observed in any animal (502).

Dermal. TGEE can be absorbed through the skin in toxic concentrations. The dermal LD_{50} was reportedly 8 mL/kg (8.2 g/kg) for rabbits (415).

Eye Irritation. TGEE caused mild irritation to the eyes of rabbits (415). TGEE was not judged an eye irritant when tested in rabbits (501).

Skin Irritation. When TGEE was tested on the skin of rabbits, it was not a skin irritant (501).

30.4.1.2 Chronic and Subchronic Toxicity. Drinking Water. The subchronic toxicity of TGEE in drinking water was evaluated in 50 male rats for 30 days at concentrations of 0, 0.12, 0.5, 2, or 8% (0, 0.18, 0.75, 3.30, and 3.29 g/kg/day). Mortality occurred in all 10 rats at the high dose. Clinical observations included lowered water consumption in the high-dose group and lowered body weight gain in the 2% group. Necropsy revealed liver and kidney injury in the high-dose group, high blood urea concentrations in 4 of 10 rats in the 2% group, and kidney damage and liver abnormalities in one of six rats, respectively, in the 2% group. No abnormalities were detected in rats in the 0.5 and 0.12% groups.

Dermal. TGEE was administered dermally to rabbits (five males and five females) at 1000 mg/kg, 5 days/week for 3 weeks (20). The results indicated very slight erythema and edema; by day 21 all animals were clear of edema. All other parameters monitored, including macroscopic and microscopic pathology, were considered within normal variations. These data indicate that TGEE does not share toxicity profiles similar to those of methyl and ethyl ethers of ethylene glycols.

30.4.1.3 Pharmacokinetics, Metabolism, and Mechanisms. The main metabolic pathway is oxidation via alcohol dehydrogenase that leads to the formation of an alkoxy acid. A second important route of metabolism is oxidation by microsomal P-450 mixed-function oxidases (O-dealkylation) that lead to the formation of triethylene glycol (TEG). TEG may be oxidized to a carboxylic acid.

30.4.1.4 Reproductive and Developmental. *Developmental Toxicity/Oral.* TGEE's potential to induce developmental toxicity was examined in a rat screening study (503, 20). Pregnant rats were given TGEE by gavage at dose levels of 0, 250, or 1000 mg/kg body weight on days 7 through 16 of gestation. No adverse effects on the litters, maternal body weight, or clinical condition resulted from oral administration of TGEE at the doses tested. The NOEL for oral administration under the conditions of this study was 1000 mg/kg body weight for rats. Kimmel (426) reviewed the study of Leber et al. (20) and considered the data inadequate to derive an NOAEL (see Table 86.17) for developmental toxicity.

30.4.2 Human Experience

30.4.2.1 General Information. No adverse human experiences have been reported for TGEE.

30.4.2.2.3 Pharmacokinetics, Metabolism, and Mechanisms. *In Vitro Testing.* The absorption rate of TGEE was determined in an *in vitro* procedure designed to determine the ability of materials to penetrate human skin. Under the conditions of the study, the mean rate of absorption of TGEE through the skin was 24.1 µg/cm^2/h (SD±0.91, n = 5) (504). The rate of absorption reached a steady state after 1 to 2 h.

30.5 Standards, Regulations, or Guidelines of Exposure

No occupational exposure standards or regulations have been established for TGEE.

31.0 Triethylene Glycol Mono-*n*-butyl Ether

31.0.1 CAS Number: [143-22-6]

31.0.2 Synonyms: 2-(2-(2-Butoxyethoxy) ethoxy) ethanol; butoxytriglycol; TGBE

31.0.3 Trade Names: Poly-Solve® TB

31.0.4 Molecular Weight: 206.28

31.0.5 Molecular Formula: C$_{10}$H$_{22}$O$_4$

31.0.6 Molecular Structure: HO~~O~~O~~O~~

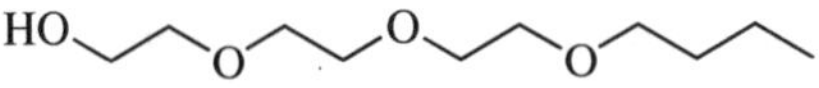

31.1 Chemical and Physical Properties

31.1.1 General

See Table 86.13.

ETHERS OF ETHYLENE GLYCOL AND DERIVATIVES

31.2 Production and Use

Triethylene glycol mono-n-butyl ether (TGBE) is used primarily as a brake fluid component, solvent, and a plasticizer intermediate (505). Refer to Table 86.2 for additional production and use information.

31.3 Exposure Assessment

No occupational standards have been established for exposure to TGBE.

31.4 Toxic Effects

TGBE has a low order of acute toxicity by oral and dermal routes and was not toxic by the inhalation route, possibly due in part to the low vapor concentration that could be generated. TGBE may cause marked eye irritation. The dermal route is the most likely route for human exposure because of the use of TGBE in brake fluids. TGBE showed no evidence of systemic toxicity in a 3-week dermal study in rabbits but did cause skin irritation at the site of repeated application. TGBE showed no evidence for potential developmental toxicity in a screening assay in rats after oral administration. The data indicate little to no potential for hazard to humans exposed to TGBE or TGBE-containing products.

31.4.1 Experimental Studies

31.4.1.1 Acute Toxicity. *Oral.* The acute oral LD$_{50}$ in rats was 6700 mg/kg (348). In more recent studies, the acute oral LD$_{50}$ in rats was 5300 mg/kg (506); clinical signs at high doses included loss of righting reflex, flaccid muscles, and coma.

Inhalation. A single 8-h exposure of rats to essentially saturated vapors (348) or for 1 h to 200 ppm (nominal concentration) (506) caused no significant adverse effects.

Dermal. The dermal LD$_{50}$ in rabbits was 3540 mg/kg (348). No effects were reported after 24 h dermal exposure of rabbits to 2000 mg/kg (507). The systemic toxicity of TGBE has been determined after repeated occluded dermal application to male and female rabbits at 1000 mg/kg, 6 h/day, 5 days/week for 3 weeks in a "limit test" (20). There was no evidence of any systemic toxicity, but TGBE did cause skin irritation, beginning in the second week of application, which progressed to include desquamation and fissuring in some rabbits.

Skin and Eye Irritation. Acute dermal exposure was not irritating, but TGBE was a marked eye irritant in the rabbit (506).

31.4.1.4 Reproductive and Developmental. The potential for developmental toxicity of TGBE has been evaluated in an *in vivo* developmental toxicity screening assay. TGBE was administered daily by gavage to pregnant rats on gestational days 7 to 16 at either 250 or 1000 mg/kg; the dams were allowed to deliver and rear the young to postpartum day 5, and maternal and neonatal parameters were determined (20). There were no changes in any parameter after treatment with TGBE, although under these conditions the methyl ether of ethylene glycol (EGME) administered at a lower dose (50 mg/kg/day) as a positive control was clearly developmentally toxic. The data indicate that TGBE has low to no potential for embryo-fetal toxicity or teratogenicity.

31.4.2 Human Experience

There were no reports of adverse health effects from normal use of TGBE or TGBE-containing products. The rate of dermal penetration of undiluted TGBE through whole human skin has been determined *in vitro* (20). The rate of penetration was 22 µg/cm^2/h, about 100 times less than that of EGME. The low rate of dermal penetration, the low vapor pressure, plus the low intrinsic toxicity of TGBE, indicate little to no potential for hazard to humans exposed to TGBE or TGBE-containing products.

31.5 Standards, Regulations, or Guidelines of Exposure

No occupational exposure standards have been established for TGBE.

32.0 Tetraethylene Glycol Monovinylethyl Ether

32.0.1 CAS Number: None found

32.0.2 Synonyms: 3,6,9,12,15-Pentaoxaheptadec-1-ene; 2-vinylethoxytetraethylene glycol

32.0.4 Molecular Weight: 248.3

32.0.5 Molecular Formula: $C_{12}H_{24}O_5$

32.0.6 Molecular Structure:

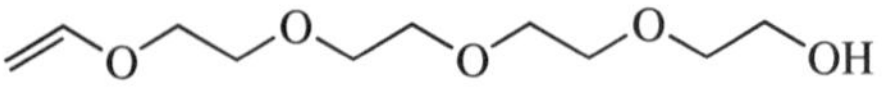

32.1 Chemical and Physical Properties

32.1.1 General

Boiling point: 295.6°C
Freezing point: −21°C

32.4 Toxic Effects

32.4.1 Experimental Studies

32.4.1.1 Acute Toxicity. Tetraethylene glycol monovinylethyl ether has low single-dose oral toxicity; the LD$_{50}$ value for rats is 6.17 mL/kg. It may cause mild eye irritation and only minor skin irritation. It is listed 3 and 2, respectively, on a scale of 10 in rabbit tests. It is not likely to pose a health hazard from skin absorption because the LD$_{50}$ value for rabbits is 6.35 mL/kg. None of the rats exposed for 8 h to the vapors of this material generated at room temperature died. This may be due in part to its low volatility and to its low degree of toxicity (348). There are no data reported on the effects of repeated exposure to vapor.

ETHERS OF ETHYLENE GLYCOL AND DERIVATIVES

32.5 Standards, Regulations, or Guidelines of Exposure

Reasonable safety precautions should be adequate in most industrial operations. No occupational exposure standard has been established for this material.

33.0 Tetraethylene Glycol Monophenyl Ether

33.0.1 **CAS Number:** *[36366-93-5]*

33.0.2 **Synonyms:** 2-(2-(2-(2-Phenoxyethoxy)ethoxy)ethoxy) ethanol

33.0.3 **Trade Names:** Dowanol® T 4Ph; Bellacide 3062 (Ciba-Geigy Corp.)

33.0.4 **Molecular Weight:** 270.2

33.0.5 **Molecular Formula:** $C_{14}H_{22}O_5$

33.0.6 **Molecular Structure:**

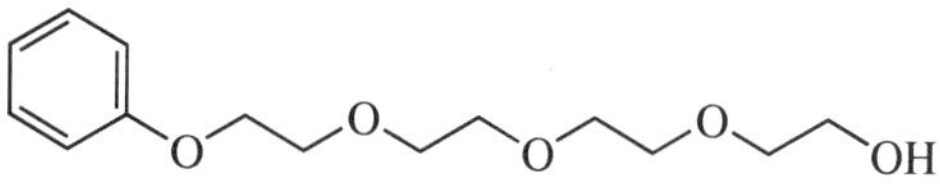

33.1 Chemical and Physical Properties

33.1.1 **General**

Physical state: light yellow liquid

33.4 Toxic Effects

Tetraethylene glycol monophenyl ether has low single-dose oral toxicity; the LD_{50} for rats is approximately 1.26 to 5.0 g/kg. It was not significantly irritating to skin nor did it produce systemic toxicity following skin contact. However, eye contact produced irritation, moderate corneal damage, and some impairment of vision in rabbits. Vapor inhalation studies were not conducted because of the material's very low volatility. Based on these limited data, tetraethylene glycol monophenyl ether should be handled with care to prevent eye contact. However, skin contact and ingestion should not pose a problem in industrial handling (508). The tri-, tetra-, and pentaethyleneglycol ethers of phenol, *p*-chlorophenol, and 2,4,5-trichlorophenol inhibit the growth of *Staphylococcus aureus* 209P and *E. coli*.

34.0 Tetraethylene Glycol Diethyl Ether

34.0.1 **CAS Number:** *[4353-28-0]*

34.0.2 **Synonyms:** 3,6,9,12,15-Pentaoxyheptadecane; diethoxytetraethylene glycol

34.0.4 **Molecular Weight:** 250.38

34.0.5 **Molecular Formula:** $C_{12}H_{26}O_5$

34.0.6 **Molecular Structure:**

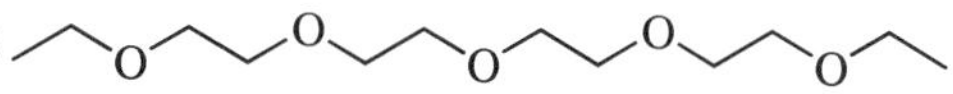

34.1 Chemical and Physical Properties

Physical state: colorless liquid

34.4 Toxic Effects

34.4.1 Experimental Studies

34.4.1.1 Acute Toxicity. Tetraethylene glycol diethyl ether has low single-dose oral toxicity; the LD_{50} for rats is 4.29 mL/kg. The material is only slightly irritating to the skin and eyes and is not likely to be absorbed through the skin in acutely toxic amounts; the percutaneous LD_{50} value for rabbits is 6.35 mL/kg. A single 8-h exposure of rats to essentially saturated vapors resulted in no significant adverse effects (205).

34.5 Standards, Regulations, or Guidelines of Exposure

Tetraethylene glycol diethyl ether should pose no significant health hazard from ordinary industrial handling. No occupational exposure standard has been established for this material.

35.0 Polyethylene Glycol Methyl Ethers

35.0.1 CAS Number: [9004-74-4]

35.0.2 Synonyms: Carbowax; methoxypolyethylene glycols; MPEGS; Poly(ethylene glycol)(350) monomethyl ether; Poly(ethylene glycol) (550) monomethyl ether; Poly(ethylene glycol) (750) monomethyl ether; Poly(ethylene glycol) (1900) monomethyl ether; Poly(ethylene glycol) (5000) monomethyl ether; GLYCOLS 350, 500, 550, 750

35.4 Toxic Effects

Studies of polyethylene glycol methyl ethers whose average molecular weights are 350, 550, and 750 indicate that they have low single-dose oral toxicity; the LD_{50} value for rats are 22 mL/kg, 39.8 mL/kg, and 39.8 mL/kg, respectively. All caused only minor irritation and possibly minor transient corneal injury when tested in the eyes of rabbits. Skin irritation tests using rabbits resulted in no more than a trace of irritation. None of these products was absorbed through the skin to any appreciable extent; the LD_{50} by absorption was higher than 20 mL/kg for all three (509). Because of their very low degree of toxicity and their low volatility, it is reasonable to assume that these products are not likely to present a health hazard from inhalation in anticipated industrial handling.

35.4.1 Experimental Studies

35.4.1.1 Acute Toxicity. Hermansky and Leung (510) conducted cutaneous toxicity studies with methoxy polyethylene glycol 350 (MPEG-350) that contained less than 1.5% EGME, DGEE, and TGME as impurities. Studies were conducted for 14 or 28 days in rats and 9 or 90 days in rabbits. Rats were treated with 1.0 mL/animal of undiluted MPEG-350 for 5 days/week and rabbits with either undiluted or a 50% solution of MPEG-350 using the same exposure regimen. No animals died, but slight body weight decreases were noted for all groups of treated male rats. Slight skin irritation was noted for both species. Slight decreases in absolute testes, spleen, and thymus weights were observed in rats treated for 14 days; similar changes were not noted after 28 days of treatment. A single rat had moderate to high aspermatogenesis and multinucleated spermatids; no microscopic changes were seen in testes of other rats. It was concluded that repeated cutaneous exposure to this material would not be expected to result in significant toxicological effects.

35.4.1.3 Pharmacokinetics, Metabolism, and Mechanisms. *Pharmacokinetics.* Intravenous administration of I^{125}-labeled poly(ethylene glycol) (PEG) of differing molecular weights to mice resulted in differing pharmacokinetics depending on molecular weight (511). Higher molecular weight PEGs were retained longer in blood circulation. The half-life for elimination extended from 18 min to 1 day as the molecular weight of the PEG increased from 6000 to 190,000 Da. PEGs tended to accumulate in muscle, skin, bone, and liver to a higher extent than other organs, regardless of molecular weight. Urinary clearance decreased with increasing molecular weights, whereas liver clearance increased.

36.0 2,2′-[1,4-Phenylenebis(oxy)]bisethanol

36.0.1 CAS Number: *[104-38-1]*

36.0.2 Synonyms: Hydroquinone, bis(2-hydroxyethyl) ether; 2,2′-(1,4-phenylenedioxy)diethanol; hydroquinone di-(beta-hydroxyethyl) ether; *p*-bis(2-hydroxyethoxy)-benzene; 1,4-bis(2-hydroxyethoxy)benzene

36.0.3 Trade Names: NA

36.0.4 Molecular Weight: 198.22

36.0.5 Molecular Formula: $C_{10}H_{14}O_4$

36.0.6 Molecular Structure:

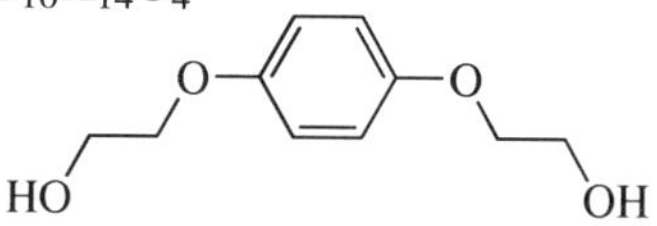

36.1 Chemical and Physical Properties

Specific gravity: 1.15
Boiling point: 185–200 at 0.3 mmHg
Vapor pressure: negligible
Flash point: 224°C (Cleveland open cup)
Autoignition temperature: 468°C

36.4 Toxic Effects

36.4.1 Experimental Studies

36.4.1.1 Acute Toxicity. The acute oral toxicity of 2,2′-[1,4-phenylenebis(oxy)]bisethanol is low; the LD_{50} is in excess of 3.2 g/kg (the highest dose tested) for both rats and mice.

Intraperitoneally, the LD_{50} is in the range of 1.6 to 3.2 mg/kg for these same species (512). Oral administration to male and female rats at levels up to 1% in the diet caused no adverse effects in the females at the highest dose administered (851 mg/kg/day) (513). The only significant compound-related effect reported in the male rats at the highest dose tested (848 mg/kg/day), was a slight decrease in blood platelets (513). The solid compound moistened with water and held in contact with the shaved skin of guinea pigs for 24 h produced only slight irritation. The dermal LD_{50} was in excess of 1.0 g/kg. Skin sensitization tests in guinea pigs were negative (512).

36.5 Standards, Regulations, or Guidelines of Exposure

No occupational exposure standard has been established for 2,2'-[phenylenebis(oxy)] bisethanol.

37.0 Ethylene Glycol Monomethyl Ether Acetate

37.0.1 CAS Number: [110-49-6] and [32718-56-2]

37.0.2 Synonyms: 2-Methoxyethanol acetate; EGMEA

37.0.3 Trade Names: Methyl Cellosolve® Acetate; Glycol Ether EM acetate

37.0.4 Molecular Weight: 118.13

37.0.5 Molecular Formula: $C_5H_{10}O_3$

37.0.6 Molecular Structure:

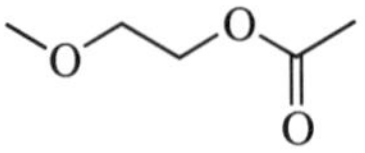

37.1 Chemical and Physical Properties

37.1.1 General

See Table 86.20.

37.3 Exposure Assessment

37.3.3 Workplace Methods

NIOSH Method 1451 is recommended for determining workplace exposure to this chemical (14).

37.4 Toxic Effects

Ethylene glycol monomethyl ether acetate (EGMEA) has low single-dose oral toxicity, is not significantly irritating to the eyes or skin, is poorly absorbed through the skin, and is moderately toxic when inhaled. Its effects are similar to those of ethylene glycol monomethyl ether and are centered in the blood, kidneys, brain, and testes. Ethylene glycol monomethyl ether and its acetate are embryo toxic and teratogenic by all routes tested.

Table 86.20. Physical and Chemical Properties of Some Ether Acetates of Certain Glycols

	Ethylene Glycol			Diethylene Glycol		
Property	Methyl	Ethyl	Butyl	Methyl	Ethyl	Butyl
CAS Number	110-49-6 3271/8-56-2	111-15-9	112-07-2	629-38-9	112-15-2	124-17-4
Molecular formula	$C_5H_{10}O_3$	$C_6H_{12}O_3$	$C_8H_{16}O_3$	$C_7H_{14}O_4$	$C_8H_{16}O_4$	$C_{10}H_{20}O_4$
Molecular weight	118.13	132.16	160.21	162.19	176.21	204.27
Specific gravity 25/4°C	1.007	0.975 at 20°C	0.94	1.04	1.01	0.98
Boiling point (°C at 760 mmHg)	144.5	156.4	192	194.2	210–220	246
Vapor pressure (mmHg) (25°C)	2.0–3.72 torr at 20°C	2.8	0.29	0.12	0.05	<0.01
Refractive index (25°C)	1.402	1.406				
Flash point (°F) (open cup)	60°C	49°C (closed cup)	76°C	180	100°C	105°C
Vapor density (air = 1)	4.1	4.7	5.5		6.07	
Percent in saturated air (25°C)	0.31–0.60	0.21–0.27			0.01	<0.002
1 ppm≈mg/m³ at 25°C, 760 mmHg	4.83	5.40	6.55	6.63	7.20	8.34
1 mg/l≈ppm at 25°C, 760 mmHg	207	185	153	151	139	120

37.4.1 Experimental Studies

37.4.1.1 Acute Toxicity. *Oral.* The LD$_{50}$s of EGMEA of 1.25 g/kg for guinea pigs and 3.93 g/kg for rats reported by Smyth et al. (434) were determined by feeding 50% aqueous solutions. Later, Smyth and Carpenter (415) reported that the LD$_{50}$ for rats is 3.39 g/kg but did not specify in what form it was fed.

Gross (49) reported that rabbits died after receiving three daily doses of either 0.5 or 1.0 mL/kg. All showed kidney injury and had albumin and granular casts in the urine.

Nagano et al. (32) fed mice doses that ranged from 62.5 to 4000 mg/kg, 5 days/week for 5 weeks and observed testicular atrophy and leukopenia, whose intensity was dose-related. The response was essentially equivalent to that observed with ethylene glycol monomethyl ether.

Injection. Browning (59) reported that the lethal dose for mice following subcutaneous injections was 5.04 and 4.63 g/kg. Seven injections of 0.5 mL or four injections of 1.0 mL in guinea pigs caused their deaths 1 to 5 days after treatment. Kidney injury was apparent (49).

Inhalation. Gross (49) reported the results of several inhalation studies that may be summarized as follows: (1) In an essentially saturated atmosphere that contained 22 mg/L (about 4554 ppm), mice and rabbits tolerated single exposures for 3 h and had only irritation of the mucous membranes; guinea pigs survived exposure for 1 h but succumbed days later. (2) Cats died after receiving one 9-h exposure to 2500 ppm and survived one 7-h exposure to 1500 ppm but did show an increase in blood clotting time and changes in the brain. (3) Cats, guinea pigs, rabbits, and mice were given repeated 8-h exposures to 500 or 1000 ppm; at 500 ppm the cats showed slight narcosis and died, but the other species lived. Deaths occurred in all species at 1000 ppm. Kidney injury occurred at both concentrations. (4) Cats tolerated repeated 4- to 6-h exposures to 200 ppm, but decreases in blood pigments and of red cell numbers were noted. Smyth and Carpenter (415) reported that a single 4-h exposure to 7000 ppm killed two of six rats.

Dermal. EGMEA can be absorbed through the skin if exposure is prolonged. Using the "cuff" procedure on rabbits, Smyth and Carpenter (415) estimated that the LD$_{50}$ is 5.29 g/kg.

Eye Irritation. EGMEA is mildly irritating to the eyes (66).

Skin Irritation. EGMEA is not significantly irritating to the skin (49).

37.4.1.2 Chronic and Subchronic Toxicity. Refer to the discussion of EGME in Section 1.

37.4.1.3 Pharmacokinetics, Metabolism, and Mechanisms. *Metabolism and Pharmacokinetics.* Browning (59) was the first to state that this material must be hydrolyzed by the body because it gives rise to toxic signs similar to those caused by ethylene glycol methyl ether. Studies by Stott and McKenna (514) and others showed that EGMEA is hydrolyzed to EGME by carboxylesterases in nasal mucosa, blood, and liver. Enzyme kinetics were determined using 0.4–0.7 mg of protein incubated with 2–115 mM EGMEA. The extent of hydrolysis was determined by measuring EGME levels in the supernatant by gas chromatography. The V_{max} (mmoles/min), K_m (mM), and V_{max}/K_m values were 0.870, 13.4, and 0.065 for nasal mucosa, respectively. The metabolic rate values show that EGMEA is readily metabolized in mouse nasal tissues to produce EGME. EGME is further metabolized by alcohol dehydrogenase to 2-methoxyacetic acid, a known reproductive and developmental toxicant. The metabolism, pharmacokinetics, and mechanism of action of EGME are described in Section 1.

37.4.1.4 Reproductive and Developmental. EGMEA is metabolized to EGME by tissue carboxylesterases. The reproductive and developmental toxicities of EGME are described in Section 1.

37.4.1.6 Genetic and Related Cellular Effects Studies. Zeiger et al. (515) studied the mutagenicity of EGMEA using the Ames tester strains (TA100, TA1535, TA98, TA97, TA1537) with and without S9 activation (rat and hamster liver, AROCLOR 1254) at concentrations of 100–10,000 µg/plate. The results were negative in all cases. These investigators also studied the mutagenicity of the metabolites of EGMEA, EGME and 2-MAA. The results of these and other studies on mutagenicity are discussed in Section 1.

37.4.1.7 Other: Neurological, Pulmonary, Skin Sensitization. EGMEA is metabolized to EGME and 2-MAA in mammalian tissues. Oral dosing of adult male F344 rats with EGME or its principal metabolite 2-MAA resulted in suppressing the primary plaque-forming cell (PFC) response to trinitrophenyl-lipopolysaccharide (TNP-LPS) (35). Oral dosing of rats with EGME or 2-MAA for 10 days (50 to 200 mg/kg/day) resulted in thymic atrophy in the absence of body or spleen weight loss, reduced lymphoproliferative responses to B and T cell mitogens, reductions in interleukin 2 (IL-2) production, and alterations in the primary antibody response to sheep red blood cells (SRBC) and TNP-LPS. A more extensive review of the immunotoxicities of EGME and 2-MAA is given in Section 1.

37.4.2 Human Experience

37.4.2.1 General Information. Jordan and Dahl (516) reported that a 58-year-old woman developed dermatitis on the nose from her eyeglasses. Patch testing revealed that the causative agent could have been ethylene glycol monomethyl ether acetate and/or ethyl acetate. They state that ethylene glycol monomethyl ether acetate should be considered only as a rare cause of allergic dermatitis because this was the first incident recorded in the literature.

37.4.2.2 Clinical Cases. Dermal exposure to EGMEA is a suspected cause of hypospadia in two boys born to a female lacquer laboratory worker who was exposed to EGMEA during pregnancy. The malformations were corrected surgically and by treatment with chlorionic gonadotropin (517).

37.4.2.3 Epidemiology Studies. A number of epidemiology studies were carried out that involved spontaneous abortions among workers who were exposed to glycol ethers in the semiconductor industry. The glycol ethers and their acetates were identified as EGME, EGEE, EGMEA, and EGEEA. A detailed discussion of these studies is presented in Section 1 and 3 as they pertain to EGME and EGEE.

37.5 Standards, Regulations, or Guidelines of Exposure

OSHA proposed a reduction in exposure limits to EGME and EGMEA. The new limits would reduce permissible exposure limits for EGME and EGMEA to 0.1 parts per million (ppm) (518, 519). Worldwide standards for these solvents are presented in Table 86.9 and are as follows: ACGIH TLV TWA 5 ppm (24 mg/m^3); OSHA PEL TWA 25 ppm (120 mg/m^3); NIOSH REL TWA 0.1 ppm (0.5 mg/m^3); DFG MAG (Germany) TWA 5 ppm (25 mg/m^3); HSE MEL (United Kingdom) TWA 5 ppm (25 mg/m^3); The Netherlands TWA 0.3 ppm (1.5 mg/m^3).

38.0 Ethylene Glycol Monoethyl Ether Acetate

38.0.1 CAS Number: [111-15-9]

38.0.2 Synonyms: 2-Ethoxyethanol acetate; EGEEA

38.0.3 Trade Names: Cellosolve® Acetate; Poly-solv® EE Acetate

38.0.4 Molecular Weight 132.16

38.0.5 Molecular Formula: C$_6$H$_{12}$O$_3$

38.0.6 Molecular Structure:

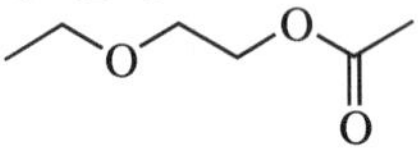

38.1 Chemical and Physical Properties

38.1.1 General

See Table 86.20.

38.2 Production and Use

Ethylene glycol monoethyl ether acetate (EGEEA) is used as a solvent for nitrocellulose, oils, and resins, in lacquers, varnish removers, wood stains, textiles, and leather, and as a chemical intermediate. Refer to Table 86.2 for additional production and use information.

38.3 Exposure Assessment

38.3.1 Air

Piacitelli et al. (520) collected air samples from several manufacturing sites that used EGEEA. Air samples from the aerospace industry and electronic industry contained EGEEA that was less than the limit of detection. Detectable amounts were found in the airline industry (0.29–2.69 ppm), coatings manufacturing (0.07–0.35 ppm), and automotive industry (< 0.02 to 0.05 ppm).

38.3.3 Workplace Methods

NIOSH Method 1450 is recommended for determining workplace exposures to EGEEA (14).

ETHERS OF ETHYLENE GLYCOL AND DERIVATIVES

38.3.5.2 Urine. A study of urinary biomarkers of occupational exposure to glycol ethers was conducted by Laitinen et al. (521). Over-shift urine samples were collected from five silk printing workers who were occupationally exposed to EGEEA and other glycol ether acetates and ten nonexposed office workers. The urine samples were analyzed for 2-EEA. EGEEA breathing zone samples contained from 4.9 to 32.3 mg/m^3. Excretion of 2-EAA was significantly correlated with breathing zone exposures.

38.4 Toxic Effects

Ethylene glycol monoethyl ether acetate (EGEEA) has fairly low single-dose oral toxicity, is somewhat irritating to the eyes, is not appreciably irritating to the skin, is poorly absorbed through the skin, and is not especially toxic when inhaled in amounts likely to be encountered under ordinary conditions. It can cause central nervous system depression, blood changes, and lung and kidney injury. EGEEA is a developmental toxicant and a teratogen in rats and rabbits when exposed by oral, dermal, or inhalation routes of exposure. Similarly to EGEE, EGEEA causes testicular effects in mice.

38.4.1 Experimental Studies

38.4.1.1 Acute Toxicity. Oral. Smyth et al. (354) reported LD$_{50}$ values of 5.10 g/kg for rats and 1.91 g/kg for guinea pigs when the material was fed as a 50% aqueous suspension in 1% Tergitol 7 surfactant solution. Carpenter (522) reported an oral LD$_{50}$ of 1.95 g/kg for rabbits. The toxic effects seen were some degree of gastrointestinal tract irritation and kidney injury and to a lesser degree, injury to the liver. Bloody urine was rarely seen, but the bile was often orange or red. Narcosis occurred mostly in those animals at or higher than the LD$_{50}$ dose level.

Truhaut et al. (523) reported that the oral LD$_{50}$ is 2.9 g/kg for female rats and 3.9 g/kg for male rats. Toxic effects reported were blood in the urine and hypertrophic kidneys dilated with blood.

Nagano et al. (32) treated mice by gavage with doses that ranged from 62.5 to 4000 mg/kg 5 days/week for 5 weeks and observed testicular atrophy and leukopenia, whose intensity was dose-related. They concluded that the response was similar to that of the unesterified ethyl ether.

Injection. Von Oettingen and Jirouch (157) reported that the lethal dose for mice was 5.0 mL/kg. Gross (49) reported that single doses of 0.5 or 1.0 mL/kg were well tolerated when injected intraperitoneally in guinea pigs. When injected subcutaneously seven times, doses of 0.5 and 1.0 mL/kg caused temporary ill effects but no deaths.

Inhalation. According to Gross (49), guinea pigs survived a 1-h exposure to an atmosphere essentially saturated with vapor (< 4000 ppm). Cats exposed once for 2 to 6 h to an atmosphere laden with fog survived, but two such exposures caused vomiting, paralysis, albumin in the urine, and death. Mice, guinea pigs, and a rabbit were unaffected by twelve 8-h exposures to a concentration of 450 ppm, but another rabbit and two cats died before the end of the exposure period. Albumin occurred in the urine, and the kidneys of the animals that died were injured.

Carpenter (522) stated that rats died from a 2-h exposure to a saturated vapor concentration generated by bubbling air through boiling liquid but survived a 4-h exposure when the air was bubbled through the liquid at room temperature (1500 ppm). Under the latter conditions, deaths resulted from an 8-h exposure.

Truhaut et al. (523) reported that rats and rabbits survived a single 4-h exposure to 2000 ppm of EGEEA. In rabbits only, slight transient hemaglobinuria and/or hematuria were observed that lasted 24 to 48 h. Upon necropsy, no gross pathological lesions were observed.

Dermal. Absorption of the material through intact skin can occur, but the lethal dose is large. Carpenter (522) reported an LD_{50} of 10.3 g/kg for rabbits. Other researchers reported that the dermal LD_{50} for EGEEA is 10.5 g/kg for rabbits (523).

Eye Irritation. EGEEA is somewhat irritating to the eyes of rabbits (66). Union Carbide (524) reported that it was moderately irritating to the eyes of rabbits. Truhaut et al. (523) exposed six rabbits to EGEEA to evaluate it for eye irritation using the method of Draize. The authors concluded that it was nonirritating to the eyes of rabbits.

Skin Irritation. The liquid is not significantly irritating to the skin unless exposure is prolonged or frequently repeated (522). Union Carbide (524) reported that EGEEA is only mildly irritating when applied to the skin of rabbits. More recent testing confirmed these results. EGEEA produced very mild irritation when applied to the skin of rabbits (523).

38.4.1.2 Chronic and Subchronic Toxicity. *Inhalation.*

Dogs survived 120 daily 7-h exposures to a concentration of 600 ppm without apparent injury. Carpenter (522) was unable to detect methemoglobin in the blood, other hematologic changes, any effects upon numerous clinical tests, or any histopathological changes in the tissues. This is a bit unexpected when one considers the effects caused by EGEE and EGEEA on rats. It may be explained by the difference in susceptibility of the two species and perhaps the dosage. Carpenter concluded that EGEEA is in the same range of toxicity as methyl ethyl ketone, propylene dichloride, and tetrachloroethylene, but its hazards are believed to be less because its vapor pressure is substantially lower.

38.4.1.3 Pharmacokinetics, Metabolism, and Mechanisms. *Metabolism and Pharmacokinetics.*

Guest et al. (525) administered ^{14}C-EGEEA intravenously to beagle dogs (1 mg/kg) and found 20 and 61% of the radio-label in the urine at 4 and 24 h, respectively. Approximately 1.6% of the doses were recovered in exhaled air 8 h postdosing. Based on the decrease of radioactivity in the blood, the half-life was 7.9 h. Stott and McKenna (514) studied the *in vitro* hydrolysis of EGEEA by carboxylesterase from nasal mucosal tissue of B6C3F1/CrlBr mice. Enzyme kinetics were determined using 0.4–0.7 mg of protein incubated with 2–115 mM EGEEA. The extent of hydrolysis was determined by measuring EGEE levels in the supernatant by gas chromatography. The V_{max} (mmoles/min), K_m (mM), and V_{max}/K_m values were 0.730, 9.52, and 0.077, respectively. The metabolic rates show that EGEEA is readily metabolized in mouse nasal tissues to produce EGEE. The metabolism, pharmacokinetics, and mechanism of action of EGEE are presented in Section 3 of this Chapter.

38.4.1.4 Reproductive and Developmental.

Reproductive. Nagano et al. (206) administered EGEEA by gavage to male mice for 5 days/week for 5 weeks at daily dose levels of 500, 1000, 2000, or 4000 mg/kg body weight. The results demonstrated a dose-dependent decrease in the weight of the testes; significant decreases occurred at dose levels of 1000 mg/kg and higher. Histopathological examination revealed atrophy of the seminiferous tubular epithelium at dose levels of 1000 mg/kg and higher. A significant decrease in white blood cells at doses of 2000 and 4000 mg/kg was observed. Other results include a decrease in hematocrit and a slight reduction in hemoglobin and erythrocytes at 4000 mg/kg. The NOEL for effects on the testes in this study was 500 mg/kg and 1000 mg/kg for blood effects by gavage in male mice.

Developmental. Hardin et al. (179) applied 0.35 mL EGEEA to the shaved skin of pregnant rats four times a day on days 7 through 16 of gestation. The results from this study were reduced maternal body weight, embryo toxicity (significant increase in resorptions), fetotoxicity (delayed skeletal ossification), and malformations (significant increase in the number of cardiovascular defects) compared to the negative control animals.

Doe (180) exposed rabbits to EGEEA by inhalation at 25, 100, or 400 ppm concentrations. The results included reduced maternal body weight and malformations of the vertebral column at 400 ppm. There were increased resorptions and minimal fetotoxicity (delayed skeletal ossification) at 100 and 400 ppm. The NOEL for developmental toxicity was 25 ppm, and for teratogenicity it was 100 ppm.

Nelson et al. (101) exposed rats by inhalation of EGEEA at 130, 390, or 600 ppm concentrations for 7 h/day on days 7 through 15 of gestation. At 600 ppm all offspring from the dams exposed were resorbed. Malformations of the vertebral column, heart, and umbilicus were observed at 390 ppm. There was a significant increase in resorptions at 390 ppm, and fetal body weights were reduced at 130 and 390 ppm. At 130 ppm, one fetus had a malformation of the heart, but no other abnormalities or significant fetotoxicity were observed at this level. However, because of the rarity of heart defects and because none were observed in the control animals, the authors concluded that the inhalation of 130 or 390 ppm of EGEEA was teratogenic in the rat.

Tyl et al. (526) exposed rats and rabbits to EGEEA via inhalation for 6 h/day on days 6 through 15 of gestation (rats) and days 6 through 18 of gestation (rabbits). Exposure concentrations were 0, 50, 100, 200, or 300 ppm of EGEEA. Maternal toxicity was apparent in both species at concentrations of 100, 200, and 300 ppm EGEEA, and developmental toxicity was also observed in both rats and rabbits at these concentrations. Teratogenicity (visceral and skeletal malformations) was observed at 200 and 300 ppm in both species. The NOEL for developmental toxicity was 50 ppm for both species.

38.4.2 Human Experience

38.4.2.1 General Information.
No reports of adverse human experience have been attributed to EGEEA. The major routes of occupational exposure to EGEEA are inhalation and skin absorption. A total of 357 personal and workplace air samples were collected between 1982 and 1985 at seven U.S. semiconductor manufacturing plants that used EGEEA. EGEEA concentrations ranged from 0.001–18.0 ppm and the mean concentrations were 0.05 ppm for TWA exposure and 1.56–2.82 ppm for short-term exposure (527).

Sohnlein et al. (151) measured exposure and urine levels of metabolites in workers occupationally exposed to EGEEA in the production of varnishes. In the varnish production area, the air concentration of EGEEA varied between 0.1 and 0.5 ppm. The mean urinary 2-EAA concentration varied between 53.2 and 53.8 ppm. After studying 17 persons for their excretion of 2-EAA in urine during an exposure-free weekend, calculated half-lives were of the order of 60 h.

Vincent et al. (528) conducted a study on the exposure of paint stripping workers in the aeronautical industry to EGEEA. During painting operations, air concentrations were higher than 0.1 ppb. Ethoxyacetic acid was detected in urine samples from EGEEA exposed workers who had an average preshift value of 108.4 mg/g creatinine and an average postshift value of 139.4 mg/g creatinine. The results suggested that the main route of exposure was via skin because respiratory protection was provided.

Low levels of EGEEA (TWA of 0.51 ppm) had no effect on the menstrual cycle (period) and duration (days) of the menses of women who worked in the crystal display (LCD) manufacturing industry (529).

38.4.2.2 Clinical Cases

38.4.2.2.3 Pharmacokinetics, Metabolism, and Mechanisms. In Vitro Skin Absorption. Evidence that EGEEA can penetrate the skin is provided by Dugard et al. (21), who evaluated EGEEA in an *in vitro* procedure using human cadaver abdominal epidermis. Under the conditions of the study, the mean rate of absorption through the skin was 0.80 mg/cm^2/h (21). The rate of absorption reached a steady state after a lag time of just under 1 h. In a separate study, Barber et al. (208) determined that the *in vitro* absorption rate in human stratum corneum is 1.41 mg/cm^2/h.

38.4.2.3 Epidemiology Studies. A number of epidemiology studies were carried out that involved spontaneous abortions among workers in the semiconductor industry who were exposed to glycol ethers. The glycol ethers and their acetates were identified as EGME, EGEE, EGMEA, and EGEEA. A detailed discussion of these studies is presented in Section 3 as they pertain to EGEE.

38.5 Standards, Regulations, or Guidelines of Exposure

The ACGIH (143) has recommended a TLV of 5 ppm (27 mg/m^3) as an 8-h TWA for EGEEA. OSHA adopted 100 ppm (540 mg/m^3) as a PEL; rule-making is in progress to adopt a new workplace exposure standard. In 1991, NIOSH (144) recommended limiting exposure to EGEEA in the workplace to 0.5 ppm (2.7 mg/m^3) as a 10-h TWA.

OSHA recently proposed reducing the current 8-h time-weighted average limits for EGEEA to 0.5 ppm (518). Table 86.9 gives the international standards and regulations for inhalation and skin exposure to EGEEA.

39.0 Ethlyene Glycol Mono-*n*-butyl Ether Acetate

39.0.1 CAS Number: [112-07-2]

39.0.2 Synonyms: 2-*n*-Butoxyethanol acetate; EGBEA; BGA

39.0.3 Trade Names: Butyl Cellosolve® Acetate; Eastman® EB Acetate; Butyl Glycol Acetate

39.0.4 Molecular Weight: 160.21

39.0.5 Molecular Formula: $C_8H_{16}O_3$

39.0.6 Molecular Structure:

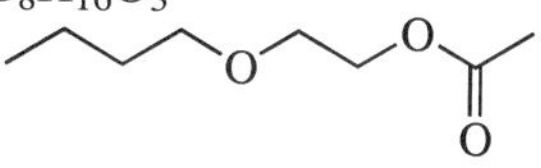

39.1 Chemical and Physical Properties

39.1.1 General

Ethylene glycol mono-*n*-butyl ether acetate (EGBEA) is a colorless liquid that has a fruity odor. It is slightly soluble in water (1.5% w/w at 20°C), and is soluble in hydrocarbons and organic solvents. See Table 86.20.

39.2 Production and Use

EGBA is used primarily as a slowly evaporating solvent and coalescing agent in latex paints and lacquers at concentrations of 1 to 5% (530). It is also used in some ink and spot remover formulations. Refer to Table 86.2 for additional production and use information.

39.3 Exposure Assessment

OSHA Method 83 can be used to determine concentrations of EGBEA in air (245). The quantification limit is 0.024 ppm. NIOSH recommends analyzing all samples by gas chromatography-mass spectrometry to correct for potential interference. NIOSH method 1403 may also be adapted for EGBEA (245).

39.4 Toxic Effects

EGBEA has a low order of acute oral toxicity. Both rats and mice survived a 4-h inhalation exposure to a saturated atmosphere of about 400 ppm and showed no signs of toxicity. EGBEA was practically nonirritating by the dermal route. EGBEA is rapidly and quantitatively metabolized to ethylene glycol butyl ether (EGBE); consequently, its systemic toxicity is expected to be similar to that of EGBE. Administration of large oral doses of EGBEA to rats, resulted in hemolysis and nephropathy, which may have been secondary to the hemolytic effect. No adverse effects from human exposure have been reported. The toxicity of EGBEA was recently summarized (245).

39.4.1 Experimental Studies

39.4.1.1 Acute Toxicity. Oral. The acute oral LD_{50} for rats was reportedly 7000 mg/kg (348). In a more recent study, it was reportedly 3000 mg/kg for male rats and 2400 mg/kg for females (523). The acute oral LD_{50} for mice was 3200 mg/kg (531). When administered by gavage to rats at near-lethal doses, EGBEA caused hemolysis, hemoglobinuria, and hypertrophic kidneys congested with blood (523).

Inhalation Exposure. Rats and rabbits were exposed acutely for 4 h to a saturated concentration (approximately 400 ppm) of EGBEA; transient hemoglobinuria and/or hematuria were observed only in rabbits, and no gross pathological lesions were observed 2 weeks later at necropsy (523).

Dermal. The acute dermal LD_{50} in rabbits was 1500 mg/kg (523).

Eye Irritation. EGBEA produced slight conjunctival redness in the rabbit eye (523).

Skin Irritation. A single dermal exposure to EGBEA was practically nonirritating and produced only very slight erythema in rabbit skin (523) and in human skin, even when applied under occlusion for 4 h (532). Prolonged or repeated exposure could result in more pronounced irritation. Zissu (233) reports that EGBEA is a nonirritant when applied dermally to rabbits under the EEC testing protocol and is a moderate irritant under the Draize method. EGBEA was tested with cultured human epidermal keratinocytes in an *in vitro* study of irritant potential (533). As a nonirritant, it stimulated release of prostaglandin E-2 only at concentrations that compromised cellular integrity.

39.4.1.2 Chronic and Subchronic Toxicity.

Inhalation. Rats and rabbits were exposed by inhalation to 400 ppm EGBEA for 4 h/day, 5 days/week for 1 month, or to approximately 100 ppm for 10 months. After 2 weeks at 400 ppm, hemoglobinuria became apparent and was more pronounced in rabbits than rats. Two rabbits died and showed histological evidence of kidney damage, whereas no gross pathological lesions were reported in surviving rabbits or rats. No adverse effects were reported during or after 10 months daily exposure to 100 ppm (523).

39.4.1.3 Pharmacokinetics, Metabolism, and Mechanisms.

Metabolism and Pharmacokinetics. Although no studies of metabolism of EGBEA have been reported, the analogous diethylene glycol butyl ether acetate (DGBEA) was studied. DGBEA was rapidly metabolized by serum and other esterases to the parent glycol ether (463). The hemolytic effects of EGBEA are identical to those reported for EGBE, indicating similar rapid metabolism of EGBEA to EGBE. See Section 6 for the metabolism and pharmacokinetics of EGBE.

39.4.2 Human Experience

39.4.2.1 General Information.
There are no human studies of the effects of exposure to EGBEA. However, it was concluded there were no health hazards from solvents, including EGBEA, in a plant where exposures to EGBEA ranged from 0.8 to 3.9 ppm TWA (245).

39.5 Standards, Regulations, or Guidelines of Exposure

NIOSH (245) recommends a 5 ppm (31 mg/m^3) occupational exposure standard (TWA) for EGBEA. A number of other countries have also established occupational exposure limits (Table 86.9): DFG MAK (Germany) TWA 20 ppm (130 mg/m^3); The Netherlands TWA 20 ppm (135 mg/m^3), STEL 40 ppm (270 mg/m^3).

40.0 Diethylene Glycol Monomethyl Ether Acetate

40.0.1 CAS Number: [629-38-9]

ETHERS OF ETHYLENE GLYCOL AND DERIVATIVES

40.0.2 Synonyms: 2-(2-Methoxyethoxy) ethanol acetate; DGMEA

40.0.3 Trade Names: Methyl Carbitol® Acetate

40.0.4 Molecular Weight: 162.19

40.0.5 Molecular Formula: $C_7H_{14}O_4$

40.0.6 Molecular Structure:

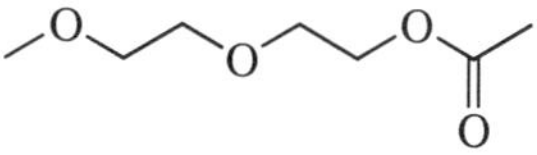

40.1 Chemical and Physical Properties

40.1.1 General

See Table 86.20 for chemical and physical properties.

40.4 Toxic Effects

40.4.1 Experimental Studies

40.4.1.1 Acute Toxicity. Diethylene glycol monomethyl ether acetate (DGMEA) has a low single-dose oral toxicity. The LD_{50} values reported by Smyth et al. (354) are 11.96 g/kg for rats and 3.46 g/kg for guinea pigs when the material is fed as a 50% aqueous solution. Carpenter and Smyth (66) found that the material is appreciably irritating to the rabbit eye. DGMEA is believed to be metabolized in the body to diethylene glycol monomethyl ether (DGME). The toxicity of DGME is presented in Section 19. DGME exerts its principal toxicological action upon the brain, kidney, liver, thymus, and testes. DGME is a developmental toxicant in experimental animals by the oral route.

40.4.1.2 Chronic and Subchronic Toxicity. Refer to Section 19, DGME.

40.4.1.3 Pharmacokinetics, Metabolism, and Mechanisms. Metabolism. DGMEA is believed to be metabolized in animals to DGME by tissue carboxylesterases. The further metabolism of DGME to 2-(2-methoxyethoxy) acetic acid and 2-methoxyacetic acid is discussed in Section 19. The toxicological properties of DGMEA are expected to be similar to those of DGME (see Section 19).

40.5 Standards, Regulations, or Guidelines of Exposure

No occupational exposure standard has been established for DGMEA.

41.0 Diethylene Glycol Monoethyl Ether Acetate

41.0.1 CAS Number: [112-15-2]

41.0.2 Synonyms: 2-(2-Ethoxyethoxy) ethanol acetate: DGEEA

41.0.3 Trade Names: Carbitol® Acetate; Eastman® DE Acetate

41.0.4 Molecular Weight: 176.21

41.0.5 Molecular Formula: $C_8H_{16}O_4$

41.0.6 Molecular Structure:

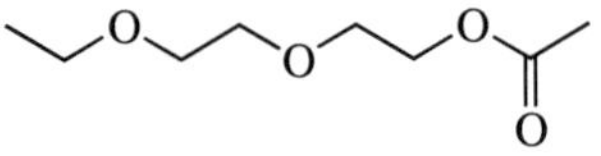

41.1 Chemical and Physical Properties

41.1.1 General

See Table 86.20 for chemical and physical properties.

41.2 Production and Use

Diethylene glycol monoethyl acetate (DGEEA) is listed by Johanson and Rick (534) among the top 15 most important glycol ethers used in Sweden from 1986 to 1993. DGEEA is used in paints and lacquers as a solvent. Refer to Table 86.2 for additional production and use information.

41.4 Toxic Effects

Diethylene glycol monoethyl ether acetate (DGEEA) has a low order of toxicity through oral, dermal, and inhalation routes of exposure. DGEEA is slightly irritating to the eyes and skin of laboratory animals and produces only a mild irritation when tested on humans.

41.4.1 Experimental Studies

41.4.1.1 Acute Toxicity. *Oral.* The oral $LD_{50}s$ reported by Smyth et al. (354) are 11.0 g/kg for rats and 3.39 g/kg for guinea pigs when DGEEA is administered as a 50% aqueous solution. Union Carbide (355) reported an LD_{50} of 4.4 g/kg for rabbits.

Inhalation. Rats and guinea pigs exposed to essentially saturated atmospheres of DGEEA at room temperature for 8 h all survived, although injury to the lungs and kidneys was observed at necropsy (535).

Dermal. The dermal LD_{50} for rabbits was reportedly 15.0 g/kg (355).

Eye Irritation. Carpenter and Smyth (66) reported that DGEEA produced slight irritation of the rabbit eye.

Skin Irritation. DGEEA produced slight skin irritation in rabbits and when tested using human patch tests (undiluted material), only a limited number of people developed mild skin irritation (536).

41.4.1.2 Chronic and Subchronic Toxicity. Refer to Section 20, DGEE, for a discussion of subchronic toxicity.

41.4.1.3 Pharmacokinetics, Metabolism, and Mechanisms. DGEEA is believed to be metabolized in animals to DGEE by tissue carboxylesterases. The toxicological properties of DGEEA are expected to be similar to those of DGEE (see Section 20).

41.5 Standards, Regulations, or Guidelines of Exposure

No occupational exposure standard has been established for DGEEA.

ETHERS OF ETHYLENE GLYCOL AND DERIVATIVES

42.0 Diethylene Glycol Mono-n-butyl Ether Acetate

42.0.1 CAS Number: [124-17-4]

42.0.2 Synonyms: DGBEA; 2(2-butoxyethoxy) ethanol acetate; butyl diglycol acetate; BDGA

42.0.3 Trade Names: Butyl Carbitol® Acetate; Eastman® DB Acetate

42.0.4 Molecular Weight: 204.27

42.0.5 Molecular Formula: $C_{10}H_{20}O_4$

42.0.6 Molecular Structure:

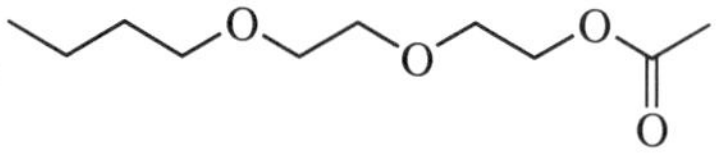

42.1 Chemical and Physical Properties

42.1.1 General

See Table 86.20.

42.2 Production and Use

Diethylene glycol mono-n-butyl ether acetate (DGBEA) is used primarily as a solvent and coalescing agent in paints and other coatings, usually at concentrations of 2 to 5%. Refer to Table 86.2 for additional production and use information.

42.3 Exposure Assessment

Concentrations of DGBEA in room air, simulating use of a paint containing 2% DGBEA, were determined by using a modification of NIOSH method 127 (537). The limit of quantitative detection was about 0.03 ppm for a 1-h sampling time.

42.4 Toxic Effects

Diethylene glycol monobutyl ether acetate (DGBEA) has a low acute toxicity by the oral, dermal, and inhalation routes of exposure. The dermal and inhalation routes are the most likely routes for human exposure because of the use of DGBEA in paints and coatings. DGBEA is absorbed dermally to an extent similar to DGBE, and it is rapidly and quantitatively metabolized to DGBE. Therefore, its toxicity is likely to be similar to that of DGBE, which has been extensively tested. The available data suggest that, like DGBE, DGBEA presents little risk of systemic toxicity from occupational exposure to undiluted material or from exposure to DGBEA in formulated consumer products.

42.4.1 Experimental Studies

42.4.1.1 Acute Toxicity. Oral. The acute oral toxicity of DGBEA is summarized in Table 86.21. At high oral doses, DGBEA causes narcosis in rats (354).

Table 86.21. Acute Toxicity of DGBEA Following Oral Administration

Species	LD$_{50}$(mg/kg)	Ref.
Rat	11,920	354
Rat	7,100	539
Guinea pig	2,700	539
Guinea pig	2,340	354
Mouse	6,600	539
Rabbit	2,800	539

Inhalation. When male rats were exposed to high concentrations of DGBEA for 4 h, the LC$_{50}$ calculated was 8700 ppm (538). Clinical signs included inactivity and respiratory irritation, and lung congestion was reported at necropsy.

Dermal. The dermal LD$_{50}$ for rabbits was 5500 mg/kg, which compared with the oral LD$_{50}$ of 2800 mg/kg, implies that a significant proportion of the dermally applied dose can be absorbed under the conditions used (539).

42.4.1.2 Chronic and Subchronic Toxicity. Subchronic Toxicity, Dermal. In a 13-week dermal application study of rabbits treated repeatedly with DGBEA, hemolysis, hemoglobinuria, and degenerative changes in the kidney were reported (539). It is possible that these effects were not due to DGBEA itself, but rather to the butyl ether of monoethylene glycol acetate (EGBEA), which may have been an impurity.

Eye Irritation. DGBEA was only slightly irritating when applied to the rabbit eye (66).

Skin Irritation. Repeated and prolonged skin contact of DGBEA caused mild erythema and exfoliation in rabbits and humans (539).

42.4.1.3 Pharmacokinetics, Metabolism, and Mechanisms. Metabolism and Pharmacokinetics. When incubated with whole rat blood *in vitro*, DGBEA was hydrolyzed to DGBE and had a half-life of less than 3 min, suggesting that the subsequent metabolism and toxicity of DGBEA would be identical to that of DGBE (463).

DGBEA was rapidly absorbed from the gastrointestinal tract, metabolized, and excreted primarily (85% of the dose) in the urine. The major urinary metabolite was 2-butoxyethoxyacetic acid, together with diethylene glycol, hydroxybutoxyethoxyacetic acid, two unidentified metabolites, and exhaled CO$_2$ (5% of the dose). No unchanged DGBEA was detected, and neither were the potential metabolites DGBE nor 2-butoxyacetic acid (463). Following dermal applications (462), the major urinary metabolite detected was 2-(2-butoxyethoxy) acetic acid (50–60%) with small amounts of the glucuronide conjugate of DGBE (5–8%).

Skin Absorption. When DGBE and DGBEA were applied to the skin for 5 min then washed, most (90%) of the material was recovered (462). When applied under occlusion for 24 h, the calculated absorption rates for DGBEA were similar (1.58, 1.28 mg/cm^2/h for males and females, respectively; mean = 1.43) to those for DGBE (0.73, 1.46; mean = 1.10 mg/cm^2/h).

42.4.2 Human Experience

42.4.2.1 General Information. One case of possible skin sensitization has been reported in an individual exposed to DGBEA and other materials for many years (540). At one time, a formulation containing 50% DGBEA was used as an insect repellent; one case of nephrosis was reported in a child who had repeated and prolonged dermal exposure to the repellent (541). There were no reports of adverse health effects from normal use of DGBEA-containing products. The concentration of DGBEA in room air has been determined under conditions simulating the use of a paint that contained 2% DGBEA. The maximum concentration of DGBEA was 0.11 ppm, and the average breathing concentration was less than 0.05 ppm (537). Adverse effects are considered unlikely under these conditions (455).

42.5 Standards, Regulations, or Guidelines of Exposure

No occupational exposure standard has been established for DGBEA.

BIBLIOGRAPHY

1. H. Chinn with E. Anderson, and M. Tashiro, CEH marketing research report: glycol ethers. *Chemical Economics Handbook*, SRI International, 1996.
2. R. L. Smith, Review of glycol ether and glycol ether ester solvents used in the coating industry. *Environ. Health Perspect.* **57**, 1–4 (1984).
3. D. Norback et al., House painters exposure to glycol and glycol ethers from water based paints. *Occup. Hyg.* **2**, 111–117 (1996).
4. H. Pastides et al., Spontaneous abortion and general illness among semiconductor manufacturers. *J. Occup. Med.* **30**, 543–551 (1988).
5. R. H. Gray et al., *Final report: The Johns Hopkins University Retrospective and Prospective Studies of Reproductive Health Among IBM Employees in Semiconductor Manufacturing*, Baltimore, The Johns Hopkins University, 1993.
6. R. H. Gray et al., Ethylene glycol ethers and reproductive health in semiconductor workers *Occup. Hyg.* **2**, 331–338 (1996).
7. S. H. Lamm, J. S. Kutcher, and C. B. Morris, Spontaneous abortions and glycol ethers used in the semiconductor industry: an epidemiologic review. *Occup. Hyg.* **2**, 339–354 (1996).
8. S. K. Hammond et al., Exposures to glycol ethers in the semiconductor industry. *Occup. Hyg.* **2**, 355–366 (1996).
9. M. Schenker et al., *Epidemiologic Study of Reproductive and Other Health Effects Among Workers Employed in the Manufacture of Semiconductors, Final Report to the Semiconductor Industry Association*, 1992.
10. M. B. Schenker, Reproductive health effects of glycol ether exposure in the semiconductor industry. *Occup. Hyg.* **2**, 367–372 (1996).
11. S. H. Swan and W. Forest, Reproductive risks of glycol ethers and other agents used in semiconductor manufacturing. *Occup. Hyg.* **2**, 373–385 (1996).
12. S. M. Pinney and G. K. Lemasters, Spontaneous abortions and stillbirths in the semiconductor employees. *Occup. Hyg.* **2**, 387–401 (1996).

13. ATSDR, *Toxicological Profile for 2-Butoxyethanol and 2-Butoxyethanol Acetate*, U.S. Department of Health and Human Services, Public Health Service, Agency for Toxic Substances and Disease Registry, Aug., 1998.
14. NIOSH, *NIOSH Manual of Analytical Methods*, 4th ed., National Institute for Occupational Safety and Health, U.S. Govt. Printing Office, Supt. of Docs. Washington, DC, 1994.
15. OSHA, *2-Butoxyethanol (Butyl Cellosolve), 2-Butoxyethyl Acetate (Butyl Cellosolve Acetate), Method 83*, Salt Lake City, Utah: Organic Methods Evaluation Branch, OSHA Analytical Laboratory, Occupational Safety and Health Administration, 1990.
16. Dow Chemical Company, *Industrial Hygiene Monitoring Method: Monitoring Nine Dowanol Glycol Ethers in Air using Adsorbent Tube/Pump Methods*, HEH-IMM-82-28, 1982.
17. Supelco, Inc., New capillary column eliminates peak tailing for many acidic compounds. *The Supelco Reporter* **6**(5), 1–2 (1987).
18. W. Eckel, G. Foster, and B. Ross, Glycol ethers as ground water contaminants. *Occup. Hyg.* **2**, 97–104 (1996).
19. S. Kezic et al., Dermal absorption of vaporous and liquid 2-methoxyethanol and 2-ethoxyethanol in volunteers. *Occup. Environ. Med.* **54**, 38–43 (1997).
20. A. P. Leber et al., Triethylene glycol ethers. Evaluation of *in vitro* absorption through human epidermis, 21-day dermal toxicity in rabbits and a developmental toxicity screen in rats. *J. Am. Coll. Toxicol.* **9**(5), 507–515 (1990).
21. P. H. Dugard et al., Absorption of some glycol ethers through human skin *in vitro*. *Environ. Health Perspect.* **57**, 193–198 (1984).
22. F. Welsch et al., Linking embryo dosimetry and teratogenic responses to 2-methoxyethanol at different stages of gestation in mice. *Occup. Hyg.* **2**, 121–130 (1996).
23. R. A. Corley, G. A. Bormett, and B. I. Ghanayem, Physiologically-based pharmacokinetics of 2-butoxyethanol and its major metabolite 2-butoxyacetic acid, in rats and humans. *Toxicol. Appl. Pharmacol.* **129**, 61–79 (1994).
24. S. Jenkins-Sumner et al., Characterization of urinary metabolites produced following administration of [1,2, methoxy-^{13}C]-2-methoxyethanol to male F-344 rats and pregnant CD-1 mice. *Occup. Hyg.* **2**, 25–31 (1996).
25. S. C. J. Sumner et al., Characterization of urinary metabolites from [1,2-methoxy-^{13}C]2-methoxyethanol in mice using ^{13}C nuclear magnetic resonance spectroscopy. *Chem. Res. Toxicol.* **5**, 553–560 (1992).
26. S. J. Sumner et al., Dose effects on the excretion of urinary metabolites of 2-[1,2-methoxy-^{13}C]methoxyethanol in rats and mice. *Toxicol. Appl. Pharmacol.* **134**, 139–147 (1995).
27. R. B. Sleet, J. A. Greene, and F. Welsch, The relationship of embryotoxicity to disposition of 2-methoxyethanol in mice. *Toxicol. Appl. Pharmacol.* **93**, 195–207 (1988).
28. B. I. Ghanayem, An overview of the hematoxicity of ethylene glycol ethers. *Occup. Hyg.* **2**, 253–268 (1996).
29. D. O. Clarke et al., Protection against 2-methoxyethanol-induced teratogenesis by serine enantiomers: studies of potential alteration of 2-methoxyethanol pharmacokinetics. *Toxicol. Appl. Pharmacol.* **110**, 514–526 (1991).
30. F. H. Wiley et al., The formation of oxalic acid from ethylene glycol and related solvents. *J. Ind. Hyg. Toxicol.* **20**, 269–277 (1938).
31. E. G. Stenger et al., Toxikologie des athylenglykol-monoathyiathers. *Thomann, Arzneim.-Forsch.* **21**, 880–885 (1971).

32. K. Nagano et al., Testicular atrophy of mice induced by ethylene glycol monoalkyl ethers. *Jpn. J. Ind. Health* **21**, 29–35 (1979) (in Japanese, English summary).

33. V. L. Horton et al., Developmental phase-specific and dose-related teratogenic effects of ethylene glycol monomethyl ether in CD-1 mice. *Toxicol. Appl. Pharmacol.* **80**, 108–118 (1985).

34. K. Nagano et al., Embryotoxic effects of ethylene glycol monomethyl ether in mice. *Toxicology* **20**, 335–343 (1981).

35. R. J. Smialowicz, The immunotoxicity of 2-methoxyethanol and its metabolites. *Occup. Hyg.* **2**, 269–274 (1996).

36. B. J. Chou et al., *Two-Year Chronic Inhalation Toxicity and Carcinogenicity Study of 2-Butoxyethanol in F344/N Rats*, Final report prepared for the National Toxicology Program, Research Triangle Park, NC, Battelle Pacific Northwest National Laboratories, Richland, WA, 1996.

37. B. J. Chou et al., *Two-year Chronic Inhalation Toxicity and Carcinogenicity Study of 2-Butoxyethanol in B6C3F1 Mice*, Final report prepared for the National Toxicology Program, Research Triangle Park, NC, Battelle Pacific Northwest National Laboratories, Richland, WA, 1996.

38. H. J. Morris, A. A. Nelson, and H. O. Calvery, Observations on the chronic toxicities of propylene glycol, ethylene glycol, diethylene glycol, ethylene glycol mono-ethyl-ether, and diethylene glycol mono-ethyl-ether. *J. Pharmacol. Exp. Therap.* **74**, 266–273 (1942).

39. D. McGregor, A review of some properties of ethylene glycol ethers relevant to their carcinogenic evaluation. *Occup. Hyg.* **2**, 213–235 (1996).

40. D. O. Clarke et al., Scale-up of a physiologically-based pharmacokinetic model for 2-methoxyethanol and 2-methoxyacetic acid from pregnant mice to rats. *Toxicologist* **12**, 101 (1992).

41. D. O. Clarke et al., Pharmacokinetcis of 2-methoxyethanol and 2-methoxyacetic acid in the pregnant mouse: a physiologically based mathematical model. *Toxicol. Appl. Pharmacol.* **121**, 239–252 (1993).

42. M. L. Gargas et al., A toxicokinetic study of inhaled ethylene glycol ethyl ether acetate and validation of a physiologically based pharmacokinetic model for rat and human. *Toxicol. Appl. Pharmacol.* **165**, 63–73 (2000).

43. D. O. Clarke, J. M. Duignan, and F. Welsch, 2-Methoxyacetic acid dosimetry-teratogenicity relationships in CD-1 mice exposed to 2-methoxyethanol. *Toxicol. Appl. Pharmacol.* **114**, 77–87 (1992).

44. C. A. Staples, R. J. Boatman, and M. L. Cano, Ethylene glycol ethers: an environmental risk assessment. *Chemosphere* **36**(7), 1585–1613 (1998).

45. J. May, *Staub* **26**, 385 (1966).

46. The Dow Chemical Company, unpublished data.

47. D. Groeseneken et al., Gas chromatographic determination of methoxyacetic and ethoxyacetic acid in urine. *Br. J. Ind. Med.* **43**, 62–65 (1986).

48. T.-S. Shih et al., Improved method to measure urinary alkoxyacetic acids. *Occup. Environ. Med.* **56**, 460–467 (1999).

49. E. Gross, In K. B. Lehmann and F. Flury, Eds., *Toxicology and Hygiene of Industrial Solvents*, transl. by E. King and H. F. Smyth, Jr., Springer, Berlin, 1938.

50. C. P. Carpenter et al., The toxicity of butyl cellosolve solvent. *AMA Arch. Ind. Health* **14**, 114–131 (1956).

51. Union Carbide Corporation, Bushy Run Research Center, *Special Report 8-3-38*.
52. E. Saparmamedov, *Zdravookhr. Turkm.* **18**, 26 (1974).
53. W. J. Krasavage and C. J. Terhaar, Comparative toxicity of nine glycol ethers I. acute oral LD_{50}, unpublished data, Eastman Kodak Company, Corporate Health and Environment Laboratories, Report No. TX-81-16, Feb.17, 1981.
54. D. Grant et al., Acute toxicity and recovery in the hemopoietic system of rats after treatment with ethylene glycol monomethyl and monobutyl ethers. *Toxicol. Appl. Pharmacol.* **77**, 187–200 (1985).
55. H. W. Werner et al., The acute toxicity of vapours of several monoalkyl ethers of ethylene glycol. *J. Ind. Hyg. Toxicol.* **25**, 157–163 (1943).
56. H. W. Werner et al., Effects of repeated exposures of rats to vapors of monoalkyl ethylene glycol ethers. *J. Ind. Hyg. Toxicol.* **25**, 374–379 (1943).
57. H. W. Werner et al., effects of repeated exposure of dogs to monoalkyl ethylene glycol ether vapors. *J. Ind. Hyg. Toxicol.* **25**, 409–414 (1943).
58. H. F. Smyth, Jr., In G. O. Curme, Jr., and F. Johnson, Eds., *Glycols*, American Chemical Society Monograph Series 114, Reinhold, New York, 1952.
59. E. Browning, *Toxicity and Metabolism of Industrial Solvents*, Elsevier, New York, 1965.
60. M. E. Goldberg, C. Haun, and H. F. Smyth, Jr., Toxicologic implications of altered behavior induced by an industrial vapor. *Toxicol. Appl. Pharmacol.* **4**, 148–164 (1962).
61. E. Z. Dajani, *Diss. Abst. Int. B.* **30**, 1819 (1969).
62. J. H. Draize, G. Woodward, and H. O. Calvery, Methods for the study of irritation and toxicity of substances applied to the skin and mucous membranes. *J. Pharmacol. Exp. Therap.* **82**, 377–390 (1944).
63. V. K. Rowe, The Dow Chemical Company, unpublished data, 1947.
64. W. J. Krasavage and C. J. Terhaar, *Comparative toxicity of nine glycol ethers II. acute dermal LD_{50}*. Unpublished data, Eastman Kodak Company, Corporate Health and Environment Laboratories, Report No. TX-81-38, July 1, 1981.
65. M. A. Wolf, The Dow Chemical Company, unpublished data, 1954.
66. C. P. Carpenter and H. F. Smyth, Jr., Chemical burns of the rabbit cornea. *Am. J. Ophthalmol.* **29**, 1363–1372 (1946).
67. W. M. Grant, *Toxicology of the Eye*, 2nd ed., Charles C. Thomas, Springfield, IL, 1974.
68. M. P. Dieter, NTP, *Technical Report on Toxicity Studies of Ethylene Glycol Ethers 2-Methoxyethanol, 2-Ethoxyethanol, and 2-Butoxyethanol (CAS Nos. 109-86-4, 110-80-5, and 111-76-2) Administered in Drinking Water to F344/N Rats and B6C3F1 Mice*, NIH Publication 93-3349, Research Triangle Park, NTP, 1993.
69. R. R. Miller et al., Ethylene glycol monomethyl ether I. subchronic vapor inhalation study with rats and rabbits. *Fundam. Appl. Toxicol.* **3**, 49–54 (1983).
70. R. R. Miller, L. L. Calhoun, and B. L. Yano, The Dow Chemical Company USA, Toxicology Research Laboratory, Report to the Chemical Manufacturers Association, 1982.
71. M. R. Zavon, Methyl cellosolve intoxication. *Am. Ind. Hyg. Assoc. J.* **24**, 38–41 (1963).
72. R. R. Miller et al., Comparative metabolism and disposition of ethylene glycol monomethyl ether and proplyene glycol monomethyl ether in male rats. *Toxicol. Appl. Pharmacol.* **67**, 229–237 (1983).
73. J. Moss et al., The role of metabolism in 2-methoxyethanol-induced testicular toxicity. *Toxicol. Appl. Pharmacol.* **79**, 480–489 (1985).

74. C. A. Mebus et al., 2-Methoxyethanol metabolism in pregnant CD-1 mice and embryos. *Toxicol. Appl. Pharmacol.* **112**, 87–94 (1992).

75. M. T. Moslen et al., Species differences in testicular and hepatic biotransformations of 2-methoxyethanol. *Toxicology* **96**, 217–224 (1995).

76. L. Aasmoe and J. Aarbakke, Gender difference in the elimination of 2-methoxyethanol, methoxyacetic acid and ethoxyacetic acid. *Xenobiotica* **27**, 1237–1244 (1997).

77. F. Welsch, G. M. Blumenthal, and R. B. Conolly, Physiologically based pharmacokinetic models applicable to organogenesis: extrapolation between species and potential use in prenatal toxicity risk assessments. *Toxicol. Lett.* **82/83**, 539–547(1995), (*Proceedings of the International Congress of Toxicology-VII*, Seattle, WA, July 2–6).

78. K. K. Terry et al., Developmental phase alters dosimetry-teratogenicity relationship for 2-ME in CD-1 mice. *Teratology* **49**, 218–227 (1994).

79. K. K. Terry et al., Development of a physiologically model describing 2-methoxyacetic acid disposition in the pregnant mouse. *Toxicol. Appl. Pharmacol.* **132**, 103–114 (1995).

80. K. K. Terry et al., A physiologically-based pharmacokinetic model describing 2-methoxyacetic acid disposition in the pregnant mouse. *Occup. Hyg.* **2**, 57–65 (1996).

81. C. E. Green et al., Comparative metabolism of glycol ethers in rat and human hepatocytes. *Toxicologist* **9**, 239 (1989).

82. G. Johanson and B. Dynesius, Liquid/air partition coefficients of six commonly used glycol ethers. *Br. J. Ind. Med.* **45**, 561–564 (1988).

83. C. E. Green et al., In vitro metabolism of glycol ethers by human and rat hepatocytes. *Occup. Hyg.* **2**, 67–75 (1996).

84. L. Aasmoe, J.-O. Winberg, and J. Aarbakke, The role of liver alcohol dehydrogenase isoenzymes in the oxidation of glycol ethers in male and female rats. *Toxicol. Appl. Pharmacol.* **150**, 86–90 (1998).

85. C. A. Mebus and F. Welsch, The possible role of one-carbon moieties in 2-methoxyethanol and 2-methoxyacetic acid-induced developmental toxicity. *Toxicol. Appl. Pharmacol.* **99**, 98–109 (1989).

86. P. J. Beattie, M. J. Welsch, and M. J. Brabec, The effect of 2-methoxyethanol and methoxyacetic acid on sertoli Cell lactate production and protein synthesis *in vitro*. *Toxicol. Appl. Pharmacol.* **76**, 56–61 (1984).

87. P. J. Beattie and M. J. Brabec, Methoxyacetic acid and ethoxyacetic acid inhibit mitochondiral function *in vitro*. *J. Biochem. Toxicol.* **1**, 61–70 (1986).

88. E. Kirsten, M. L. Sharma, and E. Kun, Molecular toxicology of (-)-erythro-fluorocitrate: selective inhibition of citrate transport in mitochondria and the binding of fluorocitrate to mitochondrial proteins. *Mol. Pharmacol.* **14**, 172–184 (1978).

89. F. Welsch, R. B. Sleet, and J. A. Greene, Attenuation of 2-methoxyethanol and methoxyacetic acid-induced digit malformations in mice by simple physiological compounds: Implications for the role of further metabolism of methoxyacetic acid in developmental toxicity. *J. Biochem. Toxicol.* **2**, 225–240 (1987).

90. C. A. Mebus, F. Welsch, and P. K. Working, Attenuation of 2-methoxyethanol-induced testicular toxicity in the rat by simple physiological compounds. *Toxicol. Appl. Pharmacol.* **99**, 110–121 (1989).

91. D. B. Stedman and F. Welsch, Inhibition of DNA synthesis in mouse whole embryo culture by 2-methoxyacetic acid and attenuation of the effects by simple physiological compounds. *Toxicol. Lett.* **45**, 111–117 (1989).

92. J. L. Ambroso et al., The amino acid L-serine protects against 2-methoxyethanol-induced neural tube defects in CD-1 mice. *The Toxicologist* **36**, Abstract No. 506 (1997).
93. B. D. Hardin et al., Testing of selected workplace chemicals for teratogenic potential. *Scand. J. Work Environ. Health* **7**(4), 66–75 (1981).
94. T. R. Hanley et al., Comparison of the teratogenic potential of inhaled ethylene glycol monomethyl ether in rats, mice, and rabbits. *Toxicol. Appl. Pharmacol.* **75**, 409–422 (1984).
95. B. D. Hardin and J. P. Lyon, Summary and overview: NIOSH symposium on toxic effects of glycol ethers. *Environ. Health Perspect.* **57**, 273–278 (1984).
96. P. M. D. Foster et al., Testicular toxicity of ethylene glycol monomethyl and monoethyl ethers in the rat. *Toxicol. Appl. Pharmacol.* **69**, 385–399 (1983).
97. P. M. D. Foster et al., Testicular toxicity produced by ethylene glycol monomethyl and monoethyl ethers in the rat. *Environ. Health Perspect.* **57**, 207–218 (1984).
98. R. E. Chapin and J. C. Lamb, IV, Effects of ethylene glycol monomethyl ether on various parameters of testicular function in the F344 rat. *Environ. Health Perspect.* **57**, 219–224 (1984).
99. M. H. Feuston et al., Reproductive toxicity of 2-methoxyethanol applied dermally to occluded and nonoccluded sites in male rats. *Toxicol. Appl. Pharmacol.* **100**, 145–161 (1989).
100. W. J. Scott et al., Teratologic potential of 2-methoxyethanol and transplacental distribution of its metabolite, 2-methoxyacetic acid, in non-human primates. *Teratology* **39**, 363–373 (1989).
101. B. K. Nelson et al., Comparative inhalation teratogenicity of four glycol ether solvents and an amino derivative in rats. *Environ. Health Perspect.* **57**, 261–272 (1984).
102. G. A. Wickramaratne, The teratogenic potential and dose-response of dermally administered ethylene glycol monomethyl ether (EGME) estimated in rats with the Chernoff-Kavlock assay. *J. Appl. Toxicol.* **6**(3), 165–166 (1986).
103. M. H. Feuston, S. L. Kerstetter, and P. D. Wilson, Teratogenicity of 2-methoxyethanol applied as a single dermal dose to rats. *Fundam. Appl. Toxicol.* **15**, 448–456 (1990).
104. R. W. Tyl et al., The developmental toxicity of 2-ethylhexanol applied dermally to pregnant Fischer 344 rats. *Fundam. Appl. Toxicol.* **19**, 176–185 (1992).
105. R. B. Sleet, J. A. John-Greene, and F. Welsch, Localization of radioactivity from 2-methoxy[1,2-14C]ethanol in maternal and conceptus compartments of CD-1 mice. *Toxicol. Appl. Pharmacol.* **84**, 25–35 (1986).
106. B. K. Nelson and W. S. Brightwell, Behavioral teratology of ethylene glycol monomethyl and monoethyl ethers. *Environ. Health Perspect.* **57**, 43–46 (1984).
107. R. B. Sleet, J. A. Greene, and F. Welsch, Teratogenicity and disposition of the glycol ether 2-methoxyethanol and their relationship in CD-1 mice. In F. Welsch, ed., *Approaches to Elucidate Mechanisms of Teratogenesis* [8th CIIT Conference on Toxicology], Hemisphere Publishing Corp. New York, 1987, pp. 33–57.
108. R. B. Sleet, J. A. Greene, and F. Welsch, The relationship of embryotoxicity to disposition of 2-methoxyethanol. *Toxicol. Appl. Pharmacol.* **93**, 195–207 (1988).
109. R. B. Sleet et al., Developmental phase specificity and dose-response effects of 2-methoxyethanol in rats. *Fundam. Appl. Toxicol.* **29**, 131–139 (1996).
110. R. B. Sleet et al., Developmental phase specificity and dose response effects of teratogenicity relationships in CD-1 mice exposed to 2-methoxyethanol. *Fund. Appl. Toxicol.* in press.

111. J. A. Pitt and E. W. Carney, Evaluation of various toxicants in rabbit whole-embryo culture using a new morphologically-based evaluation system. *Teratology* **59**, 102–109 (1999).

112. D. McGregor, A review of some properties of ethylene glycol ethers relevant to their carcinogenic evaluation. *Occup. Hyg.* **2**, 213–235 (1996).

113. D. B. McGregor, Genotoxicity of glycol ethers. *Environ. Health Perspect.* **57**, 97–104 (1984).

114. D. B. McGregor et al., Genetic effects of 2-methoxyethanol and bis (2-methoxy) ether. *Toxicol. Appl. Pharmacol.* **70**, 303–316 (1983).

115. K. S. Rao et al., Ethylene glycol monomethyl ether II. reproductive and dominant lethal studies in rats. *Fundam. Appl. Toxicol.* **3**, 80–85 (1983).

116. D. B. McGregor et al., Genetic effects of 2-methoxyethanol and bis (2-methoxyethyl) ether. *Toxicol. Appl. Pharmacol.* **70**, 303–316 (1983).

117. H. Shimizu et al., The results of microbial mutation test for forty-three industrial chemicals. *Sangyo Igaku* **27**, 400–419 (1985).

118. E. Zeiger et al., Mutagenicity testing of di (2-ethylhexyl) phthalate and related chemicals in Salmonella. *Environ. Mutagen.* **7**, 213–232 (1985).

119. E. Zeiger et al., Salmonella mutagenicity tests: V. Results from the testing of 311 chemicals. *Environ. Mol. Mutagen.* **19**(21), 2–141 (1992).

120. W. W. Au, D. L. Morris, and M. S. Legator, Evaluation of the clastogenic effects of 2-methoxyethanol in mice. *Mutat. Res.* **300**, 273–279 (1993).

121. T. Chiewchanwit and W. W. Au, Cytogenetic effects of 2-methoxyethanol and its metabolite, methoxyacetaldehyde, in mammalian cells *in vitro*. *Mutat. Res.* **320**, 125–132 (1994).

122. Z. Elias et al., Genotoxic and/or epigenetic effects of some glycol ethers: Results of different short-term tests. *Occup. Hyg.* **2**, 187–212 (1996).

123. S. Dhalluin et al., Apoptosis inhibition and ornithine decarboxylase superinduction as early epigenetic events in morphological transformation of syrian hamster embryo cells exposed to 2-methoxyacetaldehyde, a metabolite of 2-methoxyethanol. *Toxicol. Lett.* **105**, 163–175 (1999).

124. G. H. Henningsen et al., Comparative acute toxicity of three glycol ethers in rodents exposed via drinking water. *The Toxicologist* **9**, 248 (Abstract 994) (1989).

125. R. V. House et al., Immunological studies in B6C3F$_1$ mice following exposure to ethylene glycol monomethyl ether and its principal metabolite methoxyacetic acid. *Toxicol. Appl. Pharmacol.* **77**, 358–362 (1985).

126. D. P. Houchens, A. A. Ovejera, and R. W. Niemeier, Effects of ethylene glycol monomethyl (EGME) and monoethyl (EGEE) ethers on the immunocompetence of allogeneic and syngeneic mice bearing L1210 mouse leukemia. *Environ. Health Perspect.* **57**, 113–118 (1984).

127. M. P. Dieter et al., *Leuk. Res.* **13**(9), 841 (1989).

128. J. H. Exon et al., Effects of subchronic exposure of rats to 2-methoxyethanol or 2-butoxyethanol. Thymic atrophy and immunotoxicity. *Fundam. Appl. Toxicol.* **16**, 830–840 (1991).

129. R. J. Smialowicz, M. M. Riddle, and W. C. Williams, Methoxyacetaldehyde, an intermediate metabolite of 2-methoxyethanol, is immunosuppressive in the rat. *Fundam. Appl. Toxicol.* **21**, 1–7 (1993).

130. G. Young and L. B. Woolner, A case of fatal poisoning from 2-methoxyethanol. *J. Ind. Hyg. Toxicol.* **28**, 267–268 (1946).

131. E. Donley, Toxic encephalopathy and volatile solvents in industry. Report of a case. *J. Ind. Hyg. Toxicol.* **18**, 571–577 (1936).
132. C. E. Parsons and M. E. M. Parsons, Toxic encephalopathy and granulopenic anemia due to solvents in industry. Report of two cases. *J. Ind. Hyg. Toxicol.* **20**, 124–133 (1938).
133. L. Greenburg et al., Health hazards in the manufacture of "fused collars" I. Exposure to ethylene glycol monomethyl ether. *J. Ind. Hyg. Toxicol.* **20**, 134–147 (1938).
134. S. Nitter-Hauge, Poisoning with ethylene glycol mono-methyl ether. *Acta Med. Scand.* **188**, 277–280 (1970).
135. G. Ohi and D. H. Wegman, Transcutaneous ethylene glycol monomethyl ether poisoning in the work setting. *J. Occup. Med.* **20**, 675–676 (1978).
136. D. R. Mattie et al., Determination of skin: air partition coefficients for volatile chemicals: Experimental method and applications. *Fundam. Appl. Toxicol.* **22**, 51–57 (1994).
137. A. Boman and H. Maibach, Influence of evaporation and repeated exposure on the percutaneous absorption of organic solvents. In P. Elsner, Ed., *Current Problems in Dermatology (Basel)*, Vol 25, *Prevention of Contact Dermatitis*, Zurich, Switzerland, October 4–7, 1995, S. Karger AG, New York, 1996.
138. NIOSH, Symposium on the toxic effects of glycol ethers. *Environ. Health Persps.* **57**, 1–275 (1984).
139. D. J. Paustenbach, Assessment of the developmental risks resulting from occupational exposure to select glycol ethers within the semiconductor industry. *J. Toxicol. Environ. Health* **23**, 29–75 (1988).
140. M. B. Schenker et al., The association of spontaneous abortion and other reproductive effects with work in the semiconductor industry. *Am. J. Ind. Med.* **28**, 635–638 (1995).
141. S. H. Swan and W. Forest, Reproductive risks of glycol ethers and other agents used in semiconductor manufacturing. *Occup. Hyg.* **2**, 373–385 (1996).
142. R. Elliot et al., *Spontaneous Abortions in the UK Semiconductor Industry: an HSE Investigation*, Health Safety Executive, March 1998.
143. American Conference of Governmental Industrial Hygienists, *Threshold Limit Values for Chemical Substances and Physical Agents and Biological Exposure Indices for 1992–1993*, Cincinnati, OH, 1992, TLVS and other occupational exposure values, 1999.
144. *NIOSH Criteria for a Recommended Standard, Occupational Exposure to EGME, EGEE, and their Acetates*, DHHS (NIOSH) Publication No. 91–119, 1991.
145. A. Edens, Proposed regulation for occupational exposure to select glycol ethers in the United States. *Occup. Hyg.* **2**, 445–449 (1996).
146. F. A. Rudi, Hygienic standardization of vinyl ethers of glycols. *Gig. Sanit.* **11**, 94–97 (1974); *Chem Abstr.* **82**, 133649H (1974).
147. L. D. Gadaskina and F. A. Rudi, On the mechanism of toxic action of vinyl glycol ethers. *Gig. Tr. Prof. Zabol.* **2**, 31–33 (1976).
148. ECETOC, *The Toxicology of Glycol Ethers and Its Relevance to Man*, Technical Report No. 64, Brussels, Aug., 1995.
149. A. V. Bitkina and L. A. Mal'Kova, Toxicity of vinyloxyethanol. *Gig. Tr. Prof. Zabol.* **12**, 60 (1989).
150. D. E. Clapp et al., Workplace assessment of exposure to 2-ethoxyethanol. *Applied Industrial Hygiene* **2**, 183–187 (1987).
151. S. Sohnlein et al., Occupational chronic exposure to organic solvents: XIV. Examinations concerning the evaluation of a limit value for 2-ethoxyethanol and 2-ethoxyethylacetate and

the genotoxic effects of these glycol ethers. *Int. Arch. Occup. Environ. Health* **64**, 479–484 (1993).

152. L. K. Lowry, The biological exposure index: Its use in assessing chemical exposures in the workplace. *Toxicology* **47**, 55–69 (1987).

153. L. K. Lowry et al., Applications of biological monitoring in occupational health practice: practical application of urinary 2-ethoxyacetic acid to assess exposure to 2-ethoxyethyl acetate in large format silk-screening operations. *Int. Arch. Occup. Environ. Health* **65**, S47–S51 (1993).

154. H. F. Smyth, Jr., C. P. Carpenter, and C. S. Weil, Range-finding toxicity data. List IV. *Arch. Ind. Hyg. Occup. Med.* **4**, 119–122 (1951).

155. E. P. Laug et al., The toxicology of some glycols and derivatives. *J. Ind. Hyg. Toxicol.* **21**, 173–201 (1939).

156. G. L. Sparschu, The Dow Chemical Company, unpublished data, 1982.

157. W. F. von Oettingen and E. A. Jirouch, The pharmacology of ethylene glycol and some of its derivatives in relation to their chemical constitution and physical properties. *J. Pharmacol. Exp. Therap.* **42**, 355–382 (1931).

158. C. P. Waite, F. A. Patty, and W. P. Yant, *U.S. Public Health Rep.* **45**, 1459 (1930).

159. M. E. Goldberg et al., Effect of repeated inhalation of vapors of industrial solvents on animal behavior. 1. evaluation of nine solvent vapors on pole-climb performance in rats. *Am. Ind. Hyg. Assoc. J.* **25**, 369–375 (1964).

160. A. A. Kasparov, N. V. Marinenko, and E. I. Marienko, *Nauch. Tr. Irkutsk. Med. Inst.* **115**, 26 (1972).

161. C. S. Weil and R. Scala, Study of intra- and interlaboratory variability in results of rabbit eye and skin irritation tests. *Toxicol. Appl. Pharmacol.* **19**, 276–360 (1971).

162. P. T. Kan, M. A. Simetskii, and V. I. Tlyashohanko, *Vses. Nauchn. Issled. Inst. Vet. Sanit. Tr. Moscow* **39**, 369 (1971).

163. Union Carbide Corporation, unpublished data, 1962.

164. S. J. Barbee et al., Subchronic inhalation toxicology of ethylene glycol monoethyl ether in the rat and rabbit. *Environ. Health Perspect.* **57**, 157–164 (1984).

165. K. L. Cheever et al., Metabolism and excretion of 2-ethoxyethanol in the adult male rat. *Environ. Health Perspect.* **57**, 241–248 (1984).

166. C. H. Kennedy et al., Effect of dose on the disposition of 2-ethoxyethanol after inhalation by F344/N rats. *Fundam. Appl. Toxicol.* **21**, 486–491 (1993).

167. M. A. Medinsky et al., Disposition of three glycol ethers administered in drinking water to male F344/N rats. *Toxicol. Appl. Pharmacol.* **102**, 443–455 (1990).

168. P. J. Sabourin et al., Effect of dose on the disposition of methoxyethanol, ethoxyethanol, and butoxyethanol administered dermally to male F344/N rats. *Fundam. Appl. Toxicol.* **19**, 124–132 (1992).

169. P. J. Sabourin et al., Effect of dose on the disposition of methoxyethanol, ethoxyethanol, and butoxyethanol administered dermally to male F344/N rats. [Erratum to document]. *Fundam. Appl. Toxicol.* **20**, 508–510 (1993).

170. S. C. J. Sumner and T. R. Fennell, A possible mechanism for the formation of $^{14}CO_2$ via 2-methoxyacetic acid in mice exposed to ^{14}C-labeled 2-methoxyethanol. *Toxicol. Appl. Pharmacol.* **120**(1), 162–164 (1993).

171. D. J. Oudiz et al., Male reproductive toxicity and recovery associated with acute ethoxyethanol exposure in rats. *J. Toxicol. Environ. Health* **13**, 763–775 (1984).
172. J. C. Lamb et al., Reproductive toxicity of ethylene glycol monoethyl ether tested by continuous breeding of CD-1 mice. *Environ. Health Perspect.* **57**, 85–90 (1984).
173. H. Zenick, D. Oudiz, and R. J. Niewenhuis, Spermatotoxicity associated with acute and subchronic ethoxyethanol treatment. *Environ. Health Perspect.* **57**, 225–232 (1984).
174. K. Ninomiya et al., Analysis of rat epididymal sperm motility using the Hamilton-Thorne IVOS sperm analyzer (2nd report). *Teratology* **52**, 16B–17B (1995).
175. F. D. Andrew et al., *Final Report to NIOSH on Contract #210-79-0037*, 1981.
176. F. D. Andrew and B. D. Hardin, Developmental effects after inhalation exposure of gravid rabbits and rats to ethylene glycol monoethyl ether. *Environ. Health Perspect.* **57**, 13–24 (1984).
177. P. J. Wier, S. C. Lewis, and K. A. Traul, A comparison of developmental toxicity evident at term to postnatal growth and survival using ethylene glycol monoethyl ether, ethylene glycol monobutyl ether and ethanol. *Teratogenesis Carcinog. Mutagen.* **7**, 55–64 (1987).
178. B. D. Hardin et al., Teratogenicity of 2-ethoxyethanol by dermal application. *Drug Chem. Toxicol.* **5**(3), 277–294 (1982).
179. B. D. Hardin, P. T. Goad, and J. R. Burg, Developmental toxicity of four glycol ethers applied cutaneously to rats. *Environ. Health Perspect.* **57**, 69–74 (1984).
180. J. E. Doe, Ethylene glycol monoethyl ether and ethylene glycol monoethyl ether acetate teratology studies. *Environ. Health Perspect.* **57**, 33–42 (1984).
181. P. T. Goad and J. M. Cranmer, Gestation period sensitivity of ethylene glycol monoethyl ether in rats. *The Toxicologist* **4**, 87 (Abstract 345) (1984).
182. W. S. Webster, P. D. C. Brown-Woodman, and H. E. Ritchie, A review of the contribution of whole embryo culture to the determination of hazard and risk in teratogenicity testing. *Int. J. Develop. Biol.* **41**, 329–335 (1997).
183. P. D. Brown-Woodman et al., Induction of birth defects by exposure to solvents: an *in vitro* study. *Teratology* **51**, 288 (1995).
184. B. K. Nelson et al., Ethoxyethanol behavioral teratology in rats. *Neurotoxicology* **2**, 231–249 (1981).
185. B. K. Nelson, W. S. Brightwell, and J. V. Setzer, Prenatal interactions between ethanol and the industrial solvent 2-ethoxyethanol in rats: Maternal behavioral teratogenic effects. *Neurobehav. Toxicol. Teratol.* **4**, 387–394 (1982).
186. B. K. Nelson et al., Prenatal interactions between ethanol and the industrial solvent 2-ethoxyethanol in rats: neurochemical effects in the offspring. *Neurobehav. Toxicol. Teratol.* **4**, 395–410 (1982).
187. J. C. Hoflack et al., Mutagenicity of ethylene glycol ethers and of their metabolites in *Salmonella typhimurium* his$^-$. *Mutat. Res.* **341**, 281–287 (1995).
188. H. Fucik, *Prac. Lek.* **21**, 116 (1969).
189. J. Angerer et al., Occupational chronic exposure to organic solvents. XIII. Glycol ether exposure during the production of varnishes. *Int. Arch. Occup. Environ. Health* **62**, 123–126 (1990).
190. R. G. Browning and S. C. Curry, Clinical toxicology of ethylene glycol monoalkyl ethers. *Human and Experimental Toxicology* **13**, 325–335 (1994).

191. F. G. Bartnik et al., Percutaneous absorption, metabolism and hemolytic activity of N-butoxyethanol. *Fundam. Appl. Toxicol.* **8**, 59–70 (1987).

192. G. Johanson and P. Fernstrom, Influence of water on the percutaneous absorption of 2-butoxyethanol in the guinea pig. *Scand. J. Work Environ. Health* **14**, 95–100 (1988).

193. D. Groeseneken, H. Veulemans, and R. Masschelein, Urinary excretion of ethoxyacetic acid after experimental human exposure to ethylene glycol monoethyl ether. *Br. J. Ind. Med.* **43**(8), 544–549 (1986).

194. D. Groeseneken, H. Veulemans, and R. Masschelein, Urinary excretion of ethoxyacetic acid after experimental human exposure to ethylene glycol monoethyl ether. *Br. J. Ind. Med.* **43**, 615–619 (1986).

195. J. Laitinen et al., Urinary biochemistry in occupational exposure to glycol ethers. *Chemosphere* **29**, 781–787 (1994).

196. H. Veulemans et al., Exposure to ethylene glycol ethers and spermatogenic disorders in man: A case control study. *Brit. J Ind. Med.* **50**, 71–78 (1993).

197. L. S. Welch et al., Effects of exposure to ethylene glycol ethers on shipyard painters: II. Male reproduction. *Am. J. Ind. Med.* **14**, 509–26 (1988).

198. J. M. Ratcliffe et al., Semen quality in workers exposed to 2-ethoxyethanol. *Brit. J Ind. Med.* **46**, 399–406 (1989).

199. K. J. Olson, The Dow Chemical Company, unpublished data, 1959.

200. G. V. Katz, W. J. Krasavage, and C. J. Terhaar, Comparative acute and subchronic toxicity of ethylene glycol monopropyl ether and ethylene glycol monopropyl ether acetate. *Environ. Health Perspect.* **57**, 165–176 (1984).

201. National Institute for Occupational Safety and Health, *Doc. No. PB86-197605*, Available from the National Technical Information Service, Springfield, VA., 1984.

202. C. P. Carpenter, H. F. Smyth, Jr., and U. C. Pozzani, The assay of acute vapor toxicity and the grading and interpretation of results on 96 chemical compounds. *J. Ind. Hyg. Toxicol.* **31**, 343 (1949).

203. G. V. Katz, The basic toxicity of 2-propoxyethanol, unpublished data, Eastman Kodak Company, Report No. 104723Q, 1978.

204. D. C. Topping, Pulmonary function in animals exposed to 2-propoxyethanol by inhalation, unpublished data, Eastman Kodak Company, Report No. TX-83-16, 1983.

205. H. F. Smyth, Jr., et al., Range-finding toxicity data. List VII. *J. Am. Ind. Hyg. Assoc.* **30**, 470–476 (1969).

206. K. Nagano et al., Experimental studies on toxicity of ethylene glycol alkyl ethers in Japan. *Environ. Health Perspect.* **57**, 75–84 (1984).

207. G. V. Katz, Subchronic inhalation toxicity study for ethylene glycol monopropyl ether in the rat. Unpublished data, Eastman Kodak Company, Report #230857C (TX-86-216), Corporate Health and Environment Laboratories, June 9, 1987.

208. E. D. Barber et al., A comparative study of the rates of *in vitro* percutaneous absorption of eight chemicals using rat and human skin. *Fundam. Appl. Toxicol.* **19**, 493–497 (1992).

209. R. J. Boatman, L. G. Perry, and J. C. English, *The Disposition and Pharmacokinetics of Ethylene Glycol Mono Propyl Ether in Male Sprague-Dawley Rats After Intravenous or Oral Administration, Nose-Only Inhalation, or Dermal Exposure*, unpublished data, Eastman Chemical Company, *Report No. TX-96-190*, September 18, 1998.

210. B. I. Ghanayem et al., Effects of dose, age, inhibition of metabolism and elimination on the toxicokinetics of 2-butoxyethanol and its metabolites. *J. Pharm. Exp. Therap.* **253**, 136–143 (1990).
211. B. D. Hardin et al., Evaluation of 60 chemicals in a preliminary developmental toxicity test. *Teratogenesis, Carcinogenesis, and Mutagenesis* **7**, 29–48 (1987).
212. W. J. Krasavage, R. S. Hosenfeld, and G. V. Katz, Ethylene glycol monopropyl ether. A developmental toxicity study in rabbits. *Fundam. Appl. Toxicol.* **15**, 517–527 (1990).
213. W. J. Krasavage and G. V. Katz, Developmental toxicity of ethylene glycol monopropyl ether in the rat. *Teratology* **32**, 93–102 (1985).
214. F. Welsch and D. B. Stedman, Inhibition of metabolic cooperation between Chinese hamster V79 cells by structurally diverse teratogens. *Teratogenesis Carcinog. Mutagen.* **4**, 285–301 (1984).
215. R. Loch-Caruso, J. E. Trosko, and I. A. Corcos, Interruption of cell-cell communication in Chinese hamster V79 cells by various alkyl glycol ethers: Implications for teratogenicity. *Environ. Health Perspect.* **57**, 119–124 (1984).
216. S. J. Rawlings et al., The teratogenic potential of alkoxy acids in post-implantation rat embryo culture: structure-activity relationships. *Toxicol. Lett.* **28**, 49–58 (1985).
217. B. M. Jakobsen, *In vitro* embryotoxicity of glycol ethers and alkoxyacetic acids. *Teratology.* **51**(6), 25A (Abstract) (1995).
218. P. M. D. Foster, S. C. Lloyd, and D. M. Blackburn, Testicular toxicity of 2-methoxyacetaldehyde, a possible metabolite of ethylene glycol monomethyl ether, in the rat. *The Toxicologist.* **5**, 115 (Abstract 460) (1985).
219. T. J. B. Gray et al., Studies on the toxicity of some glycol ethers and alkoxyacetic acids in primary testicular cell cultures. *Toxicol. Appl. Pharmacol.* **79**, 490–501 (1985).
220. B. I. Ghanayem, L. T. Burka, and H. B. Matthews, Structure-activity relationships for the *in vitro* hematotoxicity of *N*-alkoxyacetic acids, the toxic metabolites of glycol ethers. *Chem. Biol. Interact.* **70**, 339–352 (1989).
221. R. J. Boatman, L. G. Perry, and V. E. Bialecki, The *in vitro* hemolytic activity of propoxyacetic acid in both rat and human blood. *The Toxicologist* **13**, 364 (Abstract 1424) (1993).
222. M. P. Dieter et al., The chemotherapeutic potential of glycol alkyl ethers: structure-activity studies of nine compounds in a fischer-rat leukemia transplant model. *Cancer Chemotherapy and Pharmacology.* **26**, 173–180 (1990).
223. K. P. Shepard, Skin sensitization study (Buehler method) of 2-propoxyethanol. Unpublished data, Eastman Kodak Company, Report TX-88-62, 1988.
224. K. P. Shepard, Skin sensitization study (footpad method) of 2-propoxyethanol. Unpublished data, Eastman Kodak Company, Report TX-88-61, 1988.
225. L. G. Bernard, A subchronic inhalation study of ethylene glycol monopropyl ether in rats using a functional observational battery and neuropathology to detect neurotoxicity. Unpublished data, Eastman Kodak Company, Report TX-89-91, 1989.
226. Dutch Expert Committee on Occupational Standards, June, 1993.
227. Deutsche Forschungsgemeinschaft, *List of MAK and BAT Values 1998: Maximum Concentrations and Biological Tolerance Values at the Workplace.* Commission for the Investigation of Health Hazards of Chemical Compounds in the Work Area, Report No. 34, Wiley-VCH Verlag GmbH, Germany, 1998.

228. M. A. Wolf, The Dow Chemical Company, unpublished data, 1959.

229. G. L. Kennedy, Jr. and G. J. Graepel, Acute toxicity in the rat following either oral or inhalation exposure. *Toxicol. Lett.* **56**, 317–326 (1991).

230. D. M. Samuels, J. E. Doe, and D. J. Tinston, The effects on the rat testis of single inhalation exposures to ethylene glycol monoalkyl ethers in particular ethylene glycol monomethylether. *Arch. Toxicol.* **7**, 167–170 (1984).

231. J. C. Gage, The subacute inhalation toxicity of 109 industrial chemicals. *Br. J. Ind. Med.* **27**, 1–18 (1970).

232. J. E. Doe, Further studies on the toxicology of the glycol ethers with emphasis on rapid screening and hazard assessment. *Environ. Health Perspect.* **57**, 199–206 (1984).

233. D. Zissu, Experimental study of cutaneous tolerence to glycol ethers. *Contact Dermatitis.* **32**, 74–77 (1995).

234. B. S. Moffett, S. Linnet, and D. Blair, Toxicology of isopropyl oxitol. Inhalation exposure of dogs, rabbits, guinea pigs and rats. Group Res. Report TLGR 0039.76., Shell Research Ltd., London, 1976.

235. A. S. Lykova et al., Data on hygienic standardization of monoisopropyl and monobutyl esters of ethylene glycol in the atmosphere. *Gig Sanit.* **11**, 7–11 (1976).

236. D. H. Hutson and B. A. Pickering, The metabolism of isopropyl oxitol in rat and dog. *Xenobiotica.* **1**, 105–119 (1971).

237. R. W. Tyl et al., *Draft Final Report on the Developmental Toxicity Evaluation of Inhaled Isopropyl Cellosolve (Ethylene Glycol Monoisopropyl Ether, EGIE) Vapor in CD (Sprague-Dawley) Rats.* Chemical Industry Institute of Toxicology, for Union Carbide Corporation, Study No. 96U1661, Aug. 20, 1997.

238. F. Welsch and D. B. Stedman, Inhibition of intercellular communication between normal human embryonal palatal mesenchyme cells by teratogenic glycol ethers. *Environ. Health Perspect.* **57**, 125–134 (1984).

239. V. O. Wagner, *Bacterial Reverse Mutation Assay with an Independent Repeat Assay.* Final Report to Union Carbide Corporation, Laboratory Study No. G96BH00.502001, Microbiological Associates, Rockville, MD, Nov. 22, 1996.

240. Chemical Manufacturers Association, *Dissolving the Myths About EGBE and Health*, 1989.

241. G. Bormett, M. J. Bartels, and D. A. Markham, Determination of 2-butoxyethanol and butoxyacetic acid in rat and human blood by gas chromatography/mass spectrometry. *J. Chromatogr. B.* **665**, 315–325 (1995).

242. G. Johanson and S. Johnsson, Gas chromatographic determination of butoxyacetic acid in human blood after exposure to 2-butoxyethanol. *Arch. Toxicol.* **65**, 433–435 (1991).

243. G. Johanson et al., Toxicokinetics of inhaled 2-butoxyethanol (ethylene glycol monobutyl ether) in man. *Scand. J. Work Environ. Health* **12**, 594–602 (1986).

244. J. Angerer et al., Occupational chronic exposure to organic solvents XIII. Glycol ether exposure during the production of varnishes. *Int. Arch. Occup. Environ. Health* **62**, 123–126 (1990).

245. *NIOSH Criteria for a Recommended Standard, Occupational Exposure to Ethylene Glycol Butyl Ether and Ethylene Glycol Butyl Ether Acetate, DHHS.* NIOSH Publication 90–118, 1990.

246. ECETOC, *Butoxyethanol Criteria Document*, Special Report No. 7, European Centre for Ecotoxicology and Toxicology of Chemicals, April, 1994.

247. R. Vincent et al., Occupational exposure to 2-butoxyethanol for workers using window cleaning agents. *Appl. Occup. Environ. Hygiene* **8**(6), 580–586 (1993).
248. V. Haufroid et al., Biological monitoring of workers exposed to low levels of 2-butoxyethanol. *Int. Arch. Occup. Environ. Health* **70**, 232–236 (1997).
249. A. W. Rettenmeier, R. Hennigs, and R. Wodarz, Determination of butoxyacetic acid and *N*-butylacetyl-glutamine in urine of lacquerers exposed to 2-butoxyethanol. *Int. Arch. Occup. Environ. Health* **65**, S151–S153 (1993).
250. T. Sakai et al., Gas chromatographic determination of butoxyacetic acid after hydrolysis of conjugated metabolites in urine from workers exposed to 2-butoxyethanol. *Int. Arch. Occup. Environ. Health* **66**, 249–254 (1994).
251. J. P. Collinot et al., Evaluation of urinary D-glucaric acid excretion in workers exposed to butyl glycol. *J. Toxicol. Environ. Health* **48**, 349–358 (1996).
252. J. Laitinen, J. Liesivuori, and H. Savolainen, Urinary NAG and GAG as biomarkers of renal effects in exposure to 2-alkoxyalcohols and their acetates. *J. Occup. Environ. Med.* **40**(7), 595–600 (1998).
253. T. R. Tyler, Acute and subchronic toxicity of ethylene glycol monobutyl ether. *Environ. Health Perspect.* **57**, 185–192 (1984).
254. European Chemical Industry Ecology and Toxicology Centre (ECETOC), Belgium, *Technical Report #17*, 1985.
255. B. I. Ghanayem and C. A. Sullivan, Assessment of the haemolytic activity of 2-butoxyethanol and its major metabolite, butoxyacetic acid, in various mammals including humans. *Human and Exper. Toxicol.* **12**, 305–311 (1993).
256. B. I. Ghanayem et al., Effect of age on the toxicity and metabolism of ethylene glycol monobutyl ether (2-butoxyethanol) in rats. *Toxicol. Appl. Pharmacol.* **91**, 222–234 (1987).
257. R. Gingell, R. J. Boatman, and S. Lewis, Acute toxicity of ethylene glycol mono-*n*-butyl ether in the guinea pig. *Food and Chem. Toxicol.* **36**(9/10), 825–829 (1998).
258. D. E. Dodd et al., Ethylene glycol monobutyl ether: Acute, 9-day and 90-day vapour inhalation studies in Fischer 344 rats. *Toxicol. Appl. Pharmacol.* **68**, 405–414 (1983).
259. D. J. Allen, *Ethyleneglycol Monobutylether: Acute Dermal Toxicity (Limit Test) in the Rat, Project No. 13/540*, Safepharm Laboratories, Ltd., Derby, U.K., Report to Mitsubishi Petrochemical Co., Ltd., Tokyo, Japan, 1993.
260. D. J. Allen, *Ethyleneglycol Monobutylether: Acute Dermal Toxicity (Limit Test) in the Rat, Project No. 13/542*, Safepharm Laboratories, Ltd., Derby, U.K., Report to Mitsubishi Petrochemical Co., Ltd., Tokyo, Japan, 1993.
261. R. I. Roudabush et al., Comparative acute effects of some chemicals on the skin of rabbits and guinea pigs. *Toxicol. Appl. Pharmacol.* **7**, 559–565 (1965).
262. D. J. Allen, *Ethyleneglycol Monobutylether: Acute Dermal Toxicity Test in the Rabbit, Project No. 13/605*, Safepharm Laboratories, Ltd., Derby, U.K., Report to Mitsubishi Petrochemical Co., Ltd., Tokyo, Japan, 1993.
263. D. J. Allen, *Ethyleneglycol Monobutylether: Acute Dermal Toxicity Test in the Rabbit, Project No. 13/606*, Safepharm Laboratories, Ltd., Derby, U.K., Report to Mitsubishi Petrochemical Co., Ltd., Tokyo, Japan, 1993.
264. H. E. Kennah et al., An objective procedure for quantitating eye irritation based upon changes of cornea thickness. *Fundam. Appl. Toxicol.* **12**, 258–268 (1989).
265. W. J. Krasavage, Subchronic oral toxicity of ethylene glycol monobutyl ether in male rats. *Fundam. Appl. Toxicol.* **6**, 349–355 (1986).

266. D. W. Hobson et al., Evaluation of the subchronic toxicity of diethylene glycol monobutyl ether administered orally to rats. *National Technical Information Service P89-1554*, 1987. Also reported in *The Toxicologist* **6**, 164 (Abstract 659) (1986).

267. *NTP Technical Report on Toxicity Studies of Ethylene Glycol Ethers: 2-Methoxyethanol, 2-Ethoxyethanol, 2-Butoxyethanol Administered in Drinking Water to F344/N Rats and B6C3F1 Mice, Toxicity Report Series No. 26*, National Toxicology Program, Research Triangle Park, NC, 1993.

268. *NTP Technical Report on Toxicology and Carcinogenesis Studies of 2-Butoxyethanol (CAS No. 111-76-2) in F344/N Rats and B6C3F1 Mice (Inhalation Studies)*, peer reviewed on Oct. 30, 1998, *NTP TR 484, NIH Publication No. 98-3974*, U.S. Department of Health and Human Services, Public Health Service, National Institutes of Health, 1998.

269. J. I. Everitt, P. W. Ross, and T. W. Davis, Urologic syndrome associated with wire caging in AKR mice. *Lab. Anim. Sci.* **38**, 609-611 (1988).

270. D. A. Mayhew, W. F. Sigier, and S. C. Pepple, *90-day Subchronic Dermal Toxicity Study in Rabbits with Ethylene Glycol Monobutyl Ether*, WIL Research Laboratories, Inc., Report to the Chemical Manufacturers Association, 1983.

271. G. Johanson, Inhalation toxicokinetics of butoxyethanol and its metabolite butoxyacetic acid in the male Sprague-Dawley rat. *Arch. Toxicol.* **68**, 588-594 (1994).

272. J. A. Dill et al., Toxicokinetics of inhaled 2-butoxyethanol and its major metabolite, 2-butoxyacetic acid, in F344 Rats and B6C3F$_1$ mice. *Toxicol. Appl. Pharmacol.* **153**, 227-242 (1998).

273. B. I. Ghanayem et al., Metabolism and disposition of ethylene glycol monobutyl ether (2-butoxyethanol) in rats. *Drug Metab. Dispos.* **15**, 478-484 (1987).

274. P. J. Sabourin et al., Effect of exposure concentrations on the disposition of inhaled butoxyethanol by F344 rats. *Toxicol. Appl. Pharmacol.* **114**, 232-238 (1992).

275. A. K. Jonsson and G. Steen, *n*-Butoxyacetic acid, a urinary metabolite from inhaled *n*-butoxyethanol (butylcellosolve). *Acta Pharmacol. Toxicol.* **42**, 354-356 (1978).

276. G. Johanson, Physiologically based pharmacokinetic modeling of inhaled 2-butoxyethanol in man. *Toxicol. Lett.* **34**, 23-31 (1986).

277. G. Johanson, M. Wallen, and M. B. Nordqvist, Elimination kinetics of 2-butoxyethanol in the perfused rat liver - dose dependence and effect of ethanol. *Toxicol. Appl. Pharmacol.* **83**, 315-320 (1986).

278. L. J. Shyr et al., Physiologically based modeling of 2-butoxyethanol disposition in rats following different routes of exposure. *Environ. Research* **63**, 202-218 (1993).

279. R. A. Corley, G. A. Bormett, and B. I. Ghanayem, Physiologically based pharmacokinetics of 2-butoxyethanol and its major metabolite, 2-butoxyacetic acid, in rats and humans. *Toxicol. Appl. Pharmacol.* **129**, 61-79 (1994).

280. K. M. Lee et al., Physiologically based pharmacokinetic model for chronic inhalation of 2-butoxyethanol. *Toxicol. Appl. Pharmacol.* **153**, 211-226 (1998).

281. L. Aasmoe, J-O. Winberg, and J. Aarbakke, The role of liver alcohol dehydrogenase isoenzymes in the oxidation of glycolethers in male and female rats. *Toxciol. Appl. Pharmacol.* **150**, 86-90 (1998).

282. B. I. Ghanayem, L. T. Burka, and H. B. Matthews, Metabolic basis of ethylene glycol monobutyl ether (2-butoxyethanol) toxicity: role of alcohol and aldehyde dehydrogenases. *J. Pharm. Exp. Therap.* **242**, 222-231 (1987).

283. B. I. Ghanayem, Metabolic and cellular basis of 2-butoxyethanol induced hemolytic anemia in rats and assessment of human risk *in vitro*. *Biochem. Pharmacol.* **38**, 1679–1684 (1989).

284. B. I. Ghanayem, I. M. Sanchez, and H. B. Matthews, Development of tolerance to 2-butoxyethanol-induced hemolytic anemia and studies to elucidate the underlying mechanisms. *Toxicol. Appl. Pharmacol.* **112**, 198–206 (1992).

285. D. V. Sivarao and H. M. Mehendale, 2-Butoxyethanol autoprotection is due to resilience of newly formed erythrocytes to hemolysis. *Arch. Toxicol.* **69**, 526–532 (1995).

286. S. Ward, C. Wall, and B. I. Ghanayem, Effects of 2-butoxyethanol (BE) and its toxic metabolite, 2-butoxyacetic acid (BAA) on blood from various mammals *in vivo* and *in vitro*. *The Toxicologist* **12**, 282 (Abstract 1079) (1992).

287. M. M. Udden, and C. S. Patton, Hemolysis and decreased deformability of erythrocytes exposed to butoxyacetic acid, a metabolite of 2-butoxyethanol. I. Sensitivity in rats and resistance in normal humans. *J. Appl. Toxicol.* **14**, 91–96 (1994).

288. M. M. Udden, Hemolysis and decreased deformability of erythrocytes exposed to butoxyacetic acid, a metabolite of 2-butoxyethanol: II. Resistance in red blood cells from humans with potential susceptibility. *J. Appl. Toxicol.* **14**, 97–102 (1994).

289. G. Johanson and P. Fernstrom, Influence of water on the percutaneous absorption of 2-butoxyethanol in guinea pigs. *Scand. J. Work Environ. Health* **14**, 95–100 (1988).

290. P. Foster, S. C. Lloyd, and D. M. Blackburn, Comparison of the *in vivo* and *in vitro* testicular effects produced by methoxy-, ethoxy- and *N*-butoxyacetic acids in the rat. *Toxicology* **43**, 17–30 (1987).

291. J. J. Heindel et al., Assessment of ethylene glycol monobutyl and monophenyl ether reproductive toxicity using a continuous breeding protocol in Swiss CD-1 mice. *Fundam. Appl. Toxicol.* **15**, 683–696 (1990).

292. R. W. Tyl et al., Teratologic evaluation of ethylene glycol monobutyl ether in Fischer 344 rats and New Zealand white rabbits following inhalation exposure. *Environ. Health Perspect.* **57**, 47–68 (1984).

293. R. B. Sleet, C. J. Price, and M. C. Marr, *National Toxicology Program, Report NTP-89-058. NTIS PB89-165849*, Springfield, VA, 1989.

294. J. R. Hailey et al., Impact of *Helicobacter hepaticus* infection in B6C3F1 mice from twelve national toxicology program two-year carcinogenesis studies. *Toxicol. Path.* **26**(5), 602–611 (1998).

295. J. G. Fox et al., Comparison of methods of identifying *Helicobacter hepaticus* in B6C3F$_1$ mice used in a carcinogenesis bioassay. *J. Clin. Microbiol.* **36**(5), 1382–1387 (1998).

296. R. W. Tennant and J. Ashby, Classification according to chemical structure, mutagenicity to salmonella and level of carcinogenicity of a further 39 chemicals tested for carcinogenicity by the U.S. National Toxicology Program. *Mutat. Res.* **257**, 209–227 (1991).

297. B. M. Elliot and J. Ashby, Review of the genotoxicity of 2-butoxyethanol. *Mutat. Res.* **387**, 89–96 (1997).

298. J. C. Hoflack et al., Mutagenicity of ethylene glycol ethers and of their metabolites in *Salmonella typhimurium* his$^-$. *Mutat. Res.* **341**, 281–287 (1995).

299. B. B. Gollapudi et al., Re-examination of the mutagenicity of ethylene glycol monobutyl ether to salmonella tester strain TA97a. *Mutat. Res.* **370**, 61–64 (1996).

300. Z. Elias et al., Genotoxic and/or epigenetic effects of some glycol ethers: results of different short-term tests. *Occup. Hygiene* **2**, 187–212 (1996).

301. J. C. Hoflack et al., Alteration in methyl-methanesulfonate-induced poly(ADP-ribosyl)ation by 2-butoxyethanol in Syrian hamster embryo cells. *Carcinogenesis* **18**(12), 2333–2338 (1997).

302. M. D. Rambourg-Schepens et al., Severe ethylene glycol butyl ether poisoning kinetics and metabolic pattern. *Human Toxicol.* **7**, 187–189 (1988).

303. F. P. Gisgenbergh et al., Acute butylglycol intoxication: a case report. *Human Toxicol.* **8**, 243–245 (1989).

304. K. G. Romer, F. Balge, and K. J. Freundt, Ethanol-induced accumulation of ethylene glycol monoalkyl ethers in rats. *Drug Chem. Toxicol.* **8**, 255–264 (1989).

305. M. M. Udden, Effects of butoxyacetic acid on human red blood cells. *Occup. Hygiene* **2**, 283–290 (1996).

306. K. K. Burkhart and J. W. Donovan, Hemodialysis following butoxyethanol ingestion. *Clin. Toxicol.* **36**(7), 723–725 (1998).

307. J. Gualtieri et al., Multiple 2-butoxyethanol intoxications in the same patient: clinical findings, pharmacokinetics, and therapy. *J. Toxicol. Clinical Toxicol.* **33**(5), 550–551 (1995).

308. P. Bauer et al., Transient non-cardiogenic pulmonary edema following massive ingestion of ethylene glycol butyl ether. *Intensive Care Med.* **18**, 250–251 (1992).

309. B. S. Dean and E. P. Krenzelok, Clinical evaluation of pediatric ethylene glycol monobutyl ether poisoning. *Vet. Hum. Toxicol.* **33**(4), 362 (abstract #41) (1991).

310. B. S. Dean and E. P. Krenzelok, Clinical evaluation of pediatric ethylene glycol monobutyl ether poisoning. *J. Toxicol. Clin. Toxicol.* **30**, 557–563 (1992).

311. G. Johanson and A. Boman, Percutanous absorption of 2-butoxyethanol vapour in human subjects. *Br. J. Ind. Med.* **45**, 788–792 (1991).

312. R. A. Corley et al., Physiologically based pharmacokinetics and the dermal absorption of 2-butoxyethanol vapor by humans. *Fundam. Appl. Toxicol.* **39**, 120–130 (1997).

313. G. Johanson and S. Johnsson, Gas chromatographic determination of butoxyacetic acid in human blood after exposure to 2-butoxyethanol. *Arch. Toxicol.* **65**, 433–435 (1991).

314. G. Johanson, A. Boman, and B. Dynesius, Percutaneous absorption of 2-butoxyethanol in man. *Scand. J. Work Environ. Health* **14**, 101–109 (1988).

315. A. H. Greespan et al., Human repeated insult patch test of 2-butoxyethanol. *Contact Dermatitis* **33**, 59–60 (1995).

316. S. Cordier et al., Congenital malformations and maternal occupational exposure to glycol ethers. *Epidemiology* **8**, 355–363 (1997).

317. M-C. Ha et al., Congenital malformations and occupational exposure to glycol ethers: a European collaborative case-control study. *Occup. Hygiene* **2**, 417–421 (1996).

318. American Conference of Governmental Industrial Hygienists, Documentation of TLVs, 1986.

319. C. A. Staples, R. J. Boatman, and M. L. Cano, Ethylene glycol ethers: An environmental risk assessment. *Chemosphere* **36**(7), 1585–1613 (1998).

320. Eastman Kodak Company, unpublished report M-165, 1969.

321. K. J. Olson, The Dow Chemical Company, unpublished data, 1961.

322. H. F. Smyth, Jr. et al., Range-finding toxicity data list V. *Arch. Ind. Hyg. Occup. Med.* **10**, 61–68 (1954).

323. B. Ballantyne and R. C. Myers, The comparative acute toxicity and primary irritancy of the monohexyl ethers of ethylene and diethylene glycol. *Vet. Hum. Toxicol.* **29**(5), 361–366 (1987).

324. Union Carbide Corporation, Bushy Run Research Center, Project Report 52-5, July 20, 1989.
325. D. R. Klonne et al., Acute, 9-day, and 13-week vapor inhalation studies of ethylene glycol monohexyl ether. *Fundam. Appl. Toxicol.* **8**, 198–206 (1987).
326. R. W. Tyl et al., Evaluation of the developmental toxicity of ethylene glycol monohexyl ether vapor in Fischer 344 rats and New Zealand white rabbits. *Fundam. Appl. Toxicol.* **12**, 269–280 (1989).
327. Union Carbide Corporation, Bushy Run Research Center, Project Report 48-82, June 28, 1985.
328. Union Carbide Corporation, Bushy Run Research Center, Project Report 48-108, Sept. 30, 1985.
329. Union Carbide Corporation, Bushy Run Research Center, Project Report 48-124, Oct. 7, 1985.
330. C. P. Carpenter, C. S. Weil, and H. F. Smyth, Jr., Range-finding toxicity data: list VIII. *Toxicol. Appl. Pharmacol.* **28**, 313–319 (1974).
331. W. J. Krasavage and M. S. Vlaovic, Comparative toxicity of nine glycol Ethers: III. six weeks repeated dose study. Unpublished data, Eastman Kodak Company, Corporate Health and Environment Laboratories, Report No. TX-82-06, March 15, 1982.
332. R .J. Lewis, Jr., *Hawley's Condensed Chemical Dictionary*, 11th ed., Van Nostrand Reinhold Co., New York, 1987.
333. J. E. F. Reynolds and A. B. Prasad, Eds., *Martindale-The Extra Pharmacopoeia*, 28th ed., The Pharmaceutical Press, London, 1982.
334. Dow Chemical Company, unpublished data, 1938.
335. P. A. Keeler and L. W. Rampy, The Dow Chemical Company, unpublished data, 1974.
336. Nipa Laboratories, unpublished data, report to The Dow Chemical Company, 1983.
337. W. J. Breslin et al., Hemolytic activity of ethylene glycol phenyl ether (EGPE) in rabbits. *Fundam. Appl. Toxicol.* **17**, 466–481 (1986).
338. K. J. Olson, The Dow Chemical Company, unpublished data, 1963.
339. H. D. Kirk et al., The Dow Chemical Company, unpublished data, 1985.
340. Nipa Laboratories, unpublished data, report to The Dow Chemical Company, 1977.
341. C. S. Roper et al., Percutaneous penetration of 2-phenoxyethanol through rat and human skin. *Food and Chemical Toxicology* **35**, 1009–1016 (1997).
342. C. S. Roper et al., A comparison of the absorption of a series of ethoxylates through rat skin *in vitro*. *Toxicol. In Vitro* **12**, 57–65 (1998).
343. B. H. Scortichini, J. F. Quast, and K. S. Rao, Teratologic evaluation of 2-phenoxyethanol in New Zealand white rabbits following dermal exposure. *Fundam. Appl. Toxicol.* **8**, 272–279 (1987).
344. Nipa Laboratories, unpublished data, report to The Dow Chemical Company, 1982.
345. V. A. Linscombe and B. B. Gollapudi, The Dow Chemical Company, unpublished data, 1987.
346. B. B. Gollapudi et al., Toxicology of diethylene glycol butyl ether 3. genotoxicity evaluation in an *in vitro* gene mutation assay and an *in vivo* cytogenetic test. *Am. Coll. Toxicol.* **12**, 155–159 (1993).
347. CIR, Final report on the safety assessment of phenoxyethanol. *J. Am. Coll. Toxicol.* **9**(2), 259–277 (1990).
348. H. F. Smyth, Jr. et al., Range finding toxicity data. *J. Am. Ind. Hyg. Assoc.* **23**, 95–107 (1962).

349. M. R. Plasterer et al., Developmental toxicity of nine selected compounds following prenatal exposure in the mouse. Naphthalene, *p*-nitrophenol, sodium selenite, dimethyl phthalate, ethylenethiourea, and four glycol ether derivatives. *J. Toxicol. Environ. Health* **15**(1), 25–38 (1985).

350. K. Uemura, The teratogenic effects of ethylene glycol dimethyl ether on the mouse. *Acta. Obstet. Gynaecol Jpn.* **32**(1), 113–121 (1980).

351. B. D. Hardin and C. J. Eisenmann, Relative potency of four ethylene glycol ethers for induction of paw malformations in the CD-1 mouse. *Teratology* **35**(3), 321–328 (1987).

352. D. E. Leonhardt, L. W. Coleman, and W. S. Bradshaw, Perinatal toxicity of ethylene glycol dimethyl ether in the rat. *Reprod. Toxicol.* **5**(2), 157–162 (1991).

353. Union Carbide Corporation, Bushy Run Research Center, Project Report 45-219, Feb. 10, 1983.

354. H. F. Smyth, Jr., J. Seaton, and L. Fischer, The single dose toxicity of some glycols and derivatives. *J. Ind. Hyg. Toxicol.* **23**, 259–268 (1941).

355. Union Carbide Corp., Bushy Run Research Center, Monthly Report 5-30-46, 1946.

356. Union Carbide Corp., Bushy Run Research Center, Monthly Report 10-31-52, 1952.

357. Union Carbide Corp., Bushy Run Research Center, Monthly Report 6-30-52, 1952.

358. Union Carbide Corp., Bushy Run Research Center, Special Report 9-11, 1946.

359. Union Carbide Corp., Bushy Run Research Center, Monthly Report 11-30-51, 1951.

360. J. D. George et al., The developmental toxicity of triethylene glycol dimethyl ether. *Fundam. Appl. Toxicol.* **19**, 15–25 (1992).

361. R. L. Schuler et al., Results of testing fifteen glycol ethers in a short-term *in vivo* reproductive toxicity assay. *Environ. Health Perspect.* **57**, 141–146 (1984).

362. L. Silverman, H. F. Schulte, and M. W. First, Further studies on sensory response to certain industrial solvent vapors. *J. Ind. Hyg. Toxicol.* **28**, 262–266 (1946).

363. W. P. Yant et al., *U.S. Public Health Rep.* **45**, 2023 (1930).

364. D. G. Laing, *Chem. Senses Flavor* **1**, 257 (1975).

365. A. M. Thiess, E. Tress, and I. Fleig, *Arbeitmed. Sozialmed. Praventivmed.* **11**, 35 (1976).

366. G. O. Curme, Jr. and J. Johnston, *Glycols*, American Chemical Society Monograph Series 114, Reinhold, New York, 1952.

367. E. W. Reid and H. E. Hoffman, *Ind. Eng. Chem.* **21**, 695 (1929).

368. *NIOSH Criteria for a Recommended Standard, Occupational Exposure to Dioxane*, DHHS (NIOSH) Publication, 1977.

369. S. de Navasquez, *J. Hyg.* **35**, 540 (1935).

370. W. Wirth and O. Klimmer, *Arch. Gewerbepathol. Gewerbehyg.* **17**, 192 (1936).

371. N. Nelson, *Med. Bull.* **11**, 226 (1951).

372. U. C. Pozzani, C. S. Weil, and C. P. Carpenter, The toxicological basis of threshold limit values. 5. the experimental inhalation of vapor mixtures by rats, with notes upon the relationship between single dose inhalation and single dose oral data. *Am. Ind. Hyg. Assoc. J.* **20**, 364–369 (1959).

373. A. Fairley, E. C. Linton, and A. H. Ford-Moore, *J. Hyg.* **34**, 486 (1934).

374. H. E. Christensen, Ed., *Registry of Toxic Effects of Chemical Substances*, U.S. Dept. Health, Education and Welfare, NIOSH, Rockville, MD, 1976.

375. W. H. Braun and J. D. Young, Identification of β-hydroxyethoxyacetic acid as the major urinary metabolite of 1,4-dioxane in the rat. *Toxicol. Appl. Pharmacol.* **39**, 33–38 (1977).
376. Y. Woo, J. C. Arcos, and M. F. Argus, Metabolism *in vivo* of dioxane: identification of *p*-dioxane-2-one as the major urinary metabolite. *Biochem. Pharmacol.* **26**, 1535–1538 (1977).
377. J. D. Young et al., 1,4-Dioxane and β-hydroxyethoxyacetic acid excretion in urine of humans exposed to dioxane vapors. *Toxicol. Appl. Pharmacol.* **38**, 643–646 (1976).
378. J. D. Young, W. H. Braun, and P. J. Gehring, Dose-dependent fate of 1,4-dioxane in rats. *J. Toxicol Environ. Health* **4**, 709–726 (1978).
379. S. S. Pawar and A. M. Mungikar, Dioxane-induced changes in mouse liver microsomal mixed function oxidase system. *Bull. Environ. Contam. Toxicol.* **15**, 762–767 (1976).
380. R. W. Lane, B. L. Riddle, and J. F. Borzelleca, Effects of 1,2-dichloroethane and 1,1,1-trichloroethane in drinking water on reproduction and development in mice. *Toxicol. Appl. Pharmacol.* **63**, 409–421 (1982).
381. E. Giavini, C. Vismara, and M. L. Broccia, Teratogenesis study of dioxane in rats. *Toxicol. Lett.* **26**, 85–88 (1985).
382. B. A. Schwetz, B. J. K. Leong, and P. J. Gehring, The effect of maternally inhaled trichloroethylene, perchloroethylene, methyl chloroform, and methylene chloride on embryonal and fetal development in mice and rats. *Toxicol. Appl. Pharmacol.* **32**, 84–96 (1975).
383. M. F. Argus, J. C. Arcos, and C. Hoch-Ligeti, Studies on the carcinogenic activity of protein-denaturing agents: hepatocarcinogenicity of dioxane. *J. Natl. Cancer Inst.* **35**, 949–958 (1965).
384. C. Hoch-Ligeti, M. F. Argus, and J. C. Arcos, Induction of carcinomas in the nasal cavity of rats by dioxane. *Br. J. Cancer* **24**, 164–167 (1970).
385. M. F. Argus et al., *Eur. J. Cancer* **9**, 231 (1973).
386. R. J. Kociba et al., 1,4-Dioxane. I. results of a 2-year ingestion study in rats. *Toxicol. Appl. Pharmacol.* **30**, 275–286 (1974).
387. National Cancer Institute, *Tech. Rep. No. 80, NC1-CG-TR-80*, U.S. Dept. HEW, 1978.
388. T. R. Torkelson et al., 1,4-Dioxane. II. Results of a 2-year inhalation study in rats. *Toxicol. Appl. Pharmacol.* **30**, 287–298 (1975).
389. V. B. Perone, L. D. Scheel, and W. P. Tolov, National Institute for Occupational Safety and Health, unpublished report, 1976.
390. M. E. King, A. M. Shefner, and R. R. Bates, Carcinogenesis bioassay of chlorinated dibenzodioxins and related chemicals. *Environ. Health Perspect.* **5**, 163–170 (DHEW Publications No. (NIH) 74-218) (1973).
391. W. T. Stott, J. F. Quast, and P. G. Watanabe, Differentiation of the mechanisms of oncogenicity of 1,4-dioxane and 1,3-hexachlorobutadiene in the rat. *Toxicol. Appl. Pharmacol.* **60**, 287–300 (1981).
392. The Dow Chemical Company, report submitted by Litton Bionetics, Inc., 1975.
393. F. K. Zimmermann et al., Acetone, methylethylketone, ethylacetate, acetonitrile and other polar aprotic solvents are strong inducers of Aneuploidy in *Saccharomyces Cerevisiae*. *Mutat. Res.* **149**, 339–351 (1985).
394. D. B. McGregor et al., Responses of the L5178Y mouse lymphoma cell forward mutation assay: V.27 Coded chemicals. *Environ. Mol. Mutagen.* **17**, 196–219 (1991).

395. T. L. Goldsworthy et al., Examination of potential mechanisms of carcinogenicity of 1,4-dioxane in rat nasal epithelial cells and hepatocytes. *Arch. Toxicol.* **65**, 1–9 (1991).
396. S. M. Galloway et al., Chromosome aberrations and sister chromatid exchanges in Chinese hamster ovary cells. Evaluations of 108 chemicals. *Environ. Mol. Mutagen.* **10**(10), 1–11, 22, 28–36, 72, 109, 143 (1987).
397. J. F. Sina et al., Evaluation of the alkaline elution/rat hepatocyte assay as a predictor of carcinogenic/mutagenic potential. *Mutat. Res.* **113**, 357–391 (1983).
398. C. W. Sheu et al., *In vitro* BALB/3T3 Cell transformation assay of nonoxynol 9 and 1,4 Dioxane. *Environ. Mol. Mutagen.* **11**, 41–48 (1988).
399. J. Heil and G. Reifferscheid, Detection of mammalian carcinogens with an immunological DNA synthesis-inhibition test. *Carcinogenesis* **13**(12), 2389–2394 (1992).
400. L. Hellmer and G. Bolcsfoldi, An evaluation of the E. *coli* K-12 *uvrB/recA* DNA repair host-mediated assay: I. *In vitro* sensitivity of the bacteria to 61 compounds. *Mutat. Res.* **272**, 145–160 (1992).
401. K. E. Appel, *Bundesgesundheitsblatt* **31**, 37 (1988).
402. K. T. Kitchin and J. L. Brown, Is 1,4-dioxane a genotoxic carcinogen?. *Cancer Lett.* **53**, 67–71 (1990).
403. Y. Uno et al., An *in vivo-in vitro* replicative DNA synthesis (RDS) test using rat hepatocytes as an early prediction assay for nongenotoxic hepatocarcinogens: screening of 22 known positives and 25 noncarcinogens. *Mutat. Res.* **320**, 189–205 (1994).
404. K. T. Kitchin, J. L. Brown, and A. P. Kulkarni, Predictive assay for rodent carcinogenicity using *in vivo* biochemical parameters: operational characteristics and complementarity. *Mutat. Res.* **266**, 253–272 (1992).
405. J. Ashby, The genotoxicity of 1,4-dioxane. *Mutat. Res.* **322**, 141–150 (1994).
406. H. Barber, *Guy's Hosp. Rep.* **84**, 267 (1934).
407. S. A. Henry, Annual Report of Chief Inspector of Factories, H. M. Stationery Office, London, 1934.
408. R. T. Johnstone, Death due to dioxane. *Arch. Ind. Health* **20**, 445–447 (1959).
409. J. D. Young et al., Pharmacokinetics of 1,4-dioxane in humans. *J. Toxicol. Environ. Health* **3**, 507–520 (1977).
410. R. H. Reitz et al., Development of a physiologically based pharmacokinetic model for risk assessment with 1,4-dioxane. *Toxicol. Appl. Pharmacol.* **105**, 37–54 (1990).
411. H. Leung and D. J. Paustenbach, Cancer risk assessment for dioxane based upon a physiologically-based pharmacokinetic approach. *Toxicol. Lett.* **51**, 147–162 (1990).
412. P. A. Buffler et al., Mortality follow-up of workers exposed to 1,4-dioxane. *J. Occup. Med.* **20**, 255–259 (1978).
413. Union Carbide Corporation, Bushy Run Research Center, Monthly Report 12-31-46.
414. Union Carbide Corp. *Methyl Carbitol: Acute Toxicity and Primary Irritancy Studies*, EPA Doc., No. 86-890000363, 1984.
415. H. F. Smyth, Jr. and C. P. Carpenter, Further experience with the range-finding test in the industrial toxicology laboratory. *J. Ind. Hyg. Toxicol.* **30**, 63–68 (1948).
416. T. Kawamoto et al., Acute oral toxicity of ethylene glycol monomethyl ether and diethylene glycol monomethyl ether. *Bull. Environ. Contam. Toxicol.* **44**, 602–608 (1990).
417. R. R. Miller, Diethylene glycol monomethyl ether 13-week vapor inhalation toxicity study in rats. *Fundam. Appl. Toxicol.* **5**, 1174–1179 (1985).

418. D. W. Hobson, A. P. D'Addario, and D. E. Uddin, A Comparative 90-day dermal exposure study of diethylene glycol monomethyl ether (DEGME) and ethylene glycol monomethyl ether (EGME) in the guinea pig. *The Toxicologist* **4**, 184 (Abstract 735) (1984).

419. D. W. Hobson et al., A subchronic dermal exposure study of diethylene glycol monomethyl ether and ethylene glycol monomethyl ether in the male guinea pig. *Fundam. Appl. Toxicol.* **6**, 339–348 (1986).

420. J. B. Knaak, L. J. Sullivan, and C. P. Carpenter, *Metabolism of UCON 50-HB-260 Lubricant in the Rat*, Mellon Institute Special Report 27-134, 1964.

421. J. B. Knaak, J. M. Eldridge, and L. J. Sullivan, Excretion of certain polyethylene glycol ether adducts of nonylphenol by the rat. *Toxicol. Appl. Pharmacol.* **9**, 331–340 (1966).

422. K. L. Cheever et al., Metabolism of bis(2-methoxyethyl)ether in the adult male rat. Evaluation of the principal metabolite as a testicular toxicant. *Toxicol. Appl. Pharmacol.* **94**, 150–159 (1988).

423. T. Kawamoto et al., Effect of ethylene glycol monomethyl ether and diethylene glycol monomethyl ether on hepatic metabolizing enzymes. *Toxicol.* **62**, 265–274 (1990).

424. Dow Chemical Company, *In Vitro Studies to Evaluate Glycol Ethers as Substrates for Alcohol Dehydrogenase*, EPA Document No. 86-890001231S, Fiche No. OTS0520741, 1982.

425. J. E. Jelnes and E. Soderlund, Health effects of selected chemicals 3.2-(2-methoxyethoxy) ethanol. *Nord.* **28**, 115–136 (1995).

426. C. A. Kimmel, Reproductive and developmental effects of diethylene and triethylene glycol (methyl-, ethyl-) ethers. *Occup. Hyg.* **2**, 131–151 (1996).

427. B. D. Hardin, P. T. Goad, and J. R. Burg, Developmental toxicity of diethylene glycol monomethyl ether (diEGME). *Fundam. Appl. Toxicol.* **6**, 430–439 (1986).

428. T. Yamano et al., Effects of diethylene glycol monomethyl ether on pregnancy and postnatal development in rats. *Arch. Environ. Contam. Toxicol.* **24**, 228–235 (1993).

429. J. E. Doe, Further studies on the toxicology of the glycol ethers with emphasis on rapid screening and hazard assessment. *Environ. Health Perspect.* **57**, 199–206 (1984).

430. B. H. Scortichini et al., Teratologic evaluation of dermally applied diethylene glycol monomethyl ether in rabbits. *Fundam. Appl. Toxicol.* **7**, 68–75 (1986).

431. BASF Corporation, *Standard Plate Test and Preincubation Test with Salmonella Typhimurium*, EPA Doc. No. 86-890000729, Fiche No. OTS0521259, 1989.

432. D. L. J. Opdyke, Monographs on fragrance raw materials. Diethylene glycol-monomethyl ether. *Food Cosmet. Toxicol.* **12**, 517–519 (1974).

433. S. V. Lucas, *GC/MS Analysis of Organics in Drinking Water Concentrations and Advanced Treatment Concentrates*, Vol 1, USEPA-600/1084-020A (NTIS PB85-128239), 1984.

434. H. F. Smyth, Jr., J. Seaton, and L. Fischer, The single dose toxicity of some glycols and derivatives. *J. Ind. Hyg. Toxicol.* **23**, 259–268 (1941).

435. Union Carbide Corporation, unpublished data, 1945.

436. P. J. Hanzlik, W. S. Lawrence, and G. L. Laqueur, Comparative chronic toxicity of diethylene glycol monoethyl ether (carbitol) and some related glycols. Results of continued drinking and feeding. *J. Ind. Hyg. Toxicol.* **29**, 233–241 (1947).

437. Union Carbide Corporation, unpublished data, 1968.

438. D. E. Hall et al., Short-term feeding study with diethylene glycol monoethyl ether in rats. *Food Cosmet. Toxicol.* **4**, 263–268 (1966).

439. I. F. Gaunt et al., Short-term toxicity of diethylene glycol monoethyl ether in the rat, mouse and pig. *Food Cosmet. Toxicol.* **6**, 689–705 (1968).

440. V. A. Drill, 90-Day skin absorption of diethylene glycol ethyl ether (D-17) in rabbits. The Dow Chemical Company, unpublished data, 1950.

441. H. F. Smyth, Jr., C. P. Carpenter, and C. B. Schaffer, Summary of toxicological data - A 2 year study of diethylene glycol monoethyl ether in rats. *Food Cosmet. Toxicol.* **2**, 641–642 (1964).

442. K. B. Butterworth, I. F. Gaunt, and P. Grasso, A nine month toxicity study of diethylene glycol monoethyl ether in the ferret. *BIBRA* **15**, 115 (1976).

443. C. J. Hardy et al., Twenty-eight-day repeated-dose inhalation exposure of rats to diethylene glycol monoethyl ether. *Fundam. Appl. Toxicol.* **38**, 143–147 (1997).

444. J. K. Fellows, F. P. Luduena, and P. J. Hanzlik, Glucuronic acid excretion after diethylene glycol monoethyl ether (carbitol) and some glycols. *J. Pharmacol. Exp. Therap.* **89**, 210–213 (1947).

445. J. P. Kamerling et al., (2-Ethoxyethoxy)acetic Acid. An unusual compound found in the gas chromatographic analysis of urinary organic acids. *Clin. Chim. Acta* **77**(3), 397–405 (1977).

446. J. Williams et al., Reproductive effects of diethylene glycol and diethylene glycol monoethyl ether in Swiss CD-1 mice assessed by a continuous breeding protocol. *Fundam. Appl. Toxicol.* **14**, 622–635 (1990).

447. R. E. Chapin, R. A. Sloane, and J. K. Haseman, The relationships among reproductive endpoints in Swiss mice, using the reproductive assessment by continuous breeding database. *Fundam. Appl. Toxicol.* **38**, 129–142 (1997).

448. R. E. Chapin and R. A. Sloane, Reproductive assessment by continuous breeding: evolving study design and summaries of ninety-one studies. *Environ. Health Perspect.* **105**(1), 199–395 (1997).

449. H. C. Bowden et al., Assessment of the toxic and potential teratogenic effects of four glycol ethers and two derivatives using the hydra regeneration assay and rat whole embryo culture. *Toxicol. In Vitro* **9**, 773–781 (1995).

450. Haskell Laboratories, *Mutagenic Activity of Carbitol in the Salmonella/Microsome Assay with Attachments and Cover Sheet*, EPA Document No. 86-890000829S, Fiche No. OTS0520945, 6/12/89.

451. A .G. Cranch, H. F. Smyth, Jr., and C. P. Carpenter, External contact with monoethyl ether of diethylene glycol (carbitol solvent). *Arch. Dermatol. Syphilol.* **45**, 553–559 (1942).

452. W. M. Meininger, External use of 'carbitol solvent', carbitol and other reagents. *Arch. Dermatol. Syphilol.* **58**, 19–25 (1948).

453. BASF Corporation, TSCA 8d Submission, dated 6/12/89, EPA/OTS, Doc. No. 86-890000720, NTIS/OTS0521226, Received 6/15/89.

454. W. B. Gibson et al., Diethylene glycol mono butyl ether concentrations in room air from application of cleaner formulations to hard surfaces. *J. Exp. Anal. Env. Epidemiol.* **1**, 369–383 (1991).

455. R. Gingell, Toxicology of diethylene glycol butyl ether. 1. exposure and risk assessment. *J. Am. Coll. Toxicol.* **12**(2), 139–144 (1993).

456. U. S. Environmental Protection Agency, Final test rule on DGBE and DGBEA. *Fed. Reg.* **53**, 5932 (1988).

457. C. T. Bedford, D. Blair, and D. E. Stevenson, Toxic fluorinated compounds as by-products of certain boron trifluoride catalyzed industrial processes. *Nature* **267**, 335 (1977).

458. V. K. Rowe and M. A. Wolfe, *Patty's Industrial Hygiene and Toxicology*, Wiley-Interscience, Vol. IIC, 3rd ed., New York, 1982, p. 3965.
459. B. Ballantyne and J. Cutan, Eye irritancy potential of diethylene glycol monobutyl ether. *J. Toxicol. Cut. Ocul. Toxicol.* **3**, 7–16 (1984).
460. C. S. Auletta et al., Toxicology of diethylene glycol butyl ether 4. dermal subchronic/reproduction study in rats. *J. Am. Coll. Toxicol.* **12**(2), 161–168 (1993).
461. The Dow Chemical Company, unpublished report, 1984.
462. R. J. Boatman et al., Toxicology of diethylene glycol butyl ether 2. disposition studies with ^{14}C-diethylene glycol butyl ether and ^{14}C-diethylene glycol butyl ether acetate after dermal application to rats. *J. Am. Coll. Toxicol.* **12**(2), 145–154 (1993).
463. P. J. Deisinger and D. Guest, Metabolic studies with diethylene glycol monobutyl ether acetate. *Xenobiotica* **19**(9), 981–989 (1989).
464. G. A. Nolen et al., Fertility and teratogenic studies of diethylene glycol monobutyl ether in rats and rabbits. *Fundam. Appl. Toxicol.* **5**, 1137–1143 (1985).
465. M. Ema, T. Itami, and H. Kawasaki, Teratology study of diethylene mono-*n*-butyl ether in rats. *Drug Chem. Toxicol.* **11**, 97–111 (1988).
466. E. D. Thompson et al., Mutagenicity testing of diethylene glycol monobutyl ether. *Environ. Health Perspect.* **57**, 105–112 (1984).
467. P. Beyrouty et al., Toxicology of diethylene glycol butyl ether 5. dermal subchronic neurotoxicity study in rats. *J. Am. Coll. Toxicol.* **12**(2), 169–174 (1993).
468. Procter and Gamble Company. *Consumer Exposure to DGBE from the Use of Hard Surface Cleaners*, with Cover Letter to EPA Dated 28 January 1985, EPA document FYI-OTS-0286-0471, Procter and Gamble, HES, Cincinnati, OH 1985.
469. K. Berlin, G. Johanson, and M. Lindberg, Hypersensitivity to 2-(2-butoxyethoxy)ethanol. *Contact Dermatitis* **32**, 54 (1995).
470. Union Carbide Corporation, Bushy Run Research Center, Report 13-36, May 9, 1950.
471. Union Carbide Corporation, Bushy Run Research Center, Project Report 50-55, May 29, 1987.
472. Union Carbide Corporation, Bushy Run Research Center, Project Report 50-79, June 23, 1987.
473. Union Carbide Corporation, Bushy Run Research Center, Project Report 50-119, Sept. 22, 1987.
474. Union Carbide Corporation, Bushy Run Research Center, Project Report 50-139, Jan. 12, 1988.
475. Union Carbide Corporation, Bushy Run Research Center, Project Report 54-62 Nov. 1, 1991.
476. K. L. Cheever et al., Testicular effects of bis(2-methoxyethyl)ether in the adult male rat. *Toxicol. Ind. Health* **5**, 1099–1109 (1989).
477. D. E. Richards et al., Comparative metabolism of bis(2-methoxyethyl)ether in isolated rat hepatocytes and in the intact rat: effects of ethanol on *in vitro* metabolism. *Arch. Toxicol.* **67**, 531–537 (1993).
478. M. A. Tirmenstein, Comparative metabolism of bis(2-methoxyethyl)ether by rat and human hepatic microsomes: formation of 2-methoxyethanol. *Toxicol. In Vitro* **7**(5), 645–652 (1993).
479. K. P. Lee, L. A. Kinney, and R. Valentine, Comparative testicular toxicity of bis(2-methoxyethyl)ether and 2-methoxyethanol in rats. *Toxicology* **59**, 239–258 (1989).

480. U. S. Environmental Protection Agency, Office of Toxic Substances, *Document No. FYI-OTS-0589-0553*, May 16, 1989.
481. C. J. Price et al., The developmental toxicity of diethylene glycol dimethyl ether in mice. *Fundam. Appl. Toxicol.* **8**, 115–125 (1987).
482. B. A. Schwetz et al., The developmental toxicity of diethylene and triethylene glycol dimethyl ethers in rabbits. *Fundam. Appl. Toxicol.* **19**, 238–245 (1992).
483. C. D. Driscoll et al., Developmental toxicity of diglyme by inhalation in the rat. *Drug and Chem. Toxicol.* **21**(2), 119–136 (1998).
484. D. W. Lynch, Developmental toxicity of bis(2-methoxyethyl)ether (DGDE), 2-(methoxyethoxy)acetic acid (MEAA), and 2-methoxyethanol (2ME) in intact *Drosophila. Teratology* **51**(3), 190 (Abstract) (1995).
485. K. Mortelmans et al., Salmonella mutagenicity tests II. results from the testing of 270 chemicals. *Environ. Mutagen.* **8**(7), 1–119 (1986).
486. A. Z. Buzina and F. A. Rudi, Substantiation of the maximum permissible concentrations of vinyl ethers of glycols in water bodies. *Gig. Sanit.* **43**, 12–15 (1977).
487. Union Carbide Corporation, Bushy Run Research Center, Report 20-100, July 31, 1957.
488. R. D. Ashford, *Ashford's Dictionary of Industrial Chemicals*, Wavelength Publications Ltd., London, England, 1994.
489. M. W. Gill and J. M. Hurley, Bushy Run Research Center, Project Report 52-606, 1990.
490. M. W. Gill and J. E. Negley, *Triethylene Glycol Monomethyl Ether. Ninety Day Subchronic Drinking Water Inclusion Neurotoxicity Study in Rats*, Report to the U.S. Chemical Manufacturers Association, Bushy Run Research Center, Project Report 52-607, 1990.
491. P. E. Losco et al., Triethylene glycol monomethyl ether (TGME): ninety-day subchronic drinking water inclusion neurotoxicity study in rats. *The Toxicologist* **12**, 234 (Abstract 869) (1992).
492. M. W. Gill et al., Subchronic dermal toxicity and oral neurotoxicity of triethylene glycol monomethyl ether in CD rats. *Int. J. Toxicol.* **17**, 1–22 (1998).
493. R. A. Corley et al., *13-Week Dermal Toxicity Study in Sprague-Dawley Rats*, Report to the US Chemical Manufacturers Association, The Toxicology Research Laboratory, Health and Environmental Sciences, The Dow Chemical Company, Midland, MI, Laboratory Project Study ID K-005610-004, 1990.
494. R. A. Corley et al., The subchronic toxicity of triethylene glycol monomethyl ether (TGME) in dermally-exposed Sprague-Dawley rats. *The Toxicologist* **12**, 233 (Abstract 868) (1992).
495. A. M. Hoberman et al., Developmental toxicity studies of triethylene glycol monomethyl ether administered orally to rats and rabbits. *J. Am. Coll. Toxicol.* **15**, 349–370 (1996).
496. H. K. Bates, Developmental neurotoxicity evaluation of triethylene glycol monomethyl ether (CAS *112-35-6*) administered by gavage to timed-mated CD® rats on gestational day 6 through postnatal day 21. Arlington, VA; Report of the Chemical Manufacturers Association, 1992.
497. Y. E. Samson and B. B. Gollapudi, Evaluation of triethylene glycol monomethyl ether (TGME) in the Ames salmonella/mammalian-microsome bacterial mutagenicity assay, Report to the Chemical Manufacturers Association, 1990.
498. M. L. McClintock and B. Gollapudi, Evaluation of triethylene glycol monomethyl ether in the mouse bone marrow micronucleus test. The Dow Chemical Company, Report to the Chemical Manufacturers Association, 1990.

499. V. A. Linscombe and B. B Gollapudi, Evaluation of triethylene glycol monomethyl ether in the Chinese hamster ovary cell/hypoxanthine-guanine-phosphoribosyl-transferase (CHO/HGPRT) forward mutation assay. The Dow Chemical Company, Report to the Chemical Manufacturers Association, 1990.

500. A Sprague, illness report involving the ingestion of Dot 3 brake fluid, Eisenhower Army Hospital, Atlanta, GA, unpublished, 1992.

501. The Olin Corporation, Acute toxicity studies. Unpublished data, Olin Study No's. 936 A, B, C, D, and E, 1977.

502. Dow Chemical Company, *Toxicological Properties and Industrial Handling Hazards of Dowanol TEH, EPA Document No. 878216072*, 1974.

503. Imperial Chemical Industries, *Triethylene Glycol Ethers: An Evaluation of Teratogenic Potential and Developmental Toxicity using an In Vivo Screen in Rats, EPA Document No. 40-8699089*, 1986.

504. R. J. Ward and R. C. Scott, Chemical Manufacturers Association, unpublished data, Oct. 31, 1986.

505. U. S. Environmental Protection Agency, testing consent order on TGME, TGEE, and TGBE. *Fed. Reg.* **54**(62), 13470 (1989).

506. The Olin Corporation, unpublished data, 1976.

507. The Olin Corporation, unpublished data, 1986.

508. D. J. Wroblewski and P. A. Keller, The Dow Chemical Company, unpublished data, 1976.

509. Union Carbide Corporation, unpublished data, 1947.

510. S. J. Hermansky and H. W. Leung, Cutaneous toxicity studies with methoxy polyethylene glycol-350 (MPEG-350) in rats and rabbits. *Food and Chem. Toxicol.* **35**(10/11), 1031–1039 (1997).

511. T. Yamaoka, Y. Tabata, and Y. Ikada, Distribution and tissue uptake of poly(ethylene glycol) with different molecular weights after intravenous administration to mice. *J. Pharm. Sci.* **83**(4), 601–606 (1994).

512. Eastman Kodak Company, unpublished data, Toxicity and Health Hazard Summary, Corporate Health and Environment Laboratories, Lab No. 59-260, Acc. No. 023646, Feb. 23, 1973.

513. R. S. Hosenfeld and G. J. Hankinson, unpublished data, Eastman Kodak Company, Corporate Health and Environment Laboratories, Lab. No. 87-0068, Acc. No. 023646, Report No. TX-88-2, March 31, 1988.

514. W. T. Stott and M. J. McKenna, Hydrolysis of several glycol ether acetates and acrylate esters by nasal mucosal carboxylesterase *in vitro. Fundam. Appl. Toxicol.* **5**, 399–404 (1985).

515. E. Zeiger et al., Salmonella mutagenicity tests. V. results from the testing of 311 chemicals. *Environ. Mol. Mutagen.* **19**(21), 2–141 (1992).

516. W. Jordan and M. Dahl, Contact dermatitis to a plastic solvent in eyeglasses. *Arch. Dermatol.* **104**, 524–528 (1971).

517. H. M. Bolt and K. Golka, Maternal exposure to ethylene glycol monomethyl ether acetate and hypospadia in offspring: a case report. *Br. J. Ind. Med.* **47**, 352–353 (1990).

518. A. Edens, Proposed regulation for occupational exposure to select glycol ethers in the United States. *Occup. Hyg.* **2**, 445–449 (1996).

519. J. A. Wess, Reproductive toxicity of ethylene glycol monomethyl ether, ethylene glycol monoethyl ether and their acetates. *Scand. J. Work Environ. Health* **18**(2), 43–45 (1992).

520. G. M. Piacitelli, D. M. Votaw, and E. R. Krishnan, An exposure assessment of industries using ethylene glycol ethers. *Appl. Occup. Environ. Hyg.* **5**(2), 107–114 (1990).

521. J. Laitinen et al., Urinary biochemistry in occupational exposure to glycol ethers. *Chemosphere* **29**, 781–787 (1994).

522. C. P. Carpenter, Queries and minor notes. *J. Am. Med. Assoc.* **135**, 880 (1947).

523. R. Truhaut et al., Comparative toxicological study of ethyl glycol acetate and butyl glycol acetate. *Toxicol. Appl. Pharmacol.* **51**, 117–127 (1979).

524. Union Carbide Corporation, unpublished data, 1967.

525. D. Guest et al., Pulmonary and percutaneous absorption of 2-propoxyethyl acetate and 2-ethoxyethyl acetate in beagle dogs. *Environ. Health Perspect.* **57**, 177–184 (1984).

526. R. W. Tyl et al., Developmental toxicity evaluation of inhaled 2-ethoxy-ethanol acetate in Fischer 344 rats and New Zealand white rabbits. *Fundam. Appl. Toxicol.* **10**, 20–39 (1988).

527. D. J. Paustenbach, Assessment of the developmental risks resulting from occupational exposure to select glycol ethers within the semiconductor industry. *Toxicol. Environ. Health* **23**, 29–75 (1988).

528. R. Vincent et al., Occupational exposure to organic solvents during paint stripping in the aeronautical industry. *Int. Arch. Occup. Environ. Health* **65**, 377–380 (1994).

529. S.-E. Chia et al., Menstrual patterns of workers exposed to low levels of 2-ethoxyethylacetate (EGEEA). *Am. J. Ind. Med.* **31**, 148–152 (1997).

530. N. I. Sax and R. J. Lewis, *Hawley's Condensed Chemical Dictionary*, 11th ed., Van Nostrand Reinhold Company, New York, 1987, pp 488–489.

530a. OSHA, *Analytical Methods Manual*, Available from ACGIH, Cincinnati, OH, 1990, 1993.

531. British Industrial Biological Research Association, *Toxicity Profile of 2-Butoxyethanol Acetate*, 1987.

532. G. A. Jacobs, A. Castellazi, and P. J. Dierick, Evaluation of a non-invasive human and an *in vitro* cytotoxicity method as alternatives to the skin irritation test on rabbits. *Contact Dermatitis* **21**, 239–244 (1989).

533. J. N. Lawrence, F. M. Dickson, and D. J. Benford, Skin irritant-induced cytotoxicity and prostaglandin E12 release in human skin keratinocyte cultures. *Toxicol. In Vitro* **11**(5), 627–631 (1997).

534. G. Johanson and U. Rick, Use and use patterns of glycol ethers in Sweden. *Occup. Hyg.* **2**, 105–110 (1996).

535. Union Carbide Corporation, unpublished data, 1939.

536. Union Carbide Corporation, unpublished data, 1941.

537. D. Guest, P. J. Deisinger, and J. E. Winter, Estimation of the atmospheric concentration of diethylene glycol monobutyl ether acetate resulting from the application of latex paint, Eastman Kodak Company, unpublished report, 1985.

538. Dupont, Acute inhalation toxicity of 2-(2-Butoxyethoxy)ethyl acetate (butyl carbitol acetate). Submission to EPA with Cover Letter, E. J Du Pont De Nemours & Co, Newark, Delaware, U.S.A., unpublished information, 1984.

539. J. H. Draize et al., Methods for the study of irritation and toxicity of substances applied to the skin and mucous membranes. *Pharmacol. Exp. Therap.* **82**, 377–390 (1948).

540. T. A. J. Dawson et al., Delayed and immediate hypersensitivity to carbitols. *Contact Dermatitis* **21**, 52–53 (1989).

541. D. Hoehn, *J. Am. Med. Assoc.* **128**, 153 (1945).

542. S. M. Hays et al., Development of a physiologically base pharmacokinetic model of 2-methoxyethanol and 2-methoxyacetic acid disposition in pregnant rats. *Toxicol. Appl. Pharmacol.* **163**, 67–74 (2000).

543. M. L. Gargas et al., A toxicokinetic study of inhaled ethylene glycol monomethyl ether (2-ME) and validation of a physiologically based pharmacokinetic model for the pregnant rat and human. *Toxicol. Appl. Pharmacol.* **165**, 53–62 (2000).

CHAPTER EIGHTY-SEVEN

Glycol Ethers: Ethers of Propylene, Butylene Glycols, and Other Glycol Derivatives

Rodney J. Boatman, Ph.D., DABT

A ETHERS OF PROPYLENE GLYCOL

A.1 Background

There are currently five U.S. manufacturers of propylene glycol ether derivatives as shown in Table 87.1 (1). This table also contains a listing of the trade names used by these manufacturers for these materials.

A.2 Production and Use

The ethers of mono-, di-, tri-, and polypropylene glycol are prepared commercially by reacting propylene oxide with the alcohol of choice in the presence of a catalyst. They also may be prepared by direct alkylation of the selected glycol with an appropriate alkylating agent such as a dialkyl sulfate in the presence of an alkali. When propylene oxide is used as the starting material, which is the case for most propylene glycol ethers, preparation under commercial conditions yields products that are mixtures of the α and β isomers. The α isomer consists of the ether linkage on the terminal hydroxyl group of propylene glycol; while the β isomer has the ether linkage on the secondary hydroxyl group, with the

---

Patty's Toxicology, Fifth Edition, Volume 7, Edited by Eula Bingham, Barbara Cohrssen, and Charles H. Powell.
ISBN 0-471-31940-6 © 2001 John Wiley & Sons, Inc.

Table 87.1. U.S. Manufacturers of Propylene Glycol Ethers and Tradenames

Manufacturer	Trade Name
Lyondell Chemical Company[a]	Arcosolv®
The Dow Chemical Company[a]	Dowanol®
Eastman Chemical Company[a]	Eastman®
Olin Chemical Corporation	Poly-Solv®
Union Carbide Corporation	Propasol®

[a]Members of the Chemical Manufacturers Association, Propylene Glycol Ethers Panel.

primary hydroxyl group unsubstituted. The α isomer is thermodynamically favored during synthesis from propylene oxide and the desired alcohol; thus, it constitutes the bulk of the resulting ether. By further manipulating the conditions of synthesis, the proportion of α isomer may be enhanced to constitute >99% of the end product. As explained later, this distinction is important regarding the toxicity of the propylene glycol ethers. The α and β isomers are shown below, where "R" represents the desired alcohol portion of the propylene glycol ether.

Major isomer	Minor isomer
OH	OR
R−O−CH$_2$−CH−CH$_3$	HO−CH$_2$−CH−CH$_3$
1-Alkoxy-2-propanol	2-Alkoxy-1-propanol
α isomer (2° alcohol)	β isomer (1° alcohol)
(ROCH$_2$CHOHCH$_3$)	(HOCH$_2$CHORCH$_3$)

The monoalkyl ethers of dipropylene glycol presumably can appear in four isomeric forms. The commercial product Dowanol® DPM Glycol Ether is believed to be a mixture of these but to consist to a very large extent of the isomer in which the alkyl group has replaced the hydrogen of the primary hydroxyl group of the dipropylene glycol; the internal ether linkage is between the 2 position of the alkyl-etherized propylene unit and the primary carbon of the other propylene unit, thus leaving the remaining secondary hydroxyl group unsubstituted. In the case of dipropylene glycol monomethyl ether, the primary isomer is 1-(2-methoxy-1-methylethoxy)-2-propanol. The monoalkyl ethers of tripropylene glycol can appear in eight isomeric forms. The commercial product Dowanol® TPM Glycol Ether, however, is believed to be a mixture of isomers consisting largely of the one in which the alkyl group displaces the hydrogen of the primary hydroxyl group of the tripropylene glycol and the internal ether linkages are between secondary and primary carbons. The known physical properties of the most common ethers are given in Tables 87.2 and 87.3.

The methyl and ethyl ethers of these propylene glycols are miscible with both water and a great variety of organic solvents. The butyl ethers have limited water solubility but are

Table 87.2. Physical and Chemical Properties of Some Common Ethers of Propylene Glycol

Property	Methyl	Ethyl	Mono Propyl	Isopropyl	n-Butyl	tert-Butyl	Butoxyethyl	Phenyl
CAS no. (1-alkyl isomer)	[107-98-2]	[1569-02-4]	[1569-01-3]	[3944-36-3]	[5131-66-8]	[57018-52-7]	[124-16-3]	[770-35-4]
CAS no. (alkoxy position unspecified)	[1320-67-8]	[52125-53-8]		[29387-84-6]	[29387-86-8]			
Molecular formula	$C_4H_{10}O_2$	$C_5H_{12}O_2$	$C_6H_{14}O_2$	$C_6H_{14}O_2$	$C_7H_{16}O_2$	$C_7H_{16}O_2$	$C_9H_{20}O_3$	$C_9H_{12}O_2$
Molecular weight	90.12	104.15	118.18	118.18	132.2	132.2	176.26	152.19
Specific gravity 25/4°C	0.919 (20°C)	0.896	0.884 (20°C)	0.875	0.879	0.87	0.931 (20°C)	1.059
Boiling point (°C) (760 mm Hg)	120.1	132.8	148.9	139–141	169–172	151	230.3	242.7
Freezing point (°C)	−97		< −70				−90	11.4
Vapor pressure (mm Hg) (25°C)	11.8 8.24 (20°C)	8.2	1.7 (20°C)	5.3	1.4	4.8 (20°C)	0.19	0.02
Refractive index (25°C)	1.402	1.404		1.405	1.415		1.429 (25°C)	1.522
Flash point (°F) (open cup)	96 38° (closed up)	130	128	140	138	44.4(°C)	250	264
Vapor density (air = 1)	3.11	3.59	4	4.07	4.56		6.08	5.25
Percent in saturated air (25°C)	1.55	1.08		0.70	0.18			
1 ppm ≈ mg/m³ at 25°C, 760 mmHg	3.68	4.25	4.83	4.83	5.41	5.30 (20°C)	7.20	6.22
1 mg/l ≈ ppm at 25°C, 760 mmHg	272	235	207	207	185	188 (20°C)	138.8	160.8

Table 87.3. Physical and Chemical Properties of Ethers of Di- and Tripropylene Glycols and Butylene Glycol Ethers

Property	Dipropylene Glycol Ethers			Tripropylene Glycol Ethers			Butylene Glycol Ethers		
	Methyl	Ethyl	Butyl	Methyl	Ethyl	Butyl	Methyl	Ethyl	n-Butyl
CAS no.	[34590-94-8]	[15764-24-6]	[29911-28-2]	[20324-33-8] [25498-49-1]	[20178-34-1]	[57499-93-1]	[53778-73-7] [111-32-0]	[111-73-9]	None found
Molecular formula	$C_7H_{16}O_3$	$C_8H_{18}O_3$	$C_{10}H_{22}O_3$	$C_{10}H_{22}O_4$	$C_{11}H_{24}O_4$	$C_{13}H_{28}O_4$	$C_5H_{12}O_2$	$C_6H_{14}O_2$	$C_8H_{18}O_2$
Molecular weight	148.2	162.23	190.28	206.28	220.3	248.3	104.15	118.2	146.2
Specific gravity 25/4°C	0.95	0.927	0.910	0.961			0.983 (25/25°C)	0.888 (25/25°C)	0.877 (25/25°C)
Boiling point (°C) (760 mmHg)	190	193–195	230	242.8	250±	255±	136	147	180–187
Freezing point (°C)	−83			−60					
Vapor pressure (mmHg)(25°C)	0.41 (0.3 torr at 20°C)	0.30	0.06	0.017	0.011	0.008	5.5	3.0	0.62
Refractive index (25°C)	1.419	1.421	1.418	1.428	1.426	1.424	1.408	1.410	1.420
Flash point (°F) (open cup)	185 85°C (closed cup)	205	212 (Setaflash)	2.60			110	145	160
Percent in saturated air (25°C)	0.051	0.04	0.008	0.003	0.0014	0.001	0.72	0.40	0.081
1 ppm ≈ mg/m³ at 25°C, 760 mmHg	6.06	6.64	7.77	8.44	9.00	10.1	4.25	4.83	5.98
1 mg/l ≈ ppm at 25°C, 760 mmHg	165	150.7	128.6	118.5	111	98.6	235	207	167

miscible with most organic solvents. This mutual solvency makes them valuable as coupling, coalescing, and dispersing agents. These glycol ethers have found applications as solvents for surface coatings, inks, lacquers, paints, resins, dyes, agricultural chemicals, and other oils and greases. The di- and tripropylene series also are used as ingredients in hydraulic brake fluids. Refer to Tables 87.4 and 87.5 for a summary of uses and production volumes of common propylene glycol ethers. Although still far less than that of the ethylene glycol ethers, the production of propylene glycol ethers has been increasing. In 1995, total U.S. production of these solvents was estimated at 217 million lb (1). Production of propylene glycol ethers in western Europe in 1995 was reported as 190,000 metric tons (1). Lesser amounts of these glycol ethers are manufactured in a number of other areas of the world, including Mexico and Brazil and in eastern Europe. Several companies in Japan also produce propylene glycol ethers; however, production figures are not available (1).

A.3 Occupational Exposures

Occupational exposure would normally be limited to dermal and/or inhalation exposure. The toxicological activity of the propylene glycol–based ethers generally indicates a low order of toxicity. Under typical conditions of exposure and use, propylene glycol ethers pose little hazard. As with many other solvents, appropriate precautions should be employed to minimize dermal and eye contact and to avoid prolonged or repeated exposures to high vapor concentrations.

Analytical methods suitable for the detection of ethylene glycol ethers may be applied to the detection of propylene glycol ethers in environmental air samples (2, 3). These methods employ adsorption onto activated charcoal or other suitable material followed by desorption and gas chromatographic analysis. NIOSH method 1403 (2) or OSHA method 83 (3) are representative.

Laitinen proposes the analysis of 2-methoxypropionic and 2-ethoxypropionic acids in human urine as a biomonitor of exposure to technical-grade 1-methoxy-2-propanol acetate and 1-ethoxy-2-propanol acetate (4). Hubner et al. (5) describe a method suitable for the detection of unmetabolized propylene glycol ethers in human urine.

A.4 Toxic Effects

The propylene glycol ethers (PGEs), even at much higher exposure levels, do not cause the types of toxicity produced by certain of the ethylene glycol ethers (EGEs). Specifically, they do not cause damage to the thymus, testes, kidneys, blood, and blood-forming tissues as seen with ethylene glycol methyl and ethyl ethers. In addition, the propylene glycol ethers do not induce the development effects of certain of the methyl- and ethyl-substituted ethylene glycol–based ethers nor the hemolysis and associated secondary effects seen in laboratory animals with ethylene glycol butyl ether.

The data in Table 87.6 show the lower toxicity of PGME and its acetate compared with either EGME or EGEE. The types of toxicities attributed to the ethylene glycol ethers are caused by ethylene glycol methyl ether (EGME) exposures as low as 25–50 ppm by inhalation (birth defects), 31 mg/kg per day orally (birth defects), and 250 mg/kg per day

Table 87.4. Propylene Glycol Ethers and Acetates: Consumer Use and Exposure Data for the United States[a]

Propylene Glycol Ether or Acetate	1993 Production Volume 10⁶ lb	Types of Commercial End Products	Percent of Production (%)	Industrial/Commercial Percentage Use	Approximate Weight Fraction in Product Types
Propylene glycol methyl ether	110–130	Surface coatings	30		2–20%
[107-98-2] (α isomer)		Cleaners	23		2–50%
[1589-47-5] (β isomer < 5%)		Inks	6		2–30%
[1320-67-8] (mixture)		Miscellaneous	7[b]	100/0	
[28677-93-2] (mixture)		PMA production	34		
Propylene glycol methyl ether acetate	72	Surface coatings	89	100/0[b]	5–30%
[108-65-6] (α isomer)		Inks	5		2–30%
[70657-70-4] (β isomer < 5%)		Cleaners	3		5–50%
[84540-57-8] (mixture)		Miscellaneous	3	100/0[b]	
Propylene glycol ethyl ether[c]	<1	Coatings	100		
[1569-02-4] (α isomer)					
[52125-53-8] (mixture)					
Propylene glycol ethyl ether acetate[c]	<1	Coatings	100		
Propylene glycol propyl ether[d]	5–10	Coatings	45		
[1569-01-3] (α isomer)		Cleaners	50		
		Inks	5		
Propylene glycol propyl ether acetate	0				
Propylene glycol butyl ether[d]	6–10	Surface coatings	50		2–20%
[5131-66-8] (α isomer)		Cleaners	50		2–10%
[10215-33-5] (β isomer <7%)					
[63716-40-5] (mixture)					
[29387-86-8] (mixture)					
Propylene glycol butyl ether acetate	0				

Propylene glycol *t*-butyl ether[c] [57018-52-7] (α isomer)	6–10	Surface coatings Cleaners Miscellaneous	0 90 10
Propylene glycol *t*-butyl ether acetate	0		20/80
Propylene glycol phenyl ether[c] [770-35-4] (α isomer) [4169-04-4] (β isomer < 5%)	1–5	Surface coatings Adhesives Inks	75 2–10% 5 2–10% 20 2–10%
Propylene glycol phenyl ether acetate	0	Cosmetics, soaps	4 0.1–1%

[a]Data collected by the Chemical Manufacturers Association, Propylene Glycol Ethers Panel. The information provided is not representative of all manufacturers.
[d]Miscellaneous includes adhesives and electronics.
[c]Manufactured by 1 company, actual production confidential.
[d]Manufactured by 2 companies, actual production confidential.

Table 87.5. Di- and Tripropylene Glycol Ethers and Acetates: Consumer Use and Exposure Data for the United States[a]

Di- and Tripropylene Glycol Ether or Acetate	1993 Production Volume 10^6 lb	Types of Commercial End Products	Percent of Production (%)	Industrial/Commercial Percentage Use	Approximate Weight Fraction in Product Types
Dipropylene glycol methyl ether[b] [20324-32-7] (α isomer) [34590-94-8] (mixture)	20–30	Surface coatings	19		2–20%
		Cleaners	63		2–50%
		Inks	5		—
		Dyes	3		2–20%
		Miscellaneous	9		2–10%
Dipropylene glycol methyl ether acetate[c] [88917-22-0] (mixture)	1–2	Surface coatings	30	100/0	
		Cleaners	60	100/0	
		Miscellaneous	10	90/0	
Dipropylene glycol ethyl ether or acetate	0				
Dipropylene glycol propyl ether or acetate	0				
Dipropylene glycol t-butyl ether [132739-31-2]					
Dipropylene glycol butyl ether[b] [29911-28-2] (α isomer) [35884-42-5] (mixture)	5–10	Surface coatings	10		2–20%
		Inks	5		2–20%
		Cleaners	85		2–10%
Dipropylene glycol butyl ether acetate	0				
Tripropylene glycol methyl ether[b] [20324-33-8] (α isomer) [25498-49-1] (mixture)	3–5	Inks	7		2–20%
		Cleaners	80	100/0	2–50%
		Functional fluids	13	100/0	—
Tripropylene glycol methyl ether acetate	0				
Tripropylene glycol butyl ether[b]	10–20 thousand	Cleaners	100		2–10%
		Miscellaneous	<1		—
Tripropylene glycol ethyl, propyl ether or acetate	0				

[a]Data collected by the Chemical Manufacturers Association, Propylene Glycol Ethers Panel. The information provided is not representative of all manufacturers.
[b]Manufactured by two companies, actual production confidential.
[c]Manufactured by one company, actual production confidential.

Table 87.6. Comparison of the Toxicities of Propylene Glycol Methyl Ether (PGME) and Its Acetate (PGMEA) with that of Ethylene Glycol Methyl Ether (EGME) and Ethylene Glycol Ethyl Ether (EGEE)[a]

		Birth Defects		Embryo/Fetal Toxicity		Testicular Toxicity		Blood Damage		Thymic Atrophy	
		NOAEL	LOAEL	NOAEL	LOAEL	NOAEL	LOAEL	NOAEL	LOAEL	NOAEL	LOAEL
PGME[b] or PGMEA[b]	Inhalation (ppm)	3000	NLF	1000	3000	3000	NLF	3000	NLF	3000	NLF
	Dermal (mg kg^{-1} day^{-1})	N/D	N/D	N/D	N/D	1000	NLF	1000	NLF	1000	NLF
	Oral (mg kg^{-1} day^{-1})	3000	NLF	370	740	3000	NLF	3000	NLF	3000	NLF
β-PGME[b] or β-PGMEA[b]	Inhalation (ppm)	145	225	145	225	560	2800	2800	NLF	2800	NLF
	Dermal (mg kg^{-1} day^{-1})	2000	NLF	2000	NLF	N/D	N/D	N/D	N/D	N/D	N/D
	Oral (mg kg^{-1} day^{-1})	N/D	N/D	N/D	N/D	N/D	N/D	NNF	1800	N/D	N/D
EGME	Inhalation (ppm)	10	50	10	50	30	100	30	100	30	100
	Dermal (mg kg^{-1} day^{-1})	NNF	250	NNF	250	NNF	650	N/D	N/D	N/D	N/D
	Oral (mg kg^{-1} day^{-1})	31	250	NNF	31	50	100	10	50	NNF	100
EGEE	Inhalation (ppm)	50	175	10	50	100	400	100	370	N/D	N/D
	Dermal (mg kg^{-1} day^{-1})	NNF	~250	NNF	~250	N/D	N/D	N/D	N/D	N/D	N/D
	Oral (mg kg^{-1} day^{-1})	NNF	200	NNF	1000	150	300	93	185	N/D	N/D

[a] *Abbreviations:* N/D = no data; NNF = no NOAEL found (i.e., lowest dose tested caused toxicity); NLF = no LOAEL found (highest dose tested did not cause toxicity).
[b] Mixture representative of commercially produced materials (primarily α isomer).

by dermal exposure (birth defects — single dose). Similar EGME exposures have also produced testicular and bone marrow damage. In contrast, inhalation exposure to 3,000 ppm propylene glycol methyl ether (PGME) did not cause birth defects, testicular damage, or injury to blood or thymus. The studies of Landry et al. (6) and of Miller et al. (7) identified inhalation no-observed-adverse-effect-levels (NOAELs) for PGME and PGME acetate (PGMEA), respectively. The similarity in toxicity (or lack thereof) of PGME and its acetate verifies the assumption that the acetates of most glycol ethers are similar in systemic toxicity to that of the parent glycol ether since the acetates are hydrolyzed to yield the parent glycols. Commercial PGME and PGMEA are much less toxic as indicated by their greater NOAELs and lowest-observed-adverse-effect levels (LOAELs) for each indicated endpoint.

An older study by Rowe et al. (8) and, more recently, that of L. L. Calhoun and K. A. Johnson (the Dow Chemical company, unpublished data, 1983) are useful for identifying the oral and dermal NOAELs and LOAELs for PGME, and, by inference, for PGMEA. Table 87.6 reveals the markedly higher NOAELs and LOAELs (and, thus, lower toxicity) of PGME and its acetate compared with that of either ethylene glycol methyl or ethyl ether. In fact, the PGME/PGMEA NOAELs are 60–100 times higher than those for EGME and 30–100 times higher than EGEE. For all except embryo or fetal toxicity, no dose of PGME or PGMEA resulted in birth defects or in testicular, blood, or thymic damage. Table 87.6 also reveals that commercially produced PGME and PGMEA (isomeric mixtures) are substantially less toxic than the corresponding pure β isomers.

Other propylene glycol ethers also exhibit a similar lack of toxicity. For example, propylene glycol ethyl ether (PGEE) and its acetate do not cause the critical toxicities of testicular, thymic, or blood injury and do not produce birth defects. Propylene glycol *tertiary*-butyl ether (PGTBE) also has been tested and fails to cause these toxicities or birth defects in rats exposed by inhalation to substantial concentrations.

A.5 Pharmacokinetics, Metabolism, and Mode of Action

The α isomers of propylene glycol ethers cannot be metabolized to a carboxylic acid, since the free hydroxyl group is a secondary rather than a primary alcohol. Miller et al. (9) showed that PGME in the rat is converted largely to propylene glycol, which, in turn, is metabolized and expired in the breath as CO_2. A small amount is excreted in the urine as the glucuronide and sulfate conjugates of the parent PGME as well as propylene glycol and free PGME. Because the α isomers are overwhelmingly preponderant in commercial preparations, PGEs do not form significant amounts of the alkoxypropionic acids, analogous to the alkoxyacetic acids seen for the methyl, ethyl, and butyl derivatives of ethylene glycol. Consequently, the toxicities attributable to these acid metabolites are not seen with the α isomers of the PGEs.

On the other hand, the β isomers of PGEs can be metabolized to the alkoxypropionic acids. The β isomers of propylene glycol methyl and ethyl ethers do show toxicities similar to their EGE counterparts. Miller et al. (10) showed that the β isomer of PGME indeed is converted predominantly to a carboxylic acid, similar to EGME and EGEE. Although the β isomers of propylene glycol ethers may cause toxicities similar to their ethylene based counterparts, they appear to be less potent. Even 100% β-PGME has not been shown to injure the testes of rats exposed orally for 10 consecutive days (11). When tested for its ability to cause birth defects, 100% β isomer caused deformities of the vertebrae in rat pups when exposed *in utero* to a concentration of 3000 ppm (12). Abnormalities were more serious and occurred at lower concentrations in rabbits as reported by Hellwig et al. (13). These investigators found that exposure of rabbits to β-PGME at concentrations of 545 ppm caused cleft palate, deformed paws, and other serious birth defects. The acetate of β-PGME also has produced similar effects (12). Conversely, when the β isomer is mixed with the α isomer at concentrations as high as 17% β isomer, these effects were not elicited (Dow/ARCO, unpublished data).

1.0 Propylene Glycol Monomethyl Ether

3-Methoxy-1-propanol is an isomer but is of little or no commercial significance.

1.0.1 CAS Numbers: [107-98-2], [1320-67-8]

1.0.2 Synonyms: 1-Methoxypropan-2-ol, 1-methoxy-2-hydroxypropane, methoxyisopropanol, 2-methoxy-1-methylethanol, propylene glycol monomethyl ether (PGME), 1-methoxy-2-propanol

1.0.3 Trade Names: Arcosolv® PM, Dowanol® PM Glycol Ether, Methyl Proxitol®, Propasol® Solvent M, Poly-Solv® MPM Solvent

1.0.4 Molecular Weight: 90.12

1.0.5 Molecular Formula: $C_4H_{10}O_2$

1.0.6 Molecular Structure: $CH_3OCH_2CHOHCH_3$

1.1 Chemical and Physical Properties

1.1.1 General

See Table 87.2.

1.1.2 Odor and Warning Properties

PGME is a colorless liquid with a mild ether-like odor. In controlled human exposures, Stewart et al. (14) found that levels of PGME above 100 ppm were objectionable because of odor. Eye, nasal, and throat irritation became objectionable prior to the first indications of central nervous system (CNS) impairment, which occurred at a level of 1000 ppm. Breath analysis data showed that PGME was rapidly excreted via the lungs. The human volunteers all experienced a rapid development of odor tolerance.

1.2 Production and Use

PGME is produced by reacting propylene oxide with methanol. In 1995, U.S. production of PGME reached 145 million lb (1). Refer to Table 87.4 for additional production and use data.

1.3 Exposure Assessment

1.3.1 Air

Refer to the Introduction for a discussion of procedures suitable for the analysis of PGME in environmental air samples.

1.3.2 Background Levels: NA

1.3.3 Workplace Methods: NA

1.3.4 Community Methods: NA

1.3.5 Biomonitoring/Biomarkers

1.3.5.1 Blood: NA

1.3.5.2 Urine. Urinary excretion of 1,2-propanediol was found to correlate with exposures to PGME measured in workers breathing zones (15). The method was applied to urine collected from 23 silkscreen printers collected at the end of a workweek. The mean concentration of diol was 2.52 mmol per mole of creatinine. Workers were exposed to a workplace atmosphere containing predominantly (90.2%) PGME.

1.4 Toxic Effects

Commercial propylene glycol monomethyl ether (PGME) is low in both single- and repeated-dose oral toxicity, transiently painful to the eyes, and not appreciably irritating to the skin, but can be absorbed through the skin in toxic amounts after repeated, high dose exposures. The vapors are low in toxicity and the hazard from inhalation also is low because acutely toxic concentrations are essentially intolerable to humans, and concentrations that might cause effects from repeated exposures are very disagreeable (irritating to the eyes and mucous membranes and nauseating to some persons). The primary response to high oral or inhalation exposures is sedation. This material did not produce cancer or teratogenic, reproductive, or mutagenic effects in animal and *in vitro* studies. The lack of toxicity may be due to its extensive metabolism to propylene glycol, a material of low toxicity.

1.4.1 Experimental Studies

1.4.1.1 Acute Toxicity

ORAL. The single-dose oral toxicity for PGME has been determined in several animal species. The LD_{50} values suggest a low oral toxicity. The LD_{50} levels for commercial-grade PGME were as follows: for the rat, 6.6 (8), 5.20 (16), 7.51 (α isomer) (17), 5.71 (β isomer) (17), and 11.9 g/kg (estimated) (18); for the mouse, 10.8 g/kg (19); for the dog, 4.6–5.5 (19) and 9.2 g/kg (estimated) (20); and for the rabbit, 5.3 g/kg (19). Rowe et al. (8) found that death in rats was associated with profound CNS depression. Shideman and Procita (20) attributed death in dogs to respiratory arrest, and pointed out that if the acute effects were tolerated there were no residual effects. It is interesting to note that the α and β isomers 1-methoxy-2-propanol and 2-methoxy-1-propanol are both similar in acute oral toxicity.

INHALATION. Acute vapor studies conducted on rats and guinea pigs (8) have shown that (a) at a concentration of approximately 5000 ppm, rats and guinea pigs survived single 7-h exposures; (b) at a concentration of approximately 10,000 ppm, time to the LD_{50} for rats was 5–6 h and for guinea pigs it was greater than 7 h; (c) at a concentration of approximately 15,000 ppm (saturated, with some mist present), time to the LD_{50} for rats was 4 h and for guinea pigs 10 h; and (d) deaths resulting from single inhalation exposures appeared to be due to CNS depression. Rats (F344) survived single, 6-h exposures to 6038 or 7559 ppm PGME but experienced general narcosis (21). Animals revived and appeared normal within 2 or 3 days. Body weights were initially depressed, however, but exceeded preexposure levels within a week.

Goldberg et al. (22) exposed rats to concentrations of 2500, 5000, and 10,000 ppm for 4 h/day, 5 days/week for 2 weeks in order to study the effect of inhalation on behavior. There was a transient, nonspecific depression of behavior for the first several exposures to 5000 and 10,000 ppm. However, there was a rapid development of tolerance. Decreased growth rate was seen at 10,000 ppm.

Miller et al. (23) exposed Fischer 344 rats and B6C3F1 mice to levels of 0, 300, 1000, or 3000 ppm PGME 6 h/day for nine exposures over an 11-day period. Exposure to 3000 ppm resulted in increased liver weights in male rats, and CNS depression and decrease in specific gravity of the urine in male and female rats. There were no gross or histological changes in either the rats or mice. Thus, at most, 3000 ppm PGME had only a minimal effect on these species. The investigators conclude that the hazard from exposure to PGME vapors is distinctly less than to EGME vapors.

DERMAL. When the ether was applied to the skin of rabbits under a cuff, the LD_{50} value was found to be in the range of 13–14 g/kg (8, 16). Depression was observed with the absorption of acutely toxic quantities of this material, especially at dosage levels above 10 mL/kg.

INJECTION. Stenger et al. (19) characterized LD_{50} values in several species following direct injection of the compound by various routes of administration. The intravenous LD_{50} levels were 4.9 (mice), 3.9 (rats), 1.1 (rabbits), and 1.8–2.3 g/kg (dogs). The intraperitoneal LD_{50} in rats was 3.9 g/kg, and the subcutaneous LD_{50} values for rats and rabbits were 7.2 and 4.6 g/kg, respectively. After intravenous injection, dogs experienced pain at the site of injection, shallow breathing, decreased blood pressure, auricular arrhythmia, and death due to convulsions. These data indicate that PGME is low in toxicity when injected.

EYE IRRITATION. Repeated application of one drop of the undiluted material into the eyes of rabbits for 5 days caused only a mild transitory irritation of the eyelids after each dose (8). Smyth et al. (16) indicated that eye irritation was mild. On the other hand, 3-methoxy-1-propanol, an isomer of PGME, was reported to be capable of causing moderate eye injury, being rated 5 on a scale of 10 (18).

SKIN IRRITATION. PGME, when tested for skin irritation on rabbits, failed to cause more than a very mild irritation, and that only after constant contact for several weeks (8). Smyth et al. (16) also reported only mild irritation. PGME was not a skin sensitizer when tested in guinea pigs (24).

1.4.1.2 Chronic and Subchronic Toxicity

SUBCHRONIC TOXICITY, ORAL. Stenger et al. (19) administered PGME 5 times a week for 13 weeks to rats and for 14 weeks to dogs. The rats received daily doses of 0.5, 1.0, 2.0, and 4.0 mL/kg and the dogs 0.5, 1.0, and 2, and 3 mL/kg. Both the dogs and rats experienced mild to severe CNS depression in a dose-related manner. In the rat, this caused a growth depression because of reduced food intake. In addition, the livers of the rats became enlarged, especially at dose levels greater than 1.0 mL/kg per day. This was accompanied by cell necrosis in the liver, mainly in the peripheral parts of the lobules. Appreciable mortality occurred at the 4.0-mL/kg dosage level. Male dogs developed numerous spermiophages in the epididymis. The meaning of this is not clearly understood. In both the rats and dogs, there was minor kidney injury at the higher doses.

Rowe et al. (8) reported that groups of rats that received 26 doses of 1.0 g/kg or less of PGME over a 35-day period showed no ill effects as judged by appearance, growth, organ weights, and histopathological examination of the organs. Under the same test conditions, 3.0 g/kg of this material produced only minor effects in the liver and kidney.

INHALATION. Rabbits and monkeys subjected to 132 exposures (5 days/week for approximately 6 months) of PGME at 800 ppm showed no evidence of adverse effects as judged by gross appearance and behavior, growth, final body and organ weights, hematology, and microscopic examination of tissues (8). In the same study, guinea pigs tolerated 3000 ppm and rats 1500 ppm without adverse effects. The effects observed at higher concentrations were slight growth depression and very slight liver and lung effects. A mild CNS depression was observed at the start of the experiments at the 3000–ppm level. However, recovery was rapid after cessation of each day's exposure. The animals developed a tolerance after several weeks so that this response was not observed later in subsequent exposure weeks.

Landry et al. (6) exposed rats and rabbits to 300, 1000, or 3000 ppm for 6 h/day, 5 days/week, for 13 weeks. Both species exhibited a slightly transient CNS depression at 3000 ppm. Rats exposed to 3000 ppm experienced mild changes in liver weight and hepatocellular swelling in the absence of any degenerative changes. No exposure-related effects were reported at the low and intermediate doses. More recently, male and female F344 rats were exposed by inhalation to 0 (control), 300 or 3000 ppm PGME for 6 h/day, 5 days/week for 13 weeks (25). The highest concentration produced sedation in both genders of rat during the first week of exposure but declined in subsequent weeks. Hepatic mixed function oxidase activity and hepatocellular proliferation were increased. Male rat–specific α-2-microglobulin nephropathy was observed at the 3000-ppm level and, to a slight extent, at 300 ppm. Mice (B6C3F1) were also exposed by inhalation to 0 (control), 300, 1000, or 3000 ppm PGME for 6 h/day, 5 days/week for 13 weeks (26). Sedation was seen in both genders of mouse for the first 3 days of exposure. Adrenal gland atrophy was seen in female mice at the 3000 ppm concentration and to a slight degree at the 1000-ppm level. Slight renal and hepatic cellular proliferation was observed with significant hepatic enzyme induction at 3000 ppm in both genders of mouse. No effects were observed at 300 ppm.

DERMAL. When PGME was repeatedly placed beneath bandages onto the clipped abdomens of rabbits over a 90-day period (65 doses), the high dose levels of 7 and 10 mL/kg caused narcosis and increased mortality. At the lower dose levels of 2 and 4 mL/kg, only mild narcosis was apparent. A slight increase in kidney weights was also noted in the highest dose animals (10 mL/kg) (8). In a more recent 21-day dermal study, rabbits received 1000 mg/kg daily 15 times. No systemic toxicity was observed; the only treatment-related response was a slight scaling and minimal inflammation associated with protective skin thickening (27).

CHRONIC. The chronic toxicity and carcinogenicity of PGME was studied in male and female F344 rats (28) and male and female B6C3F1 mice (29). Animals were exposed to PGME vapor concentrations of 0 (control), 300, 1000, or 3000 ppm for 6 h/day, 5 days/week for up to 2 years. PGME-induced sedation at 3000 ppm was initially present but resolved. This coincided with the appearance of adaptive changes in the liver, including mixed function oxidase (MFO) induction and hepatocellular proliferation. By week 52, MFO activities dropped to near control levels with a return of sedation at 3000 ppm in rats. In male rats, there was a dose-related increase in eosinophilic foci of altered hepatocytes after 2 years of exposure. Male rats also displayed kidney toxicity that was identified immunohistochemically as α-2-microglobulin nephropathy. Based on the results of these studies, 300 ppm was established as the NOEL for rats.

Similar but generally less severe effects were noted in mice in these 2-year studies. Sedation of mice was present at the highest exposure concentration during the first week of exposures but resolved, with concomitant adaptive changes in the livers (as for rats). There were chronic, but small, increases in hepatocellular proliferation in mice at 3000 ppm. However, no histopathological changes of significance were noted in any tissues. Increased mortality in the high dose male group (34% vs. 18% in controls) may have been related to the minimal liver toxicity noted. Based on the results of these studies, 1000 ppm was established as the NOEL for mice.

1.4.1.3 Pharmacokinetics, Metabolism, and Mechanisms. The metabolism of [^{14}C]propylene glycol monomethyl ether (α isomer) has been evaluated following single oral doses of 1 or 8.7 mmol/kg to rats (9). Approximately 10–20% of the administered dose was recovered in the urine, and 50–60% was excreted as $^{14}CO_2$ within 48 h. The urinary metabolites were primarily glucuronide and sulfate conjugates of the parent compound, as well as lesser amounts of unconjugated parent compound and propylene glycol. These results indicated that, unlike the metabolism of ethylene glycol ethers, the α isomer was not a good substrate for alcohol dehydrogenase. The primary metabolic pathway appeared to be a microsomal O-demethylation leading to propylene glycol, which in turn was further oxidized to carbon dioxide (see Fig. 87.1). A subsequent oral-dose study in rats revealed that, similar to the EGEs, the β isomer was extensively metabolized by alcohol

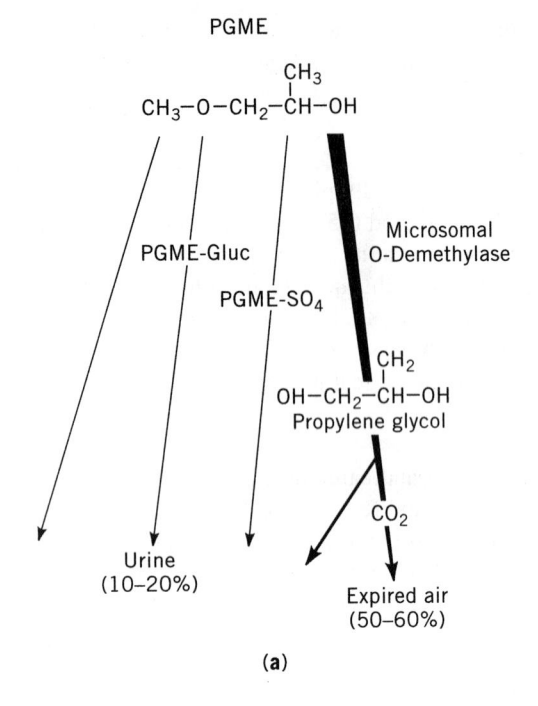

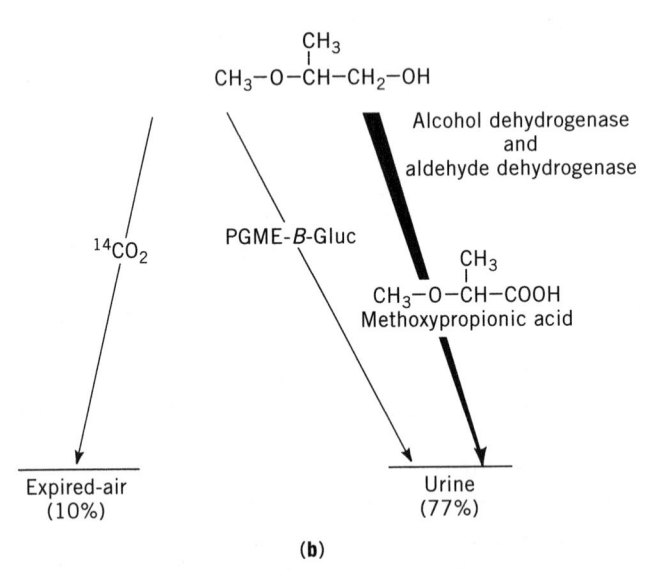

Figure 87.1. Comparative metabolism of the α (a) and β (b) isomers of propylene glycol methyl ether.

dehydrogenase to methoxyproprionic acid (see Fig. 87.1) (10). The distinct difference in the metabolism of the α isomer of PGME compared to the EGEs likely accounts for the differences in toxicity between these glycol ether series. Propylene glycol, as a metabolite of the α isomer, would be expected to have markedly less toxicity than the alkoxyacetic acid metabolites derived from ethylene glycol ethers. Ferrala et al. (30) measured PGME and its metabolite, propylene glycol (PG) in rat and mouse plasma following oral gavage administration at 90 mg/kg. Maximum plasma concentrations (C_{max}) of PGME and PG in mice were seen at 20 and 30 min, respectively, and were 76.5 µg/mL (PGME) and 18.5 µg/mL (PG).

The disposition of the α isomer has also been evaluated following inhalation exposure of rats. Blood parent compound concentrations were found to increase continually during 6-h exposures ranging from 300 to 3000 ppm (31), suggesting that absorption was limited by respiration. The end-exposure blood concentrations were not proportional to exposure concentration; the clearance of glycol ether from blood was best described as a zeroth-order process even at the low exposure concentration of 300 ppm. Repeated exposure to 3,000 ppm but not 300 ppm increased liver weight and mixed function oxidase activity. These liver changes apparently represented a metabolic compensation process in that concentrations of parent compound in blood were 50% lower after the tenth exposure compared to the first 3000-ppm exposure. This metabolic compensation may account for the rapid development of tolerance to the CNS depressant activity that is observed with repeated-inhalation exposure to high concentrations of PGME.

1.4.1.4 Reproductive and Developmental

REPRODUCTIVE TOXICITY. A continuous breeding study was conducted in CD-1 mice administered dose levels of 0.5, 1.0, or 2.0% PGME in drinking water during a 7-day premating exposure and for 14 subsequent weeks of paired cohabitation (32). Pups from the 2.0% dose group were continued on treatment through lactation and subsequent mating for a second generation. There were no effects at the 2% dose level on fertility or reproductive parameters, including litters/pair, live pups/litter, sex ratio, and sperm motility, density, or incidence of sperm abnormalities. However, the highest dose group exhibited a decrease in pup body weights in the first generation and a reduction in right epididymal and prostate gland weights in the second-generation pups. In a retrospective study, ovaries from mice exposed to PGME in an NTP continuous breeding protocol (33) were reanalyzed histopathologically (34). Changes in differential follicle counts in exposed and unexposed mice indicated that PGME did not induce ovarian toxicity.

In an inhalation study, pregnant rats were exposed to 200 or 600 ppm PGME vapor, 6 h/day on days 6–17 of gestation (35). No effects were noted on maternal body weight gain or the number, weight, or viability of pups. No alterations in testicular weight or histopathology were observed in rats exposed for 10 consecutive days, 6 h/day, to 200 or 600 ppm PGME.

A two-generation inhalation reproductive study was conducted with PGME in Sprague–Dawley rats (36). Groups of 30 male and 30 female rats were exposed to nominal levels of 0 (control), 300, 1000, or 3000 ppm of PGME vapors 6 h/day, 5 days/ week prior to mating and 6 h/day, 7 days/week during mating, gestation, and lactation for two generations. At the highest concentration of 3000 ppm, there was marked parental toxicity evidenced by sedation during and following exposures and body weights that were as much as 21% lower than controls. In female rats, the toxicity was accompanied by lengthened estrous cycles, dereased fertility, decreased ovary weights, and histological ovarian atrophy. Offspring of affected dams displayed decreased body weights, reduced survival and litter sizes, delays in puberty onset, and histological changes in the liver and kidney. The close correlation of these effects to the noted parental toxicity suggested that these effects were secondary to general toxicity or nutritional stress. No such reproductive or neonatal effects were seen at 1000 ppm, a level that also caused less marked maternal toxicity. No treatment-related effects were seen at 300 ppm. The NOEL for reproductive/neonatal effects was 1000 ppm. The NOEL for parental toxicity was 300 ppm.

DEVELOPMENTAL TOXICITY. Stenger et al. (19) fed PGME to mice, rats, and rabbits during the first 18–21 days of gestation. Only the rat fetus showed any effect, a delayed ossification of the skull, at the highest dose given (0.8 mL/kg). There were no effects on the number of pups born. In a more recent inhalation study, no teratogenic effects were observed in rats or rabbits exposed to concentrations of up to 3000 ppm (37). Mild CNS depression, decreased food consumption, and body weight gain were reported in maternal rats only at 3000 ppm. Slight fetotoxicity reflected by delayed skeletal ossification was seen in the high dose rats.

1.4.1.5 Carcinogenesis. Oncogenicity studies in rats and mice were discussed in Section 1.4.1.2 (28, 29). PGME failed to produce a dose-related increase in any tumor types in either male or female F344 rats or male or female B6C3F1 mice.

1.4.1.6 Genetic and Related Cellular Effects Studies. Mutagenicity was not detected in the Ames Salmonella assay, both with and without metabolic activation, at plate concentrations ranging from 2 to 6250 µg/plate (38). PGME also did not produce chromosomal abnormalities in the CHO metaphase analysis test (38) or induce unscheduled DNA synthesis in rat hepatocytes (A. L. Mendrala, The Dow Chemical Company, in published data, 1983). Elias et al. (39) reported on the genotoxicity of PGME in a series of short-term genetic toxicity tests. The results suggest PGME to be weakly clastogenic but at excessively high concentrations, which would potentially produce artifactual responses due to osmolarity changes in the medium.

1.4.2 Human Experience

1.4.2.1 General Information. There is no unfavorable human experience known other than from the disagreeableness of the odor and irritation associated with extremely high exposures. Groups of male volunteers were exposed to 50, 100, or 250 ppm PGME in an exposure chamber for 1–7 h (14). Tolerance to the exposure developed within 25 minutes. At 100 ppm, mild eye irritation was noted in three of six individuals. At 250 ppm, the majority of 23 subjects exposed for 1–7 h complained of eye, nose or throat irritation; several subjects developed headaches, and one was nauseated. None of the exposures resulted in changes in vision, coordination, neurological, or brake-reaction tests. Clinical studies including complete urinalysis, completed before and after exposures, showed no effects. The odor of PGME could be detected at 10 ppm.

1.4.2.2 Clinical Cases

1.4.2.2.1 Acute Toxicity: NA

1.4.2.2.2 Chronic and Subchronic Toxicity: NA

1.4.2.2.3 Pharmacokinetics, Metabolism, and Mechanisms. Dermal absorption of PGME vapors was investigated using male and female volunteers (40). Exposures were with and without a mask to determine the contribution of PGME uptake by the dermal route alone. Blood level measurements indicated that dermal absorption contributed 6.3% of the total body uptake. The elimination half-life following total body exposure averaged 1.5 h versus 2.7 h for dermal only exposure. This suggested a possible reservoir (skin) effect, giving an apparent delay in elimination.

Jones et al. (41) have developed a method for the detection of PGME in human urine. The method involves solvent extraction followed by analysis by gas chromatography/mass spectrometry. The reported detection limit for this method was 1 mµmol/L of urine. In a human volunteer study, six volunteers were exposed to 100 ppm PGME for 8 h. Postexposure levels of PGME were urine, 110 mµmol/L; blood (maximum level), 103 mµmol/L; and exhaled air, up to 252 nmol/L. PGME was excreted into urine with a half-life of less than 2.6 h.

1.4.2.2.4 Reproductive and Developmental: NA

1.4.2.2.5 Carcinogenesis: NA

1.4.2.2.6 Genetic and Related Cellular Effects Studies: NA

1.4.2.2.7 Other: Neurological, Pulmonary, Skin Sensitization, etc.

EYE IRRITATION. Studies were conducted with 12 male volunteers (42). Subjects were exposed for 2.5 h to three different exposure conditions with each exposure separated by 7 days. Diethyl ether was used as a masking agent to minimize responses caused by PGME odor. Pre- and post-exposure eye redness, corneal thickness, tear film breakup time, conjunctival epithelial damage, blinking frequency, and subjective ratings were used to determine irritating effects. Minimal subjective eye effects were noted at 150 ppm PGME; there was no impact on other objective parameters. The NOEL for eye irritation was 150 ppm.

1.5 Standards, Regulations, or Guidelines of Exposure

ACGIH in their 1998 listing of threshold limit values recommend a time-weighted average (TWA) threshold limit value (TLV) of 100 ppm (369 mg/m^3) for repeated 8-h human

Table 87.7. Occupational Exposure Guidelines for PGME[a]

Regulatory or Expert Authority (Date)	Exposure Value	Comment/Notation
ACGIH (TLV) (1998)	TWA = 100 ppm STEL = 150 ppm	
NIOSH (1994)	TWA = 100 ppm STEL = 150 ppm	
Belgium (OEL) (1993)	TWA = 100 ppm STEL = 150 ppm	
Denmark (OEL/TWA) (1993)	100 ppm	
Finland (OEL) (1993)	TWA = 100 ppm STEL = 150 ppm	Skin
France (OEL/TWA) (1993)	100 ppm	
Germany (OEL/TWA) (1993)	100 ppm	
Netherlands (OEL/TWA) Oct-97	TWA = 100 ppm	
Switzerland (OEL) (1993)	TWA = 100 ppm STEL = 200 ppm	Skin
United Kingdom (OEL) (1993)	TWA = 100 pp	Skin

[a] The OEL values set by Bulgaria, Colombia, Jordan, Korea, New Zealand, Singapore, and Vietman are based on the ACGIH value.

exposure and a short-term exposure limit (STEL) of 150 ppm (553 mg/m^3) (43). A German MAK value for the β isomer of PGME is 20 ppm. Refer to Table 87.7 for a summary of international regulatory guideline for occupational exposure to PGME.

2.0 Propylene Glycol Monoethyl Ether

2.0.1 CAS Numbers: *[1569-02-4]* (α isomer), *[52125-53-8]* (mixture of α and β isomers)

2.0.2 Synonyms: For CAS *[1569-02-4]*: 1-ethoxy-2-propanol, 1-ethoxypropan-2-ol, 1-ethoxy-2-hydroxypropane. For CAS *[52125-53-8]*: ethoxypropanol, ethyl ether of propylene glycol, PGEE

2.0.3 Trade Names: Arcosolv® PE, Dowanol® PE Glycol Ether, Propasol® Solvent

2.0.4 Molecular Weight: 104.15

2.0.5 Molecular Formula: C$_5$H$_{12}$O$_2$

2.0.6 Molecular Structure:

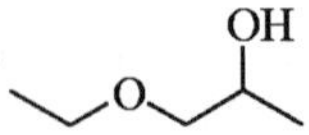

2.1 Chemical and Physical Properties

See Table 87.2.

2.2 Production and Use

Most commercial PGEE is manufactured from propylene oxide resulting in the α and β isomers, 1-ethoxy-2-hydroxypropane and 2-ethoxy-1-hydroxypropane, respectively, with the α representing at least 95% and, usually, a greater proportion of the final product. The third isomer, 3-ethoxy-1-propanol, is an isomer of PGEE manufactured from 1,3-dihydroxypropane, but is of little or no commercial importance. Refer to Table 87.4. for production and use data.

2.3 Exposure Assessment: NA

2.4 Toxic Effects

Commercial propylene glycol monoethyl ether (PGEE) is of a low order of single-dose oral toxicity, irritating to the eyes, but not appreciably irritating to the skin. This material may produce toxic effects on prolonged and widespread contact with the skin. Its vapor is only slightly toxic on short-term acute exposure; however, high concentrations are irritating and can cause narcosis. The β isomer is similar in acute toxicity to the commercial material except that it appears to be somewhat more injurious to the eyes and more toxic by percutaneous absorption (18).

2.4.1 Experimental Studies

2.4.1.1 Acute Toxicity

ORAL. Smyth et al. (17) fed the both the α and β isomers of PGEE as 50 percent aqueous solutions to rats and found the LD_{50} levels to be 7.11 and 7.00 g/kg, respectively. Unpublished studies on a commercial product yielded an oral LD_{50} value for rats of more than 5.0 g/kg. Marked narcosis and some kidney injury were observed from large doses in both studies (V. K. Rowe, The Dow Chemical Company, unpublished data, 1947).

Rats (6 per sex) administered 10 consecutive daily doses of 2 mL/kg showed little effect other than a slight decrease in weight gain and slight hematological alterations in males only and liver weight increases in both sexes (44). Smyth and Carpenter (45) administered the beta isomer to rats in their drinking water for 30 days. A daily dose of 0.68 g/kg was without adverse effect, whereas a daily dose of 2.14 g/kg caused reduced growth. In a later reference to this study (18), kidney injury, but no mortality, was noted at the 2.14 g/kg daily dosage level.

INHALATION. Gross (46) reported that a mouse, a guinea pig, and a rabbit tolerated up to 7000 ppm for 1 h without effect other than irritation of the eyes and respiratory tract. A 2-h exposure caused more severe irritation, and a rabbit showed signs of kidney injury, transient albuminuria, and red cells in the urine. All of five rats exposed for 4 h to a concentration calculated to be 10,000 ppm survived, but they showed signs of marked irritation to the eyes and nares, and were anesthetized by the end of the exposure period (V. K. Rowe, The Dow Chemical Company, unpublished data, 1948). Gross (46) also exposed cats, guinea pigs, and rabbits 8 h/day to a concentration of about 1200 ppm. One of two cats and one of two guinea pigs tolerated 12 exposures without apparent effect, but the other cat and guinea pig died after the treatment. The two rabbits succumbed after 3 and 9 days of exposure; necropsy revealed pneumonia and kidney injury. In 1983 study, rats (6 per sex per exposure level) were exposed to 1400 ppm (or 8900 mg/m^3) for 6 h/day for 9 days (47). Liver weights were slightly increased without histopathological changes in both sexes at the higher exposure level with no other effects noted.

EYE IRRITATION. Primary eye irritation tests indicate that PGEE is not severely irritating to the eyes of rabbit. Discomfort, conjunctival irritation, and corneal reactions were seen with full recovery within 7 days (48, 49). In an older study, when the commercial product was applied to the eyes of rabbits on 5 consecutive days, it caused moderate conjunctival irritation and some transient cloudiness of the cornea. Healing was essentially complete in 3–7 days (V. K. Rowe, The Dow Chemical Company, unpublished data, 1947).

DERMAL TOXICITY, SKIN IRRITATION. When PGEE was applied topically to the skin of rabbits and occluded for 24 h, all of six animals treated survived 5 mL/kg, three of five survived 7 mL/kg, and one of five survived doses of either 10 or 15 mL/kg. The LD$_{50}$ was estimated to be 9 mL/kg. The material in large doses caused marked CNS depression and deaths usually occurred within 48 h after treatment. No appreciable irritation of the skin resulted under these conditions (V. K. Rowe, The Dow Chemical Company, unpublished data, 1947). More 1980s tests explicitly designed to define skin irritation potential confirm the mild irritation potential of PGEE to skin (48, 49).

2.4.1.2 Chronic and Subchronic Toxicity

SUBCHRONIC TOXICITY/INHALATION. Rats (15 per sex per concentration) were exposed to atmospheres of PGEE at nominal concentrations of 0, 100, 300, and 2000 ppm 6 h/day, 5 days/week for 13 weeks, totaling 65 exposures (50). At 2000 ppm, rats exhibited irritation of the eyes and nose and increased urine volume during weeks 1 and 12 of exposure and only females showed increased liver weights without accompanying histopathological changes. At 300 ppm, urine volumes were increased in both sexes during the twelfth week of exposure. No effects were detected at 100 ppm. Urinalysis and hematological examination revealed no abnormalities, nor did histopathological evaluation of kidneys, testes, blood, or blood-forming organs, and a variety of other tissues examined at the end of the study (50).

2.4.1.3 Pharmacokinetics, Metabolism, and Mechanisms: NA

2.4.1.4 Reproductive and Developmental.
PGEE has been evaluated for its potential to cause developmental toxicity in two species. Pregnant rats (25 dams per exposure concentration) were exposed 6 h/day to atmospheres of PGEE at nominal concentrations

of 0, 100, 450, and 2000 ppm during days 6–15 of gestation (51). Maternal toxicity was seen at the highest exposure concentration, manifested as reduced food consumption and body weight gain. Exposure to 450 ppm also produced slightly reduced body weights, but no effect was seen at 100 ppm. No effects were detected on the number of pre- and post-implants, litter size and pup weights, or anomalies, variants, or frank malformations (51).

Pregnant rabbits (22 dams per exposure concentration) were exposed 6 h/day to PGEE atmospheres at nominal concentrations of 0, 100, 350, and 1200 ppm on days 6–18 of gestation (52). Dams exposed to 1200 ppm exhibited reduced food consumption and weight gain. No effects on the dams were noted at the lower exposure levels. No effects were found on pre- and post-implantation numbers, litter size, or pup weights. Although the incidence of malformations was slightly greater in treated versus control pups, they were not dose related and were within historical ranges (52).

2.4.1.5 Carcinogenesis: NA

2.4.1.6 Genetic and Related Cellular Effects Studies. PGEE did not cause mutations in the Ames *Salmonella* assay when tested using five histidine-requiring strains either with or without S9 rat liver microsomes at doses of ≤5000 µg/plate (53). PGEE did not damage chromosomes of cultured human lymphocytes in the presence or absence of rat liver microsomes at concentrtions of ≤5000 µg/mL (54).

2.5 Standards, Regulations, or Guidelines of Exposure

An occupational standard for exposure to PGEE has not been established.

3.0 Propylene Glycol-*n*-monopropyl Ether

3.0.1 CAS Numbers: [1569-01-3] (α isomer), [30136-13-1] (mixture)

3.0.2 Synonyms: 1-Propoxy-2-propanol, propylene glycol-*n*-monopropyl ether (PGPE)

3.0.3 Trade Names: Dowanol® PnP Glycol Ether, Propasol® Solvent P

3.0.4 Molecular Weight: 118.18

3.0.5 Molecular Formula: $C_6H_{14}O_2$

3.0.6 Molecular Structure:

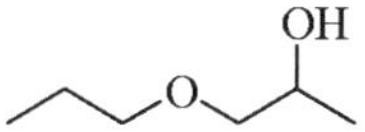

3.1 Chemical and Physical Properties

3.1.1 General

See Table 87.2.

3.2 Production and Use

See Table 87.4.

3.4 Toxic Effects

Propylene glycol-*n*-monopropyl ether (PGPE) is of low acute toxicity following either oral or dermal applications. The material is severely irritating to the eyes and can cause mild corneal injury, which may be slow in healing. It is essentially nonirritating to the skin following short-term contact; however, prolonged or occluded contact can cause severe irritation. Because of its relatively low vapor pressure, inhalation is not anticipated to present a significant hazard under conditions of ambient temperature.

3.4.1 Experimental Studies

3.4.1.1 Acute Toxicity

ORAL. PGPE has been shown to be of low toxicity. The LD_{50} for rats has been reported as 3.25 mL/kg (Union Carbide Corporation, unpublished data, 1954), 4.92 mL/kg (male), 2.83 mL/kg (female) (55), and 2.8–3.0 g/kg (K. J. Olson, The Dow Chemical Company, unpublished data, 1959). In addition, at high doses, the animals developed CNS depression and some evidence of kidney injury.

INHALATION. Union Carbide (56) reported that an 8-h exposure to substantially saturated vapor generated at 22°C killed zero of five rats (approximate concentration 11.7 mg/L or 2460 ppm). In a more recent study, Union Carbide (57) reported that an 8-h exposure to essentially saturated vapor generated at 23°C killed zero of six rats. However, the authors reported that the rats appeared to be anesthetized. Ballantyne et al. (55) reported that male and female rats exposed for 6 h to a statically generated saturated vapor at 24°C (approximate concentration 2230 ppm) produced signs of sensory irritation of the eye. However, no signs of systemic toxicity were observed during exposure nor during the 14-day post-exposure observation period. No abnormal pathological features were seen at necropsy, following the observation period. Union Carbide (58) also reported an absence of observable adverse effects in female rats exposed to a metered concentration of 600 ppm, 7 h/day for 10 days. When the exposure was increased to a metered concentration of 600 ppm, 7 h/day for 29 days, six male rats were without observable adverse effects. In a 9-day study, male and female Fischer 344 rats exposed for 6 h/day over an 11-day period to 0 (control), 503, 983, or 2000 ppm exhibited CNS depression at the highest exposure concentration. In addition, corneal lesions, depressed body weights, and increased liver and kidney weights were observed at all exposure levels (59). When groups of male Fischer 344 or Sprague–Dawley rats, Hartley guinea pigs, and New Zealand rabbits were exposed for 6 h/day for 9 days over an 11-day period to 0 (control), 105, 486, or 1824 ppm, ocular irritation, eye lesions, and CNS depression were observed in both strains of rats and in the rabbits at the highest exposure level. In addition, three of six rabbits died at the highest exposure level. However, the highest exposure level produced minimal toxicity in the guinea pig. Exposure concentrations of 105 and 486 ppm produced some potentially irreversible eye lesions in the Fischer 344 rats. However, the same concentrations produced marginal effects in the Sprague–Dawley rats and in the New Zealand rabbits. One guinea pig developed an ocular lesion at 500 ppm. The authors concluded that due to an apparent predisposition of Fischer 344 rats for mineral deposition in ocular tissue (corneal dystrophy), this strain of rat may not be an appropriate animal model for evaluating the ocular toxicity of PGPE (60). In a follow-up study, male Fischer 344 or Sprague–Dawley rats exposed for 6 h/day for 9 days over an 11-day period to 0 (control), 5.2, 48.1, or 99.1 ppm exhibited no signs of toxic effects and no apparent ocular alterations (61).

DERMAL. A number of dermal LD_{50} values have been determined for PGPE. Union Carbide (56, 57) reported an LD_{50} of 4.0 mL/kg and 3.17 mL/kg in rabbits. Ballantyne et al. (55) reported an LD_{50} of 4.29 mL/kg in male rabbits and 4.92 mL/kg in female rabbits. Signs of toxicity included comatose appearance within 15–30 min of application and prostration and lacrimation at 1 day.

EYE IRRITATION. Instillation of 0.1, 0.01, or 0.005 mL into the eyes of rabbits caused severe conjunctivitis and diffusely mild corneal injury, requiring 3–14 days for complete resolution (55).

SKIN IRRITATION. Union Carbide (56) reported that PGPE was essentially nonirritating to the skin of rabbits by short-term contact. Ballantyne et al. (55) reported that 0.5 mL of PGPE applied to the shaven dorsal skin of each of six New zealand white rabbits and covered with an occlusive dressing for 4 h produced no irritation on two animals and mild to moderate erythema and edema on the remaining animals. None of the six rabbits showed necrosis or ulceration. However, under more rigorous conditions, PGPE applied under an occlusive dressing for 24 h caused more severe irritation with persistent fissuring, desquamation, ulceration, and necrosis (55).

3.4.1.2 Chronic and Subchronic Toxicity

SUBCHRONIC INHALATION. Four groups of 20 male and 20 female Fischer 344 and Sprague–Dawley rats were exposed for 6 h/day, 5 days/week for 14 weeks to 0 (control), 30, 100, or 300 ppm. Half of the animals (10 $sex^{-1} strain^{-1} group^{-1}$) were maintained for an additional 90-day recovery period. Monitors of toxicity included clinical observation, ophthalmic examination, body and organ weights, clinical pathology, and macroscopic and microscopic examination. No exposure-related clinical signs were observed during the study. In addition, for all groups of animals, the eyes appeared normal during ophthalmic examination. At 300 ppm, body weight gains were lower for the female Fischer 344 rats, except during the recovery period. At 100 ppm, the female Fischer 344 rats had decreased body weight gains for the first 2 weeks of exposure. No additional exposure-related differences in body weights or food and water consumption were observed. No exposure-related gross lesions were observed at necropsy and organ weights were normal. In addition, no microscopic tissue lesions attributable to exposure were observed. However, corneal dystrophy was observed in rats of all groups, including the control group. The authors suggested that 300 ppm was a marginal-effects concentration and 100 ppm is a no-effect concentration (62).

3.5 Standards, Regulations, or Guidelines of Exposure

There has been no occupational exposure standard established for PGPE.

4.0 Propylene Glycol Monoisopropyl Ether

4.0.1 CAS Numbers: [3944-36-3], [29387-84-6], [110-48-5]

4.0.2 Synonyms: 1-Isopropoxy-2-propanol (α isomer), and 2-isopropoxy-1-propanol (β isomer) isopropoxypropanol, methoxy ethoxy propanol

4.0.3 Trade Names: NA

4.0.4 Molecular Weight: 118.18

4.0.5 Molecular Formula: $C_6H_{14}O_2$

4.0.6 Molecular Structure:

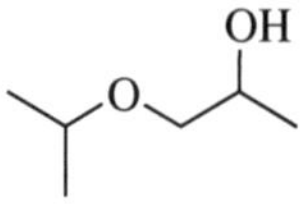

4.1 Chemical and Physical Properties

See Table 87.2.

4.2 Production and Use: NA

4.3 Exposure Assessment: NA

4.4 Toxic Effects

Propylene glycol monoisopropyl ether is low in single-dose oral toxicity for rats; the LD_{50} is greater than 2 g/kg and is typically 4 g/kg. Conjunctival irritation accompanied by some corneal injury and iritis resulted from the instillation of small quantities into the eyes of rabbits. The injury healed within a week. The ether is only slightly irritating to the intact skin even on prolonged and repeated contact. This material is at most only slightly acutely toxic by absorption through the skin. Rats that received a single 7-h exposure to an essentially saturated vapor survived but exhibited drowsiness, labored breathing, and temporary weight loss. Evidence of mild kidney injury was noted at autopsy (M. A. Wolf, The Dow Chemical Company, unpublished data, 1959). These data suggest that propylene glycol monoisopropyl ether may be considered to present no unusual health hazards in anticipated industrial operations. Precautions should be exercised to avoid eye contact. It would seem prudent to avoid repeated inhalation of vapor.

5.0 Propylene Glycol Mono-n-butyl Ether

5.0.1 CAS Numbers: [5131-66-8], [29387-86-8]

5.0.2 Synonyms: 1-Butoxy-2-propanol, *n*-butoxypropanol, PGBE

5.0.3 Trade Names: Dowanol® PnB Glycol Ether, Propasol® Solvent B

5.0.4 Molecular Weight: 132.2

5.0.5 Molecular Formula: $C_7H_{16}O_2$

5.0.6 Molecular Structure:

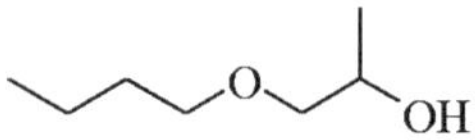

5.1 Chemical and Physical Properties
See Table 87.2.

5.2 Production and Use
See Table 87.4.

5.3 Exposure Assessment: NA

5.4 Toxic Effects

Propylene glycol mono-n-butyl ether (PGBE) is low in single-dose oral toxicity. It is markedly irritating and somewhat injurious to the eyes, and may be moderately irritating to the skin. It may be absorbed through the skin but is low in toxicity by this route. PGBE is not particularly hazardous from an occasional exposure by inhalation. However, repeated exposure may lead to liver injury, especially at high levels. It is considered to have no significant developmental toxicity in experimental animals and does not cause hemolysis in rodents.

5.4.1 Experimental Studies

5.4.1.1 Acute Toxicity

ORAL. The acute oral toxicity of PGBE has been shown to be low. The LD_{50} value for rats has been reported to be 1.9 (V. K. Rowe. The Dow Chemical Company, unpublished data, 1947), 2.5 (63), 3.3 (64), and 5.2 g/kg (18). Rats dosed by gavage with 0, 100, 200, or 400 mg/kg of PGBE daily for 14 consecutive days showed no effects (65).

INHALATION. Carpenter et al. (63) and Smyth et al. (18) have shown that rats exposed for 8 h to essentially saturated vapors of propylene glycol mono-*n*-butyl ether generated at room temperature were essentially unaffected. Rats showed no signs of toxicity after a 4-h exposure to a saturated atmosphere of PGBE (66).

In repeated exposure studies, Fischer 344 (F344) and Sprague—Dawley rats of both genders received whole-body vapor exposure at 0, 10, 100, 300, or 600 ppm for 6 h/day for 9 days over an 11-day period. The only exposure-related effects reported were increased liver weights in the 600-ppm group of F344 rats and a low incidence of mild eye lesions in the 300- and 600-ppm groups of F344 rats (67). In a second study, Fischer 344 rats exposed to 700 ppm PGBE 6 h/day for 9 days over a 2-week period showed no hematological effects or other signs of toxicity (68). In an older study, rats were exposed to nominal concentrations of 600 ppm PGBE 7 hr/day for 31 days. The only effects seen were increased liver weights in the females (69).

DERMAL. Carpenter et al. (63) determined a dermal LD_{50} of 3.1 g/kg for rabbits, and Smyth et al. (18) reported a dermal LD_{50} of 1.4 g/kg for rabbits. More recent studies in rats revealed that the dermal LD_{50} was greater than 2 g/kg, the highest dose level evaluated (70). Other dermal toxicity studies conducted essentially as described by Draize et al. (71) showed that all of five animals receiving 1.8 g/kg survived, two of five receiving 2.6 g/kg survived, and none of five receiving 4.4 g/kg survived. When the material was confined under a cuff for 24 h, severe injury to the skin occurred and the animals became deeply narcotized. Deaths from the larger doses occurred within a few hours after application of the material; all deaths occurred within 24 h after treatment or not at all (V. K. Rowe, The Dow Chemical Company, unpublished data, 1947). Further studies in rabbits have shown that application of test material for 4 h resulted in moderate to severe redness, swelling, scaling, and eschar formation. The eschar was not considered to be representative of a corrosive response (P. J. J. M. Weterings, P. A. M. Daamen, and H. G. Verschuuren, The Dow Chemical Company, unpublished data, 1987).

EYE IRRITATION. A single application of undiluted PGBE was moderately irritating to the rabbit eye with transient corneal opacity (72). PGBE was found to be appreciably irritating to the rabbit eye; one drop in an eye on five consecutive days caused marked conjuntival irritation and corneal cloudiness, which healed within a week (V. K. Rowe, The Dow Chemical Company, unpublished data, 1947). Similar finding were reported by Carpenter et al. (63) and Smyth et al. (18).

SKIN IRRITATION. PGBE was moderately irritating to skin after a 4-h single exposure when applied in pure form or as a 75% solution in water (73). A 50% solution was mildly irritating, and a 25% solution was not irritating after four hours occlusion (74). Repeated applications (10 in 14 days) to the skin of rabbits resulted in slight irritation and some evidence that toxic amounts were absorbed (V. K. Rowe, The Dow Chemical Company, unpublished data, 1947). Both Carpenter et al. (63) and Smyth et al. (18) also reported that propylene glycol mono-n-butyl ether caused only mild irritation.

5.4.1.2 Chronic and Subchronic Toxicity

ORAL. Grandjean and co-workers (75) administered PGBE in the drinking water of Fischer 344 rats at doses of 0, 100, 350, or 1000 mg/kg per day for 13 weeks. Increased liver weights were reported in the high-dose level males and increased kidney weights in the high-dose females. No histological alterations were associated with the organ weight changes. The daily NOEL dose was 350 mg/kg in both genders, and the daily NOAEL dose was 1000 mg/kg.

INHALATION. Rats (5/Sex) were exposed to atmospheres containing 600 ppm PGBE 7 h/day, 5 days/week, for 31 days. Females exhibited slight liver weight increases (Union Carbide, unpublished report, 1964).

DERMAL. Application of test material to the skin or rats 5 days/week for 13 weeks at daily dose levels of 0, 0.1, 0.3, and 1.0 mL/kg resulted in minor skin reactions ranging from redness and swelling to occasional superficial scar tissue at all dose levels (76). No systemic toxicity was noted at the high dose of 1 mL/kg (880 mg/kg), including clinical chemistries, hematology, and histopathology. Rabbits treated dermally with 0, 11.4, 114, or 1140 mg/kg daily for 13 weeks (7 h/day, 5 days/week) showed no systemic toxicity (77). The higher two dose levels produced mild to moderate redness, swelling, atony, desquamation, and fissuring of the skin.

5.4.1.3 Pharmacokinetics, Metabolism, and Mechanisms: NA

5.4.1.4 Reproductive and Developmental. Pregnant Swiss Webster mice were treated on gestation days 6–15 by either ingestion or subcutaneous injection; the highest levels were 226.4 and 151.6 mg/kg per day by ingestion and by injection, respectively (The Dow Chemical Company, unpublished data, 1972). There were no effects in any of the mice as shown by maternal body weight, mortality or behavior, reproduction, fetal body weights, or fetal development as shown by skeletal, internal, and external examinations.

Pregnant rabbits were exposed dermally to PGBE at 0, 10, 40, or 100 mg/kg per day on days 7–18 of gestation (78). No significant maternal toxicity was noted in the dams except for erythema at the site of application of the test material. There were no malformations or variations in any of the fetuses from any of the litters that could be attributed to treatment with PGBE. The NOEL for this study was 100 mg/kg/day for rabbits by dermal exposure.

Pregnant Wistar rats were treated dermally with 0, 0.3, and 1.0 mL/kg (0, 264, and 880 mg/kg) test material on gestation days 6–16. Minor skin reactions were noted in the maternal animals; no compound-related embryo or fetal toxicity or teratogenicity was seen at any dose level (79).

5.4.1.5 Carcinogenesis: NA

5.4.1.6 Genetic and Related Cellular Effects Studies. PGBE did not produce evidence of chromosomal aberrations in Chinese hamster ovary cells exposed up to 5000 µg/mL, both with and without an Aroclor-induced rat S-9 microsomal activation system (80). This material also was not mutagenic in an Ames *Salmonella* assay, with or without microsomal activation, conducted with tester strains TA98, TA100, TA1537, and TA1538 at concentrations of ≤5000 µg/plate (R. J. Bruce, B. Gollapudi, and H. G. Verschuuren, The Dow Chemical Company, unpublished data, 1988).

5.4.1.7 Other: Neurological, Pulmonary, Skin Sensitization, etc. Using a modified Bueler method, a challenge dose of 40 percent PGBE did not result in a positive sensitization response in guinea pigs (J. Vanderkom and H. G. Verschuuren, The Dow Chemical Company, unpublished data, 1987).

5.5 Standards, Regulations, or Guidelines of Exposure

No occupational exposure standard has been established for PGBE.

6.0 Propylene Glycol Mono-*tertiary*-butyl Ether

6.0.1 CAS Number: *[57018-52-7]*

6.0.2 Synonyms: Alpha isomer: 1-(1,1-dimethylethoxy)-2-propanol, 1-tertiary-butoxypropan-2-ol. Beta isomer: 2-(1,1-dimethylethoxy)-1-propanol, 2-*tertiary*-butoxypropan-1-ol. Also PGTBE, propylene glycol mono-*tert*-butyl ether; 1-*tert*-butoxy-2-propanol; 1-methyl-2-*tert*-butoxyethanol

6.0.3 Trade Names: Arcosolv® PTB

6.0.4 Molecular Weight: 132.2

6.0.5 Molecular Formula: $C_7H_{16}O_2$

6.0.6 Molecular Structure:

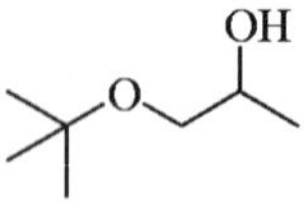

6.1 Chemical and Physical Properties

6.1.1 General

See Table 87.2.

6.2 Production and Use

Propylene glycol mono-*tertiary*-butyl ether (PGTBE) is manufactured by reacting isobutylene with excess propylene glycol in the presence of a solid resin etherification catalyst. It is then distilled to produce 99+% of the α isomer, 1-*tert*-butoxy-propan-2-ol (81). PGTBE is used in commercial cleaners, inks, adhesives, nail polish, lacquers, latex paints, and other coatings. It has properties that permit its use as a substitute for ethylene glycol butyl ether (82, 83). Refer to Table 87.4 for additional production and use information.

6.3 Exposure Assessment

Workers and consumers could be exposed to PGTBE while handling or using products containing this chemical. Relevant routes of exposure would include inhalation and dermal absorption. Workers involved in the manufacture of PGTBE may have less exposure to this chemical because the manufacturing process is largely enclosed.

6.4 Toxic Effects

PGTBE has low acute toxicity by the oral, inhalation, and dermal routes of exposure. It is slightly irritating to skin, but pure PGTBE can be severely irritating to eyes. It exhibits no sensitization potential. The subchronic toxicity of this compound also is low, and PGTBE does not damage blood, blood-forming organs, thymus, or testes. Nor does it damage the developing fetus. Chronic toxicity testing was under way at the time of writing.

6.4.1 Experimental Studies

6.4.1.1 Acute Toxicity

ORAL. In an oral LD_{50} study, Sprague–Dawley rats (8/sex per dose) were administered a single dose of 2239, 2818, 3548, or 4467 mg PGTBE/kg body weight and observed for 14 days. All animals survived the lowest dose, 75% survived 2818 mg/kg, 50% survived 3548 mg/kg, and 37% survived the high dose. Analysis of survival data using the Litchfield–Wilcoxon method resulted in an LD_{50} of 3771 mg/kg. To variable degrees, all animals showed signs of lethargy, ataxia, prostration, irregular breathing, lacrimation, crusty eyes, and yellow/brown-stained fur. Nonsurviving animals also showed crusty muzzle, salivation, red-stained fur and emaciation. Survivors gained weight throughout the study, whereas nonsurvivors lost weight. Necropsy of nonsurvivors showed red/dark-stained areas of the lungs, and damage to the stomach and liver. Survivors at necropsy revealed similar damage to lungs and mild dilation of kidney pelvises (84).

INHALATION. When Sprague–Dawley rats (5/sex per exposure level) were exposed in whole-body inhalation chambers to atmospheres of 0 or 2680 mg/m^3 ($\sim$550 ppm) PGTBE for 4 h and subsequently observed for 14 days, no deaths resulted and no abnormal clinical signs were observed. At the end of the observation period, necropsy and subsequent microscopic examination revealed mild hepatic extramedullary hematopoiesis (small foci of hematopoietic cells in some portal triads) in two males and two females exposed to PGTBE, which was considered treatment-related (85). From this study, it may be concluded that the acute LC_{50} (LC_{50} = concentration lethal to 50% of subjects tested) of PGTBE is greater than 500 ppm. As described below, subchronic exposure to concentrations in excess of these levels also is not lethal so it may be concluded further that the LC_{50} is actually greater than 700 ppm (see Section 6.4.1.2).

DERMAL. All rabbits survived a 24-h application of 2 g PGTBE per kilogram body weight. PGTBE was applied to the abraded, clipped skin of rabbits and held in contact for a period of 24 h by use of an occlusive wrap that prevented removal by evaporation or other means. After 24 h, the test material was removed from the skin and subjects were observed for 14 days. No deaths resulted from treatment and no adverse clinical signs were observed. No effects were noted on body weights. Examination of the skin at the site of application revealed transient erythema as well as persistent desquamation, atonia, erythema, fissures, discoloration, and thickened skin of some of the treated animals (86).

SKIN IRRITATION. PGTBE was applied to the clipped skin of six albino rabbits at both abraded and nonabraded sites (two each) and held in contact with the skin under an occlusive wrap for 24 h in a primary skin irritation assay following a modification of the procedure of Draize et al. (71). Transient, usually slight erythema and desquamation was observed over the 7-day observation period. The overall primary irritation index (PII) was 0.6 on a scale of 8, indicating that PGTBE has slight skin irritation potential (87).

EYE IRRITATION. PGTBE was tested in rabbits for primary eye irritation potential following a modification of the Draize procedure. In this test, 0.1 mL of pure PGTBE was instilled into the eyes of nine rabbits. Three of the rabbits had their eyes flushed with water to remove the test material immediately following instillation (washed group), and six did not (unwashed group). PGTBE was found to cause corneal opacities that were reversible after 13 days in all but one rabbit from the unwashed group and that were reversible within 3 days in all rabbits from the washed group. Other less severe signs of irritation were observed and were of a transient nature, including redness, chemosis, and discharge. After 13 days of observation, PII scores were below 10 (on a scale totaling 110 points) for both groups. However, PGTBE should be considered a severe eye irritant in pure form because of the occurrence of corneal opacities in the unwashed group, which persisted in one animal until the end of an extended 13-day observation period (88). PGTBE was slightly irritating to eyes when tested as a 20% aqueous solution (89).

6.4.1.2 Chronic and Subchronic Toxicity

SUBCHRONIC TOXICITY/INHALATION. Fischer 344 rats (30/sex per concentration) were exposed in whole-body inhalation chambers to atmospheres containing PGTBE at concentrations of 0, 28, 82, 243, or 709 ppm, 6 h/day, 5 days/week for 13 weeks (>95 α-isomer). Ten subjects/sex per exposure level were sacrificed after 4 weeks and 10/sex per level at 13 weeks and 10 more after a 3-week recovery period. During the in-life phase, subjects were monitored for clinical signs, food consumption, and body weight. Clinical chemistries and urinalysis were taken at the specified sacrifice intervals. At necropsy, tissues were taken for histopathological examination (90).

At all three sacrifice intervals, body weights were unchanged and no animals died. No abnormal behavior was observed during the course of the study. Hematology and clinical chemistries were normal with no evidence of hemolysis or injury to internal organs. Urinalysis revealed no frank or occult blood. Liver, kidney, and spleen weights were elevated in treated animals but without accompanying histopathology or clinical chemistry changes and was considered adaptive in nature. Histopathology revealed slight testicular degeneration (damage to seminiferous tubules) in 1 of 10 animals after 4 weeks of exposure and, again, in 1 of 10 animals after 13 weeks of exposure. None of the control animals showed this lesion. Because this damage (1) was slight, (2) affected only one testicle and was not bilateral, (3) did not increase in severity or extent after 13 weeks, and (4) occurs commonly in rats of this age, the testicular degeneration was not considered treatment-related. No other histopathology was noted. Specifically, no changes were found in the bone marrow, blood, or thymus. The NOAEL was 709 ppm (90).

CHRONIC TOXICITY. Currently, the National Toxicology Program (NTP) is testing PGTBE by inhalation in a 2-year study with rats and mice (50/sex per group). Exposures at nominal concentration of 0, 75, 300, or 1,200 ppm began in September 1997, and the in-life portion of the study is still in progress as of this writing. The NTP has released no interim results.

6.4.1.3 Pharmacokinetics, Metabolism, and Mechanisms. In a final report sponsored by the NTP, PGTBE was found to be rapidly absorbed, metabolized, and excreted, largely within 24 h via the urine, lungs (as CO_2), and, to a lesser extent, the feces, after administration of a single oral dose to rats (91). Urinary metabolites included sulfate and glucuronide conjugates of PGTBE and a small portion of parent compound.

6.4.1.4 Reproductive and Developmental. Pregnant New Zealand rabbits (16/group) were exposed in whole-body inhalation chambers to atmospheres containing PGTBE at actual concentrations of 0, 229, 721, or 984 ppm for 6 h/day, on days 7–19 of gestation. No toxicity to the dams was observed; no effect was seen on behavior, weight gain, or hematological parameters. Gross observation at caesarian section was normal in all treated groups. No effect was seen on fetal morphology. The NOAEL was the highest exposure level of 984 ppm (92).

Pregnant Charles River CDF® rats (25/group) were exposed to PGTBE concentrations of 0, 230, 726, or 990 ppm for 6 h/day on days 6–15 of gestation. One half of the dams from the high exposure group were pale in appearance during most of the exposure period and absolute and relative liver weights of the mid- and high-exposure level dams had statistically increased liver weights. These effects were considered exposure-related. No other effect on the dams was noted. No fetotoxicity or developmental abnormalities were noted in the pups (93).

6.4.1.5 Carcinogenesis. See Section 6.4.1.2.

6.4.1.6 Genetic and Related Cellular Effects Studies. PGTBE was tested in the Ames *Salmonella* assay with and without Aroclor-induced rat liver microsomal S-9 preparations using tester strains TA1537, TA1538, TA98, and TA100 at doses of 50, 150, 500, 1500, or 5000 µg/plate. Solvent and positive controls were within normal limits. PGTBE caused no mutagenic response in this test (94).

In a second Ames assay, PGTBE was tested in strains TA1535, TA1537, TA1538, TA98, and TA100 with and without Aroclor-induced S-9 metabolic activation at doses of 50, 167, 500, 1667, and 5000 µg/plate. Solvent and positive control agents showed revertant frequencies within normal limits. PGTBE did not induce revertants in excess of solvent controls (95).

The National Toxicology Program (NTP) has conducted a third Ames test and concluded that PGTBE was mutagenic (NTP, Ames *Salmonella* Assay of Propylene Glycol Mono-*t*-Butyl Ether, unpublished results, 1997). A data sheet of the results is available from NTP but is too cryptic for interpretation. These results are at variance with the other two Ames tests on this compound. Moreover, glycol ethers, whether ethylene- or propylene-based, generally have been found to yield negative results in the Ames assay.

PGTBE was tested in the *in vitro* Chinese hamster ovary (CHO) test to determine whether it had the potential to induce chromosome aberrations. When incubated with CHO cells at doses of 0, 0.1, 0.5, and 1% (v/v) PGTBE in the incubation medium, with and without metabolic activation, no excess incidence of aberrations were found with metabolic activation. Without metabolic activation, the incidence of excess aberrations reached borderline statistical significance which the conducting laboratory did not consider evidence of clastogenic activity because of the marginal significance and since the incidence level was within the historical negative control range for the laboratory (96).

PGTBE was also tested for its potential to induce chromosome aberrations in a cultured human cell line consisting of peripheral lymphocytes with and without metabolic activation. Doses tested included 1000, 3330, and 5000 µg PGTBE per milliliter of culture medium. Negative and positive controls produced aberration incidences within historical controls. PGTBE did not induce chromosome aberrations in this test system (97).

NTP has sponsored an *in vitro* chromosome aberration study with cultured CHO cells and reports that the results were negative (98). NTP has also sponsored an *in vitro* sister chromatid exchange assay, again, with CHO cells and reported the results as negative (99).

6.4.1.7 *Other: Neurological, Pulmonary, Skin Sensitization, etc.*

SKIN SENSITIZATION. PGTBE was tested for its potential to cause skin sensitization in the guinea pig following the OECD Guideline for Testing of Chemicals, Protocol 406 (guinea pig maximization test). During the induction phase, 20 guinea pigs received three intradermal injections at separate locations on each flank of (*1*) Freund's complete adjuvant, (*2*) 5% v/v PGTBE in water, and (*3*) 5% PGTBE in a 50:50 mixture of distilled water and Freund's adjuvant. One week later, pure PGTBE was applied covering all three injection sites on each flank for 48 h in a second induction treatment. Two weeks later for the challenge phase, treatment solutions of 20 and 30% (v/v) PGTBE in water were applied to the sites and held in place for 24 h under occlusion. Ten control animals received similar treatments where distilled water was substituted for PGTBE. No sensitization reactions were recorded in the PGTBE-treated animals (100).

6.5 Standards, Regulations, or Guidelines of Exposure

No occupational exposure limits (OELs) have been established for propylene glycol *tertiary*-butyl ether.

7.0 Propylene Glycol Monoethylbutyl Ether, Mixed Isomers

7.0.1 CAS Number: *[26447-42-7]*

7.0.2 Synonyms: 2-Ethylbutoxypropanol, mixed isomers; propylene glycol monohexyl ether, mixed isomers

7.0.3 Trade Name: NA

7.0.4 Molecular Weight: 160.2

GLYCOL ETHERS

7.0.5 Molecular Formula: $C_9H_{20}O_2$

7.0.6 Molecular Structure: $C_6H_{13}OCH_2CHOHCH_3$ mixed isomers

7.1 Chemical and Physical Properties

7.1.1 General

1 ppm $\sim$ 5.40 mg/m^3 at 760 mmHg, 25°C
1 mg/L $\sim$ 185 ppm at 760 mmHg, 25°C

7.2 Production and Use: NA

7.3 Exposure Assessment: NA

7.4 Toxic Effects

The LD$_{50}$ for rats from a single oral dose is 3.56 mL/kg. Eye tests on rabbits indicate that this chemical can cause severe irritation and damage (rated 9 on a scale of 10). Skin irritation tests on rabbits indicate that it is only very slightly irritating (rated 2 on a scale of 10). The LD$_{50}$ for rabbits by skin absorption is 6.0 mL/kg. No rats died when exposed for 8 h to essentially saturated vapors (18).

8.0 Propylene Glycol Butoxyethyl Ether

8.0.1 CAS Number: [124-16-3]

8.0.2 Synonyms: 1-Butoxyethoxy-2-propanol, 1-(2-butoxyethoxy)-2-propanol

8.0.3 Trade Names: Propasol® Solvent BEP

8.0.4 Molecular Weight: 176.26

8.0.5 Molecular Formula: $C_9H_{20}O_3$

8.0.6 Molecular Structure:

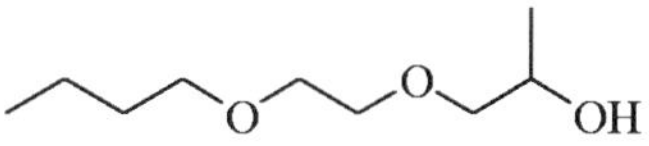

8.1 Chemical and Physical Properties

8.1.1 General

Refer to Table 87.2.

8.2 Production and Use

Propylene glycol butyoxyethyl ether (PGBEE) is a clear, colorless liquid of low volatility that is manufactured by reacting excess ethylene glycol monobutyl ether with propylene oxide. Production volume in the United States is estimated to be in excess of 2000 kg per year. PGBEE is used (*1*) in hydraulic fluids ; (*2*) as an antistalling additive for automotive fuels; (*3*) as a plasticizer intermediate; (*4*) as a solvent for waxes, and degreasing solutions; and (*5*) as a chemical intermediate (101).

8.3 Exposure Assessment: NA

8.4 Toxic Effects

The single-dose oral LD_{50} for rats has been reported to range between 4.0 and 5.66 mL/kg for PGBEE. A single drop, when applied to the eyes of a rabbit, caused moderate to severe irritation and corneal injury. Contact with the skin of rabbits resulted in minor irritation. The material is only slightly toxic on prolonged occluded contact with the skin; the LD_{50} for rabbits is in the range of 2.83–3.00 mL/kg. Concentrated vapor generated at about 19°C was not lethal to rats exposed for 8 hr. This material did not cause alteration of red blood cell fragility in rats exposed to the vapor. The estimated exposure concentration was about 0.25 mg/L or about 35 ppm (102, 103).

9.0 Propylene Glycol Phenyl Ether

9.0.1 CAS Number: [770-35-4]

9.0.2 Synonyms: 1-Phenoxy-2-propanol, phenoxyisopropanol

9.0.3 Trade Names: Dowanol® PPh Glycol Ether, Propylene Phenoxytol®

9.0.4 Molecular Weight: 152.19

9.0.5 Molecular Formula: $C_9H_{12}O_2$

9.0.6 Molecular Structure:

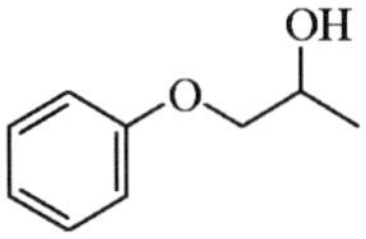

9.1 Chemical and Physical Properties

Commercial-grade propylene glycol phenyl ether (PGPhE) is a clear, colorless liquid having a slight, but disagreeable, odor. See Table 87.2 for other chemical and physical properties.

9.2 Production and Use

Refer to Table 87.4.

9.3 Exposure Assessment: NA

9.4 Toxic Effects

9.4.1 Experimental Studies

9.4.1.1 Acute Toxicity. PGPhE is low in acute toxicity. The LD_{50} was 2830 mg/kg for male rats and 3730 mg/kg for females (104). The main toxic effect noted was CNS depression and death usually occurred in 48 h or less. The rabbit dermal LD_{50} was greater than 2000 mg/kg and, via inhalation, no deaths resulted in rats after a 7-h exposure to a saturated vapor concentration generated at room temperature (104). When exposed for 7 h to a saturated atmosphere generated at 100°C, rats showed significant nasal, respiratory, and eye irritation during exposure but, again, no deaths resulted (104). When two drops of the liquid were applied to the eyes of rabbits, some initial pain, slight irritation, and slight corneal injury occurred. The eyes healed in several days to one week (104). No primary dermal irritation studies were located but when applied continuously to the skin of rabbits under a cloth covering over a 2-week period, only mild irritation resulted (105).

9.4.1.2 Chronic and Subchronic Toxicity. In the 2-week dermal study referred to previously, female rabbits exhibited no systemic toxicity when 1000 mg PGPhE was applied continuously under a cloth covering (105). In a 28-day dermal study in rabbits employing five animals per sex per dose, 19 daily applications of 0, 100, 300, and 1000 mg/kg produced no evidence of systemic toxicity. Parameters monitored included behavioral signs, organ weights, clinical endpoints, and histopatholgy. The only treatment-related effect was a mild transient dermal irritation at all dose levels (106).

9.4.1.3 Pharmacokinetics, Metabolism, and Mechanism: NA

9.4.1.4 Reproductive and Developmental: NA

9.4.1.5 Carcinognenesis: NA

9.4.1.6 Genetic and Related Cellular Effects Studies. PGPhE did not cause reverse mutations in the Ames *Salmonella typhimurium* assay when tested using strains TA98, TA100, TA1535, or TA 1537, with or without Aroclor 1254–induced rat microsomes at concentrations ranging up to 5000 µg/plate (107). PGPhE did not induce chromosome aberrations in cultured human lymphocytes in either the presence or absence of a metabolic microsomal activating system at concentrations of ≤400 µg/mL of culture medium (108).

9.5 Standards, Regulation, or Guidelines of Exposure

No occupational exposure guidelines have been established for PGPhE.

10.0 Dipropylene Glycol Monomethyl Ether

10.0.1 CAS Number: [34590-94-8]

10.0.2 Synonyms: 1-(2-Methoxy-1-methylethoxy)-2-propanol, dipropylene glycol monomethyl ether (DPGME)

10.0.3 Trade Names: Arcosolve® DPM, Dowanol® DPM Glycol Ether, Ucar® Solvent 2LM, Propasol® Solvent DM, Poly-solv® DPM Solvent

10.0.4 Molecular Weight: 148.2

10.0.5 Molecular Formula: $C_7H_{16}O_3$

10.0.6 Molecular Structure: $CH_3CHOHCH_2OCH(CH_3)CH_2OCH_3$

10.1 Chemical and Physical Properties

10.1.1 General

See Table 87.3.

10.1.2 Odor and Warning Properties

DPGME is a clear, colorless liquid with an ether-like odor. Exposure to 300 ppm was reported to be quite disagreeable to humans (8). Further studies (14) done with five unacclimated human volunteers suggested the following levels as possible guides for warning properties:

35 ppm: Probable minimum concentration that may cause minor nasal irritation.
75 ppm: Probable minimum concentration that may cause some tolerable eye, throat, and respiratory irritation.
80 ppm: Lowest concentration at which odor was rated intolerable.

10.2 Production and Use

DPGME is used as a solvent in hard-surface household cleaners and water-based coatings, as a coupling agent in water-based polishes, as a solvent for nitrocellulose and other synthetic resins, and as a solvent in hydraulic brake fluids. Refer to Table 87.5 for additional production and use data.

10.3 Exposure Assessment: NA

10.4 Toxic Effects

Dipropylene glycol monomethyl ether (DPGME) is low in single-dose oral toxicity, is transiently painful but not damaging to the eyes, and is nether appreciably irritating to the skin nor readily absorbed through the skin of rabbits in toxic amounts when exposures are prolonged and repeated. It caused neither irritation nor sensitization when tested on human subjects. It is low in toxicity by inhalation. It was not teratogenic in rats or rabbits after inhalation exposure, and was not genotoxic in several *in vitro* test systems.

GLYCOL ETHERS

10.4.1 Experimental Studies

10.4.1.1 Acute Toxicity

ORAL. The single-dose oral LD_{50} for male and female rats was found to be 5.50 and 5.45 mL/kg (5.22 and 5.18 g/kg), respectively (8). The material produced marked CNS depression. Shideman and Procita (20) estimated the single-dose oral LD_{50} for dogs to be 7.5 mL/kg. They noted that death was due to respiratory failure and usually occurred within 48 h or not at all.

INHALATION. Rats exposed for 7 h to approximately 500 ppm DPGME exhibited only mild narcosis from which they rapidly recovered (8). This exposure concentration produced an atmosphere of fog, and the animals were wet with material at the end of exposure.

DERMAL. Single applications to the skin revealed that DPGME was not absorbed through the skin in acutely toxic amounts even when massive doses (20 mL/kg, 19.0 g/kg) were held in continuous contact with a large area of the rabbit's skin for a period of 24 h (The Dow Chemical Company, unpublished data, 1950). Sufficient absorption did occur, however, to result in transient narcosis. Others have reported dermal LD_{50} values for rabbits of 13–14 g/kg (109).

INJECTION. The LD_{50} by intravenous injection to anesthesized dogs in 0.35–0.5 mL/kg (0.33–0.47 g/kg), and for artificially respired dogs 1.3 mL/kg (1.23g/kg) (109).

EYE IRRITATION. DPGME is not appreciably irritating to the eyes. When one drop of undiluted material was placed in a rabbit's eye on each of 5 consecutive days, a mild transitory irritation of the conjunctival membranes occurred. Fluorescein staining revealed no corneal damage (8).

SKIN IRRITATION. Continuous contact of DPGME with the skin of rabbits for 90 days caused only a very slight scaling (The Dow Chemical Company, unpublished data, 1950). The response was similar to that produced by water alone under the same conditions.

10.4.1.2 Chronic and Subchronic Toxicity

SUBCHRONIC TOXICITY/ORAL. Browning (109) reported that rats given DPGME at a dosage of 1.0 g/kg for 35 days showed no adverse effects.

INHALATION. Rats, rabbits, guinea pigs, and monkeys were exposed 7 h/day, 5 days/week for periods of 6–8 months in an atmosphere containing about 300–400 ppm (essentially saturated) material. Transient narcosis was observed in rats during the first few weeks of the study. The only other systemic toxicity reported was a slight increase in liver weight in all species, which was attributed to an adaptive rather than toxic response (8).

No adverse effects were observed in a 13-week inhalation study (15, 50, or 200 ppm) in rats and rabbits (110). The top concentration was approximately 40% of atmospheric vapor saturation.

DERMAL. When DPGME was applied 5 times a week for 90 days at dosage levels of 1.0, 3.0, 5.0, and 10.0 mL/kg, the following observations were made (8):

1. Mortality was high at the 10.0-mL/kg dosage level, slight at the 5.0-mL/kg level, and absent at the 1.0- and 3.0-mL/kg levels.
2. No adverse body weight changes occurred at any level, except in animals just prior to death at the high dose.
3. No hematologic changes occurred at any dosage level.
4. The effect of severe (repeated and prolonged) exposure to the skin was slight; it was similar to that caused by distilled water under similar conditions.
5. Observations for gross pathology revealed only gastric distension and occasional gastric irritation in those animals dying at the 10-mL/kg dosage level.
6. No significant organ weight changes occurred at any dose level.
7. The blood urea nitrogen (BUN) concentration was unaffected in the animals surviving the 3.0- and 5.0-mL/kg dosage levels.
8. Histopathological studies conducted on the liver, lung, spleen, adrenal, heart, testes, and stomach of those animals receiving the 5.0- and 10.0-mL/kg dosage levels revealed no changes. The kidneys of animals treated with 10.0 mL/kg showed some granular and hydropic changes, which were not apparent at lower doses.

In a more recent study, application to the skin of rats for 4 weeks, 4 h/day, 5 days/week at doses of 100 and 1000 mg/kg did not produce any evidence of systemic toxicity (111).

10.4.1.3 Pharmacokinetics, Metabolism, and Mechanisms. Male rats given a single oral dose of [^{14}C]dipropylene glycol monomethyl ether (1289 mg/kg, 8.7 mmol/kg) excreted approximately 60% of the dose in the urine and 27% as $^{14}CO_2$ within 48 h after dosing (112). Parent compound, propylene glycol monomethyl ether, dipropylene glycol, propylene glycol, and sulfate and glucuronide conjugates of dipropylene glycol monomethyl ether were all identified in the urine. These results indicated that dipropylene glycol monomethyl ether was likely metabolized by the same pathways described for propylene glycol monomethyl ether (see Fig. 87.1).

10.4.1.4 Reproductive and Developmental

DEVELOPMENTAL STUDIES. The developmental toxicity of DPGME has been evaluated in rats and rabbits following inhalation exposure to 50, 150, or 300 ppm (W. J. Breslin et al., The Dow Chemical Company, unpublished data, April and May, 1990; 113). These exposure concentrations were approximately equivalent to 0, 32, 97, and 193 mg DPGME/kg per day for rabbits and 0, 107, 322, and 644 mg DPGME/kg per day in rats. Exposure of rats for 6 h/day on days 6–5 of gestation, and rabbits on days 7–19 of gestation, did not result in any treatment-related effects in maternal, embryonal, or fetal parameters.

10.4.1.5 Carcinogenesis: NA

10.4.1.6 Genetic and Related Cellular Effects Studies

GENETIC TOXICOLOGY. DPGME was not a mutagen in CHO cells (38) or in the Ames test (114), with or without metabolic activation. In addition, a rat hepatocyte unscheduled DNA synthesis assay was negative (A. L. Mendrala, The Dow Chemical Company, unpublished data, 1983).

10.4.1.7 Other: Neurological, Pulmonary, Skin Sensitization, etc. When patch-tested on 250 human beings, DPGME produced no evidence of primary irritation or sensitization of the skin (8; V. K. Rowe, The Dow Chemical Company, unpublished data, 1951).

10.4.2 Human Experience

10.4.2.1 General Information. No injury or adverse effects have been reported from the handling and use of DPGME.

10.5 Standards, Regulations, or Guidelines of Exposure

ACGIH, in their 1998 listing of threshold limit values, recommend a time-weighted average TLV of 100 ppm (606 mg/m^3) for repeated 8-h human exposure and a short-term exposure limit (STEL) of 150 ppm (909 mg/m^3) (43). Table 87.8 summarizes recommended or regulatory guidelines of occupational exposure for DPGME.

11.0 Dipropylene Glycol Monoethyl Ether

11.0.1 CAS Numbers: [15764-24-6], [300025-38-8]

11.0.2 Synonyms: 1-(2-Ethoxy-1-methylethoxy)-2-propanol, ethoxypropoxypropanol (DPGEE)

11.0.3 Trade Name: NA

11.0.4 Molecular Weight: 162.2

11.0.5 Molecular Formula: C$_8$H$_{18}$O$_3$

11.0.6 Molecular Structure:

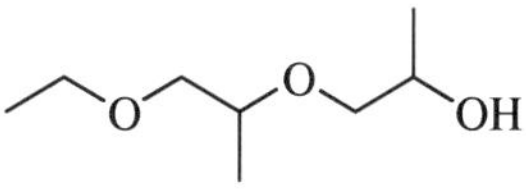

11.1 Chemical and Physical Properties

See Table 87.3.

11.2 Production and Use: NA

11.3 Exposure Assessment: NA

Table 87.8. Occupational Exposure Guidelines for DPGME[a]

Regulatory or Expert Authority (Date)	Exposure Value	Comment/Notation
ACGIH (TLV) (1998)	TWA = 100 ppm STEL = 150 ppm	Skin
NIOSH (1994)	TWA = 100 ppm STEL = 150 ppm IDLH Value = 600 ppm	Skin
OSHA (PEL) (1994)	TWA = 100 ppm	Skin
Australia (OEL) (1993)	TWA = 100 ppm STEL = 150 ppm	Skin
Austria (OEL) (1993)	TWA = 50 ppm	
Belgium (OEL) (1993)	TWA = 100 ppm STEL = 150 ppm	Skin
Denmark (OEL/TWA) (1993)	100 ppm	Skin
Finland (OEL) (1993)	TWA = 100 ppm STEL = 150 ppm	Skin
France (OEL/TWA) (1993)	100 ppm	
Germany (OEL/TWA) (1993)	50 ppm	
Netherlands (OEL/TWA) Oct-97	50 ppm	
Philippines (OEL/TWA) (1993)	100 ppm	Skin
Switzerland (OEL) (1993)	TWA = 50 ppm STEL = 100 ppm	
Turkey (OEL/TWA) (1993)	300 ppm	Skin

[a]The OEL values set by Bulgaria, Colombia, Jordan, Korea, New Zealand, Singapore, and Vietnam are based on the ACGIH value.

11.4 Toxic Effects

Dipropylene glycol monoethyl ether (DPGEE) exhibits low acute and subacute toxicity. This compound has not been evaluated for toxicity in subchronic or chronic bioassays. It has not exhibited genotoxic effects in the Ames assay or chromosome aberration tests with cultured cells.

11.4.1 Experimental Studies

11.4.1.1 Acute Toxicity

ORAL. The single-dose oral LD_{50} of DPGEE for rats is 4 mL/kg (3.71 g/kg) (V. K. Rowe, The Dow Chemical Company, unpublished data, 1947). In a more recent study, the rat oral

LD$_{50}$ was 5 mL/kg (4.64 g/kg). The lowest lethal doses were 2.76 mL/kg (2.56 g/kg) for females and 5.4 mL/kg (5.0 g/kg) for males (115).

DERMAL. Regarding acute dermal toxicity, five of five rabbits survived single doses of 5 mL/kg (4.63 g/kg) applied to the skin for 24 h under a cuff, essentially as described by Draize et al. (71); three of 10 receiving 10 mL/kg (9.27 g/kg) survived, but none of five receiving 15 mL/kg (13.9 g/kg) survived (V. K. Rowe, The Dow Chemical Company, unpublished data, 1947). More recently, no mortality or toxic signs were observed after 2000 mg/kg was applied to the shaved skin of rats (duration/occlusion unspecified) (16).

INHALATION. Via inhalation, mortality occurred in one of 12 rats exposed for 7 h to an atmosphere containing essentially saturated vapor concentrations (calculated to be approximately 400 ppm). Signs of irritation of the eyes and nares and transient weight loss was noted in the survivors, but recovery appeared complete 24 h later (V. K. Rowe, The Dow Chemical Company, unpublished data, 1947).

EYE AND SKIN IRRITATION. When one drop of the liquid was instilled in the eyes of rabbits on each of 5 consecutive days, it caused a slight transitory conjunctival irritation but no corneal injury (V. K. Rowe, The Dow Chemical Company, unpublished data, 1947). In another study, DPGEE caused transient conjunctival irritation and diffuse corneal opacity in the rabbit eye (117). It did not meet the European Community criteria of an eye irritant. When DPGEE was applied to rabbit skin under semiocclusion for 4 h, it was found to be slightly irritating (118).

11.4.1.2 Chronic and Subchronic Toxicity

DERMAL. Repeated applications (10 in 14 days) to the skin of rabbits resulted in only a very slight exfoliation with no evidence of toxicity. All animals exhibited transient weight loss. Narcosis was apparent in all animals but was profound at the higher dosages. The animals were cold to the touch, and the skin beneath the cuff was burned. In animals that succumbed, death occurred in almost all cases within 1 or 2 days. Recovery was complete in the survivors within 3 days (V. K. Rowe, The Dow Chemical Company, unpublished data, 1947).

ORAL. Sprague–Dawley rats (5/sex per group) were dosed by gavage with 0, 50, 225, or 1000 mg/kg DPGEE 5 days/week over the course of 28 days, according to OECD guidelines (119). No effects were noted on growth rate or behavior. Liver weights of both sexes were increased in the highest dose group, and male rats from the mid and high dose groups exhibited slight hyaline droplet formation in renal tubular cells.

11.4.1.3 Pharmacokinetics, Metabolism, and Mechanisms: NA

11.4.1.4 Reproductive and Developmental: NA

11.4.1.5 Carcinogenesis: NA

11.4.1.6 Genetic and Related Cellular Effects Studies

11.4.1.6.1 Genetic Toxicity. In an Ames assay, DPGEE did not increase revertant mutations in any of five strains of *Salmonella typhimurium* over a range of doses from 50 to 5000 μg/plate, with or without Aroclor 1254-induced rat liver microsomes (120). DPGEE did not cause chromosome breaks when incubated with human lymphocytes at doses of ≤ 1000 μg/mL, with or without rat liver microsome activation (121).

11.4.1.7 Other: Neurological, Pulmonary, Skin Sensitization, etc.

SKIN SENSITIZATION. DPGEE did not cause dermal sensitization in guinea pigs when tested with the Matheson–Kligman protocol (122).

11.5 Standards, Regulations, or Guidelines of Exposure

No occupational exposure limits have been developed for DPGEE.

12.0 Dipropylene Glycol Mono-n-butyl Ether

12.0.1 CAS Number: [29911-28-2]

12.0.2 Synonyms: 1-[2-Butoxy-1-methylethoxy]-2-propanol, dipropylene glycol butoxy ether

12.0.3 Trade Names: Butyl Dipropasol® Solvent, Dowanol® DPnB Glycol Ether

12.0.4 Molecular Weight: 190.28

12.0.5 Molecular Formula: $C_{10}H_{22}O_3$

12.0.6 Molecular Structure:

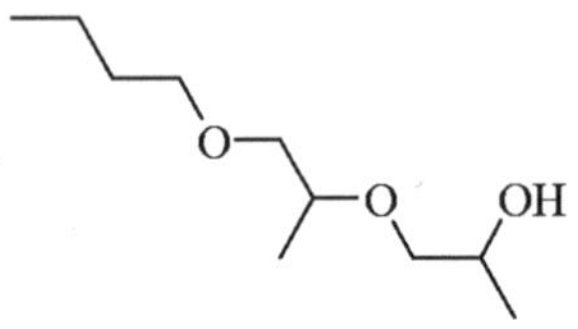

12.1 Chemical and Physical Properties

12.1.1 General

Dipropylene glycol mono-*n*-butyl ether (DPGBE) is a clear, colorless liquid with a slight odor. Its chemical and physical properties are listed in Table 87.3.

12.2 Production and Use

Refer to Table 87.5.

12.3 Exposure Assessment: NA

12.4 Toxic Effects

DPGBE exhibits low toxicity from a single exposure by any route of administration. It is slightly irritating to skin; however, the undiluted material may markedly irritate eyes, although damage was transitory when tested in rabbits. DPGBE has not caused skin sensitization when tested either in guinea pigs or humans. In experimental animals, subacute and subchronic administration of DPGBE has resulted in slight toxicity of a nonspecific nature and only at considerable dose levels. When applied to the skin of pregnant rats, no developmental effects occurred in offspring. This compound did not cause mutations or genetic damage when tested in several assays.

12.4.1 Experimental Studies

12.4.1.1 Acute Toxicity

ORAL. The single-dose LD_{50} for rats has been estimated at 2–3 g/kg. More recent studies reported an LD_{50} of 4400 mg/kg for male rats and 3700 mg/kg for female rats (64). An oral LD_{50} value in rats of 2830 µL/kg (2575 mg/kg) for males and 1620 µL/kg (1475 mg/kg) for females also has been reported (123). In a subacute study, rats (6/sex per dose) received gavage doses of 0, 100, 200, and 400 mg/kg for 14 consecutive days. No effects were noted at any dose level (124).

INHALATION. No deaths occurred in rats (5/sex) exposed for 6 h to a saturated atmosphere of DPGBE (123). DPGBE caused nasal irritation and liver enlargement in a 2-week aerosol inhalation study where Fischer 344 rats (5/sex per concentration) were exposed (nose-only) 6 h/day to aerosol concentrations of 0, 200, 810, and 2010 mg/m^3. At the highest exposure level, rats exhibited decreased activity, decreased body weight, increased liver cell size, and irritation of the membranes of the nose. The mid-concentration group showed similar but less intense effects, and 200 mg/m^3 was the no effect level (125).

DERMAL. The test material does not appear to be readily absorbed through the skin; the dermal LD_{50} in rats was reported as greater than 2000 mg/kg (70). More recently, a dermal LD_{50} of 7130 µL/kg (6490 mg/kg) for males and 5860 µL/kg (5330 mg/kg) for females was reported for rabbits (123).

SKIN IRRITATION. Repeated application (10–14 days) to the skin of rabbits resulted in only a very slight irritation (V. K. Rowe, The Dow Chemical Company, unpublished data, 1947).

EYE IRRITATION. Application for 5 days to eyes of rabbits produced conjunctival irritation not associated with corneal injury (V. K. Rowe, The Dow Chemical Company, unpublished data, 1947). More recent studies found evidence of transient corneal effects and no conjunctival irritation (P. J. J. M. Weterings, P. A. M. Daamen, and H. G. Verschuuren, The Dow Chemical Company, unpublished data, 1987).

12.4.1.2 Chronic and Subchronic Toxicity

ORAL. DPGBE caused no testicular, blood, or thymic injury in rats exposed orally for 13 weeks. In this study, rats given daily doses of 0, 200, 450, and 1000 mg/kg in the diet incurred slightly decreased body weights, minor changes in clinical chemistry, and slightly increased liver weights at the high dose but no other treatment-related effects. Minimal increase in urinary magnesium and plasma urea concentrations, not considered toxic effects, were reported in the mid-dose group. The NOEL was 200 mg/kg (P. Thevenaz et al, The Dow Chemical Company, unpublished data, 1988).

DERMAL. Dermal treatment of rats for 13 weeks also did not cause testicular, blood, or thymic damage with a NOAEL for these adverse effects of 910 mg/kg per day. In this study, rats received dermal doses 5 days/week, of 0, 91, 273, and 910 mg/kg per day. Body weights were decreased in the two highest dose groups and, in the high dose group only, rats had slightly increased liver weights and slight alterations in clinical chemistry readings. No histopathology was detected in any of the treatment groups. The NOEL for any effect, adverse or otherwise, was 91 mg/kg per day (B. A. R. Lina et al., The Dow Chemical Company, unpublished data, 1988).

12.4.1.3 *Pharmacokinetics, Metabolism, and Mechanisms:* NA

12.4.1.4 *Reproductive and Developmental Studies*

DERMAL. DPGBE was not embryotoxic, fetotoxic, or teratogenic when applied to the skin of pregnant rats on days 6–16 of gestation at doses of 0, 0.3 (273 mg/kg), or 1.0 mL/kg (910 mg/kg) (J. M. Wilmer, M. W. Marwijk, and H. G. Verschuuren, The Dow Chemical Company, unpublished data, 1988). Minor skin irritation was seen at all dose levels in the maternal animals.

12.4.1.5 *Carcinogenesis:* NA

12.4.1.6 *Genetic and Related Cellular Effects Studies*

GENETIC TOXICITY. Negative results were reported in an Ames *Salmonella* assay (tester strains TA98, TA100, TA1537, TA1538) at concentrations of $\leq$1560 µg/mL (E. J. Van de Waart, J. C. Enninga and H. G. Verschuuren, The Dow Chemical Company, unpublished data, 1987). The material did not produce chromosomal aberrations in cultured Chinese hamster ovary cells treated with concentrations of $\leq$5000 µg/mL (80). In both of these *in vitro* assays, DPGBE was tested with and without Aroclor-induced rat liver microsomal activating systems. In an *in vivo* genetic toxicity test, no treatment-related response was reported in a mouse micronucleus assay after oral doses up to 2500 mg/kg (M. L. McClintock, B. Gollapudi, and H. G. Verschuuren, The Dow Chemical Company, unpublished data, 1988).

12.4.1.7 *Other: Neurological, Pulmonary, Skin Sensitization, etc.*

SKIN SENSITIZATION. No evidence of skin sensitization was seen in a modified Buehler guinea pig assay (J. Vanderkom and H. G. Verschuuren, The Dow Chemical Company, unpublished data, 1987).

12.4.2 Human Experience

12.4.2.1 General Information. No adverse human experience has been reported. Dipropylene glycol mono-*n*-butyl ether was tested for skin sensitization in an 82-member human test panel. (A. M. Maclennon, J. Hedgecock, and H. G. Verschuuren, The Dow Chemical Company, unpublished data, 1988). Treatment consisted of nine doses of 0.4-mL test material under an occluded patch over a period of 3 weeks. No evidence of skin sensitization was seen when challenge patches were applied 17 days later.

12.5 Standards, Regulations, or Guidelines of Exposure

No occupational exposure standard has been established.

13.0 Dipropylene Glycol Mono-*tertiary*-butyl Ether

13.0.1 CAS Number: [132739-31-2]

13.0.2 Synonym: [2-(1,1-Dimethylethoxy)methylethoxy] propanol

13.0.3 Trade Name: Arcosolv™ DPTB

13.0.4 Molecular Weight: 190.3

13.0.5 Molecular Formula: $C_{10}H_{22}O_3$

13.0.6 Molecular Structure: $C_4H_9OCH_2CH(CH_3)OCH_2CH(CH_3)OH$

13.1 Chemical and Physical Properties

13.1.1 General

1 ppm ~ 7.78 mg/m^3 at 760 mmHg, 20°C
1 mg/l ~ 128 ppm at 760 mmHg, 20°C

13.2 Production and Use

Dipropylene glycol mono-*tertiary*-butyl ether (DPTB) is manufactured by reacting isobutylene with excess dipropylene glycol in the presence of a solid resin etherification catalyst. It is then distilled to produce 99+% of the α isomer, 1-*tert*-butoxypropoxy-propan-2-ol (126).

13.3 Exposure Assessment

Individuals exposed to this material would include those involved in its manufacture and use. Manufacture is mostly enclosed, precluding significant exposure.

13.4 Toxic Effects

DPTB has low acute toxicity by the oral, inhalation, and dermal routes of exposure. It is slightly irritating to skin and moderately to severely irritating to eyes and, in a guinea pig maximization test, exhibited no sensitization potential. The subchronic toxicity of this compound also is low and it does not damage blood, blood-forming organs, the thymus, or testes.

13.4.1 Experimental Studies

13.4.1.1 Acute Toxicity

ORAL. In an oral LD_{50} study, Wistar albino rats (5/sex per dose) were administered a single dose of 1600, 2000, 2600, 3200, or 5000 mg DPTB/kg body weight and observed for 14 days. All animals survived the lowest dose, five of five females survived 1600 mg/kg (no males tested), four of five males and three of five females survived 2000 mg/kg, five of five males survived 2600 mg/kg (no females tested), two of five males and one of five females survived 3200 mg/kg, and, at the high dose, one of five males and zero of five females survived 5000 mg/kg. Analysis of survival data using the Litchfield–Wilcoxon method, resulted in an estimated LD_{50} of 2600 mg/kg. To variable degrees, animals showed signs of lethargy, flaccid muscle tone, ataxia, convulsions, redness of the nose and mouth, and wet body areas. Necropsy revealed abnormal lungs, liver and gastrointestinal tract in nonsurvivors while, in survivors, these tissues appeared normal (127).

INHALATION. When Sprague–Dawley rats (5/sex per exposure level) were exposed to atmospheres of 0 or 5110 mg/m^3 (656 ppm) DPTB (nose-only) for 4 h and subsequently observed for 14 days, no deaths resulted and no abnormal clinical signs were observed. At the end of the observation period, necropsy revealed no abnormalities (128). DPTB was in the form of a respirable aerosol with a mass median aerodynamic diameter of 3.9 μm (82% respirable). The nominal concentration, calculated from the amount of DPTB consumed and the volume of air supplied, was 51,700 mg/m^3, or about 10 times greater than the measured concentration. The conducting laboratory indicated that the concentration was chosen based on OECD test guidelines. From this study, it may be concluded that the acute LC_{50} of DPTB is greater than 656 ppm.

DERMAL. Five male and five female New Zealand rabbits survived a 24-h application of 2 g DPTB per kilogram body weight. DPTB was applied to the abraded, clipped skin of rabbits and held in contact for a period of 24 h by use of a semiocclusive wrap that prevented removal by evaporation or other means. After 24 h, the test material was removed from the skin and subjects were observed for 14 days. No deaths resulted from treatment and no adverse clinical signs were observed. Examination of the skin at the site of application revealed epidermal flaking in 4 of 10 animals. At necropsy, one animal showed kidney abnormalities (129).

SKIN IRRITATION. DPTB was applied to the clipped skin of six albino rabbits (3/sex) at both abraded and non-abraded sites (two each) and held in contact with the skin under an occlusive wrap for 4 h in a primary skin irritation assay following a modification of the procedure of Draize et al. (71). A volume of 0.5 mL was applied to two intact and two abraded sites for a total dose of 2 mL/rabbit. After 4 h, the test material was removed and the application sites were observed and scored at intervals of 72 h or less. No ulceration, necrosis, or eschar was observed. Transient, usually slight erythema and edema were observed at 4, 24, and 48 h and were absent after 72 h. No abnormal behavioral signs were noted at any time interval. The modified PII (primary irritation index) was 0.13 on a scale of 0–8, indicating that DPTB has slight skin irritation potential (130).

EYE IRRITATION. DPTB was tested in rabbits for primary eye irritation potential following a modification of the Draize procedure. One tenth milliliter of pure DPTB was instilled into the eyes of nine rabbits. Three of the rabbits had their eyes flushed with water (for 1 min) to remove the test material 20–30 s following instillation (washed group), and six did not (unwashed group). In the unwashed group, DPTB caused corneal opacities in two of six eyes that cleared by day 7. Iritis was found in one of six eyes (cleared by 48 h) and conjunctival irritation occurred in six of six eyes (cleared by day 7). In the washed group, three of three eyes had corneal opacities and conjunctival irritation; both findings reversed within 7 days. No iritis occurred in the washed group (131). Undiluted DPTB is moderately to severely irritating to eyes.

13.4.1.2 Chronic and Subchronic Toxicity. In a subacute toxicity study conducted with rats, DPTB was administered by gavage to rats (5/sex per dose level) at doses of 0, 100, 300, or 1000 mg/kg for 14 consecutive days. During the course of the study, body weights and food consumption were monitored and animals were observed for clinical signs. At the end of 14 days, animals were sacrificed and necropsied, organs were weighed, blood was taken for hematology and clinical chemistries, and tissues were preserved for microscopic examination (132).

No deaths occurred during the course of treatment except for a control animal. No effects were noted upon body weights or food consumption. Salivation was observed in the 300- and 1000-mg/kg dose groups after day 7. Lymphocytes were reduced in high dose males; no changes were found in blood chemistries of either sex. Liver and kidney weights were increased in male and female rats receiving 300 and 1000 mg/kg, and females in these groups also exhibited slightly increased adrenal weights. Accompanying microscopic examination revealed centrilobular hepatocyte enlargement in livers of both sexes and hyaline droplet formation in cells of the kidney tubules of male but not female rats (the latter finding suggests α-2-microglobulin-induced kidney nephropathy). The kidney changes occurred at all dose levels in a dose-related manner in male rats while the liver changes occurred at the top two dose levels in both sexes. Females did not exhibit micropathological changes in adrenals. The NOEL was 100 mg/kg from this study, based on organ hypertrophy. A subsequent study (see below) indicates that organ hypertrophy was reversible following a 4-week recovery period, supporting the conclusion that the hypertrophy was adaptive, relating to enzyme induction, rather than pathological in nature.

Crl:CD BR rats (10/sex per dose level) were treated by gavage, 7 days/week for 13 weeks, to doses of 0, 62.5, 250, or 1000 mg DPTB/kg per day. An additional 5 rats/sex per dose were included at each dose level for neurotoxicological evaluation (including a functional observational battery, motor activity, and neurohistopathology following EPA published guidelines). In the high dose group only, another 10 rats/sex were included in order to evaluate the reversibility of any effects after a 4-week, post-treatment, recovery period. The control group also included behavioral and recovery animals. Clinical signs were monitored daily, and body weights and food consumption were assessed weekly throughout the study. Water intake was measured at week 12 and at 3 weeks post-treatment. Hematology and clinical chemistries were determined at weeks 5 and 13, and hematology was additionally evaluated at 4 weeks post-treatment. Control and high dose groups were subjected to ophthalmological examination at weeks 0 and 13. Neurological evaluations took place at weeks 4, 8, and 12 of dosing. At terminal sacrifice, necropsies were performed, organ weights determined, and tissues taken for micropathological examination (133).

No deaths occurred in the DPTB-treated groups. There was no effect on body weights or food consumption. Water consumption was increased in males from the high dose group, but this effect did not occur at the end of the recovery period. In the high dose group only, salivation, unsteady gait, collapsed posture, partially closed eyes, body spasms, and paddling of forelimbs occurred during weeks 8–13 in a small portion of animals. These signs did not occur during the recovery period. Such signs are suggestive of a neurological effect. However, no effects were noted on neurobehavioral parameters in the functional observational battery, on motor activity parameters, or on microscopic examination of the CNS. Ophthalmological examination revealed no changes. Slight but statistically significant reductions in total white cell counts and lymphocytes were observed in high dose males at week 5; at week 13, this effect was seen again in high dose males and mid and high dose females. Clinical chemistries revealed increased mean albumin and globulin in males and females at the high dose. At the 13-week sacrifice, male and female liver weights from the high dose group were increased with enlarged centrilobular hepatocytes suggesting adaptive hypertrophy. This effect reversed after a 4-week recovery period. Males and females from the high dose group and males from the mid-dose group had increased kidney weights, with, in the males only, accompanying hyaline droplet formation in tubular cells and cellular debris casts in the lumen of the kidney tubules (at all dose levels). This latter effect was partially reversible in the 4-week recovery group. The latter findings suggest the occurrence of α-2-microglobulin-induced male rat nephropathy. Adrenal weights were increased in female rats from the high dose group but not in the recovery group. Histopathology of adrenals revealed a slightly increased width of the zona fasciculata of this organ in the high dose females, which was not evident in the recovery group.

Although slight lesions were found in the kidney in the low-dose males suggesting α-2-microglobulin nephropathy, this finding is not considered relevant to humans. Therefore, a NOAEL of 62.5 mg/kg per day may be conservatively established from this study. Since other findings were minimal at 250 mg/kg per day, a less conservative NOAEL at this level may be considered. The study author concluded that the latter (250 mg/kg per day) was the move appropriate number.

13.4.1.3 Pharmacokinetics, Metabolism, and Mechanisms: NA

13.4.1.4 Reproductive and Developmental: NA

13.4.1.5 Carcinogenesis: NA

13.4.1.6 Genetic and Related Cellular Effects Studies.
DPTB was tested in the Ames *Salmonella* assay when incubated with and without Aroclor-induced rat liver microsomal S-9 preparations using tester strains TA1537, TA1538, TA98, and TA100 at doses of 50, 150, 500, 1500, or 5000 µg/plate in a preliminary assay with no toxicity evident at the top dose. In the main assay, doses of 50, 150, 500, and 1500 µg/plate were used. Solvent and positive controls were within normal limits. DPTB caused no mutagenic response in this test (134).

DPTB was tested for its ability to cause the formation of micronuclei in bone marrow cells of mice following OECD guideline 474 (135). At a low and mid-dose level, mice (15/sex per dose) were administered a single intraperitoneal injection of 200 and 400 mg/kg DPTB, respectively. At the high dose level, males were given 600 mg/kg while females were injected with 800 mg/kg. The different dose for males and females at the high dose was due to unexpectedly high mortality in males at 800 mg/kg when injected intraperitoneally. A negative control (15/sex) and a positive control (mitomycin C, 5/sex) were included in the study design. For the negative control and the DPTB-treated groups, five animals per sex per dose level were sacrificed at 24, 48, and 72 h post-injection. The mitomycin C–positive control group (5/sex) were sacrificed at 24 h post-injection. Bone marrow was collected from both femurs of all animals, processed into slides, and read blind for the incidence of micronucleated cells among polychromatic erythrocytes. DPTB did not increase the incidence of micronulei in this assay.

13.4.1.7 Other: Neurological, Pulmonary, Skin Sensitization, etc.

SKIN SENSITIZATION. DPTB was tested for its potential to cause skin sensitization in the guinea pig following the OECD Guideline for Testing of Chemicals, Protocol 406 (Magnusson/Kligman guinea pig maximization test). During the induction phase, 10 guinea pigs received three intradermal injections at separate locations on each flank of (1) Freund's complete adjuvant, (2) 1% v/v DPTB in water, and (3) 1% v/v DPTB in a 50 : 50 mixture of distilled water and Freund's adjuvant. One week later, pure DPTB was applied covering all three injection sites on each flank for 48 h in a second induction treatment. Two weeks later for the challenge phase, treatment solutions of 50 and 100% (v/v) DPTB in water were applied to the sites and held in place for 24 h under occlusion. Five control animals received similar treatments where distilled water was substituted for DPTB. No sensitization reactions were recorded in the DPTB-treated animals (136).

13.5 Standards, Regulations, or Guidelines of Exposure

No occupational exposure limits have been established for DPTB.

14.0 Tripropylene Glycol Monomethyl Ether

14.0.1 CAS Numbers: [20324-33-8], [25498-49-1]

14.0.2 Synonym: 1-(2-(2-Methoxy-1-methylethoxy)-1-methylethoxy)-2-propanol

14.0.3 Trade Names: Arcosolv® TPM, Dowanol® TPM Glycol Ether

14.0.4 Molecular Weight: 206.28

14.0.5 Molecular Formula: $C_{10}H_{22}O_4$

14.0.6 Molecular Structure:

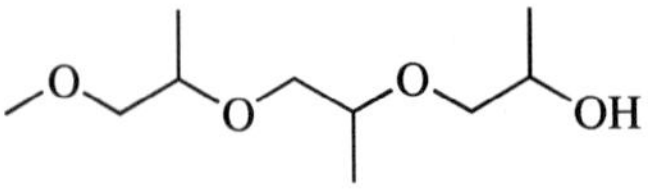

14.1 Chemical and Physical Properties

14.1.1 General

See Table 87.3.

14.2 Production and Use

Refer to Table 87.5.

14.4 Toxic Effects

Tripropylene glycol monomethyl ether (TPGME) is a colorless liquid, low in volatility with a slight, pleasant, ethereal odor and a bitter taste. It is low in single-dose oral toxicity, transiently painful but not damaging to the eye, and is not appreciably irritating to the skin. Narcosis and lethality were reported in rabbits following repeated high dose dermal treatment. It appears to have only minimal effects (liver weight increases) following inhalation exposure, and produces no developmental toxicity following this route of administration. The hazards to health from ordinary handling and use would seem to be negligible.

14.4.1 Experimental Studies

14.4.1.1 Acute Toxicity

ORAL. When administered in single doses to rats, the LD_{50} was about 3.3 g/kg (8). The primary effect of the material appeared to be narcosis. Shideman and Procita (20) have estimated the LD_{50} for dogs to be 5 mL/kg (4.8 g/kg). They asserted that the primary effect of the material was CNS depression, with death from large doses due to respiratory failure.

INHALATION. When rats were exposed once for 7 h to a saturated atmosphere at 25°C, no ill effects were observed. This indicates that the material does not present a hazard from acute vapor exposure at ordinary temperatures (8). Rats and mice exposed to an aerosol (0.15, 0.36, or 1.01 mg/L) for 6 h/day for 9 days exhibited only increased liver weights in both sexes of both species (20). The liver weight changes were not accompanied by histopathological changes. This response was considered to be an adaptive effect rather than a toxicological response.

DERMAL. The acute dermal LD$_{50}$ in rabbits was greater than 20 mL/kg, indicating very low acute dermal toxicity (8). some of the rabbits exhibited narcosis at doses of 10 and 20 mL/kg.

EYE IRRITATION. Repeated instillation of the liquid into the eye of a rabbit caused only a mild transitory conjunctival irritation. Evidence of transient pain was observed (8).

SKIN IRRITATION. TPGME applied repeatedly to the skin of rabbits caused only a very mild, simple irritation (8).

14.4.1.2 Chronic and Subchronic Toxicity

SUBCHRONIC TOXICITY/DERMAL. When TPGME was applied repeatedly to the skin of rabbits (1.0, 3.0, 4.0, and 10.0 mL/kg; 65 doses over a period of 90 days), seven out of eight animals died within 10 days at the top dose (8). Lower doses caused some narcosis that was associated with development of tolerance. Only a very mild skin irritation was observed. Some kidney toxicity was apparent even at the lowest dose administered.

At high dose levels (5–10 mL/kg, 4.8–9.65 g/kg), TGME applied repeatedly over a 90-day period to the clipped abdomen of rabbits under an occlusive wrap caused necrosis and kidney injury. At dose levels below 5 mL/kg, narcosis was not apparent, but kidney injury was still seen at the lowest dose (1 mL/kg) applied (8).

14.4.1.3 Pharmacokinetics, Metabolism, and Mechanisms

METABOLISM AND TOXICOKINETICS. Urine was the primary route of excretion after a single oral dose of 1 or 4 mmol/kg [^{14}C]TPGME (137). A majority of both doses was recovered in urine within 48 h (69 and 75%, respectively); the metabolites identified in urine were similar to those found with DPGME.

14.4.1.4 Reproductive and Developmental

DEVELOPMENTAL TOXICITY. Exposure of rats to aerosols (0.1, 0.3, or 1.0 mg/L) for 6 h/day during gestation produced maternal toxicity (muzzle staining) at the high exposure level (138). No embryo lethality, fetotoxicity, or teratogenicity was observed at any dose level.

14.4.1.5 Carcinogenesis: NA

14.4.1.6 Genetic and related Cellular Effects Studies

GENOTOXICITY. TPGME was negative when tested in the Ames bacterial tester systems (A. L. Mendrala, The Dow Chemical Company, unpublished data, 1982) and also in an *in vitro* unscheduled DNA synthesis assay in rodent hepatocytes (A. L. Mendrala, The Dow Chemical Company, unpublished data, 1982).

14.5 Standards, Regulations, or Guidelines of Exposure

No occupational exposure standards have been proposed or established for TPGME.

15.0 Tripropylene Glycol Monoethyl Ether

15.0.1 CAS Numbers: *[20178-34-1], [75899-69-3]*

15.0.2 Synonyms: 1-[2-(2-Ethoxy-1-methylethoxy)-1-methylethoxy]-2-propanol

15.0.3 Trade Name: NA

15.0.4 Molecular Weight: 220.3

15.0.5 Molecular Formula: $C_{11}H_{24}O_4$

15.0.6 Molecular Structure: $CH_3CHOHCH_2OCH(CH_3)CH_2OCH(CH_3)CH_2OCH_2CH_3$

15.1 Chemical and Physical Properties

Tripropylene glycol monoethyl ether is a colorless liquid, low in volatility, with a slight but pleasant ethereal odor and a bitter taste.

15.1.1 General

See Table 87.3.

15.2 Production and Use: NA

15.3 Exposure Assessment: NA

15.4 Toxic Effects

When single oral doses of the material were administered to rats, the LD_{50} was found to be 2 mL/kg. When one drop of the liquid was introduced into the eye of a rabbit on 5 consecutive days, it produced only a very slight conjunctival irritation and no corneal injury. Repeated applications (10 in 14 days) to the skin of rabbits resulted in slight irritation without evidence that toxic amounts were absorbed (V. K Rowe, The Dow Chemical Company, unpublished data, 1947). No quantitative skin absorption studies or inhalation studies have been reported.

15.5 Standards, Regulations, or Guidelines of Exposure

No occupational exposure standard for tripropylene glycol monoethyl ether has been established.

16.0 Tripropylene Glycol Mono-*n*-butyl Ether

16.0.1 CAS Numbers: [57499-93-1], [55934-93-5]

16.0.2 Synonyms: 1-(2-(2-Butoxy-1-methylethoxy)-1-methylethoxy)-2-propanol, 2-(2-butoxymethylethoxy)methylethoxy propanol

16.0.3 Trade Name: Dowanol® TPnB Glycol Ether

16.0.4 Molecular Weight: 248.36

16.0.5 Molecular Formula: $C_{13}H_{28}O_4$

16.0.6 Molecular Structure:

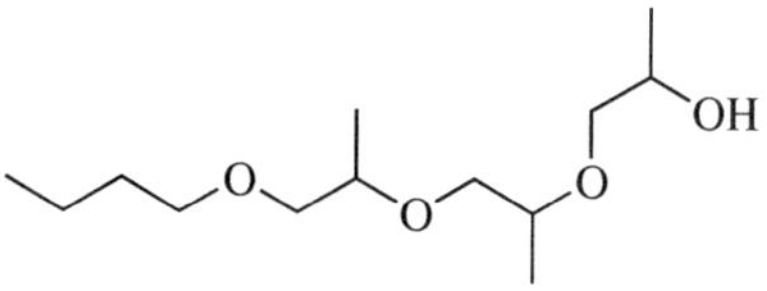

16.1 Chemical and Physical Properties

See Table 87.3.

16.2 Production and Use

Refer to Table 87.5.

16.3 Exposure Assessment: NA

16.4 Toxic Effects

16.4.1 Experimental Studies

16.4.1.1 Acute Toxicity

ORAL. The oral LD_{50} of undiluted tripropylene glycol mono-*n*-butyl ether (TPGBE) has been determined to be 3100 mg/kg in male rats and 2600 mg/kg in female rats (139). Studies conducted much earlier had estimated an LD_{50} of 1840 mg/kg (V. K. Rowe, The Dow Chemical Company, unpublished data, 1947).

Fischer 344 rats were given test material in corn oil by gavage for 4 weeks at doses of 0, 100, 350, and 1000 mg/kg per day (M. J. Mizel, L. Atkin, and B. L. Yano, The Dow Chemical Company, unpublished data, 1990). Lethargy was observed in the high dose rats during the first 4 h after dosing over the first 3 days of dosing; the high dose rats appeared normal on subsequent dose days, suggesting adaptation to this effect. Dose-dependent liver weight increases were noted in both the mid-dose and high dose males and females, which were accompanied by increases in hepatocyte size and staining in the highest dose group. Altered hepatocyte staining was reported only in the mid-dose male rats. No degenerative changes were reported in the liver. The compensatory liver hypertrophy was ascribed to metabolism of parent test material, and thus was not regarded by the authors to be of toxicological significance. The daily NOEL was reported as 100 mg/kg and the daily NOAEL as 350 mg/kg.

DERMAL. Contact with the intact skin of three rabbits for 4 h produced slight redness and swelling accompanied by slight scaling. Five applications of an 85% test material to rabbit skin produced a slight redness on intact skin or on abraded skin (P. J. J. M. Weterings, P. A. M. Daamen, and H. G. Verschuuren, The Dow Chemical Company, unpublished data, 1988; J. M. Wall, The Dow Chemical Company, unpublished data, 1988). Application of 2000 mg/kg to the skin of rats and rabbits produced no evidence of systemic toxicity, (139; J. B. J. Reijndors and H. G. Verschuuren, The Dow Chemical Company, unpublished data, 1988). This material did not produce skin sensitization when evaluated in the Buehler guinea pig assay (P. A. M. Daamen and H. G. Verschuuren, The Dow Chemical Company, unpublished data, 1989).

EYE IRRITATION. Application of pure test material to rabbit eyes resulted in slight conjunctival redness and moderate swelling, which healed in 7–14 days (P. J. J. M. Weterings, P. A. M. Daamen, and H. G. Verschuuren, The Dow Chemical Company, unpublished data, 1988). Corneal effects were noted by 24 h and had dissipated by 72 h. All dosed rabbits exhibited servere lacrimation by 1 h after application and discharge during the next 2 days.

16.4.1.6 Genetic and Related Cellular Effects Studies. Genetic toxicity TPGBE was not mutagenic when evaluated in the Ames *Salmonella* assay (tester strains TA98, TA100, TA1535, and TA1537), with or without metabolic activation, at concentrations of $\leq$5000 µg/plate (Y. E. Samson, B. Gollapudi, and H. G. Verschuuren, The Dow Chemical Company, unpublished data, 1989). A mouse micronucleus assay was negative following oral administration of doses of $\leq$1875 mg/kg (M. L. McClintock, B. Gollapudi, and H. G. Verschuuren, The Dow Chemical Company, unpublished data, 1989).

16.5 Standards, Regulations, or Guidelines of Exposure

No occupational exposure standard for tripropylene glycol monobutyl ether has been established.

17.0 Mono-, Di-, and Tripropylene Glycol Allyl Ethers

*17.0.1 **CAS Number:*** *[1331-17-5]*

*17.0.2 **Synonyms:*** Propylene glycol allyl ether, (allyloxy) propanol; propanol, 1(or 2)-(2-propenyloxy)-; (2-propenyloxy)propanol; propylene glycol monoallyl ether

*17.0.3 **Trade Names:*** NA

*17.0.4 **Molecular Weight:*** 116.16

*17.0.5 **Molecular Formula:*** $C_6H_{12}O_2$

*17.0.6 **Molecular Structure:***

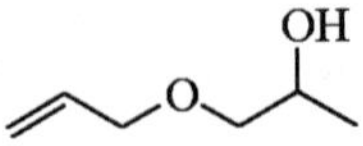

17.1 Chemical and Physical Properties

17.1.1 General

Appearance: Clear, colorless liquid
Purity: Mixture of allyl alcohol ethers of propylene glycols with less than 0.1% free allyl alcohol

17.2 Production and Use: NA

17.3 Exposure Assessment: NA

17.4 Toxic Effects

17.4.1 Experimental Studies

17.4.1.1 Acute Toxicity. Tripropylene glycol allyl ethers have low volatility at room temperature. Toxicological information is available only for a commercial-grade material. The single-dose oral toxicity is considered to be low; the LD_{50} for female rats was in the range of 1–2 g/kg (L. W. Rampy and P. A. Keeler, The Dow Chemical Company, unpublished data, 1973). Lethargy was apparent at lethal doses. Death was frequently preceded by convulsions. Gross autopsy revealed irritation of the stomach. Eye exposure using rabbits resulted in marked irritation and moderate corneal injury, which healed in 3–7 days. Continuous skin contact for 2 weeks resulted in only minor irritation in rabbits. The rabbits showed no signs of systemic toxicity. In a single, 7-h inhalation test using rats (H. O. Yakel and G. C. Jersey, The Dow Chemical Company, unpublished data, 1976), saturated vapor generated at room temperatures produced no adverse effects. However, exposure to vapors generated at 100°C resulted in complete lethality (three deaths during the exposure period and three within 24 h). Congested lungs and nasal passages were observed.

18.0 Mono-, Di-, and Tripropylene Glycol Isobutyl Ethers

18.0.1 CAS Number: None found

18.0.2 Synonyms: NA

18.0.3 Trade Names: NA

18.0.4 Molecular Weight: NA

18.0.5 Molecular Formula: NA

18.0.6 Molecular Structure: $(CH_3)_2CHCH_2[OCH_2CH(CH_3)]_xOH$

18.1 Chemical and Physical Properties

18.1.1 General

Appearance: Clear, essentially colorless liquid
Boiling point: 170–173°C
Specific gravity: 0.882 at 25°C

18.1.2 Odor and Warning Properties

A limited human volunteer study conducted with inhalation exposures ranging from 10 to 20 ppm indicated detection of odors at 10 ppm. However, intolerable concentrations likely exceed 20 ppm.

18.2 Production and Use: NA

18.3 Exposure Assessment: NA

18.4 Toxic Effects

A commercially produced mixture of these glycol ethers was low in single-dose oral toxicity. Rabbit eye tests indicate that it should be considered low in hazard from eye contact, causing only minor transient irritation. Rabbit skin tests suggested that it is not likely to cause significant skin irritation, nor is it likely to be absorbed through the skin in acutely toxic amounts. Short-term repeated inhalation exposures have a low toxicity potential. The effects of long-term inhalation exposure are not known. This material is not mutagenic.

18.4.1 Experimental Studies

18.4.1.1 Acute Toxicity

ORAL. The single-dose oral LD_{50} value for male and female rats was 4.29 g/kg (L. W. Rampy, P. A. Keeler, and H. O. Yakel, The Dow Chemical Company, unpublished data, 1974) and 2.46 g/kg (The Dow Chemical Company, unpublished data, 1967), respectively. Lethargy was the only adverse effect observed.

INHALATION. Groups of four rats were exposed for 7 h to saturated atmospheres generated at room temperature and at 100°C (nominal concentrations of 1630 and 3950 ppm, respectively). All the rats appeared drowsy and unsteady when removed; the effect was greater in rats exposed to vapor generated at 100°C. Autopsy observation indicated no adverse effects to any organs (R. E. Hefner, Jr., B. K. L. Leong, and J. F. Quast, The Dow Chemical Company, unpublished data, 1977).

Groups of 20 male rats, 4 male rabbits, and 2 male beagle dogs were subjected to twenty 7-h exposures in 28 days to 100 or 200 ppm of vapor. The rats experienced mild eye irritation during the first 5 days of exposure, which subsided to no irritation for the rest of the test period. Neither the rabbits nor the dogs showed any eye irritation. None of the animals experienced any adverse effects as reflected by hematology, clinical chemistry, pathology, or body and organ weights (R. E. Hefner, Jr., B. K. L. Leong, and J. F. Quast, The Dow Chemical Company, unpublished data, 1977).

Rats exposed for 7 h/day, 5 days a week for a total of 28 exposures to 600 ppm developed liver injury consisting of central lobular granular degeneration with occasional necrosis of parenchymal cells (R. E. Hefner, Jr., B. K. L. Leong, and J. F. Quast, The Dow Chemical Company, unpublished data, 1977).

DERMAL. The dermal LD_{50} in rabbits following application under an impervious cuff was approximately 8.0 g/kg of body weight. Slight skin irritation, edema, and necrosis occurred (L. W. Rampy, P. A. Keeler, and H. O. Yakel, The Dow Chemical Company, unpublished data, 1974).

EYE IRRITATION. The application of two drops of undiluted material into the eyes of rabbits caused only slight pain on contact followed by transient conjunctival redness in the eyes of rabbits (R. E. Hefner, Jr., B. K. L. Leong, and J. F. Quast, The Dow Chemical Company, unpublished data, 1977). Thus this product may be considered to be low in hazard from eye contact.

SKIN IRRITATION. Daily repeated application of undiluted material under a cloth covering for 2 weeks to the healthy skin of rabbits resulted in no irritation during the first week of application but slight scaling during the second week. Continuous contact under a cloth covering for 3 days to abraded skin resulted in the same response (R. E. Hefner, Jr., B. K. L. Leong, and J. F. Quast, The Dow Chemical Company, unpublished data, 1977).

18.4.1.2 Chronic and Subchronic Toxicity: NA

18.4.1.3 Pharmacokinetics: NA

18.4.1.4 Reproductive and Developmental: NA

18.4.1.5 Carcinogenesis: NA

18.4.1.6 Genetic and Related Cellular Effects Studies

GENETIC TOXICITY. Bone marrow cytogenetics was evaluated in rats following exposure to 200 ppm, 7 h/day, 5 days/week for 4 weeks. No significant effects were seen in metaphase chromosomes (J. J. Middleton et al., The Dow Chemical Company, unpublished data, 1976).

19.0 Polypropylene Glycol Butyl Ethers

19.0.1 CAS Number: [9003-13-8]

19.0.2 Synonyms: Poly[oxy(methyl-1,2-ethanediyl)], α-butyl-ω-hydroxy-; butoxypolypropylene glycol, butoxypropanediol polymer, polyoxypropylene glycol butyl ether, BPG 400, BPG 800, Crag fly repellent, Fluid-AP, LB 1145, Newpol LB3000, OPSB, polyoxypropylene glycol butyl monoether, polyoxypropylene monobutyl ether, polypropoxbutanol, PPG-14 butyl ether, PPG-16 butyl ether, PPG-33 butyl ether, Stabilene, Stabilene fly repellent

19.0.3 Trade Names: Crag Fly Repellent, Newpol® LB3000, Stabilene® Fly Repellent, PPG-14 Butyl Ether, PPG-16 Butyl Ether, PPG-33 Butyl Ether, Ucon® LB 1145, Ucon® LB1800X, Ucon® LB-250

19.0.4 Molecular Weight: NA

19.0.5 Molecular Formula: NA

19.0.6 Molecular Structure: NA

19.1 Chemical and Physical Properties

The properties for two polypropylene glycol butyl ethers having molecular weights approximating 400 (BPG 400) and 800 (BPG 800) are shown in Table 87.9.

19.2 Production and Use

A major use of this material is as an insect and fly repellent.

19.3 Exposure Assessment

A review by the Cosmetic Ingredient Review Expert Panel has concluded that the use of this material as a cosmetic agent cannot be concluded to be safe based on the lack of appropriate safety data (140).

19.4 Toxic Effects

19.4.1 Experimental Studies

19.4.1.1 Acute Toxicity. Carpenter et al. (141) have reported on toxicologic studies conducted on BPG 400 and BPG 800. Both ethers are low in toxicity when given either

Table 87.9. Physical and Chemical Properties of Representative Polypropylene Glycol Butyl Ethers

	BPG 400	BPG 800
Molecular formula	$C_4H_9O(C_3H_6O)_xH$	$C_4H_9O(C_3H_6O)_yH$
Molecular weight	~400	~800
Specific gravity (25/25°C)	~0.973	~0.990
Vapor pressure mm Hg (25°C)	<0.1	<0.1
Solubility in H_2O (g/100 g at 20°C)	0.2	0.1

orally or intraperitoneally. The oral LD$_{50}$ values for BPG 400 for male rats, male guinea pigs, and male rabbits were found to be 5.84, 2.46, and 3.30 g/kg, respectively. Similar values for these same species for BPG 800 were 9.16, 6.8, and 23.7 g/kg, respectively. The intraperitoneal LD$_{50}$ values for rats were found to be 0.32 g/kg for BPG 400 and 0.91 g/kg for BPG 800. When large, single oral doses were given, the principal effects of both materials were gastrointestinal irritation, congestion of internal organs, and death, usually within 24 h. The BPG 800 was less distressing to the animals than the BPG 400 and more likely to cause convulsions and lung hemorrhage.

Neither material is more than very slightly irritating to rabbit skin or eyes. BPG 800 is neither an irritant nor a sensitizer of human skin. Neither is readily absorbed through the skin in acutely toxic amounts but repeated inunction of the rabbit's skin for 30 days showed BPG 400 to be moderate in toxicity and BPG 800 to be low in toxicity by this route; the no-effect levels were less than 0.1 and 1.0 mL/kg per day, respectively. BPG 400 is readily absorbed from the gastrointestinal tract, whereas BPG 800 is poorly absorbed. BPG 400 was not stored in the bodies of rats fed large doeses for 30 days. Neither material presents a hazard from inhalation under reasonable conditions. Rats exposed to atmospheres essentially saturated at room temperatures for 8 h were unaffected and suffered only mild effects when exposed for 8 h to fogs of the material. It is clear that BPG 800 is distinctly less toxic than BPG 400.

19.4.1.2 Chronic and Subchronic Toxicity. Dietary studies over a 90-day period showed the daily no-effect dosage levels for BPG 400 to be between 0.16 and 0.67 g/kg and for BPG 800 to be less than 0.52 g/kg (141). The liver and/or kidney were the first organs affected when subacute dosage levels were given repeatedly.

B ETHERS OF BUTYLENE GLYCOL

B.1 Source, Uses, and Industrial Exposure

The methyl, ethyl, and *n*-butyl ethers of butylene glycol considered herein are prepared by reacting the appropriate alcohol with so-called straight-chain butylene oxide, consisting of about 80% 1,2 isomer and about 20% 2,3 isomer in the presence of a catalyst. They are colorless liquids with slight, pleasant odors. The methyl and ethyl ethers are miscible with water, but the butyl ether has limited solubility. All are miscible with many organic solvents and oils; thus they are useful as mutual solvents, dispersing agents, and solvents for inks, resins, lacquers, oils, and greases. Industrial exposure may occr by any of the common routes.

20.0 Butylene Glycol Monomethyl Ether

20.0.1 CAS Number: [111-32-0]

20.0.2 Synonyms: 4-Methoxybutanol, 4-methoxy-1-butanol

20.0.3 Trade Names: NA

20.0.4 *Molecular Weight:* 104.15

20.0.5 *Molecular Formula:* $C_5H_{12}O_2$

20.0.6 *Molecular Structure:* HO~~~O~

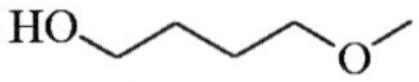

20.1 Chemical and Physical Properties

See Table 87.3.

20.2 Toxic Effects

20.2.1 Summary

Butylene glycol monomethyl ether is low in single dose oral toxicity for rats, moderately irritating and injurious to the eyes of rabbits, not appreciable irritating to the skin, and not readily absorbed through the skin of rabbits in toxic amounts. Prolonged inhalation by rats of air essentially saturated with vapors causes drowsiness, unsteadiness, and slight injury of the liver and kidneys; when some fog was present, deaths occurred (M. A. Wolf, The Dow Chemical Company, unpublished data, 1959).

20.2 Production and Use: NA

20.3 Exposure Assessment: NA

20.4 Toxic Effects

20.4.1 Experimental Studies

20.4.1.1 Actue Toxicity

ORAL. When single oral doses of butylene glycol monomethyl ether were fed by gavage to rats, doses of 2 g/kg were survived and doses of 4 g/kg were fatal. No pathology was noted on gross observation of the organs of the survivors (M. A. Wolf, The Dow Chemical Company, unpublished data, 1959).

INHALATION. An atmosphere essentially saturated (probably 6000–7000 ppm) with vapor was not lethal to any of three rats exposed for 7 h, but it did cause drowsiness, unsteadiness, temporary weight loss, and some mild kidney and liver changes. An exposure lasting 4 h caused drowsiness and unsteadiness but no histopathological changes. When six rats were exposed for 7 h to a supersaturated atmosphere (containing some fog), all animals became drowsy, were unable to stand, and breathed with difficulty, and four died. No significant pathological changes were seen in the organs of the survivors (M. A. Wolf, The Dow Chemical Company, unpublished data, 1959).

GLYCOL ETHERS

EYE IRRITATION. The undiluted material, when instilled into the eyes of rabbits, was painful and caused moderate conjunctival irritation, moderate corneal injury, and some iritis, all of which cleared within a week (M. A. Wolf, The Dow Chemical Company, unpublished data, 1959).

SKIN IRRITATION. The material caused only very slight; irritation of the skin of rabbits even though exposures were prolonged and repeated. Evidence of absorption through the skin was not apparent (M. A. Wolf, The Dow Chemical Company, unpublished data, 1959).

20.5 Standards, Regulations, or Guidelines of Exposure

No occupational exposure standard has been established for butylene glycol monomethyl ether.

21.0 Butylene Glycol Monoethyl Ether

21.0.1 CAS Number: [111-73-9]

21.0.2 Synonym: 4-Ethoxy-1-butanol

21.0.3 Trade Name: NA

21.0.4 Molecular Weight: 118.2

21.0.5 Molecular Formula: $C_6H_{14}O_2$

21.0.6 Molecular Structure: $C_2H_5OC_4H_8OH$

21.1 Chemical and Physical Properties

See Table 87.3.

21.2 Production and Use: NA

21.3 Exposure Assessment: NA

21.4 Toxic Effects

21.4.1 Experimental Studies

Butylene glycol monoethyl ether is low in single-dose oral toxicity for rats, moderately irritating and injurious to the eyes of rabbits, not appreciably irritating to the skin, and not readily absorbed through the skin in toxic amounts. Air essentially saturated with vapor was only slightly toxic to rats (M. A. Wolf, The Dow Chemical Company, unpublished data, 1959).

21.4.1.1 Acute Toxicity

ORAL. When single oral doses of butylene glycol monoethyl ether were fed to rats, doses of 1 g/kg were survived, doses of 2 g/kg caused the death of one of three, and a dose of 4 g/kg caused the death of one of two. Doses of 2 and 4 g/kg caused injury to the lungs, liver, and kidney, but lower dosage levels did not (M. A. Wolf, The Dow Chemical Company, unpublished data, 1959).

INHALATION. Rats exposed for 7 h to an atmosphere essentially saturated (probably 3000–4000 ppm) with vapor survived without serious effects, although some kidney injury was apparent on gross examination of the organs. However, when a fog was present, six rats exposed for 7 h exhibited irritation of the eye and nose, drowsiness, inability to stand, and difficulty in breathing. Although none died, all were ill and gross examination of the organs revealed injury to the lungs, kidneys, and livers (M. A. Wolf, The Dow Chemical Company, unpublished data, 1959).

EYE IRRITATION. When the material was instilled into the eyes of rabbits it caused pain, marked conjunctival and corneal injury, and slight iritis (M. A. Wolf, The Dow Chemical Company, unpublished data, 1959).

SKIN IRRITATION. The undiluted material was not appreciably irritating to the skin of rabbits even when exposures were prolonged and repeated. Evidence of absorption through the skin was not apparent (M. A. Wolf, The Dow Chemical Company, unpublished data, 1959).

21.5 Standards, Regulations, or Guidelines of Exposure

An occupational exposure standard has not been established for butylene glycol monoethyl ether.

22.0 Butylene Glycol Mono-n-butyl Ether

22.0.1 CAS Number: No number found

22.0.2 Synonyms: 4-Butoxy-1-butanol

22.0.3 Trade Names: NA

22.0.4 Molecular Weight: 146.2

22.0.5 Molecular Formula: $C_8H_{18}O_2$

22.0.6 Molecular Structure: $C_4H_9OC_4H_8OH$

22.1 Chemical and Physical Properties

See Table 87.3.

22.2 Production and Use: NA

22.3 Exposure Assessment: NA

22.4 Toxic Effects

Butylene glycol mono-*n*-butyl ether is low in single-dose oral toxicity for rats. In studies with rabbits, it was found to be moderately irritating and injurious to the eyes, not appreciably irritating to skin when unconfined but moderately injurious when confined, and single applications are not readily absorbed through the skin in toxic amounts. Acute exposure to a saturated atmosphere was low in toxicity to rats (K. J. Olson, The Dow Chemical Company, unpublished data, 1959).

22.4.1 Experimental Studies

22.4.1.1 Acute Toxicity

ORAL. When single oral doses of butylene glyucol mono-*n*-butyl ether as a 20% solution in corn oil were fed to rats, doses of 2.0 g/kg were survived and doses of 4.0 g/kg were lethal. High doses caused prostration and labored breathing. Necropsy of surviving animals revealed some kidney injury (K. J. Olson, The Dow Chemical Company, unpublished data, 1959).

INHALATION. Rats were not noticeably affected by an exposure of 7-h duration to an atmosphere saturated at 100°C and then cooled to room temperature. Necropsy, however, showed that some kidney injury had resulted (K. J. Olson, The Dow Chemical Company, unpublished data, 1959).

EYE IRRITATION. The undiluted material, when instilled into the eyes of rabbits, was painful, irritating to the conjunctival membranes, and injurious to the cornea, and caused some iritis. All these effects disappeared within 7 days after exposure (K. J. Olson, The Dow Chemical Company, unpublished data, 1959).

SKIN IRRITATION. The material was not irritating to the skin of rabbits when applied to the uncovered skin. However, when applied under a bandage for 48 h, it caused a burn. Single applications were not readily absorbed through the intact skin in toxic amounts (K. J. Olson, The Dow Chemical Company, unpublished data, 1959).

22.5 Standards, Regulations, or Guidelines of Exposure

An occupational exposure standard has not been established for butylene glycol mono-*n*-butyl ether.

C ESTERS, DIESTERS, AND ETHER ESTERS OF GLYCOLS

C.1 Sources, Uses, and Industrial Exposure

The common esters and diesters of the polyols are prepared commercially by esterifying the particular polyol with the acid, acid anhydride, or acid chloride of choice in the presence of a catalyst. Mono- or diesters result, depending on the proportions of each reactant employed. The ether esters are prepared by esterifying the glycol ether in a similar manner. Other methods can also be used (142).

The acetic acid esters have remarkable solvent properties for oils, greases, inks, adhesives, and resins. They are widely used in lacquers, enamels, dopes, and adhesives to dissolve the plastics or resins. They are also used in lacquer, paint, and varnish removers.

C.2 Physical and Chemical Properties

All the esters and ether esters of organic acids are colorless volatile liquids. Generally the odors are mild, sometimes fruity, and they all have a bitter taste. See Tables 87.10 and 87.11. for additional properties.

C.3 Determination in the Atmosphere

The choice of methods for the determination of esters, diesters, and ether esters of various glycols vary with existing conditions. Gas chromatography would seem to offer the best means not only of resolving mixtures of vapors but also of identifying components. Mass spectrometry may also be used. Chemical methods such as proposed by Morgan (143) for esters or ether esters may be useful where spectroscopic equipment is not available.

C.4 Summary of Toxicology

Generally speaking, the fatty-acid esters of the glycols and glycol ethers, in either the liquid or vapor state, are more irritating to the mucous membranes than those of the parent glycol or glycol ethers. However, once absorbed into the body, the esters are hydrolyzed and the systemic effect is quite typical of the parent glycol or glycol ethers. Lepkovski et al. (144), in studies with higher fatty acids of glycols, concluded that the fatty acids were liberated and used nutritionally. Furthermore, they observed in rats that severe injury to the tubular epithelium of the kidneys occurred when the esters of ethylene glycol and diethylene glycol were administered, but not when equivalent amounts of fatty esters of propylene glycol, glycerol, ethyl alcohol, methyl alcohol, or the free fatty acids themselves were given. Shaffer and Critchfield (145), in studies with polyethylene glycol 400 monostearate, concluded that it was low in toxicity and also could be utilized nutritionally.

It should be noted that the nitric acid esters of glycols are highly toxic and exert a physiological action quite different from that of the parent polyols. These materials are covered in Section D of this chapter.

Table 87.10. Physical and Chemical Properties of Esters of Various Glycols

Property	Ethylene Glycol Acetate	Ethylene Glycol Diacetate	Triethylene Glycol Diacetate	2-Methyl-2-propene-1,1-diol Diacetate	2,2-Dimethyl-1,3-propanediol Diacrylate	2-Methyl-2,4-pentanediol Diacetate	2,2,4-Trimethyl 1,3-pentanediol Monoisobutyrate	2,2,4-Trimethyl-1,3-pentanediol Diisobutyrate
CAS no.	[542-59-6]	[111-55-7]	[111-21-7]	[10476-95-6]	[2223-82-7]	[1637-24-7]	[25265-77-4]	[6846-50-0]
Molecular formula	$C_4H_8O_3$	$C_6H_{10}O_4$	$C_{10}H_{18}O_6$	$C_8H_{12}O_4$	$C_{11}H_{16}O_4$	$C_{10}H_{18}O_4$	$C_{12}H_{24}O_3$	$C_{16}H_{30}O_4$
Molecular weight	104.11	146.14	234.25	172.18	212.25	202.2	216.32	286.41
Specific gravity 25/4°C	1.108	1.128	1.117 (20/20°C)	1.051 (20/20°C)	1.030 (20/20°C)	1.000 (20/20°C)	0.95 (20/20°C)	0.94
Boiling point (°C) (760 mmHg)	181–182	187	300	215	253	235	244	280
Freezing point (°C)		−41	<−50					−70
Vapor pressure (mmHg) (25°C)		0.25	<0.01			0.07 (20°C)		<0.01 (25°C)
Refractive index (25°C)	1.422	1.416	1.439 (20°C)			1.423 (20°C)		
Flash point (°F)(open cup)	215	82°C (closed cup)	174	215	253	235	248	262
Vapor density (air = 1)	3.6	5.0	8.1					9.9
Percent in saturated air (25°C)		0.044	10^{-6}					
1 ppm ≈ mg/m³ at 25°C, 760 mmHg	4.25	5.98	9.58	7.03	8.68	8.25	8.85	11.68
1 mg/L ≈ ppm at 25°C, 760 mmHg	235	168	104.3	142.0	115.2	121.0	113.0	85.7

Table 87.11. Physical and Chemical Properties of Acetate Esters of Certain Propylene Glycol Ethers

Property	Propylene Glycol, Methyl	Propylene Glycol, Ethyl	Dipropylene Glycol, Methyl	Tripropylene Glycol, Methyl
CAS no.	[108-65-6]	[54839-24-6]	[88917-22-0]	None found
Molecular formula	$C_6H_{12}O_3$	$C_7H_{14}O_3$	$C_9H_{18}O_4$	$C_{12}H_{24}O_5$
Molecular weight	132.16	146.19	190.24	248.2
Melting point (°C)		−89		
Specific gravity 25/4°C	0.969			
Boiling point (°C) (760 mmHg)	145.8	158–160	209	258
Vapor pressure (mmHg) (25°C)	1.8	2.03 (20°C)		
Refractive index (25°C)	1.400			
1 ppm ≈ mg/m³ at 25°C, 760 mmHg	5.40	6.09 (at 20°C)	7.77	10.15
1 mg/L ≈ ppm at 25°C, 760 mmHg	185		128.6	98.5

23.0 Ethylene Glycol Monoacetate

23.0.1 CAS Number: [542-59-6]

23.0.2 Synonyms: 1,2-Ethanediol monoacetate, glycol monoacetate, 2-hydroxyethyl acetate, 2 glycal-monoacetate

23.0.3 Trade Names: Cellosolve® Acetate, Solvent GC

23.0.4 Molecular weight: 104.11

23.0.5 Molecular formula: $C_4H_8O_3$

23.0.6 Molecular Structure:

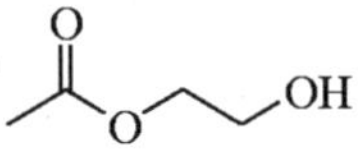

23.1 Chemical and Physical Properties

See Table 87.10.

23.4 Toxic Effects

The acute toxicity of ethylene glycol monacetate is low, but this compound may cause injury on repeated oral dosage or repeated inhalation exposure. Saturated vapors are irritating to the mucous membranes of animals and humans. However, the liquid applied to the skin of animals or humans is not appreciably irritating. The liquid is moderately irritating to the eyes. Wiley et al. (146) reported testicular degeneration in rabbits

GLYCOL ETHERS

given repeated doses of this glycol ester. This result was not confirmed in more recent studies (147).

23.4.1 Experimental Studies

23.4.1.1 Acute Toxicity

ORAL. Smyth et al. (17) reported LD_{50} values for this glycol acetate of 8.25 g/kg for rats and 3.80 g/kg for guinea pigs when fed as a 50% aqueous solution. Truhaut et al. (147) reported the actue oral toxicity (LD_{50}) of this glycol acetate when administered as a solution in olive oil as follows: male rats, 3.9 g/kg; female rats, 2.9 g/kg. Dogs were apperently unaffected by 12 feedings of 0.1 or 0.5 mL/kg (46).

INHALATION. A single 8-h exposure to near saturated vapors of this glycol acetate (1900 ppm) resulted in no deaths or after effects in cats, guinea pigs, mice, or rabbits (46). Truhaut et al. (147) reported that rats and rabbits survived a single, 4-h exposure to saturated vapors (2000 ppm). In this study, transient hemoglobinuria and/or hematuria was seen only in rabbits. No gross pathological lesions were noted in animals sacrificed following a 2-week recovery period. Studies by Rosser (148), quoted by Gross et al. (46), indicate that cats tolerated a single, 6-h exposure to an atmosphere containing mist (28 mg/L) but succumbed from two such exposures.

Twelve 8-h expsures to an atmosphere essentially sasturated with vapor at room temperature were survived by cats, guinea pigs, and mice, but caused lung irritation and slight kidney injury. One rabbit treated similarly died (46).

DERMAL. Truhaut et al. (147) reported a dermal LD_{50} of 10.5 g/kg for this glycol acetate in rabbits (modified Draize procedure, occluded, 24-h exposures). In this study, slight hematuria was seen along with a marked decrease in white blood cells.

INJECTION. The minimum lethal dose of ethylene glycol monoacetate administered by subcutaneous injection in cats was 4.5 g/kg (46). Guinea pigs showed no injury following seven subcutaneous injections of 0.1 or 0.5 mL of this material (46).

Wiley et al. (146) gave seven daily injections (route unspecified) of ethylene glycol monoacetate to dogs (8.5 mL/day) and rabbits (3.5 mL/day) and observed an increase in urinary oxalic acid similar to that caused by administration of an equivalent amount of ethylene glycol. Mild nephrosis was observed for both species. Mild to moderate degenerative lesions of the spermatogenic epithelium of the testes was reported for three of the four rabbits examined. Localized lymphoid infiltrations were noted in the meningeal membranes and meningeal, intracerebral invaginations in the treated dogs. All other organ systems were normal.

EYE IRRITATION. Ethylene glycol monoacetate is moderately irritating to the eyes of animals or humans (147).

SKIN IRRITATION. Ethylene glycol monoacetate is essentially nonirritating when applied to the skin of rabbits (147).

23.4.1.2 Subchronic Toxicity

INHALATION. Male and female rats and rabbits exposed for 10 months (4 h/day, 5 days/week) to 200 ppm ethylene glycol monoacetate displayed no signs of hematuria or ketonuria, nor were any other hematological changes or any gross pathological lesions noted (147).

23.4.1.3 Pharmacokinetics, Metabolism, and Mechanisms: NA

23.4.1.4 Reproductive and Developmental.
Gastric intubation of this glycol actetate at dose levels of 500, 1000, or 2000 mg/kg, 5 days/week for 5 weeks had no effects on the testes of mice (150).

23.5 Standards, Regulations, or Guidelines of Exposure

No occupational exposure standard has been established for ethylene glycol monoacetate.

24.0 Ethylene Glycol Diacetate

24.0.1 CAS Number: [111-55-7]

24.0.2 Synonyms: 1,2-Ethandiol diacetate, 1,2-diacetoxyethane, ethylene diacetate, ethylene glycol acetate, ethylene acetate, glycol diacetate, ethanediol diacetate

24.0.3 Trade names: NA

24.0.4 Molecular Weight: 146.14

24.0.5 Molecular: $C_6H_{10}O_4$

24.0.6 Molecular Structure:

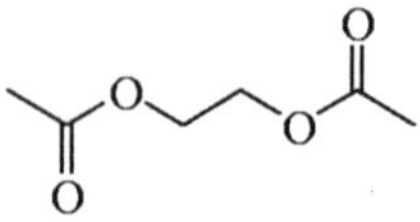

24.1 Chemical and Physical Properties

Ethylene glycol diacetate is a colorless, slightly volatile liquid with an odor similar to that of ethyl acetate and a bitter taste. See Table 87.10 for other chemical and physical properties.

24.2 Production and Use: NA

24.3 Exposure Assessment: NA

24.4 Toxic Effects

This glycol diacetate is of low acute toxicity when administered orally or intravenously. Repeated oral administration is injurious, with the deposition of calcium oxalate crystals in the kidneys.

24.4.1 Experimental Studies

24.4.1.1 Acute Toxicity

ORAL. The oral LD_{50} for ethylene glycol diacetate was 6.86 g/kg for rats and 4.94 g/kg for guinea pigs when administered as a 50% aqueous solution (17). Oral administration of large doses of this diacetate did not cause hydropic degeneration of kidney tubules in the animals studied (151).

The minimum repeated dose necessary to produce kidney damage in rats was 6 g/kg, administered at 5% in the drinking water for 7 days (109, 152). Long-term administration of 1–3% in the drinking water resulted in the deposition of calcium oxalate crystals in the kidneys of rats (152). At a concentration of 5%, crystal deposition was most rapid and resulted in death of the animal.

INJECTION. Intravenous injections (sublethal) in rabbits results in the deposition of calcium oxalate crystals in the kidneys at dose levels in excess of 2.5 mL/kg (152). In rats, acutely toxic doses had no apparent effect on the kidney and resulted in no calcium oxalate deposition (152). Intravenous administration of large doses of ethylene glycol diacetate did not cause hydropic degeneration of kidney tubules in animals (151).

24.4.1.2 Chronic and Subchronic Toxicity.
Mulinos et al. (152) reported growth retardation and a 50% mortality rate in young rats treated with 1% ethylene glycol diacetate in the drinking water for 11 weeks. Calcium oxalate fed to rabbits at 2% in the drinking water was of low toxicity, suggesting that calcium oxalate formation alone is not entirely responsible for the observed toxicity of this glycol diacetate.

24.4.1.3 Reproductive and Developmental Studies.
Oral administration of ethylene glycol diacetate by gastric intubation at dose levels of 500, 1000, or 2000 mg/kg 5 days/week for 5 weeks resulted in no significant decreases in testes weights in mice (150).

24.5 Standards, Regulations, or Guidelines of Exposure

No occupational exposure standard has been established for ethylene glycol diacetate.

25.0 Ethylene Glycol Diacrylate

25.0.1 CAS Number: [2274-11-5]

25.0.2 Synonyms:
2-Propenoic acid, 1,2-ethanediyl ester; 1,2-ethanediyl diacrylate; ethylene acrylate; 1,2-ethylene diacrylate; 1,2-diacrylyloxy ethane; ethylene diacrylate

25.0.3 Trade Names: NA

25.0.4 Molecular Weight: 170.16

25.0.5 Molecular Formula: $C_8H_{10}O_4$

25.0.6 Molecular Structure:

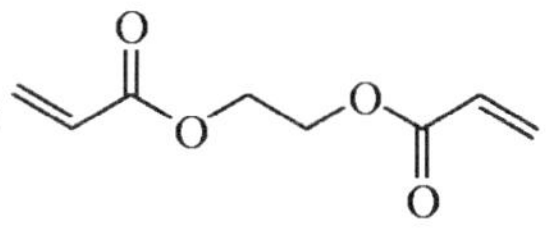

25.1 Chemical and Physical Properties

1 ppm ~ 6.95 mg/m^3 at 760 mmHg, 25°C
1 mg/L ~ 143.7 ppm at 760 mmHg, 25°

25.2 Production and Use: NA

25.3 Exposure Assessment: NA

25.4 Toxic Effects

25.4.1 Experimental Studies

25.4.1.1 Acute Toxicity. Ethylene glycol diacrylate has a low single-dose oral toxicity, the LD$_{50}$ for rats is 1.07 mL/kg. This material may be absorbed through the skin in toxic amounts, the LD$_{50}$ for rabbits is 0.57 mL/kg (63). Presumably due to the low volatility of this glycol ester, rats were not seriously affected by a single 8-h exposure to an essentially saturated atmosphere.

25.4.1.2 Chronic and Subchronic Toxicity: NA

25.4.1.3 Pharmacokinetics, Metabolism, and Mechanisms: NA

25.4.1.4 Reproductive and Developmental: NA

25.4.1.5 Carcinogenesis: NA

25.4.1.6 Genetic and Related Cellular Effects Studies. Waegemaekers and Bensink (153) reported that ethylene glycol diacrylate was negative in the standard Ames assay with TA1535, TA1537, TA1538, TA98, and TA100, either with or without the addition of Aroclor 1254 or phenobarbital-induced S9. However, this material was positive in the mouse lymphoma assay (L5178Y. TK+/TK-) either with or without metabolic activation (154). Other multifunctional acrylates respond similarly in the mouse lymphoma assay (155).

25.4.1.7 Other: Neurological, Pulmonary, Skin Sensitization. This material is capable of causing severe eye irritation and serious corneal damage, as indicated by testing in rabbits (grade 8 on a scale of 10). The material is slightly irritating to the skin of rabbits (grade 3 on a scale of 10). Van der Walle et al. (156) reported this material to be a positive skin sensitizing agent in the Freund's complete adjuvant test but nonsensitizing in the guinea pig maximization test.

25.5 Standards, Regulations, or Guidelines of Exposure

No occupational exposure standard has been established for ethylene glycol diacrylate.

26.0 Diethylene Glycol Diacrylate

26.0.1 CAS Number: [4074-88-8]

26.0.2 Synonyms: 2-Propenoic acid, oxydi-2,1-ethanediyl; DEGDA; acrylic acid, 2-ethoxyethanol diester; oxydiethylene acrylate; oxydiethylene diacrylate; TGA 2; SR 230; diethylene glycol diacrylate

26.0.3 Trade Names: NA

26.0.4 Molecular Weight: 214.22

26.0.5 Molecular Formula: $C_{10}H_{14}O_5$

26.0.6 Molecular Structure:

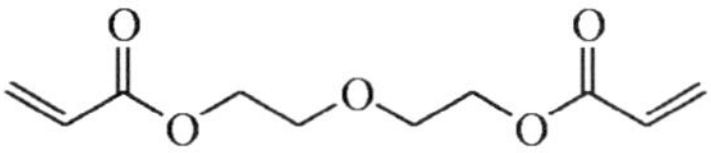

26.1 Chemical and Physical Properties

26.1.1 General Molecular formula:

1 ppm ~ 8.76 mg/m^3 at 760 mmHg, 25°C
1 mg/L ~ 114.2 ppm at 760 mmHg, 25°C

26.2 Production and Use: NA

26.3 Exposure Assessment: NA

26.4 Toxic Effects

26.4.1 Experimental Studies

26.4.1.1 Acute Toxicity. Diethylene glycol diacrylate has a low single-dose oral toxicity; the LD$_{50}$ for rats is 0.77 mL/kg. Diethylene glycol diacrylate is very toxic by skin absorption; the LD$_{50}$ for rabbits is 0.18 mL/kg. Presumably because of the low volatility of this glycol ester, rats were not seriously affected by a single 8-h exposure to an essentially saturated atmosphere.

It is severely injurious to the eyes of rabbits, causing severe corneal damage (rated 9 on a scale of 10). It is markedly irritating to rabbit skin, rated 5 on a scale of 10 (63). Andrews and Clary (155) reported this glycol ether to be corrosive (category I) in a skin irritation study.

26.4.1.2 Chronic and Subchronic Toxicity: NA

26.4.1.3 Pharmacokinetics, Metabolism, and Mechanisms: NA

26.4.1.4 Reproductive and Developmental: NA

26.4.1.5 Carcinogenesis: NA

26.4.1.6 Genetic and Related Cellular Effects Studies. Waegemaekers and Bensink (153) reported that diethylene glycol diacrylate was negative in the standard Ames assay with TA1535, TA1537, TA1538, TA98, and TA100, either with or without the addition of Aroclor 1254 or phenobarbital-induced S9.

26.4.1.7 Other: Neurological, Pulmonary, Skin Sensitization

SKIN SENSITIZATION. Van der Walle et al. (156) reported negative results in both the Freund's complete adjuvant and guinea maximization tests for skin sensitization. Bjorkner et al. (157) found it to be weakly positive (2 of 15 animals responding) in the guinea pig maximization test as well as producing a similar weak cross-reactivity to subsequent exposures to 1,4-butanediol diacrylate or tetraethylene glycol diacrylate.

26.4.2 Human Experience

26.4.2.1 General Information: NA

26.4.2.2 Clinical Cases

26.4.2.2.1 Acute Toxicity: NA

26.4.2.2.2 Chronic and Subchronic Toxicity: NA

26.4.2.2.3 Pharmacokinetics, Metabolism, and Mechanisms: NA

26.4.2.2.4 Reproductive and Developmental: NA

26.4.2.2.5 Carcinogenesis: NA

26.4.2.2.6 Genetic and Related Cellular Effects Studies: NA

26.4.2.2.7 Other: Neurological, Pulmonary, Skin Sensitization, etc. In human patch testing, this acrylate was positive when applied as a 0.1% solution (158).

27.0 Triethylene Glycol Diacetate

27.0.1 CAS Number: [111-21-7]

27.0.2 Synonyms: 2,2'-(1,2-Ethanediylbis(oxy))bisethanol diacetate; triglycol diacetate; ethanol, 2,2'-ethylenedioxydi-,diacetate; 2,2'-(ethylenedioxy)di(ethyl acetate); TDAC; 1,2-bis(acetoxyethoxy)ethane; 3,6-dioxaoctane-1,8-diyl diacetate

27.0.3 Trade Names: NA

27.0.4 Molecular Weight: 234.25

27.0.5 Molecular Formula: $C_{10}H_{18}O_6$

27.0.6 Molecular Structure:

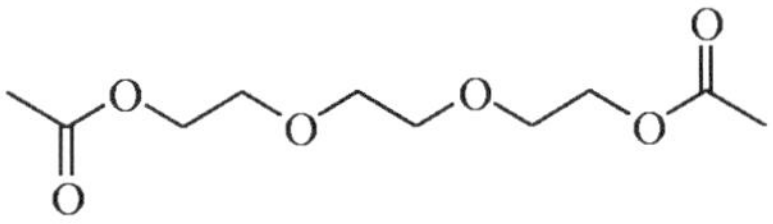

27.1 Chemical and Physical Properties

27.1.1 General

See Table 87.10.

27.4 Toxic Effects

27.4.1 Experimental Studies

27.4.1.1 Acute Toxicity. Triethylene glycol diacetate (TEGDA) is a colorless liquid with a low degree of volatility. It has a low order of single-dose oral toxicity; LD_{50} values for rats were 22.6 mL/kg (18) and 13.8 mL/kg (159). In mice, the oral LD_{50} was >3.2 g/kg and in rabbits, 3.5–7.0 g/kg. The inhalation LC_{50} in rats was >6.11 mg/L for a 6-h exposure (159). This material also has a low order of single-dose toxicity when injected intraperitoneally; the LD_{50} for rats was 1.41 mL/kg (18). When held in occluded contact with guinea pig skin for a 24-h period, there was only slight evidence of skin irritation and no evidence of toxicity caused by dermal absorption (159). The LD_{50} for rabbits by skin absorption was 8.0 mL/kg (18). In the guinea pig, the dermal LD_{50} was in excess of 20 mL/kg (159). A single, 8-h exposure of rats to essentially saturated vapors generated at approximately 20°C resulted in no significant adverse effects (18).

27.4.1.2 Chronic and Subchronic Toxicity: NA

27.4.1.3 Pharmacokinetics, Metabolism, and Mechanisms: NA

27.4.1.4 Reproductive and Developmental Studies. TEGDA was not a reproductive toxicant when evaluated in Swiss CD-1 mice in a continuous breeding assay when tested in drinking water up to 3% (160). In a retrospective study, ovaries from mice exposed to TEGDA in an NTP continuous breeding protocol (33) were reanalyzed histopathologically (34). On the basis of changes in differential follicle counts in exposed and unexposed mice, TEGDA did not induce ovarian toxicity.

28.0 Triethylene Glycol Divalerate

28.0.1 CAS Number: No CAS # Found

28.0.2 Synonyms: Pentanoic acid, 1,2-diethanylbis(oxy2,1-ethane) diyl ester; valeric acid, triethylene glycol diester

28.0.3 Trade Names: NA

28.0.4 Molecular Weight: 318.4

28.0.5 Molecular Formula: $C_{16}H_{30}O_6$

28.0.6 Molecular Structure: $(CH_2OCH_2CH_2OOC(CH_2)_3CH_3)_2$

28.1 Chemical and Physical Properties

28.1.1 General

1 ppm ~ 13.05 mg/m^3 at 760 mmHg, 25°C
1 mg/L ~ 0.999 ppm at 760 mmHg, 25°C

28.4 Toxic Effects

28.4.1 Experimental Studies

28.4.1.1 Acute Toxicity. Triethylene glycol divalerate is a liquid with a low single-dose oral toxicity; the LD$_{50}$ for rats is 14.3 mL/kg. It is essentially nonirritating to the eyes of rabbits, graded 1 on a scale of 10, and only mildly irritating to rabbit skin with a grade of 2 on a scale of 10. Rabbit skin absorption tests indicate that it is very low in toxicity by absorption; the LD$_{50}$ was > 16 mL/kg. A single, 8-h exposure of rats to an essentially saturated atmosphere of the vapor resulted in no deaths, presumably due to the low vapor pressure of this material (63).

29.0 Propylene Glycol Monoacrylate

29.0.1 CAS Numbers: *[999-61-1]* (1-acrylate), *[2918-23-2]* (2-acrylate)

29.0.2 Synonyms: 2-Hydroxy-1-propyl acrylate; 1,2-propanediol, 1(or 2)-acrylate; acrylic acid, hydroxypropyl ester

29.0.3 Trade Names: NA

29.0.4 Molecular Weight: 130.14

29.0.5 Molecular Formula: $C_6H_{10}O_3$

29.0.6 Molecular Structures: 1-acrylate,

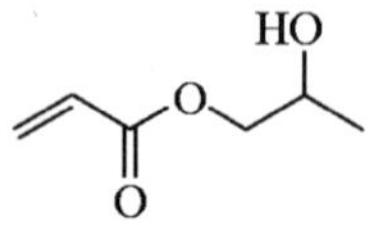

2-acrylate,

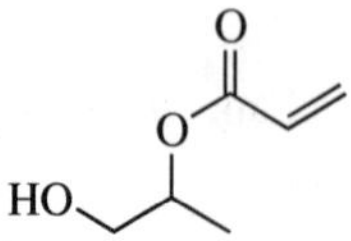

29.1 Chemical and Physical Properties

29.1.1 General

1 ppm ~ 5.32 mg/m^3 at 760 mmHg, 25°C
1 mg/L ~ 188.1 ppm at 760 mmHg, 25°C

29.1.2 Odor and Warning Properties

Propylene glycol monoacrylate is a clear to light yellow liquid with a sweetish solvent odor and is only slightly volatile. The vapor of propylene glycol monoacrylate causes eye, nasal, and respiratory irritation.

29.2 Production and Use: NA

29.3 Exposure Assessment: NA

29.4 Toxic Effects

Propylene glycol monoacrylate is moderate to low in single-dose oral toxicity. The liquid is markedly painful and injurious to the eyes, resulting in possible impairment of vision. Prolonged or repeated skin contact can cause marked irritation, even a burn. However, it is not considered to be corrosive by the Department of Transportation (DOT) test. It may be absorbed through the skin in acutely toxic amounts and should be considered a possible weak skin sensitizer.

29.4.1 Experimental Studies

29.4.1.1 Acute Toxicity

ORAL. Propylene glycol monoacrylate has an oral LD$_{50}$ in rats of 0.25–0.50 g/kg (K. J. Olson, The Dow Chemical Company, unpublished data, 1964), 0.59 g/kg (161), and 1.3 g/kg (18). The rats died overnight, and gross autopsy revealed that there were no significant pathologic effects (K. J. Olson, The Dow Chemical Company, unpublished data, 1964). Tanii and Hashimoto (162) reported the oral LD$_{50}$ in male mice to be 8.1 mmol/kg (1.06 g/kg).

INHALATION. Rats exposed for 7 or 8 h to essentially saturated vapors of propylene glycol monoacrylate generated at room temperature survived with no significant adverse effects (18). All rats exposed for 4.25 h to a vapor generated at 100°C died either during the exposure or within 1 h after termination. During the exposure, marked signs of nasal, respiratory, and eye irritation were observed (K. J. Olson, The Dow Chemical Company, unpublihsed data, 1964). Rats exposed for 1 h to a vapor generated at 100°C survived, but gross observation revealed some slight kidney and liver injury. The concentration of the vapor generated at 100°C was estimated to be in the range of 650 ppm.

Dogs, rats, rabbits, and mice were exposed to 5 ppm (the lowest concentration tested) of propylene glycol monoacrylate vapor for 6 h/day for 5 days/week for a total of 20 or 21 exposures in a month. During the exposure period, both dogs and rabbits exhibited signs of nasal and respiratory irritation. The rats, however, showed no visual signs, but histological evidence of the irritation was observed. This was seen in the dogs as well as in rabbits. Both the dogs and rabbits showed signs of eye irritation. There were no treatment-related effects in any of the animals as shown by organ to body weight ratios, hematology, clinical chemistry, and urinalysis. However, microscopic testicular changes were seen in three of the four treated dogs. These changes were not observed in the other animals. The pathological findings indicate that the changes may have been due to the respiratory system effects observed or perhaps to the sexual immaturity of the dogs (K. J. Olson, The Dow Chemical Company, unpublished data, 1964). According to these findings, the no-effect level for repeated inhalation exposure to animals must be less than 5 ppm.

DERMAL. Propylene glycol monoacrylate has been shown to be highly toxic in rabbit skin absorption tests; the LD_{50} was 0.17 g/kg (18) and in the range of 0.25 g/kg (K. J. Olson, The Dow Chemical Company, unpublished data, 1964). Death occurred overnight. Surviving animals developed severe skin irritation, moderate edema, and moderate to severe necrosis (K. J. Olson, The Dow Chemical Company, unpublished data, 1964).

EYE IRRITATION. Propylene glycol monoacrylate is markedly injurious to the eye. The application of approximately 0.1 mL to the eyes of rabbits caused marked to severe irritation and corneal injury sufficient to cause impairment of vision (18). In one report, the impairment was of such a nature that it could possibly be permanent (K. J. Olson, The Dow Chemical Company, unpublished data, 1964).

SKIN IRRITATION. A single, 24-h contact has been reported to cause marked irritation and some damage to the skin of rabbits (18). Even a 24-h exposure under a cloth covering to a 10% water solution has been observed to cause a moderate burn resulting in a scar (K. J. Olson, The Dow Chemical Company, unpublished data, 1964). However, the same dose, when applied 9 times to the uncovered skin over a 2-week period, caused no significant irritation. The application of the undiluted material to the covered skin of rabbits resulted in death within 24 h. Propylene glycol monoacrylate was tested for corrosivity as directed by the DOT and was found to be noncorrosive (K. J. Olson, The Dow Chemical Company, unpublished data, 1964).

29.4.1.2 Chronic and Subchronic Toxicity: NA

29.4.1.3 Pharmacokinetics, Metabolism, and Mechanisms: NA

29.4.1.4 Reproductive and Developmental: NA

29.4.1.5 Carcinogenesis: NA

29.4.1.6 Genetic and Related Cellular Effects Studies.
Propylene glycol monoacrylate was tested for mutagenicity in a collaborative study using *Salmonella typhimurium* strains TA102 and TA2638 (with metabolic activation) and *Escherichia coli* strains WP2/pKM101 and WP2 uvrA/pKM101 (163). It was found positive only in the *E. coli* strains.

29.4.1.7 Other: Neurological, Pulmonary, Skin Sensitization

SKIN SENSITIZATION. Male and female guinea pigs were administered six applications in 3 weeks of a 0.5% solution of this material in a solvent made of Tween 80 (1 part) and dipropylene glycol methyl ether (9 parts). A challenge dose was given after a 2-week rest period. Of the 10 guinea pigs, 4 displayed weak sensitization reactions. More recently, Rao et al. (164) reported that no sensitization occurred in guinea pigs using the contact sensitization method of Maguire (165). In contrast to these findings, Clemmensen (166) reported that 2-hydroxypropyl acrylate caused sensitization in 12 out of 12 guinea pigs using the guinea pig maximization test. In addition, Clemmensen also reported that this acrylate derivative exhibited cross-sensitization with other hydroxyalkyl acrylate esters.

29.4.2 Human Experience

29.4.2.1 General Information: NA

29.4.2.2 Clinical Cases

29.4.2.2.1 *Acute Toxicity:* NA

29.4.2.2.2 *Chronic and Subchronic Toxicity:* NA

29.4.2.2.3 *Pharmacokinetics, Metabolism, and Mechanisms:* NA

29.4.2.2.4 *Reproductive and Developmental:* NA

29.4.2.2.5 *Carcinogenesis:* NA

29.4.2.2.6 *Genetic and Related Cellular Effects Studies:* NA

29.4.2.2.7 *Other Neurological, Pulmonary, Skin Sensitization, etc.* In human patch testing, this acrylate was positive when applied as a 0.1% solution (158).

29.5 Standards, Regulations, or Guidelines of Exposure

Table 87.12 summarizes recommended or regulatory guidelines of occupational exposure for propylene glycol monoacrylate (CAS *[999-61-1]*).

Table 87.12. Occupational Exposure Guidelines for Propylene Glycol Monoacrylate[a]

Regulatory or Expert Authority (Date)	Exposure Value	Comment/Notation
ACGIH (TLV) (1997)	TWA = 0.5 ppm	Skin
NIOSH (REL) (1994)	TWA = 0.5 ppm	Skin
Australia (OEL) (1993)	TWA = 0.5 ppm	Skin
Belgium (OEL) (1993)	TWA = 0.5 ppm	Skin
Denmark (OEL) (1993)	TWA = 0.5 ppm	Skin
France (OEL) (1993)	TWA = 0.5 ppm	Skin
Netherlands (OEL) (Oct-1997)	TWA = 0.5 ppm	Skin
Russia (OEL) (1993)	STEL = 1 mg/m^3	Skin
Switzerland (OEL) (1993)	TWA = 0.5 ppm	Skin
United Kingdom (OEL) (1993)	TWA = 0.5 ppm	Skin

[a] The OEL values set by Bulgaria, Colombia, Jordan, Korea, New Zealand, Singapore, and Vietnam are based on the ACGIH value.

30.0 1,3-Butanediol Diacrylate

30.0.1 **CAS Number:** [19485-03-1]

30.0.2 **Synonyms:** 2-Propenoic acid, 1-methyl-1,3-propanediyl ester; 1,3-butylene glycol diacrylate; acrylic acid, 1,3-butylene glycol diester; 1,3-butanediol diacrylate

30.0.3 **Trade Names:** NA

30.0.4 **Molecular Weight:** 198.22

30.0.5 **Molecular Formula:** $C_{10}H_{14}O_4$

30.0.6 **Molecular Structure:**

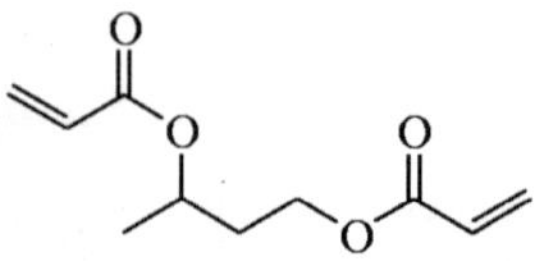

30.1 Chemical and Physical Properties

30.1.1 General

1 ppm ~ 8.10 mg/m^3 at 760 mmHg, 25°C
1 mg/L ~ 123.4 ppm at 760 mmHg, 25°C

30.2 Production and Use: NA

30.3 Exposure Assessment: NA

30.4 Toxic Effects

30.4.1 Experimental Studies

30.4.1.1 Acute Toxicity. 1,3-Butanediol diacrylate is low in single-dose oral toxicity; the LD$_{50}$ for rats is 3.54 mL/kg. It is severely irritating and damaging to the eye, rated 9 on a scale of 1–10. It is markedly irritating to the skin and may cause a burn on prolonged or repeated contact. It has moderate toxicity by skin absorption; the LD$_{50}$ for rabbits is 0.45 mL/kg. Because of its low volatility, it is not likely to be a problem due to inhalation following a single or occasional exposure. Rats exposed to essentially saturated vapors for 8 h survived (63). However, it would seem wise to avoid inhalation of the vapors of this material.

31.0 Butylene Glycol Adipic Acid Polyester

31.0.1 CAS Number: *[9080-04-0]*

31.0.2 Synonyms: Hexanedioic Acid, Polymer with butanediol

31.0.3 Trade Names: Santicizer® 344F

31.0.4 Molecular Weight: NA

31.0.5 Molecular Formula: NA

31.0.6 Molecular Structure:

$$[RCO-]_2[-O-CH(CH_3)-CH_2CH_2-O]_x[-CO-(CH_2)_4-CO-]_{x-1}$$

where R = C$_{13}$H$_{27}$, C$_{15}$H$_{31}$, C$_{17}$H$_{35}$
  x = Molecular weight in a range of 1700–2200

31.2 Production and Use

Santicizer® 334F [butylene glycol adipic polyester (BGAP)] is a polymer of alternating units of 1,3-butylene glycol and adipic acid; it is a plasticizer for polyvinyl chloride (PVC) used in food contact applications (167).

31.4 Toxic Effects

31.4.1 Experimental Studies

31.4.1.1 Acute Toxicity: NA

31.4.1.2 Chronic and Subchronic Toxicity. Rats and dogs were fed diets containing 1000, 5000, or 10,000 ppm BGAP for 2 years. No adverse effects were reported at any dose level (167).

31.4.1.3 Pharmacokinetics, Metabolism, and Mechanisms. The metabolism of BGAP radiolabeled with [^{14}C]adipic acid was studied in rats after oral and intravenous administration; in both cases, 50% was recovered as exhaled CO_2, 29% in the urine, and 11% in the feces. These data indicate that BGAP was readily absorbed, hydrolyzed, and further metabolized (168). Thus, the lack of systemic toxicity was presumably not due to lack of gastrointestinal absorption.

31.4.1.4 Reproductive and Developmental. No adverse effects were reported in rats in a three-generation reproduction study at dietary levels of 1000, 5000, or 10,000 ppm BGAP (167).

32.0 2-Methyl-2-propene-1,1-diol Diacetate

32.0.1 CAS Number: [10476-95-6]

32.0.2 Synonyms: Methacrolein diacetate, methyl allylidene diacetate

32.0.3 Trade Names: NA

32.0.4 Molecular Weight: 172.18

32.0.5 Molecular Formula: $C_8H_{12}O_4$

32.0.6 Molecular Structure:

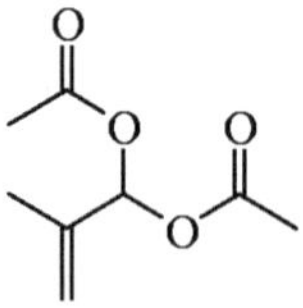

32.1 Chemical and Physical Properties

See Table 87.10.

32.4 Toxic Effects

32.4.1 Experimental Studies

32.4.1.1 Acute Toxicity. Methacrolein diacetate is an acutely toxic material. The oral LD_{50} in rats was 440 mg/kg, but the dermal LD_{50} in rabbits was 44 mg/kg, indicating that it can be readily absorbed dermally in toxic amounts (18). It was serverely irritating to both the skin and eye, causing chemical burns. When exposed by inhalation to a saturated vapor for 1 h, all rats died, and five of six rats died when exposed to 62.5 ppm for 4 h.

33.0 2,2-Dimethyl-1,3-propanediol Diacrylate

33.0.1 CAS Number: [2223-82-7]

33.0.2 Synonyms: 2-Propenoic acid, 2,2-dimethyl-1,3-propanediyl ester; 2,2-dimethylpropane diacrylate; 2,2-dimethylpropanediol diacrylate; neopentyl glycol diacrylate; neopentanediol dicacrylate; NPGDA

33.0.3 Trade Names: NA

33.0.4 Molecular Weight: 212.25

33.0.5 Molecular Formula: $C_{11}H_{16}O_4$

33.0.6 Molecular Structure:

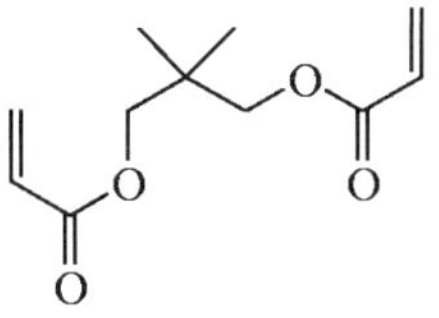

33.1 Chemical and Physical Properties

33.1.1 General

See Table 87.10.

33.4 Toxic Effects

33.4.1 Experimental Studies

33.4.1.1 Acute Toxicity. 2,2-Dimethyl-1,3-propanediol diacrylate is low in single-dose oral toxicity; the LD_{50} value for rats is 6.73 mL/kg. It is moderately irritating to the eyes and may cause some corneal injury; it is rated 4 on a scale of 0–10 in rabbit eye tests. It is also moderately injurious to the skin, rated 5 on a scale of 0–10 in rabbit skin tests. The dermal LD_{50} for rabbits was 0.40 mL/kg. However, it should be low in hazard from an occasional exposure to vapors; rats exposed for 8 h to an essentially saturated vapor did not die (63). No data involving repeated exposure are available.

33.4.1.2 Chronic and Subchronic Toxicity: NA

33.4.1.3 Pharmacokinetics, Metabolism, and Mechanisms: NA

33.4.1.4 Reproductive and Developmental: NA

33.4.1.5 Carcinogenesis. Prolonged dermal applications of NPGDA to mice gave evidence for local carcinogenicity (BIBRA, 1996).

33.4.1.6 Genetic and Related Cellular Effects Studies. NPGDA was negative as a mutagen in *Salmonella typhimurium* strains TA1535, TA1537, TA98, TA100, and TA1538 either with or without metabolic activation (153). In mammalian cell culture, NPGDA induced chromosomal damage but was not mutagenic (BIBRA, 1996).

33.4.1.7 Other: Neurological, Pulmonary, Skin Sensitization. NPGDA has been reported to be a severe skin and eye irritant in the rabbit and to have demonstrated strong sensitizing potential in guinea pigs (BIBRA, 1996).

34.0   2-Methyl-2,4-pentanediol Diacetate

34.0.1   CAS Number: [1637-24-7]

34.0.2   Synonyms: 2-Methyl-2,4-pentanediol acetic acid diester, hexylene glycol diacetate

34.0.3   Trade Names: NA

34.0.4   Molecular Weight: 202.2

34.0.5   Molecular Formula: $C_{10}H_{18}O_4$

34.0.6   Molecular Structure: $(CH_3)_2C(OOCCH_3)CH_2CH(OOCCH_3)CH_3$

34.1   Chemical and Physical Properties

34.1.1   General

See Table 87.10.

34.4   Toxic Effects

34.4.1   Experimental Studies

34.4.1.1 Acute Toxicity. 2-Methyl-2,3-pentanediol acetic acid diester has low acute toxicity; the oral LD_{50} in rats was 3360 mg/kg, and the dermal LD_{50} in rabbits was 16 g/kg, indicating that it is not absorbed significantly on dermal exposure (18). It was mildly irritating to the skin and eyes. No rats died when exposed to an essentially saturated vapor for 8 h (18).

35.0   2,2,4-Trimethyl-1,3-pentanediol Monoisobutyrate

35.0.1   CAS Numbers: [77-68-9] (3-hydroxy-2,2-dimethyl-1-(1-methylethyl)propyl 2-methylpropanoate), [77-68-9] (3-hydroxy-2,2,4-triethylpentyl 2-methylpropanoate), [25265-77-4] (mixture)

35.0.2 Synonyms:
2-Methylpropanoic acid monoester with 2,2,4-trimethylpentane-1,3-diol; 2-(2,4-trimethy-3-hydroxypentyl) isobutyrate; propanoic acid, 2-methyl-3-hydroxy-2,2,4-trimethylpenty ester

35.0.3 Trade Names: Texanol™ Ester-Alcohol [CAS *[25265-77-4]* (mixture)]

35.0.4 Molecular Weight: 216.32

35.0.5 Molecular Formula: $C_{12}H_{24}O_3$

35.0.6 Molecular Structure:

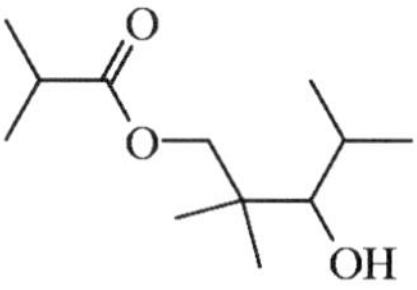

35.1 Chemical and Physical Properties

35.1.1 General
See Table 87.10.

35.2 Production and Use

Estimates of yearly production volumes of Texanol™ range from 25,000 to 50,000 tons/year (1977 TSCA inventory). It is manufactured in a continuous system by the self-condensation of isobutyraldehyde in the presence of base catalyst. Final purification is by distillation. More than 84% of this material is used in the formulation of latex paints.

35.3 Exposure Assessment

35.3.1 Air: NA

35.3.2 Background Methods: NA

35.3.3 Workplace Methods

Exposure to Texanol™ occurs primarily through its use in latex paints. In a report prepared for the National Paint and Coatings Association (169), concentrations of Texanol™ in breathing-zone samples were measured in rooms in which paints were applied using either airless spraying or roller/brush methods. The maximum concentration in breathing zones during spray applications was 0.99 ppm and during roller applications was 1.96 ppm. Within 24 h of application, Texanol™ was below limits of detection in 19 of 24 rooms treated. Ulfvarson et al. (170) reported Texanol™ at concentrations of 0.74–5.2 ppm in air following the application of a number of water-based paints. In this latter study, volatile solvents were measured in air samples by collection on activated carbon or

Amberlite® XAD-7 followed by carbon disulfide desorption and analysis by gas chromatography.

35.4 Toxic Effects

Texanol™ is of low acute toxicity by all routes of exposure. It is slightly irritating to the eyes of rabbits, but the effect is reduced with washing. The major component of this mixture is rapidly hydrolyzed in blood. The results of mutagenicity testing for this material were negative. Exposure of rats orally to levels of ≤1000 mg/kg per day resulted in no significant reproductive or developmental effects. On the basis of limited animal testing, ordinary handling and use of this glycol derivative should not present a health hazard.

35.4.1 Experimental Studies

35.4.1.1 Acute Toxicity

ORAL. The oral LD_{50} for this material in rats is 6.86 mL/kg (63). The oral LD_{50} has also been reported to be >3.2 g/kg for both male and female rats (171).

Daily administration of 100 or 1000 mg/kg of this material to groups of male or female rats (5 days/week, 11 doses over a 15-day period) resulted in slight increases in liver weights for both sexes at the highest dose level. However, liver histopathology and serum enzymes were normal (171). There was slight hyaline droplet formation in male rat kidneys. The importance of this observation is questionable since no similar hyaline droplet formation has been identified in other test species under comparable conditions.

INHALATION. A 6-h exposure of rats to a nominal concentration of 2.75 mg/L resulted in no mortality (63).

DERMAL. Acute dermal exposure of rabbits to 16 mL/kg (63) or guinea pigs to 20 mL/kg (171) resulted in no mortality.

EYE IRRITATION. This material produces slight to moderate eye irritation in rabbits; however, immediate washing of the eyes is palliative (171).

SKIN IRRITATION. Occluded application of this material to the depilated abdomens of guinea pigs for 24 h resulted in only slight skin irritation.

35.4.1.2 Chronic and Subchronic Toxicity: NA

35.4.1.3 Pharmacokinetics, Mechanisms, and Metabolism: NA

35.4.1.4 Reproductive and Devlopmental.
Female rats were administrated 0, 100, 300, or 1000 mg/kg per day of this glycol ester by gavage during the following periods: premating (14 days), mating (≤14 days), pregnancy (21–22 days), and early lactation (4 days). All male rats received 51 doses of the test material over 51 days. Absolute and

relative kidney weights were significantly higher in high dose males and absolute, and relative liver weights were significantly increased in the low, mid, and high dose male and female dose groups. Hyaline droplet formation was observed in the mid and high dose males. Centrilobular hepatocytomegaly (enlargement of hepatocytes surrounding the central vein) was observed in male and female rats in the mid and high dose groups. The liver changes noted suggest increased metabolic activity as a result of test article administration. Reproductive performance was not affected as measured by mean number of live or dead pups per litter, total implants, prenatal loss, percent survival, total litter weights, mean pup weight, pup survival, and postnatal growth. No other significant treatment-related effects were noted (172).

35.4.1.5 Carcinogenesis: NA

35.4.1.6 Genetic and Related Cellular Effects Studies

GENETIC TOXICITY. This material was negative in an Ames test using *Salmonella typhimurium* strains TA98, TA100, TA1535, TA1537, and TA1538, both with and without the addition of rat liver S9 fraction (173). In addition, it was negative in an *in vivo* mouse bone narrow micronucleus assay (174).

35.4.1.7 Other: Neurological, Pulmonary, Skin Sensitization

IN VITRO STABILITY IN BLOOD:. This glycol derivative is rapidly hydrolyzed in both rat and human blood; the half-lives for the major component of the mixture at 37°C were 19.6 and 17.3 min, respectively, for these species (175). The alcohol, 2,2,4-trimethyl-1,3-pentanediol, was the single identified product from this hydrolysis. The minor Texanol™ component (presumably the 3-substituted isomer) was resistant to hydrolysis under these conditions.

SKIN SENSITIZATION. Guinea pigs were not sensitized to this material following dermal application (171).

35.4.2 Human Experience

35.4.2.1 General Information

35.4.2.2 Clinical Cases

35.4.2.2.1 Acute Toxicity. Ulfvarson et al. (170) reported temporary health effects involving the lungs (decreased capacities) as well as increased urine volumes and increased volumes of erythrocytes among painters reporting feelings of "nuisance" from the use of water-based paints. Although Texanol™ was measured in inspired air samples, effects correlated more strongly with the presence of isothiazolinone (an antimicrobial). Because of the small study population, more definitive conclusions cannot be drawn from this study.

35.5 Standards, Regulations, or Guidelines of Exposure

No occupational exposure standard has been established for repeated vapor exposure to Texanol™.

36.0 2,2,4-Trimethyl-1,3-pentanediol Diisobutyrate

36.0.1 CAS Number: [6846-50-0]

36.0.2 Synonyms: Propanoic acid, 2-methyl-2,2-dimethyl-1-(1-methylethyl)-1,3-propanediyl ester

36.0.3 Trade Names: Eastman® TXIB™ Plasticizer

36.0.4 Molecular Weight: 286.41

36.0.5 Molecular Formula: $C_{16}H_{30}O_4$

36.0.6 Molecular Structure: $RCOOCH_2C(CH_3)_2CH(COOR)(R)$, where R = isopropyl

36.1 Chemical and Physical Properties

36.1.1 General

See Table 87.10.

36.2 Production and Use

TXIB™ is prepared in a single step from isobutyraldehyde. It is used as a viscosity control agent in various plastisol, rotomolding, and rotocasting operations, for which the major product is rolled sheet vinyl flooring. A minor use of TXIB™ is as a coalescing aid in latex paints.

36.3 Exposure Assessment

An analytic detection method has been developed (176). More recently, Norback et al. (177) reported that TXIB™ may be trapped from air with activated charcoal or other resins, desorbed with carbon disulfide or methylene chloride, and analyzed using gas chromatography and/or mass spectrometry.

36.4 Toxic Effects

TXIB™ has a low potential for acute or subchronic toxicity; no toxic effects were detected in animal studies. At very high oral doses, the only effect was an adaptive response of the liver.

36.4.1 Experimental Studies

36.4.1.1 Acute Toxicity

ORAL. The acute oral LD_{50} was reported to be greater than 3200 mg/kg in rats, and greater than 6400 mg/kg in mice, because no deaths occurred at these doses (178).

INHALATION. Inhalation exposure of rats to 5.3 mg/L for 6 h caused only vasodilation (178).

36.4.1.2 Chronic and Subchronic Toxicity

ORAL. Rats were fed diets containing 0.1 or 1% TXIB™ for 103 days; the only effect noted was a slight increase in liver weight (178). This effect occurred only at the 1% concentration, and was reversible when rats were returned to control food (179). This increased liver weight was associated with increased liver mixed function oxidase activity, and was considered to be an adaptive response of the liver, not a toxic effect (179).

Dogs administered up to 1% TXIB™ in the diet for 90 days showed no toxicologically significant effects (179).

36.4.1.3 Pharmacokinetics, Metabolism, and Mechanisms

METABOLISM AND TOXICOKINETICS. After oral administration to rats, more than half the dose was eliminated in the urine, indicating that the lack of systemic toxicity was not due to a lack of gastrointestinal absorption. The urine contained unchanged TXIB™ and its metabolites the O-glucuronide, the O-sulfate, and 2,2,4-trimethyl-3-hydroxyvaleric acid and its glucuronide, as the result of partial or complete hydrolysis of the diester bonds. The detection of the monoester in feces suggested that some hydrolysis occurred before absorption (178).

36.4.1.4 Reproductive and Developmental: NA

36.4.1.5 Carcinogenesis: NA

36.4.1.6 Genetic and Related Cellular Effects Studies: NA

36.4.1.7 Other: Neurological, Pulmonary, Skin Sensitization

SKIN IRRITATION AND SENSITIZATION. TXIB™ was slightly irritating when held under occlusion in contact with guinea pig skin, and was not a potent skin sensitizer (178).

36.4.2 Human Experience

36.4.2.1 General Information: NA

36.4.2.2 Clinical Cases

36.4.2.2.1 Acute Toxicity: NA

36.4.2.2.2 Chronic and Subchronic Toxicity.
Exposure to emissions from water-based paints has been proposed to contribute to reports of asthma and asthma-like symptoms. Indoor air concentrations of a number of water-based paint components, including TXIB™ were increased in recently painted houses (180). In these same dwellings, asthma

symptoms and other clinical changes were correlated with the use of water-based paints but the etiologic agent responsible was not identified.

36.4.2.2.3 Pharmacokinetics, Metabolism, and Mechanisms: NA

36.4.2.2.4 Reproductive and Developmental: NA

36.4.2.2.5 Carcinogenesis: NA

36.4.2.2.6 Genetic and Related Cellular Effects Studies: NA

36.4.2.2.7 Other: Neurological,Pulmonary, Skin Sensitization. A statistical correlation has been reported between air levels of TXIB™, emitted from water-based paints, and increased serum levels of a protein marker of inflammation or allergic reaction (181).

36.5 Standards, Regulations, or Guidelines of Exposure

No occupational exposure standards have been established for TXIB™.

37.0 Propylene Glycol Monomethyl Ether Acetate (a), Dipropylene Glycol Monomethyl Ether Acetate (b), Tripropylene Glycol Monomethyl Ether Acetate (c)

37.0.1 CAS Numbers: (a) *[108-65-6]*, (b) *[88917-22-0]*, (c) no CAS number found

37.0.2 Synonyms: 2-Methoxy-1-methylethyl acetate *[108-65-6]*, 1(or 2)-(2-methoxymethylethoxy) propanol acetate *[88917-22-0]*, methoxypropyl acetate (MPA)

37.0.3 Trade Names: Dowanol® PMA Glycol Ether Acetate *[108-65-6]*, Eastman® PM Acetate *[108-65-6]*, Dowanol® DPMA *[88917-22-0]*, Arcosolv® DPMA *[88917-22-0]*

37.0.4 Molecular Weight: 132.16 *[108-65-6]*

37.0.5 Molecular Formula: $C_6H_{12}O_3$ *[108-65-6]*

37.0.6 Molecular Structure:

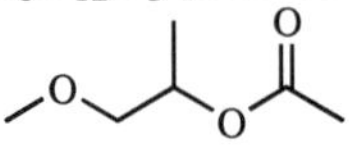

37.1 Chemical and Physical Properties

37.1.1 General

See Table 87.11.

37.2 Production and Use

Refer to Table 87.4.

37.3 Exposure Assessment

37.3.1 Air: NA

37.3.2 Background Levels: NA

37.3.3   Workplace Methods: NA

37.3.4   Community Method: NA

37.3.5   Biomonitoring/Biomarkers

37.3.5.1 Blood: NA

37.3.5.2 Urine. Laitinen (4) has published a biomonitoring method for exposure to technical-grade 1-methoxy-2-propanol acetate based on analysis of urinary 2-methoxypropionic acid (2MPA). The procedure involves acidification and extraction of the acid followed by conversion to the methyl ester. The methylated product is analyzed by capillary gas chromatography with flame-ionization detection. The detection limit for 2MPA was 0.1 mg/L. This method was employed to compare postshift 2MPA levels in silkscreen workers with breathing-zone measurements of the glycol acetate. Mean postshift 2MPA levels were 1.27 mmol/mol creatinine for exposures of 0–33 mL/m^3.

37.4   Toxic Effects

All these acetate esters are low in single-dose oral toxicity for rats. The propylene glycol derivative was somewhat painful and irritating to the eyes, but the others were not. These compounds were not appreciably irritating to the skin of rabbits, and did not produce evidence of systemic toxicity when applied repeatedly to skin for up to 1 week. Inhalation exposure to propylene glycol monomethyl ether acetate indicated irritation to the upper respiratory tract. The propylene glycol derivative was not teratogenic or mutagenic.

37.4.1   Experimental Studies

37.4.1.1 Acute Toxicity

ORAL. All these compounds are low in single-dose oral toxicity to rats (V. K. Rowe, The Dow Chemical Company, unpublished data, 1949). Single doses of 3 mL/kg were survived by all animals; doses of 10 mL/kg killed three of five rats fed the propylene glucol derivative, and all of five rats fed the di- and tripropylene glycol derivatives.

INHALATION. Acute 7 h inhalation exposures to saturated vapors of propylene glycol monomethyl ether acetate produced no effects other than eye and nasal irritation (R. W. Hollingsworth, unpublished data, 1949). Exposure of male and female rats and mice to 300, 1000, or 3000 ppm propylene glycol monomethyl ether acetate for 9 days produced mild systemic effects (liver weight changes) similar to those seen following inhalation exposure to propylene glycol monomethyl ether (7). However, histological evidence of irritation of the olfactory region of the nasal muscosa was reported at all exposure concentrations for mice and at the high concentration in rats.

EYE IRRITATION. The propylene glycol derivative was somewhat painful and irritating to the eyes, but the others were not.

SKIN IRRITATION. The cutaneous irritation of propylene glycol monomethyl ether acetate was determined in rabbits following the EEC (European Economic Communities) method and the Draize protocol (182).

37.4.1.2 Chronic and Subchronic Toxicity: NA

37.4.1.3 Pharmacokinetics, Metabolism, and Mechanisms

METABOLISM AND TOXICOKINETICS. The metabolism and disposition of propylene glycol monomethyl ether acetate appeared to be very similar to that of propylene glycol monomethyl ether following either single oral (8.7 mmol/kg) or single 6-h inhalation exposure (3000 ppm) to ^{14}C-labeled material (7). Approximately 53–64% of the dose was recovered as carbon dioxide from both routes of administration, and the remaining radioactivity recovered in urine exhibited a metabolic profile containing propylene glycol, propylene glycol monomethyl ether and its sulfate, and glucuronide conjugates.

37.4.1.4 Reproductive and Developmental Studies

DEVELOPMENTAL TOXICITY. No teratogenicity or developmental toxicity was reported in rats exposed to ≤4000 ppm propylene glycol monomethyl ether acetate during gestation days 6–15. Slight maternal toxicity was reported at exposure concentrations as low as 500 ppm (183).

37.4.1.5 Carcinogenesis: NA

37.4.1.6 Genetic and Related Cellular Effects Studies

GENETIC TOXICITY. Propylene glycol monomethyl ether acetate was not mutagenic in an Ames Salmonella assay (A. L. Mendrala, The Dow Chemical Company, unpublished data, 1983) and also did not produce any evidence of genotoxocity in a rat hepatocyte unscheduled DNA synthesis assay (A. L. Mendrala, The Dow Chemical Company, unpublished data 1983).

37.4.1.7 Other: Neurological, Pulmonary, Skin Sensitization

SKIN SENSITIZATION. Skin sensitization potential was determined using the Magnusson–Kligman method in the guinea pig. The glycol ether acetate was nonirritating in the EEC method and produced only slight irritation with the Draize procedure. Skin sensitization was absent.

37.5 Standards, Regulations, or Guidelines of Exposure

The German MAK (1996) value for occupational exposure to propylene glycol monomethyl ether acetate is 50 ppm (270 mg/m^3). The corresponding MAK value for the β isomer of PGMEA is 20 ppm.

38.0 Propylene Glycol Monoethyl Ether Acetate

38.0.1 CAS Number: *[54839-24-6]* (predominant α isomer), *[57350-24-0]* (2-ethoxy-1-propyl-acetate or β isomer), *[98516-30-4]* (mixed isomers)

38.0.2 Synonyms: 1-Ethoxy-2-acetoxypropanol, 2-Propylene Glycol 1-Ethyl Ether Acetate (2PG1EEA), 1-ethoxy-2-propyl acetate, 2-ethoxy-1-methylethyl acetate, ethoxypropyl acetate (EPA), PE acetate

38.0.3 Trade Names: NA

38.0.4 Molecular Weight: 146.19

38.0.5 Molecular Formula: $C_7H_{14}O_3$

38.0.6 Molecular Structure:

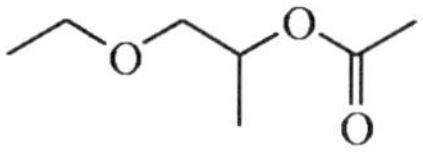

38.1 Chemical and Physical Properties

38.1.1 General

See Table 87.11.

38.3 Exposure Assessment

38.3.1 Air: NA

38.3.2 Background Levels: NA

38.3.3 Workplace Methods: NA

38.3.4 Community Methods: NA

38.3.5 Biomonitoring/Biomarkers

38.3.5.1 Blood: NA

38.3.5.2 Urine. Laitinen (4) has published a biomonitoring method for exposure to technical-grade 1-ethoxy-2-propanol acetate based on analysis of urinary 2-ethoxypropionic acid (2EPA). The procedure involves acidification and extraction of the acid followed by conversion to the methyl ester. The methylated product is analyzed by capillary gas chromatography with flame-ionization detection. This method was employed to compare postshift 2EPA levels in silkscreen workers with breathing-zone measurements of the glycol acetate.

38.4 Toxic Effects

This compound has a low single-dose oral toxicity in rats. It evoked only mild irritation in the eyes and was not appreciably irritating or sensitizing to the skin of rabbits. Exposure of rats to saturated vapor produced signs of irritation to the eyes and nose with no subacute or subchronic systemic effects. Propylene glycol monoethyl ether acetate is neither mutagenic nor fetotoxic and is not expected to be teratogenic. Studies involving prolonged or repeated exposure to this compound have not indicated any potential for neurotoxicity or immunotoxicity. It is likely that this acetate ester will readily undergo rapid metabolism to the corresponding parent glycol ether (propylene glycol monoethyl ether).

38.4.1 Experimental Studies

38.4.1.1 Acute Toxicity

ORAL. A single oral dose of 5 g/kg administered to groups of five male and five female rats caused no deaths, or any signs of ill health persisting for more than 4 days (184). It was concluded that the acute lethal oral dose to rats is greater than 5.0 g/kg body weight.

INHALATION. Acute 4-h inhalation exposure of rats to saturated (6.99 mg/L) vapor of propylene glycol monoethyl ether acetate produced no effects other than eye and respiratory irritation. It is concluded that the LC_{50} (4-h) is greater than 6.99 mg/L ($\sim$ 1150 ppm) in air. There was no histological evidence of irritation of the olfactory region or respiratory tract (185).

EYE IRRITATION. Propylene glycol monoethyl ether acetate did not evoke a response in rabbits sufficient to warrant labeling as irritating to eyes under EEC legislation (186).

SKIN IRRITATION. Single semiocclusive application of proplyene glycol monethyl ether acetate to the intact skin of rabbits evoked very slight dermal irritation to a degree insufficient to warrant labeling under EEC legislation (187).

38.4.1.2 Chronic and Subchronic Toxicity

INHALATION. Whole-body exposure of rats to propylene glycol monoethyl ether acetate vapor at $\leq$1176 ppm for 6 h per day, 5 days per week over a period of 28 days produced no signs of local or systemic toxicity (188).

38.4.1.3 Pharmacokinetics, Metabolism, and Mechanisms.
It is likely that propylene glycol monoethyl ether acetate will metabolise readily to the parent ether, therefore the subchronic, chronic, reproductive/teratological or neurological effects of this compound will be the same as for propylene glycol monoethyl ether.

38.4.1.4 Reproductive and Developmental: NA

38.4.1.5 Carcinogenesis: NA

38.4.1.6 Genetic and Related Cellular Effects Studies

GENETIC TOXICITY. Propylene glycol monoethyl ether acetate has been tested in the reverse mutation (Ames) test using five appropriate strains of *Salmonella typhimurium* (TA1535, TA1537, TA1538, TA98, and TA100) in the presence and in the absence of liver postmitochondrial fraction (S9). There was no increase in revertants in any of the strains tested at concentrations of ≤5000 µg/plate (189). Analysis of metaphase chromosomes obtained from Chinese hamster ovary (CHO) cells cultured *in vitro* with propylene glycol monoethyl ether acetate at concentrations of ≤2300 µg/mL in the presence or absence of S9 provided no evidence of potential to cause chromosome aberration (190).

38.4.1.7 Other: Neurological, Pulmonary, Skin Sensitization

SKIN SENSITIZATION. Delayed contact hypersensitivity studies in the guinea pig did not produce any evidence of potential to cause skin sensitization (191).

D NITRATE ESTERS OF ETHYLENE AND PROPYLENE GLYCOLS

The nitric acid esters of glycols are not typical of the esters or ether esters of organic acids and are considered separately in this chapter. They are used as explosives, usually in combination with nitroglycerin, to reduce the freezing point.

Industrial exposures of consequence are most likely to occur through the inhalation of vapors, but may also occur through contact with the eyes and skin. With the dinitrate, a serious hazard exists from absorption through the skin.

39.0 Ethylene Glycol Dinitrate

39.0.1 CAS Number: *[628-96-6]*

39.0.2 Synonyms: 1,2,-Ethanediol dinitrate, ethylene dinitrate, glycol dinitrate, EGDN, nitroglycol

39.0.3 Trade Names: NA

39.0.4 Molecular Weight: 152.06

39.0.5 Molecular Formula: $C_2H_4N_2O_6$

39.0.6 Molecular Structure:

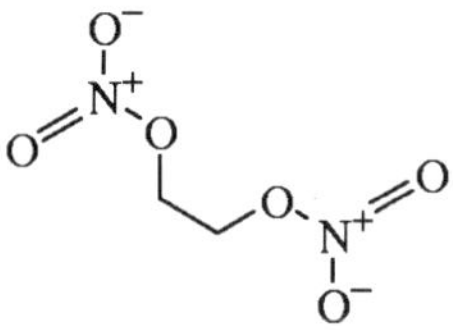

39.1 Chemical and Physical Properties

39.1.1 General

Ethylene glycol dinitrate is a yellow, oily liquid as ordinarily produced, but when pure it is colorless. It is soluble in many solvents, such as carbon tetrachloride, ether, benzene, toluene, and acetone, limited in solubility in common alcohols, and only very slightly soluble in water. It decomposes violently on heating or impact with a force similar to that of nitroglycerin. Additional physical and chemical properties are given in Table 87.14.

39.2 Production and Use

EGDN is produced by the nitration of ethylene glycol and is used as a solvent for low grade nitrocellulose in explosives.

39.3 Exposure Assessment

39.3.1 Air

In accordance with NIOSH method 2507 (192), a known volume of air is drawn through a tube packed with Tenax™ GC to trap the vapors present. The analyte is desorbed with ethyl alcohol. Analysis is made using gas chromatography employing an electron capture detector.

39.4 Toxic Effects

Ethylene glycol dinitrate is moderate in toxicity to rats when given in single oral doses. It is not likely to be significantly irritating on eye or skin contact, but it is readily absorbed

Table 87.14. Physical and Chemical Properties of Sone Glycol Dinitrates

Property	Ethylene Glycol Dinitrate	Diethylene Glycol Dinitrate	Triethylene Glycol Dinitrate	Propyl Glycol Dinitrate
CAS no.	[628-96-6]	[693-21-0]	[111-22-8]	[6423-43-4]
Common designation	EGDN, EGN	DEGDN	TEGDN	PGDN
Molecular formula	$C_2H_4N_2O_6$	$C_4H_8N_2O_7$	$C_6H_{12}N_2O_8$	$C_3H_6N_2O_6$
Molecular weight	152.1	196.1	240	166
Specific gravity (25/4°C)	1.491	—	—	1.4
Boiling point, °C (760 mm Hg)	Explodes	—	—	92 (10 mm Hg)
Vapor pressure, mm Hg (25°C)	0.072	—	—	0.07 (22.5°C)
Vapor density (air = 1)	5.24	—	—	—
Percent in saturated air (25°C)	0.0095	—	—	—
1 ppm ∼ mg/m³ at 25°C, 760 mm Hg	6.24	8.02	9.82	6.79
1 mg/L ∼ ppm at 25°C, 760 mm Hg	160.9	124.7	101.9	147.3

through the skin in toxic amounts. Even minor skin exposures (in humans) can cause adverse clinical effects. Inhalation of as little as 0.02 ppm has been reported to cause adverse effects, but absorption through the skin may have contributed significantly to the exposure.

39.4.1 Experimental Studies

39.4.1.1 Acute Toxicity

ORAL. The single-dose oral LD_{50} for rats of ethylene glycol dinitrate has been reported to be 616 mg/kg. This value is similar to that for propylene glycol dinitrate (193). The chief effect is blood vessel dilation (194).

INJECTION. Gross et al. (195) reported that the LD_{100} values for ethylene glycol dinitrate, when injected subcutaneously in the rabbit and cat, are 300 and 100 mg/kg, respectively. According to Jones et al. (196), Kylin and his co-workers reported that the toxicity of ethylene glycol dinitrate was very similar to that of propylene glycol dinitrate when given intraperitoneally; the LD_{50} values ranged from 0.4 to 1.05 g/kg, depending on the species. Sugawara (197) gave rabbits 150 mg/kg subcutaneously without significant effects.

EYE IRRITATION. There are no reports of eye tests conducted on ethylene glycol dinitrate. Extensive human experience suggests that eye contact is not likely to cause significant irritation. This is consistent with the observation that propylene glycol dinitrate causes at most only mild irritation. However, the implied toxicity of this material by absorption indicates that it would be wise to use suitable eye protection where eye exposure might occur.

SKIN IRRITATION. No animal tests for skin effects have been reported. Extensive human experience, however, indicates that ethylene glycol dinitrate is not likely to give rise to skin irritation.

39.4.1.2 Chronic and Subchronic Toxicity

INJECTION. Cavagna et al. (198) reported an increase in catecholamine in myocardial tissue reaching a maximum concentration at the end of each week during which rats were given 5 mg/kg every 12 h (11 times/week) for 15 weeks. However, there were no signs of long-term accumulation. At the end of each week and at the end of the experiment, the catecholamines peaked in 24 h but returned to normal in 48 h. Vigliani et al. (199) also saw an increase in catecholamines in the heart when rats were injected intramuscularly with 60 mg/kg daily in three divided doses for 5 days. Kalin et al. (200) reported that rats given 75 mg/kg of ethylene glycol dinitrate per day for 14 to 28 days developed no effect on the uptake, storage, release, or resynthesis of noradrenaline in the adrenergic nerves. These findings are contrary to those found by Minami et al. (201), Clark (202), and Clark and Litchfield (203), who report varying effects, such as increased catecholamines (201) at a daily dose of 65 mg/kg for 5 days/week for 8 weeks, increase in plasma corticosterone (194) at the same dose, and a fall in blood pressure in dogs given 10 mg/kg per day, 5 days/week for 5 weeks, and in rats given 65 mg/kg per day, 5 days/week for 10 weeks. The effects disappeared in 2–3 days after cessation of exposure.

INHALATION. Cats that inhaled an average concentration of 2 ppm of ethylene glycol dinitrate for a total of one thousand, 8-h exposures exhibited only temporary blood changes. If exposed to 21 ppm, there were marked blood changes. Higher levels killed the cats in a short time (195). Rats and guinea pigs exposed to 80 ppm for 6 months experienced drowsiness and Heinz body formation in the red blood cells. The pathological effects from chronic exposure were fatty changes in the liver and spleen similar to those seen in anemia.

39.4.1.3 Pharmacokinetics, Metabolism, and Mechanisms

METABOLISM AND TOXICOKINETICS. Numerous workers have conducted studies to elucidate both the metabolism and the pharmacological action of ethylene glycol dinitrate because of its use in pharmacology and its adverse human exposure effects. These studies have been summarized by Litchfield (204, 205). Basically, the studies show that when animals and humans are exposed to ethylene glycol dinitrate the material appears in the blood immediately, peaking in concentration in about 30 min. Apparently it is rapidly metabolized, for it disappears in about 4 h with very little found excreted in the urine. When it was given orally, a trace amount was found in the feces. Blood analysis shows that inorganic nitrite, inorganic nitrate, and ethylene glycol mononitrate are the main metabolites. The inorganic nitrite concentration in the blood peaked in about 2 h and returned to normal in 12 h. During this period, very little or no inorganic nitrates were found in the urine. This implies that the inorganic nitrite was either metabolized to inorganic nitrate or incorporated into the body.

Inorganic nitrate levels in the blood peaked in about 3.5 h, then returned to normal levels in 12 h. During this period, sufficient inorganic nitrate was excreted in the urine to account for approximately 60% of the amount of ethylene glycol dinitrate administered. The fate of the other 40% of inorganic nitrate is not known.

The ethylene glycol mononitrate found in the blood has been shown to be rapidly metabolized to ethylene glycol and inorganic nitrate and inorganic nitrite. Less than 0.5% of the ethylene glycol mononitrate was found to be excreted in urine.

Very little ethylene glycol has been seen in the urine after administration of either ethylene glycol dinitrate or mononitrate; however, in the blood, ethylene glycol reaches a peak level in 2–3 hr; hence it appears that it is metabolized rapidly (206). On the basis of these facts, the scheme for the metabolism of ethylene glycol dinitrate shown in Figure 87.2 has been proposed (205). The mechanism of step I is considered to proceed mainly to ethylene glycol mononitrate and inorganic nitrites via organic nitrate reductase accompanied by reduced glutathione. However, Tsuruta and Hasegawa (207) found that there are two enzymes involved: a nitrate as well as a nitrite-forming enzyme.

Step II is accomplished by the addition of –OH, but the mechanism has not been suggested. Step III occurs as a result of reductive hydrolysis, whereas step IV is the result of direct hydrolysis. Inorganic nitrites are known to be metabolized to inorganic nitrates by

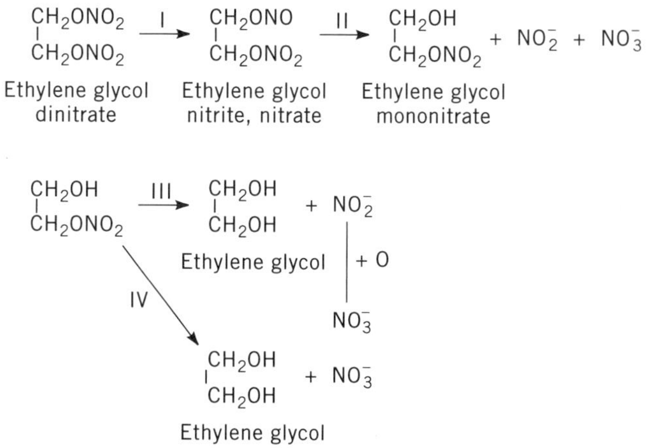

GLYCOL ETHERS

$$\begin{array}{c}CH_2ONO_2\\|\\CH_2ONO_2\end{array} \xrightarrow{I} \begin{array}{c}CH_2ONO\\|\\CH_2ONO_2\end{array} \xrightarrow{II} \begin{array}{c}CH_2OH\\|\\CH_2ONO_2\end{array} + NO_2^- + NO_3^-$$

Ethylene glycol dinitrate   Ethylene glycol nitrite, nitrate   Ethylene glycol mononitrate

$$\begin{array}{c}CH_2OH\\|\\CH_2ONO_2\end{array} \xrightarrow{III} \begin{array}{c}CH_2OH\\|\\CH_2OH\end{array} + NO_2^-$$

Ethylene glycol $+ O$

$\xrightarrow{IV}$ NO_3^-

$$\begin{array}{c}CH_2OH\\|\\CH_2OH\end{array} + NO_3^-$$

Ethylene glycol

Figure 87.2. Scheme of ethylene glycol dinitrate metabolism.

oxidation. Because of the ability of inorganic nitrites to cause the production of methemoglobinemia, it was suggested that some of the adverse effects of ethylene glycol dinitrate might be due to the presence of methemoglobin. Tests in rabbits by Hasegawa and Sato (208, 209) showed that methemoglobin is in fact produced by ethylene glycol dinitrate and that the amounts produced parallel the amount of free ethylene glycol dinitrate in the blood (208). In their study of the mechanism of production of methemoglobin, they found that it was produced only in the presence of oxygen. On this basis, they suggest that methemogbobin is formed in the following manner:

$Hb(O_2)_4 + {}_2ONOCH_2CH_2ONO_2 + H_2O \Rightarrow$
Hemoglobin, oxygenated
$Hb(OH)_4 + {}_2ONOCH_2CH_2OH + NO_2^- + 3O_2$
Methemoglobin, ethylene glycol mononitrate

This scheme, however, does not follow the postulated scheme for the metabolism of ethylene glycol dinitrate in which inorganic nitrites are formed. Hence it is possible that the production of inorganic nitrites may also be involved in the production of methemoglobin as well.

Hasegawa and Sato (209) and Hasegawa et al. (210) as well as others report that humans exposed to ethylene glycol dinitrate do not develop methemoglobin; hence the adverse effects seen apparently are not related to methemoglobin formation.

MODE OF ACTION. The clinical effects due to exposure to ethylene glycol dinitrate have been reported by many medical observers. They have been summarized by Litchfield (205) and Plunkett (211) to be lowered blood pressure because of vasodilatation, increased pulse rate, headache, dizziness, nausea and vomiting, hypotension, tachycardia, peripheral paresthesia, and chest pain. Anginal-type attacks, and in some cases deaths, that occur in workers usually result 48–60 h after cessation of exposure. Exposed workers become tolerant during the workweek but lose this tolerance over the weekend or when away from exposure. Thus, they experience the above mentioned effects when exposure occurs again. This is known as "Monday headache."

Gobbato and DeRosa (212) exposed humans to ethylene glycol dinitrate by inhalation and report that it causes severe cranioencephalic dilation as shown by an increase in systolic blood pressure, in cerebral flow, and in vasoelasticity with a reduction in local capillary resistance. The changes began almost immediately on exposure and reached a maximum in 4–8 min and disappeared in 15–20 min. This corresponds with the presence of ethylene glycol dinitrate in the blood. This suggests that the vasodilatation seen is due to the ethylene glycol dinitrate and not to its metabolites. Clark and Litchfield (206) report that the initial drop in blood pressure due to ethylene glycol dinitrate exposure returns to normal. This is then followed by a longer period of decrease in blood pressure that parallels the presence of the metabolite, ethylene glycol mononitrate.

The causes of "Monday headaches" and/or "Monday deaths" have also been explored. Vigliani et al. (199) feel that the causes of the sudden deaths associated with exposures to ethylene glycol dinitrate are very similar to those resulting from nitroglycerin exposure. Thus the causes of deaths from nitroglycerin, listed as follows, may also be considered possible causes of death from exposure to ethylene glycol dinitrate:

1. Peripheral vascular collapse due to prolonged vasodilatation
2. Coronary insufficiency due to vascular coronary spasm resulting from prolonged vasodilatation
3. Coronary insufficiency due to lowered pulse pressure increased diastolic pressure, and high pulse rate
4. Sensitization to cardioinhibitory trigeminovagal or senocarotid reflexes
5. Coronary insufficiency due to increased consumption of oxygen by the myocardium, caused by prolonged inhibition
6. Peripheral vascular collapse due to vasomotor paralysis resulting from hypothalmic alterations
7. Coronary insufficiency due to vascular sclerosis

Examination of these hypotheses reveals two major concepts: coronary insufficiency and serious changes in the dynamics of the cardiovascular system. However, the exact cause or causes have not been explained satisfactorily.

Vigliani et al. (199) attributed an increase in the catecholamine content of the heart to the biologic action of ethylene glycol dinitrate. This suggests that some of the adverse effects due to exposure to ethylene glycol dinitrate may be explained by the storage, then sudden release, of large amounts of catecholamines.

Minami et al. (201) postulate that ethylene glycol dinitrate affects the catecholamine metabolism not only of the heart but also of the brain and adrenals. Kalin et al. (200) report that they found ethylene glycol dinitrate to have no clear effect on the uptake, storage, release, or resynthesis of noradrenaline in the adrenergic nerves of the rat.

Suwa et al. (213) also studied the effect of ethylene glycol dinitrate on the heart. They found evidence that the heart continued to be affected even after ethylene glycol dinitrate

GLYCOL ETHERS

was no longer detectable in the blood. They feel that the supersensitivity of the heart might be due to adverse effects on the nervous system, not to electrolyte changes in the myocardium.

Gobbato and DeRosa (212), from studies in humans, suggested that the adverse effects from exposure were due to circulatory collapse resulting from systemic vasodilatory effects.

Clark (202), in his study of the effects of ethylene glycol dinitrate on pituitary–adrenocortical function, concluded that this material apparently increased the level of corticosterone in the plasma. This may be due to hypotension, which could stimulate the pituitary gland.

39.4.1.4 Reproductive and Developmental: NA

39.4.1.5 Carcinogenesis: NA

39.4.1.6 Genetic and Related Cellular Effects Studies

GENETIC TOXICITY. Kanonova et al. (214) evaluated a number of alkyl nitrates for mutagenic effects using the extracellular bacteriophage T4B of *Escherichia coli* and the alkylation of the DNA of the phage. Ethylene glycol dinitrate was found to cause mutagenic changes whereas nitroglycerin did not.

Tai and Tsuruta (215) reported the effects of EGDN on cardiac muscle isolated from Donryu rats. It was found that EGDN produced negative inotropic effects on left atrial muscle and positive inotropic effects on right ventricle muscle under in vitro conditions, suggesting EGDN may act directly on both cardiac muscles as well as vascular smooth muscle in cases of acute poisoning.

39.4.2 Human Experience

39.4.2.1 General Information. There are many reports in the literature of human exposure to ethylene glycol dinitrate either alone or in association with nitroglycerin, mainly in dynamite plants. Carmichael and Lieben (216) and von Oettingen (217) have summarized them.

Most authors agree that exposures in industrial plants are the result of inhalation of vapors and/or absorption through the skin. Because of the greater volatility of ethylene glycol dinitrate as compared to nitroglycerin, it is felt that the inhalation effects in operations where both are handled are due mainly to ethylene glycol dinitrate.

Repeated exposures to vapor may lead to a tolerance as long as daily exposures occur, but this is lost in 24–60 h. The major clinical effects are lowered blood pressure due to vasodilatation, increased pulse rate, headache, dizziness, nausea, vomiting, hypotension, tachycardia, peripheral paresthesis, and chest pain. Angina-type attack and, in some cases, sudden deaths have occurred on reexposure after a few days of nonexposure; a limited number have been reported in the absence of reexposure.

Human experience has indicated that those consuming alcohol are especially susceptible to ethylene glycol dinitrate poisoning (217).

39.4.2.2 Clinical Cases

39.4.2.2.1 Acute Toxicity

INHALATION. Single occasional exposures are more likely to cause the clinical signs described. Most authors agree that repeated daily exposure gives rise to a tolerance that reduces the clinical effects, and this continues during the workweek. However, over the weekend, this tolerance is lost so that on return to work and exposure the clinical effects recur. In a limited number of workers these effects can be severe, resulting in sudden death due to angina-type effects, usually in the summer months. These phenomena have come to be known as "Monday headache," or in the case of sudden death, "Monday morning death." Symanski (218) summarized the reports of sudden deaths occurring worldwide and found 44 deaths. Carmichael and Lieben (216), in a survey of such deaths in Pennsylvania in 1961, listed 12 such deaths, and Forssman et al. (219) also listed a number of such deaths.

Forssman et al. (219) reported that there were no chronic effects in workers exposed repeatedly to 5 mg/m^3 (0.5 ppm) to nitroglycerin and ethylene glycol dinitrate. Their criteria were clinical, X-ray of heart and lung, electrocardiographic studies, and blood examinations.

Einert et al. (220), in medical and industrial hygiene surveys in California, found levels of 0.013–0.44 ppm of a mixture of nitroglycerin and ethylene glycol dinitrate, and development of adverse clinical symptoms in 14—48% of the employees. The authors further studied the possible effects of skin absorption and showed that this type of exposure was a significant factor in the cause of adverse effects.

A validation study was undertaken to determine whether previous exposure chamber studies, showing headache induction in EGDN-exposed workers, could be duplicated under actual occupational conditions (221). It was found in this study that incidences of induced headaches were statistically lower than those predicted from the chamber studies.

However, Trainor and Jones (222) found that human volunteers exposed to 2 mg/m^3 (0.2 ppm) of a mixture of ethylene glycol dinitrate and nitroglycerin experienced a fall in blood pressure and marked headaches in 3 min. Even at a level of 0.7 and 0.5 mg/m^3 (0.08 and 0.05 pmm) the same effects were seen after 25 min.

Morikawa et al. (223) found that a significant number of dynamite plant workers developed slight to moderate abnormal pulse waves typical of those caused by ethylene glycol dinitrate when tested by plethysmography. These workers experienced vapor exposures exceeding 0.1 ppm of ethylene glycol dinitrate in 6 of 104 measurements. Cutaneous absorption from contaminated rubber gloves also occurred. Hematological and clinical chemical studies on 35 subjects exposed to ethylene glycol dinitrate manifested no abnormalities in blood electrolytes, liquids, cholesterol levels, and various enzyme activities (224).

Hasegawa et al. (210) report that dynamite workers exposed to 0.2–2 ppm of vapors 6 h/day repeatedly developed no detectable methemoglobinemia. They attributed this to the fact that humans are the lowest producers of methemoglobin when exposed to ethylene glycol dinitrate of the animals they studied.

DERMAL. Ethylene glycol dinitrate is readily absorbed through the skin in toxic amounts, more readily than nitroglycerin (194). The minimum dose, applied as a 10% solution in alcohol, which causes headaches in humans, was reported as 18–35 mg. Einert et al. (220), in reviewing the problems encountered in dynamite plants where nitroglycerin and ethylene glycol dinitrate are handled, quote Ebright, who states that headaches could result from merely "shaking hands with persons who handle dynamite." They also quote Gross and co-workers, who showed that practically all the ethylene glycol dinitrate was absorbed in 8 days, with up to one-third absorbed in the first 4–12 h. In their studies Einert et al. (220) showed that skin exposure contributed significantly to the clinical symptoms. They considered the risk from absorption from gloves to be low if less than 0.25 mg of the material was absorbed in 8 h, moderate if 0.25–0.75 mg was absorbed, and high if more than 0.75 mg was absorbed. These figures emphasize the high toxicity of ethylene glycol dinitrate by percutaneous absorption.

Morikawa et al. (223), in their study using plethysmographic measurements, found that nonexposed volunteers who wore protective gloves after they had been washed with detergents experienced characteristic changes in the pulse wave and a drop in systolic blood pressure, thus indicating that it is very difficult to remove ethylene glycol dinitrate from contaminated gloves. It is expected that the same would be the case with other wearing apparel, particularly shoes.

39.4.2.2.2 Chronic and Subchronic Toxicity: NA

39.4.2.2.3 Pharmacokinetics, Metabolism, and Mechanisms. Refer to Section 39.4.1.3.

39.5 Standards, Regulations, or Guidelines of Exposure

ACGIH recommends a TLV of 0.05 pmm (0.31 mg/m^3) as an 8-h TWA exposure for ethylene glycol dinitrate. A skin notation is indicated with this TLV, noting that this material can be absorbed through the skin and produce toxicity (43). Table 87.15 summarizes recommended or regulatory guidelines of occupational exposure for EGDN. OSHA DEL = 0.2 ppm (1 mg/m^3) CEIL.

40.0 Diethylene Glycol Dinitrate

40.0.1 CAS Number: [693-21-0]

40.0.2 Synonyms: 2,2′-Oxybisethanol dinitrate, digol dinitrate

40.0.3 Trade Names: NA

40.0.4 Molecular Weight: 196.12

40.0.5 Molecular Formula: C$_4$H$_8$N$_2$O$_7$

40.0.6 Molecular Structure: O=N$^+$(O$^-$)–O–CH$_2$CH$_2$–O–CH$_2$CH$_2$–O–N$^+$(=O)O$^-$

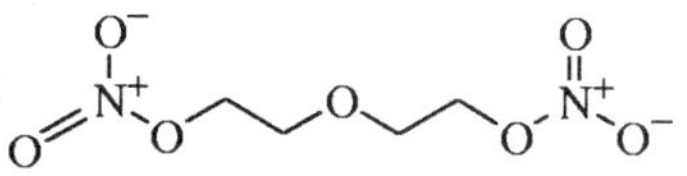

Table 87.15. Occupational Exposure Guidelines of EGDN[a]

Regulatory or Expert Authority (Date)	Exposure Value	Comment/Notation
ACGIH (TLV) (1997)	TWA = 0.05 ppm	Skin
NIOSH (1994)	STEL = 0.1 mg/m^3	Skin
OSHA (1994)	PEL = 0.2 ppm	Skin
Australia (OEL) (1993)	TWA = 0.05 ppm	Skin
Belgium (OEL) (1993)	TWA = 0.05 ppm	Skin
Denmark (OEL/TWA) (1993)	STEL = 0.02 ppm	Skin
Finland (OEL) (1993)	TWA = 0.1 ppm	Skin
	STEL = 0.3 ppm	
France (OEL/TWA) (1993)	0.25 ppm	Skin
Germany (MAK/TWA) (1993)	0.05 ppm	Skin
Japan (JSOH) (1998)	TWA = 0.05 ppm	Skin
Netherlands (OEL/TWA) (Oct. 1997)	TWA = 0.05 ppm	Skin
Philippines (OEL) (1993)	TWA = 0.2 ppm	Skin
Russia (OEL) (1993)	TWA = 0.05 mg/m^3	
Sweden (OEL) (1993)	TWA = 0.1 ppm	Skin
	STEL = 0.3 ppm	
Switzerland (OEL) (1993)	TWA = 0.05 ppm	Skin
	STEL = 0.1 ppm	
Thailand (OEL) (1993)	TWA = 0.2 ppm	
Turkey (OEL) (1993)	TWA = 0.2 ppm	Skin
United Kingdom (OEL) (1993)	TWA = 0.2 ppm	Skin
	STEL = 0.2 ppm	

[a]The OEL values set by Bulgaria, Colombia, Jordan, Korea, New Zealand, Singapore, and Vietnam are based on the ACGIH value.

40.1 Chemical and Physical Properties

40.1.1 General

See Table 87.14.

40.2 Production and Use: NA

40.3 Exposure Assessment

No official methods for determination in the atmosphere have been published, but it seems likely that NIOSH method 2507, recommended for ethylene glycol dinitrate and nitroglycerin, would be adaptable (192).

40.4 Toxic Effects

Diethylene glycol dinitrate (DEGDN) is a colorless or yellowish liquid of low volatility. Depending on purity of the product, it gives water a bitter–astringent taste at concentrations of 130 mg/L. The odor threshold in water is about 680 mg/L. It is low in single-dose oral toxicity and high in repeated-dose oral toxicity. The main adverse effects are those of CNS injury and hypotension. Although it apparently does cause the production of methemoglobinemia, it does so only at fairly high doses. Chronic exposure has been shown to cause a drop in blood pressure. No information was found concerning the effects of eye or skin contact or of inhalation.

40.4.1 Experimental Studies

40.4.1.1 Acute Toxicity

ORAL. Diethylene glycol dinitrate has a low single-dose oral toxicity; the LD_{50} levels for mice, rats, and guinea pigs are 1250, 1180, and 1250 mg/kg, respectively. The signs seen were acute cyanosis and CNS injury (225). More recently, rats and mice were administered this material by oral gavage and observed for clinical signs of toxicity and for mortality for a 2-week period (226). In both rats and mice, behavioral disturbances were the most frequently observed clinical signs. Calculated LD_{50} values for rats were 990 mg/kg (males) and 753 mg/kg (females) for mice, LD_{50} values were 1394 mg/kg (males) and 1320 mg/kg (females).

INJECTION. Intravenous injection of 0.4 mg/kg to rabbits showed that this material also reduces blood pressure and induces prolonged hypotension (225).

DERMAL. Neat diethylene glycol dinitrate (2 g/kg) was applied occluded to the clipped backs of male and female New Zealand rabbits for 24 h followed by a 14-day observation period (227). No mortality or other significant clinical signs were reported. Slight erythema was initially observed and persisted for 1–3; days, however, no gross or microscopic lesions were revealed at necropsy.

EYE IRRITATION. A dose of 0.1 mL of DEGDN was applied to the eyes of New Zealand white rabbits (227). Slight iridal vasodilation and slight conjuctival redness were observed beginning 1–4 h after dosing but resolving by 24 h. DEGDN was rated as nonirritating to the eyes.

SKIN IRRITATION. A dose of 0.5 mL DEGDN was applied to the clipped backs of New Zealand rabbits (227). After a 4-h exposure, material was removed and the sites scored for irritation by the method of Draize. All animals showed a negative response during the 72-h observation period. DEGDN was classified as nonirritating to skin.

40.4.1.2 Chronic and Subchronic Toxicity

ORAL. Rats were given diethylene glycol dinitrate by gavage in vegetable oil 6 times a week for 6 months at dose levels of 0.05, 0.5, or 5 mg/kg. These doses did not cause the production of methemoglobin. However, the rats given doses of 5 and 0.5 mg/kg showed signs of changes in conditioned reflex activity, CNS effects, and immunological condition. The 5 mg/kg dose decreased blood pressure by the fifth to the sixth months as well as changing the mitotic activity of the bone marrow and cardiovascular effects. Hematologic and histopathological studies revealed no changes. The no-effect level was determined to be 0.05 mg/kg (225).

40.4.1.3 Pharmacokinetics, Metabolism, and Mechanisms

MODE OF ACTION. The main effects seem to be those of CNS injury and hypotension, but unlike the other dinitrates, it seems to be low in its ability to cause methemoglobinemia, especially at small doses. The hypotensive effect was not accompanied by changes in electrocardiograph readings, suggesting that this material influences vascular tonus without impairing the activity of the myocardium.

40.4.1.4 Reproductive and Developmental: NA

40.4.1.5 Carcinogenesis: NA

40.4.1.6 Genetic and Related Cellular Effects Studies: NA

40.4.1.7 Other: Neurological, Pulmonary, Skin Sensitization

SKIN SENSITIZATION. DEGDN gave no response in a skin sensitization study using male guinea pigs (227).

40.5 Standards, Regulations, or Guidelines of Exposure

No occupational exposure standard has been established for diethylene glycol dinitrate.

41.0 Triethylene Glycol Dinitrate

41.0.1 CAS Number: [111-22-8]

41.0.2 Synonyms: 2,2′-Ethanediylbis(oxy)bisethanol Dinitrate, TEGDN

41.0.3 Trade Names: NA

41.0.4 Molecular Weight: 240.17

41.0.5 Molecular Formula: $C_6H_{12}N_2O_8$

41.0.6 Molecular Structure:

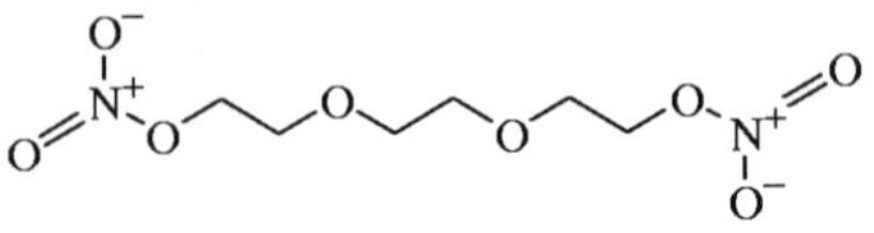

41.1 Chemical and Physical Properties

41.1.1 General

See Table 87.14.

41.2 Production and Use: NA

41.3 Exposure Assessment

No official methods for determination in the atmosphere have been published, but it seems likely that NIOSH method 2507, recommended for ethylene glycol dinitrate and nitroglycerin, would be adaptable (192).

41.4 Toxic Effects

41.4.1 Experimental Studies

41.4.1.1 Acute Toxicity. Triethylene glycol dinitrate is low in single-dose toxicity by various routes in several species, as shown in Table 87.16. The effects of exposure were those of methemoglobinemia and hypotension in rats. Unlike both ethylene and propylene glycol dinitrate-treated rats, the rats treated with triethylene glycol dinitrate developed tremors, convulsions, then death apparently due to respiratory arrest. Using a phrenic nerve diaphram preparation, triethylene glycol dinitrate blocked nerve-stimulated contraction, suggesting that this product interferes with nerve–muscle communication, which may be the cause of the tremors, convulsions, and respiratory arrest. Daily dermal applications of 21 mmol/kg (approximately 5.0 g/kg) caused death of 9 or 11 rabbits treated in 2–3 weeks. All the rabbits lost weight, and those that died developed the same signs as seen when this material was given by ingestion or by injection. As with the other glycol dinitrates, triethylene glycol dinitrate is apparently metabolized in the red cells in the presence of hemoglobin to mononitrate, inorganic nitrate, and inorganic nitrite (228).

Rats and mice were administered DEGDN by oral gavage and observed for clinical signs of toxicity and for mortality for a 2-week period (229). In both rats and mice, behavioral disturbances were the most frequently observed clinical signs. Calculated LD_{50} values for rats were 1330 mg/kg (males) and 1116 mg/kg (females). For mice, LD_{50} values were 2036 mg/kg (males) and 1866 mg/kg (females).

EYE IRRITATION. A dose of 0.1 mL of TEGDN was applied to the eyes of New Zealand white rabbits (230). Slight conjuctival redness was observed beginning 1–4 h after dosing but resolving by 24 h. TEGDN was rated as non-irritating to the eyes.

Table 87.16. Acute Toxicity of Triethylene Glycol Dinitrate in Laboratory Animals

Species	Sex	Administration Route	LD_{50} (mg/kg)
Mouse	M	IP	945
Guinea pig	M	IP	700
Rat	M	IP	796
Rat	M	SC	2520
Rat	M	Oral	1000

SKIN IRRITATION. A dose of 0.5 mL TEGDN was applied to the clipped backs of New Zealand rabbits (230). After a 4-h exposure, material was removed and the sites scored for irritation by the method of Draize. All animals displayed a slight erythema at 1–24 h after dosing, but all had returned to normal by 72 h. TEGDN was classified as nonirritating to skin.

41.4.1.2 Chronic and Subchronic Toxicity: NA

41.4.1.3 Pharmacokinetics, Metabolism, and Mechanisms: NA

41.4.1.4 Reproductive and Developmental: NA

41.4.1.5 Carcinogenesis: NA

41.4.1.6 Genetic and Related Cellular Effects Studies: NA

41.4.1.7 Other: Neurological, Pulmonary, Skin Sensitization

SKIN SENSITIZATION. TEGDN gave no response in a skin sensitization study using male guinea pigs (230).

41.5 Standards, Regulations, or Guidelines of Exposure

No occupational exposure standard has been established for TEGDN.

42.0 Propylene Glycol Dinitrate

42.0.1 CAS Number: [6423-43-4]

42.0.2 Synonyms: 1,2-Propanediol dinitrate, 1,2-propylene glycol dinitrate (PGDN), propene, 1,2-dinitrate

42.0.3 Trade Names: NA

42.0.4 Molecular Weight: 166.09

42.0.5 Molecular Formula: $C_3H_6N_2O_6$

42.0.6 Molecular Structure:

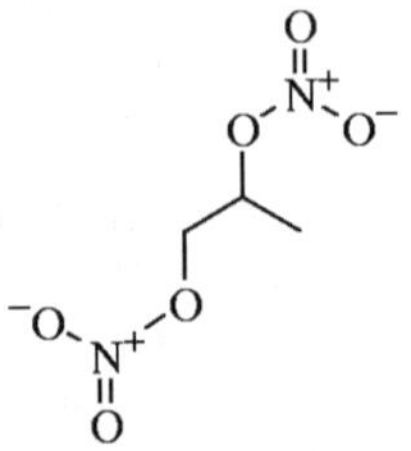

GLYCOL ETHERS

42.1 Chemical and Physical Properties

42.1.1 General

See Table 87.14.

42.2 Production and Use: NA

42.3 Exposure Assessment

No official methods for determination in the atmosphere have been published, but it seems likely that NIOSH method 2507, recommended for ethylene glycol dinitrate and nitroglycerin, would be adaptable (192).

The work of Stewart et al. (231) indicates that the detection of propylene glycol dinitrate in air by humans occurs in the range of 0.2–0.26 ppm. Repeated exposures at this level gave rise to headaches and visually evoked responses. Thus the warning properties of propylene glycol dinitrate may not be adequate to prevent some adverse responses, especially if repeated exposures occur.

42.4 Toxic Effects

Propylene glycol dinitrate (PGDN) is a red-orange or colorless liquid depending on the purity of the product. It is moderate to low in single-dose oral toxicity and moderately toxic by injection. Eye contact should cause at most only mild transient irritation, whereas skin contact may cause only mild irritation on prolonged or repeated contact. PGDN is readily absorbed through the skin. Repeated skin exposures can lead to adverse effects. A single short vapor exposure to concentrations of 1–1.5 ppm or more can lead to headaches. The main effects are circulatory collapse and the development of methemoglobinemia. The symptoms of overexposure are headaches, nasal congestion, dizziness, eye irritation, vasomotor collapse, unconsciousness, and death, depending on degree of exposure.

42.4.1 Experimental Studies

42.4.1.1 Acute Toxicity

ORAL. The single-dose oral toxicity is moderate to low for propylene glycol dinitrate; the LD_{50} values reported for female rats are 1.19 g/kg (232) and 0.86 g/kg (196) and for male rats, 0.25 g/kg (228). The adverse signs seen in the rat were prostration, lethargy, anoxia, and symptoms of methemoglobinemia (232). Jones et al. (196) reported reduced response to external stimuli and mild convulsions prior to death.

INJECTION. The reported LD_{50} values via injection and route of injection are shown in Table 87.17. The signs seen after injection were similar to those seen after ingestion. Methemoglobinemia was shown to be directly related to the dose administered. Rats receiving doses that proved fatal to a significant number experienced an increase in methemoglobin levels of up to 70–85 percent (228). Male Fischer F344 rats administered a rapid intravenous dose of PGDN demonstrated a maximum fall in systolic blood pressure

Table 87.17. Acute Toxicity of Propylene Glycol Dinitrate in Laboratory Animals Following Injection

Species	Sex	Route of Injection	LD$_{50}$ (g/kg)	Ref.
Rat	F	SC	0.463	232
	M	SC	0.524	232
	M	SC	0.53	228
Mouse	F	SC	1.21	232
Cat	F	SC	0.2–0.3	232
Mouse		IP	0.93	232
Mouse	M	IP	1.05	228
Guinea pig	M	IP	0.40	228
Rat	M	IP	0.48	228

within 1 min of dosing (233). This effect demonstrated a dose response over the dose range of 0.1–30 mg/kg

Clark and Litchfield (232) reported that the hemoglobin in rats given an LD$_{50}$ dose subcutaneously was nearly all converted to methemoglobin. They postulate that death is the result of this methemoglobin formation.

Clark and Litchfield (232) also followed the blood pressure of rats given PGDN subcutaneously and found again that the drop in blood pressure reflected the dose administered. A dose of 0.16 g given subcutaneously caused an average drop of about 60%.

INHALATION. Jones et al. (196) exposed six rats for 4 h to a mist of PGDN (estimated concentration 1350 mg/m^3) and found that the rats showed no toxic signs and there was no mortality. However, the mean methemoglobin level immediately after exposure had risen to 23.5%. Both a rabbit and a squirrel monkey were exposed 23 h/day to concentrations of 240 and 415 mg/m^3, respectively. The rabbit died on the fourth day; the monkey, on the third day. Both experienced increased methemoglobin levels, the rabbit 18.2% and the monkey 40.2% on death. Both animals became cyanotic during the experiment.

Jones et al. (196) also exposed rats for 7 h/day, 5 days/week for a total of 30 exposures at a level of 65 mg/m^3 (about 10 ppm). These rats experienced neither mortality nor toxic effects as shown by hematology and histopathological examination.

Mattsson et al. (234) exposed rhesus monkeys to the vapor of propylene glycol dinitrate for 4 h at concentrations of 2–33 ppm. At the 2-ppm level the visually evoked response was altered (increased by 20%) but behavioral response was unaltered even at the 33-ppm level.

Young et al. (235) found that primates (*Macaca mulatta*) when exposed to 28.2 mg/m^3 (4 ppm) of propylene glycol dinitrate for 23 h/day for 14 days showed no significant disruption of avoidance behavoir, motor coordination, or sensory function.

Stewart et al. (231) exposed human volunteers to levels of PGDN in air of 0.03–1.5 ppm for up to 8 h. In addition, some were exposed 8 h/day for 5 days to 0.2 ppm. The detectable level of PGDN vapor in air was found to be in the range of 0.2–0.26 ppm by most of the subjects. A single 8-h exposure to 0.2 ppm or more caused disruption of the organization of the visually evoked response and headache in a majority of the subjects. Marked impairment of balance occurred in those exposed for 6.5 h to 0.5 ppm. Those exposed for 40 min to 1.5 ppm developed eye irritation in addition. Repeated 8-h exposures for 5 days to 0.2 ppm resulted in tolerance to the induction of headaches; however, the alteration in the visually evoked response appeared to be cumulative. These findings led the authors (231) to suggest that the guide for the control to 0.2 ppm may not be adequate to prevent adverse responses, especially where repeated exposures may occur.

DERMAL. Clark and Litchfield (232) applied a 10% solution of PGDN to the skin of rats at a dose of 50 mg/kg and found that the rats experienced a detectable fall in blood pressure. This suggests that this material is absorbed through the skin.

Jones et al. (196) administered this material to the skin of rabbits for 2 h/day for 20 days at dose levels of 1.0, 2.0, and 4.0 g/kg. At the 2-g/kg level the rabbits all initially developed signs of weakness, cyanosis, and shallow, rapid breathing. One of the five animals died. Those that survived gradually improved so that at the end of the experiment they appeared normal. At the 4-g/kg level, 13 of 14 rabbits died after the 50th application. Autopsy revealed that the internal organs were dark blue-gray in color and the urinary bladder was markedly distended. The hemoglobin and hematocrit values were depressed and urinary nitrates accounted for approximately 7% of the propylene glycol dinitrate given at the 4-g/kg level.

Godin et al. (236) concluded that measurement of blood pressure by tail cuff in rats was not a sufficiently sensitive or specific biomarker of effect for exposure to PGDN. Thus, following subcutaneous injection, the most marked fall in systolic blood pressure occured over the dose range of 5–20 mg/kg. At doses of ≥ 160 mg/kg, dilation of the abdominal vasculature with cyanosis was noted. Following dermal administration, no correlation was found between amount of PGDN absorbed and blood pressure. Rats exposed to a nearly saturated vapor of PGDN showed no effects.

EYE IRRITATION. Jones et al. (196) found that PGDN is only very slightly irritating to the eyes of rabbits. Application of 0.1 mL to the eye caused only mild irritation but no corneal response. The irritation disappeared entirely within 24 h.

SKIN IRRITATION. Skin tests on rabbits showed that this material caused no observable skin irritation when the method similar to that of Draize was used (232). Jones et al. (196) reported that the application of 1 g/kg to the skin of rabbits repeatedly caused slight irritation.

42.4.1.2 Chronic and Subchronic Toxicity

INHALATION. In further studies (196), rats of both sexes, guinea pigs of both sexes, male squirrel monkeys, and male beagle dogs were exposed to 67 ± 8, 108 ± 11, or

236 ± 24 mg/m^3 of propylene glycol dinitrate continuously for 90 days. These levels are estimated to be 10, 16, and 35 ppm, respectively. None of the animals died as a result of exposures. The adverse effects seen were anemia, pigment deposition in various organs, fatty changes in the liver, methemoglobin formation, and increased serum and urinary nitrates at the highest exposure level. Dogs and monkeys exposed to about 10 ppm exhibited fatty changes and pigment deposits, and the guinea pigs developed some minor lung hemorrhages at the 16-ppm level. In these tests three of the monkeys at the 35-ppm level were removed from exposure for 2 h once a week to evaluate avoidance behavior. Their behavior was unaffected. On the basis of these data the authors recommend an industrial hygiene standard of 1.2 mg/m^3 (0.2 ppm) for a 40 h week exposure and a zero level for those in confined spaces.

42.4.1.3 Pharmacokinetics, Metabolism, and Mechanisms

METABOLISM AND TOXICOKINETICS. Clark and Litchfield (232) studied the metabolism of PGDN in rats. After a subcutaneous dose of 65 mg/kg it was found that PGDN rapidly enters the bloodstream, where it is readily metabolized to PGDN inorganic nitrate, and a small amount of inorganic nitrite. The PGDN apparently is readily metabolized, for the blood level decreases with time and none appears in the urine. Urinanalysis showed that inorganic nitrates are the major product; they represented about 56% of the dose administered. Essentially no PGDN or propylene glycol mononitrate and very little inorganic nitrites were excreted. Studies using whole blood and serum indicate that the metabolism of PGDN occurs in the red blood cells (237). Considering these findings, the metabolism of PGDN is presumed to be similar to that of ethylene glycol dinitrate.

MODE OF ACTION. The presence of propylene glycol dinitrate in the blood has been shown to lead to a correlatable drop in blood pressure. As the material is metabolized, there is a correlatable increase in the formation of methemoglobin. In rats it has been shown that the hemoglobin is essentially completely converted to methemoglobin at doses that result in the death of the animals (232). On this basis it may be postulated that death is the result of circulatory collapse and anoxia due to the high level of methemoglobin in the blood.

42.4.2 Human Experience

42.4.2.1 General Information.
Stewart et al. (231) reported that human exposures to Otto fuel II, which is largely propylene glycol dinitrate, caused headaches, nasal congestion, dizziness, eye irritation, vasomotor collapse, and unconsciousness. However, the number of people involved was not given.

42.5 Standards, Regulations, or Guidelines of Exposure

AGGIH recommends a TLV of 0.05 ppm (0.34 mg/m^3) as an 8-h TWA exposure for PGDN. A skin notation is indicated with this TLV, noting that this material can be absorbed

Table 87.18. Occupational Exposure Guidelines for Exposure to PGDN[a]

Regulatory or Expert Authority	Exposure Value	Comment/Notation
ACGIH (TLV) (1997)	TWA = 0.05 ppm	Skin
NIOSH (1994)	TWA = 0.05 ppm	Skin
Australia (OEL) (1993)	TWA = 0.05 ppm	Skin
Belgium (OEL) (1993)	TWA = 0.05 ppm	Skin
Denmark (OEL) (1993)	TWA = 0.02 ppm	Skin
Finland (OEL) (1993)	TWA = 0.02 ppm STEL = 0.06 ppm	Skin
France (OEL/TWA) (1993)	0.05 ppm	Skin
Germany (MAK/TWA) (1993)	0.05 ppm	Skin
Netherlands (OEL/TWA) (1993)	TWA = 0.05 ppm	Skin
Sweden (OEL) (1993)	TWA = 0.1 ppm STEL = 0.3 ppm	Skin
Switzerland (OEL) (1993)	TWA = 0.05 ppm	Skin
United Kingdom (OEL) (1993)	TWA = 0.2 ppm STEL = 0.2 ppm	Skin

[a]The OEL values set by Bulgaria, Colombia, Jordan, Korea, New Zealand, Singapore, and Vietnam are based on the ACGIH value.

through the skin and produce toxicity (43). Table 87.18 summarizes recommended or regulatory guidelines of occupational exposure for PGDN.

BIBLIOGRAPHY

1. H. Chinn, with E. Anderson and M. Tashiro, *CEH Marketing Research Report: Glycol Ethers*, Chemical Economics Handbook, SRI International, 1996.
2. National Institute for Occupational Safety and Health (NIOSH), *NIOSH Manual of Analytical Methods*, 4th ed., Issue 2, Method 1403, Public Health Service, U.S. Department of Health, Education, and Welfare, NIOSH, Cincinnati, OH, 1994.
3. Occupational Safety and Health Administration (OSHA), *Method 83, 2-Butoxyethanol (Butyl Cellosolve), 2-Butoxyethyl Acetate (Butyl Cellosolve Acetate)*, Organic Methods Evaluation Branch, OSHA Analytical Laboratory, Salt Lake City, UT, 1990.
4. J. Laitinen, Biomonitoring of technical grade 1-alkoxy-2-propanol acetates by analyzing urinary 2-alkoxypropionic acids. *Sci. Total Environ.* **199** (1/2), 31–39 (1997).
5. B. Hubner, K. Geibel, and J. Angerer, Gas chromatographic determination of propylene- and diethylene glycol ethers in urine. *Fresenius' J. Anal. Chem.* **342**(9), 746–748 (1992).
6. T. D. Landry, T. S. Gushow, and B. L. Yano, Propylene glycol monomethyl ether. A 13-week vapor inhalation toxicity study in rats and rabbits. *Fundam. Appl. Toxicol.* **3**, 627–630 (1983).
7. R. A. Miller et al., Propylene glycol monomethyl ether acetate (PGMEA) metabolism, disposition, and short-term vapor inhalation toxicity studies. *Toxicol. Appl. Pharmacol.* **75**, 521–530 (1984).
8. V. K. Rowe et al., Toxicology of mono-, di- and tripropylene glycol methyl ethers. *Arch. Ind. Hyg. Occup. Med.* **9**, 509–525 (1954).

9. R. R. Miller. et al., Comparative metabolism and disposition of ethylene glycol monomethyl ether and propylene glycol monomethyl ether in male rats. *Toxicol. Appl. Pharmacol.* **67**, 229–237 (1983).
10. R. R. Miller et al., Metabolism and disposition of propylene glycol monomethyl ether (PGME) beta isomer in male rats. *Toxicol. Appl. Pharmacol.* **83**, 170–177 (1986).
11. ECETOC, *The Toxicology of Glycol Ethers and Its Relevance to Man*, Tech. Rep. No. 64, ECETOC, Brussels, 1995.
12. J. Merkle, H. J. Klimisch, and R. Jackh, Prenatal toxicity of 2- methoxypropylacetate-1 in rats and rabbits. *Fundam. Appl. Toxicol.* **8**, 71–79 (1987).
13. J. Hellwig, H. J. Klimisch, and R. Jackh, Prenatal toxicity of inhalation exposure to 2-methoxypropanol-1 in rabbits. *Fundam. Appl. Toxicol.* **23**, 608–613 (1994).
14. R. D. Stewart et al., Experimental human exposure to vapor of propylene glycol monomethyl ether. *Arch. Environ. Health.* **20**, 218–223 (1970).
15. J. Laitinen, J. Liesivuori, and H. Savolainen, Biological monitoring of occupational exposure to 1-methoxy-2-propanol. *J. Chromatogr. B.* **694**, 93–98 (1997).
16. H. F. Smyth, Jr. et al., Range finding toxicity data. *Am. Ind. Hyg. Assoc. J.* **23**, 95–107 (1962).
17. H. F. Smyth, Jr., J. Seaton, and L. Fischer, The single dose toxicity of some glycols and derivatives. *J. Ind. Hyg. Toxicol.* **23**, 259–268 (1941).
18. H. F. Smyth, Jr. et al., Range-finding toxicity data. List VII. *J. Am. Ind. Hyg. Assoc.* **30**, 470–476 (1969).
19. E. G. Stenger et al., Zur toxizitat des propylenglykol-Monomethylaethers. *Arzneim.-Forsch.* **22**, 569–574 (1972).
20. F. E. Shideman and L. Procita, The pharmacology of the mono methyl ethers of mono-, di-, and tripropylene glycol in the dog with observations on the auricular fibrillation produced by these compounds. *J. Pharmacol. Exp. Ther.* **102,** 79–87 (1951).
21. F. S. Cieszlak and J. W. Crissman, *Dowanol PM Glycol Ether: An Acute Vapor Inhalation Study in Fischer 344 Rats*, unpublished report, The Dow Chemical Company, 1991.
22. M. E. Goldberg et al., Effect of repeated inhalation of vapors of industrial solvents on animal behavior. 1. Evaluation of nine solvent vapors on pole-climb performance in rats. *J. Am. Ind. Hyg. Assoc.* **25**, 369–375 (1964).
23. R. R. Miller et al., Comparative short-term inhalation toxicity of ethylene glycol monomethyl ether and propylene glycol monomethyl ether in rats and mice. *Toxicol. Appl. Pharmacol.* **61**, 368–377 (1981).
24. R. E. Carreon and J. M. Wall, *Propylene Glycol Monomethyl Ether: Skin Sensitization Potential in the Guinea Pig*, unpublished report, The Dow Chemical Company, 1984.
25. F. S. Cieszlak et al., *Propylene Glycol Monomethyl Ether: A 13-Week Vapor Inhalation Study to Evaluate Hepatic and Renal Cellular Proliferation, P450 Enzyme Induction and Protein Droplet Nephropathy in Fischer 344 Rats*, unpublished report, The Dow Chemical Company, 1996.
26. F. S. Cieszlak et al., *Propylene Glycol Monomethyl Ether: A 13-Week Vapor Inhalation Study to Evaluate Hepatic and Renal Cellular Proliferation and P450 Enzyme Induction in B6C3F1 Rats*, unpublished report, The Dow Chemical Company, 1996.
27. L. L. Calhoun and K. A. Johnson, *Propylene Glycol Monomethyl Ether (PGME). 21-Day Dermal Study in New Zealand White Rabbits*, unpublished data, Toxicology Research Laboratory Report, The Dow Chemical Company, 1984.

28. F. S. Cieszlak et al., *Propylene Glycol Monomethyl Ether: A Two-Year Vapor Inhalation Chronic Toxicity/Oncogenicity Study and Evaluation of Hepatic and Renal Cellular Proliferation, P450 Enzyme Induction and Protein Droplet Nephropathy in Fischer 344 Rats*, unpublished results, The Dow Chemical Company, 1999.

29. F. S. Cieszlak et al., *Propylene Glycol Monomethyl Ether: A Two-Year Vapor Inhalation Oncogenicity Study and Evaluation of Hepatic Cellular Proliferation and P450 Enzyme Induction in B6C3F1 Mice*, unpublished results, The Dow Chemical Company, 1999.

30. N. F. Ferrala, B. I. Ghanayem, and A. A. Nomeir, Determination of 1-methoxy-2-propanol and its metabolite 1,2-propanediol in rat and mouse plasma by gas chromatography. *J. Chromatogr. B* **660**(2), 291–296 (1994).

31. D. A. Morgott and R. J. Nolan, Nonlinear kinetics of inhaled propylene glycol monomethyl ether in Fischer 344 rats following single and repeated exposures. *Toxicol. Appl. Pharmacol.* **89**, 19–28 (1987).

32. National Toxicology Program (NTP), *Reproduction and Fertility Assessment in CD-1 Mice when Administered in the Drinking Water*, Rep. No. NTP-86-062, Research Triangle Institute, Research Triangle Park, NC, 1986.

33. R. E. Morrissey et al., Results and evaluations of 48 continuous breeding reproduction studies in mice. *Fundam. Appl. Toxicol.* **13**, 747–777 (1989).

34. B. Bolon et al., Differential follicle counts as a screen for chemically induced ovarian toxicity in mice: Results from continuous breeding assays. *Fundam. Appl. Toxicol.* **39**, 1–10 (1997).

35. J. E. Doe et al., Comparative aspects of the reproductive toxicology by inhalation in rats of ethylene glycol monomethyl ether and propylene glycol monomethyl ether. *Toxicol. Appl. Pharmacol.* **69**, 43–47 (1983).

36. E. W. Carney et al., Two-generation inhalation reproduction study with propylene glycol monomethyl ether in Sprague–Dawley rats. *Toxicol. Sci.* (1999) (to be published).

37. T. R. Hanley et al., Teratologic evaluation of inhaled propylene glycol mono-methyl ether in rats and rabbits. *Fundam. Appl. Toxicol.* **4**, 784–794 (1984).

38. D. J. Kirkland, *Metaphase Analysts of Chinese Hamster Ovary Cells Treated with Dowanol DPM*, unpublished data, Dow Chemical Europe, Horgen, Switzerland, 1983.

39. Z. Elias et al., Genotoxic and/or epigenetic effects of some glycol ethers: Results of different short-term tests. *Occup. Hyg.* **2**, 187–212 (1996).

40. K. Jones, *Dermal Absorption of Vapours I-Final Report*, unpublished data, Health and Safety Laboratory, Health and Safety Executive, UK, 1997.

41. K. Jones et al., A biological monitoring study of 1-methoxy-2-propanol: Analytical method development and a human volunteer study. *Sci. Total Environ.* **199**(1/2), 23–30 (1997).

42. H. H. Emmen, *Human Volunteer Study with Propylene Glycol Monomethyl Ether: Potential Eye Irritation During Vapor Exposure*, unpublished report, TNO Nutrition and Food Research Laboratory, 1997.

43. American Conference of Governmental Industrial Hygienists (ACGIH), *Threshold Limit Values for Chemical Substances and Physical Agents and Biological Exposure Indices for 1998–1999*, ACGIH Cincinnati, OH, 1999.

44. BP Chemicals, *A Study of the Ten Day Repeat Dose Oral Toxicity of Ethoxypropanol in Rats*, B.P. GOHC Expt. No. 83T18, BP Chemicals, UK, 1983.

45. H. F. Smyth, Jr. and C. P. Carpenter, Further experience with the range-finding test in the industrial toxicology laboratory. *J. Ind. Hyg. Toxicol.* **30**, 63–68 (1948).

46. E. Gross, in K. B. Lehmann and F. Flury, eds.; E. King and H. F. Smyth, Jr., transl., *Toxicology and Hygiene of Industrial Solvents*, Springer, Berlin, 1938.
47. BP Chemicals, *The Nine Day Repeated Exposure Inhalation Toxicity of Ethoxypropanol in Rats*, B.P. GOHC Expt. No. 83T20, BP Chemicals, UK, 1983.
48. BP Chemical, *Napsol PE1 (Ethoxypropanol). Test de Tolérance Locale Chez le Lapin*, IFREB Rep. 110336, BP Chemicals, UK, 1981.
49. BP Chemicals, *A Study of the Eye Irritancy of Ethoxypropanol in Rabbits*, B.P. GOHC Expt. No. 84T027, BP Chemicals, UK, 1984.
50. BP Chemicals, *Ethoxypropanol 90 Day Inhalation Study in Rats*, Rep. BPC 49/86965, Huntingdon Research Centre, BP Chemicals, UK, 1986.
51. BP Chemicals, *Effect of Ethoxypropanol on Pregnancy in the Rat (Inhalation Exposure)*, Report BPC 47/86852, Huntingdon Research Centre, BP Chemicals, UK, 1986.
52. BP Chemicals, *Effect of Ethoxypropanol on Pregnancy in the Rabbit (Inhalation Exposure)*, Report BPC 49/86965, Huntingdon Research Centre, BP Chemicals, UK, 1986.
53. BP Chemicals, *Study to Determine the Ability of Ethoxypropanol to Induce Mutation in Five Histidine-Requiring Strains of* Salmonella typhimurium, Report BOH 1/S, Microtest Research Ltd., UK, 1988.
54. BP Chemicals, *Study to Evaluate the Chromosome Damaging Potential of Ethoxypropanol by Its Effects on Cultured Human Lymphocytes Using an* In Vitro *Cytogenetics Assay*, Report BOH 1/HLC, Microtest Research Ltd., UK, 1988.
55. B. Ballantyne, R. C. Myers, and P. E. Losco, The acute toxicity and primary irritancy of 1-propoxy-2-propanol. *Vet. Hum. Toxicol.* **30**(2), 126–129 (1988).
56. Union Carbide Corporation, Project Report 27-45, April 15, Bushy Run Research Center, 1964.
57. Union Carbide Corporation, Project Report 40-6, December 27, Bushy Run Research Center, 1977.
58. Union Carbide Corporation, Project Report 27-153, November 25, Bushy Run Research Center, 1964.
59. Union Carbide Corporation, Project Report 49-120, January 15, Bushy Run Research Center, 1987.
60. Union Carbide Corporation, Project Report 50-8, September 30, Bushy Run Research Center, 1987.
61. Union Carbide Corporation, Project Report 52-28, April 4, Bushy Run Research Center, 1989.
62. Union Carbide Corporation, Project Report 53-44, July 19, Bushy Run Research Center, 1990.
63. C. P. Carpenter, C. S. Weil, and H. F. Smyth, Jr., Range-finding toxicity data: List VIII. *Toxicol. Appl. Pharmacol.* **28**, 313–319 (1974).
64. J. B. J. Reijnders, A. M. Zucker-Keizer, and H. G. Verschuuren, *Evaluation of the Acute Oral Toxicity of Dowanol PnB in the Rat*, Toxicology Research Laboratory Report, unpublished data, Dow Europe, Horgen, Switzerland, 1987.
65. F. M. H. Debets, and H. G. Verschuuren, *Assessment of Oral Toxicity, Including the Haemolytic Activity of Dowanol PnB in the Rat. 14-Day Study*, Toxicology Research Laboratory Report, Dow Europe, Horgen, Switzerland, 1987.

66. R. A. Corley et al., *Propylene Glycol n-Butyl Ether. An Acute Vapour Inhalation Study in Fischer 344 Rats*, Toxicology Research Laboratory Report, The Dow Chemical Company, 1989.
67. Union Carbide Corporation, Project Report 51-5, January 5, Bushy Run Research Center, 1989.
68. R. A. Corley et al., *Propylene Glycol n-Butyl Ether. Two-Week Vapour Inhalation Study with Fischer 344 Rats*, Toxicology Research Laboratory Report, The Dow Chemical Company, 1989.
69. Union Carbide Corporation, Project Report 28-11, February 1, Bushy Run Research Center, 1965.
70. J. B. J. Reijnders and H. G. Verschuuren, *Evaluation of the Acute Oral Toxicity of Dowanol PnB in the Rat*, Toxicology Research Laboratory Report, unpublished data, Dow Europe, Horgan, Switzerland, 1987.
71. J. H. Draize, G. Woodward, and H. O. Calvery, Methods for the study of irritation and toxicity of substances applied to the skin and mucous membranes. *J. Pharmacol. Exp. Ther.* **82**, 377–390 (1944).
72. P. J. J. M. Weterings, P. A. M. Daamen, and H. G. Verschuurenn, *Assessment of Acute Eye Irritation/Corrosion by Dowanol PnB in the Rabbit*, Toxicology Research Laboratory Report, Dow Europe, Horgan, Switzerland, 1987.
73. P. J. J. M. Weterings, P. A. M. Daamen, and H. G. Verschuurenn, *Assessment of Primary Skin Irritation/Corrosion by Dowanol PnB in the Rabbit*, Toxicology Research Laboratory Report, Dow Europe, Horgan, Switzerland, 1987.
74. P. J. J. M. Weterings, P. A. M. Daamen, and H. G. Verschuurenn, *Assessment of Primary Skin Irritation/Corrosion by Dowanol PnB Diluted to 75%, 50%, and 25% (w/w) in the Rabbit*, Toxicology Research Laboratory Report, Dow Europe, Horgan, Switzerland, 1987.
75. M. Grandjean, J. R. Szabo, and H. G. Verschuuren, *Propylene Glycol n-Butyl Ether. 13-Week Drinking Water Study in Fischer 344 Rats*, Toxicology Research Laboratory Report, unpublished data, The Dow Chemical Company, 1992.
76. D. Jonkers, B. A. R. Lina, and H. G. Verschuuren, *Sub-Chronic (13-Week) Dermal Toxicity Study with Propylene Glycol n-Butyl Ether in Rats*, Toxicology Research Laboratory Report, unpublished data, Dow Europe, Horgan, Switzerland, 1988.
77. J. D. Innis and G. A. Nixon, No evidence of toxicity associated with subchronic dermal exposure of rabbits to butoxypropanol. *Toxicologist* **8**, 213 (Abstr. 849) (1988).
78. W. B. Gibson, G. A. Nolen, and M. S. Christian, Determination of the developmental toxicity potential of butoxypropanol in rabbits after topical administration. *Fundam. Appl. Toxicol.* **13**(3), 359–365 (1989).
79. D. H. Waalkens-Berendsen et al., *Dermal Embryotoxicity/Teratogenicity Study with Propylene Glycol n-Butyl Ether (PnB) in Rats*, Toxicology Research Laboratory Report, unpublished data, Dow Europe, Horgan, Switzerland, 1989.
80. B. B. Gollapudi, V. A. Linscombe, and H. G. Verschuuren, *Evaluation of Propylene Glycol n-Butyl Ether in an* In Vitro *Chromosomal Aberration Assay Utilizing Chinese Hamster Ovary (CHO) Cells*, Toxicology Research Laboratory Report, unpublished data, The Dow Chemical Company, 1988.
81. V. P. Gupta, Recovery of propylene glycol mono T-butyoxy ether. U.S.P. 4,675,082 (1987).
82. L. R. Nudy and W. A. Johnston, PTB: A safe, effective solvent for hard surface cleaners. *Household Personal Prod. Ind.*, April, pp. 90–92 (1990).

83. R. A. Heckman, Using P-series glycol ethers in water-reducible coatings. *Mod. Paint Coat.* **76**(6), 36–42 (1986).
84. ARCO Chemical Company, *Acute Oral LD50 Study in Rats of A 209429 (PTB)*, Conducted by Toxgenics, Inc., Study 410-0935, unpublished report, June 4, ARCO Chem. Co., 1982.
85. ARCO Chemical Company, *Acute Inhalation Toxicity Test in Sprague-Dawley Rats Using A 209429 (PTB)*, Conducted by Midwest Research Institute, MRI Project No. 7450-B, unpublished report, November 12, ARCO Chem. Co., 1982.
86. ARCO Chemical Company, *Acute Dermal Toxicity Study in Rabbits of A 209429 (PTB) at a Dose Level of 2 Grams per Kilogram of Body Weight*, Conducted by Toxgenics, Inc., Study 410-0936, unpublished report, May 27, ARCO Chem. Co., 1982.
87. ARCO Chemical Company, *Primary Dermal Irritation Study in Rabbits of A 209429 (PTB)*, Conducted by Toxgenics, Inc., Study 410-0938, unpublished report, May 19, ARCO Chem. Co., 1982.
88. ARCO Chemical Company, *Primary Eye Irritation Study in Rabbits of A 209429 (PTB)*, Conducted by Toxgenics, Inc., Study 410-0937, unpublished report, May 26, ARCO Chem. Co., 1982.
89. ARCO Chemical Company, *Primary Eye Irritation Study in Albino Rabbits Administered Test Article PTB (20%)*, Conducted by Bio-Research Laboratories LTD., Project No. 51158, unpublished report, January 10, ARCO Chem. Co., 1985.
90. ARCO Chemical Company, *A 4- and 13-Week Inhalation Toxicity Study (with 3-Week Regression) of ARCOSOLV PTB in the Albino Rat*, Conducted by Bio-Research Laboratories LTD., Project No. 81910, unpublished report, July 31, ARCO Chem. Co., 1985.
91. I. G. Sipes and D. E. Carter, *The Metabolism and disposition of Propylene Glycol t-Butyl ether in the male Fischer 344 Rat* (Final Report), NIEHS Contract NO1-ES-85320, unpublished report, National Toxicology Program, Washington, DC, 1994.
92. ARCO Chemical Company, *Inhalation Developmental Toxicity Study in Rabbits*, Conducted by International Research and Development Corporation (IRDC), IRDC Report 419-031, unpublished report, June 3, ARCO Chem. Co., 1988.
93. ARCO Chemical Company, *Inhalation Developmental Toxicity Study in Rats*, Conducted by International Research and Development Corporation (IRDC), IRDC Report 419-029, unpublished report, June 3, ARCO Chem. Co., 1988.
94. ARCO Chemical Company, *Ames Metabolic Activation Test to Assess the Potential Mutagenic Effect of ARCOSOLV PTB*, Conducted by Huntingdon Research Centre, Ltd., HRC Rep. No. ARO 9/871215, unpublished report, October 27, ARCO Chem. Co., 1988.
95. ARCO Chemical Company, *Ames Salmonella/Microsome Plate Test (EPA/OECD)*, Conducted by Pharmakon Research International, Inc., Pharmkon Report PH 301-AR-002-86, unpublished report, April 24, ARCO Chem. Co., 1986.
96. ARCO Chemical Company *Analysis of Metaphase Chromosomes Obtained from CHO Cells Cultured* In Vitro *and Treated with 1-(1,1-Dimethylethoxy)-2-propanol CAS RN 57018-52-7* (Commercially available from Arco Chemical Co. as ARCOSOLV® PTB Solvent), Conducted by Huntingdon Research Centre, Report ARO 6/871093, unpublished report, November 10, ARCO Chem. Co., 1987.
97. ARCO Chemical Company, *Evaluation of the Ability of ARCOSOLV PTB to Induce Chromosome Aberrations in Cultured Peripheral Human Lymphocytes*, Conducted by RCC NOTOX B.V., Project 043291, unpublished report, January 23, ARCO Chem. Co., 1988.

98. National Toxicology Program (NTP), In Vitro *Chromosomal Aberration Study using CHO Cells with Propylene Glycol mono-t-butyl Ether*, unpublished results, NTP, Washington, DC, 1997.
99. National Toxicology Program (NTP), In Vitro *Sister Chromatid Exchange Assay with Propylene Glycol Mono-t-butyl Ether*, unpublished results, NTP, Washington, DC, 1997.
100. ARCO Chemical Company, *Delayed Contact Hypersensitivity in the Guinea-Pig with ARCOSOLV PTB*, Conducted by Huntingdon Research Centre, Ltd., HRC Rep. No. 87718D/ARO 5/SS, unpublished report, June 22, ARCO Chem. Co., 1988.
101. HSDB (Hazardous Substance Databank), Database Generated by the National Occupational Institute for Safety and Health (NIOSH) and Distributed by the National Library of Medicine (NLM), Bethesda, MD, 1999.
102. Union Carbide Corporation, Project Report 16-2, December 15, Bushy Run Research Center, 1952.
103. Union Carbide Corporation, Project Report 28-78, June 15, Bushy Run Research Center, 1965.
104. J. M. Norris and K. J. Olson, *Toxicological Properties and Industrial Handling Hazards of Dowanol PPh (1-Phenoxy-2-Propanol)*, Toxicology Research Laboratory Report, unpublished data, The Dow Chemical Company, 1968.
105. J. E. Phillips et al., *Ethylene Glycol Phenyl Ether and Propylene Glycol Phenyl Ethers. Comparative 2-Week Dermal Toxicity Study in Female Rabbits*, Toxicology Research Laboratory Report, unpublished data, The Dow Chemical Company, 1985.
106. L. L. Calhoun et al., *Propylene Glycol Phenyl Ether 28-Day Dermal Toxicity Study in Rabbits*, Toxicology Research Laboratory Report, unpublished data, The Dow Chemical Company, 1986.
107. J. Bootman and K. May, *Mutagenicity Studies*, Ames Test Report, Life Sciences Res. Ltd., UK, 1985.
108. J. Bootman, *Mutagenicity Test, Metaphase Analysis, Human Peripheral Lymphocytes, Chromosome Aberration*, Confid. Report NIPA Lab. Ltd., Life Science Res. Mid Clamorgan, GB-CF38-25N, UK, 1986.
109. E. Browning, *Toxicity and Metabolism of Industrial Solvents*, Elsevier, New York, 1965.
110. T. D. Landry and B. L. Yano, Dipropylene glycol monomethyl ether. A 13-week inhalation toxicity study in rats and rabbits. *Fundam. Appl. Toxicol.*, **4**, 612–617 (1984).
111. S. Fairhurst et al., Percutaneous toxicity of ethylene glycol monomethyl ether and of dipropylene glycol monomethyl ether in the rat. *Toxicology*, **57**, 209–215 (1989).
112. R. R. Miller et al., Metabolism and disposition of dipropylene glycol monomethyl ether (DPGME) in male rats. *Fundam. Appl. Toxicol.*, **5**, 721–726 (1985).
113. W. J. Breslin et al., Evaluation of the developmental toxicity of inhaled dipropylene glycol monomethyl ether (DPGME) in rabbits and rats. *Occup. Hyg.* **2**, 161–170 (1996).
114. D. J. Kirkland and R. Varley, *Bacterial Mutagenicity Test on Dowanol DPM*, Toxicology Research Laboratory Report, unpublished data, The Dow Chemical Company, 1983.
115. BP Chemicals, *Dipropylene Glycol Monoethyl Ether. Acute Oral Toxicity in the Rat*. Life Science Research, Rep. No. 86/BP0033/06, BP Chemicals, UK, 1986.
116. BP Chemicals. *Dipropylene Glycol Monoethyl Ether. Test to Evaluate the Acute Toxicity Following a Single Cutaneous Application (Limit Test) in the Rat*. Report 15590, BP Chemicals, Hazleton, UK, 1990.

117. BP Chemicals, *Dipropylene Glycol Monoethyl Ether. Acute Eye Irritation/Corrosion Test in the Rabbit,* Life Science Research, Rep. No. 85/BP0035/795, BP Chemicals, UK, 1985.

118. BP Chemicals, *Dipropylene Glycol Monoethyl Ether. Acute Dermal Irritation/Corrosion Test in the Rabbit,* Life Science Research, Rep. No. 85/BP0034/786, BP Chemicals, UK, 1985.

119. BP Chemicals, *Dipropylene Glycol Monoethyl Ether. 4 Week Oral (Gavage) Study in the Rat,* Report 15590, BP Chemicals, Hazleton, UK, 1990.

120. BP Chemicals, *Dipropylene Glycol Monoethyl Ether. Assessment of Its Mutagenic Potential in Histidine Auxotrophs of Salmonella typhimurium.* Life Science Research, Rep. No. 85/BP0036/698, BP Chemicals, UK, 1985.

121. BP Chemicals, *Dipropylene Glycol Monoethyl Ether. Test to Evaluate the Ability to Induce Chromosome Aberrations in Human Lymphocytes.* Report 16490D, BP Chemicals, Hazleton, UK, 1990.

122. BP Chemicals *Dipropylene Glycol Monoethyl Ether. Delayed Contact Hypersensitivity Study in Guinea Pigs.* Life Science Research, Rep. No. 85/BP0032/788. BP Chemicals, UK, 1985.

123. ATDAEI, Acute toxicity data. *J. Am. Coll. Toxicol., Part B* **1**, 172 (1992).

124. F. M. H. Debets, Study conducted at NOTOX Laboratories, unpublished report, sponsored by The Dow Chemical Company, 1987.

125. F. S. Cieszlak, R. J. Stebbins, and H. G. Verschuuren, Report K-0054-010, unpublished report, The Dow Chemical Company, 1990.

126. ARCO Chemical Company, Summary Information for DPTB, unpublished report, ARCO Chem. Co., 1997.

127. ARCO Chemical Company, *Single Dose Oral Toxicity in Rats/LD50 in Rats,* Conducted by MB Research Laboratories, Inc., Study MB92-1741 A, unpublished report, ARCO Chem. Co., 1992.

128. ARCO Chemical Company, *DPTB Acute Inhalation Toxicity in Rats, 4-Hour Exposure,* Conducted by Huntingdon Research Laboratory, ARO 48/970681, unpublished report, ARCO Chem. Co., 1997.

129. ARCO Chemical Company, *Acute Dermal Toxicity in Rabbits/LD50 in Rabbits,* Conducted by MB Research Laboratories, Inc., Study MB92-1741 B, unpublished report, ARCO Chem. Co., 1992.

130. ARCO Chemical Company, *Primary Dermal Irritation in Albino Rabbits; Dipropylene Glycol Tertiary Butyl Ether,* Conducted by MB Research Laboratories, Inc., Study MB 92-1741C, ARCO Chem. Co., 1992.

131. ARCO Chemical Company, *Primary Eye Irritation and/or Corrosion in Rabbits; Dipropylene Glycol Tertiary Butyl Ether,* Conducted by MB Research Laboratories, Inc., MB 92-1741 D, unpublished report, ARCO Chem. Co., 1992.

132. ARCO Chemical Company, *DPTB: Toxicity to Rats by Repeated Oral Administration for 14 Days,* Conducted by Huntingdon Life Sciences, Ltd., ARO 25/960403, unpublished report, ARCO Chem. Co., 1996.

133. ARCO Chemical Company, *Dipropylene Glycol t-Butyl Ether (DPTB): Toxicity to Rats by Repeated Oral Administration for 13 Weeks Incorporating a Neurotoxicity Screen and Followed by a 4-Week Recovery Period,* Conducted by Huntingdon Life Sciences, Ltd., ARO 26/962148, unpublished report, ARCO Chem. Co., 1996.

134. ARCO Chemical Company, *Dipropylene Glycol t-Butyl Ether: Bacterial Mutation Assay,* Conducted by Huntingdon Life Sciences, Ltd., ARO 27/960041, unpublished report, ARCO Chem. Co., 1996.

135. ARCO Chemical Company, *DPTB: Mouse Micronucleus Test,* Conducted by Huntingdon Life Sciences, Ltd., ARO 28/960732, unpublished report ARCO Chem. Co., 1996.

136. ARCO Chemical Company, *Dipropylene Glycol t-Butyl Ether: Skin Sensitization in the Guinea Pig,* Conducted by Huntingdon Life Sciences, Ltd., ARO 24/960421/SS, unpublished report, ARCO Chem. Co., 1996.

137. L. L. Calhoun et al., *Metabolism and Disposition of Tripropylene Glycol Monomethyl Ether (TPGME) in Male Rats,* Toxicology Research Laboratory Report, unpublished data, The Dow Chemical Company, 1986.

138. C. Breckenridge et al., *A Teratological Study of Inhaled Dowanol TPM in the Albino Rat,* Toxicology Research Laboratory Report, unpublished data, The Dow Chemical Company, 1985.

139. F. M. H. Debets and H. G. Verschuuren, *Assessment of Oral Toxicity, Including the Haemolytic Activity of Dowanol PnB in the Rat. 14-Day Study,* Toxicology Research Laboratory Report, unpublished data, Dow Europe, Horgan, Switzerland, 1988.

140. S. N. J. Pang, Final report on the safety assessment of PPG-40 butyl ether. *J. Am. Coll. Toxicol.* **12**(3), 257–259 (1993).

141. C. P. Carpenter et al., Toxicology of two butoxypolypropylene glycol fly repellents. *Arch. Ind. Hyg. Occup. Med.* **4**, 261–269 (1951).

142. A. B. Boese, C. K. Kink, and H. G. Goodman, Jr., in G. O. Curme, Jr. and F. Johnston, eds., *Glycols,* Am. Chem. Soc. Monogr. Ser. 114, Reinhold, New York, 1952.

143. P. W. Morgan, *Ind. Eng. Chem. Anal. Ed.* **18**, 500 (1946).

144. S. Lepkovski, R. A. Over, and H. M. Evans, *J. Biol. Chem.* **108**, 431 (1935).

145. C. B. Shaffer and F. H. Critchfield, *Fed. Proc., Fed. Am. Soc. Exp. Biol.* **7**, 254 (1948).

146. F. H. Wiley et al., The formation of oxalic acid from ethylene glycol and related solvents. *J. Ind. Hyg. Toxicol.* **20**, 269–277 (1938).

147. R. Truhaut, et al., Comparative toxicological study of ethyl glycol acetate and butyl glycol acetate. *Toxicol. Appl. Pharmacol.* **51**, 117–127 (1979).

148. E. Rosser, in K. B. Lehmann and F. Flury, eds., E. King and F. F. Smyth, Jr., transl., *Toxicology and Hygiene of Industrial Solvents,* Springer, Berlin, 1938.

149. C. P. Carpenter and H. F. Smyth, Jr., Chemical burns of the rabbit cornea. *Am. J. Ophthalmol.* **29**, 1363–1372 (1946).

150. K. Nagano et al., Experimental studies on toxicity of ethylene glycol alkyl ethers in Japan. *Environ. Health Perspect.* **57**, 75–84 (1984).

151. H. D. Kesten, M. G. Mulinos, and L. Pomerantz, Pathological effects of certain glycols and related compounds. *Arch. Pathol.* **27**, 447–465 (1939); *Chem. Abstr.* **33**, 4659 (1939).

152. M. G. Mulinos, L. Pomerantz, and M. E. Lojkin, The metabolism and toxicology of ethylene glycol and ethylene glycol diacetate. *Am. J. Pharm.* **115**, 51-63 (1943); *Chem. Abstr.* **37**, 4136 (1943).

153. T. H. J. M. Waegemaekers and M. P. M. Bensink, Non-mutagenicity of 27 aliphatic acrylate esters in the *Salmonella*-microsome test. *Mutat. Res.* **137**, 95–102 (1984).

154. T. P. Cameron et al., Genotoxicity of multifunctional acrylates in the *Salmonella*/mammalian-microsome assay and mouse lymphoma TK+/-assay. *Environ. Mol. Mutagen.* **17**(4), 264–271 (1991).

155. L. S. Andrews and J. J. Clary, Review of the toxicity of multifunctional acrylates. *J. Toxicol. Environ. Health,* **19**, 149–164 (1986).

156. H. B. Van der Walle, T. H. Waegemaekers, and T. Bensink, Sensitizing potential of 12 di(methyl)acrylates in the guinea pig. *Contact Dermatitis* **9**, 10–20 (1983).

157. B. Bjorkner, I. Dahlquist, and S. Fregert, Allergic contact dermatitis from acrylates in ultraviolet curing inks. *Contact Dermatitis* **6**, 405–409 (1980).

158. L. Kanerva et al., Fingertip paresthesia and occupational allergic contact dermatitis caused by acrylates in a dental nurse. *Contact Dermatitis* **38**(2), 114–116 (1998).

159. Eastman Kodak Company, *Toxicity and Health Hazard Summary*, Lab. No. 65-563, Acc. No. 902855, unpublished data, Corporate Health and Environment Laboratories, Eastman Kodak Co., Rochester, NY, 1966.

160. N. L. Bossert et al., Reproductive toxicity of triethylene glycol and its diacetate and dimethyl ether derivatives in a continuous breeding protocol in Swiss CD-1 mice. *Fundam. Appl. Toxicol.* **18**(4), 602–608 (1992).

161. W. E. Rinehart, M. Kaschak, and E. A. Pfitzer, *Chem-Toxicol. Ser. Bull.*, 6, Industrial Hygiene Foundation of America, 1967.

162. H. Tanii and K. Hashimoto, Structure-toxicity relationship of acrylates and methacrylates. *Toxicol. Lett.* **11**, 125–129 (1982).

163. K. Watanabe, K. Sakamoto, and T. Sasaki, Comparisons on chemically-induced mutagenicity among four bacterial strains, *Salmonella typhimurium* TA102 and TA2638, and *Escherichia coli* WP2/pKM101 and WP2 uvrA/pKM101: Collaborative Study I. *Mutat. Res.* **361**, 143–155 (1996).

164. K. S. Rao, J. E. Betso, and K. J. Olson, A collection of guinea pig sensitization test results-grouped by chemical class. *Drug and Chem. Toxicol.* **4**, 331–351 (1981).

165. H. C. Maguire, *J. Soc. Cosmet. Chem.* **24**, 151 (1973).

166. S. Clemmensen, Cross-reaction patterns in guinea pig sensitized to acrylic monomers. *Drug Chem. Toxicol.* **7**, 527–540 (1984).

167. O. E. Fancher et al., Toxicology of a butylene glycol adipic acid polyster. *Toxicol. Appl. Pharmacol.* **26**, 58-62 (1973).

168. P. L. Wright et al., Metabolism of intravenously or orally administered butylene glycol adipic acid polyester. *Toxicol. Appl. Pharmacol.* **29**, 115 (abstr.) (1974).

169. J. R. Kominsky and R. W. Freyberg, *Exposure to Volatile Components of Polyvinyl Acetate (PVA) Emulsion Paints During Application and Drying*, Report Prepared by International Technology Corporation, Cincinnati, OH for the National Paint and Coatings Association, 1992.

170. U. Ulfvarson et al., Temporary health effects from exposure to water-borne paints. *Scand. J. Work, Environ. Health* **18**, 376-387 (1992).

171. J. L. O'Donoghue, Lab. No. 84-0079, Acc. No. 000701, Rep. No. TX-84-35, unpublished data, Corporate Health and Environment Laboratories, Eastman Kodak Company, Rochester, NY, 1984.

172. W. D. Faber and R. S. Hosenfeld, Lab. No. 91-0053, Acc. No. 000701, Rep. No. TX-92-57, unpublished data, Corporate Health and Environment Laboratories, Eastman Kodak Company, 3, Rochester, NY, 1992.

173. Eastman Kodak Company, Lab. No. 84-0079, Acc. No. 000701, Rep. No. TX-85-5, unpublished data, Corporate Health and Environment Laboratories, Eastman Kodak Company, Rochester, NY, 1985.

174. E. D. Barber et al., Rep. No. TX-91-309, unpublished data, Corporate Health and Environment Laboratories, Eastman Kodak Company, Rochester, NY, 1992.

175. R. J. Boatman, Rep. No. TX-89-28, unpublished data, Corporate Health and Environment Laboratories, Eastman Kodak Company, Rochester, NY, 1989.
176. H. C. Shields and C. J. Weschler, Analysis of ambient concentrations of organic vapors with a passive sampler. *J. Air Pollut. Control Assoc.* **37,** 1039–1045 (1987).
177. D. Norback, G. Wieslander, and C. Edling, Occupational exposure to volatile organic compounds (VOCs), and other air pollutants from the indoor application of water-based paints. *Ann. Occup. Hyg.* **39**(6), 783–794 (1995).
178. B. D. Astill, C. J. Terhaar, and D. W. Fassett, The toxicology and fate of 2,2,4-trimethyl-1,3-pentanediol diisobutyrate. *Toxicol. Appl. Pharmacol.* **22,** 387–399 (1972).
179. W. J. Krasavage, K. S. Tischer, and R. L. Roudabush, The reversibility of increased rat liver weights and microsomal processing enzymes after feeding high levels of 2,2,4-trimethyl-1,3-pentanediol diisobutyrate. *Toxicol. Appl. Pharmacol.* **22,** 400–408 (1972).
180. G. Wieslander et al., Asthma and the indoor environment: The significance of emission of formaldehyde and volatile organic compounds from newly painted indoor surfaces. *Int. Arch. Occup. Environ. Health* **69,** 115–124 (1997).
181. D. Norback et al., Asthma symptoms in relation to volatile oganic compounds (VOC) and bacteria in dwellings. In G. Leslie and R. Perry, eds., *Volatile Organic Compounds in the Environment,* IAI, Rothenfluh, Switzerland, 1993, pp. 377–386.
182. D. Zissu, Experimental study of cutaneous tolerance to glycol ethers. *Contact Dermatitis* **32**(2), 74–77 (1995).
183. A. E. Asaki and J. T. Houpt, *Asssessment of the Developmental Toxicity of Propylene Glycol Monomethyl Ether Acetate (PM Acetate) in Rats,* Rep. No. USAEHA-75-51-0753-90, Army Environmental Hygiene Agency, Aberdeen Proving Ground, MD, 1990.
184. BP Chemicals, *Acute Oral Toxicity to Rats of Ethoxypropyl Acetate,* Report 851078D/BPC 51/AC, Huntingdon Research Centre, BP Chemicals, UK, 1985.
185. BP Chemicals, *Acute Inhalation Toxicity in Rats 4-hour Exposure-Ethoxypropyl Acetate,* Report BPC 55/851141, Huntingdon Research Centre, BP Chemicals, UK, 1985.
186. BP Chemicals, *Irritant Effects on the Rabbit Eye of Ethoxypropyl Acetate,* Report BPC 85938D/BPC 52/SE, Huntingdon Research Centre, BP Chemicals, UK, 1986.
187. BP Chemicals, *Irritant Effects on the Rabbit Skin of Ethoxypropyl Acetate,* Report BPC 85937D/BPC 52/SE, Huntingdon Research Centre, BP Chemicals, UK, 1986.
188. BP Chemicals, *Ethoxypropyl Acetate 28-Day Inhalation Toxicity Study in Rats by Administration on 5 Days each Week,* Report BPC 56/8655, Huntingdon Research Centre, BP Chemicals, UK, 1986.
189. BP Chemicals, *Ames Metabolic Activation Test to Assess the Potential Mutagenic Effect of Ethoxypropyl Acetate,* Report BPC 57/851010, Huntingdon Research Centre, BP Chemicals, UK, 1985.
190. BP Chemicals, *Analysis of Metaphase Chromosomes Obtained from CHO Cells Cultured* In Vitro *and Treated with Ethoxypropyl Acetate,* Report BPC 58/851160, Huntingdon Research Centre, BP Chemicals, UK, 1985.
191. BP Chemicals, *A Study of Delayed Contact Hypersensitivity in the Guinea Pig with Ethoxypropyl Acetate,* Report BPC 851091D/BPC 54/SS, Huntingdon Research Centre, BP Chemicals, UK, 1986.
192. National Institute of Occupational Safety and Health (NIOSH), *NIOSH Manual of Analytical Methods,* 3rd ed., Vol. 2, Method 2507 (1) Nitroglycerine and (2) Ethylene Glycol Dinitrate,

U.S. Department of Health, Education and Welfare (NIOSH), Cincinnati, OH 1990, pp. 2507–2511.

193. H. E. Christensen, ed., *Registry of Toxic Effects of Chemical Substances*, U.S. Department Health, Education and Welfare, NIOSH, Rockville, MD, 1976.

194. Anonymous, Ethylene glycol dinitrate (nitroglycol, dinitroglycol, EGDN, EGN, ethylene nitrate). *Am. Ind. Hyg. Assoc. J.* **27**, 574–577 (1966).

195. E. Gross, M. Bock, and F. Hellrung, *Naunyn-Schmiedebergs, Arch. Exp. Pathol. Pharmakol.* **200**, 271 (1942).

196. R. A. Jones, J. A. Strickland, and J. Siegel, Toxicity of propylene glycol 1,2-dinitrate in experimental animals. *Toxicol. Appl. Pharmacol.* **22**, 128–137 (1972).

197. N. Sugawara, *Bull. Yamaguchi Med. Sch.* **20**, 99 (1973).

198. G. Cavagna, G. Locati, and M. Capizzi, *Med. Lav.* **59**, 772 (1968).

199. E. C. Vigliani et al., Biological effects of nitroglycol on the metabolism of catecholamines. *Arch. Environ. Health* **16**, 477–484 (1968).

200. M. Kalin, B. Kylin, and T. Malmfors, The effect of nitrate explosives on adrenergic nerves. *Arch. Environ. Health.* **19**, 32–35 (1969).

201. M. Minami et al., Effect of ethylene glycol dinitrate on metabolism of catecholamines and on blood pressure reaction to re-exposure. *Br. J. Ind. Med.* **29**, 321–327 (1972).

202. D. G. Clark, Effects of ethylene glycol dinitrate on pituitary-adrenocortical function in the rat. *Toxicol. Appl. Pharmacol.* **21**, 355–360 (1972).

203. D. G. Clark, and M. H. Litchfield, Metabolism of ethylene glycol dinitrate (ethylene dinitrate) in the rat following repeated administration. *Br. J. Ind. Med.* **26**, 150–155 (1969).

204. M. H. Litchfield, *J. Pharm. Sci.* **60**, 1599 (1971).

205. M. H. Litchfield, *Drug Metab. Rev.* **2**, 239 (1973).

206. D. G. Clark and M. H. Litchfield, Metabolism of ethylene glycol dinitrate and its influence on the blood pressure of the rat. *Br. J. Ind. Med.* **24**, 320–329 (1967).

207. H. Tsuruta and H. Hasegawa, *Ind. Health* **8**, 119 (1970).

208. H. Hasegawa and M. Sato, *Ind. Health* **1**, 20 (1963).

209. H. Hasegawa and M. Sato, *Ind. Health* **8**, 88 (1970).

210. H. Hasegawa, M. Sato, and H. Tsuruta, *Ind. Health* **8**, 153 (1970).

211. E. R. Plunkett, *Handbook of Industrial Toxicology*, Chem. Publ. Co., New York, 1976.

212. F. Gobbato and E. DeRosa, *Folia Med.* **52**, 537 (1969).

213. K. Suwa, M. Sato, and H. Hasegawa, *Ind. Health* **2**, 80 (1964).

214. S. D. Kanonova et al., *Genetika* **8**, 101 (1972).

215. T. Tai and H. Tsuruta, The effects of nitroglycol on rat isolated cardiac muscles. *Ind. Health* **35**(4), 515–518 (1997).

216. P. Carmichael and J. Lieben, Sudden death in explosives workers. *Arch. Environ. Health* **7**, 424–439 (1963).

217. W. F. von Oettingen, *Natl. Inst. Health Bull.* **186** (1964).

218. H. Symanski, *Arch. Hyg. Bakteriol.* **136**, 139 (1952).

219. S. Forssman et al., Untersuchungen des Gesundheitszustandes von Nitroarbeitern bei drei schwedischen Sprengstoffabriken *Arch. Gewerbepathol. Gewerbehyg.* **16**, 157–177 (1958).

220. C. Einert et al., Exposure to mixtures of nitroglycerin and ethylene glycol dinitrate *J. Am. Ind. Hyg. Assoc.* **24**, 435–447 (1963).

221. S. H. Lamm, K. S. Grumski, and R. F. Fethers, Short-term exposures to ethylene glycol dinitrate concentrations greater than 0.4 mg/m3 show no evidence of nitrate-induced headaches *Am. J. Epidemiol.* **138**(8), 653 (abstr.) (1993).

222. D. C. Trainor and R. C. Jones, Headaches in explosive magazine workers. *Arch. Environ. Health* **12**, 231–234 (1966).

223. Y. Morikawa et al., Organic nitrate poisoning at an explosives factory. *Arch. Environ. Heath* **14**, 614–621 (1967).

224. P. Martini and L. Massari, *Med. Lav.* **56**, 62 (1965).

225. G. N. Krasovsky, A. A. Korolav, and S. A. Shigan, *J. Hyg. Epidemiol. Microbiol. Immunol.* **17**, 114 (1973).

226. L. D. Brown et al., Acute oral toxicological evaluation of diethyleneglycol dinitrate. *J. Am. Coll. Toxicol.* **12**(6), 602–603 (1993).

227. D. W. Korte, Jr. et al., Ocular and dermal toxicity of diethyleneglycol dinitrate. *J. Am. Coll. Toxicol.* **12**(6), 600–601 (1993).

228. M. E. Andersen and R. G. Mehl, A comparison of the toxicology of triethylene glycol dinitrate and propylene glycol dinitrate. *J. Am. Ind. Hyg. Assoc.* **34**, 526–532 (1973).

229. D. W. Korte, Jr. et al., Acute oral toxicological evaluation of triethyleneglycol dinitrate. *J. Am. Coll. Toxicol.* **12**(6), 606–607 (1993).

230. D. W. Korte, Jr. et al., Ocular and dermal toxicity of triethyleneglycol dinitrate *J. Am. Coll. Toxicol.* **12**(6), 604–605 (1993).

231. R. D. Stewart et al., Experimental human exposure to propylene glycol dinitrate[1,2] *Toxicol. Appl. Pharmacol.* **30**, 377–395 (1974).

232. D. G. Clark and M. H. Litchfield, The toxicity, metabolism, and pharmacologic properties of propylene glycol 1,2-dinitrate. *Toxicol. Appl. Pharmacol.* **15**, 175–184 (1969).

233. C. S. Godin et al., Effect of propylene glycol 1,2-dinitrate on cerebral blood flow in rats: A potential biomarker for vascular headache? *Toxicol. Lett.* **75**(1–3), 59–68 (1985).

234. J. L. Mattsson, J. W. Crock, Jr., and L. J. Jenkins, Jr., Report AD-A023770, U.S. Department of Commerce, National Technical Information Service, Washington, DC, 1975.

235. R. W. Young et al., Report AD-A0266841, U.S. Department of Commerce, National Technical Information Service, Washington, DC, 1976.

236. C. S. Godin et al., Effect of exposure route on measurement of blood pressure by tail cuff in F-344 rats. *Toxicol. Lett.* **66**(2), 147–155 (1993).

237. M. E. Andersen and R. A. Smith, On the mechanism of the oxidation of human and rat hemoglobin by propylene glycol dinitrate. *Biochem. Pharmacol.* **22**, 3247–3256 (1973).

CHAPTER EIGHTY-EIGHT

Synthetic Polymers

Bailus Walker, Jr., Ph.D., MPH, and Laureen Burton, MPH

INTRODUCTION

Polymers are among the oldest chemicals known to humans. Since ancient times, natural polymers, including silk, wool, rubber, and cellulose, have been known and exploited. For a century and a half now, synthetic fibers, plastics, and modified natural polymers have been part of everyday life. The early history of these materials is being documented, and new generations are being taught about their production and fabrication. In the United States, a number of institutions have been established to preserve and transmit knowledge of polymer technology. Natural polymers such as proteins occupy a central position in the architecture and function of living matter. Nucleic acid polymer, called deoxyribonucleic acid (DNA), automatically controls the formation of another nucleic acid, ribonucleic acid (RNA), that controls the formation of a specific protein. These natural polymers are increasingly important, especially in agriculture and medicine, and much has been written about them. They are not considered here.

Many of the general considerations discussed in the chapters on synthetic polymers in the fourth edition of *Patty's Industrial Hygiene and Toxicology* are still applicable. To update those portions that need alteration, literature since 1994 when the fourth edition was published has been searched using primarily the National Library of Medicine's Hazardous Substances Data Bank (HSDB) and ToxLine. Other sources including the polymer science literature were reviewed for useful information. However, there has been no intent to compile an exhaustive bibliography. References to the multiple dimensions of synthetic polymers are so numerous that it has been necessary to limit citations to those that are most useful to understanding and controlling the toxicological problems of polymers in the occupational environment.

Patty's Toxicology, Fifth Edition, Volume 7, Edited by Eula Bingham, Barbara Cohrssen, and Charles H. Powell.
ISBN 0-471-31940-6 © 2001 John Wiley & Sons, Inc.

The Polymer Industry

Perhaps no other aspect of the chemical industry has grown as rapidly as synthetic polymers. Each year since 1868 when cellulose nitrate was first introduced, new commercial polymers have been produced for a broad range of industrial and nonindustrial applications. Prominent examples include polyvinyl chloride which was introduced in 1927, ethyl cellulose in 1931, fluoropolymers in 1943, acrylonitrile-butadiene-styrene (ABS) in 1948, polypropylene in 1957, and polyisoprene in 1962.

In the 1970s and 1980s, engineering polymers made their mark by replacing traditional materials such as metals, rubber, automotive glass, electronics, and other materials. During the decade of the 1980s, the polymer industry's focus shifted to developing blends and alloys of existing polymers. In the 1990s, new products were rapidly developed, as manufacturers capitalized on developments in metallocene catalysis and other new chemistry. A manufacturer using metallocene catalysis transformed polystyrene — an amorphous, low melting point, relatively brittle plastic — into a tough crystalline material that has a melting point of 270°C, and employs the same styrene monomer. Medical and diagnostic containers are expected to benefit from another promising new cycloolefin copolymer, which is expected to be produced in large quantities by mid-2000. The copolymer's clarity, stiffness, and moisture and shatter resistance should make it attractive for pharmaceutical packaging, where moisture resistance is needed. It will also be useful in cut-produce packaging where the copolymer's high oxygen permeability is a desirable property.

In 1997, production in most sectors of the chemical industry grew. Production in overall chemical industries rose 4% from the previous year, up from the 3% gain in 1996. Output of synthetic material posted a 5% gain in 1997 continuing a six-year pattern of steady growth. Within that category, plastic material output rose 5%, and synthetic fiber output increased 4%.

The Society of Plastics Industry reports production gains of 5% for thermosetting resins and 6% for thermoplastic resins in 1997. The largest gain was for polypropylene, which was up 11% to 13.3 billion 1b. Nearly 1,000 new plastic-related products and services were on display at the 1997 National Plastics Exposition. A broad array of recent advances in resins and additives were also exhibited. Synthetic rubber shipment grew 3% in 1997, marking the sixth year of continuous growth. Outpacing the average for the group were both styrene-butadiene rubber and polybutadiene, up 4% to 939,006 and 555,000 metric tons, respectively. There is also evidence that more stringent air pollution regulations are fueling the demand for polymer products. For example, the demand for emulsion polymers is expected to reach 4.5 billion lb by 2000, representing a 3.1% average annual increase. The switch from solvent-borne to waterborne emulsion polymers and other improved formulations will help reduce a source of air pollution.

In 1998, several major chemical producers announced plans to expand capacity for polymeric materials in response to U.S. and global demands, and there is already substantial evidence of emerging new waves of polymers which could spur further growth in the industry. Another measure of the expanse and vibrance of the synthetic polymer industry is the size of its labor force. U.S. Department of Labor figures show that 60% of the workers in the chemical industry work in some way with synthetic polymers, and the future of this market appears to be bright. The U.S. Department of Labor expects the number of all jobs in the chemical industry to rise about 38,000, nearly 40%, to 1.1 million between 1996 and 2006.

Indeed, the hiring wave predicted in 1997 for workers in the chemical industry materialized in the spring of 1998, as reflected in the space devoted to help-wanted ads in chemical journals. Such ads, in *Chemical and Engineering News* for example, accounted for 170 pages in the first 16 weeks of 1998 compared with 136 pages in the same period in 1997.

Polymer Characteristics

Although considerable portions of textbooks in organic chemistry are devoted to polymers, polymer chemistry is not simply a subdivision of organic chemistry. It has its own methods of synthesis and its own methods of characterization.

A number of specific characteristics are defined by polymeric compounds:

- Their molecular weight is high and ranges from a few thousands to several millions (or even to infinity) in cross-linked polymeric materials.
- They are constituted of repeat units (or *mers*) generally named after the low molecular weight constituents (monomers) from which they originate.
- Polymer synthesis always involves a large number of steps; each of them results in forming a bond and implies extension of the chain.
- Polymers are usually polydisperse, which means that macromolecules that exhibit the same chemical composition but different chain length coexist in a sample.

Classification of Polymers

There are a number of ways to classify polymers. One of the oldest methods is based on the material's response to heat. In this system there are two types: thermoplastic and thermosets. Thermoplastic, the polymers melt when heated and solidify when cooling. This cycle of heating and cooling can be applied several times without affecting properties. Thermoset polymers melt only the first time they are heated. A more important classification is based on molecular structure. The polymers could be (1) linear-chain polymers, (2) branched-chain polymers, and (3) network or gel polymers. The repeat units in linear-chain polymers are joined by strong covalent bonds, and different molecules are held together by weaker secondary forces. A branched-chain polymer contains molecules that have a linear backbones, and branches emanate randomly from it.

Whenever monomers are polymerized, the polymer evolves through a collection of linear chains to a collection of branched chains that ultimately form a network, or a *gel*, polymer. These variations in chemical structure give polymeric compounds their distinctive properties: high strength, elasticity, high viscosity as melts and solutions. Such properties have made polymers desirable for use in food packaging, clothing, home and office furnishing, transportation, medical equipment, and information technology. To make polymers suitable for specific end uses, various additives are incorporated in varying amounts. Some of the additives are antioxidants, antistatic agents, flame retardants, inhibitors, and plasticizers. Polymers may also require fillers, stabilizers, and reinforcing agents to modify mechanical and electrical properties. Additives occupy a major niche in polymer production. In 1997, the global plastic additive business was valued at $16 billion. This figure does not include colorants or reinforcements. The dividing line between polymers of low molecular weight (oligomers) and high is usually placed in the molecular range of 10,000, but sometimes as low as 5,000. Individual materials discussed in this review are essentially prototype polymers. The space allotted in this volume precludes discussion of the infinite array of products presently marketed.

Two types of polymerization mechanisms that result in polymer molecules (macromolecules) should be distinguished. *Step-growth* polymerization, also referred to as polycondensation, is a process in which two chemicals of different type interact, frequently by eliminating fragments or small molecules such as water or formaldehyde. Typical examples of polymers that arise from step-growth polymerization include polyester from a diacid and a dialcohol, polyamides from a diamine and a diacid, and polyurethane from a diisocyanate and a diacid. *Chain-growth polymerization* includes several events: initiation that results in the formation of an active site on the monomer molecule and chain growth (or propagation) that comprises a large number of steps. Each implies the addition of a monomer molecule to an active site; termination involves the destruction of the site where the growth of the chain stops, and transfer reactions imply the removal of the active site from a macromolecule that stops growing. Polyethylene and other olefin polymers are examples of polymers that are manufactured by chain-growth polymerization.

Synthetic polymers characteristically are a mixture of many chains of different weights. Different manufacturing processes result in trace amounts of different ingredients that can have a major effect on product behavior. Polymers that have the same fundamental unit structure but significantly different molecular weight, may have quite different properties.

The polymer's molecular weight or chain conformation is essentially an average that varies somewhat with different measurement methods. Therefore, one should not speak of molecular weight, but rather the average molecular weight. This weight is generally expressed as the number (M_N), the weight (M_W), or the viscosity (M_V) average. When all molecules are within a narrow weight range, $M_W = M_N$, and the system is monodisperse; otherwise, and more commonly, M_W exceeds M_N. The ratio of M_W and M_N defines the polydispersity index (Q), which is a measure of the molecular weight distribution. Q of a monodisperse polymer is unity. Commercial polymers may have a Q between 2 and 20.

Kumar and Gupta (1) summarized the importance of measuring the molecular weight. The molecular weight and its distribution determine the viscous and elastic properties of molten polymers. This affects the processibility of the melt and also the behavior of the resulting solid material. Differences in molecular weight distribution also influence the extent of polymer chain entanglement and the amount of melt elasticity, as measured by phenomena such as extrudate swell. The effects of swell show up during processing, wherein flow results in different amounts of chain extension and orientation that remain frozen in the solidified part. Consequently, two chemically similar polymers that are processed identically and have the same molecular weight, but different molecular weight distribution, may result in a product that shows significantly different shrinkage, tensile properties, and failure properties.

The regions of dramatically different polymer properties are separated by the glass transition temperature (T_g) defined as the temperature at which hard glassy polymer becomes a rubbery material. The T_g of a polymer can often characterize its potential use. The T_g of fibers should be well above room temperature and that of rubber well below room temperature for typical "plastics."

Toxicological Characteristics

Our understanding of the toxicological characteristics of synthetic polymers has changed little since the fourth edition of *Patty's Industrial Hygiene and Toxicology*. Many of the properties already discussed reduce the toxic potential of polymeric materials. For example, the high molecular weight of the synthetic polymer itself impedes its ability to penetrate membranes and reach blood capillaries by diffusion through cells, where the effects of toxic natural or synthetic chemicals unfold. Consequently, it is rare for exposure to a fully reacted or "cured" polymer to increase the risk of toxic injury. The risks are most often attributable to residual monomers, unreacted or underreacted constituents, products of degradation, or of thermal decomposition. Decomposition products depend on the chemical composition of the intact polymer and also on the condition by which they are decomposed. The temperature to which the polymer is subjected, the atmosphere in which decomposition occurs, and the material of the vessel used can alter the kinds and quantities of the decomposition products formed.

Toxic injury to the respiratory system may be provoked by exposure to polymer dust, volatile chemical species generated during processing of polymers, and thermal degradation products. Assessing the nature and extent of human contact with these agents is within the purview of industrial hygiene. Although the industrial hygiene techniques for identifying, evaluating, and controlling products of polymer processing are often tailored for each type of risk, the pinciples of evaluation can be generalized. These principles are discussed (in detail) elsewhere in this book.

Of relevance and concern are the high toxicity of many monomers, variations in the quantities of complex catalysts and additives used in the polymerization process, the presence of extractable residual monomers in the final polymer product, and the toxicological properties of low molecular weight polymer that consist of only a few monomer units (dimers, trimers). Also of toxicological concern is the biotransformation of polymeric materials that may reach systemic circulation and could result in forming less toxic or more toxic metabolites. The biotransformation may be at a fast or slow rate with obvious consequences for the concentration and thus the toxicity at the target site. Biotransformation, along with other interrelated processes of disposition — absorption, distribution, and excretion — may increase the rate of elimination of polymeric materials from an organism. This in turn, will lower the concentration and hence its toxicity in a target tissue or organ.

It is also important to know that there is the potential for "solid-state carcinogenesis" in which solid material that has an uninterrupted, smooth, round surface and critical size can cause local tissue sarcoma when implanted. The chemical nature of the polymer implant is not the critical factor in its ability to transform normal cells to neoplastic cells.

The biocompatibility of polymers, specifically breast implants that contain silicone gel, has been a controversial issue in recent years. FDA estimates that 1 million to 2 million

U.S. women have received various types of implants since they were introduced for cosmetic surgery. By 1992, widespread reports of adverse reactions and insufficient evidence of the safety of implants led the agency to pull silicone gel implants off the market. These implants consist of silicone rubber envelopes filled with silicone gel. Litigation and settlements tied to claims of implanted-related health impairment have been going on for years. In 1998, a panel of immunologists, epidemiologists, and rheumatologists issued a report concluding that some of the research was flawed and that scientific evidence has so far failed to show that silicone breast implants cause disease.

A National Cancer Institute study of 13,500 women who have had implants is examining potential long-term health impacts. A panel convened by the Institute of Medicine is also continuing to review the medical evidence to determine what is not yet known.

Dose–Response Relationship

In toxicology, the quantal dose response is used extensively. Determination of the median lethal dose (LD_{50}) is among the first measures of acute toxicity for new chemicals. For synthetic polymers, the values of the LD_{50} are often above the limits of practical experimental testing. The sheer bulk of the required dose of a relatively inert polymer for dose–response measurement may overshadow the toxic potential.

Although by itself the LD_{50} value has limited significance, acute lethality studies are essential to characterize the toxic effects of chemicals and their risk for human health. The most meaningful risk assessment information comes from clinical observations and postmortem examination of experimental animals rather than from a specific LD_{50} value. Toxicologists do not, however, have convincing evidence that every type of toxic response to a chemical observed in species of laboratory animals will also be expected to develop in similarly exposed humans. This issue is further complicated by the many examples of different species of laboratory animals that exhibit different responses to the same chemical exposure. The LD_{50} has come under sharp criticism from animal welfare groups for its use of 30–100 experimental animals per test. In 1999, these groups objected to the large number of animal tests that would be needed by an EPA program in which companies would voluntarily conduct toxicity tests for some 3000 high-production volume chemicals sold or imported in amounts exceeding 1 million lb per year. In response to these concerns, FDA and Proctor & Gamble recently proposed a new test method called the "up-and-down" procedure. It uses a fraction of the number of animals used in the LD_{50} and produces enough information to evaluate the consequences of a single chemical exposure and to serve as a basis for hazard classification and labeling. The development of other methods for testing chemicals is on the rise, including research aimed at developing innovative skin and eye analogs for *in vitro* chemical testing that may eventually replace traditional animal testing. Some synthetic polymers intended as direct food additives have been assessed for acceptable daily intake (ADI). The ADI defines the daily intake of chemicals which, during an entire lifetime, is without appreciable risk on the basis of all facts known at that time. For polymers intended for food contact application, such as wrapping or packaging, the toxic potential is more often assessed in terms of any possible migration of low molecular weight components into foods.

Estrogenic Effects

In recent years, there has been much discussion about environmental and industrial chemicals, such as monomers and polymers, that may mimic or interfere with endocrine functions. The so-called endocrine disrupters have been the subject of a plethora of publications, conferences symposia, and seminars. Reports have hypothesized a link between endocrine disrupters and testicular tumors, lower sperm count, breast cancer, and birth defects. Human exposure to estrogenic chemicals is not confined to industrial products. Naturally occurring compounds have been identified. Besides the human effects suggested, there are reported estrogenic effects in wildlife populations that are related to fertility and sexual development.

This information raises the question whether estrogenic effects identified pose a risk for human health. There is no clear evidence now that polymers contribute to the pool of suggested environmental estrogens. The Environmental Protection Agency has begun screening thousands of chemicals to determine which may be endocrine disrupters and need further study. Such studies will require considering bioavailability, lipophobicity, metabolism, and toxicokinetics. These characteristics are not totally understood for many monomers, polymers, and their thermal decomposition products.

The level, time, and duration of exposure to these products is also important in establishing any biological significance for polymers. Therefore, future studies of the effects of polymers on normal endocrine function must be pursued at several levels — basic, clinical, and epidemiological.

Clearly, toxicological interests in polymers include both acute and chronic effects. Any situation in which a worker may be exposed to high concentrations of vapors or dust should be controlled by best industrial hygiene practices. Even these do not always preclude overexposure to toxic levels of chemicals. Thus, it is necessary for industrial health professionals to have guidelines for judging the potential health hazards of industrial chemicals. The guidelines — derived from experimental and epidemiological investigations in industrial toxicology — are expressed in terms of maximum allowable concentrations (MAC), occupational exposure limits (OEL), or permissible exposure limits (PEL). Limits been defined and published for finished synthetic polymers in only a very few cases. The problems and philosophical dimensions of setting limits have been reviewed in numerous publications and discussed in a range of conferences and symposia.

Fire and Polymers

As indicated previously, fully "cured" polymers pose few risks of toxic injury. However, almost all polymers can be pyrolyzed and/or burned. Pyrolysis is defined as "the irreversible chemical decomposition caused by heat, usually without oxidation;" combustion is defined as the "chemical process of oxidation that occurs at a rate fast enough to produce a temperature rise and usually light either as a glow or flame" (2). These processes have both physical and chemical elements (3). This section discusses features of polymer pyrolysis and thermal decomposition and illustrates current environmental and health considerations.

With sufficient heat, polymeric materials often first undergo phase change, such as melting in the case of thermoplastics, followed by chemical decomposition. Chemical decomposition processes may involve a number of mechanisms, such as chain scission, chain stripping, and cross-linking. These are endothermic processes that produce volatile low molecular weight products that may in turn burn.

Smoke from pyrolysis often contains gases which, if sufficiently heated, then burn, in the presence of an oxidizing agent (air) and an ignition source. For combustion to be self-sustaining, it is necessary for the burning gases to release sufficient heat energy to the material to continue producing gaseous fuel vapors or volatiles. The process is a continuous feedback loop: heat transferred to the polymer generates flammable volatiles, these volatiles react with the oxygen in the air above the polymer to generate heat, and part of this heat is transferred back to the polymer to continue the process. Chemical oxidation for the combustion of organic polymers is generally quite exothermic, and more than ample energy is produced to continue the pyrolysis and bond-breaking processes.

It is difficult to generalize on the combustibility of polymers because fire performance is influenced by a number of factors, including the chemical composition and structure of the polymer, the use of additives in formulated polymer systems, and even the conditions of the combustibility tests themselves. Highly cross-linked thermoset polymers normally burn less readily than thermoplastics. Cellular plastics (foams) generally burn quite readily because of their high surface area and good thermal insulating properties, which prevent dissipation of heat.

Polymer systems that contain halogens (e.g., polyvinyl chloride) burn with difficulty; however, the addition of plasticizers increases their propensity to burn. The fire properties of polymer systems can be modified by the addition of fire retardants that include organic halogen and phosphorus compounds, as well as a number of metal oxides and hydrates. Some are used in combinations. Although the use of fire retardants introduces additional potential toxicants to a combustion atmosphere, this is more than offset by the reduction of the hazard through increased resistance to ignition and/or lower burning rates (4). However, polymer systems that contain fire-retardant additives and even inherently fire-resistant polymers still burn, given sufficient heat input or oxidative conditions.

There are numerous tests for the various aspects of combustibility, including those for ignition, flame spread, heat release, visible smoke, and fire endurance. Fire test response depends greatly on the conditions under which the tests are conducted; thus materials often produce results that point to different and sometimes contradictory conclusions, even when the tests measure the same characteristic. Furthermore, few laboratory fire tests relate well to real-scale fires, so that predicting actual fire performance from laboratory data is difficult at best. This applies particularly to the assessment of the toxicity of smoke produced in actual fires.

Smoke

Smoke is defined as the airborne solid and liquid particulates and gases evolved when a material undergoes pyrolysis or combustion (2). Knowledge of the toxicity of the combustion products of polymeric materials is important to understanding smoke toxicity. Therefore, it is important to identify the fate of the various chemical elements present during the combustion of polymer systems. Because it is an oxidation process, combustion results in transforming each chemical element to some oxidized state, the extent of which often depends on the supply of oxygen (generally dependent on ventilation conditions) and the energy of the fire. Hydrogen is oxidized to form water, and carbon is oxidized to carbon monoxide (CO) and to carbon dioxide (CO_2). The ratio of CO_2/CO, often used as a descriptive characteristic of a fire, depends more on the ventilation conditions of the fire than on the nature of the materials being burned (5).

Combustion of materials whose polymer structure contains nitrogen bonds results in the formation of hydrogen cyanide (HCN) and nitriles (6). Production of HCN is also temperature-dependent; it increases with temperature and contributes to the toxicity of the smoke generated by the combustion of such polymers. For example, data on modacrylic fibers show that the 30-minute LC_{50} value decreases as the temperature of decomposition increases. The increase in toxicity (i.e., lower LC_{50}) is attributable to the increasing production of HCN. With sufficient oxygen at higher temperatures, HCN is further oxidized to oxides of nitrogen (NO_X).

Both pyrolysis and combustion of polymer systems that contain halogens (fluorine, chlorine, or bromine) form halogen acids (HF, HCl, and HBr). Halogen acids form during the pyrolysis phase, whether or not flaming combustion occurs, and are not oxidized further.

Because oxygen is consumed in oxidation, oxygen deficiency must also be considered an important component of smoke toxicity. Careful investigation is necessary when evaluating the toxicological impact of combustion and pyrolytic products of polymer systems to account for and distinguish the effects of oxygen deficiency.

In addition to combustion chemistry, the physics of combustion can explain the rapidity with which smoke atmospheres are generated in a fire. Fire size is generally described in terms of its rate of energy (heat) release in kilowatts (kw). A small fire confined to an object of origin produces 30 to 100 kw, whereas a large fire that progresses beyond its compartment of origin (flashover) generates in excess of 1000 kw (1 molecular weight (M_w)). The rate of heat release of a fire (and the rate at which smoke is produced) is related to its mass loss rate through the fuel's effective heat of combustion. Using a typical effective heat of combustion for many polymers of about 25 megajoules (MJ)/kilogram (kg), a fire size of about 1000 kw translates to a mass loss rate of the order of 0.04 kg/s. This mass loss rate represents approximately 7200 g of smoke evolved from the 1000-kw fire in approximately 3 minutes.

Simple calculations based on the mass of smoke released illustrate why smoke is rapidly fatal to humans exposed without respiratory protection, even if the smoke is assumed to be of only "average" toxicity. In a typical $15 \times 18 \times 8$-ft room, the smoke concentration in the room with the fire described would exceed 100,000 mg/m^3. Incapacitation might well occur within only 2 to 3 minutes. The toxic effects of smoke from fires usually result from inhaling extraordinarily high concentrations of toxicants during quite short periods of time. Thus many values commonly used in toxicology to describe the onset of hazardous conditions, such as TLVs, are often improperly applied in evaluating situations in fire toxicology.

Similar calculations can be made for the rise in temperature due to the heat released by combustion. In real fires, heat becomes an equally significant and sometimes overpowering stress for an extensive area around a fire center. There is some evidence that heat stress affects the toxicity of fire gases.

Toxicity of Smoke

Fire gas toxicants usually produce three basic kinds of toxic effects: (1) chemical asphyxiation; (2) irritation, both sensory (upper respiratory, eye, and mucous membrane) and pulmonary; and (3) other unusual effects.

In combustion toxicology, the concentration of chemical asphyxiants is often so high that severe central nervous system depression occurs, often followed rapidly by loss of consciousness and ultimately death. The acute effects of these toxicants depend on the accumulated dose, a simple product of concentration and the duration of the exposure. The severity of the effects increases with increasing dose. Although a number of asphyxiants may be produced by combustion of materials, only CO and HCN have been measured in sufficient concentrations in fire gases to cause significant acute toxic effects. Inhalation of CO and HCN is manifested by accumulation of carboxyhemoglobin (COHb) and cyanide, respectively, in the blood of those exposed.

Carbon dioxide is quite low in toxicological potency and is not, by itself, normally considered a significant factor in the severity of the toxic response to combustion products. However, it does stimulate respiration that leads to an increased rate of loading of blood COHb from inhaling CO. The same equilibrium COHb saturation is reached as in the absence of CO_2. Increased incidence of death (particularly postexposure) of rats has been observed with certain combinations of CO and CO_2 (7). This effect may be associated with the combined insult of respiratory acidosis (caused by CO_2) and metabolic acidosis (caused by CO), a condition from which the rodent has difficulty recovering postexposure. Whether this effect occurs in primates and the consequent importance of the effect for exposed humans has not been determined.

Irritants are produced in essentially all fire atmospheres and produce effects of two distinct types: (1) upper respiratory tract and mucous membrane (or sensory) irritation and (2) pulmonary irritation. Most fire irritants produce signs and symptoms characteristic of both sensory and pulmonary irritation. Halogen acids are among the irritants produced in fires. The most important is HCl from the decomposition of polyvinyl chloride. The irritants have received attention in recent years because of the perception that they contribute to significantly increased hazards in fires. Fire-retardant additives based on chlorine or bromine are also sources of halogen acids in fires.

Eye irritation, an immediate effect that depends primarily on the concentration of an irritant, may significantly hinder the escape of a victim from a fire. Nerve endings in the cornea are stimulated, causing pain, reflex blinking, and tearing. Severe irritation may produce permanent eye damage. Victims can alleviate these effects somewhat by closing their eyes, but doing so may also hinder their escape from a fire.

Airborne irritants that enter the upper respiratory tract cause burning sensations in the nose, mouth, and throat, along with the secretion of mucus. These sensory effects are related primarily to the concentration of the irritant and do not normally increase in severity as exposure time is increased.

Following signs of initial sensory irritation, significant amounts of inhaled irritants may be quickly taken into the lungs, and the symptoms of pulmonary or lung irritation are exhibited. Lung irritation is often characterized by coughing, bronchoconstriction, and increased pulmonary flow resistance. Tissue inflammation and damage, pulmonary edema, and subsequent death may follow exposure to high concentrations, usually after 6 to 48 hours. Exposure to pulmonary irritants also increases susceptibility to bacterial infection. Unlike sensory irritation, the effects of pulmonary irritation depend on both the concentration of the irritant and the duration of the exposure.

Although individual fire gas toxicants may exert quite different physiological effects through different mechanisms, in a mixture each may result in a certain degree of compromise in an exposed subject. It should not be unexpected that varying degrees of a partially compromised condition may be roughly additive in contributing to incapacitation or death. This has been demonstrated in a number of studies using rodents and has led to the use of the fractional exposure dose (FED) in combustion toxicology (8–18). The FED of a fire gas toxicant is the ratio of the Ct product (concentration × time) for a gaseous toxicant produced in a given test to that Ct product of the toxicant, which based on statistical determination from to independent experimental data to produces lethality in 50% of test animals within a specified exposure and postexposure time. Because time values in this ratio numerically cancel, the FED is also simply the ratio of the average concentration of a gaseous toxicant to its LC_{50} value for the same exposure time.

For example, it is fairly well agreed that carbon monoxide and hydrogen cyanide are additive when expressed as the FED required to cause an effect (11, 16). In experiments using mixtures of hydrogen chloride with either carbon monoxide or hydrogen cyanide, empirical analysis of the toxicological data also suggests that the contribution of HCl to the FED may be additive on the basis of lethality in rats (17, 18).

Tests for Toxicity of Smoke

Documented concern over the toxicity of smoke produced in fires dates at least as far back as the 1930s — long before the widespread use of synthetic polymers in structural building components and furnishings (19). However, it was not until the 1970s, when synthetic materials were widely used, that concerted efforts were made to understand the toxicity of smoke better and to develop laboratory tests for evaluating materials and products.

Assuming that the toxicity of smoke produced from the burning of materials could be as exotic as its chemical composition, literally hundreds of constituents may be present. Early efforts in the 1970s were directed toward developing "laboratory tests" for "smoke toxicity" (20). It was originally felt that only a bioassay would be a reliable way of evaluating the combined effects of gases in smoke and that every polymeric material and every formulation would have to be tested. Studies eventually demonstrated, however, that smoke toxicity could, to a large extent, be explained both qualitatively and quantitatively in terms of a small number of important toxic gases: CO, HCN, HCl, O_2 deficiency, and NO_2.

Early tests for smoke toxicity concentrated primarily on simple thermal decomposition of materials, subjecting rodents to the emissions produced, and observing of the effects on the animals. There was little attempt to determine concentration–response relationships or to establish testing protocols. Data produced by investigators in different laboratories were difficult, if not impossible, to compare.

Tests for the toxicity of smoke produced by a burning material now involve a quantitative measurement of the lethal toxic potency in the laboratory. A concentration–response relationship is determined by measuring the response of rodents exposed over a fixed time to different concentrations of a combustion atmosphere. This is accomplished by a series of experiments in which the quantity of combusted material or the flow rate of diluting air is varied to produce the different concentrations. The number of animals that show a response, such as lethality, increases as the exposure concentration is increased.

Smoke toxicity test data are generally reported as 30-minute LC_{50} values in units of grams of material per unit exposure chamber volume or total air flow (g/m^3). The weight of material may be expressed either as that subjected to the test conditions or as that consumed in the test. These values are identical if the specimen is completely combusted and has no residue. One commonly used test (the University of Pittsburgh, or UPITT method) expresses the LC_{50} only as grams of material subjected to the test protocol. Thus the UPITT LC_{50} is not a "lethal concentration, 50%" as traditionally defined in industrial toxicology, nor do the concentrations reported necessarily reflect the concentrations to which the animals were actually exposed. The term LCUPITT is used here to refer to values generated in these tests; concentrations calculated by dividing the total mass consumed by the total airflow through the chamber are designated LC_{50}Pitt. Care must be taken in comparing LC_{50} values obtained from different test methods to ensure that the basis for reporting data is the same.

Most of the interest in testing the toxicity of smoke has centered on those materials used in construction, interior finishes, or furnishings. Polymers that are used primarily as fibers (except those for carpeting), films, coatings, and so on have received little attention. The major test methods for assessing materials for the toxic potency of combustion products can be found in several sources.

Significance of Smoke Toxicity

Combustion toxicology in the 1970s was clouded with considerable controversy over issues such as the possibility of "supertoxicants," the need to assess incapacitation; the validity of data obtained from tests using rodents in assessing toxicity to humans; the precision and accuracy of smoke toxicity data; the relevance of laboratory scale tests to real fires; and, ultimately, how such data should be used.

Based on more experience in testing materials, the issue of potential "supertoxicants" in fire atmospheres has been generally dismissed because there have been few documented examples. One involved the formation of a neurotoxin from the thermal decomposition of a noncommercial rigid polyurethane foam (21), and another the unusual toxic potency exhibited by polytetrafluoroethylene in certain laboratory tests. The latter case now appears to be largely an artifact of the test method (22).

Furthermore, as noted previously, it is widely accepted that smoke toxicity can, to a large extent, be explained both qualitatively and quantitatively on the basis of a small number of toxic gases. Experience has shown that the LC_{50} values for most polymeric materials fall within a range of approximately one order of magnitude from about 5 to 60 g/m^3 (23). Values for most varieties of wood and other cellulosic materials also fall within that range.

Calculations based only on yields of carbon monoxide under oxygen-deficient, postflashover conditions suggest LC$_{50}$ values of 25 g/m^3 for most polymeric materials (24).

Although considerable attention was once directed toward incapacitation studies, current smoke toxicity tests make no such assessment. Incapacitation is simply inferred from lethality data. Combustion toxicologists generally consider that incapacitating exposure doses are about one-third to one-half of those required for lethality. Incapacitation due to sensory irritants has been proposed as a criterion for assessing the hazard of construction materials, and tests have been developed to assess smoke irritation (25). Such tests have not been used significantly, however, due, at least in part, to lack of validation between the tests and actual effects on humans. The propensity for smoke to have the same effects on humans in fires can be inferred only to the extent that the test animal model (rat or mouse) predicts the response of humans as a biological system. As a result of such comparisons, smoke toxicologists generally endorse the use of rats as an acceptable model for human exposure (26).

Some consider that the 30-minute exposure time for animals in laboratory smoke toxicity tests is rather long compared to most human exposures to smoke in fires. However, toxic potency test data can be extrapolated to shorter exposures by using Haber's rule (concentration × time = constant), which is reasonably applicable over the range of concentrations of interest.

It was once felt that every polymeric material and formulation would have to be tested because the toxicity of its smoke could well be unique to the material, that is, it is highly material-dependent. When studies demonstrated that smoke toxicity could be explained in terms of a small number of important toxic gases noted previously, it became possible to study the combustion characteristics of polymeric materials more systematically. Moreover, it was recognized that the formation of these major toxicants more often depended on the conditions of the fire, that is, they are highly fire-dependent.

Some toxicants are both material-dependent and fire-dependent. For example, CO production is somewhat material-dependent, but is highly fire-dependent with regard to the ventilation or air supply. The formation of HCN is highly material-dependent because only nitrogen-containing materials produce HCN, but it is also fire-dependent because HCN evolution increases with increasing temperatures (production of HCN may also be influenced by ventilation, but this has not been sufficiently studied). Formation of halogen acids, for example, HCl, is clearly material-dependent, but it is fire-dependent only to the extent that the temperature must be sufficiently high to evolve the halogen acid.

Because the production of common toxicants is both material- and fire-dependent, some doubt is cast on the significance of data obtained using laboratory tests that may not simulate either real or full-scale test fire conditions. It is important to recognize that the results obtained from smoke toxicity tests relate only to the specific conditions of the test itself and may not predict conditions in any real-scale fire. The term real-scale refers to both controlled, experimental, full-scale fires, as well as unwanted, real-life fires.

A particular problem exists because CO is the major toxicant in most real-scale fires and confounds the prediction of conditions in real fires compared to laboratory-scale tests. The cause is the changing ventilation (air supply) throughout the course of a real fire. Combustion consumes oxygen from the air, causing increasingly oxygen-deficient conditions that favor increased production of CO. These conditions are generally not attained in small-scale laboratory tests. A correction for CO production under postflashover conditions has been proposed to enable using laboratory LC_{50} values in fire hazard assessment (27).

Hazards in Fires

Smoke toxicity test data do not, in themselves, indicate toxic fire hazard, or even relative toxic fire hazard. Thus they should not be used in the absence of a toxic fire hazard assessment. "Toxic fire hazard" is most succinctly defined as the potential for harm from the toxic products of combustion. Elements such as smoke toxicity potency, rate of smoke production, smoke transport, conditions and quantity of materials used, and factors intrinsic to humans exposed can all affect the potential toxic fire hazard. Furthermore, toxic fire hazard is but one element of fire hazard. Other hazards include ease of ignition, smoke density, flame spread, rate of heat release, and ease of suppression. Toxic fire hazard could, at times, be the primary concern in fire hazard or at other times it could be irrelevant. For this reason, toxic fire hazard should be integrated into overall fire hazard assessment. Mathematical computer models, such as Hazard I developed at NIST, are available for such analyses (28).

Exposure Assessment

Exposure is the key element in the chain of events that leads from release of pollutant into the occupational or ambient environment to actual human exposure to disease or injury. But, exposure should not be equated with toxicity. The fundamental principles of toxicology that have evolved over the years require that there be exposure *and* toxic effect. Exposure to polymers or its decomposition products can occur from contact with a variety of elements (air, water, food) that in turn influences the pathway of exposure (inhalation, ingestion, dermal). Individual interactions with these elements, even in the workplace, are complex, and therefore, it is not surprising that exposure assessment and the related area of dose estimation create challenges to efforts to manage the risk of toxic injury.

The objective of exposure assessment is to determine the source, type, magnitude, and duration of contact with the polymeric material of interest. It is the key phase of risk assessment because hazard does not occur in the absence of exposure. As indicated previously, exposure in the production of polymers may be to the raw materials, the chemical intermediates, the catalysts and additives, or the polymer products. Thus exposure to airborne polymeric materials at work is variable composition and concentration, depending on the materials (monomers, additives, catalysts) handled, the polymerization process, the degree of engineering controls applied to minimize release to the air, and the work practices followed.

There are techniques for measuring exposures in the workplace over the range of types of exposures (gases, vapors, aerosols or dust) to polymeric material. Although the approaches are tailored for assessing each type of exposure, the general principles of evaluation can be generalized. Most measurement methods assess airborne contaminants

SYNTHETIC POLYMERS

because of the importance of inhalation exposure in the workplace. One class of measurement is the extractive method in which the polymeric material (monomers, additives, catalyst) is removed from the air for laboratory analysis. Direct reading instruments detect the instantaneous air concentration and may produce a reading on a dial. These monitoring methods can measure continuously and report results immediately, which allows examining patterns of exposure over time. Personal sampling is an increasingly common approach to accurate and precise measurement of worker exposure to particulate or gaseous air contaminant. The characteristics of equipment for air sampling in industry are described in detail in *Air Sampling Instruments* (ACGIH, 1995). The specific methods are described in the NIOSH's *Manual of Analytical Methods* and in other NIOSH and OSHA procedures manuals.

In addition to these publications, there is a growing literature on the advances in the technology of workplace exposure assessment and epidemiological evaluations. The developments mean that improved scientific criteria can be brought to bear on questions of occupational exposure and related issues of dose–response. For example, biological markers — measurements conducted in biological samples such as blood or urine that evaluate an exposure or biological effect — integrate all routes and sources of exposure and can provide measures of the actual dose of a polymeric material an individual received when appropriate metabolic data are available. Typical biomarkers of exposure include measurement of the toxic agent or its specific metabolite. Detection of a product of interaction between polymeric material and an endogenous component represents an exposure biomarker. A clear definition of the relationship between time of exposure and sample collection can aid in the interpretation of biological marker information. Utilizing biomarker measurements has the potential to provide exposure information far superior to present methods for exposure assessment.

Methods for exposure assessments outside the workplace are included in EPA's *Method for Assessing Exposure to Chemicals*, and *Guidelines for Exposure Assessment*. In *Risk Assessment Guidance for Superfund*, EPA describes its model for estimating the magnitude of actual or potential human exposure to chemicals in environmental media at Superfund sites.

The technology for sampling air and water polymeric materials is relatively well developed, as is the technology for sample separation of mixtures of pollutants or co-pollutants. However, questions of when, where, how long, and at which rate and frequency to collect data relevant to polymeric exposure pose significant challenges for ambient exposure assessment.

Recognizing the need for further improvements in exposure assessment, the National Institute of Environmental Health Sciences, in collaboration with other federal agencies (e.g., EPA and NIOSH), began a new interagency initiative in exposure assessment in 1998. The work builds on the interagency pilot study led by EPA, the National Human Exposure Assessment Survey (NHEXAS). Much has already been learned through a series of EPA studies about individual exposures to airborne contaminants. Through total exposure assessment methodology (TEAM), research has developed a number of novel approaches to assessing human exposures to airborne contaminants in the ambient environment. These approaches include miniaturized sampling and analytical techniques suitable for personal monitoring; demographic sampling techniques to choose individual representatives of community exposure; and, most importantly, approaches allowing integration of indoor and outdoor exposure data for individuals. The approaches pioneered by the TEAM study are continuing to be adapted and improved.

Environmental Impact

The life cycle of a polymer—manufacturing, packaging, transportation, use, recovery, recycling, and disposal—is important in designing a health and environmental protection program. The life cycles of specific polymeric materials may take divergent patterns. Some unpolymerized monomers may become widely dispersed in the environment, whereas monomers combined in the manufacture of a polymer may be released to the environment in very low or negligible quantities. Some of the additives and catalysts may have only limited exposure potential and only to the workforce involved in the processing.

The concept of product stewardship practiced by many polymer manufacturers includes protection of people and the environment. It also emphasizes protection of the polymer from the environment—an environment in which the material could be misused, abused, and disposed of improperly, all of which could impact the ambient environment. The disposal end of the life cycle of polymers has attracted increased attention because they make up a growing fraction of municipal solid waste streams and pose environmental challenges. For example, there are more than two billion waste tires stockpiled in the United States, and another 250 million tires are added each year. Products made of polystyrene, polypropylene, and polyvinyl chloride also make up a significantly large volume of municipal waste.

Once in the waste stream, polymers are dealt with in one of three ways: incineration, burial, or recycling. Recycling is the least practiced of the three primarily because it is not economically attractive. Incineration is used to dispose of more than 16% of all municipal waste. Potentially hazardous emissions from incinerating some wastes have been the focus of much attention, including regulatory measures. Even with stricter air pollution standards, there is considerable public opposition to incineration. Some of this opposition is based on experience with antiquated incineration technology or combustors not properly designed or operated for efficient destruction of wastes. There is ample evidence that incineration technology now available can be operated with no adverse impacts on humans or the environment. Shy's group investigated the effects of these three types of incinerators on respiratory disease: a biomedical incinerator, a municipal incinerator, and a liquid hazardous waste-burning industrial furnace. The data suggest that even though an incinerator may be a point source of air pollution in a community, its contribution to the total mass of air pollution in that community may be relatively small and nearly undetectable by standard air monitoring techniques. The study results also indicate that if a particular emission can cause respiratory effects in the exposed community, standard measures of air pollution may fail to detect relevant differences in human exposure (29).

Landfilling polymers is generally a benign practice because polymers are chemically inert. Some additives, which may not be chemically bound in the polymer matrix, do cause concern if they should migrate from the polymer into the leachate. For example, heat stabilizers may be toxic, but their stability depends in part on the types of polymer and heat

SYNTHETIC POLYMERS

stabilizer. Heavy metals are being phased out of packaging materials and are a diminishing problem. Some phthalate plasticizers are hazardous substances and have been found in a number of leachate analyses at various concentrations.

A more significant problem for landfilling some polymer waste, (plastics, for example) is that they now constitute about 10% by weight and approximately 20% by volume of the municipal waste stream. Because polymers are essentially nonbiodegradable, they may consume a disproportionate amount of landfill space. EPA reports that plastics, for example, comprised an estimated 400,000 tons of municipal solid waste in 1960. By the mid-1990s, the figure was 19.8 million tons. New polymer recycling technology, including bacterial degradation of natural and synthetic rubbers, may make recycling an economically attractive alternative and thereby reduce the potential impact of polymers on the environment.

INTRODUCTION TO CLASS OF CHEMICALS

Olefin Resins

Olefin is a class of unsaturated hydrocarbon, obtained by cracking naphtha or other petroleum fractions at high temperatures. The simplest members are ethylene, propylene, and butylene which are the starting points for certain resins.

The commercial development of polyethylene began in the 1930s and was followed by full-scale production of the major olefin resins during the 1940s and through the 1960s. Polyethylene and other olefins are closely related to paraffins. As a class, these resins have very low water absorption, moderate to high gas permeability, good toughness and flexibility at low temperatures, and relatively low heat resistance. Environmental stress cracking is greater with polyethylene, whereas polypropylene is more susceptible to oxidation. Polyethylene cross-links on oxidation, whereas polypropylene degrades to form lower molecular weight products. Typical properties are given in Table 88.1.

Toxicological Potential

Monomer residue has not been considered a problem. Ethylene, propylene, and 1-butene monomers act toxicologically as asphyxiant gases. Ethylene is a natural constituent of apples and other fruit and is used commercially to accelerate ripening of fruit. Low molecular weight additives or impurities may be of concern if significant migration of these materials into food or water is possible.

1.0 Polyethylene

1.0.1 CAS Number: [9002-88-4]

1.0.2 Synonyms: PE; polyethene; polythene; chlorinated polyethylene (CMS), ethene, homopolymer; ethylene polymer; polyethylene as; agilene; alathon; alcowax &; amybethene; courlene x 3; diothene; Dylan; dylan-super; etherin; etherol; irrathene r; lupolen; lupolen n; orizon; polythene; suprathen; tenaplas; epolene; flothene; fortiflex a 60/500;

Table 88.1. Physical Constants of Selected Polymers

Compound	CAS	Molecular Formula	Molecular Weight	Boiling Point (°C)	Specific Gravity	Melting Pt (°C)	Sol in Water (at 68F) mg/ml	Refractive Index (20°c)	Vapor Pressure (mmHg)	UEL % by Vol. (Room Temp)	LEL % by Vol. (Room Temp)
Polyethylene	[9002-88-4]	$(C_2H_4)_n$	28.054	—	0.92	130–145	—	1.51–1.54	—	—	—
Polypropylene	[9003-07-0]	$(C_3H_6)_n$	42.080	—	0.89–0.94	160–171	—	1.49	—	—	—
Polybutylene	[9003-29-6]	$(C_4H_8)_n$	56.107	—	0.87–0.95	124–135	—	—	—	—	—

SYNTHETIC POLYMERS

grisolen; hoechst wax pa 520; hostalen; marlex 50; microthene f; petrothene lc 731; poly-em 40; PY 100; rigidex 35; sholex 6000c; stamylan 1700; yukalon lk 30; Polyethylene wax; ethylene resin; ethylene latex; poly(ethylene), low density, average M.W. 50.000

1.0.3 Trade Names

The *Dictionary of Chemical Names and Synonyms* lists more than 100 trade names for polyethylene. Examples are AC 394, AC 680, AC 1220, Bakelite DFD 330, Bareco Polywax, Bicolene C, BPE-1, Chem C and Chemiplex 3006.

1.0.4 Molecular Weight

Low molecular weight polyethylene: 2000–5000; extra-high-molecular weight materials range from 150,000 to 1,500,000; and ultra-high molecular weight materials are in the 1,500,000 to 6,000,000 range.

1.0.5 Molecular Formula: $[C_2H_4]_n$

1.0.6 Molecular Structure:

1.1 Chemical and Physical Properties

Chemical and physical properties are markedly affected by increasing density which is affected by the shape and spacing of the molecular chain. Low-density materials have highly branched and widely spaced chains, whereas high-density materials have comparatively straight and closely aligned chains. Low-density polyethylene (0.926 to 0.9409/cm^3) is soluble in organic solvents at temperatures higher than 200°F. It is insoluble at room temperature. High density polyethylene (0.041 to 0.965) is hydrophobic, permeable to gas, and has high electrical resistivity.

Low molecular weight (2,000–5,000) polyethylenes have excellent electrical resistance and resistance to water and to most chemicals. They are fluids used as lubricants. Medium molecular weight polymers are waxes miscible with paraffin wax; and polyethylene polymers whose molecular weights are higher than 10,000 are the familiar tough and strong resins that are flexible or stiff. By varying the catalyst and methods of polymerization, properties such as density, crystallinity, molecular weight, and polydispersity can be regulated over wide ranges. Polymers whose densities range from approximately 0.910 to 0.925 g/cm^3 are low-density polyethylene; those whose densities range from 0.926 to 0.940 are medium density; and those whose densities range from 0.941 to 0.965 and higher are high-density polyethylene. As the crystallinity or density increases, the products generally become stiffer and stronger and have a high softening temperature and high resistance to penetration by liquid and gases. At the same time, they lose some of their resistance to tear and environmental stress cracking.

Polyethylene can be cross-linked by irradiation (electron beam, gamma, or X radiation) or by free-radical catalysts such as peroxides. Controlled levels of cross-linking result in improved resistance to stress cracking without sacrificing good electrical or mechanical properties. Such cross-linked resins are especially useful as electric cable insulation.

1.1.1 General

Polyethylenes are large family of resins. The available forms range in crystallinity from 50–90%. As crystallinity increases, the products generally become stiffer and stronger, have high softening temperatures, and high resistance to penetration by liquids or gases. A key property of polyethylene is its chemical inertness. Strong oxidizing agents eventually cause some oxidation and some solvents cause softening or swelling, but there is no known solvent for polyethylene at room temperature. The brittleness temperature is $-100°C$.

Two newer types of polyethylene are (1) extra-high molecular weight (EHMWPE) and (2) ultra-high molecular weight (UHMWPE) in the molecular weight range of 150,000 to 1,500,000. These do not melt or flow. When fully cross-linked by chemical additives electron beam, or gamma radiation, polyethylene is no longer a plastic, and has superior strength, impact resistance, and electrical properties. A newer member of the polyethylene family is linear low-density polyethylene used in grocery bags. Another new subfamily is very low density polyethylene.

1.2 Production and Use

At least five distinct processes are used for producing PE (30, 31). Linear PE is produced by a low-pressure solution or gas-phase process that is initiated by a variety of transition-metal catalyst. The most common catalysts are Ziegler titanium compounds with aluminum alkyls and Phillips chromium oxide-based catalysts. The gas-phase and slurry processes are used to produce high molecular weight, high-density (HMW-HDPE) products. The highest density linear PEs can be made from an α-olefin comonomer, typically octene for the solution process and butene or hexene for the gas-phase process. Linear PE does not have long-chain branches and is therefore more crystalline. The short-chain branches found in linear PE serve as tie molecules, which give the higher α-olefin copolymers improved puncture and tear properties. Included in the linear PE family are ultra-low-density PE (ULDPE), linear low-density PE (LLDPE), and high-density PE (HDPE).

The major uses of polyethylene are in packaging films, containers and bottles, molded articles, electrical insulation, coatings, monofilament, and pipes. Some foamed products are also made from polyethylene.

1.4 Toxic Effects

The major concerns for toxicity are those from acute exposure to highly active chemicals that are potentially irritating to tissues. The systemic toxicity of monomers is species specific and is linked to the metabolism of the monomer.

1.4.1 Experimental Studies

Rats survived acute doses of 7.95 g/kg of "Marlex" 50, homopolymer whose density is 0.96, without evidence of adverse effects (32). Mice that received a single oral dose of

2500 mg/kg powdered unstabilized and stabilized high-density PE also showed no toxic effects (33).

Dietary levels of 1.25, 2.5, or 5% "Marlex" 50 for 90 days produced no adverse effects in rats (32). In other 90-day tests (34, 35), rats and dogs were fed an extract of low molecular weight PE film; the film had been extracted with isooctane to yield 568 mg extract/100 g of film. Rats fed at a level of 13,500 ppm film extract showed liver changes (fat droplets, cloudy swelling, and increased liver weight) that were considered reversible in all cases. Rats fed at levels of 2700 and 540 ppm and dogs fed 2700 ppm showed no adverse effects.

Swine fed 20% shredded PE for 65 or 77 days in a program to control weight gain showed no adverse effects on the digestive system (36). Feeding animals aqueous extracts of stabilized PE for 16 to 19 months produced only insignificant, generally transient effects on body weight and conditioned reflexes, but no pathological changes (32).

A biochemical correlation with physical factors comes from the tests of Matlaga et al. (37) on six "medical-grade polymers," including PE. They found that triangular-shaped implants of the respective polymers showed the highest level of cellular lysosomal phosphatase enzymatic activity, whereas pentagonal shapes showed less activity and circular shapes had the lowest enzymatic activity.

1.4.1.5 Carcinogenesis. Tumors in animals from implanted PE are generally considered consistent with solid-state carcinogenesis. Tumors observed in women who used PE intrauterine contraceptive devices are also consistent with this rationale. In an apparent exception, intraperitoneal implants of PE particles < 0.4 cm did produce tumors in rats, but it was "suggested that the critical size of particles implanted in the abdominal cavity may be much less for solid-state carcinogenesis to be operative" (38).

1.4.2 Human Experience

Asthma was reported by a worker who heated polyethylene film as part of food-wrapping activity: the film used reportedly contained no additives. However, asthma is much more often reported by food wrappers who are exposed to the pyrolytic products of polyvinyl chloride film.

1.4.2.1 General Information. When heated to decomposition, PE emits acrid fumes and smoke that irritate mucous membranes and the upper respiratory tract.

1.4.2.2.5 Carcinogenesis. As indicated in Section 1.4.1.5, tumors have been observed in women who used intrauterine contraceptive devices made of polyethylene.

2.0 Polypropylene

2.0.1 CAS Number: [9003-07-0]

Isotatic PP is produced for commercial application and has the specific CAS Number [25085-53-4]. The amorphous (atactic) forms of polypropylene are unwanted by-products in some manufacturing processes.

2.0.2 Synonyms: PP; polypropene; 1-propene polymer

2.0.3 Trade Names: Appryl, Carlona P, Daplen, El-rex, Eltex P, Eref, Es con, Es corene, PP, Finapro, Fortilene, Hifax, Hostalen PP, Lacqtene P, Luparen Moplan, Napryl Novolen, Polypro, Profax, Refene, Royalene, Snialene, Stamylan P, Tenite, Valtec, and Vestolen

2.0.4 Molecular Weight

The number-average molecular weight of commercial polypropylene ranges from 38,000 to 60,000, and average molecular weights are from 200,000 to 700,000.

2.0.5 Molecular Formula: $[C_3H_6]_n$

2.0.6 Molecular Structure: ⟋⟍

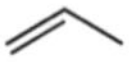

2.1 Chemical and Physical Properties

Polypropylene is a low-density resin (0.896 to 0.94) that offers a good balance of thermal, chemical, and electrical properties, along with moderate strength. Strength can be significantly increased by using reinforcing agents such as glass fiber. Polypropylene has limited heat resistance, but it can be used in applications that must withstand boiling water or steam sterilization.

Polypropylenes can resist chemical attack and are unaffected by aqueous solutions of inorganic salts or mineral acids and bases, even at high temperatures. They are not attacked by most organic chemicals, and there is no solvent for these resins at room temperature. The resins are attacked, however, by halogens, fuming nitric acid, other active oxidizing agents, and by aromatic and chlorinated hydrocarbons at high temperatures (39).

2.1.1 General

Three forms of polypropylene are possible.

> Isotactic (fiber-forming): methyl groups are all on the same side of a plane of a zigzag carbon atom chain.
> Syndiotactic methyl groups are on alternate sides of the carbon atom chain.
> Atactic (not fiber-forming, amorphous): methyl groups are in a random arrangement with respect to the plan of the carbon atom chain.
> The desired form is the isotactic arrangement (at least 93% is required to give desired properties). Commercial polypropylene is 90–95% isotatic.

Polypropylene is translucent and autoclavable. Properties can be improved by compounding with fillers, by blending with synthetic elastomers, and by copolymerzing with small amounts of other monomers.

2.1.2 Odor and Warning Properties

When heated to decomposition, polypropylene emits acid smoke and irritating fumes.

2.2 Production and Use

In PP production, propylene monomer is polymerized to make the homopolymer by using a Ziegler–Natta type coordination catalyst. This catalyst results from the reaction and interaction of a transition-metal compound and an organometallic compound, usually an alkylaluminum compound. Halide atoms are involved in most such catalyst systems (40).

Polypropylene can be made by solution, slurry (or solvent), bulk (or liquid propylene) or gas-phase polymerization, or a combination of these processes (41). The most widely used is the slurry process; however, the current trend is toward the gas-phase process. In the solution, slurry, and bulk processes, the catalyst system is mixed with propylene and a hydrocarbon diluent (usually hexane, heptane, or liquid propylene) in a reactor. After polymerization, the reaction mixture enters a flash tank where unreacted propylene is removed and recycled. Propylene-ethylene copolymers CAS Number [*9010-79-1*] can be manufactured when ethylene is fed along with propylene to the polymerization reactor or by adding ethylene and propylene to a postpolymerization reactor that contains PP. This mixture may then be purified to remove low molecular weight and atactic fractions and washed to remove catalyst residues. The polypropylene resin is then dried and pelletized. During this time additives may be incorporated (42) in the gas-phase process; no liquid diluent is used (40).

Polypropylene products may be manufactured by blow molding (< 2%), extrusion (36%), or injection molding (43). The major categories of unreinforced PP products include fibers that are used in carpet and fabrics, film, packaging (containers and closures), and medical applications such as hypodermic syringes and surgical mesh. Both reinforced and unreinforced PPs are used in automotive, appliance, and electrical applications.

2.4 Toxic Effects

Mice given an acute oral dose of 8 g/kg showed no noticeable toxic effects. Feeding rats and mice aqueous extracts of PP (extraction temperatures 20 and 60°C) for 15 months produced no significant changes (44). Rats fed edible-oil solutions of extractable material from a typical but ^{14}C-tagged PP excreted radioactive material in the feces within 2 days. "This accounted for about 93 percent of the amount fed" (45). Two-year tests with oligomeric material of molecular weight 800 showed no effects on rats fed a diet as high as 20,000 ppm and dogs fed up to 1 g/kg (46).

Polypropylene microfoam sheeting produced no evidence of skin irritation or sensitization when tested on 20 humans (47).

2.4.1.5 Carcinogenesis.
The data from one experiment with subcutaneous implants in rats (48) suggest that the response of local fibrosarcomas observed with disks and to a relatively small extent, with powder is essentially similar to that observed with polyethylene. As with polyethylene, PP implants of triangular, pentagonal, or circular shape provoked varying levels of cellular lysosomal acid phosphatase enzymatic activity (49). In a 6-month test of subcutaneous implants in hamsters (50), sequential examination of PP

filaments with and without an antioxidant showed that *in vivo* degradation could be effectively retarded by the using an antioxidant.

Tissue reactivity to 10 different suture materials, including PP, was evaluated in the rat by conventional microscopic assessment. At 7 days postsurgery, there were no system differences among tissue reaction to the sutures (51).

Histological reaction to nylon, PP, polyglactin-910, and polydioxanone microsutures was assessed in the uterine horn of the rabbit. At 24 days after insertion of the microsuture, a marked infiltration of histocytes was seen around the nylon, PP, and polydioxanone microsutures. Giant cells were seen around the polyglactin-910. At 80 days after insertion of the microsutures, polydioxanone was almost entirely absorbed, and the reaction to polyglactin-910 was minimal. Moderate histiocytic infiltration persisted around the nylon and PP sutures. Fibrosis was also detected around the nylon and PP sutures at 90 days (52).

2.4.2 Human Experience

There is little clinical or epidemiological literature on the health effects of human exposure to polypropylene, primarily due to the low risk of exposure in the manufacturing process.

2.4.2.1 General Information. Polypropylene is moderately toxic by ingestion and by the intraperitoneal route. When heated to decomposition, it emits acrid smoke and irritating fumes that may increase the risk of localized inflammation of the respiratory system. PP microfoam produced no skin irritation or sensitization when tested on 20 humans (47).

2.4.2.2.5 Carcinogenesis. Following a report of an excess of colorectal cancers among workers involved in manufacturing polypropylene by a heavy diluent process, Exxon initiated a study of colorectal cancer incidence among its PP pilot-plant workers. Overall, there were three observed colorectal cases versus 3.3 expected (standardized incidence ratio = 0.9, 90% confidence interval 0.3 to 2.3). Analyses by duration of employment and latency did not show patterns consistent with the colorectal cancer excess previously reported (53). The feasibility of a European study to see if there is any link between stages in propylene polymerization and colorectal cancer is being evaluated by the European Chemical Industry Ecology and Toxicology Centre (ECOTOC) (54).

2.4.2.2.7 Other: Neurological, Pulmonary, Skin Sensitization, etc. Polypropylene microform sheets produced no evidence of skin sensitization or irritation when tested on 20 humans. No information was found on neurological effects.

3.0 Polybutylene

3.0.1 CAS Number: [9003-29-6]

The homopolymer exists in isotactic (crystalline) [CAS#25036-29-7] and atactic (amorphous) forms.

3.0.2 Synonyms: PB; poly-1-butene; polyisobutylene; polyisobutene, polybutylene resins; Amoco 15H; butene, polymers; Chevron 12; Chevron 16; Chevron 18; Chevron 6;

HV 1900; HYVIS 07; Indapot H-100; Indopol; LV 50; Oktol; OKTOL 600; Oronite 128; Oronite 32

3.0.3 Trade Names: Butuf, Vestolen BT

3.0.4 Molecular Weight

The weight-average molecular weight is 10^5 to 10^6.

3.1 Chemical and Physical Properties

Polybutylene resins resist most acids and bases at temperatures less than 90°C and many chemical solvents and detergents. Resistance is considered poor to strong oxidizing acids and to aromatic and chlorinated hydrocarbon solvents at temperatures higher than 60°C. Oxidation induced by heat and ultraviolet radiation is prevented by stabilizers. Polybutylene exhibits high tear, impact, and puncture resistance. It is combustible.

3.1.1 General

Polybutylenes are a family of polymers that consist of isotactic, stereoregular, highly crystalline polymers based on 1-butene. The general properties are similar to those of polypropylene and linear polyethylene.

3.1.2 Odor and Warning Properties

No published data were found. Based on the properties of 1-butene, acid smoke and fumes emitted when the polymer is heated may serve as a warning to prevent excessive chronic exposure.

3.2 Production and Use

Polybutylene is synthesized from 1-butene by a Ziegler–Natta polymerization. As indicated in 3.1.1, the commercial outlet for PB resin is pipes for hot and cold water plumbing. PB is also used in specialty films and film coatings for packaging dry foods, meat, and medical parts. Other uses are in adhesives and sealant formulations.

3.4 Toxic Effects

Rats fed a diet of 10 or 1% polybutene-1 showed no adverse effect.

4.0 Other Olefins

At high molecular weights, polyisobutylene becomes a rather sluggish, unusable rubber. Polyisobutylene is used as an oil additive. The copolymer with 2% isoprene is the widely used butyl rubber.

Polymethylpentene is a predominantly isotactic polymer distinguished by unusual transparency, exceptionally low density (0.83), and high softening point (Vicat softening point 179°C). Its limitations are high permeability and environmental stress cracking (55).

Ionomers are ionic polymers that contain metal or other inorganic counterions and are intended for film and plastic applications (56). Most commercial ionomers (such as Surlyn®) are derived from ethylene copolymers. These copolymers contain anionic pendent groups introduced from an unsaturated carboxylic acid such as methacrylic acid. The copolymers are then treated with a metal derivative (typically of zinc or sodium) to make the carboxylic group ionize reversibly with heat. The products can resemble thermosetting resins at ambient temperatures and linear thermoplastics at elevated temperatures (30).

BIBLIOGRAPHY

1. A. Kumar and R. K. Gupta, *Fundamental of Polymers*, McGraw Hill Inc., New York, 1998, pp. 258–259.
2. *Standard Terminology of Fire Standards, ASTM E176-89a*, American Society for Testing and Materials, Philadelphia, PA, 1989.
3. C. Beyler, Thermal decomposition of polymers. In P. J. DiNenno, Ed., *The SFPE Handbook of Fire Protection Engineering*, Chapts. 1–12, National Fire Protection Association, Quincy, MA, 1988.
4. V. Babrauskas et al., *Fire Hazard Comparison of Fire-Retarded and Non-Fire-Retarded Products*, NBS Special Publication 749, National Bureau and Standards, U.S. Department of Commerce, Gaithersburg, MD, 1998.
5. V. Babrauskas et al., Large-scale validation of bench-scale fire toxicity tests. *J. Fire Sci.* **9**(2), 125–148 (1991).
6. G. D. Clayton and F. E. Clayton, Eds., *Patty's Industrial Hygiene and Toxicology*, 3rd rev. ed., Vol 2B, Wiley, New York, 1983.
7. B. C. Levin et al., Toxicological interactions between carbon monoxide and carbon dioxide. *Proceeding of 16th Conference of Toxicology*, Air Force Medical Research Laboratory, Dayton, OH, 1986; *Toxicology* **47**, 135–164 (1987).
8. B. C. Levin et al., Toxicological effects of the interactions of fire gases and their use in a toxic hazard assessment computer model. *The Toxicologist* **5**, 127 (1985).
9. V. Babrauskas, B. C. Levin, and R. G. Gann, A new approach to fire toxicity data for hazard evaluation. *Fire J.* **81**, 22–71 (1987); also in *ASTM Stand. News* **14**, 28–33 (1986).
10. B. C. Levin et al., Effects of exposure to single or multiple combinations of the predominant toxic gases and low oxygen atmospheres products in fires. *Fundam. Appl. Toxicol.* **9**, 236–250 (1987).
11. B. C. Levin et al., Further studies of the toxicological effects of different time exposures to the individual and combined fire gases: Carbon monoxide, hydrogen cyanide, carbon dioxide, and reduced oxygen. In *Polyurethanes'88, Proceedings of the 31st Society of Plastics Industry*, Polyurethane Division, Meeting Technomic, Lancaster, PA, 1988, pp. 249–252.
12. B. C. Levin and R. G. Gann, Toxic potency of fire smoke: Measurement and use. In G. L. Nelson, Ed., *Fire and Polymers: Hazards Identification and Prevention (ACS Symp. Series 425)*, American Chemical Society, Washington, DC, 1990, pp. 3–11.
13. G. E. Hartzell, D. N. Priest, and W. G. Switzer, Modeling of toxicological effects of fire gases: II. Mathematical modeling of intoxication of rats by carbon monoxide and hydrogen cyanide. *J. Fire Sci.* **3**(2), 115–128 (1985).
14. G. E. Hartzell et al., Modeling of toxicological of fire gases: III quantification of post-exposure lethality of rats from exposure to HCl atmospheres. *J. Fire Sci.* **3**(3), 195–207 (1985).

15. G. E. Hartzell et al., Modeling of toxicological effects of fire gases: V. Mathematical modeling of intoxication of rats by combined carbon monoxide and hydrogen cyanide atmospheres. *J. Fire Sci.* **3**(5), 330–342 (1985).
16. G. E. Hartzell, A. F. Grand, and W. G. Switzer, Modeling of toxicological effects of fire gases: VI. further studies on the toxicity of smoke containing hydrogen chloride. *J. Fire Sci.* **5**(6), 368–391 (1987).
17. G. E. Hartzell, A. F. Grand, and W. G. Switzer, Modeling of toxicological effects of fire gases: VII. Studies on evaluation of animal models in combustion toxicology. *J. Fire Sci.* **6**(6), 411–431 (1988).
18. G. E. Hartzell, A. F. Grand, and W. G. Switzer, Toxicity of smoke containing hydrogen chloride, In G. L. Nelson, Ed., *Fire and Polymers*, American Chemical Society, Washington, DC, 1990, pp. 12–20.
19. G. E. Ferguson, Fire gases. *Qu. Natl. Fire Prot. Assoc.* **27**(2), 110 (1933).
20. H. L. Kaplan, A. F. Grand, and G. E. Hartzell, *Combustion Toxicology: Principles and Test Methods*, Technomic, Lancaster, PA, 1983.
21. J. H. Petajan et al., Extreme toxicity for combustion products of a fire-retarded polyurethane foam. *Science* **187**, 542–544 (1975).
22. B. B. Baker, Jr. and M. A. Kaiser, Understanding what happens in a fire. *Anal. Chem.* **63**, 79–83 (1991).
23. D. A. Purser, Toxicity assessment of combustion products and modeling of toxic and thermal hazards in fire. *SFPE Handbook of Fire Protection Engineering*, National Fire Protection Association, Quincy, MA, Section 1, 1988, pp. 200–245.
24. V. Babrauskas et al., *Toxic Potency Measurement for Fire Hazard Analysis*, NIST Special Publication 827, National Institute of Standards and Technology, Gaithersburg, MD, 1991.
25. C. S. Barrow, Y. Alarie, and M. F. Stock, Sensory irritation evoked by the thermal decomposition products of plasticized Poly vinyl chloride. *J. Fire Sci.* **1**, 147–153 (1976).
26. *Prediction of Toxic Effects of Fire Effluents*, TR 9122-Part 5, International Standards Organization, Geneva, Switzerland, 1993.
27. V. Babrauskas et al., *Toxic Potency Measurement for Fire Hazard Analysis*, NIST Special Publication 827, National Institute of Standards and Technology, Gaithersburg, MD, 1991.
28. R. W. Bukowski et al., *Hazard 1*, Vol. 1, *Fire Hazard Assessment Method*, NBSIR 87-3602, U.S. Department of Commerce, National Bureau of Standards, 1987.
29. C. M. Shy, D. Degman, D. L. Fox, S. Mukerjee, et al., Do waste incinerators induce adverse respiratory effects? an air quality and epidemiological study in six communities. *Environ Health Perspect.* **105**(3), 714–724 (1995).
30. J. A. Brydson, *Plastics Materials*, 3rd ed., Whitefriars Press, London, 1975.
31. J. R. Flesher, Polyethylene. In *Modern Plastics Encyclopedia, 1978–79 ed.*, Vol. 55, No. 10A, McGraw-Hill, New York, 1978–1979, pp. 59, 60, 63, 64.
32. M. L. Westrick, P. Gross, and H. H. Schrenk, unpublished data, Industrial Hygiene Foundation of America, Pittsburgh, PA, March 1956, courtesy of The Society of the Plastics Industry, Inc., New York.
33. B. Y. Kalinin and L. P. Zimnitskaya, Toxicology of low-pressure (high density) polyethylene. *Toksikol. Vysokomol. Mater. Khim. Syr'ya Ikh Sin., Gas. Nauch.-Issled. Inst. Polim. Plast. Mass.* 21–40 (1966); *Chem. Abstr.* **66**, 103638.

34. C. S. Weil and P. E. Palm, unpublished data, Mellon Institute of Industrial Research, Pittsburgh, PA, October 1960, courtesy of The Society of the Plastics Industry, Inc., New York.
35. C. S. Weil and H. F. Smyth, Jr., unpublished data, Mellon Institute of Industrial Research, Pittsburgh, PA, March 1966, courtesy of The Society of the Plastics Industry, Inc., New York.
36. J. A. Boling, R. H. Grummer, and E. R. Hauser, A comparison of plastic dilution of diets with full and limit-fed diets for growing swine. *College of Agricultural and Life Sciences, Research Report 55*, University of Wisconsin, Madison, WI, April 1970, pp. 1–4.
37. B. F. Matlaga, L. P. Yasenchak, and T. N. Salthouse, Tissue response to implanted polymers: The significance of sample shape. *U. Biomed. Mater. Res.* **10**, 391–397 (1976).
38. J. Autian, Film carcinogenesis. In P. Bucalossi, V. Veronesi, and N. Cascinelli, Eds., *Chemical and Viral Oncogenesis*, Vol. 2, American Elsevier, New York, 1975, pp. 94–101.
39. Plastics, *Machine Design*, **57**, 143 (April 18, 1985).
40. R. L. MacGovern, *Linear Polyethylene and Polypropylene, Process Economics Program*, Sanford Research Institute, Nov. 1966, pp. 49–50.
41. *Polypropylene Globalization Markets, Technology and Who's Who*, Phillip Townsend and Associates, 1990, p. 3-1.
42. International Agency for Research on Cancer, *IARC Monographs on the Evaluation of the Carcinogenic Risk of Chemicals of Humans; Some Monomers, Plastics and Synthetic Elastomers and Acrolein*, Vol. 19, Lyon, France, 1979.
43. *Mod. Plastics*, 89 (Jan. 1992).
44. B. Y. Kalinin, Toxicity of polypropylene. *Toksikol, Vysokomol, Mater, Khim, Syr'ka Ikh, Sin, Gos, Nauch-Issled. Inst. Polim. Plasti. Mass.* **55–63** (1966); *Chem. Abstr.* **66**, 103640a.
45. F. K. Kinoshita, personal communication, Hercules, Inc., Wilmington, DE, July 1979, tests conducted at Hercules 1958–1959.
46. P. J. Garvin, personal communication, Standard Oil Company, Chicago, IL, Aug. 1979, tests conducted at Industrial Bio-Tests Laboratories 1964–1966.
47. D. F. Edwards, unpublished data, E. I. duPont de Nemours and Co., Inc. Haskell Laboratory, Wilmington, DE, Sept. 1975.
48. J. Vollmar and G. OH-Heidelberg, Experimentelle geschwulstauslosung durch kunststoffe aus chirurgischer sicht. *Langenbecks Arch. Klin. Chir.* **298**, 729–736 (1961).
49. R. H. Rigdon, Plastics and carcinogenesis. *South. Med. J.* **67**, 1454–1465 (1974).
50. T. C. Liebert et al., Subcutaneous implants of polypropylene filaments. *J. Biomed. Mater. Res.* **10**, 939–951 (1976).
51. I. B. Smit et al., Tissue reactions to suture materials revisited: Is there argument to change our views? *Eur. Surg. Res.* **23**, 347–354 (1991).
52. L. D. Delbeke, V. Gomel, P. F. McComb, and N. Jetha, Histologic reaction to four synthetic micro suture in the rabbit. *Fetil. Steril.* **40**(2), 248–252 (Aug 1983).
53. J. F. Acquavella and C. V. Owens, *J. Occup. Med.* **32**(2), 127–130 (1990).
54. S. Robinson, Euro-study to look again at possible PP cancer link. *Plastics & Rubber Weekly* (1398) (Aug. 17, 1991).
55. W. J. Roff and J. R. Scott, with J. Pacitti, *Handbook of Common Polymers*, CRC Press, Cleveland, OH, 1971.
56. J. J. Throne, *Plastic Process Engineering*, Marcel Dekker, New York, 1979.

CHAPTER EIGHTY-NINE

Synthetic Polymers — Olefin, Diene Elastomers, and Vinyl Halides

Bailus Walker, Jr., Ph.D., MPH, and Laureen Burton, MPH

INTRODUCTION

Elastomers, also called rubber, can withstand considerably greater deformation than other materials and uniquely return essentially to their original shape even after substantial deformation. A familiar example is the behavior of a stretched rubber band after its release. All elastomers are composed of long macromolecular chains that assume a random coil conformation when undeformed (1). Deformation causes these coils to straighten out. Upon being allowed to relax, an elastomer returns essentially to its original shape because the chains reassume their random conformation.

The first elastomer identified, natural rubber, was described by Columbus as a ball that bounced. More practical early applications included primitive waterproof clothing and the rubber tires made for the carriage of Queen Victoria in 1846. Although the first specialty elastomers, polysulfides and polychloroprene, were commercialized in the 1930s, natural rubber was the major industry product until World War II, when styrene-butadiene rubber (SBR) and acrylonitrile-butadiene rubber (NBR) were established as important synthetic rubbers.

From these early beginnings, the elastomer industry grew rapidly to a global elastomer demand of 15 million metric tons in 1990 (2). The range and diversity of synthetic rubber becomes evident upon reviewing the Synthetic Rubber Manual (3) that describes both thermosetting elastomers (TSE) and thermoplastic elastomers (TPE).

Patty's Toxicology, Fifth Edition, Volume 7, Edited by Eula Bingham, Barbara Cohrssen, and Charles H. Powell.
ISBN 0-471-31940-6 © 2001 John Wiley & Sons, Inc.

TSE and TPE exhibit important similarities. The most useful properties are the result of their long molecular chains linking to one another to form a three-dimensional network. In TSE this network is linked together with essentially irreversible cross-links. Vulcanization is the process of forming these cross-links, most typically using sulfur as the cross-linking agent.

In contrast, the attachments between chains in TPE can be reversibly broken and reformed by heating the TPE. This feature permits the direct recycling of scrap TPE by molding and other shaping processes. This is the principal difference in properties between TSE and TPE; the irreversible sulfur links of TSE cause the polymer to break down when sufficiently heated, whereas the TPE can be reliably reprocessed.

TPEs are generally provided as pellets to rubber product manufacturers. Although additives might be incorporated into the pellets by the manufacturer (a process called compounding) of the final product, the TPE pellets are usually converted directly into products. This differs from typical practice with TSE.

TSE generally arrives at the rubber fabricators in bales. Ten or more ingredients might be added to the bale in heavy mixers before the compounded elastomer is shaped into a product and vulcanized. Schunk (4) has characterized the health hazards of many of these ingredients, including carbon blacks, mineral fillers, plasticizers, protective and cross-linking agents, and accelerators. Broadly considered, these health hazards can be considered in terms of the following:

1. monomers, solvents, and other materials used to prepare elastomers
2. storage and handling of elastomer (bales, pellets, and powder)
3. processing of elastomers, generally at high temperatures
4. finished rubber product

Health hazards in processing, and storage and handling elastomers are the dominant focus of this section; limited references will be made to the other two areas where appropriate.

Certain portions of the material refer to monomer toxicology and epidemiology because some of the monomers used in manufacturing elastomers remain at low levels in the polymer. A full discussion of the toxicity of monomers is beyond the scope of this chapter.

Table 89.1 lists some typical basic properties of certain elastomers. Properties within a given class of elastomers can vary significantly. For example, increasing acrylonitrile content in NBR reduces swelling of the NBR caused by some oils and solvents. Figure 89.1 gives comparative properties of various elastomers.

Most rubber is sold raw or uncured as a solid or liquid latex. The basic steps in the manufacture of some types of dry synthetic rubber are polymerization, coagulation, washing, and drying. The basic steps in producing a latex are polymerization, stabilization, and usually, concentration. A latex is defined as a stable aqueous dispersion that contains discrete polymer particles about 0.05 to 5 mm in diameter (5).

Emulsion polymerization systems contain water, monomer(s), initiator, and anionic or cationic surfactants. Solution polymerization with stereospecific catalysts involves reacting one or more monomers in an inert solvent; system conditions can be controlled to maximize a desired isomer arrangement in the polymer.

Table 89.1. Physical Properties of Selected Polymers

Compound	CAS	Molecular Formula	Molecular Weight 68.118	Boiling Point (°C)	Specific Gravity	Melting Point (°C)	Sol. in Water (at 68°F) mg/mL	Refractive Index (20°C)	Vapor Pressure (mmHg)	UEL % by Vol. Room Temp.	LEL % by Vol. Room Temp.
Polyisoprene	[9003-31-0]	$(C_5H_8)_n$	NR <100,000–4,000,000 IR 700,000– >1,600,000	—	0.90–0.93	—	—	—	—	—	—
Styrene-butadiene	[9003-55-8]	$(C_{12}H_{14})_n$	158.24	—	0.94	−59	—	—	—	—	—
cis-Polybutadiene	[9003-17-2]		420,000	—	0.89–0.94	—	—	1.52	—	—	—
Ethylene propylene copolymer	[9010-79-1]	$[(C_2H_4)_x–(C_3H_6)_y]_n$	100,000–500,000	—	—	—	—	—	—	—	—
Acrylonitrile butadiene copolymer	[9003-18-3]	$[(CNC_2H_3)_x–(C_4H_6)_y]_n$	20,000–10^6	—	—	—	—	—	—	—	—
Vinylidene fluoride-hexafluoro propylene	[9011-17-0]	$[(CH_2CF_2)_x–(CF_2CF\text{-}CF_3)_y]_n$	—	—	1.8	—	—	—	—	—	—
Chlorosulfonated polyethylene	[68037-39-8]	$[(C_5H_9ClC_2H_4)_x–(CH_2CHSO_2–Cl)]_n$	—	—	1.28	—	—	—	—	—	—
Butyl rubber	[9010-85-9]	$[(C_4H_8)_x–(C_5H_8)_y]_n$	400,000–600,000	—	0.92	—	—	—	—	—	—
Perfluoro rubber	—	$[(C_2F_4)_x–(CF_2CF\text{-}OCF_3)_y–(CF_2CF\text{-}ORx)_z]_a$ Rx=$(CF_2)_n$CN; CO_2H; SO_2F; $CF_2(CFO\text{-}CF_3)$–C_6F_5	—	—	2.02	—	—	—	—	—	—
Acrylic elastomer	[9003-32-1]	$[(C_2H_4)_x–(CH_2CHO\text{-}OR)_y]_n$	95,000	—	—	—	—	—	—	—	—
Chlorinated polyethylene	—	$(C_2H_3\text{-}Cl\text{-}C_2H_4)_n$	—	—	—	—	—	—	—	—	—
Epichlorohydrin	[106-89-8]	$(C_3H_5ClO)_n$	92.525	117.9	1.181–1.183	−57	50–100	1.4361	—	21.0	—
Silicone	[63148-62-9]	$(C_4H_{12}O_3Si_3)$	—	—	0.963	—	—	—	—	—	—
Polyurethane elastomers	[9009-54-5]	$[C_3H_8N_2O]_n$	88.1	—	—	—	—	—	—	—	—
Polyvinyl chloride	[9002-86-2]	$(C_2H_3Cl)_n$	62,499	—	1.4	—	<1	1.54	—	—	—
Polyvinylidene chloride	[9002-85-1]	$(C_2H_3Cl_2)_n$	20,000–50,000	—	—	—	—	—	—	—	—
Vinylidene chloride vinyl chloride	[9011-06-7]	$(C_4H_5Cl_3)_n$	−159.44	—	—	−183	—	—	—	—	—
Vinylidene chloride-methyl acrylate	[25038-72-6]	—	—	—	—	—	—	—	—	—	—

427

Table 89.1. (Continued)

Compound	CAS	Molecular Formula	Molecular Weight	Boiling Point (°C)	Specific Gravity	Melting Point (°C)	Sol. in Water (at 68°F) mg/mL	Refractive Index (20°C)	Vapor Pressure (mmHg)	UEL % by Vol. Room Temp.	LEL % by Vol. Room Temp.
Vinylidene chloride acrylonitrile	[9010-76-8]	—	68.118	—	—	—	—	—	—	—	—
Polyvinyl fluoride	[24981-14-4]	$(C_2H_3F)_n$	46.044	—	1.38–1.44	200–230	—	1.46	—	—	—
Polytetrafluoroethylene	[9002-84-0]	$(C_2F_4)_n$	—, 100.02	—	2	327	—	1.35	—	—	—
Fluorinated ethylene propylene copolymer	[25067-11-2]	$[(C_2F_4)_x–(C_2H_4)_y]_n$	—	—	—	245–280	—	—	—	—	—
Perfluoroalkoxy copolymer	[26655-00-5]	$[(CF[OR_f]CF_2)_x–(C_2F_4)_y]_n$ OR_f = perfluoroalkoxy group	—	—	—	300	—	—	—	—	—
Ethylene tetrafluoro-ethylene copolymer	[54302-05-5]	$[(C_2F_4)_x–(C_2H_4)_y]_n$	—	—	1.70	250	—	1.40	—	—	—
Polyvinylidene fluoride	[24937-79-9]	—	300,000–600,000	—	1.75–2.02	156–220	—	1.42	—	—	—
Polychloro-trifluoro-ethylene	[9002-83-9]	$(CF_2–CFCl)_n$	—	—	—	210–220	—	—	—	—	—
Ethylene chloro-tri-fluorethylene copolymer	[25101-45-5]	—	—	—	—	220–245	—	—	—	—	—

SYNTHETIC POLYMERS—OLEFIN, DIENE ELASTOMERS, AND VINYL HALIDES

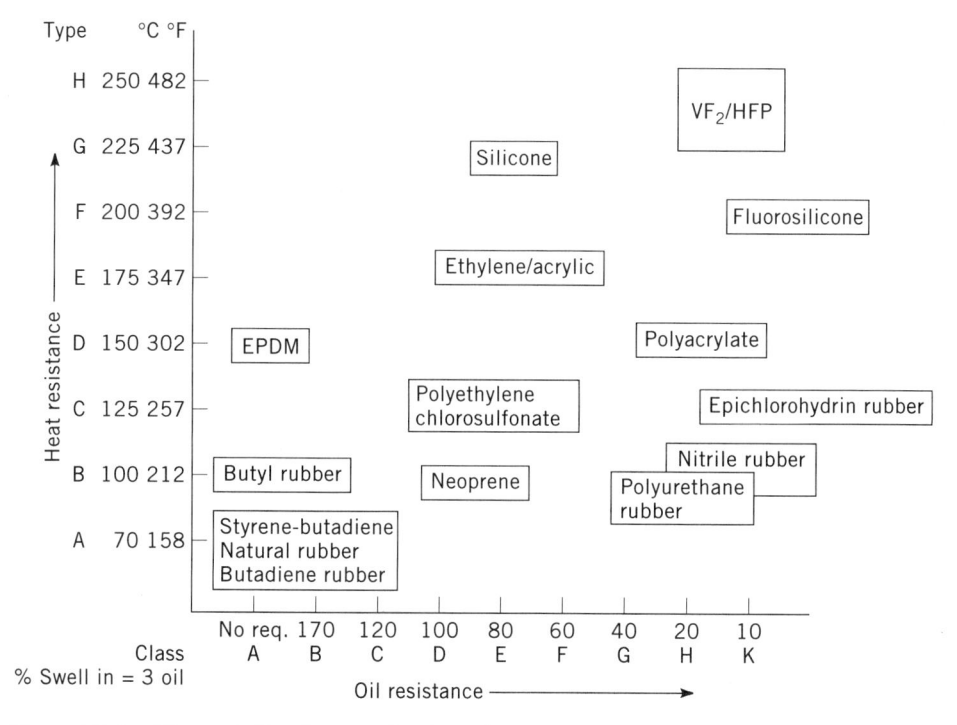

Figure 89.1. Heat and oil resistance of various elastomers (courtesy of D. H. Geschwind, DuPont Company, Inc., Polymer Products Department, Wilmington, Delaware.

Antioxidants are generally added for shelf, processing, and in-service stability. Unsaturation of the components in the polymer chain correlates with sensitivity to oxidation. For example, among synthetic rubbers, butadiene and isoprene polymers are much more sensitive to oxidation than ethylene-propylene polymers. Natural rubber contains some natural antioxidants from rubber trees.

Vulcanization is usually done with sulfur, sulfur-containing compounds, or peroxides, but it may also be accomplished with other compounds that yield free radicals at curing temperature or by radiation. Various supplementary materials such as cure accelerators, cure retarders, or reinforcing agents are commonly part of the compounding recipe. Vulcanization ideally begins when the elastomer assumes its final shape in a mold. The elastomer type and its viscosity significantly affect molding behavior (6).

Analysis and Specifications

Measurements to determine the amounts of residual monomer have been of particular interest (7–9). Residual butadiene, acrylonitrile, and styrene can be determined in solid rubber to <1 ppm (by weight, w/w) by headspace gas chromatography (9). Butadiene monomer in 0.1 g samples of latex can be measured by gas chromatography with a detection limit of 50 ppm (10). A gas chromatographic/mass spectrometric method to identify volatile materials released during simulated vulcanization at 160 to 200°C has been reported (11); the polymers tested were *cis*-polybutadiene, styrene-butadiene, and a blend of these two. Pyrolysis/mass spectrometry (samples <5 mg) has been used to distinguish among adhesives based on natural rubber, styrene-butadiene, and polychloroprene (12). More recent work has helped to unravel some of the complexities of ingredient–emission relationships through computerized analysis of emissions from vulcanized rubber (13).

Residual acrylonitrile in nitrile rubber reportedly varies from "nondetectable to something less than 100 ppm" (14). The concentration of chloroprene monomer is less than 1 ppm (w/w) in solid polychloroprene, but amounts as high as 5000 ppm have been reported in some latex samples (8).

Accepted practice in raw rubber manufacturing calls for quality control of the polymer within various specifications (usually +10%). Standards for these measurements are generally formulated by ASTM (15) or product-oriented organizations such as the Society of Automotive Engineers (16).

Elastomers that meet certain specifications are permitted in specific food-additive applications (17). Both natural rubber and polyisoprene are specifically identified among the elastomers permitted in rubber articles intended for repeated use. Natural rubber, natural latex, *cis*-1,4-polyisoprene, synthetic rubber, and rubber hydrochloride are specifically identified among the elastomers permitted in adhesives, sealing gaskets, paper, paperboards, and coatings. Styrene-butadiene, ethylene-propylene copolymer (and certain diene-containing terpolymers), acrylonitrile-butadiene copolymer, and chloroprene polymers are similarly permitted in specific applications. The components permitted in rubber articles intended for repeated use also include vinylidene fluoride-hexafluoropropylene copolymer that has a minimum number-average molecular weight of 70,000 and tetrafluoroethylene terpolymer that has a minimum number-average molecular weight of 100,000. Chlorosulfonated polyethylene is also permitted in certain food-contact or drinking water applications; specifications require that the chlorine content and the sulfur content not exceed 25 and 1.15% by weight, respectively, and that the molecular weight is in the range of 95,000 to 125,000. Permissible direct additive applications include the use of natural rubber, styrene-butadiene, and butyl rubber as masticatory substances in chewing gum base.

Reviews of analytical methods useful in assessing health and safety aspects of rubber processing and handling have been published biennially (18–21). These methods focus on many modern analytic techniques.

Toxicological Potential

Dry solid polymers usually contain less residual monomer (or solvent) than latex materials. The processing necessary to produce the dry product drives the residual monomer or solvent out of the resin, usually by heat.

Several reports address worker health problems in the rubber fabrication industry. For example, one study (22) suggests an association between the mortality risk of lung cancer and employment in operations involving reclaim, chemicals, and special products. Another study (23) showed that processing workers had increased mortality from leukemia, emphysema, and cancers of the stomach, large intestine, biliary passages, and liver. Other findings (24) demonstrated that men who were employed for at least 10 years experienced small increases in deaths from cancers of the large intestine, pancreas, and lung. Studies of cancer mortality in the British rubber industry found (25) an absence of any excess mortality from bladder cancer among men who entered the industry after January 1, 1951, possibly the result of removing putative bladder carcinogens from production processes in July 1949.

Industrial dermatitis from finished rubber products due to the various chemicals added during polymerization, curing, and processing is not uncommon (26–30). Mercaptobenzothiazole, tetramethylthiuram disulfide, N-isopropyl-N'-p-phenylenediamine (IPPD), and related compounds are the most common offenders.

Elastomers degraded at high temperatures around 800°C can yield more toxic products than elastomers degraded at smoldering temperatures or gradually rising temperatures. This is to be especially so with nitrile-butadiene.

Workplace Practice and Standards

At ambient temperatures, the main concern with vulcanized polymer is possible effects from inhalation of dust generated in grinding operations and dermatitis in "rubber-sensitive" individuals. Operations that may generate dust should, as a minimum, be controlled within recommended limits for nuisance dusts (31). Dermatitis can usually be controlled only by avoiding either the specific materials that provoked the reaction or "rubber" in general. In some cases, avoiding occluded contact with the skin may eliminate dermatitis (32).

Depending on the particular polymer and type, unvulcanized rubber may require special handling because of the residual monomer or solvent content. Uncured chlorosulfonated polyethylene or other elastomers similarly prepared by dissolving an existing polymer generally contain residual solvent. Monitoring carbon tetrachloride from chlorosulfonated polyethylene, is desirable if bulk polymer is stored in warehouses with limited ventilation or if ventilation in the processing area is questionable. Ventilation adequate to protect against diffusion of the residual carbon tetrachloride should be adequate for any other vapors that may develop during normal curing operations.

Ammonia is added to NR (polyisoprene) latex to preserve it (33), and it may be released during processing. Handling of latex requires consideration of the toxic potential of the monomer(s), cosolvents, other ingredients, and also of the particular operation(s) involved to determine if monitoring the working environment is indicated. Standard operating procedures should include provisions for appropriate personal protective equipment and monitoring to ensure that exposures are well controlled where gross exposure is unavoidable, as in tank entry.

A OLEFIN AND DIENE ELASTOMERS

1.0 Polyisoprene

1.0.1 CAS Number: [9003-31-0]

1.0.2 Synonyms: Rubber; 1,3-butadiene, 2 methyl-homopolymer; poly (2 methyl-1,3 butadiene)

1.0.3 Trade Names: Nalsyn 220; Nalsyn 2210; Nalsyn 2205

1.0.4 Molecular Weight

Natural rubber comprises a range of polymers with estimated molecular weight from < 100,000 to 4,000,000 (34).

1.0.5 Molecular Formula: $[C_5H_8]_n$

1.0.6 Molecular Structure:

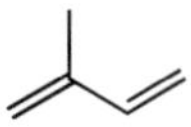

1.1 Chemical and Physical Properties

Natural rubber has been partially replaced by synthetics, particularly styrene-butadiene, as a general-purpose rubber. High resilience, low heat build-up, and easy processing are particular advantages of natural rubber when it is often used in blends with synthetic polyisoprene and other elastomers. Natural rubber, alone and in combination with neoprene, received high ratings for resistance to water, dimethyl sulfoxide and some alcohols in a comparative test of glove materials; resistance to other solvents varied from good to poor. Polyisoprene is thermoplastic until mixed with sulfur and vulcanized. It supports combustion.

1.2 Production and Use

The latex of natural rubber is obtained from trees (*Hevea brasiliensis*); the actual monomer is isopentenyl pyrophosphate that has been formed by biosynthesis. Natural rubber contains low molecular weight impurities; small amounts of sugar, fatty acids, proteins, and trace metals all play an important part in processing.

Depending on the catalyst and conditions, rubber may undergo 1,2-; 3,4-; or 1,4- addition polymerization that leads to several isomeric structures.

Almost all commercial synthetic polyisoprenes are prepared from purified isoprene monomer by a solution process. A stereospecific catalyst, such as an Al-Ti Ziegler type, is required to direct polymerization to the *cis*-1,4 isomer.

The production of the finished polymer requires two separate manufacturing processes: (1) formation of the raw polymer and (2) conversion of the polymer to the finished rubber product. The first step is similar to that of plastic production. Large-scale operations use bulk materials in an enclosed system.

cis-1,4-Polyisoprene is used in tires, tire products, molded and mechanical goods, foam rubber, rubber sheeting, rubber bands, bottle nipples, footwear, and sporting goods. *trans*-1,4-Polyisoprene, is used mainly in golf ball covers, orthopedic devices, and splints. *trans*-1,4-Polyisoprene is occasionally used in cable covering and adhesives.

1.4 Toxic Effects

Table 89.2 summarizes toxicity test data on selected olefin resin. Information on the toxic effects of natural (NR) and isoprene rubber (IR) has changed little since the fourth edition

Table 89.2. Summary of Toxicity Tests of Selected Polymers

Chemical Name	CAS	Species	Exposure Route	Approximate Dose g/kg	Treatment Regimen	Observed Effects	Reference
Polyethylene	[9002-88-4]	Rat	ig	7.95	Single dose	No effect	37
Polypropylene	[9003-07-0]	Mice	ig	8.0	Single dose	No effect	38
Chlorosulfonated polyethylene	[68037-39-8]	Rat	ig	20.0	Single dose/5 days	No effect	77
Polyvinyl chloride	[9002-86-2]	Dog	ig	25.0	Daily dose	No effect	100
Polyvinylidene chloride	[9002-85-1]	Rats/mice	ig	25.0	Single oral dose	Tissue damage	
Polyvinyl acetate	[9003-20-7]	Rats/mice	ig	2.5	Daily dose 12-months	Tissue damage	6
Polyvinyl alcohol	[9002-89-5]	Rats	ig	50g/kg	30 single doses	Minimal changes in liver and myocardial cells	14
Polyacrylonitrile	[25014-41-9]	Rat	ig	0.25 to 0.5	Daily doses for 6 months	Reversible changes in liver, kidney and thyroid	42
		Rats	ig	4.0			
Polyacrylamide	[9003-05-8]	Rats/dogs	ig	MTD[a]	Repeated oral administration for 2 years	No effect	71

[a]MTD = Maximum tolerated dose.

of this book. NR has long been recognized as a low toxicity material suitable for a wide range of food-contact and body-contact applications.

1.4.2 Human Experience

In 1991, the Food and Drug Administration issued a medical alert concerning "Allergic reaction to latex-containing medical devices" (35). The alert discussed reports of allergic reactions including anaphylaxis in some exposed individuals who used various medical devices manufactured from natural latex rubber. Recent studies found that latex allergies affect up to 14% of healthcare workers. Chronic hand dermatitis in surgeons may be produced or made worse by the use of certain rubber gloves. The allergies in rubber gloves are usually mercaptobenzothiazole or tetramethylthuriam. The Elastyren glove, which is free of rubber chemicals, is usually well tolerated by surgeons who have dermatitis due to rubber (36).

The number of latex allergies has increased from a single case report in 1979 to 6% of Americans in the mid-1990s. A report in the January 1995 issue of *Journal of Allergies and Clinical Immunology* found that urban air contained latex particles shed into the environment by normal automobile tire wear. One study collected particulate air pollution samples, which included black rubber fragments containing latex, which were recognized in tests by human antibodies to latex proteins. More than half (58%) of the airborne debris was small enough to be inhaled into the lungs. Although historically this health risk has been elevated in hospital personnel and patients, there is now evidence of a significant risk for the general population (37, 38). A survey of blood bank sera from the general adult population in Detroit, Michigan, revealed the presence of anti-latex IgE antibody in 6.5% of the volunteers, thus indicating the potential for latex allergy in roughly 17 million individuals in the United States (39). Symptoms of latex allergy resulting from IgE-mediated reactions are those customarily diagnosed for atopic allergic disease, including sneezing, watery eyes, asthma, decrease in blood pressure, and circulatory collapse. Severity can range from mild to fatal.

1.4.2.2 Clinical Cases. Cases of allergies to latex are cited in Section 1.4.2.

2.0 Styrene-Butadiene

2.0.1 CAS Number: [9003-55-8]

2.0.2 Synonyms: Benzene, ethenyl-, polymer with 1,3-butadiene; styrene, 1,3-butadiene polymer; styrene-butadiene copolymer; butadiene-styrene latex; butadiene-styrene resin; 5% styrene, SB butadiene-styrene copolymer; poly-(styrene-co-butadiene); poly-(butadiene-co-styrene)

2.0.3 Trade Names: Some 47 trade names are listed in the literature. Examples are Afcolac.B101; Andrez; Base 661; butadiene styrene resin; Butakon 85-71; Diarex 600; Dienol S. Europrene; finaprene; Tufprene; Uridene

2.0.4 Molecular Weight

The molecular weight for "hot" rubber is given (34) as 50,000 to 400,000 (viscosity average) or 30,000 to 100,000 (number average); for "cold" rubber the corresponding values are 280,000 or 110,000 to 260,000 (or 500,000 as weight-average molecular weight).

2.0.5 Molecular Formula: $[C_{12}H_{14}]_n$

2.0.6 Molecular Structure:

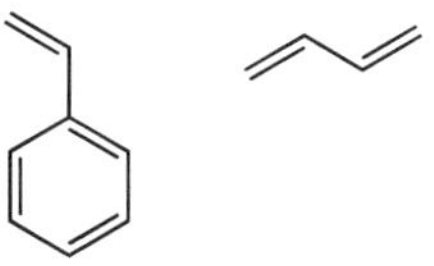

2.1 Chemical and Physical Properties

The properties of "cold" SBR offers certain advantages over those of hot SBR. Solution SBR offers some advantages over emulsion SBR, and it is replacing it in some applications. In general, mechanical properties are slightly inferior to those of natural rubber and vary according to composition. "Cold" styrene-butadiene has improved abrasion resistance and better tensile strength. Solution SBR is essentially pure rubber with a trace of antioxidant added because it does not require all of the additives needed to provide a stable emulsion as in the emulsion SBR process. Molecular weights of 100,000 to 200,000 (number average) are supplied, and the higher molecular weight polymers are oil-extended. Weight-average molecular weights of more than 500,000 are available. Solution SBR is heavily used in tire applications where it provides an excellent balance of wear, wet traction, and rolling resistance. It also is used in adhesives. One advantage of SBR over natural rubber is that it can tolerate much larger quantities of oil extenders without deterioration of vulcanizate properties. SBR has better abrasion resistance and aging resistance than natural rubber, but poorer dynamic mechanical properties.

2.1.1 General

SBR elastomers are often characterized on the basis of the percentage of styrene coreacted with the polymer. SBRs are available with a wide range of styrene levels; a level of 23.5% styrene is used in many applications. There has long been a concern for the toxicity of synthetic elastomers like SBR, as evidenced by early patch tests on the skin of rabbits (40).

2.1.2 Odor and Warning Properties

No information was found in the literature except about its irritating fumes and acrid smoke during thermal decomposition.

2.2 Production and Use

SBR is prepared mainly by emulsion and solution polymerization in a free-radical process with a mercaptan, such as dodecylmercaptan, which is used as a chain transfer agent to control molecular weight. In the emulsion process, a soap-stabilized water emulsion of styrene and butadiene is polymerized in continuous reactors to form SBR elastomer. "Cold" SBR is produced at a temperature of 5 to 10°C (41), and "hot" SBR at about 50°C (42). Solution SBR is typically made in hydrocarbon solution with an n-butyllithium initiator. The major differences between emulsion and solution styrene-butadiene are in the linearity and molecular weight distribution of the polymer chains. In nonpolar solvents, the tendency is to produce block copolymer, but random copolymers are formed in polar solvents. These polymers are linear and may suffer from cold flow. This can be eliminated by linking the polymer chains together to form a star polymer. SBR can be vulcanized similarly to natural rubber but generally requires less sulfur and more accelerator because of its low unsaturation. SBR is by far the most widely used synthetic rubber for tires, footwear, mechanical goods, coatings, adhesives, and carpet backing.

2.3.5 Biomonitoring/Biomarkers

Styrene-butadiene is not in the published list of chemicals for which biological determination may be useful for evaluating internal dose.

2.4 Toxic Effects

The tire industry has given considerable attention to SBR because of its treads and related uses. Several studies discuss toxicological factors affecting SBR developments (43), as well as those associated with rubber and rubber cements (44, 45). The prevalence of occupational dermatitis among tire and cement workers has been studied (44). Solvents were the main cause of contact dermatitis in another study (45). Extensive literature is available on the toxic properties of both monomers, styrene, a colorless to yellowish liquid, and 1,3-butadiene, the gaseous hydrocarbon referred to as butadiene. In the early 1990s, there was much controversy surrounding the health effects of occupational inhalation exposure to butadiene focused largely on its potential as a human carcinogen. The contention over this issue is revealed in a number of editorial comments and scientific reports since 1989.

2.4.2 Human Experience

Contact dermatitis and leukemia have been associated with solvents and additives used, but not with the polymer itself (45a,b).

2.4.2.3 Epidemiology Studies. Controversy surrounding the health effects of exposure to styrene-butadiene has focused increased attention on exposure to butadiene — 4,107 million pounds of which were produced in 1997. Of concern is the multiple sources of exposure in the workplace (i.e., decontamination and maintenance of processing equipment, product and transport of the monomer or its chemical mixture) and industrial emissions to the ambient environment. The concern for the health effects is focused largely on its potential as a human carcinogen.

The relevant epidemiological studies of butadiene involve one cohort of workers at a butadiene monomer production facility (46, 47), and two cohorts involved in SBR production (48–52). Studies of workers involved in tire or rubber products manufacturing (53, 54), although often cited for relevance in assessing butadiene exposure, are not germane because (1) butadiene is not released from synthetic rubber during processing and (2) the study cohort was not adequately controlled for exposures to known and suspected leukemogenic agents (55). One study of rubber manufacturing workers (54) did report findings for a subgroup of synthetic plant workers, that produced at least some butadiene-related elastomers, but the lack of detail about the amount and duration of butadiene usage in these operations makes the findings about these workers of questionable relevance. Mortality rates for one subgroup of SB latex workers have been reported twice (56, 57) but involved few workers and showed no elevation in rates of lymphopoietic cancers.

The largest butadiene epidemiological study evaluated mortality between 1943 and 1982 for 12,110 workers at eight SBR plants in the United States and Canada (50, 52). Eligible workers in U.S. plants, were employed for at least 1 year between 1943 and 1976, whereas eligible employees in the one Canadian plant worked at least 10 years or reached age 45 during employment. Records were complete from 1943 at four of the plants, but mortality follow-up at the remaining plants started when records were judged complete (1953, 1958, 1964, 1970).

For the total cohort, mortality rates from all causes (observed/expected, O/E = 2441/3000.8, standardized mortality ratio, SMR = 0.81, 95 % confidence interval = CI 0.78 to 0.85) and all cancers (O/E 518/609.4, SMR = 0.85, 95% CI 0.78 to 0.92) were significantly lower than U.S. rates. Mortality for the lymphopoietic cancers was consistent with expected values (O/E = 55/56.7, SMR = 0.97, 95% CI 0.73 to 1.26), including lymphosarcoma (O/E = 7/11.5, SMR = 0.61, 95% 0.24 to 1.26) and leukemia (O/E = 22/22.8, SMR = 0.96, 95% CI 0.60 to 1.46). There were no mortality trends by duration of employment for any cause of death.

Mortality rates by job category showed variable findings by racial group. White production workers showed lymphopoietic cancer mortality consistent with expected values (O/E = 13/11.9, SMR = 1.10, 95% CI 0.58 to 1.87), no excess for leukemia (O/E = 4/4.8, SMR = 0.84, 0.22 to 2.13) or lymphosarcoma (O/E = 0/2.4, SMR = 0, 95% CI 0 to 1.54), and an elevated SMR for other lymphatic cancers (O/E = 7/3.1, SMR = 2.30, 95% CI 0.90 to 4.65).

Black production workers showed a significant excess of lymphopoietic cancers (O/E = 6/1.2, SMR = 5.07, 95% CI 1.83 to 10.88) due to elevated mortality from leukemia (O/E = 3/0.5, SMR = 6.56, 95% CI 1.24 to 17.53) and other lymphatic cancers (O/E = 2/0.4, SMR = 4.82, 95% CI 0.61 to 18.06). Interpretation of the findings for black workers was hindered by inaccurate enumeration of former workers because of missing racial information at several of the plants. The findings for other lymphatic cancers were the only consistent finding for both racial groups. Five of the nine deaths from other lymphatic cancers were classified as multiple myelomas [versus an estimated 1.7 expected (58), SMR = 2.94, 95% CI 0.95 to 6.86], and the remainder were unspecified non-Hodgkins lymphomas. Mortality from non-Hodgkins lymphomas was consistent with expected values (O/E = 5/4.3) (58) (the five included four other lymphatic cancers and one lymphosarcoma). SMRs for maintenance workers showed lower mortality than expected for all lymphopoietic cancer subgroups for both black and white workers.

The SBR workers' job category analyses omitted 2391 workers who had incomplete work histories. Roughly 75% of these workers were active employees at one plant (50). As such, SMRs by job category overestimate worker mortality (59). For example, Acquavella (58) estimated leukemia SMRs correcting for omitted production and maintenance workers at 1.11 (O/E = 7/6.3) and 0.70 (O/E = 6/8.6), respectively, versus reported SMRs of 1.34 (O/E = 7/5.2) and 0.85 (O/E = 6/7.1).

The most recent epidemiological study is a lymphopoietic cancer case control study "nested" within the largest butadiene workers cohort study (50). Case control studies yield two measures of importance: (1) an estimate of the ratio of disease rates for exposed versus unexposed workers (called the odds ratio (OR) and (2) the exposure prevalence among controls — an estimate of exposure prevalence in the base study population. The findings from case control and cohort studies can be related from the OR and the exposure prevalence (namely, the OR and SMR).

The OR for butadiene exposure was reportedly 7.6 (95% CI 1.6 to 35.6) for leukemia, 0.5 (95% CI 0.1 to 4.2) for lymphosarcoma, and 1.5 (95% CI 0.5 to 4.8) for the other lymphatic cancers. The exposure prevalence was estimated at approximately 60% for the controls of the leukemia cases. These leukemia findings (OR = 7.6, exposure prevalence of 60%) conflict markedly with the results of the base cohort study (60), which found no leukemia excess overall (O/E = 22/22.9) and no excess for production (O/E = 7/6.3) and maintenance workers 0.70 (O/E = 6/8.6) — job categories that had the greatest potential for butadiene exposure (61).

Three possible explanations for these conflicting results have been suggested (59): bias in the cohort mortality study, bias in the case control study, or an extremely low leukemia rate among unexposed workers. Until these possibilities are evaluated further, proper evaluation of the case control study can be done only in concert with the results of the base cohort study.

None of the studies to date has had quantitative exposure data or quantitative estimates of exposure. However, an update of the largest SBR workers cohort study at this time was ongoing, and the authors planned to incorporate quantitative butadiene exposure estimates in it (47). Information from this study should resolve the conflicting leukemia findings from the SBR workers case control study and further clarify the lymphopoietic cancer findings reported among SBR workers.

Several consistent themes emerge from the butadiene epidemiological literature. First, worker mortality rates from all causes and all cancers are lower than those of the general population. Second, there are no trends of increasing mortality with increasing duration of employment for any cause of death. Third, findings for worker subgroups for individual lymphopoietic cancers are quite variable across studies and show elevations for leukemia, lymphosarcoma, and multiple myeloma in subgroups in the individual studies, respectively. Fourth, long-term workers in each cohort show favorable mortality patterns for all individual causes of death.

2.5 Standards, Regulations, or Guidelines of Exposure

Some types of SBR are extended with oil. Threshold limit values (TLVs) are listed for some types of oil, for example, solvent-refined heavy naphthenic oil (62). Production processes for mineral oils have changed over time, and more recent manufacturing methods produce highly refined products that contain smaller amounts of contaminants, such as polycyclic aromatic hydrocarbons (63).

3.0 Polybutadiene

3.0.1 CAS Number: *[9003-17-2]*

3.0.2 Synonyms: BR, PB, PBD

Polybutadiene is a homopolymer of 1,3-butadiene.

3.0.3 Trade Names: Ameripol CB; Budene; Buna; Buna CB; Cariflex BR; *Cis*; Cisdene; Diene; Europrene *Cis*; Europrene Sol P; Finaprene Cis; Finaprene PB; Intene; Intolene; JSR-BR; Neocis; Nipol; Phillips-cis-4; Solprene Synpol-EBR; Taktene

SYNTHETIC POLYMERS—OLEFIN, DIENE ELASTOMERS, AND VINYL HALIDES

3.0.4 Molecular Weight

The catalyst in the polymerization process determines the structure and molecular weight. Polybutadiene produced by lithium catalysts has a linear microstructure in a narrow molecular weight distribution, and limited long-chain branching. Ziegler–Natta type catalysts produce polybutadiene that has wide molecular weight distributions and various degrees of branching, depending on the type of metal used. *cis*- and *trans*-Polybutadiene have an average molecular weight of 420,000.

3.1 Chemical and Physical Properties

Polybutadiene typically contains a mixture of isomers, including large amounts of the 1,2-isomers. It has high gel content, wide molecular mass distribution, and a high glass transition (T_g) temperature (that temperature at which a hard glassy polymer becomes rubbery). Lithium-produced polybutadienes contain approximately 35% *cis*-1,4, whereas those produced by Ziegler–Natta catalysts contain more than 90% *cis*-1,4.

In the polymerization of butadiene, both 1,2- and 1,4-polymers can occur. In the case of 1,4-, both *cis* and *trans* isomers are possible. Only *cis*-1,4-polymer is a useful rubber; all other stereoregular polymers crystallize too readily. Only when *cis*-1,4-polymer has very high *cis* content will it crystallize and then only at very low temperatures. Thus, polybutadiene rubbers do not exhibit high tensile strength due to crystallization on large deformation. Polybutadiene is known for its excellent abrasion resistance and good low temperature flexibility. It is combustible.

3.1.1 General

Polybutadiene contains 36% *cis*, 55% *trans* 1,4-, and 95% vinyl 1,2-. Other general properties are as follows:

Specific gravity: 0.93
Service Temperature (°C): minimum: −62   maximum: 79–100
Tensile strength (lbs·in^{-2}) (23°C): 2500–3000
Solubility parameter: 6.3

3.1.2 Odor and Warning Properties

No information was found in the literature except for the mildly aromatic odor of the monomer; butadiene-styrene has a reported odor threshold of 0.01 ppm.

3.2 Production and Use

The production of polybutadiene involves copolymerizing three parts butadiene and one part styrene. Also present in small amounts is an initiator or catalyst, which is usually a peroxide, and a chain-modifying agent such as dodecyl mercaptan. It is used in tires, footwear, mechanical goods, coating adhesives, and carpet backing.

3.3.4 Community Methods

The EPA reports that butadiene emission samples have been collected from the vent stream of a plant that manufactured synthetic polymer from styrene and butadiene. Samples were analyzed on-site by using a gas chromatograph equipped with a flame ionization detector. The precision of butadiene concentration was determined from simultaneous samples collected at a nominal sample rate of 0.050 L/min rather than at the recommended sampling rate of 0.5 L/min. Acceptable precision was observed at both sampling rates.

3.3.5 Biomonitoring/Biomarkers

No information was found in the literature. Polybutadiene is not in the list of chemicals for which biological measurements are available to assess internal dose.

3.4 Toxic Effects

Both BR as well as SBR are extended with oils. The processing oils in the stock could be released from the surface and result in exposure through skin contact. Prolonged contact with untreated naphthenic/paraffinic and aromatic oils caused skin cancer in laboratory animals, when applied over a 2-year period. Untreated naphthenic/paraffinic and aromatic oils are classified as carcinogenic to humans by the IARC (63).

3.4.1 Experimental Studies

See Section 3.4.

3.4.1.5 Carcinogenesis. Skin cancer occurs in experimental animals, as cited in Section 3.4.

3.4.2.1 General Information. General information about human experience with polybutadiene is sparse. By contrast, there is substantial information about the monomers, as previously indicated.

3.4.2.2.5 Carcinogenesis. The individual oil extenders are classified as suspected human carcinogens.

3.4.2.3.2 Chronic and Subchronic Toxicity. Liquid polybutadiene may cause skin irritation.

3.5 Standards, Regulations, or Guidelines of Exposure

Standard operating procedure should include appropriate industrial hygiene procedures to reduce the risk of gross exposure.

4.0 Ethylene-Propylene

4.0.1 CAS Number: *[9010-79-1]*

The terpolymer has CAS Number *[25038-37-3]*.

4.0.2 Synonyms

Ethylene-propylene is referred to as EPM or EPDM. EPM and EPDM, respectively, are preferred abbreviations for ethylene-propylene copolymer and for terpolymer containing ethylene and propylene in the backbone and a diene that has the residual unsaturated portion of the diene in the side chain (63). EPDM is also used as a collective term for both the copolymer and terpolymer (65).

4.0.3 Trade Names: Buna AP; Dutral; EPCar; EPM; Inolan; Keltan Nordel; Polysar Royalene; Vistalon.

4.0.4 Molecular Weight: M_w 100,000 to 500,000 (34)

4.1 Chemical and Physical Properties

The outstanding feature of these rubbers is their inherent resistance to oxidation and ozone as a result of their essentially saturated molecular backbone. They are noted for excellent weather, heat, aging, and electrical characteristics, as well as low permeability to gases.

At atmosphere pressure and temperature, EPM is soluble in aliphatic hydrocarbons of C_5 and higher. EPM is not soluble in aldehydes, ketones, carboxylic acids, esters, or dioxane. The polymer is a saturated hydrocarbon except for a small mole fraction of olefin. The amount of ethylene and propylene in the polymer is among the key determinants of its properties. For example, oxidative stability and resistance to scission increase as ethylene increases and the number of methine hydrogens decreases. Crystallinity and glass-transition temperature vary with ethylene content. In most commercial grades, the glass-transition temperature is between -55 and $-60°C$. At high diene and low ethylene content, it can be 10–20°C higher.

4.1.1 General

Ethylene-propylene copolymer and ethylene-propylene terpolymer are specialty elastomers. They possess particular qualities, such as chemical resistance, especially to ozone that suit them to a number of applications. They tolerate high concentrations of filler and oil and retain good physical properties.

4.2 Production and Use

EPM is made by copolymerizing ethylene and propylene. The monomers may or may not be combined with a third of even fourth monomer to provide olefinic sites along the backbone. Olefinic sites improve cross-linking response with peroxides and permit direct sulfur cure. EPM is used in making hoses, belts, gaskets, and footwear.

4.4 Toxic Effects

Ethylene-propylene elastomers generally have low toxicity. Certain grades have received FDA approval for contact with skin and in pharmaceutical applications. Residual monomers, polymerization diluents, catalysts residue, slurry, antifoam or deashing aids, scale from polymerization vessels, and antioxidants are the most likely contaminants; all have low to moderate toxicity.

4.4.2 Human Experience

EPDM is available as a blend with silicone rubber (66) and in a compounded version for a hose cover (67). A trace bloom which contains minute quantities of sulfur-containing curing chemicals may develop on the hose surface. Workers who have been sensitized previously to thiazole or thiocarbamate type materials may experience allergic reactions when handling the hose. Health effects may include skin irritation (itching, swelling, cracking and/or blistering, and/or a burning sensation), and/or respiratory responses, including breathing difficulties and chest tightening, may also be experienced (67).

5.0 Acrylonitrile-Butadiene

5.0.1 CAS Number: *[9003-18-3]*

5.0.2 Synonyms

Acrylonitrile-butadiene; it is also called nitrile rubber and NBR; originally it was called Perbunan or Buna N. NBR is a preferred abbreviation for nitrile-butadiene rubber (63). The term "nitrile rubber" is generic and refers to any copolymers of an unsaturated nitrile and a diene; in practice these are usually acrylonitrile and butadiene, respectively.

5.0.3 Trade Names: Breon; Buna N; Butracil; Butakon A; Butaprene; Chemigum N; Elaprim; Europrene N; FR-N; GR-A; Hycar; JSR-N; Krymac; Nipol N; Nysin; Paracil; Perbunan N; SiR

5.0.4 Molecular Weight

Molecular weight determined by fractionation ranges from 20,000 to 1,000,000 (34).

5.1 Physical and Chemical Properties

NBR is important for its high oil and temperature resistance. Flexibility at very low temperatures is obtained with a lower acrylonitrile content; in turn, this sacrifices some oil resistance. Low-temperature flexibility for some applications can be modified by plasticizers or blends. Additional properties are resistance to heat, abrasion, and water, as well as resistance to gas permeation. NBR is frequently plasticized with organic phosphate, phthalic ester, or dibenzyl ether.

5.2 Production and Use

NBR is produced by polymerizing acrylonitrile and butadiene in varying proportions. The basic steps are the emulsion polymerization techniques used for styrene-butadiene. The process yields both "hot" and "cold" rubber. Hot rubbers are branched and contain some cross-linked gel. The more linear cold rubbers are slightly cross-linked by incorporating divinyl benzene to improve processibility and compression set resistance. Applications include fuel lines, hoses, automobile parts, structural adhesives, oil-resistant clothing or articles, gloves, and shoe soles and heels. Gloves made of nitrile rubber are comparable to neoprene gloves in their resistance to organic solvents; nitrile rubber gloves were superior in resistance to *n*-hexane and inferior in resistance to phenol. NBR is used for hoses, gaskets, and protective clothing.

5.4 Toxic Effects

Residual monomer and exposure to monomers during polymerization are the most common toxicological hazards. The toxicological properties and health risk of exposure to butadiene and acrylonitrile were identified during the decades of nitrile rubber production. A 1995 study of workers in a butadiene production plant demonstrated an excess of mortality from lymphosarcoma and reticulosarcoma. The data were consistent with other butadiene production cohort previously studied. The findings were also consistent with a mouse bioassay. The investigators concluded that the results of the study added to the weight of evidence that butadiene is carcinogenic to humans (67). Acrylonitrile is an irritant to the mucous membrane and the skin. Experimental studies have shown that acrylonitrile is a cyanide-releasing compound and its immediate toxic effects were thought to be cyanide intoxication. Later studies, however, have shown that the toxicity was largely due to the compound itself, not to liberated cyanide.

The IARC classifies acrylonitrile as a class 2A carcinogen because of limited evidence in humans but sufficient evidence in animals.

Special consideration is given to powdered NBR. For example, one manufacturer's material safety data sheet (MSDS) states, "Processing operations may produce vapors or dust that may cause eye, skin, and respiratory tract irritation. Contains talc which may cause lung damage if inhaled. Contains quartz which may cause cancer based on tests with laboratory animals. May form flammable/explosive air-dust mixtures (68)."

Hydrogenated NBR (HNBR) is a relatively new elastomer, and the Goodyear Tire and Rubber Company supplies an HNBR latex that contains less than 0.01% hydrazine (69).

5.4.1 Experimental Studies

Experimental studies have focused on the monomers 1,3-butadiene and acrylonitrile. No studies of the polymer acrylonitrile-butadiene or nitrile rubber itself were found.

5.4.2 Human Experience

Risk has been assessed on the monomers.

5.5 Standards, Regulations, or Guidelines of Exposure

By simulating manufacturing conditions, objective data have been generated indicating that related acrylonitrile-butadiene elastomers cannot release acrylonitrile, resulting in airborne concentrations above 1 ppm under normally expected conditions of handling, processing, and use. Exaggerated conditions, including open mill mixing at elevated temperatures, indicate that the maximum acrylonitrile level to which a mill operator would be exposed is less than 0.5 ppm. Based on these data, these NBR elastomers do not require warning labels under OSHA' acrylonitrile standard (70) according to a Zeon Chemical's MSDS because exposures above the action level of 1 ppm are not likely to result from handling and use. However, there may be conditions of use, storage, processing, or handling that were not anticipated by these studies. Therefore, because these products contain trace amounts of acrylonitrile, users of NBR should not rely solely on these data but should do sufficient in-plant testing of acrylonitrile levels to ensure that their operations comply.

NIOSH recommends an 8-hour TWA of 1 ppm for the acrylonitrile monomer, and the ACGIH TLV of 2 ppm (skin) which was adopted by OSHA as the PEL. OSHA also adopted the NIOSH recommended 15-minute ceiling value of 10 ppm. The OSHA exposure limit for the monomer 1,3-butadiene, OSHA PEL-TWA (8-hours), is 1.0 ppm and the STEL is 5 ppm.

6.0 Vinylidene Fluoride-Hexafluoropropylene

6.0.1 CAS Number: *[9011-17-0]*

6.0.2 Synonyms

Also known as FPM and FKM. FPM is an abbreviation formerly recommended for vinylidene fluoride-hexafluoropropylene copolymer. FKM is now recommended to indicate fluororubber of the polymethylene type that has substituent fluorine and perfluoroalkyl or perfluoroalkoxy groups on the polymer chain. FFKM designates a separate subdivision for fully fluorinated elastomers.

6.0.3 *Trade Names:* Dai-El; Fluorel; SKF; Technoflon FOOR; Viton A; Viton B

6.0.4 Molecular Weight

The molecular weight of some fluoropropene copolymers has been reported as 100,000 to 200,000 (71) and as low as 60,000 (34). Fluoroalkoxy rubbers may have a molecular weight as high as 1,000,000 (71).

6.1 Chemical and Physical Properties

A primary advantage of fluororubber is its exceptional heat, chemical, and solvent resistance. The copolymer is biologically inert. The decomposition products of FPM depend on the chemical composition of the intact polymer and also on the conditions under which it is decomposed. The temperature to which the polymer is subjected, the atmosphere in which decomposition occurs, and the material of the vessel used, can alter the kinds and quantities of the decomposition products. When a copolymer of vinylidene fluoride and hexafluoropropylene decomposed at 550°C and 800°C, carbon monoxide and carbon dioxide were produced. Hydrogen fluorides and other flurocarbons have been reported as decomposition products.

6.2 Production and Use

Vinylidene fluoride-based elastomers have generally been prepared by free-radical, emulsion polymerization of the monomers with organic or inorganic peroxide initiators.

Low molecular weight fluid or semifluid polymers have been made with various chain transfer agents, special initiators, or dehydrofluorination oxidation methods. Applications include seals and protective coatings on paper, wood, rubber, and leather.

6.3 Exposure Assessment

No one analytical technique detects all of the potential decomposition products of FPM, but gas chromatography has been used to measure some of the pyrolytic products. Sampling and analytical procedures are discussed in NIOSH manuals of analytical methods and sampling procedures (71a,b).

Pelosi et al. (72) reported quantitative determinations of volatile emissions evolved during press cure (typically 10 minutes at 193°C) and the usual 24-hour post-cure. Stepwise analysis of several stocks showed the greatest weight loss, 1.32 to 2.16%, during the first 12 hours of oven cure. During both curing operations, the emissions were primarily water and carbon dioxide, and secondarily, curing agent fragments; no fluorocarbon decomposition products were detected. Small amounts of hydrogen fluoride were evolved during press cure, and even small amounts during postcure.

6.4 Toxic Effects

The oral LD_{50} for rats dosed intragastrically with a latex dispersion that contained 60 to 65% terpolymer (vinylidene fluoride-hexafluoropropylene-tetrafluoroethylene) was >40,000 mg/kg (73). Rats fed the uncured copolymer crumb (vinylidene fluoride-hexafluoropropylene) for 2 weeks at 25% in the diet showed no clinical or nutritional signs of toxicity; enlarged livers were seen at the end of the feeding period, but there was some evidence of remission after a 14-day recovery period (74).

Uncured sheets and latex dispersions of both the copolymer and this terpolymer produced only minimal erythema when tested with 24-hour occluded contact on rabbit skin; eye tests with the dispersions showed only very mild, temporary conjunctival irritation (75). A subsequent skin test (76) was conducted to determine the irritancy potential of the copolymer when cured with 2-dodecyltetramethylguanidine (present at 1.33% in the compounded polymer). Rabbits showed erythema 2 days after a series of five 6-hour exposures. Reactions were somewhat more pronounced after four additional exposures but resolved completely within a 14-day recovery period.

Thermal degradation starts at 245 to 250°C. As might be expected of these fluorinated elastomers, the toxicity of the effluent depends strongly on the degradation temperature and the curing recipe.

6.4.1 Experimental Studies

See Section 6.4.

6.4.1.1 Acute Toxicity. See Section 6.4.

6.4.2.2.7 Other: Neurological, Pulmonary, Skin Sensitization, etc. Toxic and corrosive hydrogen fluoride may be liberated during processing above 200°C or from smoking tobacco or cigarettes contaminated with resin dust. These vapors can irritate the eyes, nose, throat, and lungs. Lung effects may be delayed for several hours (77). Prolonged skin contact with vinylidene fluoride copolymer may produce skin irritation (78).

6.5 Standards, Regulations, or Guidelines of Exposure

Recommended exposure limits have been set for some of the decomposition products. These limits are in the 2000 chemical substances thresholds limit values of the American Conference of Governmental Industrial Hygenists (ACGIH) and the 1997 OSHA permissible exposure levels (PEL). Since the values are reviewed and updated frequently, interested readers should contact the sponsoring organizations to obtain current information and supporting documentation for the values.

7.0 Chlorosulfonated Polyethylene

7.0.1 CAS Number: [68037-39-8]

7.0.2 Synonyms: Ethylene resin, chlorosulfonated; poly(ethylene), chlorosulfonated, contains 43% Cl, 1.1% S; CSM

7.0.3 Trade Names: Hypalon

7.1 Chemical and Physical Properties

CSM contains 1–5% sulfur as the sulfonyl chloride group and 25–43% chlorine. The substitution of chlorine and sulfur dioxide onto the polyethylene molecules destroys the crystalline segments and changes the thermoplastic materials into an amorphous polymer that behaves as an elastomer. The addition of chlorine also increases oil resistance. CSM is also more resistant to heat and ozone than synthetic rubber. Other properties are toughness, weatherability, and colorability.

7.2 Production and Use

This type of elastomer is made by dissolving polyethylene in carbon tetrachloride and then treating it with chlorine and sulfonyl chloride in the presence of a catalyst. After the desired degree of chlorosulfonation has been attained, the residual chlorine–sulfur dioxide mixture is stripped off, a stabilizer is added, and the commercial product is isolated as raw rubber in crumb or film form. Major applications are in hose and wire covering and in sheet form for uses such as swimming pool liners or reservoir containers.

7.4 Toxic Effects

Rats fed 20 g/kg of a chlorosulfonated polyethylene in ground chow showed no clinical effects (79). A limited skin sensitization test with two commercial solutions (U.S. and Japanese) produced one instance of sensitization; the causative agent was not identified

(76). More extensive testing with four formulations that contained 45 to 50% chlorosulfonated polyethylene, cured and uncured, produced no reactions, either irritation or sensitization (80).

7.4.2 Human Experience

No human health effects have been attributable to CSM during its several decades of use.

7.4.2.2.5 Carcinogenesis. Carbon tetrachloride is liberated from CSM on standing; evolution is accelerated by heat. Decomposition products include carbon monoxide, hydrogen chloride, sulfur dioxide, and hydrocarbon oxidation products, including organic acids, aldehydes, and alcohols (77). Carbon tetrachloride and chloroform have been classified by IARC and NTP as carcinogens on the basis of tests with laboratory animals. Exposure to carbon tetrachloride at concentrations above the permissible exposure level (PEL) may cause liver damage (77). Lymphatic leukemia was especially strongly related to carbon tetrachloride (81).

7.5 Standards, Regulations, or Guidelines of Exposure

PELs have been established for known decomposition products, including carbon tetrachloride, but not for the polymer itself.

8.0 Butyl Rubber

8.0.1 CAS Number: [9010-85-9]

Butyl (IIR) and halobutyl rubbers are copolymers of isobutylene with small amounts of isoprene.

Butyl rubber may be halogenated with either chlorine or bromine to form chlorobutyl rubber *[68081-82-3]* or bromobutyl rubber *[68441-14-5]*.

8.0.2 Synonyms: 2 Methyl-1-propane, poly (isobutylene)-co-isoprene

8.0.3 Trade Names: Esso Butyl; Polysar Butyl; Total Butyl

8.0.4 Molecular Weight

The M_w for butyl and halobutyl rubber ranges from 400,000 to 600,000.

8.1 Chemical and Physical Properties

Butyl rubber has the chemical resistance characteristic of saturated hydrocarbons. Oxidative degradation is slow, and the rubber may be further protected by incorporating antioxidants. Isobutylene is the major component of bulk rubber. It provides good aging resistance and low gas permeability. Isoprene, a minor component, enhances vulcanizability into flexible stable rubber products.

8.1.1 General

Density (g/cm^3): 0.917

Glass-transition temperature: -75 to $-67°C$.

The structures of both chlorinated and brominated butyl rubber are not what would be expected based on halogenation of multisubstituted alkenes. The methyl group adjacent to the unsaturation in butyl rubber prevents halogenation across the double bond. Thus the properties of polyisobutylene and chlorinated and brominated butyl rubber are similar. The solubility of polyisobutylene butyl rubber and halogenated butyl rubber are similar.

8.2 Production and Use

The bulk of butyl rubber is made by a slurry process using aluminum chloride at -98 to $-9°C$ and methyl chloride as a diluent. The extremely rapid reaction is unique, and proceeds via cationic polymerization to completion at $-100°C$ in less than a second. Butyl rubber may be vulcanized by three basic methods: accelerated sulfur vulcanization, cross-linking with dioxime and related dinitroso compounds, and polymethylol-phenol resin cure. Halogenated butyl rubber allows broadened vulcanization latitude and rate, and enhanced co-vulcanization to general-purpose elastomers while it maintains the unique attributes of the basic butyl molecule. Applications include tires, automotive parts, adhesives and sealants, and pharmaceutical closures.

8.5 Standards, Regulations, or Guidelines of Exposure

Chlorobutyl rubber withstands 170°C for 5 to 8 minutes before decomposing and releasing HCl (up to 2% on rubber) and low molecular weight halogenated compounds. Bromobutyl rubber contains trace amounts of bromoform, dibromochloromethylpropane (DBCMP), tribromomethylpropane, and other low molecular weight halogenated compounds that may be emitted in concentrations in the parts per billion range when heated above 150°C. Adequate ventilation and good industrial hygiene practices should be used to control exposures within the recommended occupational exposure levels for the decomposition products.

9.0 Perfluoro Rubber

The literature on perfluorubber is vague. It is a fluoroelastomer in which all substituents on the carbon chain are completely fluorinated.

9.0.2 Synonyms: FKKM; tetrafluorethylene-perfluromethylvinylethercopolymer

9.0.3 Trade Names: Tecmaflon; Kalrez; Viton

9.1 Chemical and Physical Properties

FFKM requires difficult and complex techniques to produce finished parts. It has one of the best thermo-oxidative stabilities of all fluoroelastomers. It has excellent resistance to automotive fuels and oils, hydrocarbon solvents, and certain chlorinated solvents.

SYNTHETIC POLYMERS—OLEFIN, DIENE ELASTOMERS, AND VINYL HALIDES 449

9.2 Production and Use

It is typically prepared by high pressure, free-radical, aqueous emulsion polymerization. All of the substituents on the carbon chain are completely florinated; fluorine atoms replace all hydrogen atoms. FFKM is used for seals of all kinds in the chemical, aircraft, aerospace, and automotive industries.

9.4 Toxic Effects

It is biologically inert and generally has low toxicity.

9.4.2.3.7 Other: Neurological, Pulmonary, Skin Sensitization, etc. No information was found in the literature, but residual monomers, primarily acrylate monomer vapors, may increase the risk of eye and or skin irritation.

10.0 Acrylic Elastomer

10.0.1 CAS Number: [9003-32-1]

10.0.2 Synonyms: ACM; acrylate rubber; acrylic rubber; polyacrylate rubber; polyacrylic elastomer

10.0.3 Trade Names: Acralen; Cyanacryl; Elaprin AR; Europrene AR; Hycar; Krymac Lactoprene BN; Lactoprene EV; Nipol AR; Paracril; Thiacryl

10.0.4 Molecular Weight

Commercial polyacrylates are generally high M_w (> 200,000) thermoset type polymers.

10.1 Chemical and Physical Properties

Depending on the specific choice of backbone monomer, they have glass-transition temperatures that vary from about − 16 to − 40°C. The low-temperature types have somewhat poorer oil resistance than the high T_g types. They are most often filled with carbon black at a total loading of no more than 100 parts per hundred parts rubber (phr).

10.1.1 General

Acrylate elastomers are generally stable and not reactive with water. Above 300°C, these elastomers may pyrolize to release ethyl acrylate and other alkyl acrylates. Otherwise, thermal decomposition or combustion may produce carbon monoxide, carbon dioxide, and hydrogen chloride.

10.2 Production and Use

Acrylates can be polymerized in a variety of ways including bulk, solution, suspension, and emulsion polymerization. The most common industrial methods are suspension and emulsion polymerization. The three major backbone monomers used in acrylate elastomers are ethyl acrylate, butyl acrylate, and methoxyethyl acrylate.

Acrylic elastomers are widely used in the automotive industry for seals in automatic transmissions, valve stems, crankshafts, oil pans, and packings. Other uses include hose, tubes, boots, spark plugs, and fabric coatings.

10.5 Standards, Regulations, or Guidelines of Exposure

Occupational exposure levels for ethyl acrylate and butyl acrylate monomers are listed in the recommendations of the 2000 chemical substances threshold limit values (TLV) of ACGIH and the 1997 OSHA permissible exposure limit (PEL). Because of the varying toxicity of monomers, polymerization should not be undertaken without first consulting appropriate material data safety sheets.

11.0 Chlorinated Polyethylene

11.0.1 CAS Number: [9002-86-2]

11.0.2 Synonyms: CM; CPE

11.0.3 Trade Names: Alcryn; Bayer CM; Daisolac DOW CPE; Elaslen; Holothene; Hostrapren; Kelrinal; Tyrin

11.0.4 Molecular Weight

The number-average molecular weight of CPE ranges from 30,000 to about 120,000.

11.1 Chemical and Physical Properties

Peroxide- or thiadiazole-cured CPE exhibits good thermal stability up to 150°C and is much more oil resistant than nonpolar elastomers such as natural rubber or EPDM. Commercial products are soft when the chlorine content is 28–38%. At more than 45% chlorine content, the material resembles polyvinyl chloride. Higher molecular weight polyethylene yields a chlorinated polyethylene that has high viscosity and tensile strength.

11.1.1 General

The general properties of CPE depend on the properties of the starting material and the amount and distribution of chlorine introduced. Chlorination destroys the structural regularity of the polyethylene and hence crystallinity. Cured CPE exhibits good thermal stability up to 150°C.

11.2 Production and Use

In chlorinating polyethylene, chlorine atoms substitute for hydrogen atoms of the polyethylene chain in both crystalline and amorphous regions. The most common chlorination method is treating polyethylene powder in an aqueous suspension that contains hydrochloric acid and a free-radical initiator with chlorine gas. After the desired level of chlorination is obtained, the CPE is water washed and dried, and an antiblocking agent is then added. Major applications are in wire and cable jacketing, sheeting applications in

SYNTHETIC POLYMERS—OLEFIN, DIENE ELASTOMERS, AND VINYL HALIDES 451

automotive under-the-hood applications such as hoses, mechanical goods, and sponge, and as an impact modifier in PVC.

11.4 Toxic Effects

CPE polymers are classified as low toxicity compounds.

11.4.1.7 Other: Neurological, Pulmonary, Skin Sensitization. Skin contact, especially with hot polymer, may irritate the skin of some people and result in redness and itching. Skin irritation tests in rabbits showed that the material is a mild irritant. The material is not a skin sensitizer as determined in tests with guinea pigs (82).

12.0 Epichlorohydrin

12.0.1 CAS Number: [106-89-8]

12.0.2 Synonyms: Chloromethyloxirane rubber; DL-a-epichlorohydrin; (chloromethyl) ethylene oxide; 3-chloropropylene oxide; Ech; glycerol epichlorohydrin; gamma-chloropropylene oxide; 1-chloro-2,3-epoxypropane; (chloromethyl)oxirane; 2-(chloromethyl)oxirane; alpha-epichlorohydrin; 1,2-epoxy-3-chloropropane; 2,3-epoxypropyl chloride; skekhg; allyl chloride oxide; 3-chloro-1,2,-epoxypropane; 3-chloro-1,2-propylene oxide; 3-chloropropene-1,2-oxide; 3-chloropropyl epoxide; EPI; epoxy-3-chloropropane; epoxypropyl chloride; glycidyl chloride; (RS)-3-chloro-1,2-epoxypropane; Cardolite NC-513

12.0.3 Trade Names: Epichloromer; Herclor; Hydrin

12.0.4 Molecular Weight

The epichlorohydrins are typical rubbers whose molecular weights range from 500,000 to < 1,000,000.

12.0.5 Molecular Formula: C_3H_5ClO

12.0.6 Molecular Structure: Cl—△—O

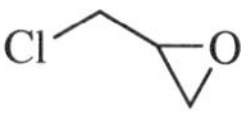

12.1 Chemical and Physical Properties

Epichlorohydrins have a unique combination of heat resistance, fuel and oil resistance, ozone resistance, and low-temperature flexibility. They provide excellent resistance to vapor penetration by hydrocarbons, fluorocarbons, and air. The homopolymer is brittle at about $-15°C$. This value is lowered to $-40°C$ in the copolymer but with loss of some oil resistance.

12.1.1 General

Epichlorohydrin elastomers are linear, amorphous polymers. All have high specific gravity. The EPI homopolymers and EPI copolymers are the most significant commercially. Like other vulcanizable elastomers, EPI elastomers are compounded with processing aids, plasticizers, stabilizers, and vulcanizing agent. Blends of EPI elastomers with other elastomers do not perform well and have not been commercially successful.

12.2 Production and Use

Polymerization is generally carried out in a solution process at 40–130°C using an aromatic solvent such as benzene or toluene. The catalyst systems are alkylaluminum–water, and akylaluminum–water–acetylacetone. A continuous commercial production process has been reported whereby solvent, catalyst and monomer are fed to a backmixed reactor maintained at desired temperature. The properties of epichlorohydrin polymers make them suitable for under-the-hood automobile applications.

12.4 Toxic Effect

The most commonly used curing agent for epichlorohhydrin elastomers is ethylenethiourea (ETU). It has been identified as carcinogenic and teratogenic to experimental animals (83, 84).

12.5 Standards, Regulations, or Guidelines of Exposure

TLVs for airborne concentrations of epichlorohydrin monomers are listed in the recommendations of the 2000 chemical substances threshold limit values from ACGIH. Interested readers should contact the sponsoring organization to obtain current information and documentation for values.

13.0 Silicone

13.0.1 CAS Number: [63148-62-9]

13.0.2 Synonyms:
Siloxane; organosiloxane silicone, all; silicones; dimethylpolysiloxane hydrolyzate; alpha-methyl-omega-methoxypolydimethylsiloxane; polydimethyl silicone oil; poly(dimethylsiloxane); polydimethylsiloxane, methyl end-blocked; polyoxy-(dimethylsilylene), alpha-(trimethylsilyl)-omega-hydroxy; poly[oxy(dimethylsilylene)], alpha-[trimethylsilyl]-omega-[(trimethylsilyl)oxy]; silicone oils; siloxane and silicones, dimethyl; siloxanes and silicones, dimethyl; alpha-(trimethylsilyl)poly[oxy(dimethylsilylene)]-omega-methyl; silicone oil; silicone oil, for oil baths; polydimethylsiloxane, trimethylsiloxy terminated; polydimethylsiloxane, trimethoxysilyl terminated; hydride terminated, 1; carboxypropyldimethyl; silicone oil, for melting point and boiling point apparatuses; Antifoam A Spray 260

13.1 Chemical and Physical Properties

Silicones are thermal and oxidatively stable. The physical properties are little affected by temperature. Other properties include high chemical inertness, resistance to weathering, good dielectric strength, and low surface tension.

13.1.1 General

Silicones were the first polymers based on organometallic chemistry. They are unique among the important commercial polymers in both basic chemistry and a wide variety of industrial applications. As suggested by the general formula ($C_4H_{12}O_3Si_3$), the molecular structure can vary considerably to include linear, branched, and cross-linked structures. The nomenclature of silicones is simplified by the use of the letters M, D, T, and Q to represent monofunctional, difunctional, trifunctional, and quadrifunctional monomers units, respectively.

13.2 Production and Use

Silicone elastomers are generally prepared from chlorosilanes. The chlorosilanes are hydrolyzed to give hydroxyl compounds that condense to form elastomers. Applications include electrical insulation, gaskets, surgical membranes and implants, and automobile engine components.

13.4 Toxic Effects

Silicone elastomers have not been identified as toxic in production or in use.

13.4.1 Experimental Studies

A number of investigations have been pursued to determine the specific action of silicones on the immune system. Two reported studies extensively evaluated the immunotoxic effects of exposure to silicone used in medical practices. In one study, mice were implanted with polydimethylsiloxane elastomers fluids and gels. No observable alterations in the immune system were reported in this study. In another study, silicone products were implanted in mice for 180 days. Modest suppression of NK cell activity was observed. The silicone product did not alter host resistance to two strains of infectious microorganisms (85, 86).

13.4.1.2 Chronic and Subchronic Toxicity. See Section 13.4.1.

13.4.2 Human Experience

As indicated previously, interest in health effects has focused on the biocompatibility of silicone and the potential for the polymer to produce immunotoxic injury. The issue has been the subject of litigation provoked by the clinical literature that includes reports of adverse reactions to silicone-containing medical materials. There were also claims from women who had silicone breast implants. No unequivocal link has been established between exposure to silicone and human disease. Research in this area is continuing.

13.5 Standards, Regulations, or Guidelines of Exposure

Precautions are needed for handling residual unreacted diisocyanates. Certain aromatic diamines, for example, 4,4 methylene bis- (2 chloroaniline) used to cure the polymer are suspected cancer agents and require safety measures.

13.6 Studies on Environmental Impact

Although silicone polymers are considered to have low toxicity, raw materials and intermediates may have adverse effects on the environment if not properly managed. Most solid and liquid waste from silicone production can be incinerated in accordance with state and local regulations.

14.0 Polyurethane Elastomers

14.0.1 CAS Number: [9009-54-5]

14.0.2 Synonyms: Polyurethane rubber; urethane rubber

14.0.3 Trade Names: NA

14.0.4 Molecular Weight: 88.109

14.0.5 Molecular Formula: $[C_3H_8N_2O]_n$

14.0.6 Molecular Structure:

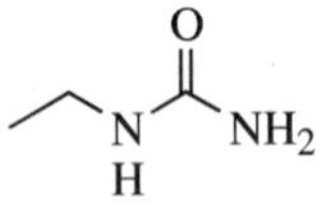

14.1 Chemical and Physical Properties

Specific properties vary widely with the composition of the polymer. The major components of polyurethane elastomers are polyester and polyether diols, diisocyanates, and chain extenders. The segmented molecular chain and two-phase domain morphology of polyurethane elastomers are responsible for their excellent tear, abrasion, impact, and wear resistance. Resistance to hydrocarbons and aromatic oils is another important property of polyurethane elastomers.

14.1.1 General

Polyurethane elastomers are block copolymers that consist of alternating polyurethane and polyol segments. The segments are designated hard and soft segments, respectively, because they are below and above their softening temperatures at the normal use temperatures. Studies of these properties and others and their relationship to molecular structures have been spurred by the growth in the commercial importance of polyurethane products formed by reaction injection molding (RIM).

14.2 Production and Use

Castable polyurethane elastomers are fabricated from polyurethane prepolymers, which are obtained by reacting an excess of diisocyanate with high molecular weight (500 to 3000) diols. These NCO-terminated oligomers (prepolymers) are commercially available under a variety of trade names and in numerous types depending upon the types of diisocyanates and polyols that are used to synthesize them. The NCO content of these prepolymers can vary from less than 3% to as much as 20%. A cast polyurethane fabricator mixes these liquid prepolymers with approximately stoichiometric quantities of a curing agent (or a blend of curing agents) such as an appropriate low molecular weight diol or diamine.

Generally a prepolymer is heated to reduce its viscosity before mixing it with a liquid or a molten curing agent. The prepolymer curing agent blends have a limited working time (pot life) during which they are still liquid and can be poured into molds. The liquid prepolymer/curing agent blend is degassed and then poured into molds, which are often heated to expedite curing. After curing, the solid polyurethane elastomer articles are removed from the mold and are sometimes finished by keeping them at elevated temperature to complete the cure and to maximize mechanical properties.

The numerous and varied uses of polyurethane elastomers include roller-skate wheels, elevator wheels, snowplow blades, bumper pads, hydraulic seals, brake diaphragms, helicopter-blade sleeves, grain buckets, and grain and coal chutes.

14.4 Toxic Effects

Polyurethane prepolymers contain residual unreacted diisocyanates in amounts that vary from less than 0.1 to 20% or more. Inhalation is the primary route of exposure to most isocyanates, whose vapors or aerosols can irritate mucous membranes in the respiratory tract. The toxic effects of isocyanate are discussed in detail in other sections of this work.

14.5 Standards, Regulations, or Guidelines of Exposure

The American Conference of Governmental Industrial Hygienists has recommended threshold limit values (TLVs) for isocyanates.

B VINYL HALIDES

The commercial polymers in this group contain chlorine atoms, fluorine atoms, or both in a few cases. In very diverse ways these halogens can be used to produce vinyl polymers that have such characteristics as increased resistance to water, oils, and solvents, plus other distinctive properties. The prototypes are polyvinyl chloride and polyvinylidene chloride. Polyvinyl chloride and its copolymers rank first in production/consumption volume among polymers in the United States and abroad. Their key attribute is low-cost versatility. Polyvinylidene chloride resins have an extremely regular, closely packed molecular structure that results in outstanding impermeability to water, oils, and gases (87, 88).

15.0 Polyvinyl Chloride

15.0.1 CAS Number: [9002-86-2]

15.0.2 Synonyms: PVC; chloroethene homopolymer; chloroethylene polymer; vinyl chloride homopolymer; vinyl chloride polymer; polychloroethene; poly-1-chloroethylene

15.0.3 Trade Names: Afcodur; Afcoplst; Afcovyl; Breon; Carina; Corvic; Dacon; Ekavyl; Gedevyl; Geon; Hostalit; Lacovyl; Lacqvyl; Laddene; Lakavyl; Lonzalit; Lonzavyl; Lucolene; Lucorex; Lucovyl; Mavlan; Marvylan; Mipolam; Norvyl; Opalon; Pekevic; Pevikon; Quirvil; Ravinil; Rhodapas; Rhovylite; Scon; Sicron; Solvic; Trosiplast;

Ultryl; Varlan; Vestolit; Vinatex; Vinnol; Vinflex; Vionylite; Vipla; Viplast; Viplavi; Vybak (plastic materials); and Fibravyl; Isovyl; Khlorin; Leavil; PCU; Retractyl; Rhovyl; Tevilon; Thermovyl (fibers)

15.0.4 Molecular Weight

Molecular masses for average molecular weights of commercial polymers are 60,000 to 150,000. The average molecular weight of the polymer varies greatly depending on the method of production, but is generally in the range of 45,000 to 150,000.

15.0.5 Molecular Formula: $[C_2H_3Cl]_n$

15.0.6 Molecular Structure: $\diagup\!\!\diagdown_{Cl}$

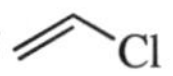

15.1 Chemical and Physical Properties

PVC resin is white or colorless. Depending on the polymerization techniques, the resin particles can range from 0.05 to 150 μm in diameter. PVC resin has a specific gravity of 1.4 and a refractive index of 1.54. Solvents of unmodified PVC of high molecular weight are cyclohexanone, methylcyclohexanone, dimethylformamide, nitrobenzene, tetrahydrofuran and mesityl oxide. Solvents of lower molecular weight PVC include methyl ethyl ketone, dioxane and methyl chloride. In the absence of added stabilizers, PVC is unstable to heat and ultraviolet light.

15.2 Production and Use

PVC resins are manufactured by a variety of polymerization processes to produce materials that have a wide range of chemical and physical properties. A resin suitable for an intended application is selected and then formulated with one or more additives such as thermal stabilizers, flame retardants, lubricants, processing aids, impact modifiers, and plasticizers to produce a PVC compound (89). These compounds are offered in many forms, including liquids, dry powder blends, cubes, and pellets, as required by the processing and fabrication techniques employed. These techniques include coating, calendering, molding, and extruding. To address the possible hazards to humans and the environment from PVC or its production, the chemical composition of the polymerization ingredients and the types of additives used to compound the resin must be known.

The manufacture of PVC resins from VCM involves reacting the monomer in agitated pressure vessels in the presence of catalysts and converting these liquids and/or gases to solid resins. A considerable amount of heat that is generated by this reaction, and is removed by cooling the vessel. As pointed out by Wheeler (89), as the monomer is converted to polymer during the reaction, the rate of reaction slows down; thus after some optimal reaction time and 80 to 95% conversion, the unconverted remaining monomer is removed, and the PVC resin is recovered as a white powder, a liquid latex, or a solution. This polymerization reaction takes place in pure monomer, in a solution, in a water-monomer emulsion, or in an aqueous suspension of monomer, depending on the polymerization process used. The nature of the polymerization process determines the nature of the subsequent recovery process and the nature of the resin particles produced (89).

Four basic VCM polymerization techniques exist; these are, in order of frequency of use, (*1*) suspension polymerization, (*2*) emulsion polymerization, (*3*) bulk polymerization, and (*4*) solution polymerization (2, 87).

Suspension polymerization is the major process for manufacturing PVC resins. One or two parts water and one part VCM are charged to an agitated reactor along with initiator and suspending agents (e.g., polyvinyl alcohol). The mass is reacted at 35 to 80°C until 75 to 90% of the VCM is converted to resin. The resin–water mixture is heated and steam-stripped until the unconverted VCM is essentially removed. Then, the resin is separated from the water and dried. The dried resin is transferred to storage silos, then shipped to fabrication plants in hopper rail cars, bulk containers, or paper bags. Suspension polymerization produces relatively large resin particles, that range from 50 to 200 mm. The particles have controlled internal morphology that facilitates monomer removal in the monomer recovery operation. In the past, residual VCM in suspension PVC resins was as high as 2,000 ppm by weight; however, improved methods were developed in response to the recognized need for lower residual levels (89). Current manufacturing technology produces general PVC that contains less than 1 ppm residual VCM in virgin PVC powder, and modern medical-grade PVC contains less than 10 ppb VCM.

The second most widely used process for manufacturing PVC resins is emulsion polymerization. This is not a single process but rather a large family of processes, and each produces specialized products. The two major process families use a water-soluble initiator system and an oil-soluble initiator system. Using the water-soluble initiator method, the VCM, water, a surfactant, and water-soluble initiator are mixed in an agitated reactor and reacted at 30 to 60°C to form a synthetic latex. When 80 to 95% of the monomer is converted to PVC, unreacted VCM can be removed or it may be subjected to a second initiator treatment (89). Although the polymer may be simply filtered and distributed as a latex, it may also be recovered as a dry resin by using a number of different techniques.

In the oil-soluble initiator system, a mixture of monomer and an organic peroxide is emulsified in water with a surfactant. The resulting emulsion is reacted at 30 to 60°C to form a synthetic latex, and unreacted monomer is removed. Once again, the product may be sold as a latex or processed to yield a dry powder. The basic difference between the oil-soluble initiator and the water-soluble initiator processes is that the size of the emulsified monomer particle determines the resin particle size in the oil-soluble initiator process, whereas the polymerization technique determines the resin particle size in the water-soluble initiator process. The resin particles formed during emulsion polymerization range in size from 0.05 to 2.0 mm in diameter. In the course of recovery and drying, there may be agglomeration that produces particles as large as 30 mm.

Bulk polymerization is the third major process for manufacturing PVC resins. The process involves charging VCM and initiator to a first-stage polymerizer, where approximately 10% of the monomer is converted to polymer. Then, this batch is transferred to a second-stage polymerizer where additional monomer, and sometimes initiator, are added. When the reaction has progressed to approximately 80 to 95% completion, the unreacted VCM is removed by heat, vacuum, and an air or steam sweep; then the finished resin product is transferred to bins for subsequent shipment to fabrication plants. Using this method, resin is produced that has uniform particle size (50 to 100 mm), a highly porous internal morphology, and high purity (i.e., no soaps or detergents). Before 1975, residual VCM in resin made by this process was of the order of 1000 ppm when produced. Today, because of advances in technology, VCM levels of less than 1 ppm are common.

Solution polymerization is the fourth method for PVC resin production. Monomer, solvent, and initiators are fed to a continuous reactor system. The polymer formed is soluble in the reacting mass so that the reactor product is a viscous resin solution. This solution is distilled to remove the unconverted VCM, and the resin product is recovered by treating the resin solution with heat and drying the product (89). The median particle size is approximately 75 mm and has less than 0.2 ppm residual VCM (89).

In 1987, the production of polyvinyl chloride and copolymers amounted to 7,971 million lb. By 1997, it was 14,084 million lb. In the United States, significant volumes of the PVC are used in the form of rigid product: films, sheets, flooring wire, and cables. A minor, but important, use of PVC is in medical devices, including blood and blood-component storage bags, cannulas, drip chambers, enema packs, endotracheal tubes, hemodialysis sets, and intravenous tubing (90).

15.3 Exposure Assessment

Concerns about the health effects of VCM have necessitated exposure assessment techniques to detect VCM at low concentrations in polymers and in air and water. Headspace gas chromatography has become a principal technique in measuring VCM. The EPA specifies the use of dynamic headspace gas chromatographic methods.

15.3.1 Air

See Section 15.3.

15.4 Toxic Effects

As indicated, a large number of additives can be used as components of PVC formulations. Given the extensive use of many of these, a brief review of the main additives is warranted.

Typical plasticizers can be classified as follows: phthalates used for general-purpose compounding; esters of aliphatic dibasic acids such as adipates, azelates, and sebacates used for low-temperature and food-contact applications; polyesters and trimellitates used for high permanency and low migration; organophosphates used to impart flame retardancy; epoxides used as secondary or co-thermal stabilizers; and extenders such as aliphatic hydrocarbons, aromatic hydrocarbons, and alkylated aromatics.

Of the plasticizers, di(2-ethylhexyl) phthalate (DEHP) is the most widely used and has become the standard plasticizer for flexible PVC used in medical devices (90, 91). It is also employed in certain nonfatty food-contact applications, although di(2-ethylhexyl) adipate (DEHA) is the major food-contact film plasticizer because of its desirable properties in cling wrap (90, 91). It has been estimated that more 3000 articles have been published regarding the toxicology of DEHP (90, 92). Although the acute toxicity of DEHP is considered low, repeated administration to rodents has resulted in testicular damage, liver damage, and liver tumors. These effects are probably the result of one or more of the metabolites of DEHP, not the parent compound itself (90, 91, 93–99). The hepatocarcinogenicity in rodents is believed due to peroxisome proliferation rather than the direct genetic effects of DEHP, a response that may not occur in humans (100, 101).

Flame retardancy is required for high-performance thermoplastic resins because of their use in electrical and high-temperature applications. Rigid PVC is inherently flame retardant due to its high (56%) chlorine content. Flexible PVC products plasticized with organic-based esters, however, will burn, and flame retardants are needed to protect them from fire. The most common classes of flame retardants include inorganic fillers (e.g., aluminum trihydrate and calcium carbonate); inorganic agents (e.g., antimony oxide, zinc, and copper compounds); borates; chlorine-containing aliphatic, cycloaliphatic, and aromatic compounds (e.g., chlorinated paraffins); phosphate esters (e.g., tricresyl phosphate); chlorinated/brominated phosphates [e.g., tris(2,3-dichloropropyl) phosphate and tris(2,3-dibromopropyl) phosphate] (102).

15.4.1.2 Chronic and Subchronic Toxicity. Dogs fed 250 mg of PVC acrylic sheeting/kg body weight for 5 days experienced no adverse effects (103), and no adverse effects were noted in a 2-year dietary study in which rats received a copolymer of 95% vinyl chloride and 5% vinyl acetate at levels of 1.5 or 12% of the diet (104). In a similar study, rats and dogs received a copolymer of vinyl chloride and vinylidene chloride in the diet for 2 years at a level of 5% by weight. No signs of toxicity were noted (105). Two other studies treated rats orally with PVC dusts. Approximately 550 days following a mixed treatment of PVC orally and intraperitoneally, 2 out of 30 animals developed subcutaneous fibrosarcomas; one of the animals also showed a tumor of the lung (106). In the second study, rats were treated with PVC dust in distilled water for 15 days or in the diet for approximately 10 months; findings included hyperplasia of the gastroenteric mucosa, pulmonary effects, and one case of histiocytic lymphoma (107). The lack of both statistics and details regarding the PVC used make the significance of these studies unclear.

When rats received daily intravenous injections of PVC particles (up to 5 mm with 80% < 1 mm) for a period of 28 days (3×10^9 particles in total), histological examination revealed the presence of particulates in liver, spleen, and lung tissue, but no adverse histopathological changes were reported (108). Peritoneal macrophages harvested from these animals contained particles, presumably from engulfing the particulates. No macrophage toxicity was reported, although there was a statistically significant increase in the release of interleukin-1.

PVC powders, administered to rats by intraperitoneal injection do not lead to tissue fibrosis (109–111). In one of the studies, a representative range of PVC powder formulations was included. These included particles of various sizes, porosity, and method of production (110). In these studies and others discussed previously (108, 112), PVC was engulfed by macrophages, and there was no evidence of toxicity to this cell type. This is unlike the situation for fibrogenic agents such as silica, where toxicity and necrosis of the phagocytic cells is implicated in the etiology of fibrosis (111).

15.4.1.3 Pharmacokinetics, Metabolism, and Mechanisms

15.4.1.3.1 Absorption. Following a single dose of intratracheally instilled PVC dust (< 0.5 mm) in rats, particles were reportedly cleared through the lymphatic circulation and progressively accumulated in the tracheobronchial lymph nodes (113). Similar clearance results have been suggested or reported previously (114, 115). Ingested or rectally administered PVC particles were reportedly within the lymph vessels associated with the intestinal wall of rats, guinea pigs, rabbits, chickens, dogs, and pigs (112). Particles were reportedly found in the blood, bile, urine, and cerebrospinal fluid of dogs and in the blood of other species, but details of the experimental methods and results are lacking in this report (112). Throughout the current literature there is no evidence that PVC is metabolized, thus implying that PVC is an inert polymer.

15.4.1.3 Carcinogenesis. A number of implantation studies conducted in experimental animals using PVC products have shown equivocal results. Earlier studies were reviewed by the IARC (116) and include subcutaneous (s.c.) implantation of PVC into the abdominal walls of rats (117, 118), Implantation of PVC film (s.c.) into the abdomens of rats (119), implantation of PVC surrounding the kidneys of rats (120, 121), implantation of PVC particles into the muscles of rabbits (122), and s.c. implantation of PVC sponge into the abdominal walls of dogs (123) and rats (124). In the majority of these studies, adverse effects were noted, including fibrosis at the site of implantation and/or local tumors. As reported in the 4th edition, it is difficult to draw conclusions regarding these reports, which were published between 1952 and 1968. In fact, the IARC has determined that the data are inadequate for drawing conclusions regarding carcinogenicity (125). In many cases, the PVC formulation was known to contain additives, but the compositions of the PVC product in the remainder of the studies was not reported or was simply not known. In addition, PVC produced in the 1970s and earlier contained much higher levels of residual VCM than PVC today due to advances in manufacturing technology (125–128).

15.4.1.7 Other: Neurological, Pulmonary, Skin Sensitization. In 1970, Szende et al. (129) first described radiological and histopathological evidence of diffuse pulmonary fibrosis in a worker exposed to PVC dust. Since that time, much effort has gone into studying potential adverse pulmonary effects due to PVC. In the case of human epidemiological data, which are discussed in the following section, exposure to PVC dusts is confounded by the potential for concomitant exposure of workers to VCM and to additives, which can be present as respirable dusts themselves (130). The weight of evidence from data generated by inhalation and intratracheal exposure to PVC particles in experimental animals indicates that PVC itself possesses little or no biological activity and its physical presence produces benign pneumoconiosis at elevated dust concentrations.

In a study conducted by the National Institute for Occupational Safety and Health (NIOSH) in the United States, rats, guinea pigs, and monkeys were exposed by inhalation (6 hour/day, 5 days/week) for up to 22 months to a 13 mg/m^3 concentration of PVC dust. No fibrosis or significant cellular infiltrates were noted in the lungs of animals, except for aggregates of alveolar macrophages that contained PVC particles, and no significant effects on pulmonary function could be demonstrated in monkeys exposed to PVC. These results led the authors to conclude that PVC produced a benign pneumoconiosis (131). Similar weak reactions to inhaled PVC were noted in rats exposed to inhalation of PVC dust at a concentration of 10 mg/m^3 for 6 hour/day, 5 days/week for 15 weeks (132). In this study, no evidence of overt fibrosis was observed, although randomly scattered lung lesions were noted. These were characterized by hypercellularity of the interstitium of the alveolar walls with particle-laden macrophages. As noted by the authors, the weak biochemical and histological effects seen in animals exposed to PVC during this study contrast with those conducted using dusts of high biological reactivity such as chrysotile asbestos and amosite asbestos (132). In a comparable study, in which rats were exposed by inhalation to 12 mg PVC dust/m^3 for 7 hour/day, 5 days/week for 5 months, no evidence of significant pulmonary disease was observed; the authors considered that particle deposition in the lungs was simply evidence of exposure (133).

In an early study, rats and guinea pigs were placed in same work environment as PVC workers (near bagging operations) for 24 hour/day for 2 to 7 months (134). Animals of both species developed fibrosis after exposure to total dusts at the level of approximately 10,000 particles/cm^3. Although fibrosis was observed, the lack of control animals and the small number of experimental animals used make it difficult to draw conclusions. Interpretation is further complicated by the observation of pulmonary fibrosis induced in guinea pigs under controlled conditions by inhaling of VCM (135). Similarly, when rats were exposed to PVC dust (97 g/m^3; 92% of the particles were < 5 mm in diameter) for 1-hour exposures daily for 1 to 12 months, evidence of pulmonary toxicity was noted (136). This included bronchiectasis, emphysema, pneumonia, lung abscesses, and squamous metaplasia in bronchial epithelium, but no fibrosis. As pointed out in the NIOSH discussion (131), although they were free of disease, only 10 control animals were examined in this study, and conditions of animal housing were not included. This last point was considered significant, considering the propensity for untreated rodents to contract acute and chronic respiratory diseases (131).

Intratracheal instillation of PVC dusts in experimental animals also has been performed. The results indicate that such exposure results in acute pulmonary response to the particles that is self-limiting and is characterized by the presence of phagocytic cells that do not pose a serious threat to the biochemical and histological structure of the lungs. The lung of rats displayed enzymatic and microscopic alterations following a single dose of PVC dust (25 mg/rat in saline) (113). These effects indicated a brief inflammatory response after the initial insult, yet decreased in severity with time. Rats that were treated intratracheally with a single 2- or 25-mg dose of PVC dust in saline and were followed for 5 to 12 months following treatment displayed weak acute reactions that gradually declined to normal levels as the masses of PVC dust were engulfed by inflammatory cells and/or cleared (presumably through lymphatic drainage) (110, 112, 137). In these and similar studies, acute exposure to PVC particles did not produce severe histopathological effects on lung structure (110, 113, 112, 137–139).

Exposure to PVC in the workplace has been identified as a risk factor for respiratory irritation and asthma — commonly known as meat wrappers' asthma. This name, descriptive if imprecise, refers to clinical effects associated with fumes — particulates formed from condensing solids, vaporized by heat — from thermal cutting and sealing of PVC film generally used to package meat in retail outlets. Particulates that have extremely small particle size and plasticizer products have been identified as the responsible agents.

15.4.2 Human Experience

This discussion and that dealing with human experience with PVC are limited to the toxicology and potential hazards of PVC itself. Many of the studies that have investigated PVC toxicity are difficult to interpret because often the actual formulation of the PVC is not known. In many cases, it is unclear whether "pure" PVC resin was used or formulated PVC using a variety of additives.

15.4.2.2 Clinical Cases. Since the first report of pulmonary effects from exposure to PVC dust (129, 140), a number of epidemiological studies have investigated the pulmonary effects of PVC and VCM (141–154), and several additional case reports have been published that describe pulmonary effects in exposed workers (155–158). Case reports have shown profound pulmonary effects on the individual workers, yet they are of limited use for understanding the effect of PVC alone because of the small number of individuals involved and co-exposure to other chemicals. Epidemiological studies have often shown an increased prevalence of radiological, lung function, or subjective abnormalities (141–146, 148, 149, 152, 153), although in a number of the studies, there were conflicting results regarding impairment of lung function (143, 145, 146, 150, 153). These pulmonary abnormalities cannot be ascribed to PVC dust alone because it was known that there were concomitant exposures to VCM, PVC thermal decomposition products, and other air contaminants. For example, VCM concentrations in the workplace have been estimated at approximately 1000 ppm between 1945 and 1955, 400 to 500 ppm from 1955 to 1965, 300 to 400 ppm from 1966 to 1972, 150 ppm by 1973, and 5 ppm after 1975 (159).

15.4.2.2.5 Carcinogenesis. No information was found in the literature. PVC is not listed as a carcinogen or potential carcinogen by the National Toxicology Program (NTP).

15.4.2.2.6 Genetic and Related Cellular Effects Studies. A number of investigations have been published regarding the *in vitro* toxicity of PVC and PVC products. The results of these studies suggest that PVC itself has negligible effects on cultured mammalian cells; apparently conflicting cytotoxicity results are attributable to the presence of surfactant material or other formulation additives, such as organotin stabilizers (108–110, 122, 160–164).

15.4.2.2.7 Other: Neurological, Pulmonary, Skin Sensitization, etc. Following an outbreak of acneiform eruptions on the skins of workers in a PVC manufacturing plant, a follow-up study showed that PVC resin powder was not comedogenic (causing blackheads) on rabbit ear (165). No confirmation of such dermal effects due to PVC has been reported. Skin reactions may develop from low molecular weight additives, stabilizers, and plasticizers, but very few have been reported (166–168).

15.4.2.3 Epidemiology Studies. Based on epidemiological evidence, exposure to high levels of PVC dusts in the workplace can increase the risk of low-grade pneumoconiosis, similar to that experienced from inhaling other unreactive dusts. Chest X-rays films of workers have often shown abnormalities, yet these changes have not always been associated with decreased lung function. A 1983 critical review of the epidemiological data on the respiratory effects of PVC dusts in humans concluded that such exposure did not produce neoplastic effects (169).

15.4.2.3.4 Reproductive and Developmental. A study revealed a high rate of congenital abnormalities in communities that had PVC polymerization plants. Follow-up studies have not confirmed that the excess defects were related to PVC (170).

15.5 Standards, Regulations, or Guidelines of Exposure

The ACGIH considers occupational exposure to PVC dusts under the category "Particulates Not Otherwise Classified (PNOC)." The TLV for this category is 10 mg/m^3 time-weighted average (TWA) for inhalable particulate and 3 mg/m^3 for respirable fraction (30). OSHA has set limits for inert or nuisance dusts of 15 mg/m^3 (total) and 5 mg/m^3 (respirable) as TWA, not regulating PVC directly (171). Germany, Switzerland, and the United Kingdom have set occupational exposure limits of 5 mg/m^3 TWA for PVC (respirable dust) (172), as have Sweden and Norway (91). As part of standard industrial hygiene practices, workers who may come into contact with such "nuisance dusts" in excess of these limits are required to wear appropriate respiratory protective equipment.

In addition to occupational regulation, PVC has prior sanction for use in general food-contact applications in both the flexible and rigid forms by the U.S. FDA based on a 1951 article published by the FDA (173). In addition, PVC is also listed as an acceptable ingredient for food packaging and other food-contact articles in the United States under various sections of Title 21 of the U.S. Code of Federal Regulations.

15.6 Studies on Environmental Impact

No clear evidence shows that PVC itself is susceptible to microbiological degradation (174). In fact, rigid PVC, such as that used for manufacturing pipes, is extremely resistant to degradation in this manner, making it a material of choice for underground uses (91, 89, 174). In the case of flexible PVC products, the heavy content of plasticizers and stabilizers, which are generally biodegradable, lead to environmental degradation, embrittlement, and eventual disintegration; the biological assimilation of the polymer itself has still to be demonstrated (174). At one time it was suspected that PVC, deposited as part of the municipal solid waste stream, was the source of VCM found in landfill sites; however, this was the result of the microbial metabolism of chlorinated solvents present, although residual VCM present in older PVC products would not have contributed more than negligible amounts of monomer to the total (175). PVC does not depolymerize, even at high temperatures, to release VCM (88). The potential also exists for heavy metals used in formulating of PVC products to contribute to the solid waste burden, although data in this area are lacking.

16.0 Polyvinylidene Chloride

Saran is a generic term for polymers that have high vinylidene chloride content. The three commercially important types are vinylidene chloride-vinyl chloride copolymers, vinylidene chloride-alkyl acrylate and methyl acrylate copolymers, and vinylidene chloride-vinyl chloride-acrylonitrile copolymers. In general, the literature is sketchy on vinylidene chloride polymers, first commercialized in 1939. The most extensive coverage is given to the production and properties of vinylidene chloride monomer.

16.0.1 CAS Number: [9002-85-1]

16.0.2 Synonyms: 1,1-Dichloro-ethene homopolymer; PVDC

16.0.3 Trade Names: Ixan (homopolymer); Saran; Viclan (plastics); Krehalon; Pemalon; Tygan; Velan (fibers)

16.0.4 Molecular Weight

Molecular weights of the copolymers vary: approximately 20,000 to 50,000, 10,000 to 100,000 for vinyl chloride copolymer, and 200,000 for emulsion copolymer.

16.1 Chemical and Physical Properties

Among the valuable properties of polyvinylidene chloride (PVDC) and its copolymer are low permeability to gases and vapors. PVDC is noted for its toughness, flexibility, and clarity. The high chlorine content imparts fire resistance, but thermal instability at melt processing temperatures is a deficiency. PVDC does not dissolve in most common solvents at ambient temperatures. Copolymers, especially those of low crystallinity, are much more soluble.

16.1.1 General

Density: 1.80–1.94/g/cm^3 at 25°C experimental; 1.96 g/cm^3 calculated.

T_m,°C: 198–205

T_g,°C: −19 to −11

Density at 25°C: amorphous 1.67–1.775; unit cell 1.949–1.96; crystalline 1.80–1.97

Refractive index (crystalline): 1.63

Heat of fusion (ΔH_m) J/mol: 6275–7950

The permeability of the polymer is a function of the type and amount of comonomer. As the comonomer fraction of these semicrystalline copolymers increases, the melting temperature decreases, and the permeability increases.

16.1.2 Odor and Warning Properties

Polyvinylidene chloride—a thermoplastic polymer—is tasteless and odorless.

16.2 Production and Use

PVDC polymerizes by both ionic and free-radical reactions. Free-radical polymerization is typical. Free-radical polymerization of PVDC may be by solution, slurry, suspension, and emulsion methods. In copolymerization, usually one component is introduced to improve the processability or solubility of the polymer; others are added to modify specific properties. Properties modified by copolymerization depend on the content of components (structure, amounts, types). PVDC has a number of applications, including molding resins, extrusion resins, multilayer film, rigid barriers, containers, lacquer resin, vinylidene chloride copolymer latex and foams, resins for solvent coating, latices for coating, and pipes for chemical processing.

16.4 Toxic Effects

Most studies of toxic effects focus on the monomer, vinylidene chloride, a clear liquid that is highly flammable and reactive. The meager data available on the human health effects of vinylidene chloride indicate irritation of mucous membranes after acute exposure. Central nervous system toxicity has been associated with exposure levels above 3000 ppm (176). Experimental animal data indicate that the liver is the target organ for vinylidene chloride (176).

16.5 Standards, Regulations, or Guidelines of Exposure

Exposure limit have not been established for the monomer.

17.0 Vinylidene Chloride–Vinyl Chloride

17.0.1 CAS Number: [9011-06-7]

17.0.2 Synonyms: 1,1-Dichloroethene, polymer with chloroethene; polyvinylidene chloride-co-vinyl chloride; 1,1-dichloroethene, chloroethene polymer

17.0.3 Trade Names: Saran; Velan

17.0.4 Molecular Weight: 159.44

17.0.5 Molecular Formula: $[C_4H_5Cl_3]_n$

17.0.6 Molecular Structure:

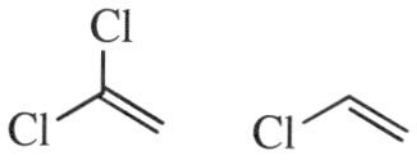

17.1 Chemical and Physical Properties

Commercial copolymers that contain about 15% vinyl chloride have properties similar to polyvinylidene chloride: they crystallize and have low gas permeability, but thermal stability is poor.

17.2 Production and Use

Useful for films with high durability and chemical resistance and as films and coatings with low gas and water vapor permeability.

18.0 Vinylidene Chloride–Methylacrylate

18.0.1 CAS Number: [25038-72-6]

18.0.2 Synonyms: 2-Propenoic acid methyl ester polymer with 1,1 dichloroethene

18.0.3 Trade Names: AMSCO

18.1 Chemical and Physical Properties

The most valuable property is low permeability to a wide range of gases and vapors. The polymer is not heat stable.

18.1.1 General

Vinylidene chloride-methyl acrylate is one of the three commercially important copolymers of the many copolymers that have been prepared by polymerizing vinylidene chloride. The other two commercially important copolymers are vinylidene chloride-vinyl chloride copolymer and vinylidene chloride-alkyl acrylate. Limited information is available on vinylidene chloride-methyl acrylate itself.

19.0 Vinylidene Chloride–Acrylonitrile

19.0.1 CAS Number: [9010-76-8]

Vinylidene chloride-acrylonitrile is one of many copolymers of vinylidene chloride. Limited information is available on this specific copolymer. Properties and uses are similar to those of other vinylidene chloride polymers and copolymers.

19.0.2 Synonyms: NA

19.0.3 Trade Names: NA

20.0 Polyvinyl Fluoride

20.0.1 CAS Number: [24981-14-4]

20.0.2 Synonyms: PVF

20.0.3 Trade Names: Tedlar

20.0.4 Molecular Weight: 60,000 to 180,000 (34)

20.1 Chemical and Physical Properties

Polyvinyl fluoride is noted for its inertness, flexibility, toughness, durability, and weather resistance. Polyvinyl fluoride is a semicrystalline polymer. The degree of crystallinity depends on the method of polymerization and the thermal history.

20.1.1 General

Density: 1.38–172 g/cm^3
Temperature Range: −70 to +107°C
Impact Strength: 6–22 kJ/m
Auto-Ignition Temperature: 390°C

Polyvinyl fluoride has better heat resistance than polyvinyl chloride but eliminates hydrogen fluoride at low temperatures.

20.1.2 Odor and Warning Properties

The literature contained warning and odor properties only on the monomer.

20.2 Production and Use

Vinyl fluoride monomer is polymerized in a free-radical initiated water suspension process. After filtration and drying, the polymer may be cast into film using an organosol based on dimethylacetamide or an equivalent solvent system. Oriented polymer film sold for adhesive bonding may contain up to 0.5% residual dimethylacetamide (177). Commercial synthesis of vinyl fluoride has not been described in the literature for proprietary reasons. Polyvinyl fluoride film is intended primarily for adhesive bonding to a variety of substrates. Construction surfaces, especially residential and commercial siding and aircraft interior surfaces, are among the major applications.

20.4 Toxic Effects

Polyvinyl fluoride produced no skin reactions when tested for irritation and sensitization on 215 human subjects (178, 179).

21.0 Polytetrafluoroethylene

21.0.1 CAS Number: [9002-84-0]

21.0.2 Synonyms: PTFE; polydifluoromethylene

21.0.3 Trade Names: Polyflon; Teflon; Hostaflon; Fluon; Algoflon; Hanlon; Fluoroplast

21.0.4 Molecular Weight

The original commercial granular type polymers were in the number-average molecular weight range of 400,000 to 10,000,000 (180).

21.0.5 Molecular Formula: [C$_2$F$_4$]$_n$

21.0.6 Molecular Structure:

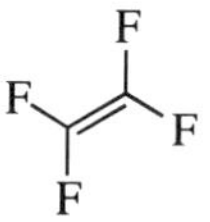

$$F_2C=CF_2$$

21.1 Chemical and Physical Properties

The thermal and chemical stability of tetrafluoroethylene polymers are well demonstrated in PTFE. It is chemically inert to industrial chemicals and solvents even at elevated temperatures and pressures. Gases and vapors penetrate PTFE more slowly than most other polymers. PTFE does not melt to form a liquid and cannot be melt extruded. When the virgin resin is heated, it forms a clear, coalescable gel. Once processed, the gel point (often referred to as the melting point) is 10°C lower than that of the virgin resin. It is sold as a granular powder, a fine powder, or an aqueous dispersion. Each is processed in a different manner.

21.2 Production and Use

Polytetrafluoroethylene is generally made from tetrafluoroethylene gas by free-radical polymerization under pressure with oxygen, peroxides, or peroxydisulfates. The "granular resins" have medium size particles that range from 30 to 600 mm. Colloidal aqueous dispersions, made by a different process, are concentrated to about 60% by weight of the polymer and have particles that average about 0.2 mm. Coagulated dispersions with agglomerates that average 450 mm are also available (180).

The use of polytetrafluoroethylene as a release agent in coatings and certain other food contact applications is permitted under FDA regulations (181). Polyvinylidene fluoride resins may be similarly used as articles or components of articles intended for repeated use in contact with food according to prescribed conditions (181). Polyvinyl fluoride resins of specified type may be used as food-contact coatings for containers whose capacity is not less than 1 gal or in formulations for wall covering and ceiling tile in meat and poultry processing plants.

Major applications include components or linings for chemical process equipment, high-temperature wire and cable insulation, molded electrical components, tape, and nonstick coatings. Polytetrafluoroethylene also has been used in a variety of medical applications (182).

Polytetrafluoroethylene fiber is among the most chemically-resistant fiber known, and it is also the most fire-resistant in oxygen-rich and high-pressure atmospheres. Bleached PTFE fiber was used for the in-flight coveralls of the Apollo astronauts.

21.3 Exposure Assessment

Sampling and analysis of decomposition products of PTFE are described in the NIOSH *Occupational Exposure Sampling Strategy Manual* and in the NIOSH *Manual for Analytical Methods*. Gas chromatography has been used to measure some of the pyrolytic products. Sampling and analytical procedures for hydrolyzable fluoride, a decomposition product of fluorocarbon polymers, are discussed in the NIOSH criteria document for inorganic fluorides.

21.3.5.2 Urine. In one study of workplace exposure, exposure to PTFE resin dust produced increases in urinary fluoride, but these were below levels considered toxic (183,

SYNTHETIC POLYMERS—OLEFIN, DIENE ELASTOMERS, AND VINYL HALIDES

184). Grossly excessive exposure to PTFE dust can elevate urinary fluoride. Urine and serum are the proposed biological materials for monitoring exposure to fluoride. In the past, biomonitoring of workers for fluorides in urine indicated that occupational exposures to these polymers did not increase the risk of illness (183).

21.4 Toxic Effects

The finished polymer is inert under ordinary conditions. Exposure to pyrolytic or decomposition products is the principal health concern.

21.4.1 Experimental Studies

No abnormalities were evident in a 7-month experiment in which rats ingested a diet containing 0.5% of a mixture that contained 21% PTFE. Rats fed one of three types of PTFE at a dietary level of 25% for 90 days "showed a shift in the number and distribution of white blood cells and, in one group fed unsintered TFE resin, an increase in the relative size of the liver, relative to body weight"; feeding did not produce any adverse effects on growth rate or behavior or any microscopic evidence of tissue change (185).

This minimal response was most likely attributable to a surfactant used in polymerization. Subsequent 2- or 3-week tests at 25 or 10% dietary levels (186) showed that PTFE resins prepared with various volatile dispersing agents produced enlarged livers, but the same resins heated to remove these volatiles did not produce enlarged livers. In one case, feeding the surfactant per se did not produce the large livers observed under test conditions similar to feeding the resins containing the surfactant; these data suggested that the "active agent" was produced in situ during the polymerization process (187). Several Russian reports (188–190) also describe effects consistent with the concept that low molecular weight additives leach out under test conditions. In one study of workplace exposure, exposure to PTFE resin dust produced increases in urinary fluoride, but these were below levels considered toxic (183, 184).

21.4.1.3 Pharmacokinetics, Metabolism, and Mechanisms. Rats exposed to PTFE pyrolytic products at 450°C developed severe epithelium damage to alveolar living cells and marked septal edema and necrosis of the tracheobronchial epithelium. Death from hemorrhagic pneumonitis of caged pet birds due to accidental overheating of empty PTFE-lined pans has been reported.

21.4.1.5 Carcinogenesis. Animal studies have not demonstrated that this polymer is carcinogenic.

21.4.2 Human Experience

Polymerization is usually conducted in a closed system, thereby reducing the risk of worker exposure to the polymers themselves. The greatest risk to workers is exposure to decomposition products through inhalation. Adverse effects could result from exposure to a single or several decomposition products. A major concern is the potential for increased risk of polymer fume fever and damage to the respiratory tract.

First described in 1951, "polymer fume fever," an influenza-like illness may be provoked by the inhalation of pyrolytic decomposition products of PTFE. Workers who smoke have experienced "polymer fume fever" as a result of PTFE-resin contaminated cigarettes. Reports indicate that the burning ember reaches a high enough temperature (875°C) to expose workers to toxic pyrolytic products (191, 192). Attacks reportedly occur several hours after exposure, often toward the end of the work shift or in the evenings after work. Long-term sequelae of polymer fume fever are poorly characterized. The pathogenesis of polymer fume fever is unknown. Unlike metal fume fever, no tolerance phenomenon has been reported, and polymer fume fever develops without regard to previous exposures.

Polytetrafluoroethylene fabricated in textile form has been effectively used for certain types of vascular grafts; thrombosis develops (193) when the prosthesis is used as a replacement for small arteries (diameter of 6 to 8 mm or less). The susceptibility of PTFE to deformation under load makes it unsuitable for prostheses such as an acetabular cup that articulates with a femoral head prosthesis of stainless steel (194).

Numerous factors, including the physical form, size, and chemical composition of the implanted material, have been implicated as important determinants of possible carcinogenic potential. Although Bischoff (195) speculated that "chemical factors were more important than the continuous surface in elicting the carcinogenic response," he noted that "neither subcutaneous implants of Teflon® mesh surgical outflow patches nor shreds of the same material produce cancer in rats. For this particular material, a chemical carcinogen was not indicated" (195). Others have not produced tumors following subcutaneous implantation with all forms of Teflon® (196), suggesting that chemical factors are not the sole determinant of carcinogenicity. This issue remains an area for additional research (197).

The IARC noted (198) a single case of fibrosarcoma without evidence of metastasis 10 years after implantation of a polyester-PTFE arterial prosthesis. This case plus the studies on mice and rats provided insufficient evidence to assess the carcinogenic risk of exposure to humans. Long-term follow-up was recommended only in the cases of patients who had medical implants.

21.5 Standards, Regulations, or Guidelines of Exposure

NIOSH recommendation for worker protection include worker education, prohibition of smoking, mandatory hand washing, and attention to airborne dust and heat sources. Properly designed and operated ventilation systems should reduce the risk of severe exposures. Figure 89.2 illustrates two modifications of industrial ventilation design recommended for processing fluoropolymer resin. Control of processing temperature should prevent exposure to decomposition products.

22.0 Fluorinated Ethylene Propylene Copolymer

22.0.1 CAS Number: [25067-11-2]

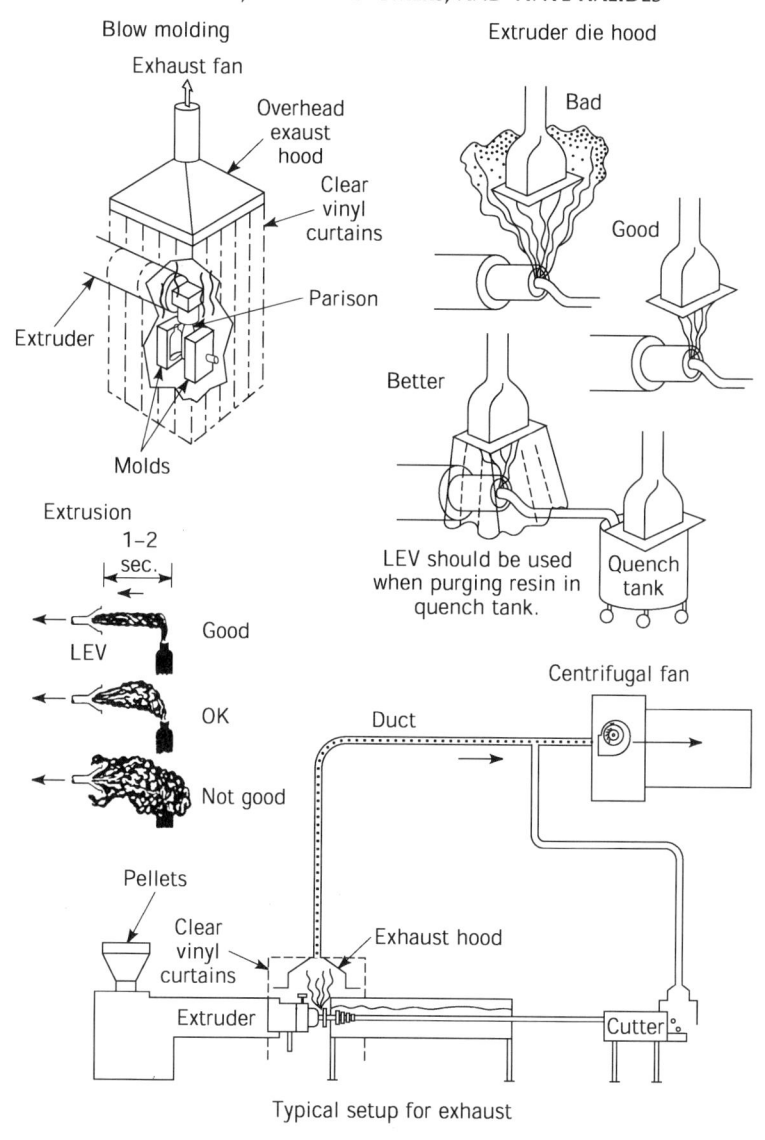

SYNTHETIC POLYMERS—OLEFIN, DIENE ELASTOMERS, AND VINYL HALIDES 471

Figure 89.2. Recommended ventilation modifications for polymer polymere processing.

22.0.2 Synonyms: Polytetrafluroethylene-co-hexaflopropylene; FEP

22.0.3 Trade Names: Fluon; FEP; Hostaflon FEP; Texflex; Teflon FEP

22.1 Chemical and Physical Properties

FEP fluoropolymer resin is a polymer of tetrafluoroethylene and hexafluoropropylene. It has a melting point range of 245 to 280°C and it is melt processable. It is supplied as translucent pellets, powder, or an aqueous dispersion. FEP has somewhat better impact strength but a lower service temperature than polytetrafluorolethylene.

22.2 Production and Use

The elastomer is prepared by high pressure, free-radical aqueous emulsion polymerization. Fluorine is substituted for hydrogen. The initiators are organic or inorganic peroxy compounds, such as ammonium persulfate, and the emulsifying agent is usually a fluorinated acid soap. Applications include tubing, hose, and gaskets.

22.4 Toxic Effects

The elastomer is biologically inert. However, extreme heat can produce harmful breakdown products: hexafluoropropylene, hydrogen fluoride, and other fluorocarbons.

22.4.2 Human Experience

Limited information is available on the human health effects of the polymer itself.

22.4.2.2.7 Other: Neurological, Pulmonary, Skin Sensitization, etc. Skin sensitization can occur after extended contact with the polymer.

23.0 Perfluoroalkoxy Copolymer

23.0.1 CAS Number: [26655-00-5]

23.0.2 Synonyms: FMVE; PFA; perfluoroalkylvinyletherpolymer; polyperfluoralkylvinylether

23.0.3 Trade Names: Teflon FPA; TFA; Hostaflon; Neoflon

23.1 Chemical and Physical Properties

PFA resin is a melt-processible copolymer that contains a fluorocarbon backbone in the main chain and perfluorinated vinylether side chains. PFA has excellent chemical and physical properties at elevated temperatures, including chemical inertness, heat-resistant toughness, flexibility, nonflammability, low moisture absorption, and good dielectric properties.

23.2 Production and Use

PFA fluropolymer resin is a polymer of tetrafluorotheylene and a perfluorinated vinyl ether. The physical and chemical properties of PFA yield a product that provides high performance service to the chemical processing and related industries. It is available as translucent pellets, powder, and an aqueous solution.

24.0 Ethylene Tetrafluoroethylene Copolymer

24.0.1 CAS Number: [54302-05-5]

24.0.2 Synonyms: Polytetrafluoroethylene-co-ethylene; tetrafluoroethylene-ethylene-copolymer; ETFE

24.0.3 Trade Names: Aflon; Hostaflon ET; Hostaflon FEP; Neoflon; Teflex; Teflon FEP; Tefzel

24.1 Chemical and Physical Properties

ETFE is an alternating copolymer of tetrafluoroethylene and ethylene similar in properties to polytetrafluoroethylene, but it has higher impact and tensile strengths. Other properties include low dielectric constant, excellent resistivity, and low dissipation factors. Thermal and cryogenic performance and chemical resistance are rated as good.

24.2 Production and Use

To prepare ETFE, ethylene and tetrafluoroethylene are copolymerized in aqueous, nonaqueous, or a mixed medium with free-radical initiators. The polymer is isolated and converted into extruded cubes, powders, and beads. ETFE is used in thin coatings, heavy-wall logging cables, insulation, tubing, and fasteners.

25.0 Polyvinylidene Fluoride

25.0.1 CAS Number: [24937-79-9]

25.0.2 Synonyms: PVDF, vinylidene fluoride; VDF or VF_2; poly-1,1,-difluoroethylene

25.0.3 Trade Names: Dyflor; Foraflon; KF; Kynar; Solef

25.0.4 Molecular Weight

Commercial grades that have molecular weights of approximately 300,000 and 600,000, respectively, have been reported.

25.1 Chemical and Physical Properties

Properties are intermediate in many respects between the monofluorinated and perfluorinated analogs. Wear and creep resistance are typically greater than for perfluorinated polymer, and weather resistance is less than that of the monofluorinated analog. The polymer science literature reports that PVDF is stable to harsh chemicals, thermal, ultraviolet, weathering and oxidizing, and high energy radiation environments. This stability approaches the essentially inert attribute of fully fluorinated polymers.

In summary, the physical and chemical properties of PVDF are typical of a strong and durable engineering thermoplastic that may be used in equipment and facilities designed for elevated temperatures. The polymer has the desirable property of low contamination because the preparatory process does not require additives for stabilization.

25.1.1 General

Tensile strength MPa at 25°C: 38–52
                 at 100°C: 17
Flammability: self-extinguishing
Thermal conductivity at 25–160°C w/(m·k): 0.17–0.19
Thermal degradation temperature (°C): 390
Specific heat J/(kg·K): 1255–1425

25.2 Production and Use

PVDF, the addition polymer of 1,1,difluoroethene, is prepared by high-pressure free-radical polymerization in aqueous systems. It has applications in electric and electronic devices, as weather-resistant binders for exterior architectural finishes, and in a number of areas of the chemical processing industry. The lack of additives and thus the exceedingly low contamination levels make the polymer useful in ultrapure water systems, for example, where high purity is necessary for construction materials.

25.3 Exposure Assessment

As indicated previously, past biomonitoring of workers involved in producing fluoropolymers revealed fluoride concentrations in the urine below levels associated with toxic exposure.

25.4 Toxic Effects

PVDF has low toxicity. It may be safely used for products intended for repeated contact with food and for human organ prostheses. The thermal decomposition of PVDF and its potential toxic effects have not been as intensively studied as those of other fluorocarbon polymers.

25.4.2 Human Experience

Operational problems may lead to the release of thermal decomposition products, including toxic hydrogen fluoride. Polymer fume fever which may result from exposure to this polymer has been previously discussed.

26.0 Polychlorotrifluoroethylene

26.0.1 CAS Number: [9002-83-9]

26.0.2 Synonyms: PCTFE

26.0.3 Trade Names: Aclar; Diaflon; Fluoroethene; Halon; hostaflon C; Kel F; Voltalef

26.0.4 Molecular Weight: $M_W = 70,000-400,000$

26.1 Chemical and Physical Properties

Unique properties include rigidity, low temperature toughness, resistance to creep, and cold flow. PCTFE reportedly has low permeability to moisture (water absorption: 0%) and gases and resistance to attack by most industrial chemicals, including strong acids and akalies.

26.2 Production and Use

PCTFE is a polymer of chlorotrifluoroethylene. It is prepared by free-radical polymerization in an aqueous emulsion using organic or water-soluble initiators or ionizing radiation. Examples of applications of PCTFE are plastic molded parts in chemical process equipment and electric parts. PCTFE films and film laminates are used in high-performance packaging.

26.3 Exposure Assessment

As indicated, past biomonitoring of workers involved in producing fluorocarbons revealed fluoride concentration in urine below levels associated with toxic exposure.

26.4 Toxic Effects

The literature focuses primarily on the toxicity of decomposition product, not on the intact polymer.

26.5 Standards, Regulations, or Guidelines of Exposure

Good industrial hygiene practices are necessary to reduce the risk of worker exposure to trace monomers. At processing temperatures, PCTFE generates thermal decomposition products that may be toxic. Local exhaust to capture contaminants near their source is recommended. Disposal of waste products by landfill is also recommended.

27.0 Ethylene Chlorotrifluoroethylene Copolymer

27.0.1 CAS Number: [25101-45-5]

27.0.2 Synonyms: Polychlorotrifluoroethylene-co-ethylene; ECTFE

27.0.3 Trade Names: Halar

27.1 Chemical and Physical Properties

ECTFE has thermal and oxidative stability to 150–175°C. It is chemically inert to a number of chemicals and solvents and has good permeability resistance. The copolymer is 50–55% crystalline, depending on the method of preparation.

27.2 Production and Use

PCTFE is a polymer of chlorotrifluoroethylene. Copolymerization of ethylene and chlorotrifluoroethylene is carried out by aqueous suspension techniques, by low temperature polymerization initiated by oxygen-activated triethylboron in dichlorotetrafluoroethane, and by radiation-induced polymerization. Wire and cable insulation is the primary application of ECTFE.

27.3 Exposure Assessment

As indicated previously, past biomonitoring of workers involved in producing fluorocarbon polymer revealed fluoride concentrations in the urine below levels associated with toxic exposure.

27.4 Human Experience

Operational problems in processing may lead to the release of thermal decomposition products, including toxic hydrogen fluoride. Polymer fume fever has been previously discussed.

BIBLIOGRAPHY

1. R. Lefaux, *Practical Toxicology of Plastics*, English ed., CRC Press, Cleveland, OH, 1968.
2. G. D. Clayton and F. E. Clayton, Eds., *Patty's Industrial Hygiene and Toxicology*, 3rd rev. ed., Vol. 2B, Wiley, New York, 1983.
3. H. F. Mark and S. Atlas, Introduction to polymer science. In H. S. Kaufman and J. J. Falcetta, Eds., *Introduction to Polymer Science and Technology: An SPE Textbook*, Wiley, New York, 1977, pp. 1–23.
4. G. G. Hawley, *The Condensed Chemical Dictionary*, 9th ed., Van Nostrand Reinhold, New York, 1976.
5. M. P. Stevens, *Polymer Chemistry; An Introduction*, Addison-Wesley, Reading, MA, 1975.
6. M. Morton, Ed., *Rubber Technology*, 2nd ed., Van Nostrand Reinhold, New York, 1973.
7. F. W. Billmeyer, Jr., *Textbook of Polymer Science*, 2nd ed., Wiley, New York, 1971.
8. R. W. Moncrieff, *Man Made Fibres*, 6th ed., Wiley, New York, 1975.
9. L. R. Whittington, *Whittington's Dictionary of Plastics*, Technomic, Lancaster, PA, 1978.
10. R. F. Boyer, Transitions and relaxations. In *Encyclopedia of Polymer Science and Technology*, 2nd ed., Suppl. 2, Wiley, New York, 1977, pp. 745–839.
11. G. T. Davis and R. K. Eby, Glass transition of polyethylene: Volume relaxation. *J. Appl. Phys.* **44**, 4274–4281 (1973).
12. J. A. Brydson, *Plastics Materials*, 3rd ed., Whitefriars Press, London, 1975.
13. J. Brandrup and E. H. Immergut, Eds., *Polymer Handbook*, 2nd ed., Wiley, New York, 1975.
14. *1981 Annual Book of ASTM Standards*, American Society for Testing and Materials, Philadelphia, PA, updated annually, see D 1418-79a for nomenclature for rubber polymers.
15. J. L. Throne, *Plastics Process Engineering*, Marcel Dekker, New York, 1979.
16. W. J. Roff and J. R. Scott, with J. Pacitti, *Handbook of Common Polymers*, CRC Press, Cleveland, OH, 1971.
17. J. Haslam, H. A. Willis, and D. C. M. Squirrell, *Identification and Analysis of Plastics*, Heyden, London, 2nd ed., 1972, reprinted 1980.
18. C. A. Harper, Ed., *Handbook of Plastics and Elastomers*, McGraw-Hill, New York, 1976.
19. H. R. Allcock and F. W. Lampe, *Contemporary Polymer Chemistry*, Prentice-Hall, Englewood Cliffs, NJ, 1981.
20. *Guide to CAS ONLINE Commands*, Chemical Abstracts Service, Columbus, OH, 1981, p. 39; See bimonthly newsletter, CAS ONLINE, for current developments.

SYNTHETIC POLYMERS—OLEFIN, DIENE ELASTOMERS, AND VINYL HALIDES

21. J. F. Rabek, *Experimental Methods in Polymer Chemistry*, Wiley, New York, 1980.
22. C. G. Smith et al., Analysis of high polymers. *Anal. Chem.* **63**, 11R–32R (1991).
23. A. Krishen, Rubber. *Anal. Chem.* **61**, 238R–243R (1989).
24. American Society for Testing and Materials, *Annual Book of Standards, Plastics*, Vol. 8.01-8.04, ASTM, Philadelphia, PA, 1992.
25. R. E. Eckardt and R. Hindin, The health hazards of plastics. *J. Occup. Med.* **15**, 808–819 (1973).
26. World Health Organization, *Toxicological Evaluation of Some Food Colours, Thickening Agents, and Certain Other Substances* (revised title), FAO Nutrition Meetings Report Series No. 55A, WHO Food Additive Ser. No. 8, Geneva, 1975 (also WHO Tech. Rep. Ser. No. 576). FAO is the Food and Agriculture Organization of the United Nations, Rome, which also issues This Nineteenth Report of the Joint FAO/WHO Expert Committee on Food Additives (subsequent reports issued by WHO in its Tech. Rep. Ser., also by FAO).
27. D. E. Till et al., *A Study of Indirect Food Additive Migration* (Second Annual Technical Progress Report October 1, 1978 to September 30, 1979), Arthur D. Little, Inc., Project No. 81166, Oct. 1979, Food and Drug Administration Contract Number 223-77-2360 (courtesy of FDA, Washington, DC).
28. J. T. Chudy and N. T. Crosby, Some observations on the determination of monomer residues in foods. *Food Cosmet. Toxicol.* **15**, 547–551 (1977).
29. Monsanto v. Kennedy et al., U.S. Court of Appeals for the District of Columbia Circuit; 613 F.2d, 947 (D.C. Cir. 1979).
30. *TLV, Threshold Limit Values for Chemical Substances in Workroom Air*, Adopted by ACGIH for 1999, American Conference of Governmental Industrial Hygienists, Cincinnati, OH, 1999.
31. *Occupational Exposure Limits for Airborne Toxic Substances*, Occupational Safety and Health Series No. 37, International Labour Office, Geneva, 1977.
32. F. Bischoff, Organic polymer biocompatibility and toxicology. *Clin. Chem.* **18**, 869–894 (1972).
33. R. D. Karp et al., Tumorigenesis by millipore filters in mice: histology and ultrastructure of tissue reactions as related to pore size. *J. Natl. Cancer Inst.* **51**, 1275–1285 (1973).
34. International Agency for Research on Cancer, *IARC Monographs on the Evaluation of the Carcinogenic Risk of Chemicals to Humans; Some Monomers, Plastics and Synthetic Elastomers and Acrolein*, Vol. 19, Lyon, France, 1979.
35. U.S. Food and Drug Administration, Allergic reactions to latex-containing medical devices. MDA 91-1 March 29, 1991.
36. A. A. Fisher, Management of dermatitis due to surgical gloves. *J. Dermatol. Surgery Oncol.* **11**(6), 628–631 (1985).
37. A. G. Miguel et al., Latex allergy in tire dust and airborne particles. *Environmental Health Perspective* **104**, 1180–1185 (1996).
38. J. E. Slater and S. K. Chabra, Latex antigens. *J. Allergy Clin. Immunol*, **93**, 644–694 (1992).
39. D. R. Ownby et al., Prevalence of anti-latex I_gE antibodies in 1,000 volunteer blood donors. *J. Allergy Clin. Immunol.* **93**. 282 (1994).
40. Anon., Toxicity of general purpose synthetic rubbers. *Rubber Age* **54**, 428 (1944).
41. R. S. Barrows, personal communications, E. I. duPont de Nemours and Co., Inc. Elastomer Chemicals Dept., Wilmington, DE, May 1979.

42. V. Babrauskas et al., Large-scale validation of bench-scale fire toxicity tests. *J. Fire Sci.* **9**(2), 125–148 (1991).
43. H. G. Haag and K. H. Nordsick, Development trends in emulsion SBR. *Kautschuk Gummi Kunst* **43**(2), 135–138 (1990).
44. G. A. Varigos and D. R. Dunt, Occupational dermatitis-an epidemiological study in the rubber and cement industries. *Contact Dermatitis* **7**(2), 105–110 (1981).
45. I. Kilpikari, Occupational contact dermatitis among rubber workers. *Contact Dermatitis* **8**(6), 359–362 (1982).
45a. R. A. Lemen, Environmental epidemiologic investigations in the styrene-butadiene industry. *Environ. Health Perspect.* **86**, 11 (1990).
45b. R. L. Melnick, J. E Huff, and M. G. Bird, Toxicology, carcinogenesis and human health aspects of 1,3 Butadiene *Environ. Health Perspect.* **86**, 3 (1990).
46. T. D. Downs, M. M. Crane, and K. W. Kim, Mortality among workers at a butadiene facility. *Am. J. Ind. Med.* **12**, 311–329 (1987).
47. B. J. Divine, An update on mortality among workers at a 1,3-butadiene facility — preleminary results. *Environ. Health Perspect.* **86**, 119–128 (1990).
48. T. J. Mainhardt et al., Environmental epidemiologic investigations of the styrene–butadiene rubber industry. *Scand. J. Work Environ. Health* **8**, 250–259 (1982).
49. G. M. Matanoski and L. Swartz, Mortality of workers in styrene–butadiene polymer production. *J. Occup. Med.* **29**, 675–680 (1987).
50. G. M. Matanoski, C. Santos-Burgoa, and L. Swartz, *Mortality of a Cohort of Workers in the Styrene–Butadiene Polymer Manufacturing Industry 1943–1982*. Final report prepared under contract to International Institute of Synthetic Rubber Producers, April 1988.
51. G. M. Matanoski et al., *Nested Case Control Study of Lymphopoietic Cancers in Workers of the Styrene–Butadiene Polymer Manufacturing Industry*, final report prepared under contract to International Institute of Synthetic Rubber Producers, April 1989.
52. G. M. Matanoski, C. Santos-Burgoa, and L. Swartz, Mortality of a cohort of workers in the styrene–butadiene polymer manufacturing industry. *Environ. Health Perspect*, **86**, 107–117 (1990).
53. D. Andjelkovich et al., Mortality of rubber workers with reference to work experience. *J. Occup. Med.* **19**, 397–405 (1977).
54. A. J. McMichael et al., Mortality among rubber workers: relationship to specific jobs. *J. Occup. Med.* **18**, 178–185 (1976).
55. H. Checkoway et al., An evaluation of the associations of leukemia and rubber industry solvent exposures. *Am. J. Ind. Med.* **5**, 239–249 (1984).
56. M. G. Ott, R. C. Kolesar, M. C. Scharnweber, et al., A mortality survey of employees in the development or manufacture of styrene-based products. *J. Occup. Med.* **22**, 445–460 (1980).
57. G. G. Bond et al., Mortality among workers engaged in the development or manufacture of styrene-based products — an update. *Scand. J. Work Environ. Health* **18**, 145–154 (1992).
58. J. F. Acquavella, Direct testimony before the Occupational Safety and Health Administration, Nov. 1990.
59. J. F. Acquavella, The paradox of butadiene epidemiology, *Exp. Pathol.* **37**, 114–118 (1989).
60. P. Cole, E. Delzell, and J. Acquavella, Exposure to butadiene and lymphatic and hematopoietic cancer, *Epidemiology* (accepted for publication).

SYNTHETIC POLYMERS—OLEFIN, DIENE ELASTOMERS, AND VINYL HALIDES

61. J. M. Fajen et al., Occupational exposure of workers to 1,3-butadiene. *Environ. Health Perspect.* **86**, 11–18 (1990).
62. MSDS No. G0027R04, The Goodyear Tire and Rubber Company, Feb. 18, 1992.
63. *IARC Monographs on the Evaluation of Carcinogenic Risks to Humans*, Supplement 7, World Health Organization, an Updating of IARC Monographs Volumes 1 to 42.
64. MSDS No. G0064R03, The Goodyear Tire and Rubber Company, Dec. 10, 1991.
65. E. V. Anderson, Downturn looms for synthetic elastomers, *Chem. Eng. News.* **57** (10), 8–9 (1979).
66. MSDS No. R578001, Uniroyal Chemical, Nov. 8, 1985.
67. E. Ward, J.M. Fajen, A.M. Ruder et al., Mortality study of workers in 1,3- butadiene production units identified from a chemical worker's cohort. *Environ. Health Perspective* **103**:525–632 (1995).
68. MSDS No. ZO1792, Zeon Chemicals, Inc., Mar. 18, 1992.
69. MSDS No. G2180R01, The Goodyear Tire and Rubber Company, Dec. 2, 1991.
70. MSDS No. ZO1591, Zeon Chemicals, Inc., Aug. 18, 1992.
71. D. A. Stivers, Fluorocarbon rubbers. In M. Morton, Ed., *Rubber Technology*, 2nd ed., Van Nostrand Reinhold, New York, 1973, pp. 407–439.
71a. P. M. Eller, Ed., *NIOSH Manual of Analytical Methods*, 4th ed., U.S. Department of Health and Human Services, 1994.
71b. N. A. Leidel and K. A Busch, Occupational exposure sampling strategy manual. *NIOSH publication 77-173*, U.S. Department of Health and Human Services, 1997.
72. L. F. Pelosi et al., The volatile products evolve from fluoroelastomer compounds during curing *Rubber Chem. Technol.* **49**, 367–374 (1976).
73. O. L. Dashiell, unpublished data, DuPont Company, Haskell Laboratory, Wilmington, DE, Aug. 1972.
74. H. Sherman, unpublished data, DuPont Company, Haskell Laboratory, Wilmington, DE Jan. 1968.
75. R. E. Reinke, unpublished data, DuPont Company, Haskell Laboratory, Wilmington, DE Sept. 1967.
76. M. E. McDonnell, unpublished data, DuPont Company, Haskell Laboratory, Wilmington, DE, Feb. 1970.
77. MSDS No. HYP001, Du Pont (Jan 31, 1992).
78. MSDS No. VIT006, Du Pont (Jan. 6, 1989).
79. O. L. Dashiell, unpublished data, E. I. duPont de Nemours and Co., Haskell Laboratory, Wilmington, DE, Sept. 1972.
80. Betro Laboratories, unpublished data, Philadelphia, PA, Nov. 1965 (work sponsored by Haskell Laboratory).
81. T. C. Wilcosky, Cancer mortality and solvent exposures in the rubber industry. *Am. Ind. Hyg. Assoc. J.* **45**(12), 809–811 (1984).
82. MSDS No. CLE001, Du Pont, Aug. 28, 1992.
83. S. L. Graham, Effects of one year administration of ethylene thiourea upon the thyroid of the rat. *J. Agric Feed Chem.* **21**. 324–331 (1975).
84. G. P. Daston, B. F. Rehnberg, B. Carver et al., Functional tetratogens of the rat kidney, *Fund. Appl Toxicol.* **11**. 401–415 (1988).

85. S.G. Bradley, A. E. Munson, J. A. McCay et al., Subchronic 10-day immunotoxicity of polydimethylsiloxane (silicone) fluid, gel, elastomer and polyurethane disks in female B6C3F$_1$ mice. *Drug and Chemical Toxicol.* **17**(3), 221–269 (1994).
86. S. E. Bradley, K. L. White, Jr., J. A. McCay et al., Immunotoxicity of 180-day exposure to polydimethylsiloxane (silicone) fluid, gel and elastomer and polyurethane disks in female B6C3F$_1$ mice, *Drug and Chemical Toxicol.* **17**(3), 221–269 (1994).
87. V. Era, *Polymer processing, Industrial Hazards of Plastics and Synthetic Elastomers*, Alan R. Liss, Inc., New York, 1984, pp. 11–17.
88. Vinyl Institute, *Polyvinyl Chloride: An Overview of the Role Chlorine Plays in this Plastic Material*, The Vinyl Institute, July 1991.
89. R. N. Wheeler, Poly(vinyl chloride) processes and products, *Environ. Health Perspect.* **41**, 123–128 (1981).
90. C. R. Blass, PVC as a biomedical polymer—plasticizer and stabilizer toxicity. *Med. Device Technol.* **3**, 32–40 (1992).
91. Norsk Hydro, *PVC and the Environment*, Ostlands-Posten, Norway, 1992.
92. M. L. Westrick, P. Gross, and H. H. Schrenk, unpublished data, Industrial Hygiene Foundation of America, Pittsburgh, PA (courtesy of The Society of the Plastics Industry, Inc., New York), March 1956.
93. O. G. Hansen, PVC in the health care sector. *Med. Device Technol.* **2**, 18–23 (1991).
94. W. F. Lawrence and S. F. Tuell, Phthalate esters: the question of safety—an update. *Clin. Toxicol.* **15**, 447–466 (1979).
95. J. A. Thomas and M. J. Thomas, Biological effects of di(2-ethylhexyl)phthalate and other phthalic acid esters. *CRC Crit. Rev. Toxicol.* **13**, 283–317 (1984).
96. S. D. Gangolli, Testicular effects of phthalate esters. *Environ. Health Perspect.* **45**, 77–84 (1984).
97. P. K. Seth, Hepatic effects of phthalate esters. *Environ. Health Perspect.* **45**, 27–34 (1984).
98. W. M. Kluwe, The carcinogenicity of dietary di(2-ethylhexyl)phthalate in the rat, *J. Toxicol. Environ. Health* **10**, 797–815 (1982).
99. B. G. Lake et al., Studies on the hepatic effects of orally administered di(2-ethylhexyl)phthalate in the rat. *Toxicol. Appl. Pharmacol.* **32**, 355–367 (1975).
100. D. E. Moody et al., Peroxisome proliferation and nongenotoxic carcinogenesis: Commentary on a symposium. *Fundam. Appl. Toxicol.* **16**, 233–248 (1991).
101. R. C. Cattley et al., Cell proliferation and promotion in peroxisome proliferator-induced rodent hepatocarcinogenicity, *CIIT Act.* **12**, 1–5 (1992).
102. L. Fishbein, Additives in synthetic polymers. In *Industrial Hazards of Plastics and Synthetic Elastomers*, Alan R. Liss, New York, 1984, pp. 19–42.
103. W. S. Johnson and R. E. Schmidt, Effects of polyvinyl chloride ingestion by dogs. *Am. J. Vet. Res.* **38**, 1891–1892 (1972).
104. H. F. Smyth and C. S. Weil, Chronic oral toxicity to rats of a vinyl chloride-vinyl acetate copolymer. *Toxicol. Appl. Pharmacol.* **9**, 501–504 (1966).
105. A. O. Seeler, M. Clinton, J. Boggs, and P. Drinker, Experiments on the chronic toxicity of vinyl and vinylidene chloride unpublished Research Report of the Department of Industrial Hygiene and Department of Pathology, Harvard Medical School, Boston, MA, 1947.
106. V. Costa and N. Frongia, Sull'attivita oncogena del cloruro di polivinile. *Pathologica* **73**, 59–67 (1981).

107. G. Faa, E. Dessy et al., Richerche sperimentali sulla tossicita' del PVC per via alimentare. *Boll. Soc. It. Biol. Sper.* **57**, 926–929 (1981).

108. J. Bommer et al., Particles from dialysis tubing stimulate Interleukin-1 secretion by macrophages. *Nephrol. Dial. Transplant*, **5**, 208–213 (1990).

109. J. A. Styles and J. Wilson, Comparison between *in vitro* toxicity of polymer and mineral dusts and their fibrogenicity. *Ann. Occup. Hyg.* **16**, 241–250 (1973).

110. G. H. Pigott and J. Ishmael. A comparison between *in vitro* toxicity of PVC powders and their tissue reaction *in vivo. Ann. Occup. Hyg.* **22**, 111–126 (1979).

111. P. Grasso et al., Tissue reaction to the intrapleural injection of polyvinyl chloride powder, a-quartz and titanium dioxide. *Ann. Occup. Hyg.* **27**, 415–425 (1983).

112. G. Volkheimer, Hematogenous dissemination of ingested polyvinyl chloride particles. *Ann. NY. Acad. Sci.* **246**, 164–171 (1975).

113. D. K. Agarwal, R. K. S. Dogra, and R. Shanker, Pathobiochemical response to tracheobronchial lymph nodes following intratracheal instillation of polyvinylchloride dust in rats. *Arch. Toxicol.* **65**, 510–510 (1991).

114. D. K. Agarwal et al., Some biochemical and histological changes induced by polyvinyl chloride dust in rat lung. *Environ. Res.* **16**, 333–341 (1978).

115. S. Takenaka et al., Morphological effects of nuisance dusts on the respiratory system in rats. *J. Aerosol. Sci.* **18**, 717–720 (1987).

116. IARC, Vinyl chloride, polyvinyl chloride, and vinyl chloride-vinyl acetate copolymers *IARC Monographs on the Evaluation of the Carcinogenic Risk of Chemicals to Humans* **19**, 377–438 (1979).

117. B. S. Oppenheimer, E. T. Oppenheimer, and A. P. Stout, Sarcomas induced in rodents by imbedding various plastic films. *Proc. Soc. Exp. Biol. Med.* **79**, 366–369 (1952).

118. B. S. Oppenheimer et al., Further studies of polymers as carcinogenic agents. *Cancer Res.* **15**, 333–340 (1955).

119. F. E. Russell et al., Tumors associated with embedded polymers. *J. Natl. Cancer Inst.* **23**, 305–315 (1959).

120. N. T. Raikhlin and A. H. Kogan, On the development and malignization of connective tissue capsules around plastic implants (Russian). *Vopr. Onkol.* **7**, 13–17 (1961).

121. A. K. Kogan and V. N. Tugarinova, On the blastomogenic action of polyvinyl chloride. (Russian). *Vopr. Onkol.* **5**, 540–545 (1959).

122. W. L. Guess and J. B. Stetson, Tissue reactions to organotin-stabilized polyvinyl chloride (PVC) catheters. *J. Am. Med. Assoc.* **24**, 118–122 (1968).

123. J. H. Harrison, D. S. Swanson, and A. F. Lincoln, A Comparison of the tissue reactions to plastic materials. *AMA Arch. Surg.* **74**, 139–144 (1957).

124. J. Calnan, The use of inert plastic material in reconstructive surgery. I. A biological test for tissue acceptance. II. Tissue reactions to commonly used materials. *Br. J. Plast. Surg.* **16**, 1–22 (1963).

125. IARC, Vinyl chloride, polyvinyl chloride, and vinyl chloride-vinyl acetate copolymers, *IARC Monographs on the Evaluation of the Carcinogenic Risk of Chemicals to Humans* **19**, 377–438 (1979).

126. R. N. Wheeler, Poly(vinyl chloride) processes and products. *Environ. Health Perspect.* **41**, 123–128 (1981).

127. V. N. Thomas and T. Ramstad, A dynamic headspace GC method for the determination of vinyl chloride monomer in stored solutions of cefmetazole sodium in PVC bags. *Acta Pharm. Nord.* **4**, 97–104 (1992).

128. A. A. Van Dooren, PVC as a pharmaceutical packaging material. *Pharm. Weekbl. (Sci.)*, **13**, 109–118 (1991).

129. B. Szende et al., Pneumoconiosis caused by the inhalation of polyvinyl chloride dust. *Med. Lav.* **61**, 433–436 (1970).

130. S. Vainiotalo and P. Pfaffli, Air impurities in the PVC plastics processing industry. *Ind. Occup. Hyg.* **34**, 585–590 (1990).

131. D. H. Groth et al., Pneumoconiosis in animals exposed to poly(vinyl chloride) dust. *Environ. Health Perspect.* **41**, 73–81 (1981).

132. R. J. Richards et al., Effects in the rat of inhaling PVC dust at the nuisance dust level (10 mg/m^3). *Arch. Environ. Health* **36**, 14–19 (1981).

133. J. C. Wagner and N. F. Johnson, Preliminary observations of the effects of inhalation of PVC in man and experimental effects. *Environ. Health Perspect.* **41**, 83–84 (1981).

134. N. Frogia, A. Spinazzola, and A. Bucarelli, Lesioni polmonari sperimentali da inalazione prolungata di polveri di PVC in ambiente di lavoro. *Med. Lav.* **65**, 321–342 (1974).

135. L. Prodan et al., Experimental chronic poisoning with vinyl chloride (monochloroethene). *Ann. NY. Acad. Sci.* **246**, 159–163 (1975).

136. J. Popow, Effect of poly(vinyl chloride) (PVC) dust on the respiratory system of the rat. *Rocz. Akad. Med. Bialymstoku.* **24**, 5–48 (1969).

137. D. K. Agarwal, Biochemical assessment of the bioreactivity of intratracheally administered polyvinyl chloride dust in rat lung. *Chem. Biol. Interact.* **44**, 195–201 (1983).

138. I. Yu, I. Vertkin, and A. Nikonov, Nature of the action of poly(vinyl chloride) dust on lungs. *Gig. Tr. Prof. Zabol.* **7**, 48 (1981).

139. T. D. Tetley, F. A. Rose, and R. J. Richards, Biochemical and cellular reaction of PVC paste polymer and latex following intratracheal instillation into rats. *Inflammation* **5**, 137 (1981).

140. C. J. Hilado and N. V. Huttlinger, Toxicity of pyrolysis gases from some synthetic polymers. *J. Combust. Toxicol.* **5**, 361–369 (1978).

141. A. Miller et al., Changes in pulmonary function in workers exposed to vinyl chloride and polyvinyl chloride. *Ann. NY Acad. Sci.* **246**, 42–52 (1975).

142. R. Lilis et al., Pulmonary changes among vinyl chloride polymerization workers. *Chest* **69**, 299–303 (1976).

143. J. Gamble et al., Effects of occupational and non-occupational factors on the respiratory system of vinyl chloride and other workers. *J. Occup. Med.* **18**, 659–670 (1975).

144. G. Mastrangelo et al., Polyvinyl chloride Pneumoconiosis: epidemiological study of exposed workers. *J. Occup. Med.* **21**, 540–542 (1979).

145. C. A. Soutar et al., Epidemiological study of respiratory disease in workers exposed to polyvinyl chloride dust. *Thorax* **35**, 644–652 (1980).

146. C. P. Chivers, C. Lawrence-Jones, and G. M. Paddle, Lung function in workers exposed to polyvinyl chloride dust. *Br. J. Ind. Med*, **37**, 147–151 (1980).

147. G. Mastrangelo, G. Marcer, and G. Piazza, Epidemiological study of pneumoconiosis in the italian poly(vinyl chloride) industry. *Environ. Health Perspect.* **41**, 153–157 (1981).

148. M. E. Baser, M. S. Tockman, and T. P. Kennedy, Pulmonary function and respiratory symptoms in polyvinylchloride fabrication workers. *Am. Rev. Respir. Dis.* **131**, 203–208 (1985).

149. P. Ernst et al., Obstructive and restrictive ventilatory impairment in polyvinylchloride fabrication workers. *Am. J. Ind. Med.* **14**, 273–279 (1988).

150. A. Siracusa et al., An 11-year longitudinal study of the occupational dust exposure and lung function of polyvinyl chloride, cement and asbestos cement factory workers. *Scand. J. Work Environ. Health* **14**, 181–188 (1988).

151. J. Nielsen et al., Small airways function in workers processing polyvinylchloride. *Int. Arch. Occup. Environ. Health* **61**, 427–430 (1989).

152. H. S. Lee, T. P. Ng, and W. H. Phoon, Diurnal variation in peak expiratory flow rate among polyvinylchloride compounding workers. *Br. J. Ind. Med.* **48**, 275–278 (1991).

153. T. P. Ng et al., Pulmonary effects of polyvinyl chloride dust exposure on compounding workers, *Scand. J. Work Environ. Health*, **17**, 53–59 (1991).

154. C. A. Soutar and S. Gauld, Clinical studies of workers exposed to polyvinylchloride dust. *Thorax* **38**, 834–839 (1983).

155. A. Arnaud et al., Polyvinyl chloride pneumoconiosis. *Thorax* **33**, 19–25 (1978).

156. F. Sulotto et al., A case of PVC pneumoconiosis in man, *Med. Lav.* **76**, 304–308 (1985).

157. M. Antti-Poika et al., Lung disease after exposure to polyvinyl chloride dust. *Thorax* **41**, 566–567 (1986).

158. H. S. Lee et al., Occupational asthma due to unheated polyvinylchloride resin dust. *Br. J. Ind. Med.* **46**, 820–822 (1989).

159. Anon., A scientific basis for the risk assessment of vinyl chloride. *Reg. Toxicol. Pharmacol.* **7**, 120–127 (1987).

160. R. J. Richards et al., Biological reactivity of PVC dust. *Nature*, **256**, 664–665 (1975).

161. Y. Kotouro et al., A method for toxicological evaluation of biomaterials based on colony formation of V79 cells. *Arch. Orthop. Trauma Surg.* **104**, 15–19 (1985).

162. V. Habermann and D. Waitzova, On the safety evaluation of extracts from synthetic polymers used in medicine. *Arch. Toxicol.* (Suppl 8) 458–560 (1985).

163. R. E. Marchant, J. M. Anderson, and E. O. Dillingham, *In vivo* biocompatibility studies. VII. Inflammatory responses to polyethylene and to a cytotoxic polyvinylchloride. *J. Biomed. Mat. Res.* **20**, 37–50 (1986).

164. R. J. Richards, R. Desai, and F. A. Rose, A surface-active agent involved in PVC-induced hemolysis. *Nature* **260**, 53–54 (1976).

165. C. L. Goh and S. F. Ho, An outbreak of acneiform eruption in a polyvinyl chloride manufacturing factory. *Dermatosen* **2**, 53–57 (1988).

166. MSDS No. GO379RO8, The Goodyear Tire and Rubber Company, Aug. 26, 1991.

167. T. Nguyen and J. W. Burnett, Local skin reaction caused by the plastic catheter tubing of the continuous subcutaneous insulin infusion system. *Cutis* **41**, 355–356 (1988).

168. V. Di Lernia, N. Cameli, and A. Patrizi, Irritant contact dermatitis in a child caused by the plastic tube of an infusion system. *Contact Dermatitis.* **21**, 339–353 (1989).

169. J. K. Wagoner, Toxicity of vinyl chloride and poly(vinyl chloride): A critical review. *Environ. Health Perspect.* **52**, 61–66 (1983).

170. P. F. Infante, J. K. Wagoner, A. J. McMichael, et. al, Genetic risk of vinyl chloride Lancet **23**(1), 734–735 (1976).

171. U.S. Code of Federal Regulations, Title 29 Part 1910.1000 (29 *CFR*. 1910.1000), Table 1 (1992).

172. W. M. Saltman, Styrene-butadiene rubbers. In M. Morton, Ed., *Rubber Technology*, 2nd ed., Van Nostrand Reinhold, New York, 1973, pp. 178–198.
173. A. J. Lehman, *J. Assoc. Food Drug Off.* (July 1951).
174. R. B. Cain, Microbial degradation of synthetic polymers, *Microbial Control of Pollution*, Cambridge University Press, 1992, pp. 293–338.
175. P. M. Molton, R. T. Hallen, and J. W. Pyne, *Study of Vinyl Chloride Formation at Landfill Sites in California.*, prepared for the California State Air Resources Board, Sacramento CA, January NTIS PB 87-161279, Springfield, VA, 1987.
176. Agency for Toxic Substances and Disease Registry (ATSDR), *Toxicological Profiles for 1,1-Dichloroethene*, U.S. Department of Health and Human Services, 1992, p. 123.
177. P. J. Vanderhorst, personal communication, DuPont Company, Plastics Products and Resins Dept., Wilmington, DE, May 1979.
178. R. W. Morrow, unpublished data, DuPont Company, Haskell Laboratory, Wilmington, DE, Oct. 1972.
179. C. H. Hine, unpublished data, Hine Laboratories, Inc., Research and Development, San Francisco, CA, Dec. 1972.
180. D. I. McCane, Tetrafluoroethylene Polymers. In *Encyclopedia of Polymer Science and Technology*, Vol. 13, Wiley, New York, 1970, pp. 623–671.
181. U.S. Code of Federal Regulations, 21 *CFR*. 170–199, April 1, 1981, also 21 CFR 500–599. Revised annually. Guide available from *Food Chemical News*.
182. B. D. Halpern and W. Karo, Medical applications. In *Encyclopedia of Polymer Science and Technology*, 2nd ed., Suppl. 2, Wiley, New York, 1977, pp. 368–403.
183. P. L. Polakoff, K. A. Busch, and M. T. Okawa, Urinary fluoride levels in polytetrafluoroethylene fabricators. *Am. Ind. Hyg. Assoc. J.* **35**, 99–106 (1974).
184. M. T. Okawa and P. L. Polakoff, Occupational health case reports — No. 7. *J. Occup. Med.* **16**, 350–355 (1974).
185. A. J. McMichael et al., Mortality among rubber workers: relationship to specific jobs. *J. Occup. Med.* **18**, 178–185 (1976).
186. S. B. Fretz and H. Sherman, unpublished data, DuPont Company, Haskell Laboratory, Wilmington, DE, Oct. 1968.
187. S. B. Fretz and H. Sherman, unpublished data, DuPont Company, Haskell Laboratory, Wilmington, DE, Sept. 1969.
188. R. S. Khamidullin and G. A. Petrova, Hygienic evaluation of fluoroplast-4 films. *Cig. Vop. Proizvod. Primen. Polim. Mater.* 242–248 (1969); *Chem. Abstr.* **75**, 107796q.
189. R. S. Khamidullin and G. A. Petrova, Experimental hygienic assessment of teflon film. *Gig. Sanit.*, **34**, 247–248 (1969).
190. R. S. Khamidullin et al., Hygienic assessment of a fluoroplast 4 film in connection with its use in the food industry. *Gig. Sanit.* **1**, 38–41 (1977).
191. C. E. Lewis and G. R. Kerby, An epidemic of polymer-fume fever, *J. Am. Med. Assoc.* **191**, 103–106 (1965).
192. D. H. Wegman and J. M. Peters, Polymer fume fever and cigarette smoking. *Ann. Int. Med.* **81**, 55–57 (1974).
193. R. Preiss, Properties and toxicology of polytetrafluoroethylene and its uses in industry and medicine. *Pharmazie* **28**, 281–284 (1973).

194. R. A. Elson and J. Charnley, The direction and resultant force in total prosthetic replacement of the hip joint. *Med. Biol. Eng.* **6**, 19–27 (1968).
195. F. Bischoff, Organic polymer biocompatibility and toxicology. *Clin. Chem.* **18**, 869–894 (1972).
196. R. L. Davidson, Ed., *Handbook of Water-Soluble Gums and Resins*, McGraw-Hill, New York, 1980.
197. R. B. Pedley, G. Meachim, and D. F. Williams, Tumor induction by implant materials. In D. F. Williams, Ed., *Fundamental Aspects of Biocompatibility*, Vol. 2, CRC Press, Boca Raton, FL, 1981, pp. 175–202.
198. International Agency for Research on Cancer, *IARC Monographs on the Evaluation of the Carcinogenic Risk of Chemicals to Humans; Some Monomers, Plastics and Synthetic Elastomers and Acrolein*, Vol. 19, Lyon, France, 1979.

CHAPTER NINETY

Polyvinyl Acetate, Alcohol, and Derivatives, Polystyrene, and Acrylics

Bailus Walker, Jr., Ph.D., MPH, and Laureen Burton, MPH

A POLYVINYL ACETATE, ALCOHOL AND DERIVATIVE POLYMERS

Polyvinyl acetate, the most widely used vinyl ester, is noted for its adhesion to substrates and high cold flow. Polyvinyl acetate serves as the precursor for polyvinyl alcohol and, directly or indirectly, the polyvinyl acetals. Both polyvinyl acetate and polyvinyl alcohol are insoluble in many organic solvents but water sensitive. Polyvinyl acetate absorbs from 1 to 3% water, up to 8% on prolonged immersion (1). Polyvinyl alcohol absorbs 6–9% water when humidity conditioned and can usually be dissolved completely in water above 90°C, but it can also be insolubilized by chemical treatment. Table 90.1 gives data on properties for typical polymers.

Production and Processing

U.S. manufacturers currently sell polyvinyl acetate in emulsion form and polyvinyl alcohol as granules. Polyvinyl alcohol is processed into films and formulated with other materials into emulsion intermediates. Both polymers are typically used in aqueous systems.

Both polyvinyl acetate and polyvinyl alcohol meeting certain specifications are permitted in stated food contact applications such as packaging, coatings, and adhesives.

Patty's Toxicology, Fifth Edition, Volume 7, Edited by Eula Bingham, Barbara Cohrssen, and Charles H. Powell.
ISBN 0-471-31940-6

Table 90.1. Physical Properties of Selected Polymers

Compound	CAS	Molecular Formula	Molecular Weight	Boiling Point (°C)	Specific Gravity	Melting Point (°C)	Sol. in water (at 68°F) mg/ml	Refractive Index (20°C)	Vapor Pressure (mm. Hg)	UEL % by Vol. Room Temp	LEL % by Vol. Room Temp
Polyvinyl acetate	[9003-20-7]	$(C_4H_6O_2)_n$	86,090	—	1.18	35	—	1.46–1.47	—	—	—
Polyvinyl acetate-butyl acrylate	[25067-01-0]	—	—	—	—	—	—	—	—	—	—
Polyvinyl acetate-1-ethylhexyl acrylate	[25067-02-1]	—	—	—	—	—	—	—	—	—	—
Polyvinyl acetate-ethylene	[24937-78-8]	—	—	—	—	—	—	—	—	—	—
Polyvinyl alcohol	[9002-89-5]	$(C_2H_3OH)_n$	37,000–185053	—	1.3	228	—	1.49–1.53	—	—	—
Polystyrene	[9003-53-6]	$(C_8H_8)_n$	104.15	—	1.05	240	—	1.57–1.60	—	—	—
Polyacrylo-nitrile	[25014-41-9]	$(C_2H_3CN)_n$	100,000–150,000	—	1.07–1.53	317–340	—	1.52	—	—	—
Styrene-Acrylonitrile	[9003-54-7]	$(C_8H_8\text{—}C_3H_5N)_n$	100,000–400,000	—	1.06–1.08	—	—	—	—	—	—
Acrylonitrile Butadiene Styrene	[9003-56-9]	—	60,000–200,000	—	1.02–1.07	—	—	—	—	—	—
Polymethyl methacrylate	[9011-14-7]	$(C_5H_8O_2)_n$	100.13	—	1.188	160– >200	—	1.49	—	—	—
Polyacryl amide	[9003-05-8]	$(C_3H_5NO)_n$	91.009	—	1.302	—	—	—	—	—	—
Polyacrylic acid	[9003-01-4]	$(C_3H_4O_2)_n$	72.063	—	1.09	106	—	1.53	—	—	—
Poly (2-hydroxy-ethyl methacrylate) hydrogel	—	—	—	—	—	—	—	—	—	—	—
Methyl-methacrylate	[80-62-6]	$(C_5H_8O_2)$	100.1	100	0.943	−48	1.5	1.41	40	12.5	2.1
Ethyl-acrylate polymer	[9003-32-1]	$(C_5H_8O_2)_n$	—	—	—	—	—	—	—	—	—

POLYVINYL ACETATE, ALCOHOL, AND DERIVATIVES

Ethylene–vinyl acetate copolymers and ethylene–vinyl acetate–vinyl alcohol terpolymers [CAS # 26221-27-2] are similarly permitted in certain food contact applications. Polyvinyl acetate with a minimum molecular weight of 2000 is permitted as a synthetic masticatory substance in chewing gum base (2).

Toxicologic Potential

Monomer residue has not been considered a problem in end-use products. Latexes or solutions of polyvinyl acetate that are essentially intermediates may contain residual vinyl acetate, essential emulsifiers, or initiators (3). No detailed information is available on the amount of unreacted monomer in either polyvinyl acetate or polyvinyl alcohol resins (3).

Local sarcomas have been produced in rats with polyvinyl alcohol sponges, but implants of both polyvinyl alcohol and polyvinyl acetate in powder form did not produce tumors. IARC considered that additional studies would be required prior to evaluation of carcinogenic potential (3).

Inhalation and combustion toxicity have not been considered problems. This may be attributed to polymer structure and degradation characteristics as well as the nature of ordinary intermediate and end-use products.

Work Practices

With products formulated as solutions or emulsions, the potential for inhalation toxicity or skin reactions from residual monomer or additives may in some cases require evaluation. Recommended industrial hygiene procedures for dealing with nuisance dusts should provide adequate control for any dust hazards from solid forms of polymer. Spillage of these polymers can produce slipping hazards; spills should be cleaned up immediately to prevent falls.

1.0a   Polyvinyl Acetate

1.0b   Polyvinyl Acetate 1-Ethylhexyl Acetate

1.0c   Polyvinyl Acetate Ethylene

1.0.1a   CAS Number: [9003-20-7]

1.0.1b   CAS Number: [25067-02-7]

1.0.1c   CAS Number: [24937-78-8]

Commercial copolymers containing at least 60% vinyl acetate units are generally termed "polyvinyl acetate."

1.0.2   Synonyms: PVA, PVAC, poly-1-acetoxyethylene, acetic acid ethenyl ester, homopolymer; vinyl acetate homopolymer; vinyl acetate resin, vinyl acetate latex; and poly(vinyl acetate), sec. stand., typical M.W. 194800, typical M.N. 63600

1.0.3 Trade Names:
Catalac, Elvacet, Emultex, Epok V, Gelva, Mowilith, Texicote V, Texilac, Vandike, Vinalk, Vinavil, Vinamul, and Vinnapas

1.0.4 Molecular Weight

The molecular weight has been described as varying from 5000 to more than 500,000 (4) and from 11,000 to 1,500,000.

1.0.5 Molecular Formula: $[C_4H_6O_2]_n$

1.0.6 Molecular Structure:

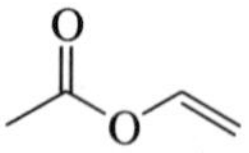

1.1 Chemical and Physical Properties

The chemical properties of polyvinyl acetate are primarily those of an aliphatic ester. Thus acidic and basic hydrolysis produce polyvinyl alcohol and acetic acid. Copolymerization can modify these chemical properties. A comonomer containing a carboxy group can increase the solubility of the copolymer in aqueous alkali. These copolymers adhere better to many types of surfaces than homopolymers or neutral copolymers. This "adherence property" is due to the interaction between the acid group and the surface. Polyvinyl acetate resins are soluble in organic solvents such as halogenated hydrocarbons and carboxylic acids. they are insoluble in glycols, water and nonpolar liquids such as ether, aliphatic hydrocarbons, oil, and fats.

Commercial polymers are atactic and therefore, if free from emulsifier, transparent (5). Polyvinyl acetate is noted for its high tensile strength and toughness. In general, copolymerization results in a particular hardening or softening effect desirable in the intended application. Properties vary with humidity. It is resistant to weathering.

1.1.1 General

PVA is a thermoplastic polymer, a colorless solid, and combustible. Physical constants are:

> Density (g/cm^3) at 20°C: 1.19
> Density (g/cm^3) at 200°C: 1.05
> Decomposition temperature (°C): 150
> Softening temperature (°C): 35–50
> Tensile strength (MPa): 29.4–49.0

1.1.2 Odor and Warning Properties

PVA is odorless and tasteless.

1.2 Production and Use

Polyvinyl acetate is derived from the polymerization of vinyl acetate. the catalysts used in polymerization may include hydrogen peroxide, peroxy sulfates, or various redox combinations. The polymerization process is described as being carried out by charging all ingredients to the reactor, heating to reflux, and stirring until the reaction is over. Typically, only a part of the monomer and catalyst is initially charged; the remainder is added during the course of the reaction.

The applications of PVAC include adhesives for paper, wood, glass, metal, porcelain. It is also used in book binding, textile finishing, and as a strengthening agent for cement.

1.3 Exposure Assessment

No information was found in the literature.

1.4 Toxic Effects

The limited data available indicate very low toxicity. Rats and mice apparently survived a single oral administration of 25 g/kg polyvinyl acetate as well as 12-m administration at 250 mg/kg that produced some tissue damage (6). Animals fed various doses of several modifications of polyvinyl acetate showed effects ranging from increased "work capacity" to hepatocyte necrosis. Weeks and Pope reported that the oral lethal dose of polyvinyl acetate formulated as an emulsion dust control material was > 9.7 g/kg for rats (7). Skin irritation tests on rabbits with this emulsion or the base latex produced moderate to severe irritation on intact and abraded skin. (The irritation potential received the most emphasis in medical surveillance recommendations.) However, Hood (8) found no evidence of primary irritation or skin sensitization in any of 210 human subjects exposed to film made from two different polyvinyl acetate emulsions or cotton cloth impregnated with 40% polyvinyl acetate resin. Her subjects wore 1-in. squares of the test material for 6 d and then again for 1 d after a 10-d rest period. IARC reviewed (3) data on subcutaneous or intraperitoneal implantation of polyvinyl acetate powder in mice and rats. No local sarcomas were found. No data are available in humans.

1.4.1 Experimental Studies

No information was found in the literature. Only IARC data are available.

1.4.1.5 Carcinogenesis. IARC reviewed data on subcutaneous or intraperitoneal implantation of polymer vinyl acetate powder in rats and mice. No local sarcomas were found.

1.4.2 Human Experience

No information was found in the literature.

1.5 Standards, Regulations, or Guidelines of Exposure

The FDA permits the use of PVAC homopolymer and copolymer as components of adhesives, resinous and polymeric coating when they are intended for use in contact with food. In many processes, significant quantities of solvent, monomer, or additive vapors may evolve. Recommended ventilation and related good industrial hygiene procedures should reduce exposures and risks of toxic injuries.

1.6 Studies on Environmental Impact

Polyvinyl acetate emulsions have been found nontoxic to pike, carp, daphnia, and other species of aquatic life at concentrations ranging from 0.3 to 80 mg/L (9–11).

Griffin and Mivetchi (12) report that polyvinyl acetate is biodegradable, and ethylene–vinyl acetate copolymer shows the effect to a degree, depending on vinyl acetate content. Starch filler in polyvinyl acetate stimulates biodegradability.

2.0a Polyvinyl Acetate-Butyl Acrylate

2.0b Polyvinyl Acetate-2-ethylhexyl Acrylate

2.0c Polyvinyl Acetate Ethylene

2.0.1a CAS Number: [25067-01-0]

2.0.1b CAS Number: [25067-02-1]

2.0.1c CAS Number: [24937-78-8]

Commercial copolymers containing at least 60% vinyl acetate units are generally termed "polyvinyl acetate." Polyvinyl acetate-butyl acrylate is one of the copolymers. The others are polyvinyl acetate-2-ethylhexyl acrylate. It is reported that many manufacturers have not taken properly into account the differences in rates of copolymerization of acrylic esters and vinyl acetate. As a result, many "acrylic copolymers" are mixtures with composition ranging from almost pure acrylic polymer to almost pure vinyl acetate homopolymer. These copolymer blends produce apparently clear films when dried as the various polymer components tend to homogenize each other. The physical properties of these films are different from those of "true" copolymers. No other relevant information on these chemicals was found in the literature.

3.0 Polyvinyl Alcohol

3.0.1 CAS Number: [9002-89-5]

3.0.2 Synonyms: PVA, PVAL, Elvanol; Polyviol; Vinol; Alvyl; Polyvinyl Alcohol 500; PVOH; Ethenol, homopolymer; vinyl alcohol polymer; liquifilm; alcotex 88/05; alcotex 88/10; alkotex; aracet apv; cipoviol w 72; Covol; covol 971; elvanol 50-42; elvanol 52-22; elvanol 70-05; elvanol 71-30; elvanol 90-50; elvanol 522-22; elvanol 73125g; EP 160; galvatol 1-60; gelvatol 1-30; gelvatol 1-60; gelvatol 1-90; gelvatol 3-91; gelvatol 20-30; gelvatol 20-30; gelvatol 2090; GH 20; GL 02; GL 03; GLO 5; GM 14; gohsenol; gohsenol ah 22; gohsenol gh; gohsenol gh 17; gohsenol gh 20; gohsenol gh 23; gohsenol gl 02; gohsenol gl 03; gohsenol gl 05; gohsenol gl 08; gohsenol gm 14; gohsenol gm 94; gohsenol kh 17; gohsenol nh 05; gohsenol nh 17; gohsenol nh 18; gohsenol nh 20;

gohsenol nh 26; gohsenol nk 114; gohsenol nl 05; gohsenol nm 14; gohsenol nm 114; ivalon; kurare poval 1700; kurare pva 205; kurate poval 120; Lemol; lemol 5-88; lemol 5-98; lemol 12-88; lemol 16-98; lemol 24-98; lemol 30-98; lemol 51-98; lemol 60-98; lemol 75-98; lemol gf-60; M 13/20; mowiol; mowiol n 30-88; mowiol n 50-98; mowiol n 70-98; NH 18; NM 11; NM 14; polydesis; polysizer 173; polyvinol; polyviol m 13/140; polyviol mo 5/140; polyviol w 25/140; polyviol w 40/140; poval 117; poval 120; poval 203; poval 205; poval 217; poval 1700; poval c 17; pva 008; PVS 4; resistoflex; rhodoviol; rhodoviol 4/125; rhodoviol 16/200; rhodoviol 4-125p; rhodoviol r 16/20; solvar; sumitex H 10; vibatex s; vinacol mh; vinalak; vinarol dt; vinarole; vinarol st; vinavilol 2-98; vinnarol; vinol 125; vinol 205; vinol 351; vinol 523; vinol unisize; vinylon film 2000; akwa tears; moviol; sno tears; PVA1; vinylon; Polyvinyl Alcohol 15000; POLYVINYL ALCOHOL (POLYMER MIXTURE)

3.0.3 Trade Names: There are over 50 trade names for PVA. Examples are: Alcotex 88/05, Alcotex 88/10, Alvyl, Aracet APV, Cipoviolw 72 Covol, Covol 971, EL Vanol 70-05, Elvanol 71-30, Elvanol 90-50, Elvanol 90-50, Elvanol 522-22, Elvanol 73125g, EP160, C-Oshenol NH05, and Goshenol NH17

3.0.4 Molecular Weight

Molecular weight depends on and is approximately half that of the polyvinyl acetate from which the polyvinyl alcohol was derived. The average molecular weight is 120,000.

3.0.5 Molecular Formula: $[C_2H_4O]_n$

3.0.6 Molecular Structure: ⌇OH

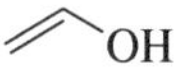

3.1 Chemical and Physical Properties

The physical properties vary, depending on the degree of alcoholysis. Polyvinyl alcohol is amorphous to polycrystalline, depending on mechanical and heat treatment. When stretched as fibers the polymer becomes up to 60% crystalline (4). PVA is largely unaffected by hydrocarbons, chlorinated hydrocarbons, carboxylic acid esters, animal grease, or vegetable oils. As hydrolysis increases, resistance to organic solvents increases. Data on chemical properties indicate that polyvinyl alcohol undergoes chemical reactions similar to that of other secondary alcohols. At the commercial level, reactions with aldehydes to form acetals, such as polyvinyl butyral, are of great importance. Chemical data on PVA also indicate that its oxygen-barrier properties at low humidity are the best of any resin.

The thermal decomposition of PVA in the absence of oxygen is reported as a two-stage process. The first stage begins at approximately 200°C and is mainly dehydration, accompanied by the formation of volatile products. The residue is predominately macromolecules of polyene structure. Further heating to 400–500°C yields carbon and hydrocarbons. The physical and chemical properties of polyvinyl alcohol make it one of the most readily biodegradable synthetic polymers.

3.1.1 General

The degree of crystallinity affects solubility, water sensitivity, tensile strength, oxygen barrier properties, and thermoplastic properties. Some physical constants for PVA are:

> Appearance: White to ivory white
> Specific gravity: 1.25–1.35
> Melting point: 212–267°C
> Specific heat J (g K): 1.67.

3.1.2 Odor and Warning Properties

When heated to decomposition, it emits acid smoke and irritating fumes that may be adequate to prevent overexposure.

3.2 Production and Use

Various indirect methods that involve alcoholysis (also called saponification or hydrolysis) of polyvinyl acetate are used to prepare polyvinyl alcohol. Preparation can be carried out by dissolving polyvinyl acetate in methanol or ethanol with an alkaline or acid catalyst and heating to precipitate the polyvinyl alcohol from the solution.

Polyvinyl acetate is converted to the corresponding polyvinyl alcohol by hydrolysis or catalyzed alcoholysis. The process can be catalyzed by strong acids or strong bases. Alkaline alcoholysis is generally used large-scale industrial productions. Batch and continuous processes are used in commercial production. The continuous process is the most frequently used techniques.

The chemical and physical properties of PVA have attracted broad use: textile sizing, adhesives, paper coating, polymerization stabilizers, water-soluble films, and nonwoven fabric binders.

3.3 Exposure Assessment

No information was found in the literature.

3.4 Toxic Effects

Oral tests on animals date back to 1939, when Hueper reported (13) on a test with four rats conducted in connection with more extensive parenteral tests. The rats were fed a dietary level of 4% polyvinyl alcohol increased to 8 and then 29% at 2-wk intervals; two were sacrificed at 4 wk. The two surviving rats that received the higher dose level showed microscopic tissue changes in the liver, stomach, and sternum. Sections stained with iodine solution did not show the characteristic blue color indicative of polyvinyl alcohol deposits. Zaeva et al. (14) describe only minimal changes in the liver and myocardial cells of rats receiving 30 doses of 500 mg/kg. Yamatani and Ishikawa (15) found no polyvinyl alcohol in the urine or blood of rats dosed orally with 2 mL of 2% polyvinyl alcohol (molecular weight = 22,000).

In eye drops polyvinyl alcohol in soluble form increases viscosity; a 1.4% neutral solution of polymer with molecular weight > 100,000 has been used successfully in tests on rabbit eyes and on human eyes (16, 17). Other reports suggest continuing interest in solutions of polyvinyl alcohol as a component of formulations for topical treatment of "dry eye" (18) and intraocular pressure (19).

3.4.1 Experimental Studies

See Section 3.4.

3.4.1.1 Acute Toxicity. See Section 3.4.

3.4.1.2 Chronic and Subchronic Toxicity. Polyvinyl alcohol released into circulation may cause serious dysfunction, particularly in the kidneys. This was first demonstrated by Hueper (13), who showed that polyvinyl alcohol injected subcutaneously into rats diffused throughout major body organs. Most of the kidneys showed degeneration or necrosis of the tubular epithelium, which was sometimes accompanied by other changes. During the 1960s Hall and Hall (20) also conducted an extensive series of subcutaneous tests with various grades of polyvinyl alcohol. They confirmed that injections of polyvinyl alcohol may affect elements of the reticuloendothelial system and cause a nephrotic syndrome. Initial experiments suggested a correlation with molecular weight; solutions of a medium-size molecule (average molecular weight = 133,000) appeared more injurious than those of a larger molecule (average molecular weight = 185,000) and those of a relatively small molecule (average molecular weight = 37,000). The last caused only a slight elevation of arterial pressure in a small percentage of rats and its inertness was presumed due to ready filtration of its small size through renal glomeruli. Additional work showed that the hypertensive syndrome found after injection of some but not all types of polyvinyl alcohol did not correlate with infiltration of the kidney; infiltration and varying degrees of anemia were considered the only common sequelae with various types of polymer (21).

3.4.1.5 Carcinogenesis. IARC (3) summarizes various animal studies, some but not all of which produced tumors. All these studies were parenteral. Reference is also made to one case of hemangiopericytoma of the bladder in an occupationally exposed worker, which the authors "speculated" might be analogous to that of angiosarcoma of the liver and vinyl chloride. Polyvinyl alcohol (5) was not mutagenic to any of five strains of *salmonella typhimurium* (Ames test) when tested with or without activation (22).

3.4.2 Human Experience

No information was found in the literature, except as summarized in Section 3.4.2.3.5.

3.4.2.3.5 Carcinogenesis. In IARC review of PVA, reference is made to one case of hemangiopericytoma of the bladder in an occupationally exposed worker, which the authors "speculated" might be analogous to that of angiosarcoma of the liver and vinyl chloride (3).

3.5 Standards, Regulations, or Guidelines of Exposure

The American Standard for Precautionary Labeling of Hazardous Industrial Chemicals classifies PVA as a nonhazardous material.

3.6 Studies on Environmental Impact

Bluegill sunfish exposed for 96 h to 10,000 mg/L of polyvinyl alcohol showed no mortality or other evidence of physiological response (23). Biodegradation of polyvinyl alcohol can be accomplished in properly designed and operated waste treatment systems where > 90% of the polyvinyl alcohol in the influent may be removed by acclimated microoraganisms (24, 25). The chemical oxygen demand rate coefficient was found to be h = 1.2 mg oxygen/mg polyvinyl alcohol/d in a domestic (municipal) activated waste system (24) and k = 0.50 in a textile waste treatment facility (25). Organisms degrading polyvinyl alcohol in mixed culture (26) have remained viable under hydrogen peroxide treatment conditions that were toxic to filamentous and nitrifying organisms. Product differences, particularly the number and location of residual acetate groups (27), may have to be considered. Ozonization (28) may lower polymer molecular weight and improve biodegradability.

B POLYSTYRENE AND ACRYLICS POLYMERS

Since the 1700s when Newman first isolated styrene by stream distillation from liquid ambar, a solid resin obtained directly from a family of trees native to the Far East and California, a substantial industry has developed for styrene-based products. Today, "styrene-based" plastics most commonly are polystyrene, successfully commercialized in 1938, plus the derivatives containing butadiene, acrylonitrile, or both. The derivatives containing acrylonitrile are also called "acrylonitrile polymers" or "nitrile polymers." Polystyrene is made in three different forms: crystal, impact, and expandable. Producers generally refer to the polystyrene market as including only crystal and impact grade. Expandable polystyrene — a foam product, with primary markets in construction and packaging — is a separate specialty product.

Structurally the acrylic polymers include those containing repeating units of acrylonitrile, acrylic acid, acrylates, methacrylates, and all the various derivatives. "Acrylic plastics" may imply only polymers of acrylic or methacrylic acid ester (29), among which the prototype is polymethyl methacrylate. The demand for sheet polymethyl methacrylate dates from World War II, when it was used for aircraft glazing (5).

Polyacrylonitrile is used primarily as fibers, commonly called "acrylic,"that have been formulated with varying amounts of comonomer. Processing may involve blending with other fibers. Dyeing may be done with either the yarn or woven fabric. "Polymeric acrylontrile" is a powder that can be hydrolyzed to form water-soluble polyacrylamides or cast as plastic or film from solvent.

The main copolymer types derived from both styrene and acrylonitrile are (*1*) styrene–acrylonitrile (SAN) copolymer resin and (*2*) acrylonitrile–butadiene–styrene (ABS), in which discrete butandiene particles are dispersed in a SAN copolymer matrix and

POLYVINYL ACETATE, ALCOHOL, AND DERIVATIVES

then sold as pellets or powder. Temperatures described for ABS processing are in the 190–275°C range (30).

Acrylate and methacrylate esters are generally available from the manufacrturer in granules or powder. Dyes, pigments, plasticizers, or ultraviolet absorbers may be added during processing. Commercial processing of polymethyl methacrylate per se uses three intermediate types of approach: the melt state for injection molding and extrusion; sheets, rods, and tubes that are mechined or welded; and monomer–polymer dough, primarily for dentures (5).

4.0 Polystyrene

4.0.1 CAS Number: [9000-53-6]

4.0.2 Synonyms:
PS, HIPS, crystal polystyrene, impact polystyrene, benzene, ethenyl-, homopolymer, styrene, homopolymer; polystyrene beads; A 3–80; afcolene; atactic polystyrene; bactolatex; bakelite smd 3500; basf iii; bdh 29–790; bextrene xl 750; bicolastic a 75; bicolene h; bio-beads s-s 2; bp-klp bsb-s; bustren k; bustren u 825; bustreny 825; cadco 0115; carinex gp; copal z; cosden 550; denka qp3; diarex 43g; dorvon; dow 456; dylene 8; dylite f; esbrite 2; estyrene g 15; foster grant 834; Gedex; hostyren n; kb polymer; lacqren 506; lustrex hp 77; poligostyrene; owispol gf; pelaspan 333; piccolastic a; polyflex; polystrol d; polystrene beads; printels; rexolite 1422; rhodoline; shell 300; sternite 30; styrafoil; styragel; styrene polymers; styrex; styrocell pm; styrofoam; toporex 500; trolitul; trycite 1000; ubatol u 2001; UP 1; UP 2; vestolen p 5232g; vestyron; vinamul n 710; vinylbenzene polymer; X 600; styrene latex; Poly(styrene)

4.0.3 Trade Names:
Benzene, Ethenyl-, Homopolymer. A 3-80, Afcolene, Atactic Polystyrene, Bactolatex, Bakelite SMD 3500, BASF III, BDH 29– 790, Bextrene X 750, Bicolastic A 75, Bicolene H, Bio-Beads S-S 2, BP- KLP BSB-S, Bustren K %00, Bustren U 825, Cadco 0115, Carinex GP, Copal Z, Cosden 550, Denka QP3, Diarex 43G, Dorvon, Dow 456, Dylene 8, Dylite F, Esbrite 2, Esstyrene G 15, Foster Grant 874, Gedex, Hostryren N, KB Polymer, Lacqren 506, Lustrex HP 77, Poligostyrene, Owispol GF, Pelaspan 333, Piccolasstic A, Polyflex, Polystrol D, Polystrene Beads, Printels, Rexolite 1422, Rhodoline, Shell 300, Sternite 30, Styrafoil, Styragel, Styrene Polymers, Styrex C, Styrocell PM, Styrofoam, Styron, Toporex 500, Trolitul, Trycite 1000, Ubatol U 2001, UP 1, UP2, Vestolen P 5232G, Vestyron, Vinamul N 710, Vinylbenzene Polymer, X 600

4.0.4 Molecular Weight
The molecular weight ranges from less than 60,000 to more than 300,000.

4.0.5 Molecular Formula: $[C_8H_8]_n$

4.0.6 Molecular Structure:

4.1 Chemical and Physical Properties

Molecular weight, molecular weight distribution, and additives affects the properties of polystyrene. The commercial success of polystyrene is attributed to its transparency, lack of color, ease of fabrication, thermal stability, low specific gravity, and good electrical properties. The three common commercial grades of general purpose polystyrene are described as: easy flow, medium flow and high heat. The choice of resin grade depends primarily on the fabrication but also on end use. In addition to the common grades of polystyrene, several specialty grades are produced. These have unique properties. They are designated as fast-cycle resins, low-molecular-weight resins, and food-contact-grade reins.

"Rubber-modified polystyrene" or high-impact polystyrene (HIPS) describes a product in which a rubber, usually polybutadiene, has been added to reduce brittleness but which keeps most of the characteristic properties of polystyrene (31). High-impact polystyrene may contain 5–15% polybutadiene. Polystyrene foam has many very desirable properties. Closed-cell foams have excellent insulating properties, low weight, and good cushioning.

Industry data often refer to general-purpose polystyrene as crystal styrene. This designation refers to the clarity of the resin and not to its molecular structure.

4.1.1 General

No information was found in the literature.

4.1.2 Odor and Warning Properties

When heated to decomposition, it emits acrid smoke and irritating fumes.

4.2 Production and Use

Most general-purpose polystyrene is produced by solution polymerization in a continuous process. The volatiles are removed at high temperature and vacuum, and the molten polymer is then cooled and pelletized. Suspension polymerization is used for products for which a small spherical form is desirable, such as foam-in-place beads and ion-exchange resins. Rubber-modified polystyrene is manufactured in a very similar manner (32; 31 and citations therein). The principal challenges to polymerization of polystyrene by solution process are control of the large amount of heat generated in the polymerization reaction and handling of viscous materials. The solution process produces polystyrene and rubber-modified polystyrene of high purity with low residual monomer concentrations. The system may consist of several reactors in series held at progressively higher temperatures. Styrene monomer and, optionally, solvent are fed into the first reactor. At the last reactor, the effluent is fed into a devolatilizer, which removes the remaining monomer and solvent. The hot melt is extruded and pelletized. The monomer and solvent are recovered and recycled back in to the process.

Major applications for general-purpose polystyrene include packaging products, disposal medical ware, toys, tumblers, cutlery, tape reels, storm windows, and consumer electronics. Polystyrene foam application includes fast-food packaging, cushioning meterial for packaging, and building insulation.

4.3 Exposure Assessment

No information was found in the literature.

4.4 Toxic Effects

As reported in the fourth edition of *Patty's Industrial Hygiene and Toxicology*, long-term feeding tests with polystyrene are known to have been conducted, but few details are available in the published literature, Rats fed a diet of 4% polystyrene-based plastic for an unspecified length of time showed no abnormal effects (33). A 1976 French trade report (34) mentions two earlier 2-yr feeding tests with rats, neither of which revealed any toxic effects. The Huntingdon Reasearch Centre conducted a study at a dietary level of 5% polystyrene with a sample that contained 120 ppm monomer; a BASF study used a 10% dietary level.

4.4.1 Experimental Studies

See Section 4.4.

4.4.1.5 Carcinogenesis. Implant studies with polystyrene have produced tumors consistent with carcinogenesis caused by physical rather than chemical faactors. As summarized by IARC (3), "subcutaneous implantation of polystyrene discs, rods, spheres, or powder in rats induced local sarcomas, the incidence of which varied with the size and form of the implant." Additionally, a study with mice given interperitoneal injections of polystyrene latex showed no carcinogenic effect.

4.4.2 Human Experience

Occupational health risks associated with the production of polystyrene are associated with the monomer, styrene.

Polystyrene is considered physiologically inert. Solids or dusts may cause eye irritation or corneal injury due to mechanical action. Dusts also may cause irritation to the upper respiratory tract. Skin absorption and vapor inhalation are unlikely because of the physical state of polystyrene.

4.5 Standards, Regulations, or Guidelines of Exposure

Exposure guidelines have been established for the monomer, styrene, but not for polystyrene. Inhalation of fine dust, aerosol, or fumes should be minimized. Solvent and/or residual monomer released during processing should be controlled by exhaust ventilation or other appropriate means. Combustion of polymers or copolymers of styrene may yield unusually dense smoke. "Smoldering" combustion in environments containing limited amounts of oxygen may be particularly hazardous with acrylic polymers.

4.6 Studies on Environmental Impact

Disposal of water-soluble polymers may be accomplished by burial in sanitary landfill or by passage through water-treatment plants, depending on federal, state, or local regulations. Biodegradation may occur in oligomers (35), but the polymer is considered highly resistant to biologic decay (36–40). Canadian investigators (38, 39) found degraded polystyrene more resistant to microbial attack than degraded polyethylene or degraded polypropylene. Radioactive tracer studies with a photodegraded polystyrene–viny ketone copolymer indicated a slow rate of biodegradation by bacteria. This slow rate was compared to " the known slow rates of biological degradation of lignin and other natural products containing armoatic residues."

5.0 Polyacrylonitrile

5.0.1 CAS Number: [25014-41-9]

5.0.2 Synonyms: Acrylonitrile homopolymer; acrylonitrile polymer, Dralon T,2-propenenitrile, acrylonitrile resin, and bulana

5.0.3 Trade Names: NA

5.0.4 Molecular Weight

Polycrylonitrile is the basic material in acrylic and modacrylic fibers, the molecular weight of which, is reported to be 100,000–150,000. "Continuous filament" contains molecules of different molecular weight; one calculation (weight basis) has shown about 45% of the molecules over 100,000,33% from 50,000 to 100,000, and the remainder less than 50,000 (41).

5.1 Chemical and Physical Properties

The system of polyacrylonitrile includes crystalline, quasicrystalline, and amorphous phases. The characteristics of polyacrylonitrile include: hardness and rigidity; resistance to most oils, chemicals, and solvents, sunlight heat and microorganisms; slow burning, and charring, and low permeability to gases such as oxygen and carbon dioxide. These properties are due to the prevailing polar nature of polyacrylonitrile.

On heating polyacrylonitrile, it first becomes yellow, progressively red, and then black. Above 400°C the decomposition products include nitriles and HCN.

5.2 Production and Use

Acrylonitrile monomer is generally copolymerized with other ingredients by free-radical addition methods, primarily in suspension or solution with water or organic solvents. The exact composition of "acrylic fibers" varies widely. Materials identified in patent literature as used with acrylonitrile include among others, vinyl acetate, vinyl chloride, styrene, isobutylene, acrylic esters, and acrylamide (41). A typical acrylic fiber may contain 85–91% acrylonitrile, 7–8% of a comonomer such as methyl acrylate or methyl methacrylate, plus 2–5% other comonomers and additives such as dyeing acids, pigments, optical brighteners, stabilizers, and fire retardants (3). In modacrylic fibers, the nonacrylic moiety. consists of vinylidene chloride, vinyl chloride, vinyl bromide and other vinyl monomers. The modacrylic fiber produced in the largest quantity is composed of 37 percent acrylonitrile, 40 percent vinylidene chloride, 20 percent isopropyl acrylamide and 3 percent methyl acrylate.

Industrial applications of acrylic fibers are as filters for dry substances, in air pollution control, and in dye nets. Other pertinent pertinent uses are fiberfill batting, sandbags, and bagging. Industrial uses of modacrylic fibers include paint roller covers, filters for aqueous substances, and chemical-resistant twine. Both of these fiber types are used for flets, webbing, protective fabrics, carpets, upholstery, and draperies (3).

5.3 Exposure Assessment

No information was found in the literature.

5.4 Toxic Effects

A 1962 Russian report (42) indicates that rats survived acute oral doses of 2–3 g polyacrylonitrile (molecular weight = 25,000–50,000) with little effect. Daily doses of 0.25–0.5 g/kg to rats for 6 mo were reported to produce changes in the liver, kidneys, and thyroid that were reversible within the next 2 mo. Skin and eye test in rabbits produced no irritation.

Representative experimental tests, using methods described by Fleming (43) and Hood and Ivanova-Binova (44), have produced isolated reactions form a proposed surface modifier (45) or other additives, but reactions from commercial control materials were consistent with effects from occlusion.

According to a 1974 report (46), none of eight samples of 100% acrylic fabric had a free formaldehyde content >30 ppm. This study involved 112 American-made clothing samples of 39 types; 18 samples had a free formaldehyde >750 ppm (formaldehyde was not associated with the intrinsic nature of any fabric but with finishing agents and, very briefly, "manufacturing techniques").

The one available animal inhalation study on polyacrylonitrile "dust" (42) describes respiratory irritation and some pathological changes in rats exposed to concentrations of 1.3 mg/L of the same relatively low-molecular-weight polymer for oral and irritation tests. Workers (47) exposed to polyacrylonitrile fiber dust at mean concentrations of 0.42 or 1.04 mg/m^3 showed mild acute reductions of ventilatory capacity, but this response did not appear to be dose related.

5.4.1 Experimental Studies

See Section 5.4.

5.4.1.5 Carcinogenesis. Given the avialable experimental animal data on acrylonitrile, the relevance to the carcinogenicity of the polymer depends on the level of residual monomer and on how firmly fixed the acrylonitrile is in the polymer matrix.

5.4.2 Human Experience

General clinical experience in humans indicate that acrylic fibers are not associated with allergic contact dermatitis. There are reports of dermal toxicity resulting from prolonged contact with the monomer. Several clinical reports refer to acrylic and other synthetic fibers but lack data on exposure concentrations (46, 48), or related evidence of internal dose absorbed by the groups (49, 50).

5.4.2.2.5 Carcinogenesis. Arcylonitrile has been regarded as a suspected human carcinogen based on consistent production of tumors in experimental animals. But questions have been raised about human studies, with some investigators expressing the view that epidemiological studies have not provided evidence of an increased cancer risk in occupationally exposed workers (51).

The practical significance of the acrylonitrile carcinogenicity studies, and its classification as a cancer-causing agent, depends on the level of residual acrylonitrile monomer and its stability in the polyacrylonitrile matrix.

The amount of residual acrylonitrile in finished resins or products appears to be low (1 ppm in acrylic and modacrylic), and the risk of acrylonitrile migration or release under intended use of the polymer has not been a major concern.

5.5 Standards, Regulations, or Guidelines of Exposure

No information was found in the literature.

5.6 Studies on Environmental Impact

No information was found in the literature.

6.0 Styrene–Acrylonitrile (SAN)

6.0.1 CAS Number: *[9003-54-7]*

6.0.2 Synonyms: Acrilafil, acrylonitrile–styrene co-polymer, acrylonitrile–styrene polymer, acrylonitrile–styrene resin, ACS, AS 61CL, Bakelite RMD 4511, Ceviaan HL, Dialux, Estyrene AS, Kostil, Litac, Luran, Lustran, polystyrene–acrylonitrile, 2-Pro-Penenitrile Polymer with ethenylbenzene, Rexene 106, Sanrex, SN 20, Styren-Acrylonitrilepolymer, styrene–acrylonitrile copolymer, Terulan KP 2540, and Tyril

6.0.3 Trade Names: Bexan, Fostacryl, Kostil, Lacqsan, Lustran A, Norvodur W Restil, Sniasar, Tyril, and Vestroran

6.0.4 Molecular Weight

The molecular weight of SAN has been reported as 100,000 to 400,000 (3).

6.0.5 Molecular Formula: $[C_{11}H_{13}N]_n$

6.0.6 Molecular Structure:

6.1 Chemical and Physical Properties

The acrylonitrile content is a primary determinant of SAN's properties. Commercial SAN has an acrylonitrile content of 10–35%. It has better solvent resistance, higher impact strength, and a higher softening point than polystyrene. As the acrylonitrile content increases, these properties improve. For example, at 5.5% acrylonitrile content SAN properties are: tensile strength (MPa) is 42.27, impact strength (J/m notch) is 26.6, and heat distortion is 72°C. At 27.0% acrylonitrile tensile strength is 72.42, impact strength is 27.1, and heat distortion is 88°C.

6.1.1 General

See Table 90.1.

6.1.2 Odor and Warning Properties

No information was found in the literature.

6.2 Production and Use

Emulsion, suspensions and continuous bulk are the three processes by which SAN may be produced. Fabrication methods include injection molding, blow molding extrusion, and thermoforming.

SAN are used in appliances, packaging meterials, instrument covers, medical apparatuses, and custom molding products.

6.3 Exposure Assessment

No information was found in the literature.

6.4 Toxic Effects

No information was found in the literature.

6.4.2 Human Experience

No information was found in the literature.

6.5 Standards, Regulations, or Guidelines of Exposure

No information was found in the literature.

6.6 Studies on Enviromental Impact

No information was found in the literature.

7.0 Acrylonitrile–Butadiene–Styrene (ABS)

7.0.1 CAS Number: [9003-56-9]

7.0.2 Synonyms: ABS

7.0.3 Trade Names: NA

7.0.4 Molecular Weight

The molecular weight of ABS has been given as 60,000–200,000 (34).

7.1 Chemical and Physical Properties

ABS is resistant to attack by mineral oils, waxes, and related commercial material because of the polar charcater of the nitrile group from acrylonitrile component. This property reduces interaction of the polymer with hydrocarbon solvent. ABS will undergo stress cracking when brought into contact with certain chemical agents under stress. Grades of ABS vary widely in flow and tensile strength, allowing flexibility in part design and fabrication methods. The electrical properties are constant over a wide range of frequencies and are unaffected by temperature or humidity. A broad range of flame-retardant grades are available. The increase susceptibility of ABS to oxidation, halogenation, and sulfonation by reactive chemicals is attributable to the polybutadiene component of the polymer.

7.1.1 General

ABS polymers are elastomers and thermoplastic (qv) that exhibit toughness and stability, a general property that attracts users to ABS products. ABS is structured effectively to dissipate the energy of an impact blow.

7.1.2 Odor and Warning Properties

No information was found in the literature.

7.2 Production and Use

The ABS production process is the free radical polymerization of styrene and acrylonitrile in the presence of polybutadiene or butadiene copolymers. Commerical production utilizes bulk, emulsion, or suspension processes. The bulk process is a single-phase reaction in which all polymerization is completed in a monomer–polymer medium starting with a linear rubber dissolved in styrene and acrylonitrile. Both the emulsion and suspension reactions involve two steps.

In the emulsion process the rubber substrate is produced, followed by the grafting of styrene and acrylonitrile. The first step of the suspension reaction involves the making of a bulk polymerized prepolymer using styrene and acrylonitrile monomers. The second step is polymerization in a suspension reaction.

In the United States the acrylonitrile–butadiene–styrene market includes pipe, automotive trim application (structural parts, grilles, dashboards), and appliance parts. Among large appliances, the largest applications are in refrigerator liners, vegetable crisper pans, door shelves, and so on. Smaller markets include business machines, telephones, electrical and electronic equipment, luggage, and furniture (52). ABS is used infrequently in food packaging.

7.3 Exposure Assessment

No information was found in the literature.

7.4 Toxic Effects

ABS is considered physiologically inert with no toxic effects reported in experimental animals or in human studies. The monomers have biological properties that appear to be diminished in the finished product of ABS. Particulates or dust of the material may increase the risk of mechanical irritation of eyes and of upper respiratory tract. Skin absorption and vapor inhalation are unlikely because of the physical state of the polymer.

7.4.1 Experimental Studies

No information was found in the literature.

7.4.2 Human Experience

No information was found in the literature.

7.5 Standards, Regulations, or Guidelines of Exposure

No information was found in the literature.

7.6 Studies on Environmental Impact

No information was found in the literature.

8.0 Polymethyl Methacrylate

8.0.1 CAS Number: [9011-14-7]

8.0.2 Synonyms: Acrylite, Acrypet, Alutor M 70, CMW Bone Cement, Crinothene, Degalan S 85, Delpet 50M, Diakon, Dispasol M, DV 400, Elvacite, Kallocryl K, Kallodent Clear, Korad, LPT, Lucite, Metaplex NO, methacrylic acid methyl ester polymers, methyl methacrylate homopolymer, methyl methacrylate polymer, methyl methacrylate resin, 2-methyl-2-propenoic acid methyl ester homopolymer, Organic Glass E 2, Osteobond Surgical Bone Cement, Palacos, Paraglas, Paraplex P 543, Perspex, Plexi-glas, Plexigum

M 920, PMMA, Pontalite, Repairsin, Resarit 4000, Rhoplex B 85, Romacryl, Shinkolite, SOL, Stellon Pink, Sumiplex LG, Superacryle AE, Surgical Simplex, Tensol 7, and Vedril

8.0.3 Trade Names:
Acryl-ace, Altuglas, and, in sheet form, Asterite, Implex, Lucryl, Oroglas, Perspex, and Plexiglas

8.0.4 Molecular Weight

The molecular weight would appear to be in the range of 50,000 to 60,000 for molding compounds and about 1,000,000 for acrylic sheet (1, 5).

8.0.5 Molecular Formula: $[C_5H_8O_2]_n$

8.0.6 Molecular Structure:

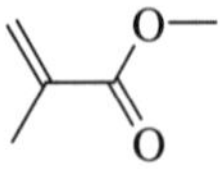

8.1 Chemical and Physical Properties

Polymethyl methacrylate is an important thermoplastic material. Polymerization of methyl methacrylate produces two forms, the isotatic [CAS # *25188-98-1*] and syndiotactic [CAS # *25188-97-0*]. The isotatic polymer has a T_g value of 45°C and a T_m value of 115°C. Syndiotactic polymer has a T_g value of 115°C.

Polymethyl methacrylate is typically tougher than polystrene but less tough than cellulose acetate or acrylonitrile–butadiene–styrene polymers (5). Polymethyl methacrylate is particularly noted for light transmission, plus shatter resistance and light weight, as compared to glass. Polymethyl methacrylate cracks on impact, but fragments are normally less sharp and jagged than those of glass; correspondingly, polymethyl methacrylate is inferior to glass in scratch resistance. Light transmission through unblemished sheet can exceed that of plate glass 92–93% versus 89%.

8.1.1 General

No information was found in the literature.

8.1.2 Odor and Warning Properties

When heated to decomposition, it emits acrid smoke and irritating fumes. This property may be adequate to prevent excessive repeated exposure.

8.2 Production and Use

Methyl methacrylate monomer can polymerize readily during storage unless an inhibitor, such as 0.1% hydroquinone, is added; the inhibitor is then removed prior to commercial polymerization. To avoid the side development of methacrylate peroxides, polymerization is usually done in the absence of oxygen (either by bulk polymerization in a full cell or blanketing with inert gas). Bulk polymerization is extensively used to manufacture sheets and, to a lesser extent, rods and tubes. Suspension polymerization, which avoids the potential serious exotherm problem found with bulk polymerization, gives a lower-molecular-weight polymer suitable for molding powder. Additives may include buffering agents, chain transfer agents, lubricants, and emulsifiers.

Polymethyl methacrylate and its copolymers are widely used in molding resins, exterior and interior paints, enamels, polishes, adhesives, and miscellaneous coatings. Polymethyl methacrylate and other acrylic resins have been used for specified food contact items (2), dentures, and medical applications such as hard contact lenses, intraocular lenses, or bone cements.

8.3 Exposure Assessment

No information was found in the literature.

8.4 Toxic Effects

The toxic effects are attributable to the monomer; the polymer is characterized as inert. Mice inhaling decomposition products from PMMA showed effects somewhat sooner than mice comparably exposed to decomposition products from a series of other polymers. Elevated temperatures produce a classical unzipping of the polymer and can yield 99% monomer at 425°C; at higher temperatures monomer content decreases and aldehydes, particularly formaldehyde, may be evolved. Some manufacturing processes such as sawing or molding the polymer may, under certain circumstances, release fine dust or monomer-containing vapors (53); comparatively small amounts of methanol may also be released (54). The heating, lathing, molding, and grinding processes associated with the processing of PMMA lenses are capable of degrading the polymer sufficiently to cause a measurable, although not persistent, increase in monomer content of the polymer (55).

8.4.1 Experimental Studies

The literature includes some data on carcinogenicity studies, cited in Section 8.4.1.5.

8.4.1.5 Carcinogenesis. Experimental data on animals plus various types of data on human exposures all indicate no evidence of metastatic cancer but reveal that newly polymerized or implanted resin may not be entirely inert. IARC (3) called for experimental and epidemiologic studies on the basis of animal studies that essentially showed sarcomas at the site of implantation and the evident human exposure arising from widespread industrial or medical use. Available data were considered inadequate to evaluate either the monomer or the polymer.

Stinson (56) implanted polymethyl methacrylate particles > 76 mm in the muscle and knee joint of guinea pigs. These particles produced "an extremely mild reaction," without evidence of erosion over 3-yr period. The author contrasts these results with his earlier results from disks in guinea pigs and also rats when tumors did occur. Barvic (57) conducted an extensive study on 450 rats with both "small flat pieces of pure [poly]methyl-methacrylate and neutral specimens of [poly]hydrocolloidalmethacrylate." Perforated specimens caused no tumors and all observed neoplasms were induced by specimens prepared from a solid sheet. "Small specimens" rarely caused tumors, whereas "larger implants" produced a 12–30% incidence of tumors with a shorter latency.

Tomatis (58) reports that mice given subcutaneous implants of ^{14}C- and ^{3}H-labeled PMMA films showed increased radioactivity in the urine until the seventh to eighth week after implantation and sharply decreased radioactivity thereafter. Very low levels of radioactivity were detected in saline in which labeled implants were immersed.

8.4.1.7 Other: Neurological, Pulmonary, Skin Sensitization. No information was found in the literature.

8.4.2 Human Experience

Although the monomer is a recognized contact allergen, heat-polymerized PMMA appears essentially inert in this respect. The level of residual monomer in cold-cured dentures may be as much as 50% (59) but can be reduced by heat curing to a level not detected by a person strongly sensitive to 0.1% methyl methacrylate (60). The monomer is more commonly "a cause of hand dermatitis in dentists and dental mechanics than of denture sore mouth" (61). Dermal sensitivity to additives such as accelerators may also occur. Residual methyl methacrylate appears to be a weaker allergen than residual acrylates.

8.4.2.2 Clinical Cases. One case report described a chondrosarcoma intimately associated with the fibrous capsule surrounding PMMA spheres used as plombage (62). The spheres had been implanted for compression of a tuberculous cavity 18 yr prior to terminal hospital admission when the tumor was detected.

8.4.2.3 Epidemiology Studies. Polymethyl methacrylate has been widely utilized as a cement in bone surgery, particularly for hip replacements in elderly patients and in malignant neoplastic fractures. Acute hypotension, cardiovascular collapse, and other misadventures have been associated with the use of acrylic cement in hip replacements, but the reported incidence of these events varies widely. Charnley's survey of some 3700 hip replacements (63) lists four patients with cardiac arrest, but Lipecz et al. (64) describe hypotension in 58% of 300 patients. Factors considered significant include operative technique, age and medical fitness of the patient, and release of the monomer from the implantation directly into the circulation. The literature discussing these aspects is intensive (3, 63–69).

8.5 Standards, Regulations, or Guidelines of Exposure

No information was found in the literature.

8.6 Studies on Environmental Impact

No information was found in the literature.

9.0 Polyacrylamide

9.0.1 CAS Number: [9063-05-8]

Polyacrylamide is available in many modifications. Polymers with < 4.0% hydrolysis are usually considered nonionic (70).

POLYVINYL ACETATE, ALCOHOL, AND DERIVATIVES

9.0.2 Synonyms: 2 propenamide, homopolymer, J-100; polyacrylamide; polyacrylamides; acrylamide resin; poly(acrylamide), granular (nonionic), average M.W. 5 to 6.000.000

9.0.3 Trade Names: NA

9.0.4 Molecular Weight

The moclecular weights for commercial polyacrylamides are typically very high. They range from at least 1,000,000 to 3,000,000.

9.0.5 Molecular Formula: $[C_3H_5NO]_n$

9.0.6 Molecular Structure:

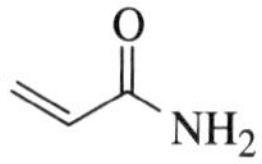

$$\underset{NH_2}{\overset{O}{\|}}$$

9.1 Chemical and Physical Properties

Polyacrylamide is soluble in water over a wide range of concentrations and temperatures. Hydrolyzed, it interacts with many metal cations in solution. Solvents for polyacrylamide include hydrazine, ethylene glycol, and morpholine. Many other organic liquids are nonsolvents.

Polyacrylamide resembles a gel in appearance and in physical properties when in aqueous solutions. Polyacrylamide is a useful precursor to a variety of derivatives because it undergoes reactions typical of low-molecular-weight amides.

9.1.1 General

Some physical properties for solid acrylamide are:

Chain structure: Predominantly Heterotactic
Density 23°C (g/cm³): 1.302
Softening temperature (°C): 210

Polyacrylamide retains water during ordinary drying and, when dry, rapidly absorbs moisture from the environment. Therefore, reported properties of the solid must be viewed with caution.

9.1.2 Odor and Warning Properties

No information was found in the literature.

9.2 Production and Use

The production and marketing literature on the industrial use of acrylamide polymers has been growing significantly in recent years. Development of the catalytic process for production of the monomer, acrylamide, and the introduction of easily handled emulsion polymers are among the significant advances in the acrylamide industry.

The earliest, and still widely used, method of producing polyacrylamide employs a dilute aqueous solution (9–20%) and a conventional 9z0, peroxy or redox initiator. Batch reactor or continuous stirring reactor may be used. A reported method for preparing low- to moderate-molecular-weight polymer is to polymerize acrylamide dissolved in a mixture of water and a water-miscible liquid. Photopolymerization (q_v) processes have also been recommended. Major applications of polyacrylamides are in water treatment, paper production, oil recovery, and mining.

9.3 Exposure Assessment

No information was found in the literature.

9.4 Toxic Effects

Various mammalian toxicity tests with polyacrylamide polymers all indicate a very low order of toxicity.

9.4.1 Experimental Studies

The single-dose LD_{50} for rats by oral admnistration, and for rabbits by skin penetration, has been described as > 8.2 g/kg; 20% solutions were reported as nonirritating to the rabbit eye (71). Only a very mild dermal inflammatory reaction developed in rabbits after 24-h contact with pure moistened product. Cationic polyacrylamides may be somewhat more irritating (70). A brief Russian report states that "prolonged exposure" (animal model unspecified) to technical-grade polyacrylamide at doses up to 70 mg/kg was not toxic, and histological examination revealed no toxic effects (72).

McCollister and his colleagues (73) found that rats survived the maximum feasible oral dose, 4 g/kg, of a nonionic and anionic polyacrylamide, and conducted three separate 2-yr feeding studies with rats. Two were done with nonionic product containing 0.02 or 0.08% residual acrylamide, respectively; in the third, anionic product containing 0.07% residual monomer was fed to both rats and dogs for 2 yr. Dogs were also fed nonionic product (0.01% monomer) for 1 yr. "The laboratory animals tolerated 5–10% in their total diet without effects other than those believed to be attributable indirectly to the large, hydrophilic, non-nutritive bulkiness of the materials. The unequivocal 'no ill-effect' levels were 1% in the diet of rats and 5–6% in the diet of dogs." Additional ^{14}C radioactive tracer studies with both types of product indicated that "negligible amounts of polymer, if any, pass through the walls of the intestinal tract of the rat."

Christofano et al. (74) administered "strongly cationic, high molecular weight, water-soluble polyacrylamide resins" to rats and dogs. Tests included maximum single-dose, 90-d feeding, and 2-yr feeding with three-generation reproduction studies. Animals fed $> 1\%$ dietary levels showed, in some cases, changes such as depressed weight gain and altered liver weight, whereas levels of $< 0.2\%$ produced no effect.

9.4.2 Human Experience

Environmental studies have been conducted in polyacrylamide production plants of the Dow Chemical Company over a 5-yr period (73). Airborne concentrations averaged approximately 1 mg/m³ with almost all particles greater than 50 mm in size. "It was postulated that about 5 mg/day might be ingested by human subjects working in such areas. The physicians found no indication of any more pathology than one would expect to find in a similar group of men in the general population."

Polyacrylamides and polyacrylic acids may decompose at temperatures greater than 200–300°C. Thermal degradation of polyacrylamide in air at 700°C has yielded relatively small amounts of HCN and ammonia.

9.4.2.1 General Information. There is substantial literature on the health risk associatd with exposure to acrylamide monomer (75–77).

9.5 Standards, Regulations, or Guidelines of Exposure

Standards, regulations and guidelines of exposure have been established for the monomer, acrylamide, not polyacrylamide.

9.6 Studies on Environmental Impact

Studies with the nonionic and anionic polyacrylamides discussed in Section 9.4.1 (73) also included several species of fish, such as Lake Emerald shiners, yellow perch, fathead minnows, rainbow trout, and blue gills. The anionic products at 2500 ppm caused 100% mortality: "The solution was so viscous that the fish had difficulty in swimming. Fish were maintained in water containing 1000 ppm of these resins for 5 days and 100 ppm for 90 days without apparent adverse effect."

Two Japanese reports (78, 79) indicate that some types of polyacrylamide or their derivatives can be toxic to fish. The 48-h median tolerance limit (TL_m) values in one bioassay (78) appeared to vary from > 80 to < 1 ppm.

Polyelectrolytes may act to restrict microbial growth by chelating essential metals or they may flocculate microbial cells. Certain polyacrylamides can actually serve as a nitrogen source in a chemically defined medium and stimulate bacterial growth (*Pseudomonas aeruginosa*), but they do not appear to be biodegradable in the usual sense of the word (80). The breakage of polymer chains by ozone has been reported to be strongly accelerated by ultraviolet radiation under acidic and neutral conditions (81). Formaldehyde was produced in the ozonization process.

10.0 Polyacrylic Acid

10.0.1 CAS Number: [9003-01-4]

10.0.2 Synonyms: Acrylic acid polymer; 2-propenoic acid, homopolymer; propenoic acid polymer; propenoic acid, polymers, homopolymer; carbopol 940; Carbomer; acrylic polymer resins; propenoic acid, homopolymer; poly(acrylic acid), sec. stand

10.0.3 Trade Names: Acrysol, Carbopol, Carboset, Plexileim, Syncol, Texigel, Texipol, Versicol

10.0.4 Molecular Weight: Aver. M.W. 1.080.000, aver. M.N. 135.000

10.0.5 Molecular Formula: [C₃H₄O₂]ₙ

10.0.6 Molecular Structure:

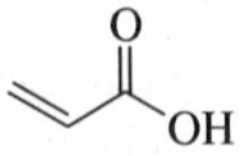

10.1 Chemical and Physical Properties

The polymer is readily soluble in waters and in alcohols. Polyacrylic acid, also noted for its flocculent action, is a weaker acid than its monomeric counterpart. The polyacid acts as an excellent buffer in the range of pH 4–6.4. Precise titration in water alone is diffcult, but titration in 0.01 to 1 N solutions of neutral salts gives precise end points (82). Copolymers with only a few cross-links are gels; more cross-linking results in a brittle polymer (82).

10.2 Production and Use

Polyacrylic acid may be polymerized in aqueous solution at concentrations of 25% or less (82). Polymerization can be accomplished with a peroxydisulfate initiator at 90–100°C or a redox initiator activated by trace of ferric iron at 0–10°C. Polymerization of undiluted monomer is hazardous. Polymers may contain at least 5% moisture to facilitate re-solution (70). Polymerization can also be accomplished in nonaqueous solvents that dissolve the monomer but not the polymer. Application includes textile finishes and paint as well as being useful as suspension and flocculating agent.

10.3 Exposure Assessment

No information was found in the literature.

10.4 Toxic Effects

No information was found in the literature.

10.4.2 Human Experience

No information was found in the literature.

10.5 Standards, Regulations, or Guidelines of Exposure

No information was found in the literature.

10.6 Studies on Environmental Impact

No information was found in the literature.

11.0 Poly(2-hydroxethyl methacrylate) Hydrogel

11.0.1 CAS Number: [25249-16-5]

11.0.2 Synonyms: Poly-HEMA

11.0.3 Trade Names: Hydron

11.1 Chemical and Physical Properties

The polymer is characterized by hydrophilicity and insolubility in water. These properties are due to the presence of groups such –OH, –COOHH, –CONH–, –SO$_2$H. The polymer has a T_g value of 55 or 86°C (conflicting data).

11.1.1 General

No information was found in the literature.

11.1.2 Odor and Warning Properties

No information was found in the literature.

11.2 Production and Use

Poly-HEMA is produced by free radical polymerization of 2-hydroxyethylmethacrylate. It has been used extensively for hydrophilic contact lenses ("soft lenses"). Applications such as blood-compatible prosthetic materials are limited by thrombogenic capacity recently associated with radiation-grafted hydrogels. Another use is in membranes for the controlled drug release because of their compatibility.

11.3 Exposure Assessment

No information was found in the literature.

11.4 Toxic Effects

No information was found in the literature.

11.4.1 Experimental Studies

No information was found in the literature.

11.4.2 Human Experience

No information was found in the literature.

11.5 Standards, Regulations, or Guidelines of Exposure

No information was found in the literature.

11.6 Studies on Environmental Impact

No information was found in the literature.

12.0 Methyl Methacrylate

12.0.1 CAS Number: [80-62-6]

12.0.2 Synonyms: Methylacrylic acid methyl ester, methyl-2-methyl-2-propenoate, MME, MMA, 2-methylacrylic acid methyl ester, methyl methyacrylate, methyl α-methylacrylate, diakon, and methyl 2-methylpropenoate

12.0.3 Trade Names: NA

12.0.4 Molecular Weight: 100.12

12.0.5 Molecular Formula: $C_5H_8O_2$

12.0.6 Molecular Structure:

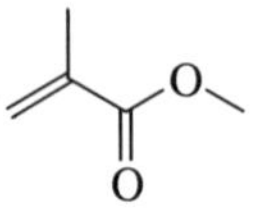

12.1 Chemical and Physical Properties

Key properties of MMA polymer are resistance to heat, light, and weathering. It is unaffected by inorganic acids, alkalis, aliphatic hydrocarbon. The resin varies from hard brittle solid, to elastomeric structure, depending on the method.

12.2 Production and Use

The resin is made by free-radical polymerization of the monomer, polymethylmethacrylate, initiated by peroxide or azo catalysts. The polymer is used in dental and orthodontic devices, glass substitutes, furniture components, and illuminated signs.

12.3 Exposure Assessment

No information was found in the literature.

12.4 Toxic Effects

The toxic effects are due to the monomer MMA; the polymer appears to be biologically inert. The inhalation of MMA resin powder may increase the risk of interstitial pneumonitis. In one case report the chest x-ray cleared after exposure ceased. A common occupational health concern is exposure of MMA through skin absorption. This route of exposure has been most probable in health care facilities where dermal contact with medical devices containing residual amounts of monomers may occur. Resin containing acrylic monomers tends to present a more serious dermatitis than those containing methylacrylic monomers. Reports indicate that patch testing to methyl methacrylate has generally been negative in dermatitis associated with light-sensitive acrylate. In one case where acrylonitrile was used with methyl methacrylate in a plastic finger splint, acrylonitrile was identified as the sensitizer (83–85).

12.4.1 Experimental Studies

No information was found in the literature.

12.4.2 Human Experience

See Section 12.4.

12.5 Standards, Regulations, or Guidelines of Exposure

The ACGIH TLV for this chemical is 50 ppm and marked as a skin sensitizer (86).

12.6 Studies on Environment Impact

No information was found in the literature.

13.0 Ethyl Acrylate Polymer

13.0.1 CAS Number: [900-32-1]

The limited information in the literature focuses primarily on properties of the monomer, ethyl acrylate, described as a colorless liquid widely used in the production of polymers and copolymers. These polymers are for manufacturing textiles, latex paints, paper coatings, and specialty plastics. The monomer is an irritant of the skin, eyes, respiratory tract, and mucus membrane, and it is carcinogenic in experimental animals. The 2000 ACGIH threshold limit value (TLV) is 5 ppm, with a short-term exposure limit of 15 ppm (86).

BIBLIOGRAPHY

1. W. J. Roff and J. R. Scott, with J. Pacitti, *Handbook of Common Polymers*, CRC Press, Cleveland, OH, 1971.
2. U.S. Code of Federal Regulations, 21 C.F.R. **296**, 170–199, 21 C.F.R. **296**, 500–599, Superintendent of Documents, Washington, DC, 1981 (revised annually. Guide available from Food Chemical News).
3. International Agency for Research on Cancer (IARC), *Monographs on the Evaluation of the Carcinogenic Risk of Chemicals to Humans; Somes Monomers, Plastics and Synthetic Elastomers and Acrolein*, Vol. 19, IARC, Lyon, France, 1979.
4. M. P. Stevens, *Polymer Chemistry: An Introduction*, Addison-Wesley, Reading, MA, 1975.
5. J. A. Brydson, *Plastics Meterials,* 3rd ed., Whitefriars Press, London 1975.
6. B. I. Shcherbak et al., Toxicological characteristics of some poly(vinyl acetate) dispersions (PVAD) *Uch. Zap.—Mosk. Nauchno-Issled. Inst. Gig.* **22**, 74–80 (1975); *Chem. Abstr* **86**, 66421v (also 66422w, 66423x).
7. M. H. Weeks and C. R. Pope, *Toxicological Evaluation of Polyvinyl Acetate (PVA) Emulsion Dust Control Material*, AD 784603, U.S. Army, Environmental Hygiene Agency, Aberdeen, MD, 1973–1974.
8. D. B. Hood, DuPont Company, Haskell Laboratory, Wilmington, DE, 1962 (unpublished data).

9. V. A. Goreva, Study of the toxic effect of poly(vinyl acetate) emulsion on fish. *Tr. Sarat. Otd. GosNIORKh* **13**, 89–92 (1975); *Chem. Abstr.* **87**, 96940b.
10. G. V. Gurova, Effect of poly(vinyl acetate) emulsion on phytophilic fish in early ontogenesis. *Tr. Sarat. Otd. GosNIORKh* **13**, 92–95 (1975); *Chem. Abstr.* **87**, 96941c.
11. A. E. Shapshal and N. I. Berlyakova, Effect of poly(vinyl acetate) emulsion on chironomus dorsalis meig. and *Daphnia Magna* Straus. *Tr. Sarat. Otd Gos.NIORKh* **13**, 85–89 (1975); *Chem. abstr.*, **87**, 96939h.
12. G. J. l. Griffin and H. Mivetchi, Biodegradation of ethylene/vinyl acetate co-polymers. *Proc. Int. Biodegradation Symp., 3rd, 1975* (1976), pp. 807–817.
13. W. C. Hueper, Organic lesions produced by polyvinyl alcohol in rats and rabbits. *Arch. Pathol.* **28**, 510–531 (1939).
14. G. N. Zaeva et al., Toxicological charcteristics of poly(vinyl alcohol), polyethylene, and polypropylene. *Toksikol. Nov. Prom. Khim. Veshchestv* **5**, 136–149 (1963); *Chem. Abstr.* **61**, 6250h.
15. Y. Yamatani and S. Ishikawa, Polyvinyl alcohol as a water-soluble marker, Part 1. Absorption of excretion polyvinyl alcohol from the gastro-intestinal tract of adult rat. *Agric. Biol. Chem.* **32**(4), 474–478 (1968).
16. W. M. Grant, Polyvinyl alcohol. *Toxicology of the Eye*, 2nd ed., Thomas, Springfield, IL, 1974, pp. 849–851.
17. N. Krishna and B. Mitchell, Polyviny alcohol as an ophthalmic vehicle. *Am. J. Ophthalmol.* **59**, 860–864 (1965).
18. D. O. Shah and M. J. Sibley, Treatment of dry eye. *U.S. Pat.* 4,131,651 (Cl. 424–78; A61K31/74). (December 26, 1978) (Appl. 844,555, October 25, 1977); *Chem. Abstr.* **90**, 110015z.
19. V. I. Trautmann, Poly(vinyl alcohol) as vehicle for eye drops. *Dtsch. Gesundheitswes.* **22**, 317–320 (1967).
20. C. E. Hall and O. Hall, Polyvinyl alcohol nephrosis: Relationship of degree of polymerization to pathophysiologic effects. *Proc. Soc. Exp. Biol. Med.* **112**, 86–91 (1963).
21. C. E. Hall and O. Hall, Polyvinyl alcohol: Relationship of physicochemical properties to hypertension and other pathophysiologic sequelae. *Lab. Invest.* **12**, 721–736 (1963).
22. M. E. Sipple, DuPont Company, Haskell Laboratory, Wilmington, DE, 1977 (unpublished data).
23. B. H. Sleight, Bionomics, Inc. Wareham, MA, 1971 (unpublished data).
24. Q. D. Wheatley and F. C. Baines, Biodegradation of polyvinyl alcohol in wastewater. *Text. Chem. Color.* **8**, 23–28 (1976).
25. W. H. Hahn, E. L. Barnhart, and R. B. Meighan, The biodegradability of synthetic size material used in textile processing. *Proc. Ind. Waste Conf.* **30**, 530–539 (1977).
26. J. P. Casey and D. G. Manly, Polyvinyl alcohol biodegradation by oxygen activated sludge. *Proc. Int. Biodegradation Symp., 3rd, 1975* (1976), pp. 33, 819.
27. J. J. Porter and E. H. Snider, Long-term biodegradability of textile chemicals. *J. Water Pollut Control Fed.*, **48**, 2198–2210 (1976).
28. J. Su-uki, K. Hukushima, and S. Su-uki, Effect of ozone treatment upon biodegradability of water-soluble polymers. *Environ. Sci. Technol.* **12**, 1180–1183 (1978).
29. D. B. Bivens and R. B. Rector, Acrylics. In *Modern Plastics Encyclopedia*, Vol. 55, No. 10A, McGraw-Hill, New York, 1978–1979, pp. 7–9.
30. Dow Chemical Company, Styron Polystyrene Resins, Midland, MI, 1977.

31. E. R. Moore, ed., Styrene polymers. In *Encyclopedia of Polymer Science and Engineering*, 2nd ed., Vol. 16, Wiley, New York, 1989.
32. P. C. Hiemenz, *Polymer Chemistry*, Dekker, New York, 1984.
33. I. Phillips and G. C. Marks, as cited in R. Lefaux, ed., *Practical Toxicology of Plastics*, CRC Press, Cleveland, OH, 1968.
34. Les emballages en polystyrene et la santé-du Consommateur. *Rev. Gen. Caoutch. Plast.* **562** (1976).
35. M. Sielicki, Microbial degradation of styrene and styrene polymers. *Diss. Abstr. Int. B* **38** 4076-B (1978).
36. J. E. Potts et al., The biodegradability of synthetic polymers. *Polym. Sci. Techno.* **3**, 61–79 (1973).
37. J. E. Potts, R. A. Clendinning, and W. B. Ackart, The effect of chemical structure on the biodegradability of plastics. PB 213488, EPA-R2-72-0-46, *Degradability of Packaging Plastics*, Union Carbide Corporation, Bound Brook, NJ, 1972 (sponsored by U.S. Environmental Protection Agency, Washington, DC).
38. P. H. Jones et al., Biodegradabilty of photodegraded polymers. *Environ. Sci. Technol.* **8**, 919–923 (1974).
39. J. E. Guillet, T. W. Regulski, and T. B. McAneney, Biodegradability of photodegraded polymers. *Environ. Sci. Technol.* **8**, 923–925 (1974).
40. D. L. Kaplan, R. Hartenstein, and J. Sutter, Biodegradation of polystyrene, poly(methylmethacrylate), and phenol formaldehyde. *Appl. Environ. Microbiol.* **38**, 551–553 (1979).
41. R. W. Moncrieff, *Man Made Fibres*, 6th ed., Wiely, New York, 1975.
42. G. V. Lomonova, Toxic properties of acrylonitrile polymer. *Gig. Tr. Prof. Zabol.* **6**(6), 54–57 (1962).
43. A. J. Fleming, The provocative test for assaying the dermatitis hazards of dyes and finishes used on nylon. *J. Invest. Dermatol.* **10**, 281–291 (1948).
44. D. B. Hood and A. Ivanova-Binova, The skin as a portal of entry and the effects of chemicals on the skin, mucous membranes and eye. In *Principles and Methods for Evaluation the Toxicity of Chemicals*, Part II, Environ. Health Criteria Ser. No. 6, World Health Organization, Geneva.
45. L. A. Wells, DuPont Company, Haskell Laboratory, Wilmington, DE, 1968 (unpublished data).
46. W. F. Schorr, E. Keran, and E. Poltka, Formaldehyde allergy. *Arch. Dermatol.* **110**, 73–76 (1974).
47. F. Valic and E. Zuskin, Respiratory-function changes in textile workers exposed to synthetic fibers. *Arch. Environ. Health*, **32**, 283–287 (1977).
48. J. C. Pimentel, R. Avila, and A. G. Lourenco, Respiratory disease caused by synthetic fibres: A new occupational desease. *Thorax* **30**, 204–219 (1975).
49. A. Muittari and T. Veneskoski, Natural and synthetic fibers as causes of asthma and rhinitis. *Ann. Allergy* **41**, 49–50 (1978).
50. A. Bouhuys, Fibers and fibrosis. *Ann. Intern. Med.* **83**, 898–899 (1975).
51. C. E. Ward. T. B. Starr, Comparison of cancer risks projected from animal boiassay to epidemiological studies of acrylonitrile-exposed workers *Regul. Toxicol. Pharmacol.* **18**, 24–232 (1993).
52. U.S. Consumer Product Safety Commission, *Assessment of Arcylonitrile Contained in Consumer Products*, Final Report, Contract No. CPSC-C-77-0009, U.S. Government Printing Office, Washington, DC, 1978.
53. K. E. Malten and R. L. Zielhuis, *Industrial Toxicology and Dermatology in the Production and Processing of Plastics*, Elsevier Monographs on Toxic Agents, Elsevier, New York, 1964.

54. W. H. Martin, DuPont Company, Polymer Products Dept., Wilmington, DE, 1980 (unpublished data).
55. D. G. Anderson and J. T. Vandeberg, Coatings. *Anal. Chem.* **51**, 80R–90R (1979).
56. N. E. Stinson, Tissue reaction induced in guinea-pigs by particulate polymethylmethacrylate, polythene and nylon of the same size range. *Br. J. Exp. Pathol.* **46**, 135–146 (1965).
57. M. Barvic, Reaction of the body to the presence of acrylic allografts and possible carcinogenic effects of such grafts, *Acta Univ. Carol. Med.* **8**, 707–753 (1962).
58. L. Tomatis, Subcutaneous carcinogenesis by 14C- and 3H-labeled poly(methyl methacrylate) films. *Tumori*, **52** 165–172 (1966).
59. M. A. Guill and R. B. Odom. Hearing aid dermatitis, *Arch. Dermatol.* **114**, 1050–1051 (1978).
60. A. I. Fernstrom and G. Oquist, Location of the allergenic monomer in warm-polymerized acrylic dentures. *Swed. Dent. J.* **4** (part 11), 253–260; see also part 1, pp. 241–252.
61. R. J. G. Rycroft, Contact dermatitis from acrylic compounds. *Br. J. Dermatol.* **96**, 685–687 (1977).
62. J. R. Thompson and S. D. Entin, primary extraskeletal chondrosarcoma. *Cancer (Philadelphia)* **23**, 936–939 (1969).
63. Acrylic cement and the cardiovascular system. *Lancet* **2**, 1002–1004 (1974).
64. J. Lipecz et al., Kreislaufkomplikationen bei Alloarthroplastiken des Huftgelenkes. *Anaesthesist* **23**, 382–388 (1974).
65. E. O. Dillingham et al., Biological evaluation of polymers 1. Poly(methyl methacrylate). *J. Biomed. Mater. Res.* **9**, 569–596 (1975).
66. R. H. Ellis, Hypotension and methylmethacrylate cement. *Br. Med. J.* **1**, 236 (1973).
67. F. T. Schuh et al., Circulatory changes following implantation of methyl-methacrylate bone cement. *Anesthesiology* **39**, 455–457 (1973).
68. G. Schlag et al., Does methylmethacrylate Induce cardiovascular complications during alloarthroplastic surgery of the hip joint? *Anaesthesist* **25**, 60–67 (1976); (authors' English abstract).
69. J. Kraft, Polymethylmethacrylate — A review. *J. Foot Surg.* **16**, 66–68 (1977).
70. J. D. Morris and R. J. Penzenstadler, Acrylamide polymers. In *Encyclopedia of Chemical Technology*, 3rd ed., Vol. 1, Wiely, New York, 1978, pp. 312–330.
71. American Cyanamid Company, *Cyanamer P26 Acrylamide Copolymer; Cyanamer P250 Polyacrylamide*, American Cyanamid Brochure, American Cyanamid Company, Process Chemicals Department, Wayne, NJ, 1957.
72. N. A. Rakhmanina, Toxicological characteristics of technical grade polyacryl-amide. *Nauchn. Tr. — Akad. Kommun. Khoz.* **22**, 56–59 (1963); *Chem. Abstr.* **61**, 15249d.
73. D. D. McCollister et al., Toxicologic investigations of polyacrylamides. *Toxicol. Appl. Pharmacol.* **7**, 639–651 (1965).
74. E. E. Christofano et al., The toxicology of modified polyacrylamide resin. *Toxicol. Appl. Pharmacol.* **14**, 616 (1969).
75. A. S. Kuperman, effects of acrylamide on the central nervous system of the cat. *J. Pharmacol. Exp. Ther.* **123**, 180–182 (1958).
76. D. D. McCollister et al., Toxicology of acrylamide. *Toxicol. Appl. Pharmacol.* **6**, 172–181 (1964).
77. M. S. Miller and P. S. Spencer, The mechanism of acrylamide axonopathy. *Annu. rev. Pharmacol. Toxicol.* **25**, 643–666 (1985).

78. Y. Matsuo, Bioassay of the acute toxicity of condensing agents of high molecular weight. *Osaka-shi Suidokyoku Komubu Suishitsu Chosa Hokoku narabini Shiken Seiseki* **24**, 7–11 (1972); *Chem. Abstr.* **85**, 138141h.
79. N. Miyanaga et al., Toxicity of polymer flocculent in water. *Mizu Shori Gijutsu* **18**(4), 333–342 (1977); *Chem. Abstr.* **88**, 65557y.
80. M. M. Grula and M. Huang, Interactions of polyacrylamides with certain soil pseudomonads, *Dev. Ind. Microbiol.* 451–457 (1981).
81. J. Su-uki, H. Harada, and S. Su-uki, Ozone treatment of water-soluble polymers. V. Ultraviolet irradiation effects on the ozonization of polyacrylamide. *J. Appl. Polym.* **24**, 999–1006 (1979).
82. M. L. Miller, Arcylic acid polymers. In *Encycolpedia of Polymer Science and Technology*, 3rd ed., Vol. 1, Wiley, New York, 1964, pp. 197–226.
83. K. Mizunuma, Biological monitoring of possible health effects in worker's occupationally exposed to methyl methacrylate. *Int. Arch. Occup. Environ. Health* **65**, 227–232 (1993).
84. M. Farli, Occupational contact dermatitis in two dental technicians. *Contact Dermatitis*, **22**, 282–287 (1990).
85. T. H. Waegemaekers, Permeability of surgeon's gloves for methyl methacrylate. *Acta Orthop. Scand.* **66**, 790–795 (1983).
86. American Conference of Governmental and Industrial Hygienists (ACGIH), *TLV's and BEIS Threshold Limit Values for Chemical Substances and Physical Agents*, ACGIH, Cincinnati, OH, 2000.

CHAPTER NINETY-ONE

Synthetic Polymers — Cellulosics, Other Polysaccharides, Polyamides, and Polyimides

Finis L. Cavender, Ph.D., DABT, CIH

INTRODUCTION

Natural polymers are biological macromolecules and are as old as life itself. The natural polymers include such diverse materials as proteins, polypeptides, polysaccharides, DNA, wood, wool, and silk. The word *polymer* is derived from the Greek words *poly*, or many, and *meros*, or parts. This chapter will focus on cellulose and it derivatives, polyamides such as nylon, and polyimides.

A CELLULOSICS

"Cellulose plastics are produced by the chemical modification of cellulose. Raw cellulose is not a thermoplastic: it does not melt. Cellulose is a substance that forms the cell walls of many trees and plants. Raw cellulose can be made into a fiber or film, but it must be chemically modified to produce a thermoplastic" (1).

1.0 Cellulose

1.0.1 CAS Number: *[9004-34-6]*

Patty's Toxicology, Fifth Edition, Volume 7, Edited by Eula Bingham, Barbara Cohrssen, and Charles H. Powell.
ISBN 0-471-31940-6 © 2001 John Wiley & Sons, Inc.

1.0.2 Synonyms: Regenerated cellulose, rayon, cellophane, and cellulose, microcrystalline

1.1 Chemical and Physical Properties

1.1.1 General

Cellulose is a white substance that is practically insoluble in water and other solvents. It can be dissolved in zinc chloride, ammoniacal copper hydroxide, or caustic alkali with carbon disulfide (2). The specific properties of the many types of rayon vary widely (3). Rayon is readily blended with other fibers. Rayon can contribute pleasing texture and touch quality, moisture absorbency, or strength. Cellophane is noted for clarity, crisp hand, and dimensional stability. Coatings of nitrocellulose or saran reduce moisture and oxygen permeability; polyethylene or ionomer coatings reduce heat loss. Both rayon and cellophane can be highly flammable (4). See Tables 91.1–91.3.

1.1.2 Odor and Warning Properties

Dry, untreated cellulosic products are often highly flammable (4).

1.2 Production and Use

Regenerating cellulose to yield the products rayon and cellophane removes the natural impurities. These regenerated products are essentially inert unless new toxicants such as finishes and plasticizers are added in sufficient quantities to cause injury. Many cellulose derivatives would appear to be similarly inert. "Rayon" is, by definition, established by the Federal Trade Commission, the "generic name for a manufactured fiber composed of regenerated cellulose as well as manufactured fibers composed of regenerated cellulose in which substitutes have replaced not more than 15% of the hydrogens of the hydroxyl groups."

Table 91.1. Limiting Oxygen Indices and Ignition Temperatures for Various Polymers[a]

Polymer	Range of Limiting Oxygen Indices[b]	Ignition Temperature (°C) Flash Ignition	Self-Ignition
Wood	22.4–25.4	228–264 (white pine shavings)	260
Cotton	18.6–27.3	230–266	254
Cayon	18.7–19.7		
Cellulose acetate	16.8–27	305	475
Cellulose nitrate		141	141
Wool	23.8–25.0		
Nylon 66 fabric	20–21.5		
Nylon 6 fabric	20–21.5		
Polyethylene terephthalate fabric	20–21		

[a]Summarized from Ref. 4. Additional ignition temperatures are summarized in Table 91.2.
[b]The limiting oxygen index can also be expressed in terms of the volume of oxygen in the atmosphere, where n is a decimal.

Table 91.2. Explosion Parameters of Polymeric Dusts

A. Explosion Parameters of Selected Polymeric Dusts as Compiled by the National Fire Protection Association[a]

Type of Dust	Explosibility Index	Ignition Sensitivity	Explosion Severity	Maximum Explosion Pressure (psig)	Maximum Rate of Pressure Rise (psi/sec)	Ignition Cloud Temperature Cloud (°C)	Ignition Cloud Temperature Layer (°C)	Minimum Ignition Energy (joules)	Minimum Explosion Concentration (oz/ft^3)
Agricultural dusts									
Cellulose	2.8	1.0	2.8	130	4,500	480	700	0.080	0.055
Cork dust	>10	3.6	3.3	96	7,500	460	100	0.035	0.035
Cotton linter, raw	<0.1	<0.1	<0.1	73	400	50	—	1.92	0.50
Wood, birch bark ground	6.7	3.7	1.8	103	7,500	450	50	0.060	0.020
Wood, flour, white pine	9.9	3.1	3.2	113	5,500	470	260	0.040	0.035
Thermo plastic resins and molding compounds									
Cellulose acetate	>10	8.0	1.6	85	3,600	40	—	0.015	0.040
Cellulose triacetate	7.4	3.9	1.9	107	4,300	430	—	0.030	0.040
Cellulose acetate butyrate	5.6	4.7	1.2	85	2,700	410	—	0.030	0.035
Nylon (polyhexamethylene adipamide) polymer	>10	4.7	1.8	95	4,000	500	430	0.020	0.030
Rayon (viscose) flock, 1.5 denier, 0.00 in maroon	0.2	0.3	0.8	107	1,700	−520	250	0.240	0.055

B. Explosion Parameters of Selected Polymetric Dusts as Tested at the Joint Fire Research[b]

Type of Dust	Minimum Ignition Temperature (°C)	Minimum Explosible Concentration (kg/m^3)	Minimum Ignition Energy (mJ)	Maximum Explosion Pressure lbf/in^2	Maximum Explosion Pressure kN/m^2	Maximum Rate of Pressure Rise lbf/in.^{2} sec	Maximum Rate of Pressure Rise kN/m^2 sec	Maximum Oxygen Concentration to Prevent Ignition (% by vol)
Rayon, viscose	40	—	—	—	—	—	—	—
Sodium carboxymethyl cellulose	320	1.1	440	49	340	400	2800	5

[a] Summarized from Ref. 4. The potential hazard of a dust is related to its ignition sensitivity and to the severity of the subsequent explosion.... The index of explosibility is the product of the ignition sensitivity and the explosion severity. This index is a dimensionless quantity having a numerical value of 1.0 for a dust equivalent in explosibility to the standard Pittsburgh coal. An index greater than 1.0 indicates a hazard greater than that for the coal dust. The notation ≪1.0 designates materials presenting primarily a fire hazard as ignition of the dust cloud is not obtained by spark or flame, but by a surface heated to a relatively high temperature.... The relative explosion hazard of a dust may be further classified by ratings of weak, moderate, strong, or severe. These ratings are correlated with the empirical index as follows:

Relative Explosion Hazard Rating	Ignition Sensitivity	Explosion Severity	Index of Explosibility
Weak <0.2	<0.5	<0.1	
Moderate	0–1.0	0.5–1.0	0.1–1.0
Strong	1.0–5.0	1.0–2.0	1.0–10
Severe	>5.0	>2.0	>10

[b] See Ref. 4. These data are reproduced by permission from Fire Research Technical paper No. 1, *Explosibility Tests for Industrial Dusts*, ©Crown copyright 1975, Borehamwood, Hertfordshire, England.

The classification of explosibility is as follows:

(a) Dusts which, when tested, ignited and propagated flame in the test apparatus, and
(b) Dusts which, when tested, did not ignite and propagate flame in the test apparatus.

Dusts are classified as Group (*a*) if they ignite and propagate flame in any of the three tests with a small source of ignition, either "as received" or after sieving and drying the sample. The explosibility of dusts may be modified by factors such as large particle size, or high moisture content which is not removed by the usual drying methods, or if the sample is a mixture and the explosible proportion is inerted by the other ingredients. With some dusts, although flame propagation has been observed in the tests, and they are therefore classified as explosible, the explosion pressure that the dust can develop under industrial conditions may be very small.

The classification only applied to conditions where a dust is dispersed at or near ordinary atmospheric temperatures. Where a dust cloud is in a heated environment other considerations may apply.

Table 91.3. Some Typical Properties of Cellulose Derivatives

Property	Cellulose Acetate	Cellulose Triacetate	Cellulose Nitrate	Sodium Carboxymethyl Cellulose[a]	Hydroxyethyl Cellulose	Ethyl Cellulose
Melting temperature (°C)	230–260 (with decomposition)	290–396	Flow temperature ~150. Plastics serviceable to 60[b]			
Charring temperature (°C)				252		
Browning temperature (°C)				227	205–210	
Softening temperature (°C)	75–120	220–225	80–90		145–140	152–162
Density	1.28–1.32 (sheeting), 1.22–1.34 (molding)	1.30	1.58–1.65 (cast film, Ref. 550). 1.35–1.5 (trinitrate; plastics, ~1.38)	0.75 (bulk). 1.59 (film). 1.007 (2% solution)	0.6 (bulk). 1.34 (film; 50% relative humidity). 1.003 (2% solution)	
Refractive index, n_D	1.46–1.50	1.47–1.48	1.51	1.515 (film). 1.336 (2% solution)	1.51 (film). 1.336 (2% solution)	1.479
Moisture absorption	3% (depends on plasticizer)		1% (24 h at 80% RH)			2% (24 h at 80% RH)
Glass-transition temperature (°C)		49–478 ("Conflicting data; depends on acetate and water content and degree of crystallinity," 180 (2nd-order transition, fibers)	53–66 ("conflicting data")			
Solvents/nonsolvents/relatively unaffected by	Solubility varies with degree of substitution; water and 2-methoxyethanol can act as both solvent and nonsolvent	Soluble in methylene chloride, chloroform; acetone may act as solvent and nonsolvent. Relatively unaffected by hydrocarbons, most oils, and greases		Solubility varies with degree of substitution; generally, carboxymethyl cellulose and hydroxyethyl cellulose are soluble in water, ethyl cellulose insoluble; see		

Table 91.3. (*Continued*)

Property	Cellulose Acetate	Cellulose Triacetate	Cellulose Nitrate	Sodium Carboxymethyl Cellulose[a]	Hydroxyethyl Cellulose	Ethyl Cellulose
Decomposed by	Less resistant than cellulose triacetate	Moderately concentrated acids, alkalis with pH >9.5				
Biologic oxygen demand				Reported values after 5 days of incubation in the range of 7000–18,000 ppm depending on substitution of polymer; comparable value for cornstarch is over 800,000 ppm		

[a]Summarized from Ref. 4.
[b]Ignition temperature depends on sample size and composition, heating conditions, time, and test method.

Cellophane is "regenerated cellulose, chemically similar to rayon, made by mixing cellulose xanthate with a dilute sodium hydroxide solution to form a viscose, then extruding this viscose into an acid for regeneration. The term rayon is used when the material is in fibrous form" (5). Rayon is made from regenerated cellulose by forcing it through holes into the coagulating acid bath at the end of the process, while cellophane is the film form of regenerated cellulose that has been forced through a slit (6).

All methods of preparation essentially depend upon solubilizing short-fibered forms of natural cellulose, reshaping it into long fibers or film by extrusion through a spinneret or slit aperture, then immediately converting the extruded product back into solid cellulose. Rayon was first commercialized in the nineteenth century by the now discarded Chardonnet process that used highly flammable cellulose nitrate. The cuprammonium process replaced the Chardonnet process and is still used to a limited extent to produce extremely fine, silk-like filaments. Today the most widely used process is the xanthate or viscose process (3, 7).

Generally, alkali cellulose is prepared by reacting wood pulp with excess sodium hydroxide (or other alkali), followed by aging to permit partial depolymerization. The alkali cellulose is reacted with carbon disulfide to form sodium xanthate, which is then dissolved in alkali and extruded into an acid bath that converts the filaments or film into cellulose. These filaments may be stretched, desulfurized, washed, dried, or otherwise finished.

Extruding the viscose through a slit into an acid bath yields cellophane. Cellophane can be plasticized by washing the product with glycerol, propylene glycol, or polyethylene glycol (8, 9). Regenerated cellulose may also be prepared by saponification of cellulose acetate. Industrial uses of rayon include reinforcing cords for tiers, belts, and hoses, as well as in "disposable," nonwoven fabrics. It is also widely used in textiles, although less extensively today. Cellophane films are widely used for sausage casings (10).

1.3 Exposure Assessment

These products are so diverse that the references given here are primarily limited to reviews (11–14).

1.3.3 Workplace Methods

The NIOSH Method 0500 for total particulate not otherwise regulated and Method 0600 for respirable particles are recommended for determining workplace exposures to cellulose (14a).

1.4 Toxic Effects

Unprocessed rayon does not cause dermatitis (15). Commercial fabrics may contain free formaldehyde or formaldehyde resins. Analyses of 12 samples of 100% rayon clothing showed free formaldehyde levels ranging from 15 to 3517 ppm (16); formaldehyde is present in the finishing agents for fabrics.

1.4.1 Animal Studies

1.4.1.1 Acute Toxicity. The oral LD_{50} of cellulose is >5000 mg/kg in rats; the dermal LD_{50} in rabbits is >2000 mg/kg; and the 4-h LC_{50} in rats is >5800 mg/m^3 (17). Male

Fischer 344 rats were given 0, 0.25, 1.0, or 4.0 mg via intratrachael instillation on four consecutive days. Bleached cellulose produced no effects at any dosage level, whereas cellulose insulation and microcrystalline cotton caused cytotoxicity and inflammation at the highest dosage (18). Table 91.4 gives an overview of inhalation toxicity data pertaining to the major types of cellulose products.

1.4.1.5 Carcinogenesis. Regenerated cellulose per se is not known to be associated with the development of neoplasms. Subcutaneous implants of cellophane characteristically produce tumors at the site of the injection (19–21). The incidence of these tumors varies with the physical shape of the implant and is not generally recognized as consistent with solid-state carcinogenesis. A study with surgical cotton or cotton linters (from which cellophane had been manufactured) showed no sarcomas (19).

1.4.1.7 Other: Neurological, Pulmonary, Skin Sensitization. Although cellulosic and other polysaccharide materials constitute a very large and diverse class, only wood has been well studied for its smoke toxicity. In fact, wood has often been chosen as a reference for comparative purposes. This is unfortunate, because no material is as dependent as wood on the conditions of combustion for the composition of its smoke. Furthermore, the toxicity of smoke produced from wood also depends on the density of the wood. The more dense woods yield smoke that is somewhat less toxic than the smoke from those varieties having lower density.

The thermal decomposition of cellulose is highly complex, involving desorption of water, followed by a variety of cross-linking, unzipping, repolymerization, and volatilization processes (22). A great many volatile products can be formed, among which aldehydes, especially acrolein, are of particular significance in considering the toxicity of the smoke. Acrolein is a potent sensory irritant, as well as a strong pulmonary irritant. It causes severe sensory irritancy in humans in the range of only 1–5.5 ppm (23). Although it has been reported that a baboon could escape from a chamber after a 5-min exposure to 2780 ppm, the animal later died from pulmonary effects (24). The same results were observed at 1025 ppm. No signs of pulmonary effects were observed following exposure to 505 ppm and below. In general, smoke from wood and other cellulosics is quite irritating and toxic.

Depending on the rate of heating, various dehydration and repolymerization mechanisms lead to the formation of a char structure from the combustion of wood. The char serves to insulate undecomposed material, thus slowing its rate of combustion. However, the solid-phase combustion of char can cause sustained smoldering, a characteristic fire property of cellulosics. Once ignited, either from an external source or from autoignition, wood and other cellulosics generally burn freely with the production of CO_2 and CO, usually accompanied by irritants and other products of thermal decomposition. The composition of the smoke, including the CO produced, depends strongly on the ventilation or air supply during combustion.

In laboratory smoke toxicity tests, wood produces smoke of average toxicity. Using the NBS method, Douglas fir was reported to have 30-min LC_{50}s of about 40 g/m^3 (flaming) and approximately 20 g/m^3 (nonflaming) (25). The latter value probably reflects increased toxicity due to the presence of irritants in the smoke produced under nonflaming

SYNTHETIC POLYMERS

Table 91.4. Overview of Inhalation and Thermal Degradation Data on Cellulosic Polymers[a]

Polymer: Species Exposed	Description of Material	Procedure[b]/Observations/Conclusions
Rayon: mouse	100% rayon viscose fibers with and without flame retardant	Time to incapacitation and death longer from products generated in the rising temperature program longer than the median among a series of fabrics comparably tested. Death attributed to CO. Addition of an "organic phosphorus–sulfur based" flame retardant "appeared to increase char yield but had slight if any effect in reducing time to death"
		"Aldehydes are present in very small amounts, if any, in the combustion and decomposition products of... acetate rayon." "Formaldehyde in appreciable concentrations is found in the fumes evolved by burning and smoldering cotton fabric under conditions of oxygen deficiency"
Cellulose acetate		Very thin films or bulk material degraded under vacuum between 230 and 320°C: main volatile product from films was acetic acid, some water also formed; tar from bulk degradation contained acetyl derivatives of D-glucose
		Change in sample weight at burning temperature < 200°C
Cellulose nitrate Human		*Review, long-term survivorship study: on May 15, 1929, "gases from the the decomposition of an estimated 50,000 nitrocellulose x-ray films ignited and three explosions shook the Cleveland clinic.... [The investigating committee] concluded that both CO and HCN contributed to causing the death of the 97 acutely fatal cases"
Rat	Celluloid	*DIN 53436 protocol 200°C degradation products lethal, caused 60% COHb (data ≥300°C degradation)
Rat	"Nitrocellulose"	*Comparison of toxicity of pyrolysis products generated from various polymers in standardized test; deaths from products generated at <200°C (only material so determined); PVC and several wood products were "≤300")
Ethyl cellulose		"Volatile products from ethyl cellulose include H_2O, CO, CO_2, C_2H_4, C_2H_6, C_2H_3OH, CH_3CHO, unsaturated aliphatic compounds, and furan derivatives." Products identified from degradation of thin films are mainly water, ethanol, and acetaldehyde; products of bulk degradation separated into yellow-brown tar, liquid, and gas fractions (procedure, see above this table)

[a]Summarized from Ref. 4.
[b]* = no effluent analysis.

conditions. Other 30-min LC_{50} values reported for wood include 56 g/m^3 for the NIST test (26), 20–50 g/m^3 for the DIN test (27), and an average for woods of several different densities of about 33 g/m^3 for the UPITT test. An overall average toxicity for wood smoke compiled from several test methods has been reported to be about 42 g/m^3, with the range of materials tested being from about 5 to 60 g/m^3 (28).

1.4.2 Human Experience

In the general population, harm is more likely to result from the flammability of these cellulosics or the means used to retard flammability than any other factor associated with their use (4). Untreated cellulosic materials exposed to smoldering flames readily generate serious or lethal amounts of carbon monoxide. Analysis of New York City autopsy records from a 2-yr period (1966 to 1967) showed that 79% of fire victims who died during the first 12 h had carboxyhemoglobin poisoning (29, 30).

1.5 Standards, Regulations, or Guidelines of Exposure

Practices for handling regenerated cellulose and process waste depend primarily on the specific alkalis, acids, or other materials involved. Rayon, cellophane, and cellulose derivatives may be handled without specific precautions or, in particular cases, may require special precaution (generally due to flammability). The ACGIH 8-h TLV–TWA for cellulose is 10 mg/m^3 (31), the 10-h NIOSH REL is 10 mg/m^3 for total cellulose dust and is 5 mg/m^3 based on the respirable fraction (32), and the 8-h OSHA PEL for total cellulose dust is 15 mg/m^3 and is 5 mg/m^3 based on the respirable fraction (33).

2.0 Cellulose Acetate

2.0.1 CAS Number: *[9004-35-7]*

2.0.2 Synonyms: Secondary acetate, acetate cotton, acetyl cellulose, and acetose

2.1 Chemical and Physical Properties

2.1.1 General

Cellulose acetate is a partly acetylated cellulose that exists as white flakes or powder with a melting point of approximately 260°C. Its density varies between 1.27 and 1.34, and it is soluble in acetone, ethyl acetate, cyclohexanol, nitropropane, and ethylene dichloride (17). Properties of the plastics vary with acetic acid content and molecular weight. Acetyl content of less than 13% results in a relatively insoluble product, whereas products with higher acetyl content may have a narrow range of solubility in certain solvents (11). The plastics are noted for their clarity. Acetate fibers have a bright, lustrous appearance and lower strength and abrasion resistance compared to most other manmade fibers. They are weakened by prolonged exposure to elevated temperatures in air. See Tables 91.1–91.3.

SYNTHETIC POLYMERS

2.1.2 Odor and Warning Properties

No odor data were located.

2.2 Production and Use

Cellulose acetate is prepared by hydrolyzing the triacetate to remove some of the acetyl groups. The fibers are used primarily in textiles and household applications where aesthetic appeal is desirable. The plastic is used in photographic film, sheeting, lacquers, and for molded or extruded products (17).

2.3 Exposure Assessment

No analytical methods were located.

2.4 Toxic Effects

2.4.1 Fire and Thermal Degradation

Thermal decomposition data indicate that cellulose acetate would be expected to yield carbon monoxide, water, acetic acid, and related materials, depending on specific conditions. See Table 91.4 for details.

2.4.2 Human Experience

Cellulose acetate and other cellulose esters may be used for plastic film attached to adhesive tape (34), eyeglasses, hearing aids, and other products that are in contact with the skin for prolonged periods (35). Dermatitis from the final plastic product is uncommon but "may result from pressure, chemical irritation, or allergy" (35). Identified allergens include resorcinol monobenzoate, dyes for plastic colorant, and triphenyl phosphate (15).

2.5 Standards, Regulations, or Guidelines of Exposure

No specific standards or guidelines were identified.

3.0 Cellulose Triacetate

3.0.1 CAS Number: *[9012-09-3]*

3.0.2 Synonyms: Primary acetate and acetylated cellulose

3.1 Chemical and Physical Properties

3.1.1 General

Properties of the plastics vary with acetic acid content and molecular weight. Acetyl content of less than 13% results in a relatively insoluble product, whereas products with higher acetyl content may have a narrow range of solubility in certain solvents (11). The plastics are noted for their clarity. Triacetate fibers have a bright, lustrous appearance and lower strength and abrasion resistance compared to most other manmade fibers. They are weakened by prolonged exposure to elevated temperatures in air. See Tables 91.1–91.3.

3.1.2 Odor and Warning Properties

No data on odor threshold or recognition were located.

3.2 Production and Use

Cellulose triacetate plastic is customarily prepared by reacting cellulose with acetic anhydride in the presence of a catalyst, often sulfuric acid (36). Usually wood pulp is the source of the cellulose, although acetate produced from cotton linters may have better color and solution clarity. Fibers are most commonly prepared with the same reactants. Fibers can also be prepared by a solvent process with perchloric acid catalyst that dissolves the acetate as formed or by a nonsolvent process that yields a product physically similar to the original cellulose (13). Acetic acid is the most common solvent.

The fibers are used primarily in textiles and household applications where aesthetic appeal is desirable. The plastic is used in photographic film, sheeting, lacquers, and for molded or extruded products.

3.3 Exposure Assessment

No specific analytical methods were located.

3.4 Toxic Effects

3.4.1 Experimental Studies

Cellulose triacetate may be used for plastic film attached to adhesive tape (34), eyeglasses, hearing aids, and other products that are in contact with the skin for prolonged periods (35). Dermatitis from the final plastic product is uncommon but "may result from pressure, chemical irritation, or allergy" (35). Identified allergens include resorcinol monobenzoate, dyes for plastic colorant, and triphenyl phosphate (15).

3.4.2 Fire and Thermal Degradation

Thermal decomposition data indicate that cellulose acetate would be expected to yield carbon monoxide, water, acetic acid, and related materials, depending on specific conditions. See Table 91.4 for details.

3.5 Standards, Regulations, or Guidelines of Exposure

No specific standards or guidelines were identified.

SYNTHETIC POLYMERS

4.0 Cellulose Acetate Butyrate

4.0.1 CAS Number: [9004-36-8]

4.0.2 Synonyms: Cellulose acetate butanoate

4.1 Chemical and Physical Properties

4.1.1 General

Compared to cellulose acetate, polymers of cellulose acetate butyrate are typically of lower density, slightly softer, and may have slightly lower water absorption values (37). The properties of cellulose acetate butyrate may be considerably modified by plasticizer, chain length, or degree of substitution (37).

4.1.2 Odor and Warning Properties

No data on odor threshold or recognition were located.

4.2 Production and Use

Cellulose acetate butyrate is noted for toughness, clarity, and ease of processing (11). It is extensively used in lacquers, such as with acrylics on automobiles.

4.3 Exposure Assessment

No specific analytical methods were located.

4.4 Toxic Effects

No specific toxicity data were located.

4.5 Standards, Regulations, or Guidelines of Exposure

No specific standards or guidelines were identified.

5.0 Cellulose Nitrate

5.0.1 CAS Number: [9004-70-0]

5.0.2 Synonyms: Collodion, nitrocellulose, pyroxylin, nitrocotton, guncotton; flexible collodion, Synpor; Parlodion; Parlodion strips; and Pyroxilene

5.1 Chemical and Physical Properties

5.1.1 General

Cellulose nitrate is a colorless to slightly yellow, clear or slightly opalescent, syrupy liquid. It has a density of 0.765–0.775 and is slightly soluble in methanol, acetone, and glacial acetic acid (17). Cellulose nitrate is known for its extreme flammability, which limits its commercial use. Dry cellulose nitrate may explode when subjected to heat or sudden shock (11); cellulose nitrate is therefore generally shipped or handled wet with water or alcohol, usually ethanol. Physical properties vary with the degree of nitration, which, in turn, is usually directed toward the proposed end use. Cellulose nitrate with about 12% nitrogen is compatible with many synthetic resins, soluble in ketones and esters, and is used in lacquer coatings. Cellulose nitrate with about 11% nitrogen is more thermoplastic, is soluble in ethyl or isopropyl alcohol, and is used widely in flexographic inks for paper and foil (11). See Tables 91.1–91.3.

The term *celluloid* [CAS No. *8050-88-2*] is now used as a general term for plasticized cellulose nitrate compositions, which are extremely flammable and have fair to poor chemical resistance. Celluloid is a reasonably tough thermoplastic that can be molded for balls with "bounce" and small articles requiring pleasant appearance (37).

5.1.2 Odor and Warning Properties

It has the odor of ether (17).

5.2 Production and Use

Cellulose nitrate is prepared from cellulose and nitric acid, usually in the presence of sulfuric acid (7). Completely nitrated cellulose contains 14.4% nitrogen. The three main types (11) contain approximately 12, 11.5, or 11% nitrogen, respectively. Celluloid is cellulose nitrate plasticized with camphor.

5.3 Exposure Assessment

Several analytical methods for nitrocellulose were located. To quantify it in the presence of nitroglycerin, Peak (38) used TLC and Lloyd (39) used a reductive mode electrochemical detection. To detect cellulose nitrate in wastewater, Epstein et al. (40) used gas chromatography and Barkley and Rosenblatt (41) used an automated procedure.

5.4 Toxic Effects

5.4.1 Experimental Studies

The oral LD_{50} in rats and in mice is >5000 mg/kg (17).

5.4.1.1 Fire and Thermal Degradation. Toxic combustion products, mainly carbon monoxide and oxides of nitrogen, are readily generated at relatively low temperatures (<200°C). See Table 91.4 for details.

5.4.2 Human Experience

In a matched case-control study of workers from a plastics producing plant (42), possible association between rectal cancer and cellulose nitrate workers was identified and warrants further surveillance. A man without allergies injured his foot, which was wrapped in collodian-soaked lint (43). Twelve days later, the skin beneath the wrap was erythematous, vescular, and scaly. This suggests that occulusion can lead to sensitization.

5.5 Standards, Regulations, or Guidelines of Exposure

No specific standards or guidelines were identified.

5.6 Studies on Environmental Impact

No acutely toxic effects were observed among several species of fish, algae, or invertebrates exposed to nominal concentrations of 1000 mg/L, except in the case of the green alga *Selenastrum capricornutum* (44). Based on the data obtained, a water quality criteria of 50 mg/L was proposed. A combined chemical–biologic treatment has been suggested for disposal (41).

6.0 Ethyl Cellulose

6.0.1 CAS Number: [9004-57-3]

6.0.2 Synonyms: Cellulose ether, cellulose, ethyl ether, and Azo Dye N5

6.1 Chemical and Physical Properties

6.1.1 General

Ethyl cellulose resins have been used in foods and pharmaceuticals as diluents, binders, and fillers. They are insoluble in water and are most frequently used in solvent mixtures containing 60–80% aromatic hydrocarbon and 20–40% alcohol; this percentage of alcohol results in minimum viscosity. The ethoxyl content of ethyl cellulose affects thermal behavior. The melting point of approximately 210°C with an ethoxyl content of 44% may decrease to approximately 170°C with an ethoxyl content of 48–49% (12).

6.1.2 Odor and Warning Properties

No odor threshold or recognition data were located.

6.2 Production and Use

Ethyl cellulose is manufactured by reacting ethyl chloride with cellulose in the presence of sodium hydroxide or by reacting cellulose and ethanol in the presence of a dehydrating agent. It is used in printing inks, surface coatings, thermoplastic molds, and in controlled-release products (17).

6.3 Exposure Assessment

One analytical was identified for the determination of ethylcellulose in hair fixatives (46).

6.4 Toxic Effects

Ethylcellulose is a mild dermal irritant and the oral LD_{50} in rats is >5000 mg/kg, while the dermal LD_{50} in rabbits is >5000 mg/kg (17).

6.5 Standards, Regulations, or Guidelines of Exposure

No specific standards or guidelines were identified.

7.0 Sodium Carboxymethyl Cellulose

7.0.1 CAS Number: [9004-32-4]

7.0.2 Synonyms: Cellulose sodium glycolate; cellulose carboxymethyl ether, sodium; CMC Sodium Salt; Carboxymethyl Cellulose, Sodium Salt; Cellex; Aquacide I, Calbiochem; Aquacide II, Calbiochem and cellulose gum

7.1 Chemical and Physical Properties

7.1.1 General

Sodium carboxymethyl cellulose is soluble in water (4).

7.1.2 Odor and Warning Properties

No odor threshold or recognition data were located.

7.2 Production and Use

Sodium carboxymethyl cellulose is prepared by reacting alkali cellulose with sodium acetate (12). Major applications include food uses such as a thickening, emulsifying, or bulk additive (46).

7.4 Toxic Effects

Cellulose gums such as sodium carboxymethyl cellulose have been considered inert for several decades. A number of negative dermal irritation studies have been reported (12, 14).

7.5 Standards, Regulations, and Guidelines of Exposure

No specific standards or guidelines were identified.

8.0 Methyl Cellulose

8.0.1 CAS Number: [9004-67-5]

8.0.2 Synonyms: Cellulose methylate, cellulose methyl ether; adulsin; bagolax; bufapto methalose; bulkaloid; celacol m; celacol m20; celacol m450; celacol mm; celacol mm

SYNTHETIC POLYMERS

10p; celacol m 20p; cellapret; cellogran; cellothyl; cellulose methyl; cellumeth; cethylose; cethytin; culminal k 42; edisol m; hydrolose; mapolose m25; mapolose 60sh50; mco 8000; mc 4000 cp; mc 20000s; mellose; methocel 10; methocel 15; methocel 181; methocel 400; methocel 4000; methocel a; methocel chg; methocel 400cps; methocel 4000cps; methocel mc; methocel mc25; methocel mc4000; methocel mc 8000; methocel sm 100; methulose; methyl cellulose-a; methyl cellulose ether; metolose mc 8000; metolose 60sh; metolose 60sh400; metolose sm 15; metolose sm 100; metolose sm 4000; mmts-btr; napolone; Nicel; rhomellose; syncelose; tylose 444; tylose A4s; tylose mf; tylose mh; tylose mh20; tylose mh50; tylose mh300; tylose mh1000; tylose mh2000; tylose mh4000; tylose mh300p; tylose sap; tylose sl; tylose sl 100; tylose sl 400; tylose sl 600; tylose Iwa; methylcellulose; viscol; viscontran 152; viscosol; walsroder mc 20000s; Celevac; Cologel; Citrucel; and Methylcel MC

8.1 Chemical and Physical Properties

8.1.1 General

Methyl cellulose is nonionic and can function as a polymeric surfactant (12). It is soluble in hot and cold water (47).

8.1.2 Odor and Warning Properties

Methyl cellulose is odorless (2).

8.2 Production and Use

Methyl cellulose is manufactured by reacting methyl chloride and cellulose in the presence of sodium hydroxide. It can also be produced using dimethyl sulfide or methanol (17). Methyl cellulose and its modifications have the unusual property that they are more soluble in cold than in hot water (12). This thermal gelation property is widely utilized in adhesives, where it provides a quick set and control of penetration into the substrate. Applications include pharmaceuticals, cosmetics, agricultural products, and cement and plaster formulations. It is also used as an insecticide carrier in veterinary medicine (2). Bachmann et al. (48) recommend 0.5% methyl cellulose as a suspending medium of choice for oral administration of water-insoluble components in experimental studies. This dose was not observed to alter mitochondrial function or activity of mixed-function oxidases, whereas these effects were observed at a higher dose level of methyl cellulose and with various levels of sodium carboxymethyl cellulose, gum arabic, or gum tragacanth.

8.3 Exposure Assessment

Methyl cellulose has been determined using ASTM Method DI347-56 (50) and by colorimetric methods (49).

8.4 Toxic Effects

Methyl cellulose has also been considered essentially innocuous (14) by the oral route on the basis of extensive feeding tests extending back to at least 1942. McCollister and associates (50) found no evidence of oral toxicity in either rats or dogs at concentrations up to 6% (methyl cellulose, 10 or 4000 cP) or 10% (hydroxypropyl methyl cellulose, 10 cP) for 90 d. Two-yr studies in rats with methyl cellulose (15, 400, or 4000 cP) at dietary levels of 1 or 5% were also without effect. Braun and his colleagues (51) found no significant absorption of a single dose of 500 mg/kg of methyl cellulose (3300 cP) or five daily doses. After administration of the single dose, the measured amount of ^{14}C activity in the feces was 102%. No accumulation of ^{14}C activity was detected in the body or in selected tissues after multiple dosing. This suggests that these substances are not absorbed from the gastrointestinal tract.

Tumors have been observed at the site of injection in some animals given massive, repeated subcutaneous injections. Hematologic and biochemical changes are more characteristic; death in dogs given repeated intravenous injections of methyl cellulose was attributed to renal failure. Repeated ingestion of a short-chain methyl cellulose was not tolerated in two patients with an erratic history of fluid retention and obesity (52).

8.5 Standards, Regulations, and Guidelines of Exposure

No specific standards or guidelines were identified.

8.6 Studies on Environmental Impact

Testing of methyl cellulose (53) with activated sludge indicates that 96% can be dissipated in 20 d. Reported values for the biologic oxygen demand of two samples of hydroxyethyl cellulose are 7000 and 18,000 ppm, respectively, after 5 d of incubation (12).

9.0 Hydroxypropyl Methyl Cellulose

9.0.1 CAS Number: [9004-65-3]

9.0.2 Synonyms: Cellulose 2-hydroxypropyl ester; cellulose, 2-hydroxypropyl methyl ether; Hypromellose; Isopto-Tears; Methopt; Poly-Tears; and Tears Naturale

9.1 Chemical and Physical Properties

9.1.1 General

Hydroxypropyl methyl cellulose is nonionic and can function as a polymeric surfactant (12). It is soluble in hot and cold water (47).

9.1.2 Odor and Warning Properties

No odor threshold or recognition data were located.

9.2 Production and Use

Hydroxypropyl methyl cellulose is manufactured by reacting methyl chloride, chloropropanol, and cellulose in the presence of sodium hydroxide (17).

SYNTHETIC POLYMERS

9.3 Exposure Assessment

No specific analytical methods were located.

9.4 Toxic Effects

Hydroxypropyl methyl cellulose has been considered essentially innocuous (14) by the oral route on the basis of extensive feeding tests extending back to at least 1942. McCollister and associates (52) found no evidence of toxicity in either rats or dogs fed these materials at concentrations up to 10% (hydroxypropyl methyl cellulose, 10 cP) for 90 d.

9.5 Standards, Regulations, and Guidelines of Exposure

No specific standards or guidelines were identified.

10.0 Hydroxyethyl Cellulose

10.0.1 CAS Number: [9004-62-0]

10.0.2 Synonyms: 2-Hydroxyethyl cellulose; Cellulose, 2-hydroxyethyl ether; Alcogum 5499-R; BL 15; Cellosize; Cellosize UT 40; Cellulose, ethylene oxide-grafted; Glutofix 600; HEC; HEC-AL 5000; Hetastarch; Hydroxyethyl cellulose ether; Hydroxyethyl ether cellulose; Natrosol; Natrosol 240JR; Natrosol 250 H; Natrosol 250 HHR; Natrosol 250 M; Natrosol L 250; and Natrosol LR

10.1 Chemical and Physical Properties

10.1.1 General

Hydroxyethyl cellulose is nonionic and soluble in both cold and hot water. Solution viscosity decreases with increasing temperature (12). It is insoluble in most organic solvents and partially soluble in acetic acid (17).

10.1.2 Odor and Warning Properties

Hydroxyethyl cellulose is odorless (56).

10.2 Production and Use

Hydroxyethyl cellulose is manufactured by reacting ethylene oxide and cellulose in the presence of sodium hydroxide (17). Major applications include coatings and petroleum production; others are pharmaceuticals, agricultural products, textile processing, and construction (12). It is also used as a thickener, protective colloid, binder stabilizer, and suspending agent and in artificial tear solutions (17).

10.3 Exposure Assessment

No analytical methods were identified in the literature (17).

10.4 Toxic Effects

Hydroxyethyl cellulose has been administered to rats in single oral doses as high as 23,000 mg/kg (50% in corn oil) without observed toxic effects (55). Rats maintained for 2 yr on diets containing approximately 5% hydroxyethyl cellulose showed no adverse effects (14). Hydroxypropyl cellulose has been fed to rats for 90 d at dietary levels of 5, 1, or 0.2%. The only differences observed between any groups were an increase in food consumption and decrease in food utilization among rats fed both the hydroxyethyl cellulose and the unmodified cellulose control (56). Repeated intraperitoneal injections in mice resulted in a marked increase in mortality (57). Mice injected with 1% or 4% on gestation days 3–7 showed an increased incidence in fetal resorptions and an increase in macrophage granulomas in the intestinal cavity (58).

10.5 Standards, Regulations, and Guidelines of Exposure

No specific standards or guidelines were identified.

11.0 Hydroxypropyl Cellulose

11.0.1 CAS Number: [9004-64-2]

11.0.2 Synonyms: 2-Hydroxypropyl cellulose ester; cellulose, 2-hydroxypropyl ether; and hydroxypropyl cellulose, Average M.W. 100.000

11.1 Chemical and Physical Properties

11.1.1 General

Hydroxypropyl cellulose is nonionic but has a thermal gel point and other properties quite different from the hydroxyethyl derivative (4).

11.1.2 Odor and Warning Properties

No odor threshold or recognition data were located.

11.2 Production and Use

It functions as a protective colloid in suspension polymerization, in paint removers, cosmetics, and pharmaceuticals, and also as a protective coating and glaze for food items (4).

11.3 Exposure Assessment

No analytical methods for hydroxypropyl cellulose were located.

11.4 Toxic Effects

The oral LD_{50} in rats is 10,200 mg/kg and in mice is >5000 mg/kg (17). The cellulose gums have been considered innocuous for several decades (4).

SYNTHETIC POLYMERS

11.5 Standards, Regulations, and Guidelines of Exposure

No specific standards or guidelines were identified.

12.0 Carboxymethyl Cellulose

12.0.1 CAS Number: *[9000-11-7]*

12.0.2 Synonyms: Cellulose carboxymethyl ether; Cellulose CM; Cellulose, carboxy methyl ether; Almolose; Cm Cellulose; Carboxymethyl cellulose ether; CMC; Carmellose; and Croscarmellose

12.1 Chemical and Physical Properties

12.1.1 General

Carboxymethyl cellulose is an anionic cellulose with exceptionally high water-binding capacity (46). It is soluble in both hot and cold water.

12.1.2 Odor and Warning Properties

No odor threshold or recognition data were located.

12.2 Production and Use

Carboxymethyl cellulose is used as a thickening, emulsifying, and bulk additive in food products, pharmaceuticals, and detergents, and in sizing of paper and textiles (17).

12.3 Exposure Assessment

No analytical methods for carboxymethyl cellulose were located.

12.4 Toxic Effects

Data on the very low oral toxicity of all these products extend back for several decades. Hake and Rowe (14) reviewed a number of tests with carboxymethyl cellulose, including a 6-mo study with dogs, a 1-yr test with guinea pigs, and a 25-mo study with rats. Three humans fed 20–30 g/d showed a depression of protein digestion and an increase in fat digestion. A 1968 report (59) of 2-yr feeding tests indicates that rats and mice showed little adverse response from dietary levels of 1 or 10% sodium carboxymethyl cellulose. Rats showed an increase in food consumption and slight retardation of growth consistent with the non-nutritive value of the diet. The cellulose gums have been considered nonirritating and nonsensitizing for several decades. A number of negative tests on animals and humans are cited by Hake and Rowe (14) and Greminger (12). In a rare exception, eczema of the hands of 8 year's duration in a Japanese baker was attributed (100) to carboxymethyl cellulose after a positive closed-patch test (2% in petrolatum).

12.5 Standards, Regulations, and Guidelines of Exposure

No specific standards or guidelines were identified.

12.6 Studies on Environmental Impact

Cellulose gums are generally slowly biodegraded. Wirick (61) review some general principles and notes that the biologic oxygen demand for carboxymethyl cellulose is about 3% of the theoretical amount of oxygen required for complete oxidation during the normal 5-d test period. Acclimated bacteria may, however, cause some 55–75% biodegradation, depending on the percentage of saturated groups originally present in the two samples; approximately 90% of the unsaturated groups of both samples degraded during the standard test period. Reported 5-d values for two samples with different properties were 11,000 and 17,300 ppm, respectively, compared to a value of 800,000 ppm for starch (12).

B POLYAMIDES AND POLYIMIDES

The synthetic polyamides and polyimides are all step-growth or condensation polymers. As a group, they are considered performance polymers, whereas the chain growth or addition polymers include the typical commodity polymers of polyethylene, polyvinyl chloride, and polystyrene as well as the high-performance fluoropolymers. Polyamides are linked with the word nylon, the first major synthetic polyamide. Nylon was developed as a fiber in the 1930s and as a plastic in the 1940s. Polyamide polymers also include protein fibers such as wool and silk that have been commercially important for several millenia (4). These natural protein fibers are not discussed in this section.

Nylon is a generic term for a synthetic aliphatic polyamide of well-defined structure and certain typical properties either as a fiber (62) or as a plastic (63) The name system reflects the chemical structure and preparation. Nylons 66, 610, and 612 are all prepared from a six-carbon diamine and a 6-, 10-, or 12-carbon dibasic acid, respectively. [The names can also be written in the style nylon 6/6, nylon 6.6, or nylon 6,6 to reflect the two-monomer origin; the simpler style of nylon 66 — always "six six," or "six ten" for nylon 610 — is usually preferred (28).] Nylons 6, 11, and 12 are prepared from an amino acid or derivative thereof with 6, 11, or 12 carbons, respectively. Nylon 66 was developed in the United States and nylon 6 was developed abroad (37), but both are now found worldwide. As a group, the nylons are tough, strong, abrasion resistant, and resistant to alkalies, hydrocarbons, ketones, and esters (63, 64).

Aromatic polyamides such as Nomex were formerly called nylon (65), but aramid is now the official generic classification of the U.S. Federal Trade Commission and the International Standards Organization. Aramid denotes a long-chain synthetic polyamide fiber in which at least 85% of the amide linkages are attached directly to two aromatic rings, whereas nylon now indicates that less than 85% of the amide linkages are so attached (66). Aromatic polyamide fibers typically have many desirable properties of nylon fibers plus improved heat resistance and strength (67).

Polyimides are a completely synthetic class of polymers developed as a variation on polyamides to provide increased resistance to high temperature (68). Aromatic polyimides have exceptional heat resistance. Conventional tensile strength has been measured up to 500°C (69). Thermoplastic varieties, or those that become rubbery rather than melt at the glass-transition temperature of approximately 310°C (70), retain high strength at almost 300°C. Table 91.5 summarizes some typical physical properties. Figure 91.1 compares the performance of selected aramid and nylon fibers with other types in a standardized test.

Production and Use

Nylon and aramid are processed as fibers. Nylon can also be readily processed as molded or extruded plastic or as film. In the United States "nylon" as a plastic usually refers to nylon 66, the first commercial product and still dominant in the U.S. market (63). Both nylon 66 and nylon 6 are used extensively for textiles in the United States (71). Nylon 6 is more prominent abroad. Polyimides are used as molded plastic, plastic film, and fiber.

Analysis and Specifications

Anton (72) discusses the identification of nylons by solubility tests, color tests, thermal methods, and infrared spectra. Many basic types of nylon can meet the requirements of the Food and Drug Administration for food contact in particular applications (73).

Toxicologic Potential

Available data concerning residual reactions or solvents in the synthetic polymers of this group are meager. Negligible amounts of residual reactants would be expected on a stoichiometric basis in nylons formed by polymerization of a nylon salt and the known aromatic polyamides or polyimides. In the case of nylon 6, residual caprolactam has been present in the polymer, but little concern on this point appears to have developed (71). Residual solvent might be a concern in view of the powerful solvent systems required for polymerization and processing of the high-melting aromatic polymers. Examples of such solvents for aromatic polyamides are dimethylacetamide, *N*-methylpyrrolidone, hexamethylphosphoramide, tetramethylurea, and mixtures of these solvents, which may be used with inorganic salts to increase solvating power (74). Subsequent processing as a textile would generally remove most or all polymerization solvent.

Allergic dermatitis from currently manufactured, commercial nylon fabric would appear to be rare and associated with the dyed products (4).

Thermal degradation of polyamides and polyimides can yield toxic gases, particularly carbon monoxide, hydrogen cyanide, or ammonia. The temperatures at which these gaseous products are released can vary appreciably (4). Inhalation and thermal degradation data are presented in Table 91.6.

Figure 91.2 compares the evolution of ammonia from wool and nylon 6 with that of several other polymers when the materials were heated under nitrogen.

Workplace Practices and Standards

Inhalation of excessive dust or fume should be avoided. In the event of fire, precaution should be taken against inhalation of toxic gases and smoke (4).

Table 91.5. Physical Properties of Selected Polyamide and Polyimide Polymers[a]

A. Nylons

Property	Nylon 66	Nylon 612	Nylon 6	Nylon 11	Nylon 12
Melting temperature (°C)	250–270	217	200–250	182–220	179
Glass-transition temperature (°C)	~40 (conflicting data, most values ~40)	46	40–87 (conflicting data, most values ~40–52)	43, 92 (affected by thermal history and relaxation effects)	41
Thermal deterioration	Fairly rapid oxidative degradation when heated >200°C in air; degradative discoloration may occur at 150°C (depending on type)	As nylon 66			
Density	1.20–1.25; 1.09, amorphous moldings; 1.14–1.45, crystalling moldings		1.220–1.25; 1.10, amorphous moldings; 1.12–1.14, crystalline moldings 1.52–1.58	1.04–1.05[b]	1.034–1.10
Refractive index, n_D	1.52–1.58		1.52–1.58	1.51–1.55	
Water absorption (%) (plastics)	8.5 ± 0.5 (moldings, 20–90°C); 1.5		9.5 ± 0.5 (moldings, 20–90°C); ~2.5 (can approach 100% on prolonged immersion)	"Slightly less" than nylon 610 (can approach 2–3% on prolonged immersion)	
Water retention (%) (fibers)	~25		15	3	
Moisture regain (%) (fibers)	4.5, commercial value; 3.9–4.2		~15	~1	

544

Solvents[d]	*Room temperature:* trichloroethanol, phenol, cresols formic acids, halogenated acids, halogenated acetic acids, sulfuric acids, saturated solutions of alcohol-soluble salts (as magnesium chloride in methanol). At 120–180°C: benzyl alcohol, diethylene glycol, acetic acid, formamide, DMSO	
Nonsolvents/ relatively unaffected by	Hydrocarbons, aliphatic alcohols, chloroform, dieethyl ether, aliphatic ketones and esters	Hydrocarbons, chloroform, alcohols, ethers, ketones, esters
Decomposed by	Nylon in general: concentrated acids, hot dilute mineral acids, oxidizing agents, halogens (including chlorine-containing bleaching agents); hot concentrated alkalies cause loss in weight	

B. Several Aromatic Types

	Aromatic Polyamide		SP Polyimide
	Poly-*m*-Phenylene–Isophthalamide	Poly-*p*-Phenylene–Terephthalamide	Poly(oxy-di-*p*-phenylene) pyromellitimide
Melting temperature (°C)	350–400 (softening and extension)	500–560	None (begins to char >800°C)
Glass-transition temperature (°C)	~275	~300–345	
Decomposition temperature (°C)	400	>362 in air; > 489 in nitrogen (differential scanning calorimetry); 590	
Zero-strength temperature (°C)			815 (20 psi load for 5 sec)
Cut-through temperature (°C)			435 (1 mil) 525 (2.5 mil)
Density	1.14		1.42

Refractive index n_D	1.78 (1 mil film, Becke line)
Moisture absorption	1.3%, 50% RH; 2.9%, immersion for 24 h
	2.9%, 5 days at 100% RH, 23°C (absorbs water reversibly)
Solvents	No known organic solvent[c]
Nonsolvents/relatively unaffected by	Some changes in mechanical properties after 1-yr exposure to benzene, toluene, methanol, acetone at room temperature, also 100°C water for 4 d; no obvious change after 6 month exposure to 150°C transformer oil
Decomposed by	10% sodium hydroxide (5 d)

[a] Summarized from Ref. 4.
[b] The density of 1.04–1.05 for nylon 11 is from Ref. 16, which also gives values of 1.13–1.14 for nylon 6, 1.14 for nylon 66, and ≈0.98 for nonfibrous polyamides.
[c] ASTM Standard D1909-77.
[d] Solubility of nylon in hydrochloric acid at room temperature varies with type; proceeding from most soluble to least solube the order is nylon 6, 66, 610, and 11. Nylon 6 is soluble in 4 N HCl; nylon 11 is insoluble in 6 N acid.
[e] Some derivatives have been structurally modified to provide solubility.

SYNTHETIC POLYMERS

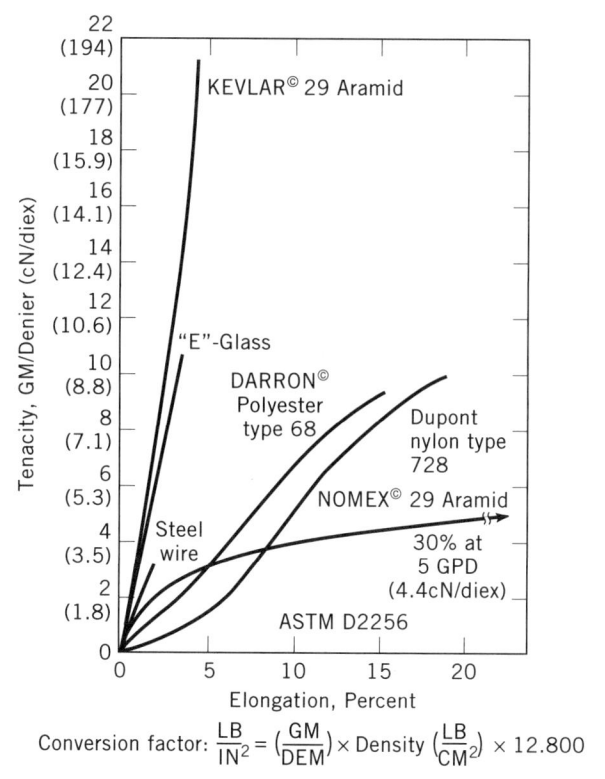

Figure 91.1. Stress–strain behavior of yarns. From Wardle. Reproduced by permission of the *Journal of Coated Fabrics*, Technomic Publishing Co., Inc., Westport, Connecticut.

Fire and Toxicity of Smoke

Of the polyamides, only nylon 6 and wool, used extensively in carpeting, have been studied with regard to the toxicity of smoke produced. Being nitrogen-containing polymers, they produce HCN, along with CO and CO_2, upon combustion. The yields of these toxicants, as with essentially all polymers, depend heavily upon the thermal conditions and ventilation or air supply (4).

Use of the University of Pittsburgh (UPITT) smoke toxicity test method yielded LC_{50Pitt} values for both nylon 6 and wool, which would place them on the highly toxic end of the overall range of test results. The value reported for the LC_{50} of nylon 6 was 6 g, and that for wool was 3 g (75). Adjusting for total air flow during the UPITT test, these LC_{50} values would be the equivalent of about 10 and 5 g/m^3, respectively.

Data from use of the NBS test yielded an LC_{50} for wool of 25–30 g/m^3, which is considerably less toxic than that reported with the UPITT test (25). This difference may be due to the higher temperatures attained under the UPITT test conditions, which favor the production of HCN. The data suggest that fires involving polyamides, both natural and synthetic, produce relatively toxic smoke atmospheres, particularly under conditions of

Table 91.6. Overview of Inhalation and Thermal Degradation Data on Polyamide and Polyimide Polymers[a]

Polymer: Species Exposed	Description of Material	Procedure[b]/Observations/Conclusions
Nylon		
Human		*As summarized in Ref. 4
Rat	Nylons 66 and 6	*Degradation products formed at 500–600°C lethal, caused 22–53% COHb (procedure, Ref. 4)
Rat	Nylon fabric	*Fuel load of 7 g produced 50% mortality, compared to 8–11 g for several other materials; see Ref. 4
Guinea pig	Nylons 66, 6, 610, and 611	Decomposition at 500°C in air current: all 4 materials produced HCN, ammonia, CH$_4$, C$_2$H$_6$, C$_3$H$_2$, and C$_4$H$_{10}$; vol % of ammonia greatest in effluent from nylon 66. Guinea pigs exposed to gases of nylon 6 died when cyanide precipitate observed in effluent trap
Mouse	Nylon fiber/fabric Polycaprolactam	Products of 100% nylon (all 3 types) among the least toxic as judged by time to death among a series of polymers so tested. CO considered an important toxicant in the pyrolysis of polycaprolactam (procedure, Ref. 4)
Mouse	Nylon 66	Rapid deaths from products generated at 750°C attributed to HCN poisoning; no deaths in similar tests with 350 or 500°C decomposition (procedure, Ref. 4)
Mouse	Nylon 66; polyamide	Products generated at 750 but not 350 or 500°C lethal; death attributed to CO and HCN. Product generated at 850°C considered about twice as toxic as those generated at 550°C (procedure, Ref. 4)
Mouse	"Polyamide" film, analyzed as nylons 6 and 66	Exposure 1 h, film decomposed in airstream at 550°C. ALC = 79.6 g/polyamide per 1000 L of air. Main contents of effluent = 2700 ppm CO, 700 ppm ammonia, small amount of propylene. Some mice died during exposure showed >55% COHb; some died at 1–2 d with pulmonary edema
Mouse	Nylon 6 with inorganic additives	Decomposition in air ≥400°C; NaHCO$_3$, Na$_2$CO$_3$, Na borate additives decreased formation of HCN, and increased ammonia formation at 400°C
	Nylon 66	*In mg/g sample, at 50 or 100 L/h air flow, respectively: 194 or 205 mg CO, 563 or 590 mg CO$_2$, 4 or 10 mg NH$_3$, 26 or 31 mg HCN, 39 or 40 mg CH$_4$, 82 or 94 mg C$_2$H$_4$, 7 or 15 mg C$_2$H$_2$ (procedure Ref. 4)

Commercial nylon 6 contains 12% nitrogen	*Increased air flow results in decreased HCN evolved at temperatures 500–900°C. Evolution of HCN under nitrogen generally increases with temperature and is approximately proportional to nitrogen content at 900°C. Some ammonia, little formaldehyde evolved. See Ref. 4
Nylon 6	Degradation, 800°C in air: nylon generally evolved less HCN per weight of sample than acrylic fiber but more than wool Significant quantities of ammonia evolved; amounts increased as decomposition increased from 400 to 1000°C. Other materials detected included alkanes, aldehydes, alkenes, alkylamines; HCN not listed (procedure, Ref. 4)
Nylons 4, 7, 12, 66, 610 (also nylon 4, 10, nylon 11, 6, nylon 12, 12	Cyclic oligomers detected below 200°C thermal decomposition above 350°C. Mass spectra recorded at 170°C and ~400°C are characteristic
Nylons 66 and 610	Degradation of nylon 66 under vacuum ≤1000°C yielded mostly NH_3, H_2O, CO, CO_2, cyclopentanone, and several hydrocarbons; similar procedure with nylon 610 showed H_2O, CO, CO_2, 1,5-hexadiene and other hydrocarbons as the major products, no ammonia detected
Nylon 66	Escape of volatile material from heating nylon above its melting point results in rapid gelation and color formation, even in the absence of oxygen.
Nylon 66	HCN, CO, NH_3 evolved at 350°C in air, nitrogen oxides evolved at 600°C. Comparable temperatures for release of these gases in nitrogoen identical or within 50°C (polyacrylonitrile yielded HCN at 250°C in air and nitrogen)
Nylon 66	Degradation 310–380°C: "H_2O, CO_2, cyclopentanone, traces of saturated and unsaturated hydrocarbons; purification from water and acid polymerization catalysts increases stability and decreases yield of CO_2"
Nylon 6	Major products found during oxidative degradation to 1000°C were CO_2, H_2O, ε-caprolactam, methane, oligomers, also propene, propenenitrile, ethylene, acet-onitrile, HCN; pyrolysis in helium gave high yield of oligomers, less ε-caprolactam, as main products
Nylon 6 and 66	Slightly more ammonia, slightly less cyanide and nitrogen oxide from nylon 66 compared to nylon 6
Nylon	Molten droplets contained considerable energy but inhibited burning
Material proposed for aircraft use	
Chlorinated and fluorinated polya-mide: mouse	30-min exposure to products from pyrolysis at 700°C; LC_{50} values lowest in series of 7 polymers — approximately 1/3 that of chlorinated polyamide, 1/3 that of polyamide, and 1/10 that of polyvinyl fluoride based on weight of sample charged LC_{50} for fluorinated polyamide

549

Table 91.6. (*Continued*)

Polymer: Species Exposed	Description of Material	Procedure[b]/Observations/Conclusions
Aromatic polyamidie		
Rat	Poly(*p*-phenylene) terephthalamide (dust representative of workroom atmosphere)	4-h acute exposure, also 4 h/d, 5 d/wk × 2 wk; 150 and 130 mg/m^3 in acute and repeated exposures, respectively.[c] Inactivity, shallow respiration during exposures; slow weight gain during repeated exposure series, normal weight gain in 2-wk recovery period. Slight phagocytosis of foreign particles in lung found after tenth exposure, persistent to same degree 14 d later
Mouse	Aromatic polyamide	*Time to death longer than median among products of a group of polymers similarly tested (procedure, Ref. 4). Thermogravimetric analysis showed less weight loss than with polybenzimidazole until ~450°C
	Aramid fiber	Fast pyrolysis of 2.8 mg by introduction into 650°C furnace yielded 63% CO, 35% benzene, 2%.toluene, plus 0.6 mg residue. Slow pyrolysis at 40°C/min for 15 min yielded "at least 15 compounds and a series of nitrogen containing cyclic components [but] no ammonia or hydrogen cyanide was observed"
	Two samples of aromatic polyamide	Based on decomposition of wool and other materials in combustion tubes and under flaming conditions; gas yields more reproducible for gases such as CO and HCN; "the utility of measuring NO$_2$, SO$_2$, and HCHO for the purpose of ranking interior materials is questionable..."
	"Polyaramide" as aramid fiber woven blanket and as a resin system	Impingement under atmospheric O$_2$ conditions, also thermal radiation under atmospheric and low O$_2$ conditions. Polyaramide itself resistant to stress, few toxic combustion products formed. Addition of resin for structural strengthening "resulted in a rapidly deteriorating environment under all three stresses"
Rat	SP polymer (polyimide film)	Dust levels as high as 15 mg/L (nominal) produced signs of discomfort, inactivity, and deep difficult respiration. Pyrolysis products produced death in 20–140 min at 450–500°C, did not produce death in 4-h exposures ≤400°C. CO detected in effluent with pyrolysis temperature of 300°C. Rats dying after exposure to products generated at 450°C showed microscopic evidence of edema, congestion, and occasional hemorrhage in lungs, no effects in liver or kidneys; death considered due to either or both CO and lung irritant(s)

Mouse	Polyimide flexible foam[d]	Products of polyimide foams generally associated with relatively short times to death in rising- and fixed-temperature programs (see Ref. 4). With polyimide flexible foam, "the toxicants causing death were evolved at temperatures below 600°C;... the principal cause of death was not carbon monoxide, but probably hydrogen cyanide and/or nitrogen dioxide." Thermogravimetric analysis in air showed 1-step degradation starting at 500°C, with complete combustion at 650°C. CO, CO_2, ammonia, cyanide identified in effluent
	H-film	Degradation 300–510°C; thermal oxidation occurred at 426°C in air, degradation at 510°C in nitrogen. Vacuum stability of film exposed to air at elevated temperatures significantly less than that of film as received
	Cured polyimide resin	Decomposition in vacuum using simultaneous mass spectral and differential thermal analysis. "The volatile decomposition products were H_2, CO, CO_2, H_2O, HCN, and minor quantities of benzonitrile, benzene, methane, and ammonia"
	Polypyromellitimide H film	Decomposition in vacuum pyrolysis at 540°C yielded CO and CO_2 plus smaller or trace amounts of hydrogen, water hydrogen cyanide, benzene, benzonitrile
	Polypyromellitimide H film	Data "suggest that the primary scission occurs at the imide bonds, most likely followed by a secondary cleavage resulting in the elimination of CO groups"

[a] Summarized from Ref. 4. ALC = approximate lethal concentration.
[b] * = no effluent analysis.
[c] A low but unspecified concentration of particles was <10 μm in diameter. The concentration for these exposures was calculated from an increase in filter weight (obtained by drawing a known volume of sample) per liter of air.
[d] Polymethacrylimide and polybismaleimide rigid foams were also tested.

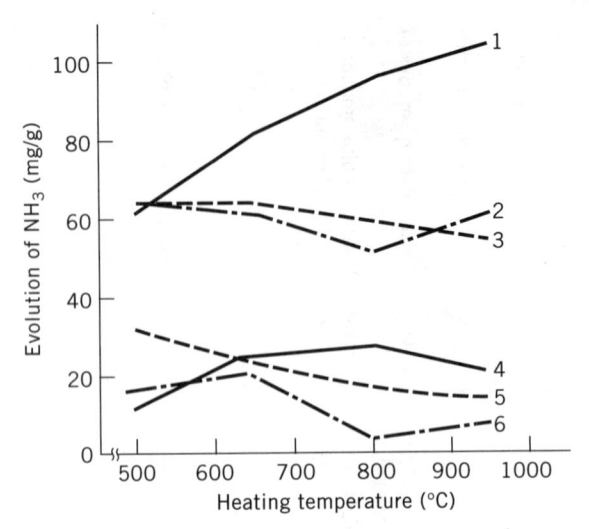

Figure 91.2. Ammonia from nylon. Materials: 1, melamine resin; 2, urea resin; 3, wool; 4, nylon 6; 5, polyacrylonitrile; 6, polyurethane flexible foam. Reproduced by permission of the *Journal of Combustion Toxicology*, Technomic Publishing Co., Inc. Westport, Connecticut.

high thermal energy (4). This does not necessarily result in an untoward toxic hazard. Only a full hazard assessment can ascertain that for a specific fire scenario (4).

13.0 Nylon

13.0.1 CAS Number: [63428-83-1]

13.0.2 Synonyms: Polyamide and polyamide resins

13.1 Chemical and Physical Properties

13.1.1 General

For useful properties as polyamide fibers or plastics the number-average molecular weight must be above 10,000 (75). Nylons are tough, strong, and have good abrasion resistance but are notch sensitive (63, 64, 77, 78). They have a low coefficient of friction and are suitable for use where lubricants are undesirable. Nylons show no appreciable change in aging properties at room temperature in indoor service or when protected from sunlight (37). When heat stabilized, their continuous use temperature is about 120°C (63). Nylons absorb a varying percentage of water; fibers are considered hydrophobic compared to cotton, whereas plastics are hygroscopic compared to many other synthetic polymers (4). Figure 91.3 shows the generic structures for some common nylons.

13.1.2 Odor and Warning Properties

No odor threshold or recognition data were located.

SYNTHETIC POLYMERS

$$H-[HN(CH_2)_6NH-CO(CH_2)_4CO]_n-OH$$

$$H-[HN(CH_2)_6NH-CO(CH_2)_8CO]_n-OH$$

$$H-[HN(CH_2)_6NH-CO(CH_2)_{10}CO]_n-OH$$

$$H-[HN(CH_2)_5CO]_n-OH$$

$$H-[HN(CH_2)_{10}CO]_n-OH$$

$$H-[HN(CH_2)_{11}CO]_n-OH$$

Figure 91.3. Structures of various nylon resins.

13.2 Production and Use

Nylons of one-monomer origin are commercially prepared either by self-condensation of an omega-amino acid or by opening of a lactam ring. The automotive industry is the single largest user of nylon, in uses ranging from nylon tire cord to emission control canisters, fuel filter bowls, and door strikers. Speedometer and window-lift gears are molded from glass-reinforced nylon. Industrially, nylon is used in gears for electric drills, cams and slides, and bearing housings. Plumbing, electrical, personal, and household applications include devices such as pipe fittings, plugs, sockets, hairbrushes, toothbrushes, and vacuum cleaner brushes (63).

Many other types of nylon have been developed (7, 28, 37, 77, 78). Copolymers such as nylon 66/610/6, with an irregular structure that limits interchain bonding and crystallization, are soluble in alcohols and many other common polar solvents. Methyl methoxy nylons can be prepared by dissolving a nylon (as in 90% formic acid) and then treating it with formaldehyde and an alcohol in the presence of an acidic catalyst. Polymers with about 33% of the –NH– groups substituted are soluble in lower aliphatic alcohols and phenols and can absorb up to 21% moisture when immersed in water (4).

13.3 Exposure Assessment

No specific analytical methodology was identified.

13.4 Toxic Effects

Experimental tests in rabbit corneas have shown that nylon thread (Tubinger nylon) is well tolerated in microsurgical sutures (79). Nylon has for some years been considered unsuitable as an endothesis (80). Fragmentation and crazing result in total loss of serviceability within a 1-yr period (81).

13.4.1 Experimental Studies

No known reports attribute any metastatic carcinogenic potential to nylon. IARC (71) considered that a definite assessment of the carcinogenicity of caprolactam and its polymer could not be made in view of the limited data available. Local sarcomas in rats have been reported after intraperitoneal implantation of nylon films, about 10 mm in diameter (71). A local tumorigenic response was also observed (20) when "nylon" was embedded as a flexible plain film, and to a lesser extent as a flexible perforated film, but not as a soft textile (7 of 26 rats, 2 of 31 rats, and 0 of 33 rats, respectively). The critical factor would most likely appear to be the physical form of the material (4).

The toxicity of the thermal degradation products of nylon is often attributed to carbon monoxide and/or hydrogen cyanide. Ammonia can be evolved in significant quantities and accounts for as much as half the nitrogen eliminated (77). The role of ammonia in bioassay tests does not appear to have been defined. Concentrations of the pyrolysis/combustion products are generally not lethal until decomposition occurs, which varies from approximately 400 to 750°C, depending on test method (4).

13.4.2 Human Experience

Clinical dermatitis due to unprocessed nylon appears to be rare, although it was reported in 1955 and 1960 (15). Nylon does not absorb perspiration readily and is lipophilic. Materials that contain stiff bristles sometimes produce temporary irritation. One instance of very mild erythema, at the initial 2-d reading only, was observed in one of 176 subjects exposed by the Schwartz–Peck technique to a sample of an untreated nylon carpet fiber (82). A recent modified Draize–Shelanski test (83), 10 24-h applications plus a 24-h challenge after a 2-wk rest period, showed no reaction in 100 subjects exposed to a sample of knitted nylon (84).

Dermatitis associated with the dyes and finishes on nylon (85), sometimes called "nylon stocking dermatitis," is now rare because of improved dyeing technique (15). However, a 1979 case report describes a patient who showed stronger skin reactions to nylon blanks dyed with the component colors than to the dyes alone (86). The patient showed no reaction to undyed nylon. Several reports from abroad describe inhalant reactions in humans in connection with occupational exposure (type of nylon unspecified). Human studies show that nylon sutures were well tolerated when examined at intervals from less than 1 yr to more than 5 yr after surgery (87).

13.5 Standards, Regulations, or Guidelines of Exposure

No specific standards or guidelines were identified.

13.6 Studies on Environmental Impact

Nylon polymers per se are generally not considered susceptible to growth of microorganisms. Growth on nylon may occur if a supply of extraneous nutrients is available

SYNTHETIC POLYMERS

(as from lubricants, plasticizers, or contaminating organic matter). Forty-two-day incubation of three strains of *Penicillium janthinellum* on agar containing strips of nylon 66 yielded an abundance of pink pigment in the agar and stains on the nylon strips (88). No significant changes in the tensile strength of nylon 66 occurred after samples were cultured on agar for 42 d with *P. janthinellum* and then buried in soil for 42 additional days.

14.0 Nylon 6

14.0.1 CAS Number: [25038-54-4]

14.0.2 Synonyms: Polyamide 6, Perlon, poly[imino(1-oxo-1,6-hexanediyl)], poly(caprolactam), Poly-epsilon-caprolactam, and policapram

14.0.3 Trade Names: NA

14.0.4 Molecular Weight: 131.17

14.0.5 Molecular Formula: $[C_6H_{13}NO_2]_n$

14.0.6 Molecular Structure: $H_2N-(CH_2)_5-C(=O)-OH$

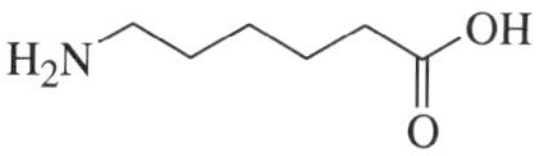

14.1 Chemical and Physical Properties

14.1.1 General

Nylons are tough, strong, and have good abrasion resistance but are notch-sensitive (63, 64, 77, 78). They have a low coefficient of friction and are suitable for use where lubricants are undesirable. Nylons show no appreciable change in aging properties at room temperature in indoor service or when protected from sunlight (37). When heat stabilized, their continuous use temperature is about 120°C (63). Nylons absorb a varying percentage of water; fibers are considered hydrophobic compared to cotton, whereas plastics are hygroscopic compared to many other synthetic polymers (4).

14.1.2 Odor and Warning Properties

No odor threshold or recognition data were located.

14.2 Production and Use

Nylon 6 is one of the main products in the American and European markets (4). Nylon 6 has been polymerized from epsilon-caprolactam by continuous and batch processes run at about 250°C. In a batch process, caprolactam, water that serves as catalyst, and a molecular weight regulator such as acetic acid may be reacted under a nitrogen blanket for about 12 h. Low-molecular-weight materials can be removed by leaching and/or vacuum distillation to achieve desired physical properties (37). Alternatively, residual caprolactam may serve as an effective plasticizer and not be extracted after polymerization has reached equilibrium (65). Casting in the mold with anionic polymerization has also been described (37).

The automotive industry is the single largest user of nylon, in uses ranging from nylon tire cord to emission control canisters, fuel filter bowls, and door strikers. Speedometer and window-lift gears are molded from glass-reinforced nylon. Industrially, nylon is used in gears for electric drills, cams and slides, and bearing housings. Plumbing, electrical, personal, and household applications include devices such as pipe fittings, plugs, sockets, hairbrushes, toothbrushes, and vacuum cleaner brushes (63).

14.3 Exposure Assessment

No analytical methodology was identified in the literature.

14.4 Toxic Effects

Nylon 6 was fed to rats at a dietary level of 25% for 2 wk. In the 2-wk test average weight and food consumption was normal except in the group fed nylon 6; these animals showed a slower rate of weight gain, a lowered food consumption, and a decreased ability to utilize their food effectively. No anatomic injury attributable to any of these resins was observed after the 2-wk study.

14.5 Standards, Regulations, or Guidelines of Exposure

No specific standards or guidelines were identified.

15.0 Nylon 11

15.0.1 CAS Number:

15.0.2 Synonyms: Polyamide 11

15.1 Chemical and Physical Properties

15.1.1 General

Nylon 11 is a speciality nylon noted particularly for low water absorption. Nylon 11 is a commercially important polymer prepared from an amino acid or derivative thereof (4). Table 91.5 provides some comparative data.

15.1.2 Odor and Warning Properties

No odor threshold or recognition data were located.

15.2 Production and Use

Nylon 11 is polymerized from omega-aminoundecanoic acid, a derivative of castor oil, at temperatures slightly above 200°C (4).

15.3 Exposure Assessment

No specific analytical methodology was identified.

SYNTHETIC POLYMERS

15.4 Toxic Effects

No toxicity data were located in the literature.

15.5 Standards, Regulations, or Guidelines of Exposure

No specific standards or guidelines were located.

16.0 Nylon 12

16.0.1 CAS Number: [24937-16-4]

16.0.2 Synonyms: Polyamide 12

16.1 Chemical and Physical Properties

16.1.1 General

Nylon 12 is a speciality nylon noted particularly for low water absorption. Nylon 12 is a commercially important polymer prepared from an amino acid or derivative thereof (4). Table 91.5 provides some comparative data.

16.1.2 Odor and Warning Properties

No odor threshold or recognition data were located.

16.2 Production and Use

Nylon 12 is prepared from lauryl lactam at somewhat higher polymerization temperatures, above those used for polymerization of caprolactam. Both reactions are normally catalyzed. These polymerization reactions essentially go to completion and a washing step is not considered necessary (92).

16.3 Exposure Assessment

No specific analytical methodology was identified.

16.4 Toxic Effects

No toxicity data for nylon 12 was located.

16.5 Standards, Regulations, or Guidelines of Exposure

No specific standards or guidelines were identified.

17.0 Nylon 66

17.0.1 CAS Number: [32131-17-2]

17.0.2 Synonyms: Polyamide 66, poly(hexamethylene adipamide), and poly(hexamethylene dodecanediamide)

17.1 Chemical and Physical Properties

17.1.1 General

The main products in the American and European markets are nylon 66. Nylon 66 is a commercially important polymer prepared from a diamine and a diacid. Nylon 66 has a melting point of 264°C and it starts to gel (cross-link) after prolonged steam heating (77).

17.2 Production and Use

Nylon 66 can be manufactured (37, 63) by mixing equimolar amounts of hexamethylene diamine and adipic acid in alcoholic or aqueous solution to form hexamethylene diammonium adipate (nylon salt). This salt, which has a melting point approximately 190°C, precipitates out of solution and is then subjected to heat and pressure to condense it to a high-molecular-weight polyamide. Polymers of this type can also be prepared by interfacial polymerization (76). The polymer forms almost instantaneously when an aqueous solution of one reactant (as hexamethylenediamine) contacts a nonmiscible solution of the other reactant (as adipyl chloride dissolved in a chlorinated hydrocarbon) (4).

17.3 Exposure Assessment

No analytical methodology for nylon 66 was located.

17.4 Toxic Effects

17.4.1 Experimental Studies

Sherman and his associates (90) fed nylon 66 at 10% of the diet to male and female rats for 2 yr and male dogs for 1 yr. Figure 91.4 shows a slight retardation in the rate of weight gain in male rats compared to controls, but no differences were observed in tibia lengths, hematologic or biochemical measurements, anatomic or histopathological lesions, or incidence and types of tumors (4).

17.4.2 Human Experience

Human volunteers exposed to unfinished nylon 66 fibers or fabrics by the Schwartz–Peck technique (83) occasionally show erythematous or papular reactions after 6-d occluded contact. Reactions other than at 6 d are uncommon and are not typical of sensitization (91).

17.5 Standards, Regulations, and Guidelines of Exposure

No specific standards or guidelines were identified.

18.0 Nylon 610

18.0.1 CAS Number: [9008-66-6]

18.0.2 Synonyms: Polyamide 610

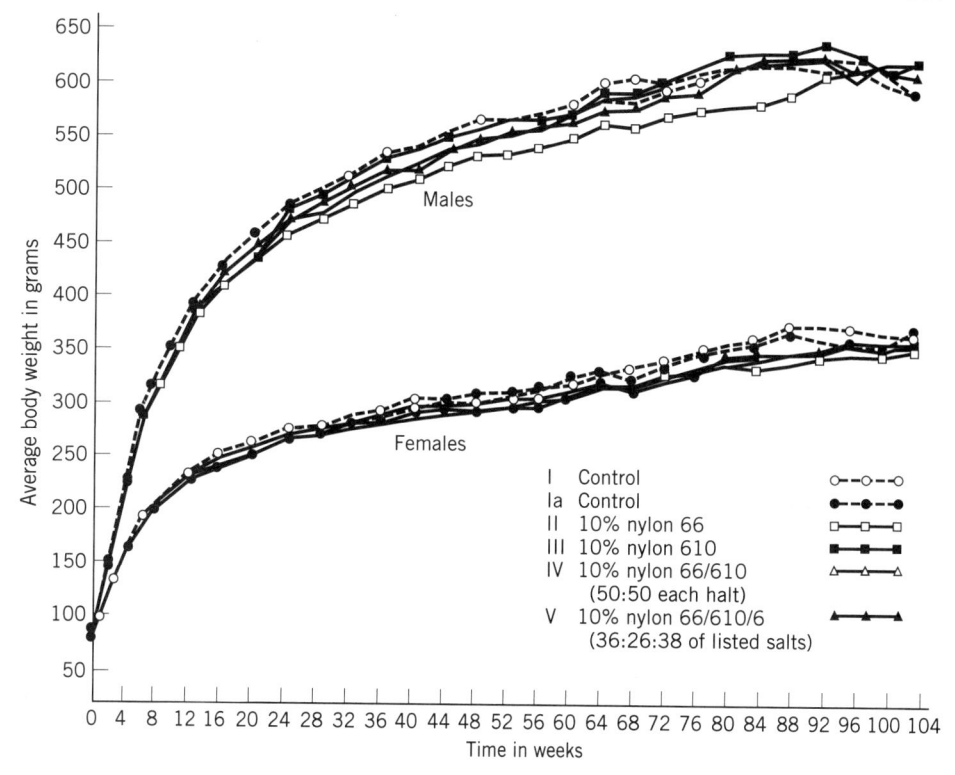

Figure 91.4. Growth response of male and female rats fed various nylon resins. Adapted with permission from H. Sherman et al. (90).

Legend:
- I   Control
- Ia Control
- II 10% nylon 66
- III 10% nylon 610
- IV 10% nylon 66/610 (50:50 each halt)
- V   10% nylon 66/610/6 (36:26:38 of listed salts)

18.1 Chemical and Physical Properties

18.1.1 General

Nylons 610 is considered a special-purpose nylon (63). Although the melting point of nylon 610 is lower than that of nylon 66 (215 and 264°C, respectively), nylon 610 has greater thermal stability (37).

18.1.2 Odor and Warning Properties

No threshold or recognition data were located.

18.2 Production and Use

Nylons 610 is prepared from hexamethylenediamine and sebacic acid by processes similar to those for nylon 66. Polymers of this type can also be prepared by interfacial polymerization (76). The polymer forms almost instantaneously when an aqueous solution of one reactant (as hexamethylenediamine) contacts a nonmiscible solution of the other reactant (as adipyl chloride dissolved in a chlorinated hydrocarbon).

18.3 Exposure Assessment

No analytical methodology for nylon 610 was located.

18.4 Toxic Effects

Dogs showed no toxic effects or tumor development. Similar tests with nylon 610 and two copolymers were devoid of any observed effect (4). The copolymers were composed, respectively, of (*1*) 36% nylon 66, 26% nylon 610, and 38% nylon 6, or (2) 50% each of nylon 66 and nylon 610.

18.5 Standards, Regulations, and Guidelines of Exposure

No specific standards or guidelines were identified.

19.0 Nylon 612

19.0.1 CAS Number:

19.0.2 Synonyms: Polyamide 612

19.1 Chemical and Physical Properties

19.1.1 General

Nylon 612 is considered a special-purpose nylon (63).

19.1.2 Odor and Warning Properties

No odor threshold or recognition data were located.

19.2 Production and Use

Nylon 612 is prepared from hexamethylenediamine and dodecanedoic acid by processes similar to those for nylon 66. Polymers of this type can also be prepared by interfacial polymerization (76). The polymer forms almost instantaneously when an aqueous solution of one reactant (as hexamethylenediamine) contacts a nonmiscible solution of the other reactant (as adipyl chloride dissolved in a chlorinated hydrocarbon).

19.3 Exposure Assessment

No analytical methodology foe nylon 612 was located.

19.4 Toxic Effects

Nylon 612 was fed to rats and dogs of both sexes for 90 d at a dietary level of 10% without any clinical, nutritional, hematologic, urinary, biochemical, or pathological evidence of toxicity (92).

SYNTHETIC POLYMERS

19.5 Standards, Regulations, and Guidelines of Exposure

No specific standards or guidelines were identified.

20.0a Fatty Acid Polyamides

These materials, sometimes simply called "fatty polyamides," are obtained by prepolymerizing fatty acids to rather variable structure (37). These dimers are then treated with amines (93), typically ethylenediamine (37), to form viscous liquids or brittle resins of low molecular weight (2,000–15,000). They are useful as hardeners or flexibilizers in epoxide resins, thixotropic paints, and adhesives (1).

20.0b Aromatic Polyamides

The current American products are most frequently aramid-type fibers, a polyamide fiber containing at least 85% of the specified aromatic amide structure. The polymers are generally formed by treating aromatic diamines and aromatic diacid chlorides in a solution polymerization process with an amide-type solvent (74). Several types of aramid fibers have now been developed (74); the basic structures appear to be isophthalamide or terephthalamide with variations such as amide hydrazide. The condensation product of m-phenylenediamine and isophthalic acid (37, 94) is

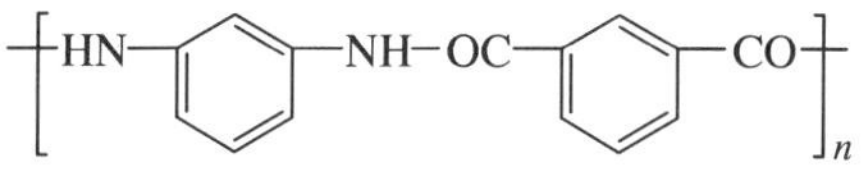

$$\left[HN-\underset{}{\bigcirc}-NH-OC-\underset{}{\bigcirc}-CO \right]_n$$

The condensation product of p-phenylenediamine and terephthaloyl chloride (74, 95) is

$$\left[HN-\underset{}{\bigcirc}-NH-OC-\underset{}{\bigcirc}-CO \right]_n$$

and the condensation products of p-aminobenzhydrazide and terephthaloyl chloride (74, 94) are described by

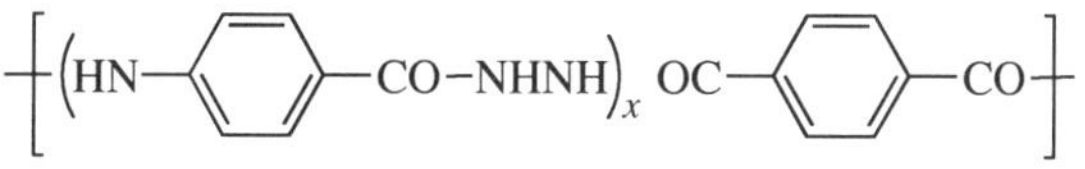

$$\left[\left(HN-\underset{}{\bigcirc}-CO-NHNH \right)_x OC-\underset{}{\bigcirc}-CO \right]$$

Other types of aromatic polyamide (94) include poly(quinazolinedione) fiber and transparent polyamide plastics that are mostly European in origin. The best known of the plastics is probably poly(trimethylhexamethylene terephthalamide). This polymer is obtained by treating terephthalic acid with an isomeric mixture of trimethylhexamethylenediamines (37, 94, 96).

20.1 Chemical and Physical Properties

Preston (74) notes that the molecular weight "probably should be 60,000 or greater" for the best balance of tensile properties. The aramid fibers have been developed for (1) heat and flame resistance or (2) ultrahigh strength and high modulus. Polymers of the first type usually contain primarily *m*-oriented phenylene rings, whereas those of the second type contain primarily *p*-oriented phenylene rings (588). The basic isophthalamide and terephthalamide structures shown have been modified with chlorine, fluorine, or phosphorus groups to provide particular properties. Aramid fibers characteristically burn with difficulty and produce a thick char. The limiting oxygen index of poly(*p*-phenylene terephthalamide) can be raised from approximately 28–30 up to 40–42 with the incorporation of 1% phosphorus (74). Table 91.5 and Figure 91.1 give some comparative data on properties.

20.2 Production and Use

The heat and flame resistance of the aramid fibers makes them suitable in such varied applications as filter bags for hot stack gases, insulation paper for electrical transformers, welder's clothing, and jump suits for forest fire fighters. Aramid pajamas and robes have been quite useful for certain nonambulatory patients. Aramid drapes are specified on some aircraft and all ships of the U.S. Navy. Aramid fibers of moderately high modulus are used in the reinforcement of fire hose and V-belts. Ultrahigh strength, high modulus fibers are used in tire cord, V-belts, cables, and body armor, and as reinforcement in aircraft and other vehicles of transportation.

20.4 Toxic Effects

20.4.1 Experimental Studies

When administered by intragastric intubation to rats, the approximate lethal dose of poly(*p*-phenylene terephthalamide) was greater than 7500 mg/kg (97). This was considered the maximum feasible acute dose.

Intratracheal insufflation tests in rats have been conducted to assess the respiratory tract reaction to particles of two aromatic polyamides. A single 0.5-mL injection of a 5% suspension of poly(*p*-phenylene terephthalamide) in 0.9% saline (98) was given under mild ether anesthesia. Rats were sacrificed serially at 2, 7, and 49 d, and also at 3, 6, and 12 mo; the last group of 10 rats was killed at 21 mo. Respirable-sized dust particles (<10 mm) caused only the minimal lung tissue reaction seen with inert-type dusts. Large nonrespirable dust particles (up to 150 mm) produced mild foreign body granulomas. After 21 mo the lung dust content, granulomatous reaction, and bronchiolar polyploid protrusions (considered a result of mechanical trauma) were decreased compared to those seen at 1 yr.

Rats similarly given a dose of 0.25 mL of a 1% suspension of poly(*m*-phenylene isophthalamide) (paper dust) were sacrificed at 6 mo, 1 yr, and 2 yr (99). The small-sized dust particles were mostly phagocytized within 2 yr. Large dust particles (up to 100 mm long and 20 mm wide) provoked giant cell reactions without any fibrosis or collagenization.

Rats inhaling the dust of poly(*p*-phenylene terephthalamide) at a maximum practicable concentration showed a depressed rate of weight gain, which was followed by normal gain when exposures were discontinued (see Table 91.7). The listed thermal degradation data for aromatic polyamides do not indicate that they release hydrogen cyanide or ammonia, but these eventualities should be anticipated if thermal decomposition occurs.

20.4.2 Human Experience

Skin tests conducted with two types of poly(*m*-phenylene isophthalamide) showed no sensitization reactions and no irritation reactions after 48-h contact. Moderate erythema was observed after 6 d of occluded contact in one of 207 persons tested (100). This isolated reaction was not considered sensitization and may well have been related to occlusion. No reactions were observed in the panel of 200 people tested with another sample (101).

20.5 Standards, Regulations, and Guidelines of Exposure

No specific standards or guidelines were identified.

21.0 Polyimides and Polyetherimides

Polyimides in general are used where high heat resistance is required. Polyimides may be found as either thermoplastic or thermosetting materials. Thermosetting polyimides may be processed by injection, transfer, extrusion, and compression methods. Many polyimides do not melt but must be fabricated by machining or other forming methods. True thermoplastic materials can be processed by traditional methods of thermoplastic processes.

21.1 Chemical and Physical Properties

The basic polyimide structure has a carbon–nitrogen linkage of the type:

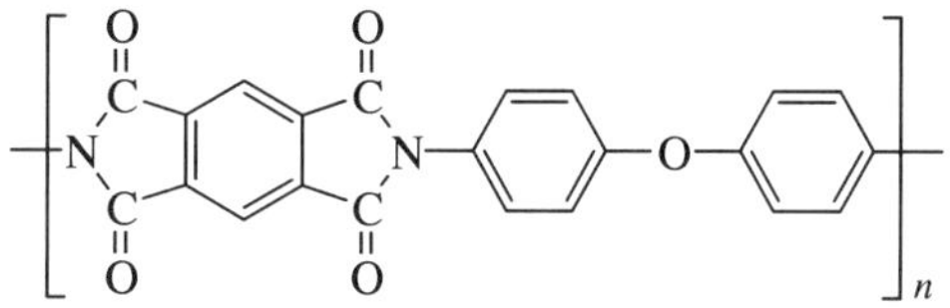

One variation of the polymer is the polyamide–imide structure:

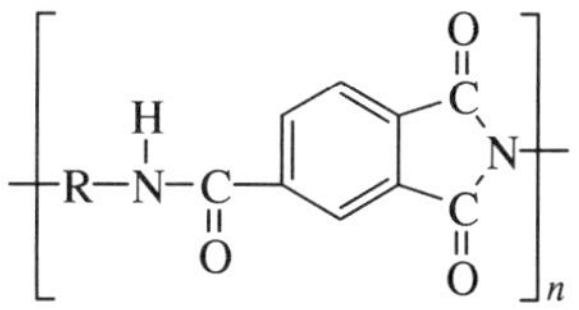

A group of polymers also contains polymers having the etherimide linkage:

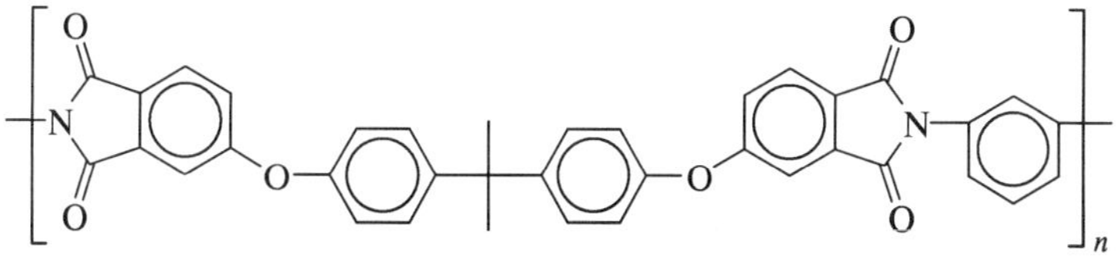

Finally, the formula for bismaleimide (BMI) is

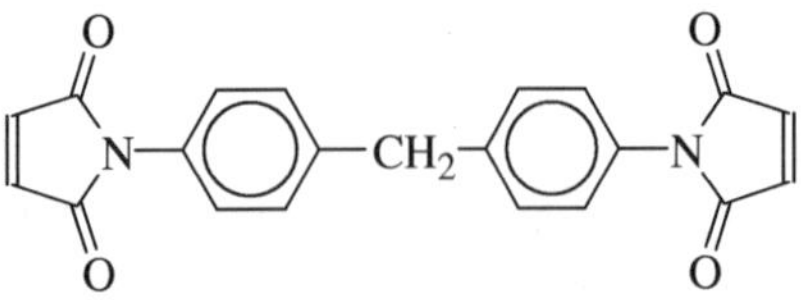

Polyimides exhibit good chemical resistance and are hydrolytically stable to acidic or neutral aqueous environments. However, almost all polyimides undergo hydrolytic degradation. Aromatic polypyromellitimides are insoluble in most organic solvents, but are soluble in strong acids like sulfuric or nitric acid.

In contrast, aromatic PEI displays a variety of solubilities in some solvents including chlorinated hydrocarbons (methylene chloride, chlorobenzene), phenols (*m*-cresol, chlorophenol), and dipolar aprotic solvents such as dimethylformamide, dimethyl acetate, and certain ketones (i.e., methyl ethyl ketone). Crystalline PEI resins behave very much like polypyromellitimides and are normally only soluble in strong acids. The solvent resistance of polyimides can be increased by cross-linking.

Polyimides offer excellent mechanical, physical, and electrical properties. The outstanding mechanical properties make them excellent candidates for replacement of metal, glass, and other materials in high-performance applications. Polyimides are tough and offer high impact strength, exhibiting high flexural modulus below the glass-transition temperature (T_g). Tensile, flexural, and compressive strength are excellent. Mechanical properties can be enhanced by inert additives such as glass, carbon powder, and minerals, which increase stiffness. Tensile elongation is normally modest (between 2 and 100%). Resistance to degradation by ultraviolet light is adequate for most applications. Polyimides are yellow to brown in color depending on the method of polymerization. If derived from highly purified monomers they range in color from yellow to nearly colorless, depending on the structure.

PEI resins are amorphous, amber in color with a high heat distortion temperature ($T_g = 215-230°C$) and high modulus. They can be processed by traditional thermoplastic processes, such as injection molding, extrusion, and thermoforming.

PEI resins are inherently flame resistant, with low smoke generation. Chemical resistance is high for amorphous resins. They are essentially unaffected by organic chemicals, automotive and aircraft fluids, fully halogenated hydrocarbons, and alcohols. Hydrolytic stability is excellent; however, exposure to partially chlorinated hydrocarbons and strong alkaline environments should be avoided.

21.2 Production and Use

Early synthesis of polyimides involved the reaction of aromatic tetracarboxylic acids, such as pyromellitic acid, with aromatic or aliphatic diamines to form a salt (67). Using aromatic monomers, the polyimides derived from long-chain aliphatic diamines were usually of high molecular weight (64, 68). This process was later improved by the use of an aromatic dianhydride and an aromatic diamine to form a high-molecular-weight polyamic acid.

This polyamide acid could then be cyclodehydrated by chemical (102) or thermal (103) means to yield a polyimide. Excess anhydride in these processes is needed to limit the molecular weight of the polymer, and moisture must be excluded from the reaction to prevent hydrolytic cleavage of the polyamide acid (104, 105). The evolution of solvent and condensation by-products during the thermal conversion can result in severe processing problems (37, 106). This has led to the development of other synthetic routes.

Polyetherimide resins are a new type of engineering thermoplastics. They can be produced by a nucleophillic substitution process or by condensation of diamines and dianhydrides (107). Aromatic polypyromellitimides are not processible in the molten state by conventional methods but are usually processed from polyamic acid precursors in dipolar aprotic solvents to yield thin films. Processing of plastic materials requires thermal stability at temperatures of at least $50-100°C$ above the normal processing temperatures. Polypyromellitimides in general do not meet this criteria for normal processing techniques. For nonmelt processes, polyimides are handled as powder or polyamic acid solutions. A combination of heat and pressure or vacuum is then used to form the polymer into a desired shape in the fully imidized form. The choice of processing method depends on the properties of the plastic and the desired end product.

Aromatic PEIs on the other hand are easily processed by conventional thermoplastic processing methods: extrusion, and injection and compression molding. Thermoplastic PEIs offer the additional advantage of being easily and readily recyclable into new products. Polypyroimides (thermosetting materials), however, are very difficult to recycle or are not recyclable at all.

The materials derived from polyamide acid are used as film and moldings. Film derived from polyamide acid has excellent mechanical properties and is used as an insulator and as substrate and packaging materials in the electronic industry, and as packaging materials for high-temperature materials.

PEI resins can be used in food packaging for microwave ovens. Certain PEI resins are used in many engineering applications where high heat deflection, high heat resistance, and chemical and flame resistance are needed. These resins can be blended with other polymers for applications having special requirements. For example, materials with high heat resistance, abrasion resistance, and improved high impact values can be obtained by blending a high heat resistance polyetherimide with a high impact polycarbonate. Blends with polyphenylene sulfide and polyetherimide are being developed to offer even higher chemical resistance properties. New technological developments have contributed to the increased use of high-performance engineering plastic materials, including polyimides. These materials find wide applications in the marketplace including ground and aerospace transportation industries, appliances, construction, cookware, food packaging, and electronics. Polyimides are widely used in load-bearing applications such as chassis and brackets in automotive and aircraft structures because of their high flexural modulus and compressive strength.

The high heat resistance and relatively high chemical resistance make polyimides excellent candidate materials for under-the-hood applications in the automotive industry. Polyimides are also excellent materials for cookware and food packaging because of their chemical resistance to oils, greases, and fats, their microwave transparency, and their thermal resistance. Polyimides can be used as films in high-temperature insulation materials and passivation layers in integrated circuits and flexible circuitry. Other applications include fibers, advance composites, and insulating tapes. Additionally, the low smoke and low flammability make polyimides excellent materials for aircraft interiors, furnishings, and wire insulation.

Typical uses for PEI include aircraft interior components, high-temperature switches, connectors, circuit boards, fuse bodies, gears, fans for computers and business machines, sterilizable surgical equipment, iron skirts, and microwave oven components.

21.4 Toxic Effects

21.4.1 Experimental Effects

Results from a 14-d study of rats fed a diet containing 25% polyimide resin showed that rats fed this diet had a slightly lower food consumption and a lower rate of weight gain compared to control rats fed 25% Alphacel *f* cellulose (108). Gross pathological evaluation showed chronic murine pneumonitis present at a slightly greater degree than in the control animals, but no microscopic changes were attributable to the polymer. After a 14-d recovery period food consumption and body weight returned to normal. Pathological examinations did not find significant changes, and microscopic examination of tissues showed no effect due to the test material.

Skin tests on guinea pigs showed no allergic reactions and generally no irritation. Mild transient erythema has been occasionally observed in more susceptible animals (108).

Toxicological aspects of plastic combustion products are very complex. Depending on a variety of factors, the qualitative and quantitative nature of chemicals present in the combustion product mixtures will vary. Factors contributing to toxicity of real-life fires include increased carbon monoxide and carbon dioxide, decrease of oxygen, and the presence of irritant gases. The primary effects from fires are to tissues and irritation of the respiratory tract by combustion gases. A study using animal models has shown that inhalation of combustion products generated at temperatures above 450°C can be lethal (109). Additional studies using a plastic pyrolysis approach to simulate plastic combustion have also shown that combustion products from imide-type polymers can be very toxic (110, 111).

The estimated acute oral (rat) LD_{50} of PEI resins is greater than 10 g/kg (112). PEI resins are not considered to be skin irritants. The Draize Skin Primary Irritation Score (rabbit) for this product in finely divided form, for a 24-h response, is zero. It is not expected to be a skin sensitizer based on results of the Modified Buehler Guinea Pig Sensitization Test. The estimated acute dermal toxicity LD_{50} (rabbit) is greater than 2 g/kg (113).

PEI resins are not considered to be primary eye irritants. When finely divided PEI resins were placed into the eyes of rabbits, slight transient redness or discharged occurred, consistent with the expected slightly abrasive nature of the resin particles (624).

PEI resin is not a mutagen by Ames (*Salmonella*) assay with and without activation (114). In acute inhalation tests, laboratory rats were exposed to processing fumes at concentrations exaggerating those that would likely occur in workplace situations. No deaths or signs of toxicity, except transient irritancy in some cases, were noted during the 6-h fume exposure tests.

There were no distinct or consistent treatment related tissue or organ changes noted in gross necropsies (115).

21.4.2 Human Experience

Dianhydride monomers are skin and eye irritants. Some aromatic diamines used in polyimide synthesis are suspect carcinogens (4).

21.5 Standards, Regulations, and Guidelines of Exposure

No specific standards or regulations were identified. Worker protection should be designed to control thermal decomposition products. Thermal decomposition at high temperatures can result in volatile products such as carbon monoxide, carbon dioxide, water, and small amounts of aromatic compounds such as benzene, phenolics, and aniline. The nature and type of volatile by-products depend on the polymer structure and the process used in its manufacture and conditioning preceding the decomposition process. The presence of aliphatic hydrocarbon segments reduces thermal and oxidative stabilities, decomposing at lower temperatures and at a higher rate than purely aromatic polymers. However, these same aliphatic hydrocarbons increase flexural strength and impact properties. When the aliphatic carbons are attached directly to the imide nitrogen, the resins tend to decompose more quickly and at lower temperatures than those in which the aliphatic carbons are attached elsewhere in the molecule (4).

Polyimides that decompose at higher rates tend to do so over a narrow temperature range, whereas those with lower decomposition rates tend to decompose over a broader temperature range. Oxidative stability of most polyimides drops fairly quickly above the glass-transition temperature. Most probably this is caused by increased migration of molecular oxygen in the plastic bulk (4).

Processing fumes of PEI, like any other plastic materials, highly depend on product formulation and processing conditions. In general, processing fumes from polyether-imide processed at recommended processing conditions may include trace levels of o-dichlorobenzene and phenol. In some instances additives containing certain heavy metal compounds may be present. These ingredients are essentially bound in the plastic matrix and are unlikely to contribute to workplace exposure under recommended processing conditions (4).

A continuous supply of fresh air to the workplace together with removal of processing fumes through exhaust systems is recommended. Processing fume condensate may be a fire hazard and toxic; it should be removed periodically from exhaust hoods, duct work, and other surfaces using appropriate personal protection. Ventilation requirements must be locally determined to limit exposure to materials at their point of use (4).

Processing fumes may cause irritation to the eyes, skin, and respiratory tract, and in cases of severe overexposure, nausea and headaches. Fumes from plastic thermal processes may condense on processing equipment, ventilation systems, and other surfaces. These grease like processing fume condensates can cause irritation and injury to skin. Use of proper personal protection, including gloves, eye protection, and appropriate respiratory protection where the condensate is heated and/or solvents are used, should be observed while cleaning or handling these condensates (4).

For thermoset-type materials, safe handling of all monomers, prepolymers, solvents, and additives used in the processes of polyimides should be practiced. Dianhydride monomers are skin and eye irritants. Appropriate personal protective equipment, including clothing, gloves, and eye protection should be used. Some aromatic diamines used in polyimide synthesis are suspect carcinogens and should be handled with proper protection. Inhalation or ingestion of these substances should be avoided. Before using any new material, one is advisable to check the MSDS from the product manufacturer for specific instructions and precautions (4).

BIBLIOGRAPHY

1. R. E. Scales, Cellulosics. In *Modern Plastics Encyclopedia*, McGraw-Hill, New York, 1990, p. 23.
2. S. Budvari et al., *The Merck Index*. 12th ed., Whitestation, NJ, 1996.
3. R. L. Mitchell and G. C. Daul, Rayon. In *Encyclopedia of Polymer Science and Technology*, Vol. 11, Wiley, New York, 1969, pp. 810–847.
4. L. R. Harris and D. G. Sarvadi, Synthetic Polymers. In G. D. Clayton and F. E. Clayton, Eds., *Patty's Industrial Hygiene and Toxicology*, 4th rev. ed., Vol. 2B, Chapter 37, Wiley, New York, 1994.
5. L. R. Whittington, *Whittington's Dictionary of Plastics*, Technomic, Lancaster, PA, 1978.
6. *Kirk-Othmer Encyclopedia of Chemical Technology*, Vol. 10, Wiley, New York, 1980, p. 151.
7. M. P. Stevens, *Polymer Chemistry; An Introduction*, Addison-Wesley, Reading, MA, 1975.
8. P. J. Vanderhorst, DuPont Company, Plastics Products and Resins Dept., Wilmington, DE, May 1979, personal communication.
9. H. I. Bolker, *Natural and Synthetic Polymers*, Marcel Dekker, New York, 1974.
10. A. F. Turbak et al., Cellulose. In *Encyclopedia of Chemical Technology*, Vol. 5, 1979, pp. 70–88.
11. R. T. Bogan, C. M. Kuo, and R. J. Brewer, Cellulose derivatives, esters. In *Encyclopedia of Chemical Technology*, 3rd ed., Vol. 5, Wiley, New York, 1979, pp. 118–143.
12. G. K. Greminger, Cellulose derivates, ethers. In *Encyclopedia of Chemical Technology*, 3rd ed., Vol. 5, Wiley, New York, 1979, pp. 143–163.
13. G. A. Serad and J. R. Sanders, Cellulose acetate and triacetate fibers. In *Encyclopedia of Chemical Technology*, 3rd ed., Vol. 5, Wiley, New York, 1979, pp. 89–117.
14. C. L. Hake and V. K. Rowe, Ethers. In F. A. Patty, Ed., *Industrial Hygiene and Toxicology*, 2nd ed., Vol. 2, Wiley, New York, 1963, pp. 1655–1718.
14a. NIOSH, *Manual of Analytical Methods*, 4th ed., US Supt of Docs, Govt Printing Office, 1994.
15. A. A. Fisher, *Contact Dermatitis*, 2nd ed., Lea & Febiger, Philadelphia, PA, 1973.
16. W. F. Schorr, E. Keran, and E. Poltka, Formaldehyde Allergy. *Arch. Dermatol.* **110**, 73–76 (1974).
17. National Library of Medicine, TOXNET Databases, as of Feb. 1999.
18. G. M. Adamson et al., Effects of different cellulose-containing respirable samples in the lung of Fischer 344 Rats. *Toxicol. Sci.* **48**, 117 (1999).
19. B. S. Oppenheimer, E. T. Oppenheimer, and A. P. Stout, Sarcomas induced in rodents by imbedding various plastic films. *Proc. Soc. Exp. Biol. Med.* **79**, 366–369 (1952).

20. B. S. Oppenheimer et al., Further studies of polymers as carcinogenic agents in animals. *Proc. Am. Assoc. Cancer Res.* **2**, 333–340 (1955).
21. B. S. Oppenheimer et al., The latent period in carcinogenesis by plastics in rats and its relation to the presarcomatous stage. *Cancer* **11**, 204–213 (1958).
22. C. Beyler, Thermal decomposition of polymers. In P. J. DiNenno, Ed., *The SFPE Handbook of Fire Protection Engineering*, National Fire Protection Association, Quincy, MA, 1988, Chapts. 1–12.
23. *Documentation of the Threshold Limit Values for Substances in Workroom Air*, American Conference of Governmental Industrial Hygienists, Cincinnati, OH, 1980.
24. H. L. Kaplan et al., Effects of combustion gases on escape performance of the baboon and the rat. *J. Fire Sci.* **3**(4), 228–244 (1985).
25. B. C. Levin et al., *Further Development of a Test Method for Assessment of the Acute Inhalation Toxicity of Combustion Products*, NBSIR 82-2532, National Institute of Standards and Technology, Gaithersburg, MD, 1982.
26. V. Babrauskas et al., Toxic potency measurement for fire hazard analysis. *NIST Special Publication 827*, National Institute of Standards and Technology, Gaithersburg, MD, 1991.
27. G. Kimmerle and F. H. Prager, The relative toxicity of pyrolysis products. Part II. polyisocyanate-based foam materials. *J. Comb. Toxicol.* **7**, 54–68 (1980).
28. M. I. Kohan, Ed., *Nylon Plastics*, Wiley, New York, 1973.
29. B. A. Zikria et al., Smoke and carbon monoxide poisoning in fire victims. *J. Trauma* **12**, 641–645 (1972).
30. J. B. Terrill, R. R. Montgomery, and C. F. Reinhardt, Toxic gases from fires. *Science* **200**, 1343–1347 (1978).
31. *Threshold Limit Values for Chemical Substances and Physical Agents and Biological Exposure Indices*, American Conference of Governmental Industrial Hygienists, 2000.
32. National Institute for Occupational Safety and Health, *Recommendations for Occupational Health Standards (Recommended Exposure Limits)*, Cincinnati, OH, 1998.
33. Occupational Safety and Health Standards (Permissible Exposure Levels). *Code of Federal Regulations*, 29, §1910.1000, Table Z-1A, 1998.
34. W. P. Jordan and M. V. Dahl, Contact dermatitis from cellulose ester plastics. *Arch. Dermatol.* **105**, 880–885 (1972).
35. N. Hjorth, Contact dermatitis from cellulose acetate film. *Berufsdermatosen* **12**, 86–100 (1964).
36. R. E. Scales, Cellulosic. In *Modern Plastics Encyclopedia*, 1978–1979 ed., Vol. 55, No. 10A, McGraw-Hill, New York, pp. 13–14.
37. J. A. Brydson, *Plastics Materials*, 3rd ed., Whitefriars Press, London, 1975.
38. S. A. Peak, *Forensic Sci.* **25**, 679–681 (1980).
39. J. B. F. Lloyd, *Anal. Chem.* **56**, 1907–1912 (1984).
40. J. Epstein et al., Environmental Quality Standards, ISS ARLCSL-TR-77025, AD-E410040, 1978.
41. J. J. Barkley and D. H. Rosenblatt, *Automated Nitrocellulose Analysis*, USAMBRDL-TR-7807, 1978.
42. G. M. Marsh, Mortality among workers from a plastics producing plant: a matched case-control study nested in a retrospective cohort study. *J. Occup. Med.* **25**, 219–230 (1983).
43. J. H. Barth, Colophony sensitivity — a regional variant. *Contact Dermatitis* **7**, 165–166 (1981).

44. R. E. Bentley et al., Laboratory evaluation of the toxicity of nitrocellulose to aquatic organisms. U. S. Army Medical Research and Development Command, Washington, DC, AD A037749, 1977.
45. T. M. Wendt and A. M. Kaplan, A chemical-biological treatment process for cellulose nitrate disposal. *J. Water Pollut. Contr. Fed.* **48**, 660–668 (1976).
46. S. Strutsman, in *Newberger's Manual of Cosmetic Analysis*, 2nd ed., 1977, p 72.
47. R. E. Klose and M. Glicksman, Gums. in T. E. Furia, Ed., *Handbook of Food Additives*, 2nd ed., CRC Press, Cleveland, OH, 1972, pp. 295–359.
48. E. Bachmann et al., Biochemical effects of gum arabic, gum tragacanth, methylcellulose and carboxymethylcellulose-Na in rat heart and liver. *Pharmacology* **17**, 39–49 (1978).
49. E. P. Samsel and R. A. Delap, *Anal. Chem.* **23**, 1795 (1951).
50. S. B. McCollister, R. J. Kociba, and D. D. McCollister, Dietary feeding studies of methylcellulose and hydroxypropylmethylcellulose in rats and dogs. *Food Cosmet. Toxicol.* **11**, 943–953 (1973).
51. W. H. Braun, J. C. Ramsey, and P. J. Gehring, The lack of significant absorption of methylcellulose, viscosity 3300cP, from the gastrointestinal tract following single and multiple oral doses to the rat. *Food Cosmet. Toxicol.* **12**, 373–376 (1974).
52. M. G. Crane et al., Excessive Fluid retention related to cellulose ingestion: studies on two Patients. *Metabolism* **18**, 945–960 (1969).
53. F. A. Blanchard, I. T. Takahashi, and H. C. Alexander, Biodegradability of [^{14}C] methylcellulose by activated sludge. *Appl. Environ. Microbiol.* **32**, 557–560 (1976).
54. A. Osol et al. *Remington's Pharmaceutical Sciences*. 15th ed., Easton, PA, 1975.
55. Hercules Inc., *Natrosol f 250*; Summary of Toxicological Investigations, Coatings & Specialty Products Dept., Bulletin T-101B, Wilmington, DE.
56. Hercules Inc., *Klucel f Hydroxypropyl Cellulose- Summary of Toxicological Investigations*, Coatings & Specialty Products Dept. Bulletin T-122, Wilmington, DE.
57. J. Guettner et al. *Zwierzeta Lab.* **12**, 5 (1975).
58. J. Guettner et al. *Anat. Anz.* **149**, 282 (1981).
59. T. F. McElligott and E. W. Hurst, Long-term feeding studies of methyl ethyl cellulose ('Edifas' A) and sodium carboxymethyl cellulose ('Edifas' B) in rats and mice. *Food Cosmet. Toxicol.* **6**, 449–460 (1968).
60. T. Hamada and S. Horiguchi, Allergic contact dermatitis due to sodium carboxymethyl cellulose. *Contact Dermatitis* **4**, 244 (1978).
61. M. G. Wirick, Aerobic biodegradation of carboxymethylcellulose. *J. Water Pollut. Contr. Fed.* **46**, 512–521 (1974).
62. R. W. Moncrieff, *Man Made Fibres*, 6th ed., Wiley, New York, 1975.
63. M. Meisters, Nylon, in *Modern Plastics Encyclopedia*, 1978–1979 ed., Vol. 55, No. 10A, McGraw-Hill, New York, pp. 28–30.
64. M. I. Besenov et al., *Polyimides Thermally Stable Polymers*, Consultant Bureau, New York, 1987.
65. E. M. Hicks et al., The production of synthetic-polymer fibres. *Text. Progr.* **3**, 1–113 (1971).
66. U. S. Federal Trade Commission, Rules and Regulations Under the Textile Fiber Products Identification Act, Washington, DC, effective March 3, 1960, as amended to November 1, 1974.
67. C. E. Sroog, *Encyclopedia of Science and Technology*, Vol. 11, 1969, p. 247.

68. K. L. Mittal, *Polyimide Synthesis, Characterization, and Applications*, Plenum Press, New York, 1987.
69. C. E. Sroog, Polyimides. *J. Polym. Sci. Part C* **16**, 1191–1209 (1967).
70. F. P. Recchia and W. J. Farrissey, Jr., Polyimide; thermoplastic polyimide. In *Modern Plastics Encyclopedia*, Vol. 55, No. 10A, McGraw-Hill, New York, 1978–1979 ed., pp. 66.
71. International Agency for Research on Cancer, *IARC Monographs on the Evaluation of the Carcinogenic Risk of Chemicals to Humans*; Some Monomers, Plastics and Synthetic Elastomers and Acrolein, Vol. 19, Lyon, France, 1979.
72. A. Anton, Polyamides. In *Encyclopedia of Industrial Chemical Analysis*, Vol. 17, Wiley, 1973, pp. 275–306.
73. *U.S. Code of Federal Regulations*, US Supt. Doc., Washington, DC, 21 CFR §170–199, April 1, 1981; 21 CFR §500–599. Revised annually. Guide available from Food Chemical News.
74. J. Preston, Aromatic polyamide fibers. In *Encyclopedia of Polymer Science and Technology*, Vol. 2, Wiley, New York, 1977, pp. 84–112.
75. F. M. Esposito and Y. Alarie, Inhalation toxicity of carbon monoxide and hydrogen cyanide gases released during the thermal decomposition of polymers. *J. Fire Sci.* **6**(3), 195–242 (1988).
76. W. J. Roff and J. R. Scott, with J. Pacitti, *Handbook of Common Polymers*, CRC Press, Cleveland, OH, 1971.
77. W. Sweeny and J. Zimmerman, Polyamides. In *Encyclopedia of Polymer Science and Technology*, Vol. 10, Wiley, New York, 1969, pp. 483–597.
78. O. E. Snider and R. J. Richardson, Polyamide Fibers. In *Encyclopedia of Polymer Science and Technology*, Vol. 10, Wiley, New York, 1969, pp. 347–460.
79. J. Faulborn, Studies on the tolerance of silk, nylon, dacron, and collagen suture material in the cornea of the rabbit. *Adv. Ophthalmol.* **30**, 50–54 (1975).
80. R. I. Leininger, Changes in properties of plastics during implantation. In *American Society for Testing and Materials*. Plastics in Surgical Implants, ASTM Special Technical Publication No. 386, Philadelphia, PA, 1965, pp. 71–76.
81. E. Roggendorf, The biostability of silicone rubbers, a polyamide, and a polyester. *J. Biomed. Mater. Res.* **10**, 123–143 (1976).
82. N. C. Goodman, DuPont Company, Haskell Laboratory, Wilmington, DE, June 1976, unpublished data.
83. D. B. Hood and A. Ivanova-Binova, The skin as a portal of entry and the effects of chemicals on the skin, mucous membranes and eye. In *Principles and Methods for Evaluating the Toxicity of Chemicals*, Part II, Environmental Health Criteria Series 6, World Health Organization, Geneva, 1996.
84. J. F. Wilson and K. L. Gabriel, Biosearch, Inc., Philadelphia, PA, Oct. 1979, unpublished data.
85. A. J. Fleming, The provocative test for assaying the dermatitis hazards of dyes and finishes used on nylon. *J. Invest. Dermatol.* **10**, 281–291 (1948).
86. S. A. Imbeau and C. E. Reed, Nylon stocking dermatitis; an unusual example. *Contact Dermatitis* **5**, 163–164 (1979).
87. R. W. Postlethwait, D. A. Willigan, and A. W. Ulin, Human tissue reaction to sutures. *Ann. Surg.* **181**, 144–150 (1975).
88. M. R. Rogers and A. M. Kaplan, Effects of penicillium janthinellum on parachute nylon—is there microbial deterioration? *Int. Biodeterior. Bull.* **7**, 15–24 (1971).

89. E. C. Schule, Polyamide Plastics. In *Encyclopedia of Polymer Science and Technology*, Vol. 10, Wiley, New York, 1969, pp. 460–482.
90. H. Sherman et al., DuPont Company, Haskell Laboratory, Wilmington, DE, March 1959, unpublished data.
91. R. W. Morrow, DuPont Company, Haskell Laboratory, Wilmington, DE, February 1980, personal communication.
92. H. Sherman, DuPont Company, Haskell Laboratory, Wilmington, DE, April 1971, unpublished data.
93. D. E. Peerman, Polyamides from fatty acids. In *Encyclopedia of Polymer Science and Technology*, Vol. 10, Wiley, New York, 1969, pp. 597–615.
94. H. G. Elias, *New Commercial Polymers 1969–1975*, Gordon and Breach Science Publishers, New York, 1977.
95. T. I. Bair, P. W. Morgan, and F. L. Killian, Poly(1,4-phenyleneterephthalamides). polymerization and novel liquid-crystalline solutions. *Macromolecules* **10**, 1396–1400 (1977).
96. J. B. Titus, *New Plastics: Properties, Processing, and Potential Uses*, U. S. Army Armament Research and Development Command, Dover, NJ, 1977 (AD A056990).
97. N. C. Goodman, DuPont Company, Haskell Laboratory, Wilmington, DE, Nov. 1974, unpublished data.
98. D. P. Kelly, E. I. Du Pont de Nemours and Company, Inc., Haskell Laboratory, Wilmington, DE, Sept. 1979, unpublished data.
99. J. B. Terrill, DuPont Company, Haskell Laboratory, Wilmington, DE, June 1977, unpublished data.
100. J. W. McAlack, DuPont Company, Haskell Laboratory, Wilmington, DE, June 1973, unpublished data.
101. A. M. Kligman, Ivy Research Laboratories, Inc., Philadelphia, PA, April 1977, unpublished data.
102. J. A. Kreuz et al., *J. Polym. Sci.* Part A-1, **4**, 260 (1966).
103. U. S. Pat. 3073785 (1963); U. S. Pat. 3282878 (1967), R. J. Angelo.
104. C. E. Sroog et al., *J. Polym. Sci. Part A-3*, 1374 (1965).
105. S. R. Sandler and W. Karo, *Polymer Synthesis*, Academic Press, New York, Vol. 1, 1974, p. 216.
106. S. A. Zakoshchikov et al., *Sov. Plast.* **4**, 13 (1967).
107. U. S. Pat. 3875116 (April 1, 1975), D. R. Heath and R. G. Wirth, (to General Electric Co.)
108. R. S. Waritz and S. D. Morrison, DuPont Company, Haskell Laboratory, Wilmington, DE, Feb. 1965, unpublished data.
109. C. J. Hilado, The practical use of the USF toxicity screening test method. *J. Combust. Toxicol.* **5**, 331–338 (1978).
110. C. Arnold, Jr. and L. K. Borgman, Chemistry and kinetics of polyimide degradation. *Ind. Eng. Chem. Prod. Res. Develop.* **11**, 322–325 (1972).
111. T. H. Johnson and C. A. Gaulin, Thermal decomposition of polyimides in vacuum. *J. Macromol. Sci. Chem.* **A**3, 1161–1182 (1969).
112. R. H. Cox, for GE Plastics, TPS Inc., Nov., 1980, unpublished data.
113. S. M. Rosemberger, for GE Plastics, TPS Inc., May, 1981, unpublished data.
114. S. M. Rosemberger, for GE Plastics, TPS Inc., Jan., 1982, unpublished data.
115. D. J. Brusick, for GE Plastics, Litton Bionetics, April, 1982, unpublished data.

CHAPTER NINETY-TWO

Synthetic Polymers, Polyesters, Polyethers, and Related Polymers

Steven T. Cragg, Ph.D., DABT

1 OVERVIEW

Commercial use of polyester resins dates from the early twentieth century, when alkyd resins were first used in surface coatings (1). The polyesters are found today as fibers, films, laminating resins, molding resins, and engineering plastics. The high-molecular-weight polyethers are known primarily as engineering plastics (an exception is high-molecular-weight polyoxyethylene, a water-soluble packaging polymer), as are the polysulfides and the polysulfones. Table 92.1 gives some basic properties.

1.1 Production and Processing

Production data for the general categories are listed in Table 92.2. Processing techniques vary widely and are discussed in the appropriate subsection.

1.2 Specifications and Test Methods

Clearance for certain food contact applications with many of these materials has been obtained under Title 21 of the Code of Federal Regulations (2). Those specifically listed include polyethylene terephthalate, polybutylene terephthalate, polycarbonate, polyoxymethylene, polyoxyethylene derivatives, polyphenylene sulfide, and polysulfone. Specifications concerning mechanical and other types of performance depend on end use.

Patty's Toxicology, Fifth Edition, Volume 7, Edited by Eula Bingham, Barbara Cohrssen, and Charles H. Powell.
ISBN 0-471-31940-6 © 2001 John Wiley & Sons, Inc.

Table 92.1. Selected Properties of Some Polyesters, Polyethers, and Related Polymers[a]

A. Polyesters

Property	Polyethylene Terephthalate	Polybutylene Terephthalate	Polycarbonate of Bisphenol A
Melting temperature (°C)	254–284; 256 (commercial PET); 271 (highly crystalline PET)	221–232	215–230
Glass-transition temperature (°C)	69, 67, amorphous; 81, crystalline; 125, crystalline and oriented	17–80; 22–43; unfilled, 50	145–149
Service temperature (°C)	−60 to +150	Can be used for prolonged periods of time at 120–140°C. Embrittlement by hydrolysis on long-term aging at 60–85°C under humid conditions	135 maximum
Density	1.33 (amorphous); 1.45 (crystalline)	1.31–1.32	1.20
Refractive index, n_D	Film: amorphous, 1.5760; crystalline and biaxially oriented, 1.64		1.585
Moisture absorption 0.8%, immersion	0.55%, 24-h immersion of commercial films; 0.8%, immersion in water at 25°C for 1 wk	0.08%, 24-h immersion at 23°C	0.2% in air at 60% relative humidity
Moisture regain (fiber)	0.4%, commercial value and normal conditions		
Solvents	Crystalline: choralhydrate, phenol, phenol tetrachloroethane, (1:1 vol.), nitrobenzene, DMSO (hot)	"Complex phenols" such as o-chlorophenol	Methylene chloride and chloroform; less soluble in tetrachloroethane, trichloroethylene, dichloroethane, tetrahydrofuran, dioxane, cyclohexanone, dimethylformamide. Swells in benzene, chlorobenzene, acetone, ethyl acetate, carbon tetrachloride

Nonsolvents/relatively unaffected by	Crystalline: hydrocarbons, chlorinated hydrocarbons, aliphatic alcohols, ketones, carboxylic esters, ethers.	Very resistant to most chemicals	Hydrocarbons, styrene, carbon tetrachloride, acetone, lower esters.
Decomposes	Strong acids and bases, particularly when hot. Hydrolyzes slowly in water at elevated temperature (fibers with ~20% loss in strength after 1 week at 100°C, without measurable loss in strength after several weeks at 70°C		Hot alcoholic alkalies, amines, and other organic bases; surface attack by aqueous alkali. Hydrolyzes in water > 600°C (can withstand relatively short exposures)

B. Polyethers

Property	Polyoxymethylene	Polyoxyethylene	Polyphenylene Oxide
Melting point (°C)	175–200		298 (as poly-p-phenylene oxide); 261–272 [as poly(2,6-dimethyl-p-phenylene oxide]
Glass-transition temperature			105–120 (phenylene oxide-based resin)
Density	1.40–1.42 for molded parts, range to 1.56 for special grades of resin		1.314 (as poly-2,6-dimethyl-p-phenylene oxide); 1.408 (as poly-p-phenylene oxide)
Refractive index, n_D	1.489–1.553	1.51–1.54	
Water absorption, 24 h (%)	0.02–0.3		0.006
Solvents	At elevated, temperature; benzyl alcohol, phenol, chlorophenols, aniline, formamide, DMF, γ-butyrolactone, bromobenzene, diphenylether	Benzene, chloroform, carbon tetrachloride, alcohols, cyclohexanone esters, DMF, water (cold), aqueous, K_2SO_4 (0.45 M above 35°C); (swells in dioxane).	Poly(oxy-2,6-dimethyl-1,4-phenylene)[b] amorphous, α-pinine (hot); crystalline, benzene, toluene, chloroform chlorobenzene poly(oxy-1,3-phenylene)[c] benzene, biphenyl, 3-pentanol, phenyl ether, pyridine, benzophenone, nitrobenzene, DMF, DMSO

Table 92.1. (*Continued*)

B. Polyethers

Property	Polyoxymethylene	Polyoxyethylene	Polyphenylene Oxide
Nonsolvents/relatively unaffected by	Aliphatic hydrocarbons, lower alcohols, diethyl ether, lower esters	Aliphatic hydrocarbons, ethers, water (hot)	poly(oxy-2,6-dimethyl-1,4-phenylene).[d] Amorphous, α-pinene (cold), methanol, ethanol; crystalline, α-pinene (hot), methanol, ethanol, nitromethane poly(oxy-1,3-phenylene).[b] methanol
Decomposes	Alkalies with pH > 9; acids with pH < 1 (extended contact)		

C. Polyphenylene Sulfide and Polysulfones

Property	Polyphenylene Sulfide	Polysulfone of Bisphenol A	Polyethersulfone 200P
Melting point (°C)	~285–295		
Glass-transition temperature (°C)	97	~190	230 (other polysulfones ranging to 315)
Density	1.440; 1.34, unfilled; 1.64, 40% glass	1.24	
Refractive index, n_D		1.633	
Water absorption	0.02, unfilled; 0.01, 40% glass		
Solvents	Biphenyl, dimethyl-p-terphenyl, chloronaphthalene, some other solvents at elevated temperature	Chlorinated hydrocarbons, dimethylformamide, N-methylpyrrolidone; swells in dimethyl sulfoxide	
Nonsolvents/relatively unaffected by	At reflux temperature:toluene, pyridine, phenyl oxide, phenyl sulfide	Inorganic acids, alkalies, aliphatic alcohols	
Decomposes		Concentrated sulfuric acid (dissolves with degradation)	

[a] From Ref. 2a
[b] "Phenylene-oxide based resin" may soften or dissolve in certain halogenated or aromatic hydrocarbons. If an application requires such exposure, stressed samples should be tested under operating conditions (10).

SYNTHETIC POLYMERS, POLYESTERS, POLYETHERS, AND RELATED POLYMERS

Table 92.2. Production of Certain Synthetic and Natural Polymers

Polymer	Dry Weight Basis 1000 lb	1000 kg
A. U.S. Production of Plastics, Resins, and Elastomers (1990)		
Thermoplastic resins[a]		
Low-density polyethylene	7,254,891	3,291,693
Linear low-density polyethylene	3,893,252	1,766,448
High-density polyethylene	8,339,360	3,783,739
Polypropylene	8,310,409	3,770,603
Acrylonitrile–butadiene–styrene (ABS)	1,161,855	527,157
Styrene–acrylonitrile (SAN)	135,173	61,331
Polystyrene	5,021,090	2,278,172
All other styrene-based resin	1,189,204	539,566
Nylon	558,307	253,315
Polyvinyl chloride	9,095,534	4,126,830
Polyvinyl acetate	987,190	447,908
Other vinyl/vinylidene resins[c]	211,560	95,989
Thermoplastic polyester[d]	1,878,629	852,373
Engineering resins[e]	1,382,806	627,407
Acrylic[b]	1,507,031	683,771
Other thermoplastic resins	781,856	354,744
Total thermoplastic resins	51,708,147	23,461,046
Thermosetting resins[a]		
Epoxy	499,321	226,552
Polyester (unsaturated)	1,221,160	554,065
Urea–formaldehyde	1,495,531	678,553
Melamine–formaldehyde	201,986	91,645
Phenolic and other tar acid resins[f]	2,946,276	1,336,786
Polyurethane[g]	2,951,818	1,339,300
Alkyd[b]	769,238	349,019
Furfuryl type resins[b]	14,035	6,368
All other thermosetting resins[b]	152,488	69,187
Total thermosetting resins	10,251,853	4,651,475
Synthetic elastomers[h] (include latex)		Metric tons
Styrene–butadiene		852,851
Polybutadiene		403,388
Nitrile		562,261
Ethylene–propylene		255,537
Other synthetic elastomers (includes polychloroprene, polyisoprene, and butyl rubber but excludes polyurethane rubber)		
Total		2,620,535
B. World Production of Certain Textile Fibers (1990)[i]		
		1000 Metric tons
Natural fibers		
Raw cotton		18,714

Table 92.2. (*Continued*)

	Dry Weight Basis	
Polymer	1000 lb	1000 kg
Raw wool		1,964
Raw silk		66
Total		20,744
Percent of world total = 54		
Man-made fibers except olefin and textile glass		
Rayon and acetate		2,846
Acrylic and modacrylic		2,326
Nylon and aramid		3,765
Polyester		8,621
Certain other noncellulosic fibers		157
Total		17,715
Percent of world total = 46		

[a] Adapted from data in Society of the Plastics Industry, "Monthly Statistical Report—Resins, Full Year 1990, Production and Sales & Captive Use of Thermosetting & Thermoplastic Resins" issued by the SPI Committee on Resin Statistics as compiled by Ernst & Young; other data reported in SPI *Facts and Figures of the U.S. Plastics Industry*, 1991 Edition; and data in *Synthetic Organic Chemicals*, USITC Publ. 2470, U.S. International Trade Commission, Washington, DC, 1991. SPI data are used where available. Data reported to the USITC do not necessarily coincide with those reported to the SPI because of differences in both the reporting instructions and the coverage of certain resins.
[b] As reported in *Synthetic Organic Chemicals*, USITC Publication 2470, December 1991.
[c] Includes only polyvinyl butyral, polyvinyl formal, and polyvinylidene.
[d] Does not include polyester for film and tape.
[e] Engineering resins include acetal, granular fluoropolymers, polyamide-imide, polycarbonate, modified polyphenylene oxide, polyphenylene sulfide, polysulfone, polyether imide, and liquid crystal polymers, (ABS and nylon resins are listed separately.)
[f] Material is reported on a "gross weight" or "as sold" basis including the weight of water and other liquid diluents.
[g] Polyurethanes are derived from starting materials isocyanates (TDI/MDI) and polyols (polyether polyester). Raw materials reported to the SPI sold in the United States for conversion into polyurethanes provide the basis for these data.
[h] The term *elastomers* may be defined as substances in bale, crumb, powder, latex, and other crude forms that can be vulcanized or similarly processed into materials that can be stretched at 68°F to at least twice their original length and, after having been stretched and the stress removed, return with force to approximately their original length (*Synthetic Organic Chemicals*, USITC Publ. 2470, U.S. International Trade Commission, Washington, DC, 1991). Production figures are based on the *RMA Industry Rubber Report*, December 1990, and are reprinted by permission of the Rubber Manufacturers Association, Washington, DC.
[i] As listed in *Fiber Organon*, 62(6), June 1991. Data are reprinted by permission of the Fiber Economics Bureau, Inc., Roseland, NJ. "The silk and man-made fiber data are on a calendar-year basis, while the figures for cotton and wool are on a seasonal basis."

1.3 Toxicologic Potential

Industrial hygiene concerns with these polymers depend on the type of resin in use. Chemicals released in significant quantities may include (*1*) styrene during the fabrication of unsaturated polyester resins and in any subsequent release of styrene, (*2*) other volatile products generated at elevated temperatures or in fires, and (3) dust or particulate generated in the manufacturing and processing of polyester fibers. The use of styrene in polyester resins that are fabricated into glass-reinforced plastics can provide the greatest intensity of exposure to styrene in workplace situations (3–5). Toxic vapors can also be released by solvent systems, particularly at elevated temperatures, and during fabrication of engineering resins. For example, the processing of polyoxymethylene (acetals) in a poorly ventilated space may release biologically significant amounts of formaldehyde into the adjacent atmosphere.

Sulfur-containing engineering resins typically have a high resistance to thermal deterioration but may yield hydrogen sulfide or sulfur dioxide if heated to decomposition temperatures.

Inhalation tests in rats indicate no particular hazard to humans from particles of commercial polyester fiber but suggest that sufficient inhalation of cyclic ethylene terephthalate trimer dust may produce a granulomatous tissue response. Both types of particles produced a nonspecific inflammatory tissue response when first introduced into the lungs (6). This initial response was transitory and was not followed by collagen formation or any development of fibrosis. However, small particles of crystalline cyclic ethylene terephthalate trimer dust produced more prominent foreign body giant cell reactions than small polyester fiber particles. The latter, formed during high-speed processing by abrasive fracture of the surface of commercial fiber, produced a modest response consisting mostly of macrophages; the large nonrespirable particles produced artifactual foreign body granulomas.

1.4 Synthetic Fibers and Human Responses

Prior to 1950, "textile dust" was essentially the dust of cotton and other natural fibers. Harmful health effects were particularly identified with the inhalation of the dusts of raw cotton, flax, and hemp in the workplace. Today, the production and use of synthetic fibers is comparable to that of natural fibers (Table 92.2). No conclusive evidence links the manufacturing or processing of synthetic fibers with any serious health effects.

Except for cotton and other byssinogenic dusts, relatively few epidemiologic studies that involve exposure to organic fibers have been conducted. Three studies that consider exposure to polyacrylonitrile fiber, rayon, or an unidentified synthetic fiber, respectively, do so in the context of a comparison to cotton. Valic and Zuskin (7) report a mild reduction of ventilatory capacity in 1975 Yugoslav textile workers exposed to polyacrylonitrile fibers, including 30 with previous exposure to cotton and 77 with previous exposure to hemp; however, this response did not appear to be dose related. Tiller and Schilling (8) found an insignificant change in ventilatory capacity (less than 1%) among 26 English rayon workers, 13 of whom had been previously exposed to cotton. Another British report (9) describes a decreased prevalence of bronchitis among workers in two spinning mills using "man-made" fiber compared to workers in 14 cotton mills.

Several clinical reports record the symptoms or test results of affected textile workers but do little to clarify the origin of these observations. The significance of these reports outside the immediate environment is difficult to determine and quite likely limited. Pimentel et al. (10) describe seven patients in Portugal with a history of exposure to various textiles during the manufacturing process. The seven patients were variously affected with asthma, extrinsic allergic alveolitis, chronic bronchitis with bronchiectasis, spontaneous pneumothorax, or pneumonia. All had worked with one or more synthetic fibers; five had also worked with wool and/or cotton. The terminal patient identified as Case 3 had been exposed to the dust of "wool, cotton, synthetic fibers" over approximately a quarter-century and had apparently continued work with fibers for some 3 years after a hospital admission for progressive breathlessness upon exertion 8 years earlier. Many fibers were found in fibrotic tissue and some were identified as polyester. However, the mere presence of particles in the respiratory tract cannot be considered pathological. Each day, the average person inhales some 20,000 L of air laden with particles, many of which are deposited on the alveolar surface (11).

Bouhuys (12) points out that this report of case histories (13) suffers from lack of control biopsy samples, specific exposure data, and information on the nature and size of the populations at risk. A later report by Pimentel et al. (14) describes two patients with unusual "sarcoid-like granulomas" of the skin that the authors attributed to acrylic or nylon fibers, respectively; one of the patients also had respiratory tract lesions that were considered similar to those described in the first report. Unfortunately, this second report does not remedy the deficiencies of the first report.

Two Finnish investigators (13) cite the first Pimentel et al. study and also fail to give exposure data when reporting the results of several inhalation challenge tests in textile workers. Neither group provides specific evidence to support its implication of an immunogenic phenomenon.

These clinical reports pay little or no attention to the dimensional characteristics of the inhaled fibers as distinct from the chemical nature of the polymer in question. Both size and shape are important factors in biologic responses to durable fibers—particularly carcinogenicity (see Section 1.4). Work with fibrous glass indicates that short fibers (< 8 μm long) have negligible carcinogenic potential, but fine-diameter fibers (< 1.5 μm) that are long (> 8 μm) appear to increase in carcinogenic potential as their length increases (15) The synthetic organic fibers used in textiles are generally larger in diameter than this apparently critical 1.5-μm diameter (16).

1.5 Workplace Practices and Standards

No specific standards are known that pertain to ordinary industrial use of the finished polymeric products. Fabrication of the raw polymers or use of the prepolymers may release volatile materials and require appropriate industrial hygiene measures. Possible fire hazards should be considered when large amounts of these polymers are present in a given area, particularly with prepolymers of unsaturated polyester that contain appreciable amounts of flammable solvent or styrene.

In general, exposure to particles of polymeric fibers should be minimized by appropriate industrial hygiene measures for the control of exposure to insoluble organic dust. Particles of fiber dust should not be considered harmless but rather as particles that can—like particles of any dust—have biologically significant consequences if inhaled in gross concentrations. The presence of oligomers or other impurities may require evaluation in specific workplace situations.

SYNTHETIC POLYMERS, POLYESTERS, POLYETHERS, AND RELATED POLYMERS

The Textile Research Institute has provided a current review (16) on methods for collecting and measuring the particle sizes of respirable dust with special emphasis on dusts of importance to the fiber and textile industries.

1.6 Fire and Toxicity of Smoke

These polymers generally are not used in materials of construction, interior finish, or furnishings. They have had little or no study with respect to their smoke toxicity upon burning.

2 LINEAR TEREPHTHALATE POLYESTERS

All high-molecular-weight polyester polymers of commercial significance as fibers or films are derived from dimethyl terephthalate (or terephthalic acid). The basic structure is polyethylene terephthalate or its copolymers. Poly-1,4-cyclohexylenedimethylene terephthalate has also been used, but to a much more limited extent.

Polybutylene terephthalate is known as an engineering resin. The two fiber-forming terephthalates can also be adapted for this purpose as homopolymers or copolymers (1).

2.0 Polyethylene Terephthalate and Polyester Fibers

2.0.1 CAS Number: [25038-59-9]

2.0.2 Synonyms: NA

2.0.3 Trade Names: NA

2.0.4 Molecular Weight

The molecular weight of commercial polymers has been given as 15,000–20,000 M_n and 20,000–30,000 M_w (24). Farrow and Hill (25) describe M_n in the range of 10,000–50,000, the latter figure representing fibers used particularly for industrial applications.

2.0.5 Molecular Formula: NA

2.0.6 Molecular Structure:

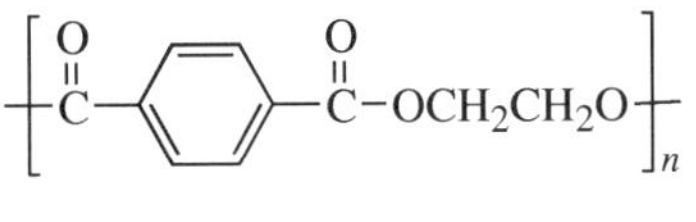

$$\left[\begin{matrix} O \\ \| \\ C \end{matrix} - \bigcirc - \begin{matrix} O \\ \| \\ C \end{matrix} - OCH_2CH_2O \right]_n$$

2.2 Production and Use

2.2.1 Structure, Synthesis, and Processing of Polyester Fiber

As noted above, the basic structure in most polyester fiber is polyethylene terephthalate.

The structure of the end groups has been described as mainly hydroxyethyl ester with a small number of carboxyl end groups (17). The Federal Trade Commission defines a polyester fiber as "a manufactured fiber in which the fiber-forming substance is any long chain synthetic polymer composed of at least 85 percent by weight of an ester of a dihydric alcohol and terephthalic acid" (18).

Most polyester fiber is composed primarily of a polyethylene terephthalate. Other fibers are derived from polycyclohexylenedimethylene terephthalate (17). Other variations include terephthalate-substituted isophthalate copolymers that provide enhanced dyeability (19, 20).

Polymerization has two basic steps (1, 17, 19): (*1*) reaction of ethylene glycol with terephthalic acid (esterification) or dimethyl terephthalate (alcoholysis) to produce oligomeric hydroxyethyl terephthalate, ranging from dimer to pentamer; (2) polycondensation of this oligomeric mixture to the desired molecular weight and removal of the excess glycol and by-products.

Commercial polyester fibers are prepared by melt spinning. The extruded melt is forced into filamentary streams that are formed into spun filaments and then drawn to fibers, either in a separate drawing step or in combination with spinning (19).

Two impurities are "normally" present in polyester fibers (17). Ethylene glycol used in the synthesis may be converted to diethylene glycol, and the corresponding ether group may be present to a slight degree (1–3 mol%) in the polyester. Cyclic trimer is present to the extent of $< 1.5\%$ in the polymer or in the fiber derived from it. Part of this trimer may be removed during dyeing or reprecipitate on the fiber.

"Snow" deposits that can develop on the surfaces of high-speed friction twist texturing machines (21) are aggregates of irregular polymeric particles ("skin" particles that describe the outermost layer of the individual fibers) rather loosely held together by finish oils. Analyses of these deposits average approximately half polymer, half finish. Generally the polymeric particles have the melting point and molecular weight characteristics of polyethylene terephthalate and should not be confused with cyclic trimer. "Snow" appears to be the result of local yarn heating caused by friction. Optimal control of yarn and process variables can reduce the level of snow generation.

2.2.2 Structure and Processing of Film

The basic structure in most of these films is again polyethylene terephthalate. Polycyclohexylenedimethylene terephthalate can also be used (22, 23). These films can be prepared by quenching (solidifying) extruded polymer to the amorphous state and then reheating and stretching the sheet approximately threefold in each direction at 80–100°C. Orienting the film and then annealing it under restraint at 180–210°C can raise the crystallinity to 40–42% (1).

2.2.3 Properties and Applications

Fabrics made of polyester fibers are noted for their strength, wrinkle resistance, and resistance to moisture at ordinary temperatures. Weathering resistance is good and superior to that of the polyamides (24); resistance to sunlight is inferior to that of the acrylics. Polyester fibers blend well with cotton or wool, and blends with cotton for clothing are easily the largest single end use (25). Industrial uses of polyester fibers include rubber reinforcing material, filter cloths, sieve cloths, and marine applications such as fishing nets or tarpaulins.

Polyethylene terephthalate film in its oriented crystalline state ranks among the strongest of the thermoplastics. Properties vary widely among the different types and subtypes,

SYNTHETIC POLYMERS, POLYESTERS, POLYETHERS, AND RELATED POLYMERS 583

but its basic advantages are its toughness, durability, excellent flex life, resistance to most organic solvents and mineral acids, very low moisture retention, and general absence of plasticizers (26). Applications include magnetic tape, X-ray and other photographic film, electrical insulation (metallized for capacitors), food packaging, and boil-in-bag food pouches. Polyethylene terephthalate can also be blow-molded to prepare bottles (1). Additional properties are listed in Table 92.1.

2.4 Toxic Effects

2.4.1 Oral Toxicity

Several 90-d feeding tests have been conducted. In a series of tests with films intended for food packaging, polyethylene terephthalate (27) and heat-treated polyethylene terephthalate (28) were fed to rats and dogs. The dietary level was 10% in all cases. No clinical, nutritional, hematologic, urinary, biochemical, or pathological evidence of toxicity was observed. Chloroform extracts of powdered resin produced no adverse effects when administered to rats in oral doses as high as 10 g/kg in an acute test and 400 mg/kg in a 90-d test (29).

2.4.2 Skin Contact

As with acrylic and nylon fabrics, clinical experience over the past several decades indicates that the basic polyethylene terephthalate fiber is essentially innocuous when applied to the skin. Tests conducted by the Schwartz–Peck procedure with a 2-wk rest period (30, 31) showed no sensitization but occasional irritation reactions consistent with occlusion. Fabric made from textured yarn (false twist) has also caused some slight reactions that appeared to be related to mechanical irritation from the rough edges formed during draw texturing and to increasing denier per filament (32). Several recent tests conducted by a modified Shelanski repeated insult patch test procedure (33, 34) or a modified Draize repeated insult patch test procedure (35) were entirely negative. These procedures are described in Ref. 36.

Like acrylic fiber, 100% polyester knit fabric is low in free formaldehyde content; 15 samples of American clothing contained ≤30 ppm free formaldehyde (37), formaldehyde not attributable to the intrinsic fiber. Samples of polyester/cotton fabrics showed in some cases a relatively high level of free formaldehyde.

2.4.3 Inhalation Toxicity, Thermal Degradation, and Related Data

Insufflation tests have been conducted with the cyclic ethylene terephthalate trimer dust (38) and also with the polyester "skin" particles (39) released from some types of filament yarns during high-speed processing (see preceding). Under examination by a light microscope, the particles of trimer dust were refractile and varied in size from 1 to 7 µm; many particles were 1 to 3 µm. The dust was brilliantly birefringent under polarized light. Forty rats were given intratracheal injections of 0.25 mL of a 1% trimer dust suspension in 0.9% saline; three other groups of rats received injections of the same volume of a 10% suspension of quartz dust in saline, a 1% suspension of quartz dust in saline, or the saline diluent alone. Rats were killed serially in groups of five at 2, 7, 28, 91, 183, and 371 d after dosing. The remaining 10 rats of each group (or less if some died) were killed at 2 yr after treatment.

The injected trimer dust was found to be scattered in the alveoli adjacent to the respiratory bronchioles. Rats that were killed 2 d after exposure showed extensive acute peribronchiolar pneumonia from the dust accumulation. This reaction was much more intense than that observed after a 2-wk series of 10 4-h exposures to 0.4 mg/L trimer dust (mass median diameter = 5.9 ± 1.1 μm, histologically visible particles mostly < 2 μm). The lungs of rats so inhaling trimer dust showed a nuisance dust cell reaction after the tenth exposure with significant reduction of dust-laden macrophage cells 2 wk later (6).

At 1 wk after exposure the inflammatory exudate had disappeared and a small number of minute dust-laden granulomas developed. The granulomas were readily detected under polarized microscopic examination. Dust particles a few micrometers or less in diameter were directly surrounded by foreign body giant cells and lymphocytes. Some dust particles had been transported from the lung to the tracheal lymph nodes. The number of granulomas appeared to be somewhat less at 1 yr and was further diminished at 2 yr, although a few active foreign body granulomas were still evident at this time. No fibrogenic activity was evident in the granulomas or in the dust-laden macrophages of the tracheobronchial lymph nodes.

The pulmonary reactions from the quartz-treated groups were quite different. Two years after treatment at the 2.5-mg dose level, the dust was almost entirely eliminated from the lungs and normal architecture was observed. At the 25-mg dose level, quartz lesions were characterized by silicotic nodule formation with progressive collagenization.

The polyester fiber "skin" particles appeared irregularly constricted and sausagelike when examined under a light microscope and were brilliantly birefringent in polarized light (39). Size varied from 1 to 1000 μm in length and 1 to 40 μm in width.

Forty rats given an intratracheal injection of 0.25 mL of a 1% suspension of these particles were maintained on the sacrifice schedule described for the trimer dust. The long nonrespirable particles were found in the small bronchi and terminal bronchioles, whereas shorter fibers were trapped in the respiratory bronchioles and adjoining alveoli. On the second day after exposure, the inflammatory reaction consisted of bronchitis, bronchiolitis, and peribronchial pneumonia. At 1 wk, the inflammatory reaction had disappeared; the test material was retained in the foreign body giant cells or macrophages, but no significant tissue reaction was observed. The intensity of the reaction progressively decreased at 1-mo, 6-mo, 1-yr, and 2-yr observations. At 2 years, large nonrespirable particles were retained in the terminal air passages while small respirable particles had been mostly removed by the lung clearing mechanism. No evidence of collagen formation, fibrosis, or significant alteration of the lung stromal architecture was observed.

Combustion toxicity tests (Table 92.1) show no unusual hazard from the products evolved from polyester compared to those from other fabrics similarly tested. Thermal degradation at approximately 300°C yields primarily acetaldehyde, whereas somewhat higher temperatures yield carbon monoxide (40–51).

2.4.4 Carcinogenic Potential/Cytotoxicity/Implant Studies

Many studies have been conducted with implants of polyethylene terephthalate, mostly in a form described as mesh or velour but also as sutures or powder. Polyethylene terephthalates have often been favored on the basis of minimal toxicologic response, durability, and mechanical properties (52; see also 53–55). Sutures of this fiber have been preferred by some investigators (20) but not others (59). Athough relatively inert, polyethylene terephthalate fiber is subject to slow degradation in body fluids (58, 59).

Particulate polyethylene terephthalate showed little cytotoxicity when tested in rats (60, 61). Two instances have been reported in which a prosthetic graft made from this polymer was associated with tumor development in humans (62, 63); pore size between the polymeric strands was identified as a determining factor in the second case. Vascular prostheses should be knitted with the largest possible pores to promote connective tissue organization and blood supply within the knitted structure of the prosthesis (59). A Russian report states that subcutaneous implants of polyethylene terephthalate "fibers" were resorbed in humans after an average interval of 30 yr (64).

Polyethylene terephthalate has been fabricated into sutures in filamentary form (e.g., for corneal surgery) and into supporting ligature in sheet form (e.g., for hernias) in biomedical implants. Several studies have examined the types and rates of complications resulting from use of polyethylene terephthalate as opposed to other prosthetic materials (65–69). The polyesters made from polyethylene terephthalate polyesters had higher rates of infection, fistula formation, and hernia failure than materials made from polypropylene mesh (68). Biomedical implants made from polyethylene terephthalate also may cause an inflammatory response due to increased adhesion of fibrinogen to the material; this may be related to the type of weave used in the material as well as any intrinsic property of the polyester (69). Tissue compatibility may be improved to reduce the likelihood of thrombosis and inflammation by impregnating the polyester with agents that reduce fibrinogen or lymphocyte adhesion (70). The continuing use of polyethylene terephthalate as permanent prosthetic implants points to the low toxicity of these materials.

2.4.5 Biodegradation

Human data as cited indicate that polyethylene terephthalate is slowly biodegradable.

3.0 Polybutylene Terephthalate

3.0.1 CAS Number: [24968-12-5], [26062-94-2], [30965-26-5]

3.0.2 Synonyms: Polytetramethylene terephthalate and PBT

3.0.3 Trade Names: NA

3.0.4 Molecular Weight

The low-molecular-weight polymers generally have M_n of 23,000–30,000 and M_w of 36,000–50,000. High-molecular-weight polybutylene terephthalate resins, which have preferable mechanical properties, have M_n of 36,000–50,000 and M_w of 60,000–90,000 (71).

3.0.5 Molecular Formula: NA

3.0.6 Molecular Structure:

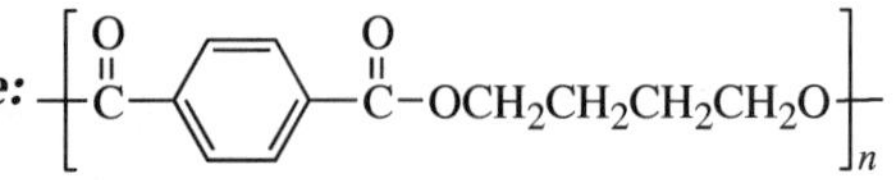

$$\left[\overset{O}{\underset{\|}{C}} - \underset{}{\bigcirc} - \overset{O}{\underset{\|}{C}} - OCH_2CH_2CH_2CH_2O \right]_n$$

3.1 Synthesis, Properties, and Applications

Polybutylene terephthalate can be made by the catalyzed condensation of 1,4-butanediol with either terephthalic acid or more frequently dimethyl terephthalate (71, 72).

Polybutylene terephthalate has a relatively low glass-transition temperature (see Table 92.1) and crystallizes at a rapid rate; molding cycles are very short. These resins are noted for their low coefficients of friction and are resistant to abrasion. They can be readily glass reinforced. Typical applications include exterior parts in the automotive and related fields and connectors and fuse cases in the electrical and electronic industries. Polybutylene terephthalate materials are also used in appliances, pump housings, and impellers (71). Polybutylene terephthalate has been approved for use by the FDA as a material that may contact food (21CFR 177.1655) and it is used in biomedical implants.

3.4 Toxic Effects

3.4.1 Oral Toxicity

Polytetramethylene terephthalate (i.e., polybutyleneterephthalate) with a M_w of 45,000–85,000 has been fed at dietary levels up to 5%, to rats for 148 d, and to dogs for 90 d (73). At the 5% level both male and female dogs showed a somewhat enhanced food intake, but otherwise no untoward effects were observed. Analysis of urine from rats and dogs fed the 5% dietary level revealed no evidence of free or combined terephthalic acid at the detection limit of 1 μg/mL; it was therefore concluded that < 0.003% of the daily intake of polytetramethylene terephthalate was absorbed from the gastrointestinal tract and then eliminated via the urine. The investigators reasoned that if the 2.5% level of polymer intake was considered an acceptable no-effect level, and the likely maximum was 1.8 mg extractable material from the polymer migrating into the total food intake, the safety factor would be in excess of 10,000.

3.4.2 Inhalation Toxicity and Thermal Degradation Data

Cartier et al. (74) report a case of asthma in a 39-year-old male who developed cough, dyspnea, sweating, and tremors 2–3 wk after starting a job where he was exposed to a variety of chemicals. Challenge with heated polyester fiber would initiate an asthmatic episode.

Reported data indicate no unusual inhalation toxicity of the pyrolysis/combustion products of polybutylene terephthalate per se (75–78). Flame-retarded samples may exhibit additional toxic properties; rats inhaling vapors from some but not all brominated samples heated to simulate processing temperatures developed slightly enlarged livers.

3.4.3 Implant Studies

Polybutylene terephthatate is used in a variety of biomedical implants similar to polyethylene terephthalate. It seems to have fewer coagulation, inflammatory, or rejection complications than the latter polyester. The plasma coagulation system, as measured by thrombin activation markers such as fragment 1 + 2 and fibrinopeptide A, was activated when polybutylene terephthalate was incubated *in vitro* with the coagulation system but not as great as polyethylene terephthalate (79). When used as an artificial ligament in Wistar rats, polybutylene terephthalate was less rejected than polyethylene terephthalate (Teflon was most inert) (80). As a copolymer with poly(ethylene oxide hydantoin), polybutylene terephthalate, used as an alloplastic tympanic membrane substitute, was assessed in rats for periods up to 1 yr (81). This copolymer was better accepted than a similar implant fabricated from Estane 5712 F1 polyether urethane, routinely used for this procedure, and much better than an implant made of polypropylene oxide. Minimal macrophages and foreign-body giant cells were present, but fibrous tissue and bone growth into the implant were acceptable. The continued use of polybutylene terephthalate in biomedical implants highlights the low toxicity of this polymer.

3.5 Elastomeric Polyester

This term has been used to describe a random copolymer of polybutylene terephthalate and polybutylene ether glycol. These polyester elastomers have strength comparable to many thermoplastics plus a rubberlike extensibility (82). A variety of polymers have been used in biomedical implants, indicating low toxicity, including biodegradable nerve guidance channels (83), replacement arteries (84), and biodegradable artificial skin (85).

4 UNSATURATED POLYESTER RESINS

Polyester resins comprise a large group of polymers and may include both these resins and the alkyds discussed in Section 4.

4.1 Properties

4.1.1 Molecular Weight, Properties, and Applications

The molecular weight for uncured resins has been given as 7,000–40,000 (24). When used in polyester–glass laminates, the cured resins are noted for their good strength and rigidity, low density, toughness, and translucency. They can be formulated to be fire retardant and generally have superior heat resistance compared to most rigid thermoplastics available in sheet form.

Propylene and diethylene glycol are often used to influence the rigidity or flexibility of the resin. With phthalate–fumarate resins, increasing propylene glycol content increases hardness but may cause a reduction in tensile and flexural strength. Diethylene glycol can make the polyester more flexible and also more susceptible to water absorption. Dibromoneopentyl glycol, tetrabromophthalic anhydride, and chlorendic acid (anhydride) are used for fire retardancy. Chlorendic acid (anhydride) and propoxylated bisphenol A are used for chemical resistance.

Generally, unsaturated polyester resins are classified as either general purpose or specialty polyester resins (88). General-purpose resins are widely used in the so-called "open mold" processes (hand lay-up/spray-up) to produce a wide range of products including boats, truck components, furniture, and applications that do not require the premium performance of higher-cost grades of polyester resin. Specialty polyesters are chemically tailored to meet the requirements of a wide range of applications including flexibilized polyesters, electrical-grade polyesters, heat-resistant polyesters, low flame/low smoke, translucent polyesters, and low shrink/low profile.

4.2 Production and Use

4.2.1 Structure, Synthesis, and Processing

Thermosetting, unsaturated polyester resins are made in several steps (1, 24, 86). A saturated dihydric alcohol is generally condensed with both a saturated and an unsaturated dicarboxylic acid. The alcohol used in the prepolymer is almost always a glycol such as ethylene, propylene, butylene, diethylene, or neopentyl or dipropylene glycol. The use of ethylene glycol is limited because unsaturated polyester resins made from it have poor solubility in styrene. Polyhydric alcohols are sometimes used to provide strength and chemical resistance. Maleic acid (anhydride) and fumaric acid are the usual unsaturated acids, although itaconic or mesaconic acid may be used to give flexibility. Phthalic anhydride is the most widely used saturated acid component; however, isophthalic and adipic acids have common use.

The prepolymer (or first-stage resin) is dissolved in a vinyl monomer, usually styrene, with appropriate inhibitors; the product may be supplied as a syrup containing 20–50% monomer (vinyl toluene, methyl methacrylate, diallyl phthalate, or other monomers can be used) (87). To add strength to the cured resin, glass fiber can be added as reinforcement (other fibers such as aramid, boron, carbon, nylon, polyester, polyethylene can also be used) (88). Curing is by free-radical polymerization of styrene monomer with the unsaturated acid residues of the resin. An organic peroxide is used as an initiator. Benzoyl peroxide is frequently used for elevated temperature curing; methyl ethyl ketone peroxide or cyclohexane peroxide can be used with a cobalt accelerator at room temperature.

Products of high quality often require a relatively high percentage of unsaturation in the polymeric chain. This can result in resins that during the curing process produce high temperatures that must be controlled to prevent explosions. Styrene-containing unsaturated polyesters polymerize with time at ambient temperatures.

Maximum mechanical strength may not be attained until more than a week after curing. Unsaturated polyester may remain undercured, soft, and, in some cases, tacky, if freely exposed to air during this period.

4.4 Toxic Effects

4.4.1 Skin Reactions

The finished, completely polymerized products are not considered dermatologic hazards (89), although there has been one case report of an 8-month-old child who developed contact sensitivity to cured unsaturated polyester resin in a limb prosthesis (90). Exposure to the uncured unsaturated polyester resin systems used in manufacturing have been associated with outbreaks of dermatitis during the 1960s (91, 92), but this was less common during the 1970s (89, 93). However, dermatitis can readily become a major problem in plants with poor industrial hygiene practices (94). The incompletely hardened macromolecular resin was considered the main causative agent in 17 dermatitis cases cited in a 1962 Czech report (95).

The dermatitis is reported as caused mostly by primary irritation but occasionally by sensitizing agents (92, 96–98; see also Ref. 89). Reactions are eczematous and more frequent on the backs of the hands, wrists, and forearms. The suggested patch test concentration for the unsaturated polyester resin is 10% in acetone. Malten (99) discusses several earlier studies (including Ref. 94), rates the sensitizing capacity of the polyester resin system low, and calls attention in this connection to a test with volunteers exposed to benzoyl peroxide that induced a 40% incidence of sensitization.

Kanerva et al. (100), reported two cases of hand and face contact sensitization in two car repair workers using an unsaturated polyester resin filler putty. The sensitizing agent was determined to be diethylene glycol maleate. Tarvainen et al. (101) found that some workers may react to the hardened glue of unsaturated polyester resins after becoming sensitized. These researchers also identified diethylene glycol maleate as the putative agent responsible for sensitization with this unsaturated polyester system. Irritant dermatosis of the hands in workers using unsaturated polyester resins may also be of a nonallergic origin (102).

4.4.2 Inhalation and Thermal Degradation

As indicated earlier, styrene is frequently used as a cross-linking monomer in the preparation of unsaturated polyester products. Potential exposure to styrene during processing is generally considered to present the most serious inhalation hazard associated with polyester resin. Styrene invariably evaporates when the resin surface is exposed to the atmosphere during molding operations, and without proper control, operators may be exposed to unacceptable levels of vapors (103). When products are manufactured in open molds typical of boat manufacturing, as much as 10% of the styrene can volatilize into the workplace air (104). Overall exposure to styrene monomer in seven representative U.S. plants in the fiberglass plastic boat industry ranged from 2 to 183 ppm; mean exposures for the primary job categories ranged from 44 to 78 ppm (3); see also Refs. 4, 105, 106, and 107. A Scandinavian report indicates that workers can be exposed "to a styrene concentration typically ranging from 20 to 300 ppm" (5). A French study (108) conducted under model laboratory conditions indicates that a standard-type resin released the equivalent of 4.3 mg styrene/cm^2 surface in 4 h, whereas a resin with reduced potential for evaporation of styrene yielded 0.7 mg/cm^2.

Other studies of boat manufacturing facilities report average exposures during work shifts typically did not exceed the then current OSHA PEL of 100 ppm (99, 109); however, those exposures did exceed the current PEL of 50 ppm. Studies of other hand lay-up and spray-up operations also report typical exposure levels below the then-current 100 ppm PEL (110–112). In 1989, OSHA reduced the 8-h TWA PEL for styrene from 100 to 50 ppm. In the rule making, OSHA noted that, with the exception of hand lay-up and spray-up operations in the boat building industry, the 50 ppm PEL could be achieved with engineering and work practice controls. OSHA found, for these boat-building and similar large scale lay-up operations, that employers must use respiratory protection in combination with engineering controls and work practices to achieve these limits (113). The PEL for styrene has since been increased to 100 ppm.

Fiberglass particles released during processing may be coated with a mixture of resins and finishes. Lim et al. (94) report finding minute amounts of trivalent and hexavalent chromium on fiberglass samples collected during walk through plant surveys.

When exposed to fire, both conventional and fire-retarded formulations of unsaturated polyester resins typically yield copious amounts of smoke because the major decomposition product is usually styrene, which burns with a very smoky flame (86). Resins based on alkyl and particularly acrylate monomers may produce less smoke (114). Tests conducted by German investigators indicate that products evolved from pyrolysis in the 330–400°C range or above may be lethal to rats. Carbon monoxide can be released in biologically significant amounts but is not necessarily the cause of death. Additional details can be found in Refs. 114–117.

4.4.3 Cyto-/Genotoxicity

Furniture workers exposed to unsaturated polyester resins showed increased frequencies of sister chromatid exchanges in their circulating lymphocytes compared to controls (118). Frequencies were higher than for nonexposed smokers. The specific agent responsible for the increase was not identified.

5 ALKYD RESINS

Commercial production of alkyd resins was begun in the 1920s, and by the 1940s they had become the backbone of the coatings industry, widely used in paints, varnishes, lacquers, baking enamels, and other surface coatings. In 1973, consumption of alkyd resins peaked — they constituted about one-third of all synthetic resins used in coatings. Now they comprise about one-fourth (119).

Because of environmental concerns for the solvents typically associated with these resins and the better performance of latex paints aimed at the consumer architectural coatings market, their use has declined and will continue to do so. However, they are one of the more versatile binders with potential for modification for use in high-solids coatings, an attribute that will make them important for some time to come.

5.1 Production and Chemical Characteristics

Alkyd resins are condensation polymers formed from esterification reactions. However, they differ from other polyesters that are unsaturated and thermosetting in that they are modified with a triglyceride oil or the fatty acids of that oil. They are the reaction products of a polybasic acid, a polyhydric alcohol, and a monobasic fatty acid or oil. The name alkyd is derived from a combination of alcohol and acid (120).

The most commonly used polybasic acids are the anhydrides of dicarboxylic acids, with phthalic anhydride being by far the most important. Pentaerythritol is the most commonly used polyhydric alcohol; others used frequently include glycerol, trimethylolpropane, and trimethylolethane. When oils are used as starting materials, alcoholysis from a polyol, glycerol being very common, must first be carried out to form the monobasic fatty acid. This reaction is often catalyzed by a very small quantity of basic compounds such as lithium hydroxide. Currently, most alkyd resins are manufactured using the solvent process that refluxes the solvent and the reactants until the completion of the reaction. The final product is usually let down or thinned with a solvent or diluent to a 50% or more solids content in solution.

The fatty acid moiety determines whether the alkyd is drying or nondrying. Unsaturated fatty acids in the alkyd undergo oxidation in air and cross-link or "dry." The cross-linking is catalyzed by "driers," usually metallic soaps added during the paint formulation stage. Alkyds without these unsaturated groups are nondrying and are usually used as plasticizers with other resins. Sometimes monobasic acids other than fatty acids are used as starting materials to make "oil-free" alkyds (e.g., rosin acids) or to terminate an alkyd chain [e.g., p(tert-butyl)benzoic acid].

Alkyds may be modified chemically by combining them with other molecules. Vinyl modification is common and occurs by grafting such molecules as styrene, vinyltoluene, and acrylic esters to the unsaturated groups in the resin chain via free-radical initiation. Alkyd resins may also be modified by alkoxypolysiloxanes, phenolics, polyamides, and diisocyanates. Alkyd resins may also be mixed with other resins to obtain the desired properties. They are commonly added to nitrocellulose lacquers.

By varying the starting materials, the modifiers, and/or the polymers with which they are blended, they may be "designed" for a wide variety of uses. Vapor emissions and hazards from solvents may also be minimized by making an alkyd resin water reducible. This is accomplished by creating an emulsion or introducing groups into the resin to make it water soluble.

Exposure to the ingredient dicarboxyic acid precursors, phthalic, trimellitic, and maleic anhydrides, was evaluated in a retrospective cohort study of alkyd resin workers (121). This study found that exposures in alkyd resin production facilities were generally low in recent years at levels below the TLVs for these anhydrides capable of causing occupational asthma. Another similar study in a cushioned flooring manufacturing plant showed similar low exposures (122).

5.4 Toxic Effects

Little information exists on the toxicity of the alkyd resins themselves. It is quite common to perform no toxicity tests at all on new alkyd resins for two reasons. First, they are almost always encountered as part of formulations that contain other chemicals assumed to present greater hazards. Thus exposure is controlled by limiting the exposure to the other chemicals. Second, the precursors to alkyd resins and any residual monomers are considered to be far more toxic than these large molecules, most of which exceed 1000 molecular weight.

Aside from the solvents, the dibasic acids used as starting materials for the basic alkyd resins present the greatest hazards during the manufacturing process. Phthalic anhydride, along with other low-molecular-weight anhydrides, is a known irritant and has been identified as a sensitizer, with a potential for respiratory sensitization. The EPA has identified carboxylic acid anhydrides as a category of concern for potential pulmonary sensitization and may require testing for new chemicals containing this group during the premanufacture notification review process (123). (These groups are not present in the resin owing to esterification.) Yokota et al. (124) recently have reviewed prevention of occupational allergy due to these anhydrides. The chain terminator, p-(tert-butyl)benzoic acid, has been identified as a potential reproductive toxin. The toxicity of such ingredient monomers, stabilizers, plasticizers, catalysts, and other performance enhancing chemicals that may be present as precursors or residuals, is dealt with in other chapters of this treatise.

Chemical modifiers such as styrene, acrylic esters, and diisocyanates present health hazards, including irritation and sensitization potential, unique to these chemicals during the manufacturing process (125). The hazards associated with the solvents used to manufacture the resin are also associated with the processing and use of the resin and the coatings formed from it. Small amounts of residual monomers (usually well below 1% by weight) remain in the resin and may present a potential for exposure during their use. Release of monomers (e.g., phthalic anhydride) may be of concern during heat-accelerated curing processes such as baking.

Toxic combustion products usually are not a problem. Most of these resins contain only carbon, hydrogen, and oxygen, and they are contained, owing to their customary use, in thin films, making less of a contribution to the combustion products than the substrates and other materials in the environs. Protective measures aimed at solvents and other attendant chemicals encountered during the manufacturing, processing, and use of alkyd resins should protect against any hazards from the resins themselves, residual monomers, or low-molecular-weight species. Additional protection may be required when handling raw materials such as acid anhydrides, styrene, and acrylic esters, for the manufacturing operation.

5.5 Additives

Other substances may be used as modifiers or additional ingredients, including phenolic resins, epoxy resins, styrene, cobalt naphthenate, lead soaps, and fire retardants. Additives are used to resist mildew and ultraviolet light. Curing or cross-linking is accomplished by air oxidation of the unsaturated groups (24, 86).

6 ALLYL POLYMERS

The term *allyl resin* typically refers to unsaturated polyester cross-linked with an allyl-type monomer, such as allyl alcohol, rather than styrene. Such resins are notable for their retention of electrical properties under conditions of high temperature and high humidity. "Allyl molding compounds" may refer to "nonpolymeric" substances used in the preparation of thermoset moldings. Those in widest commercial use are the monomers and prepolymers of diallyl phthalate and diallyl isophthalate (126). Another self-polymerizing monomer, CR-39 (CAS# [*142-22-3*]; a.k.a., diethylene glycol *bis*(allyl carbonate) or diallyl diglycol carbonate), is used for lenses and optical devices because of its light weight and resistance to impact, scratch, and abrasion.

CR-39 monomer is a colorless, slightly volatile liquid (molecular weight 274) prepared from diethylene glycol chloroformate and allyl alcohol (127). A peroxide catalyst (typically, benzoyl peroxide, isopropyl percarbonate, cyclohexyl percarbonate) is dissolved at about 3% by weight in the diallyl glycol carbonate and then the liquid is polymerized first to a gel and then to a fusible solid.

The monomer is known to be an irritant. Skin contact during the polymerization process may result in a rapidly developing, irritant dermatitis with an incidence as high as 70%. Dermatitis frequently appears to result from direct contact with the liquid monomer. Some cases would appear to be traceable to residual monomer or other ingredient(s) in partially polymerized polymer because "adherence of the polymer to the molds is often associated by the workers with the irritant nature of the resin" (127). The National Fire Protection Association ranks the health hazard for this monomer as a 1 (slightly toxic) (128).

7 POLYCARBONATE RESINS

7.1 Chemical and Physical Properties

Polycarbonate resins are characterized by water-white clarity and high impact resistance. They are extremely tough and rigid and have good electrical insulation characteristics (1, 86).

7.1.1 Structure

Thousands of polycarbonates have been prepared since polycarbonate was first introduced in the late 1950s. Approximately 99% of the polycarbonate sold is the polycarbonate of bisphenol A with the basic structure

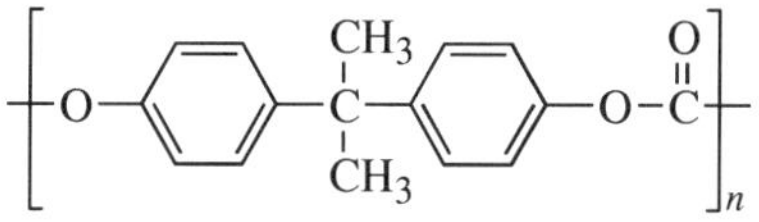

$$\left[O - \!\!\left\langle\!\!\!\bigcirc\!\!\!\right\rangle\!\! - \underset{\underset{CH_3}{|}}{\overset{\overset{CH_3}{|}}{C}} - \!\!\left\langle\!\!\!\bigcirc\!\!\!\right\rangle\!\! - O - \overset{\overset{O}{\|}}{C} \right]_n$$

7.1.2 Properties

The properties of polycarbonates are dramatically affected by the comonomers, molecular weight, end groups and branching agents. BPA polycarbonate homopolymer has a T_g of 145–150°C. Polycarbonate is typically amorphous. The polymer has a polymer decomposition temperature as measured by thermal gravimetric analysis in nitrogen of ~510°C. The polymer is soluble in solvents such as chloroform, methylene chloride, and tetrachloroethane. Other nonchlorinated solvents include pyridine, m-cresol, and phenol. Polycarbonates are insoluble in water, alcohol, organic acids, and hydrocarbons (130). The resin swells in various aromatic solvents, ketones, and esters.

7.2 Production and Use

7.2.1 Preparation

Aromatic polycarbonates are prepared by the condensation of a bisphenol with a carbonic acid derivative. Commercially, the bisphenol is typically 4,4'-hydroxyphenylpropane

(BPA), and the carbonic acid derivative is normally phosgene. Other comonomers can be added to tailor the properties of the resin to specific applications. In addition, various monofunctional compounds are added to control the molecular weight of the resin.

The interfacial polymerization process is the most widely used method in the manufacture of polycarbonates and polycarbonate copolymers. Two variations of the process have been described in the literature (129).

In the standard interfacial process, the bisphenol, chain terminator, catalyst, organic solvent, water, and caustic are charged into a stirred reactor followed by the addition of phosgene. The sodium salt of BPA is generated in the aqueous phase and allowed to react with phosgene in the organic phase in the presence of a catalyst. The catalyst is typically a tertiary amine. In this process, the phosgene reacts with the sodium salt of BPA to form BPA chloroformates. Concurrently, the catalyst facilitates the reaction of phenoxide with the chloroformates to form high polymer. The molecular weight of the resin is controlled by chain terminators, typically monofunctional phenols, carboxylic acids, or acid chlorides.

In the second process the tertiary amine catalyst, and in some cases the terminator, are not added in the first step. Without the catalyst present, the reaction of phenoxide with the chloroformate end group is suppressed resulting in the formation of BPA chloroformates and chloroformate oligomers. After the chloroformates are formed, the terminator and the tertiary amine are added. The amine catalyzes the condensation of the phenoxides with the chloroformates affording high polymer. This process has been claimed to produce resin with less carbonate functionality.

In both processes, after the reaction is complete, the organic phase is separated from the aqueous phase and washed free of catalysts and salts. Isolation of the polymer can be accomplished by antisolvent precipitation, spray drying, and devolatization using an extruder (129).

7.2.2 Additives

A variety of additives are added to polycarbonate to enhance its properties, its retention of properties, and/or its processability. These include the following:

- UV stabilizers (typically benzotriazoles)
- Thermal stabilizers [such as phosphonites, phosphites, and organosilicon compounds (131–133)]
- Epoxides for increased hydrolytic stability (134)
- Release agents [typically, long-chain carboxylic acids and esters (135, 136)]

Other commercially significant additives and fillers include the following:

- Fire-retardant agents such as brominated polycarbonate copolymers (137) organic sulfonates (138–141), and finely dispersed fluoropolymer
- Glass fiber
- Blowing agents, such as 5-phenyltetrazole (142) or henyldihydrooxadiazinone (143), for use in foamed polycarbonate

7.2.3 Processing

7.2.3.1 Injection Molding. Although polycarbonate has been processed in most common injection molding machines, in-line reciprocating screw machines are preferred. Melt temperatures of 280–350°C and mold temperatures of 80–120°C are typically recommended for general-purpose grades.

7.2.3.2 Extrusion. Film, sheet, multilayer sheet, profiles, tubes, and rod stock can be manufactured using high-viscosity polycarbonate resin in extrusion processes. Melt temperatures of 230–260°C are typically recommended.

7.2.3.3 Thermoforms. Polycarbonate sheet can be thermoformed. The sheet must be dried at 80–110°C prior to forming. Typical forming temperatures are 175–205°C.

7.2.3.4 Blow Molding. Blow molding of branched polycarbonate has been extensively used to manufacture a wide variety of items from baby bottles to 5-gal water bottles. Typically, the non-Newtonian rheological properties obtained with branched resin are needed for blow-molding applications.

7.2.3.5 Applications. Applications include medical equipment, beer pitchers, automotive lenses and trim extrusions, drapery fixtures, door and window components, furniture, and plumbing. Foamed polycarbonate is used in major automotive components such as a one-piece bus seat frame. Sheet glazing products that conform to specifications for safety glazing and burglar resistance may be used in schools, off-highway installations, and security facilities. Coated thin-gauge sheet has been used for protective eyewear and business machines (144). Registration of fast-neutron-induced recoil and (n, a) tracks in polycarbonate foils provides a sensitive means of dosimetry (145). Polycarbonate resins are approved for use with foods by the FDA (21 CFR 177.1580).

7.4 Toxic Effects

7.4.1 Toxicologic Information

Polycarbonates in general are not considered primary eye irritants. When polycarbonate products, in finely divided form, were placed into the eyes of rabbits, slight transient redness or discharge occurred, consistent with the expected slightly abrasive nature of the resin particles (146).

Polycarbonates are not considered primary skin irritants. The Draize skin primary irritation score (rabbit) for polycarbonate resins, in finely divided form, for a 24-h exposure is 0. These resins are not expected to be skin sensitizers based on results of modified Buehler guinea pig sensitization. The dermal LD_{50} (rabbit) is estimated to be greater than 2 g/kg (146).

The estimated acute oral LD_{50}(rat) is greater than 5 g/kg (147). Bisphenol-A polycarbonate has been found to be physiologically inert when fed to rats at a level of 6% in food (148).

In acute inhalation tests, laboratory rats were exposed to processing fumes at concentrations exaggerating those that would likely occur in workplace situations. No deaths or signs of toxicity, except transient irritancy in some cases, were noted during the 6-h fume exposure tests. There were no distinct or consistent treatment-related tissue or organ changes noted in gross necropsies (149).

Recently, bisphenol A has been implicated as an environmental "endocrine disruptor" or chemical that interferes with normal endocrine metabolism by mimicking estrogen. Brotons et al. (150) reports finding the "xenoestrogen" bisphenol A in foods from cans lined with a polycarbonate resin. However, other studies that have evaluated the migration potential of bisphenol A from polycarbonate indicate that degradation of polycarbonate and consequent migration of bisphenol A is minimal. Specifically, a study using FDA worst-case extraction procedures found no migration/extraction of bisphenol A from several polycarbonate formulations (151). Bisphenol A was found not to migrate from 24 different brands of polycarbonate baby bottles following sterilization by steam, alkaline hypochlorite, or washing in an automatic diswasher at 65°C with detergent (152).

Toxicologic aspects of plastic combustion products are very complex. This is due mainly to the fact that it is very difficult to simulate real-world environments under controlled conditions. Depending upon a variety of factors, the qualitative and quantitative nature of chemicals present in the combustion product mixtures will vary. Factors contributing to toxicity of real-life fires include increase of carbon monoxide and carbon dioxide, decrease of oxygen, and the presence of irritant gases. The primary effects from fires are asphyxia due to oxygen deficiency, poisoning from carbon monoxide, heat damage to tissues, and irritation of the respiratory tract by combustion gases (see Section 1.5). A study using rats as models found that combustion products from polycarbonate resins induce rapid sensory irritation of the upper respiratory tract. Rapid recovery also occurred following termination of exposure (153). Another study on toxicity of plastic products using eight different plastic resins found that combustion products from polycarbonate were among the least toxic relative to the other resins (154). Polycarbonates manufactured with the flame retardant tetrabromobisphenol A should not present an additional toxicological hazard during manufacturing since this compound has low acute or chronic toxicity (155). Its combustion toxicity has not been assessed. When polycarbonate products were manufactured with a flame-retardant metallic covering film, combustion toxicity products were no worse than for noncoated polycarbonate (156).

7.4.2 Plastic Processing Fumes

The chemical composition of processing fumes of polycarbonates, like any other plastic materials, depends greatly on product formulation. In general, processing fumes from polycarbonates processed at recommended processing conditions may include trace levels of phenol, alkylphenols, and diarylcarbonates. Polycarbonate treated with brominated fire retardants may also evolve small traces of hydrogen bromide. Products formulated with polyfluorocarbon additives may evolve hydrogen fluoride and fluorocarbon compounds in the resulting processing fumes. Thermal emissions from impact-modified polycarbonate resins may also include styrene and acrylates, depending on the impact modifier used in its formulation.

In some instances, additives containing certain heavy metal compounds may be present. These ingredients are essentially bound in the plastic matrix and are unlikely to contribute to workplace exposure under recommended processing conditions.

Processing fumes may cause irritation to the eyes, skin, and respiratory tract, and in cases of severe overexposure, nausea and headaches. Fumes from plastic thermal processes may condense on processing equipment, ventilation systems, and other surfaces. These greaselike processing fume condensates can cause irritation and injury to skin. Use of proper personal protection while cleaning or handling these condensates should be observed.

When polycarbonate was machined using either hot-gas welding or laser cutting techniques, significant particulate was formed compared to polymethyl methacrylate polymer or polyethylene terephthalate (157, 158).

7.4.3 Exposure Controls and Personal Protection

A continuous supply of fresh air to the workplace sufficient to replace that used to remove processing fumes through local exhaust systems is necessary for proper operation of these systems. Processing fume condensate may be a fire hazard and toxic; it must be removed periodically from exhaust hoods, duct work, and other surfaces. Persons performing such tasks must be provided appropriate personal protection including face protection and respirators. Ventilation requirements must be locally determined to limit exposure to materials at their point of use.

8 POLYETHERS

8.0 Acetal Resins — Polyoxymethylene

8.0.4 Molecular Weight

Most manufacturers supply a number of different molecular-weight resins. The Mw of some commercial polyoxymethylenes has been reported in the range from 20,000 to 110,000 (1). In general, increasing molecular weight leads to increased toughness, and increased melt viscosity. Table 92.3 provides an example of the effect of molecular weight on properties. Coupled with low density (as compared to metal) and ease of fabrication, acetals are used as replacements for metal in many applications (1).

Table 92.3. Effect of Molecular Weight on Impact Properties of Acetal Homopolymer

Approximate M_n	Notched Izod (ft-lb/in)[a]	Tensile Impact (ft-lb/in^2)[b]
65,000	2.4	170
40,000	1.4	100
30,000	1.3	70

[a]Method: ASTM D256.
[b]Method: ASTM D1822.
Source: Ref. 2a.

8.0.6 Molecular Structure

Acetal resins [9002-81-7] are also known as polyformaldehyde and polyoxymethylene. Polyoxymethylene heart valve biopolymeric materials may be found in the biomedical literature with the name Delrin [9085-38-5]. Although paraformaldehyde [30525-89-4] also is a polymer of formaldehyde, this material has repeating units of 8–100 according to Hawley's Condensed Chemical Dictionary (159) and should not be considered the same as the larger polyformaldehyde resins discussed here. Unlike the larger, more refined polyoxymethylene polymers, paraformaldehyde may be highly irritating and an allergic sensitizer, perhaps due to large amounts of residual free formaldehyde (160, 161). High-molecular-weight acetal resins are highly crystalline, resulting in excellent retention of properties and dimensions and predictable behavior. Because of these notable features of metal-like stiffness, resistance to fatigue, and resistance to organic solvents, they are referred to as engineering resins.

The dominant structural repeat unit of acetal resins is

$$-[CH_2-O]_n-$$

or formaldehyde. Acetal homopolymers consist entirely of these repeating units, with acetate and caps to provide stability. In acetal copolymers, the structural regularity of —OCH_{2-n} is interrupted after approximately 65 units ($n = 65$) by a comonomer unit derived from ethylene glycol, 1,4-butylene glycol, or diethylene glycol. The purpose of including comonomer units is to provide depolymerization stoppers. The inclusion of the comonomer units also reduces the inherent crystallinity, affecting the strength and rigidity of the resin. Reduction of properties is generally compensated to some extent by addition of nucleating agents.

8.1 Properties

Typical properties of the intermediate molecular weight, probably the most widely sold grades of acetal resins, are listed in Table 92.4. Again, owing to the highly crystalline nature, acetals have excellent solvent resistance. However, all acetals are attacked by strong acids; copolymer resins in general are resistant to strong bases.

Table 92.4. Properties of Typical Acetal Resins at 23°C

Property	Method	Acetal[a]
Tensile strength	ASTM D638	8,800–10,000 psi
Elongation at break	ASTM D638	40–50%
Flexural modulus	ASTM D790	375,000–400,000 psi
Melting point		165–175°C
Specific gravity	ASTM D792	1.41–1.42 g/ml

[a]Range accounts for both homopolymer and copolymer values.

Acetals in contact with steel generally produce less friction than acetals with other metals, and acetals used with nylon are usually preferable to either alone (1). Most manufacturers also sell resins modified with various additives, including fluoropolymer fibers for reduced friction and wear, glass fiber for increased modulus and strength, and toughener to increase impact resistance.

8.2 Production and Use

Acetal resins are manufactured from monomeric formaldehyde or from its cyclic trimer, trioxane. Homopolymers are manufactured from highly purified anhydrous formaldehyde (1, 162). One process for making these resins starts with aqueous formaldehyde. The manufacture of polymers with useful molecular weights (typical M_w are 25,000–80,000 with $M_w/M_n\sim2$) requires very high-purity monomer. Thus the concentration of chain transfer agents such as water and methanol must be reduced to very low levels. All acetal homopolymers are believed to be manufactured by anionic polymerization. To ensure good thermal stability, the homopolymers are end capped with acetate end groups. Esterification with acetic anhydride or other anhydrides can be conducted at 130–200°C with sodium acetate as a catalyst.

Most copolymers are manufactured from trioxane by cationic polymerization. Like the formaldehyde used for making homopolymers, the trioxane must also be highly purified to be able to produce copolymer resins of useful molecular weights. (M_n of 15,000–40,000, with $M_w/M_n\sim3$–80). Because the "as-polymerized" copolymer resins contain unstable ends, the polymer is typically stabilized by thermal or hydrolytic degradation back to depolymerization-stopping comonomer units.

Acetal resins may be readily processed on conventional injection molding, blow molding, and extrusion equipment, provided that overheating is avoided. Overheating can lead to the production of formaldehyde gas and a serious or even dangerous buildup of pressure. Gas pressure created by decomposition can rapidly become extremely high when processing machines are not properly vented. A shotgun-like reaction (blow-back) from a pellet hopper was reported to develop within a few minutes after failure of one of three heating zone circuits (163). Recommended control measures (164, 165) include (*1*) the use of vented feed screw and proper design of injection nozzle, (*2*) preventive maintenance and replacement of worn gaskets and valves, (*3*) automatic heat regulation and monitoring of excess temperature conditions by an audible alarm or warning device in direct view of the operator, (*4*) a standby water cooling device not dependent on the main electrical system, and (*5*) personnel instruction.

8.4 Toxic Effects

8.4.1 Oral Toxicity and Skin Contact

Ninety-day feeding tests have been conducted with two different types of polyoxymethylene (96+ and 99% active ingredient, respectively, both with M_w of 25,000–30,000). Rats in groups of 40 and dogs in groups of 4 were fed a 10% dietary level of each test polymer with Alphacel® cellulose as a control (28). One or two rats died in each of the two test groups; one death was also observed in the corresponding control group. No deaths occurred among any of the dogs. No changes attributable to the test polymer were evident in behavior and appearance, body weight, food consumption, ophthalmoscopy, and hematologic, biochemical, or urinalysis studies. No gross or microscopic lesions or variation in organ weight attributable to treatment were observed in any of the animals.

Skin irritation and sensitization tests with 1-in disks of polyoxymethylene extruded sheeting were conducted by the Schwartz–Peck procedure (36). In another study with a 10-d rest period, on a panel of 212 volunteers (166), some mild or moderate erythema consistent with occlusion was observed at the 6-d reading, but otherwise no reactions were observed. The panel included one subject known to be sensitive to urea–formaldehyde who showed no reaction through 4 d after removal of the test polymer.

8.4.2 Inhalation Toxicity and Thermal Degradation Data

A massive single 6-h dust inhalation exposure (47 mg/L nominal concentration) caused emphysema and atelectasis in rats. Fumes evolved at temperatures as low as 150°C, in sufficient quantity, can be lethal to animals that are exposed for 4 h. Both formaldehyde and carbon monoxide have been identified in the effluent evolved at elevated temperatures. More details can be found in Refs. 168–171.

8.4.3 Cytotoxicity/Implant Studies

Many studies have evaluated the long-term efficacy and biocompatability of heart valves manufactured from Delrin, a highly refined polyoxyethylene, in humans (172–180). Other than mechanical wear, sometimes manifesting as distortions in the shape of the valve as a result of cavitation and other hemodynamic stresses, no toxicity and little evidence of rejection have been reported in heart-valve patients for periods as long as 20 yr. Delrin, used as a bone substitute in a rabbit model, was less inert than titanium inducing foreign body reactions (181). Polyoxyethylene (Delrin) was more cytotoxic than polyvinyl chloride, inhibiting cell growth in the continuous cell line, L-929, as measured by cell density, total protein, total per cell, protein fraction of cells in mitosis and other parameters (182).

8.4.4 Injection Studies

Rats receiving intraperitoneal injections of an unstated concentration of dust suspension showed granulomas with some scarring when examined 90 d later (183). In another study 40 mg/kg was administered to male guinea pigs. One animal died due to pneumonia and infection. The other four animals survived the 32-d observation period. At sacrifice, peritoneal nodules consisting of the polymer surrounded by grant cells were found in all animals. The nodules showed no inflammation and were typical of the reaction caused by an inert, insoluble material (184).

9.0 Polyoxyethylene

Polyoxyethylene is available as high- or low-molecular-weight polymer.

SYNTHETIC POLYMERS, POLYESTERS, POLYETHERS, AND RELATED POLYMERS

9.0.4 Molecular Weight

One group of low-molecular-weight polyoxyethylenes range from liquids to waxy solids, with a molecular weight up to 10,000. High-molecular-weight resins have a molecular weight described as ≥100,000 (16) and ranging up to 4,000,000 (185).

9.0.6 Molecular Structure: The generic structure is

$$\pm CH_2-CH_2-O\pm_n$$

9.1 Properties and Applications

Even the high-molecular-weight resins are water-soluble thermoplastic polymers that can be extruded as films. They can be used in packaging for food, as textile sizes, and for the reduction of hydrodynamic friction.

9.4 Toxic Effects

9.4.1 Acute Toxicity

Polyoxyethylene polymers (a.k.a, polyethylene glycols or PEGs) with molecular weights below 10,000 have very low acute oral toxicity with LD_{50}s generally above 5,000 mg/kg and often above 10,000 or even 20,000 mg/kg (160). Even intraperitoneal and intravenous LD_{50}s often are above 5,000 mg/kg (160). In an older report (185), solutions of 0.1% commercial material killed rats when first administered intravenously at a dosage of 3 mg/kg (0.3 mL). This may have been due to clumping of cells and death from embolism. Subsequent injection of this concentration after shearing the polymer in an Osterizer did not kill rats receiving 40 mg/kg (4 mL). Few dermal LD_{50}s were found for polyethylene glycols of any molecular weight. Those found generally were greater than 5,000 mg/kg with no deaths reported at the highest dose tested (160).

Primary irritation testing of a variety of low-molecular-weight PEGs in rabbits reveals minimal irritation of eyes or skin (160, 161). PEG 400 (molecular weight = 400) has been used to wash human eyes as a 50% aqueous solution (160). Low-molecular-weight polyoxyethylenes were found to be nonsensitizing in patch tests. Low-molecular-weight PEGs have parasympathomimetic properties (160).

9.4.2 Repeat-Dose Toxicity

Polyoxyethylene (PEG 400) was administered to Fischer 344 rats by daily gavage 5 d/wk for 13 wk at doses of 0, 1, 2.5, or 5 mL/kg/d (186). No changes were noted in mortality, clinical chemistries, or hematology. High-dose animals showed a slight decrease in body weights. Mid- and high-dose groups exhibited loose stools. All treatment groups consumed water in excess of controls in a dose-related manner. Kidney weights were slightly increased, but histopatholgy revealed no lesions. A functional difference was seen in kidney function evidenced by increased urinary N-acetyl-β-D-glucosaminidase, vascular cell findings, and bilirubin levels. Animals allowed to recover for 6 wk showed no kidney changes.

Rats were exposed to aerosols of polyoxyethylene (PEG 3350) at concentrations of 0, 109, 567 or 1008 mg/m^3, 6 h/d, 5 d/wk for 2 wk (187). No effect was noted on clinical signs, ophthalmology, clinical chemistries, urinalyses, or gross pathology. Body weight gains in males of both groups were statistically decreased but with no associated dose–response effect. Lung weights of both sexes were increased, and histopathology revealed some macrophage-filled alveoli. No histological lesions were found in other tissues.

Polyoxyethylene with a mean molecular weight of 4,000,000 was fed to rats for 90 d at dietary levels averaging 8.0 and 18.4 g/kg/d (188). These rats showed changes, rated minor, in the liver and kidney. Two-year tests with rats fed at dietary levels averaging up to 2.76 g/kg/d and dogs fed at dietary levels in the range of 0.6 g/kg/d showed no detectable effect. Testing with a ^{14}C sample revealed no significant absorption of the polymers *per se* from the gastrointestinal tract of the rat or dog. The content of glycols and polyglycols in the tagged sample was considered sufficient to account for all the radioactivity found in the urine (0.72 and 1.1% of the dose administered to 20 rats and 1 dog, respectively).

9.4.5 Reproductive Toxicity

Facial malformations have been reported in the offspring of mice but not rats where pregnant mice were dosed orally with 0.5 and 0.7 mL/kg/d on days 6–17 of gestation, and rats were dosed orally with 1.5–5 mL/kg/d on days 6–14 or 11–16 (188). Maternal toxicity was evident in rats but not mice. Offspring of mice exhibited marked facial and thoracic skeletal defects. In the rabbit, PEG 300 and 400 caused maternal toxicity at daily oral doses of 2 mL/kg during organogenesis but no adverse effects on the fetus (189). Gupta et al. (190) reported no malformations in soft or skeletal tissue in either rats or rabbits (10 dams/group) treated daily during organogenesis with oral doses of 1 mL/kg (rats) or 2 mL/kg (rabbits).

9.4.6 Human Clinical Use

Review of the biomedical literature reveals many studies where polyoxyethylene polymers have been evaluated for efficacy and adverse effects when used as vehicles for medications and as oral or rectal colonic cleansers prior to colonoscopies. The continued use of polyoxyethylene polymers of diverse molecular weights points toward the low toxicity of this family of polymers.

10.0 Polyphenylene Oxide-Based Resin

Polyphenylene oxide can be blended with crystal polystyrene (PS) in all proportions, forming an alloy with a single glass-transition temperature. The higher the PS content, the lower the heat distortion temperature of the material and the easier the alloy to process. These alloys can be brittle and must be further modified to improve toughness by the addition of rubber impact modifiers. In practice, polyphenylene oxide–PS resin blends are also compounded with many additives, including stabilizers and fire retardants. The resulting resins are cream colored and have excellent melt flow. Alloys of polyphenylene oxide can be modified with glass or carbon fibers and a variety of mineral fillers to improve rigidity, tensile strength, and static dissipative properties. This resin family can be

SYNTHETIC POLYMERS, POLYESTERS, POLYETHERS, AND RELATED POLYMERS

filled to high levels with combinations of fiber and mineral fillers to provide high-modulus, controlled shrinkage resins.

10.0.1 CAS Numbers: *[9041-80-9], [25134-01-4]*

10.0.6 Molecular Structure:

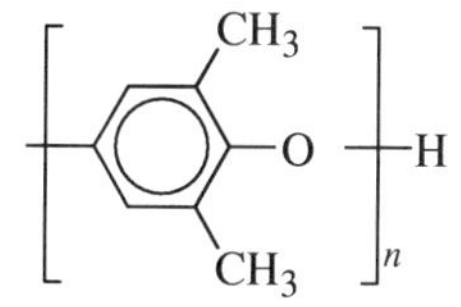

$$\left[\begin{array}{c} CH_3 \\ \\ \\ CH_3 \end{array} \left\langle \bigcirc \right\rangle - O \right]_n - H$$

10.1 Physical and Chemical Properties

Polyphenylene oxide is a thermoplastic resin made from phenolic monomers. Commercially available resins are based on 2,6-dimethylphenol (a.k.a., 2,6-xylenol) and are produced by oxidative polymerization in the presence of an amine catalyst.

Polyphenylene oxide homopolymers have a relatively high T_g (above 212°C) and are very difficult to process (191). At normal processing conditions (300–350°C), melt flow is very stiff, and special precautions must be taken to minimize oxidation.

The blendability of polyphenylene oxide and PS in all proportions provides the capability of producing materials with a wide variety of physical properties and heat resistance. Products are available over the 75–175°C heat distortion range (192). Because both polyphenylene oxide and polystyrene are hydrophobic in nature, the resulting alloys are characterized by low moisture absorption; this, in turn, results in very good electrical properties over a wide range of temperature and humidity.

These resin blends have relatively low coefficients of thermal expansion. In fact, glass-reinforced resins have nearly the same coefficient of thermal expansion as that of metals, such as aluminum.

Alloys of polyphenylene oxide and PS are amorphous materials and can be chemically attacked by a number of organic solvents. These resins will soften or dissolve in many halogenated and aromatic hydrocarbons. Several alloys, however, have been developed to provide improved chemical resistance. These blends are based on polyphenylene oxide and crystalline polymers, such as nylon. Polyphenylene oxide provides excellent heat resistance and toughness, and the crystalline resin provides chemical resistance. These types of blends have utility in the automotive industry, where chemical resistance to compounds such as oils and gasoline is needed in addition to good heat resistance at relatively high temperatures.

10.2 Production And Use

Polyphenylene oxide–PS resins can be processed by standard injection molding, extrusion, blow molding, and structural foam molding. Processing temperatures vary, depending on resin grade, from 230 to 320°C. These resins are easily foamed either by chemical heat-activated blowing agents or by physical injection of inert volatile gases.

Major applications of polyphenylene oxide–PS blends include automotive interiors, such as instrument panels and seat backs; automotive exteriors, such as wheel covers, mirror housings, and rear spoilers; and electrical applications such as fuse boxes and connectors. Other applications include telecommunications and business machines. For telecommunications and computer and business equipment applications, these resins are available with varying levels of flame retardants to meet a wide range of applications: computer housings, card frames, keyboard bases, printer housings, and so on. High-modulus grade resins also find applications in chassis for copiers and laser printers. Examples of home appliance applications include power tools, hair dryers, pump housings, and portable mixers (193). These resins also find extensive use in food contact applications in cookware and food packaging. In food contact applications, these resins offer great improvements over polystyrene in heat resistance for microwave food packaging.

10.4 Toxic Effects

Polyphenylene oxide is not considered a primary eye irritant. In a primary eye irritation study, polyphenylene oxide produced mild and transient irritation in albino rabbits (194). Within 1 h of instillation, redness of the conjunctivae and chemosis was observed, but had resolved by 48 h.

Polyphenylene oxide is not considered a primary skin irritant or a skin sensitizer. Polyphenylene oxide was nonirritating when applied dermally (500 mg/site) to rabbits for 4 h and examined for 3 d following removal (194). Polyphenylene oxide did not cause delayed contact hypersensitivity in guinea pigs given three weekly 6-h epicutaneous applications of a 100% concentration of polyphenylene oxide and challenged 2 wk later with a final application to a naive site (195).

Estimated acute oral exposure LD_{50} (rat) is greater than 5 g/kg. No mortality or signs of toxicity were reported in a group of five male and five female Sprague–Dawley rats given polyphenylene oxide acutely at 5000 mg/kg (146). In two other acute oral toxicity studies in rats, up to 15 g/kg produced no mortality or signs of toxicity (196, 197).

A no observed adverse effect level (NOAEL) for repeated inhalation of polyphenylene oxide powder is estimated to be 7 mg/m^3. In a 13-wk repeated inhalation study, a group of Fischer 344 rats (10 of each sex/group) receiving 1, 7, or 50 mg/m^3 of polyphenylene oxide dust for 6 h/d, 5 d/wk for 2 wk had no treatment-related mortality, clinical signs of toxicity, significant body weight or food consumption differences, or changes in hematology and clinical chemistry parameters (198). Gross and microscopic examination in the 50-mg/m^3 group revealed an exposure-related localized toxicity in the lungs and regional lymph nodes, consisting of a significant increase in lung weights and lung to body weight ratio and a mild, acute/chronic pulmonary inflammation and thymic/peribronchial lymph node histiocytosis. These symptoms were still apparent in the high-exposure group following a 13-wk recovery period.

Groups of 30 male and 30 female rats fed 1.0, 5.0, and 10% of polyphenylene oxide in the diet for 22 mo had no treatment-related mortality, clinical signs of toxicity, body weight differences, changes in clinical chemistry parameters, or apparent gross or microscopic alterations (12). Beagle dogs fed 1.0, 5.0, and 10.0% of polyphenylene oxide in the diet for 2 yr developed no treatment-related changes in mortality or physical parameters, and there were no alterations noted upon gross or histopathological examination (199).

Polyphenylene oxide is not a mutagen by the Ames assay. Several samples of polyphenylene oxide were negative in the *in vitro* Ames *Salmonella*/microsome plate test with and without activation at concentrations up to 2500 µg/plate (200, 201).

10.4.1 Exposure to Processing Fumes

In acute inhalation tests, laboratory rats were exposed to processing fumes at concentrations exaggerating those that would likely occur in workplace situations. During the exposure periods (6-h duration) signs of eye and nasal irritation were observed. These signs of irritation disappeared shortly after the animals were removed from the exposure chamber. No deaths or signs of toxicity were noted during the fume exposure period. There were no distinct or consistent treatment-related tissue or organ changes noted in gross necropsies (202).

10.4.2 Exposure to Combustion Products

The toxicologic aspects of plastic combustion products are very complex. This is mainly because it is very difficult to simulate real-world environments under controlled conditions. Depending on a variety of factors, the qualitative and quantitative nature of chemicals present in the combustion product mixtures will vary. Factors contributing to toxicity of real-life fires include increase of carbon monoxide and carbon dioxide, decrease of oxygen, and the presence of irritant gases. The primary effects from fires are asphyxia due to oxygen deficiency, poisoning from carbon monoxide, heat damage to tissues, and irritation of the respiratory tract by combustion gases.

Some older studies have attempted to elucidate the pyrolysis products of polyphenylene oxide and compare the toxicity of combustion products from this polymer and natural fibers. Under varying fire generation conditions, carbon monoxide, carbon dioxide, and particulates were major combustion products of polyphenylene oxide (203, 204). Bucci et al. attempted to characterize the lung histopathology of mice exposed to the pyrolysis products of several synthetic polymers, including polyphenylene oxide (205). Except for the lungs, no histopathology was evident in major organs from survivors 2 wk after exposure. In the lungs, extensive damage was found 2 wk post-exposure, but damage was too extensive to distinguish differences among polymers. Hilado et al. found that a modified polyphenylene oxide were less toxic than cellulosic materials, based on time of death (206). Based on time to incapacitation, however, synthetic polymers were more toxic. A fire-retarded polyphenylene oxide produced carbon monoxide less rapidly than non-fire-retarded synthetic polymers or natural materials (207). The authors of this study concluded that qualitatively identifying chemicals produced from pyrolysis should be the goal of such studies rather than ranking the toxicity of various polymers or natural materials. These results should be interpreted with caution since, as stated previously, the conditions of fire generation may profoundly influence the formation rates of toxic combustion species.

10.4.3 Plastic Processing Fumes

Processing fumes from polyphenylene oxide and its blends, like any other plastic materials, depend highly on product formulation. In general, processing fumes from polyphenylene oxide processed under recommended conditions may include trace levels of alkylphenols, aliphatic amines and aldehydes, dimethylcyclohexanone, and toluene. Polyphenylene oxide/polystyrene blends may evolve styrene, styrene dimers, toluene, aliphatic amines, aldehydes and alcohols, ethylbenzene, and 4-vinylcyclohexene. Polyphenylene oxide/polystyrene blends flame retarded with triarylphosphate esters also evolve triarylphosphate esters and phenol. Products formulated with polyfluorocarbon additives may evolve hydrogen fluoride and fluorocarbon compounds in the resulting processing fumes. In some instances, additives containing certain heavy metal compounds may be present. These ingredients are essentially bound in the plastic matrix and are unlikely to contribute to workplace exposure under recommended processing conditions.

Processing fumes may cause irritation to the eyes, skin, and respiratory tract, and in cases of severe overexposure, nausea and headaches. Fumes from plastic thermal processes may condense on processing equipment, ventilation systems, and other surfaces. These grease-like processing fume condensates can cause irritation and injury to skin. Use of proper personal protection while cleaning or handling these condensates should be observed.

10.4.4 Exposure Controls and Personal Protection

A continuous supply of fresh air to the workplace together with removal of processing fumes through exhaust systems is recommended. Processing fume condensates may be a fire hazard and toxic; it should be removed periodically from exhaust hoods, duct work, and other surfaces using appropriate personal protection. Ventilation requirements must be locally determined to limit exposure to materials at their point of use.

11 SULFUR POLYMERS

The main commercial polymers are polyphenylene sulfide and the polysulfones.

11.0a Poly(*p*-phenylene sulfide)

11.0.1a CAS Number: [53027-72-8], [9016-75-5]

11.0.2a Synonyms: Poly(thio-1,4-phenylene)

11.0.6a Molecular Structure:

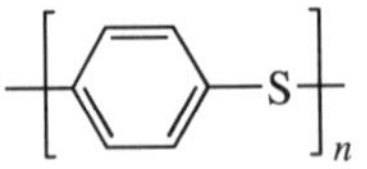

11.0.b Polysulfone of Bisphenol A

11.0.1b CAS Number: [25135-51-7]

11.0.6b Molecular Structure:

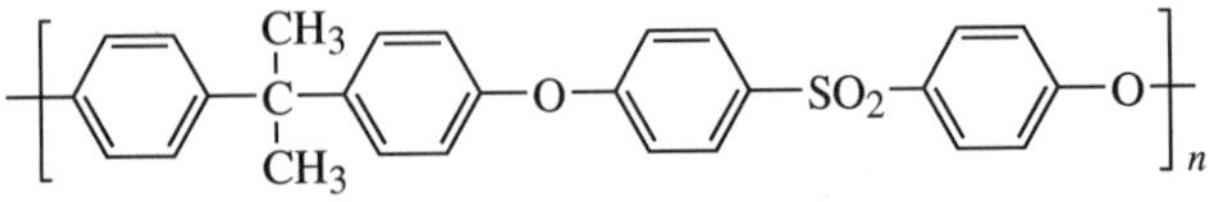

Other polysulfones, such as polyether sulfone and polyaryl sulfone, that exist as powders differ in the types of linkage between aromatic rings and vary in their properties accordingly.

11.1a Synthesis, Properties, and Applications of Polyphenylene Sulfide

This polymer has been manufactured commercially by reacting *p*-dichlorobenzene and sodium sulfide, apparently in *N*-methylpyrrolidone (as cited in Ref. 208). Heating under oxygen cures the product to form a fine white powder, apparently linear. The white, lightly crystalline polymer discolors on heating in air to form a brown product, presumably crosslinked. Polyphenylene sulfide is known for thermal and chemical resistance, toughness, and flexibility. Additional properties are summarized in Table 92.1 (209).

Polyphenylene sulfide molded parts can be used in submersible and centrifugal pumps, computer housings, and telecommunications. Coatings of this polymer often utilized for chemical resistance in industrial process operations (210).

11.1b Synthesis, Properties, and Applications of Polysulfone

For commercial products an aryl ether can be treated with sulfonyl chloride in the presence of catalysts to give sulfone groups (polysulfonylation) or sulfones can be treated with phenolates to give ether groups (polyetherification). The structures made by one route are usually not made by the other, although some can be made by either process. All polysulfones have excellent creep resistance. They are stable to oxygen and thermal degradation and are flame resistant. Their toughness is affected by the inclusion of bulky side groups in the polymer structure or deviations from the all-*para* orientation of groups linking the aromatic rings (208, 211). See Table 92.1 for additional data.

11.4 Toxic Effects

Several experimental tests indicate that products evolved at elevated temperatures can be rapidly toxic to mice. Thermal degradation of polyphenylene sulfide may yield hydrogen sulfide; the sulfur of polysulfone is typically evolved as sulfur dioxide. See Table 92.1. Additional details may be found in Refs. 40 and 212–216.

Ground fiberglass-reinforced polyphenylene sulfide (CAS# [9016-75-5]) was fed to rats in their diets at concentrations of 0, 0.5, 2.75, or 5% for 6 mo. A small number of high-dose rats exhibited yellow staining of the anogenital region. Red or black staining was noted in some rats around the eyes and masses on the neck. No changes were noted in behavior or hematology, and clinical chemistries revealed a transient elevation in blood urea nitrogen and a decrease in SGOT in the two highest-dose groups. No gross lesions were evident at necropsy, and organ weights were normal (217).

Highly refined polysulfone has been used as membranes in kidney dialysis machines and other biomedical applications with little toxicity.

BIBLIOGRAPHY

1. J. A. Brydson, *Plastics Materials*, 3rd ed., Whitefriars Press, London, 1975.
2. U. S. Code of Federal Regulations, Supt. Doc., Washington, DC, 21 C.F.R. 170–199, April 1, 1981, also 21 *CFR* 500–599, revised annually, guide available from *Food Chemical News*.

2a. G. P. Clayton and F. E. Clayton, Eds., *Patty's Industrial Hygiene and Toxicology*, 4th ed., vol. 11, Part E, Wiley, New York, 1994, Chapt. 37.

3. M. S. Crandall, Worker exposure to styrene monomer in the reinforced plastic boat-making industry. *Am. Ind. Hyg. Assoc. J.*, **42**, 499–502 (1981).

4. R. L. Schumacher et al., Styrene exposure in the fiberglass fabrication industry in Washington state. *Am. Ind. Hyg. Assoc. J.* **42**, 143–149 (1981).

5. A. Tossavainen, Styrene use and occupational exposure in the plastics industry. *Scand. J. Work Environ. Health* **4**, 7–13 (1978).

6. K. P. Lee, personal communication, Haskell Laboratory, DuPont Company, Wilmington, DE, July 1980.

7. F. Valic and E. Zuskin, *Respiratory-function changes in textile workers exposed to synthetic fibers*. *Arch. Environ. Health* **32**, 283–287 (1977).

8. J. R. Tiller and R. S. F. Schilling, Respiratory function during the day in rayon workers: a study in byssinosis. *Trans. Assoc. Ind. Md. Offic.* **7**, 161–162 (1958).

9. G. Berry, M. K. B. Molyneux, and J. B. L. Tombleson, Relationships between dust level and byssinosis and bronchitis in Lancashire cotton Mills. *Br. J. Ind. Med.* **31**, 18–27 (1974).

10. J. C. Pimentel, R. Avila, and A. G. Lourenco, Respiratory disease caused by synthetic fibres: a new occupational disease, *Thorax* **30**, 204–219 (1975).

11. J. D. Brain et al., Pulmonary distribution of particles given by intratracheal instillation or by aerosol inhalation. *Environ. Res.* **11**, 13–33 (1976).

12. A. Bouhuys, Fibers and fibrosis. *Ann. Int. Med.* **83**, 898–899 (1975).

13. A. Muittari and T. Veneskoski, Natural and synthetic fibers as causes of asthma and rhinitis. *Ann. Allergy* **41**, 48–50 (1978).

14. J. C. Pimentel, Sarcoid granulomas of the skin produced by acrylic and nylon fibres. *Br. J. Dermatol.* **96**, 673–677 (1977).

15. M. F. Stanton et al., Carcinogenicity of fibrous glass: pleural response in the rat in relation to fiber dimension. *J. Natl. Cancer Inst.* **58**, 587–597 (1977).

16. Textile Research Institute, *Methods of Measuring the Particle Sizes of Respirable Dusts; A Review of the Literature with Special Emphasis on Fibrous Dusts*, Report No. 5, Princeton, NJ., 1980.

17. R. W. Moncrieff, *Man Made Fibres*, 6th ed., Wiley, New York, 1975.

18. U.S. Federal Trade Commission, Rules and Regulations Under the Textile Fiber Products Identification Act, Washington, DC, effective March 3, 1960, as amended to November 1, 1974.

19. E. M. Hicks et al., The production of synthetic-polymer fibres. *Text. Progr.* **3**, 1–113 (1971).

20. H. I. Bolker, *Natural and Synthetic Polymers*, Marcel Dekker, New York, 1974.

21. R. C. Knowlton, N. C. Pierce, and P. Popper, Snow deposits in friction twisting. *Fiber Prod.* **6**(6), 18–30 (1978).

22. G. G. Hawley, *The Condensed Chemical Dictionary*, 9th ed., Van Nostrand Reinhold, New York, 1976.

23. J. B. Titus, *Trade Designations of Plastics and Related Materials*, (Revised), Plastec Note N9C, U.S. Army Material Development and Readiness Command (Plastics Technical Evaluation Center), Dover, NJ, 1978 (ADA 058395).

24. W. J. Roff and J. R. Scott, with J. Pacitti, *Handbook of Common Polymers*, CRC Press, Cleveland, OH, 1971.

25. G. Farrow and E. S. Hill, Polyester fibers. *In Encyclopedia of Polymer Science and Technology*, Vol. 11, Wiley, New York, 1969, pp. 1–41.
26. J. M. Hawthorne and C. J. Heffelfinger, Polyester films. *In Encyclopedia of Polymer Science and Technology*, Vol. 11, Wiley, New York, 1969, pp. 42–61.
27. H. Sherman, unpublished data, DuPont Company, Haskell Laboratory, Wilmington, DE, Jan. 1968.
28. F. X. Wazeter and E. I. Goldenthal, unpublished data, International Research and Development Corp., Mattawan, MI, June 1973 (sponsored by Haskell Laboratory).
29. T. Otaka et al., Safety evaluation of polyethylene terephthalate resins. 1. studies on acute and subacute (3 months) toxicities by oral administration in rats. *Shokuhin Eiseigaku Zasshi; Chem. Abstr.*, **91**, 14755r.
30. K. M. Frank, unpublished data, DuPont Company, Haskell Laboratory, Wilmington, DE, July 1972.
31. R. W. Morrow, unpublished data, DuPont Company, Haskell Laboratory, Wilmington, DE, Nov. 1972.
32. J. W. McAlack and N. C. Goodman, unpublished data, DuPont Company, Haskell Laboratory, Wilmington, DE, Dec. 1976.
33. M. V. Shelanski, unpublished data, Product Investigations, Inc., Conshohocken, PA, Oct. 1979 (sponsored by Haskell Laboratory).
34. M. V. Shelanski, unpublished data, Product Investigations, Inc., Conshohocken, PA, Dec. 1979 (sponsored by Haskell Laboratory).
35. J. F. Wilson and K. L. Gabriel, unpublished data, Biosearch, Inc., Philadelphia, PA, Oct. 1979 (sponsored by Haskell Laboratory).
36. D. B. Hood and A. Ivanova-Binova, The skin as a portal of entry and the effects of chemicals on the skin, mucous membranes and eye. *In Principles and Methods for Evaluating the Toxicity of Chemicals, Part II, Environmental Health Criteria Series 6*, World Health Organization, Geneva, in press.
37. W. F. Schorr, E. Keran, and E. Poltka, Formaldehyde allergy. *Arch. Dermatol.* **110**, 73–76 (1974).
38. J. B. Terrill and K. P. Lee, personal communication, Haskell Laboratory, DuPont Company, Wilmington, DE, May 1977.
39. R. W. Hartgrove and K. P. Lee, personal communication, Haskell Laboratory, DuPont Company, Wilmington, DE, July 1978.
40. C. J. Hilado, J. J. Cumming, and C. J. Casey, Toxicity of pyrolysis gases from natural and synthetic materials. *Fire Technol.* **14**, 136–146 (1978).
41. K. Yamamoto, Acute toxicity of the combustion products from various kinds of fibers. *Z. Rechtsmed.* **76**, 11–26 (1975); *Chem. Abstr.* **84**, 922m.
42. R. L. Schumacher and P. A. Breysse, Combustion and pyrolysis products from synthetic textiles. *J. Combust. Toxicol.* **3**, 393–424 (1976).
43. C. J. Hilado and H. J. Cumming, Relative toxicity of pyrolysis gases from materials: effects of chemical composition and test conditions. *Fire Mater.* **2**, 68–79 (1978).
44. J. R. Bercaw, The melt-drip phenomena of apparel. *Fire Technol.* **9**, 24–45 (1973).
45. Y. Yoshida et al., Toxicity of pyrolysis products of thermal-resistant plastics including polyamide and polyester. *Jpn. J. Hyg.* **33**, 450–458 (1978).
46. N. Igarashi, The thermal degradation of nylon 6, polyethylene terephthalate and polycarbonate polymers. *Diss. Abstr. Int. B.* **39**, 3458 (1979).

47. J. B. Terrill, unpublished data, DuPont Company, Haskell Laboratory, Wilmington, DE, Sept. 1975.
48. E. P. Goodings, Thermal degradation of polyethylene terephthalate. In Society of Chemical Industry, *High Temperature Resistance and Thermal Degradation of Polymers*, S. C. 1. Monograph No. 13, MacMillan, New York, 1961, pp. 211–228.
49. L. H. Buxbaum, The degradation of poly(ethylene terephthalate). *Angew. Chem. Int. Ed.* **7**, 182–190 (1968).
50. W. L. Hergenrother, Influence of copolymeric poly(diethylene glycol) terephthalate on the thermal stability of poly(ethylene terephthalate). *J. Polym. Sci.* **12**, 875–883 (1974).
51. W. T. Langstaff and L. C. Trent, The effect of polyester fiber content on the burn injury potential of polyester/cotton blend fabrics. *J. Consum. Prod. Flammability* **7**, 26–39 (1980).
52. R. N. King and D. J. Lyman, Polymers in contact with the body. *Environ. Health Perspect.* **11**, 71–74 (1975).
53. J. H. Harrison, D. S. Swanson, and A. F. Lincoln, A comparison of the tissue reactions to plastic materials. *AMA Arch. Surg.* **74**, 139–144 (1957).
54. H. Kus, K. Kawecki, and E. Szewczak, Studies on carcinogenesis and the usefulness of polyester yarn in alloplasty. *Arch. Immunol. Ther. Exp.* **12**, 730–739 (1964).
55. H. C. Amstutz, W. F. Coulson, and E. David, Reconstruction of the canine achilles and patellar tendons using dacron mesh silicone prosthesis. 1. clinical and biocompatibility evaluation. *J. Biomed. Mater. Res.* **10**, 47–59 (1976).
56. Y. M. Tardif, C. L. Schepens, and F. I. Tolentino, Vitreous Surgery. *Arch. Ophthalmol.* **95**, 229–234 (1977).
57. NIOSH, *Manual of Analytical Methods*, 4th ed., 1994.
58. R. H. Hayward and F. L. Korompai, Degeneration of knitted dacron grafts. *Surg.* **79**, 581–583 (1976).
59. H. Kulenkampff and G. Simonis, Biological tolerance of vascular prostheses made of dacron and synthetic suture material. *Chirurg*, **47**, 189–192 (1976).
60. J. A. Styles and J. Wilson, Comparison between *in vitro* toxicity of polymer and mineral dusts and their fibrogenicity. *Ann. Occup. Hyg.* **16**, 241–250 (1973).
61. D. M. Conning et al., Comparison between *in vitro* toxicity of dusts of certain polymers and minerals and their fibrogenicity. In W. H. Walton, Ed., *Inhaled Particles III*, Vol. I, Unwin Brothers Limited, Surrey, England, 1971, pp. 499–506.
62. W. A. Burnes et al., Fibrosarcoma occurring at the site of a plastic vascular graft. *Cancer* **29**, 66–72 (1972).
63. T. X. O'Connell, H. J. Fee, and A. Golding, Sarcoma associated with dacron prosthetic material. *J. Thorac. Cardiov. Surg.* **72**, 94–96 (1972).
64. T. T. Daurova et al., Kinetics of the degradation of poly(ethylene terephthalate) in body tissues. *Dokl. Akad. Nauk SSSR* **231**(4), 919–920 (1976); *Chem. Abstr.* **87**, 15639k.
65. D. L. Myers and C. A. LaSala, Conservative surgical management of Mersilene mesh suburethral sling erosion. *Am. J. Obstet. Gynecol.* **179**(6, Part 1), 1424–1428 (1998).
66. F. M. Mutlu, K. Tuncer, and C. Can, Extrusion and granuloma formation with mersilene mesh brow suspension. *Ophthalmic Surg. Lasers* **30**(1), 47–51 (1999).
67. J. Frucht-Pery, Mersilene sutures for corneal surgery. *Ophthalmic Surg.* **26**(2), 117–120 (1995).
68. G. E. Leber et al., Long-term complications associated with prosthetic repair of incisional hernias. *Arch. Surg.* **133**(4), 378–382 (1998).

69. L. Tang, Mechanisms of fibrinogen domains: biomaterial interactions. *J. Biomater. Sci. Polym. Ed.* **9**(12), 1257–1266 (1998).

70. J. A. Chinn et al., Blood and tissue compatibility of modified polyester: thrombosis, inflammation, and healing. *J. Biomed. Res.* **39**(1), 130–140 (1998).

71. D. P. Wyman, Thermoplastic polyester: PBT. *In Modern Plastics Encyclopedia, 1978–79 ed.*, Vol. 55, No. 10A, McGraw-Hill, New York 1978–1979, p. 49.

72. W. F. H. Borman and M. Kramer, Poly(butylene-terephthalate) chemical and physical properties affecting its processing and usage. *Am. Chem. Soc. Div. Org. Coatings Plast. Chem.* U, 77–85 (1974).

73. B. D. Astill and R. L. Raleigh, unpublished data, Health, Safety and Human Factors Laboratory, Rochester, NY, May 1974 (courtesy of The Society of the Plastics Industry, Inc., New York).

74. A. Cartier et al., Respiratory and systemic reaction following exposure to heated electrostatic polyester paint. *Eur. Respir. J.* **7**(3), 608–611 (1994).

75. G. L. Nelson, E. J. Hixson, and E. P. Denine, Combustion product toxicity studies of engineering products. *J. Combust. Toxicol.* **5**, 222–238 (1978).

76. J. W. Sarver, unpublished data, DuPont Company, Haskell Laboratory, Wilmington, DE, Oct. 1974.

77. O. L. Dashiell, unpublished data, DuPont Company, Haskell Laboratory, Wilmington, DE, June 1975.

78. V. Passalacqua et al., Thermal degradation of poly(butylene terephthalate). *Polymer* **17**, 1044–1048 (1976).

79. E. Cenni et al., Activation of the plasma coagulation system induced by some biomaterials. *J. Biomed. Mater. Res.* **31**(1), 145–148 (1996).

80. A. Weckbach, E. Kunz, and T. Kirchner. Alloplastic ligament replacement. A study of the biological fixation of 5 non-resorbable materials. *Unfallchirurg.* **93**(8), 380–383 (1990).

81. D. Bakker et al., Biocompatibility of a polyether urethane, polypropylene oxide, and a polyether polyester copolymer. A qualitative and quantitative study of three alloplastic tympanic membrane materials in the rat middle ear. *J. Biomed. Mater. Res.* **24**(4), 489–515 (1990).

82. H. E. Schroeder, Thermoplastic or elastomer? a new engineering polymer. *Shell Polym.* **3**, 70–73 (1979).

83. M. Borkenhagen et al., *In vivo* performance of a new biodegradable polyester urethane system used as a nerve guidance channel. *Biomaterials* **19**(23), 2155–2165 (1998).

84. B. S. Gupta and V. A. Kasyanov. Biomechanics of human common carotid artery and design of novel hybrid textile compliant vascular grafts. *J. Biomed Mater. Res.* **34**(3), 341–349 (1997).

85. G. J. Beumer et al., A new biodegradable matrix as part of a cell seeded skin substitute for the treatment of deep skin defects: a physico-chemical characterisation. *Clin. Mater.* **14**(1), 31–37 (1993).

86. National Materials Advisory Board, Materials: state of the art. In *Fire Safety Aspects of Polymeric Materials*, Vol. 1, Technomic, Westport, CT, 1977.

87. H. P. Cordts and J. A. Bauer, Unsaturated polyester. In *Modern Plastics Encyclopedia, 1978–79 ed.*, Vol. 55, No. 10A, McGraw-Hill, New York, 1978–1979, pp. 54–59.

88. SPI Composites Institute, *Introduction to Composites*, The Society of Plastics Industry, Inc., Washington, DC, 1992.

89. A. A. Fisher, *Contact Dermatitis*, 2nd ed., Lea & Febiger, Philadelphia, PA, 1973.
90. A. W. MacFarland, R. K. Curley, and C. M. King, Contact sensitivity to unsaturated polyester resin in a limb prosthesis. *Contact Dermatitis*, **15**(5), 301–303 (1986).
91. L. B. Bourne and F. J. M. Milner, Polyester resin hazards. *Br. J. Ind. Med.* **20**, 100–109 (1963).
92. M. M. Key and D. P. Discher, Polyester resins—their dermatologic aspects in industry. *Cutis* **2**, 27–29 (1966).
93. K. E. Malten, Occupational dermatoses in the processing of plastics. *Trans. St. John's Hosp. Dermatol. Soc.* **59**, 78–113 (1973).
94. J. Lim et al., Fiber glass reinforced plastics. *Arch. Environ. Health*, **20**, 540–544 (1970).
95. L. Jirasek, Polyester resins and glass laminates. *Prac. Lek.* **14**, 120–124, 1962; *Chem. Abstr.* **57**, 7560d.
96. A. Dooms-Goossens and G. DeJong, *Letter to the Editor. Contact Dermatitis* **12**(4), 238 (1985).
97. C. Linden, A. Löfström, and Storgårds-Hatam, Contact allergy to unsaturated polyester in a boatbuilder. *Contact Dermatitis* **11**(4), 262–264 (1984).
98. M. R. Kantz, Advanced polymer matrix resins and constituents: an overview of manufacturing, composition and handling. *Appl. Ind. Hyg.* **4**, 1–8 (1989).
99. A. H. Okum et al., Mortality study of workers exposed to styrene in the reinforced plastic boat building industry. *NIOSH PB84 242171*, NIIS, Springfield, VA, 1984, 57 pp.
100. L. Kanerva et al., Occupational allergic contact dermatitis from unsaturated polyester resin in a car repair putty. *Int. J. Dermatol.* **38**(6), 447–452 (1999).
101. K. Tarvainen, R. Jolanki, and T. Estlander, Occupational contact allergy to unsaturated polyester resin cements. *Contact Dermatitis* **28**(4), 220–224 (1993).
102. K. Tarvainen et al., Exposure, skin protection and occupational skin diseases in the glass fibre-reinforced plastics industry. *Contact Dermatitis* **29**(3), 119–127 (1993).
103. Health & Safety Executive, *A Guide to Health & Safety in GRP Fabrication*, ISBN 07176 0294X (112/87).
104. P. Kalliokoski, Procedures and uses of styrene containing polymers. In *Industrial Hazards of Plastics and Synthetic Elastomers*, Alan R. Liss, New York, 1984, pp. 193–202.
105. H. F. Mark and S. Atlas, Introduction to polymer science. In H. S. Kaufman and J. J. Falcetta, Eds., *Introduction to Polymer Science and Technology: An SPE Textbook*, Wiley, New York, 1977, pp. 1–23.
106. International Agency for Research on Cancer, *IARC Monographs on the Evaluation of the Carcinogenic Risk of Chemicals to Humans; Some Monomers, Plastics and Synthetic Elastomers and Acrolein*, Vol. 19, Lyon, France, 1979.
107. J. S. McDermott, Health and safety in reinforced plastics fabrication. In Society of Plastics Engineers, Inc., *Safety Health with Plastics, National Technical Conference*, Denver, CO, Nov. 8–10, 1977, pp. 187–191.
108. H. L. Boiteau and F. Rossel-Renac, Etude experimentale de l'evaporation du styrene au cours de la polymerisation des resines polyesters insaturees. *Arch. Mal. Pros. Med. Trav. Secur. Soc.* **39**, 52–59 (1977).
109. P. Roper, Health hazard evaluation report monark boat company. *NIOSH No. HETA-86-441-1913, PB89-130140*, Springfield, VA, 1988, 21 pp.
110. M. S. Crandall and R. W. Hartle, Health hazard evaluation report. Henry R. Hinckley & Co., *NIOSH No. HETA 83-128-1485 (PB85-208866)*, Springfield, VA, 1985, 35 pp.

111. G. L. Swinehart, Survey report on control technology for FRP tank manufacture at FMC corporation, Jonesboro, Arkansas. *PB86-144649*, 1985, 12 pp.
112. W. F. Todd, Control of styrene vapor during the manufacture of fiber reinforced plastics small parts. *Report No. CT-107-06 (PB85 221455/AS)*, 1984, 58 pp.
113. Unsaturated polyester resins. *Fed. Reg.* **54**, 2429 (Jan. 19, 1989).
114. D. P. Miller, R. V. Petrella, and A. Manca, An evaluation of some factors affecting the smoke and toxic gas emission from burning unsaturated polyester resins. *Am. Chem. Soc. Div. Org. Coatings Plast. Chem.* **36**, 576–581 (1976).
115. H. T. Hofmann and H. Sand, Further investigations into the relative toxicity of decomposition products given off from smoldering plastics. *J. Combust. Toxicol.* **1**, 250–258 (1974).
116. G. Kimmerle, Aspects and methodology for the evaluation of toxicological parameters during fire exposure. *J. Combust. Toxicol.* **1**, 4–51 (1974).
117. C. S. Barrow, Y. Alarie, and M. F. Stock, Sensory irritation and incapacitation evoked by thermal decomposition products of polymers and comparisons with known sensory irritants. *Arch. Environ. Health* **33**, 79–88 (1978).
118. A. E. Karakaya, S. Sardas, and M. Sun, Sister chromatid exchanges in furniture workers exposed to unsaturated polester resins. *Arch. Toxicol.* (Suppl. 14), 307–310 (1991).
119. K. F. Lin, Alkyd resins, In J. Kroscnwitz, Ed., *Encyclopedia of Chemical Technology*, 4th ed., Vol. 2, Wiley, New York, 1992, pp. 53–85.
120. J. R. Blegen, Alkyd resins. In W. R. Fuller, Ed., *Federation Series on Coatings Technology, Unit 5*, Federation of Societies for Paint Technology, Philadelphia, 1967, pp. 7–31.
121. M. J. van Tongeren et al., Retrospective exposure assessment for a cohort study into respiratory effects of acid anhydrides. *Occup. Environ. Med.* **55**(10), 692–696 (1998).
122. M. J. van Tongeren et al., Exposure to acid anhydrides in three resin and one cushioned flooring manufacturing plants. *Ann. Occup. Hyg.* **39**(5), 559–571 (1995).
123. New Chemicals Branch, Carboxylic acid anhydrides. In *New Chemical Program-Categories of Concern*, Office of Pollution Prevention and Toxics, U. S. Environmental Protection Agency, 1992.
124. K. Yokota, T. Takeshita, K. Morimoto, Prevention of occupational allergy caused by exposure to acid anhydrides. *Ind. Health* **37**(3), 281–288 (1999).
125. K. E. Malten, Old and new, mainly occupational dermatological problems in the production and processing of plastics. In H. S. Maibach, Ed., *Occupational and Industrial Dermatology*, 2nd ed., Year Book Medical Publishers, Chicago, 1987, pp. 290–340.
126. J. L. Thomas, Allyl. In *Modern Plastics Encyclopedia, 1978–79 ed.*, Vol. 55, No. 10A, McGraw-Hill New York, pp. 9–10.
127. M. Lacroix, et al., Irritant dermatitis from diallyglycol carbonate monomer in the optical industry. *Contact Dermatitis* **2**, 183–195 (1976).
128. HSDB (Hazardous Substances Databank), Toxicological database generated and maintained by the National Occupational Institute for Safety and Health (NIOSH), Bethesda, 1999.
129. D. C. Clagett and S. J. Shafer, Polycarbonates. *Comp. Polym. Sci.* **5**, 345 (1989).
130. D. Freitag, U. Grigo, P. R. Muller, and W. Nouvertne, Polycarbonates. In *Encyclopedia of Polymer Science and Engineering*, Vol. 11, Wiley. New York, 1988, p. 648.
131. Ger. Pat. (DOS) 2,117,509 (Apr. 10, 1971), F. N. Liberti (to General Electric Co.); U.S. Patent 4,092,888 (July 1, 1978), K. J. Wilson (to General Electric Co.); Ger. Pats. (DOS) 2,658,849

(Dec. 24, 1976), R. J. Axelrod (to General Electric Co.) and 2,726,662 (June 14, 1977), R. J. Axelrod and C. A. Bialous (to General Electric Co.).

132. Ger. Pat. (DAS) 2,140,207 (Aug. 21, 1971), E. Eimers and D. Margotte (to Farbenfabriken Bayer AG); U.S. Pat. 3,794,629 (Aug. 11, 1971), E. Eimers, D. Margotte, R. Dhein, and H. Schmid (to Farbenfabriken Bayer AG); Ger. Pats. (DOS) 1,769,823 (July 19, 1968), H. L. Rawlings (to Mobay Chemical Co.) and (DOS) 3,011,125 (Mar. 22, 1979), R. J. Axelrod and C. A. Bialous (to General Electric Co.).

133. U.S. Pat. 3,679,629 (Apr. 1, 1972), A. J. Chalk and A. Factor (to General Electric Co.); Ger. Pats. (DOS) 2,920,450 (May 21, 1979), K. Idel and W. Cohnen (to Farbenfabriken Bayer AG) and 2,920,451 (May 21, 1979) (to Farbenfabriken Bayer AG).

134. Ger. Pat. (DOS) 2,400,045 (Jan. 2, 1974), C. A. Bialous and G. F. Macke (to General Electric Co.); U.S. Pat. 3,679,629 (Apr. 1, 1972), A. J. Chalk and A. Factor (to General Electric Co.); Ger. Pats. (DOS) 2,920,450 (May 21, 1979), K. Idel et al., (to Farbenfabriken Bayer AG) and 2,920,451 (May 21, 1979), H. Vernaleken et al., (to Farbenfabriken Bayer AG).

135. Ger. Pats. (DOS) 2,507,748 (Feb. 22, 1975), S. Adelmann et al., (to Farbenfabriken Bayer AG) and 2,729,485 (June 30, 1977), S. W. Scott (to General Electric Co.).

136. Ger. Pats. (DOS) 2,064,095 (Dec. 28, 1970), H. Schirmer and G. Peilstocker (to Farbenfabriken Bayer AG).

137. U.S. Pat. 4,831,100 (May 16, 1989), T. Komatsu and E. Terada (to Idemitsu Petrochemical Co., Ltd.); U.S. Pat. 4,888,410 (Dec. 19, 1989), T. Komatsu and E. Terada (to Idemitsu Petrochemical Co., Ltd.); U.S. Pat. 4,918,155 (Apr. 17, 1990), T. Komatsu and E. Terada (to Idemitsu Petrochemical Co., Ltd.).

138. V. Mark, *Am. Chem. Soc. Div. Org. Coat. Plast. Chem. Pap.* **43**, 71 (1980).

139. J. L. Webb, *Am. Chem. Soc. Div. Org. Coat. Plast. Chem. Pap.* **43**, 79 (1980).

140. U.S. Pat. 3,775,367 (Nov. 27, 1973), W. Nouvertne (to Farbenfabriken Bayer AG).

141. I. S. Thomas and S. A. Ogoe, *Tech. Pap. Reg. Tech. Conf. Soc. Plast. Eng.* Baltimore-Washington Sect. 100 (1985).

142. U.S. Pat. 3,442,829 (May 6, 1969), L. D. Moore and J. J. Randal (to Borg-Warner Corp.).

143. U.S. Pat. 4,097,425 (June 27, 1978), G. E. Niznik (to General Electric Co.).

144. R. O. Carhart, R. L. Stadterman, and R. O. L. Lynn, Polycarbonate. In *Modern Plastics Encyclopedia, 1978-79 ed.*, Vol. 55, No. 10A, McGraw-Hill, New York 1978–1979, p. 46.

145. M. Sohrabi and K. Z. Morgan, A New polycarbonate fast neutron personnel dosimeter. *Am. Ind. Hyg. Assoc. J.* **39**, 438–447 (1978).

146. V. T. Mallory, unpublished data, for General Electric Co., Pharmakon Research International, July, 1986.

147. G. E. Parke, unpublished data, for GE Plastics, Cannon Laboratories Inc., January, 1978.

148. G. Bornmann and A. Loeser, *Arzneim.-forsch.* **9**(9), 9–13 (1959); *Chem. Abstr.* 53, 11662f.

149. S. M. Rosemberger, unpublished data, for GE Plastics, TPS Inc., July, 1982.

150. J. A. Brotons et al., Xenoestrogens released from lacquer coatings in food cans. *Environ. Health Perspect.* **103**(6), 608–612 (1995).

151. S. R. Howe, L. Borodinsky, 1998. Potential exposure to bisphenol A from food-contact use of polycarbonate resins. *Food Addit. Contam.* **15**(3), 370–375 (1998).

152. K. A. Mountfort et al., Investigations into the potential degradation of polycarbonate baby bottles during sterilization with consequent release of bisphenol A. *Food Addit. Contam.* **14**(6–7), 737–740 (1997).

153. Y. Alarie, C. K. Lin, and D. L. Geary, Sensory irritation evoked by plastic decomposition products. *Am. Ind. Hyg. Assoc. J.* **35**, 654–661 (1974).
154. G. L. Moan and M. Chaigneau, *Ann. Pharm. Fr.* **35**, 641, (1977).
155. Anonymous, *Tetrabromobisphenol A and derivatives, Environmental Health Criteria 172*, Office of Publications, World Health Organization, Geneva, 1995.
156. J. Larsen et al., Effects of metallic coatings on the combustion toxicity of engineering plastics. *Fire and Materials* **18**(2), 121–130 (1994).
157. H. J. Taylor, and M. J. Troughton, Products evolved during hot gas welding of plastics. *Health and Safety Executive*, HSE Books, Sudbury, Suffolk, U.K., 1995.
158. H. J. Taylor, and M. J. Troughton, Products evolved during laser cutting of plastics. *Health and Safety Executive*, HSE Books, Sudbury, Suffolk, U.K., (1995).
159. N. I. Sax and R. J. Lewis, Eds. *Hawley's Condensed Chemical Dictionary*, 12th ed. Van Nostrand Reinhold, New York, 1994.
160. *RTECS (Registry of the Toxic Effects for Chemical Substances)*, toxicology database compiled and maintained by the National Institute for Occupational Safety and Health (NIOSH) available through the National Library of Medicine. Bethesda, MD, 1999.
161. *HSDB (Hazardous Substances Data Base)*, a computerized toxicology database generated by the National Institute for Occupational Safety and Health (NIOSH) available through the National Library of Medicine. Bethesda, MD, 1999.
162. K. J. Persak and L. M. Blair, Acetal resin. In *Encyclopedia of Chemical Technology*, Vol. 1, Wiley, New York, 1979, pp. 112–123.
163. Thermo plastic resin decomposition. *Mich. Occup. Health* **16**(2), 8 (1970–1971).
164. W. M. Cleary, Thermalplastic resin decomposition. *Ind. Med.* **39**(3), 34–36 (1970).
165. L. Bedouin, Acetal resins, injection molding: The risks and their prevention, *Cah. Notes DOC.* **85**, 545–552, (1976) *Chem. Abstr.* **87**, 17218b.
166. R. J. Neher, unpublished data, DuPont Company, Haskell Laboratory, Wilmington, DE, July 1964.
167. C. J. Hilado and N. V. Huttlinger, Toxicity of pyrolysis gases from some synthetic polymers. *J. Combust. Toxicol.* **5**, 361–369 (1978).
168. J. W. Clayton and G. Limperos, unpublished data, DuPont Company, Haskell Laboratory, Wilmington, DE, Aug. 1956–Aug. 1957.
169. R. G. McKee et al., Effects of combustion mode on the smoke toxicity of polyoxymethylene (delrin), *Natl. SAMPE Tech. Conf.* **11**, 582–592 (1979); *Chem. Abstr.* **92**, 70553f.
170. C. J. Hilado, J. E. Schneider, and D. P. Brauer, Toxicity of pyrolysis gases from polyoxymethylene. *J. Combust. Toxicol.* **6**, 30–36 (1979).
171. C. E. Schweitzer, R. N. MacDonald, and J. O. Punderson, Thermally stable high molecular weight polyoxymethylenes. *J. Appl. Polym. Sci.* **1**, 158–163 (1959).
172. C. M. Zapanta et al., A comparison of the cavitation of potential of prosthetic heart valves based on valve closing dynamics. *J. Heart Valve Dis.* **7**(6), 665–667 (1998).
173. M. Triggiani et al., Six-and-half years experience with the St. Jude BioImplant porcine prosthesis. *J. Heart Valve Dis.* **6**(2), 138–144 (1997).
174. M. Hirai et al., Long-term results of cardiac valve replacement with a Delrin-disk model of the Bjork-Shiley valve prosthesis - comparative analysis with spherical-disk model. *Nippon Kyobu Genka Gakkai Zasshi* **44**(11), 1986–1992 (1996).

175. P. D. Stein and K. C. Dellsperger, Summary and recommendations. *J. Heart Valve Dis.* (Suppl. 2), S246–248 (1996).
176. K. Thyagarajan, C. H. Conlin, and D. W. Wieting, Evaluation of disc retention of Bjork-Shiley Delrin heart valves. *J. Heart Valve Dis.* (Suppl. 2), S243–245 (1996).
177. H. A. McKellop, H. L. Milligan, and T. Rostlund, Long-term biostability of polyacetal (Delrin) implants. *J. Heart Valve Dis.* (Suppl. 2), S238–242 (1996).
178. K. C. Dellsperger, G. L. Grunkemeier, and P. D. Stein, Clinical experience with the Bjork-Shiley Delrin tilting disc heart valve. *J. Heart Valve Dis.* (Suppl. 2) S169–177 (1996).
179. D. W. Wieting, The Bjork-Shiley Delrin tilting disc heart valve: historical perspective, design and need for scientific analyses after 25 years. *J. Heart Valve Dis.* (Suppl. 2), S157–168 (1996).
180. M. Abe et al., Report of a case with re-mitral valve replacement for Bjork-Shiley Delrin disc valve prosthesis—disc wear of the Delrin disc 20 years after implantation. *Nippon Kyobu Genka Gakkai Zasshi* **44**(4), 545–547 (1996).
181. A. Ohlin and L. Linder, Biocompatibility of polyoxymethylene (Delrin) in bone. *Biomaterials* **14**(4), 285–289 (1993).
182. A. P. Weislander et al., *In vitro* toxicity of biomaterials determined with cell density, total protein, cell cycle distribution and adenine nucleotides. *Biometer. Artif. Cells Immobilization Biotechnol.* **21**(1), 63–70 (1993).
183. J. Kopecny, E. Cerny, and D. Ambroz, Effect of polyformaldehyde dust in rat tissues. *Scr. Med.* **41**, 405–409 (1968).
184. Unpublished data, E. I. du Pont de Nemours and Company, Inc., Haskell Laboratory, Wilmington, DE, (1957).
185. H. F. Smyth, Jr., et al., Experimental toxicity of a high molecular weight poly(ethylene oxide). *Toxicol. Appl. Pharmacol.* **16**, 442–445 (1970).
186. S. J. Hermansky et al., Effects of polyethylene glycol 400 (PEG 400) following 13 weeks of gavage treatment in Fischer-344 rats. *Food Chem. Toxicol.* **33**(2), 139–149 (1995).
187. D. R. Klonne et al., Two-week aerosol inhalation study on polyethylene glycol (PEG) 3350 in F-344 rats. *Drug Chem. Toxicol.* **12**(1), 39–48 (1989).
188. B. Vannier et al., Teratogenic effects of polyethylene glycol 200 in the mouse but not in the rat. *Teratology* **40**(3), 302(1989).
189. R. W. Lewis, M. E. Moxon, and P. A. Botham, Evaluation of oral dosing vehicles for use in developmental toxicity studies in the rat and rabbit. *Toxicologist* **36**(1 Pt 2): 259–260 (1997).
190. U. Gupta et al., *Teratology* **53**(2), 111 (1996).
191. R. I. Warren, *Polyphenylene Ethers and the Alloys*, SPE-RETEC, Mississauga, Ontario, Canada, Oct. 16, 1984, pp. 1–7.
192. *NORYL Resin Design Brochure CDX-83*, General Electric Co., Pittsfield, MA, 1983, p. 6–7.
193. D. Aycock, Polyphenylene oxide, modified. In R. Juran, Ed., *Modern Plastics Encyclopedia–90*, McGraw-Hill, New York, 1989, (mid-Oct. issue,) pg. 90.
194. V. T. Mallory, unpublished data, for General Electric Co., Pharmakon Research International, June 25, 1986.
195. V. T. Mallory, unpublished data, for General Electric Co., Pharmakon Research International, June 19, 1986.
196. M. B. Powers, unpublished data, for General Electric Co., Hazleton Laboratories, Inc., Nov. 1964.

197. M. B. Powers, unpublished data, for General Electric Co., Hazleton Laboratories, Inc., March 1965.
198. M. E. Placke, unpublished data, for General Electric Co., Battelle Institute, Columbus, OH, Nov. 1987.
199. M. B. Powers, unpublished data, for General Electric Co., Hazleton Laboratories, 1968.
200. L. F. Stankowski, unpublished data, for General Electric Co., Pharmakon Research International, Sept. 1986.
201. L. F. Stankowski, unpublished data, for General Electric Co., Pharmakon Research International, Jan. 1987.
202. R. H. Cox, unpublished data, for GE Plastics, TPS Inc., Nov. 1980.
203. G. Ball, B. Weiss, and E. A. Boettner, Analysis of the volatile combustion products of polyphenylene oxide plastics. *Am. Ind. Hyg. Assoc. J.* **31**, 572–578 (1970).
204. E. A. Boettner, G. L. Ball, and B. Weiss, Combustion products from the incineration of plastics. Office of Research and Development, National Environmental Research Center, U.S. Environmental Protection Agency, Cincinnati, OH, *Report No. EPA-670/2-73-049*, 1973.
205. T. J. Bucci et al., Two-week studies of survivors from exposures to pyrolysis gases. *J. Combus. Toxicol.* **5**(3), 278–289 (1978).
206. C. J. Hilado et al., Relative toxicity of pyrolysis products of some synthetic polymers. *J. Combus. Toxicol.* **3**(3), 270–283 (1976).
207. G. L. Nelson, Combustion product toxicity studies of engineering plastics. *J. Combus. Toxicol.* **5**(2), 222–237 (1978).
208. H. G. Elias, *New Commercial Polymers 1969–1975*, Gordon and Breach Science Publishers, New York, 1977
209. G. C. Bailey and H. W. Hill, Jr., Polyphenylene sulfide: a new industrial resin. *Am. Chem. Soc. Div. Org. Coatings Plast. Chem.* **34**, 156–161 (1974).
210. P. J. Boeke, Polyphenylene sulfide. In *Modern Plastics Encyclopedia, 1978–1979 ed.*, Vol. 55, No. 10A, McGraw-Hill, New York, 1978–1979 pp. 73–74.
211. V. J. Leslie et al., Polyethersulphone — new high temperature engineering thermoplastic. *Am. Chem. Soc. Div. Org. Coatings Plast. Chem.* **34**, 142–155 (1974).
212. E. A. Boettner, G. L. Ball, and B. Weiss, *Combustion Products from the Incineration of Plastics*, U.S. Environmental Protection Agency, Cincinnati, Ohio, 1970 (PB 222001).
213. C. J. Hilado and N. V. Huttlinger, Concentration-response data on toxicity of pyrolysis cases from some natural and synthetic polymers. *J. Combust. Toxicol.* **5**, 196–213 (1978).
214. G. F. L. Ehlers, K. R. Fisch, and W. R. Powell, Thermal degradation of polymers with phenylene units in the chain. II. sulfur-containing polyarylenes. *J. Polym. Sci.* **7**, 2955–2967 (1969).
215. N. S. J. Christopher, et al., Thermal degradation of poly(phenylene sulfide) and perfluoropoly(phenylene sulfide). *J. Appl. Polym. Sci.* **12**, 863–870 (1968).
216. C. J. Hilado and E. M. Olcomendy, Toxicity of pyrolysis gases from polyether sulfone. *J. Combust. Toxicol.* **6**, 117–123 (1979).
217. W. C. Thomas et al., The subchronic toxicity of polyphenylene sulfide. *J. Appl. Toxicol.* **4**(1), 8–11 (1984).

CHAPTER NINETY-THREE

Polyurethanes, Miscellaneous Organic Polymers, and Silicones

Steven T. Cragg, Ph.D., DABT

OVERVIEW

The toxicity of the polymers discussed in this chapter may be generally attributed to the residual monomers, catalysts, and other additives present rather than the polymer *per se*. The cured polymer itself may be of high molecular weight and, consequently, more or less toxicologically inert. Carefully manufactured, highly refined polymers contain few residual toxic chemicals. However, some of the polymers discussed in this chapter, at least in some applications, go through an intermediate stage consisting of "prepolymers" (sometimes referred to as "resins") that react further to achieve their final, cured form. An example is a polyurethane system for making foam cushions. To manufacture polyurethane foam for cushions, workers combine diisocyanate molecules with a polyol prepolymer. Such "systems" inherently have more potential for exposure of workers if not the general public to toxic monomers or other reactive chemicals. The exposure potential of glues, paints, and coatings may extend more broadly to the consumer. Thus, examination of the toxicity of the polymers discussed in this chapter focuses on monomers and prepolymers. This is not always so. Some of polymers in this chapter are used in biomedical devices or in a way that puts them in intimate contact with humans. Here, the issue of biodegradation becomes important because of potential toxicity from breakdown products of the polymer, or rejection may ensue if the polymer is incompatible with the surrounding tissues.

Patty's Toxicology, Fifth Edition, Volume 7, Edited by Eula Bingham, Barbara Cohrssen, and Charles H. Powell.
ISBN 0-471-31940-6 © 2001 John Wiley & Sons, Inc.

A POLYURETHANES

Polyurethanes are an extremely complex class of polymers (1) that are essentially ester-amide derivatives of carbonic acids (2). They were first developed commercially about 1937–1941 as nylon-like fibers (3, 4). Today they exist as foams, elastomers, coatings, adhesives, and elastomeric fibers. A major concern of the polyurethanes, polymers and silicones is the toxic health effects not only during manufacture, but also during installation, use and combustion of these materials. The exposed populations, the chemicals and types of exposure may vary considerably. This information is provided when it is available.

Polyurethanes, also called polyurethans, urethanes (these compounds should not be confused with urethane gas, a chemical used as an anesthetic), and polycarbamates, are polymers that have a urethane linkage in the polymer backbone, which is derived from the condensation of an isocyanate group and an alcohol group. A typical polyurethane may contain aliphatic and aromatic hydrocarbon moieties, linked as esters, ethers, or amides (5). Polyurethanes can be prepared by reacting bischloroformates with diamines or by other techniques, but are most commonly made by reacting isocyanates with polyhydroxy compounds. These reactions characteristically involve polycondensation by the addition of hydrogen across the carbon–nitrogen double bond of the isocyanate group (6). The term "polyurethane" now generally includes all of the complex polymers formed from diisocyanates and polyols (3).

1.0 Polyurethane

1.0.1 CAS Number: *[9009-54-5]*

1.0.2 Synonyms: Polyurethane; polyurethane foam

1.0.3 Trade Names: NA

1.0.4 Molecular Weight: 88.109

1.0.5 Molecular Formula: $[C_3H_8N_2O]_n$

1.0.6 Molecular Structure: CH₃–NH–C(=O)–NH₂

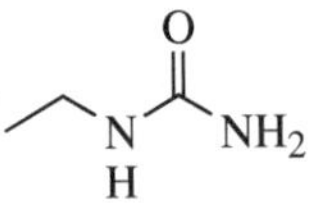

1.2 Production and Use

Polyurethane compounds are formed by reacting polyisocyanates with polyalcohols (or "polyols").

The use of polyurethane is expanding worldwide and exceeded 11 billion lb in 1990. The United States accounted for about 29% of world use (11–13). In 1991, 51% of the U.S. market was in flexible foam, 26% in rigid foam, and 23% in elastomerics and other types. Production data for polyether and polyester polyols are given in Table 93.1. Toluene diisocyanate (TDI) and 4,4'-methylenebisphenyl diisocyanate (MDI) are the principal commercial aromatic diisocyanates. Commercial aliphatic diisocyanates include iso-

POLYURETHANES, MISCELLANEOUS ORGANIC POLYMERS, AND SILICONES

Table 93.1. Production of Selected Synthetic and Natural Polymers

Polymer	Dry Weight Basis 1000 lb	1000 kg
A. U.S. Production of Plastics, Resins, and Elastomers (1990)		
Thermoplastic resins[a]		
Low-density polyethylene	7,254,891	3,291,693
Linear low-density polyethylene	3,893,252	1,766,448
High-density polyethylene	8,339,360	3,783,739
Polypropylene	8,310,409	3,770,603
Acrylonitrile–butadiene–styrene (ABS)	1,161,855	527,157
Styrene–acrylonitrile (SAN)	135,173	61,331
Polystyrene	5,021,090	2,278,172
All other styrene-based resin	1,189,204	539,566
Nylon	558,307	253,315
Polyvinyl chloride	9,095,534	4,126,830
Polyvinyl acetate	987,190	447,908
Other vinyl/vinylidene resins[c]	211,560	95,989
Thermoplastic polyester[d]	1,878,629	852,373
Engineering resins[e]	1,382,806	627,407
Acrylic[b]	1,507,031	683,771
Other thermoplastic resins	781,856	354,744
Total thermoplastic resins	51,708,147	23,461,046
Thermosetting resins[a]		
Epoxy	499,321	226,552
Polyester (unsaturated)	1,221,160	554,065
Urea–formaldehyde	1,495,531	678,553
Melamine–formaldehyde	201,986	91,645
Phenolic and other tar acid resins[f]	2,946,276	1,336,786
Polyurethane[g]	2,951,818	1,339,300
Alkyd[b]	769,238	349,019
Furfuryl type resins[b]	14,035	6,368
All other thermosetting resins[b]	152,488	69,187
Total thermosetting resins	10,251,853	4,651,475
Synthetic Elastomers[h] (includes latex)		Metric tons
Styrene–butadiene		852,851
Polybutadiene		403,388
Nitrile		562,261
Ethylene–propylene		255,537
Other synthetic elastomers (includes polychloroprene, polyisoprene, and butyl rubber but excludes polyurethane rubber)		546,498
Total		2,620,535

Table 93.1. Production of Selected Synthetic and Natural Polymers

	Dry Weight Basis	
Polymer	1000 lb	1000 kg
B. World Production of Certain Textile Fibers (1990)[i]		
		1000 metric ton
Natural fibers		
Raw cotton		18,714
Raw wool		1,964
Raw silk		66
Total		20,744
Percent of world total = 54		
Man-made fibers except olefin and textile glass		
Rayon and acetate		2,846
Acrylic and modacrylic		2,326
Nylon and aramid		3,765
Polyester		8,621
Certain other noncellulosic fibers		157
Total		17,715
Percent of World total = 46		

[a] Adapted from data in Society of the Plastics Industry, "Monthly Statistical Report—Resins, Full Year 1990, Production and Sales & Captive Use of Thermosetting & Thermoplastic Resins" issued by the SPI Committee on Resin Statistics as compiled by Ernst & Young; other data reported in SPI *Facts and Figures of the U.S. Plastic Industry*, 1991 Edition; and data in *Synthetic Organic Chemicals*, USITC Publ. 2470, U.S. International Trade Commission, Washington, DC, 1991. SPI data are used where available. Data reported to the USITC do not necessarily coincide with those reported to the SPI because of differences in both reporting instructions and the coverage of certain resins.
[b] As reported in *Synthetic Organic Chemicals*, USITC Publication 2470, December 1991.
[c] Includes only polyvinyl butyral, polyvinyl formal, and polyvinylidene.
[d] Does not include polyester for film and tape.
[e] Engineering resins include acetal, granular fluoropolymers, polymde-imide, polycarbonate, modified polyphenylene oxide, polyphenylene sulfide, polysulfone, polyether imide, and liquid crystal polymers. (ABS and nylon resins are listed separately.)
[f] Material is reported on a "gross weight" or "as sold" basis including the weight for water and other liquid diluents.
[g] Polyurethanes are derived from starting isocyanates (TDI/MDI) and polyols (polyether/polyester). Raw materials reported to the SPI sold in the United States for conversion into polyurethanes provide the basis for these data.
[h] The term "elastomer" may be defined as substances in bale, crumb, powder, latex, and other crude forms that can be vulcanized or similarly processed into materials that can be stretched at 68°F to at least twice their original length and, after having been stretched and the stress removed, return with force to approximately their original length (*Synthetic Organic Chemicals*, USITC Publ. 2470, U.S. International Trade Commission, Washington, DC, 1991). Production figures are based on the *RMA Industry Rubber Report*. December 1990, and are reprinted by permission of the Rubber Manufacturers Association, Washington, DC.
[i] As listed in *Fiber Organon*, 62(6), June 1991. Data are reprinted by permission of the Fiber Economics Bureau, Inc., Roseland, NJ. "The silk and man-made fiber are on a calendar-year basis, while the figure for cotton and wool are on a seasonal basis."

phorone diisocyanate (IPDI), hexamethylene diisocyanate (HDI), methylene cyclohexyl diisocyanate, and naphthalene diisocyanate (NDI).

1.2a Polyurethane Foams

Polyurethane foams are typically made by reacting of diisocyanates, polyester or polyether resins, a blowing agent, water, catalysts, and surfactants (80). They also may contain flame retardants, fillers, extenders, bacteriostats, and dyes. They can be classed as flexible, semirigid, or rigid. Flexible foams probably are the best known and most common type of polyurethane.

Flexible Foams. Flexible polyurethane foams are used widely in automotive seating and trim, in furniture and bedding, acoustic insulation, footwear, carpet underlay, and specialized packaging applications. Other uses include stuffed toys, dolls, and sports equipment.

Flexible foams are prepared from liquid monomer streams by mixing-activated polymerization. One of these streams is a polyisocyanate, usually an aromatic type such as TDI, MDI, or chemically modified variants thereof. Foaming and polymerization are initiated by mixing the polyisocyanate with at least one additional stream that contains a "softblock" resin and a blowing agent, usually water.

The softblock resin is generally an aliphatic polyether or polyester of molecular weight 1000 to 8000, containing two to four, usually primary, hydroxyl groups. These "polyols" are most commonly oligomers of propylene oxide that have minor amounts of ethylene oxide located at the chain ends, thus providing primary hydroxyl termini. The polyol resin provides elasticity to the foam. It normally constitutes 40 to 80% by weight of the polymer. Portions of this resin, occasionally as much as 100%, are often pre-reacted with the isocyanate stream to form isocyanate-terminated prepolymers.

Water is by far the most important blowing agent. Each mole of water reacts with two equivalents of isocyanate (−NCO) and produces a mole of carbon dioxide and a urea linkage. Water functions as both an expanding agent and as a monomer. This "chain extending" reaction produces a polymeric urea phase that imparts strength and resiliency to the foam.

The final polymer is actually a block copolymer, in which the rigid urea "hardblock" links together the elastic polyol (softblock) phase. By changing the formulation, the weight ratio of these phases can be varied over a wide range of properties to provide enormous product versatility.

Physical blowing agents, such as inert gases or halocarbons, are also extensively used in flexible foam production. These nonreactive species may be incorporated into the liquid streams or injected separately during the mixing process.

The isocyanate reactive stream(s) may also contain low molecular weight (MW < 500) polyols or polyamines. These species are chain extenders or cross-linkers and are useful for fine-tuning key foam properties such as resiliency. Foam formulations almost invariably contain catalysts for the urethane/urea reactions. Other typical additives are fire retardants, fillers, cell-opening agents, surfactants, pigments, and antioxidants.

The most important processing methods are molding and continuous pouring ("slabstock" production). Foam slabs, prepared according to the latter process, are subsequently cut into useful shapes.

Table 93.2. Typical Ranges of Foam Properties for Major Applications[a]

Property[b]	Product Class[c]				
	I	II	III	IV	V
Density(lb/ft^3)	2–3.5	4–8	0.25–1.10	0.40–1.10	1.2–3.5
Tensile strength (psi)	12–20	20–45			12–20
Shore A hardness			10–80	30–80	
Ultimate elongation (%)	100–200	100–150	40–700	300–500	70–250
Resilience (% ball rebound)	30–70				30–50
Flex fatigue resistance			50 to > 100		
75% compression set (%)	5–12				
90% compression set (%)					3–14
75% humid aged compression set (%)	7–14				
90% humid aged compression set (%)					4–31
Hardness; ILD-25% deflection (lb)	20–50				25–50
Hardness; ILD-65% deflection (lb)	55–110				50–110
Tear (pli)	1.5–3	2–4			1–3
Abrasion resistance			50–150		

[a]The Society of the Plastics Industry, Inc., Polyurethane Division, private communication, New York, 1992.
[b]Method: Abrasion resistance: DIN 53516, cubic milimeters; Flex fatigue resistance: Ross Flex, kilocyles at 15°C.
[c]Product class: I: Molded seating (MDI; polyether polyol); II: Molded automotive trim (MDI; polyether polyol); III: Shoe soles (MDI; polyester polyol); IV: Shoe soles (MDI; polyesther polyol); V: Slabstock (TDI; polyether polyol).

The range of flexible foam properties is vast; Table 93.2 lists properties for five representative foam types. Density can vary from less than 1 to 70 lb/ft^3. Flexural moduli (bulk polymer) ranging from less than 4000 psi (soft rubber) to nearly 100,000 psi (semi-rigid elastomer). Ultimate elongation ranges from less than 50% to more than 1000%. Flexible foams can be formulated for very high resiliency (as in automobile seats) to essentially zero resilience (energy-absorbing elastomers). As a class, flexible polyurethane foams are appreciated for their toughness and abrasion (as in high-quality shoe soles).

Rigid Foams. The most common type of rigid foam is formed by reacting polymeric MDI (functionality = 2.7) with highly branched polyols (functionality = 3 to 8). An alternative to polyurethane foam (PUR) is polyisocyanurate foam (PIR), which is produced by trimerizing isocyanate groups by using a trimerization catalyst. High functionality polymeric MDI (functionality = 2.9) is often used for polyisocyanurate foam.

Typical foam additives are tertiary amine catalysts to modify reactivity, silicone surfactants to optimize cell size, flame retardants to reduce flammability, and blowing agents to modify foam density. There are two types of blowing agents, chemical and mechanical.

The primary chemical blowing agent is water. The water and isocyanate reaction generates carbon dioxide gas, which is used to expand the polyurethane mass.

Mechanical blowing agents expand as gas bubbles, once the reaction temperature reaches the boiling point of the blowing agent. CFC-11 has been the mechanical blowing agent of choice because it is easy to use in processing and has superior thermal conductivity. CFC-11 is being replaced by HCFCs and other more exotic blowing agents because of the effect of CFCs on the stratospheric ozone layer.

The closed cells of low-density rigid polyurethane foam retain most of the low thermal conductivity blowing agents in the cells until the foam is destroyed. This property makes polyurethane and polyisocyanurate foams the most efficient thermal insulating materials commercially available. The foams also have excellent adhesion to surfaces with which they come into contact during the foaming process, thereby creating a sandwich structure that possesses superior rigidity.

The primary applications for rigid foam use its excellent insulating characteristics. Rigid polyurethane and polyisocyanurate foam laminated with a facer such as aluminum are used in new home construction. In appliances, rigid foam is poured in place during the manufacture of refrigerators. The foam provides both insulation and structural strength.

Industrial applications include spraying polyurethane foam onto chemical tanks and piping to provide thermal insulation. Low-density foam is used as a packaging material for delicate objects such as computers, and rigid foam is added to the core of surfboards and boats to provide buoyancy.

1.2b Polyurethane Elastomers

Depending on composition, versatile polyurethane elastomers can be processed by liquid casting, as vulcanizable millable rubbers, by thermoplastic techniques, or by spraying. Typical of other forms of polyurethanes, the isocyanates used to manufacture elastomers include MDI, TDI, hexamethylene diisocyanate (HDI), naphthalene diisocyanate (NDI), dimethyl diphenyl diisocyanate, and methylene dicyclohexyl diisocyanate. The isocyanate is usually reacted with a hydroxyl-terminated polymer and a low molecular weight diol, triol, or amine (often called a chain extender). The hydroxyl-terminated polymer can be an adipic acid-ethylene glycol such as polybutylene glycol or polypropylene glycol, or a hydroxy-terminated polybutadiene (100). Polyurethane elastomers have the special feature of segmented or block structure added to the basic characteristics of rubbery behavior. Figure 93.1 shows in simplified form the regions of "hard" segments derived from the isocyanate component and the "soft" segments derived from the hydroxyl-terminated polymer.

Polyurethane elastomers are noted for their exceptional abrasion resistance at moderate temperatures, high tensile and tear strength, and high hardness with mechanical strength. Polyester urethane elastomers show less swelling in oils and fuels than polyether elastomers but are more susceptible to deterioration by moisture at elevated temperatures. Polyurethane elastomers are used in solid tires for industrial trucks, potting and sealing of electronic components, shoe heels and soles, and elastic thread (100, 101). Specially formulated elastomers are considered promising for use in vascular repair and cardiovascular devices (102); see also Reference 103.

1.2c Polyurethane Coatings and Adhesives

Polyurethane coatings are typically tough, flexible, abrasion resistant, impact resistant, and fast curing, and adhere well to a variety of substrates. Solvent resistance may vary

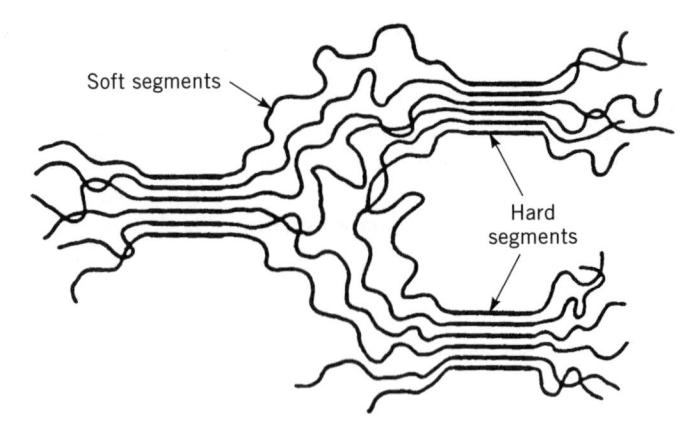

Figure 93.1. Segregation of hard segments in polyurethane elastomers (100) (from *Rubber Technology*, 2nd ed., Maurice Morton, Ed., copyright © 1973 by Van Nostrand Reinhold, reprinted by permission of Van Nostrand Reinhold).

from fair to excellent (106). The coatings are often sensitive to extreme temperatures, humidity, and reactive contaminants. Isocyanate prepolymers are based on TDI, but MDI, hexamethylene diisocyanate HDI), isophorone diisocyanate (IPDI), and other aliphatic isocyanates are used for clear or color-stable pigmented coatings. Polyols may be polyesters or polyethers. The traditional types defined by ASTM are the alkyd or oil urethanes (type 1), the isocyanate prepolymers that cure with atmospheric moisture or by baking (types 2 and 3, respectively), and the more cumbersome two-part prepolymer catalyst or prepolymer polyol systems (types 4 and 5, respectively) that provide maximum properties (107). Organic solvents may also be present.

The coatings are used in wood and machine finishes and in marine enamels. Applications of the other types include floor finishes; coatings for auto and machine parts; and finishes for nylon rainwear, leather, and rubber (107). Polyurethane adhesives are used in boots and shoes (24).

1.2d Polyurethane Fibers

Fibers that have an extension at break in excess of 200% as well as the property of rapid recovery when tension is released are classed as elastomers. Spandex, used as a generic term for polyurethane elastomeric fibers, is defined as a "manufactured fiber in which the fiber-forming substance is a long chain synthetic polymer comprised of at least 85% of a segmented polyurethane" (112).

The first polyurethane fiber was not elastomeric but was a nylon-like yarn spun in Germany during World War II. Unlike spandex, this fiber was a completely polyurethane fiber made by treating 1,4-butanediol with hexamethylene diisocyanate at 195°C (113).

Spandex fibers are block polymers (see Fig. 93.1). Relatively long "soft" segments that are very flexible, rubbery, and noncrystalline alternate in the molecular chain of the fiber with short "hard" segments that are usually cross-linked, crystalline, and polar. The soft segments, generally a polyester or a polyether, are easily deformed to give high extensions

POLYURETHANES, MISCELLANEOUS ORGANIC POLYMERS, AND SILICONES

in contrast to the hard isocyanate segments that are not deformed. The use of preformed soft-segment polymer blocks results in a more regular spacing of tie points along the molecular chain than occurs in randomly vulcanized rubber (114).

Four stages of synthesis have been described (113). A low molecular weight linear polymer is made with terminal hydroxyl groups; this may be either polybutylene glycol or polyester made with excess glycols. Then, this prepolymer is treated with excess diisocyanate (either TDI or MDI) to give a polyurethane that has terminal isocyanate groups. Third, water is added, just sufficient to convert some of the terminal isocyanate groups to amine groups. Heat then causes the amine and isocyanate groups to cure and form urea cross-linkages, which provide the snapback property. Elastomeric fibers are commonly formed by dry spinning into a heated gas or by wet spinning into a dilute solvent bath (115).

The outstanding feature of these polyurethane fibers is their stretch and recovery. The stretch of spandex yarn before break may range from 520 to 610%, compared to 540 to 760% for rubber (see Refs. 113 and 114). Spandex is relatively stronger and lighter than rubber and can be used uncovered, whereas rubber fiber is generally covered with rayon or nylon. Spandex has a melting point of about 250°C, its specific gravity is about unity, its and moisture regain is about 0.3%. The fiber is soluble in boiling dimethylformamide and accepts dye readily. Spandex is used mainly in clothing. The technical importance of the yarns is considerably greater than may appear from total production figures because very small amounts can give textiles elastic properties (116).

1.3 Exposure Assessment

The hazards involved in manufacturing and processing vary with the type of product, but basically result from diisocyanate exposures in handling the monomers, prepolymers, or uncured products. In addition, fabrication of cured products may release isocyanates when heated, ground, milled, or otherwise processed. Adverse health effects in individuals exposed to diisocyanates include increased incidence of respiratory illnesses, mucous membrane irritation, allergic dermatitis, and allergic occupational asthma. Studies of nonallergic workers in polyurethane production facilities have shown that exposures to toluene diisocyante at concentrations of 5 ppb or less and excursions that do not exceed 20 ppb do not cause excessive respiratory illness (7–9). Concentrations slightly above these may decrease pulmonary function and irritate the eyes, nose and respiratory system even in nonallergic workers. Similar effects at similar concentrations have been reported among polyurethane workers exposed to diphenylmethane diisocyanate (MDI) (10).

Several analytical methods are available to determine free toluene diisocyanate (TDI) (14–16) and aromatic diisocyanates (17–19) monomer in various polyurethane products. Procedures also are available for determining residual isocyanate groups in polyurethane foams (20) and in thermoplastic polyurethanes (21).

1.3.1 Air

Methods have been developed for determining isocyanates in air as well, including those for TDI (22), 4,4'-diphenylmethane diisocyanate (MDI) (23), and others (24, 25) and also several aliphatic diisocyanate procedures. The methods include sampling (26) and determination by gas chromatography (27), HPLC (28–31), and continuous monitoring (32). (These methods could be adapted for determining aliphatic monomers in polyurethane products.) Air analysis has also been extended to the corresponding aromatic and aliphatic diamines (33, 34). The analysis of amines in biological materials (35, 36), water (37), and polyurethane foams (16, 38, 39), films (40, 41), thermoplastics (42), and sterilized polyurethanes (43) is also available. In addition, several researchers have developed methods to study the thermal decomposition/pyrolysis of various polyurethane materials to characterize polymers (44) and to analyze resulting compounds (45–47), including isocyanates and amines (48).

1.3.3 Workplace Methods

More recently, methods have been developed to monitor exposure of polyurethane workers by measuring their isocyanate precursors (or metabolites thereof) in plasma and urine (49–51). In addition to monitoring isocyanate monomers and metabolites of monomers in plasma and urine, immunoglobulin serum levels and lung function also have been employed as indexes of isocyanate exposure because these chemicals are potent sensitizing agents (52).

1.4 Toxic Effects

Polyurethanes are a large family of polymeric materials made by reacting a wide choice of isocyanates with polyols. Polyurethane products can be produced that cover a broad range of properties and applications, including foams, elastomers, adhesives, and coatings.

1.4.1 Experimental Studies

1.4.1.1 Acute Toxicity—Smoke Inhalation Toxicity. Most studies of flammability and smoke toxicity have been conducted on polyurethane foams (63). Foams may be flexible, semirigid, or rigid, and have many variations for specific uses. Many additives are used in formulating polyurethane foams to control the chemistry and the processing of the foams and to modify the ultimate performance (including flammability) of the product.

Using the DIN test method, exposure of test animals to pyrolytic products from flexible polyurethane foam caused narcosis that resulted in incapacitation and death due to chemical asphyxia (64). At decomposition temperatures in the range of 300 to 800°C, carboxyhemoglobin (COHb) levels indicated that CO was the primary toxicant, although some HCN was produced. Above 800°C, HCN became the dominant toxicant.

The NBS test method also showed that with flaming combustion, deaths occurred during exposure and were attributable to CO and HCN in combination. However, in the case of nonflaming thermal decomposition, there was significant postexposure mortality with pulmonary edema, indicating pulmonary irritation (65). Pulmonary irritation was also observed as the predominant toxic effect in other laboratory tests (66). One study, using monkeys, evaluated the irritation produced by smoke generated from flexible polyurethane foams in the DIN 53436 method (67). At 300°C, irritation was observed at a mass loss concentration of 16 g/m^3. A concentration of 10 g/m^3 produced irritation at 600°C, whereas irritation and semiconsciousness resulted with only 3 g/m^3 at 900°C. The small amounts of isocyanates that have been detected analytically cannot account for the irritant response, which is more likely due to the presence of aldehydes and other irritants common to the decomposition products of many materials.

The effects of pulmonary irritation have not predominated under true smoldering conditions in large-scale tests. Overall, the toxic effects of the combustion products of flexible polyurethane foams are typical of many materials used in construction and furnishings. Chemical asphyxia attributed to carbon monoxide and hydrogen cyanide generally predominates.

Rigid polyurethane foams also have been studied extensively (68) and exhibit greater toxic potency than the flexible foams under conditions of flaming combustion. The greater toxic potency of rigid polyurethane and isocyanurate foams may be due to their greater thermal stability. Higher temperatures required for combustion favor HCN formation. Char structures produced during combustion of rigid foams also generate larger amounts of CO (flexible foams decompose at lower temperatures and do not lead to char formation). Deaths during exposure have predominated with rigid foams, and it has been calculated that the CO and HCN produced account for the asphyxiant effects noted.

Evidence for respiratory irritation has been provided from some studies in the form of a minimal number of postexposure deaths with pulmonary involvement. No particular chemical has been detected in smoke to account for the irritation, apart from the usual mixture of aldehydes and other chemicals often found in the combustion products of many materials.

The toxicity of the combustion products of polyurethane foam was also examined in one laboratory using a "physiogramme" method (69, 70). The method exposed conscious rabbits immobilized with curare and subjected to forced positive pressure ventilation. A variety of physiological measurements were made, including electrocardiograms, electroencephalograms, and blood pressure. These three parameters were plotted on a three-dimensional plot to form a "physiogramme." The products of combustion of flexible and rigid polyurethane foams caused "physiogramme" patterns similar to those produced from exposure to carbon monoxide and hydrogen cyanide, indicating that these were the major toxicants.

The only exception to the typical toxicity profiles normally observed was from a noncommercial rigid polyurethane foam formulation based on trimethylolpropane which contained a phosphorus fire retardant (71). Nonflaming combustion gave rise to the formation of a bicyclic phosphate ester, 4-ethyl-1-phospha-2,6,7-trioxabicyclo[2.2.2] octane 1-oxide (TMPP) that caused myoclonic seizures in rats exposed to the smoke. Large numbers of other fire retardants have been tested in rigid polyurethane foam, but no additional unusual toxic effects have been observed.

In summary, both flexible and rigid polyurethane foams display typical toxicity profiles when burned, predominantly chemical asphyxia, accompanied by irritation. The effects are attributable to combinations of CO and HCN, and the irritation is probably caused by aldehydes and other reactive species.

The three smoke toxicity tests most frequently used to quantify the toxic effects of polyurethane foams are the National Bureau of Standards (NBS, now the National Institute of Standards and Technology, or NIST), the University of Pittsburgh (UPITT), and the German Standards Institution (DIN 53436 Test). The data in Table 93.3 (72, 73) show some of the results. Most data on conventional flexible polyurethane foams obtained using

Table 93.3. Combustion: Acute Toxicity of Polyurethane Foam

Fire Model	Animal Model	Comments	LC$_{50}$ Values[a] (g/m^3)
Flexible			
UPITT	Mice (head-only) 30 min exposure	LC$_{50}$ calculated on total airflow even when material no longer decomposes. Thus LC$_{50}$ may appear too low	22
			18
			14
			24
DIN 53436	Rats (restrained, but whole body) 30-min exposure	Significant lethality only at 450°C and above. LC$_{50}$ at 600°C	20.6
			24
			24
DIN 53436	Rats	425 or 490°C	21–24
NSB	Rats	Flaming	> 38
			> 50
		Nonflaming	28
			40
			27
Rigid			
UPITT	Mice	LC$_{50}$s calculated from total airflow	17
			14
			13
			13
DIN 53436	Rats	600°C	7
		500°C	8
		400°C	29
NBS	Rats	Flaming	13
			11
		Nonflaming	> 40
			> 34
			> 35
Utah (similar to NBS)	Rats	Flaming	11
			14
			12
			11
			17
		Nonflaming	> 40
			> 37
			> 37

[a]Experience with the NBS test has indicated that an uncertainty of about 15–40% is associated with determination of LC$_{50}$ values. This includes variability in the measurement process and in animal responses. Comparable uncertainties would be expected for other test methods as well.

Source: Chapt 37 of Patty's Industrial Hygene and Toxicology, 4th ed.

the NBS test place these materials in the same range as Douglas fir for both nonflaming and flaming combustion. Results obtained using the UPITT, the DIN, and the NIST test methods generally support this conclusion. In the UPITT test, flexible polyurethane foams, including high-resilience foams, appear at about the middle of the range of toxic potencies where 69% of the materials tested reportedly fell. Data on melamine-modified (for improved flammability properties) flexible polyurethane foam show that the smoke produced had a 30-min LC_{50} of about 13 g/m^3 (74). This fourfold increase in smoke toxicity over that for conventional flexible polyurethane foams is likely due to significant amounts of HCN produced from decomposition of the melamine.

There are no laboratory toxicity test data to suggest that smoke from flexible polyurethane foams is unusually toxic under the conditions of the tests used. The situation may be more complicated than simple laboratory testing suggest, however. HCN formation is favored by high temperatures; About 600°C and even higher is required for significant generation rates. This is readily attainable in a fire at flashover. Laboratory studies of flexible polyurethane foams under the conditions of the NBS test are conducted well below the temperatures required to produce significant concentrations of HCN. The UPITT test does reach the required temperatures but probably only after most of a test sample has been consumed. Independent studies at the Fire Research Station (75) and at the NIST (76) have shown that heating flexible polyurethane foam at a relatively low, nonflaming temperature to produce char and an intermediate "yellow smoke," will yield, upon heating to higher temperatures, as much as 10 times the HCN normally produced from virgin foam. Laboratory tests may not be relevant to the fire scenario of a multistage decomposition, perhaps one of low-temperature charring followed by the temperatures reached at flashover. Some large-scale studies have confirmed that significant quantities of HCN are generated when flexible polyurethane slabs or upholstered chairs burst into flame following a period of smoldering (76). Fully furnished room fires have produced 1200 to 1500 ppm HCN at the doorway at flashover, although the toxic contribution from the CO produced still far exceeded that for the HCN (77). Laboratory test results may be misleading in that they do not simulate some of the possible events during the progress of a real fire.

Rigid polyurethane foams generally exhibit a low order of smoke toxicity, even less toxic than wood, when thermally decomposed, nonflaming, in the NBS test (78). Flaming combustion in this test results in LC_{50} values that show significantly increased toxicity, although the values are still not alarming. In the UPITT test procedure, rigid polyurethane and isocyanurate foams yield LC_{50} values of 6 to 8 g (79). Thus these foams place on the low (more toxic) end of the 4.5- to 23-g spread that encompasses 69% of materials tested for the State of New York (these values for UPITT LC_{50}s should not be confused with those shown in Table 93.3 where LC_{50}s have been converted to g/m^3 units). The increased toxicity that results from the UPITT test and with the flaming NBS test may be rationalized in that the temperatures required for significant HCN production are reached due to the thermal stability of the rigid foams.

1.4.1.2 Chronic and Subchronic Toxicity—Dust Inhalation Toxicity. Laskin et al. (89) exposed hamsters and Sprague–Dawley rats to freshly generated polyurethane foam dust at concentrations of 3.6 or 20 mg/m^3, 6 h/day for 30 exposure days. The test material was described as a commercial isotropic fluorocarbon-blown urethane foam used for structural insulation.

The mass median diameter determined by electron microscopy was 1.8 mm and the geometric standard deviation was 1.8 mm. A high incidence of macrophages that contained particles was observed in rats exposed to 20 mg/m^3. Three of 15 rats observed from the 20-mg/m^3 group showed evidence of centrilobular emphysema. One rat exposed at each concentration developed squamous cell carcinoma of bronchogenic origin late in the life-span observation period, but no neoplasms of this type were seen in hamsters.

Although Laskin et al. were experienced and thorough with the technology that was available, respiratory disease in the experimental animals makes interpretation difficult. Pneumonitis incidence in the control rats was 89% (67% in 20-mg/m^3 exposed rats). The authors attributed the variation in median life span of the experimental groups to the endemic disease normally present in these animals.

Thyssen et al. (90) also exposed Sprague–Dawley rats (50 per sex per exposure group, 6 h/day for 12 weeks) to respirable dust concentrations averaging 8.65 mg/m^3. This study evaluated a freshly ground polyurethane dust synthesized from a sucrose polyether-based polyol and an MDI-based isocyanate. Control rats were exposed to titanium dioxide or to air alone. Rats were held for observation for up to 140 weeks. Necropsies were conducted on all rats, and surviving rats were sacrificed after 140 weeks.

Intra-alveolar and interstitial macrophages that contained dust particles were found in rats exposed to polyurethane and titanium dioxide dust. Peritracheal and peribronchial lymphocytic infiltration were observed, but were considered within the usual physiological range. No indications of a carcinogenic effect of the inhaled dusts on the respiratory tract could be established. Although a standard design oncogenicity study is not available, the polyurethane dust as tested did not demonstrate a carcinogenic potential.

An intratracheal instillation test in rats was conducted by Stemmer et al. (91) using foam samples of the same composition used by Laskin et al. (89). The samples for intubation were either freshly prepared or aged from the identical block, and 94% of the particles had diameters of 10 mm or less. The initial inflammatory response was followed by the development of areas of fibrosis, then nodular scars and perifocal emphysema; the authors thought that the latter changes were probably below the level that would cause functional impairment. Four rats had benign intrabronchial adenomas at 18 months after intubation, and one rat developed a hyperplastic subpleural lesion without evident dust particles.

Intratracheal instillation bypasses normal pulmonary defense mechanisms; Stemmer et al. (91) noted that their intratracheal administration studies were not intended to replace inhalation exposures and that intubation introduces large particles into the lung that would probably not occur by inhalation.

Styles and Wilson (92) evaluated polymer and mineral dusts by intraperitoneal administration; 50 mg/kg of fine polyurethane dust particles resulted in a persistent granulomatous reaction. This contrasted with silica dusts and asbestos, which caused progressive fibrosis. However, based on *in vitro* macrophage cytotoxicity, the polyurethane dust was ranked in an intermediate group (which included silica).

Monkeys that were exposed to 10 mg/m^3 of predominantly respirable polyurethane dust for 6 h/day for 18 months showed no remarkable histopathological changes (93).

Sporadic changes in lung mechanical properties were reported: reduced flow maxima at 75% vital capacity after 4 months, reduced end tidal forced expiratory volume at 14 months, and elevated phase IV of closing volume at 18 months. The authors considered that these represent small airway obstructive impairment. However, there was not a consistently affected parameter in measurements taken at 4, 14, and 18 months. The reported effects were based on physiological measurements; there were no remarkable effects based on pathology. The inconsistency of the physiological effects and the lack of corroborative pathological effects suggest that the potential toxicity of polyurethane dust exposure in this study was not definitively established.

1.4.1.6 Genetic and Related Cellular Effects Studies—Release of Monomers and Additives. Rats implanted with TDI-based polyurethane foams did not form tumors at the site of implant or elsewhere after 42 weeks (88). No genetic damage was detected in these rats from measured DNA adduct formation or HPRT lymphocyte mutations. The authors concluded that 2,4-TDI biodegraded from polyurethane implants poses little risk for genetic damage or tumor formation.

1.4.2 Human Experience

Toxicity from exposure to polyurethane foams, as for all forms of polyurethanes, results from release of monomeric precursors or additives before curing is complete or from degradation as a result of heating, fabrication processing (e.g., grinding, milling), or biodegradation. Again, as for all forms of polyurethanes, inhalation and dermal contact are the relevant routes of exposure in industry. Because polyurethanes may be fabricated into biomedical devices or coat biomedical devices, direct absorption of biodegraded monomers may also be a relevant route of exposure.

1.4.2.2 Clinical Experience

1.4.2.2.2 Chronic and Subchronic Toxicity. Toxicity data are limited regarding coatings and adhesives but the considerable literature pertaining to other forms of polyurethane is relevant because adverse health effects may be attributed to their common monomeric constituents and additives. Cases of "intoxication" have occurred during processing; thermal decomposition at elevated temperatures frequently yields isocyanates (108) and may also yield aromatic amines (109).

Toxicity data for cured and uncured elastomers are limited. Contact dermatitis in polyurethane molding is relatively uncommon but has been reported in a plant using an aliphatic diisocyanate prepolymer with poor hygienic practices (104). "Polyurethane rubber" in shoes has been implicated as a cause of contact dermatitis (105), but this dermatitis may be attributable at least in part to polyurethane adhesives and coatings.

Exposures to monomers and additives during production—foams. Workers may be exposed to monomers of any of the several commonly used diisocyanates during the initial manufacture of polyurethane polymers, the use of polyurethane prepolymers (e.g., in glues), or subsequent fabrication of the finished polyurethane into a specific product. Often during fabrication, polyurethanes may be heated short of actual combustion, and, consequently, constituent diisocyanates may be released into the air from incompletely cured resins or from thermal degradation. Likewise, the use of solvents or other processing techniques during fabrication may also free diisocyanates from polyurethanes.

Jakobsson et al. (55) found decreased lung function among pipelayers exposed to degradation products of MDI-based polyurethane pipes by measuring forced expiratory volume. These workers also had higher incidences of eye irritation, nose congestion, throat irritation, productive cough, and dyspnea or wheezing. None of the workers exhibited positive responses when challenged in patch tests with MDI or other isocyanates, but two of 50 workers had elevated plasma levels of MDI-specific immunoglobulin. In another study, workers in a car parts manufacturing plant using heated isocyanate-based polyurethane glue had elevated plasma and serum concentrations of methylenediphenyl diisocyanate (MDI) (52). These workers also had higher rates than normal of respiratory symptoms and specific antibodies against MDI. Fuortes et al. (56) reported naphthalene diisocyanate-induced asthma in workers from a polyurethane plastic factory where heated polyurethane glue was used to coat metal surfaces. In this study, increased respiratory symptoms consistent with asthma were reported in 17 of 26 production workers.

Isocyanates such as methylenebisphenyl diisocyanate (MDI), the toluene diisocyanates (TDI), and other aromatic and nonaromatic diisocyanates may cause allergic sensitization reactions of the skin and also of the respiratory tract. In fact, diisocyanates are well known respiratory sensitizers as reflected by some of the quite low ACGIH TLVs (e.g., the TLV TWA for TDI is 5 ppb). Rates of respiratory sensitization may exceed 10% of the exposed population. Occupational asthma may result from excessive exposure to diisocyanates in polyurethane production facilities (57, 58). Rather than base the diagnosis on respiratory symptomatology alone, the diagnosis of occupational asthma also requires (*1*) establishing the presence of specific antibodies to the suspected diisocyanate and (*2*) positive challenge tests (59). Pulmonary function tests also may be used to diagnose respiratory dysfunction and may reveal impaired lung function even when no symptoms are present (60). Case control studies also have been used to establish a link between isocyanate exposure and occupational asthma (61). Using this method, investigators found that the relative risk of occupational asthma in polyurethane workers is 22 times normal and ranges up to 100 times normal for certain jobs (milling and baking).

Dust inhalation—foams. Industrial hygiene surveys were conducted in three flexible polyurethane foam manufacturing facilities (94). Respirable dust exposure in even the dustiest operations, for example, buffing and sanding, were all substantially less than the current guidelines of 10 mg/m^3 for "particulates not otherwise classified" by the ACGIH (95). In one facility, the respirable dust concentrations determined for seven finishing operators ranged from 0.19 to 0.72 mg/m^3 (mean 0.45 mg/m^3). In the second facility, the respirable dust exposures to six workers ranged from 0.05 to 0.66 mg/m^3 (mean = 0.21 mg/m^3). Personal respirable dust exposures among six finishing workers in the third facility ranged from 0.14 to 0.57 mg/m^3 (mean = 0.34 mg/m^3).

An assessment of spray foam applicators reported reddening of the eyes and lacrimation in numerous workers (96). In all cases, sprayers had polyurethane particles deposited on the neck, chin, nose, and mouth. The assistants had deposits on their clothes, hands, faces, and hair. Discolored spray deposits on the analytical sampling tape were reported, indicating that airborne polyurethane mist contained unreacted isocyanate liquid. The authors reported that diisocyanate vapor exposures were higher than 0.02 ppm and that proper protective equipment was not used. The particles that were found on the workers apparently originated from the spray aerosol rather than as a dust from a solid material.

Available information suggests that polyurethane dust(s) have some potential for biological reactivity, although the dusts studied by Thyssen et al. (90) and Moorman et al. (93) had limited toxic potency. Aerodynamic size is the predominant determinant of deposition site within the respiratory tract and therefore is a useful consideration for potential effects.

A moderate amount of inert polymeric material deposited in the upper respiratory tract is normally cleared rapidly by mucociliary action. Clearance of fine "respirable" dust from the lung is much slower, and in general, respirable dust is more likely to have adverse consequences. Boeniger measured less than 1 mg/m^3 of respirable particles in the "dustiest" operations of a flexible foam facility. Freshly reacting polyurethane systems, for example, spray foaming, create significant potential diisocyanate exposure (96). Exposure control and appropriate skin and respiratory protection is warranted in these and similar situations, regardless of particle size.

Biodegradation. Polyurethane materials have been used as coverings for implants. As with silicone breast implants, the safety of the products for this use is under investigation by the FDA and is unresolved at this time. The relevance of this and significance of the reported test conditions of the implantations are unclear for workplace safety and health.

Polyurethanes are unique among synthetic polymers in that the polymer itself provides a direct contribution toward the overall resistance or susceptibility of the formulated product to microbial growth. Polyester-based urethanes are characteristically less resistant than polyether-based urethanes, which can be quite resistant to microbial attack (97, 98). Reference 97 (preliminary report) specifically refers to rigid foams. Polyurethane based on polyester diol (no other specific detail) showed the only growth among 30 generic types of commercial "plastics" tested according to ASTM D1924-63 (99).

1.4.2.2.5 Carcinogenesis. The 2,6-toluene diisocyanate isomer has been implicated by both the International Agency for Research on Cancer (IARC) and the U.S. National Toxicology Program (NTP) as a suspect human carcinogen. An epidemiological study by Schnorr et al. (62) in 1996 did not find any statistically significant excesses of cancer mortality in a cohort of men and women numbering more than 4500 workers in the polyurethane foam industry who used toluene diisocyanate. Estimated TDI exposures were below 0.04 mg/m^3 during the mid-1980's but were above this standard prior to 1980. Although not statistically significant, the authors did find a slight elevation in Hodgkin's lymphoma and non-Hodgkin's disease (both increased with employment time) as well as rectal cancer (did not increase with employment time). The authors also noted that the relatively young age of the cohort would require further follow-up to clarify possible associations between TDI exposure and cancer and noncancer mortality/morbidity.

1.4.2.2.7 Other—urethane coatings. Fully cured polyurethane coatings are quite inert. Four thin coats of polyurethane applied to the skin contact surfaces of costume jewelry and nickel coins completely prevented dermatitis in 8 of 11 nickel-sensitive patients (110). Some polyurethane resins of the type for coatings and adhesives reportedly caused contact dermatitis from shoes (105). The urethane prepolymer coating used in the Letterflex® photoprepolymer mixture has also been connected with contact dermatitis (111); the causative agent in at least some of these cases is a polythiol cross-linking component, pentaerythritol tetrakis-3-mercaptopropionate.

1.4.2.2.7b Other—urethane fibers. During the mid-1960s, a number of investigators here and abroad described allergic contact dermatitis with itching and erythema that was attributed to contact with spandex. This dermatitis typically developed under spandex panels in brassieres and in some cases resembled "classical rubber dermatitis" (117). In many cases sensitivity was clearly related to sensitivity to the 2-mercaptobenzothiazole present as a trace contaminant in some, but not all, brands of spandex. However, three patients in the Netherlands all reacted to both N-phenyl- N'-isopropyl-p-phenylenediamine and zinc diethyl dithiocarbamate as well as "Wonderlastic"(118). All three had positive reactions to one or more additional substances. Two reacted to both rubber cloth and 2-mercaptobenzothiazole, but none reacted to "Lycra."

Several human volunteer tests showed no evidence of chemical irritation or sensitization from commercial spandex materials without 2-mercaptobenzothiazole. Panels of 200 subjects (119), 151 subjects (120), or 203 subjects (121) showed no reactions when tested by the Schwartz–Peck technique with fabrics that contained three different types of "Lycra."

A rare case of vitiligo (122) observed after the wearing of an elasticized brassiere containing spandex may have involved a "depigmenting agent" in the yarn as well as friction and pressure. Depigmentation from contact with rubber-containing antioxidants has been reported previously (123).

1.4.2.3 Epidemiology Studies—Release of Monomers and Additives during Production. In an epidemiological study of the toluene diisocyanate polyurethane foam industry within the United Kingdom, the pulmonary function and respiratory symptomatology was evaluated in 780 workers from 12 factories during a 5-year period (81). Results indicated that at concentrations near the U.K. occupational exposure limit (i.e., 8 h TWA of 5.8 ppb), declines in pulmonary function were not evident from forced vital capacity and forced expiratory volume measurements. Reported respiratory symptoms did increase with time. Forty-four of 96 West African workers who produced diisocyanate-based polyurethane paints and foam were diagnosed with occupational asthma (82). Diagnosis was based on questionnaires, pulmonary function tests, immunoglobulin E levels, and radioallergosorbent tests (RAST). Specific antibodies to isocyanates were found in two of 20 subjects tested. The findings of these two studies contrast markedly, presumably as a consequence of good industrial hygiene practices in the former instance.

A hypersensitive pneumonitis-like reaction and occupational asthma were found among polyurethane foam workers who used a 1,3-bis (isocyanatomethyl) cyclohexane-based prepolymer (83). Either of these reactions was found in 23 of 34 workers employed in an injection molding operation (pneumonitis-like reaction–27%; hyperresponsiveness–

38%). Symptoms included dyspnea, cough, chest tightness, chills, wheezing, myalgias, arthralgias, and nausea. Use of this prepolymer was discontinued. This study shows that vigilance is required in monitoring the respiratory health of workers in this industry because new monomers and prepolymers are introduced continuously. Elevated levels of IgG and IgE antibodies specific for one or more diisocyanates were found in 10 of 113 residents who lived near a polyurethane foam production plant (84). Coexposure of the affected residents to other isocyanate sources could not be ruled out in this study.

In humans, highly refined polyurethanes have been used in biomedical implants such as coverings for silicone breast implants and skin wound dressings. 2,4-Toluenediamine (TDA) was measured in the serum and urine of 61 women who had polyurethane foam covers for breast implants (85). TDA presumably forms from the biodegradation of the implant. No TDA was found in sera, but 30 had quantifiable levels, and 18 more had detectable levels (detection limit was 10 pg/mL). Seven of 61 control subjects had detectable but non-quantifiable levels in their urine. The authors concluded that small quantities of TDA escape from polyurethane coverings of breast implants, but that the amounts do not pose a significant risk to health. Sepai et al. (86) found an initial steep drop in TDA urinary and plasma concentrations soon after implantation of polyurethane-covered breast implants which leveled-off. Elevated levels were found 2 years after surgery. Brinton et al. (87) discuss several ongoing studies evaluating the risk from polyurethane coatings for breast implants.

1.5 Standards, Regulations, or Guidelines of Exposure

Uncured resins that may involve respiratory exposure or skin contact to isocyanates should be evaluated before use to assess the types and severity of hazard (53). The discussion of spills of all kinds of partially polymerized polyurethane (53) illustrates the range of advisable practices rather than serving as a didactic code. Users of polyisocyanate materials in the manufacture of polyurethane products should consult with manufacturers of the isocyanates for more detailed information.

Skin or eye contact requires prompt attention. In the event of skin contact, immediately but gently remove excess polymer, then flush with copious amounts of water. If eye contact occurs, irrigate eyes with water and seek medical attention without delay. Butyl rubber or comparable overshoes should be worn in areas susceptible to floor contamination. Acid-suited personnel should be swabbed down with decontamination solution by a helper before showering and removing the suit. Chem suits, which additionally offer the respiratory protection of an air supply, are recommended where the resin temperature exceeds 70°C or when an unusually volatile material is present.

Contaminated personal clothing should be swabbed or immersed in decontaminated solution before laundering or else should be discarded. A neutralizing solution of aqueous ammonia, about 1 to 2% by weight, or 5 to 10% sodium carbonate, plus a heavy-duty detergent is used effectively (54). Personnel responsible for cleanup of minor spills should be provided with personal protective equipment, including impervious clothing, footwear, and gloves. Positive-pressure, self-contained respirators, with full facepiece, are recommended.

Small spills should be absorbed with vermiculite or other absorbent and shoveled into open-top containers. The containers should not be sealed. After transporting the containers outdoors, neutralizing solution should be added at the rate of 10 parts neutralizer to one part isocyanate. The mixture should be allowed to stand for 48 h or until all evolved carbon dioxide has escaped (54). The waste must be discarded in accordance with Resource Conservation and Recovery Act (RCRA) requirements.

Ammonia–isocyanate mixtures can volatilize most unpleasantly in confined spaces. Respirators must be worn in areas of poor ventilation.

Processing of polyurethanes at elevated temperatures should be conducted under exhaust ventilation or preferably in a sealed system. Bulk quantities of polyurethane foam should be stored separately from other flammable substances.

B MISCELLANEOUS ORGANIC POLYMERS

Miscellaneous organic polymers discussed in this section include phenolics, aminoplastics including urea formaldehyde and melamine-formaldehyde; furan polymers and resins; polyvinylpyrrolidone and polybenzimiadiazole.

Phenolics

The phenolics were the first completely synthetic class of resins developed commercially and are still a major product line (Tables 93.1 and 93.4). The term phenolic describes a resinous material produced by reacting a phenol or a mixture of phenols with an aldehyde. Phenol itself, or sometimes cresols, are usually combined with formaldehyde and occasionally with other aldehydes such as furfural (3). The resin has sometimes been associated with dermatitis that is traceable to the phenolic monomer component, the resin itself, or less frequently formaldehyde. Inhalant effects from dust or fumes during processing have also been reported. No toxicologically significant emission of volatile substances is has been connected with the cured resin. Phenol-formaldehyde is highly resistant to biological decay (124). Other common resins of this class include (*1*) *p-tert*-butylphenol-formaldehyde [*25085-50-1*], (*2*) bisphenol A/epichlorohydrin/formaldehyde [*28906-96-9*], and (*3*) aniline/phenol/formaldehyde [*24937-74-4*] copolymers.

2.0 Phenol–Formaldehyde Resin

2.0.1 CAS Number: [*9003-35-4*]

2.0.2 Synonyms: Phenol formaldehyde resin; P-F-R-2; phenol formaldehyde resin, resol; phenol formaldehyde; phenol, polymer with formaldehyde; formaldehyde, phenol polymer; paraformaldehyde, formaldehyde, phenol polymer; paraformaldehyde, phenol polymer, phenol, formaldehyde polymer; novolac; resole

2.2 Production and Use

The polymerization of phenol and formaldehyde readily becomes exothermic and may present an explosion hazard (3, 125). The reaction is generally conducted only to the

Table 93.4. Some Commercial Polymers and Approximate Year of Introduction[a]

Date	Material	Typical Application
1868	Cellulose nitrate	Eyeglass frames, table tennis balls
1900	Viscose rayon	Lining in clothing, curtains, tablecloths
1909	Phenol–formaldehyde	Telephone handset, electrical insulators
1919	Casein	Knitting needles
1926	Alkyd	Exterior paint, electrical insulators, distributor caps
1927	Cellulose acetate	Toothbrushes, packaging film, lacquers
1927	Polyvinyl chloride	Wall coverings, pipe, siding, flooring
1929	Urea–formaldehyde	Lighting fixtures, wood adhesives, electrical fixtures and parts
1931	Polychloroprene	Industrial hoses, wire and cable, footwear
1935	Ethyl cellulose	Flashlight cases, coatings
1936	Acrylic	Display signs, brush backs
1936	Polyvinyl acetate	Flashbulb lining, adhesives
1937	Styrene–butadiene copolymers	Tires, footwear, molded items
1938	Cellulose acetate butyrate	Packaging, tubing, lacquers
1938	Polystyrene	Kitchenware, toys
1939	Nylon	Fibers, films, gears
1939	Polyvinylidene chloride	Packaging film, paper coating
1939	Melamine–formaldehyde	Tableware
1942	Unsaturated polyester	Boat hulls
1942	Low-density polyethylene	Packaging film, squeeze bottles
1943	Silicone	Rubber goods, motor insulation
1943	Fluoropolymers	Industrial gaskets, coatings
1943	Polyurethane	Foam cushions, insulation, adhesives
1943	Butyl rubber	Tubeless tire liner, inner tubes
1945	Cellulose propionate	Pens and pencils
1947	Epoxy	Coatings, industrial equipment
1948	Acrylonitrile–butadiene–styrene (ABS)	Pipe and fittings, luggage, appliances
1950	Polyacrylonitrile	Sweaters, knitwear, blankets
1950	Chlorosulfonated polyethylene	Automotive hoses, wire and cable
1952	Polyethylene terephthalate (fiber grade)	Clothing, fiberfill, sailcloth
1954		Housewares, automotive trim
1956	Styrene–acrylonitrile (SAN)	Auto parts
1956	Acetal	High-temperature seals and gaskets
1957	Fluoroelastomers	Milk bottles
1957	High-density polyethylene	Safety helmets, carpet fiber, battery cases
1957	Polypropylene	Appliance parts
1958	Polycarbonate	Wire insulation, film
1959	Fluorinated ethylene propylene	Valves and fittings
1961	Chlorinated polyether	Tires, footwear, medical items
1962	Polyisoprene (high cis)	Tires, general-purpose rubber
1962	Ethylene propylene rubber	Bottles
1964	Phenoxy	Skin packaging moldings

Table 93.4. (*Continued*)

Date	Material	Typical Application
1964	Ionomer	High-temperature moldings
1964	Polyphenylene oxide	High-temperature films, wire coatings
1964	Polyimide	Adhesives and coatings
1965	Ethylene–vinyl acetate	Electrical/electronic parts
1965	Polysulfone	Clear moldings
1965	Poly (4-methylpentene-1)	Electrical/electronic parts
1970	Thermoplastic polyester	Piping, film
1973	Polybutylene	Telecommunications
1974	Polyphenylene sulfide	Printed circuit boards
1974	Polyethersulfone	Clothing, reinforcement/on tires and plastic composites
1975	Aromatic polyamides	Non-food packaging
1978	Nitrile barries resins	Extruded film
1982	Linear low-density polyethylene	Electrical/electronic parts
1983	Polyetherimide	Wire and cable
1984	Polyetheretherketone	Electrical/electronic parts
1985	Aromatic copolyester	Electrical/electronic parts
1988	Liquid crystal polymers	High temperature transparent
1988	Polymethylpentene	Packaging, electrical electronic, medical products

[a]Adapted with permission from a compilation by H. G. Mark and S. Atlas, "Introduction to Polymer Science," in H. S. Kaufman and J. J. Falcetta, Eds., *Introduction to Polymer Science and Technology*, Copyright © 1977 by John Wiley and Sons, Inc., New York, and from *Facts and Figures of the U.S. Plastics Industry*, (1991 Edition) prepared annually by The Society of the Plastics Industry, Inc., Washington, D.C.

intermediate condensate product (126, 127), which can be either a one-step resol (or resole) or a two-stage novolak (or novolac). Resols are prepared with an alkaline catalyst and sufficient formaldehyde (about 1.5:1 mole ratio with phenol) so that the resins can later cross-link and cure simply by heat. Their molecular weights are reportedly 300 to 700 (127).

The novolaks are produced by the controlled reaction of phenol and formaldehyde in the presence of an acid catalyst to form a brittle resin. The amount of formaldehyde, about 0.75 to 0.90 mole formaldehyde/mole phenol, is insufficient to permit cross-linking; this is accomplished by adding of hexamethylenetetramine (or "hexa"). The novolaks have molecular weights as high as 1200 to 1500 (127).

Both types of resins are available in granular, powdered, or liquid forms (see Refs. 3, 126, 128, 129). Wood flour is the general-purpose filler; special-purpose fillers include cotton flock, hydrated alumina, glass, mica, and other minerals. The molecular weight of a cured (or thermoset) phenolic resin may be in the range of several hundred thousand. The common property that forms the basis of the varied applications of phenolic resins is first that they are low molecular weight, fusible, soluble resins that can be easily handled and then subsequently polymerized to high molecular weight, strong, heat-resistant products (126). They are also noted for surface hardness, chemical resistance, and useful electric properties. Comparing the two types, resols offer better resistance to cracking and essentially no ammonia emission, whereas novolaks offer greater molding latitude, better dimensional stability, and better long-term storage properties (129). There has been an extensive increase in the use of phenolics in the aerospace, mass transit, marine, ducting, piping, and construction markets because of the development of low-viscosity liquid phenolics that are easy to process. This resulted from the inherent fire retardant and excellent low smoke and low smoke toxicity of phenolics (130).

The cured resins are probably most widely known as electrical, automotive, and appliance components, for example, automotive distributor caps; water pumps; appliance bases, handles, and knobs; electrical switch gears, circuit breakers, and wiring devices. Some phenolic resins based on substituted phenols are permanently soluble and fusible. These resins are used in varnish coatings and as rubber tackifiers (126). Phenolic fibers provide flame resistance and thermal insulation, but the upper temperature limit in air for long-term use is around 150°C (131). Certain food-contact applications are permitted (132).

2.3 Exposure Assessment

Free phenol in phenol resole resins has been determined by gel permeation chromatography (133).

2.4 Toxic Effects

2.4.1 Experimental Studies

2.4.1.1 Acute Toxicity—Fire and Smoke Toxicity. Because phenolics are not generally used in materials of construction, interior finishes, or furnishings, these materials have been little studied for their smoke toxicity when burning. One study reported that phenol-formaldehyde foam produced rather large amounts of carbon monoxide (166). The resulting LC_{50} values for both flaming and nonflaming combustion were in the range of 6 to 8 g/m^3. Compared with other cellular plastics, therefore, phenol-formaldehyde foam produces quite toxic smoke.

2.4.1.4 Reproductive and Developmental. Two investigators evaluated the oral teratogenic potential of *p-tert*-butylphenol-formaldehyde resin by gavage and dietary feeding in rats. Kanaka et al. (162) administered this resin by stomach tube to pregnant Wistar rats at doses of 250, 500, and 1000 mg/kg on days 7–17 of gestation. No maternal or embryo or fetotoxicity was reported, and no malformations were noted in either skeletal or soft tissues of pups. In a second study, Itami et al. (163) administered diets that contained this resin at concentrations of 0, 2.5, 5, or 10% to pregnant Wistar rats on days 6 through 15 of pregnancy. No maternal toxicity was noted, but dams exhibited lower food consumption and body weights during the treatment period. No reductions in pup body weights or placental weights were noted in any of the treated groups, and no other signs of embryo or fetotoxicity were found. Pups in controls and some of the treated groups exhibited variations including short tail, dilatation of the cerebral ventricle and renal pelvis, dextrocardia, and some skeletal variations (wavy ribs and split or asymmetric sternebra) that did not significantly increase in treated groups. The authors considered the variations anomalous rather than frank defects and concluded that *p-tert*-butylphenol-formaldehyde resin was not a developmental toxin under the conditions of the test. Solokhima (164), as reported in Schardein (165), tested a phenol-formaldehyde resin, called viam-B, for teratogenic potential in rats and obtained negative results.

2.4.2 Human Experience

2.4.2.2 Clinical Experience

2.4.2.2.2 Chronic and Subchronic Toxicity—Inhalation of Dusts and Fumes. Emissions from solid finished commercial articles have presented relatively few if any problems due to the stable nature of the phenol-formaldehyde bond. Formaldehyde may be detected by gel permeation chromotography as described in Section 3.3. Ammonia is produced in the curing process of the "hexa"-cured novolaks and "over a period of time from the molded parts" (129). Some resols contain virtually no "hexa" and therefore are used in applications where ammonia emissions would be undesirable.

Several reports from abroad associate inhalation of "bakelite" dust with pulmonary changes in workers (151–153). In one case, the dust concentration at a cutting machine was reportedly 10.5 mg/m^3; however, the "comparatively large size of the particles appeared to be related to the nasal symptoms in 4 cases and history of nasal diseases in 4 cases" among 16 workers examined (151). "Slight pneumoconiotic opacities" have also been described (152). Extrinsic allergic alveolitis, which the author identifies as "caused by the inhalation of vegetable, animal, or chemical dust" (153), was reported in two patients who had histories of occupational exposure to "Bakelite" dust. Pulmonary changes have also been described in rats (154) and guinea pigs (151) exposed to unstated atmospheric concentrations of phenol-formaldehyde or "Bakelite" dust.

In the United States, vapors and particulates from phenol-formaldehyde resins have been associated with effects on the respiratory tract. Long-term exposure to phenol-formaldehyde resin fumes reportedly caused airway obstruction as evidenced by pulmonary function tests (155). The resin emissions came from a production line where acrylic wool filters were dipped into a vat of liquid phenol-formaldehyde and then placed into a curing oven at 160°C. Levels of phenol found in breathing zone samples ranged from 7 to 10 mg/m^3 (about 1.6 to 2.6 ppm). Comparable levels of formaldehyde varied from an atypical high of 16.3 mg/m^3 (14 ppm) to an estimated 0.5 to 1 mg/m^3 (0.4 to 0.8 ppm).

Workers in a tire manufacturing plant reported an excess of acute irritant symptoms and significant reductions in expiratory flow rates at low volumes (156). This plant used an adhesive system whereby resorcinol (as the phenol) and hexamethylenetetramine (as both a formaldehyde donor and catalyst) were added directly into the rubber stock during compounding. No significant association was found between changes in pulmonary function and environmental levels of resorcinol, formaldehyde, ammonia, or area sample particulates. Some associations were found between respirable particulate (personal samples) and decrements in lung function; the investigators suggest that the particulate may have carried an adsorbed agent deep into the lung. Although only small decreases in pulmonary function were found, the incidence of acute symptoms was high among the 40 current workers. These symptoms included eye irritation (98%), cough (72%), rash (35%), and phlegm (20%).

Phenolic resin paints can release appreciable quantities of phenol plus smaller amounts of ammonia and formaldehyde during industrial drying processes, according to a German study (157). Alcohols and phosphoric acid were also detected in the drying oven air. In one particular case, the emission of phenol was in the range of 700 g/h (157).

Phenolic resins show an appreciable rate of decomposition at temperatures above 300°C, particularly under oxidative conditions (158–160). The primary products are phenol and methyl phenols, plus carbon dioxide and/or carbon monoxide. Formaldehyde can be formed as a minor product under oxidative conditions at 400°C (160) or its yield may be "negligible" (159). At temperatures > 600°C, ring scission occurs; the amounts of phenol and methyl-substituted phenols decrease as benzene, toluene, benzaldehyde, and benzyl alcohol are formed.

Lemiere et al. (161) reported three cases of occupational asthma: one in a lab researcher, another in a worker from a phenol-formaldehyde resin plant, and a third in a carpenter who used phenol-formaldehyde processed pine boards. All three cases responded positively to inhalation challenges either with formaldehyde gas or phenol-formaldehyde resin. The authors concluded that inhalation challenge should be used to verify a diagnosis of occupational asthma.

2.4.2.2.7 Other—Skin Contact. Foussereau et al. (134) associated allergy to *p-(tert-*butyl) phenol resins with their presence in polychloroprene adhesives. In Europe, occupational allergy to these resins is reportedly most common in the shoemaking industry and to a lesser extent in the automobile industry. Both dermatitis (135) and vitiligo (136) have been reported in the United States; vitiligo is the more common complaint.

Skin irritation is presumed to be a consequence of contact with phenolic resins, particularly phenol-formaldehyde resins, because of (*1*) the potential for dermal irritation of residual formaldehyde or phenol in the resin and (*2*) reports of free phenol in resole resins. Observations drawn from occupational exposure, however, are problematic in attempting to establish a cause and effect relationship because of possible mixed chemical exposures. For the same reason, this conclusion may be extended outside an occupational setting to the end consumer (e.g., of glues, paints). In the case of shoe manufacturing and synthetic fiber used for clothing, particularly in Europe, a number of chemical exposures have been associated with dermatitis in workers engaged in manufacture and in the public that uses synthetic fibers. In the absence of well-controlled laboratory studies to demonstrate the potential for effects and to isolate noncontributors to toxicity, observations of dermatitis or dermal sensitization in workers are compromised and should not be relied on to ascribe toxicity to specific compounds (137).

It is difficult to establish causal relationships, but it is not impossible. For example, human patch tests and other immunological assays are quite specific in implicating particular allergens. Kanerva et al. (138, 139) found allergic skin reactions during patch testing to novolac epoxy resin and phenol-formaldehyde resin in workers who had dermatitis and were followed over a 6 year period. Patch testing has established allergy to phenol-formaldehyde resins and benzoyl peroxide in such consumer items as swimming goggles (140). Several other investigators have found similar results (141–147). Graa-Thomsen and Soderlund report that 1–3% of the patients who have skin disease reacted positively to challenge of phenol-formaldehyde resin (148). Moreover, Zimerson et al. (149, 150) reported dermal contact allergy to phenolic resins in the guinea pig that corroborated the effect in humans.

Aminoplastics

The term aminoplastic describes a group of resinous polymers that are produced by the interaction of amines or amides with aldehydes. The two polymers of major commercial importance are urea-formaldehyde and melamine-formaldehyde. Other types include coreacted polymers such as melamine-phenol-formaldehyde molding powders and benzoguanamine-based resins for coating applications.

Aminoplastics are generally prepared as liquid or dry resins of relatively low molecular weight that are then converted to insoluble and infusible end use items. Cure is ordinarily effected with heat or heat plus acidic catalysts and also without heat in the case of catalyzed urea resins (3, 167).

3.0 Urea-Formaldehyde

3.0.1 CAS Number: [9011-05-6]

3.0.2 Synonyms: Kaurit S; dimethylol urea; formaldehyde, urea polymer urea, paraformaldehyde polymer; urea-formaldehyde resin; polynoxylin

3.0.3 Trade Names: NA

3.0.4 Molecular Weight: 120.11

3.0.5 Molecular Formula: $[C_3H_8N_2O_3]_n$

3.0.6 Molecular Structure:

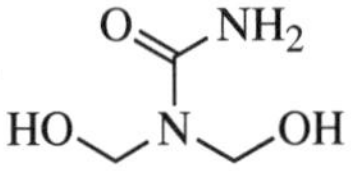

3.2 Production and Use

These resins, first commercialized in the 1920s, were developed for adhesives and resins in uses such as glues for wood, wet-strength paper applications, and crease-resistant finishes for clothing (see Table 93.4). Complaints from the consuming public that have been relatively infrequent, considering their extensive use, were primarily limited to odor and dermatitis. Complaints of both odor and some systemic effects, principally irritation of the respiratory tract, were more common from occupational exposure. Urea-formaldehyde foam for thermal insulation was first used extensively in Europe; extensive use in the United States resulted from increased interest in residential energy conservation during the 1970s.

Urea-formaldehyde resins are usually prepared in a two-stage reaction. Urea and formaldehyde are first partially reacted under neutral or mildly alkaline conditions to form mono- and dimethylol ureas, whose ratio depends on the urea-formaldehyde ratio. In a second stage, the resulting intermediate is then subjected to acid conditions and heat to form a solution that converts to a gel; eventual evolution of water and formaldehyde yields a hard, colorless, transparent, infusible mass. The reaction is often arrested before gelation by changing to a slightly alkaline pH and removing some or all of the water; then it is continued when desired by changing again to an acid pH. The powdery condensation products are normally insoluble but can be dissolved in aqueous solutions of lithium bromide, lithium iodide, and magnesium perchlorate (170).

Urea and formaldehyde are heated to give dimethylol urea and other low molecular weight products that are acidified for further reaction and then stabilized with alkali. The resins are hardened by adding an aqueous solution of an acid donor (3) or mixed with fillers that contain an acid catalyst for use in cold pressing (167).

Powders. Molding powders based on urea-formaldehyde usually contain a number of additional ingredients: filler, pigment, hardener or accelerator, stabilizer, plasticizer, and lubricant. The filler is usually bleached wood pulp, although wood flour may be used. The cellulose-filled materials are used in wiring devices, cosmetic closures, buttons, toilet seats, knobs, and handles (167).

Adhesives and coatings. Urea-formaldehyde glues are commonly used with wood, particularly particle board and plywood. Particle board is a mixture of sawdust or wood shavings held together with a urea-formaldehyde resin (3, 129). Cationic urea-formaldehyde coatings are used to improve the wet strength, dry tensile strength, and bursting strength of paper (167). Certain types of urea-formaldehyde resins are permitted in food-contact applications (132).

Foams. Urea-formaldehyde foams can be generated from three types of formulations: aqueous solution, powder mixed on-site in water, and concentrated solution diluted with water. The aqueous solution is the most common in the United States (171). A typical foam formulation (167) contains 54% urea-formaldehyde resin, 38% water, 7% phosphoric acid (75%), and 1% foaming agent (quaternary ammonium salt).

Urea-formaldehyde foam insulation burns with difficulty (3, 171, 172). The air-dried foam reportedly weighs 0.75 lb/ft^3(167), although a literature survey shows considerable variation in this parameter (171). The foam is reportedly friable and has negligible load-bearing characteristics. It is also hydrophobic (167) but can absorb some water (171). In many respects, data on properties are contradictory or have not been determined (171). Much of this variation is undoubtedly due to variation in formulation and application under conditions of relatively poor quality control. Shrinkage of the foams and lack of resistance to temperature and humidity have been identified as problem areas in its major application of home insulation use. Other reported applications of urea-formaldehyde foam include floral decorations, artificial snow in television, fire lighters made by saturating urea-formaldehyde foam with paraffin, and arrester beds to stop aircraft on runways (3).

3.4 Toxic Effects

3.4.1 Experimental Studies

3.4.1.1 Acute Toxicity. *Oral and Dermal Toxicity.* Urea-formaldehyde resin [*9011-05-6*] has low acute oral and dermal toxicity. LD$_{50}$s in rats, mice, and rabbits all exceed 2 g/kg

(173). In an acute inhalation test conducted with rats, the 4-hour LC_{50} reportedly exceeded 167 mg/m^3 (174). Urea-formaldehyde (presumably uncured) resin, was severely irritating to the eyes and skin in primary irritation tests with rabbits (173).

Several brief reports (204, 205) indicate that thermal degradation products from urea-formaldehyde resins are toxic to rats. If sufficiently heated, carbon monoxide, carbon dioxide, ammonia, and cyanide have been identified as (or in) effluent gases (206–209).

Inhalation Toxicity—Smoke. When assayed by the UPITT test, urea-formaldehyde foam produced smoke that has rather high toxicity due to the production of HCN (79). An LC_{50} value 3of 3.5 g was reported that corresponds to about 6 g/m^3 when total airflow is accounted for during the test.

3.4.2 Human Experience

The acute toxicity and irritation potential of urea-formaldehyde resins are closely linked to the amount of free formaldehyde present in either uncured or cured forms (175). Consequently, methods have been devised to determine the amount of free formaldehyde in the resins, and classification systems based on formaldehyde content have been devised to rank the quality of these resins (175).

3.4.2.2.2 Chronic and Subchronic Toxicity. Inhalation—Off gassing of formaldehyde. The emission of formaldehyde from urea-formaldehyde foam used in building construction received widespread public attention in the late 1970s (169). Earlier, a 1975 Danish report (186) described 23 dwellings where the concentration (range 0.08 to 2.24 mg/m^3; mean 0.62 mg/m^3 or about 0.5 ppm) of formaldehyde was associated with its emission from chipboard (particle board, wood shavings held together with urea-formaldehyde glue). Specific data on the contribution of formaldehyde from these resins in relation to other contributing sources are generally lacking. Other sources may include cigarette smoke (supported by a measurement of 0.23 ppm formaldehyde from five cigarettes smoked in a 30-m^3 chamber, as cited in Ref. 168) and combustion products released by cooking and burning fuels (no known measurements but undoubtedly a wide variation, depending on the type of residence and personal idiosyncrasy). Reference 180 reviews the analytical methods.

Most of the levels in residential dwellings that contain urea-formaldehyde foam insulation are less than 5 ppm, although levels as high as 10 ppm were reported in a Connecticut survey (168; and Sardinas et al., as cited therein). In a retrospective list (187), clinical symptoms commonly included eye and respiratory tract irritation; headaches, vomiting, and diarrhea were mentioned less frequently. As for particle board, other sources that may contribute to measured formaldehyde concentrations have not been rigorously evaluated.

Long et al. (188) attribute short-term emission of formaldehyde to free formaldehyde and associate longer term emission with hydrolysis of the polymer. Increased temperature and decreased humidity result in higher initial and lower subsequent concentrations of formaldehyde in the adjacent atmosphere. Varying intrinsic quality of the foam from using inappropriate materials or failing to adhere to recommended instructions during installation can also affect the emission of formaldehyde. Residual hot-water-soluble, low molecular weight acidic components affect the hydrolytic breakdown of urea-formaldehyde foams. The total quantity of formaldehyde emitted from a commercial product during a 30 day period (55°C, 90% relative humidity) was reduced approximately sixfold when acidic components were removed by washing before testing (189). Increasingly, houses are constructed to eliminate air exchanges between the outside and inside environments to reduce energy for heating and cooling. As a downside consequence, however, indoor air pollution has increased. Thus, the "airtightness" of the construction also contributes to the final concentration of formaldehyde in a dwelling that has urea-formaldehyde foam insulation and to any resulting health impacts.

According to a Japanese study (190), four commercial soil-consolidating agents contained up to 192 mg/mL formaldehyde and up to 21 mg/mL formic acid in the urea-formaldehyde condensate before gelation. After gelation of the resin with sand, up to 14,900 mg/m^3 formaldehyde and up to 96 mg/m^3 formic acid evaporated from the gel to air at equilibrium.

Laboratory tests conducted in France to determine the evolution of formaldehyde during a 90-min oxidative pyrolysis showed a steady increase with increasing temperature to approximately 140°C (191). The emissions (in units of g/100 g) were 0.19 at 5°C, 0.25 at 55°C, 0.58 at 88°C, 0.65 at 111°C, and 1.29 at 140°C.

The half-life for formaldehyde in a commonly used Scandinavian particle board was reportedly about 2 years with a ventilation rate of 0.3 changes/h (192). The level of formaldehyde can be reduced to approximately half the original value by coating the board with a formaldehyde-absorbent paint. Such painting may be less effective with urea-formaldehyde foam (187). The quality of foam manufacturing process or installation may also profoundly influence off-gassing of formaldehyde.

Whether formaldehyde from off-gassed urea-formaldehyde or similar resins causes occupational asthma has been investigated over a long period of time but the issue has not been completely resolved (193). Some recent studies conclude that a cause and effect relationship does exist between formaldehyde exposure and occupational asthma (194) or, at least, between formaldehyde-containing resins and asthma (161, 195–197), whereas others do not (199–201), and still others are equivocal (202). In 1997, an attempt was made to answer this question definitively. To resolve this issue, a comprehensive review of the literature in 1997 by a panel of experts concluded that a cause and effect relationship between formaldehyde exposure and occupational asthma could not be established (203).

3.4.2.2.7 Other—Skin Sensitization. Sensitivity to urea-formaldehyde resin does not necessarily involve sensitivity to formaldehyde, but it may be because formaldehyde causes allergic dermatitis. Allergic or non allergic dermal reactions also may be due to curing agents or other components of urea-formaldehyde resin systems. The issue was extensively explored during the 1960s (176–179), during the 1970s by Hsiao and Villaume (180), and more recently by Jolanki et al. (181), Sommer et al. (182), and Marks et al. (183). Hypersensitivity to formaldehyde has occurred in some of the human subjects of these investigations but is not a factor in their reactions to textiles finished with formaldehyde resins. Historically, dermatitis associated with industrial exposure to urea-formaldehyde molding powder was reported as early as the 1940s and 1950s (184, 185).

3.5 Standards, Regulations, or Guidelines of Exposure

Formaldehyde may be emitted by the degassing of these foams, particularly if installation or manufacture is poor and curing is incomplete. The qualitative data—prolonged odor and also adverse health effects—reported as a consequence of installation of these foams resulted in a series of investigations (168 and references cited therein) and banning of further installation of these foams in some areas (169). The current TLV for exposure to formaldehyde is a ceiling of 0.3 ppm. The present OSHA standard is a TWA of 0.75 and a short-term exposure limit (STEL) of 2 ppm.

4.0 Melamine-Formaldehyde Resins

Many features of these resins are similar to those of urea-formaldehyde resins. In general, melamine-formaldehyde resins offer more desirable properties than urea-formaldehyde materials (3). Emissions of volatile substances from these resins are less in consumer use because they are cured at a higher temperature than urea-formaldehyde resins, particularly the foam. Melamine-formaldehyde is biodegradable (210, 211).

4.0.1 CAS Number: [9003-08-1]

4.0.2 Synonyms: 1,3,5,-Triazine-2,4,6-triamine, polymer with formaldehyde; melamine, polymer with formaldehyde

4.0.3 Trade Names: NA

4.0.4 Molecular Weight: NA

4.2 Production and Use

Melamine-formaldehyde resins are synthesized by treating melamine with formaldehyde to yield a product that has up to six methylol groups per molecule. The methylol content depends on the melamine-formaldehyde ratio and reaction conditions. Compared to urea-formaldehyde, melamine-formaldehyde moldings have lower water absorption, better staining resistance to aqueous solutions, better electrical properties in damp weather, better heat resistance, and greater hardness. Like the phenolics and urea-formaldehyde resins, these resins may be sold as a low molecular weight product intended for further processing. These resins may be formulated for use as molding powders, laminates, adhesives, and for treating of textiles. The molding powders and laminates are probably most familiar as dinnerware, the largest single use, and as laminates. Subject to certain specifications, melamine-formaldehyde resins are permitted in food-contact applications such as coatings, components of packaging, and molded articles (132).

4.3 Exposure Assessment

A Taiwanese report (212) describes a formaldehyde content of 25 ppm in melamine bowls, compared to 20 to 42 ppm in urea resin in or coated on chopsticks.

Because the melamine-formaldehyde resins contain significant nitrogen, they produce hydrogen cyanide, along with carbon monoxide and carbon dioxide, upon burning or thermal degradation.

4.4 Toxic Effects

4.4.1 Experimental Studies

4.4.1.1 Acute Toxicity—Oral and Dermal. As reported in RTECS (173), the acute oral (rats) and dermal (rabbit) LD_{50}s of melamine-formaldehyde resin are higher than 10 gm/kg. In a review of the literature on this resin published in 1995 by Pang (213), a 1% aqueous solution of melamine did not cause sensitization in guinea pigs. The resin reportedly caused skin sensitization not attributed to formaldehyde (213). Several investigators reported allergic reactions to the resin in skin patch tests (214–216), but see the results from exposure to highly cured resins provided below. One report claims an occupational asthma from exposure to wood dusts where melamine-formaldehyde resin was used to create laminates. If real, however, this may be due to other environmental allergens (217).

4.4.1.2 Chronic and Subchronic Toxicity. The cured product was fed to dogs for 2 years at dietary levels of 2.5 and 5% (two males and two females per group). Overall health and appearance were good, except for a slightly increased incidence of diarrhea compared to control dogs. Pathogenic examination showed that the testes weight was reduced in both males at the 5% level and in one at the 2.5% level. Microscopic examination at the higher level showed depression of spermiogenesis and low-grade inflammation of the alimentary tract. See next section for additional information.

4.4.1.4 Reproductive and Developmental. A melamine–formaldehyde resin sold as a low molecular weight, wet-strength additive was treated to simulate the conditions of cure that this resin generally meets during its use in papermaking (218). Accordingly, a commercial batch of resin was dried as a thin layer in an oven for 3 days at 115°C. The cured product was fed to rats, 40 males and 40 females per group, at dietary levels of 2.5, 5, and 10% for 2 years (219). Some decreases in mean weight gain were observed at the two higher levels; no effects were evident in hematologic studies or mortality patterns. Pathological examination indicated no effects attributable to feeding the resin except at the 10% level, where mean tests weight was significantly reduced. No evidence of reproductive toxicity was observed in a separate test where rats were fed a 10% dietary level of resin and mated three times. See previous section for additional information on reproductive effects.

4.4.2 Human Experience

Overall, dermatitis is less frequent than with urea-formaldehyde resin but may occur during processing. A 1963 Italian report (220) describes an incidence of 7% dermatitis in the production of phenolic, urea, and melamine resins; the relative importance of the melamine resins is unclear. A series of human patch tests with resin-treated paper and textiles indicates that these resins have "no significant sensitizing potential in their ordinary and usual uses in paper and textiles" (218).

Furan Polymers/Furan Resins

5.0 Furfuryl Alcohol Resin

5.0.1 CAS Number: [25212-86-6]

5.0.2 Synonyms: Furfuryl alcohol resin

5.0.3 Trade Names: NA

5.0.4 Molecular Weight: NA

The expressions "furan polymers" and "furan resins" often refer to polymers whose starting monomer is furfuryl alcohol or furfural (the aldehyde). This is often misleading because furan is a distinct chemical compound whose properties are quite unlike those of furfural and furfuryl alcohol. Following is a brief summary of the more commonly known furfuryl alcohol and furfural polymers and resins.

5.2 Production and Use

Furfuryl alcohol is readily polymerized in the presence of acidic catalysts. The reaction proceeds via intermolecular dehydration, in which the hydroxyl group of one furfuryl alcohol molecule reacts with the alpha hydrogen of another molecule, resulting in the formation of a 5-furfuryl alcohol dimer. Further reaction in this manner results in higher molecular weight linear polymers of this structure:

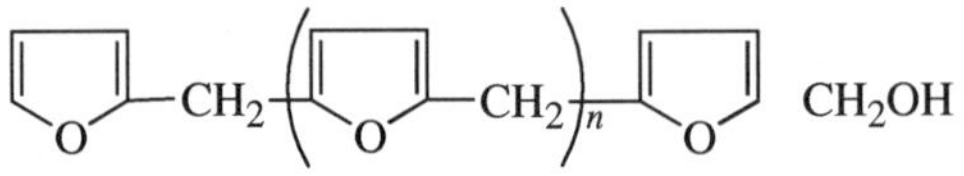

$$\text{[furan]}-CH_2-\left(\text{[furan]}-CH_2\right)_n-\text{[furan]}-CH_2OH$$

where n is usually limited to one, two, or three units before competing side reactions occur. Typical of these side reactions are the formation of both furan- and hydroxymethyl-terminated homologues of difurfuryl ether and difurylmethane. Small amounts of transient formaldehyde have been identified as a by-product of polymerization via decomposition of internal ether linkages and terminal hydroxymethyl groups. Evidence also exists that some opening of the heterocyclic furan ring leads to carboxylic and ketonic groups, suggesting that levulinic acid has been formed during the reaction.

Furfuryl alcohol resins frequently are modified with coreactants such as formaldehyde, urea, and furfural to improve handling, curing, and end-use properties. Such resins are dark colored liquids whose viscosities range from 50 to 15,000 cp. They are stable for extended periods in closed containers at normal room temperatures. Furfuryl alcohol resins are cured by adding mineral, organic, or Lewis acids as catalysts. The rate and degree of cure depend on the amount and activity of the acid catalyst used and the cure temperature.

The furfuryl alcohol resins described are used to produce cements, grouts, and mortars for chemically resistant brick and tile floors, vats, towers, and other masonry structures; as impregnants for carbon, graphite, sandstone, and other substrates to impart imperviousness and resistance to harsh chemical environments; and in down-well resin treatments used in oil fields to consolidate, impregnate, or otherwise alter the producing zones to improve and prolong output.

They are also used to fabricate corrosion- and fire-resistant fiberglass-reinforced tanks, pipe, ductwork, hoods, sumps and other such equipment for handling a broad range of acids, bases, solvents, and combinations of these aggressive chemicals.

The resins are used in the foundry industry to produce rapid production automotive castings and very large, limited production metal castings. The large castings are made with molds and cores that utilize furfuryl alcohol no-bake binders. This results in high-quality iron and steel castings and allows ease of sand reclamation for further use. Aside from providing the high-speed core production rates required in the automotive industry, furfuryl alcohol binders produce excellent quality castings and very low scrap rates.

As with the alcohol, the aldehyde, furfural, can be polymerized with acidic catalysts to produce furan resins, but few practical uses have been developed for such products. More frequently, furfural is used as a modifying agent for phenolic novolac resins and furfuryl alcohol resins. Phenol-formaldehyde novolac resins are prepared by using acidic catalysts; phenol-furfural novolacs are readily produced by using alkaline catalysts. The phenol-furfural novolacs have long flow times and are used in molding compounds alone or with phenol-formaldehyde resins to modify molding and end-use properties.

5.4 Toxic Effects

In the past, there have been reports of sensitization and provoked late asthmatic responses from exposure to the emissions from some "furan"resin systems (221–223). Whether copolymers, catalysts, or other additives are responsible for such reactions is still uncertain. Some published literature implicates furfuryl alcohol (224, 225) and furfural (193, 226, 227) as dermal and/or respiratory sensitizers.

6.0 Polyvinylpyrrolidone

Polyvinylpyrrolidone (PVP), the water-soluble homopolymer of 1-ethenyl-2-pyrrolidinone (9CI), known generically as PVP, is sold in a variety of grades. A water-insoluble cross-linked form is known as polyvinyl-polypyrrolidone (PVPP) or crospovidone.

6.0.1 CAS Number: [9003-39-8]

6.0.2 Synonyms: Polyvinylpyrrolidone K-30; Povidone; PVP; poly(vinyl pyrrolidone); poly(1-vinyl-2-pyrrolidinone); hueper's polymer no. 5; poly-(1-vinyl-2-pyrrolidone); 1-ethenyl-2-pyrrolidinone, homopolymer; polyvidone; poly(N-vinylbutyrolactam); poly-(vinylpyrrolidinone); poly(N-vinylpyrrolidone); N-vinylbutyrolactam polymer; vinylpyrrolidinone polymer; 1-vinyl-2-pyrrolidinone polymer; 1-vinyl-2-pyrrolidone polymer; pyrrolidinone, 1-ethenyl-, homopolymer; poly(N-vinyl-2-pyrrolidone); polyvinylpyrrolidone, K15, average M.W. 10.000

6.0.3 Trade Names: NA

6.0.4 Molecular Weight: 111.14

6.0.5 Molecular Formula: $(C_6H_9NO)_n$

6.0.6 Molecular Structure:

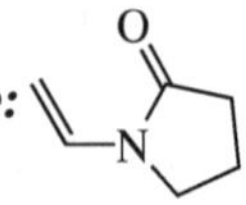

6.1 Chemical and Physical Properties

Physical state: white powder, hygroscopic and light sensitive
Melting point (°C): 225
Water sloubility: ≥ 10 g/100 mL at 20°C

6.1.1 General

The molecular weight of PVP is not a discrete figure, but rather the average molecular weight of a mixture that contains a range of lower and higher molecular weight species. The nomenclature of PVP polymers is expressed as K values, which are related to viscosity-average molecular weight. Commonly used PVP grades are K-12 ($M_v \sim 3,900$), K-17 ($M_v \sim 9,200$), K-25 ($M_v \sim 26,000$), K-30 ($M_v \sim 42,000$), and K-90 ($M_v \sim 1,100,000$).

PVP is extremely soluble in water and also in many organic solvents. PVP forms tight complexes with many substances; insoluble complexes formed by the addition of polybasic acids such as polyacrylic acid or tannic acid can be reversed by neutralizing with a base. Complexes of PVP with iodine retain an antimicrobial effect with reduced toxicity and staining tendency (229).

6.2 Production and Use

The monomer *N*-vinylpyrrolidone is typically polymerized in bulk or in aqueous solution with heat and free-radical catalysts, for example, hydrogen peroxide and ammonia (4, 228, 229).

PVP is used as a thickening agent, stabilizer, and complexing agent, in human and veterinary drugs, foods, cosmetics and toiletries, cleaners, plastics, adhesives, paints, and coatings. In the textile and dye industries, the complexing and surfactant properties of PVP improve the dye receptivity of hydrophobic fibers such as polyacrylonitrile or polypropylene (229). PVP is used to clarify and stabilize color in fruit juice, beer, and wine. Use as a plasma expander in the treatment of shock has been generally discontinued. It is still used today as a cryogenic preserving agent for mammalian cells.

6.3 Exposure Assessment

PVP can be detected in foods, beverages, and in other media at concentrations as low as 0.1 ppm $\pm 5\%$ by a chromatographic-colorimetric method (234). An infrared method permits detecting as little as 0.1 mg with $7 \pm 1\%$ accuracy in solutions as dilute as 0.1% (235). See Reference 34 for additional methods.

6.4 Toxic Effects

Extensive toxicological testing has been performed to support the pharmaceutical, food, and cosmetic uses of PVP, and several detailed reviews of the data (230–233) have appeared during the past 30 years. In many cases, the authors of the reviews had access to unpublished reports. Much of the information summarized in subsequent sections is extracted from these reviews.

6.4.1 Experimental Studies

6.4.1.1 Acute Toxicity. The acute oral LD_{50} of PVP that has a molecular weight of 10,000 to 30,000 is 40 g/kg (236). That of PVP K-30 (MW 40,000) is 100 g/kg in the rat and the guinea pig (237). RTECS reports the oral LD_{50}s of a variety of PVPs, CAS Number *[9003-39-8]*, (presumably differing by molecular weight) all of which show similar low oral acute toxicity (173). No dermal LD_{50}s were reported, but absorption of this polymer should be negligible, rendering this polymer low in dermal acute toxicity (173). No reports of primary dermal irritation results were in this database or others, however, medical use of this material implies low irritation potential. In a 1991 review of the toxicity of polyvinylpyrrolidone by a BIBRA working group (238), this polymer did not induce sensitization reactions in humans. However, in a case report by Gonzalo-Garijo et al. (239), an anaphylactic reaction was reported in a 37-year-old male that was attributed to "providone," a medical name for PVP, after challenge tests.

A shock reaction has been reported following intravenous injection of large amounts of PVP in dogs. The observed response is similar to systemic anaphylaxis but is not immunologically mediated. The associated increased plasma histamine level has been ascribed to the release of histamine from mast cells. Similar reactions have not been reported in rooster, rat, or humans (233, 255).

A PVP of MW 50,000 administered orally to rabbits impeded weight gain in the highest dose group of 2700 mg/kg (240). Effects noted in intravenous studies using small numbers of dogs or monkeys were attributed to the osmotic imbalance caused by the hypertonic dosing solutions (241, 242). When used as a blood tonic in Taiwan, PVP did not exhibit acute toxicity (but, as discussed later, very serious chronic toxicity may result from repeated exposure).

6.4.1.2 Chronic and Subchronic Toxicity. Several studies addressed the short-term repeated administration of PVP polymers but many reports lack details critical for comparative assessments. Those described in sufficient detail are summarized. The effect of repeated oral administration of PVP on the gastrointestinal tract of rats and dogs was described in a doctoral thesis. Severe desquamation of the gastric mucosa and excipient bronchial pneumonia were noted in 7 of 11 rats that survived 48 doses of 10 g/kg PVP administered by gavage during 2 months. No abnormal findings were noted in two dogs that received oral doses of 5 g/kg PVP (MW 220,000) for 1.5 weeks or PVP (MW 1,500,000) for 2 weeks (236).

PVP K-90 was administered in the diet to rats for 28 days at 0, 2.5, or 5% by weight (243) or for 90 days at 0, 2, 5, or 10% (244) without eliciting any clinical signs or histological changes that could be attributed to PVP administration. Similarly, PVP K-90 failed to induce toxic effects or pathological changes other than increased spleen weight in the high-dose females when administered in the diet to beagles for 28 days at 0, 2.5, 5, or 10% by weight (243). No treatment-related pathology was described in dogs that received 0, 2, 5, or 10% (244) PVP K-90 in the diet for 90 days. The high-dose group did not gain weight compared to controls, and evidence of PVP accumulation in the mesenteric lymph nodes was confirmed in all dose groups.

6.4.1.3 Pharmacokinetics, Metabolism, and Mechanisms. The pattern of absorption and excretion of PVP polymers varies with the molecular weight or molecular size, route of administration, and the species of interest. Most PVP products contain some fraction of low molecular weight polymers that are more readily absorbed from the gastrointestinal tract. Neonatal animals can absorb more of an orally administered dose, probably by pinocytosis (introduction of fluids into a cell that results in forming vesicles in the cell). The efficiency of this mechanism, an active mechanism that absorbs colostrum, decreases but does not disappear completely as an animal matures. Orally administered PVP is primarily excreted in the feces. PVP clearance after injection depends on the molecular weight of the polymer and the degree of vascularization at the injection site. Low molecular weight PVP is excreted by glomerular filtration into the urine or is retained in the draining lymph nodes. Excretion by the kidney is limited by the maximal pore size in the glomerulus.

6.4.1.4 Reproductive and Developmental. Rabbits administered intravenous doses of 50, 250, or 1250 mg/kg PVP K-12 on days 6 to 18 of gestation were sacrificed on day 28. No treatment-related effects on pregnancy rate or fetal parameters were reported in any dose group. Transient clinical signs (trembling, rapid breathing, convulsions) on treatment day 2 were noted only in the dams of the high-dose group; food intake was also slightly reduced in this group (249).

PVP K-25 or PVP K-90 was administered in the diet (10% by weight) to Sprague–Dawley rats for 20 days after mating; the dams were then sacrificed, and the fetuses were evaluated. Neither PVP affected the pregnancy rate; fetal parameters were comparable to those of controls that received a normal diet (250, 251).

6.4.1.5 Carcinogenesis. Early chronic studies of multiple PVP products (K-20, K-22, K-30, K-62, and others) by subcutaneous (s.c.), intraperitoneal (i.p.), and intravenous (i.v.) injection suggested increased incidence of reticuloendothelial system sarcomas and total carcinomas in mice and/or rats. These effects were not seen in rabbits, but the number of animals in the rabbit studies was very low. The incidence rates in the rodent studies were difficult to interpret because controls consisted of pooled treated and untreated animals from several studies that used a variety of treatment regimens (261–263).

Bethesda black rats that received three i.p. injections of a 25% solution of PVP K-17 and PVP K-25 at monthly intervals and observed for their lifetimes did not exhibit any difference in tumor incidence compared to controls (264). Sprague–Dawley rats fed diets containing 0, 5, or 10% PVP K-25 for 24 months did not show signs of toxicity. Body weights, food intake, blood and urine parameters and gross and histopathological findings were comparable to controls (265).

PVP K-30 was administered to rats, rabbits, and dogs to assess chronic effects. In a combined chronic toxicity and oncogenicity protocol, Sherman-Wistar rats (50/sex/group) received 0, 1, or 10% in the diet for 24 months. Clinical signs included soft stools and reduced body weight gain in the high-dose group. Albumin was noted in the urine at 18 months in the high-dose group and in all groups at 21 months. No other signs of treatment-related toxicity, no differences in gross or histopathological findings, and no evidence of carcinogenicity relative to controls were reported (266). Groups of female Sprague–Dawley rats tolerated single s.c. injections of 0.5 g PVP K-30 powder without evidence of injection site sarcomas or differences in tumor incidence. Average survival time was 20 months (266).

Dogs that received 0, 2, 5, or 10% PVP K-30 and cellulose in the diet (two of each/sex/group) did not exhibit any effects on body weights, clinical signs, or increased tumor incidence. Slight swelling of the reticuloendothelial cells was somewhat dose-related. No other gross or histopathological abnormalities were noted (267). In another 1-year dietary study in dogs, the only effect reported was evidence of polymer storage in the lymph nodes surrounding the gastrointestinal tract at doses greater than 5% in the diet (230).

PVP K-30, K-20, and K-60 polymers were administered i.v. to rabbits in doses of 0.4 to 3.0 g/kg 2 to 14 times at monthly intervals in studies from 21 to 89 months long. The number of tumors in the K-20 group was similar to that in the controls. No tumors were found in the K-30 and K-60 treated rabbits, although histological evidence of PVP storage in the liver and spleen of these animals was reported (268).

In another combination chronic toxicity and oncogenicity study, PVP was administered in the diet to Sprague–Dawley rats at levels of 0, 1, 2.5, or 5% until 70% mortality occurred in the control group (129 to 138 weeks). Interim sacrifice groups at 26, 52, and 104 weeks did not show any evidence of abnormal storage of PVP in liver, kidney, heart, or lymph nodes. No increases in tumor incidence compared to the cellulose control were reported (269).

Other studies in which the K-value designations and/or molecular weights of the materials tested were not specified have been reported for PVP polymers but are not summarized here.

6.4.1.6 Genetic and Related Cellular Effects Studies. PVP K-30 was evaluated in several test systems and did not possess genotoxic or clastogenic potential. Ames tests with and without metabolic activation were negative (256, 257), as was the mouse lymphoma forward mutation assay (258). The cell transformation assay using the Balb/c 3T3 cultures showed no increase in transformation frequency compared to controls. A single intraperitoneal administration of 3.16 g/kg PVP K-30 failed to elicit clinical signs or to influence the rate of conception, average number of implantations, percentage of live fetuses, or the mutagenic index of treated animals compared to controls in the mouse dominant lethal test (259) or to induce chromosomal aberrations in the bone marrow of Chinese hamsters (260).

6.4.2 Human Experience

6.4.2.2 Clinical Cases

6.4.2.2.3 Phamacokinetics, Metabolism, and Mechanisms. The term foam cells describes macrophages that contain large, clear pinocytized inclusions which contribute to a foamy appearance. PVP storage is characterized by the appearance of foam cells in liver, spleen, bone marrow, or lymph nodes.

Thesaurosis, the excessive storage of foreign substances in the body, has been associated with PVP administered by inhalation or injection. Granulomas typical of PVP storage disease have been misdiagnosed as signet-ring cell carcinomas. Caution should be exercised to ensure correct diagnosis after treatment with polymers. Pseudotumors have been identified in patients who received large quantities (70 g or more) of injectable preparations that contained PVP for extended periods of time. (See Ref. 233 for an extensive review of PVP uptake excretion and storage).

PVP was first related to thesaurosis in beauticians in 1958 (252, 253). The nature of the factors involved has subsequently been disputed (254). Increased bronchovascular markings, but no thesaurosis, were noted in 11 of 227 subjects during a survey of beauticians. Hair spray exposure was considered the most likely origin of the pulmonary infiltrate in five patients who had major exposure to aerosols. In 1977, 32 cases of thesaurosis had been identified, 10 of which were in beauticians. Discher (253) concluded that all these cases "seem to bear a remarkable resemblance to a recognized chemical condition, sarcoidosis, which existed prior to the use of PVP hairspray."

PVP was once used as a plasma expander in humans and has been administered to at least 500,000 individuals without incidence of adverse effects from PVP storage.

Several instances of people who used intravenously injected PVP as a "blood tonic" were reported in Taiwan. This use may stem from PVP's prior but now discontinued use as a plasma expander in individuals who required plasma for traumatic blood loss. Because the polymer is too large to excrete, accumulation from repeated infusion leads to potentially extremely serious consequences. Kuo et al. (245) reported five cases of "PVP storage disease" where the repeated infusion of PVP caused massive infiltration of PVP-containing macrophages into the bone marrow and consequent destruction of bone marrow, anemia, and destruction of bone. Dunn et al. (246) reported a case where PVP injection had been repeated over the course of several years and resulted in severe bone marrow failure and bone degeneration that led to multiple fractures and arthritis. The deposition of PVP in histiocytes stimulates their proliferation, further uptake of PVP, and infiltration into all parts of the reticuloendothelial system. Symptoms are similar to hereditary storage disease, osteomyelitis, or signet-ring cell carcinoma. It may be diagnosed by skin biopsy. Other manifestations of this syndrome may include frank skin lesions and polyneuropathy. Other cases have also been reported from Taiwan (247) and from Eastern Europe (248). PVP storage disease, or thesaurosis, is discussed further in Section 7.4.2.2.2.

6.4.2.2.7 Other: Skin Sensitization

7.0 Polybenzimidazole

7.0.1 CAS Number: [26985-65-9]

Polybenzimidazole is known as PBI and has this structure:

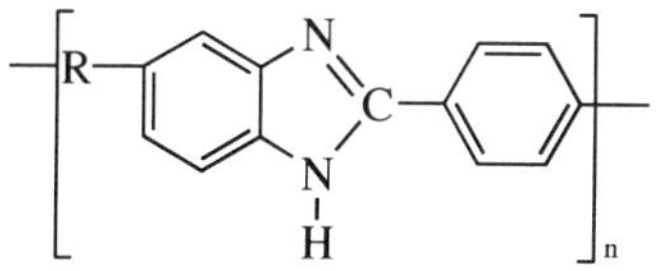

$$\left[R \underset{\underset{H}{N}}{\overset{N}{\diagdown}} C - \right]_n$$

7.2 Production and use

The most common polymer of PBI, poly(2,2'-*m*-phenylene)-5, 5'-bisbenzimidazole, can be prepared by heating tetraamino biphenyl and diphenyl isophthalate under nitrogen, typically at about 200°C, to yield a low molecular weight, glassy, friable foam, which is generally ground and cured by heating to a typical temperature of 385°C. This resulting amorphous powder can be prepared as a heat-resistant fiber by dry spinning from a solution of dimethylacetamide, as an adhesive, or as sheet film (270). Foam can be made by heating the low molecular weight, first-stage product.

As a fiber, polybenzimidazole is noted for exceptional high-temperature performance. The useful temperature limit is reportedly about 560°C (270).

7.4 Toxic Effects

Thermal decomposition of polybenzimidazole foams in air begins slowly at 375°C and becomes more rapid above 500°C (198). Studies conducted by various investigators in inert environments and in air at temperatures up to 1000°C indicate that volatile pyrolytic products include hydrogen, carbon monoxide, carbon dioxide, methane, water, hydrogen cyanide, ammonia, propene, acrylonitrile, phthalonitrile, benzene, aniline, and benzonitrile. Water is the only significant product released at less than 550°C (271). The pyrolytic products formed during combustion depend on the burning temperature. More toxic products are formed at higher temperatures (272, 273). Polybenzimidazole may be considered "less toxic" than other materials under fire conditions because it requires more time to decompose (if it reaches decomposition temperature at all), rather than resulting from an intrinsic difference in the pyrolytic products eventually formed.

C SILICONES

Silicones or silicon hydrides of increasing chain length up to about Si_6H_{14} are known. Above this length however, the Si–Si chain becomes thermally unstable. Commercial silicone polymers contain the siloxane link Si–O–Si.

8.0 Silicone

8.0.1 CAS Number: [63148-62-9]

8.0.2 Synonyms: Dimethylpolysiloxane hydrolyzate; alpha-methyl-omega-methoxypolydimethylsiloxane; polydimethylsiloxane, methy end-blocked, silicone, all; silicones; dimethylpolysiloxane hydrolyzate; polydimethyl silicone oil; poly(dimethylsiloxane);

polydimethylsiloxane, methyl end-blocked; Polyoxy(dimethylsilylene), alpha-(timethylsilyl)-omega-hydroxy; Poly[oxy(dimethylsilylene)], alpha-[trimethylsilyl]-omega-[(trimethylsilyl)oxy]; silicone oils; siloxane and silicones, dimethyl; siloxanes adn silicones, dimethyl, alpha-(trimethylsilyl)poly[oxy(dimethylsilylene)]-omega-methyl; silicone oil; silicone oil, for oil baths; polydimethylsiloxane, trimethylsiloxy terminated; polydimethylsiloxane, trimethoxysilyl terminated; hydride terminated, 1; carboxypropyldimethyl; silicone oil, for melting point and boiling point apparatuses; antifoam spray 260

8.0.3 Trade Names: NA

8.0.4 Molecular Weight: 111.14

8.1 Chemical and Physical Properties

Specific Gravity: 0.963
Flash Point: 300°C

8.2 Production and Use

Polyorganosiloxanes are generally prepared by reacting organochlorosilanes with water and then condensing the intermediate material to form a finished polymer. A broad variety of products can be made from relatively few monomers and intermediates (274). PDMS materials are the most important commercially available silicones from the standpoint of sales volume. Cross-linking these materials converts the fluid polymer chain into a gel, resin, gum, or elastomer, depending on the molecular weight and degree of cross-linking.

9.0 Dimethyl Silicone

9.0.1 CAS Number: [9016-00-6]

9.0.2 Synonyms: Demethicone; silicone rubber; latex; silicone rubber, latex; dimethylpoly-siloxane; simethicone; dimethyl silicone; Dermafilm; Dimethicream; Silbar

9.0.3 Trade Names: NA

9.0.4 Molecular Weight: NA

9.0.5 Molecular Formula:

The linear polydimethylsiloxanes (or PDMS) may be represented by CAS Number [9016-00-6], [63148-62-9], and [63394-02-5], and their molecular structure by the generic formula

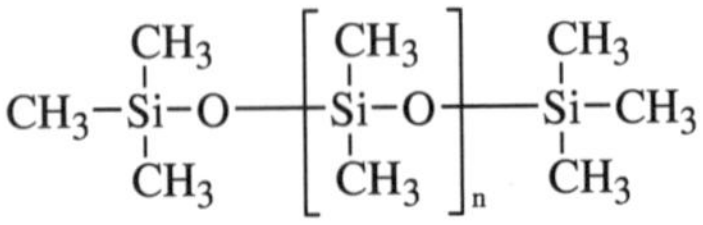

$$CH_3-\underset{\underset{CH_3}{|}}{\overset{\overset{CH_3}{|}}{Si}}-O\left[\underset{\underset{CH_3}{|}}{\overset{\overset{CH_3}{|}}{Si}}-O\right]_n\underset{\underset{CH_3}{|}}{\overset{\overset{CH_3}{|}}{Si}}-CH_3$$

where n may range from 0 to $> 10,000$. In the cyclic oligomers, the terminal end groups are lacking, n may be 3 to 8, and $n = 4$ and 5 predominate (275).

9.1 Chemical and Physical Properties

Physical state: Clear, viscous liquid
Specific Gravity: 0.98

9.2 Production and Use

Silicon Fluids and Resins

The most common liquid silicones are polydimethylsiloxanes (PDMS), which occur as oligomers up to polymers of high molecular weight ($>500,000$). Commercially useful properties are retained over a wide temperature range ($-70°C$ to $>200°C$). PDMS has prolonged stability at 150°C, but oxidizes at 250°C unless inhibited by antioxidants (3). PDMS is extremely water repellent and has strong antiadhesive properties. The liquid silicones are excellent heat-transfer agents and dielectric fluids and are also used in polishes, waxes, release coatings, high performance paints, and in packaging, paper, and textile treatments. There are significant applications in personal care shampoo and skin creams and also in the controling foam in widely diverse applications that range from food preparation to amusement park waterfalls.

Silicone resins have excellent heat resistance but are mechanically weaker than corresponding organic resin polymers (292). High phenyl content silicone resins are compatible with phenol-formaldehyde, urea–formaldehyde, melamine–formaldehyde, and oil-modified alkyd resins (not nonmodified alkyds). Silicone resins may be used to prepare heat-resistant glass-cloth laminates, particularly for electric motors, printed circuits, and transformers. Lesser uses include molding powders, water-repellent treatments, and release agents. The use of silicone resins, as heat-cured coatings on commercial bakery pans, aids in the release of bread and other foods after baking.

Silicone Elastomers

Polydimethylsiloxane (PDMS) rubber is prepared primarily from dimethyl cyclic tetramer. The most useful property of silicone elastomers is their wide service temperature, which can range from $-65°C$ to 315°C, and their stability to long-term exposure to environmental agents, such as ultraviolet light, ozone, and rain.

Silicone rubber is widely used in aerospace, automotive, and electrical applications. Consumer uses include roof coatings, household appliances, toys, pharmaceutical devices, and adhesives and gaskets for structural-glass building applications. Repair of damage due to failed organic elastomers that provide protection from weather is an increasing commercial application of PDMS elastomers.

The household consumer has found hundreds of uses of silicone adhesives that cure in the presence of water vapor to form a silicone rubber. These applications commonly include weather sealing around window frames/doors, aquarium building, ceramic tile repair, sealants for bathroom tub and shower to wall joints, and general-purpose adhesives.

Public Highway departments have used these silicone rubber adhesives to seal joints between road sections to prevent entry of water and reduce freeze/thaw damage to highways.

9.3 Exposure Assessment

Programmed heating of fluid PDMS at 10°C/min under vacuum causes the evolution of volatile products that are detectable at 343°C and reach a maximum at 443°C. Cyclic trimer, cyclic tetramer, and higher cyclics were detected. When phenyl-containing PDMS was studied under these conditions, benzene was also detected (285–287). Unpublished data indicates that PDMS heated in air to temperatures greater than 150°C can evolve trace levels of formaldehyde. There is a direct correlation between temperature and the amount of evolved formaldehyde and an indirect correlation between the degree of cross-linking (molecular weight) and evolved formaldehyde (288).

9.4 Toxic Effects

The FDA sponsored a monograph (278) on the safety of PDMS. This monograph summarizes various individual items found in the scientific literature from 1920 through September 1978.

9.4.1 Experimental Studies

In October 1992 a Significant New Alternative Policy (SNAP) report was submitted to the EPA (293). This is a review of published and nonpublished data on low molecular weight PDMS materials that have both linear and cyclic structures. Another recent toxicological data submission on medium to high molecular weight PDMS materials was made to the European Chemical Industry Ecology and Toxicology Center for publication (294).

These reviews indicate that acute ingestion of massive doses up to 50 mL/kg (guinea pigs) or 30 mL/kg (rats) had no significant adverse effects. These materials caused, at most, mild, transient, eye discomfort from direct contact, possibly from upsetting the normally hydrophilic eye membrane, that lasted several hours. These materials are generally nonirritant to skin; however, hexamethyldisiloxane can cause slight to moderate irritation if prolonged occlusion occurs. There are no indications of dermal absorption in acutely toxic amounts. Inhalation is not a likely route of exposure because of the very low vapor pressure; the most volatile species (hexamethyldisiloxane) caused no adverse effects from exposure to 25,000 ppm for 30 min. However, saturated vapor concentrations (39,000 to 40,000 ppm at 21°C), usually caused death in 15 to 30 min from respiratory failure. Aerosol exposure of rats to 10,000 centistoke PDMS fluid for 4 h at concentrations up to 695 mg/m^3 caused no adverse effects.

Subchronic studies that consisted of oral, dermal, and inhalation exposure indicate minimal adverse effects. High dose levels via the oral or inhalation route have resulted in increased liver weights in rats, hamsters, and mice but not among rabbits, guinea pigs, or dogs. The increased liver weight effect seen in rodents did not show any histopathological injury, and the liver weights of satellite groups returned to normal after about 4 weeks postexposure. It is believed this effect is the result of acclimation to increased metabolic activity of the rodent liver. These theories are supported by measured increases in rodent liver enzymatic activity.

Chronic feeding studies of rats at dietary levels up to 0.28% PDMS (1000 centistokes) for 2 years did not show any evidence of adverse effects. In another study, mice were fed PDMS at 2.35% of the diet for 80 weeks. Macro- and microscopic examination of these animals found no significant increase in malignant or benign tumors, nor was there any significant increase in mortality. Body weight data were not given. A 26-month dermal application to monkeys dosed at 200 mg/kg (5 times a week for 26 months) revealed no carcinogenic effects nor any other indication of adverse effects.

9.4.1.1 Acute Toxicity. Early toxicological studies were conducted by Rowe et al. (276, additional work is cited in Ref. 277). The initial series included acute and short-term tests with hexamethyldisiloxane and dodecamethylpentasiloxane; two dimethylpolysiloxanes of grease-like consistency, one a sealing compound and the other an antifoam (Dow Corning Antifoam A); and three silicone resins, one a methylpolysiloxane and two methylphenylpolysiloxanes. The short-term tests showed low toxicity. The Registry for the Toxic Effects of Chemical Substances (RTECS), a NIOSH-generated database, shows the acute oral, dermal, and other LD_{50} values for the polydimethylsiloxanes with the three CAS Numbers enumerated earlier (173). All have low acute toxicity. The Hazardous Substances Databank (HSDB) indicates low irritation potential to eyes and skin for this family of polymers (278).

9.4.1.2 Chronic and Subchronic Toxicity—Oral. Rats fed a dietary level of 0.3% Antifoam A for 2 years showed no significant toxic effects. Two additional 2 year tests with rats fed a polymethyl silicone fluid showed no direct evidence of toxic effects although the data were less conclusive (Gloxhuber and Hecht, as cited in Ref. 279).

Cutler and co-workers (280) reported that mice given a lifetime diet containing 0.25 and 2.5% of a silicone antifoam agent from weaning showed no significant toxic effects. The antifoam agent contained 94% dimethylsiloxane silicone oil and 6% finely divided silicon dioxide. Analysis of the whole animals revealed no silicone component in 10 mice that had been fed a diet containing 2.5% silicone for 75 weeks. Two other groups of mice received a single subcutaneous injection of 0.2 mL antifoam and 0.2 mL liquid paraffin, respectively, at weaning. The mice that received the paraffin had an increased incidence and earlier appearance of subcutaneous fibromas at the injection site in males, whereas the silicone-injected mice showed a greater incidence of cysts.

Inhalation—Thermal degradation products. Hilado et al. (284) exposed mice to the pyrolytic gases evolved from samples of 11 silicon polymers: a transformer liquid, four elastomeric products, and six silicone resins. As judged by time to death or incapacitation, these polymers were among the least toxic of approximately 300 materials tested in the rising temperature program. Carbon monoxide was identified as an important toxicant.

9.4.1.3 Pharmacokinetics, Metabolism, and Mechanisms. Dermal tests (rat, monkey, human) with a mixture of cyclic oligomers (specified as 70% tetramer octamethylcyclotetrasiloxane and 20% pentamer heptamethylcyclopentasiloxane) showed no evidence of absorption (276). No increase in silicon levels was detected in expired air by atomic absorption analysis or in urine by emission spectroscopy.

9.4.1.4 Reproductive and Developmental. Several grades of polydimethylsiloxane caused no evident change when tested for reproductive and teratogenic effects in rats and rabbits (281) and testicular effects in rabbits (282). Lack of effect on male reproductive function was also observed (283) for polydimethylsiloxane, a trimethyl end-blocked dimethylphenyl-ethylpolysiloxane, tris(trimethylsiloxy)phenylsilane, and trifluoropropylmethylpolysiloxane. However, a testicular effect was observed (282) in each of a series of 19 tests of rabbits given different batches of an equilibrated copolymer of mixed polymethyl and dimethyl cyclic siloxanes.

Reproductive and developmental toxicity studies in which 350-centistoke PDMS was injected subcutaneously into pregnant rabbits at 20, 200, or 1000 mg/kg and a parallel study in which pregnant rabbits were dermally dosed at 200 mg/kg indicated an increase in resorption sites in direct proportion to the dose. (The sesame oil control groups also showed significant resorption sites.) No fetal abnormalities nor maternal body weight effects were observed at the high dose, but 3.8% of the fetuses had gross abnormalities at the intermediate dose level (synophthalmia, umbilical hernia, clubbing of extremities). This preliminary work was followed by a larger study using both rats and rabbits. Twenty pregnant rats were dosed at 20, 200, or 1000 mg/kg via subcutaneous injection and 1000-mg/kg sesame oil control groups were used. These results indicate a slight increase in delayed ossification to sternebra and cranial bones only at the high dose. A parallel rabbit study (same doses/route) indicated no adverse effects other than slight variations in the number of ribs (a common finding that is not considered a significant adverse effect).

9.4.1.6 Genetic and Related Cellular Effects Studies. Immunotoxicity studies conducted on 1000-centistoke PDMS using a *Listeria* host-resistance assay indicated that 1 mL of PDMS fluid implanted subcutaneously in the mid-back region of female B6C3F$_1$ mice had no effect on immune competence.

The NTP completed a series of immunotoxicity tests on polydimethylsiloxane. In a skin sensitization test, PDMS was applied to the backs of B6C3F1 female mice for 5 days in an induction phase. Seven days later, mice were challenged with a similar application and immunological responses were monitored by measuring (*1*) the migration of ^{125}I-deoxyuridine labeled cells into the challenge site and (2) the amount of ear swelling. No allergic dermal reactions were found in this test (295). In two other NTP studies that evaluated the immunotoxicity of PDMS in B6C3F1 female mice implanted with silicone gel or fluid, no effects were found on "B cell number, proliferative ability, or ability to differentiate into antibody-producing cells to a T-dependent antigen." No effect on cell-mediated immunity was found (296, 297).

9.4.2.2.5 Carcinogenesis. Bischoff (289) presented an extensive review of the literature pertaining to medical uses and carcinogenic potential through 1972. This review indicates that silicone fluids can migrate from an injection site and cause granulomatous reactions.

Malignancies have been observed under certain test conditions and with certain species. These tumors are thought to be associated with rodents and have minimal impact on human health.

Implants. Concern about the safety of silicone breast implants has led to a debate that is still ongoing. A search of the available scientific literature reveals many epidemiological studies of women who have silicone breast implants where the frequency of autoimmune disease and (perhaps related) connective tissue and neurological disorders, neoplasia, and other adverse health effects are compared with the general population. Many of these studies are ongoing, and their purpose is to determine whether a causal relationship exists between silicone breast implants and these adverse health effects. Other studies of animals are intended to investigate whether the effects suspected in humans who have silicone implants may be reproduced in an animal model, and if so, whether their mechanisms may be elucidated. Only selected studies are discussed here because the volume of literature is too large to review in a single chapter. The reader may obtain more information by searching the National Library of Medicine's (NLM) MedLine/ToxLine databases for other citations using MESH terms such as (1) "silicones—adverse effects," (2) "connective tissue diseases—immunology," (3) "connective tissue diseases—epidemiology," and other variations (rather than searching solely by CAS numbers and subject heading terms).

9.4.2.3 Epidemiology Studies—Implants.

Wong (298) did not find a relationship between silicone implants and connective tissue disease based on a meta-analysis of 15 epidemiological studies that involved more than 4000 cases. In an epidemiological study that evaluated the incidence of systemic sclerosis in more than 2500 augmentation mammoplasty patients, Hochberg et al. (299) did not find an association. In a study of more than 500 women, Englert et al. (300) did not find an increase in scleroderma. This result was confirmed in a study of 274 breast implant patients by Burns et al. (301). In a study of 4229 women, Goldman et al. (302) found no increased risk of rheumatoid arthritis or other diffuse connective tissue disease. With regard to neurological disease, Winther et al. (303) studied a cohort of 1135 women who received silicone breast implants, and found no increased incidence. Similarly, in a cohort of 7433 women who had breast implants, Nyren et al. found no increased incidence of neurological disease (304).

No epidemiological studies were found claiming a positive association between silicone implants and the adverse affects listed before from a fairly comprehensive search of the literature using the above-mentioned NLM MESH terms, CAS numbers, and the subject term "polydimethylsiloxane" (and restricting the search to recent years). The only review of the literature found that reported a positive association between silicone breast implantation and immunological disease, claimed by the authors as causative, is that by Brautbar et al. (305). The authors base this conclusion on a review of the literature and on the bases that silicone is immunogenic in animals and humans and that symptoms of women who have implants have reversed upon removal of the implant. Two case reports support an association between immunological and neurological disorders and silicone breast implants. A case study by Yoshida (306) reports a neurosarcoidosis in a women that developed 22 years after silicone breast implantation. Cambell and Brautbar (307) reported a case where a silicone Norplant device resulted in "systemic immunological complications." Case reports have the obvious limitation of being classified as anecdotal in that they cannot define whether the reported effect occurs at a rate higher than in the general population.

9.5 Standards, Regulations, or Guidelines of Exposure

The FDA permits adding 10 ppm PDMS (> 100 centistokes) as a direct food additive to nonstandardized foodstuffs. In addition, the World Health Organization has established an acceptable daily intake of 1.5 mg/kg for PDMS (290).

The Cosmetic Ingredient Review Board published a review of the toxicity of cyclomethicone as part of its responsibility to the Cosmetic, Toiletry, and Fragrance Association. This publication deals primarily with dimethyl cyclic tetramer and dimethyl cyclic pentamer (291).

9.6 Studies on Environmental Impact

Several polydimethylsiloxane fluids and formulations exhibited very low toxicity (283) in tests with *Daphnia* water fleas, freshwater and marine fish, mallards, bobwhite quail, and domestic chickens. Radioactive tracer studies ($[^{14}C]$polydimethylsiloxane) showed no detectable degradation by sewage microorganisms or bioaccumulation in bluegill sunfish. Silicon residues in eggs and tissue samples of white Leghorn chickens were below the detection limit of 2 to 4 ppm after the chickens had been maintained on diets that contained 200 to 5000 ppm polydimethylsiloxane.

BIBLIOGRAPHY

1. H. J. Fabris, Thermal and oxidative stability of urethanes, in K. C. Frisch and S. L. Reegen, Eds., *Advances in Urethane Science and Technology*, Vol. 6, Technomic, Westport, CT, 1978, pp. 173–196.
2. M. P. Stevens, *Polymer Chemistry; An Introduction*, Addison-Wesley, Reading, MA, 1975.
3. J. A. Brydson, *Plastics Materials*, 3rd ed., Whitefriars Press, London, 1975.
4. W. J. Roff and J. R. Scott, with J. Pacitti, *Handbook of Common Polymers*, CRC Press, Cleveland, OH, 1971.
5. K. A. Pigott, Polyurethanes, in *Encyclopedia of Polymer Science and Technology*, Vol. 11, Wiley, New York, 1969, pp. 506–563.
6. G. G. Hawley, *The Condensed Chemical Dictionary*, 9th ed., Van Nostrand Reinhold, New York, 1976.
7. K. Omae et al., Four-year follow-up of effects of toluene diisocyanate exposure on the respiratory system in polyurethane foam manufacturing workers. I. Study design and results of the first cross-sectional observation. *Int. Arch. Occup. Environ. Health* **63**(8), 559–564 (1992).
8. K. Omae et al., Four-year follow-up of effects of toluene diisocyanate exposure on the respiratory system in polyurethane foam manufacturing workers. II. Four-year changes in the effects on the respiratory system. *Int. Arch. Occup. Environ. Health* **63**(8), 565–569 (1992).
9. R. N. Jones et al., Abnormal lung function in polyurethane foam producers. Weak relationship to toluene diisocyanate exposure. *Am. Rev. Respir. Dis.* **146**(4); 871–877 (1992).

10. F. Sulotto et al., Short and long-term respiratory effects of exposure to low concentrations of diphenylmethane diisocyanate in the production of expanded polyurethane. *Prevenzione oggi* **6**(4), 163–175 (1994).
11. World polyurethane use heads for 7 billion lb. *Chem. Eng. News* **57**(42), 12–13 (1979).
12. R. R. Reen, *2000 and Beyond: Succeeding in the Next Century, Polyurethanes World Congress, 1991 Proceedings*, Technomic Publishing Inc., 1991, pp. 7–9.
13. The Society of the Plastics Industry, Inc. Polyurethane Division, *End Use Market Survey on the Polyurethane Industry in the U.S. and Canada*, 1991.
14. R. G. Ridgeway et al., *Microchem. J.* **42**(1), 138–145 (1990); *Chem. Abstr.* **113**, 60456q.
15. Q. Zhang and Z. Yong, *Sepu* **5**(5), 305–308 (1987); *Chem. Abstr.* **108**, 76202m.
16. *Proposition 65 Compliance Workbook for TDI Exposure*, published by The Society of the Plastics Industry, Inc., New York, 1990.
17. Q. Zhang and Z. Yong, *Tuliao Gongye* **3**, 46–50 (1988); *Chem. Abstr.* 110, 76556w.
18. I. Padovani and L. Trevisan, *Chim. Ind.* **69**(4), 30–31 (1987); *Chem. Abstr.* **107**, 78548u.
19. S. C. Rastogi, *Chromatographia* **5**, 73–74 (1989); *Chem. Abstr.* **111**, 175058n.
20. K. C. Cole et al., *J. Appl. Polym. Sci.* **34**(1), 395–407 (1987); *Chem. Abstr.* **107**, 116218j.
21. I. V. Leikin and V. V. Zharkov, *Plast.* **5**, 73–74 (1989); *Chem. Abstr.* **111**, 234216n.
22. G. Balle, M. Kuck, and W. Wellner, *Staub-Reinhalt. Luft* **51**(6), 231–236 (1991); *Chem. Abstr.* **115**, 286140k.
23. K. Andersson et al., *Am. Ind. Hyg. Assoc. J.* **44**(11), 802–808 (1983); *Chem. Abstr.* **100**, 38820p.
24. C. Rosenberg and T. Tuomi, *Am. Ind. Hyg. Assoc. J.* **45**(2), 117–121 (1984); *Chem. Abstr.* **100**, 179381h.
25. M. E. Krzymien, *Int. J. Environ. Anal. Chem.* **36**(4), 193–207 (1989); *Chem. Abstr.* **111**, 238770m.
26. G. E. Podolak et al., *ASTM Spec. Tech. Publ.* **957**, 203–212 (1987); *Chem. Abstr.* **108**, 10534q.
27. G. G. Esposito and T. W. Dolzine, *Anal. Chem.* **54**(9), 1572–1575 (1982); *Chem. Abstr.* **97**, 43329j.
28. F. Schmidtke, B. Seifert, and J. Fresenius, *Anal. Chem.* **336**(8), 647–654 (1990); *Chem. Abstr.* **113**, 64418b.
29. W. S. Wu, L. K. Huang, and V. S. Gaind, *Am. Ind. Hyg. Assoc. J.* **47**(8), 482–487 (1982); *Chem. Abstr.* **105**, 138892k.
30. S. P. Levine et al., *Anal. Chem.* **51**(8), 1106–1109 (1979); *Chem. Abstr.* **91**, 43896t.
31. K. Samejima, *J. Chromatogr.* **96**(2), 250–254 (1974); *Chem. Abstr.* **82**, 28034f.
32. V. Dharmarajan and R. J. Rando, *Am. Ind. Hyg. Assoc. J.* **41**(6), 437–441 (1980); *Chem. Abstr.* **93**, 191281v.
33. C. J. Warwick, D. A. Bagon, and C. J. Purnell, *Analyst (London)* **106**(1263), 676–685 (1981); *Chem. Abstr.* **95**, 85457p.
34. G. Skarping, M. Dalene, and L. Mathiasson, *J. Chromatogr.* **435**(3), 435–468 (1988); *Chem. Abstr.* **108**, 136881u.
35. M. Dalene, G. Skarping, and T. Brorson, *J. Chromatogr.* **516**(2), 405–413 (1990); *Chem. Abstr.* **114**, 37243e.
36. K. Samejima et. al., *Anal. Biochem.* **76**(2), 392–406 (1976); *Chem. Abstr.* **85**, 188610f.
37. Y. Nishikawa, *J. Chromatogr.* **392**, 349–359 (1987); *Chem. Abstr.* **106**, 201435b.

38. C. J. Hull et al., *J. Chromatogr.* **477**(2), 387–395 (1989); *Chem. Abstr.* **111**, 195757t.
39. T. Hirayama et al., *J. Assoc. Offic. Anal. Chem.* **68**(4), 746–748 (1985); *Chem. Abstr.* **103**, 88497w.
40. T. Inoue, H. Ishiwata, and A. Tanimura, *Shokuhin Eiseigaku Zasshi* **26**(4), 326330 (1985); *Chem. Abstr.* **104**, 110875n.
41. M. M. O'Mara, D. A. Ernes, and D. T. Hanshumaker, *Prog. Biomed. Eng.* **1**, 83–92 (1984); *Chem. Abstr.* **103**, 76297t.
42. A. L. Mazzu and C. P. Smith, *J. Biomed. Mater. Res.* **18**(8), 961–968 (1984); *Chem. Abstr.* **101**, 235546a.
43. H. Shintani, *J. Chromatogr.* **600**(1), 93–97 (1992); *Chem. Abstr.* 117, 76422u.
44. N. Yoshitake, M. Maruo, and M. Furukawa, *Ariake Kogyo Koto Senmon Gakko Kiyo* **25**, 61–66 (1989); *Chem. Abstr.* **111**, 79647h.
45. C. Rosenberg, *Analyst (London)* **109**(7), 859–866 (1984); *Chem. Abstr.* **101**, 215618x.
46. A. Kolbrecki, *Polimery (Warsaw)* **35**(3), 81 (1990); *Chem. Abstr.* **116**, 215814t.
47. C. C. Greenwalt, J. H. Futrell, and D. J. Lyman, *J. Polym. Sci., Part A* **27**(1), 301–315 (1989); *Chem. Abstr.* **110**, 173895p.
48. L. Renman, C. Sangoe, and G. Skarping, *Am. Ind. Hyg. Assoc. J.* **47**(10), 621–628 (1986); *Chem. Abstr.* **105**, 213571k.
49. M. Dalene, G. Skarping, and P. Lind, Workers exposed to thermal degradation products of TDI- and MDI-based polyurethane: biomonitoring of 2,4-TDA, 2,6-TDA, and 4,4-MDA in hydrolyzed urine and plasma. *Am. Indus. Hyg. Assoc. J.* **58**(8), 587–591 (1997).
50. P. Lind, G. Skarping, and M. Dalene, Biomarkers of toluene diisocyanate and thermal degradation products of polyurethane, with special reference to the sample preparation. *Analytica Chemica Acta* **333**(3), 277–283 (1996).
51. H. Tinnerberg, M. Dalene, and G. Skarping, Air and biological monitoring of toluene diisocyanate in a flexible foam plant. *Am. Indus. Hyg. Assoc. J.* **58**(3), 229–235 (1997).
52. G. Skarping et al., Biomarkers of exposure, antibodies, and respiratory symptoms in workers heating polyurethane glue. *Occ. Environ. Med.* **53**(3), 180–187 (1996).
53. B. R. Grant, personal communication, DuPont Company, Polymer Products Dept., Wilmington, DE, March 1980.
54. *Polyurethanes*, The Society of the Plastics Industry, Inc., Polyurethanes Division, New York, 1992, p. 12.
55. K. Jakobsson et al., Airway symptoms and lung function in pipelayers exposed to thermal degradation products from MDI-based polyurethane. *Occ. Environ. Med.* **54**(12), 873–879 (1997).
56. L. J. Fuortes, S. Kiken, and M. Makowsky, An outbreak of naphthalene di-isocyanate-induced asthma in a plastics factory. *Arch. Environ. Health* **50**(5), 337–340 (1995).
57. X. Baur et al., Respiratory and other hazards of isocyanates. *Int. Arch. Occup. Environ. Health* **66**(3), 141–152.
58. X. Baur, Occupational asthma due to isocyanates. *Lung* **174**(1), 23–30 (1996).
59. G. M. Liss, S. M. Tarlo, and D. E. Banks, Evidence for occupational asthma among compensation claimants at a polyurethane utilizing facility. *Canadian J. Public Health* **87**(6), 401–403 (1996).

60. F. Akbar-Khanzadeh and R. D. Rivas, Exposure to isocyanates and organic solvents, and pulmonary-function changes in workers in a polyurethane molding process. *J. Occ. Environ. Med.* **18**(12), 4205–1212 (1996).

61. G. Mastrangelo et al., Repeated case-control studies as a method of surveillance for asthma in occupations. *J. Occ. Environ. Med.* **39**(1), 51–57 (1997).

62. T. M. Schnorr, Mortality of workers exposed to toluene diisoyanate in the polyurethane foam industry. *Occ. Environ. Med.* **53**(10), 703–707 (1996).

63. G. E. Hartzell, The combustion toxicology of polyurethane foams. *J. Cellular Plastics* **28**, 330–358 (1992).

64. S. Womble, Analytical characterization of flexible polyurethane foams. In R. Orzel, Ed., *Appendix to Memorandum on Literature Review of the Combustion Toxicity of Flexible Polyurethane Foam*, U.S. Consumer Products Safety Commission, Washington, DC, 1984.

65. B. C. Levin et al., *Further Development of a Test Method for Assessment of the Acute Inhalation Toxicity of Combustion Products*, NBSIR 82–2532, National Institute of Standards and Technology, Gaithersburg, MD, 1982.

66. D. A. Purser and W. D. Woolley, Biological studies of combustion atmospheres. *J. Fire Sciences* **1**(2), 118–144 (1983).

67. D. A. Purser and W. D. Woolley, Biological studies of combustion atmospheres. *Conference on Smoke and Toxic Products from Burning Polymers*, London, England, 1982.

68. M. Paabo and B. C. Levin, *A Review of the Literature on the Gaseous Products and Toxicity Generated from the Pyrolysis and Combustion of Rigid Polyurethane Foams*, NBSIR 85–3224, Center for Fire Research, National Bureau of Standards, Gaithersburg, MD, 1985; In *Fire Mater.* **11**, 1–29 (1987).

69. J. Boudene and J. M. Jouany, Study of the toxicity of air-borne combustion and pyrolysis products of polyurethanes. Report to the International Isocyanate Institute, 1978.

70. R. Truhaut, J. Boudene, and J. M. Jouany, Etude de la toxicite des produits de combustion et de pyrolyse de materiaux utilise dans le batiment, I etude de la toxicite aigue par voie aerienne, des toxiques major pouvant etre liberes lors d'incednies. *Arch. Mal. Prof.* **36**, 707–238 (1975).

71. J. H. Petajan et al., Extreme toxicity for combustion products of a fire-retarded polyurethane foam. *Science* **187**, 542–544 (1975).

72. G. E. Hartzell, D. N. Priest, and W. G. Switzer, Modeling of toxicological effects of fire gases: II. Mathematical modeling of intoxication of rats by carbon monoxide and hydrogen cyanide. *J. Fire Sci.* **3**(2), 115–128 (1985).

73. G. E. Hartzell et al., Modeling of Toxicological of fire gases: III. Quantification of post-exposure lethality of rats from exposure to HCl atmospheres. *J. Fire Sci.* **3**(3), 195–207 (1985).

74. V. Babrauskas et al., *Toxic Potency Measurement for fire Hazard Analysis*, NIST Special Publication 827, National Institute of Standards and Technology, Gaithersburg, MD, 1991.

75. W. D. Woolley and P. J. Fardell, The prediction of combustion products. *Fire Res.* **1**, 11–21 (1977).

76. B. C. Levin et al., Generation of hydrogen cyanide from flexible polyurethane foams under different combustion condition. *Fire Mater.* **9**(3), 125–134 (1985).

77. A. F. Grand et al., An evaluation of toxic hazards from full-scale furnished room fire studies. In T. Z. Harmathy, Ed., *Fire Safety: Science and Engineering*, ASTM STP 882, Philadelphia, PA, 1985, pp. 330–353.

78. B. C. Levin, M. Paabo, and M. M. Birky, *An Interlaboratory Evaluation of the National Bureau of Standards Test Method for Assessing the Acute Inhalation Toxicity of Combustion Products*, NBSIR 83–2678, U.S. National Bureau of Standards, Gaithersburg, MD, 1983.
79. F. M. Esposito and Y. Alarie, Inhalation toxicity of carbon monoxide and hydrogen cyanide gases released during the thermal decomposition of polymers. *J. Fire Sci.* **6**(3), 195–242 (1988).
80. R. R. Beard, Polyurethanes; Hazards & control. *Mich. Occup. Health* **16**(1), 2–7 (1970).
81. R. L. Clark et al., *Int. Arch. Occup. Environ. Health* **71**(3), 169–179 (1998).
82. F. Deschamps et al., Prevalence of respiratory symptoms and increased specific IgE levels in West-African workers exposed to isocyanates. *J. Toxicol. Environ. Health* **54**(5), 335–342 (1998).
83. C. Simpson et al., Hypersensitivity pneumonitis-like reaction and occupational asthma associated with 1,3-bis(isocyanatomethyl) cyclohexane pre-polymer. *Am. J. Ind. Med.* **30**(1), 48–55 (1996).
84. K. G. Orloff, Antibodies to toluene diisocyanate in an environmentally exposed population. *Environ. Health Perspect.* **106**(10), 665–666 (1998).
85. T. R. Hester, Jr., et al., Measurement of 2,4-toluenediamine in urine and serum samples from women with Meme or Replicon breast implants. *Plast. Reconstr. Surg.* **100**(5), 1291–1298 (1997).
86. O. Sepai et al., Exposure to toluenediamines from polyurethane-covered breast implants. *Toxicol. Lett.* **77**(1–3), 371–378 (1995).
87. L. A. Brinton, P. Toniolo, and B. S. Pasternack, Epidemiologic follow-up studies of breast augmentation patients. *J. Clin. Epidemiol.* **48**(4), 557–563 (1995).
88. K. B. Delclos et al., Assessment of DNA adducts and the frequency of 6-thioguanine resistant T-lymhocytes in F344 rats fed 2,4-toluenediamine or implanted with a toluenediisocyanate-containing polyester polyurethane foam. *Mutat. Res.* **367**(4), 210–218 (1996).
89. S. Laskin et al., Inhalation studies with freshly generated polyurethane foam dust. In T. T. Mercer, P. E. Morrow, and W. Stober, Eds., *Assessment of Airborne Particles*, Springfield, IL, 1972, pp. 382–404.
90. J. Thyssen et al., Inhalation studies with polyurethane foam dust in relation to respiratory tract carcinogenesis. *J. Environ. Pathol. Toxicol.* **1**, 501–508 (1978).
91. K. L. Stemmer, E. Bingham, and W. Barkley, Pulmonary response to polyurethane dust. *Environ. Health Perspect.* **11**, 109–113 (1975).
92. J. A. Styles and J. Wilson, Comparison between *in vitro* toxicity of polymer and mineral dusts and their fibrogenicity. *Ann. Occup. Hyg.* **16**, 241–250 (1973).
93. W. J. Moorman et al., Pulmonary effects of chronic exposure to polyurethane foam (PUF), *The Toxicologist* **1**(1), abstract #290 (1981).
94. M. F. Boeniger, Nonisocyanate exposure in three flexible polyurethane manufacturing facilities. *Appl. Occup. Environ. Hyg.* **6**(11), 945–952 (1991).
95. *TLV's, Threshold Limit Values for Chemical Substances in Workroom Air*, Adopted by ACGIH for 1998–99, American Conference of Governmental Industrial Hygienists, Cincinnati, OH, 1999.
96. H. R. Hosein and S. Farkas, Risk associated with the spray application of polyurethane foam. *Am. Ind. Hyg. Assoc. J.* **42**, 663–665 (1981).

97. A. M. Kaplan et al., Microbial deterioration of polyurethane systems. In Society for Industrial Microbiology, *Dev. Ind. Microbial.* Vol. 9, American Institute of Biological Sciences, Washington, DC, 1968, pp. 201–217.
98. R. Martens and K. N. Domsch, Microbial degradation of polyurethane foams and isocyanate based polyureas to different media. *Water, Air, and Soil Pollution* **15**, 503–509 (1981).
99. J. E. Potts et al., The biodegradability of synthetic polymers. In J. Guillet, Ed., Polymer Science and Technology, Vol. 3, *Polymers and Ecological Problems*, Plenum Press, New York, 1973, pp. 61–79.
100. D. A. Meyer, Urethane elastomers. In M. Morton, Ed., *Rubber Technology*, Reinhold, New York, 1973, pp. 440–458.
101. D. A. Meyer, Polyurethane elastomers, in G. G. Winspear, Ed., *Rubber Handbook*, Vanderbilt, New York, 1968, pp. 208–220.
102. D. J. Lyman et al., Polyurethane Elastomers in surgery. *Int. J. Polym. Mater.* **5**, 211–229 (1977).
103. J. Smahel, Tissue reactions to breast implants coated with polyurethane. *Plast. Reconstr. Surg.* **61**, 80–85 (1987).
104. E. A. Emmett, Allergic Contact dermatitis in polyurethane plastic moulders. *J. Occup. Med.* **18**, 802–804 (1976).
105. F. Grimalt and C. Romaguera, New Resin allergens in shoe contact dermatitis. *Contact Dermatitis* **1**, 169–174 (1975).
106. J. A. Mock, Plastic and elastomeric coatings. In C. A. Harper, Ed., *Handbook of Plastics and Elastomers*, McGraw-Hill, New York, 1975, pp. 9-1 to 9–51.
107. H. Ulrich, Polyurethane. In *Modern Plastics Encyclopedia, 1978–79 ed.*, Vol. 55, No. 10A, McGraw-Hill, New York, 1978–1979, pp. 88, 90, 96, 97.
108. V. J. Seemann and U. Wolcke, Šber die Bildung toxischer Isocyanatd, mpfe bei der thermischen Zersetzung von Polyurethanlacken und ihren polyfunktionellen Hartern. *Zentralbl. Arbeitsschutz* **26**, 2–9 (1976); from author's English abstract.
109. O. H. Bullitt, personal communication, DuPont Company, Fabrics & Finishes Dept., Wilmington, DE, Jan. 1980.
110. J. C. Moseley and H. J. Allen, Jr., Polyurethane coating in the prevention of nickel dermatitis. *Arch. Dermatol.* **103**, 58–60 (1971).
111. K. E. Malten, Contact sensitization to letterflex urethane photoprepolymer mixture used in printing. *Contact Dermatitis* **3**, 115–121 (1977).
112. U. S. Federal Trade Commission, Rules and Regulations Under the Textile Fiber Products Identification Act, Washington, DC, effective March 3, 1960, as amended to Nov. 1, 1974.
113. R. W. Moncrieff, *Man Made Fibres*, 6th ed., Wiley, New York, 1975.
114. E. M. Hicks, Jr., A. J. Ultee, and J. Drougas, Spandex elastic fibers. *Science* **147**, 373–379 (1965).
115. E. M. Hicks et al., The production of synthetic-polymer fibres. *Text. Progr.* **3**, 1–113 (1971).
116. H. Oertel, Structure, modification potential and properties of segmented polyurethane elastomer filament yarns. *Chemiefasern /Textildustire* **29/80**, E10–E12 (1978).
117. H. L. Joseph and H. I. Maibach, Contact dermatitis from spandex brassieres. *J. Am. Med. Assoc.* **201**, 880–882 (1967).
118. E. Van Dijk, Contact dermatitis due to spandex. *Acta Dermato-Venereol.* **48**, 589–591 (1968).

119. A. M. Kligman, unpublished data, Ivy Research Laboratories, Inc., Philadelphia, PA, April 1977 (sponsored by Haskell Laboratory).
120. O. L. Dashiell, unpublished data, DuPont Company, Haskell Laboratory, Wilmington, DE, Aug. 1974.
121. C. W. Colburn, unpublished data, DuPont Company, Haskell Laboratory, Wilmington, DE, Jan. 1970.
122. S. S. Bleehen and P. Hall-Smith, Brassier depigmentation: Light and electron microscope studies. *Br. J. Dermatol.* **83**, 157–160 (1970).
123. A. A. Fisher, *Contact Dermatitis*, 2nd ed., Lea & Febiger, Philadelphia, PA, 1973.
124. D. L. Kaplan, R. Hartenstein, and J. Sutter, Biodegradation of polystyrene, poly(methylmethacrylate), and phenol formaldehyde. *Appl. Environ. Microbio.* **38**, 551–553 (1979).
125. P. A. Waitkus and G. R. Griffiths, Explosion venting of phenolic reactors-toward understanding optimum explosion vent diameters. In *Society of Plastics Engineers, Safety & Health with Plastics, National Technical Conference*, Denver, Colo., Nov. 8–10, 1977, pp. 181–186.
126. W. A. Keutgen, Phenolic resins, in *Encyclopedia of Polymer Science and Technology*, Vol. 10, Wiley, New York, 1969, pp. 1–73.
127. W. R. Sorensen and T. W. Campbell, *Preparative Methods of Polymer Chemistry*, 2nd ed., Wiley, New York, 1968.
128. *1981 Annual Book of ASTM Standards*, American Society for Testing and Materials, Philadelphia, PA, updated annually, see D 1418–79a for nomenclature for rubber polymers.
129. T. E. Steiner, Phenolic, in *Modern Plastics Encyclopedia, 1978–79 ed.*, Vol. 55, No. 10A, McGraw-Hill, New York, 1978–1979, pp. 34–36.
130. BP Chemicals, private communication, Aug. 21, 1992.
131. J. Economy and L. Wohrer, Phenolic fibers, in *Encyclopedia of Polymer Science and Technology*, Vol. 15, Wiley, New York, 1971, pp. 365–375.
132. U. S. Code of Federal Regulations, Supt. Doc., Washington, DC, 21 *CFR*-170–199, April 1, 1981, also 21 *CFR*-500–599, revised annually, guide available from *Food Chemical News*.
133. M. Tsuge, T. Miyabayashi, and S. Tanaka, Determination of free phenol in phenol resole resin by gel permeation chromatography. *Chem. Lett.* **3**, 275–278 (1973).
134. J. Foussereau, C. Cavelier, and D. Selig, Occupational eczema from *p-tertiary*-butylphenol formaldehyde resins: a review of the sensitizing resins. *Contact Dermatitis* **2**, 254–258 (1976).
135. L. E. Gaul, Absence of formaldehyde sensitivity in phenol-formaldehyde resin dermatitis. *J. Invest. Dermatol.* **48**, 485–486 (1967).
136. A. A. Fisher, Vitiligo due to contactants. *Cutis* **17**, 431–448 (1976).
137. J. H. Butala, Private communication, 1992.
138. L. Kanerva, R. Jolanki, and T. Estlander, Allergic and irritant patch test reactions for plastic and glue allergens. *Contact Dermat.* **37**(6), 301–302 (1997).
139. L. Kanerva et al., Patch-test reactions to plastic and glue allergens. *Acta Derm. Venereol.* **79**(4), 296–300 (1999).
140. R. M. Azurdia and C. M. King, Allergic contact dermatitis due to phenol-formaldedhyde resin and benzyl peroxide in swimming goggles. *Contact Dermat.* **38**(4), 234–235 (1998).
141. A. M. Downs and J. E. Sansom, Palmoplantar dermatitis may be due to phenol-formaldehyde resin contact dermatitis. *Contact Dermat.* **39**(3), 147–148 (1998).

142. L. Massone et al., Sensitization to *para-tertiary*-butylphenolformaldehyde resin. *Int. J. Dermatol.* **35**(3), 177–180 (1996).

143. M. Kiec-Swierczydska and W. Szymczk, The effects of the working environment on occupational skin disease development in workers processing rockwool. *Int. J. Occup. Med. Environ. Health* **8**(1), 17–22 (1995).

144. C. Vincenzi et al., Allergic contact dermatitis due to phenol-formaldehyde resins in a knee-guard. *Contact Dermat.* **27**(1), 54 (1992).

145. R. Hayakawa et al., Allergic contact dermatitis from *para-tertiary*-butylphenol-formaldehyde resin. *Contact Dermat.* **30**(3), 187–188 (1994).

146. M. Bruze, Diagnostic pearls in occupational dermatology. *Dermatologic Clinics* **12**(3), 485–489 (1994).

147. T. Estlander et al., Active sensitization and occupational allergic contact dermatitis caused by *para-tertiary*-butylcatechol. *Contact Dermat.* **38**(2), 96–100 (1998).

148. K. Graa Thomsen and E. Soderlund, Health effects of selected chemicals 2. Phenol formaldehyde resin. *Nord* **28**, 153–173 (1995).

149. E. Zimerson and M. Bruze, Contact allergy to the monomers of *p-tert*-butylphenol-formaldehyde resin in the guinea pig. *Contact Dermat.* **39**(5), 222–226 (1998).

150. E. Zimerson, M. Bruze, and A. Goossens, Simultaneous *p-tert*-butylphenol-formaldhyde resin and *p-tert*-butylcatechol contact allergies in man and sensitizing capacities of *p-tert*-butylphenol and *p-tert*-butylcatechol in guinea pigs. *J. Occup. Environ. Med.* **41**(1), 23–28 (1999).

151. A. Abe and T. Ishikawa, Studies on pneumoconiosis caused by organic dusts. *J. Sci. Labor* **43**, 19–41 (1967); from author's English abstract.

152. T. Sano, Pathology and pathogenesis of organic dust pneumoconiosis. *J. Sci. Labor* **43**, 3–18 (1967); from author's English abstract.

153. J. C. Pimentel, A granulomatous lung disease produced by bakelite. *Am. Rev. Resp. Dis.* **108**, 1303–1310 (1973).

154. O. M. Ratner and G. N. Tkachuk, Toxicological-hygienic study of the dust of glass-fiber reinforced plastic from phenol–formaldehyde resin. *Gig. Tr. Prof. Zabol.* **10**, 52–53 (1976); *Chem. Abstr.* **86**, 12429k.

155. J. B. Schoenberg and C. A. Mitchell, Airway disease caused by phenolic (phenol-formaldehyde) resin exposure. *Arch. Environ. Health* **30**, 574–577 (1975).

156. J. F. Gamble et al., Respiratory function and symptoms: an environmental-epidemiological study of rubber workers exposed to a phenol-formaldehyde type resin. *Am. Ind. Hyg. Assoc. J.* **37**, 499–513 (1976).

157. H. Schulz and R. Gunther, Paint processing and prevention of air pollution–investigations concerning organic substances from paint binders appearing in the drying air during paint drying in continuous ovens. *Staub-Reinhalt. Luft* **32**(12), 1–11 (1972) (English translation, U.S. Environmental Protection Agency).

158. G. F. Heron, Pyrolysis of a phenol-formaldehyde polycondensate. In Society of Chemical Industry, *High Temperature Resistance and Thermal Degradation of Polymers, S.C.I. Monograph No. 13*, MacMillan, New York, 1961, pp. 475–498.

159. Y. Tsuchiya and K. Sumi, *Toxicity of Decomposition Products-Phenolic Resin*, Division of Building Research (Building Research Note No. 106), National Research Council of Canada, Ottawa, Dec. 1975.

160. J. Q. Walker and M. A. Grayson, "Thermal Oxidative Degradation Characteristics of a Phenolic Resin," in *Society of Plastics Engineers, Safety & Health with Plastics*, National Technical Conference Denver, CO, Nov. 8–10, 1977, p. 171–173.

161. C. Lemiere et al., Occupational asthma due to formaldehyde resin dust with and without reaction to formaldehyde gas. *Eur. Resp. J.* **8**(5), 861–865 (1995).

162. R. Kanaka et al., Studies on the teratogenic potential of *p-tert*-butylphenolformaldehyde resin in rats. *Eisei Shikenjo Hokoku* **110**, 22–26 (1992).

163. T. Itami, M. Ema, and H. Kawasaki, Teratogenic evaluation of *p-tert*-butylphenol formaldehyde resin (novolak type) in rats following oral exposure. *Drug Chem. Toxicol.* **16**(4), 369–382 (1993).

164. T. A. Solokima, Embryotoxic effects of the polymer materials. Phenylon-2s and phenol formaldehyde resin viam-B. *Sb. Tr. Nauchno-issled Inst. Gig. Tr. Polfzabol, Tiflis* **15**, 202–205 (1976).

165. J. L. Schardein, Chemically induced birth defects, Second ed., Rev. and Expanded Chapter 29. In *Plastics*. (J. L. Schardein, Ed.), Marcel Dekker, Inc., New York, 1993, Chapt. 29.

166. D. G. Farrar et al., *Development of a Protocol for the Assessment of the Toxicity of Combustion Products Resulting from the Burning of Cellular Plastics*. Final Report to the Products Research Committee, Vol. 1, UTEC 79/130, Flammability Research Center, University of Utah, Salt Lake City, UT, 1979.

167. W. H. Fried, Amino, in *Modern Plastics Encyclopedia, 1978–79 ed.*, Vol. 55, No. 10A, McGraw-Hill, New York, 1978–1979, pp. 10–13.

168. National Research Council, Assembly of Life Sciences, Board on Toxicology and Environmental Health Hazards, Committee on Toxicology, *Formaldehyde–An Assessment of Its Health Effects*. National Academy of Sciences, Washington, DC, January 1980; prepared for the Consumer Product Safety Commission.

169. U.S. Consumer Product Safety Commission, Public hearing concerning safety and health problems that may be associated with release of formaldehyde gas from urea formaldehyde (UF) foam insulation. *Fed. Reg.* **44**, 69578–69583 (1979).

170. G. Widmer, Amino resins, in *Encyclopedia of Polymer Science and Technology*, Vol. 2, Wiley, New York, 1965, pp. 1–94.

171. W. J. Rossiter, Jr. et al., *Urea-Formaldehyde Based Foam Insulation: An Assessment of Their Properties and Performance*, National Bureau of Standards, Washington, DC, 1977.

172. National Materials Advisory Board, Materials: state of the art. In *Fire Safety Aspects of Polymeric Materials*, Vol. 1, Technomic, Westport, CT, 1977.

173. *RTECS (Registry of the Toxic Effects for Chemical Substances)*. Toxicology database compiled and maintained by the National Institute for Occupational Safety and Health (NIOSH) available CCOHS, 1999.

174. S. N. J. Pang, Final report on the safety assessment of polyoxymethylene urea. *J. Am. Coll. Toxicol.* **14**(3), 204–220 (1995).

175. V. Vargha, Urea-formaldehyde resins and free formaldehyde content. *Acta Biol. Hung.* **49**(2–4), 463–475 (1998).

176. A. A. Fisher, N. B. Kanof, and E. M. Biondi, Free formaldehyde in textiles and paper. *Arch. Dermatol.* **86**, 753–756 (1962).

177. K. E. Malten, Textile finish contact hypersensitivity. *Arch. Dermatol.* **89**, 215–221 (1964).

178. S. E. O'Quinn and C. B. Kennedy, Contact dermatitis due to formaldehyde in clothing textiles. *J. Am. Med. Assoc.* **194**, 123–126 (1965).

179. H. Shellow and A. T. Altman, Dermatitis from formaldehyde resin textiles. *Arch. Dermatol.* **94**, 799–801 (1966).

180. S. Hsiao and J. E. Villaume, *Occupational Health and Safety and Environmental Aspect of Formaldehyde Resins*, Science Information Services Department, Franklin Institute Research Laboratories, Philadelphia, PA, April 1978; supported by U.S. Army Medical Research and Development Command, Fort Detrick, Frederick, MD (AD A054991/5ST).

181. R. Jolanki et al., Occupational dermatoses from exposure to epoxy resin compounds in a ski factory. *Contact Dermat.* **34**(6), 390–396 (1996).

182. S. Sommer, S. M. Wilkinson, and B. Dodman, Contact dermatitis due to urea-formaldehyde resin in shin-pads. *Contact Dermat.* **40**(3), 159–160 (1999).

183. J. G. Marks et al., *J. Am. Acad. Dermatol.* **38**(6 Pt 1), 911–918 (1998).

184. D. K. Harris, Health problems in the manufacture and use of plastics. *Br. J. Ind. Med.* **10**, 255–268 (1953).

185. L. Schwartz, Dermatitis from synthetic resins. *J. Invest. Dermatol.* **6**, 239–255 (1945).

186. I. Anderson, G. R. Lundqvist, and L. Molhave, Indoor air pollution due to chipboard used as a construction material. *Atmos. Environ.* **9**, 1121–1127 (1975).

187. U.S. Consumer Product Safety Commission, *Summary of In-Depth Investigations Urea Formaldehyde Foam Home Insulation Table*, Washington, DC, July 1978.

188. K. R. Long et al., *Problems Associated with the Use of Urea–Formaldehyde Foam for Residential Insulation, Part I: The Effects of Temperature and Humidity on Formaldehyde Release from Urea–Formaldehyde Foam Insulation*, Oak Ridge National Laboratory, Oak Ridge, TN (operated by Union Carbide Corporation for the Department of Energy), Sept. 1979; ORNL/SUB-7559/I, Contract No. W-75004-eng.26.

189. G. G. Allan, J. Dutkiewicz, and E. J. Gilmartin, Long-term stability of urea–formaldehyde foam insulation. *Environ. Sci. Technol.* **14**, 1235–1240 (1980).

190. Y. Matsumura and H. Arito, Toxic volatile components of organic soil consolidating agents. *Ind. Health* **13**, 135–149 (1975).

191. M. Chaigneau, G. Le Moan, and C. Agneray, Study of the pyrolysis of plastic materials. IX. Urea–formaldehyde resins. *Ann. Pharm. Fr.* **36**, 551–554 (1978).

192. C. D. Hollowell, J. V. Berk, and G. W. Traynor, *Impact of Reduced Infiltration and Ventilation on Indoor Air Quality in Residential Buildings*, ASHRAM Trans., 85, Part 1, 1979, preprint.

193. ACGIH (American Conference of Governmental Industrial Hygienists). *Documentation of the Threshold Limit Values-1999*, ACGIH, Cincinnati, 1999.

194. P. F. G. Gannon, Occupational asthma due to glutaraldehyde and formaldehyde in endoscopy and X ray departments. *Thorax* **50**(2), 456–459 (1995).

195. D. W. Cockcroft, V. H. Hoeppner, and J. Dolovich, Occupational asthma caused by cedar urea formaldehyde particle board. *Chest* **82**(1), 49–53 (1982).

196. G. Weislander et al., Asthma and the indoor environment: The significance of emission of formaldehyde and volatile organic compounds from newly painted indoor surfaces. *Int. Arch. Occ. Environ. Health* **69**(2), 115–124 (1997).

197. R. Becher et al., Environmental chemicals relevant for respiratory hypersensitivity: The indoor environment. *Toxicol. Lett.* **86**(2/3), 155–162 (1996).

198. T. Morikawa, Evaluation of hydrogen cyanide during combustion and pyrolysis. *J. Combust. Toxicol.* **5**, 315–330 (1978).

199. D. K. Milton et al., Endotoxin exposure-response in a fiberglass manufacturing facility. *Am. J. Indus. Med.* **29**(1), 3–13 (1996).

200. A. Krakowiak, Airway response to formaldehyde inhalation in asthmatic subjects with suspected respiratory formaldehyde sensitization. *Am. J. Indus. Med.* **33**(3), 274–281 (1998).

201. J. Smedley and D. Coggon, Health surveillance for hospital employees exposed to respiratory sensitizers. *Occ. Med.* **46**(1), 33–36 (1996).

202. F. Akbar-Khanzadeh and J. S. Mlynek, Changes in respiratory function after one and trhee hours of exposure to formaldehyde in non-smoking subjects. *Occ. Environ. Med.* **54**(5), 296–300 (1997).

203. D. Paustenbach et al., A recommended occupational exposure limit for formaldehyde based on irritation. *J. Toxicol. Environ. Health* **50**(3), 217–263.

204. B. F. Matlaga, L. P. Yasenchak, and T. N. Salthouse, Tissue Response to implanted polymers: the significance of sample shape. *U. Biomed. Mater. Res.* **10**, 391–397 (1976).

205. R. H. Moss, C. F. Jackson, and J. Seiberlich, Toxicity of Carbon monoxide and hydrogen cyanide gas mixtures, a preliminary report. *AMA Arch. Ind. Hyg. Occup. Med.* **4**, 53–64, (1951).

206. E. A. Boettner, G. L. Ball, and B. Weiss, *Combustion Products from the Incineration of Plastics*. U.S. Environmental Protection Agency, Cincinnati, Ohio, 1970, PB 222001. Available from NTIS, Springfield, VA.

207. K. Sumi and Y. Tsuchiya, Combustion products of polymeric materials containing nitrogen in their chemical structure. *J. Fire Flammability* **4**, 15–22 (1973).

208. G. Ball and E. A. Boettner, Combustion products of nitrogen-containing polymers. *Am. Chem. Soc. Div. Org. Coatings Plast. Chem.* **33**, 431–437 (1973).

209. K. Hiramatsu, Mass-spectrometric analysis of pyrolysis products of melamine–formaldehyde and urea–formaldehyde resins. *Osaka Furitsu Koghy-Shoreikan Hokoku* **43**, 28–33 (1967); *Chem. Abstr.* **69**, 28147b.

210. J. Mills, The Biodeterioration of synthetic polymers and plasticizers. *CRC Crit. Rev. Environ. Control* **4**, 341–351 (1974).

211. A. M. Kaplan, Microbial decomposition of synthetic polymeric materials. In *Proceedings of the First Interjectional Congress of AIMS*, Vol. 2, T. Hasegawa, 1975, pp. 535–545.

212. T. Kuo, J. Lai, and W. Lin, Formaldehyde content of some plastic dinnerwares in Taiwan. *Taiwan I Hsueh Hui Tsa Chih* **77**, 218–225; *Chem. Abstr.* **89**, 71749z.

213. S. N. J. Pang, Final report on the safety assessment of melamine/formaldehyde resin. *J. Am. Coll. Toxicol.* **14**(5), 373–385 (1995).

214. D. L. Holness and J. R. Nethercottdagger, Results of patch testing with a specialized collection of plastic and glue allergens. *Am. J. Contact Dermat.* **8**(2), 121–124 (1997).

215. J. F. Fowler, Jr., S. M. Skinner, and D. V. Belsito, Allergic contact dermatitis from formaldehyde resins in permanent press clothing: an underdiagnosed cause of generalized dermatitis. *J. Am. Acad. Dermatol.* **27**(6 pt 1), 962–968 (1992).

216. A. K. Srivastava et al., Clinical evaluation of workers handling melamine formaldehyde resin. *J. Toxicol - Clin. Toxicol.* **30**(4), 677–681 (1992).

217. J. L. Malo et al., Occupational asthma caused by Oak wood dust. *Chest* **108**(3), 856–858.

218. A. E. Sherr, personal communication, American Cyanamid Company, Toxicity Dept., Wayne, NJ, Feb. 1980.

219. G. J. Levinskas, L. B. Vidone, and C. B. Shaffer, unpublished data, American Cyanamid Company, Toxicity Dept., Wayne, NJ, Nov. 1960-Feb. 1961.

220. G. Armeli and F. Azimonti, Observations and prevention of occupational diseases in the production of phenolic, ureic, and melamine resins. *Med. Lav.* **59**, 534–539 (1968).

221. G. N. Mazyrov and B. Ju. Jampol'skaja, The effects of furan resins on the skin and allergic reactions of workers employed in the production of these resins. *Gigiena truda I professional'nye zabolevanija* **13**(10), 19–22 (1969).

222. D. W. Cockroft et al., Asthma caused by occupational exposure to a furan-based binder system. *J. Allergy Clin. Immunol.* **66**(6), 458–463 (1980).

223. M. Ahman et al., Impeded lung function in moulders and coremakers handling furan resin sand. *Int. Arch. Occup. Environ. Health* **63**(3), 175–180 (1991).

224. NIOSH Working Group, *Occupational Exposure to Furfuryl Alcohol, Criteria for a Recommended Standard*, NIOSH 79–133, 1979.

225. BIBRA Working Group, *Furfuryl Alcohol, Toxicity Profile*, The British Industrial Bioloical Research Association, 1989.

226. S. Fregert, Contact allergy to phenoplastics. *Contact Dermat.* **7**(3), 170 (1981).

227. P. J. Nigro and W. B. Bunn, Petroleum. In P. Harper, M.B., Schenker, and J.R. Balmes, Eds., *Occupational Environmental and Respiratory Disease*, Mosby-Year Book, Inc. St Louis, 1996, pp. 688–696.

228. M. Windholz, Ed., *The Merck Index*, 9th ed., Merck & Company, NJ, 1976.

229. D. H. Lorenz, *N*-vinyl amide polymers, in *Encyclopedia of Polymer Science and Technology*, Vol. 14, Wiley, New York, 1971, pp. 239–251.

230. L. W. Burnette, A review of the physiological properties of polyvinylpyrrolidinone. *Proc. Sci. Sect. Toilet Goods Assoc.* **38**, 1–4 (1962).

231. J. F. Borzelleca and S. L. Schwartz, unpublished critique for acceptable daily intake, Medical College of Virginia, Richmond, Virginia; critique submitted to World Health Organization by L. Blecher, GAF Corporation, Wayne, NJ, December 1979; see also S. L. Schwartz, Evaluation of the safety of providone and crospovidone. *Yakuzaijaku* **41**, 205–217 (1981).

232. W. Wessel, M. Schoog, and E. Winkler, Poly(vinylpyrrolidone)(PVP), its diagnostic, therapeutic, and technical application and consequences thereof. *Arzneim-Forsch.* **21**, 1468–1482 (1971).

233. B. V. Robinson et al., *PVP, A Critical Review of the Kinetics and Toxicology of Polyvinylpyrrolidone (Povidone)*, Lewis Publishers, Chelsea, MI, 1990.

234. L. J. Frauenfelder, Universal chromatographic-colorimetric method for the determination of trace amounts of polyvinylpyrrolidone and its copolymers in foods, beverages, laundry products, and cosmetics. *J. Assoc. Offic. Anal. Chem.* **57**, 796–800 (1974).

235. K. Ridgway and M. H. Rubinstein, The quantitative analysis of polyvinylpyrrolidinone by infrared spectrophotometry. *J. Pharm. Pharmacol.* **23**, 587–589 (1971).

236. D. Scheffner, *Tolerance and Side Effects of Various Kollidons Administered by Mouth and Their Behaviour in the Gastrointestinal Tract*, Doctoral thesis, University of Heidelberg, 1955.

237. A. A. Shelanski, M. V. Shelanski, and A. Cantor, Polyvinylpyrrolidone (PVP) as a useful adjunct in cosmetics. *J. Soc. Cosm. Chem.* **5**, 129–132 (1954).

238. BIBRA Working Group, *Polyvinylpyrrolidone, Toxicity Profile*, BIBRA Toxicology International, 1991.
239. M. A. Gonzalo-Garijo et al., Anaphylatic shock following providone. *Ann. Pharmacother.* **30**(1), 37–40 (1996).
240. A. Neumann et al., Study on the acute oral toxicity of PVP (MW 50,000) in rabbits. unpublished report to BASF, 1979.
241. R. P. Zendzian and W. R. Teeters, Acute intravenous administration of PVP in beagle dogs, unpublished report by Hazleton Laboratories for National Cancer Institute, 1970.
242. R. P. Zendzian, W. R. Teeters, and R. P. Kwapien, Acute intravenous administration of polyvinylpyrrolidone (PVP) in Rhesus monkeys, unpublished report by Hazleton Laboratories for National Cancer Institute, 1981.
243. P. Kirsch et al., *Report on a Study of the Effects of Kollidon 90 When Applied Orally to Dogs Over a 28 Day Period*, BASF Gewerbehygeine und Toxikologie, submitted to WHO by BASF, 1975.
244. M. V. Shelanski, Ninety-day feeding study with PVP K-90 in rats, unpublished report from the Industrial Biological Research and Testing Laboratories (USA) for GAF, submitted to WHO by BASF, 1959a.
245. T. T. Kuo et al., Cutaneous involvement in polyvinylpyrrolidone storage disease: a clinicopathologic study of five patients, including two patients with severe anemia. *Am. J. Surg. Pathol.* **21**(11), 1361–1367 (1997).
246. P. Dunn, Bone marrow failure and myelofibrosis in a case of PVP storage disease. *Am J. Hematol.* **57**(1), 68–71 (1998).
247. J. J. Kepes, W. Y. Chen, and Y. F. Jim, 'Mucoid dissolution' of bones and multiple pathologic fractures in a patient with past history of intravenous administration of polyvinylpyrrolidone (PVP). A case report. *Bone Miner.* **22**(1), 33–41 (1993).
248. E. D. Cherstovoi et al., Tissue reactions in children treated with hemodynamic dextran and detoxication polyvinylpyrrolidone plasma substitutes. *Arkh. Patol.* **54**(11), 21–27 (1992).
249. H. T. Hofmann and J. Peh, unpublished report for BASF, Gewerbehygeine und Toxikologie, report on testing of Kollidon CE5080 K12 (Compound No. XXV1/17-2) for prenatal toxicity in rabbits, 1977.
250. H. Zeller and J. Peh, Report on a study on the effects of Kollidon 25, batch 1229, on the prenatal toxicity with rats, unpublished report from BASF Gewerbehygeine und Toxikologie, submitted to WHO by BASF, 1976.
251. H. Zeller and J. Peh, Report on a Study on the Effects of Kollidon 90, Batch 5, on the Prenatal Toxicity with Rats, unpublished report from BASF Gewerbehygeine und Toxikologie, submitted to WHO by BASF, 1976.
252. International Agency for Research on Cancer, *IARC Monographs on the Evaluation of the Carcinogenic Risk of Chemicals to Humans; Some Monomers, Plastics and Synthetic Elastomers and Acrolein*, Vol. 19, Lyon, France, 1979.
253. D. P. Discher, Inhalation of hairspray resin-does it cause pulmonary disease? in Society of Plastics Engineers, *Safety & Health with Plastics*, National Technical Conference, Denver, CO, Nov. 8–10, 1977, pp. 21–24.
254. J. M. Gowdy and M. J. Wagstaff, Pulmonary infiltration due to aerosol thesaurosis. *Arch. Environ. Health* **25**, 101–108 (1972).

255. M. Adant, Quelques effets de l'injection intraveineuse de polyvinyl-pyrrolidone. *Arch. Int. Physiol.* **62**, 145–146 (1954).
256. Clairol Laboratories, Ames test performed on six samples of hair spray resins, unpublished report for GAF, 1978.
257. R. Bruce, unpublished report on polyvinylpyrrolidone in the Ames test, carried out in the Ontario Cancer Institute for Dr. G. Ege of the Department of Nuclear Medicine, Princess Margaret Hospital, Toronto, personal communication to GAF.
258. F. K. Kessler et al., Assessment of somatogenotoxicity of providone-iodine using two *in vitro* assays. *J. Environ. Pathol. Toxicol.* **4**, 327–335 (1980).
259. H. Zeller and C. Englehardt, Testing of Kollidon 30 for mutagenic effects in male mice after a single intraperitoneal application, dominant lethal test, unpublished report from BASF Gewerbehygiene and Toxikologie, submitted to WHO by BASF, 1977.
260. BASF, Effect of Kollidon 30 on bone marrow chromosomal abberation in Chinese hamsters. unpublished report submitted to WHO, 1980.
261. W. C. Hueper, Experimental carcinogenicity studies in macromolecular chemicals, I. neoplastic reactions in rats and mice after parenteral introduction of polyvinylpyrrolidone. *Cancer* **10**, 8–18 (1957).
262. W. C. Hueper, Carcinogenic studies on water-soluble and insoluble macromolecules. *Am. Med. Assoc. Arch. Pathol.* **67**, 589–617 (1959).
263. W. C. Hueper, Bioassay of polyvinylpyrrolidones with limited molecular weight range. *J. Natl. Cancer Inst.* **26**, 229–237 (1961).
264. BASF, Bericht uber die Prufung von Kollidon K12 and K25 auf etwaige cancerogene Wirkung, unpublished report VII/72–73, 1960.
265. M. V. Shelanski, Two year chronic oral toxicity study with PVP K-30 in rats, unpublished report from the Industrial Biological Research and Testing Laboratories (USA) for GAF. Submitted to WHO by BASF, 1957.
266. BASF, Vorlaufiger Bericht uber die Prufung von Kollidon 30, Typ K26 und Dextran auf etwaige cancerogene Wirkung, unpublished report (V-406, V-408), 1958.
267. J. V. Princiotto, E. P. Rubbacky, and V. J. Dardin, Two year feeding study in dogs with polyvinylpyrrolidone (Plasone C), unpublished report from Chemo Medical Consultants (USA) for GAF. Submitted to WHO by BASF, 1954.
268. BASF, unpublished report.
269. BASF, Chronic oral toxicity of Kollidon-90 USP XIX VERS NR 77–244 in Sprague-Dawley rats–repeated dosage over 129/138 weeks, unpublished report submitted to JECFA 1983, 1980.
270. R. H. Jackson, PBI fiber and fabric–properties and performance. *Text. Res. J.* **48**, 314–319 (1978).
271. I. N. Einhorn, D. A. Chatfield, and D. J. Wendel, Thermochemistry of polybenzimidazole foams, paper Presented at the *Third International Symposium on Analytical Pyrolysis*, Amsterdam, Sept. 1976 (Flammability Research Center, University of Utah, FRC/UU-065, UTEC 76126, March, 1976).
272. C. J. Hilado and E. M. Olcomendy, Toxicity of pyrolysis gases from some flame-resistant fabrics. *J. Combust. Toxicol.* **7**(1), 69–72 (1980).
273. C. J. Hilado, Relative toxicity of pyrolysis products of some foams and fabrics. *J. Combust. Toxicol.* **3**(1), 32–60 (1976).

274. T. J. Gair and R. J. Thimineur, Silicone. In *Modern Plastics Encyclopedia, 1978–79 ed.*, Vol. 55, No. 10A, McGraw-Hill, New York, 1978–1979, pp. 102–108.
275. J. C. Calandra et al., Health and environmental aspects of polydimethylsiloxane fluids. *Polym. Preps., Am. Chem. Soc., Div. Polym. Chem.* **17**, 1–4 (1976).
276. V. K. Rowe, H. C. Spencer, and S. L. Bass, Toxicological studies on certain commercial silicones. *U. Ind. Hyg. Toxicol.* **30**, 332–352 (1948).
277. R. Dailey, *Methylpolysilicones*, Informatics, Inc. Rockville, MD, prepared for Food and Drug Administration, Washington, DC, Bureau of Foods, PB 289396, 1978.
278. *HSDB (Hazardous Substances Databank)*, Toxicology database generated and maintained and distributed by National Library of Medicine, 1999.
279. Gloxhuber and Hecht, in ref 277.
280. M. G. Cutler et al., A lifespan study of a polydimethylsiloxane in the mouse. *Food Cosmet. Toxicol.* **12**, 443–450 (1974).
281. G. L. Kennedy, Jr., et al., Reproductive, teratologic, and mutagenic studies with some polydimethylsiloxanes. *J. Toxicol. Environ. Health* **1**, 909–920 (1976).
282. E. J. Hobbs, O. E. Fancher, and J. C. Calandra, Effect of selected organopolysiloxanes on male rat and rabbit reproductive organs. *Toxicol. Appl. Pharmacol.* **21**, 45–54 (1972).
283. E. J. Hobbs, M. L. Keplinger, and J. C. Calandra, Toxicity of polydimethylsiloxanes in certain environmental systems. *Environ. Res.* **10**, 397–406 (1975).
284. C. J. Hilado et al., Toxicity of pyrolysis gases from silicone polymers. *J. Combust. Toxicol.* **5**, 130–140 (1978).
285. N. Grassie and I. G. Macfarlane, The thermal degradation of polysiloxanes I; poly(dimethylsiloxane). *Eur. Polym. J.* **14**, 875–884 (1978).
286. N. Grassie and I. G. Macfarlane, *Degradation Reactions in Silicone Polymers*, Department of Chemistry, Glasgow, Scotland, 1976 (AD A050366).
287. N. Grassie, I. G. Macfarlane, and K. F. Francez, The thermal degradation of polysiloxanes II; poly(methylphenylsiloxane). *Eur. Polym. J.* **15**, 415–422 (1979).
288. Dow Corning Corp, private communication, 1993.
289. F. Bischoff, Organic polymer biocompatibility and toxicology. *Clin. Chem.* **18**, 869–894 (1972).
290. World Health Organization, *ADI, Silicones, 200-300 MW*, Technical Report Series, 648-XXIII/31, 1979.
291. Director, Cosmetic Ingredient Review, Final report on the safety assessment of cyclomethicone. *J. Am. College Toxicol.* **10**(1) (1991).
292. C. A. Harper, Ed., *Handbook of Plastics and Elastomers*, McGraw-Hill, New York, 1976.
293. M. E. Thelen, Submission to the Environmental Protection Agency (EPA) Office of Stratospheric Ozone Protection, Significant New Alternatives Policy (SNAP) Submission, Oct. 13, 1992, Washington, DC.
294. European Chemical Industry Ecology and Toxicology Center (ECETOX), Joint Assessment of Commodity Chemicals (JACC), unpublished data (submitted 1993 for publication), Dr. Wolfgang Haebler, ECETOX, Av. E. VanNieuwenhuyse 4 (BTC 6) B-1160 Brussels, Belgium.
295. Anonymous, NTP report on the immunotoxicity of polydimethylsiloxane (CAS No. *9016-00-6* or *63394-02-5*) in female B6C3F1 mice (contact hypersensitivity studies). (IMM90006), NTIS Report PB94-121449 (PB92-140383). Available NTIS, Springfield, VA.

296. Anonymous, NTP report on the immunotoxicity of silicone (CAS No. *9016-00-6*) in female B6C3F1 mice. (IMM89050), NTIS Report PB94-121456. Available NTIS, Springfield, VA.

297. Anonymous, NTP report on the immunotoxicity of silicone (CAS No. *9016-00-6*) in female B6C3F1 mice. (IMM89051), NTIS Report PB94-121365. Available NTIS, Springfield, VA.

298. O. Wong, A critical assessment of the relationship between silicone brest implants and connective tissue diseases. *Regul. Toxicol. Pharmacol.* **23**(1 pt 1), 74–85 (1996).

299. M. C. Hochberg et al., Lack of association between augmentation mammoplasty and systemic sclerosis (scleroderma). *Arthritis Rheum.* **39**(7), 1125–1131 (1996).

300. H. Englert, D. Morris, and L. March, Scleroderma and silicone gel breast prostheses - the Sydney study revisited. *Aust. N. Z. J. Med.* **26**(3) 349–355 (1996).

301. C. J. Burns et al., The epidemiology of scleroderma among women: Assessment of risk from exposure to silicone and silica. *J. Rheumatol.* **23**(11), 1904–1911 (1996).

302. J. A. Goldman et al., Breast implants, rheumatoid arthritis, and connective tissue diseases in a clinical practice. *J. Clin. Epidemiol.* **48**(4), 571–582 (1995).

303. J. F. Winther et al., Neurologic disease among women with breast implants. *Neurology* **50**(4), 951–955 (1998).

304. O. Nyren et al., Breast implants and risk of neurologic disease: a population-based cohort study in Sweden. *Neurology* **50**(4), 956–961 (1998).

305. N. Brautbar, A. Compbell, and A. Vojdani, Silicone breast implants and autoimmunity: causation, association, or myth? *J. Biomater. Sci. Polym. Ed.* **7**(2), 133–145 (1995).

306. T. Yoshida et al., Neurosarcoidosis following augmentation mammoplasty with silicone. *Neurol. Res.* 18(4), 319–320 (1996).

307. A. Campbell and N. Brautbar, Norplant: Systemic immunological complications - case report. *Toxicol. Ind. Health* **11**(1), 41–47 (1995).

Table 94.1. Physical and Chemical Properties of Organic Sulfur Compounds

Compound	CAS	Empirical Formula	Mol Wt.	Boiling Point (°C)	Melting Point (°C)	Specific Gravity	Water Sol[a] (g/100 mL at room temp)	Vapor Pressure, mmHg (at °C)	UEL (vol % at room temp)	LEL (vol % at room temp)	Flash Pt. (°C)	Odor/Threshold (ppb)	Hazard Rating	Classification
Methyl mercaptan	[74-93-1]	CH_4S	48.1	6	−123	0.87	2.3	1520 (26)	21.8	3.9	−18	2	3	Flammable gas
Ethyl mercaptan	[75-08-1]	C_2H_6S	62.1	36	−148	0.84	<0.1	442 (20)	18.2	2.8	−48	0.1	3	Flammable liquid
n-Propyl mercaptan	[107-03-9]	C_3H_8S	76.2	68	−113	0.84	<0.1	155 (25)			−21	1.6	3	Flammable liquid
Isopropyl mercaptan	[75-33-2]	C_3H_8S	76.2	53	−131	0.81	<0.1	454 (38)			−34		3	Flammable liquid
n-Butyl mercaptan	[109-79-5]	$C_4H_{10}S$	90.2	98	−116	0.83	<0.1	83 (38)			3	0.1–1	3	Flammable liquid
Isobutyl mercaptan	[513-44-0]	$C_4H_{10}S$	90.2	85–95	−79	0.84	<0.1				−9	0.8	3	Flammable liquid
s-Butyl mercaptan	[513-53-1]	$C_4H_{10}S$	90.2	85	−165	0.83	<0.1				−23		3	Liquid
t-Butyl mercaptan	[75-66-1]	$C_4H_{10}S$	90.2	64	−0.5	0.81	<0.1	305 (38)			−26	0.1	3	Liquid
n-Pentyl mercaptan	[110-66-7]	$C_5H_{12}S$	104.2	127 (460 mmHg)	−76	0.84	i	14 (25)			18	0.8	3	Flammable liquid
Isoamyl mercaptan	[541-31-1]	$C_5H_{12}S$	104.2	116–118	−111	0.84	i				18	8.3	3	Flammable liquid
t-Amyl mercaptan	[1679-09-0]	$C_5H_{12}S$	104.2	95–119		0.83	i				−1		3	Flammable liquid
n-Hexyl mercaptan	[111-31-9]	$C_6H_{14}S$	118.2	149–151	−81	0.84	i				20		3	Flammable liquid
n-Heptyl mercaptan	[1639-09-4]	$C_7H_{16}S$	132.3	175	−43	0.84	i				46		3	Flammable liquid
n-Octyl mercaptan	[111-88-6]	$C_8H_{18}S$	146.3	199	−49	0.84	i				69–79		3	Flammable liquid
t-Octyl mercaptan	[141-59-3]	$C_8H_{18}S$	146.3	154–166		0.85	i				43		3	Flammable liquid
n-Nonyl mercaptan	[1455-21-6]	$C_9H_{20}S$	160.2	220		0.84	i				78		3	Flammable liquid
t-Nonyl mercaptan	[25360-10-5]	$C_9H_{20}S$	160.2	188–196		0.86	i	1.2 (26)			40		3	Flammable liquid
n-Decyl mercaptan	[143-10-2]	$C_{10}H_{22}S$	174.4	241	−26	0.84	i				98		3	Flammable liquid

CHAPTER NINETY-FOUR

Organic Sulfur Compounds

Howard G. Shertzer, Ph.D.

Sulfur appears in group VI of the periodic table, just below oxygen. Similar to other second row elements, sulfur is more versatile as a reaction center than first-row counterparts. Both oxygen and sulfur tend to alter the reactivity of an adjacent carbon atom, but sulfur itself is more reactive than oxygen toward nucleophilic, electrophilic, and radical reagents. Therefore, chemical reactions that involve organic sulfur compounds tend to occur directly at sulfur. Organic sulfur compounds are economically important in every type of industry, including manufacturing, agriculture, the environment, and pharmaceuticals. The chemicals included in this chapter were selected on the basis of their common industrial uses and commercial importance, although some compounds have multiple types of use. The chemicals have been grouped according to chemical classes. The major focus is on mercaptans (also called thiols or sulfhydryls), and sulfides, chemicals with both sulfur and oxygen atoms (sulfoxides, sulfones, sulfonates, and sulfonyl chlorides), and a miscellaneous group (thiophenes and benzothiazoles). There is a paucity of toxicological data in the open literature for many of these compounds, and a wealth of scientific information that is contained in unpublished industrial documents. Many of these unpublished reports have been referred to in previous editions of Patty's. So that the important information contained in these reports is not lost to the scientific and industrial communities, they have been maintained in the current chapter and referred to in the text, rather than in the list of references.

The basic chemical and physical properties of industrially important organic sulfur compounds are summarized in Table 94.1. The distinctive odor associated with sulfur is particularly important for many of these compounds. Often thought of as pleasant at low concentrations, higher subtoxic concentrations are often extremely offensive. It is

Patty's Toxicology, Fifth Edition, Volume 7, Edited by Eula Bingham, Barbara Cohrssen, and Charles H. Powell.
ISBN 0-471-31940-6 © 2001 John Wiley & Sons, Inc.

Name	CAS	Formula	MW	BP	MP	Density	Sol.	VP	FP			State
n-Dodecyl mercaptan	[112-55-0]	C$_{12}$H$_{26}$S	202.4	267–278	−7	0.85	i	2.5 (25)	88–128		3	Flammable Liquid
t-Dodecyl mercaptan	[25103-58-6]	C$_{12}$H$_{26}$S	202.4	230–247	−8	0.85	i	<0.1 (24)	110		3	Liquid
n-Tetradecyl mercaptan	[2079-95-0]	C$_{14}$H$_{30}$S	230	176–180	6.5	0.84	i				3	Liquid
n-Hexadecyl mercaptan	[2917-26-2]	C$_{16}$H$_{34}$S	258.2	123–128	18–20		i		135		3	Liquid
n-Octadecyl mercaptan	[2885-00-9]	C$_{18}$H$_{38}$S	286.6	188	24–26	0.85	i	0.1	185		3	Solid
Allyl mercaptan	[870-23-5]	C$_3$H$_6$S	74	67–68		0.93	i		21		3	Flammable liquid
Cyclohexyl mercaptan	[1569-69-3]	C$_6$H$_{12}$S	116.2	159		0.98	i	10	43–49		3	Flammable liquid
Benzyl mercaptan	[100-53-8]	C$_7$H$_8$S	124.2	195		1.06	i		70		3	Flammable oil
Phenyl mercaptan	[108-98-5]	C$_6$H$_6$S	110.2	169	−15	1.08	i	2.0 (25)	56	0.25	3	Flammable liquid
1,2-Ethanedithiol	[540-63-6]	C$_2$H$_6$S$_2$	94.2	144–146	−41	1.12	i	4.0 (23)	40		3	Liquid
Ethylcyclohexane-dithiol	[28679-10-9]	C$_8$H$_{16}$S$_2$	176			1.06	i					
d-Limonene dimercaptan	[4802-20-4]	C$_{10}$H$_{20}$S$_2$	204	154		1.03	i					
Vinylcyclohexane-dimercaptan	[37241-32-0]	C$_8$H$_{16}$S$_2$	176			1.05	i					
Perchloromethyl mercaptan	[594-42-3]	CC$_{14}$S	185.9	147–148		1.7	i	65 (70)	None	1	3	Corrosive oil
Dimethyl sulfide	[75-18-3]	C$_2$H$_6$S	62.1	37	−83	0.85	d	15		1	3	Liquid
Dimethyl disulfide	[624-92-0]	C$_2$H$_6$S$_2$	94.2	110	−85	1.06	i	29 (25) 19.7	24		3	Flammable liquid
Diethyl sulfide	[352-93-2]	C$_4$H$_{10}$S	90.2	92–93	−102	0.84	i		−10		3	Flammable Liquid
Dipropyl disulfide	[629-19-6]	C$_6$H$_{14}$S$_2$	150.3	195	−86	0.96	i		66		3	Liquid
Dibutyl sulfide	[544-40-1]	C$_8$H$_{18}$S	146.3			0.89	i				2	Liquid
Ethylene sulfide	[420-12-2]	C$_2$H$_4$S	60.1	55–56		1.02	i	375 (25) 2.2	10		3	Liquid
Propylene sulfide	[1072-43-1]	C$_3$H$_6$S	74.2	72–75		0.95	i		10		3	Flammable liquid
Propylallyl disulfide	[2179-59-1]	C$_6$H$_{12}$S$_2$	148.3		71–72		i				1	Liquid
Dibenzyl disulfide	[150-60-7]	C$_{14}$S$_{14}$S$_2$	246.4				i				1	Solid
Dimethyl sulfone	[67-71-0]	C$_2$H$_6$O$_2$S	94.1	238	109		>1		143		3	Solid
Dibutyl sulfone	[598-04-9]	C$_8$H$_{18}$O$_2$S	178.3	287–295	43–45		>0.1		143		3	Solid
Divinyl sulfone	[77-77-0]	C$_4$H$_6$O$_2$S	118.2	90–92	−26		>0.1		102		3	Liquid
Diphenyl sulfone	[127-63-9]	C$_{12}$H$_{10}$O$_2$S	218.3	379	128	1.18	i				3	Solid

683

Table 94.1. (*Continued*)

Compound	CAS	Empirical Formula	Mol wt.	Boiling point (°C)	Melting point (°C)	Specific gravity	Water sol (g/100 mL at room temp)[a]	Vapor pressure, mmHg (at °C)	UEL (vol % at room temp)	LEL (vol % at room temp)	Flash pt. (°C)	Odor/Threshold (ppb)	Hazard rating	Classification
Diethyl sulfoxide	[70-29-1]	$C_4H_{10}OS$	106.2	89		1.015	>1				93		2	Liquid
3-Chloropropyl n-octyl sulfoxide	[3569-57-1]	$C_{11}H_{23}COS$	238.9		37-39		i						3	Solid
Allyl n-octyl sulfoxide	[3868-44-8]	$C_{11}H_{22}OS$	202.1				i							
2-methallyl n-octyl sulfoxide	[4886-36-6]	$C_{12}H_{24}OS$	216.1				i							
Methanesulfonic acid	[75-75-2]	CH_4O_3S	96.1	167	20	1.48	>1	<1 (20)			>110		3	Corrosive Liquid/solid
Methanesulfonyl chloride	[124-63-0]	CH_3ClO_2S	114.6	161	−33	1.48	d	2.1 (20)			110		D	Corrosive liquid
Ethanesulfonyl chloride	[594-44-5]	$C_2H_5ClO_2S$	128.6	177	−39	1.36	d	13 (65)			83		3	Corrosive oil
Propanesulfonyl chloride	[10147-36-1]	$C_3H_7ClO_2S$	142.6	66 (8 mmHg)	−46	1.28	d	0.5 (20)			91		3	Corrosive oil
Benzothiazole	[95-16-9]	C_7H_5NS	135.2	224	2	1.25	i				>110		3	Liquid
2-Mercaptobenzothiazole	[149-30-4]	$C_7H_5NS_2$	167.3		177–181	1.42	i						3	Solid
n-Isopropyl-2-benzothiazolesulfenamide	[10220-34-5]	$C_{10}H_{12}N_2S_2$	224		96		i						3	Solid
N,N-Diisopropyl-2-benzothiazolsulfenamide	[95-29-4]	$C_{13}H_{18}N_2S_2$	265		57								2	Solid
N-tert-Butyl-2-benzothiazolesulfenamide	[95-31-8]	$C_{11}H_{14}N_2S_2$	238.4		108		i						3	Solid
n-Cyclohexyl-2-benzothiazolesulfenamide	[95-33-0]	$C_{13}H_{16}N_2S_2$	264.4		103–104	1.27	i						3	Solid
N,N-Dicyclohexylbenzothiazolesulfenamide	[4979-32-2]	$C_{19}H_{26}N_2S_2$	346.6		95		d						1	Solid

[a] i (essentially insoluble); d (decomposes in water)

important to note that all chemicals that contain the thiol moiety emit toxic sulfur oxide fumes when heated to decomposition.

Industrial uses of organic sulfur compounds include intermediates in chemical synthesis, gas odorants (especially C_1 to C_4 mercaptans), chelating or complexing agents, catalysts, solvents, and accelerators or components of synthetic rubber products. The sulfonates are used to produce anionic detergents, surfactants, and wetting agents, and in some cases as lubricant additives. Industrial exposure limits to some important organic sulfur compounds are listed in Table 94.2.

Sulfur-containing materials are widespread in the environment and are essential for the survival of mammals and other species. Bacterial decomposition of vegetative matter contributes significantly to the atmospheric content of organic sulfur compounds, along with their release from and occurrence in natural gases and oil deposits. Validated atmospheric monitoring methods exist for several volatile organosulfur compounds, including hydrogen sulfide, carbon disulfide, methyl and ethyl mercaptans, butyl mercaptan, mercaptobenzothiazole, and tetrahydrothiophene. New methods are being developed, especially field analyses using gas chromatography-mass spectrometry, to assay mixtures of sulfur-containing compounds.

Methyl and ethyl mercaptans can be used as sources of sulfur or carbon fragments in mammalian biosynthetic pathways. Excretion of excess amounts of these compounds occurs either as unchanged material through the lungs or in urine or feces as sulfate or sulfate conjugates generated through sulfone and/or sulfoxide intermediates. Higher mercaptan homologues are not incorporated into biosynthetic pathways, although similar hepatic oxidative biotransformations result in the excretion of sulfides, sulfates, and other oxidized materials in an intermediate oxidation state. Methyl and ethyl sulfides can be excreted unchanged through the lungs but are also readily metabolized. For example, methyl sulfide is a product of the reductive metabolism of dimethyl sulfoxide.

A MERCAPTANS

The toxicological mechanisms for organic mercaptans (RSH, also known as thiols or sulfhydryls) are based on their chemical and biochemical properties. They are best known for their strong odors, most notably in *Allium* vegetables (garlic, onion, leek, shallot), rotting foodstuff (cabbage, eggs) and as the characteristic odorant for natural gas.

Mercaptans are the essential functional chemical moiety for synthesizing other organosulfur compounds. They are more acidic than the corresponding alcohols and readily form the reactive soft-nucleophilic thiolate anion (RS^-). The most significant difference between the chemistry of alcohols and mercaptans is the relative ease of mercaptan oxidation that produces (in order of oxidant strength) disulfides ($RSSR'$), sulfoxides ($RSOR'$), and sulfones (RSO_2R'). Thus, mercaptan oxidation tends to occur at the sulfur atom itself, whereas the oxidation of an alcohol usually increases the oxidation state of an adjacent carbon (as in the oxidation of organic alcohols to produce aldehydes or acids). Mercaptans are extremely important in fundamental biological processes. The thiol-containing amino acid cysteine is important in maintaining proper protein structure via disulfide bonds between two cysteine residues. Cysteine is also important in regulating

Table 94.2. Exposure Limits

Compound	CAS	OSHA PEL	NIOSH Exposure Limit	ACGIH TLV	MSHA standard	OEL-Germany	OEL-United Kingdom	OEL-Russia	OEL-Australia	OEL-The Netherlands	OEL-Finland
Methyl mercaptan	[74-93-1]	TWA 0.5 ppm	0.5 ppm	0.5 ppm	0.5 ppm	0.5 ppm	0.5 ppm	0.45 ppm	0.5 ppm	0.5 ppm	0.5 ppm
Ethyl mercaptan	[75-08-1]	TWA 0.5 ppm	0.5 ppm	0.5 ppm	0.5 ppm	0.5 ppm	0.5 ppm	0.5 ppm	0.5 ppm	0.5 ppm	0.5 ppm
Butyl mercaptan	[109-79-5]	TWA 0.5 ppm	0.5 ppm	0.5 ppm	0.5 ppm	0.5 ppm				0.5 ppm	0.5 ppm
Phenyl mercaptan	[108-98-5]		0.1 ppm	0.5 ppm			0.5 ppm		0.5 ppm	0.5 ppm	0.5 ppm
Perchloromethyl mercaptan	[594-42-3]	0.1 ppm	0.1 ppm	0.1 ppm	0.1 ppm				0.1 ppm	0.1 ppm	0.1 ppm
Propylallyl disulfide	[2179-59-1]	2 ppm	2 ppm	2 ppm	2 ppm	2 ppm			2 ppm	2 ppm	2 ppm

genetic events in cells, including DNA replication, gene transcription, and protein translation. In addition, cysteine forms the reactive moiety of the intracellular antioxidant glutathione that helps maintain the highly reduced state of cells and protects cells from potentially toxic electrophiles and radicals that may originate from metabolism or from foreign sources (xenobiotics).

Mercaptans, it has long been known, react with several heavy metals, including mercury. The formation of insoluble mercury complexes is characteristic of this class of compounds and is the source of the name *mercaptan* (*mercurium captans*, capturing mercury). Although not in current use, the dithiol 2,3-dimercapto-1-propanol (BAL, dimercaprol, British anti-Lewisite CAS Number [*59-52-9*]) was developed to protect against arsenical war gases (including Lewisite) and has been used as an antidote for exposure to arsenic, mercury, and other heavy metals. Interestingly, mercaptans such as the intracellular antioxidant glutathione activate the environmentally and occupationally important heavy metal Cr^{6+} and also the economically important solvent methylene chloride to genotoxic products.

1.0 Methyl Mercaptan

1.0.1 CAS Number: *[74-93-1]*

1.0.2 Synonyms: Mercaptomethane; methanethiol; methylthioalcohol; methyl sulfhydrate; thiomethanol

1.0.3 Trade Names: NA

1.0.4 Molecular Weight: 48.10

1.0.5 Molecular Formula: CH_4S

1.0.6 Molecular Structure:

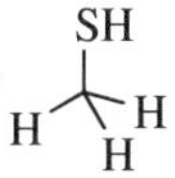

1.1 Chemical and Physical Properties

Methyl mercaptan is a flammable (flash point $-18°C$), slightly water-soluble gas that has a disagreeable odor described as rotten cabbage. It has a boiling point of 5.96°C, a vapor pressure of 1520 mmHg (26.1°C), and a vapor density of 1.66 relative to air. An extensive literature exists for methyl mercaptan, including a Public Health Service Toxicological Profile (1).

1.1.2 Odor and Warning Properties

Methyl mercaptan is a powerful odorant by stimulating adenyl cyclase activity or by stimulating the inositol phosphate second messenger system in the olfactory epithelium (26). Even though methyl mercaptan has an extremely unpleasant odor, at high concentrations, olfactory desensitization or fatigue occurs. Therefore, odor and symptoms of irritation may not adequately provide warning of high concentrations of methyl mercaptan. The human odor threshold is 2 ppt.

1.2 Production and Use

Methyl mercaptan is used as a gas odorant; an intermediate in the production of pesticides, jet fuels, and plastics, in the synthesis of methionine; and as a catalyst. Methyl mercaptan occurs naturally in a wide variety of vegetables such as garlic and onion, in "sour" gas in West Texas oil fields, and in coal tar and petroleum distillates. Methyl mercaptan can be prepared from sodium methyl sulfate and potassium hydrosulfide, catalytically from methanol and hydrogen sulfide, and from methyl chloride and sodium hydrosulfide (2, 3).

1.3 Exposure assessment

Sampling for personal or area air monitoring can be accomplished with glass fiber filters impregnated with mercuric acetate, and quantification is by gas chromatography and flame photometric detection (4, 5). There is a concern for methyl mercaptan exposure in workers at paper pulp manufacturing plants using the kraft process (5). In pulp mills, methyl mercaptan was found at levels of 0 to 15 ppm. The highest levels were at chip chutes and evaporation vacuum pumps (6).

1.3.3 Workplace Methods

NIOSH Method 2542 is recommended for determining workplace exposures to methyl mercaptan (6a).

1.4 Toxic Effects

1.4.1 Experimental Studies

1.4.1.1 Acute Toxicity. Studies have been designed to assess the concerns for high concentrations of sulfur compounds, (especially methyl mercaptan, dimethyl sulfide, and dimethyl disulfide), in the air in the production areas of paper and pulp plants (7, 8). Methyl mercaptan concentration in pulp mills increased in the winter due to restricted ventilation and was associated with worker headaches and increased frequency of sick leaves (6). An oral LD_{50} for mice of 61 mg/kg has been reported (7). The acute 4-h inhalation LC_{50} value in mice is 1664 ppm. For rats, inhalation LC_{50} values of 675 and 1680 have been reported for 4-h and 1-h exposure periods, respectively (9; Elf Atochem, Pharmacology Research Incorporated, 1977). Clinical signs following acute exposure included initial hyperactivity, tachypnea, cyanosis, muscular weakness, convulsions, respiratory depression, narcosis, skeletal muscular paralysis, and paralysis of the respiratory musculature. Increasing concentrations caused inflammation in the nasal mucosa and lungs. Further acute studies (10) reported that concentrations of 1600 to 2200 ppm produced coma in 15 min in male rats and similar effects following 30-min exposures of rats to 500, 700, and 1500 ppm, and 1-min exposures to 10,000 ppm. At 1500 ppm, the righting reflex was abolished and histopathological effects (hemorrhage and wall thickening) were noted in the lung alveoli. The 1-min exposure to 10,000 ppm caused convulsions, CNS depression, paralysis of locomotor muscles, and mucous membrane irritation; death occurred within 14 min. Inhalation exposure (7 h/day, 5 days/week) of male rats for 13 weeks at 2, 17, and 57 ppm reduced body weight increments at 57 ppm and equivocal evidence for hepatic toxicity (9).

ORGANIC SULFUR COMPOUNDS

Fish are highly sensitive to methyl mercaptan. Haydu et al. (19) determined that the no-effect level is 0.5 ppm and the minimum lethal concentration is 0.9 ppm for the coastal cutthroat trout and king and silver salmon. Methyl mercaptan was lethal at 1 ppm to white bass, yellow perch, large- and smallmouth bass, bluegills, and rock bass (20). The median lethal concentration (LC_{50}) for *Daphnia* was 7.86 ppm (21). In fish, the offensive odor of the flathead *Calliurichthys doryssus* was due to methyl mercaptan, as well as dimethyl disulfide [CAS Number 624-92-0] in the skin (22).

Methyl mercaptan is a central nervous system (CNS) depressant and acts like hydrogen sulfide on the respiratory center to produce death by respiratory paralysis. Lower concentrations produce pulmonary edema. Chief signs and symptoms are eye and mucous membrane irritation, headache, dizziness, staggering gait, nausea, and vomiting, and more or less pronounced paralysis of the locomotor muscles and respiration (1).

1.4.1.3 Pharmacokinetics, Metabolism, and Mechanisms. Methyl mercaptan is a normal mammalian metabolite formed primarily by the degradation of methionine and related substances (11–13). It is introduced exogenously into the mammalian system primarily by inhalation. Skin and eye absorption are minimal. The gas is absorbed rapidly through the respiratory system and directly translocated to the vascular system. Methyl mercaptan binds to protein and erythrocytes and is extremely effective in stabilizing erythrocyte membranes against hypotonic hemolysis (14). The compound reacts directly with collagen (15). Methyl mercaptan can be metabolized by serving as a methyl, sulfur, or methylthio donor for synthesizing amino acids and proteins. The compound is readily oxidized to carbon dioxide and inorganic sulfates (13). Intraperitoneal administration of radiolabeled materials to male rats resulted in approximately 40% excretion as carbon dioxide in expired air within 6 h of administration, and 6% in expired air as unchanged mercaptan in the first hour. Urinary excretion over an 8-h period accounted for approximately 30% of the administered dose. Tissue levels were notable in plasma protein, liver, kidney, lung, spleen, and testis (13). Methyl mercaptan reportedly inhibits mitochondrial respiration by interfering with cytochrome c oxidase, to depress hepatic, splenic, and erythrocyte catalase and to inhibit carbonic anhydrase, Na^+, K^+-ATPase, and Mg^{2+}-ATPase (14, 16–18).

1.4.2 Human Experience

1.4.2.2 Clinical Cases

1.4.2.2.1 Acute Toxicity. A human fatality attributed to exposure to methyl mercaptan involved a laborer handling tanks containing methyl mercaptan. The man was hospitalized in a coma, with acute hemolytic anemia and methemoglobinemia, and later found to have a deficiency of glucose-6-phosphate dehydrogenase (23). It was once thought that methyl mercaptan caused encephalopathy (24), but since children with hypermethioninaemia and higher blood levels of methyl mercaptan did not develop encephalopathy (25), methyl mercaptan alone is insufficient to produce this condition.

1.4.2.2.2 Chronic and Subchronic Toxicity. Oral malodor (halitosis) is a condition attributable in large part to methyl mercaptan, a major chemical contributor to this condition (27, 28). Methyl mercaptan also acts as a major mediator in promulgating periodontal disease by a variety of mechanisms (11, 28–30). High concentrations of methyl mercaptan were associated with deep and inflamed gingival crevicular sites in humans (28). The compound was implicated as a contributing factor in periodontal tissue degradation by increasing cyclic-AMP and procollagenase levels in human gingival fibroblasts (30) and by decreasing the intracellular pH of periodontal ligament cells (29). In human gingival fibroblast cultures, 10 ng/mL methyl mercaptan produced a 39% decrease in collagen synthesis and increased the degradation of newly-synthesized collagen (31). The inhibition of collagen synthesis was inhibited irreversibly (32).

1.5 Standards, Regulations, or Guidelines of Exposure

The human odor threshold is 2 ppb, and the level immediately dangerous to life or health (IDLH) is 150 ppm. The Occupational Safety and Health Administration (OSHA) permissible exposure level is a ceiling of 10 ppm/15 min and the American Conference of Governmental Industrial Hygienists (ACGIH) threshold limit value is 0.5 ppm with a STEL/C of 100 ppm (32a).

2.0 Ethyl Mercaptan

2.0.1 CAS Number: [75-08-1]

2.0.2 Synonyms: Ethanethiol; Mercaptoethane; ethyl sulfhydrate; ethyl hydrosulfide; ethyl thioalcohol; mercaptan

2.0.3 Trade Names: NA

2.0.4 Molecular Weight: 62.13

2.0.5 Molecular Formula: C_2H_6S

2.0.6 Molecular Structure: ⌒SH

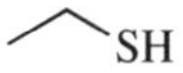

2.1 Chemical and Physical Properties

Ethyl mercaptan is a colorless, highly flammable liquid (flash point $-55°F$) and has a strong, penetrating, leek-like or garlic-like odor. It occurs naturally in illuminating gas, in "sour" Texas gas wells, and in petroleum distillates and coal tar. The boiling point is 36°C, the vapor pressure is 442 mmHg at 68°F and 838 mmHg at 100°F, and the vapor density is 2.14 relative to air.

2.1.2 Odor and Warning Properties

The human odor recognition in air appears at a concentration of 0.1 to 1.0 ppb; the odor detection in air ranges from about $2.1–6.6 \times 10^{-10}$ to 4.6×10^{-4} mg/m^3 and in water from 1.9×10^{-4} mg/L to 43.5 ppm.

2.2 Production and Use

Ethyl mercaptan is used as an intermediate and starting material in the manufacture of plastics, insecticides, and antioxidants and as an odorant for natural gas and propane. It can be detected in human blood after propane exposure (33). It can be prepared from sodium ethyl sulfate and potassium hydrosulfide or catalytically from ethanol and hydrogen sulfide (2).

2.3 Exposure Assessment

2.3.1 Air

Air can be sampled by collection in an aqueous solution of mercuric acetate–acetic acid (34), followed by extraction with ether (35) or by a chemisorbent method based on silica gel. Gas chromatographic methods have been the most effective analytical methods for the lower molecular weight thiols.

2.3.3 Workplace Methods

NIOSH Method 2542 is recommended for determining workplace exposures to ethyl mercaptan (6a).

2.4 Toxic Effects

2.4.1 Experimental Studies

2.4.1.1 Acute Toxicity. Acute LD_{50} and LC_{50} values in rats were 682 mg/kg (oral), 226 mg/kg (intraperitoneal), and 4420 ppm (4-h inhalation exposure). Mice were more susceptible to inhalation exposure and the LC_{50} was 2770 ppm (36). Exposure of male rats to a concentration of 33,000 ppm for 15 min caused the loss of the righting reflex in 50% of the animals tested (10). Respiratory function was measured in rabbits exposed to 10, 100, or 1000 ppm for 20 min. Breathing rate and expiratory volume were decreased at 1000 ppm. Other acute 4-h inhalation studies in rats found no mortality at 991 ppm (head-only) (Phillips Petroleum, Hazelton, UK, 1987) or 27 ppm (whole body) (Phillips Petroleum, Hazelton Laboratories, 1983). Male rats survived a 1-h exposure to 28,400 ppm; three of five female rats died during a 1-h exposure to 27,700 ppm (37). The acute dermal LD_{50} (24-h occluded contact) in rats was higher than 2000 mg/kg (Elf Atochem, Pharmacology Research Inc., 1977).

Instillation of 0.1 mL of undiluted material into the conjunctival sac of rabbits caused slight irritation (36). Dermal application in rats and rabbits caused pain and skin discoloration in rats but no persistent irritation was noted. Moderate erythema (redness/inflammation) lasting less than 24-h was noted in rabbits after a 4-h contact period. Mice exposed to 35 ppm in two 1-min exposures showed no upper airway irritation measured by whole-body plethysmography.

Rabbits that inhaled 30 ppm for 25 min had trace amounts of ethyl mercaptan in the blood. Inhalation of 10,000 ppm for 60 min resulted in significant amounts in the blood. After the exposure ended, the amount present in the blood was very small and diminished very rapidly. Ethyl methyl sulfide and diethyl sulfide were detected in the blood and expired air.

2.4.1.2 Chronic and Subchronic Toxicity. Subcutaneous administration of 10 and 90 mg/kg to rats and rabbits daily or every other day for 1 year caused localized necrosis at the site of injection. Rabbits showed reductions in red blood cells and hemoglobin and increases in leukocytes and reticulocytes. The most marked changes were noted in the spleen and included hyperemia, dilation of the sinusoids, deposits of hemosiderin, fibrosis, erythrocyte destruction, and increased hematopoiesis. Some of the histopathological changes were also seen in control animals and may have been due to the localized injection site deterioration. Minor microscopic changes were also noted in the liver, lungs, kidney, and testes, and were also possibly incidental to the generalized debilitation caused by chronic subcutaneous injection. However, some of the hemolysis may have been due to thiol-disulfide redox cycling with free-radical formation and specific toxicity to the hemoglobin in erythrocytes (38).

2.4.1.3 Pharmacokinetics, Metabolism, and Mechanisms. A study of the metabolism of ethyl mercaptan in mice and guinea pigs showed that the sulfur was excreted mainly as inorganic sulfate. Organic metabolites, ethyl methyl sulfone, and an unidentified product constituted 10 to 20% of the sulfur excreted in the urine. It was postulated that oxidation converted the thiol to the sulfide and then to the sulfone (39). Ethyl mercaptan decreases Na,K-ATPase in rat brain (40).

2.4.1.6 Genetic and Related Cellular Effects Studies. Ethyl mercaptan was negative in the Ames *Salmonella typhimurium* assay, produced equivocal results in the mouse lymphoma forward mutation assay, and elicited a positive response in the Chinese hamster ovary sister chromatid exchange assay (Phillips Petroleum, Hazelton Laboratories, 1983 and 1984).

2.5 Standards, Regulations, or Guidelines of Exposure

The OSHA permissible exposure limit is a ceiling of 10 ppm/15 min and the ACGIH time-weighted average (TWA) TLV is 0.5 ppm. The NIOSH IDLH is 500 ppm and the NIOSH REL is 0.5 ppm (32a).

3.0 Propyl Mercaptan

3.0.1 CAS Number: *[107-03-9]*

3.0.2 Synonyms: 1-Propanethiol; 1-mercaptopropane; *n*-propyl mercaptan; propanethiol

3.0.3 Trade Name: NA

3.0.4 Molecular Weight: 76.16

3.0.5 Molecular Formula: C_3H_8S

3.0.6 Molecular Structure: ⁀⁀SH

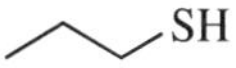

3.1 Chemical and Physical Properties

Propyl mercaptan is a colorless, flammable, offensive smelling liquid with a characteristic odor of cabbage. The flash point and boiling point are $-5°F$ and 67 to 68°C, respectively; the vapor pressure is 265 mmHg at 100°F. Its specific gravity is 0.842, and it is soluble in water.

3.1.2 Odor and Warning Properties

The odor of propyl mercaptan is very faint. It has an odor threshold close to 0.0016 ppm, and it is easily noticeable at 0.36 ppm.

3.2 Production and Use

Propyl mercaptan is used as a synthetic flavoring agent, a chemical intermediate, and a gas odorant. Methods of synthesis include the reaction of propyl alcohol and hydrogen sulfide, propyl alcohol or propyl disulfide and naphthalene, or *n*-propyl chloride and potassium hydrosulfide.

3.3 Exposure Assessment

Methods of sample collection include absorption in mercuric cyanide solution and direct gas chromatography from a gas sampling bottle (41). Detection and quantitation are best optimized by gas chromatography (42).

3.4 Toxic Effects

3.4.1 Experimental Studies

3.4.1.1 Acute Toxicity. Fairchild and Stokinger (36) determined the acute toxicity of propyl mercaptan. They reported an oral LD_{50} of 1790 mg/kg for the rat, an intraperitoneal LD_{50} of 515 mg/kg for the rat, and an estimated inhalation LC_{50} of about 7300 ppm for the rat, and 4010 ppm for the mouse for 4-h exposures. Other investigators (Phillips Petroleum, Utah Biomedical Testing Laboratory, 1981 and 54) reported oral LD_{50} values of 3000 mg/kg and 2.22 mL/kg for the rat, which were somewhat higher than the earlier study, although the results indicate relatively low oral toxicity.

An acute inhalation study in rats found that the lethal concentration is higher than 5.66 mg/L (approximately 1818 ppm), although some aerosol was present in the exposure chamber for a 4-h exposure. No deaths occurred, but signs of respiratory and ocular irritation were evident during the exposure (Elf Atochem, Huntingdon Research Centre Ltd., 1987). In another study the 4-h inhalation LC_{50} in rats was 8170 ppm (Phillips Petroleum, Utah Biomedical Testing Laboratory, 1981). During exposures, clinical signs included depressed activity, squinting, ataxia, shaking movements, and labored breathing. A depressed appearance persisted during the 14-day observation period.

Three acute dermal studies in rabbits all resulted in a 24-h dermal LD$_{50}$ larger than 2000 mg/kg (Elf Atochem, Pharmacology Research Inc., 1976; Elf Atochem, Product Safety Laboratories, 1985; Phillips Petroleum, Utah Biomedical Testing Laboratories, 1982). The application sites were typically erythematous, with thickened skin, and in some cases exhibited eschar formation. Propyl mercaptan produced severe eye irritation when instilled into rabbit eyes (36). Severe eye irritation occurred in rabbit eyes rinsed either 10 or 60 s following instillation (Elf Atochem, Pharmacology Research Incorporated, 1958). In another rabbit eye irritation study, both washed and unwashed eyes showed dark redness, swelling, and discharge at 1 h. Propyl mercaptan is a slight eye irritant (Phillips Petroleum, Utah Biomedical Testing Laboratory, 1981). Exposure of animals to high concentrations of propyl mercaptan vapor induced rubbing and closing of the eyes in about 15 min.

A 4-h skin test in rabbits was negative for corrosivity (Elf Atochem, Pharmacology Research Incorporated, 1976). In the 24-h occluded dermal toxicity test, 2000 mg/kg produced gray mottled skin which became dry, thick, and corrugated after 4 days. By day 7, cracks appeared which were followed by exfoliation. Animals reportedly experienced pain immediately after application (Elf Atochem, Pharmacology Research Incorporated, 1976). In another skin irritation study, 0.5 mL applied to rabbit skin for 24 h was a slight irritant and caused erythema in three of six rabbits on both abraded and nonabraded sites. All sites were normal by day 7 (Phillips Petroleum, Utah Biomedical Testing Laboratory, 1981).

3.4.2 Human Experience

Propyl mercaptan is released from freshly chopped onions. As with the lower molecular weight mercaptans, the strong, offensive smell of propyl mercaptan can cause headache and nausea in humans. Contact with the liquid or vapor may irritate the skin, eyes, and mucous membranes of the upper respiratory tract. Acute respiratory effects were associated with propyl mercaptan exposure from potato fields treated with the pesticide ethoprop (Mocap) (43). High concentrations of vapor can cause a sense of coldness at the extremities, tachycardia, pulmonary irritation, cyanosis, respiratory paralysis, and unconsciousness.

3.5 Standards, Regulations, or Guidelines Exposure

The National Institute for Occupational Safety and Health (NIOSH) recommended exposure limit for n-propyl mercaptan in air is a 0.5 ppm ceiling for 15 min (32a).

4.0 Isopropyl Mercaptan

4.0.1 CAS Number: [75-33-2]

4.0.2 Synonyms: 2-propanethiol; 2-mercapto propane

4.0.3 Trade Names: NA

4.0.4 Molecular Weight: 76.16

4.0.5 Molecular Formula: C_3H_8S

4.0.6 Molecular Structure:

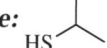

$HS-CH(CH_3)_2$

4.1 Chemical and Physical Properties

Isopropyl mercaptan is a colorless liquid with an extremely repulsive, unpleasant odor of skunk. It is highly flammable, has a flash point of $-30°F$, a boiling point of 52.6°C, vapor pressure of 454 mmHg at 100°F, and a vapor density of 2.62 relative to air a melting point of $-131°C$, and its specific gravity is 0.814.

4.2 Production and Use

Isopropyl mercaptan is used as a chemical intermediate, an odorant, and as a standard for petroleum analysis. It is manufactured by reacting of propylene and hydrogen sulfide.

4.4 Toxic Effects

4.4.1 Experimental Studies

4.4.1.1 Acute Toxicity. The acute oral median lethal dosage (LD_{50}) in rats was higher than 2000 but less than 5000 mg/kg. Clinical signs included hypotonia, ataxia, loss of righting reflex, and body weight loss. Recovery was complete in 4 days (Phillips Petroleum, Utah Biomedical Testing Laboratory, 1981). A single 2000-mg/kg dermal application to rabbit skin (24-h occluded contact) caused initial vocalization and weight losses lasting up to 4 days. No deaths occurred and the LD_{50} it was concluded was in excess of 2000 mg/kg (Elf Atochem, Pharmacology Research Inc., 1977).

Four-hour acute inhalation studies in the rat showed that the acute LC_{50} is higher than 1792 mg/kg and 5917 ppm. No deaths occurred in either study. Signs noted during exposure were typical of ocular, nasal, and pulmonary irritation. Weight gain in the exposed rats as slightly reduced 2 to 5 days (Elf Atochem, Huntingdon Research Centre, Ltd., 1987). A third 4-h inhalation study was conducted in rats at a concentration of 3899 ppm. Clinical signs noted during exposure included hyperactivity and ataxia, labored respiration, prostration, and squinted eyes. Reduced body weights were noted for 4 to 7 days postexposure. No deaths occurred and no gross pathological changes were noted at necropsy (Phillips Petroleum, Hazelton Laboratories, 1983).

Liquid isopropyl mercaptan produced slight irritation of the skin and eyes of rabbits (Elf Atochem, Pharmacology Research Inc., 1977), which recovered within 48 h. Head-only exposure of mice to a concentration of 4362 ppm for two 1-min periods did not produce upper airway irritation as measured by individual plethysmographs (Phillips Petroleum, Hazelton Laboratories, 1983).

4.4.2 Human Experience

The strong and offensive smell of mercaptans can cause headache and nausea in humans. High concentrations of mercaptan vapors can produce unconsciousness, cyanosis, a sense of coldness at the extremities, quickening of the pulse, and pulmonary edema.

Butyl Mercaptans

The butyl mercaptans exist in four isomeric forms; the n, $CH_3(CH_2)_3SH$; the *sec-*, $CH_3CH_2CH(CH_3)SH$; the iso-, $CH_3CH(CH_3)CH_2SH$; and the *tert-*, $CH_3C(CH_3)_2SH$. They all possess strong disagreeable odors similar to that of the skunk.

5.0 n-Butyl Mercaptan

5.0.1 CAS Number: [109-79-5]

5.0.2 Synonyms: 1-Butanethiol; *n*-butanethiol; 1-mercaptobutane; butanethiol; *n*-butyl thioalcohol; butane-1-thiol; bear skunk

5.0.3 Trade Names: NA

5.0.4 Molecular Weight: 90.18

5.0.5 Molecular Formula: $C_4H_{10}S$

5.0.6 Molecular Structure: ∕∖∕∖SH

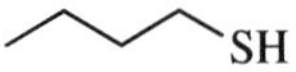

5.1 Chemical and Physical Properties

n-Butyl mercaptan is a colorless, flammable liquid (flash point 38°F) and has a strong, skunk-like odor. Its boiling point is 98.5°C, its vapor pressure is 83 mmHg, and its vapor density is 3.1 relative to air, its specific gravity 0.842 and it is soluble in water.

5.1.2 Odor and Warning Properties

n-Butyl mercaptan has a skunk-like odor. Its threshold ranges from 0.0001 to 0.001 ppm. The readily noticeable level is about 0.1 to 1 ppm.

5.2 Production and Use

n-Butyl mercaptan is used as a solvent, an intermediate in the production of pesticides, and an odorant. It can be prepared in a number of ways; two of them are reacting *n*-butylene with hydrogen sulfide in the presence of a catalyst or passing vapors of butanol and hydrogen sulfide over a thorium oxide catalyst.

5.3 Exposure Assessment

5.3.1 Workplace Methods

NIOSH Method 2542 is recommended for determining workplace exposures to n-butyl mercaptan (6a).

5.4 Toxic Effects

5.4.1 Experimental Studies

5.4.1.1 Acute Toxicity. Acute oral LD_{50} values in rats of 1500 mg/kg (36) and 1800 mg/kg (Phillips Petroleum, Industrial Biotest, 1960) have been reported, and clinical signs include sedation, ataxia, occasional muscular tremors, and labored respiration. Necropsy did not reveal any gross pathological changes. The acute intraperitoneal LD_{50} in rats was 399 mg/kg (36). The acute dermal LD_{50} in rabbits was considered higher than 34.6 g/kg. Skin reactions consisted of mild to moderate erythema that subsided in 72 to 96 h. Signs of toxicity included inactivity, weakness, lassitude, and loss of appetite (Phillips Petroleum, Industrial Biotest, 1960). Fairchild and Stokinger (36) reported 4-h LC_{50} values of 4020 ppm for rats and 2500 ppm for mice. The higher toxicity in mice is probably due to the larger respiratory minute volume per unit body weight compared to the rat. Clinical signs were similar in both species; initial stimulant effects were on respiration and activity and were followed by CNS depression, weakness, muscular paralysis, incoordination, and cyanosis. An acute 4-h LC_{50} of 6060 ppm for rats was also reported (Elf Atochem, Bio/Dynamics Inc., 1986). Signs of toxicity included respiratory abnormalities, lacrimation, prostration, and tremors. Acute inhalation exposure of rats to 200 mg/L (approximately 54,200 ppm) caused the deaths of all animals within 97 min (Phillips Petroleum, Industrial Biotest, 1960).

Slight ocular irritation was reported in rabbits by Fairchild and Stokinger (36), and marked irritation that lasted up to 4 days has been reported (Elf Atochem, Pharmacology Research Inc., 1958). In another study, iridial irritation was noted during the first 24 h and moderate to slight conjunctival irritation through 72 h (Phillips Petroleum, Industrial Biotest, 1960). Skin irritation reports vary from no irritation to slight irritation (Elf Atochem, Pharmacology Research Inc., 1958 and 1978; Phillips Petroleum, Industrial Biotest, 1960). Mucous membrane irritation occurred in rats and mice exposed by inhalation (36; Elf Atochem, International Research and Development Corporation, 1981 and 1982). *n*-Butyl mercaptan is not a sensitizer in the guinea pig (48).

The 96-h LC_{50} in channel catfish was 3600 mg/L. In 21-day studies, concentrations greater than 80 mg/L caused significant and persistent increases in methemoglobin in the catfish starting on day 2 (52). The concentration in water that can cause tainting of fish and other aquatic organisms is 60 mg/L (53). The 96-h EC_{50} for algae growth inhibition was more than 1000 mg/L.

5.4.1.2 Chronic and Subchronic Toxicity. A 90-day inhalation study of rats at levels of 9, 70, and 150 ppm (6 h/day, 5 days/week) produced no toxic effects apart from increased alveolar macrophage numbers in both sexes at 150 ppm (Phillips Petroleum, International Research and Development Corporation, 1982). Repeat dose experiments by Szabo and Reynolds (49) showed that rats given 20 mg/100 g body weight three times daily for 2 days and then 40 mg/100 g body weight once daily for days 3 and 4 did not develop duodenal ulcers but did develop some necrosis in their adrenal glands. *n*-Butyl mercaptan may affect biotransformation reactions because it binds to microsomal cytochrome P450 (50).

5.4.1.3 Pharmacokinetics, Metabolism, and Mechanisms. *n*-Butyl mercaptan is a metabolite of the pesticides *S,S,S*-tri-*n*-butylphosphorotrithioate (DEF) and *S,S,S*-tri-*n*-butylphosphorotrithioite (merphos) (45), and is released from cotton defoliants (46). In metabolism studies with DEF, single oral doses of *n*-butyl mercaptan were given to hens at dosages of 400 or 1000 mg/kg (45, 47). These dosages produced weakness, malaise, and hemolysis of erythrocytes for 1 to 2 days, but no histopathological lesions in the central or peripheral nerve tissue. A single oral dose of 500 mg/kg was given to hens and produced similar effects. Additional hematologic investigations showed that Heinz bodies and erythrocyte deformation were present in blood smears taken from hens 24 and 48 h after treatment. Hemoglobin concentration, hematocrit, and erythrocyte counts were lower than in controls, whereas methemoglobin concentration was increased. As the hens improved, the hematologic effects disappeared (45).

5.4.1.4 Reproductive and Developmental. Inhalation exposure (6 h/day) of pregnant rats (gestation days 6 to 19) and mice (gestation days 6 to 16) at levels of 10, 68, and 152 ppm elicited slight maternal effects with no teratogenicity at 152 ppm in rats. Maternal and embryo toxicity were seen in mice at both 68 and 152 ppm but no teratogenic effects at 68 ppm; mortality precluded fetal evaluation at 152 ppm (51).

5.4.1.6 Genetic and Related Cellular Effects Studies. *n*-Butyl mercaptan tested negative in the Ames *S. typhimurium* test (Phillips Petroleum, Hazelton Laboratories, 1982) and in the Chinese hamster ovary sister *in vitro* chromatid exchange assay (Phillips Petroleum, Hazelton Laboratories, 1983). It was weakly mutagenic in the mouse lymphoma forward mutational assay (Phillips Petroleum, Hazelton Laboratories, 1982).

5.4.2 Human Experience

Accidental exposure of seven workers at concentrations thought to be in the range of 50 to 500 ppm for 1 h caused symptoms of CNS toxicity. All workers experienced muscular weakness and malaise; six of the individuals experienced sweating, nausea, vomiting, and headache. Three experienced confusion, and one lapsed into a coma for 20 min. Six recovered within a day, but the most seriously affected worker experienced more persistent weakness, dizziness, vomiting, drowsiness, and depression (54).

5.5 Standards, Regulations, or Guidelines of Exposure

The NIOSH IDLH is 500 ppm. The ACGIH TWA TLV is 0.5 ppm and the OSHA permissible exposure limit (PEL) is set at 10 ppm (32a). An approved NIOSH-validated method for butyl mercaptan uses Chromosorb 104 and gas chromatography (44).

6.0 Isobutyl Mercaptan

6.0.1 CAS Number: [513-44-0]

6.0.2 Synonyms: 2-Methyl-1-propanethiol

6.0.3 Trade Names: NA

6.0.4 Molecular Weight: 90.18

6.0.5 Molecular Formula: $C_4H_{10}S$

6.0.6 Molecular Structure:

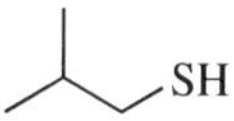

6.1 Chemical and Physical Properties

Isobutyl mercaptan (2-methyl, 1-propanethiol) is a flammable liquid (flash point 15°F) and has a heavy skunk odor. Its boiling points is 85 to 95°C.

6.1.2 Odor and Warning Properties

Isobutyl mercaptan has a skunk-like odor and its median odor threshold is 0.84 ppb.

6.3 Exposure Assessment

Air is sampled by Tenax trapping and gas chromatography (54a).

6.4 Toxic Effects

6.4.1 Experimental Studies

The oral LD_{50} in rats after 48 h observation was 7168 mg/kg, and the inhalation LC_{50} was more than 2.5% in air for the rat and the mouse. The intraperitoneal LD_{50} was 917 mg/kg for the rat after a 15-day observation period. The compound was slightly irritating to rabbit eyes (36).

7.0 sec-Butyl Mercaptan

7.0.1 CAS Number: [513-53-1]

7.0.2 Synonyms: 2-Butanethiol

7.0.3 Trade names: NA

7.0.4 Molecular Weight: 90.18

7.0.5 Molecular Formula: $C_4H_{10}S$

7.0.6 Molecular Structure:

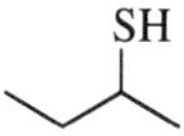

7.1 Chemical and Physical Properties

sec-Butyl mercaptan (2-butanethiol) is a flammable (flash point −10°F), colorless, obnoxious smelling liquid that has a boiling point of 84 to 85°C and a vapor density of 3.11. Its specific gravity is 2.831.

7.1.2 Odor and Warning Properties

No odor threshold data were found.

7.2 Production and Use

sec-Butyl mercaptan is used as an odorant for natural gas. It is manufactured by reacting 2-butene with hydrogen sulfide in the presence of a catalyst or from 2-iodobutene by an alcoholic KSH solution.

7.3 Exposure Assessment

Sampling methods include Tenax trapping and gas chromatography or gas chromatography with photoionization detection (55), which are identical to the *tert*-butyl mercaptan methods.

7.4 Toxic Effects

7.4.1 Experimental Studies

7.4.1.1 Acute Toxicity. The oral LD_{50} in the rat was 5176 mg/kg, the dermal lethal dose in the rat was more than 2000 mg/kg, and no skin irritation was noted after 4-h contact in rabbits. Symptoms after oral exposure included ataxia and body weight loss. Instillation into rabbit eyes caused acute pain and moderate conjunctival irritation for 2 days. Slight iridial irritation was noted for 1 day (Elf Atochem, Pharmacology Research Inc., 1981).

8.0 *tert*-Butyl Mercaptan

8.0.1 CAS Number: [75-66-1]

8.0.2 Synonyms: 2-Methyl-2-propanethiol; tert-butanethiol

8.0.3 Trade Names: NA

8.0.4 Molecular Weight: 90.18

8.0.5 Molecular Formula: $C_4H_{10}S$

8.0.6 Molecular Structure:

8.1 Chemical and Physical Properties

It is a colorless, flammable (flash point-15°F), obnoxious smelling liquid. Its boiling point is 63.7 to 64.2°C, its vapor pressure is 305 mmHg (100°F), and its vapor density is 3.1 relative to air.

8.1.2 Odor and Warning Properties

Its odor threshold is 0.08 ppb.

8.2 Production and Use

tert-Butyl mercaptan is used as a gas odorant and a chemical intermediate. It is prepared by reacting isobutylene and hydrogen sulfide or from tert-butyl iodide, zinc sulfide, and alcohol.

8.3 Exposure Assessment

Sampling methods include Tenax trapping and gas chromatography and gas chromatography with photoionization detection (55).

8.4 Toxic Effects

8.4.1 Experimental Studies

8.4.1.1 Acute Toxicity. Acute oral LD_{50} values of 4729 mg/kg (36) and 8400 mg/kg (Phillips Petroleum, Industrial Biotest, 1960) have been reported; clinical signs of toxicity included general inactivity and sedation. No gross pathological changes were noted at necropsy. Fairchild and Stokinger (36) also reported an acute intraperitoneal LD_{50} of 590 mg/kg. The acute percutaneous LD_{50} in rabbits was 20.8 g/kg. Skin reactions consisted of mild erythema and discoloration of the skin at the application site. Moderate to severe inactivity and weakness were noted during the first 3 days after skin contact ceased. Necropsy did not reveal any gross pathological changes (Phillips Petroleum, Industrial Biotest, 1960). The median lethal level by the inhalation route was 22,200 ppm for rats and 16,500 ppm for mice (36). The higher toxicity in mice was probably due to the larger respiratory minute volume per unit body weight compared to the rat. Clinical signs in both species included initial stimulant effects on respiration and activity followed by CNS depression, weakness, muscular paralysis, lack of coordination, and cyanosis. An acute inhalation study determined the 4-h LC_{50} in rats at 26,432 ppm (97.5 mg/L). Generalized inactivity and sedation were noted during the exposure period. Labored respiration and convulsions were noted in the highest exposure level groups (62.5 and 126.1 mg/L). Necropsy did not reveal any gross pathological changes (Phillips Petroleum, Industrial Biotest, 1960). Another investigator reported an acute 4-h inhalation LC_{50} for rats of 26,643 ppm. Clinical signs included those described plus ataxia and tremors (Elf Atochem, Bio/dynamics Inc., 1986).

Slight to moderate conjunctival irritation that subsided by the seventh day after instillation into the eye had been reported. Slight skin irritation to rabbit skin was also reported (Phillips Petroleum, Industrial Biotest, 1960). Mucous membrane irritation, particularly involving the eyes, was seen in mice, rats and rabbits exposed by inhalation (36).

8.4.1.2 Chronic and Subchronic Toxicity. A 90-day inhalation study in rats at levels of 9, 97, and 196 ppm (6 h/day, 5 days/week) produced no toxic effects other than increased alveolar macrophage numbers in both sexes at 97 and 196 ppm and kidney effects (renal tubular nephrosis) at levels as low as 9 ppm only in male rats (Phillips Petroleum, International Research and Development Corporation, 1982).

8.4.1.4 Reproductive and Developmental. Inhalation exposure (6 h/day) of pregnant rats (gestation days 6 to 19) and mice (gestation days 6 to 16) at levels of 11, 99, and 195 ppm elicited slight maternal effects at 195 ppm but no teratogenic effects at any level (51).

8.4.1.6 Genetic and Related Cellular Effects Studies. *tert*-Butyl mercaptan tested negative in the Ames *S. typhimurium* assay (Phillips Petroleum, Hazelton Laboratories, 1982) and in the Chinese hamster ovary *in vitro* sister chromatid exchange assay (Phillips Petroleum, Hazelton Laboratories, 1983). It tested positive in the mouse lymphoma forward mutational assay (Phillips Petroleum, Hazelton Laboratories, 1982).

9.0 Triisobutyl Mercaptan

9.0.1 CAS Number: *[25103-58-6]*

9.0.4 Molecular Weight: 200.1

9.1 Chemical and Physical Properties

Triisobutyl mercaptan has boiling range of 153 to 186°F at 5 mmHg and a specific gravity of 0.8649.

9.1.2 Odor and Warning Properties

It has a pungent irritating odor at 0.1 ppm.

9.3 Exposure Assessment: NA

9.4 Toxic Effects

9.4.1 Experimental Studies

9.4.1.1 Acute Toxicity. The acute oral LD$_{50}$ in rats was 7.6 g/kg. Clinical signs included hypoactivity, ruffled fur, and severe weight loss. No gross pathological changes were noted at necropsy (Phillips Petroleum, Industrial Biotest, 1965). No deaths occurred after dermal application of doses up to 10.2 g/kg to rabbits, although moderate to severe erythema and edema were noted at the application site. Rats exposed to a saturated air concentration (nominal concentration of 2.1 mg/L) for 4 h experienced convulsions and severe hyperpnea 3 min after exposure started. Hemorrhagic rhinitis and salivation were also noted but no deaths occurred. Instillation into rabbit eyes caused iritis and slight conjunctivitis for 24 h. All eyes were normal by 48 h. Application of undiluted material to rabbit skin caused severe irritation (Phillips Petroleum, Industrial Biotest, 1965).

10.0 *n*-Pentyl Mercaptan

10.0.1 CAS Number: *[110-66-7]*

10.0.2 Synonyms: Amyl mercaptan; 1-pentanethiol; amyl thioalcohol; n-amyl mercaptan; pentylmercaptan; pentanethiol

10.0.3 Trade Names: NA

10.0.4 Molecular Weight: 104.2

10.0.5 Molecular Formula: $C_5H_{12}S$

10.0.6 Molecular Structure: ⁓⁓SH

10.1 Chemical and Physical Properties

n-Pentyl mercaptan (1-pentanethiol, *n*-amyl mercaptan) is a flammable (flash point 65°F), offensive smelling liquid. Its boiling point is 123 to 124°C, vapor pressure is 13.8 mmHg (25°C), and vapor density is 3.59 relative to air.

10.1.2 Odor and Warning Properties

The odor threshold of *n*-pentyl mercaptan has been reported as 0.8 ppb or 0.5 to 0.4 µg/m^3.

10.2 Production and Use

n-Pentyl mercaptan is used as intermediate in the synthesis of organic sulfur compounds and as a synthetic flavoring agent. It is the chief constituent of the odorant used in gas lines to locate leaks. It is manufactured either from 2-methy-1-butylisothiourea picrate by conversion to (*S*)-2-methylbutyl disulfide, followed by reduction to the corresponding thiol using sodium metal in liquid ammonia or by mixing amyl bromide and potassium hydrosulfide in alcohol.

10.3 Exposure Assessment

For sampling, collection is via the mercuric acetate-acetic acid method, and gas chromatography with flame photometric detection is used for analysis (56).

10.4 Toxic Effects

10.4.1 Experimental Studies

Limited toxicological data are available. A range finding study found inhalation of 2000 ppm for 4 hr killed two to four rats out of six exposed (57). Sandor et al. (58) produced stomach ulcers and adrenal gland damage with *n*-pentyl mercaptan at dosages that caused 70–100% mortality in 4 to 5 days. No skin irritation was reported in rabbits after a 24-h exposure, but marked eye irritation in rabbits lasted for 3 days (Elf Atochem, Pharmacology Research Inc., 1958).

11.0 Isoamyl Mercaptan

11.0.1 CAS Number: [541-31-1]

11.0.2 Synonyms: 3-Methyl-1-butanethiol

11.0.3 Trade Names: NA

11.0.4 Molecular Weight: 104.2

11.0.5 Molecular Formula: $C_5H_{12}S$

11.0.6 Molecular Structure:

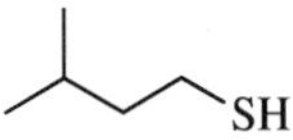

11.1 Chemical and Physical Properties

Isoamyl mercaptan (3-methyl-1-butanethiol) has been detected in the secretion gland of the skunk and in petroleum crude oils. Its boiling range is 116 to 118°C.

11.1.2 Odor and Warning Properties

Its odor is faint at 8.3 ppb, easily noticeable at 0.16 ppm, and strong at 3.1 ppm.

10.3 Exposure Assessment: NA

11.4 Toxic Effects

11.4.1 Experimental Studies

Lethal inhalation concentrations in mice were 27 mg/L (6334 ppm) in 30 min, 18 mg/L (4223 ppm) in 2 h, and 10 to 12 mg/L (2346 to 2815 ppm) in 4 h. Clinical signs included narcosis and convulsive jerking. Changes noted in the internal organs of the 18 mg/L exposure level were congestion in the liver and spleen, blood in the lung bronchi and alveoli, swollen heart fibers, cloudy swelling and partial necrosis in the kidney tubules, and hemorrhages and edema in the cerebrum and cerebellum (59).

12.0 *tert*-Amyl Mercaptan

12.0.1 CAS Number: [1679-09-0]

12.0.2 Synonyms: 2-Methyl-2-butanethiol; 1,1-dimethyl-1-propanethiol

12.0.3 Trade Names: NA

12.0.4 Molecular Weight: 104.2

12.0.5 Molecular Formula: $C_5H_{12}S$

12.0.6 Molecular Structure:

12.1 Chemical and Physical Properties

It is a flammable (flash point 30°F) liquid with a boiling range of 95 to 119°C and a strong, offensive odor.

12.2 Production and Use

tert-Amyl mercaptan is used as an odorant, a chemical intermediate, and a bacterial nutrient.

12.3 Exposure Assessment: NA

12.4 Toxic Effects

12.4.1 Experimental Studies

12.4.1.1 Acute Toxicity. An acute oral LD_{50} of 7.56 g/kg for rats has been reported. Findings noted after oral dosing consisted of generalized weakness, narcosis, decreased respiratory rate, and diarrhea. Gross pathological findings included kidney and liver ischemia plus gastrointestinal tract irritation. The acute dermal LD_{50} in rats was more than 2000 mg/kg. Rats exposed by inhalation to a nominal concentration of 238 mg/L died within 18 min. Clinical signs included narcosis, lacrimation, and convulsions. Gross pathological findings consisted of ischemia of the kidneys and liver, plus hyperemia of the lungs. Moderate eye irritation was noted in rabbits, and conjunctival irritation, persisted for 72 h postinstillation. Four-hour exposure on rabbit skin caused no signs of irritation, but 24-h contact caused slight irritation (Phillips Petroleum, Lifestream Laboratories, 1968; Elf Atochem, Pharmacology Research Inc., 1977).

13.0 Hexyl Mercaptan

13.0.1 CAS Number: [111-31-9]

13.0.2 Synonyms: Hexyl mercaptan; 1-hexanethiol

13.0.3 Trade Names: NA

13.0.4 Molecular Weight: 118.2

13.0.5 Molecular Formula: $C_6H_{14}S$

13.0.6 Molecular Structure: HS~~~

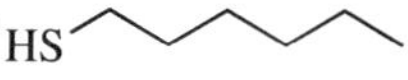

13.1 Chemical and Physical Properties

Hexyl mercaptan is a colorless liquid with an unpleasant odor and a boiling point of 149 to 150°C.

13.2 Production and Use

Hexyl mercaptan (1-hexanethiol) is used as a chemical intermediate, as an antioxidant in white oil, and in synthetic rubber processing.

13.3 Exposure Assessment

Sampling and analysis are possible with gas chromatography and flame photometric detection methods (56).

13.4 Toxic Effects

13.4.1 Experimental Studies

13.4.1.1 Acute Toxicity. Fairchild and Stokinger (36) determined the oral rat LC_{50} at 1254 mg/kg, the 4-h inhalation LC_{50} at 1080 ppm in rats and 528 ppm in mice, and the intraperitoneal LC_{50} in rats at 396 mg/kg. The predominant effects after acute dosing were CNS depression and respiratory paralysis. Inhalation study in rats reported that the 4-h LC_{50} was 1994 ppm. Eye and respiratory tract irritation, decreased motor activity, and lethargy occurred at the higher dose levels. Decedent rats showed gross lung changes upon necropsy (Elf Atochem, Pharmacology Research Inc., 1978). It was not irritating when instilled into rabbit eyes. No toxicity was reported after 24-h dermal exposure of rats to 2000 mg/kg. Four-hour skin contact on rabbits caused moderate erythema that resolved by the 24-h observation point (Elf Atochem, Huntingdon Research Centre Ltd., 1989).

13.5 Standards, Regulations, or Guidelines of Exposure

The NIOSH recommended exposure ceiling limit is 0.5 ppm/15 min (32a).

14.0 Heptyl Mercaptan

14.0.1 CAS Number: [1639-09-4]

14.0.2 Synonyms: Heptanethiol; 1-heptanethiol

14.0.3 Trade Names: NA

14.0.4 Molecular Weight: 132.3

14.0.5 Molecular Formula: $C_7H_{16}S$

14.0.6 Molecular Structure: HS~~~~

14.1 Chemical and Physical Properties

It is a highly odorous liquid with a boiling point of 177°C.

14.2 Production and Use

Heptyl mercaptan (1-heptanethiol) is used in froth flotation.

14.3 Exposure Assessment

Sampling and analysis are best accomplished by gas chromatography-flame photometric methods (56).

ORGANIC SULFUR COMPOUNDS

14.4 Toxic Effects

14.4.1 Experimental Studies

The acute intraperitoneal LD_{50} in mice, was 200 mg/kg, and the acute intravenous LD_{50} was more than 316 mg/kg. In food repellency tests with deer mice, an average amount of 1113 mg/kg/day was ingested during 3 days without killing more than 50% of the test mice (60).

14.5 Standards, Regulations, or Guidelines of Exposure

The NIOSH recommended exposure ceiling limit is 0.5 ppm/15 min (32a).

15.0 n-Octyl Mercaptan

15.0.1 CAS Number: [111-88-6]

15.0.2 Synonyms: Octanethiol; n-octanethiol; octyl mercaptan

15.0.3 Trade Names: NA

15.0.4 Molecular Weight: 146.3

15.0.5 Molecular Formula: $C_8H_{18}S$

15.0.6 Molecular Structure: ~~~~~SH

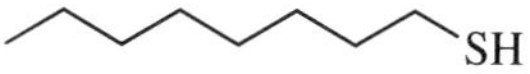

15.1 Chemical and Physical Properties

It is a colorless, combustible (flash point 69–79°C). It has a boiling point of 199°C, a vapor pressure of 1.2 mmHg at 100°F, and a vapor density greater than 1.

15.2 Production and Use

n-Octyl mercaptan (1-octanethiol) is used as a polymerization conditioner and as an intermediate in organic synthesis.

15.3 Exposure Assessment: NA

15.3.3 Workplace Methods

NIOSH Method 2510 is recommended for determining workplace exposures to n-octyl mercaptan (6a).

15.4 Toxic Effects

15.4.1 Experimental Studies

15.4.1.1 Acute Toxicity. Reported acute oral LD_{50} values in rats were 2.90 mL/kg and 2000 mg/kg. Clinical signs included lethargy, narcosis, and body weight losses. The dermal LD$_{50}$ in rats and rabbits was more than 2000 mg/kg and 2.0 mL/kg, respectively. Four-hour inhalation exposures to 40 ppm (saturated vapor concentration) and 508 ppm caused no deaths in rats. Instillation into rabbit eyes caused moderate or slight irritation of 4 days duration. This material was a slight skin irritant (score of 0.3) in rabbits (Elf Atochem, Pharmacology Research Inc., 1978; Elf Atochem, Industrial Biotest, 1974; Elf Atochem, Huntingdon Research Centre Ltd., 1987; Elf Atochem, Pharmacology Research Inc., 1958) and a moderate skin sensitizer in guinea pigs (48). Brooks et al. (61) observed that direct application to mouse skin three times a week for 2 weeks caused an epidermal hyperplasia.

15.5 Standards, Regulations, or Guidelines of Exposure

The NIOSH recommended exposure level is a ceiling of 0.5 ppm/15 min (32a).

16.0 tert-Octyl Mercaptan

16.0.1 CAS Number: [141-59-3]

16.0.2 Synonyms: 2-Pentanethiol,2,4,4-trimethyl-

16.0.3 Trade Names: NA

16.0.4 Molecular Weight: 146.3

16.0.5 Molecular Formula: C$_8$H$_{18}$S

16.0.6 Molecular Structure:

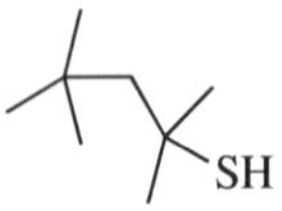

16.1 Chemical and Physical Properties

It is a colorless, combustible liquid (flash point 109°F) and has a boiling range of 154 to 166°C and a vapor density of 5.0.

16.2 Production and Use

tert-Octyl mercaptan is used in polymer modification and as a lubricant additive.

16.3 Exposure Assessment: NA

16.4 Toxic Effects

16.4.1 Experimental Studies

tert-Octyl mercaptan is one of the more highly toxic alkythiols. Reported oral LD$_{50}$ values for rats range from 52 mg/kg (Elf Atochem, MB Research Lab Inc., 1982) and 64.5 mg/kg (Phillips Petroleum, Industrial Biotest, 1961) to 85.3 mg/kg (36). The oral LD$_{50}$ for mice was 50 mg/kg (Elf Atochem, Pharmacology Research Inc., 1969 and 1978). Clinical signs noted after dosing included tremors, muscle spasms, convulsions, and seizures. In contrast to the other mercaptans, *tert*-octyl mercaptan is a potent CNS stimulant initially after dosing. Death is usually preceded, however, by CNS depression and respiratory failure (36). The acute intraperitoneal LD_{50} dosage for rats was 12.9 mg/kg. Repeated intraperitoneal (14 doses) dosing of 4.3 mg/kg during a month caused convulsions in rats, but only one death. Acute dermal LD_{50} of 1954 mg/kg for rats (36) and more than 2000 mg/kg for rabbits (Elf Atochem, MB Research Lab Inc., 1982) were reported. Signs noted in the rabbits included nasal discharge, diarrhea, lethargy, gross abnormalities in the lungs and gastrointestinal tract, and severe irritation at the site of application. Reported 4-h inhalation LD_{50} values in rats were 33 ppm (combined sexes), 59 ppm (males), and 17 ppm (females) (Elf Atochem, Temple University School of Dentistry, 1982). Other LD reports include 50 ppm and 51 ppm for rats and 47 ppm in mice (36). A 1-h LD_{50} of 61 ppm for mice was reported. Clinical signs included tremors, muscular twitching, and convulsions. Congestion or petechial hemorrhages were noted in the lungs. *tert*-Octyl mercaptan caused slight conjunctival irritation (reversible in 24 to 48 h) when instilled into rabbit eyes. It was not corrosive after 4-h contact on rabbit skin (36); (Elf Atochem, Pharmacology Research Inc., 1969 and 1978; Phillips Petroleum, Industrial Biotest, 1961).

16.5 Standards, Regulations, or Guidelines of Exposure: NA

17.0 *n*-Nonyl Mercaptan

17.0.1 CAS Number: [1455-21-6]

*17.0.2 Synonyms: n-*Nonanethiol; 1-nonanethiol

17.0.3 Trade Names: NA

17.0.4 Molecular Weight: 160.3

17.0.5 Molecular Formula: $C_9H_{20}S$

17.0.6 Molecular Structure: HS~~~~~

17.3 Exposure Assessment: NA

17.4 Toxic Effects

17.4.1 Experimental Studies

The intravenous LD_{50} in mice was more than 316 mg/kg. In food repellency studies, deer mice ingested an average amount of more than 1150 mg/kg/day over 3 days without causing death in 50% of the animals (60).

17.5 Standards, Regulations, or Guidelines of Exposure

n-Nonyl mercaptan (1-nonanethiol) has a NIOSH recommended exposure ceiling limit of 0.5 ppm/15 min (32a).

18.0 tert-Nonyl Mercaptan

18.0.1 CAS Number: [25360-10-5]

18.0.2 Synonyms: tert-Nonanethiol, Sulfole 90

18.0.3 Trade Names: NA

18.0.4 Molecular Weight: 160.3

18.0.5 Molecular Formula: $C_9H_{20}S$

18.0.6 Molecular Structure:

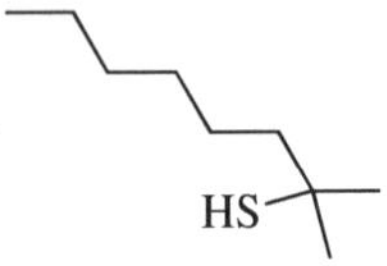

18.1 Chemical and Physical Properties

tert-Nonyl mercaptan is a colorless, combustible (flash point 147°F) liquid and has a repulsive odor. Its boiling range is 188 to 196°C, its vapor pressure is 1.2 mmHg at 78°F, and its vapor density is more than 1.

18.3 Exposure Assessment: NA

18.4 Toxic Effects

18.4.1 Experimental Studies

Reported oral LD_{50} values in rats were 3700 mg/kg and 5550 mg/kg. Symptoms included hypothermia, loss of righting reflex, disorientation, incoordination, and motor inactivity. Dermal LD_{50} values were more than 2000 mg/kg for rats and more than 10.2 g/kg for rabbits. Rabbits showed moderate erythema and edema at the application sites. Two 4-h inhalation studies have been conducted on rats. No deaths occurred at 1.97 mg/L (300 ppm: vapor only) or at a nominal vapor concentration of 7.9 mg/L. At the 300-ppm concentration, signs noted during exposure included irritation of the eyes and respiratory tract. Animals were ataxic immediately after the exposure. Slight reversible eye irritation was reported after instillation into rabbit eyes. This material was not corrosive after 4-h contact on rabbit skin, but another investigator reported severe irritation (primary irritation index of 6.1/8.0) after a 24-h occluded contact on rabbit skin (Phillips Petroleum, Industrial Biotest, 1965; Elf Atochem, Pharmacology Research Inc., 1977; Elf Atochem, Huntingdon Research Centre Ltd., 1987). The acute intraperitoneal LD_{50} was 1152 mg/kg for rats. Effects included general CNS depression, head tremors, and lack of coordination.

ORGANIC SULFUR COMPOUNDS 711

19.0 n-Decyl Mercaptan

19.0.1 CAS Number: [143-10-2]

19.0.2 Synonyms: 1-Decanethiol

19.0.3 Trade Names: NA

19.0.4 Molecular Weight: 174.3

19.0.5 Molecular Formula: $C_{10}H_{22}S$

19.0.6 Molecular Structure: ⌒⌒⌒⌒SH

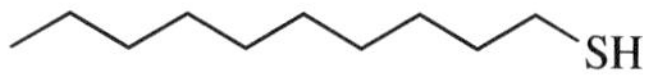

19.1 Chemical and Physical Properties

It is a combustible liquid that has a strong odor and a boiling point of 240.6°C (760 mmHg).

19.2 Production and Use

n-Decyl mercaptan is used as a chemical intermediate and in synthetic rubber processing.

19.3 Exposure Assessment

Analysis may be by gas chromatography-flame photometric methodology (56).

19.4 Toxic Effects

19.4.1 Experimental Studies

The acute oral median lethal dosage in rats was estimated at 2300 mg/kg. Symptoms observed were motor depression, disorientation, hypotonia, ataxia, and loss of righting reflex (Elf Atochem, Pharmacology Research Inc., 1978). The dermal LD_{50} in rats was more than 2000 mg/kg. n-Decyl mercaptan was corrosive to rabbit eyes and caused conjunctival, corneal, and iridial damage that did not reverse after 7 days. It caused no irritation to rabbit skin when applied under an occlusive patch for 24 h (Elf Atochem, Pharmacology Research Inc., 1958).

19.5 Standards, Regulations, or Guidelines of Exposure

The NIOSH recommended exposure level is a ceiling of 0.5 ppm/15 min (32a).

20.0 n-Dodecyl Mercaptan

20.0.1 CAS Number: [112-55-0]

20.0.2 Synonyms: Lauryl mercaptan; 1-dodecanethiol; n-dodecylmercaptan; n-lauryl mercaptan; 1-mercaptododecane

20.0.3 Trade Names: NA

20.0.4 Molecular Weight: 202.4

20.0.5 Molecular Formula: $C_{12}H_{26}S$

20.0.6 Molecular Structure: ∿∿∿∿SH

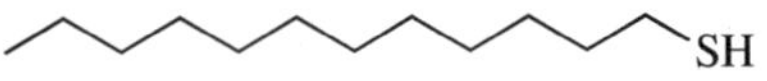

20.1 Chemical and Physical Properties

It is a colorless, repulsive smelling liquid that has boiling points of 143°C at 15 mmHg and 267 to 268°C at 760 mmHg, a flash point of 262°F, and a vapor pressure of less than 1 mmHg. The saturated vapor concentration is approximately 8 to 9 ppm. Aerosols have been used in some inhalation studies owing to the extremely low volatility of this material. The technical material is a mixture of isomers that has a flash point of 210°F.

20.2 Production and Use

n-Dodecyl mercaptan is used in pharmaceuticals, insecticides, nonionic detergents, synthetic rubber processing, and as a froth flotation agent for metal refining.

20.3 Exposure Assessment: NA

20.4 Toxic Effects

20.4.1 Experimental Studies

20.4.1.1 Acute Toxicity. A single oral dosage of 5000 mg/kg to male rats did not cause any mortality (Elf Atochem, Pharmacology Research Inc., 1977), whereas an oral LD_{50} of 4225 mg/kg was reported for mice (62). Ingestion of 138 mg/kg/day by wild deer mice for 3 days did not kill more than half of the test group (60). The acute intravenous LD_{50} for mice was more than 316 mg/kg. A 24-h occluded application of 2000 mg/kg to rat skin did not cause mortality or clinical signs during a 7-day observation period (Elf Atochem, Pharmacology Research Inc., 1977).

No irritation of abraded or non-abraded skin was seen in rabbits following 24-h occluded contact (Elf Atochem, Pharmacology Research Inc., 1958). Application of 3 mg of ether to mouse skin three times during a period of 5 days caused epidermal hyperplasia and elongation of hair follicles but no effect on sebaceous glands (61). Instillation of the undiluted material into rabbit eyes severely irritated the conjunctiva and the iris. The rabbits had not recovered from the iridial changes within 7 days. Removal of the material after 1 min or 10 sec did not moderate the responses. The material should be regarded as potentially corrosive.

Induction by the intradermal route and cutaneous application to guinea pigs has demonstrated the potential for allergic responses (48; Phillips Petroleum, Hazelton Laboratories, 1983). However, a recent Buehler sensitization test in guinea pigs gave negative results (Elf Atochem, Product Safety Labs, 1988). Delayed contact sensitization has been described in humans (63).

20.4.1.2 Chronic and Subchronic Toxicity.

A series of inhalation studies as conducted on various animal species for periods of 5 days up to 4 weeks. Rats and mice of both sexes were exposed to aerosol concentration of 0.21, 0.42, and 0.83 mg/L for 6 h/day for 5 days. All animals died at all three concentrations either on the last exposure day (day 5) or within 2 days thereafter. Death was preceded by marked body weight losses usually commencing on or about day 4, although this was not so notable at 0.83 mg/L, where a more gradual loss was noted from the first day of exposure. Food and water intake were reduced in all groups (Elf Atochem, Temple University School of Medicine, 1982).

Rats of both sexes were exposed for 6 h/day, 5 days/week for 2 weeks using a near-saturated vapor (9.4 ppm, 0.078 mg/L) and two higher levels of mixtures of vapor and aerosol. The aerosol levels were 0.7 and 3.7 mg/L. The 0.7 and 3.7 mg/L levels were reduced on days 5 and 3, respectively, to 0.32 and 2.1 mg/L. The aerodynamic mass-median diameter of the aerosol particles was 3.4 to 3.8 mm. Ocular and nasal irritation are seen at all levels and were accompanied by respiratory abnormalities at the two higher levels. All animals at 0.3 and 2.1 mg/L died or were sacrificed in extremis. Most animals at the 0.3-mg/L exposure concentration died on days 4 to 6 and those at 2.1 mg/L died on days 3 or 4. Reduced body weight gain was noted in males at the 9.4-ppm level during both weeks of treatment, whereas only a trend to reduced body weight increments was seen in females. Necropsy findings were confined to pulmonary hemorrhage and/or congestion in animals that died. No similar signs were noted in survivors to termination at 9.4 ppm. There were no organ weight differences from controls that were associated with treatment (Phillips Chemical, International Research and Development Corporation, 1983).

Groups of mice, rats, and dogs were exposed to vapors of *n*-dodecyl mercaptan for 6 h/day, 5 days/week for 4 weeks. The levels used were 0.44, 1.9, and 8.0 ppm for mice and dogs and 0.43, 1.6, and 7.3 ppm for rats. All mice at 8.0 ppm died or were sacrificed at the point of death on days 18 or 19, except for two mice that died on days 8 and 17. All rats and dogs at the high level survived. Ocular, nasal, and skin irritation and skin loss and respiratory abnormalities were noted in rats during weeks 2 through 4. Few clinical signs were noted in mice, although occasional hypoactivity and tremors were noted in five mice. Findings in dogs were predominantly skin irritation, which included peeling, cracking, and redness.

Closed/squinting eyes were also noted in all species. No treatment-related signs were noted in any species at 1.6 ppm and less. Reduced body weight increments were noted in rats at the high level. Body weight effects were not seen at lower levels in the rats or at any level in mice or dogs. Food consumption measurements indicated reductions during the first week or two for rats and mice exposed to the high level and increases in food consumption relative to control values in the last 2 weeks of the study for rats. Total leukocyte and segmented neutrophil numbers increased in male rats at the high level but not in female rats or dogs. These increases were attributed to the inflammatory response to skin irritation. All other hematologic parameters for rats and dogs were within the limits of normal expectation. Aspartate and alanine aminotransferase activities were elevated in rats of both sexes at the high level. Blood urea nitrogen was elevated in male rats at the high level. No other clinical chemistry changes were noted in rats or dogs. Skin irritation and secondary lymph node enlargement were noted in rats and dogs at the high level. Histologically, these lesions were characterized as acanthosis, hyperkeratosis, and/or inflammation in rats, and dermatitis, hyperkeratosis, or parakeratosis, and/or acanthosis in dogs. The enlarged lymph nodes microscopically showed lymphoid hyperplasia (increase in number of cells) in rats and hemorrhage or edema in dogs (Phillips Chemical, International Research and Development Corporation, 1985).

Rats exposed to 3.4 mg/L, 4 h/day for 5.5 months showed reduced growth, reduced liver and adrenal function, general congestion in internal organs, and microscopic changes in lungs, liver, kidney, heart, and brain (62).

20.4.1.4 Reproductive and Developmental. Pregnant rats and mice were exposed during gestation for 6 h/day to 7.4 ppm. Mice were exposed on days 6 to 16 and rats on days 6 to 19. All mice died or were sacrificed at the point of death during days 13 to 16. One rat was sacrificed in extremis on day 15. Overt signs of toxicity were noted in both mice (ungroomed appearance, closed/squinting eyes, decreased activity, weight loss, and red vaginal discharge) and rats (ocular and nasal irritation, unkempt appearance, ventral staining, red vaginal discharge, peeling skin on the ears, and hair loss). Uncoordinated motor activity, loss of reflexes, and loss of the use of hind limbs were noted in a few mice. These latter observations were made when the animals were in a moribund condition and are not necessarily indications of CNS toxicity. Fetal examinations were restricted for mice owing to premature mortality. Fourteen mice had implants that were developing normally. The remaining 12 dams were not pregnant. No evidence of teratogenic effects was found in rats. There was marked reduction of maternal body weight gain and overt signs of maternal toxicity. The study at and above the maternally toxic level did not show any evidence of embryo toxicity fetal toxicity, or teratogenic effects (Phillips Chemical, International Research and Development Corporation, 1985).

20.4.1.6 Genetic and Related Cellular Effects Studies. n-Dodecyl mercaptan tested negative in the Ames. *S. typhimurium* assay, in the mouse lymphoma forward mutation assay, and in the *in vitro* sister chromatid exchange assay (Phillips Petroleum, Hazelton Laboratories, 1983 and 1984).

20.4.2 Human Experience

Skin irritation has been reported in humans (Elf Atochem, Pharmacology Research Inc., 1958).

20.5 Standards, Regulations, or Guidelines of Exposure

The NIOSH recommended exposure limit is a ceiling of 0.5 ppm/15 min (32a).

21.0 *tert*-Dodecyl Mercaptan

21.0.1 CAS Number: [25103-58-6]

21.0.2 Synonyms: t-Dodecanethiol; *tert*-dodecylthiol; 2,3,3,4,4,5-hexamethyl-2-hexanethiol; Sulfole 120

21.0.3 Trade Names: NA

21.0.4 Molecular Weight: 202.4

21.0.5 Molecular Formula: $C_{12}H_{26}S$

21.0.6 Molecular Structure:

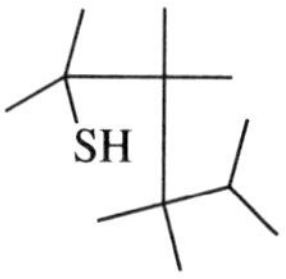

21.1 Chemical and Physical Properties

It is a colorless liquid that has a repulsive odor, a boiling range of 230 to 247°C, a flash point of 230°F, and a vapor pressure of less than 0.1 mmHg at 75°F. It has negligible solubility in water.

21.2 Production and Use

tert-Dodecyl mercaptan is used as a polymerization modifier, in ion-exchange resins, and as a plastic intermediate for the manufacture of oil additives.

21.3 Exposure Assessment

Its concentration in air is determined by GC with plasma photometric detection (64).

21.4 Toxic Effects

21.4.1 Experimental Studies

21.4.1.1 Acute Toxicity. The acute oral LD_{50} for mice was 12 g/kg, and for rats was 4.38 g/kg (Elf Atochem, Pharmacology Research Inc., 1961); the acute oral LD_{50} for rats was 6.8 g/kg for a 50% solution. Clinical signs included moderate sedation, ataxia, mild tremors, and diuresis (Phillips Petroleum, Industrial Biotest, 1961). The acute dermal LD_{50} for rabbits was 12.6 g/kg; clinical signs included inactivity, loss of appetite, weakness, and moderate erythema and skin discoloration at the application site. No deaths occurred among rats exposed by inhalation to a nominal atmospheric concentration of 12 mg/L for 4 h, although labored breathing, exophthalmos, and signs of semiconsciousness were noted. The acute intraperitoneal LD_{50} for rats was 1833 mg/kg. Clinical signs included CNS depression that sometimes lasted for 2 to 3 days (65).

When applied to the skin of a rabbit for 4 h, *tert*-dodecyl mercaptan caused local irritation in one animal but was not corrosive (Elf Atochem, Pharmacology Research Inc., 1979). In other reports, 24-h occlusive patch testing on intact and abraded sites on rabbits produced no irritation (Elf Atochem, Pharmacology Research Inc., 1961), and slight irritation was reported by another investigator (Phillips Petroleum, Industrial Biotest, 1961). Daily application to guinea pig skin caused dermatitis in 7 to 8 days. Application of a 5% preparation to rabbit skin for 20 days caused inflammation, hyperemia, edema, and desquamation. Instillation into rabbit eyes produced slight conjunctival irritation that reversed in 3 days.

tert-Dodecyl mercaptan caused sensitization reactions in animals and may have contributed to allergic reactions in humans when used as a component of certain rubber polymers. It was a slight dermal sensitizer in guinea pigs when tested in a Buehler sensitization assay (Phillips Petroleum, Hazelton Research Laboratories, 1983).

21.4.1.2 Chronic and Subchronic Toxicity. Rats exposed by inhalation (6 h/day for 2 weeks) at levels of 11, 22, and 55 ppm did not show any signs of toxicity. Two groups of these animals were then exposed for 2 additional weeks to aerosol concentrations of 3.4 and 1.3 mg/L. Mortality, respiratory abnormalities, increased liver weights, and stomach lesions were noted at both levels. Liver discoloration was noted in the 3.4 mg/L group (Phillips Chemical, International Research and Development Corporation, 1983).

Mice and dogs were exposed to 25 or 109 ppm for 6 h/day, 5 days/week for 4 weeks. Rats were similarly exposed to levels of 26 or 98 ppm. Reduced ovarian weights and reduced numbers or absence of corpora lutea were seen at the high level in mice. Increased liver weights in all species, and concomitant hepatic cell hypertrophy seen microscopically in mice and dogs were noted at the high level. Male rats at both levels exhibited renal tubular degeneration similar to that seen with branched-chain hydrocarbons. This lesion is species- and sex-specific and is not believed relevant to humans. Skin irritation and accompanying histopathological changes were noted in mice at the high level (Phillips Chemical, International Research and Development Corporation, 1985).

21.4.1.4 Reproductive and Developmental. Pregnant rats and mice were exposed to vapor concentrations of 23 and 89 ppm for 6 h/day on days 6 to 16 of gestation (mice) and days 6 to 19 of gestation (rats). Rat dams exposed to 89 ppm showed a treatment-related decrease in mean body weight gain. *tert*-Dodecyl mercaptan was not teratogenic in rats or mice (Phillips Chemical, International Research and Development Corporation, 1983).

21.4.1.6 Genetic and Related Cellular Effects Studies. *tert*-Dodecyl mercaptan tested negative in the Ames *S. typhimurium* assay and in the mouse lymphoma forward mutation assay (Phillips Petroleum, Hazelton Laboratories, 1983), as well as in the *in vitro* sister chromatid exchange assay (Phillips Petroleum, Hazelton Laboratories, 1984).

21.6 Studies on Environmental Impact

Static exposure of rainbow trout to 0.06 mg/L caused the deaths of all fish in 48 h. These fish exhibited darkened pigmentation and loss of coordination before death. No LC_{50} could be calculated because mortalities were not concentration-related (the material is poorly soluble in water). The no-observable-effect concentration was 0.01 mg/L. In a *Daphnia magna* 48 h static test, the highest concentration that did not cause immobility was 0.05 mg/L, the lowest concentration that caused 100% immobility was 1.1 mg/L, and the EC_{50} was estimated at 0.29 mg/L. Another researcher determined a static 48-h LC_{50} of 0.598 mg/L for *Daphnia magna*. In *Ceriodaphnia dubia* the LC_{50} was 0.45 mg/L, and reproductive capabilities were diminished (66). *tert*-Dodecyl mercaptan at concentrations up to 10 g/L did not inhibit the respiration of activated sludge. It was not readily biodegradable in the closed bottle test (Elf Atochem, Life Science Research, 1990; Dow Chemical, TSCA Section 8E submission, TSCATS Accession No. 45955, Fiche No. 0538045, 1992).

22.0 Tetradecyl Mercaptan

22.0.1 CAS Number: [2079-95-0]

22.1 Chemical and Physical Properties

It is a liquid with a strong odor and a boiling range of 176 to 180°C at 22 mmHg.

22.2 Production and Use

Tetradecyl mercaptan (myristyl mercaptan) is used as an organic intermediate and in synthetic rubber processing.

22.4 Toxic Effects

Limited toxicological information is available. Instillation into rabbit eyes produced mild conjunctival irritation that persisted for 5 days (Elf Atochem, Pharmacology Research Inc., 1975). A 20% solution in acetone caused intense contact sensitization in a repeat application guinea pig study (48).

23.0 *n*-Hexadecyl Mercaptan

23.0.1 CAS Number: [2917-26-2]

23.0.2 Synonyms: Cetyl mercaptan; hexadecanethiol; 1-hexadecanethiol; *n*-hexadecyl mercaptan

23.0.3 Trade Names: NA

23.0.4 Molecular Weight: 258.5

23.0.5 Molecular Formula: $C_{16}H_{34}S$

23.0.6 Molecular Structure: /\/\/\/\/\/\/\/\SH

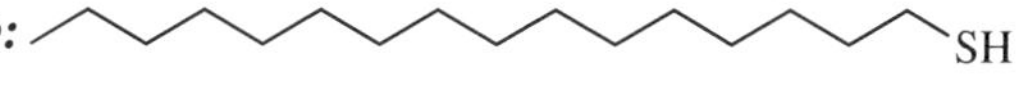

23.1 Chemical and Physical Properties

It is a liquid that has a boiling range of 185 to 190°C at 7 mmHg, a flash point of 275°F, and a vapor pressure of 0.1 mmHg.

23.2 Production and Use

n-Hexadecylmercaptan (1-hexadecanethiol, cetyl mercaptan) is used as a chemical intermediate, in synthetic rubber processing, in surface-active agents, and as a corrosion inhibitor.

23.3 Exposure Assessment

Analysis can be accomplished by gas chromatography with flame photometric detection (56).

23.4 Toxic Effects

23.4.1 Experimental Studies

The acute oral LD_{50} for rats was more than 5000 mg/kg, and the acute dermal LD_{50} for rats was more than 2000 mg/kg. Instillation into rabbit eyes caused mild conjunctival and eyelid inflammation that dissipated within 24 h. Erythema at the application site was noted following a 4-h occluded dermal application; the erythema persisted for 2 days. The intravenous LD_{50} for mice was more than 316 mg/kg (Elf Atochem, Pharmacology Research Inc., 1975 and 1978).

23.5 Standards, Regulations, or Guidelines of Exposure

The NIOSH recommended exposure limit is a ceiling of 0.5 ppm/15 min (32a).

24.0 n-Octadecyl Mercaptan

24.0.1 CAS Number: [2885-00-9]

24.0.2 Synonyms: Stearyl mercaptan; octadecyl mercaptan; 1-octadecanethiol

24.0.3 Trade Names: NA

24.0.4 Molecular Weight: 286.6

24.0.5 Molecular Formula: $C_{18}H_{38}S$

24.0.6 Molecular Structure: /\/\/\/\/\/\/\/\/\SH

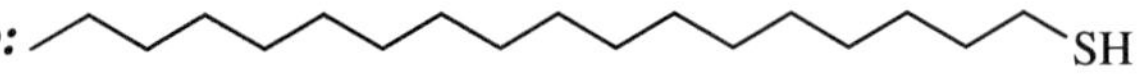

24.1 Chemical and Physical Properties

It is a greasy, white semisolid with a melting point of 25°C and a boiling range of 205 to 209°C at 11 mmHg.

24.2 Production and Use

n-Octadecyl mercaptan is used as an organic intermediate and in synthetic rubber processing.

24.3 Exposure Assessment

Analysis can be accomplished by gas chromatography with either flame photometric detection or chemiluminescence (56).

ORGANIC SULFUR COMPOUNDS

24.4 Toxic Effects

24.4.1 Experimental Studies

The acute oral LD_{50} for rats was more than 5000 mg/kg, and the acute dermal LD_{50} for rats was more than 2000 mg/kg. Four-hour contact on rabbit skin was neither corrosive nor irritating. Instillation into rabbit eyes caused only mild conjunctival irritation that dissipated within 24 h (Elf Atochem, Pharmacology Research Inc., 1977). The acute intravenous LD_{50} in mice was more than 316 mg/kg. Application of 500 mg of the undiluted material to mouse skin three times during 5 days caused an increase in epidermal weight and dermal cholesterol and a decrease in dermal D7-cholestenal. The skin was hyperkeratinized with extreme hyperplasia of the epidermis and hair follicles, and the sebaceous glands were no longer visible (61).

24.5 Standards, Regulations, or Guidelines of Exposure

The NIOSH recommended exposure limit is a ceiling of 0.5 ppm/15 min (32a).

25.0 Allyl Mercaptan

25.0.1 CAS Number: [870-23-5]

25.0.2 Synonyms: 2-Propene-1-thiol

25.0.3 Trade Names: NA

25.0.4 Molecular Weight: 74.14

25.0.5 Molecular Formula: C_3H_6S

25.0.6 Molecular Structure: HS⌒⫽

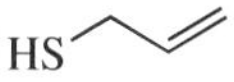

25.1 Chemical and Physical Properties

It is a water-white liquid that has a strong garlic odor and a boiling point of 67–68°C.

25.1.2 Odor and Warning Properties

Its odor is very faint at 1.5 ppb and easily noticeable at 15 ppb.

25.2 Production and Use

Allyl mercaptan is used as a pharmaceutical and rubber accelerator intermediate.

25.4 Toxic Effects

25.4.1 Experimental Studies

25.4.1.3 Pharmacokinetics, Metabolism, and Mechanisms. Allyl mercaptan is partially responsible for the odor and flavor of garlic. Some of the beneficial effects of garlic have been attributed to allyl sulfur components, such as allyl mercaptan, diallyl sulfide, CAS Number [592-88-1], and diallyl disulfide, CAS Number [2179-57-9]. These compounds decrease DNA strand breakage and mutagenicity from exposure to aflatoxin B_1 (67). In the isolated perfuse rat liver, diallyl disulfide is reduced to allyl mercaptan (68). Sandor et al. (58) reported that subcutaneous dosing three times daily for 4 days for a total dose of 89 mmol/kg (equivalent to the LD_{70-100} dose) caused moderate duodenal ulcerogenic and adrenocorticolytic effects. This regimen and a single subcutaneous LD_{50-90} dose also caused severe degenerative changes and necrosis in the thyroid. Allyl mercaptan inhibited ^{14}C-acetate incorporation into cholesterol in primary rat hepatocytes, and the IC_{50} was 450 µM (69).

25.4.1.6 Genetic and Related Cellular Effects Studies. Allyl mercaptan tested negative in the Ames *S. typhimurium* mutagenicity test and had no alkylating activity in the nitrobenzylpyridine test (70). Introduction of allyl mercaptan by stomach gavage protects C57BL/6 mice against benzo[*a*]pyrene-induced nuclear aberrations in the intestine (71).

26.0 Cyclohexyl Mercaptan

26.0.1 CAS Number: [1569-69-3]

26.0.2 Synonyms: Cyclohexanethiol

26.0.3 Trade Names: NA

26.0.4 Molecular Weight: 116.2

26.0.5 Molecular Formula: $C_6H_{12}S$

26.0.6 Molecular Structure: HS—⬡

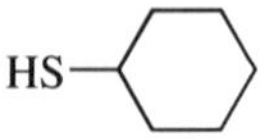

26.1 Chemical and Physical Properties

It is a clear, repulsive-smelling liquid that has a boiling range of 157 to 159°C, a flash point of 110 to 120°F, a relatively low vapor pressure of 10 mmHg, and a vapor density of 4.0 relative to air.

26.2 Production and Use

Cyclohexyl mercaptan (cyclohexanethiol) is used as a chemical intermediate and in pesticides, flavoring agents, and synthetic rubber processing.

26.3 Exposure Assessment: NA

26.4 Toxic Effects

26.4.1 Experimental Studies

26.4.1.1 Acute Toxicity. The acute oral LD_{50} in rats is reportedly between 0.56 g/kg and 1.22 g/kg (Phillips Petroleum, Lifestream Laboratories Inc., 1970). Signs noted in a second study included general weakness, lethargy, and diarrhea. Gross pathological findings included hyperemia of the lungs, intestinal mucosa, and pyloric region of the stomach. Blood was present in the gastrointestinal tract. The acute oral LD_{50} for mice was 1.9 g/kg (Elf Atochem, Pharmacology Research Inc., 1968), whereas the acute intravenous LD_{50} for mice was estimated at 316 mg/kg. In acute dermal studies, 1.0 g/kg applied to rabbit skin for 24 h caused no deaths. The acute dermal LD_{50} for rabbits was 7.82 g/kg. Gross signs included generalized weakness, plus moderate to severe erythema and edema at the application site. Skin drying was followed by necrosis and skin sloughing. Gross pathological findings consisted of hyperemia of the lungs and liver. Four hour inhalation studies of rats found that a nominal concentration of 23.5 mg/L (4943 ppm) saturated atmosphere was lethal to all rats exposed within 16 h. Clinical signs included lethargy, general weakness, ataxia, lacrimation, narcosis, and hyperventilation. Gross pathological findings included hyperemia of the lungs and liver. A lower nominal concentration of 13.3 mg/L (2798 ppm) was lethal to all rats exposed within 25 h, and signs were similar to those seen at the higher concentration. Exposure to a nominal concentration of 5.6 mg/L (1178 ppm) caused no deaths and no clinical signs. Exposure to a saturated atmosphere for 1 h was lethal to 50% of the mice exposed. Signs included sensory irritation, ataxia, and loss of the righting reflex. Instillation into rabbit eyes caused iridial irritation which resolved in 24 h and slight conjunctival irritation that cleared in 72 h. Application of cyclohexyl mercaptan to rabbit skin caused immediate pain. Cyclohexyl mercaptan is a severe skin irritant (Elf Atochem, Pharmacology Research Inc., 1968; Phillips Petroleum, Lifestream Laboratories, 1969 and 1970).

26.5 Standards, Regulations, or Guidelines of Exposure

The NIOSH recommended exposure level is a ceiling of 0.5 ppm/15 min (32a).

27.0 Benzyl Mercaptan

27.0.1 CAS Number: [100-53-8]

27.0.2 Synonyms: Benzenemethanethiol; alpha-toluenethiol; thiobenzyl alcohol; Phenylmethyl mercaptan

27.0.3 Trade Names: NA

27.0.4 Molecular Weight: 124.2

27.0.5 Molecular Formula: C_7H_8S

27.0.6 Molecular Structure: HS–CH₂–C₆H₅

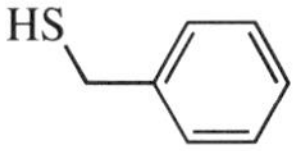

27.1 Chemical and Physical Properties

It is a colorless, combustible (flash point of 158°F) liquid with a repulsive garlic-like odor. Its boiling point is 194 to 195°C, and its vapor density is 4.28 relative to air. It can react vigorously with oxidizing materials and oxidizes in air to dibenzyl disulfide.

27.1.2 Odor and Warning Properties

Its odor threshold is 2.6 ppb; human nasal irritation starts at 4.5 ppm and eye irritation at 7.5 ppm.

27.2 Production and Use

Benzyl mercaptan is used as an odorant and a flavoring agent.

27.4 Toxic Effects

27.4.1 Experimental Studies

27.4.1.1 Acute Toxicity. The oral LD_{50} for rats was 493 mg/kg, the intraperitoneal LD_{50} 373 mg/kg, and the 4 h inhalation LC_{50} more than 235 ppm (highest concentration tested). The 4 h LC_{50} in mice was 178 ppm (36). Benzyl mercaptan administered by inhalation caused a prolonged narcotic action. In other inhalation studies, a concentration of 17 mg/L (3347 ppm) was lethal to mice in 30 min, and a concentration of 6.3 mg/L (1240 ppm) was lethal in 2 h (59). Instillation of benzyl mercaptan into rabbit eyes caused slight irritation (36); the rating was 1 on a scale of 1–10 for eye irritation.

27.4.1.2 Chronic and Subchronic Toxicity. Repeated dermal application of 5% benzyl mercaptan in ethanol and glycerin or petrolatum to mice three times weekly for 6 months produced histological abnormalities in differentiation and organization of the skin sites. However, no malignancies developed at that time or 6 months later (72). Introduction of benzyl mercaptan by stomach gavage protects C57BL/6 mice against benzo[*a*]pyrene-induced nuclear aberrations in the intestine (71).

27.4.1.6 Genetic and Related Cellular Effects Studies. Benzyl mercaptan tested negative in the Ames *S. typhimurium* mutagenicity test (73). It is readily metabolized to benzyl disulfide, and then to benzyl alcohol by the basidiomycete fungi *Coriolus versicolor* and *Tyromyces palustris* (74).

28.0 Phenyl Mercaptan

28.0.1 CAS Number: *[108-98-5]*

28.0.2 Synonyms: Thiophenol; benzenethiol; mercaptobenzene

28.0.3 Trade Names: NA

28.0.4 Molecular Weight: 110.2

28.0.5 Molecular Formula: C_6H_6S

28.0.6 Molecular Structure: HS—⟨phenyl ring⟩

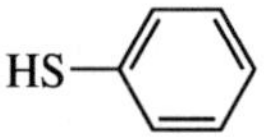

28.1 Chemical and Physical Properties

Its boiling point is 168.7°C, its melting point −14.8°C, and its flash point 132°F. Its vapor pressure is 2 mmHg at 77°F.

28.1.2 Odor and Warning Properties

Its odor is very faint at 0.26 ppb and easily noticeable at 0.72 ppm.

28.2 Production and Use

Phenyl mercaptan is used as a chemical intermediate for pesticides, pharmaceuticals, and amber dyes. It is a colorless liquid with a repulsive, penetrating, garlic-like odor. It is manufactured by reducing benzenesulfonyl chloride with zinc dust in sulfuric acid or by reacting hydrogen sulfide with chlorobenzene to produce thiophenol and diaryl sulfides.

28.3 Exposure Assessment

Two analytic methods have been reported: a gas-liquid chromatographic procedure and a gas chromatography-flame ionization detection method after derivatization with pentafluorylbenzyl bromide (75–77).

28.4 Toxic Effects

28.4.1 Experimental Studies

28.4.1.1 Acute Toxicity. Fairchild and Stokinger (36) reported an oral LD_{50} of 46.2 mg/kg for the rat, an intraperitoneal LD_{50} of 9.8 mg/kg for the rat, and a dermal LD_{50} of 300 mg/kg for the rat and 134 mg/kg for the rabbit. Dosages of 2 mL/kg applied dermally to guinea pigs were lethal in 24 h, whereas 0.65 cm^3/kg caused no mortality. Oral LD_{50} values reported for birds include 32 mg/kg for the starling and 24 mg/kg for the redwing blackbird. The 4-h LC_{50} was 33 ppm for the rat and 28 ppm for the mouse. Clinical signs after all routes of administration and in all species included increased respiration, incoordination, muscular weakness, partial skeletal muscle paralysis, cyanosis, lethargy, and mild sedation. Sedation quickly terminated upon exposure to a normal atmosphere. Histologically, changes included mild degenerative changes in the liver and kidneys, and capillary engorgement, patchy edema, and occasional hemorrhage in the lungs after inhalation exposure. A considerable amount of latent mortality was noted after inhalation exposure, and the investigators thought that the phenyl mercaptan exposure might be exacerbating latent respiratory infections in the rats.

Phenyl mercaptan was a severe eye irritant in rabbits and caused conjunctival irritation and corneal injury (36). Instillation into rabbit eyes caused moderate to severe redness, chemosis, and discharge for 3 to 4 days. The conjunctivas cleared in all rabbits by day 16. The corneal opacity increased through days 16–19 and then gradually resolved in 1.5 to 2 months. Flushing the eye with water after the exposure worsened the injury. Flushing with a dilute 0.5% silver nitrate solution first, and then flushing with large amounts of water alleviated some of the injury. Skin areas around and below the eyes were depilated for 2 to 3.5 weeks. Treatment of rabbit skin with phenyl mercaptan caused an inflammatory reaction that disappeared in 24 to 48 h (36). Skin irritation studies in guinea pigs indicated strong skin irritation after 24-h contact (Eastman Kodak, TSCA Section 8E submission, TSCATS Accession No. 19571, Fiche No. 0510336, 1968).

28.4.1.2 Chronic and Subchronic Toxicity. Repeated intraperitoneal injections of 3.5 mg/kg in nine doses to rats during a 3-week period caused fibrous thickening of the splenic capsule, enlargement of the spleen, and mild degenerative changes in the kidneys (36).

28.4.1.3 Pharmacokinetics, Metabolism, and Mechanisms. Rats dosed orally with ^{35}S-labeled phenyl mercaptan at 6 mg/kg excreted benzene-soluble and water-soluble metabolites in the urine. One identified benzene-soluble metabolite was methyl phenyl sulfone. Some water-soluble metabolites were identified as *para-* and *ortho-*hydroxylated methyl phenyl sulfone. This information suggests that phenyl mercaptan undergoes S-methylation *in vivo*, followed by oxidation of the sulfide to methyl phenyl sulfone. This is subsequently hydroxylated and further conjugated for excretion into the urine (78).

28.4.1.6 Genetic and Related Cellular Effects Studies. The EC_{50} in the Microtox system using *Photobacterium phosphorium* was 4.8 mg/mL (79). No increases in revertants over control values were noted in strains TA100 and TA98 in the Ames *Salmonella typhimurium* assay. However, concentrations greater than 25 mg/plate caused poor survival in the TA100 strain (80).

Munday and associates (38, 81) have performed *in vitro* studies with human red blood cells. Phenyl mercaptan can cause oxidant damage in the presence of hemoglobin that causes conversion of oxyhemoglobin to methemoglobin and Heinz body formation and resultant cell lysis.

28.4.2 Human Experience

In odor threshold studies using human volunteers, phenyl mercaptan caused a choking sensation in the throat, eye and nose irritation, and headache.

28.5 Standards, Regulations, or Guidelines of Exposure

The NIOSH recommended exposure level is a ceiling of 0.1 ppm.

29.0 1,2-Ethanedithiol

29.0.1 CAS Number: [540-63-6]

29.0.2 Synonyms: Dithioglycol; 1,2-ethanedithiol; ethylene dimercaptan

29.0.3 Trade Names: NA

29.0.4 Molecular Weight: 94.19

29.0.5 Molecular Formula: $C_2H_6S_2$

29.0.6 Molecular Structure: HS~~SH

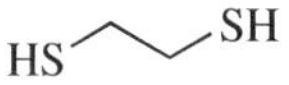

29.1 Chemical and Physical Properties

It is a clear liquid with a repulsive odor, a boiling point of 146°C, a flash point of 122°F, and a vapor pressure of 4 mmHg at 73°F. It is negligibly soluble in water.

29.1.2 Odor and Warning Properties

Its odor is very faint at 31 ppb and easily noticeable at 5.6 ppm.

29.2 Production and Use

1,2-Ethanedithiol is used as a metal chelating agent. It is manufactured by reacting ethanol, thiourea, and ethylene dibromide, and subsequent alkaline hydrolysis of the ethylene diisothiuronium bromide.

29.3 Exposure Assessment

This compound has been separated and quantitatively measured by permeation tubes/exponential dilution flasks, a solid sodium bicarbonate filter, and a flame photometric detector (81a).

29.4 Toxic Effects

29.4.1 Experimental Studies

29.4.1.1 Acute Toxicity. Several oral LD_{50} determinations in rats gave estimates that ranged between 120 mg/kg and 342 mg/kg (82; Elf Atochem, Pharmacology Research Inc., 1963; Phillips Petroleum, Hazelton Laboratories, 1982; Phillips Petroleum, Lifestream Laboratories, 1968). In general, toxicological signs noted after oral dosing included decreased activity, depression, ataxia, labored respiration, loss of righting reflex, tremors, and convulsions. Reported dermal LD_{50} values in rabbits after 24-h contact were 197 mg/kg, more than 500 mg/kg, and 1189 mg/kg. Toxicological signs included depression, weakness, labored respiration, tremors, and erythema, edema, and sloughing of the skin at the application site. A nominal 1-h LC_{50} of 583 ppm for mice was reported, and clinical signs included lack of muscular coordination, lacrimation, ataxia, labored respiration, and progressive cyanosis. A respiratory tract irritation study of mice found depressions in respiratory rates at 87 ppm (1-min exposure intervals). Acute 4-h LC_{50} values for rats ranged from 84 ppm, to 299 ppm, to 728 ppm. Toxicological signs generally included respiratory distress, signs of sensory irritation, ataxia, tremors, and convulsions. Instillation into rabbit eyes caused mild to moderate irritation that persisted through 48 h. Washing the eye after exposure shortened the time for recovery. Skin irritation studies of rabbits indicated none to slight irritation after 24-h exposure. In one primary dermal irritation study, four of six of the rabbits tested died during the 24-h exposure (Elf Atochem, Pharmacology Research Inc., 1963; Phillips Petroleum, Hazelton Laboratories, 1968, 1982 and 1983; Phillips Petroleum, Lifestream Laboratories, 1968 and 1982).

29.4.1.6 Genetic and Related Cellular Effects Studies. 1,2-Ethanedithiol tested negative in the Ames *S. typhimurium* assay, but positive in the mouse lymphoma forward mutation assay and in the *in vitro* sister chromatid exchange assay in Chinese hamster ovary cells (Phillips Petroleum, Hazelton Laboratories, 1982 and 1983).

29.6 Studies on Environmental Impact

Static aquatic assays were conducted for 24 h of three species of fish. No mortality or change in equilibrium resulted from 5 ppm in northern squawfish, and 10 ppm caused the death of coho salmon, steelhead trout, and northern squawfish in 2 to 14 h (high alkalinity and hardness). In different water conditions (low alkalinity and hardness), mortality was observed in northern squawfish at 10 ppm in 7 to 19 h.

30.0 Ethylcyclohexyldithiol

30.0.1 CAS Number: [28679-10-9]

30.0.2 Synonyms: Ethylcyclohexanedithiol; ethylcyclo hexyldimercaptan

30.1 Chemical and Physical Properties

Ethylcyclohexanedithiol is a water-white liquid with a repulsive odor. Its boiling point is 290°C, its flash point is 250°F, and its vapor pressure is less than 0.1 mmHg at 70°F.

30.4 Toxic Effects

30.4.1 Experimental Studies

30.4.1.1 Acute Toxicity. The acute oral LD$_{50}$ was 1.0 g/kg for rats, and clinical signs comprised muscular weakness, hypoactivity, ruffled fur, and emaciation. The acute dermal LD$_{50}$ for rabbits was 5.6 g/kg. Severe erythema and edema were noted at the application sites. Deaths were delayed for 3 to 5 days in the higher dose groups and up to 10 days in the lower dose groups. No deaths and no abnormal findings were noted in rats exposed to a saturated atmosphere of ethylcyclohexanedithiol for 4 h. Instillation into rabbit eyes caused moderate conjunctival and iridial irritation that persisted for 96 h. This compound is a severe skin irritant. Ethylcyclohexanedithiol tested negative in the Draize (intracutaneous injection) guinea pig sensitization test (Phillips Petroleum, Industrial Biotest, 1965).

31.0 D-Limonene Dimercaptan

31.0.1 CAS Number: [4802-20-4]

31.0.2 Synonyms: 3-Mercapto-beta-4-dimethyl-cyclohexaneethanethiol; p-menthane-2,9-dithiol; dipentenedimercaptan; dipentene; p-menthane-2,9-dithiol

31.0.3 Trade Names: NA

31.0.4 Molecular Weight: 204.4

31.0.5 Molecular Formula: $C_{10}H_{20}S_2$

31.0.6 Molecular Structure:

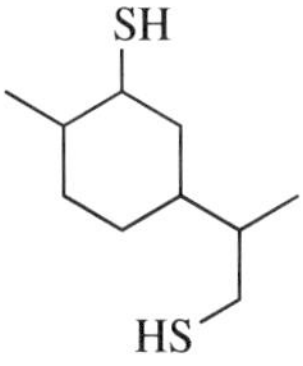

31.1 Chemical and Physical Properties

D-Limonene dimercaptan is a clear, colorless liquid with a specific gravity of 1.025.

31.2 Production and Use

It is used as a marker for *Cannabis* fields in Mexico that have been sprayed with herbicides.

31.3 Exposure Assessment

It is detected by GC by flame photometric detection (83).

31.4 Toxic Effects

31.4.1 Experimental Studies

31.4.1.1 Acute Toxicity. The acute oral LD_{50} was 2.06 g/kg for mice. Clinical signs included motor and sensory depression, ataxia, and loss of righting reflex. Dermal application of 1.0 g/kg on rabbits for a 24-h contact period caused no deaths. Skin irritation after dermal contact was delayed, and blanching of the skin appeared after 24 h, mild erythema at 5 days, and lack of hair growth for at least 16 days after treatment. Instillation into rabbit eyes caused mild con-junctival inflammation that persisted for 24 h (Elf Atochem, Pharmacology Research Inc., 1970).

31.4.1.6 Genetic and Related Cellular Effects Studies. D-Limonene dimercaptan tested negative in the Ames *S. typhimurium* mutagenicity assay.

32.0 Vinyl Cyclohexene-Derived Dimercaptan

32.0.1 CAS Number: [37241-32-0]

32.0.2 Synonyms: 3 (or 4)-mercaptocyclohexaneethanethiol

32.1 Chemical and Physical Properties

It is a colorless, malodorous liquid that has a boiling range of 97 to 99°C (at 1.25 mmHg).

32.2 Production and Use

Vinyl cyclohexene-derived dimercaptan is used as a pharmaceutical intermediate, in rubber and plastics manufacture, and as an agricultural chemical.

32.4 Toxic Effects

32.4.1 Experimental Studies

32.4.1.1 Acute Toxicity. The acute oral LD_{50} for mice was 0.71 g/kg, and there were signs of motor depression, incoordination, and tremor noted. The acute dermal lethal dose was more than 2000 mg/kg. Signs included hypertonia, coarse tremor, and increased sensitivity to stimuli. Instillation into rabbit eyes caused slight irritation of the conjunctiva that resolved within 24 h. Mild to moderate erythema was noted at treated sites in a 4-h Department of Transportation skin corrosivity test in rabbits (Elf Atochem, Pharmacology Research Inc., 1973 and 1979). Severe skin irritation, but no contact sensitization, was noted in a guinea pig Buehler sensitization test (Elf Atochem, Product Safety Labs, 1988).

32.4.1.6 Genetic and Related Cellular Effects Studies. Vinyl cyclohexene-derived dimercaptan tested negative in the Ames *S. typhimurium* mutagenicity test (Elf Atochem, Life Science Research Ltd., 1987).

33.0 Perchloromethyl Mercaptan

33.0.1 CAS Number: [594-42-3]

33.0.2 Synonyms:
Perchloromethanethiol; PMM; trichloromethyl sulfur chloride; thiocarbonyl tetrachloride; trichloromethane sulfurylchloride; trichloromethanesulfenyl chloride; clairsit; PCM; perchloro-methyl-mercaptan; thrichloromethanesulfenyl chloride; trichloromethylsulfenyl chloride

33.0.3 Trade Names: NA

33.0.4 Molecular Weight: 185.88

33.0.5 Molecular Formula: CCl_4S

33.0.6 Molecular Structure:

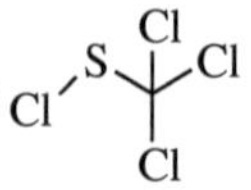

33.1 Chemical and Physical Properties

It is an oily, yellow liquid with a strong unpleasant odor. Its boiling point is 147 to 148°C, its vapor pressure is 3 mmHg at 20°C and 65 torr at 70°C, and its vapor density is 6.414 relative to air. It is neither flammable nor a serious fire hazard.

ORGANIC SULFUR COMPOUNDS

33.1.2 Odor and Warning Properties

The human odor threshold is 1 ppb and the threshold for irritation 0.22 ppm.

33.2 Production and Use

Perchloromethyl mercaptan is used as an intermediate for the synthesis of dyes and fungicides. It is manufactured by chlorinating carbon disulfide, thiophosgene, or methyl thiocyanate.

33.3 Exposure Assessment

When heated to decomposition, perchloromethyl mercaptan emits very toxic fumes of chlorine and sulfur oxides, as well as chlorine gas. Two analytic methods have been reported; one uses gas chromatography (85), and another uses a reaction with resorcinol and spectrophotometric measurement of the resulting color at 434 nm (86).

33.4 Toxic Effects

33.4.1 Experimental Studies

33.4.1.1 Acute Toxicity. Perchloromethyl mercaptan is a severe pulmonary irritant and a lacrimating agent. It has tear gas properties that have been of military interest. Eye irritation is noted at 1.3 ppm, and nausea, eye, throat, and respiratory irritation at 8.8 ppm.

The acute oral LD_{50} in rats was 83 mg/kg (1); the acute dermal LD_{50} in rabbits was 1410 mg/kg (37). Kodak reported an oral LD_{50} for mice and rats of 400 and 800 mg/kg, respectively. Clinical signs included weakness, darkening of the eyes, cyanosis, labored respiration, diarrhea, tremors, and prostration. No methemoglobin was found in rats treated orally with 200 mg/kg (Eastman Kodak, TSCA Section 8E submission, TSCATS Accession No. 42513, Fiche No. 0533569, 1961). The 1-h LC_{50} was 11 ppm for male rats and 16 ppm for female rats (37). Mice and cats exposed to 45 ppm for 15 min died from pulmonary edema within 1 or 2 days. The 3-h LC_{50} in the mouse was 9 ppm (87).

Skin irritation studies in guinea pigs indicated severe skin irritation and absorption of the material through the skin; animals that received 2.5 mL/kg died within 2 days (Eastman Kodak, TSCA Section 8E submission, TSCATS Accession No. 42513, Fiche No. 0533569, 1961). It was severely irritating to the skin and eyes of rabbits.

Perchloromethyl mercaptan was lethal to fish at 4.5 to 5.0 mg/L (86).

33.4.1.2 Chronic and Subchronic Toxicity. In a repeated-exposure inhalation study, Sprague–Dawley rats were exposed to 0.13, 1.0, and 8.7 mg/m^3 (1 ppm) for 6h/day, 5 days/week for 2 weeks. The high-dose rats showed labored breathing, tremors, and mild nasal irritation, and pulmonary edema was noted by microscopic examination. The rats exposed at the two lower concentrations showed no clinical signs. Rats exposed to 2 ppm for 6 h/day, 5 days/week for 4 weeks, showed initial respiratory distress, and congested lungs were noted at necropsy. Exposure to 0.5 ppm under the same exposure regimen caused no effects (88).

33.4.1.6 Genetic and Related Cellular Effects Studies.
Perchloromethyl mercaptan was mutagenic in DNA polymerase-deficient *E. coli* without metabolic activation (89) and inhibited DNA polymerase activity in isolated bovine liver nuclei (90).

33.4.2 Human Experience

Higher concentrations due to accidental exposure caused pulmonary edema in two of three workers exposed and death in the third worker. The fatality resulted from a spill of the liquid on the clothing and floor, and caused exposure by inhalation and direct skin contact. Pathological evaluation of the victim revealed necrosis in the trachea, hemorrhagic pulmonary edema, marked nephrosis in the kidneys, and vacuolization of centrilobular cells in the liver (87).

33.5 Standards, Regulations, or Guidelines of Exposure

The OSHA PEL, NIOSH REL and the ACGIH TLV 8-hr TWA are 0.1 ppm. The IDLH is 10 ppm (32a).

B SULFIDES

Sulfides (RSR', also known as thioethers) are weak Lewis bases and are highly nucleophilic due to the lone pairs of electrons on sulfur. Sulfides are readily oxidized to sulfoxides (RSOR') and then to sulfones (RSO$_2$R'). Sulfides tend to form insoluble complexes with heavy metal salts, such as mercuric chloride. Sulfides also react with alkyl halides to form the corresponding sulfonium halide salts (R$_3$S$^+$X$^-$). Sulfides readily form disulfides (RSSR') under mild oxidizing conditions, such as in the presence of iodine. The amino acid cysteine forms intramolecular and intermolecular protein disulfide bonds to maintain proper conformation essential for activity, such as enzyme-mediated catalysis. Protein cysteinyl residues also form mixed disulfides with glutathione (protein-cysteine-*SS*-cysteine-glutathione) during episodes of cellular oxidative stress and for gene regulation. A number of cellular enzymes (thiol isomerases and reductases) are involved in regulating the oxidation state of protein disulfides.

1.0 Dimethyl Sulfide

1.0.1 CAS Number: *[75-18-3]*

1.0.2 Synonyms: Methylthiomethane; 2-thiapropane; methyl sulfide; thiobismethane; DMS; Methyl thioether; Thiopropane

1.0.3 Trade Names: NA

1.0.4 Molecular Weight: 62.13

1.0.5 Molecular Formula: C$_2$H$_6$S

1.0.6 Molecular Structure: ╱S╲

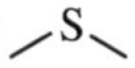

ORGANIC SULFUR COMPOUNDS

1.1 Chemical and Physical Properties

Dimethyl sulfide is a liquid at room temperature and has boiling point of 37.3°C, a vapor pressure of 15 mmHg, and a vapor density of 2.14 relative to air.

1.1.2 Odor and Warning Properties

The human odor threshold is approximately 1 ppb.

1.2 Production and Use

Dimethyl sulfide is used as a gas odorant, catalyst impregnator, and food flavoring agent.

1.3 Exposure Assessment

Analytic procedures for detecting dimethyl sulfide include chemiluminescence and gas-liquid chromatography either in the vapor phase or after trapping on a suitable matrix. There is concern for dimethyl sulfide exposure in workers at paper pulp manufacturing plants that use the kraft process (5). Using gas chromatography, concentrations of 0.04 to 0.69 mg/m^3 of dimethyl sulfide were found in the breath of workers in sulfate paper mills (91).

In pulp mills, exposure to dimethyl disulfide was associated with an increased frequency of headaches and sick leaves in workers (6).

1.4 Toxic Effects

Oral LD$_{50}$ values in rats and mice were 535 and 3700 mg/kg, respectively. The acute dermal LD$_{50}$ value in rabbits is > 5000 mg/kg. Inhalation exposure of rats for 4 h yielded an LC$_{50}$ of 40,250 ppm (9). Dimethyl sulfide is a slight skin irritant and severe eye irritant (92).

Repeated daily oral administration to rats for 14 weeks at dosages of 2.5, 25, and 250 mg/kg/day did not cause any adverse effects on behavior, body weight gain, hematology, blood chemistry, or microscopic pathology (93).

2.0 Dimethyl Disulfide

2.2 CAS Number: [624-92-0]

2.0.2 Synonyms: 2,3-Dithiabutane; (methyldithio)methane; sulfa-hitech; DMDS; methyl disulfide

2.0.3 Trade Names: NA

2.0.4 Molecular Weight: 94.19

2.0.5 Molecular Formula: C$_2$H$_6$S$_2$

2.0.6 Molecular Structure: /S\S/

2.1 Chemical and Physical Properties

Dimethyl disulfide is a liquid that has a boiling point of 109.8°C, a density of 1.0625 at 20°C, a vapor density of 3.24 (relative to air), and a vapor pressure of 28.6 mmHg at 25°C.

2.2 Production and Use

It occurs naturally in *Brassica* vegetables at levels between 0.5 and 1.5% and is a volatile constituent arising from cooked potatoes. In pulp mills, dimethyl disulfide was found at levels of 1.5 ppm, and the highest levels were at chip chutes and evaporation vacuum pumps (6).

2.4 Toxic Effects

2.4.1 Experimental Studies

2.4.1.1 Acute Toxicity. The acute median lethal concentration (LC_{50}) for rats exposed by inhalation for 4 h is 805 ppm, and the acute oral LD_{50} for rats is 190 mg/kg. No mortality was seen in rabbits after dermal application of 2000 mg/kg, although rapid onset of ocular, central nervous system, and respiratory signs were noted. Conjunctival hyperemia, chemosis, discharge, and slight corneal opacity were noted after ocular instillation. Irrigation of the eye diminished these responses, and recovery was complete in 4 to 7 days. Application to intact and abraded skin caused moderate to severe erythema accompanied by edema at 24 h. Erythema disappeared within 7 to 14 days at abraded sites and in 7 to 10 days at intact sites. No delayed contact sensitization was detected in the Buehler guinea pig assay using undiluted material (Elf Atochem, Product Safety Labs, 1985 and 1986).

Acute median lethal concentrations (LC_{50}) for *Daphnia* were 4 ppm (48 h), 15 ppm (24 h), and 21.4 ppm (4 h). The acute median lethal concentration (120 h) for trout was 1.75 ppm and for the guppy (*Lebistes reticulatus*) 50 mg/L (96 h) (Elf Atochem, Krachtwerktuigen Laboratorium, 1988).

2.4.1.2 Chronic and Subchronic Toxicity. A 90-day inhalation study in rats (6 h/day, 5 days/week) at 10, 50, and 250 ppm indicated marginal irritation of the eyes, nose, and respiratory tract during the first few weeks of exposure at 500 ppm. Body-weight increments were reduced at 50 and 250 ppm, predominantly in the first few weeks of exposure. Microscopic pathology changes were confined to the nasal turbinates and included squamous metaplasia of the respiratory mucosa at all exposure levels, accompanied by atrophy and microcavitation of the olfactory epithelium of the anterior turbinates. Recovery was noted after 4 weeks of nonexposure but was not complete at 50 and 250 ppm concentration. The no-observable-effect level (NOEL) for nasal irritation was judged to be slightly lower than 10 ppm (Elf Atochem, Hazelton Laboratories UK, 1992).

Repeated dermal exposure of rabbits (6 h/day, 5 days/week) to dosages in the range of 10.6 to 1063 mg/kg/day for 28 days caused skin irritation (erythema, edema, and necrosis) at all dosages. The severity of the responses was related to dosage and increased with time. Microscopic evaluations demonstrated acanthosis, hyper- and/or parakeratosis,

ORGANIC SULFUR COMPOUNDS 733

and inflammatory cell infiltration. Transient lethargy during the exposure period was seen at 106 and 1063 mg/kg/day. During the third week, spasms were noted in rabbits at 1063 mg/kg/day, but this may have related to impending morbidity. Erythrocyte numbers and hemoglobin concentrations were reduced in males at 1063 mg/kg/day, and there were concomitant increases in reticulocytes and hematopoiesis. Approximately half the rabbits at the high dosage died; myocarditis and myocardial degeneration together with the severe debilitation may have been the presumptive cause(s) of death. The NOEL for systemic toxicity was 10.63 mg/kg/day (Elf Atochem, CIVO Institutes TNO, 1989).

2.4.1.4 Reproductive and Developmental. A rat inhalation developmental toxicity study at 5, 15, and 50 ppm (6 h/day on days 6 to 15 of gestation) caused marked maternal toxicity at 50 ppm and secondary effects on fetal development (reduced fetal weight and delayed ossification). The no-effect level for fetal effects was 15 ppm, and the NOEL for maternal effects was 5 ppm (Elf Atochem, Hazelton Laboratories UK, 1991).

2.4.1.6 Genetic and Related Cellular Effects Studies. A battery of mutagenicity studies indicates that dimethyl disulfide does not pose a genotoxic or clastogenic hazard. It was negative in the Ames *S. typhimurium* mutagenicity assay, *in vitro* and *in vivo* mammalian cell DNA repair assays, a mouse inhalation micronucleus assay at 500 ppm, and a Chinese hamster ovary cell HPRT assay. Increases in chromosomal aberrations were noted in cultured human lymphocytes incubated with cytotoxic concentrations, but the relevance of this is doubtful in the light of the negative micronucleus test result (94); Elf Atochem, Pharmakon Research International, 1985; Elf Atochem, Huntingdon Research Centre Ltd., 1985 Elf Atochem, Sanofi Recherche Service Commun de Toxicologie, 1990; Elf Atochem, CIVO Institutes TNO, 1989 and 1990). Dimethyl disulfide, as well as diallyl disulfide CAS Number [*2179-57-9*] and allyl mercaptan (Section 25.0) decreased forestomach tumors and pulmonary adenomas produced by *N,N*-diethylnitrosamine in female Strain A mice (95).

3.0 Diethyl Sulfide

3.0.1 CAS Number: [352-93-2]

3.0.2 Synonyms: Ethyl sulfide; ethyl thioether; 1,1′-thiobisethane; diethyl thioether

3.0.3 Trade Names: NA

3.0.4 Molecular Weight: 90.18

3.0.5 Molecular Formula: $C_4H_{10}S$

3.0.6 Molecular Structure: ⌒S⌒

3.1 Chemical and Physical Properties

It is a liquid at room temperature and has a vapor density of 3.11 relative to air.

3.2 Production and Use

Diethyl sulfide is used as a chemical intermediate and in electroplating. It occurs naturally in petroleum deposits and can be formed by bacterial decomposition of plant and animal matter.

3.3 Exposure Assessment

Analytic measurement uses either gas–liquid chromatography or potentiometric titration.

3.4 Toxic Effects

The oral LD_{50} in rats was 3415 mg/kg, and reactions to treatment included mydriasis, reduced reaction to stimuli, diminished coordination, and diarrhea (Elf Atochem, Pharmacology Research Inc., 1978). No effects of treatment were reported after a 24-h occluded application of 2000 mg/kg on a rabbit. The 4-h inhalation LC_{50} for rats of a mixture of vapor and aerosol was in excess of 17.9 mg/L. Clinical signs during exposure included slowed respiration, evidence of ocular and mucous membrane irritation, and decreased activity (Elf Atochem, Huntingdon Research Centre, 1990). Slight eye irritation was noted in rabbits; two studies reported slight or moderate irritation of rabbit skin. No evidence of delayed cutaneous sensitization was seen in a guinea pig Buehler sensitization assay (Elf Atochem, Product Safety Labs, 1988).

The 96-h LC_{50} of the neutralized material at pH 7.0 for fish (*Lebistes reticulatus*) was 500 mg/L (Elf Atochem, Krachtwerktuigen Laboratory, 1988).

4.0 Di-*n*-propyl Disulfide

4.0.1 CAS Number: [629-19-6]

4.0.2 Synonyms: Propyl disulfide; di-*n*-propyl disulfide; dithiaoctane; PDS

4.0.3 Trade Names:

4.0.4 Molecular Weight: 150.3

4.0.5 Molecular Formula: $C_6H_{14}S_2$

4.0.6 Molecular Structure: ∼∼S−S∼∼

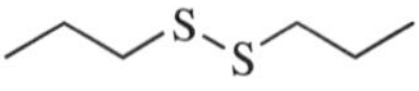

4.1 Chemical and Physical Properties

Di-*n*-propyl disulfide is a clear liquid that has a boiling point of 195°C. It is a natural component of garlic and onion.

4.4 Toxic Effects

4.4.1.1 Acute Toxicity. The acute oral LD_{50} in rats was in excess of 2000 mg/kg. The compound is a mild to moderate skin irritant after application to rabbit skin for 24 h and a very mild eye irritant. A weak sensitization response was detected in a Buehler guinea pig assay (Elf Atochem, Centre International de Toxicologie, 1992).

4.4.1.6 Genetic and Related Cellular Effects Studies. Dipropyl disulfide, and the related compound diallyl disulfide CAS Number [*2179-57-91*], were not mutagenic; rather, they reduced the mutagenicity associated with heterocyclic aromatic amines (found in boiled pork juice), as shown by using the *S. typhimurium* tester strain T98 (96).

5.0 Dibutyl Sulfide

5.0.1 CAS Number: [544-40-1]

5.0.2 Synonyms: n-Dibutyl sulfide; 1,1′-thiobisbutane; *n*-butyl sulfide; dibutyl sulfide; di-*n*-butyl sulfide; thiobisbutane

5.0.3 Trade Names: NA

5.0.4 Molecular Weight: 146.3

5.0.5 Molecular Formula: $C_8H_{18}S$

5.0.6 Molecular Structure: ⌇⌇⌇S⌇⌇⌇

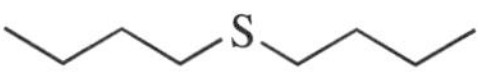

5.1 Chemical and Physical Properties

Melting point (°C): −80
Boiling point (°C): 189
Flash point (°C): 76
Specific gravity: 0.838

5.2 Production and Use

Dibutyl sulfide is prepared by refluxing sodium sulfide and sodium *n*-butyl sulfate.

5.4 Toxic Effects

It is reportedly a moderate skin irritant and has an acute oral LD_{50} of 2200 mg/kg in the rat (97).

6.0 Ethylene Sulfide

6.0.1 CAS Number: [420-12-2]

6.0.2 Synonyms: Thiirane; thiacyclopropane; 2,3-dihydrothiirene; ethylene episulfide

6.0.3 Trade Names: NA

6.0.4 Molecular Weight: 60.11

6.0.5 Molecular Formula: C_2H_4S

6.0.6 Molecular Structure:

6.1 Chemical and Physical Properties

It is a colorless liquid that has a boiling point of 55 to 56°C (with decomposition) and a vapor pressure of 375 mmHg at 25°C.

6.2 Production and Use

Ethylene sulfide is synthesized by the reaction of 2-haloethylthiocyanates with sodium sulfide, ethylene carbonate with sodium or potassium thiocyanate, or ethylene monothiocarbonate with sodium carbonate. It is used as a chemical intermediate.

6.4 Toxic Effects

6.4.1 Experimental Studies

6.4.1.1 Acute Toxicity. It has an oral LD_{50} of 178 mg/kg in the rat. Predominant signs of toxicity include CNS depression and unconsciousness (98). The oral LD_{50} in mice is 35.6 mg/kg, and inhalation LC_{50} values in rats are 4000 ppm (30 min), 2800 ppm (1 h), and 690 ppm (6 h). Ethylene sulfide is a mild skin and eye irritant with recovery within 24 h.

6.4.1.5 Carcinogenesis. Weekly subcutaneous injections of 8 or 16 mg/kg to rats for 1 year followed by observation until death was associated with the development of local sarcomas in 4 of 15 rats at the high dosage. A local fibroma in one rat was also noted at the low dosage (99). On the basis of this study, the International Agency for Cancer Research (IARC, 1976) considered that there was limited evidence for carcinogenicity, and ethylene sulfide was assigned Class 3 (not classifiable as to its carcinogenicity potential for humans).

7.0 Propylene Sulfide

7.0.1 CAS Number: *[1072-43-1]*

7.0.2 Synonyms: 1,2-Epithiopropane; 2-methylthiirane; thiirane, methyl

7.0.3 Trade Names: NA

7.0.4 Molecular Weight: 74.14

7.0.5 Molecular Formula: C_3H_6S

7.0.6 Molecular Structure:

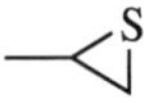

7.1 Chemical and Physical Properties

Boiling point (°C): 72–75
Flash point (°C): 10
Specific gravity: 0.946

7.2 Production and Use

Propylene sulfide is used in polymer technology for its hardening and antiozonant properties.

7.4 Toxic Effects

Acute oral and intraperitoneal LD_{50} values in rats are 254 and 44 mg/kg (98), respectively. Inhalation exposure of rats for 6 h yielded an LC_{50} of 660 ppm; responses to exposure were confined to the lungs and included congestion, edema, hemorrhage, and fibrosis. Inhalation exposure of male rats for 30 min to nominal vapor concentrations of 2400 to 9600 ppm yielded an approximate LC_{50} of 500 ppm. Respiratory irritation and pulmonary edema were the major responses to exposure and the probable cause of death (98).

7.4.1.6 Genetic and Related Cellular Effects Studies. Propylene sulfide was negative in an Ames assay with *S. typhimurium* with or without metabolic activation. It was positive in the mouse lymphoma assay [L5178Y(TK+/TK−)] without metabolic activation (99a).

8.0 Propyl Allyl Disulfide

8.0.1 CAS Number: [2179-59-1]

8.0.2 Synonyms: 2-Propenyl propyl disulfide; onion oil; 4,5-dithia-1-octene; allyl propyl disulfide

8.0.3 Trade Names: NA

8.0.4 Molecular Weight: 148.3

8.0.5 Molecular Formula: $C_6H_{12}S_2$

8.0.6 Molecular Structure:

8.1 Chemical and Physical Properties

Melting point (°C): −15
Physical state: Pale yellow liquid
Water solubility: insoluble; < 0.1 g/100 mL at 20°C

8.2 Production and Use

Propyl allyl disulfide has a pungent irritating odor of cooked onions and is used as a flavoring agent in baked goods, condiments, pickles, meat, meat sauces, and soups at concentrations up to 2 ppm. Propyl allyl disulfide is included in the FDA generally regarded as safe (GRAS) listing.

8.3 Exposure Assessment

Evaluation of airborne concentrations in an onion processing plant indicated levels up to 3.4 ppm; irritation of the eyes, nose, and throat occurred at this and lower levels. No analytical methods were located.

8.4 Toxic Effects

Guinea pigs were sensitized with garlic water-soluble extracts and tested (open epicutaneous tests) with several fractions. The presence of diallyl disulfide was detected in the sensitizing chromatographic fractions. Guinea pigs were successfully sensitized to this product and cross-reacted to garlic; animals sensitized to garlic extracts cross-reacted to diallyl disulfide. Both groups reacted to allicin, an oxidized derivative of diallyl disulfide present in garlic.

8.4.2 Human Experience

It is reportedly a lacrimator and causes immediate ocular burning, blepharospasm, lacrimation, and pain, but no tissue damage. However, in higher concentrations, chemical burns and loss of the corneal epithelium may be noted. Initial symptoms may be followed by chest tightness and coughing, burning of the tongue and mouth, salivation, and vomiting. Burning of the skin, followed by erythema, may occur. Symptoms subside rapidly 15 min after exposure stops in most individuals. The oral administration of allyl propyl disulfide to six normal, fasting human volunteers caused a significant increase in serum insulin and a concomitant fall in blood glucose during the 4-h period after dosing (100). Papageorgiou et al. (101) investigated the role of mono-, di-, and trisulfides in garlic (*Allium sativum*). Water- and ethanol- soluble extracts were prepared and purified by column chromatography and tested on a group of garlic-sensitive patients. The separations and human studies showed that the allergenic fraction was well located in a few column chromatographic fractions. Garlic-sensitive patients showed positive reactions to diallyl disulfide, allyl propyl disulfide, allylmercaptan, and allicin.

8.5 Standards, Regulations, or Guidelines of Exposure

The OSHA PEL, NIOSH REL is 2 ppm; the ACGIH TWA TLV is 2 ppm; and the short-term exposure limit is 3 ppm, for both NIOSH and ACGIH (32a).

9.0 Dibenzyl Disulfide

9.0.1 CAS Number: *[150-60-7]*

9.0.2 Synonyms: Benzyldisulfide; bis(phenylmethyl) disulfide

9.0.3 Trade Names: NA

9.0.4 Molecular Weight: 246.4

9.0.5 Molecular Formula: $C_{14}H_{14}S_2$

9.0.6 Molecular Structure:

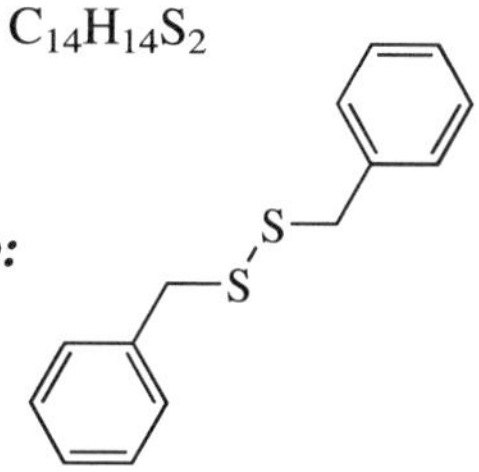

9.4 Toxic Effects

Dibenzyl disulfide caused mild irritation to both rabbit skin and eyes.

C SULFONES, SULFOXIDES, SULFONIUM CHLORIDES, SULFONATES

The formation of sulfoxides, sulfones, and sulfonium chlorides from sulfides was discussed previously. Because sulfoxides contain sulfur in an intermediate oxidation state, they are readily oxidized to sulfones or reduced to the corresponding sulfides by mercaptans, hydroiodic acid, or tertiary phosphorus (R_3P^{3+}) compounds. Sulfones are not very reactive but can be reduced to sulfides by lithium aluminum hydride or sodium borohydride. Because of electron withdrawal by the sulfone substituent group, most reactions involving sulfones occur at the electropositive α-carbon.

Sulfonic acids (RSO_3H) are stronger acids than the corresponding sulfuric acids ($ROSO_3H$), but are weaker oxidizing agents. Therefore, these compounds are used industrially as acid catalysts. Aromatic sulfonic acids are typically produced by *p*-sulfonation of the corresponding hydrocarbon (e.g., toluene + H_2SO_4 → *p*-toluenesulfonic acid). Aliphatic sulfonic acids and their salts (sulfonates) can be produced by a variety of methods, including oxidation of the corresponding mercaptan (RSH), nucleophilic displacement of the corresponding alkyl halide (RX), or a radical-mediated addition to an alkene ($RCHCH_2$). An important industrial synthesis of phenols involves reacting benzenesulfonyl chloride or other aryl sulfonates with molten sodium hydroxide.

Combustion or contact with strong oxidizing materials (including high oxygen concentrations) may yield hazardous decomposition or oxidation products.

1.0 Dimethyl Sulfone

1.0.1 CAS Number: [67-71-0]

1.0.2 Synonyms: Sulfonylbismethane; methyl sulfone

1.0.3 Trade Names: NA

1.0.4 Molecular Weight: 94.13

1.0.5 Molecular Formula: $C_2H_6O_2S$

1.0.6 *Molecular Structure:*

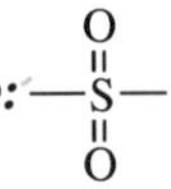

$$-\underset{\underset{O}{\overset{\overset{O}{\|}}{S}}}{}-$$

1.1 Chemical and Physical Properties

It is a solid at room temperature and has a melting point of 109°C.

Melting point (°C): 108–110
Boiling point (°C): 238
Flash point (°C): 143

1.2 Production and Use

Dimethyl sulfone is used as a high-temperature solvent for inorganic and organic compounds.

1.4 Toxic Effects

1.4.1 Experimental Studies

1.4.1.1 Acute Toxicity. Oral and dermal LD_{50} values of more than 5000 mg/kg were observed for rats and rabbits in a series of range-finding studies that investigated the acute toxicity and irritation potential of dimethyl sulfone (Rohm & Haas, TSCA Section 8D submission, TSCATS Fiche No. OTS0533525, 1991). The test material was reportedly slightly irritating to rabbit skin (24-h exposure) and moderately irritating to rabbit eyes. Repeated ingestion of dimethyl sulfone reportedly diminished autoimmune-lymphoproliferative disease in autoimmune strains of mice (102) and prolonged the time to appearance of tumors induced by dimethylbenzanthracene or 1,2- dimethylhydrazine in rats (103, 104).

Dimethyl sulfone was nontoxic to algae, fish, and invertebrates. The 96-h LC_{50} values for invertebrates (fiddler crabs, mysid shrimp) and fish (minnows) were more than 1000 ppm. No effect on cell growth of a marine alga occurred from 96-h exposure to 1000 ppm. Effective degradation (95%) was demonstrated in an activated sludge reactor (Rohm & Haas, TSCA Section 8D submission, TSCATS Fiche Nos. OTS053326-OTS053329 and OTS0533777, 1991).

1.4.1.6 Genetic and Related Cellular Effects Studies. Dimethyl sulfone was negative in the Ames test using *S. typhimurium* strains TA98, TA100 and TA102, with and without metabolic activation (94).

2.0 n-Dibutyl Sulfone

2.0.1 CAS Number: [598-04-9]

2.0.2 Synonyms: n-Butyl sulfone; di-n-butyl sulfone; butane, 1,1'-sulfonylbis-

2.0.3 Trade Names: NA

2.0.4 Molecular Weight: 178.3

2.0.5 Molecular Formula: $C_{18}H_{18}O_2S$

2.0.6 Molecular Structure:

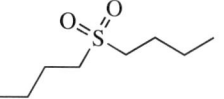

2.1 Chemical and Physical Properties

Melting point (°C): 43–45
Boiling point (°C): 287–295
Flash point (°C): 143

2.4 Toxic Effects

2.4.1.1 Acute Toxicity. The acute oral LD_{50} of *n*-dibutyl sulfone in male rats was 1870 mg/kg. Toxic reactions included tremors, heightened irritability, and loss of equilibrium. The acute dermal LD_{50} value in male albino rabbits was more than 3980 mg/kg. No deaths and no signs of toxicity occurred in male rats exposed for 8 h to a stream of dried air that had been passed through a cylinder packed with glass beads and sulfone crystals. *n*-Dibutyl sulfone caused slight iritis and injection of the blood vessels of the eyelids and conjunctivae 24 h after instillation into eyes of albino rabbits. No signs of irritation were observed after application to intact skin of albino rabbits (Phillips Chemical, 1993).

2.4.1.6 Genetic and Related Cellular Effects Studies. No evidence of mutagenicity was observed in an early microbial assay conducted by the paper-disk method using *E. coli* strain Sd-4-73 and measuring reversion from streptomycin dependence to independence (105).

3.0 Divinyl Sulfone

3.0.1 CAS Number: [77-77-0]

3.0.2 Synonyms: Vinyl sulfone; 1,1′-sulfonylbisethene; TL 797

3.0.3 Trade Names: NA

3.0.4 Molecular Weight: 118.2

3.0.5 Molecular Formula: $C_4H_6O_2S$

3.0.6 Molecular Structure:

3.1 Chemical and Physical Properties

Melting point (°C): −26
Boiling point (°C): 234.3
Flash point (°C): 102
Physical state: liquid
Specific gravity: 1.177
Water solubility: ≥10 g/100 mL at 17°C

3.4 Toxic Effects

A series of range-finding studies conducted with vinyl sulfone found that the oral LD_{50} in male rats is 25 to 43 mg/kg and the dermal LD_{50} in male rabbits is 14 to 36 mg/kg; no rats died after an 8-h exposure to a stream of air that was passed through the test material. Application to rabbit skin and eyes produced skin necrosis and severe eye burns (106). Divinyl sulfone produced ATP depletion and cell death in cultured murine spleen lymphocytes, presumably by inhibiting glyceraldehyde-3-phosphate dehydrogenase (107).

3.4.1.6 Genetic and Related Cellular Effects Studies. No mutagenic or genotoxic activity was observed for vinyl sulfone in microbial assays with *S. typhimurium* strain TA100 (108), with and without metabolic activation, in an *in vivo* mouse dominant-lethal assay (109), or in single and multiple exposure *in vivo* mouse bone-marrow micronucleus assays (109). Increases in chromosomal aberrations and micronuclei were found in cultured mouse lymphoma cells (110), but the relevance of this is doubtful in light of the negative *in vivo* results.

4.0 Diphenyl Sulfone

4.0.1 CAS Number: *[127-63-9]*

4.0.2 Synonyms: Phenyl sulfone; 1,1′-sulfonylbisbenzene

4.0.3 Trade Names: NA

4.0.4 Molecular Weight: 218.3

4.0.5 Molecular Formula: $C_{12}H_{10}O_2S$

4.0.6 Molecular Structure:

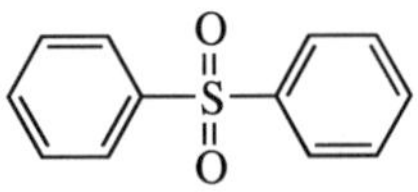

4.1 Chemical and Physical Properties

Physical state: White crystalline solid
Melting point (°C): 128
Boiling point (°C): 379

ORGANIC SULFUR COMPOUNDS

4.2 Production and Use

Diphenyl sulfone has been used as a pesticide, especially as a mite ovicide.

4.4 Toxic Effects

An approximate lethal dose of 2250 mg/kg was reported for male rats. The toxicological targets included liver, kidney, stomach, and the central nervous system. No rats died after a 4-h exposure to diphenyl sulfone dust or after a 4- or 8-h exposure to products released by heating diphenyl sulfone to 150°C, although labored breathing and pulmonary injury were noted. Application of a 50% suspension in 1% aqueous Duponol-PT to intact and abraded skin of guinea pigs resulted in mild irritation. The material did not exhibit the potential to produce dermal sensitization in male guinea pigs. In a repeated dose study, rats were administered diphenyl sulfone, by gavage, at a dosage of 450 mg/kg/day, 5 days/week for 2 weeks. As reported for acutely exposed animals, repeated administration of test material to rats produced effects on the central nervous system, liver, kidney, and stomach; one animal died after the fifth dose (E.I. Du Pont de Nemours, TSCA Section 8D submission, TSCATS Fiche No. OTS0533717, 1991).

5.0 Diethyl Sulfoxide

5.0.1 CAS Number: [70-29-1]

5.0.2 Synonyms: 1,1'-Sulfinylbisethane

5.0.3 Trade Names: NA

5.0.4 Molecular Weight: 106.2

5.0.5 Molecular Formula: $C_4H_{10}OS$

5.0.6 Molecular Structure:

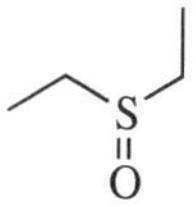

5.4 Toxic Effects

The acute intraperitoneal LD_{50} value in mice was between 2500 mg/kg and 4130 mg/kg, and in rats 4370 mg/kg. Oral LD_{50} values in rats and mice were 5650 and 3610 mg/kg, respectively. Intravenous injection yielded LD_{50} values in rats and mice of 4990 and 4370 mg/kg, respectively. No teratogenic effects were reported in early studies of rodents following administration by various routes. Malformations were observed in an early study with chicken embryos (109a).

6.0 Chloropropyl *n*-Octyl Sulfoxide

6.0.1 CAS Number: [3569-57-1]

6.0.2 Synonyms: Chloropropyl octyl sulfoxide; Repellent 1207; 1-[(3-chloropropyl)-sulfinyl]octane

6.0.3 Trade Names: NA

6.0.4 Molecular Weight: 238.8

6.0.5 Molecular Formula: $C_{11}H_{23}ClOS$

6.0.6 Molecular Structure:

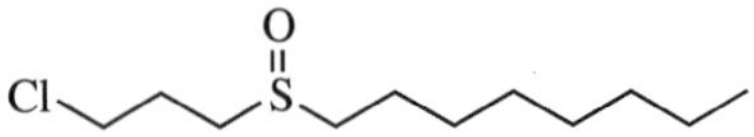

6.1 Chemical and Physical Properties

It is a crystalline solid at room temperature and has a melting point of 37 to 39°C.

6.2 Production and Use

3-Chloropropyl *n*-octyl sulfoxide has been used as an insect repellent.

6.4 Toxic Effects

The oral LD_{50} in male rats was 5660 mg/kg. An acute dermal limit test with male albino rabbits resulted in areas of necrosis and ulceration; the LD_{50} was 8000 mg/kg. Test material applied to rabbit eyes (0.5 mL) resulted in marked iritis, congestion and edema of the lids and conjunctiva, and corneal necrosis. In a skin irritation test with albino rabbits, redness was observed at 24 h. Rats fed 3-chloropropyl *n*-octyl sulfoxide for 90 days at dietary levels of 0, 1, or 2% exhibited reduced body weights. No adverse effects on food consumption, organ weights, or histopathology were observed (Phillips Chemical, 1993).

7.0 Allyl *n*-Octyl Sulfoxide

7.0.1 CAS Number: *[3868-44-8]*

7.4 Toxic Effects

Allyl *n*-octyl sulfoxide has an acute oral LD_{50} in male rats of 3730 mg/kg. In an acute dermal limit test with male albino rabbits, it was corrosive to the skin; the LD_{50} was 5.6 mL/kg. Instillation into rabbit eyes resulted in marked iritis, congestion and edema of the lids and conjunctivas, and corneal necrosis. Redness and scaling were observed with application of 0.01 mL test material to albino rabbit skin for 24 h (Phillips Chemical, 1993).

8.0 Methallyl *n*-Octyl Sulfoxide

8.0.1 CAS Number: *[4886-36-6]*

8.4 Toxic Effects

The acute oral LD$_{50}$ of methallyl *n*-octyl sulfoxide in male rats was 5660 mg/kg. In an acute dermal limit test with male albino rabbits, it was corrosive to the skin and produced ulcerated areas; the LD$_{50}$ was more than 7.95 mL/kg. No signs of toxicity and no mortality occurred in male rats following an 8-h inhalation exposure to a saturated vapor of methallyl *n*-octyl sulfoxide. Instillation into rabbit eyes resulted in marked iritis, congestion and edema of the lids and conjunctivas, and corneal necrosis. Redness and scaling were observed from application of 0.01 mL test material to albino rabbit skin (Phillips Chemical, 1993).

9.0 Methanesulfonic Acid

9.0.1 CAS Number: *[75-75-2]*

9.0.2 Synonyms: MSA; methylsulfonic acid; methylsulfonate

9.0.3 Trade Names: NA

9.0.4 Molecular Weight: 96.10

9.0.5 Molecular Formula: CH$_4$O$_3$S

9.0.6 Molecular Structure:

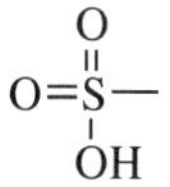

$$O=\overset{O}{\underset{OH}{\overset{\|}{S}}}-$$

9.1 Chemical and Physical Properties

It is a clear, colorless moisture-sensitive liquid at room temperature and has a boiling point of 167°C, a vapor pressure of less than 1 mmHg at 20°C, a vapor density of 3.3 relative to air.

9.2 Production and Use

Methanesulfonic acid is used as an intermediate and as a catalyst for esterification condensation reactions, for alkylation and olefinic polymerization, and for curing of conversion coatings, Use as a solubilizing agent or as a combination catalyst-solvent for reaction mediums is also prevalent.

9.3 Exposure Assessment

Analytic measurement uses the reaction of sulfonic acids with diazoalkanes and subsequent separation via gas chromatography (111).

9.4 Toxic Effects

9.4.1 Experimental Studies

9.4.1.1 Acute Toxicity. The oral LD$_{50}$ of neutralized 70% methanesulfonic acid in mice was 6200 mg/kg. For rats, acute oral LD$_{50}$ values in the 200–400 mg/kg range have been reported. Contradictory results were reported in acute dermal limit tests. The LD$_{50}$ for guinea pigs was more than 2000 mg/kg and more than 200 mg/kg when administered as an aqueous 10% w/v solution; 2000 mg/kg (undiluted) resulted in 100% mortality in rabbits because of extensive skin necrosis. Application of 200 mg/kg (10% aqueous solution), a nonlethal dose, resulted in well-defined erythema and numerous small lesions resembling acid burns. The 6-h inhalation LC$_{50}$ in rats was 330 ppm. Methanesulfonic acid was corrosive to rabbit eyes and caused immediate pain and necrosis of all ocular tissue. Both anhydrous and 70% methanesulfonic acid were corrosive to mouse skin by the anesthetized tail method after a 1-h exposure. No adverse effects on body weight, liver, or kidney weights or gross pathology were observed in rats fed methanesulfonic acid or its potassium salt in the diet for 1 week at dosage levels of up to 2000 mg/kg (Elf Atochem, Pharmacology Research Inc., 1969, 1976 and 1978; Union Carbide, TSCA Section 8E submission, Document 8EHQ-0392-2681).

9.4.1.6 Genetic and Related Cellular Effects Studies. No evidence of mutagenic or genotoxic potential was observed for methanesulfonic acid in the *E. coli* DNA polymerase test (113), in microbial mutagenicity tests using *S. typhimurium* and *E. coli* strains, with and without metabolic activation, and in the mouse micronucleus test after single oral doses of up to 500 mg/kg (114, 115; Elf Atochem, Life Science Research Ltd., 1989; Elf Atochem, Sanofi Recherche, 1990).

9.6 Environmental Impact

The 96-h LC$_{50}$ for guppies at pH 7 was more than 1100 mg/L and for 70% methanesulfonic acid (also pH 7) was more than 770 mg active ingredient per liter (Elf Atochem, Krachtwerktuigen Laboratorium, 1986 and 1988). When neutralized, methanesulfonic acid is considered readily biodegradable on the basis of an OECD 301D Closed Bottle Test and BOD testing (Elf Atochem, CIVO Institutes TNO, 1986). Following intraperitoneal injection of radiolabeled methanesulfonic acid to rats, the label was rapidly excreted (100%) in the urine (112).

10.0 Methanesulfonyl Chloride

10.0.1 CAS Number: [124-63-0]

10.0.2 Synonyms: Mesyl chloride; MsCl

10.0.3 Trade Names: NA

10.0.4 Molecular Weight: 114.5

10.0.5 Molecular Formula: CH$_3$ClO$_2$S

10.0.6 Molecular Structure:
$$-\overset{\overset{O}{\|}}{\underset{\underset{O}{\|}}{S}}-Cl$$

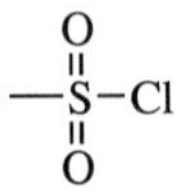

10.1 Chemical and Physical Properties

Methanesulfonyl chloride is a light yellow liquid that has an unpleasant odor at room temperature. It has a boiling point higher than 320°F, a vapor pressure of 2.1 mmHg at 20°C, and a vapor density of 3.9 relative to air. When heated to decomposition, methanesulfonyl chloride emits very toxic fumes of chlorine and sulfur oxides, as well as chlorine gas.

10.2 Production and Use

Most applications of methanesulfonyl chloride are as intermediates in the photographic, fiber dye, agricultural, and pharmaceutical industries. There are also some miscellaneous uses as a stabilizer, catalyst, curing agent, and chlorinating agent.

10.4 Toxic Effects

The oral LD_{50} in rats was 255 mg/kg and there were clinical signs of hypertonia, general distress, and stomach irritation. A single dermal application of 2000 mg/kg to rabbits killed all animals within 24 h. No deaths occurred following a dermal application of 200 mg/kg. The 1-h and 4-h inhalation LC_{50} values in rats were approximately 200 ppm and 25 ppm, respectively. Clinical signs during exposure included salivation and marked eye and respiratory tract irritation. Effects noted postexposure included lethargy, persistent disturbances in respiratory pattern, decreases in body weight and food and water intake, lung congestion, and damage to the corneal surface of the eyes. Methanesulfonyl chloride is corrosive to eyes, skin, and mucous membranes. In rabbits, irrigation of the eye 20 to 30 s after instillation reduced the response, but corneal opacity and congestion of the iris with no pupillary light reaction were still noted. Methanesulfonyl chloride was corrosive to mouse skin by the anesthetized tail method after a 1-h exposure (Elf Atochem, Pharmacology Research Inc., 1976 and 1977; Elf Atochem, Bio/dynamics, 1986; Elf Atochem, Huntingdon Research Centre Ltd., 1987).

10.6 Environmental Impact

The 96-h LC_{50} values of methanesulfonyl chloride for freshwater fish (bluegill sunfish) and saltwater fish (tidewater silverside) using static protocols were 11 and 15 mg/L, respectively (116). For neutralized material, the 96-h LC_{50} for guppies was more than 1200 mg/L (Elf Atochem, Krachtwerktuigen Laboratorium, 1988).

11.0 Ethanesulfonyl Chloride

11.0.1 CAS Number: [594-44-5]

11.0.2 Synonyms: NA

11.0.3 Trade Names: NA

11.0.4 Molecular Weight: 128.6

11.0.5 Molecular Formula: $C_2H_5ClO_2S$

11.0.6 Molecular Structure:

[Structure: CH₃CH₂-S(=O)(=O)-Cl]

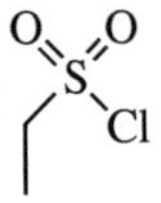

11.1 Chemical and Physical Properties

Boiling point (°C): 177
Flash point (°C): 83
Comments: Lachrymator/moisture sensitive
Specific Gravity: 1.357

11.4 Toxic Effects

11.4.1 Experimental Studies

11.4.1.1 Acute Toxicity. The acute oral LD_{50} in rats was 360 mg/kg. A single dermal application of 2000 mg/kg to rabbits killed two of three animals. No deaths occurred following a dermal application of 200 mg/kg to rabbits. Acute (4-h) inhalation studies indicated an LC_{50} of approximately 86 ppm in the rat. Clinical signs included eye irritation, lacrimation, salivation, and gasping. Decedents exhibited pulmonary hemorrhage. All rats died within 30 to 130 min during an exposure to 1273 ppm. Ethanesulfonyl chloride is corrosive to rabbit skin and eyes, and the vapor caused irritation of the eyes and respiratory tract. In rabbits, irritation of the eye 20 to 30 s after instillation did not reduce the conjunctival necrosis, corneal opacity, or iridial changes. Corneal opacity persisted through 7 days (Elf Atochem, MB Research, 1983). When heated to decomposition, ethanesulfonyl chloride emits very toxic fumes of chlorine and sulfur oxides, as well as chlorine gas.

12.0 Propanesulfonyl Chloride

12.0.1 CAS Number: [10147-36-1]

12.0.2 Synonyms: NA

12.0.3 Trade Names: NA

12.0.4 Molecular Weight: 142.6

12.0.5 Molecular Formula: $C_3H_7ClO_2S$

12.0.6 Molecular Structure:

[Structure: Cl-S(=O)(=O)-CH₂CH₂CH₃]

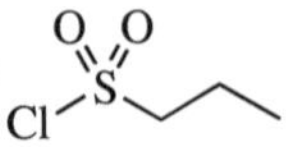

12.1 Chemical and Physical Properties

Boiling point (°C): 66 at 8 mmHg
Flash point (°C): 91
Comments: Moisture Sensitive; Lachrymator
Specific gravity: 1.28

12.4 Toxic Effects

12.4.1 Experimental Studies

12.4.1.1 Acute Toxicity. The acute oral LD$_{50}$ in rats was 320 mg/kg, and responses to treatment included lethargy, piloerection, chromorhinorrhea, diarrhea, body weight losses, and gastrointestinal irritation. The dermal LD$_{50}$ in rabbits was more than 2000 mg/kg. Severe skin responses (erythema, edema, skin discolorations, eschar, necrosis) were observed at the application site. Acute (4-h) inhalation studies indicated an LC$_{50}$ of approximately 100 ppm for the rat. Signs included eye closing, lacrimation, salivation, and quick shallow breathing. All rats died in 130 min during an exposure to 1006 ppm. Propanesulfonyl chloride is corrosive to rabbit skin and eyes, although skin necrosis was delayed (Elf Atochem, MB Research, 1982 and 1983). When heated to decomposition, propanesulfonyl chloride emits very toxic fumes of chlorine and sulfur oxides, as well as chlorine gas.

D BENZOTHIAZOLE DERIVATIVES

When heated to decomposition, compounds containing the benzothiazole moiety emit very toxic fumes of nitrogen and sulfur oxides, as well as cyanide gas.

1.0 Benzothiazole

1.0.1 CAS Number: [95-16-9]

1.0.2 Synonyms: Benzosulfonazole; 1-thia-3-azaindene

1.0.3 Trade Names: NA

1.0.4 Molecular Weight: 135.2

1.0.5 Molecular Formula: C$_7$H$_5$NS

1.0.6 Molecular Structure:

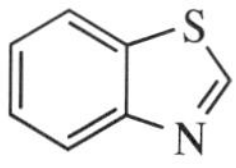

1.1 Chemical and Physical Properties

Melting point (°C): 2
Boiling point: 231
Specific gravity: 1.238

1.2 Production and Use

Benzothiazole is an isolated intermediate used in synthesizing mercaptobenzothiazoles and substituted benzothiazole sulfenamides.

1.4 Toxic Effects

1.4.1 Experimental Studies

1.4.1.1 Acute Toxicity. Following acute single doses benzothiazole is slightly to moderately toxic by both oral and dermal routes of administration. Oral LD$_{50}$ values in the rat are 380 to 492 mg/kg. An oral LD$_{50}$ for mice of 900 mg/kg has also been reported (82). Two dermal LD$_{50}$ values in the rabbit have been reported: between 630 and 1000 mg/kg and between 126 and 200 mg/kg. Exposure of rats to a saturated vapor for 6 h did not cause toxicity or mortality. It is slightly irritating to rabbit skin and eye (Monsanto, Younger Laboratories, 1964 and 1976).

1.4.1.6 Genetic and Related Cellular Effects Studies. No mutagenic activity was observed with or without metabolic activation in the Ames *S. typhimurium* or *S. cerevisiae* assays (Monsanto, Bionetics, 1976). A conflicting positive with *S. typhimurium* strain TA 1537 was reported with metabolic activation (117).

2.0 2-Mercaptobenzothiazole

2.0.1 CAS Number: *[149-30-4]*

2.0.2 Synonyms: 2(3H)-Benzothiazolethione; 2-benzothiazolyl mercaptan; benzothiazole-2-thione; MBT

2.2 Production and Use

2-Mercaptobenzothiazole is a cream to light yellow solid used as an accelerator in rubber vulcanization. It does not measurably hydrolyze over a 7-day period (Monsanto, ABC Laboratories, 1984).

2.4 Toxic Effects

2.4.1 Experimental Studies

2.4.1.1 Acute Toxicity. The acute oral and dermal toxicities are low; LD$_{50}$ values are > 3800 mg/kg (rat) and > 7940 mg/kg (rabbit), respectively (Monsanto, Younger Laboratories, 1975). An oral LD$_{50}$ was also determined in rabbits and was in the range 7500 to 8750 mg/kg (Monsanto, Scientific Associates, 1974).

It was practically nonirritating in rabbit eye and skin irritation studies (Monsanto, Younger Laboratories, 1975). In a delayed contact sensitization assay in guinea pigs (modified Buehler), challenges from 0.5% and 2% petrolatum formulations elicited positive responses (118). A strong positive response was also reported in a Magnusson–Kligman guinea pig sensitization assay (119).

Dietary administration for 4 weeks to rats at levels of up to 2500 ppm reduced body weight increments in males at levels above 1500 ppm and in females at 2000 ppm and above. Increased liver weights were observed in all treated groups. 2-Mercaptobenzothiazole was administered by gavage (diluted in corn oil) to rats and mice of both sexes for 13 weeks at dosages up to 3000 mg/kg in rats and 1500 mg/kg in mice. Mortality was noted in rats at 3000 mg/kg and in mice at dosages of 750 mg/kg and above. Reduced body weight gains were noted in rats and mice at 750 mg/kg and above. Necrosis of the renal distal convoluted tubular epithelium was seen in rats at 3000 mg/kg (Monsanto, Monsanto Environmental Health Laboratory, 1988).

2.4.1.4 Reproductive and Developmental. In a rat teratology study 2-mercaptobenzothiazole was administered to pregnant rats on days 6 to 15 of gestation at dosages of 300, 1200, and 1800 mg/kg. Signs of maternal toxicity were seen at 1200 mg/kg and above. There were some equivocal intergroup differences in postimplantation loss, but these are unlikely to be of toxicological significance. 2-Mercaptobenzothiazole was also administered to pregnant rabbits on days 6 to 18 of gestation at dosages of 50, 150, and 300 mg/kg. Signs of maternal toxicity were seen at 300 mg/kg. There were no indications of fetal or developmental toxicity at any dosage (Monsanto, Springborn Laboratories, 1991). In a two-generation reproduction study in rats, parental animals received diets that contained 2500, 8750, or 15,000 ppm from the premating period through gestation and weaning. Body weights of F1 pups were reduced at 8750 ppm and higher; similar effects were noted in all groups from day 14 of lactation in all treatment groups in the F2 generation. Reproductive indexes were unaffected by treatment at any level. Renal changes (pigmentation of proximal convoluted tubules and kidney weight increases) were seen at 8750 ppm and above in both the F0 and F1 animals. Hepatocyte hypertrophy in both sexes was seen at 8750 ppm and above in the F1 generation, which was accompanied by hepatomegaly at 8750 ppm and higher in males and at 15,000 ppm in females (Monsanto, Springborn Laboratories, 1981).

2.4.1.5 Carcinogenesis. Some evidence of carcinogenicity was reported in male (increased incidences of preputial and adrenal tumors) and female F344 rats (increased pituitary and adrenal tumors). No increased tumor incidences were seen in either male or female B6C3F1 mice (Monsanto, 1993).

2.4.1.6 Genetic and Related Cellular Effects Studies. 2-Mercaptobenzothiazole was not mutagenic in Ames *S. typhimurium*, *S. cerevisiae*, L5178Y mouse lymphoma, CHO/HPRT, mouse micronucleus, or dominant lethal assays (Monsanto, Bionetics, 1976; Monsanto, Pharmakon Research International, 1986; Monsanto, Litton Bionetics, 1986; Monsanto, Springborn Laboratories, 1991).

2.4.1.7 Other: Neurological, Pulmonary, Skin Sensitization. In an acute neurotoxicity evaluation in rats at dosages of 500, 1250, and 2750 mg/kg, responses to treatment were confined to nonspecific effects such as decreased motor activity, salivation, and decreased vocalization. Motor activity testing and a functional observation battery indicated that the responses were likely to be related to nonspecific effects (Monsanto, Bio-Research Laboratories, 1991).

2.4.2 Human Experience

There was no evidence of primary skin irritation or delayed-contact skin sensitization following a repeated insult patch test in 50 human volunteers (Monsanto, Product

Investigations, 1976). Skin sensitization reactions have been reported in humans repeatedly exposed to rubber articles containing 2-mercaptobenzothiazole (120). Positive reactions to 2-mercaptobenzothiazole or 2-mercaptobenzothiazole mixtures containing other sulfur rubber accelerators have also been reported (121, 122). Individuals that responded to 2-mercaptobenzothiazole often responded to other rubber additives, although similar cross-reactions were not seen by other investigators (123, 124).

3.0 N-Isopropyl-2-benzothiazolesulfenamide

3.0.1 CAS Number: *[10220-34-5]*

3.0.2 Synonyms: *N*-(1-methylethyl)-2-benzothiazolesulfenamide; *N*-isopropylbenzothiazol-2-sulfenamide.

3.2 Production and Use

N-Isopropyl-2-benzothiazolesulfenamide is a solid material used as an accelerator in rubber vulcanization.

3.4 Toxic Effects

3.4.1 Experimental Studies

3.4.1.1 Acute Toxicity. The acute oral and dermal toxicity is low; LD_{50} values are >7940 mg/kg (rat) and >7940 mg/kg (rabbit), respectively. Exposure of a group of six rats for 1 h to a nominal concentration of 200 mg/L did not cause any mortality (Monsanto, Younger Laboratories, 1975).

Slight eye irritation but no skin irritation were noted in rabbit irritation studies (Monsanto, Younger Laboratories, 1975).

Oral administration for 3 months to rats of both sexes at 0, 10, 50, 125, or 500 mg/kg/day reduced body weight gains in males at 500 mg/kg and increased liver weights in females at the same level. There was no other evidence of systemic toxicity, and microscopic examination of the livers from female animals did not provide any corroborative evidence for an adverse effect on the liver of females at the high dosage (Monsanto, Monsanto Environmental Health Laboratory, 1988).

3.4.1.6 Genetic and Related Cellular Effects Studies. No evidence of genotoxicity was found in an Ames *S. typhimurium* assay with and without metabolic activation, an *in vitro* cytogenetics assay in Chinese hamster ovary cells with or without metabolic activation, or in an *in vitro* unscheduled hepatocyte DNA synthesis (UDS) assay (Monsanto, Stanford, Research Institute, 1986).

3.4.2 Human Experience

Patch testing of a panel of 54 human volunteers using a 1% preparation demonstrated the material was not a primary skin irritant but one individual demonstrated a positive response for delayed contact sensitization. The material is considered to possess weak sensitization potential at this concentration (Monsanto, Product Investigations, 1986).

4.0 N,N-Diisopropyl-2-benzothiazolesulfenamide

4.0.1 CAS Number: [95-29-4]

4.0.2 Synonyms: N,N-Bis(1-methylethyl)-2-benzothiazolesulfenamide

4.1 Chemical and Physical Properties

It is completely hydrolyzed to mercaptobenzothiazole and isopropylamine during an 8-day period at pH 8.0; hydrolysis occurs more readily at lower pH values (Monsanto, ABC Laboratories, 1984).

4.2 Production and Use

It is a solid material used as an accelerator in rubber vulcanization.

4.4 Toxic Effects

4.4.1 Experimental Studies

4.4.1.1 Acute Toxicity. The acute oral and dermal toxicity is low; LD_{50} values are 5700 mg/kg (rat) and >2000 mg/kg (rabbit), respectively. An oral LD_{50} value in mice was 3892 mg/kg (Monsanto, Bio/dynamics Inc., 1982).

Slight eye (mild, transient irritation of the conjunctiva) and slight skin (slight erythema and edema) irritation were observed in rabbit irritation studies (Monsanto, Bio/dynamics Inc., 1982).

Oral administration to rats of both sexes for 3 months at 0, 250, 500, or 100 mg/kg/day reduced body weight increments in males at 500 mg/kg and higher. Increased absolute and relative liver weights in both sexes at all dosages were related to hepatocellular hypertrophy. There was an accumulation of hyaline droplets in the renal tubular epithelium in males at all dosages. This finding was not present in females, but a brown pigment was present in the same cells in females at dosages of 500 mg/kg and above (Monsanto, Monsanto Environmental Health Laboratory, 1987).

4.4.1.6 Genetic and Related Cellular Effects Studies. No evidence of mutagenicity was found in an Ames *S. typhimurium* assay or in *S. cerevisiae* (Monsanto, Bionetics, 1976). It was also negative in an *in vitro* unscheduled heptocyte DNA synthesis assay. An *in vitro* chromosomal aberration assay in Chinese hamster ovary cells was positive without metabolic activity but not when a source of metabolic activation was included (Monsanto, SRI International, 1986). This positive result is not considered indicative of a clastogenic hazard when compared to a negative *in vivo* rat bone marrow cytogenetics assay following oral administration of 2850 mg/kg and bone marrow cell harvests at 6, 18, and 30 h after administration (Monsanto, Pharmakon Research International, 1987).

4.4.2 Human Experience

Patch testing of a panel of 54 human volunteers using a 1% preparation in petrolatum indicated that the material was neither a primary skin irritant nor a delayed contact sensitizer (Monsanto, Product Investigations, 1986). Patch testing of humans previously demonstrated as sensitized to 2-mercaptobenzothiazole gave positive results, although the likely presence of impurities and inadequate documentation of sample quality and source render the validity of this result questionable (123).

5.0 N-*tert*-Butyl-2-benzothiazolesulfenamide

5.0.1 CAS Number: [95-31-8]

5.0.2 Synonyms: *N*-(1,1'-dimethylethyl)-2-benzothiazolesulfenamide

5.1 Chemical and Physical Properties

It is completely hydrolyzed to 2-mercaptobenzothiazole and *tert*-butylamine within a 25-h period at pH 7.0; the hydrolysis rate is slower at pH 9.0 (Monsanto, ABC Laboratories, 1984).

5.2 Production and Use

It is a solid material used as an accelerator in rubber vulcanization.

5.4 Toxic Effects

5.4.1 Experimental Studies

5.4.1.1 Acute Toxicity. The acute oral and dermal toxicities are low; LD_{50} values are >6310 mg/kg (rat) and >7940 mg/kg (rabbit), respectively. Slight eye irritation and minimal skin irritation were defined in rabbit irritation studies (Monsanto, Younger Laboratories, 1973). A 25% solution in ethanol did not induce any delayed-contact sensitization reactions in a guinea pig sensitization (Buehler) assay (Monsanto, Pharmakon Research International, 1982).

Oral administration for 3 months to rats of both sexes at 0, 100, 300, or 1000 mg/kg/day reduced body weight increments in males at dosages of 300 mg/kg and higher. Liver and kidney weights were also increased in females at 1000 mg/kg, but there was no microscopic evidence of morphological change in these organs and the findings were not considered of toxicological significance (Monsanto, Monsanto Environmental Health Laboratory, 1982). A dust inhalation study of rats (6 h/day, 5 days/week) for 4 weeks at 0, 2.4, 29, and 84 mg/m^3 elicited morphological effects on the liver and lymph nodes at 84 mg/m^3 that were detected during microscopic examination (Monsanto, International Research and Development Corporation, 1978). Repeated daily dermal applications of 0, 125, 500, or 2000 mg/kg/day caused slight dermal irritation but no evidence of systemic toxicity (Monsanto, International Research and Development Corporation, 1979).

5.4.1.4 Reproductive and Developmental. No maternal toxicity, fetal toxicity, or developmental toxicity were seen in rats following oral administration of 0, 50, 150, or 500 mg/kg/day to pregnant female rats on days 6 to 15 of gestation (Monsanto, International Research and Development Corporation, 1978).

5.4.1.6 Genetic and Related Cellular Effects Studies. No evidence of mutagenicity was found in Ames *S. typhimurium, E. coli, S. cerevisiae,* or Chinese hamster ovary HPRT assays (125; Monsanto, Bionetics, 1978). Positive results were obtained in mouse lymphoma assays of L5178Y cells with metabolic activation (Monsanto, Bionetics, 1978). Cell transformation was noted at the highest concentration tested in BALB/3T3 cells, but significant cytotoxicity may confound the relevance of this finding (Monsanto, Monsanto Environmental Health Laboratory, 1993).

5.4.2 Human Experience

Patch testing of a panel of 54 human volunteers using a 60% preparation in petrolatum indicated the material was not a primary skin irritant, but a strong delayed contact sensitization response occurred (Monsanto, Product Investigations, 1982). Patch testing of humans previously demonstrated to be sensitized to 2-mercaptobenzothiazole gave positive results, although the likely presence of impurities and inadequate documentation of sample quality and source render the validity of this result questionable (123).

6.0 N-Cyclohexyl-2-benzothiazolesulfenamide

6.0.1 CAS Number: *[95-33-0]*

6.1 Chemical and Physical Properties

It is a pale, buff-colored powder or granule completely hydrolyzed within 25 h at pH 7.0 (Monsanto, ABC Laboratories, 1984).

6.2 Production and Use

N-Cyclohexyl-2-benzothiazolesulfenamide is used as an accelerator in rubber vulcanization.

6.4 Toxic Effects

6.4.1.1 Acute Toxicity. The acute oral and dermal toxicities are low; LD_{50} values are 5300 mg/kg (rabbit) and > 7940 mg/kg (rat), respectively (Monsanto, Younger Laboratories, 1973). It is slightly irritating to the rabbit eye (Monsanto, Younger Laboratories, 1973) and practically nonirritating to rabbit skin (Monsanto, Pharmakon Research International, 1982).

In a guinea pig delayed-contact skin sensitization assay (Buehler) using a 25% preparation in ethanol, no evidence of sensitization was found after challenge with the 25% solution. No evidence of comedogenicity was detected in rabbits when solutions in chloroform up to 10% were applied to the inner surface of the ears for 4 weeks (Monsanto, Pharmakon Research International, 1982).

Dermal application of up to 2000 mg/kg/day to intact and abraded skin of rabbits for 21 days did not elicit any evidence of toxicity (Monsanto, International Research and Development Corporation, 1979). Inhalation exposure of rats for 4 weeks (5 days/week) to dust concentrations of 4.3, 14.4, and 48.0 mg/m^3 caused microscopic lesions of the conjunctivas, lymph nodes, and spleen at 48 mg/m^3. Elevated aspartate aminotransferase activities at 14.4 mg/m^3 and above were not associated with any morphological abnormalities (Monsanto, International Research and Development Corporation, 1978). Administration to rats of diets that contained levels approximating 100, 250, 500, 1000, or 3000 mg/kg for 4 weeks reduced food intake and body weight gain at 500 mg/kg and higher; there was no other evidence of systemic toxicity (Monsanto, International Research and Development Corporation, 1979). Administration to rats by gavage for 5 weeks (5 days/week) of dosages up to 1.25 mg/kg/day reduced body weight gain and increased thyroid weights relative to body weight. There was no evidence of pathological change, so the thyroid weight finding may reflect reduced body weight gain (Monsanto, 1993).

6.4.1.3.3 Excretion. After a single oral dosage of 250 mg/kg of ^{14}C-labeled material 65 and 24% of the dose was excreted with in 3 days in the urine and feces, respectively. Biliary excretion over this period accounted for approximately 5% of the dose. There was no evidence of selective accumulation or concentration in any organs. The urinary metabolites were identified as cyclohexylamine and 2-mercaptobenzothiazole (129).

6.4.1.4 Reproductive and Developmental. Oral administration to pregnant rats of dosages of up to 500 mg/kg on days 6 to 15 of gestation caused maternal toxicity at 500 mg/kg and concomitant reduction in fetal body weight; there were no teratological responses at this dosage. No maternal or fetal effects were noted at 300 mg/kg (Monsanto, International Research and Development Corporation, 1978). In a separate study, oral administration, of dosages of 50, 150 and 450 mg/kg to pregnant rats during organogenesis, produced maternal toxicity at the highest dosage and dose-dependent fetal toxicity (fetuses/litter and hydrocephalus) (126). These investigators found that the dosage of 450 mg/kg produced embryotoxicity, shown by increases in late resorptions and post implantation losses, decreased fetal body weight and length, and subcutaneous hemorrhage (127). In another study, dietary administration at dietary levels approximating dosages of 0.7, 7.1, 69.6, and 288.8 mg/kg to pregnant female rats on days 0 to 20 of gestation caused reduced maternal food intake and body weight gains at 288.8 mg/kg and a concomitant reduction of fetal body weight. Maternal body weight gain was also reduced at 69.6 mg/kg, although there were no effects on fetal body weight. No effects on pre- or postimplantation loss, litter size, or the incidence of malformations or visceral or skeletal variations were reported (128).

6.4.1.6 Genetic and Related Cellular Effects Studies. No evidence of mutagenicity either with or without metabolic activation was found in the Ames *S. typhimurium*, *E. coli*, or Chinese hamster ovary HPRT assays (Monsanto, Bionetics, 1976; Monsanto, 1993), the *S. cerevisiae* assay (Monsanto, Bionetics, 1976), or the L5178 mouse lymphoma assay (Monsanto, Bionetics, 1978).

6.4.2 Human Experience

Occupational exposure reportedly causes irritation of the eyes, skin and upper respiratory tract. Patch testing of a panel of 51 human volunteers using a 70% preparation in petrolatum did not cause primary or cumulative irritation, but sensitization responses occurred in 5 of the 51 subjects (Monsanto, Product Investigations, 1982). The literature contains many reports of skin sensitization, but in most cases the patients were also sensitized to other components of the "mercapto mix" used for skin testing. Industrial experience indicates that this material is a weak sensitizer.

7.0 N,N-Dicyclohexylbenzothiazolesulfenamide

7.0.1 CAS Number: [4979-32-2]

7.0.2 Synonyms: *N,N*-dicyclohexyl-1-benzothiazolesulfenamide.

7.2 Production and Use

It is an off-white solid, a powder or pellet, that is used as an accelerator in rubber vulcanizing.

7.4 Toxic Effects

7.4.1 Experimental Studies

7.4.1.1 Acute Toxicity. The material has low acute oral and dermal toxicities; LD_{50} values are > 5000 mg/kg (rat) and > 2000 mg/kg (rabbit), respectively. It was practically nonirritating in rabbit skin and eye assays (130); (Monsanto, Bio/Dynamics Inc., 1984). Using the Magnusson–Kligman maximization assay for delayed-contact cutaneous sensitization, it was concluded that it is not a skin sensitizer (Monsanto, International Research and Developmental Corporation, 1984).

7.4.1.2 Chronic and Subchronic Toxicity. Administration to rats in the diet for 4 weeks at 2000 to 10,000 ppm caused a dose-related decrease in food consumption and body weight increments at all dietary concentrations. There was no evidence of any systemic toxicity based on evaluations of the cellular and chemical constituents of blood, organ weight analysis, or macroscopic pathology examinations (130); (Monsanto, Bio/Dynamics Inc., 1987). Administration to rats in the diet for 13 weeks at 2500 or 5000 ppm decreased food intake and reduced body weight increments at both dietary concentrations. There was no evidence of any systemic toxicity or microscopic pathology changes at either dietary concentration. The no-effect level on subchronic administration in the diet was 500 mg/kg diet (130; Monsanto, Monsanto Environmental Health Laboratory, 1988).

7.4.1.5 Carcinogenesis. An increased incidence of sarcomas occurred at the site of subcutaneous administration of 20 g/kg body weight, administered as 1g/kg body weight at irregular intervals during 413 days (130). However, there was no evidence for mutagenic or clastogenic effects in the Ames *S. typhimurium* assay, the Chinese hamster ovary HPRT assay, and an *in vitro* hepatocyte DNA repair (UDS) assay, or in an *in vivo* rat bone marrow chromosomal aberration assay after oral administration of a single 1000-mg/kg dose (130); (Monsanto, Pharmakon International, 1984; Monsanto, Hazelton Laboratories, 1984; Monsanto, 1993; Monsanto, SRI International, 1984).

General

Internet site references with chemical and safety information:

http://ecdin.etomep.net	Environmental Chemicals Data Information Network
http://ecphin.etomep.net	European Community Pharmaceutical Information Network
http://www.jrc.org.isis	Institute for Systems Information and Safety
http://www.msdsonline.com	MSDS Information
http://www.safety.utoledo.edu/safety/msds.htm	MSDS links
http://www.chem.uky.edu/resources/msds.html	One of the better MSDS and chemistry information sites
http://www.aitl.uc.edu	Academic Information Technology and Libraries at the University of Cincinnati
http://www.vmi.edu/~chem/ind-home.html	Chemical Industries Homepages
http://www.chemweb.com	Chemistry information site

BIBLIOGRAPHY

1. U. S. Department of Health & Human Services, *Toxicological Profile for Methyl Mercaptan.*
2. R. L. R. Kramer, Catalytic preparation of mercaptans. *J. Am. Chem. Soc.* **43**, 880–890 (1921).
3. C. B. Scott, W. S. Dorsey, and H. C. Huffman, Methyl mercaptan from methyl chloride. *Ind. Eng. Chem.* **47**, 876–877 (1955).
4. R. Knarr and S. M. Rappaport, Determination of methanethiol at parts-per-million air concentrations by gas chromatography. *Analyt. Chem.* **52**, 733–736 (1980).
5. N. Goyer, Evaluation of occupational exposure to sulfur compounds in paper pulp kraft mills. *Am. Indus. Hyg. Assoc. J.* **51**, 390–394 (1990).
6. J. Kangas, P. Jappinen, and H. Savolainen, Exposure to hydrogen sulfide, mercaptans and sulfur dioxide in pulp industry. *Am. Indus. Hyg. Assoc. J.* **45**, 787–790 (1984).
6a. NIOSH, *Manual of Analytical Methods*, 4th ed., U.S. Gov't Printing Office, Supt of Docs, 1994.

7. G. V. Selyuzhitskii, Experimental data used to determine the maximum permissible concentration of methyl mercaptan, dimethyl sulfide, and dimethyl disulfide in the air of the production area of paper and pulp plants (Russ.). *Gig. Tr. Prof. Zabol.* **16**, 46–47 (1972).

8. G. V. Selyuzhitskii and V. P. Timofeev, Sanitary-toxicological study of the sulfur-containing components from sulfate pulp manufacture emissions (Russ.). *Khim. Seraorg. Soedin. Soderzh. Neftyakh Nefeprod.* **9**, 587–590 (1972).

9. M. F. Tansy et al., Acute and subchronic toxicity studies of rats exposed to vapors of methyl mercaptan and other reduced-sulfur compounds. *J. Toxicol. Environ. Health* **8**, 71–88 (1981).

10. L. Zieve, W. M. Doizaki, and J. Zieve, Synergism between mercaptans and ammonia or fatty acids in the production of coma: A possible role for mercaptans in the pathogenesis of hepatic coma. *J. Lab. Clin. Med.* **83**, 16–28 (1974).

11. K. Yaegaki and K. Sanada, Biochemical and clinical factors influencing oral malodor in periodontal patients. [Review][37 refs]. *Journal of Periodontology* **63**, 783–789 (1992).

12. S. Persson et al., The formation of hydrogen sulfide and methyl mercaptan by oral bacteria. *Oral Microbiology & Immunology* **5**, 195–201 (1990).

13. E. S. Canellakis and H. Tarver, The metabolism of methyl mercaptan in the intact animal. *Arch. Biochem. Biophys.* **42**, 446-455 (1953).

14. K. Ahmed, L. Zieve, and G. Quarfoth, Effects of methanethiol on erythrocyte membrane stabilization and on Na^+, K^+-adenosine triphosphatase: relevance to hepatic coma. *J. Pharmacol. Exper. Ther.* **228**, 103–108 (1984).

15. P. W. Johnson and J. Tonzetich, Sulfur uptake by type I collagen from methyl mercaptan/dimethyl disulfide air mixtures. *Journal of Dental Research* **64**, 1361–1364 (1985).

16. R. L. Waller, Methanethiol inhibition of mitochondrial respiration. *Toxicol. Appl. Pharmacol.* **42**, 111–117 (1977).

17. G. Quarfoth et al., Action of methanethiol on membrane (Na^+, K^+)-ATPase of rat brain. *Biochem. Pharmacol.* **25**, 1039–1044 (1976).

18. S. Schwimmer, Inhibition of carbonic anhydrase by mercaptans. *Enzymologia* **37**, 163–173 (1969).

19. E. P. Haydu, H. R. Amberg, and R. E. Dimick, The effect of kraft-mill-waste components on certain salmonoid fishes of the Pacific Northwest. *Tappi* **35**, 545–549 (1952).

20. A. E. Cole, The toxicity of methyl mercaptan for fresh-water fish. *J. Pharmacol.* **54**, 448–453 (1935).

21. A. E. Werner, Sulfur compounds in kraft-mill effluents. *Can. Pulp Paper Ind.* **16**, 35–43 (1963).

22. K. Shiomi et al., Volatile sulfur compounds responsible for an offensive odor of the flat-head, Calliurichthys Doryssus. *Comparative Biochemistry Physiology—B: Comparative Biochemistry* **71**, 29–31 (1982).

23. W. T. Shults, E. N. Fountain, and E. C. Lynch, Methanethiol poisoning. Irreversible coma and hemolytic anemia following inhalation. *J. Am. Med. Assoc.* **211**, 2153–2154 (1970).

24. L. Zieve et al., Ammonia, octanoate and a mercaptan depress regeneration of normal rat liver after partial hepatectomy, *Hepatology* **5**, 28–31 (1985).

25. H. Al Mardini et al., Effect of methionine loading and endogenous hypermethioninaemia on blood mercaptans in man. *Clinica Chimica Acta* **176**, 83–89 (1988).

26. E. Fabbri et al., Olfactory transduction mechanisms in sheep. *Neurochem. Res.* **20**, 719–725 (1995).

27. A. Bosy, Oral malodor: Philosophical and practical aspects. [Review] [73 refs]. *Journal/ Canadian Dental Association. Journal de l Association Dentaire Canadienne* **63**, 196–201 (1997).
28. J. M. Coli and J. Tonzetich, Characterization of volatile sulphur compounds production at individual gingival crevicular sites in humans. *Journal of Clinical Dentistry* **3**, 97–103 (1992).
29. H. Lancero, J. Niu, and P. W. Johnson, Exposure of periodontal ligament cells to methyl mercaptan reduces intracellular pH and inhibits cell migration. *Journal of Dental Research* **75**, 1994–2002 (1996).
30. L. G. Ratkay, J. D. Waterfield, and J. Tonzetich, Stimulation of enzyme and cytokine production by methyl mercaptan in human gingival fibroblast and monocyte cell cultures. *Archives of Oral Biology* **40**, 337–344 (1995).
31. P. Johnson, K. Yaegaki, and J. Tonzetich, Effect of methyl mercaptan on synthesis and degradation of collagen. *Journal of Periodontal Research* **31**, 323–329 (1996).
32. P. W. Johnson, W. Ng, and J. Tonzetich, Modulation of human gingival fibroblast cell metabolism by methyl mercaptan. *Journal of Periodontal Research* **27**, 476–483 (1992).
32a. National Institute for Occupational Safety and Health (NIOSH), *Pocket Guide to Chemical Hazards, DHHS (NIOSH) Publ. No. 85-114*, U.S. Department of Health and Human Services, Washington, DC, 1997.
33. W. T. Lowry et al., Toxicological investigation of liquid petroleum gas explosion: human model for propane/ethyl mercaptan exposures. *Journal of Forensic Sciences* **36**, 386–396 (1991).
34. H. Moore, H. L. Helwig, and R. J. Graul, A spectrophotometric method for the determination of mercaptans in air. *Am. Indus. Hyg. Assoc. J.* **21**, 466–470 (1960).
35. L. J. Priestley Jr. et al., Determination of subtoxic concentrations of phosgene in air by electron capture gas chromatography. *Anal. Chem.* **37**, 70–71 (1965).
36. E. J. Fairchild II and H. E. Stokinger II, Toxicologic studies on organic sulfur compounds I. Acute toxicity of some aliphatic and aromatic thiols (mercaptans). *Am. Indus. Hyg. Assoc. J.* **19**, 171–189 (1958).
37. E. H. Vernot et al., Acute toxicity and skin corrosion data for some organic and inorganic compounds and aqueous solutions. *Toxicol. Appl. Pharmacol.* **42**, 417–423 (1977).
38. R. Munday, Toxicity of thiols and disulphides: involvement of free-radical species. *Free Radic. Biol. Med.* **7**, 659–673 (1989).
39. G. A. Snow, The metabolism of compounds related to ethanethiol. *Biochem. J.* **65**, 77–82 (1957).
40. D. Foster, K. Ahmed, and L. Zieve, Action of methanethiol on Na^+, K^+-ATPase: implications for hepatic coma. *Ann. N.Y. Acad. Sci.* **242**, 573–576 (1974).
41. E. W. Thomas, Direct determination of hydrocarbon sulfides in kraft gases by gas-liquid chromatography. *Tappi* **47**, 587–588 (1964).
42. H. A. Mardini, K. Bartlett, and C. O. Record, An improved gas chromatographic method for the detection and quantitation of mercaptans in blood. *Clinica Chimica Acta* **113**, 35–41 (1981).
43. R. G. Ames and J. W. Stratton, Acute health effects from community exposure to *N*-propyl mercaptan from an ethoprop (Mocap)-treated potato field in Siskiyou County, California [see comments]. *Arch. Environ. Health* **46**, 213–217 (1991).

44. National Institute for Occupational Safety and Health, *n*-Butyl mercaptan standards completion program validated method No. S350. *NIOSH Manual of Analytical Methods*, 1978, pp. 1–9 (See Ref. 6a for update).

45. K. M. Abdo et al., Heinz body production and hematological changes in the hen after administration of a single oral dose of *n*-butyl mercaptan and *n*-butyl disulfide. *Fundam. Appl. Toxicol.* **3**, 69–74 (1983).

46. R. G. Ames and J. Gregson, Mortality following cotton defoliation: San Joaquin Valley, California, 1970–1990. *Journal of Occupational & Environmental Medicine* **37**, 812–819 (1995).

47. M. B. Abou-Donia, D. G. Graham, K. M. Abdo, and A. A. Komeil, Delayed neurotoxic, late acute and cholinergic effects of *S,S,S*-tributyl phosphorotrithioate (DEF): subchronic (90 days) administration in hens. *Toxicology* **14**, 229–243 (1979).

48. M. Cirstea, Studies on the relation between chemical structure and contact sensitizing capacity of some thiol compounds. *Revue Roumaine de Physiologie* **9**, 485–491 (1972).

49. S. Szabo and E. S. Reynolds, Structure-activity relationships for ulcerogenic and adrenocorticolytic effects of alkyl nitriles, amines, and thiols. *Environ. Health. Perspect.* **11**, 135–140 (1975).

50. W. Nastainczyk, H. H. Ruf, and V. Ullrich, Binding of thiols to microsomal cytochrome P-450. *Chem.-Biol. Interact.* **14**, 251– 263 (1976).

51. W. C. Thomas et al., Inhalation teratology studies of *n*-butyl mercaptan mercaptan in rats and mice. *Fundam. Appl. Toxicol.* **8**, 170–178 (1987).

52. E. Mather-Mihaich and R. T. Di Giulio, Antioxidant enzyme activities and malondialdehyde, glutathione and methemoglobin concentrations in channel catfish exposed to DEF and *n*-butyl mercaptan. *Comparative Biochemistry & Physiology—C: Comparative Pharmacology & Toxicology* **85**, 427–432 (1986).

53. D. W. Connell and G. J. Miller, Petroleum hydrocarbons in aquatic ecosystems - Behavior and effects of sublethal concentrations: Part II. *CRC Crit. Rev. Environ. Control* **11**, 105–162 (1981).

54. F. Gobbato and P. M. Terribile, Toxicological properties of mercaptans. *Folia Medica* **51**, 329–341 (1968).

54a. A. Jangerman, *J. Chromtag.* **366**, 205–216 (1986).

55. V. B. Stein and R. S. Narang, Determination of mercaptans at microgram-per-cubic-meter levels in air by gas chromatography with photoionization detection. *Analyt. Chem.* **54**, 991–992 (1982).

56. C. Bradley and D. J. Schiller, Determination of sulfur compound distribution in petroleum by gas chromatography with a flame photometric detector. *Analyt. Chem.* **58**, 3017–3021 (1986).

57. C. P. Carpenter, H. F. Smyth Jr., and U. P. Pozzani, The assay of acute vapor toxicity, and the grading and interpretation of results on ninety-six chemical compounds. *J. Ind. Hyg. Toxicol.* **31**, 343–346 (1949).

58. S. Sandor, E. S. Reynolds, and S. H. Unger, Structure-activity relations between alkyl nucleophilic chemicals causing duodenal ulcer and adrenocortical necrosis. *J. Pharmacol. Exper. Ther.* **223**, 68–76 (1982).

59. N. J. Bikbulatov, The toxicology of some organic sulfur compounds of the type found in petroleum (Rus). *Khim. Seraorg. Soedin. Soderzh. Neftyakh Nefeprod.* 369–374 (1959).

60. E. W. Schafer Jr. and W. A. Bowles Jr., Acute oral toxicity and repellency of 933 chemicals to house and deer mice. *Arch. Environ. Contam. Toxicol.* **14**, 111–129 (1985).
61. S. C. Brooks, J. J. Lalich, and C. A. Baumann, Skin sterols XII. Contrasting effects of certain mercaptans, amines, and related compounds on sterols and sebaceous glands. *Cancer Res.* **17**, 148–152 (1957).
62. M. S. Gizhlaryan, The toxicity of sulfenamide BT (Rus). *Khim. Seraorg. Soedin. Soderzh. Neftyakh Nefeprod.* **19**, 41–46 (1966).
63. F. Grimalt and C. Romaguera, New resin allergies in shoe contact dermatitis. *Contact Derm.* **1**, 169–174 (1975).
64. A. S. Adzhibekian and K. O. Kazarian, [Selective method of gas chromatographic determination of higher mercaptans in the air]. [Russian]. *Gig. Tr. Prof. Zabol.* 49–50 (1990).
65. B. B. Shugaev, Comparative toxicity of a series of higher mercaptans (Rus). *Khim. Seraorg. Soedin. Soderzh. Neftyakh Nefeprod.* **8**, 681–686 (1968).
66. U. M. Cowgill and D. P. Milazzo, The response of the three brood Ceriodaphnia test to fifteen formulations and pure compounds in common use. *Arch. Environ. Contam. Toxicol.* **21**, 35–40 (1991).
67. A. M. Le Bon et al., *In vivo* antigenotoxic effects of dietary allyl sulfides in the rat. *Cancer Lett.* **114**, 131–134 (1997).
68. C. Egen-Schwind, R. Eckard, and F. H. Kemper, Metabolism of garlic constituents in the isolated perfused rat liver. *Planta Medica* **58**, 301–305 (1992).
69. R. Gebhardt and H. Beck, Differential inhibitory effects of garlic-derived organosulfur compounds on cholesterol biosynthesis in primary rat hepatocyte cultures. *Lipids* **31**, 1269–1276 (1996).
70. E. Eder et al., Mutagenic potential of allyl and allylic compounds. Structure-activity relationship as determined by alkylating and direct *in vitro* mutagenic properties. *Biochem. Pharmacol.* **29**, 993–998 (1980).
71. M. J. Wargovich and V. W. Eng, Rapid screening of organosulfur agents for potential chemopreventive activity using the murine NA assay. *Nutr. Cancer* **12**, 189–193 (1989).
72. S. P. Reimann, The hyperplastic reaction of the skin to sulfhydryl and its significance in neoplasia, *Am. J. Cancer* **15**, 2149–2168 (1931).
73. D. Wild et al., Study of artificial flavouring substances for mutagenicity in the Salmonella/microsome, Basc and micronucleus tests. *Fd. Chem. Toxicol.* **21**, 707–719 (1983).
74. N. Itoh et al., Fungal cleavage of thioether bond found in Yperite. *FEBS Letters* **412**, 281–284 (1997).
75. H.-L. Wu et al., Derivatization-gas chromatographic determination of mercaptans. *Anal. Lett.* **14**, 1625–1635 (1981).
76. H. B. Lee and A. S. Y. Chau, Determination of trifluralin, diallate, triallate, atrazine, barlan, diclofop-methyl, and benzoylprop-ethyl in natural waters at parts per trillion levels. *J. Assoc. Off. Anal. Chem.* **66**, 651–658 (1983).
77. H. B. Lee and A. S. Y. Chau, Gas chromatographic determination of trifluralin, diallate, triallate, atrazine, barlan, diclofop-methyl, and benzoylprop-ethyl in sediments at parts per billion levels. *J. Assoc. Off. Anal. Chem.* **66**, 1322–1326 (1983).
78. J. B. McBain and J. J. Menn, *S*-Methylation, oxidation, hydroxylation and conjugation of thiophenol in the rat. *Biochem. Pharmacol.* **18**, 2282–2285 (1969).

79. L. Somasundaram et al., Application of the Microtox system to assess the toxicity of pesticides and their hydrolysis metabolites. *Bull. Environ. Contam. Toxicol.* **44**, 254–259 (1990).

80. E. LaVoie et al., Mutagenicity of aminophenyl and nitrophenyl ethers, sulfides, and disulfides. *Mutation Res.* **67**, 123–131 (1979).

81. P. Amrolia et al., Toxicity of aromatic thiols in the human red blood cell. *J. Appl. Toxicol.* **9**, 113–118 (1989).

81a. T. De Sauza and S. P. Bhatra, *Anal. Chem.* **48**(19), 2234–2240 (1976).

82. E. J. Moran, O. D. Easterday, and B. L. Oser, Acute oral toxicity of selected flavor chemicals. *Drug Chem. Toxicol.* **3**, 249–258 (1980).

83. C. E. Turner et al., Analysis of micro-encapsulated D-limonene dimercaptan, a possible herbicide marker for Cannabis sprayed with paraquat, using gas chromatography. *Bulletin on Narcotics* **33**, 43–54 (1981).

84. S. Haworth et al., Salmonella mutagenicity test results for 250 chemicals. *Environ. Mutagenesis* **5**, 1–142 (1983).

85. L. Kremer and L. D. Spicer, Gas chromatographic separation of hydrogen sulfide, carbonyl sulfide, and higher sulfur compounds with a single pass system. *Analt. Chem.* **45**, 1963–1964 (1973).

86. A. Hellwig and D. Hempel, Quantitative determination of trichloromethyl-sulfenyl chloride in water and air (Ger). *Z. Chem.* **7**, 315–316 (1967).

87. H. Althoff, [Fatal perchloromethyl mercaptan intoxication (author's transl)]. [German]. *Archiv fur Toxikologie* **31**, 121–135 (1973).

88. J. C. Gage, The subacute inhalation toxicity of 109 industrial chemicals. *British Journal of Industrial Medicine* **27**, 1–18 (1970).

89. Z. Leifer et al., An evaluation of tests using dna repair-deficient bacteria for predicting genotoxicity and carcinogenicity. A report of the U. S. EPA's Gene-TOX Program. *Mutat. Res.* **87**, 211–297 (1981).

90. J. W. Dillwith and R. A. Lewis, Inhibition of DNA polymerase β activity in isolated bovine liver nuclei by captan and related compounds. *Pestic. Biochem. Physiol.* **14**, 208–216 (1980).

91. P. Jappinen et al., Volatile metabolites in occupational exposure to organic sulfur compounds. *Arch. Toxicol.* **67**, 104–106 (1993).

92. D. L. Opdyke, Monographs on fragrance raw materials. *Fd. Cosmet. Toxicol.* **17**, 357–390 (1979).

93. K. R. Butterworth et al., Short-term toxicity of dimethyl sulphide in the rat. *Fd. Cosmet. Toxicol.* **13**, 15–22 (1975).

94. H. U. Aeschbacher et al., Contribution of coffee aroma constituents to the mutagenicity of coffee. *Fd. Toxicol.* **27**, 227–232 (1989).

95. L. W. Wattenberg, V. L. Sparnins, and G. Barany, Inhibition of *N*-nitrosodiethylamine carcinogenesis in mice by naturally occurring organosulfur compounds and monoterpenes. *Cancer Res.* **49**, 2689–2692 (1989).

96. S. J. Tsai, S. N. Jenq, and H. Lee, Naturally occurring diallyl disulfide inhibits the formation of carcinogenic heterocyclic aromatic amines in boiled pork juice. *Mutagenesis* **11**, 235–240 (1996).

97. D. L. Opdyke, Monographs on fragrance raw materials. *Fd. Cosmet. Toxicol.* **17**, 695–923 (1979).

98. J. R. Brown and E. Mastromatteo, Acute toxicity of three episulfide compounds in experimental animals. *Am. Indus. Hyg. Assoc. Jo.* **25**, 560–563 (1964).

99. H. Druckrey et al., Carcerogene alkylierende Substanzen. 3. Alkyl-halogenide, -sulfate, -sulfonate und ringgespannte Heterocyclen (German). *Z. Krebsforschung* **74**, 241–273 (1970).

99a. National Cancer Institute Report, 1998.

100. K. T. Augusti and M. E. Benaim, Effect of essential oil of onion (allyl propyl disulphide) on blood glucose, free fatty acid and insulin levels of normal subjects. *Clinica Chimica Acta* **60**, 121–123 (1975).

101. C. Papageorgiou et al., Allergic contact dermatitis to garlic (Allium Sativum L.). Identification of the allergens: the role of mono-, di-, and trisulfides present in garlic. A comparative study in man and animal (guinea-pig). *Arch. Dermatol. Res.* **275**, 229–234 (1983).

102. J. I. Morton and B. V. Siegel, Effects of oral dimethyl sulfoxide and dimethyl sulfone on murine autoimmune lymphoproliferative disease. *Proc. Soc. Exptl. Biol. Med.* **183**, 227–230 (1986).

103. D. McCabe et al., Polar solvents in the chemoprevention of dimethylbenzanthracene-induced rat mammary cancer. *Archives of Surgery* **121**, 1455–1459 (1986).

104. P. J. O'Dwyer et al., Use of polar solvents in chemoprevention of 1,2-dimethylhydrazine-induced colon cancer. *Cancer* **62**, 944–948 (1988).

105. W. Szybalski, special microbiological systems II. Observations on chemical mutagenesis in microorganisms. *Ann. N. Y. Acad. Sci.* **76**, 475–489 (1958).

106. H. F. Smyth Jr. et al., Range-finding toxicity data; List VI. *Am. Indus. Hyg. Assoc. J.* **23**, 95–108 (1977).

107. D. S. Choi et al., Glyceraldehyde-3-phosphate dehydrogenase as a biochemical marker of cytotoxicity by vinyl sulfones in cultured murine spleen lymphocytes. *Cell Biol. Toxicol.* **11**, 23–28 (1995).

108. W. R. Leopold, J. A. Miller, and E. C. Miller, Comparison of some carcinogenic, mutagenic, and biochemical properties of S-vinylhomocysteine and ethionine. *Cancer Res.* **42**, 4364–4374 (1982).

109. M. D. Shelby et al., Mouse dominant lethal and bone marrow micronucleus studies on methyl vinyl sulfone and divinyl sulfone. *Mutat. Res.* **250**, 431–437 (1991).

109a. F. Caujolle et al., *C. R. Hebd Acad. Sci. Ser. D.* **260**, 327–330 (1965).

110. K. L. Dearfield et al., Genotoxicity in mouse lymphoma cells of chemicals capable of Michael addition. *Mutagenesis* **6**, 519–525 (1991).

111. C. S. Aaron et al., Detection of methanesulfonic acid in ethyl methanesulfonate. *Anal. Biochem.* **54**, 307–309 (1973).

112. A. R. Jones and K. Edwards, Alkylating esters. VII. The metabolism of iso-propyl methanesulphonate and iso-propyl iodide in the rat. *Experientia* **29**, 538–539 (1973).

113. E. R. Fluck, L. A. Poirer, and H. W. Ruelius, Evaluation of a DNA polymerase-deficient mutant of *E. coli* for the rapid detection of carcinogens. *Chem.-Biol. Interact.* **15**, 219–231 (1976).

114. R. E. McMahon, J. C. Cline, and C. Z. Thompson, Assay of 855 test chemicals in ten tester strains using a new modification of the Ames test for bacterial mutagens. *Cancer Res.* **39**, 682–693 (1979).

115. E. Zieger and D. A. Pagano, Mutagenicity of the human carcinogen treosulphan in *Salmonella. Environmental & Molecular Mutagenesis* **13**, 343–346 (1989).

116. G. W. Dawson et al., The acute toxicity of 47 industrial chemicals to fresh and salt water fishes. *J. Hazard. Mater.* **1**, 303–318 (1977).

117. N. Kinae et al., Studies on the toxicity of pulp and paper mill effluents II. Mutagenicity of the extracts of the liver from spotted sea trout. *Water Res.* **15**, 25–30 (1981).

118. X. S. Wang and R. R. Suskind, Comparative studies of the sensitization potential of morpholine, 2-mercaptobenzothiazole and 2 of their derivatives in guinea pigs. *Contact Derm* **19**, 11–15 (1988).

119. B. Magnusson and A. M. Kligman, The identification of contact allergens by animal assay. The guinea pig maximization test. *J. Invest. Dermatol.* **52**, 268–276 (1969).

120. T. Estlander, R. Jolanki, and L. Kanerva, Dermatitis and urticaria from rubber and plastic gloves. *Contact Derm.* **14**, 20–25 (1986).

121. A. K. Bajaj et al., Shoe dermatitis in India. *Contact Derm.* **19**, 372–375 (1988).

122. D. S. Wilkinson, M. G. Budden, and E. M. Hambly, A 10-year review of an industrial dermatitis clinic. *Contact Derm.* **6**, 11–17 (1980).

123. J. Foussereau et al., Allergy to MBT and its derivatives. *Contact Derm.* **9**, 514–516 (1983).

124. C. W. Lynde et al., Patch testing with mercaptobenzothiazole and mercapto-mixes. *Contact Derm.* **8**, 273–274 (1982).

125. R. K. Hinderer et al., Mutagenic evaluations of four rubber accelerators in a battery of *in vitro* mutagenic assays. *Environ. Mutagenesis* **5**, 193–215 (1983).

126. K. Sitarek, B. Berlinska, and B. Baranski, Effect of oral sulfenamide TS administration on prenatal development in rats. *Teratog. Carcinog. Mutag.* **16**, 1–6 (1996).

127. B. Berlinska, K. Sitarek, and B. Baranski, Evaluation of the teratogenic potential of sulfenamide TS in rats. *Teratology* **53**, 36A (1996).

128. M. Ema et al., Evaluation of the teratogenic potential of the rubber accelerator *N*-cyclohexyl-2-benzothiazylsulfenamide in rats. *J. Appl. Toxicol.* **9**, 187–190 (1989).

129. T. Adachi, A. Tanaka, and T. Yamaha, Absorption, distribution, metabolism and excretion of *N*-cyclohexyl-2-benzothiazyl sulfenamide (CBS), a vulcanizing accelerator, in rats (Japanese). *Radioisotopes* **38**, 255–258 (1989).

130. Anonymous, *N-N*-Dicyclohexyl-2-benzothiazolsulfenamid. *Berufsgenossenschaft der Chemischen Industrie* **242**, 1–13 (1994).

CHAPTER NINETY-FIVE

Organophosphorus Compounds

Jan E. Storm, Ph.D

1.0 Introduction

Organophosphate pesticides are a highly diverse group of chemicals to which workers may be exposed during manufacture and formulation and during or after application for their intended uses (1, 2). They are all characterized by their ability to inhibit the enzyme acetylcholinesterase (AChE) that deactivates the neurotransmitter acetylcholine (ACh).

Compounds in this class are numerous and have been categorized in many ways according to the nature of the substituents. Gallo and Lawryk, for example, categorized them into four main categories (Groups I-IV) based on the characteristics of the leaving group (X) (1). Group I compounds, phosphorylcholines, have a leaving group that contains a quaternary nitrogen and are among the most potent organophosphates (e.g., Shradan). Group II compounds, fluorophosphates, have a fluoride leaving group and are also generally highly toxic (e.g., diisopropyl fluorophosphate). Group III compounds have leaving groups that contain cyanide or a halogen other than fluoride and are generally less potent than Groups I or II (e.g., Parathion). Group IV contains most of the organophosphates used as insecticides today. These compounds have alkoxy, alkylthio, aryloxy, arylthio or heterocyclic leaving groups and a wide variety of other substituents.

Another classification scheme is based on the nature of the atoms that immediately surround the central phosphorus atom and results in 14 different categories (2). According to this scheme, phosphates are the prototype for the entire class and are those compounds where all four atoms that surround the phosphorus atom are oxygen (e.g., dichlorvos,

---

Patty's Toxicology, Fifth Edition, Volume 7, Edited by Eula Bingham, Barbara Cohrssen, and Charles H. Powell.
ISBN 0-471-31940-6 © 2001 John Wiley & Sons, Inc.

mevinphos). Sulfur-containing organophosphate compounds (phosphorothioates; phosphorothiolates; phosphorodithioates; phosphorodithiolates) are far more numerous than phosphates and include well recognized organophosphate insecticides such as parathion, diazinon, chlorpyrifos, etc. Other groups contain nitrogen (phosphoramides and phosphorodiamides), nitrogen and sulfur (phosphoramidothionates and phosphoramidothiolates), carbon (phosphonates and phosphinates), or carbon and sulfur (phosphonothionates, phosphonothionothiolates and phosphinothionates).

All aspects of organophosphate chemistry, toxicity, analysis, and exposure potential have been previously reviewed (1–9). Additionally information regarding the toxicity of this class of compounds has expanded greatly in recent years as a result of toxicity data supplied by registrants to the U.S. EPA's Office of Pesticides to support reregistration. These data are being made publically available by the U.S. EPA on their internet web site (9a). The following discussion draws heavily from recent reviews but also includes summaries of relevant toxicity data submitted to the U.S. EPA available when this chapter was completed. Due to space limitations detailed data reviews are included here for only 30 organophosphate pesticides. Information on other pesticides registered or undergoing reregistration in the United States can be readily obtained from the previously mentioned website.

1.1 Production and Use

Organophosphates are the most widely used insecticides today; more than 40 are currently registered for use in the United States (3). They were the number one cause of symptomatic illnesses reported in 1996, according to the American Association of Poison Control Centers (3). The earliest organophosphates (e.g., schradan) were initially developed as war gases, and they have also been used as therapeutic agents, gasoline additives, hydraulic fluids, cotton defoliants, fire retardants, plastic components, growth regulators, and industrial intermediates (363). Their use as insecticides far exceed these uses.

1.2 Acute Toxicity

1.2.1 Mechanism of Action

All organophosphate pesticides exert toxicity via a common mechanism of action — binding to and phosphorylation of the enzyme acetylcholinesterase (10, 11). This causes its inhibition and a buildup of the neurotransmitter acetylcholine at central and peripheral nervous system synapses that result in overstimulation at muscarinic and nicotinic cholinergic synapses. Typical manifestations of overstimulation at muscarinic receptors include hypersalivation, excess lacrimation, hyperhydrosis, miosis, intestinal cramps, vomiting, diarrhea, urinary and fecal incontinence, bronchorrhoea, and bronchoconstriction. Overstimulation at nicotinic synapses results in muscle cramps, fasciculation, weakness, paralysis and pallor. Central nervous system effects include anxiety, restlessness, dizziness, confusion, ataxia, convulsions, and respiratory and circulatory depression (12). Symptoms are more or less severe, depending on the compound, the dose, the route, frequency and duration of exposure, and the time of observation relative to the time of peak toxic effect. Effects may be immediate or delayed by hours or even days.

Acetylcholinesterase inhibition is not necessarily irreversible. Recovery of acetylcholinesterase activity occurs in some cases via hydrolytic removal of the inhibiting phosphoryl moiety. The rate of recovery, however, like the rate of binding, varies considerably, depending on the specific compound. Additionally, some organophosphate–acetylcholinesterase complexes are characterized by a process termed "aging" in which the rate of reactivation declines soon after exposure due to dealkylation of the phosphorlyated acetylcholinesterase (2). When this happens, reactivation of inhibited acetylcholinesterase is not possible. *De novo* synthesis of the acetylcholinesterase enzyme itself also leads to recovery of acetylcholinesterase activity. The rate of acetylcholinesterase synthesis in the nervous system varies between about 12 and 24 hours, depending on the anatomical location, although longer times have been reported (2).

The acute toxicities of organophosphate compounds generally vary over two orders of magnitude. The World Health Organization (WHO) has defined the following hazard classes for organophosphate pesticides based on their acute toxicity: extremely hazardous, highly hazardous, moderately hazardous, or slightly hazardous (8). The acute toxicity and WHO hazard classification of the 30 organophosphate compounds included in this review are summarized in Table 95.1.

Binding affinity of the parent compound and/or its metabolic products for acetylcholinesterase is a major determinant of organophosphate potency. Structural characteristics such as the lability of the P=X bond and the overall hydrophobicity and steric characteristics of the molecule determine binding affinity (1, 12a). For example, P=O compounds are generally more potent than P=S compounds because the greater electronegativity of O weakens the P=X bond, thereby enhancing binding to acetylcholinesterase; compounds that have longer n-alkyl R groups are generally more potent than those that have shorter n-alkyl groups because their greater hydrophobicity enhances binding to acetylcholinesterase. Compounds that have less bulky R or X groups are generally more potent than those that have bulkier groups which hinder interaction with acetylcholinesterase (1, 12a). Differences in binding affinity for other esterases relative to acetylcholinesterase and in the rate and degree of binding reversibility and aging of phosphorylated acetylcholinesterase also impact potency (1, 2).

Toxicodynamic processes alone, however, do not determine organophosphate potency. Equally important are their toxicokinetic characteristics. For most organophosphates, absorption is rapid and complete, or nearly complete, via either oral or inhalation exposure, and distribution is usually extensive. Metabolic reactions that detoxify and toxify these compounds, however, are most often highly complex and variable. The overall balance between metabolic generation and elimination of more versus less toxic derivatives is a major factor that determines potency (13–18). For example, differences in the overall rates of detoxification (rather than toxification) contribute to the greater sensitivity of female rats to some organophosphates (parathion, methyl parathion, EPN, and chlorpyrifos) compared to male rats (16, 19, 20) and to the greater sensitivity of young animals to some organophosphates (parathion, methyl parathion, and EPN) compared to adults (18, 20, 21). Additionally, differences in the overall balance between generating and eliminating more versus less toxic derivatives following oral exposures compared to

Table 95.1. Summary of Acute Toxicity of 30 Organophosphates

Compound	Rat 1-Hour LC$_{50}$ (mg/m^3) M	Rat 1-Hour LC$_{50}$ (mg/m^3) F	Rat 4-Hour LC$_{50}$ (mg/m^3) M	Rat 4-Hour LC$_{50}$ (mg/m^3) F	Rat Oral LD$_{50}$ (mg/kg) M	Rat Oral LD$_{50}$ (mg/kg) F	WHO Hazard Classification[b]
Azinphos-methyl	396	310	155	132	5	4	High
Chlorpyrifos			> 2500[a]		155	82	Moderate
Coumaphos	1081	341			41	16	Extreme
Demeton	175		47		6	3	Extreme
Demeton-S-methyl			310	210	40	63	High
Diazinon			3500[a]		285	250	Moderate
Dichlorvos	340		455		80	56	High
Dicrotophos	610		90[a]		21	16	High
Dioxathion	1398[a]				43	23	High
Disulfoton	290	63	60	15	7	2	Extreme
EPN	106[a]				36	8	Extreme
Ethion			2310	450	191	21	Moderate
Fenamiphos	110[a]		91[a]		3	3	Extreme
Fenthion	1838	1637	507	454	245	215	Moderate
Fonofos	900[a]		460[a]		7	3	Extreme
Malathion			> 5299[a]		1375	1000	Slight
Methyl parathion	257	287	135[a]				Extreme
Mevinphos	74				6	3	Extreme
Monocrotophos	163	176	100		17	20	High
Naled					191	92	Moderate
Parathion	115		32		3	7	Extreme
Phorate	60	11			4	1	Extreme
Ronnel					1250	2630	Moderate
Sulfotepp	330	160	59	38	14	10	Extreme
Sulprofos	> 3840[a]		> 4130[a]		107	65	Moderate
Temephos			> 1300[a]		444[a]		Unlikely
TEPP	24		7		1		Extreme
Terbufos					2	2	Extreme
Trichlorphon			533[a]		173	136	Slight

[a]Sex not specified.
[b]International Programme on Chemical Safety (IPCS). The WHO (World Health Organization) Recommended Classification of Pesticides by Hazard and Guidelines to Classification 1996–1997. WHO/PCS/94.2.

inhalation exposures, due to first-pass hepatic metabolism, can contribute to differences in potency for the same compound when administered via these different routes.

1.2.2 Symptoms

In acute exposures, signs of organophosphate poisoning reflect stimulation of the autonomic and central nervous system that results from accumulation of acetylcholine.

Signs may occur within minutes or hours, depending on the compound and the route of exposure. Initially, signs are consistent with stimulation of the muscarinic receptors of the parasympathetic nervous system and include increased secretions (salivation, lacrimation, urination, defecation), bronchoconstriction, and miosis (6). This is followed by signs consistent with stimulation and subsequent blockage of nicotinic receptors and include tachycardia, hypertension, muscle fasiculation, tremors, muscle weakness, and/or flaccid paralysis. The most commonly reported symptoms in humans include headache, blurred vision, nausea, cramps, diarrhea, and tightness in the chest. Death may ensue as a direct result of respiratory failure.

If smaller doses of an organophosphate are received over a longer period of time, there may be no sign of toxicity despite a total dose that may equal or exceed a toxic acute dose. In these cases, the correlation of overt toxicity and inhibition of AChE is not clear. Plasma, brain, and RBC AChE activity may be completely inhibited, yet individuals appear normal. This may occur as a result of absorption and metabolic and/or distribution processes that slow the rate of AChE binding and/or as a result of other processes such as muscarinic receptor down-regulation that may tend to accommodate the effect of elevated ACh concentration.

1.2.3 Treatment

Treatment of organophosphate poisoning should consist of immediate removal from the contaminated environment, removal of any contaminated clothing, and washing of any contaminated skin. Oxygen and positive pressure respiration may be required. Atropine, a muscarinic cholinergic blocking agent that counteracts the effects of excessive acetylcholine at muscarinic receptors is administered. Pyridinium oxides (e.g., pralidoxime) that reactivates acetylcholinesterase are also frequently administered. In some cases, a CNS depressant such as diazepam may be given to reduce agitation, fasciculation, and convulsions (4). The following course of treatment and recommendations has recently been summarized in the "Recognition and Management of Pesticide Poisonings" (3). The reader should consult this volume for detailed information on dosing regimens.

1. Ensure that a clear airway exists. Intubate the patient and aspirate secretions if necessary. Administer oxygen by mechanically assisted pulmonary ventilation if respiration is depressed. Tissue oxygenation should be increased as much as possible before administering atropine to minimize the risk of ventricular fibrillation.
2. Administer atropine sulfate intravenously or intramuscularly if intravenous injection is not possible. Atropine antagonizes the effects of excessive ACh by competing with ACh at muscarinic receptors. It is effective, then, against muscarinic manifestations of poisoning but is ineffective against nicotinic actions, specifically muscle weakness and twitching and respiratory depression.
3. Consider the use of glycopyrolate infusion as an alternative or supplement to atropine sulfate. This compound acts similarly to atropine.
4. Administer pralidoxime in cases of severe poisoning in which respiratory depression, muscle weakness, and/or twitching are severe. Pralidoxime is a cholinesterase

reactivator and also slows the "aging" process of phosphorylated chlinesterase to a nonreactivatable form. When administered less than 48 hours after poisoning, pralidoxime relieves the nicotinic and the muscarinic effects of poisoning.

5. Decontamination must proceed concurrently with whatever resuscitative and antidotal measures are taken. Eyes should be flushed with copious amounts of clean water. Contaminated clothing should be promptly removed, and skin and hair should be washed with copious amounts of water and soap.
6. If organophosphate has been ingested, consideration should be given to gastric lavage, charcoal adsorption, and/or catharsis.
7. The poisoned individual should be closely observed for at least 72 hours to ensure that symptoms do not recur as atropinization is withdrawn. In very severe poisonings, especially with lipophilic and slowly metabolized compounds, elimination may require 5–14 days.
8. If pulmonary edema persists, administration of furosemide, a diuretic, may be considered.
9. Pulmonary ventilation should be monitored carefully, even after recovery from muscarinic symptomology.
10. Hydrocarbon aspiration that occurs as a result of ingesting liquid organophosphate concentrates should be treated as acute respiratory distress syndrome. In this case, pulmonary edema and poor oxygenation will not respond to atropine.
11. In cases of severe poisoning, cardiopulmonary function should be monitored by continuous ECG recording.
12. In some cases, despite atropine and pralidoxime therapy, convulsions may occur that can be treated with benzodiazepines.
13. The following are contraindicated in all cases of organophosphate poisoning: morphine, succinylcholine, theophylline, phenothiazines, and reserpine.
14. Individuals who have been clinically poisoned by organophosphate pesticide should not be reexposed until blood ChE levels return to normal levels.
15. Neither atropine nor pralidoxime should be administered prophylactically to workers.

1.3 Intermediate Syndrome

In addition to acute poisoning episodes, an "intermediate" syndrome has been described in some individuals beginning generally 24–96 hours after an acute poisoning episode (6). The intermediate syndrome is characterized by acute respiratory paresis and muscular weakness, primarily in the facial, neck and proximal limb muscles. Cranial nerve palsies are common, and a requirement for mechanical respiratory ventilation is common. Symptoms do not respond well to atropine and oximes; therefore treatment is mainly supportive. Artificial ventilation is frequently required in these cases to ensure survival. Cases of intermediate syndrome have been reported in individuals poisoned by monocrotophos, methamidophos, parathion, fenthion, methyl parathion, dimethoate, and fenthion.

1.4 Organophosphate-Induced Delayed Neuropathy

Some phosphate, phosphonate, and phosphoramidate organophosphates produce a delayed paralysis in animals and humans termed organophosphate-induced delayed neuropathy (OPIDN). This syndrome was first described in individuals who ingested an alcoholic extract of Jamaican ginger contaminated with the ortho-isomer of tri-*o*-tolyl phosphate (TOTP) during the years of Prohibition. It was characterized by initial flaccidity and muscle weakness in the arms and legs that gives rise to a clumsy, shuffling gait. These symptoms were replaced by a spasticity, hypertonicity, hyperreflexia, clonus and abnormal reflexes (6). Similar syndromes were described in individuals accidentally exposed to mipafox and leptophos (6). Overall, though, only these few organophosphates have been implicated as causes of delayed neuropathy in humans.

This syndrome has been intensively studied since its discovery, but its etiology has been only partially described. Hens and cats are the most sensitive species for this effect. In hens, exposure to agents known to cause OPIDN (TOTP, DFP, mipafox, and leptophos) is associated with a Wallerian "dying back" degeneration of large diameter axons and their myelinic sheaths in distal parts of the peripheral nerves and long spinal cord tracts. Biochemical studies have shown that OPIDN agents inhibit a neuronal, nonspecific carboxylesterase termed neuropathy target esterase (NTE), but the specific relationship between this binding and subsequent neuropathy is not understood.

Some epidemiological studies have suggested that individuals who are acutely poisoned by organophosphates can suffer long-term neuropsychiatric sequelae (22–24). Effects reported include deficits in performance on neurobehavioral tests that include memory, concentration and mood, and evidence of peripheral neuropathy in some cases. Case studies have sometimes indicated persistent headaches, blurred vision, muscle weakness, depression, memory and concentration problems, and irritability following acute exposures (25–28).

1.5 Cancer, Reproductive, and Developmental Effects

With few exceptions organophosphate pesticides are not carcinogenic in animal bioassays. Of the thirty compounds reviewed here positive results were obtained only when mice and/or rats were exposed to dichlorvos or methyl parathion (Table 95.2). However, in both cases, the relevance of observed responses to humans has been questioned. There are no human data that relate specific organophosphate exposures to cancer. Some studies have related the occurrence of cancer in farmers to pesticide use in general, but this connection has been challenged by expert panel reviews of the data (29, 30).

Reproductive effects following chronic exposures to organophosphates reviewed here have been more frequently observed. Teratogenic and/or developmental effects following gestational exposures have been less frequently detected. Both reproductive and developmental effects, however, occur at levels greater than those associated with significant AChE inhibition. Hence, protection against this adverse effect, it is anticipated, protects against the possibility of these effects as well. Due to space limitations, these reproductive/developmental effects are not discussed further. The reader is encouraged to consult the references in Table 95.3 for more detailed information.

Table 95.2. Summary of Genotoxicity and Carcinogenicity Data for Organophosphate Pesticides

Compound	Genotoxicity	Species	Result	References
Azinphos-methyl	Negative	Mice	Negative	584
		Rats	Negative	65
Chlorpyrifos	Positive	Dogs	Negative	202
Coumaphos	Negative	Rats	Negative	584
Demeton	Positive and negative	No data		
Demeton-S-methyl	Positive and negative	Rats	Negative	262
		Mice	Negative	262
Diazinon	Positive and negative	Rats	Negative	338
		Mice	Negative	338
Dichlorvos	Positive in vitro; negative in vivo	Rats	Negative	45, 130, 136
		Rats	Equivocal	131
		Mice	Negative	95, 130
		Mice	Equivocal	131
Dicrotophos	Positive	Rats	Negative	95
		Mice	Negative	95
Dioxathion	Positive and negative	Rats	Negative	585
		Mice	Negative	585
Disulfoton	Positive and negative	Rats	Negative	584
		Mice	Negative	584
EPN	No data	Rats	Negative	277
Ethion	Positive and negative	Rats	Negative	584
		Mice	Negative	584
Fenamiphos	Negative	Dogs	Negative	
		Rats	Negative	307
		Mice	Negative	147
Fensulfothion	Negative	Dogs	Negative	316
Fenthion	Negative and positive	Dogs	Negative	584
		Rats	Negative	336, 338, 586
		Mice	Negative	338, 584
Fonofos	No data	Rats	Negative	95
		Dogs	Negative	95
Malathion	Positive and negative	Rats	Negative	338
		Mice	Negative	338
Methyl parathion	Positive and negative	Rats	Negative	338, 584
		Mice	Negative	338
Mevinphos	Positive and negative	Rats	Negative	448
		Mice	Negative	448
Monocrotophos	Positive	Dogs	Negative	458
		Rats	Negative	458
		Mice	Negative	458

Table 95.2. (*Continued*)

Compound	Genotoxicity	Carcinogenicity Species	Carcinogenicity Result	References
Naled	Positive and negative	Rats	Negative	584
		Mice	Negative	584
Parathion	Negative	Rats	Negative	370, 384, 584
		Mice	Negative	338
		Rats	Equivocal	338, 485
Phoratae	Negative	Rats	Negative	584
		Mice	Negative	584
Ronnel	No data	Dogs	Negative	587
		Rats	Negative	41
Sulfotepp	Positive and negative	Rats	Negative	
		Mice	Negative	
Sulprofos	Negative	Dogs	Negative	539
		Rats	Negative	539
		Mice	Negative	539
Temephos	Negative	Rats	Negative	584
TEPP	No data	No data		
Terbufos	Positive and negative	Rats	Negative	584
		Mice	Negative	
Trichlorfon	Positive and negative	Rats	Negative	584, 588
		Mice	Negative	584

1.6 Exposure Assessment

1.6.1 Worker

The most common and most important route of industrial exposure to organophosphates is by dermal contact. Almost without exception, organophosphate compounds are well absorbed across the skin, and when exposures are both dermal and respiratory, dermal intake frequently exceeds inhalation intake (31–33). For example, in a pesticide formulating plant, dermal intake was 184 mg/h and respiratory intake was 0.03 mg/h, and in orchard spraymen, mean dermal and respiratory exposures were 19 mg/h and 0.02 mg/h, respectively (34) Significant dermal exposure of workers may also occur when they enter a previously sprayed crop area for further cultivation or hand-harvesting. Because of this, the EPA frequently establishes reentry times after a crop is sprayed during which workers are not allowed to work in the field. Inhalation of pesticide dusts, vapors, mists, and gases may also present a hazard, especially for the more volatile organophosphates such as dichlorvos. Oral exposure is rarely a problem, except for accidental ingestion by children and in the case of suicide.

Air monitoring has been unsatisfactory for assessing agricultural worker exposure to organophosphates because of the variability in pesticide particulate characteristics and their dispersion over treated areas. More successful approaches have included monitoring

Table 95.3. Reproductive and Developmental Effects of Organophosphates

Reproductive/Developmental Toxicity

Compound	Study Type	Species	Exposure Route	Results	Reference
Azimphos-methyl	Developmental/teratogenic	Rats	Gavage	Negative	584
	Two-generation	Rats	Diet	Positive; NOAEL 5 ppm; LOAEL 15 ppm ($\downarrow$ pup viability, lactation index, litter weight)	584
Chlorpyrifos	Developmental/teratogenic	Rats	Gavage	Negative	211
		Mice	Gavage	Negative	589
	Two generation	Rats	Diet	Negative	211
Coumaphos	Developmental/teratogenic	Rats	Gavage	Negative	584
	Two-generation	Rats	Gavage	Negative	584
Demeton	Developmental/teratogenic	Mice	i.p. injection	Positive; NOAEL not identified; at 7 mg/kg/day $\downarrow$ fetal weight; $\uparrow$ mortality	254
Demeton-S-methyl	Developmental/teratogenic	Rats	Gavage	Negative	262
		Rabbits	Gavage	Negative	262
	Two-generation	Rats	Diet	Positive; NOAEL 5 ppm; LOAEL 25 ppm ($\downarrow$ pup viability, lactation index, body weight gain)	262
Diazinon	Developmental/teratogenic	Rats	Gavage	Negative	94
		Mice	Gavage	Positive; NOAEL not identified; at 0.2 mg/kg/day neuromuscular performance deficits	590
		Rabbits	Gavage	Negative	91, 92
		Hamsters	Gavage	Negative	85
		Dogs	Capsule	Positive; NOAEL not identified; at 1.2 mg/kg/day $\uparrow$ stillbirths	85
	Reproductive	Rats	Diet	Negative	83, 86
		Dogs	Capsule	Positive; NOAEL not identified; at 10 mg/kg/day testicular atrophy, arrested spermatogenesis	85
	Four-generation	Rats	Diet	Negative	89

Dichlorvos	Developmental/teratogenic	Rats	i.p. injection	Negative	379
		Rats	Gavage	Negative	209
		Rats	Inhalation	Negative	37
		Mice	Gavage	Negative	126
		Mice	Inhalation	Negative	126
		Rabbits	Gavage	Negative	584
		Rabbits	Inhalation	Negative	37
	Reproductive	Mice	Inhalation	Negative	37
		Rats	Inhalation	Positive; NOAEL not identified; at 2.4 mg/m^3 estrus delay	132
	Three-generation	Swine	Diet	Negative	133
		Rats	Diet	Negative	95
Dicrotophos	Developmental/teratogenic	Mice	i.p. injection	Negative	157
	Two-generation	Rats	Diet	Positive; NOAEL 2 ppm; LOAEL 5 ppm ($\downarrow$ pup survival)	95
Dioxathion	Three-generation	Rats	Diet	Negative	167
Disulfoton	Developmental/teratogenic	Rats	Gavage	Negative	171
		Rabbits	Gavage	Negative	45
	Reproduction	Rats	Gavage	Positive; NOAEL not identified; at 0.5 mg/kg/day $\downarrow$ pregnancy rate	182
	Two-generation	Rats	Diet	Positive; NOAEL 2 PPM; LOAEL 3 ppm ($\downarrow$ litter size, weights, viability)	584
EPN	Developmental/teratogenic	Mice	Gavage	Negative	591
Ethion	Developmental/teratogenic	Rats	Gavage	Negative	584
		Rabbits	Gavage	Negative	584
	Three-generation	Rats	Diet	Negative	584
Fenamiphos	Developmental/teratogenic	Rats	Gavage	Negative	584
		Rabbits	Gavage	Negative	584
	Two-generation	Rats	Diet	Positive NOAEL 30 ppm; LOAEL 40 ppm ($\downarrow$ pub weight gain)	584
Fensulfothion	Developmental/teratogenic	Rabbits	Gavage	Negative	316
	Three generation	Mice	Diet	Positive; NOAEL 1 ppm; LOAEL 5 ppm ($\downarrow$ lactation index)	316

Table 95.3. (*Continued*)

Compound	Study Type	Reproductive/Developmental Toxicity Species	Exposure Route	Results	Reference
Fenthion	Developmental/teratogenic	Rats	Gavage	Positive; NOAEL 4.2 mg/kg/day; LOAEL 18 mg/kg/day (↑ resorptions)	584
		Rabbits	Gavage	Positive; NOAEL 1 mg/kg/day; LOAEL 2.7 mg/kg/day (↑ resorptions)	584
	Two-generation	Rats	Diet	Positive; NOAEL 2 ppm; LOAEL 14 ppm (cytoplasmic vacuolation of epithelial ductal cells)	584
	Five-generation	Mice	Water	Positive; NOAEL not identified; at 9.5–10.5 mg/kg/day longer time to first litters; ↓ pup survival, growth	254
Fonofos	Developmental/teratogenic	Rats	Gavage	Negative	95
		Mice	Gavage	Positive; NOAEL 2 mg/kg/day; LOAEL 6 mg/kg/day (fetotoxicity)	95
	Three-generation	Rats	Diet	Negative	95
Malathion	Developmental/teratogenic	Rats	i.p. injection	Negative	379
		Rats	Diet	Positive; NOAEL not identified; at 500 mg/kg/day↓ implantation, live fetuses	592
	Reproduction	Rabbits	Gavage	Negative	378
		Rats	Diet	Positive; NOAEL not identified; at 4000 ppm ↓ pup viability, growth	377
Methyl parathion	Developmental/teratogenic	Rats	i.p. injection	Negative	592
		Rats	Gavage	Negative	584
		Mice	i.p. injection	Negative	593
		Rabbits	Gavage	Negative	
	Reproductive	Rats	i.p. injection	Positive; NOAEL 3.5 mg/kg/day; LOAEL 4.0 mg/kg/day (↓ ovarian weight, number of health follicles)	423

		Mice	Gavage	Positive; NOAEL not identified; at 9.4 mg/kg/day ↑ percentage of abnormal sperm	422
Mevinphos	Two-generation	Rats	Diet	Negative	584
	Developmental/teratogenic	Rats	Gavage	Negative	448
		Rabbits	Gavage	Negative	448
Monocrotophos	Two-generation	Rats	Diet	Negative	458
	Developmental/teratogenic	Rats	Gavage	Negative	458
		Rabbits	Gavage	Negative	458
	Two-generation	Rats	Diet	Positive; NOAEL 0.1 ppm; LOAEL 3 ppm (↓ pup weights)	458
Naled	Developmental/teratogenic	Rats	Gavage	Negative	255, 584
		Rabbits	Gavage	Negative	584
	Two-generation	Rats	Diet	Positive; NOAEL 6 mg/kg/day; LOAEL 18 mg/kg/day (↓ pup survival, body weight)	584
Parathion	Developmental/teratogenic	Rats	Gavage	Negative	45
		Mice	i.p. injection	Positive; NOAEL 4 mg/kg/day; LOAEL 8 mg/kg/day (↑ resorptions; ↓ fetal weight)	488
	Reproductive	Rats	Diet (1 × /day)	Positive; NOAEL not identified; at 10 ppm ↓ fertility, pup survival	486
	Developmental	Rats	s.c. injection	Positive; NOAEL 0.5 mg/kg/day; LOAEL 1.0 mg/kg/day (↓ body weight gain; tremor)	492
	Two-generation	Rats	Diet	Positive; NOAEL 10 ppm; LOAEL 20 ppm (↓ pup weight gain)	45
Phorate	Developmental/teratogenic	Rats	Gavage	Negative	584
		Rabbits	Gavage	Negative	584
	Three-generation	Rats	Diet	Positive; NOAEL 2 ppm; LOAEL 3 ppm (↓ pup survival, body weight; lactation indexes, viability indexes)	584
Ronnel	Developmental/teratogenic	Rats	Gavage	Positive; NOAEL 400 ppm; LOAEL 600 ppm (↑ incidence of extra ribs)	255
		Rabbits	Gavage	Positive; NOAEL not identified; at 12.5 mg/kg/day ↑ malformed fetuses	526

Table 95.3. (*Continued*)

Reproductive/Developmental Toxicity

Compound	Study Type	Species	Exposure Route	Results	Reference
Sulfotepp	Developmental/teratogenic	Rats	Gavage	Negative	593
		Rabbits	Gavage	Negative	533
Sulprofos	Developmental/teratogenic	Rats	Gavage	Negative	539
		Rabbits	Gavage	Negative	539
	Three-generation	Rats	Diet	Negative	539
Temephos	Developmental/teratogenic	Rabbits	Gavage	Negative	584
		Rabbits	Dermal	Negative	584
	Reproductive	Rats	Diet	Negative	584
	Three-generation	Rats	Diet	Negative	584
TEPP	No data				
Terbufos	Developmental/teratogenic	Rats	Gavage	Positive; NOAEL 0.1 mg/kg/day; LOAEL 0.2 mg/kg/day (↑ resorption, post implantation losses)	584
		Rabbits	Gavage	Positive; NOAEL 0.25 mg/kg/day; LOAEL 0.5 mg/kg/day (↓ fetal weight, ↑ resorptions)	584
	Three-generation	Rats	Diet	Positive; NOAEL 1 ppm; LOAEL 2.5 ppm (↓ pregnancy rate, male fertility, pup body weight gain)	584
Trichlorphon	Developmental/teratogenic	Rats	Diet	Positive; NOAEL 1125 ppm; LOAEL 2500 ppm (↓ ossification of skulls, vertebrae, sternebrae)	584
		Rats	Gavage	Positive; NOAEL not identified; at 480 mg/kg/day teratogenicity	584
		Mice	Gavage	Positive; NOAEL not identified; at 600 mg/kg/day ↑ cleft palate	567
		Rabbits	Gavage	Positive; NOAEL not identified; at 35 mg/kg/day ↑ abortions	584

Hamsters	Gavage	Positive; NOAEL 200 mg/kg/day; LOAEL 300 mg/kg/day (↑ fetal deaths, stunted and malformed fetuses)	584
Guinea pigs	Gavage	Positive; NOAEL not identified; at 100 mg/kg/day brain hypoplasia, ataxia, tremors	569
Guinea pigs	Gavage	Positive; NOAEL not identified; at 125 mg/kg/day ↓ brain weight	570
Pigs	Gavage	Positive; NOAEL not identified; at 56 mg/kg/day congenital ataxia, tremors, cerebellar and spinal hypoplasia	1
Two-generation Rats	Diet	Positive; NOAEL 500 ppm; LOAEL 1750 ppm (pulmonary, renal lestion, ↓ pup body weight	584
Three-generation Rats	Diet	Positive; NOAEL 300 ppm; LOAEL 1000 ppm (↓ pups/litter, pup body weight)	584

residues on foliage soil, and clothing. Absorbent pads attached to clothing have also been used. Alternatively, pesticide residues can be removed from the skin by an appropriate solvent and measured (35).

1.6.2 Community

Exposure of the general population to organophosphorus pesticides is through ingestion of treated food crops treated incorrectly or harvested prematurely before residues have declined, through contact with treated areas, or through domestic use of pesticide products.

Exposure from food should be minimal because tolerances for all pesticides registered in the United States are established for food crops on the basis of health considerations (6).

Because all organophosphates are subject to degradation by hydrolysis that yields water-soluble products that are believed nontoxic, any toxic hazard from their use in the environment and their subsequent dispersion to air and/or water is essentially short-term. Most organophosphates have comparatively low volatility (except dichlorvos) and contamination of areas beyond limits of 1–2 km from the spraying source have not been documented (15). A comprehensive review of spray drift that resulted from aerial application of pesticides has been published (36).

1.7 Biomonitoring

1.7.1 Red Blood Cell Acetylcholinesterase

For all organophosphate pesticides, neuronal acetylcholinesterase inhibition is widely regarded as a direct biomarker for toxicity, even though significant acetylcholinesterase inhibition is frequently observed in the absence of overt toxicity (18, 37–43a). Acetylcholinesterase is also present in red blood cells (RBCs) where, it has been shown, it is inhibited in parallel fashion to neuronal acetylcholinesterase after exposure to several organophosphates (10, 44, 45). Inhibition of RBC acetylcholinesterase itself has no known neurotoxic consequences. Therefore, the relationship between RBC acetylcholinesterase inhibition and overt toxicity is indirect. The rate of RBC acetylcholinesterase inhibition varies, depending on the specific organophosphate. Additionally, rates of RBC acetylcholinesterase recovery frequently differ from rates of neuronal acetylcholinesterase recovery for any particular organophosphate because RBCs cannot synthesize acetylcholinesterase. Recovery of RBC acetylcholinesterase activity is limited by the production of new RBCs which takes about 120 days (46, 47). Despite these factors, significant inhibition of RBC acetylcholinesterase is considered adverse because it involves the same molecular target that is responsible for the neurotoxic effects of organophosphates (10, 44, 45).

Organophosphates also bind to a wide variety of other esterases in blood and/or tissues — the specific esterase(s) involved and the magnitude of binding depend on the species and specific organophosphate involved (13, 14, 21, 48–50). Such binding, however, is not directly related to the neurotoxic consequences of these compounds. And, in some cases (e.g., parathion), it actually protects against neurotoxicity because it prevents compounds from reacting with acetylcholinesterase. Therefore, binding to these esterases is not considered adverse but is important to consider in relation to acetylcholinesterase binding.

For these reasons, monitoring of both plasma and RBC AChE activity is an accepted means of monitoring organophosphate exposure. Both interindividual and intraindividual variability is high, however, for both types of esterases, so comparisons are best made between an individual's pre- and postexposure values. Based on an analysis of intraindividual variation and analytical imprecision of cholinesterase measurements, it has been shown that a fall of 15% in RBC AChE and 7.5% in plasma ChE between two successive samples suggests significant inhibition, if the laboratory's analytical performance is such that the coefficients of variation for the two assays are 3.5 and 2.5%, respectively (8).

In some cases, such as unexpected poisonings, preexposure values are not available. In such instances, blood AChE and ChE activities can be compared to the following lower limits of normal activities in humans published recently by the U.S. EPA (3).

Method	Plasma	RBC	Blood	Whole Units
pH (Michel)	0.45	0.55		ΔpH per mL per h
pH Stat (Nabb-Whitfield)	2.3	8.0		µm per mL per min
BMC Reagent Set (Ellman–Boehringer)	1875		3000	mU per mL per min
Dupont ACA	< 8			Units per mL
Garry–Routh (Micro)			Male 7.8 Female 5.8	µM-SH per 3 mL per min
Technicon	2.0	8.0		

Lower levels than those listed generally indicate excessive absorption of a cholinesterase-inhibiting chemical. The WHO recommends using the Ellman method for monitoring plasma and blood ChE activities. In this method, the substrate, acetylthiocholine, is hydrolyzed by AChE. The thiocholine released reacts with 5,5-dithiobis (2-nitrobenzoic acid) to produce the yellow anion of 5-thio-2-nitrobenzoic acid. The rate of color production is measured at a wavelength of 412 nm and is proportional to AChE activity. Quinidine sulphate is added to suppress any plasma cholinesterase activity that originates from trapped plasma. Quinidine is omitted for measuring plasma cholinesterase.

For any exposed individuals', a fall in their RBC AChE activity to 70% of the preexposure value is an indication of overexposure. In these circumstances, the workers concerned should be removed from any further exposure until their cholinesterase activity returns to at least 80% of preexposure activity (8). A 70% action level for cholinesterase monitoring is also laid down by the American Conference of Governmental Industrial Hygienists (ACGIH) Biological Exposure Indexes (154) and the Deutsche Forschungsgemeinschaft (DFG), Maximum Concentrations at the Workplace and Biological Tolerance Values for Working Materials.

1.7.2 Urinary Alkyl Phosphates

Alkyl phosphates and phenols to which some organophosphates are hydrolyzed in the body can often be detected in the urine during pesticide absorption and up to about 48 hours thereafter. These analyses are sometimes useful in identifying and quantifying the actual pesticide to which workers have been exposed. Urinary alkyl phosphate and phenol analysis can demonstrate organophosphate absorption at doses lower than those required to depress cholinesterase activities and at doses much lower than those required to produce symptoms and signs (3). However, technical difficulties with analytical methods required to extract dialkyl phosphates, phosphorothioates, and phosphorodithioates from urine have so far prevented the widespread use of urinary analyses in biomonitoring organophosphate exposure. Moreover, few studies correlate excretion of alkyl phosphates with overt cholinergic toxicity.

Nevertheless, the WHO noted that 75 to 100% of most orally administered organophosphates are metabolized to water soluble metabolites which appear in the urine in the first 24–48 hours. The rate of appearance of urinary metabolites following dermal exposure generally occurs during several days. Twenty-four to 48 hour full urine collections are needed for near quantitative recovery. The WHO recommended an analytical method modified from methods published by Nutley and Cocker (51) and Reid and Watts (52) that involves concentrating alkyl phosphate metabolites from urine by removing water by azeotropic distillation with acetonitrile, treating the dry residue with pentafluorobenzylbromide, and analyzing derivatized alkyl phosphates by high-resolution gas chromatography with flame photometric detection (8).

1.8 Analytical Methods

Analytical methods for various media (air, water, food, etc.) for organophosphates are too numerous to describe. Useful references are Thompson, ATSDR Toxicological Profiles (53–57) NIOSH (1994), and NIOSH Pocket Guide to Chemical Hazards (available on-line at http://www.cdc.gov/npg/npg.html).

2.0 Azinphos-Methyl

2.0.1 CAS Number: *[86-50-0]*

2.0.2 Synonyms: Phosphorodithioic acid *O,O*-dimethyl-*S*-((4-oxo-1,2,3,-benzotirazin-3 (4*H*)-yl)methyl)ester; metiltriazotion guthion; Guthion, Gusation; S-(3,4-dihydro-4-oxobenzo[d][1,2,3]triazin-3-ylmethyl) *O,O*-dimethyl phosphorodithioate; Azinphos-Me; Gusathion; Methyl guthion; Guthion(R); Phosphorodithioic acid *O,O*-dimethyl *S*-[4-oxo-1,2,3-benzotriazin-3(4*H*)-yl)methyl] ester; phosphorodithioic acid *O,O*-dimethyl ester, *S*-ester with 3-mercaptomethyl-1,2,3-benzotriazin-4(3*H*)-one; Bayer 17147; Cotnion-methyl; Gusathion M; Azimil; Bay; R 1582; Gusthion M; Gution; *O,O*-dimethyl *S*-(4-oxobenzotriazino-3-methyl) phosphorodithioate; S-(3,4-dihydro-4-oxo-1,2,3-benzotriazin-3-ylmethyl) *O,O*-dimethyl phosphorodithioate; azinophos-methyl; bay 9027; bay 17147; bayer 9027; 3-(mercaptomethyl)-1,2,3-benzotriazin-4(3*H*)-one *O,O*-dimethyl phosphorodithioate; carfene; cotneon; crysthion 21; crysthyon; DBD; *O,O*-dimethyl *S*-(benzaziminomethyl) dithiophosphaate; *O,O*-dimethyl *S*-(1,2,3-benzotriazinyl-4-keto)-methyl phosphorodithioate; *O,O*-dimethyl *S*-(3,4-dihydro-4-keto-1,2,3-benzotriazinyl-3-methyl) dithiophosphate; dimethyl dithiophosphoric acid *N*-methylbenzazimidyl ester;

O,O-dimethyl *S*-(4-oxo-3*H*-1,2,3-benzotriazine-3-methyl) phosphorodithioate; *O,O*-dimethyl *S*-(4-oxo-1,2,3-benzotriazino (3)-methyl) thiothionophosphatae; *O,O*-dimethyl *S*-4-oxo-1,2,3-benzotriazin-3(4*H*)-ylmethyl phosphorodithioate; gothnion; gusathion-20; gusathion 25; gusathion k; gusathion methyl; 3-(mercaptomethyl)-1,2,3-benzotriazin-4(3*H*)-one *O,O*-dimethyl phosphorodithioate *S*-ester; methylazinphos; *N*-methylbenzazimide, dimethyl dithiophosphoric acid ester; metiltriazotion; Beetle Buster; Ketokil No. 52; Crysthyon 2L; Dimethoxy ester of (4-oxo-1,2,3-benzotriazin-3(4*H*)-yl) methyl ester of dithiophosphoric acid; dimethyl *S*-((4-oxo-1,2,3-benzotriazin-3(4*H*)-yl)methyl) phosphorodithioate; dimethyl *S*-(3-(mercaptomethyl)-1,2,3-benzotriazin-4(3*H*)-one) phosphorodithioate; Methyl gusathion; Guthion (Azinphos-Methyl)

2.0.3 Trade Names: Guthion®, Gusathion®, methyl Guthion, Azimil, Bay 9027, Bay 17147, Carfene

2.0.4 Molecular Weight: 317.34

2.0.5 Molecular Formula: $C_{10}H_{12}N_3O_3PS_2$

2.0.6 Molecular Structure:

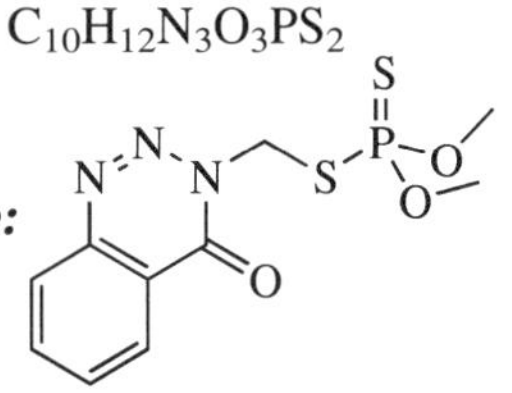

2.1 Chemical and Physical Properties

Pure azinphos-methyl is a white crystalline solid; the technical material is a brown, waxy solid.

Specific gravity: 1.44 at 20°C

Melting point: pure 73–74°C; technical 65–68°C; unstable at temperatures higher than 200°C

Solubility: soluble in methanol, ethanol, propylene glycol, xylene, and other organic solvents: slightly soluble in water (33 mg/L)

Vapor pressure: $< 3.8 \times 10^{-4}$ mmHg at 20°C

2.2 Production and Use

Azinphos-methyl is an insecticide used for control of pests on various fruits, melons, nuts, vegetables, field crops, ornamentals and shade trees. It is formulated as a liquid (20% active ingredient) and a wettable powder (20–50% active ingredient). Azinphos methyl is a restricted use pesticide.

2.3 Exposure Assessment

Absorption after dermal exposure constitutes the primary source of azinphos-methyl exposure (57a). Urinary alkyl phosphate metabolites of azinphos-methyl have been proposed as reliable markers of exposure regardless of the route of exposure (58–66).

2.4 Toxic Effects

2.4.1.1 Acute Toxicity.
Azinphos-methyl is a highly acutely toxic organophosphate that has oral LD$_{50}$s of 4–20 mg/kg, (242). The dermal LD$_{50}$ for azinphos-methyl is 220 mg/kg in both males and females (64a). One-hour LC$_{50}$ values in rats for azinphos-methyl are 310 to 396 mg/m^3. Four-hour LC$_{50}$ values in rats are 107–155 mg/m^3 (61). An oral LD$_{50}$ of 7 mg/kg and an LD$_{01}$ of 4–5 mg/kg in male and female mice, respectively, demonstrated that the dose–lethality curve is very steep (62). At the LD$_{50}$, cholinergic signs occur within 4–6 minutes; death usually occurs within 10–30 minutes. Rats tolerated cumulative oral doses of up to five to ten times the LD$_{50}$ given as daily doses of 0.5–2 mg/kg for 2–8 weeks demonstrating the lack of cumulative toxicity.

When 2, 6, or 12 mg/kg (males) and 1, 3, or 6 mg/kg (females) azinphos-methyl was given by gavage to rats, plasma and RBC cholinesterase inhibition occurred at the lowest dose tested (63). Brain cholinesterase inhibition and cholinergic effects occurred in males and females at 6 and 3 mg/kg, respectively. A high incidence of mortality (28% males and 83% females) was observed at the 6- and 12-mg/kg dose.

Azinphos-methyl did not cause acute delayed neurotoxicity in atropinized hens administered 0.1, 1.0, 10.0, 100.0 or 300 mg/kg in corn oil followed by a second dose at 21 days.

Azinphos methyl was judged to be a mild eye irritant, but nonirritating to the skin, and it did produce dermal sensitization in guinea pigs (242).

2.4.1.2 Chronic and Subchronic Toxicity.
A 90-dose LD$_{50}$ was 10.5 mg/kg for rats fed ground chow that was contaminated with azinphos-methyl (196). Using the 90-dose LD$_{50}$ and the single dose LD$_{50}$ of 11.0 mg/kg (196), a "chronicity factor" (single-dose LD$_{50/90}$ dose LD$_{50}$) of about 1 was calculated, indicating that azinphos-methyl did not have a cumulative effect.

When rabbits were treated dermally with 2, 4, 8 or 20 mg/kg azinphos-methyl for 6 h/day, 5 days/week, 15 times over three weeks, RBC cholinesterase activity was inhibited at 4 mg/kg/day and higher (242). Spleen and kidney weights increased in males and body weight gain decreased in females at this exposure level.

In rats fed diets that contained 15, 45, 90 (females), or 120 (males) ppm azinphos-methyl (about 0.9, 2.8, and 7.9 mg/kg/day (males) and 1.1, 3,2, and 7.0 mg/kg/day (females)) for 13 weeks, body weight gain decreased in both males and females fed 120 or 90 ppm, respectively (63). In males, 120 ppm (7.9 mg/kg/day) was associated with increased reactivity and autonomic cholinergic; no cholinergic signs were associated with 45 or 15 ppm. In females, 90 ppm (7.0 mg/kg/day) was associated with neuromuscular cholinergic signs and perianal staining; 45 ppm (3.2 mg/kg/day) was associated with increased reactivity and urine stains; no cholinergic effects were associated with the 15-ppm diet. Brain and RBC cholinesterase was inhibited at all dietary levels. In dogs given dietary levels of 0, 20, 50, 100, 200, or 400 ppm azinphos-methyl for 19 weeks, dose-related cholinesterase inhibition (whole blood) was observed at all dose levels. No information was provided on the occurrence of cholinergic effects.

In rats exposed to 0.195, 1.24, or 4.72 mg/m^3 azinphos-methyl aerosol for 6 h/day 5 days/wk, for 12 weeks, no effects occurred at the two lower exposure levels (242). Male rats exposed to 4.72 mg/m^3 showed lower body weight gain, and plasma and RBC cholinesterase inhibition (30–40%) occurred in both sexes exposed to 4.72 mg/m^3. Brain cholinesterase activity was not affected at any dose.

2.4.1.3 Pharmacokinetics, Metabolism, and Mechanisms. Studies that showed that oral, intraperitoneal, inhalation, and intravenous LD$_{50}$s and LC$_{50}$s are equivalent on a milligram per kilogram basis indicate that azinphos-methyl is equally well absorbed following oral or inhalation exposures. Neither oral nor inhalation absorption has been quantified. However, a study in humans showed that about 70% of an intravenous dose was absorbed and excreted within 5 days (66). About 30% of the dose was excreted in the urine in the first 24 hours after exposure; 20% was excreted in the second 24 hours, and 11% was excreted in the third 24 hours. An i.v. elimination half-life of 30 hours was calculated.

Dermal absorption of azinphos-methyl can be considerable. Dermal absorption of azinphos-methyl was determined in rats exposed to 0.93, 9.3 and 93 μg azinphos-methyl/cm^2 as the wettable powder (equivalent to 0.056, 0.56, or 5.6 mg/kg) on their clipped dorsal skin (242). By 10 hours, 32%, 22%, and 24% of the applied doses respectively, remained on the skin, suggesting that absorption had been 68%, 78%, and 76% of the applied dose (uncorrected using intravenous absorption), respectively. To simulate worker exposure, the test site of rats exposed for 24, 72 and 168 hours was wiped with a moistened gauze pad after 10 hours of exposure. Maximum systemic absorption occurred 168 hours after exposure, and 42% 22% and 18% of the applied dose was recovered in urine, feces, carcass, and cage wash combined for the 0.056, 0.56 and 5.6 mg/kg doses, respectively. In humans, about 16% of an applied dermal dose (4 μg/cm^2 on ventral forearm) was absorbed within 5 days. About 5.5% of the applied dose was excreted in urine within the first 24 hours, 5% was excreted during the second 24 hours, and 3% was excreted during the third 24 hours (corrected using intravenous excretion).

By 72 hours after oral dosing of rats with azinphos-methyl, 92–109% of the dose had been excreted. Between 63–79% of the dose was eliminated in urine, and between 20–27% was eliminated in feces. The highest residual concentrations of dose occurred in the blood, kidney, live, lung, and brain.

Azinphos methyl requires metabolic activation to azinphos-methyl oxon by microsomal mixed-function oxidases to inhibit cholinesterase (67). However, hydrolytic deactivation of the active metabolic is also very rapid (67). *In vitro* studies of azinphos-methyl metabolism indicate that metabolism of azinphos-methyl in rats proceeds largely through the actions of glutathione *S*-transferase and a mixed-function oxidase (242).

Cysteinyl methyl benzazimide sulfone (13–20% of the dose) and methyl sulfonly methyl benzazimide (14–20% of the dose) were the major urinary metabolites in rats given oral doses of azinphos-methyl (67). In feces, the methyl sulfonly methyl benzazimide, cysteinyl methyl benzazimide sulfoxide, desmethyl isoazinphos-methyl, azinphos-methyl oxygen analog, and methyl thiomethyl benzazimide were identified, but did not comprise greater than 5% of the administered dose. No azinphos-methyl or glucuronic or sulfate conjugates were found in urine or feces.

2.4.1.4 Reproductive and Developmental. Azinphos-methyl did not cause developmental effects or fetotoxicity at maternally nontoxic doses when given to rats (0.5–2 mg/kg/

day) or rabbits (1–6 mg/kg/day) during gestation, at maternally nontoxic doses (242).

In a two-generation reproductive study, rats were given diets that contained 5, 15, or 45 ppm azinphos-methyl (0.25, 0.75 or 2.25 mg/kg/day) (242). Increased dam mortality, decreased body weight of parental males and F1 males and females, and cholinergic signs ("poor condition", convulsions) occurred among rats given 45 ppm. Reduced pup viability and lactation indexes (death between postnatal days 0–5 and 5–28) and decreased mean total litter weights at weaning on postnatal day 28 occurred among rats given 15 and 45 ppm. No adverse effects occurred among rats given 5 ppm. In a supplementary study, rats were given diets that contained 15 or 45 ppm azinphos-methyl (0. 43, 1.30 or 3.73 mg/kg/day (males); 0.55, 1.54, or 4.87 mg/kg/day (females)) for one year. The pup viability index (death of offspring during postnatal days 0–5) was reduced, and pup weights decreased at postnatal days 14 and 21 among the 15- and 45-ppm groups, but not the 5-ppm group. At 45 ppm, reduction in brain cholinesterase activity occurred in pups on postnatal days 5 and 28; a reduction in brain weight occurred on postnatal day 5 but not day 28.

2.4.1.5 Carcinogenesis. In a 52-week study, dogs were given diets containing 5, 25, or 125 ppm azinphos-methyl (equivalent to 0.149, 0.688, or 3.844 mg/kg/day (males); 0.157, 0.775, or 4.333 (females)) (242). Mucoid diarrhea occurred in males given 25 ppm and in males and females that received 125 ppm. Both sexes given 125 ppm exhibited inhibition of plasma, RBC, and brain cholinesterase that began at week 4 of treatment and continued until week 52. In an earlier study, dogs given diets that contained 20 ppm azinphos-methyl for 36 weeks exhibited "irregular, slight" RBC cholinesterase depression, dogs given diets that contained 50 ppm for 15 months exhibited "slight to moderate" RBC cholinesterase inhibition, dogs given diets that contained 100 ppm exhibited "moderate" RBC cholinesterase inhibition; and dogs given diets that contained 150 ppm for 27 weeks followed by a diet that contained 300 ppm for 21 weeks exhibited "severe" (>75% inhibition) RBC cholinesterase inhibition (64). Cholinergic signs occurred only in dogs given 300 ppm.

When rats were given diets that contained 5, 15, and 45 ppm azinphos-methyl (equivalent to 0, 0.25, 075, and 2.33 mg/kg/day (males); 0.31, 0.96 and 3.11 mg/kg/day (females)) for up to 24 months, no cholinergic signs occurred, nor was clinical chemistry affected (242). There was, however, a marked inhibition of RBC, plasma and brain cholinesterase at 45 ppm, inhibition of RBC cholinesterase in males at 15 ppm, and, inhibition of RBC and plasma cholinesterase in females at 15 ppm. At 5 ppm, RBC cholinesterase decreased by 12% in male rats. When rats were fed a diet that had 50 ppm azinphos-methyl for 47 weeks followed by 100 ppm for 49 weeks, consistently depressed brain, plasma, and RBC cholinesterase activity occurred. When they were given a diet of 20 ppm for 97 weeks, depression of plasma and RBC cholinesterase activity occurred that decreased as the study progressed; exposure to diets of 2.5 or 5 ppm had no effect (64). Convulsions were seen in a "small number of female rats" (*sic*) (5/40) after increasing their dietary exposure from 50 to 100 ppm.

When azinphos-methyl was fed to mice at 0, 5, 20, or 40 ppm (about 0.79, 3.49, 11.33 (males); 0.98, 4.12, 14.30 (females)) for two years, no compound-related toxicity or evidence of cancer occurred (242). However, up to 80% inhibition of plasma, RBC, and brain cholinesterase activity occurred in mice given 20 and 40 ppm. At 5 ppm, RBC cholinesterase was still slightly inhibited.

In a chronic bioassay, rats were fed diets that contained 78 or 156 ppm (males) or 62.5 or 125 pm (females) azinphos-methyl for 80 weeks, then observed for 34-35 weeks, and mice were fed diets that contained 31.3 or 62.5 ppm (males) or 62.5 or 125 ppm (females) for 80 weeks, then observed for 12–13 weeks (65). Body weight gain decreased in male rats and mice given either diet and in female rats or mice given the high dose diet. Signs of organophosphate intoxication (hyperactivity, tremors, dyspnea) occurred "in a few animals of both species" (*sic*) fed the higher dose diets. During the second year, cholinergic toxicity occurred at both dietary levels. An equivocal increase in the incidence of tumors of the pancreatic islets and follicular cells of the thyroid suggested, but did not provide sufficient evidence of, carcinogenicity in male rats. There was no increased incidence of tumors of any kind among mice.

2.4.1.6 Genetic and Related Cellular Effects Studies. Azinphos-methyl with and without metabolic activation showed no evidence of mutagenicity in *Salmonella typhirium* (242). Azinphos-methyl was also negative in the reverse mutation induction assay with *Sacharomyces cerevisiae* and in the primary rat hepatocyte unscheduled DNA synthesis assay (242). In an *in vitro* cytogenetics assay using human lymphocytes, azinphos-methyl was clastogenic only with a metabolic activating system (242). However, a clastogenic effect was not observed in femoral marrow prepared from mice treated intraperitoneally with 5 mg/kg azinphos-methyl (242).

2.3.5   Biomonitoring/Biomarkers

Among rats exposed dermally to 100, 200 or 400 µg azinphos-methyl on the shaved dorsal skin in the scapular region, there was a good correlation between dose and amount of dimethylthiophosphate in urine (69).

Studies of agricultural field workers exposed to azinphos-methyl have shown that they excrete dimethyl phosphate (DMP), dimethyl thiophosphate (DMTP), and dimethyl dithiophosphate (DMDTP) in urine (59, 60, 69–71). However, urinary metabolite excretion is not always well correlated with serum or RBC cholinesterase inhibition (60, 71).

2.4.2   Human Experience

2.4.2.2.1 Acute Toxicity. A pilot who had spilled azinphos-methyl concentrate on his hands experienced visual disturbances, headache, tightness in the chest, abdominal cramps, nausea, vomiting, weakness and some excessive salivation (72).

2.4.2.2.2 Chronic and Subchronic Toxicity. Work in orchards treated with azinphos methyl is associated with blood cholinesterase activity reductions of up to 70% (30% of pre exposure baseline) and that intake is primarily due to dermal exposures (70). It was estimated that workers who sprayed apple orchards (with solutions of 0.5 to 6 lb of 25% wettable powder per 100 gallons of water) were exposed to 0.05 to 2.55 mg/m^3 (average 0.64 mg/m^3) azinphos-methyl for periods of 15 to 45 minutes; workers responsible for filling tanks were exposed to 0.26 to 6.20 mg/m^3 (average 2.76 mg/m^3) azinphos-methyl for brief periods; and formulators were exposed to 1.07 to 9.64 mg/m³, presumably for 8 hours/day. In all cases, dermal exposures exceeded inhalation exposure when quantified on a milligram per day basis. Serum cholinesterase activities were slightly depressed (78–91% of preexposure levels) after exposure, but no other effects were noted (73).

Orchardists who sprayed azinphos-methyl (wettable powder) were potentially exposed to an average concentration of 0.05 mg/m³ (range 0.02–0.11 mg/m³) based on air concentrations "measured near the spraymen during spraying operations," although inhalation exposures were probably significantly less than this because workers reportedly wore respirators. Dermal exposures were estimated at from 9 to 43 µg/kg. No evidence of serum or RBC cholinesterase inhibition was observed (57a). However, in another study of orchard workers, where exposures were primarily dermal (ranging up to 8,315 µg/hand wipe sample in thinners; up to 14,498 µg/sample on shirts of harvesters), median RBC cholinesterase activity declined by about 19% during a 6-week spraying season, and median plasma activity fell by about 12% (60).

Workers engaged for 5 days in thinning a peach orchard after it was treated with azinphos-methyl, showed very slightly inhibited RBC cholinesterase activity (mean decrease of about 6% compared to baseline) and essentially no change in plasma cholinesterase activity. No symptoms associated with cholinesterase inhibition occurred. Daily urinary excretion of dimethyl phosphate (DMP) and dimethyl phosphorothionate (DMPT) increased in exposed workers, and daily mean DMP and DMTP excretion was highly correlated with the mean percentage decline in RBC cholinesterase activity from baseline (74).

No changes in plasma or RBC cholinesterase activities occurred in human volunteers given oral doses 4, 4.5 or 6 mg/day (0.06, 0.06 or 0.09 mg/kg/day) (75); 7, 8 or 9 mg/day (0.10, 0.11 or 0.13 mg/kg/day) (76); or, 10, 12, 14 or 16 mg/day (0.14, 0.17, 0.20 or 0.23 mg/kg/day) for 30 days (77).

2.5 Standards, Regulations, or Guidelines of Exposure

All azinphos-methyl liquids with a concentration greater than 13.5% are classified as restricted use pesticides by the U.S. EPA because of their high toxicity.

The ACGIH TLV for azinphos-methyl is 0.2 mg/m³ and is associated with a skin notation (154). OSHA has established a PEL-TWA of 0.2 mg/m³ with a skin notation, NIOSH has established a REL-TWA of 0.2 mg/m³ with a skin notation, and most other countries have also established Occupational Exposure Limits of 0.2 mg/m³ (e.g., Australia, Austria, Belgium, Denmark, Federal Republic of Germany, Netherlands, Philippines, Switzerland, Thailand, United Kingdom). Finland has established an OEL of 0.02 mg/m³.

3.0 Chlorpyrifos

3.0.1 CAS Number: *[2921-88-2]*

3.0.2 Synonyms: (*O,O*-diethyl *O*-(3,5,6-trichloro-2-pyridinyl) phosphorothioate, Dursban; Lorsban; Dursban(R); chlorpyrifos-ethyl; Dowco 179; Pyrinex; phosphorothioic acid

O,O-diethyl *O*-(3,5,6-trichloro-2-pyridinyl) ester; Brodan; Chloropyrifos; *O,O*-diethyl *O*-(3,5,6-trichloro-2-pyridyl) phosphorothioic acid; *O,O*-diethyl *O*-3,5,6-trichloro-2-pyridyl phosphorothioate; *O,O*-diethyl *O*-(3,5,6-trichloro-2-pyridinyl) phosphorothioic acid; 3,5,6-trichloro-2-pyridinol *O*-ester with *O,O*-diethyl phosphorothioate; chlorpyriphos-ethyl; detmol u.a.; dursban f; oms-0971; stipend; dursban 4E; chloropyriphos; killmaster; dursban 10cr; suscon; lorsban 50sl; coroban; terial 401; terial; danusban; durmet; *O,O*-diethyl *O*-(3,5,6-trichloro-2-pyridinyl) phosphorothioate; Dursban/Lorsban; Lorsban 4E-SG; Chlorpyrifos 4E-AG-SG; diethyl *O*-(3,5,6-trichloro-2-pyridyl) phosphorothioatae; diethyl *O*-(3,5,6-trichloro-2-pyridinyl) phosphorothioate; Dursban HF; Eradex; Pyrindol. 3,5,6-trichloro-, *O*-ester with *O,O*-diethyl phosphorothioate; Super I.Q.A.P.T.; Trichlorpyriphos

3.0.3 Trade Names: Dursban®; Dowco 179®; ENT-27,311; Eradex®; Lorsban®; Pyrinex®

3.0.4 Molecular Weight: 350.57

3.0.5 Molecular Formula: $C_9H_{11}Cl_3NO_3PS$

3.0.6 Molecular Structure:

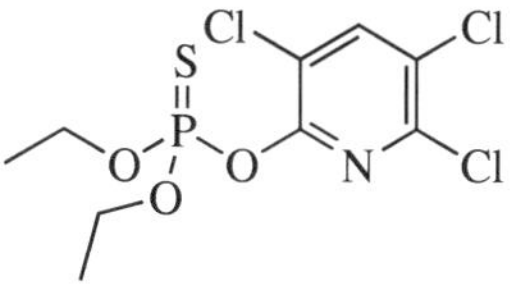

3.1 Chemical and Physical properties

Pure chlorpyrifos is a white crystalline powder; technical chlopyrifos is an amber to white crystalline powder.

Specific gravity: 1.398 g/cm^3 at 43.5°C
Melting point: 42.5–43°C
Boiling point: decomposes at approximately 200°C
Vapor pressure: 1.87×10^{-5} mmHg at 20°C
Solubility: soluble in most organic solvents; 0.00013 g/100 mL in water

3.1.2 Odor and Warning Properties

Mild mercaptan odor due to diethyl disulfide in technical product.

3.2 Production and Use

Chlorpyrifos is a broad spectrum pesticide and acaracide that acts as a contact poison. Originally used to control mosquito larvae, it is no longer registered for this use. It has a wide range of applications on crops, lawns, ornamental plants, domestic animals, and a variety of building structures. Chlorpyrifos is available as 25% wettable powders, 1–10% granules, and emulsifiable concentrates of 2 and 4 lb/gal.

3.3 Exposure Assessment

Most exposures to chlorpyrifos are anticipated to be via inhalation of aerosols during its application; dermal absorption in humans is not significant (56, 199). Chlorpyrifos metabolism yields a unique urinary metabolite, 3,5,6-trichloro 2 pyridinol (TCP), which has been correlated to chorpyrifos exposure if analyzed within 48 hours of exposure (200, 201).

3.4 Toxic Effects

3.4.1.1 Acute Toxicity. Chlorpyrifos is an organophosphate compound that has moderate toxicity and oral LD_{50}s of 80–250 mg/kg (64a, 202). Inhalation LD_{50}s of 78 and 94 mg/kg were calculated for female mice and rats, respectively, based on lethality observed after 60- to 180-minute exposures to aerosol concentrations of 5900 mg/m^3–7900 mg/m^3 (203). In female and male rats, 20 and 80% mortality occurred following 4-hour exposure to 5300 mg/m^3 chlorpyrifos, respectively, whereas no mortality occurred after exposure to 2500 mg/m^3 (56). A dermal LD_{50} of 202 mg/kg was reported in rats (64a). There is evidence that young animals are more sensitive than older animals; and that females are more sensitive than males to the lethal effects of chlorpyrifos (64a, 203a, 204).

Signs typically associated with organophosphate poisoning do not necessarily precede death caused by chlorpyrifos (203). "Apparent" (*sic*) tremors occasionally occurred in neonatal rats treated with a maximum tolerated dose (MTD) (45 mg/kg), but typical signs of a "cholinergic crisis" were not noted. Adult mice treated with a MTD (279 mg/kg) showed only slight to moderate signs of toxicity (e.g., diarrhea, fasciculations, lacrimation, slight tremors) (204). Cholinergic toxicity associated with sublethal exposure to chlorpyrifos was reflected only as mild diarrhea and hypoactivity evident only for the first two days following single s.c. injection of 279 mg/kg and not evident at all following repeated s.c. injections of 40 mg/kg (1×, 4 days, 16 days) (205, 206). Brain cholinesterase activity, however, was markedly depressed for at least 6 weeks following a single subcutaneous dose of 279 mg/kg and for at least 4 days following repeated subcutaneous doses of 40 mg/kg.

Prolonged cholinesterase inhibition occurred in rats given single subcutaneous injections of 0, 60, 125, or 250 mg/kg and examined up to 53 days later (207). Cholinergic toxicity was evident as a fine tremor in rats given 250 mg/kg but not 60 mg/kg (tremors in the 125 mg/kg group were not reported). Tremor intensity reached a peak in 9 days and returned to normal by 14 days after dosing. No overt signs of cholinergic toxicity occurred in rats treated with 0, 30, 60, or 125 mg/kg subcutaneously at any time from 1 to 35 days after dosing, although RBC cholinesterase activity was significantly inhibited in all groups (44). However, deficits in conditioned behavior occurred 2–16 days after dosing rats with 60, 125, or 250 mg/kg (207), and single oral doses of 12.5, 25, 37.5, or 50 mg/kg chlorpyrifos (or repeated doses of 12.5 mg/kg/day 5 days/week for eight weeks) caused significant deficits in response acquistition and performance measurements in a conditioned behavior task (208).

In rats given single oral doses of 20, 50, or 100 mg/kg chlorpyrifos by gavage, cholinergic signs peaked at $3\frac{1}{2}$ hours in the 100-mg/kg group, were still present at 24 hours, and had disappeared by 72 hours after dosing (324). Only hypoactivity occurred in the

20-mg/kg group. When rats were given single oral doses of 10, 30, 60, or 100 mg/kg chlorpyrifos, no cholinergic effects occurred at 10 mg/kg, rats exhibited cholinergic signs at 30 mg/kg, tremors were also evident at 60 mg/kg, and cholinergic symptoms were present and severe at 100 mg/kg (43). The time of peak effect was always $3\frac{1}{2}$ hours after dosing, and by 24 hours all symptoms had disappeared in the 30 mg/kg, but ataxia and hypoactivity were still evident in the 60- and 100-mg/kg group. Brain, plasma, and RBC cholinesterase inhibition were significantly and dose-dependently depressed at all dose including the 10 mg/kg group at both $3\frac{1}{2}$ and 24 h. Thus, behavioral and biochemical effects of chlorpyrifos exposure were poorly correlated; and, behavioral signs showed recovery at 24 h whereas cholinesterase activity did not (43).

When rats were given 10, 50, or 100 mg/kg chlorpyrifos orally and observed 1, 8, and 15 hours later, cholinergic effects occurred only on day 1 in 100-mg/kg treated female rats (209). One female rat treated with 50 mg/kg had tremors, two exhibited incoordination, and one showed pronounced lacrimation. One male rat given 100 mg/kg exhibited only minimal tremor, and one male exhibited incoordination and lacrimation. Rats treated with 50 or 100 mg/kg were significantly hypoactive only on day 1. Thus, cholinergic effects were widespread at 100 mg/kg, minor at 50 mg/kg, moderated over a few days, and were more severe in females than males.

Responses of adult rats given single oral doses of 80 mg/kg were compared with those of 17-day-old rats treated with single doses of 15 mg/kg; these doses were equally effective in inhibiting cholinesterase (203a). Compared to adults, young rats showed similar behavioral changes and cholinesterase inhibition although at a fivefold lower dose. The onset of maximal effects was somewhat delayed, cholinesterase activity recovered more quickly, more extensive muscarcinic receptor down-regulation occurred, and no gender-related difference in sensitivity was noted.

Chlorpyrifos caused neurotoxicity, indicated by leg weakness. Onset was 3-18 days after dosing and lasted from 10–20 days when given subcutaneously to atropinized chickens at a dose of 200 mg/kg but not 100 mg/kg (64a).

3.4.1.2 Chronic and Subchronic Toxicity. No treatment-related signs of toxicity, changes in body weight; or, in plasma, RBC, or brain cholinesterase activities occurred among rats exposed for 6 h/day, 5 days/week, 13 weeks, nose-only, to 72, 143, or 287 µg/m^3 chlorpyrifos (210). Nor was there any adverse effect at any exposure level based on urinalysis, clinical chemistry, hematology, gross pathological or histopathological evaluation.

When rats were exposed to 0.1, 1, 5, 15 mg/kg/day chlorpyrifos via the diet for 13 weeks, mild clinical effects (perineal soiling) were noted in females given 5 and 15 mg/kg/day (209). Plasma cholinesterase was significantly inhibited in males and females given 1 mg/kg or more, RBC cholinesterase was significantly inhibited in females given 1 mg/kg or more, and, brain cholinesterase was inhibited in both males and females given 5 or 15 mg/kg. Motor activity of males and females given 15 mg/kg mildly decreased only at week 4 of exposure, and no treatment related differences were apparent subsequently, consistent with the occurrence of tolerance upon repeated exposure to chlorpyrifos.

Oral treatment of hens with 1, 5, or 10 mg/kg/day chlorpyrifos for 13 weeks did not induce neurotoxicity (1237).

3.4.1.3 Pharmacokinetics, Metabolism, and Mechanisms.

There are no studies available that describe the pharmacokinetics of chlorpyrifos following inhalation exposure. However, absorption via inhalation was demonstrated indirectly in mice and rats who experienced lethality when exposed to 5900–7900 mg/m^3 aerosolized chlorpyrifos for 27–180 minutes (203). Indirect evidence of inhalation absorption in humans was provided in a study showing urinary excretion of diethyl phosphate chlorpyrifos metabolities by pesticide workers involved in treating structures with a spray emulsion of chlorpyrifos and vaponite (212).

Chlorpyrifos is well absorbed orally. Nearly 90% of orally administered chlorpyrifos was eliminated by rats in urine by 48–66 hours after dosing (201, 200a). The remaining 10% was eliminated in the feces. Chlorpyrifos was distributed especially to tissues involved in metabolism and excretion (liver and kidney), to tissues with considerable blood circulation (e.g., muscle, heart, lungs, spleen, testes, and bone), and to tissues high in lipids (e.g., fat and skin). It was eliminated nearly exclusively in urine (primarily as 3,5,6-trichloro-2-pyridinol phosphate). Elimination half-lives for liver, kidney, muscle, and fat were 10, 12, 16, and 62 hours, respectively, indicating considerable storage in fat.

About 72% of an oral dose of chlorpyrifos was absorbed (following a 1–2 hour delay), metabolized, and nearly completely eliminated (primarily as 3,5,6-trichloro-2-pyridinol (TCP)) in urine by male volunteers within 120 hours (199). Following either exposure route, chlorpyrifos was widely distributed and was eliminated from blood with a half-life of about 27 hours. In three individuals who ingested a concentrated solution of chlorpyrifos, its elimination was biphasic (276). An average elimination half-life for chlorpyrifos of 6 ± 2 hours in the initial elimination phase and of 80 ± 25 hours in a slower elimination phase was calculated.

Skin absorption of chlorpyrifos by humans is limited (56). In humans, less than 3% of a dermal dose of 5 mg/kg was absorbed (following a 22 hour delay) and was eliminated in urine within 180 hours (199). Considerable dermal absorption of chlorpyrifos has been reported in animals, although irritation and/or blistering from some high doses may have compromised the skin barrier. For example, 80–96% of a 22-mg/kg dermal dose was absorbed by goats by 12–16 hours after dosing and was distributed primarily to blood, liver, and fat, and lesser amounts were distributed to heart, gastrointestinal tract, and skeletal tissue (56, 57). About 60% of a dermal dose of chlorpyrifos (3–15 µM/5.6 cm^2) applied to rats was absorbed by 72 hours after dosing; young rats absorbed up to 90% of the applied dose. Eight hours after dermal treatment of mice with 1 mg/kg chlorpyrifos, about 74% of the dose was found primarily in urine and feces, carcass, blood, intestine, liver, and kidney, and an elimination half-life of 21 hours was estimated (213).

Chlorpyrifos is metabolized to chlorpyrifos oxon via cytochrome p450-dependent desulfuration (16, 215, 216). The oxon is rapidly hydrolyzed to 3,5,6-trichloro-2-pyridinol (TCP) via microsomal esterase (including paraoxonase and chlorpyrifos oxonase) (49, 217, 218) or via a nonenzymatic process (14, 215, 216). Alternatively, chlorpyrifos is dearylated to form diethyl thiophosphoric acid and TCP in a reaction also catalyzed by microsomal enzymes. TCP is a relatively unique metabolite of chlorpyrifos and it (or one of its conjugates) is almost exclusively (90%) excreted in the urine (200a, 201).

Chlorpyrifos oxon binds to and irreversibly inhibits acetylcholinesterase. However, the relative affinity of chlorpyrifos oxon for plasma and hepatic esterase exceeds that for acetylcholinesterase (13, 14, 219). Moreover, chlorpyrifos oxon causes relatively greater and longer lasting inhibition of hepatic esterase *in vivo* compared to brain acetyklcholinesterase (13, 14, 219). Noncatalytic binding of chlorpyrifos oxon to hepatic and plasma esterase represents a significant detoxication mechanism because it prevents much hepatically generated chlorpyrifos oxon from entering the general circulation and target tissues (56). High rates of hepatic dearylation and esterase binding may represent protective factors but it should also be recognized that chlorpyrifos can be activated in extrahepatic tissues such as brain (219).

Comparative differences in the rates of hepatic esterase binding and rates of dearylation have been implicated as contributors to the greater sensitivity of female rats to chlorpyrifos toxicity compared to male rats, to the greater sensitivity of some tissues (e.g., brain) to chlorpyrifos compared to other tissues (17), to the greater sensitivity of young animals to chlorpyrifos toxicity compared to adults (18), and to the greater toxicity of parathion compared to chlorpyrifos (13, 15, 19, 221).

Although human chlorpyrifos oxonase has not been shown to exhibit clear genetic polymorphism as has been shown for paraoxonase (218, 222), a 13-fold variation in chlorpyrifos oxonase activity has been found in human serum (223–225).

3.4.1.4 Reproductive and Developmental. No effects on reproduction, fertility indexes, or neonatal development were noted in two generations of rats fed diets that contained 0.1, 1.0, and 5 mg/kg/day chlorpyriphos (211). Parental toxicity at the highest dose was accompained by a decrease in pup body weight and increased pup mortality in the F1 litters.

Chlorpyrifos was not fetotoxic to rats given 0, 0.1, 3.0, or 15 mg/kg/day chlorpyriphos by gavage on days 6 through 15 of gestation (211). When pregnant mice were given 0,1,10, or 25 mg/kg/day chlorpyriphos by gavage on gestation days 6 through 15, minor fetotoxic responses and skeletal variations were noted at 25 mg/kg/day, a dose that also caused severe maternal toxicity (589). Teratogenicity (exencephaly) was observed in one pup at 1 mg/kg/day, but was not noted at 10 or 25 mg/kg/day.

3.4.1.5 Carcinogenesis. Neither rats maintained for 2 years nor dogs maintained for 1 or 2 years on diets that contained 0.01, 0.03, 0.1, 1, or 3 mg/kg chlorpyrifos showed signs of cholinergic toxicity or carcinogenicity at 12, 18, or 24 months (202). However, RBC cholinesterase activity was intermittently depressed among female rats given 0.1 mg/kg (at 30 and 365 days only) and was consistently depressed among male and female rats given 1 mg/kg and 3 mg/kg (202). Brain cholinesterase activity in rats was intermittently depressed in both sexes given 1 mg/kg (at 180 days and 547 days in females; at 365 and 730 days in males), and was consistently depressed in both sexes given 3 mg/kg. RBC and brain cholinesterase returned to normal levels within 7–8 weeks in a subset of exposed rats after they were switched to control diets. RBC cholinesterase activity was significantly depressed in dogs given 1 or 3 mg/kg diets from day 30 through 730. Brain cholinesterase activity in dogs exposed for 1 or 2 years was slightly depressed in the 3-mg/kg group (81-92% of control levels).

No evidence of cholinergic toxicity occurred in rhesus monkeys given 0.08, 0.40, or 2.0 mg/kg/day chlorpyrifos orally for six months, although plasma cholinesterase activity was reduced in all dose groups and RBC cholinesterase activity was reduced in the 0.40- and 2.0-mg/kg/day groups (220).

3.4.1.6 Genetic and Related Cellular Effects Studies. Chlorpyrifos is genotoxic based on a recent review (56). Chlorpyrifos induced micronuclei in erythroblasts and caused cytogenetic effects in human lymphoid cells in a dose-related fashion (56). Chlorpyrifos produced significant increases in sister chromatid exchanges and caused X chromosome loss in *Drosophila melanogaster* and chromosomal aberrations and sister chromatid exchanges in spleen cells. Spindle poisoning and induction of micronuclei and polyploidy have been reported following chlorpyrifos exposure. Sex-linked recessive lethals have also been produced in *Drosophila melanogaster* by chlorpyrifos exposure, indicating that chlorpyrifos is genotoxic to both somatic and germ cells.

3.3.5 Biomonitoring/Biomarkers

TCP is a unique urinary metabolite of chlorpyrifos. Variation in urinary TCP was investigated in termite control workers frequently involved in spraying chemicals in closed environments (226). Variation in the urinary TCP level corresponded to the termite control season and the length of the working period.

3.4.2 Human Experience

3.4.2.2 Clinical Cases. Unconsciousness, cyanosis, wheezing, and uncontrolled urination and diarrhea occurred within 18 hours after an individual ingested an estimated dose of 300 mg/kg chlorpyrifos (227). In another suicide attempt, a 27-year-old male experienced extreme agitation, diaphoresis and excessive oral secretions, muscle weakness and fasciculations 14 hours after ingestion of an unknown amount of chlorpyrifos (228). RBC cholinesterase activity was within normal limits. He was administered atropine, pralidoxime, and ventilatory support, and symptoms resolved after 72 hours without permanent sequelae. In another case of chlorpyrifos poisoning, a woman presented with stupor, increased muscle tone, absence of superficial reflexes, and striking choreoathetotic movements of all limbs (229). Serum and RBC cholinesterase activities were markedly depressed (about 20 and 2% of normal, respectively). Atropine relieved the symptoms, and the patient was completely well one month later. A 5-year-old girl who ingested an unknown amount of Rid-A-Bug® which contains chlorpyrifos and a 3-year-old boy who experienced frothing at the mouth, coma, pinpoint pupils, nasal secretions, fasciculations of the eyelids, and twitching of the extremities after ingesting an unknown amount of Dursban®, survived through treatment with atropine and pralidoxime (230, 231). In the boy, a distal polyneuropathy developed 18 days later, but all symptoms were fully resolved by day 52.

An intermediate syndrome that required endotracheal intubation and intermittent positive pressure ventilation followed by vocal cord paralysis was reported in an individual who had recovered from a cholinergic crisis due to acute chlorpyrifos ingestion (232). Immediate and delayed toxicity was reported in eight individuals exposed to Dursban® presumably via inhalation and dermal contact as a result of its use as a commercial fumigant (233). Immediate symptoms included nausea, vomiting, and lightheadness in a worker after chlorpyrifos was inadvertently introduced into the workplace ventilating system, and headaches, nausea, and painful muscle cramps in a family of four after their house was sprayed with chlorpyrifos by an exterminator. About 4 weeks after initial exposure, the affected worker developed paresthesia in his feet, urinary frequency and pain in the suprapubic, groin, and thigh regions, and the affected family developed numbness and paresthesias especially in the legs which was accompanied by reported memory impairment. Nerve conduction studies were consistent with distal axonopathy. Delayed sensory neuropathies were also described in two women whose houses had been treated with chlorpyrifos 3–4 weeks earlier and in an exterminator who was exposed repeatedly to chlorpyrifos during a 6-month period (233). All symptoms resolved by 2 weeks–3 months after exposure stopped. No estimates of exposure or cholinesterase activity levels were provided.

A spectrum of common birth defects that involved central nervous system malformations, ventricular, eye, and palate defects, hydrocephaly, microcephaly; mental retardation; blindness; hypotonia; widely-spread nipples; and deformities of the teeth, external ears and external genitalia was described in four infants (two of whom were siblings) and attributed to maternal exposure to chlorpyrifos during the first trimester of pregnancy either in the home or at work (234, 235). However, no confirmation of exposure either through chlorpyrifos measurements or cholinesterase activities was provided nor was any estimate of dose attempted. Reports of an additional nine cases of birth defects that had the same or similar spectrum of effects associated with *in utero* chlorpyrifos exposure were provided to the EPA and reviewed by the Centers for Disease Control (CDC) (236). The EPA concluded that the cases did not support a finding of teratogenicity (236).

3.4.2.2.2 Chronic and Subchronic Toxicity. The prevalence of selected illnesses and symptoms in 175 employees involved in producing chlorpyrifos for more than one day between 1 January 1977 and 31 July 1985 and in 335 individually matched controls were compared (237). Subjects were subdivided into three exposure groups (high, moderate, and low) on the basis of job title and air monitoring data. Estimated TWA concentrations of chlorpyrifos were not provided by exposure group, but for all employees reporting ranged from 0.01 to 0.37 mg/m^3. Company medical records were examined for evidence of gastrointestinal tract and nervous symptoms. No significant differences in illness or prevalence of symptoms were observed between the exposed and unexposed groups or among the three exposure subgroups. The observation period was extended to 31 December 1994, and the study group size was increased to 496 potentially exposed employees and 911 controls in a follow-up study (238). High, moderate, low, and negligible exposures were defined as ≥ 0.2 mg/m^3 chlorpyrifos or high potential for dermal exposure, ≥ 0.03 mg/m^3 or moderate potential for dermal exposure, ≤ 0.03 mg/m^3 or low potential for dermal exposure, and ≤ 0.01 mg/m^3 or negligible dermal exposure, respectively. Employees were categorized into moderate, low or negligible exposure levels on the basis of plasma cholinesterase activities. Prevalence odds ratios for various nervous, respiratory, and digestive system symptoms were calculated by exposure as well as plasma cholinesterase inhibition, and no exposure-related effects were observed.

An increase in the frequency of blurred vision, flushing of skin, and decreased urination were reported by pet control employees who used chlorpyrifos within the previous three months (239). However, no estimates of chlorpyrifos exposure were obtained, and exposure was not confirmed with biomarkers of exposure.

Volunteers were treated with 0.014, 0.03 or 0.10 mg/kg/day chlorpyrifos by capsule for a total of 20 days at the low and mid-dose and for 9 days at the high dose (95). Treatment of the high-dose group was discontinued after 9 days due to a runny nose and blurred vision in one individual. Mean plasma cholinesterase in this group was inhibited by about 65%. No effect on RBC cholinesterase activity was apparent at any dose. No signs of toxicity were reported among human volunteers given daily oral doses of 0.014, 0.030, or 0.100 mg/kg chlorpyrifos for up to four weeks (210). Plasma cholinesterase inhibition was reported in the 0.100-mg/kg group, but RBC cholinesterase was reportedly unaffected at all exposure levels.

No signs of toxicity or depression in RBC acetylcholinesterase activity occurred among human volunteers given 0.5 mg/kg orally or 0.5 or 5 mg/kg dermally. Plasma cholinesterase activity was reduced to 15% of predose levels after the 0.5 mg/kg oral dose, but RBC cholinesterase activity was unchanged after this dose or the 5 mg/kg dermal dose (199).

3.5 Standards, Regulations, or Guidelines of Exposure

The EPA established a 24-hour reentry interval for crop areas treated with emulsifiable concentrate or wettable powder formulations of chlorpyrifos unless workers wear protective clothing. Chlorpyrifos is undergoing reregistration by the EPA (242).

The ACGIH TLV and the NIOSH REL-TWA for chlorpyrifos is 0.2 mg/m^3 with a skin notation (154). There is no OSHA TWA-PEL for chlorpyrifos. Most other countries have also established Occupational Exposure Limits of 0.2 mg/m^3 with skin notation for chlorpyrifos (e.g. Australia, Belgium, Denmark, Finland, France, Netherlands, Switzerland, and United Kingdom). NIOSH has also established a 0.6 mg/m^3 STEL for chlorpyrifos.

4.0 Coumaphos

4.0.1 CAS Number: [56-72-4]

4.0.2 Synonyms: *O,O*-diethyl *O*-(3-chloro-4-methyl-2-oxo-2*H*-1-benzopyran-7-yl) phosphorothioate; 3-chloro-7-diethoxyphosphinothioyloxy-4-methylcoumarin; Diolice; Meldane; Muscatox; Resistox; Asuntol; Bay 21/199; Bazmix; Umbethion; phosphorothioic Acid *O*-(3-chloro-4-methyl-2-oxo-2*H*-1-benzopyran-7-yl) *O,O*-diethyl ester; Asantol; Baymix; Resitox; *O,O*-diethyl *O*-(3-chloro-4-methyl-2-oxo-2*H*-1-benzopyran-7-yl)phosphorothioate; 3-chloro-7-hydroxy-4-methylcoumarin *O*-ester with *O,O*-diethyl phosphorothioate; 3-chloro-4-methylumbelliferone, *O*-ester with *O,O*-diethyl phosphorothioate; *O,O*-diethyl *O*-(3-chloro-4-methyl-7-coumarinyl) phosphorothioate; *O,O*-diethyl *O*-(3-chloro-4-methylumbelliferone); 3-chloro-7-hydroxy-4-methyl-coumarin *O,O*-diethyl phosphorothioate; 3-chloro-4-methyl-7-coumarinyl diethyl phosphorothioate; 3-chloro-4-methyl-7-hydroxycoumarin diethyl thiophosphoric acid ester; *O,O*-diethyl *O*-(3-chloro-4-methyl-2-oxo-2*H*-benzopyran-7-yl)phosphorothioate; asunthol; coumafos; agridip; *O*-(3-chloro-4-methyl-2-oxo-2*H*-1-benzopyran-7-yl) *O,O*-diethyl phosphorothioate; Coumarin, 3-chloro-7-hydroxy-4-methyl-, *O*-ester with *O,O*-diethylpyrophosphorothioate; diethyl *O*-(3-chloro-4-methyl-2-oxo-2*H*-1-benzopyran-7-yl) phosphorothioate; Umbelliferone, 3-chloro-4-methyl-, *O,O*-diethyl phosphorothioate

4.0.3 Trade Name: Agridip; Asuntol®; Bay 21; Baymix®; Co-Ral®; ENT-17957; Meldane®; Muscatox®; Negashunt; Resitox®; Suntol, Umbethion

4.0.4 Molecular Weight: 362.8

4.0.5 Molecular Formula: $C_{14}H_{16}ClO_5PS$

4.0.6 Molecular Structure:

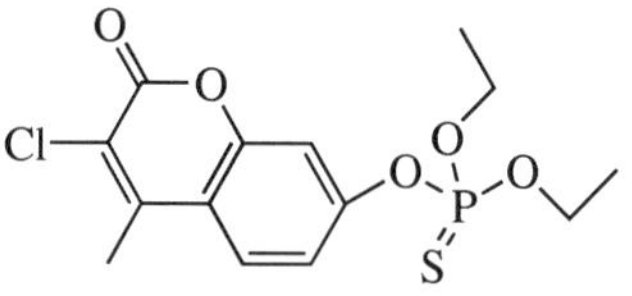

4.1 Chemical and Physical Properties

Technical coumaphos is a colorless or white and tan powder. Coumaphos is stable under normal use conditions but hydrolyzes slowly under alkaline conditions

- Specific gravity: 1.47 at 20°C
- Melting point: 91–92°C
- Boiling point: 20°C at 1×10^{-7} mmHg
- Vapor pressure: 1×10^{-7} mmHg at 20°C
- Solubility: soluble in acetone and diethyl phthalate; much less soluble in denatured alcohol and xylene; only slightly soluble in octanol, hexane and mineral spirits; insoluble in water

4.1.2 Odor and Warning Properties

Slight sulfur odor.

4.2 Production and Use

Coumaphos is an organophosphate insecticide used to control anthropoid pests on beef cattle, dairy cows, goats, horses, sheep, and swine. Formulations include wettable powders, emulsifiable liquids, flowable concentrate, ready-to-use liquids, and dusts. Coumaphos is applied by aerosol can, dust bags, hand-held dusters, dip vats, high- and low-pressure hand-held sprayers, back rubber oilers, mechanical dusters, shaker can, and squeeze applicators (241). The EPA issued a Registration Standard for coumaphos in 1989 and a Reregistration Standard in 1996 (45). Currently 26 coumaphos products are registered (243).

4.4 Toxic Effects

4.4.1.1 Acute Toxicity. Coumaphos is highly toxic and has an oral LD_{50} of 16–41 mg/kg in rats (64a). However, the EPA noted that the oral LD_{50} in male rats was >240 mg/kg and oral LD_{50} in female rats was 17 mg/kg (241). The dermal LD_{50} for coumaphos was previously reported as 860 mg/kg in rats (Gaines 1969) although EPA noted that the dermal LD$_{50}$ was >2400 mg/kg in male and female rats (241). One-hour LC$_{50}$s of 1081 mg/m^3 and 341 mg/m^3 and 341 mg/m^3 were reported for male and female rats (241).

When male rats were given single oral doses of 250 mg/kg and female rats were given single oral doses of 17.5 mg/kg, cholinergic signs occurred and lasted for 12–13 days (242). RBC cholinesterase activity was depressed at 2 mg/kg in both sexes. When sheep were given 2 or 4 mg/kg/day coumaphos orally for 6 days, 4 mg/kg/day caused RBC cholinesterase inhibition and signs of cholinergic toxicity (*sic*); 2 mg/kg/day inhibited RBC cholinesterase but caused no apparent overt cholinergic toxicity (240). Treatment with coumaphos did not significantly alter the anticholinesterase effects of the second treatment 6 weeks later, suggesting no cumulative effect. Simultaneous treatment with coumaphos (4 mg/kg/day) and an intravenous dose of trichlorfon (insufficient to cause significant inhibition of RBC cholinesterase alone) resulted in an additive effect on RBC cholinesterase inhibition (240).

When rats were treated dermally for 2 or 5 days with 0, 2.5, 5, 10, 20, or 50 mg/kg/day coumaphos, no cholinergic toxicity was noted (241). Brain, plasma, and RBC cholinesterase activity, however, were depressed at 50 mg/kg after 2 days, and RBC and brain cholinesterase activity were depressed at 20 mg/kg/day after 5 days (241).

Coumaphos is a mild eye irritant, but is not irritating to the skin and is not a skin sensitizer; nor does it produce delayed neurotoxicity in hens (241).

4.4.1.2 Chronic and Subchronic Toxicity. No signs of cholinergic toxicity were observed at any dose in rats fed diets that contained 0, 2, 5, or 10 ppm coumaphos for 13 weeks (about 0, 0.2, 0.5, or 1.0 mg/kg/day) (241). However, plasma cholinesterase was inhibited at 10 ppm (1 mg/kg/day), and RBC cholinesterase was inhibited at all dose levels. Brain cholinesterase was not inhibited at any time at any dietary level.

When rats were dermally treated with 2, 4, 20, or 100 mg/kg/day coumaphos for 21 days, cholinergic toxicity (muscle fasciculation, tremors) occurred at 20 or 100 mg/kg/day (241). RBC cholinesterase was inhibited at all doses, and brain cholinesterase was inhibited at 20 and 100 mg/kg. In another 21-day dermal study, when female rats were given 0, 0.1, 0.5, 1.1, or 2.1 mg/kg/day coumaphos, no signs of cholinergic toxicity were observed at any dose, although RBC cholinesterase was significantly inhibited at 1.1 and 2.1 mg/kg/day (242).

4.4.1.3 Pharmacokinetics, Metabolism, and Mechanisms. Coumaphos is well absorbed orally. The plasma half-life for coumaphos following oral exposure ranges from 2–3 hours at 1.0 mg/kg and 3–5 hours at 15.0 mg/kg (241). Urinary excretion is rapid; and 63–87% of an administered dose was excreted within 24 hours, and 76–96% of an administered dose was excreted within 168 hours. Tissue residues were highest in fat, kidney, liver, and muscle. Seven days after rats were given single oral doses of about 1 mg/kg coumaphos, about 55% had been excreted in urine and about 24% has been excreted in feces (246). The remaining dose was distributed primarily to the liver, abdominal fat, skin, and kidney.

The fate of dermally applied coumaphos was examined in lactating goats given about 14 mg/kg coumaphos (243). During the 7 days after treatment, an average of <0.1, 4.7, and 1% of the administered dose was eliminated in milk, urine, and feces, respectively. When goats were killed after 7 days, the highest coumaphos residues were in adipose tissue (mainly unmetabolized coumaphos) followed by the kidney and liver.

Coumaphos is extensively metabolized. The urine of rats treated orally with coumaphos contained five to eight metabolites, and the feces contained five to seven metabolites (242). The major metabolite is chlorferone (the hydroxylated leaving group). Coumaphos represented 0.1% of the urinary metabolites. Coumaphos represented 0.2% of the fecal metabolites when administered intravenously but approximately 15 to 55% of the fecal metabolites when given orally, suggesting that the process of oral absorption may slow metabolism.

4.4.1.4 Reproductive and Developmental. Reproductive toxicity did not occur in rats fed diets that contained 1, 5, or 25 ppm coumaphos (0.07, 0.30, and 1.79 mg/kg/day (F_0 males) and 0, 0.08, 0.34, or 2.02 mg/kg/day (F_0 females)) during premating for two generations (241). Dose-dependent decreases in plasma and RBC cholinesterase activity occurred among rats fed 5 or 25 ppm. Brain cholinesterase was significantly inhibited in F_0 and F_1 females. In pups, plasma and RBC cholinesterase levels were inhibited at 25 ppm on lactation day 21 but not on lactation day 4.

No developmental effects occurred in offspring of rats given 1, 5, or 25 mg/kg/day coumaphos by gavage on gestation days 6 to 15 (242). Three rats in the 25-mg/kg/day group showed tremors, and two showed additional signs of cholinergic toxicity. No developmental effects occurred in rabbits given 0.25, 2.0, or 18.0 mg/kg/day coumaphos by gavage during gestation days 7 through 19 (242). Maternal toxic signs, including death and abortion, were observed in the 18-mg/kg/day group. When pregnant bovines were dermally treated with coumaphos at various stages of gestation by pouring it along the dorsal midline, there was evidence of increasing embryonic death rates or teratogenic effects (245).

4.4.1.5 Carcinogenesis. There was no evidence of carcinogenicity at any dose in rats fed diets that contained 1, 5, or 25 ppm coumaphos (0.05, 0.25, or 1.22 mg/kg/day (males); 0.07, 0.36, or 1.70 mg/kg/day (females)) for two years (241). Body weight gain decreased in females given the 25-ppm diet (1.7 mg/kg/day). Plasma and RBC cholinesterase was inhibited in females given the 5- or 25-ppm diet (0.36 or 1.7 mg/kg/day) and in males given the 25-ppm diet (1.22 mg/kg/day). There was no evidence of carcinogenicity in another study when rats were given diets that contained 10 or 20 ppm coumaphos for 103 weeks (244). The only adverse effect observed was slightly decreased body weight gain in females given either the 10- or 20-ppm diet. No adverse effects of any kind were observed in mice given diets that contained 0, 10, or 20 ppm coumaphos for 103 weeks (241).

There were no treatment related effects other than cholinesterase inhibition in dogs given 1, 30, or 90 ppm coumaphos in the diet for one year (0.025, 0.775, or 2.295 mg/kg/day (males); 0.024, 0.7095, or 2.478 mg/kg/day (females)) (241). Plasma, RBC, brain, and ocular muscle cholinesterase activity levels were depressed in the 30- and 90-ppm fed dogs (0.775/0.7095 and 2.295/2/478 mg/kg/day).

4.4.1.6 Genetic and Related Cellular Effects Studies. Coumaphos was not mutagenic in *S. typhimurium* with or without metabolic activation, was negative in a mouse micronucleus test, and was negative in a Pol A test on *E. coli* with and without metabolic activation (241).

4.4.2 Human Experience

4.4.2.2 Clinical Cases. Six individuals suffered serious organophosphate poisoning after ingesting coumaphos mistakenly used as a food flavoring in two separate incidents (247). In the first incident, four adults developed symptoms of poisoning (nausea, vomiting, and abdominal pain) 1 hour after eating seafood flavored with coumaphos which had possibly been mistaken for monosodium glutamate. After 2 hours, obvious signs of cholinergic toxicity (sweating, urination, miosis, bronchorrhea, and hypersalivation) were apparent. One adult female died at home after about 3 hours; the remaining three adults were treated with atropine and pralidoxime and gradually recovered. In the second incident, a man and wife developed signs of organophosphate poisoning (nausea, vomiting, diarrhea, abdominal pain, and blurred vision) after eating coumaphos-contaminated fried fish and recovered with atropine and PAM.

4.5 Standards, Regulations, or Guidelines of Exposure

The EPA classifies most formulations of coumaphos as General Use Pesticides. The formulations 11.6% EC and 42% flowable concentrate end-use products have been classified as Restricted Use Pesticides because they pose a hazard of acute poisoning from ingestion. No Occupational Exposure Limits were identified for coumaphos.

5.0 Demeton

5.0.1 CAS Number: [8065-48-3]

5.0.2 Synonyms: O,O-diethyl O (and S)-2-(ethylthio)ethyl phosphorothioate mixture; Systox; Bayer 8169; Demeton-o + Demeton-s; Demox; E-1059; phosphorothioic acid O,O-diethyl O-2-(ethylthio)ethyl ester, mixed with O,O-diethyl-S-2-(ethylthio)ethyl phosphorothioate; Denox; Systemox; phosphorothioic acid O,O-diethyl O-[2-(ethylthio)ethyl] ester mixture with O,O-diethyl S-[2-(ethylthio)ethyl]phosphorothioate; phosphorothioic acid O,O-diethyl O-[2-(ethylthio)ethyl] ester, mixed. with O,O-diethyl S-[2-(ethylthio)ethyl] phosphorothioate; demeton+; Demeton (O-isomer and S-isomer); diethyl O-(and S-)(2-(ethylthio)ethyl) phosphorothioate (mixed isomers); phosphorothioic acid, O,O-diethyl O (and S)-(ethylthio)ethyl esters; Demeton O(35%)+S(56%)

5.0.3 Trade Names: Demox®; Mercaptofos®; Systox®; Bay 10756; Bayer 8169; ENT 17,295

5.0.4 Molecular Weight: 258.34

5.0.5 Molecular Formula: $C_8H_{19}O_3PS_2$

5.1 Chemical and Physical Properties

Demeton is a light brown to pale yellow, oily liquid. Demeton is the common name for a mixture of O,O-diethyl-O-2-ethylthioethyl phosphorothioate (demeton-O) and O,O-diethyl-*S*-2-ethylthioethyl phosphorothioate (demeton-S) in a ratio of approximately 2:1. Demeton decomposes to toxic gases and vapors such as sulfur dioxide, phosphoric acid mist, and carbon monoxide. Contact with strong oxidizers may cause fire and explosions.

Specific gravity: 1.18 at 20°C
Melting point: $> -25°C$
Boiling point: 134°C at 2 mmHg
Vapor pressure: 3.4×10^{-4} mmHg at 20°C (mixture)
Solubility: slightly soluble in water; soluble in most organic solvents

5.1.2 Odor and Warning Properties

Pronounced mercaptan-like odor.

5.2 Production and Use

Demeton is a systemic insecticide effective against sap-feeding insects and mites. Before 1989 when it was discontinued by the manufacturer, it was available as emulsifiable concentrates of varying active ingredient content (1). Because demeton is no longer an active ingredient in any registered pesticide product, its registration has been canceled by the U.S. EPA (45).

5.4 Toxic Effects

5.4.1.1 Acute Toxicity. Demeton is a highly toxic organophosphate compound that has oral LD_{50}s of 2–6 mg/kg (64a, 248, 249). The oral LD_{50} for the P–S isomer (demeton-O) in rats was 7.5 mg/kg and for the P–O (demeton-S) isomer was 1.5 mg/kg (250). The sulfoxide and sulfone metabolites of demeton are as lethal as demeton itself; they have oral LD_{50}s in rats of 1.9 to 2.3 mg/kg (251).

The dermal LD_{50}s are 8.2–14 mg/kg in rats, nearly equivalent to its acute oral toxicity (64a). However, formulation impacts the dermal toxicity of demeton. An equal volume of emulsifier changed the LD_{50} from less than 24 to 620 mg/kg, whereas dilution of the mixture to the strength used for spraying greatly increased the toxicity, so that a lethal dose was about 5 mg/kg (478).

A single 2-hour exposure to 18 mg/m^3 demeton was fatal within 50–90 minutes to all of a group of six rats (250). Rats exposed to 3 mg/m^3 demeton for 2 hours/day experienced "no signs of illness during the first exposure" (sic), tremors during the second exposure, lacrimation and more severe tremors during the third exposure, and mortality in 10 of 17 rats during the fourth exposure. Rats exposed to 3 mg/m^3 demeton for only 1 hour/day experienced no signs of intoxication after two days; mild tremors after 4 days; marked tremors, lacrimation, and 5% mortality (1/20) after 6 days; and 37% mortality (7/19) after 12 days. One and four-hour LC_{50} values of 175 and 47 mg/m^3 were obtained for rats (123).

Demeton was not associated with signs of organophosphate-induced delayed neuropathy when administered subcutaneously to atropinized chickens at doses that ranged from 5–80 mg/kg and were then observed for 30 days (190).

5.4.1.2 Chronic and Subchronic Toxicity. When rabbits were given greens sprayed with demeton so that intake was 2.3, 1.5, 0.5, 0.1, or 0.07 mg/kg/day for 40, 30, 100, 98, and 94 days, respectively, no effects occurred in the 0.07- or 0.1-mg/kg/day groups, one of six rabbits fed 0.5 mg/kg/day died after 64 days, four of six rabbits fed 1.5 mg/kg died, and three of six rabbits fed 2.3 mg/kg/day died (250).

Rats fed diets that contained 50 ppm demeton containing 48% of the more potent P=O isomer (equivalent to 2.6 mg/kg) for 11–16 weeks exhibited cholinergic toxicity (fasciculations, weakness, tremors, lacrimation, and salivation) at 11 weeks, but by 16 weeks they were exhibiting no signs of cholinergic toxicity despite severe brain cholinesterase inhibition (253). No cholinergic signs were observed in rats given diets that contained 1, 3, or 20 ppm demeton for 11–16 weeks. When rats were given 0.4, 0.66, 0.9, and 1.89 mg/kg/day demeton by gavage for 65 days during a 90-day period, signs of cholinergic toxicity (hyperexcitability, tremors) occurred in rats given the two highest doses after 21 days (250). There was one death in the group fed 1.89 mg/kg.

In dogs given diets that contained 1, 2, or 5 ppm demeton (0.025, 0.047, or 0.149 mg/kg/day) for 24 weeks, plasma cholinesterase activity was maximally inhibited after about 12 weeks in dogs given the 5-ppm diet and after 16 weeks in dogs given the 2-ppm diet (193). RBC cholinesterase activity was unaffected by the 1- or 2-ppm diets and was slightly inhibited by the 5-ppm diet. When demeton and parathion were in the same diet at levels necessary for cholinesterase inhibition, the effects were additive (193). In an unpublished study submitted to EPA, overt cholinergic toxicity reportedly occurred in rats fed a diet that delivered 0.9 mg/kg/day demeton but not 0.7 mg/kg/day (95). In another unpublished study, cholinsterase inhibition (*sic*) reportedly occurred in female rats fed diets that contained 3 ppm demeton for 77–112 days but not in rats fed 1 ppm). Cholinergic effects evidently occurred at 20 ppm but not at 10 ppm (95).

5.4.1.3 Pharmacokinetics, Metabolism, and Mechanisms. The principal metabolic pathway for both the O- and S-isomers is oxidation of the 2-ethylthioether to sulfoxide and sulfone. In the case of demeton-O, a secondary pathway involves oxidation of P=S to P=O and subsequent oxidation to sulfoxide and sulfone (257). These oxidation products are more potent inhibitors of acetylcholinesterase than the parent compound. Studies in mice are consistent with the notion that oxidation of the mercapto sulfur of the ethylmercaptoethyl portion of demeton and the P–O isomer to the corresponding sulfoxide and finally to the sulfone is an important metabolic pathway for demeton (258).

5.4.1.4 Reproductive and Developmental. Administration of 7 or 10 mg/kg demeton to mice as a single intraperitoneal dose or as three consecutive doses of 5 mg/kg each between days 7 and 12 of gestation was embryo toxic, as evidenced by decreased fetal weight and slightly higher mortality of the young (254). Fetuses that had intestinal hernias were found at 16 but not at 18 days. The high dose (10 mg/kg) administered on days 8, 9, or 10 of gestation had no effect on litter size at birth or on the survival rate of the young.

Ducklings hatched from eggs innoculated with demeton at the rate of 0.01 mg/egg on day 13 of incubation had partial to complete loss of voluntary control of one or both hind legs. Some were excitable and ataxic. The difficulties gradually disappeared about 1 week after hatching, but the growth of treated ducklings remained retarded. Histological examination of the skeletal muscles revealed areas of degenerative change and other areas of marked regenerative activity (255, 256).

5.4.1.5 Carcinogenesis. No published studies of the chronic toxicity or oncogenicity of demeton were identified. In an unpublished study submitted to the EPA, dogs were fed diets that contained demeton (all levels not specified) for 24 weeks (95). RBC cholinesterase was inhibited at a dietary concentration of 5 ppm (about 0.125 mg/kg/day), and plasma cholinesterase was inhibited at a dietary concentration of 2 ppm (about 0.05 mg/kg/day). A dietary level of 1 ppm (0.025 mg/kg/day) was without effect on either on either plasma or RBC cholinesterase. In another unpublished study submitted to EPA, cholinesterase (*sic*) was reportedly inhibited at 0.5 mg/kg/day but not at 0.15 mg/kg/day in rabbits that were given demeton orally for 106 days (95).

5.4.1.6 Genetic and Related Cellular Effects Studies. Demeton was reportedly both positive and negative in *in vitro* tests of bacterial mutagenicity and negative in a sex-linked lethal mutation assay using *D. melanogaster* (545).

5.4.2 Human Experience

5.4.2.2 Clinical Cases. Demeton has been associated with numerous deaths after high accidental and intentional exposures and from occupational exposures (1). Estimates of exposure are not available. A man who spilled 60 mL of concentrated liquid demeton on his thigh, then rinsed his thigh with water but continued wearing the pants, suddenly experienced nausea, vomiting and, weakness after about 9 hours, was treated with atropine, and recovered uneventfully (1).

Twelve of fourteen agricultural workers exposed to about 1 mg/m^3 demeton reportedly had lowered cholinesterase levels (*sic*) but displayed no clinical evidence of poisoning (*sic*) (259). In another study, air concentrations of up to 6 mg/m^3 were reportedly without clinical effect, although they were associated with reduced serum cholinesterase activity (259).

Eighteen different doses of demeton were evaluated in men who were given oral doses of demeton that began at 0.75 mg per day for 30 days (260). Doses of 4.5 to 6.375 mg/day (equivalent to about 0.06 to 0.09 mg/kg/day assuming a 70-kg body weight) produced average inhibition of plasma cholinesterase that was indistinguishable from normal variation. Doses of 6.75 mg/day (0.10 mg/kg/day) produced an average temporary inhibition of plasma cholinesterase, and a dose of 7.124 mg/day (0.10 mg/kg/day) produced an average of 40% inhibition by day 25. This dose was also associated with an average 16% inhibition of RBC cholinesterase inhibition. However, one of five test subjects had a marked decrease in plasma and RBC cholinesterase activities of 59% and 29%, respectively, after 24 days when given 4.125 mg/day (equivalent to 0.06 mg/kg/day). No clinical signs were observed or reported at any exposure level.

Three volunteers were exposed for two consecutive days to 9–27 mg/m^3 Metasystox (30% demeton-*S*-methyl, 70% demeton-*O*-methyl) while spraying with a hand-held nebulizer. Exposure lasted for 3 and 6 hours on the first and second days, respectively. Plasma and RBC cholinesterase activities measured up to 14 days after exposure did not show significant decreases (262).

5.5 Standards, Regulations, or Guidelines of Exposures

Demeton is not registered for use by the U.S. EPA. The ACGIH TLV for demeton is 0.11 mg/m^3 with a skin notation (154). The OSHA PEL-TWA is also 0.1 mg/m^3 with a skin notation. NIOSH established an IDLH of 10 mg/m^3 and REL-TWA of 0.1 mg/m^3 with a skin notation. Many other countries have also established Occupational Exposure Limits (OELs) of 0.1 mg/m^3 for demeton with a skin notation (Australia, Austria, Belgium, Denmark, Finland, France, Germany, India, The Netherlands, The Philippines, Switzerland, Thailand, and Turkey).

6.0 Demeton-S-methyl

6.0.1 CAS Number: [919-86-8]

6.0.2 Synonyms:
Ethanethiol, 2-(ethylthio)-S-ester with O,O-dimethyl phosphorothioate; BAY 18436; Bayer 25/154; phosphorothioic acid S-[2-(ethylthio)ethyl] O,O-dimethyl ester; Metasystox (I); metasytox thiol; S-(2-(ethylthio)ethyl) O,O-dimethyl phosphorothioate; Methyl-S-Demeton

6.0.3 Trade Names:
Demetox®; DEP 836 349; Duratox®; Isometasystox®; Isomethylsystox®; Metaisoseptox®; Metaisosytox; Metasystox (I)®

6.0.4 Molecular Weight: 230.3

6.0.5 Molecular Formula: C$_6$H$_{15}$O$_3$PS$_2$

6.0.6 Molecular Structure:

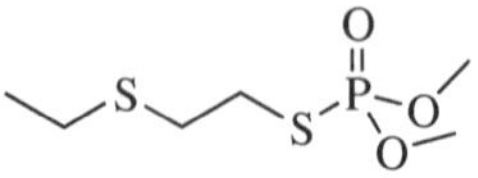

6.1 Chemical and Physical Properties

Demeton-S-methyl is an oily, colorless to pale yellow liquid. It is hydrolyzed by alkali and oxidized to the sulfoxide (oxydemeton-methyl) and sulfone (demeton-S-methylsulfone).

Specific gravity: 1.21 at 20°C
Boiling point: 74°C at 0.15 mmHg; 92°C at 0.2 mmHg; 102°C at 0.40 mmHg; 118°C at 1 mmHg
Vapor pressure: 1.6 × 10^{-4} mmHg at 10°C; 4.8 × 10^{-4} at 10°C; 1.45 × 10^{-3} at 30°C; 3.8 × 10^{-3} mm Hg at 40°C
Solubility: soluble in water (3.3 g/L); readily soluble in most organic solvents (e.g., dichloromethane, 2-propanol, toluene); limited solubility in petroleum solvents

6.1.2 Odor and Warning Properties

Unpleasant odor.

6.2 Production and Use

Demeton-S-methyl was first marketed in 1957. It replaced technical grade methyl demeton which had been introduced in 1954 and was a 70:30 mixture of *O,O*-dimethyl-*O*-ethylthioethyl phosphorothioate (demeton-*O*-methyl or O-isomer) and *O,O*-dimethyl-*S*-ethylthioethyl phosphorothioate (Demeton-*S*-methyl or S-isomer). Demeton-*S*-methyl is a systemic and contact insecticide and acaricide used to control aphids, red spider mites, whiteflies, leafhoppers, and sawflies on garden crops, fruit, and hops (155). It is applied as an emulsifiable concentrate formulation, mainly as a spray, and usually at a concentration of 0.025% active ingredient.

6.4 Toxic Effects

6.4.1.1 Acute Toxicity. Demeton-*S*-methyl is an organophosphate that has high oral toxicity with oral LD_{50}s of 33–130 mg/kg (262). The toxicity of demeton-*S*-methyl is markedly increased when it is allowed to age and forms sulphonium derivatives (263) Intravenous and oral LD_{50}s for the sulfoxide and sulfone derivatives of demeton-*S*-methyl were 22–47 mg/kg and 32–65 mg/kg, respectively (261). Intraperitoneal LD_{50}s of 7.5 and 10 mg/kg were also reported for demeton-*S*-methyl, suggesting that oral absorption may be slightly limited and/or that bypassing first-pass hepatic metabolism enhances toxicity (262). Dermal LD_{50}s from 45 to 200 mg/kg, depending on formulation and duration of exposure, were reported for rats, indicating that dermal exposures are about as potent as oral exposures on a milligram per kilogram basis (262). Dermal application of 10 mg/kg to the backs of cats caused mild signs (*sic*); application of 20 or 100 mg/kg caused death (no other details provided) (IPSC 1997). Four hour LC_{50}s for rats were 210–500 mg/m^3 (545).

Demeton-*S*-methyl applied to the shaved skin of rabbits for four hours caused mild erythema and edema that disappeared after three days (262). No signs of eye irritation occurred in rabbits whose eyes were treated with a 0.5% aqueous solution of demeton-*S*-methyl, but an undiluted formulation caused severe lacrimation and miosis. Mild corneal opacity and discrete redness and edema of conjunctivae were observed that disappeared within about 7 days (262).

Demeton-*S*-methyl had skin sensitizing potential when assessed using the guinea pig mazimization test but did not have skin sensitizing potential when assessed using the Buehler epidermal patch test on guinea pigs (262).

6.4.1.2 Chronic and Subchronic Toxicity. When groups of six rats were fed diets that contained 50, 100, or 200 ppm (aged) demeton-*S*-methyl (5, 10, and 20 mg/kg/day, respectively) for six months, cholinergic signs (slight tremors, fasciculations) occurred at 200 ppm during the first 5 weeks (263). Decreased body weight gain in the 10- and 20-mg/kg/day rats and decreased brain and RBC cholinesterase activities occurred at 5 ppm and higher.

6.4.1.3 Pharmacokinetics, Metabolism, and Mechanisms. Demeton-*S*-methyl is completely absorbed and very rapidly eliminated following either oral or intravenous administration. Blood concentration decreased, and the half-life was about 2 hours during the first 6 hours and then about 6 hours for the next 48 hours following oral administration of demeton-*S*-methyl to rats. The half-life thereafter was even longer. The half-life of urinary elimination was 2–3 hours during the first 24 hours and 1.5 days thereafter. Elimination through feces and exhaled air was minimal and accounted for 0.5–2% and about 0.2% of the dose, respectively. Except for RBCs which tended to bind the demeton-*S*-methyl, it was distributed uniformly in various body tissues and organs. At 2, 24, and 48 hours after dosing, about 60%, 1%, and 0.5% of the administered dose, respectively, remained in the body. By 10 days, demeton-*S*-methyl was almost undetectable in most organs except in the RBCs (262).

The main metabolic route of demeton-*S*-methyl is oxidation of the side chain leading to the formation of the corresponding sulfoxide, oxydemeton methyl, and to lesser extent, after further oxidation, the sulfone (262). O-demethylation also occurs. Neither glucuronide nor sulfate conjugates have been identified (262).

6.4.1.4 Reproductive and Developmental. Pup viability, lactation index, and body weight gain were reduced in F1 offspring when rats were fed a diet that contained 25 ppm demeton-*S*-methyl for two generations (202). Offspring of rats fed 1 or 5 ppm were unaffected. No compound-related malformation was found in animals of any of the treatment groups.

No alterations of physical appearance or behavior occurred in dams or fetuses from dams given 0, 0.3, 1, or 3 mg/kg demeton-*S*-methyl by gavage on days 6 to 15 of gestation (262). The numbers of live fetuses and resorptions, fetal weight, number of fetuses with malformations, and number of implants were comparable in all groups. No treatment-related visceral or skeletal abnormalities were observed (262). No abortions or increases in the numbers of implantations per day, preimplantation losses, postimplantation losses, resorptions, living and dead fetuses, or sex ratios occurred in rabbits given 3, 6, and 12 mg/kg/day demeton-*S*-methyl by gavage on gestation days 6 to 18. Diarrhea, decreased food consumption, and decreased fetal body weight occurred in the 12 mg/kg/day treated animals. There was no treatment-related increase in gross, skeletal, or visceral malformations (262).

6.4.1.5 Carcinogenesis. When dogs were fed diets containing 1, 10, or 100 ppm (day 1–36) followed by 50 ppm (day 37-termination) demeton-*S*-methyl (equivalent to 0.036, 0.36, and 4.6 followed by 1.5 mg/kg/day), diarrhea and vomiting occurred at all levels (262). Multifocal slight/moderate atrophy and/or hypertrophy of proximal renal tubules also occurred in the high dose group. Plasma and RBC cholinesterase activity was reduced at 10 and 100 ppm, and, brain cholinesterase activity was reduced at 10 ppm.

There was no evidence of carcinogenicity in mice given diets with 1, 15, or 75 ppm demeton-*S*-methyl (0.24, 3.47, or 17.81 mg/kg/day (males); 0.29, 4.18, or 20.0 mg/kg/day (females)) for two years (271). Cholinergic signs were not observed at any level nor did mortality differ among groups. Plasma, RBC, and brain cholinesterase activity decreased in mice at 15 and 75 ppm. There was no evidence of carcinogenicity in rats given 1, 7, or 50 ppm demeton-*S*-methyl in their feed (0.05, 0.31, or 2.59 mg/kg/day (males); 0.06, 0.41, or 3.09 mg/kg/day (females)) for 24 months (262). Hair loss and diarrhea occurred more frequently at 50 ppm. Body weight was reduced in males at 7 ppm and in both males and females at 50 ppm. Plasma, RBC, and brain cholinesterase activites decreased in groups given the 7- or 50-ppm diet. Increased incidence of retinal atrophy and keratitis was observed in mice given the 50-ppm diet.

6.4.1.6 Genetic and Related Cellular Effects Studies. Available information is insufficient to permit an adequate assessment of the genotoxic potential of demeton-*S*-methyl (262). Demeton-*S*-methyl did not induce DNA damage in the Pol test in *E. coli* with or without metabolic activation, but it did increase mutation rates in the Ames test and in the mouse lymphoma forward mutation assay with or without metabolic activation. In *in vivo* tests, no sister chromatid exchanges (SCEs) were found in the bone marrow of Chinese hamsters treated with high doses of demeton-*S*-methyl, and, bone marrow micronucleus and dominant lethal tests in mice treated with demeton-*S*-methyl gave negative results. However, chromosomal aberrations were found in the bone marrow of Syrian hamsters treated with a commercial formulation of demeton-*S*-methyl (262).

6.3.5 Biomonitoring/Biomarkers

Urinary levels of the metabolite dimethyl phosphorothiolated potassium salt (DMPThK) and plasma and whole blood cholinesterase activities were monitored in agricultural workers exposed to demeton-*S*-methyl for 3 consecutive days (262). Exposed subjects were identified as either mixers, sprayers, or others not directly involved in handling the pesticide. Levels of DMPThK in urine from mixers had a medium (*sic*) value of 83 µg/liter and a range of 0–822 µg/liter (neither corrected for creatinine nor for urine volume); urine from sprayers had a mean value of 30 µg/liter (limit of detection) and a range of 0–208 µg/liter, and urine from other subjects not directly exposed had a mean value of 30 µg/liter and a range of 0–100 µg/liter. Whole blood cholinesterase activity was not affected by exposure, and plasma cholinesterase activity was slightly reduced compared to preexposure levels in mixers. No correlation was found between DMPThK levels and plasma cholinesterase activity (269).

6.4.2 Human Experience

6.4.2.2 Clinical Cases. Six hundred seventy-three occupational cases of organophosphate poisoning, including three deaths, reportedly occurred in Egypt about 1 week after demeton-*S*-methyl began to be used for spraying cotton (265). Two children evidently accidentally exposed to demeton-*S*-methyl via inhalation while waiting for their father (a sprayman) to finish work, became unconscious for a few minutes. When aroused, they vomited and complained of abdominal colic. Another girl who evidently ate beans contaminated during spraying also vomited and suffered abdominal colic.

One woman attempted suicide by ingesting "1 or 2 mouthfuls" of Metasystox I (25% demeton-*S*-methyl). On admission to hospital she was comatose, sweating, salivating and had pinpoint pupils. Plasma and RBC cholinesterase activites were less than 10% of normal. She was successfully treated and released after 30 days at which time plasma cholinesterase values had returned to normal, but RBC cholinesterase activity was still below normal (262). In another suicide attempt, a 41-year-old pregnant woman was admitted to a hospital about 3.5 hrs after ingesting an estimated 12 g of methyl demeton. Upon admission, blood (*sic*) cholinesterase was 10% of normal. About 12 hours after admission, she became comatose and was treated with atropine, odoxime, haemoperfusion, and artificial ventilation. She recovered and was discharged 24 days later (262).

A man who had been an agricultural applicator for five years worked with demeton-*S*-methyl, mainly as a flagman in aerial spraying but also in preparing the spray and in cleaning containers after spraying. For about a month and half, he was potentially exposed for periods that varied from 20 minutes to $6\frac{3}{4}$ hours. Symptoms (headaches, nausea, dizziness) gradually increased in severity during the week and subsided during the weekend. Later, he developed anorexia and loss of ability to concentrate. At the end of 6 weeks, his symptoms became worse while he was driving a tractor applying disulfoton. After 2 hours, he became dissatisfied with his control of the machine and sought medical aid. Clinical findings were normal, but cholinesterase activity was low (266). The author concluded that the man had suffered from gradually worsening organophosphate poisoning primarily due to absorption of demeton-*S*-methyl through the skin (despite the use of required protective clothing) that began about 2–3 weeks after initial exposure.

Organophosphate poisoning following demeton-*S*-methyl exposure was described in six men engaged in packaging bulk loads of demeton-*S*-methyl concentrate (500 g/L) into one liter containers (267). The first poisoned worker experienced cholinergic symptoms (nausea, dizziness, weakness, difficult breathing, and diarrhea) after working 1 day filling containers; the second experienced symptoms (giddiness, nausea, weakness, sweating, and cramps) after 72 hours. The filling procedure was modified to decrease worker exposure by requiring more protective clothing and performing operations in a fume "cupboard," but a third worker experienced symptoms (nausea, abdominal cramps, and weakness) after working 2 days using the revised procedure. Plasma and RBC cholinesterase activities measured 15–30 days after exposure were below the lower limit of the normal range and did not completely recover to the normal range until 60 days after exposure. Chemical analysis of vapor in the fume hood and of residue on gloves and other clothing indicated that exposure had occurred primarily via dermal absorption that resulted from contamination of external and internal surfaces of gloves.

Six workers engaged in hop cultivation using Metasystox I (reported to contain demeton-*O*-methyl instead of demeton-*S*-methyl as the commercial name implies) were monitored. They sprayed up to 2400 liters of a 0.1% solution (in water) of the insecticide in 1 day. No significant inhibition of blood acetylcholinesterase was observed at the end of exposure or 1 or 2 days later. One subject, who was exposed twice, showed a 29% decrease in blood (sic) acetylcholinesterase after the second exposure. No cholinergic toxicity was observed in these workers (262).

In a group of men who sprayed cotton fields with demeton-*S*-methyl, signs and symptoms of cholinergic toxicity (gastrointestinal disturbances, dizziness, persistent general weakness and fatigue, respiratory manifestations, headache, sweating, salivation or lacrimation, tremors of outstretched hands, intention tremors, ataxia, exaggerated superficial and deep reflexes, hiccough, and muscular fasciculations) occurred after 1–18 days of exposure, and the mean latency period was 3 days (265). Serum cholinesterase activity estimates were performed within 24 hours after the onset of symptoms in some patients and after the cessation of symptoms in some others. In most cases, they were repeated two to three times at various intervals up to 40 days from the onset of symptoms. In general, serum cholinesterase activity underwent a marked initial fall followed by a rise above normal levels after about 30–40 days (265).

6.5 Standards, Regulations, or Guidelines of Exposure

Demeton-*S*-methyl is not registered by the EPA for use. It is anticipated that most other national registrations for demeton-*S*-methyl were probably transferred soon after 1998 to oxydemeton-methyl (262). No Occupational Exposure Limits were identified for demeton-*S*-methyl. In the 2000 ACGIH TLVs, in the notice of intended changes, ACGIH proposes a TLV of 0.05 mg/m^3, for demeton-*S*-methyl.

7.0 Diazinon

7.0.1 CAS Number: [333-41-5]

7.0.2 Synonyms: *O,O*-diethyl *O*-2-diethyl *O*-2-isopropyl-4-methyl-6-pyrimidinyl thiophosphate; phosphorothioic acid, *O,O*-diethyl-*O*-(2-isopropyl-6-methyl-4-pyrimidinyl)ester; Dimpylate; *O,O*-diethyl *O*-(2-isopropyl-6-methyl-4-pyrimidinyl), phosphorothioate; *O,O*-diethyl *O*-(6-methyl-2-(1-methylethyl)-4-pyrimidinyl) phosophorothioatae; phosphorothioic acid *O,O*-diethyl *O*-[6-methyl-2-(1-methylethyl)-4-pyrimidinyl] ester; thiophosphoric acid 2-isopropyl-4-methyl-6-pyrimidyl diethyl ester; *O,O*-diethyl *O*-2-isopropyl-4-methyl-6-pyrimidyl thiophosphate; Knox Out; dianon; gardentox; kayazinon; g-24480; diethyl 2-isopropyl-6-methyl-4-pyrimidinyl phosphorothionate; *O,O*-diethyl *O*-(6-methyl-2-(1-methylethyl)-4-pyrimidinyl) phosphorothioate; Dipofene; Diazitol; AG-500; Antigal; Dacutox; Dassitox; Dazzel; Diagran; Diaterr-fos; Diazajet; Diazide; Diazol; diethyl 2-isopropyl-4-methyl-6-pyrimidinyl phosphorothionate; diethyl 2-isopropyl-4-methyl-6-pyrimidyl thionophosphate; diethyl *O*-(2-isopropyl-6-methyl-4-pyrimidinyl) phosphorothioate; Dimpylatum; Drawizon; Dyzol; Exodin; Fezudin; Flytrol; Galesan; isopropylmethylpyrimidyl diethyl thiophosphate; Kayazol; Knox out 2FM; Neocidol; Nipsan; Nucidol; Sarolex; Dizinon; *O,O*-diethyl *O*-(2-isopropyl-4-methyl-6-pyrimidinyl) thiophosphoric acid

7.0.3 Trade Names: Spectracide®, Basudin®, Diazitol®, Dipofene®, Neocidol®, Nucidol®

7.0.4 Molecular Weight: 304.36

7.0.5 Molecular Formula: C$_{12}$H$_{21}$N$_2$O$_3$PS

7.0.6 Molecular Structure:

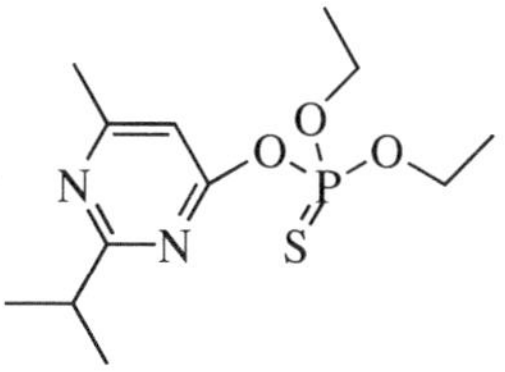

7.1 Chemical and Physical Properties

Diazinon is a colorless liquid.
Specific gravity: 1.116–1.118 at 20°C
Boling point: 83–84°C at 0.002 torr
Melting point: > 120°C (dec)
Vapor pressure: 1.4×10^{-4} mmHg at 20°C
Solubility: slightly soluble in water (0.004 g/100 mL); freely soluble in petroleum solvents; miscible with alcohol, ether, benzene, and similar hydrocarbons

7.1.2 Odor and Warning Properties

Faint ester-like odor.

7.2 Production and Use

Diazinon is a nonsystemic insecticide used on a wide variety of agricultural crops such as rice, fruit trees, corn, tobacco, and potatoes and to control fleas and ticks. Various types of formulations are available, including dusts, emulsifiable concentrates, impregnated material, granules, microencapsulated forms, pressurized sprays, soluble concentrates, and wettable powders (55).

7.4 Toxic Effects

7.4.1.1 Acute Toxicity. Diazinon is a moderately toxic organophosphate compound that has oral LD_{50}s of 250–466 mg/kg (64a). Oral LD_{50}s for an impure formulation were 76–108 mg/kg (Gaines, 1960). The dose–lethality curve is steep as illustrated by the observation that acute LD_{01}s are only 30–36% smaller than LD_{50}s (64a) and that an acute oral dose of 528 mg/kg was lethal to rats whereas a dose of 264 mg/kg was not (55). Dermal LD_{50}s were 455–900 mg/kg (64a, 268). The dermal LD_{50} for male rats of a sample allowed to completely crystallize in air for several weeks was 34 mg/kg, demonstrating that acute dermal toxicity increases significantly with an impure diazinon formulation (268). Onset of symptoms and death following acute exposures to diazinon occurs between 1 and 6 hours from oral exposures (81, 271) but may be delayed following dermal exposure by about 10 hours (272).

A 4-hour LC_{50} of 3500 mg/m^3 was reported for rats; a 4-h LC_{50} of 1600 mg/m^3 was reported for mice, and a 4-hour LC_{50} of 55,500 mg/m^3 was reported for guinea pigs (545). No deaths occurred among rats exposed to 2330 mg/m^3 diazinon for 4 hours in inhalation chambers and observed for 14 days, although decreased activity and increased salivation were noted (55).

No signs of organophosphate toxicity occurred in rats exposed to 0.05, 0.46, 1.57, or 11.6 mg/m^3 for 6 h/day, 5 days/week for three weeks (55). Serum cholinesterase activity decreased in females exposed to 0.46 mg/m^3 and higher and in males exposed to 1.57 mg/m^3

and higher; RBC acetylcholinesterase activity decreased in females exposed to 11.6 mg/m^3, and brain acetylcholinesterase activity decreased in females exposed to 1.57 and 11.6 mg/m^3.

Among male rats that were given single doses of 100, 200, or 400 mg/kg diazinon by gavage, cholinergic signs (lacrimation, salivation, miosis, hypoactivity, ataxia, increased landing foot splay, decreased tail-pinch response, tremors, chewing (smacking), and hypothermia) peaked at 4 hours in the 400-mg/kg group and were still present at 24 hours, but had disappeared by 72 hours after dosing (55). Cholinergic toxicity also occurred in the 200-mg/kg group, and only hypoactivity and decreased defecation occurred in the 100-mg/kg group. Among rats given single oral doses of 2, 132, 264, or 528 mg/kg diazinon, cholinergic signs (autonomic (smacking), neuromuscular (ataxia, abnormal gait)) occurred at 132 mg/kg and higher (55). Serum cholinesterase and RBC acetylcholinesterase activity was reduced at all levels (55).

Diazinon was not neurotoxic in atropinized hens given 11.3 mg/kg twice orally, 21 days apart (55).

7.4.1.2 Chronic and Subchronic Toxicity.
No cholinergic toxicity occurred in rats fed diets that contained up to 1000 ppm technical diazinon for four weeks (82), in rats or mice fed diets that contained up to 1600 ppm for 13 weeks (87), in rats given diets that contained up to 25 ppm diazinon for up to 92 days (79, 80) or in rats fed diets, that contained up to 125 ppm diazinon for 15–16 weeks (81). Subchronic exposures in the 3- to 100-ppm range were associated with inhibition of RBC acetylcholinesterase activity, and exposures to 1000 ppm were associated with brain acetylcholinesterase inhibition (79–82). Six week exposures of rats to up to 180 mg/kg/day via their diet caused no cholinergic toxicity, although RBC acetylcholinesterase activity was inhibited at doses of 8 or 9 mg/kg/day (83). Thirteen-week exposures of rats to 168 or 212 mg/kg/day resulted in cholinergic signs (soft stools and hypersensitivity to touch and sound), although RBC acetylcholinesterase activity was inhibited at doses as low as 15 mg/kg/day (83).

No cholinergic toxicity occurred in dogs given diazinon via their diet (0.25, 0.75, or 75 ppm) for 12 weeks, although RBC acetylcholinesterase activity was inhibited for 6 weeks after termination of dosing to dogs given the 75-ppm diet (84). When dogs were given diets that contained 0.1, 0.5, 150, and 300 ppm diazinon for 13 weeks, cholinergic signs (emesis and diarrhea) occurred but were not dose-related (90). Significant reductions in RBC and brain acetylcholinesterase levels occurred in rats fed the 150-ppm diet.

7.4.1.3 Pharmacokinetics, Metabolism, and Mechanisms.
Absorption of diazinon following oral exposures is rapid and complete. Nearly 100% of orally administered diazinon was recovered in urine (70–80%) and feces (16–24%) within 168 hours in rats (98). Diazinon was not detected later than 2 days after a final repeated dose of diazinon, demonstrating that it does not accumulate in tissues (98). The biological half-life of diazinon was 12 hours. In dogs, absorption was at least 85% after a single oral dose (99). About 68% of an oral dose of diazinon was detected in urine 60 minutes after dosing in mice. In all cases, diazinon is widely distributed to tissues, especially adipose tissue,

muscle, brain, and liver (101). Similar results were observed in dogs, guinea pigs, goats, sheep, and cows (55).

The pharmacokinetics of diazinon following inhalation exposure has not been studied. However, pharmacokinetic studies following intravenous administration of diazinon to dogs indicated rapid absorption and an elimination half-life of 6 hours (99). Recovery in urine was 58% of the administered dose after 24 hours and was primarily diethyl phosphoric and phosphorothioic acid. In rats given diazinon intravenously, plasma diazinon levels declined rapidly during the distribution phase and more slowly during the elimination phase and were characterized by an elimination half-life of about 6 hours (102). In another study that examined the toxicokinetics of diazinon in rats following oral and intravenous exposure, the elimination half-lives were 1.8 and 4.7 hours, respectively, suggesting that hepatic metabolism enhances elimination (102). In four rhesus monkeys dosed intravenously with 32 µg diazinon, about 56% was excreted in urine, and 23% was eliminated in feces after 7 days, accounting for about 70% of the total dose. Most of the dose was excreted in urine on day 1 (103).

Total dermal absorption of diazinon was estimated at about 3–4% of the dose applied to the forearm or abdomen (2 µg/cm^3) of humans during 7 days regardless of whether the vehicle was acetone or lanolin (103). Absorption measured in human skin placed in an *in vitro* diffusion cell was about 14%.

The metabolism of diazinon is complex. The primary pathways are desulfuration by cytochrome P450 enzymes to the metabolite diazoxon which then either binds to acetylcholinesterase and other esterases or is further metabolized to diethyl phosphoric acid and oxidative products such as 2-isopropyl-4-methyl-6-hydropyrimidine (104, 105). Alternatively, diazinon is dearylated and hydrolyzed by cytochrome P450 enzymes to diethyl phosphorothioic and phosphoric acid and other oxidation products (e.g. 2-isopropyl-4-methyl-6-hydropyrimidine) which are excreted mostly in urine, although minor amounts of these metabolites and some unchanged diazinon have been detected in feces (98, 100, 104).

7.4.1.4 Reproductive and Developmental. Diazinon did not cause adverse effects on fertility in rats fed 0.05 mg/kg/day diazinon in the diet for 60 days before weaning for four generations (89). No gross or histological treatment-related damage to reproductive tissues occurred in rats given up to 168 mg/kg/day (males) or 212 mg/kg/day (females) diazinon for 13 weeks (83), in rats given up to 10 mg/kg/day (males) or 12 mg/kg/day (females) for 98 weeks (86), or in dogs given up to 11 mg/kg/day for 13 weeks (90). In another study, testicular atrophy and arrested spermatogenesis were observed in one dog given 10 mg/kg/day and in all dogs given 20 mg/kg/day for 8 months via corn oil capsule (85).

No teratogenic or fetotoxic effects occurred in rabbits given 7–100 mg/kg/day diazinon by gavage during gestation (91, 92). Cholinergic signs occurred in dams given 30 mg/kg or more. In hamsters, no embryo- or teratogenicity occurred at doses up to 0.25 mg/kg/day during gestation, although cholinergic toxicity occurred in parents (diarrhea, salivation, and incoordination). In another study, however, diazinon caused an increased incidence of stillbirths in dogs given 1.2 or 5 mg/kg/day by gavage (85). In rats, doses greater than or equal to 70.6 mg/kg/day during gestation increased fetal resorptions when given on days 8 to 12 or 12 to 15; no effects occurred at doses less than 70.6 mg/kg/day (93). In another study, no differences were observed in litter size, fetal body weight, fetal brain weight, number of resorptions, or corpora lutea among rats given diazinon via peanut oil gavage at 0, 40, 50, 60, or 75 mg/kg/day on gestation days 7 through 19 (94).

Pregnant mice given 0, 0.18, or 9 mg/kg/day diazinon throughout gestation (18 days) gave birth to viable offspring, although pups exposed to 9 mg/kg grew more slowly than controls (590). Mature offspring of both treated groups displayed impaired endurance and coordination on rod cling and inclined place tests of neuromuscular function. Morphological abnormalities in brain occurred among offspring of dams exposed to 9.0 mg/kg/day.

7.4.1.5 Carcinogenesis. No signs of cholinergic toxicity occurred in rats given diets containing 10 to 1000 ppm diazinon for 72 weeks or in dogs given up to 4.6 mg/kg/day diazinon via capsule 6 days/week for up to 46 weeks (82). Doses of 4.3 to 4.6 mg/kg/day, however, caused significant RBC cholinesterase inhibition in dogs after two week (82). Cholinergic signs (soft stools) occurred in dogs given 9.3 mg/kg by 30 days and excitability and tremors occurred in one dog given 25 mg/kg/day for 6 days. In dogs given 2.5, 5.0, 10.0, or 20.0 mg/kg/day diazinon by gavage for 8 months, cholinergic signs were observed in one dog given 10 mg/kg/day and emesis, fasciculation and mortality occurred in dogs given 20.0 mg/kg/day (85). No cholinergic toxicity occurred in pigs given 1.25 mg/kg/day for eight months; but cholinergic signs were observed pigs given 2.5–10 mg/kg/day (85).

No signs of cholinergic toxicity occurred in rats given diazinon via their diet at doses of 0.004, 0.06, 5, or 10 mg/kg/day (males) and 0.005, 0.07, 6, or 12 mg/kg/day (females) for 52 or 98 weeks (86). After 1 year, RBC acetylcholinesterase activity decreased in males given 5 and 10 mg/kg/day and in females given 6 or 12 mg/kg/day; brain acetylcholinesterase activity was unchanged in males but decreased in females given 6 or 12 mg/kg/day. During a 4-week recovery period, RBC acetylcholinesterase activity returned to normal in males, whereas that of females dosed at 12 mg/kg/day remained decreased; brain acetylcholinesterase activity returned to normal in females. Results were similar at 98 weeks.

Diazinon was not carcinogenic in rats or mice when they were given diets containing 400 or 800 ppm diazinon (rats) or 100 or 200 ppm diazinon (mice) for 2 years (87). However, clinical signs of hyperactivity were noted in low- (males) and high-dose (males and females) rats and in all dosed mice (*sic*). Bloating, vaginal bleeding, and vaginal discharge were also noted in the dosed female rats.

No cholinergic, hematological, clinical chemistry, or histopathological signs occurred at any dose in monkeys given oral doses of 0, 0.05, 0.5, or 5 mg/kg diazinon 6 days/week for two years via gavage, although serum and RBC cholinesterase activities were inhibited in monkeys given 0.5 or 5 mg/kg/day (88).

7.4.1.6 Genetic and Related Cellular Effects Studies. The genotoxicity of diazinon is equivocal. *In vitro* test results showed that diazinon was positive for gene mutations in the *S. typhimurium* test assay with metabolic activation and in the mouse lymphoma cell forward mutation assay without metabolic activation (55, 96). Diazinon was also positive for chromosomal aberrations in Chinese hamster cells with metabolic activation (55). But in other tests, diazinon was negative for gene mutations in the *S. typhimurium* test assay (97) and in the rec assay utilizing strains of *Bacillus subtilis* (55) both of which were conducted without metabolic activation. Tests for sister chromatid exchange in Chinese hamster V79 cells with and without metabolic activation (55) and for chromosomal aberrations in human peripheral blood lymphocytes (55) were also negative.

7.4.2 Human Experience

7.4.2.2 Clinical Cases. Ingestion of diazinon causes typical organophosphate poisoning that varies in intensity with dose (107–109). Lethality followed adult ingestion of an estimated 293 mg/kg diazinon (110) and child ingestion of 20 mg/kg (111), although the latter estimate may have been complicated by the possible simultaneous ingestion of parathion and/or chlordane. A summary of 76 fatal cases of diazinon poisoning indicated a high incidence of miosis, froth from nose and mouth, acute pulmonary edema and congestion, acute ulcers, blood stained gastric contents, CNS hemorrhage, and evidence that death was due to asphyxiation (109). No estimates of exposure levels were provided.

Cholinergic symptoms (nausea, epigastric pain, headache, miosis and unreactive pupils, tachycardia) occurred in a woman who ingested an estimated 1.5 mg/kg diazinon (112). Severe toxicity (bradycardia, tachycardia, clonus, stupor, profuse diaphoresis, sialorrhea, miosis, hyperreflexia, weakness, dysdiadokinesis, abdominal pain, nausea, coma, twitching, restlessness, and bronchospasm) was reported in five individuals who intentionally ingested estimated doses of 240 to 986 mg/kg diazinon and recovered (113). Signs of organophosphate poisoning (profuse sweating, nausea, vomiting, and abdominal cramps) were reported in children who had eaten oatmeal contaminated with about 2.5–244 ppm diazinon (114). An earlier report indicated that oatmeal contaminated by home spraying with a 25% concentrate of diazinon caused organophosphate poisoning (nausea, vomiting, abdominal cramps, diaphoresis, muscular weakness, rolling eye movements, ataxia, and muscle cramps) in eight children from two different families (115).

A man died from cardiac arrest, despite atropine therapy, following inhalation exposure to a commercial insecticide formulation containing diazinon and malathion, but no estimate of exposure was provided (115a). Inhalation exposures to a diazinon spray used to kill cockroaches in an adjoining apartment were implicated in the organophosphate poisoning of twin infants (116).

Depressed serum cholinesterase activities were only sometimes accompanied by signs of cholinergic toxicity in individuals occupationally exposed to diazinon primarily via inhalation (117). However, cholinergic symptoms (headache, blurred vision, dizziness, fatigue, nausea, and vomiting) began within 15 minutes in mushroom workers exposed to diazinon when it was sprayed around the only entrance to a room in which they were working. Reduced serum and RBC cholinesterase activities also occurred within 48 hours, and serum cholinesterase activities remained depressed for 15 days (118). Based on comparison with stabilized cholinesterase measurements taken in affected individuals 15 days after exposure, the authors estimated that plasma cholinesterase activities had been inhibited by about 30–34% and RBC acetylcholinesterase activities had been inhibited by about 27–34%.

In another report that involved multiple routes of exposure, several family members experienced diazinon poisoning (headache, vomiting, fatigue, and chest heaviness) associated with slightly depressed serum cholinesterase activities for several months after their home had been treated with diazinon (118a). Surface concentrations in the home ranged from 126 to 1051 $\mu g/m^2$, air concentrations were between 5 and 27 $\mu g/m^3$, and some clothing showed contamination (0.5 to 07 $\mu g/g$).

Dermal exposure to diazinon caused cholinergic signs (cyanosis, forthing at the mouth, drowsiness, nausea, vomiting, abdominal colic, diarrhea, tachpnea, miosis, and sinus tachycardia) in two female gardeners (119), but estimates of exposure were not available.

Exposure to diazinon via multiple routes was estimated at an average of 0.02 mg/kg/day in 99 workers exposed to diazinon granules 8 h/day for 39 days during an insecticide application program. Slight neurological functional deficits (postshift symbol-digit speed and pattern memory accuracy) were reported among the workers, but these effects were not statistically significant (120).

7.5 Standards, Regulations, or Guidelines of Exposure

Diazinon is under reregistration by the EPA (78).

The ACGIH TLV for diazinon is 0.1 mg/m^3 with a skin notation (154). The OSHA PEL-TWA and NIOSH REL-TWA are 0.1 mgm^3 with a skin notation. Most other countries have also established an Occupational Exposure Limit of 0.1 mg/m^3 for diazinon (Australia, Belgium, Denmark, Finland, France, Germany, India, The Netherlands, and the United Kingdom).

8.0 Dichlorvos

8.0.1 CAS Number: *[62-73-7]*

8.0.2 Synonyms: *O,O*-dimethyl-*O*-2,2-dichlorvinyl dimethyl phosphate; 2,2-dichlorovinyl dimethylphosphate; DDVP; dichlorophos; Equigand; No-Pest Strip; 2,2-dichlorovinyl-*O,O*-dimethyl phosphate; phosphoric acid 2,2-dichloroethenyl dimethyl ester; phosphoric acid 2,2-dichlorovinyl dimethyl ester; SD 1750; Astrobot; Atgard; Canogard; Dedevap; Dichlorman; Divipan; Equigard; Equigel; Estrosol; Herkol; Nogos; Nuvan; 2,2-dichloroethenyl dimethyl phosphate; 2,2-dichlorovinyl dimethyl phosphoric acid ester; 2,2-dichloroethenyl phosphoric acid dimethyl ester; dimethyl 2,2-dichloroethenyl phosphate; dimethyl 2,2-dichlorovinyl phosphate; *O,O*-dimethyl dichlorovinyl phosphate; *O,O*-dimethyl *O*-2,2-dichlorovinyl phosphate; 2,2-dichlorovinyl alcohol dimethyl phosphate; apavap; atgard c; atgard v; bay-19149; benfos; bibesol; brevinyl; brevinyl e50; chlorvinphos; deriban; derribante; devikol; duo-kill;duravos; estrosesel; fecama; fly-die; fly fighter; herkal; krecalvin; MAFU; mafu strip; marvex; mopari; nerkol; nogos 50; nogos g; no-pest; NUVA; nuvan 100ec; OKO; OMS 14; phosvit; szklarniak; TASK; Tenac; task tabs; tetravos; UDVF; unifos; unifos 50 ec; vaponite; vapora ii; verdican; verdipor; vinylofos; vinylophos; bayer 19149; *O,O*-dimethyl 2,2-dichlorovinyl phosphate; fekama; insectigas d; nefrafos; nogos 50 ec; novotox; nuvan 7; panaplate; winylophos; 2,2-dichloroethenol dimethyl phosphate; Cekusan; Cypona; Delevap; Derriban; Dichloroethenyl dimethyl phosphate; Equiguard; Prentox; Verdisol; DichlorvosI [Dimethyl Dichlorovinyl Phosphate]

8.0.3 Trade Names: Vapona®

8.0.4 Molecular Weight: 220.98

8.0.5 Molecular Formula: $C_4H_7Cl_2O_4P$

8.0.6 Molecular Structure:

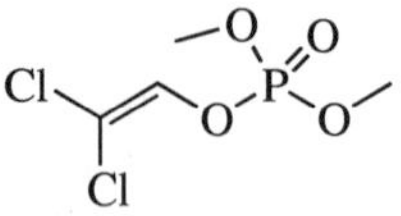

8.1 Chemical and Physical Properties

Dichlorvos is a colorless to amber, oily liquid. It hydrolyzes at a rate of 3% per day in saturated aqueous solution at room temperature; at high pH or in boiling water, it completely hydrolyzes in 1 hour. Dichlorvos decomposes to toxic gases and vapors (such as hydrogen chloride gas, phosphoric acid mist, and carbon monoxide)

- Specific gravity: 1.415 g/ml at 25°C
- Boiling point: 140°C at 20 mmHg; 221°C at 760 mmHg
- Melting point: −60°C
- Vapor pressure: 0.012 torr at 20°C
- Solubility: slightly soluble in water (1 g/100 mL at 20°C); miscible with aromatic and chlorinated hydrocarbon solvents and alcohols

8.1.2 Odor and Warning Properties

Mild chemical odor.

8.2 Production and Use

Dichlorvos is used against a wide variety of insects in greenhouses, outdoor fruit and vegetable crops, and is also used in aquaculture to rid fish of various skin parasites. In addition, it is used to control severe internal and external parasite infestations in animals and humans (57). It is a available as soluble concentrates and aerosols and is also formulated with other pesticides.

8.4 Toxic Effects

8.4.1.1 Acute Toxicity. Dichlorvos is an organophospahte that has high oral toxicity, oral LD_{50}s of 56–98 mg/kg in rats, and oral LD_{50}s 133–139 mg/kg in mice (57). Dose–lethality curves are steep; oral LD_{01}s are about one-half or less the LD_{50} value (62, 64a), and death occurs quickly within minutes. Cholinergic signs occurred within 7–15 minutes in dogs given a single oral dose of 11 or 22 mg/kg dichlorvos (62a). Three of 12 dogs given 22 mg/kg died within 10–155 minutes of treatment. Similar effects occurred when dogs were given 2–11 mg/kg dichlorvos intravenously, but death occurred slightly more rapidly, by 7 minutes in one case. When rats were given single oral doses of 0.5, 35, or 70 mg/kg dichlorvos by gavage, the 35- and 70-mg/kg groups exhibited cholinergic signs within 15 minutes after dosing (120a). Several animals in the 70-mg/kg group died. No signs of toxicity were apparent in any rats given 0.5 mg/kg or in any of the 35- or 70-mg/kg treated rats that survived 7 days after dosing (120a). Severe cholinergic signs occurred in dogs given 15–30 mg/kg/day for 12–24 days via corn oil capsule, and less severe cholinergic signs occurred in dogs given 1–10 mg/kg/day for 16–24 days (95). Doses of 0.1 mg/kg/day for up to 24 days had no effect.

Dermal LD_{50}s are 75–107 mg/kg in rats (64a). Cholinergic signs and death occurred in monkeys after a single 100 mg/kg dermal dose, after eight 50-mg/kg/day doses during 10 days, and after ten 75-mg/kg doses during 12 days, suggesting that dermal exposures may be cumulative (121).

Inhalation exposures are more potent on the basis of body weight than oral exposures (122). Four and one-hour LC_{50}s for dichlorvos in rats are 455 and 340 mg/m^3 (123). A saturated atmosphere of dichlorvos (230–341 mg/m^3) caused deaths among rats after 7 to 62 hours (121). Rabbits are more sensitive than rats or mice to dichlorvos vapor. Deaths occurred in 9 of 16 rabbits exposed to 6.25 mg/m^3 and in 6 of 20 rabbits exposed to 4 mg/m^3 for 23 h/day for 28 days during gestation (37), whereas no deaths occurred in mice or rats exposed to the same concentrations. Even exposures up to 56 mg/m^3 for 14 days did not cause deaths in rats (124). No deaths occurred in mice exposed to 30–55 mg/m^3 for 16 hours (125) or in pregnant mice or rabbits exposed to 4 mg/m^3 for 7 h/day (126). No adverse effects occurred in rhesus monkeys exposed to 0.48, 2.3, 2.6, or 12.9 mg/m^3 for 2 h/day for 4 days, although RBC cholinesterase was inhibited in monkeys exposed to 12.9 mg/m^3 (127).

Delayed neuropathy occurred in chickens after 35 days treatment with 6.1 mg/kg/day dichlorvos; 3.1- and 4.4-mg/kg/day doses were ineffective (128). Two doses of 16.5 mg/kg dichlorvos 21 days apart to hens did not cause acute delayed neurotoxicity, although signs of cholinesterase inhibition were apparent shortly after dosing (120a).

In a guinea pig mazimization test, induction with dichlorvos by intradermal injection and topical application and subsequent challenge with topical dichlorvos solutions showed sensitization (129).

8.4.1.2 Chronic and Subchronic Toxicity. A 90-day $LD_{50} > 70$ mg/kg was determined in rats fed dichlorvos in their diet for 90 days (196). Because this value was more than the single dose LD_{50} of 56 mg/kg, it was concluded that dichlorvos does not have a cumulative effect. When rats were given feed that delivered 0–360 mg/kg/day for 6 weeks, all rats that consumed 180 mg/kg or more died whereas none that consumed 90 mg/kg/day or less died (130). When mice were given feed that delivered 0–1080 mg/kg/day for six weeks, four of five females given 720 mg/kg/day died; all mice given 1080 mg/kg/day died (130).

Cholinergic signs occurred in dogs given 0.625 or 1.25 mg/kg/day dichlorvos by capsule for 70–90 days (65). RBC and brain cholinesterase activities were inhibited in dogs given 0.625 mg/kg/day. Ninety-day studies of dichlorvos in rats have shown that dietary levels up to 1000 ppm (about 70 mg/kg/day) do not result in overt cholinergic toxicity, although exposures to levels of 200 ppm or greater inhibit RBC cholinesterase

(56, 57). Oral gavage studies show lower effect levels. Daily administration of 7.5 mg/kg/day or more to rats via gavage for 13 weeks is associated with cholinergic toxicity (131) as well as significant RBC and brain acetylcholinesterase inhibition. Daily administration of 160 mg/kg/day by gavage for 13 weeks caused death in mice (131) as well as significant RBC and brain acetylcholinesterase inhibition. Gavage doses up to 40 mg/kg/day had no effect.

8.4.1.3 Pharmacokinetics, Metabolism, and Mechanisms. At least 85% of an oral dose of dichlorvos is absorbed (145). Dichlorvos is well absorbed following inhalation exposure based on the occurrence of toxic symptoms associated with inhalation exposures and the detection of specific dichlorvos metabolites (dichloroethanol and dimethyl phosphate) in urine of individuals exposed to dichlorvos (137, 138). Excretion is very rapid based on the observation that dichlorvos could not be detected in the blood of two male volunteers immediately after exposure to dichlorvos vapor (145). An elimination half-life of 13.5 minutes was estimated based on dichlorvos concentration in rat kidney after 2 or 4 hours exposure to 5 mg/m^3 (145). In mice and rats given single oral doses of dichlorvos, 59–65% was eliminated in urine, 3–7% was eliminated in feces, 14–18% was eliminated as CO_2 by 4 days after dosing, and the vast majority was eliminated by 24 hours (138). Retained dichlorvos following either oral or inhalation exposures is high because it is incorporated into intermediary metabolism (139).

Dichlorvos binds to acetylcholinesterase forming dimethoxy-phosphorylated acetycholinesterase and dichloroacetaldehyde (140). Alternatively, it is metabolized (primarily in the liver but also in the blood, adrenal, kidney, lung, and spleen) via two pathways (141). The major pathway is catalyzed by A-esterases and produces dimethyl phosphate and dichloroacetaldehyde (140). Dichloroacetaldehyde is converted to dichloroethanol which is then excreted as the glucuronide. Alternatively, dichloroacetaldehyde is dehalogenated and the carbon atoms are incorporated into normal tissue constituents via intermediary metabolism (137, 138, 141). The second minor pathway is catalyzed by glutahione-S-transferase and produced desmethyl dichlorvos and S-methyl glutathione. Subsequent degradation of desmethyl dichlorvos to dichloroacetaldehyde and monomethyl phosphate is catalyzed by A-esterases. S-methyl glutathione is broken down to methylmercapturic acid and excreted in urine. CO_2 is also the major metabolite following inhalation exposures. The major urinary metabolite following either oral or inhalation exposures is dichloroethanol glucuronide.

Dichlorvos is rapidly metabolized in human blood by A-esterases (142, 143). Unlike paraoxonse which exhibits polymorphism in the human population (144), dichlorvos A-esterase appears to be normally distributed. Half-lives for degradation of dichlorvos in whole blood after inhalation were 8.1 minutes for men and 11.2 minutes for women (145).

8.4.1.4 Reproductive and Developmental. Several studies indicated that dichlorvos is not a reproductive or developmental toxin. No impairment of male fertility occurred in mice exposed to 30 or 55 mg/m^3 dichlorvos for 16 hours or to 2.1 or 5.8 mg/m^3 for 23 hours daily for 4 weeks (37). However, estrus was delayed 10 days in female rats that were continuously exposed from birth to 2.4 mg/m^3 dichlorvos vapors from a Shell "No-pest Strip" (132). Number, viability, and growth rate were normal in offspring of swine fed up to 37 months on diets containing 200–500 ppm dichlorvos (133). No maternal or reproductive toxicity occurred in rats given dichlorvos for three generations at a dietary level of 0.1–500 ppm (about 0.005–25 mg/kg/day) (95).

No adverse effect on fetuses occurred when pregnant rats were given injections of 15 mg/kg dichlorvos on gestation day 11 (379), 0.1–21 mg/kg/day dichlorvos on gestation days 6 through 15 (95), or were exposed to 0.25–6.25 mg/m^3 dichlorvos vapor for 23 h/day on gestation day 1 through 20 (37). No adverse effect on fetuses occurred when pregnant mice were given the maximal tolerated dose of dichlorvos (60 mg/kg) on gestation days 6 through 15 or were exposed to 4 mg/m^3 dichlorvos vapor for 7 h/day (126). No adverse effect occurred on fetuses when pregnant rabbits were administered doses of 0.1 to 7.0 mg/kg/day dichlorvos by gavage on gestation days 7 through 19 (120a) or when rabbits were exposed for 23 h/day to 0.25 to 6.25 mg/m^3 dichlorvos vapor on gestation days 1 through 28 (37).

8.4.1.5 Carcinogenesis. Dichlorvos was not carcinogenic when rats were exposed to 0.05, 0.48, or 4.70 mg/m^3 dichlorvos for 23 h/day, 7 days/week for up to two years (136). Nor were any cholinergic signs observed in any group. The EPA's Carcinogenicity Peer Review Committee (CPRC) considered this study sufficient evidence that dichlorvos does not cause cancer via inhalation (68).

Carcinogenicity was not observed in rats given diets of 0, 150, or 326 ppm dichlorvos (equivalent to doses of about 8–14 and 16–29 mg/kg/day) for 80 weeks and then observed for an additional 30 weeks (130). The results of this study, however, were questioned because of an extraordinarily high mortality rate in control rats. Therefore, the chronic toxicity and carcinogenicity of dichlorvos was reevaluated in rats dosed with dichlorvos by gavage at levels of 0, 4, or 8 mg/kg/day dichlorvos for 5 days/week for 103 weeks (131). Cholinergic signs of toxicity occurred and RBC acetylcholinesterase activity decreased in both groups. Significant increase in mammary gland neoplasms in females and a significant trend (not dose related) for mononuclear cell leukemia was observed in males. Peer review panels characterized these results as "some evidence" of carcinogenic activity in males and equivocal evidence" in females (131). EPA's CPRC concluded that increased incidence of leukemia in rats may not be biologically significant (68, 120a).

Carcinogenicity was not observed in mice given diets that delivered 57 or 114 mg/kg/day dichlorvos for 80 weeks (130). However, a positive trend for squamous cell papilloma and carcinomas of the forestomach was observed in mice given 10–40 mg/kg/day dichlorvos by gavage for 5 days/week for 102 weeks (131). Significantly decreased RBC acetylcholinesterase activity was also observed at all levels. Peer review panels characterized these results as "some evidence" of carcinogenicity in male mice and "clear evidence" in female mice (131). However, the relevance of forestomach tumors to humans is questionable. Carcinogenicity was not reported in male or female mice given 58 or 95 mg/kg/day or 56 or 102 mg/kg/day, respectively, in their drinking water for 2 years (95). However, there was a dose-related decrease in absolute and relative weight of the gonads of males, and testicular atrophy increased in males given the high dose (95 mg/kg/day). The absolute and/or relative weight of the pancreas also decreased in treated females.

When dogs were given dichlorvos by capsule for 52 weeks at doses of 0, 0.1, 1.0, or 3.0 mg/kg/day, one male in the 3.0 mg/kg/day group exhibited cholinergic toxicity only at week 33 (57). RBC cholinesterase was inhibited at doses of 0.1 mg/kg/day and higher, and brain cholinesterase was inhibited in the 1.0-mg/kg/day (males only) and 3.0-mg/kg/day groups.

8.4.1.6 Genetic and Related Cellular Effect Studies.

Dichlorvos is not genotoxic when tested in *in vivo* system but is generally genotoxic or mutagenic in *in vitro* test when metabolizing enzymes are not present (57). Dichlorvos increased the frequency of chromosomal damage and micronucleus formation in Chinese hamster ovary cells; induced sister chromatid exchange, chromosomal aberrations, and transformation in cultured rat tracheal epithelial cells; induced DNA single-strand breaks in isolated rat hepatocytes; and caused increases in cell transformation of hamster embryo cells (134).

Dichlorvos was negative in the sex-linked lethal mutation test in *Drosophila*. However, increased mutations and chromosomal abnormalities occurred in flies given dichlorvos-contaminated food. Dominant lethal mutations did not occur in mice given an intraperitoneal dose or oral doses of 5 or 10 mg/kg dichlorvos or in mice exposed to dichlorvos via inhalation (30 or 55 mg/m^3). No chromosome damage occurred in mice given drinking water containing 2 mg/L dichlorvos for 7 weeks, no aberration in chromosomal structure or number occurred in bone marrow cells of mice given intraperitoneal injections of dichlorvos for 2 days, and no chromosomal abnormalities occurred in mice exposed to 64–82 mg/m^3 dichlorvos for 16 h or to 5 mg/m^3 for 21 days (134). There is a report that intraperitoneal injection of mice with lethal (LD$_{50}$, $\frac{1}{2}$ LD$_{50}$) amounts of dichlorvos causes chromosomal aberrations in bone marrow (146). However, the usefulness of this study in predicting *in vivo* genotoxicity has been challenged because of the toxic dose administered (134). In other *in vivo* studies, an increase in the percentage of hair follicles that contained nuclear aberrations occurred in mice 24 hours after a single dermal dose of dichlorvos (134a) and an increase in the incidence of micronuclei occurred in cultured skin cells from mice given a single dermal dose of dichlorvos (135).

It has been suggested that dichlorvos is not genotoxic *in vivo*, despite its methylating ability, because the phosphorus atom of the molecule is a stronger electrophile than the methyl carbons. Hence, *in vivo*, dichlorvos is much more likely to react with A-type esterases, serum cholinesterase, or acetylcholinesterase than with DNA (57, 134).

8.3.5 Biomonitoring/Biomarkers

The major metabolites of dichlorvos, dimethyl phosphate, and the glucuronide conjugate of dichloroethanol, are rapidly excreted in urine and could conceivably be used to monitor acute dichlorvos exposure. However, because other organophosphates, naled and trichlorphon, are metabolized to dichlorvos, exposure to them would have to be ruled out before a definitive exposure of dichlorvos could be made.

8.4.2 Human Experience

8.4.2.2 Clinical Cases.

Death has followed accidental or intentional ingestion of liquid dichlorvos or cake-like baits containing both malathion and dichlorvos (504). Two workers in Costa Rica died after splashing a concentration formulation of dichlorvos on their bare

ORGANOPHOSPHORUS COMPOUNDS

arms and failing to wash it off (504). Persistent contact dermatitis has also been reported following skin contact with dichlorvos (148).

RBC acetylcholinesterase activity marginally decreased in one of two dichlorvos applicators exposed to an estimated level of 0.02 mg/m^3 for about 25.5 minutes and 0.028 mg/kg/hr dermally (149). RBC acetylcholinesterase activity was reduced in some residents exposed to an estimated level of 0.2 mg/m^3 dichlorvos for about 15.8 hours, and some residents complained of headache (149). Average air concentrations of 0.13 mg/m^3 resulting from the use of resin strips in residences had no effect on RBC acetylcholinesterase activity (149a). Airbone levels of dichlorvos that caused slight to moderate RBC cholinesterase depression were 0.7 mg/m^3 averaged over 1 year in factory workers who produced dichlorvos vaporizers (149b). No significant change in RBC acetylcholinesterase activity occurred in babies exposed to an estimated 0.05 to 0.16 mg/m^3 dichlorvos for 18 h/day for 5 days (150).

Exposure of men to 0.1 to 0.3 mg/m^3 dichlorvos in 39 half-hour periods during 14 days had no effect on RBC, plasma cholinesterase, or physiological function (151). When the intensity of exposure was kept the same but the frequency of exposure increased to 96 half-hour exposures during 21 days, plasma cholinesterase activity slightly decreased, and when exposure concentration was increased to 0.4 to 0.5 mg/m^3, plasma cholinesterase activity significantly decreased. RBC cholinesterase activity was unaffected under any exposure conditions. No cholinergic signs or RBC acetylcholinesterase inhibition occurred in men exposed to average dichlorvos concentrations of 0.49 or 2.1 mg/m^3 for 1 or 2 hours on 4 consecutive days in a simulated aircraft cabin (127).

No cholinergic signs or RBC acetylcholinesterase inhibition occurred in men given 1- to 2.5-mg doses of dichlorvos via two corn oil capsules daily for up to 28 days (152) or in men given 0.9 mg dichlorvos three times a day for 21 days (152a). No adverse clinical signs or RBC acetylcholinesterase inhibition occurred in men given two oral doses of 35 mg dichlorvos (0.5 mg/kg/day), 12 or 15 daily doses of 21 mg dichlorvos (0.3 mg/kg/day), or 21 daily doses of 7 mg/kg dichlorvos (0.1 mg/kg/day) (120a). When the same individuals were given 70 mg dichlorvos for 14 days, however, RBC acetylcholinesterase was significantly inhibited. Cholinergic toxicity did not occur in volunteers given single oral doses of dichlorvos in slow release polyvinyl resin pellets ranging from 0.1 to 32 mg/kg, despite the fact that RBC acetylcholinesterase was dose-dependently inhibited because it was maximal at 24 mg/kg (153). Repeated daily administration of 8 to 38 mg/kg for 7 days caused cholinergic toxicity and dramatically decreased RBC acetylcholinesterase activity, so the experiment was terminated in most subjects in less than 7 days.

Six of 59 males and 9 of 48 females in an occupational study of flower growers showed positive reactions on patch testing of dichlorvos for an overall rate of 14%. Twelve of 18 subjects who had positive skin patch test reactions to triforine (1,4-bis (2,2,2-trichlor-1-formamidoethyl)piperazine) also showed positive reactions to dichlorvos (129).

8.5 Standards, Regulations or Guidelines of Exposure

Dichlorvos is undergoing reregistration by the EPA (9a). The ACGIH TLV for dichlorvos is 0.9 mg/m^3 with a skin notation (154). The OSHA PEL-TWA and NIOSH REL-TWA are

1 mg/m³ with a skin notation. Most other countries have established OELs of 0.1 ppm as well (Republic of Egypt, Australia, Austria, Belgium, Denmark, Finland (1 mg/m³) (Jan. 93), France, Germany 0.11 ppm (1 mg/m³), Hungary STEL 0.2 mg/m³ (Jan. 93), India, The Netherlands, The Philippines 1 mg/m³ (Jan. 93), Poland 1 mg/m³ (Jan. 93), Russia 0.2 mg/m³ (Jan. 93), Switzerland, Thailand, United Kingdom 0.1 ppm (0.92 mg/m³).

9.0 Dicrotophos

9.0.1 CAS Number: [141-66-2]

9.0.2 Synonyms: O,O-dimethyl-O-(3-dimethylamino-1-methyl-3-oxo-1-propenyl) phosphate; 3-dimethoxyphosphinyloxy-N,N-dimethylisocrotonamide; Bidirl; C709; Diapadrin; SD3562; phosphoric acid (E)-3-(dimethylamino)-1-methyl-3-oxo-1-propenyl dimethyl ester; dimethyl 1-methyl-3-(N,N-dimethylamino)-3-oxo-1-propenyl phosphate, (E)-; Penetrex; Chiles' Go-Better; Mauget Inject-A-Cide B; phosphoric acid, 3-(dimethylamino)-1-methyl-3-oxo-1-propenyl dimethyl ester, (E)-; phosphoric acid, dimethyl ester, ester with 3-hydroxy-N,N-dimethylcrotonamide, (E)-;Carbomicron; crotonamide, 3-hydroxy-N,N-dimethyl-, cis-, dimethyl phosphate; dimethyl cis-2-dimethylcarbamoyl-1-methylvinyl phosphate; dimethyl O-(N,N-dimethylcarbamoyl-1-methylvinyl) phosphate; dimethyl phosphate ester with 3-hydroxy-N,N-dimethyl-cis-crotonamide; dimethylcarbamoyl-1-methylvinyl dimethylphosphate; hydroxy-N,N-dimethyl-cis-crotonamide dimethyl phosphate; Karbicron; Oleobidrin

9.0.3 Trade Name: Bidrin®; Carbicron®; Ektafos®

9.0.4 Molecular Weight: 237.21

9.0.5 Molecular Formula: C₈H₁₆NO₅P

9.0.6 Molecular Structure:

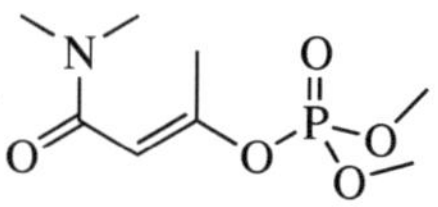

9.1 Chemical and Physical Properties

Pure dicrotophos is an amber liquid; the commercial grade which consists of 85% E-isomer is brown in color. Dicrotophos is stable when stored in glass or polyethylene containers up to 40°C but decomposes after 31 days at 75°C or after 7 days at 90°C. Dicrotophos emits toxic fumes of phosphorus and nitrogen oxides when heated to decomposition.

    Specific gravity: 1.216 at 15°C; 8.6×10^{-5} mmHg at 20°C
    Boiling point: 440°C at 760 mmHg
    Vapor pressure: 1×10^{-4} mmHg at 20°C
    Solubility: slightly soluble in xylene, kerosene, and diesel fuel; miscible with water, acetone, alcohol, 2-propanol, and other organic solvents

9.1.2 Odor and Warning Properties

A mild ester odor.

9.2 Production and Use

Dicrotophos was introduced in 1956 as a systemic and contact organophosphorus insecticide effective against sucking, boring, and chewing pests and is recommended for use on coffee, cotton, rice, pecans, and other crops. It is also used to control ticks and lice on cattle (155). Dicrotophos is available as 24% and 85% concentrates, as 40% and 50% emulsifiable concentrates, water-soluble concentrates, and ultra low volume formulations (155).

9.4 Toxic Effects

9.4.1.1 Acute Toxicity. Dicrotophos is an organophosphate that has high oral toxicity and oral LD_{50}s of 16–21 mg/kg (64a). The dermal LD_{50} of dicrotophos is 42–43 mg/kg in rats (64a) and 225 mg/kg in rabbits (155). A 4-hour LC_{50} of 90 mg/m^3 and a 1-hour LC_{50} of 610–910 mg/m^3 was reported for rats (545).

9.4.1.2 Chronic and Subchronic Toxicity. Cholinergic toxicity did not occur in rats fed diets that contained 0, 15, or 150 ppm dicrotophos for 4 weeks, although whole blood and plasma cholinesterase activites were markedly inhibited at both dose levels (156).

9.4.1.3 Pharmacokinetics, Metabolism, and Mechanisms. The oral, intraperitoneal, and intravenous LD_{50}s for dicrotophos are equivalent (ranging from 9–17 mg/kg), indicating that it is essentially completely absorbed orally. Experimental studies show it is well absorbed via other routes of exposure as well.

After 6 hours, 65% of a subcutaneously injected dose (10 mg/kg) was excreted, and after 24 hours, 83% was excreted in the urine alone (159). Similar results were obtained in other studies of rats and other species (160).

Dicrotophos is metabolized in part to monocrotophos, and the concentration of monocrotophos in tissues may be higher than that of the parent compound a few hours after administration, as reflected by analysis of rat urine and goat milk (1). Residues of both compounds are dissipated almost entirely within 24 hours, as indicated by a rapid decrease in unhydrolyzed metabolites in urine or milk.

Hydrolysis of the vinyl phosphate bond of dicrotophos and/or its oxidative metabolites (monocrotophos) to produce dimethyl phosphate is the predominant metabolic reaction (159). The proportion of dimethyl phosphate in the urine of rats increases rapidly after dosing and reaches 50% of all metabolites present in less than 4 hours and more than 80% in 20 hours. During this interval, there is a correspondingly rapid decrease in the excretion of the parent compound and its oxidation products. Desmethyl dicrotophos and inorganic phosphate are also found in minor concentrations in the urine of treated rats (159). Dimethyl phosphate has been confirmed in the urine of an individual who accidentally ingested dicrotophos (161).

9.4.1.4 Reproductive and Developmental. When rats were fed diets that contained 2, 5, 15, or 50 ppm dicrotophos (0.1, 0.25, 0.75, and 2.5 mg/kg/day) for two generations, decreased pup survival was observed at 5 ppm (95). Other effects (weakness, emaciation, and CNS effects) were seen at 50 ppm. No effects occurred at 2 ppm.

No morphological anomalies occurred in offspring of pregnant mice given intraperitoneal injections of 1, 2, 4, or 7.5 mg/kg dicrotophos on gestation day 11, 13, or days 10–12 (157). In other mice, a dose of 5 mg/kg/day on gestation days 8 though 16 did not change the developmental patterns of brain acetylcholinesterase or choline acetyltransferase in offspring through day 42, even though this dose on day 11 reduced embryonic or fetal acetylcholinesterase to 1.8% of control levels (157). The fetal brain enzyme level returned to normal by day 19 following dosing of the mother on days 8 through 16 of gestation.

9.4.1.5 Carcinogenesis. When rats were fed dicrotophos in their diets at concentrations of 0, 1, 10, or 100 ppm for two years, there were no detectable effects at the 1-ppm concentration (95). Plasma cholinesterase was inhibited at 1 ppm (95). At 10 and 100 ppm, decreased body weights and reduced cholinesterase (RBC, plasma, brain not specified) activities occurred. Dogs given dicrotophos in their diets at 0, 0.16, 1.6, or 16 ppm for two years showed some instances of slightly excessive salivation (95). At 16 ppm, both plasma and RBC cholinesterase activity was decreased.

9.4.1.6 Genetic and Related Cellular Effects Studies. Dicrotophos is considered mutagenic on the basis of its similarity in structure to monocrotophos and the observation that it induced increases in sister chromatid exchanges in cultures of Chinese hamster ovary cells (158).

9.4.2 Human Experience

An individual who inhaled a spray that contained dicrotophos and was being used to control mosquitoes in the home developed organophosphate poisoning (162). Upon hospital admission, he had abdominal cramps, nausea, vomiting, and diarrhea; the next day he exhibited increased sweating, salivation, dyspnea, coarse tremor of both legs, and generalized weakness. Plasma and RBC cholinesterase activities were nonexistent. The patient responded to atropine and pralidoxime. However, on the sixth day, respiratory paralysis occurred (typical of "intermediate syndrome"), and he required an artificial respirator for 5 days. He was discharged on day 22. In another case, a 52-year-old man accidentally drank a solution that contained dicrotophos in turpentine (163). He was brought to the hospital where he was treated effectively with atropine and pralidoxime chloride, but he required assisted respiration for more than a week.

9.5 Standards, Regulations, or Guidelines of Exposure

Dicrotophos is undergoing reregistration by the EPA (9a). Dicrotophosis is a Restricted Use Pesticide (RUP) which can be purchased and used only by certified applicators. Some specific state restrictions may apply. The ACGIH TLV for dicrotophos (intended change in the 2000 TLVS is 0.05 mg/m³) and monocrotophos is 0.25 mg/m³ with a skin notation (154). The NIOSH TWA-REL is also 0.25 mg/m³ with a skin notation.

10.0 Dioxathion

10.0.1 CAS Number: [78-34-2]

10.0.2 Synonyms: 2,3-*p*-Dioxanedithion *S,S*-bis-(*O,O*-diethyl phosphorodithioate); Hercules AC528; Ruphos; Navadel; Delnatex; Delnav; 1,4-dioxan-2,3-diyl-bis(*O,O*-diethyl phosphorothiolothionate); Delanov; Delnav(R); Phosphorodithioic acid *S,S'*-1,4-dioxane-2,3-diyl *O,O,O',O'*-tetraethyl ester; phosphorodithioic acid *S,S'*-*p*-dioxane-2,3-diyl *O,O,O',O'*-tetraethyl ester; AC 528; dioxation; 1,4-dioxan-2,3-diyl *O,O,O',O'*-tetraethyl di(phosphoromithioate); 2,3-*p*-dioxane *S,S*-bis(*O,O*-diethylphosphorodithioate); *p*-dioxane-2,3-dithiol, *S,S*-diester with *O,O*-diethyl phoshorodithioate; *p*-dioxane-2,3-diyl ethyl phosphorodithioate; dioxothion; hercules 528; kavadel; deltic; 2,3-*p*-dioxanedithiol *S,S*-bis(*O,O*-diethyl phosphorodithioate); *S,S'*-1,4-dioxane-2,3-diyl *O,O,O',O'*-tetraethyl phosphorodithioate; Cooper Del-Tox Delnav; Dextrone X; 1,4-dioxane-2,3-dithiol, *S,S*-diester with *O,O*-diethyl phosphorodithioate; 1,4-dioxane-2,3-diyl *O,O,O',O'*-tetraethyl phosphorodithioate; 1,4-dioxanedithiol *S,S*-bis(*O,O*-diethyl phosphorodithioate)

10.0.3 Trade Names: Delnav®; Hercules AC528®; Navadel®

10.0.4 Molecular Weight: 456.54

10.0.5 Molecular Formula: $C_{12}H_{26}O_6P_2S_4$

10.0.6 Molecular Structure:

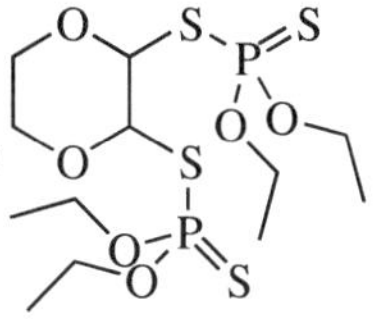

10.1 Chemical and Physical Properties

Dioxathion is a nonvolatile, chemically stable, dark amber liquid.

Specific gravity: 1.257 at 26°C
Melting point: −20°C
Boiling point: 60–68°C
Solubility: insoluble in water; soluble in aromatic hydrocarbons, alcohols, ethers, esters and ketones

10.2 Production and Use

Dioxathion is the common name for an organophosphate product that contains 70% cis and trans (1:2 ratio) isomers of 2,3-*p*-dioxanedithiol *S,S*-bis-(*O,O*-diethyl phosphorodithioate) as the principal ingredient (164). It was formely used in the United States on citrus, grapes, walnuts, and stone fruits. It was also used to control ticks, lice, and horn flies on cattle, goats, hogs, horses, and sheep when sprayed or dipped. Until 1989, when its manufacture and use in the United States was discontinued, dioxathion was available as a 25% wettable powder and a 48% emulsifiable concentrate (1, 155).

10.4 Toxic Effects

10.4.1.1 Acute Toxicity. Dioxathion is an organophosphate compound that has moderately high oral toxicity and oral LD_{50}s of 23–64 (64a, 164). An intraperitoneal LD_{50} of 30 mg/kg was obtained for rats suggesting (by comparison with the oral LD_{50} of 23 mg/kg) that dioxathion is well absorbed orally. The oral LD_{50} for dogs is 10–40 mg/kg. The dermal LD_{50} for dioxathion is 63–235 mg/kg in rats and 85 mg/kg in rabbits (64, 545). One-hour LC_{50} values of 1398 and 340 mg/m^3 were reported for rats and mice (164). The acute symptoms of dioxathion are typical of other organophosphates, but the rate of onset is "somewhat slower" (164).

When dogs were given 0.25, 0.80, 2.5, or 8.0 mg/kg/day dioxathion by capsule for 5 days/week for two weeks, those given 8.0 mg/kg/day developed signs of cholinergic toxicity (diarrhea, hypersalivation, termors, ataxia, and depression) (*sic*). Doses of 0.8 mg/kg/day and higher significantly inhibited plasma cholinesterase, and doses of 2.5 and 8.0 mg/kg/day significantly inhibited RBC cholinesterase (164).

Dioxathion showed additive or less than additive toxicity when administered in equitoxic ratios with 15 other anticholinesterase insecticides. However, when dioxathion was administered 4 hours before malathion, potentiation as great as 5.4-fold was observed (164).

When rats were given single intraperitoneal injections of 4, 8, or 16 mg/kg, liver and plasma carboxylesterase activities were 19–55%, RBC cholinesterase activity was 76%, and, brain cholinesterase activity was 96% of control in rats given 4 mg/kg/day dioxathion (165). Thus dioxathion more effectively inhibits carboxylesterases than acetylcholinesterase. A similar tendency of dioxathion to inhibit carboxylesterases to a greater extent than acetylcholinesterase was observed when enzymatic activity was examined in rats fed diets that contained 4, 10, 20 or 40 ppm dioxathion for 7 days (165). Brain acetylcholinesterase was unaffected at any level, and RBC cholinesterase was significantly inhibited at 20 and 40 ppm; however, liver carboxylesterases were significantly inhibited at all levels. Further, rats given diets that contained 4 or 10 ppm dioxathion were more susceptible than untreated rats to inhibition of brain cholinesterase by a single 100- or 200-mg/kg dose of malathion (165).

When 75 mg/kg dioxathion was given subcutaneously to rats, they displayed muscular fibrillation at 2 hours and convulsions at 4 to 8 hours (166). Symptoms of organophosphate poisoning continued several days before recovery. Oral administration of dioxathion to rats at 5 mg/kg/day for up to 21 days resulted in plasma, RBC, and brain cholinesterase inhibition within 1 day (166).

Dioxathion produced mild, transient conjunctivitis but no transient or permanent corneal damage when 0.1 mL was instilled into the eyes of rabbits (164).

Dioxathion did not produce neurotoxicity in surviving hens that received single oral doses of 10–1000 mg/kg or subcutaneous doses of 25–200 mg/kg, even though the higher rates killed some of the birds (164). However, a slightly larger subcutaneous dose,

320 mg/kg, in hens protected by atropine produced a temporary neurotoxic effect that lasted 3–31 days (164).

10.4.1.2 Chronic and Subchronic Toxicity. When rats were fed diets that contained 100 or 500 ppm dioxathion for 1–13 weeks, a dietary level of 500 ppm caused marked food refusal and loss of body weight within the first week (164). Female rats given 100 ppm (about 7.5 mg/kg/day) showed hyperexcitability and slight tremor, but males remained well. Both sexes showed marked inhibition of brain, plasma, and RBC cholinesterase activity. In another study, rats were fed diets that contained 1, 3 or 10 ppm dioxathion for 13 weeks. A dietary level of 10 ppm (0.78 mg/kg/day) produced no inhibition of brain cholinesterase but significantly reduced plasma and RBC cholinesterase activity. Dietary levels of 3 and 1 ppm (0.22 and 0.077 mg/kg/day) did not alter brain, plasma, or RBC cholinesterase activity (164).

No adverse effects occurred in dogs given 0.013, 0.025, or 0.075 mg/kg/day dioxathion for 5 days/week for 90 days via capsule (164).

10.4.1.3 Pharmacokinetics, Metabolism, and Mechanisms. Dioxathion is well absorbed orally or dermally. Rats treated orally with dioxathion for 10 consecutive days excreted dioxathion primarily in the urine (about 50% of daily administered dose) and to a lesser extent in the feces (about 20% of daily administered dose) (166). Rats given 25 mg/kg of different components of the technical product excreted about 36–45% in the urine and about 6 to 25% in the feces, depending on the component administered, by 48 hours. Hydrolytic products identified in the urine included diethyl phosphoric, phosphorothioic, and phosphorodithioic acids. When rats were given 25 mg/kg dioxathion and sacrificed after 48 hours, fat had "appreciable" (*sic*) amounts of the dose. When rats were given 5 or 10 µg/kg dioxathion for several days, the maximum level in fat (0.6 ppm) was reached within 3 days and held constant through 21 days.

The metabolism of dioxathion was examined in rats treated orally and in rat liver microsomes (168). Both the trans, and cis, isomers were rapidly and extensively metabolized by rat liver microsomes in the presence of NADPH (indicating the involvement of microsomal oxidases) and by rats *in vivo* to the corresponding oxons and dioxon. The compound also underwent oxidative O-deethylation and hydroxylation of the ring resulting in ring cleavage and the loss of both phosphorus moieties. The more toxic cis isomer was metabolized more rapidly to form oxon and dioxon and also to form CO_2 from the ethoxy group (168). Of the dose administered, 80–87% was excreted by 96 hours in urine and most of this occurred in the first 24 hours. Unmetabolized dioxathion appeared in feces.

10.4.1.4 Reproductive and Developmental. In a three-generation study, rats were fed diets containing 0, 3, or 10 ppm dioxathion (167). There were no measurable abnormalities among either parental animals or their progeny.

10.4.1.5 Carcinogenesis. No evidence of carcinogenicity was observed in rats or mice given diets that contained dioxathion for 78 weeks and then observed for an additional 33 or 12–13 weeks, respectively (585). Time-weighted average dietary concentrations were

180 and 90 ppm for male rats, 90 and 45 ppm for female rats, 567 and 284 ppm for male mice, and 935 and 467 ppm for female mice.

10.4.1.6 Genetic and Related Cellular Effects Studies. Dioxathion was positive in the *Salmonella* assay and in cultured Chinese hamster ovary (CHO) cells for the induction of sister-chromatid exchanges but was negative in the mouse lymphoma assay and in cultured CHO cells for the induction of chromosomal aberrations (65).

10.4.2   Human Experience

10.4.2.2 Clinical Cases. A 5-year-old boy, who ingested about three-quarters of a teaspoon of a 21% dioxathion formulation intended to be diluted for use as a flea dip (about 57 mg/kg) when it was mistaken for cough medicine, vomited and exhibited profuse diarrhea (169, 170). Within 2 hours, the child was mentally dull and unable to stand; he had shallow rapid respirations, muscle fasciculations, tearing, and miosis. After 12 hours of appropriate treatment, he recovered.

Volunteers were given 0.075 mg/kg/day dioxathion in divided doses three times/day, 7 days/week via capsule for 4 weeks. After 4 weeks, two of the subjects continued on this dose; other subjects received 0.150 mg/kg/day, and, the other six continued to receive 0.075 mg/kg/day dioxathion, but also received 0.150 mg/kg malathion. A dose of 0.075 mg/kg/day produced no effect on plasma or RBC cholinesterase activity. There was a slight inhibition of plasma cholinesterase activity in subjects that received 0.015 mg/kg/day. There was no effect on RBC cholinesterase activity and no clinical effect. Plasma cholinesterase measurements showed slight but statistically uncertain decreases when dioxathion was administered daily for 60 days at a rate of 0.075 mg/kg/day and malathion was given at a rate of 0.15 mg/kg/day simultaneously for the last 30 of these days (164).

10.5   Standards, Regulations, or Guidelines of Exposure

All registrations and tolerances for dioxathion have been revoked by the EPA (78). The ACGIH TLV for dioxathion is 0.2 mg/m^3 (154). NIOSH has recommended a REL-TWA of 0.2 mg/m^3 with a skin notation. Most other countries have occupational Exposure Limits of 0.2 mg/m^3 with a skin notation for dioxathion (Australia, Belgium, Denmark, France, The Netherlands, Switzerland, United Kingdom).

11.0   Disulfoton

11.0.1   CAS Number: *[298-04-4]*

11.0.2   Synonyms: *O,O*-Diethyl-*S*-ethylmercaptoethyl dithiophosphate; phosphorodithioc acid *O,O*-diethyl-*S*-(ethylthio)ethyl) ester; Thiodementon; Solvirex; Thiodemeton; Disyton(R); phosphorodithioic acid *O,O*-diethyl *S*-[2-(ethylthio)ethyl] ester; *O,O*-diethyl-*S*-ethylmercaptoethyl dithiophosphate; dithiodemeton; BAY 19639; Dithiosystox; Di-Syston; Frumin AL; Frumin G; Frumen AL; disulfoton+; *O,O*-diethyl *S*-(2-(ethylthio)ethyl) phosphorodithioate; thiometon-ethyl; Root-X; Dot-Son Brand Stand-Aid; Rigo Insyst-D; Terraclor Super-X; Diethyl S-(2-(ethylthio)ethyl) phosphorodithioate;

diethyl S-(2-ethylmercaptoethyl) dithiophosphate; Dimaz; Disipton; Disystox; Ekatin TD; ethylthiometon; Glebofos

11.0.3 Trade Names: Di-Syston®; Dithiosystox®

11.0.4 Molecular Weight: 274.38

11.0.5 Molecular Formula: $C_8H_{19}O_2PS_3$

11.0.6 Molecular Structure:

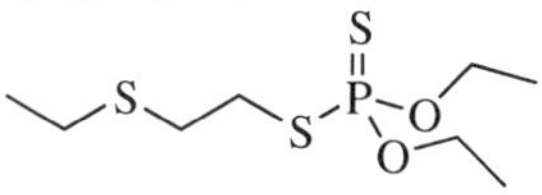

11.1 Chemical and Physical Properties

Pure disulfoton is a colorless oil that has low volatility and water solubility. The technical product is yellow.

Specific gravity: 1.144 at 20°C
Boiling point: 108°C at 0.01 mmHg
Vapor pressure: 0.00018 torr at 20°C
Solubility: insoluble in most organic solvents; slightly soluble in water (25 mg/L)

11.2 Production and Use

Disulfoton is an organophosphate insecticide effective against aphids, leafhoppers, thrips, spider mites, and coffee leaf miners. It is used on cotton, tobacco, sugar beets, corn, peanuts, wheat, ornamentals, potatoes, and cereal grains. Disulfoton is used to treat seeds and is applied to soils or plants as an emulsifiable concentrate and in granular or pelletized forms.

11.4 Toxic Effects

11.4.1.1 Acute Toxicity. Disulfoton is a highly toxic organophosphate that has oral LD_{50}s of 2.3–12.7 mg/kg for rats, mice, and guinea pigs (54, 64a). Dermal LD_{50}s are 3.6–20 mg/kg for rats, demonstrating a relatively high dermal toxicity (64a). Two of two rabbits died after dermal application of 10 mg/kg/day disulfoton that lasted 6 hours, whereas no rabbit similarly treated with 0.4 or 2.0 mg/kg/day for five days died. Treatment of rabbits five days/week for three weeks with 0.4, 0.8, 1.0, 1.6, 3.0, or 6.5 mg/kg resulted in marked cholinergic toxicity following 6.5 mg/kg, cholinergic signs and significant RBC and brain cholinesterase inhibition following 3.0 mg/kg, slight but significant RBC cholinesterase inhibition following 1.0 or 1.6 mg/kg, and, no cholinesterase inhibition following 0.4 or 0.8 mg/kg (171).

One-hour LC_{50}s values for disulfoton aerosol for rats are 290 mg/m^3 and 63 mg/m3 for males and females, respectively; four-hour LC_{50} values are 60 mg/m^3 and 15 mg/m^3 for males and females, respectively (178). Repeated inhalation exposures are more lethal than single exposures. When female rats were exposed to disulfoton 4 h/day for 5 days, the LC_{50} was between 1.8 and 9.8 mg/m^3 (sic) (178).

Symptoms caused by acutely toxic levels of disulfoton are similar to those caused by other organophosphates and develop beginning about 30 minutes after exposure, depending on dose. Time of death depends on dose; it generally occurs within 48 hours at lethal doses but is sometimes delayed by several days for doses near the LD_{50} (172).

When rats were given single gavage doses of 1.5 and 5.2 mg/kg (males) and 0.76 and 1.5 mg/kg (females) disulfoton, cholinergic toxicity developed within 0–3 days and resolved by day 4 after treatment (178). RBC cholinesterase was inhibited in mid- and high-dose females and in high-dose males. Tolerance to the cholinergic toxicity of disulfoton occurs upon repeated, subtoxic exposures. Male rats given 2.0 or 2.5 mg/kg/day disulfoton for 1–14 days exhibited cholinergic signs whose severity diminished with repeated dosing (173, 174). When rats were given 3.5 mg/kg/day for 3–4 days, clinical cholinergic signs were more severe than those exhibited by rats pretreated with 2.5 mg/kg/day for 6 days and then given 3.5 mg/kg/day for 6 more days (173, 174). Thus, rats pretreated with 2.5 mg/kg/day became tolerant to even higher doses of disulfoton. After 3 days on a diet that provided 1 mg/kg/day disulfoton, rats developed severe cholinergic signs that diminished markedly during a 62-day period despite the fact that brain and diaphragm cholinesterase activity was depressed at day 6 and remained depressed throughout the study (173).

Inhalation exposure of rats to 0.02 mg/m^3 disulfoton for 6 h/day, 5 days/week for three weeks did not cause any signs of cholinergic toxicity (171). Exposure to 0.1 to 0.5 mg/m^3 resulted in behavioral changes linked to inflammatory changes in the respiratory system, and exposure to 3.1 or 3.7 mg/m^3 caused cholinergic symptoms (muscle tremors, convulsions, increased salivation, dyspnea). Five of ten females exposed to 3.7 mg/m^3 died after three to twelve exposures, and three of twenty females exposed to 3.1 mg/m^3 died after eight to fifteen exposures. No deaths occurred in males at any exposure level. RBC cholinesterase was significantly inhibited at 0.1 mg/m^3 or more, and brain cholinesterase was significantly inhibited at 0.5 mg/m^3 in females and 3.7 mg/m^3 in males. In another 21-day study, rats exposed to 0.006, 0.07, or 0.7 mg/m^3 disulfoton showed no compound-related mortality or clinical signs of toxicity in any group. However, RBC cholinesterase was significantly inhibited at 0.7 mg/m^3. Brain cholinesterase activity was unaffected.

Neuronal degeneration was evident in some hens administered disulfoton (30 mg/kg) twice 22 days apart (95)

11.4.1.2 Chronic and Subchronic Toxicity. The mortality rate was 20% in rats given daily intraperitoneal injections of 1.0 mg/kg/day disulfoton and 100% in rats given 1.2 or 1.5 mg/kg/day for 60 days (172). After the first two doses of 1.0 mg/kg, rats displayed cholinergic signs immediately after each dose for up to 7 or 10 days. Then the rats began to recover in spite of daily treatment. Intraperitoneal doses of 0.25, 0.5, and 1.0 mg/kg produced rapid dose-related inhibition of brain and serum cholinesterase that persisted throughout the entire study period, even though cholinergic signs disappeared (175).

Disulfoton caused no cholinergic clinical signs, or adverse effects on mortality, ophthalmology, feed consumption, or body weight gain in rats exposed to 0.018, 0.16, or 1.4 mg/m^3 for 6 h/day, 5 days/week for 13 weeks (178). However, RBC, brain and plasma cholinesterase activities were significantly inhibited in rats exposed to 1.4 mg/m^3.

No adverse effects occurred in rats fed diets that contained 1, 2, 4, 5, 6, 10, or 16 ppm disulfoton (0.1, 0.2, 0.5, or 1.0 mg/kg/day) for 13 or 16 weeks, other than decreased body weight gain at 16 ppm and urine stains in females at 4 ppm (63, 171). However, RBC, brain, plasma, and tissue (submaxillary gland) cholinesterase activity was significantly inhibited in females given 2 ppm or more and in males given 5 ppm or more. In mice fed diets that provided 0.63–0.71 mg/kg/day disulfoton, cholinesterase was inhibited in all tissues, although the tissues were not specified (176). No-effect levels were 0.13–0.14 mg/kg/day.

Dogs fed diets that contained 1, 2, or 10 ppm disulfoton for 12 weeks exhibited no signs of toxicity, although plasma and RBC cholinesterase activity were significantly inhibited in dogs given 2 or 10 ppm (171).

11.4.1.3 Pharmacokinetics, Metabolism, and Mechanisms. Disulfoton is rapidly absorbed after oral exposure. An average of 80–84%, 6–8%, and 9% of a single oral dose of disulfoton was eliminated by rats in urine, feces, and expired air, respectively, in the 10 days following exposure that accounted for 96–100% of the administered dose (171). The rate of excretion was significantly lower in females—males eliminated one-half the dose in 4–6 hours, whereas females required 30–32 hours. Another experiment showed that 72 hours after an oral dose of disulfoton to rats, about 97% was eliminated in urine, 2% was eliminated in feces, and less than 1% remained in the body (171). Similar results were obtained at 12 hours in rats given 15 consecutive doses of disulfoton. There was no accumulation of disulfoton in the body. Tissue and blood levels of disulfoton peaked at 6 hours and were highest in liver followed by kidney, plasma, fat, whole blood, skin, muscle, and brain. However, on a percentage of dose basis, female livers contained a much greater amount of disulfoton than males (34% vs. 10% at 6 hours; 3% vs. 9% at 12 hours), possibly indicating slower metabolism and accounting for longer elimination in females.

Dermal absorption of 0.85, 8.5, or 85 µg/cm^2 of a disulfoton formulation ranged from 39–44% of an applied dose applied to the skin (15 cm^2) of rats (178). There was no concentration-dependent effect; the majority of absorption occurred within the first hour, and, the majority of the absorbed dose (66–91%) excreted was found in the urine.

Disulfoton is rapidly metabolized via oxidation to sulfoxides and sulfones, oxidation to oxygen analogs, and/or hydrolysis to produce a corresponding phosphorothionate or phosphate (54). Metabolism is accompanied by inhibition of microsomal enzymes (183).

In humans exposed to disulfoton, sulfones (disulfoton sulfone and demeton S-sulfone) were detected in blood (104), and diethyl phosphate (DEP), diethyl thiophosphate (DETP), diethyl dithiophosphate (DEDPT), and diethyl phosphorothiolate were detected in urine (184). Similar metabolic products were detected in the urine of rats and mice administered disulfoton intraperitoneally or orally and in liver homogenates of treated rats (54).

The oxidation reactions are toxification reactions that create metabolic products that bind to cholinesterase. In an oral study of rats, the metabolites, disulfoton sulfoxide, disulfoton sulfone, demeton-S-sulfoxide, and demeton S-sulfone caused mortality and signs of toxicity at lower doses than disulfoton itself (171a). The hydrolytic reactions create more polar products that are eliminated in the urine and therefore are detoxification reactions.

11.4.1.4 Reproductive and Developmental. Two-generation reproductive studies indicates that diets that contained 3 or 9 ppm disulfoton have adverse reproductive effects; diets that contained 0.5–2 ppm do not (242). Adverse outcomes among rats fed 9-ppm diets include decreased body weight gain during pregnancy and lactation; decreased number of implantations; decreased litter size, weights and viability; and decreased brain cholinesterase activity in F1a pups. Adverse outcomes among rats fed 3 ppm disulfoton include decreased litter size, weights, and viability in the F2 generation. Cholinergic signs were evident in rats fed 9 ppm; decreased brain cholinesterase activity occurred in rats fed a 0.5-ppm diet or more. Reproductive effects were also reported in a study in which male and female rats were given 0.5 mg/kg/day disulfoton for 60 days before and/or during mating. Two-fifths of the treated females failed to become pregnant (182).

There was no indication of a teratogenic effect in any rats given 0, 0.1, 0.3, or 1.0 mg/kg/day disulfoton via gavage on gestation days 6–15 (171). Significant depressions of RBC cholinesterase activity occurred in dams given 0.3 and 1.0 mg/kg/day. A significant increase in the incidence of incomplete ossification of the sternebrae was observed in fetuses at 1.0 mg/kg/day which was considered an effect of growth retardation due to maternal toxicity. There was no evidence of teratogenicity or embryo toxicity in rabbits given 0, 0.3, 1.0, or 3.0 mg/kg/day disulfoton (95). The highest dose was lowered to 2.0 and later to 1.5 in some but not all of the high-dose animals due to severe toxic responses and mortality.

11.4.1.5 Carcinogenesis. There was no evidence of carcinogenicity or any other adverse effect among rats fed diets that contained 1 or 2 ppm disulfoton for 104 weeks or 0.5 ppm for 80 weeks followed by 5 ppm for 24 weeks (242). Females fed 5 ppm, however, exhibited significantly inhibited plasma and brain cholinesterase. Rats fed diets that contained 10, 25, or 50 ppm disulfoton for 178 days had significantly inhibited brain cholinesterase activities but exhibited no signs of cholinergic toxicity (177). There was no evidence of carcinogenicity in rats given feed that contained 0, 1, 4, or 16 ppm disulfoton, (0.05, 0.2, and 0.1 mg/kg/day) (178). Females given the 16-ppm diet had a 40% mortality rate during the last week of the study compared with a 12% mortality in controls, and both sexes given the 16-ppm diet exhibited cholinergic signs and increased relative brain weight. There was an increased incidence of optic nerve degeneration in males given the 4-ppm diet and in females given the 4- or 16-ppm diet. Increased incidences of mucosal hyperplasia and chronic inflammation of the forestomach occurred in females given 16 ppm. Cystic degeneration of the Harderian gland occurred in male rats given 16 ppm and in female rats given 4 ppm. Corneal neovascularization was significantly increased in rats given 16 ppm. RBC and brain cholinesterase activity was inhibited in rats given 1 ppm.

There was no evidence of cancer in mice fed diets that contained 0, 1, 4, or 16 ppm disulfoton for 99 weeks (171). Nor was there any adverse effect on behavior, feed consumption, hematology, or organ weights. Significant depression of RBC, plasma, and brain cholinesterase activity occurred in mice fed diets that contained 16 ppm (equivalent to 2–2.5 mg/kg/day).

Dogs did not exhibit cholinergic signs, opthalmoscopic changes, hematological or clinical chemical changes, or any evidence of carcinogenicity when given diets that contained 0.5 or 1.0 ppm disulfoton for two years (equivalent to 0.03 or 0.14 mg/kg/day) (242). Nor was RBC cholinesterase activity inhibited. RBC cholinesterase activity was inhibited in dogs after five months exposure to ≥0.5 mg/kg/day given by capsule and when they were fed diets that contained disulfoton at a dose of 0.06 mg/kg/day for 40 weeks (242). Moderate inhibition of RBC cholinesterase occurred in dogs given diets that contained 5.0 ppm for 69 weeks (equivalent to about 0.7 mg/kg/day). Brain cholinesterase was not inhibited in dogs given a 0.5- or 1.0-ppm diet for two years but was markedly inhibited in dogs given diets that contained 2.0 ppm or more.

Ocular effects (myopia and astigmatism) associated with degenerative changes in the ciliary muscle cells occurred after 12 months in dogs given ≥0.63 mg/kg/day disulfoton for two years (128, 179). The myopia became progressively worse until dosing ceased. Necrosis and atrophy of the optic nerve and retina was observed in dogs given disulfoton (0.5–1.5 mg/kg/day) for 2 years (180). However, no ophthalmological effects occurred in dogs given diets that contained 0.5, 4, or 12 ppm disulfoton (0.015, 0.1, or 0.3 mg/kg/day) for 1 year (181). Dogs given the 4-ppm diets "demonstrated intermediate toxicity" (*sic*), and those given the 12-ppm diet "demonstrated systemic toxicity near the Maximum Tolerated Dose" (*sic*). RBC cholinesterase was significantly inhibited in females given 4 and 12 ppm, corneal cholinesterase was significantly depressed at 4 and 12 ppm in both sexes, retinal cholinesterase was significantly inhibited at 4 ppm in females in 12 ppm in males, and ciliary body cholinesterase was significantly inhibited at 12 ppm in both sexes.

11.4.1.6 Genetic and Related Cellular Effects Studies. The genotoxicity of disulfoton in *in vitro* assays has been reviewed and was mainly negative (54). Positive results for reverse mutation occurred in single assays with LT-2 or TA1535 stains of *S. typhimurium* without activation but not in several other assays with or without activation (54). Similarly, both positive and negative results for reverse mutation have been reported in *E. coli* and *S. cerevisiae* (54). Disulfoton was negative of gene conversion, mitotic crossing over and recombinants, and for DNA damage in *S. cerevisiae* with or without activation but was positive in an assay for chiasmatic frequency (genetic recombinants), mitotic index, chromosomal aberrations, and pollen fertility in barley (54).

Disulfoton was positive or weakly positive for sister chromatid exchange in Chinese hamster ovary cells in some studies, but negative in others; negative for HGPRT mutations in Chinese hamster ovary cells with or without activation; positive for forward mutations in mouse lymphoma cells for unscheduled DNA synthesis in human lung fibroblasts, and for growth inhibition and increased protein synthesis in human HeLa cells; and negative for chromosomal aberrations in human hematopoietic cell lines and for alterations of DNA or RBA synthesis in human HeLa cells (54).

11.3.5 *Biomonitoring/Biomarkers*

The presence of disulfoton and/or its metabolites in urine is a reliable biomarker for disulfoton exposure. At 2–10 days post exposure, 30–84% of an oral dose can be accounted for in the urine of animals. Although precise relationships between disulfoton exposure and urinary DEP have not been established, DEP is considered a relatively sensitive biomarker for exposure to disulfoton and other diethyl organophosphate esters (54).

11.4.2 Human Experience

11.4.2.2 Clinical Cases.
A 30-year-old man was found dead after consuming an unknown amount of disulfoton, as evidenced by the presence of disulfoton in urine and blood (186). A 75-year-old woman who ingested an unknown quantity of disulfoton as Di-Syston (5%, granular) experienced severe organophosphate poisoning from 3.5 hours to 11 days, characterized first by vomiting and diarrhea, followed by nausea, fasciculations, and then confusion, miosis, and cardiac arrhythmias. She recovered after 28 days (186a). A farmer who had worn disulfoton-contaminated gloves for several days developed signs of disulfoton toxicity (weakness, fatigue, and cyanosis) and had to be hospitalized (187).

The inhalation exposure potential of wet and dry mix procedures used to prepare disulfoton fertilizer mixtures were compared by measuring disulfoton on special filter pads used in place of the usual outer absorbent filter pads that cover the filter cartridges of respirators worn by workers (31). Dermal exposure was measured by attaching layered gauze absorbent pads to various parts of the body or clothing and allowing workers to be exposed for a timed period of work. Air exposures during dry mix operations averaged 0.633 mg/m^3 1–5 meters from the work station and, during wet mix operations, averaged 0.06 mg/m^3 1–5 meters from the work station. Dermal exposures averaged 2.0 mg/h and 0.09 mg/h during dry and wet mix operations, respectively. RBC cholinesterase values for dry mix workers were reportedly depressed by about 23% after 9 weeks of work, but it was not clear whether these measurements were from workers wearing respirator or not—which would have decreased the anticholinesterase effect.

11.5 Standards, Regulations, or Guidelines of Exposure

Disulfoton is undergoing reregistration by the EPA (178). The ACGIH TLV for disulfoton is 0.1 mg/m^3 with a skin notation (154). NIOSH has recommended a REL-TWA of 0.1 mg/m^3 with a skin notation. Most other countries also have Occupational Exposure Limits of 0.1 mg/m^3 (Australia, Belgium, Denmark, France, The Netherlands, Switzerland, and the United Kingdom).

12.0 EPN

12.0.1 CAS Number: [2104-64-5]

12.0.2 Synonyms:
O-ethyl O-p-nitrophenyl phenylphosphonothioate; PIN; EPN; phenylphosphonothioic acid, O-ethyl O-o-nitrophenyl ester; O-ethyl O-p-nitrophenylbenzenethionophosphonate; O-ethyl O-p-nitrophenyl benzenephosphonothioate; phenylphosphonothioic acid O-ethyl O-(4-nitropheny) ester; ethyl p-nitrophenyl benzenethiophosphonate; O-ethyl O-(4-nitrophenyl) phenylphosphonothioate; ethoxy-((4-nitrophenoxy) (phenyl) phosphine) sulfide; ethyl (p-nitrophenyl) phenylphosphonothioate; ethyl (p-nitrophenyl) benzenethionophosphonate; ethyl O-(4-nitrophenyl) benzenethionophosphonate; ethyl O-(p-nitrophenyl) benzenethionphosphonate; Ethyl O-(p-nitrophenyl) phenylphosphonothioate; ethyl p-nitrophenyl thiobenzene phosphonate; ethyl phenyl (p-nitrophenyl) thiophosphonate; ethyl phenylphosphonothioic acid O-(4-

nitrophenyl) ester; phenol, *p*-nitro-, *O*-ester with *O*-ethyl phenyl phosphonothioate; Santox; *O*-ethyl O-*p*-nitrophenyl phenylthiophosphonate; ethyl *p*-nitrophenyl Benzenethiophosphate

12.0.3 Trade Names: Santox

12.0.4 Molecular Weight: 323.31

12.0.5 Molecular Formula: $C_{14}H_{14}NO_4PS$

12.0.6 Molecular Structure:

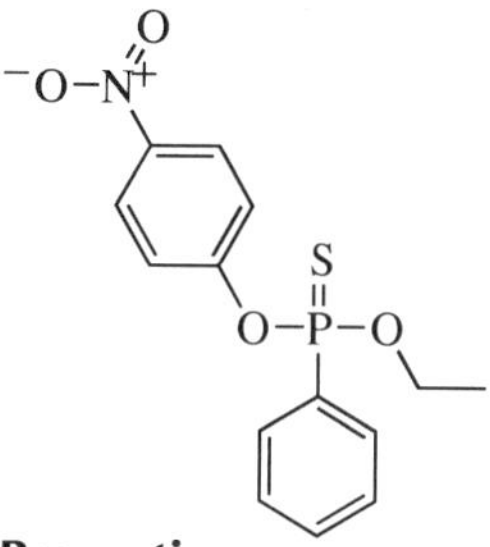

12.1 Chemical and Physical Properties

EPN is a noncombustible, light yellow solid or brown crystalline substance. Contact of EPN with strong oxidizers may cause fires and explosions. Toxic gases and vapors (e.g., oxides of sulfur and nitrogen, phosphoric acid mist, and carbon monoxide) may be released when EPN decomposes.

- Specific gravity: 1.268 at 25°C
- Melting point: 36°C
- Boiling point: 215 at 5 mmHg
- Vapor pressure: 0.0003 torr at 100°C
- Solubility: soluble in acetone, alcohols, ether, toluene; slightly soluble in water

12.2 Production and Use

EPN was introduced in 1949 for use as a nonsystemic insecticide and acaricide. It was used primarily on cotton to control the boll weevil and lepidopterous pests and was available as a 45% emulsifiable concentrate, as granules, and in combination with other insecticides (188). EPN is no longer registered for use in the United States (45).

12.4 Toxic Effects

12.4.1.1 Acute Toxicity. EPN is an organophosphate compound that has high oral toxicity and oral LD_{50}s of 14.5–91 mg/kg for rats (64a). Oral doses of 2–75 and 2–50 mg/kg technical EPN were fatal to female and male dogs, respectively (277). Oral and intraperitoneal LD_{50}s for the mouse were 12.2 and 8.4 mg/kg, respectively, and an oral LD_{50} for the dog was 20 mg/kg (545). EPN was about five times more potent in young rats compared to adult rats; the intraperitoneal LD_{50} in 23-day-old rats was 8 mg/kg, whereas it was 33 mg/kg in adults (189). Dermal LD_{50}s were 25–230 mg/kg in rats (64a). A dermal LD_{50} for the rabbit was 30 mg/kg (529), and the lethal range for dermal exposure to EPN for rabbits was 30–150 mg/kg (277). The only acute inhalation toxicity value available for EPN is a 1-hour LC_{50} of 160 mg/m^3 that was reported for rats (545).

The magnitude of brain, plasma, and RBC cholinesterase activity inhibition and its rate of recovery in rats was determined following a sublethal dose of 25 mg/kg EPN (38). Maximum brain cholinesterase inhibition occurred 4–24 hours after exposure and recovered to normal by 2 weeks, maximum plasma cholinesterase inhibition occurred at 4 hours and recovered by 72 hours, and maximum RBC inhibition occurred at 24 hours and recovered by 4 weeks (the same rate at which new rat RBCs are formed) (38).

Adult, atropinized hens treated subcutaneously with doses of EPN equivalent to the hen subcutaneous LD_{50} (60 mg/kg) developed leg weakness immediately after dosing that persisted for more than 48 hours and occurred in addition to cholinergic symptoms (140). Only 3 of 21 animals that exhibited leg weakness survived. In another study, EPN produced neurotoxicity when administered subcutaneously to atropinized hens at doses of 40 mg/kg or more, but not at doses of 20 mg/kg. Although this effect was prompt in onset and lasted as little as 6 days in some hens, it persisted for more than 330 days in others (64a). High, lethal, single oral exposures (65–100 mg/kg) also caused delayed neuropathy in atropinized hens (191). Delayed neurotoxic ataxia was also observed in mice following oral dosing. A sublethal oral dose (20 mg/kg) produced an irreversible neurotoxic ataxia in mice after 29 days (192).

EPN potentiates malathion toxicity and is itself potentiated by malathion. Oral LD_{50}s for EPN and malathion given by gavage to rats were 65 and 1400 mg/kg, respectively. When given together at approximately equitoxic doses (about a 25:1 malathion: EPN ratio), a tenfold potentiation was observed. The LD_{50} for malathion was 167 mg/kg and for EPN was about 6.6 mg/kg (193, 194). Similarly, when single doses of EPN were administered to dogs, 200 mg/kg was fatal, and 50–100 mg/kg caused moderate to severe symptoms; however, when EPN was administered simultaneously with as little as 100 mg/kg malathion (< 10% of the malathion LD_{50}), 2 mg/kg EPN caused 100% mortality.

Potentiation, it is believed, results from inhibition of the hydrolytic detoxification of malathion by EPN (194a, 195). The rate of malaoxon detoxification in liver was 10 to 80% inhibited in rats given single injections of 0.5 to 1.5 mg/kg EPN and was 29 to 95% inhibited in rats fed diets containing 5 to 100 ppm EPN for 2 weeks compared to controls. In this study, the LD_{50} of malathion was 550 mg/kg 1 hour after an intraperitoneal injection of 1.5 mg/kg EPN, compared to 1100 mg/kg in untreated controls.

Neither brain, plasma, or RBC cholinesterase inhibition occurred in rats maintained on a dietary level of 5 or 25 ppm EPN (about 0.25 and 1.25 mg/kg/day) for 8 or 2 weeks, respectively (38). When rats were given diets that contained 0, 100, 300, or 600 ppm (males) (equivalent to 5, 15, or 30 mg/kg/day) or 0, 35, 100, or 300 ppm (females) (equivalent to 2, 5, or 15 mg/kg/day) EPN for 30 days (279), "transitory nervous manifestations" (*sic*) (excitability, tremors) were noted and mortality occurred in the 600-(males) and 300-ppm (females) groups. Body weight gain decreased in the 300-(males) and 100-ppm (females) groups. The authors noted that comparable studies of rats fed the commerical 35% formulation gave consistent results.

12.4.1.2 Chronic and Subchronic Toxicity. In rats fed ground chow contaminated with EPN for 90 days and then observed until death, a calculated 90-dose LD$_{50}$ was 12 mg/kg/day (196). Using the 90-dose LD$_{50}$ and the single dose LD$_{50}$ of 7.7 mg/kg previously obtained, a "chronicity factor" (single dose LD$_{50/90}$-dose LD$_{50}$) of 0.64 mg/kg was calculated, indicating that EPN does not exhibit a cumulative toxic effect.

Cholinesterase activity declined in all organs and tissues in rats after being fed diets that contained 75 ppm EPN for 1 month (154). When rats were fed diets that contained 0.2, 1, 5, or 25 ppm EPN (0.01, 0.05, 0.25, or 1.25 mg/kg/day) for 1, 3, 6, or 13 weeks, cholinesterase activity in the brain was unaffected, although there was a dose-related inhibition of aliesterase activity in both liver and serum at dietary level of 1 ppm and above (197, 285). When rats were given diets, that contained up to at least 125 ppm, the effects noted at 125 ppm (6.25 mg/kg/day) included decreased plasma and brain cholinesterase activity, decreased female growth and decreased RBC, hemoglobin, and hematocrit in both sexes (95). At 25 ppm (1.25 mg/kg/day), RBC cholinesterase was significantly inhibited. At 5 ppm (0.25 mg/kg/day), no adverse effects were noted (95).

When dogs were treated orally with repeated doses of up to 3.0 mg/kg/day EPN, decreased RBC and brain cholinesterase activity; decreased red blood cells, hemoglobin, and hematocrit in both sexes; and pancreatic acinar cell atrophy in two males were recorded (95). Administration of 1.0 mg/kg/day had no adverse effect. No other details were provided in the study summary (95).

Potentiation of cholinesterase inhibition occurs when subchronic exposure is to EPN and malathion simultaneously (194). When rats were fed diets that contained 25 ppm EPN for 8 weeks, a small but significant inhibition of RBC cholinesterase was observed, whereas when 25 ppm EPN was fed simultaneously with 500 ppm malathion, RBC cholinesterase inhibition was marked. (500 ppm malathion alone had no effect on RBC acetylcholinesterase activity.) Similarly, when dogs were fed diets that contained 20 or 50 ppm EPN for 12 weeks, RBC cholinesterase activity marginally decreased (5 ppm had no effect), whereas diets that contained 3, 20, or 50 ppm EPN combined with 8, 100, or 250 ppm malathion (levels which did not inhibit RBC cholinesterase), respectively, markedly inhibited RBC cholinesterase (193).

12.4.1.3 Pharmacokinetics, Metabolism, and Mechanisms. EPN is readily absorbed following either oral or dermal exposure. In rats, following a single oral dose of 1.5 mg/kg EPN, about 61–68% of the dose was excreted in urine and 7–21% of the dose was excreted in the feces (279). Metabolism was essentially complete because urinary excretion of metabolites totaled 60–68% of the total dose; the corresponding fecal values were 6.7–17.2% (279). The plasma half-life was approximately 33.6 hours in males and 38.6 hours in females. In rats given 0.72 or 0.072 mg/kg EPN three times at 24-hour intervals, 14 and 26% of the dose was excreted as *p*-nitrophenol in urine (280). EPN was not detectable in blood at any time after the final dose but was detectable in adipose tissue (0.06 ppm) only on day 3.

It has been argued that the toxicokinetics and metabolism of EPN plays a major role in the development and expression of neuropathy. When a rat was preconditioned for 21 days on a diet that contained 450 ppm EPN, only traces of the dose remained in body tissues after 72 hours (280). Thus, EPN was practically completely metabolized and eliminated in rats. Two EPN-conditioned hens that received 2.5 mg/kg EPN for 21 days and two unconditioned hens excreted 94–100% of a single oral, nonneurotoxic 4-mg/kg dose within 72 hours of administration. No significant difference was apparent in the dose excreted from conditioned or unconditioned chickens. Small amounts of EPN remained in body tissues; only trace amounts were found in the brain and spinal cord. By contrast, a single oral neurotoxic dose of 50 mg/kg EPN was slowly metabolized and excreted by chickens. Only 65% of the dose was eliminated during the 72 hour test. Excreta extract contained the same five metabolites found in the rat. Thus, the 50-mg/kg dose may have exceeded the metabolic capacity of the liver to detoxify EPN (280). Based on these data, half-lives for EPN were 1.1 days for a single oral 17.3-mg/kg dose in male rats, 1.8 days for a single oral 4-mg/kg dose in chickens, and 3.5 days for a single oral dose of 5 mg/kg in chickens (280).

In cats, EPN was readily absorbed when a single dose of 20 mg/kg was dermally applied as reflected by disappearance of EPN from the application site (281). Most of the absorbed dose was excreted in urine (29.9%); 3.2% was recovered in the feces. In cats given daily dermal doses of 0.5 mg/kg EPN for 10 days, 62% of the total dose was eliminated in the urine. Thus, the metabolism and toxicokinetics of EPN in the cat are intermediate between those in the rat and in the chicken. Similarly to the rat, EPN is excreted as polar metabolite, mostly in the urine. As in the chicken, however, EPN is persistent in cat tissues, especially in nervous tissues. The finding that the cat is ten times less sensitive than the chicken to subchronic dermal exposure to EPN may be explained by the relatively rapid metabolism and elimination of EPN in the cat. (The extensive metabolism of EPN in the rat compared to (the chicken or) the cat might be explained by the presence of a higher level of cytochrome P-450 in rat liver microsomes than in the chicken or in the cat.)

EPN undergoes oxidative toxication through desulfuration that is catalyzed by microsomal oxidases (282) and detoxication through hydrolytic removal of *p*-nitrophenol (280, 283). Moreover, has been shown that the rate of hydrolytic detoxication of EPN more closely correlates with sex and age differences in EPN toxicity than the rate of toxication (20). The nitro group can be further reduced to an amino group by enzymes in the livers of mammals, birds, and fish and in the kidneys, spleens, hearts, lungs, and RBCs of mammals. The resulting amine is a weak inhibitor of cholinesterase but the importance of this reduction for detoxication *in vivo* is unknown 283. Activation and degradation of EPN is not restricted to the liver. *In vitro* studies have shown that rat brain also possesses activation and degradation capability (284).

Overall, metabolism evidently reduces the toxicity of EPN, maybe even to a greater extent than for parathion or methyl parathion. The intraperitoneal LD_{50} was 7.3 mg/kg in untreated rats and 75 mg/kg in phenobarbital-treated rats, whereas for parathion and methyl parathion, these values were 2.5 and 7.3 mg/kg and 7.0 and 8.0 mg/kg, respectively (197).

12.4.1.4 Reproductive and Developmental. EPN was administered via gavage once daily to mice (1, 3, 6, or 12 mg/kg) on days 6 through 16 of gestation (278). EPN, at dose levels up to those that were maternally lethal (12 mg/kg), did not produce fetotoxicity, fetal lethality, or teratogenicity (591).

ORGANOPHOSPHORUS COMPOUNDS

12.4.1.5 Carcinogenesis. No evidence of carcinogenicity occurred in rats fed EPN at levels of 50, 150, or 450 ppm (males) or 0, 25, 75, or 255 ppm (females) for two years (277). Dietary levels of 450 ppm (males) and 225 ppm (females) caused intermittent tremors and slight retardation of growth. Rats given the highest dietary level had depressed growth from the start of EPN feeding. Other groups of rats tolerated oral dosages up to 10 mg/kg/day. There was no indication of increasing mortality with increasing doses of EPN, although the mortality by the end of the two year study was high (86–98%).

12.4.1.7 Other: Neurological, Pulmonary, Skin Sensitization. Delayed neurotoxicity was produced in hens given daily oral doses of 0.01, 0.1, 0.5, 1.0, 0.5, 5.0, or 10.0 mg/kg EPN for 90 days and observed for 30 days (274). Hens treated with 5 mg/kg or more were atropinized to protect them from acute cholinergic toxicity. The clinical condition of most ataxic hens deteriorated during the 30-day observation period following the end of dosing, and the severity of effects dependend on the size of the daily ingested dose. Hens given small doses showed only ataxia, but those treated with large doses progressed to paralysis and died. Hens treated daily with 0.01 mg/kg EPN showed no abnormality in gait or behavior. In a study summarized by EPA, when adult hens were dosed orally at 0.01, 0.1, 0.5, 1.0, 2.5, or 5.0 mg/kg/day for 90 days, organophosphate type delayed neurotoxicity (ataxia) was seen at doses of 2.5 and 5.0 mg/kg/day but not at lower doses (95). Histopathologically observed damages to the nervous system was seen at doses of 0.1 mg/kg, but not at 0.01 mg/kg/day.

When EPN was applied dermally at doses of 0.01 to 10 mg/kg to the necks of hens for 90 days, all hens given 2.5 to 10 mg/kg EPN developed signs of cholinergic poisoning despite treatment with atropine (269). All hens given EPN, except those that received the 0.01 mg/kg dose, developed signs of delayed neurotoxicity, such as ataxia, paralysis, and death. All doses of EPN, except for the 0.01 mg/kg/day dose, caused degeneration of axons and myelin in the spinal cord.

Delayed neurotoxicity was produced in cats following the administration of a single dermal dose of 22.5, 45, 112.5, or 225 mg/kg (0.2 to 5.0 times the dermal LD_{50} of 45 mg/kg) or repeated daily doses of 0.5, 1.0, or 2.0 mg/kg EPN (269). Single dermal doses of 9.0 mg/kg and repeated dermal doses of 0.1 mg/kg were without effect. Therefore, the cat is about ten times less sensitive than the hen to EPN-induced delayed neurotoxicity.

When dogs were given 2.8 to 5.0 mg/kg EPN orally for up to a year, the first symptoms were vomiting and diarrhea, followed by gait disturbances and weight loss; histopathological changes in the nervous system were noted (275). Male sheep treated orally with 1 mg/kg/day EPN for 180 days showed no clinical signs of neurotoxicity, no histological changes in tissues, and no remarkable neurotoxic esterase (NTE) measurements (214).

12.4.2 Human Experience

12.4.2.2 Clinical Cases. In one case, ingestion of an estimated 200 mL of 50% EPN resulted in coma, miosis, sweating, bloody stools, pulmonary edema, and death despite treatment with pralidoxime and atropine (286). In another case, a 500 mL suspension that contained EPN was obtained from the gastrointestinal tract of a 46-year-old farmer who died following generalized convulsions and cardiorespiratory arrest (287). Signs and symptoms of EPN poisoning may persist for years after acute or chronic exposure due to permanent nerve damage; symptoms of nerve damage being and may be irreversible at the time of severe acute exposure (294).

Applicator personnel were monitored during aerial and ground applications of EPN to cotton in Mississippi and Arizona. Respiratory exposures based on an 8-hour work day averaged 11 µg for pilots, 15 µg for loaders, and 39 µg for ground applicators. Respiratory exposure of flagmen, monitored during a complete application cycle, averaged 317 µg/8 hours. Mean 8-hour dermal exposures were 2.1 mg for pilots, 6.3 mg for loaders, 117.7 mg for flagmen, and 7.5 mg for ground applicators (185).

Neither plasma nor RBC cholinesterase activity was affected among five volunteers given 3 mg EPN/day by capsule (about 0.04 mg/kg/day) for 32 days, nor were any clinical effects observed or reported (294). No significant effect on plasma or RBC cholinesterase activity occurred when the dose was increased to 6 mg EPN/day (0.0857 mg/kg/day) for 47 days, nor were any clinical effects observed or reported (294). When the dose was further increased to 9 mg EPN/day (about 0.13 mg/kg/day) for 56 days, plasma cholinesterase activity that began 2 weeks after first administration of EPN was depressed and continued to be depressed at 3 weeks after dosing stopped. RBC cholinesterase activity inhibition was similar to that of plasma but did not occur as soon. However, in a later study, when five other volunteers were given 6 mg EPN/day (0.086 mg/kg/day) plus 16 mg malathion/day for 44 days, both RBC and plasma cholinesterase were significantly depressed (288).

12.5 Standards, Regulations, or Guidelines of Exposure

EPN is not registered for use in the United States. The ACGIH TLV for EPN is 0.1 mg/m^3 with a skin notation (154). The OSHA PEL-TWA is 0.5 mg/m^3 with a skin notation. The NIOSH REL-TWA is 0.5 mg/m^3 with a skin notation.

13.0 Ethion

13.0.1 CAS Number: *[563-12-2]*

13.0.2 Synonyms:
O,O,O',O'-Tetraethyl-S,S'-methylene di(phosphorodithioate; O,O,O',O'-tetraethyl S,S'-methylene bis phosphorodithioate; Diethion; Ethanox; Ethiol; FMC 1240; Hylemox; Niagra 1240; Rhodiacide; Rhodocide; RP-Thion; O,O,O,O-tetraethyl S,S-methylene bisphosphorodithioate; Nialaten; Nialate(R); ethyl methylene phosphorodithioate; O,O,O',O'-tetraethyl S,S'-methylenediphosphorodithioate; bis[S-(diethoxyphosphinothioyl)mercapto]methane; S,S-methylene O,O,O',O'-tetraethyl phosphorodithioate; Ethion 8; phosphorodithioic acid, S,S'-methylene $O,O,O'O'$-tetraethyl ester; Bis(diethoxyphosphinothioylthio)methane; Embathion; Ethodan; Ethopaz; Fosfatox E; Fosfono 50; Itopaz; KWIT; methanedithiol, S,S-diester with O,O-diethyl phosphorodithioate; methylene O,O,O',O'-tetraethyl phosphorodithioate; tetraethyl S,S'-methylene bis(phosphorodithioate)

13.0.3 Trade Names:
Nialate®; Bladan; Embathion, Ethanox Ethodan, Fosfatox E; Fosfono 50; Hylemax; Hylemox Itopax, Niagara 1240; Soprathion

ORGANOPHOSPHORUS COMPOUNDS

13.0.4 Molecular Weight: 384.48

13.0.5 Molecular Formula: $C_9H_{22}O_4P_2S_4$

13.0.6 Molecular Structure:

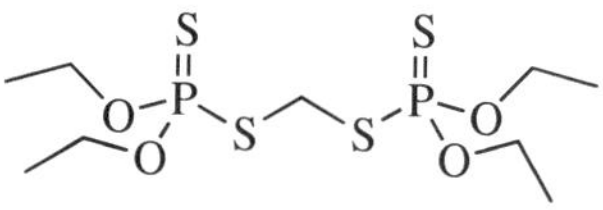

13.1 Chemical and Physical Properties

Pure ethion is a colorless liquid. Ethion emits toxic fumes of oxides of sulfur and phosphorous when heated to decomposition. It is subject to acid and base hydrolysis, and undergoes oxidation in air slowly

Specific gravity: 1.220 at 20°C
Melting point: $-12°$ to $-13°C$
Boiling point: decomposes above 150°C
Vapor pressure: 1.5×10^{-6} mm Hg at 25°C
Solubility: slightly soluble in water; readily soluable in most organic solvents, including acetone, xylene, chloroform, and methylated naphthalene

13.1.2 Odor and Warning Properties

The technical material has a very disagreeable odor.

13.2 Production and Use

Ethion is a preharvest, topical insecticide used primarily on citrus fruits, deciduous fruits, nuts, and cotton. End-use product formulations for citrus consist of emulsifiable concentrates (EC) that contained 9–82% active ingredient. Applications are made using ground boom or air blast equipment. High-pressure and low-pressure hand wands along with backpack sprayers are used for spot treatment. It is also used as a cattle dip for ticks and as a treatment for buffalo flies. Ethion is marketed as a 25% wettable powder; 2%, 3%, and 4% dusts; 5% granules; and various oil solutions and combinations with other materials (296).

13.4 Toxic Effects

13.4.1.1 Acute Toxicity. Ethion is an organophosphate that has relatively high oral toxicity and oral $LD_{50}s$ for rats of 21 to 191 mg/kg (64). Dermal $LD_{50}s$ for rats are 62 to 838 mg/kg (Gaines 1969; 242). For rabbits, the dermal LD_{50} was 915 mg/kg (IPCS 1986). A 4-hour LC_{50} of 864 mg/m^3 for ethion in rats has been reported and in studies submitted to EPA, 4-hour $LC_{50}s$ of 2310 mg/m^3 and 450 mg/m^3 were obtained for male and female rats, respectively (155, 296).

When goats were given single intravenous injections of 2, 5, or 10 mg/kg ethion, only those given 5 or 10 mg/kg displayed cholinergic toxicity (291). All doses decreased RBC cholinesterase activity that lasted for about 4 days and recovered by 8–10 days.

Ethion was slightly irritating to rabbit eye and skin but did not cause dermal sensitization in guinea pigs (296). In a study submitted to EPA, ethion was negative in an acute neurotoxicity test using hens (296).

13.4.1.2 Chronic and Subchronic Toxicity. When dogs were given diets that contained 0.5, 2.5, 25, or 300 ppm ethion for 90 days, cholinergic toxicity and body weight gain and food consumption decreased among dogs given 300 ppm (equivalent to about 6.9 or 8.25 mg/kg/day) but not 25 ppm (about 0.71 mg/kg/day). Brain and RBC cholinesterases were inhibited at 25 and 300 ppm, respectively (296).

When rabbits were treated dermally for 21 days with 0, 1.0, 3.0, 25, or 250 mg/kg/day ethion, RBC and brain cholinesterase activity were inhibited at all doses (296). Cholinergic signs were not noted. Erythema and desquamation at the application sites occurred at 25 and 250 mg/kg. Inhibition of brain cholinesterase was again observed when rabbits were treated dermally for 21 days with 0, 0.1, 0.25, 0.8, 1.0, 3.0, or 25 mg/kg/day (296).

13.4.1.3 Pharmacokinetics, Metabolism, and Mechanisms. About 80% of an oral dose to goats was excreted (in urine (64%), feces (14%), and milk (2%)) within 14 days and about 55% was excreted within the first 96 hours (291). Less than 5% of the dose was absorbed unchanged, consistent with the hydrolysis of organophosphate due to ruminal microflora. About 20% of a dermal dose of ethion (100 mg/kg) applied to goats was absorbed during 14 days. Following intravenous treatment of goats with 2 mg/kg ethion, unchanged ethion was rapidly eliminated (effective half-life of 2 hours) demonstrating the large role of ruminal microflor metabolism. Metabolized, ethion was distributed primarily to liver, kidney and fat. Tissue elimination was relatively fast the first 3 days but slow during the third and fourth weeks after dosing. The half life was about 2 weeks. Seventy-seven percent of the dose was recovered after 2 weeks in urine (55%) and feces (22%). At least five major metabolites were identified in the urine. Regardless of route of exposure, ethion was extensively bound to plasma proteins.

In humans given an intravenous dose of ethion, 38.4% of the dose was excreted in the urine by 5 days (66). About 15% of the administered dose was excreted in the first 12 hours, 9.5% in the second 12 hours, and 7.6% over the next 24 hour period. The elimination half-life was estimated at 14 hours. After topical administration of 4 µg/cm^2 to the forearm, about 3.3% of the applied dose was excreted by 5 days.

13.4.1.4 Reproductive and Developmental. In a three-generation study in which rats were given diets that contained 2, 4, or 25 ppm ethion, no adverse effects on reproduction occurred at any dose (296). In rats given 2.0, 0.6, or 2.5 mg/kg/day ethion by gavage on gestation days 6–15, maternal toxicity (hyperactivity) and developmental toxicity (delayed ossification of pubes) occurred only at 2.5 mg/kg/day (242). In rabbits given 0, 0.6, 2.4, or 9.6 mg/kg/day ethion by gavage on gestation days 6–18, maternal toxicity (weight loss, reduced food consumption, and orange colored urine) occurred at 9.6 mg/kg/day; developmental toxicity was not observed (296).

13.4.1.5 Carcinogenesis. When dogs were given diets containing 0.5, 1, 2, 20, or 100 ppm ethion (about 0.01, 0.03, 0.05, 0.5, or 2.5 mg/kg/day) for one year, cholinergic signs were not noted in any group (296). At 20 ppm, RBC cholinesterase was inhibited in females, and, at 100 ppm, both brain and RBC cholinesterase activities were (296). When rats were fed diets containing 0, 2, 4, or 40 ppm ethion (about 0.1, 0.2, and 2.0 mg/kg/day) for 24 months, the only effect was reduction in serum cholinesterase activity at 40 ppm (178). No evidence of carcinogenicity was observed. No evidence of carcinogenicity or other adverse effects occurred in mice fed diets containing 0, 0.75, 1.5, or 8.0 ppm (about 0.11, 0.22, or 1.2 mg/kg/day) ethion for 2 years (296).

13.4.1.6 Genetic and Related Cellular Effects Studies. Ethion is not mutagenic. Ethion did not cause mutations in the Ames test with or without metabolic activation, did not cause chromosomal aberrations in an *in vivo* cytogenetic test in rats, did not increase unscheduled DNA synthesis in an *in vitro* test with rat hepatocytes, and did not cause mutation in a recombinant/conversion assay using *S. cerevsiae* (296). However, there is also a report that ethion was mutagenic in an *in vivo* (mouse bone marrow) and *in vitro* (Chinese hamster lung cells) micronucleus test (289) and a report that ethion induces chromosomal aberrations when administered to chicks (290).

13.4.2 Human Experience

Six male adult human volunteers were given ethion via capsule according to the following sequential dosing regime: (1) 0.05 mg/kg/day for 21 days, (2) 0.075 mg/kg/day for 21 days, (3) 0.10 mg/kg/day for 21 days, (4) 0.15 mg/kg/day for 3 days, and (5) recovery for 19 days (296). RBC cholinesterase activity, averaged across subjects, was not inhibited at any dose level. Signs of overt cholinergic toxicity occurred in one subject on days 19–21 of 0.05 mg/kg/day dosing (headache, blurred vision, lightheadedness, and dizziness) and on day 1 of 0.075 mg/kg/day dosing. Signs of overt cholinergic toxicity (partial blindness, lightheadedness) also occurred in another subject on the first day of receiving 0.075 mg/kg/day for 0.15 mg/kg/day. EPA concluded that these results suggested a cumulative effect of ethion when administered at a dose of 0.05 mg/kg/day or higher.

13.5 Standards, Regulations, or Guidelines of Exposure

Ethion has been reregistered for use by the EPA. The ACGIH TLV for ethion is 0.4 mg/m^3 with a skin notation (154). There is no OSHA PEL-TWA. The NIOSH REL-TWA is 0.4 mg/m^3 with a skin notation. Other countries generally have the same standard (Australia, Belgium, France, The Netherlands, and Switzerland).

14.0 Fenamiphos

14.0.1 CAS Number: [22224-92-6]

14.0.2 Synonyms: Bay SRA 3886; Nemacur; ethyl 4-(methylthio)-*m*-tolyl isopropylphosphoramidate; Nemacur(R); (1-methylethyl)phosphoramidic acid ethyl 3-methyl-4-(methylthio)phenyl ester; isopropylphosphoramidic acid, 4-(methylthio)-m-tolyl ethyl

ester; ethyl 3-methyl-4-(methylthio)phenyl (1-methylethyl)phosphoramidate; isopropylamino-*O*-ethyl-(4-methylmercapto-3-methylphenyl)phosphate; 1-(methylethyl)-*O*-ethyl-*O*-(3-methyl-4-(methylthio)phenylphosphoramidate; nemacur p; phosphoroamidic acid, isopropyl-, 4-(methylthio)-*m*-tolyl ethyl ester

14.0.3   Trade Names: Bay SRA 3886; Nemacur®

14.0.4   Molecular Weight: 303.4

14.0.5   Molecular Formula: $C_{13}H_{22}NO_3PS$

14.0.6   Molecular Structure:

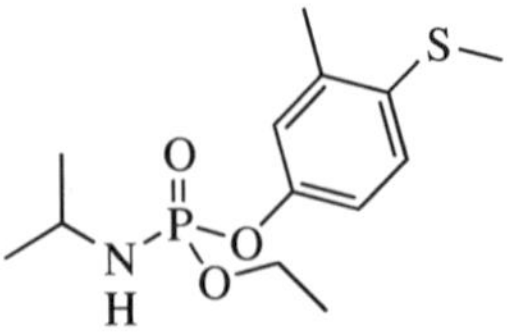

14.1   Chemical and Physical Properties

Pure fenamiphos is a colorless crystal; technical fenamiphos is an off-white to tan, waxy solid. Fenamiphos is stable under most normal use conditions but is subject to hydrolysis under alkaline conditions

>Melting point: 49.2°C (pure); 40°C (technical)
>Density: 1.15 g/cm³ at 20°C
>Vapor pressure: 7.5×10^{-7} mmHg at 30°C; 4.7×10^{-5} mmHg at 20°C
>Concentration in saturated air: 0.001 ppm (0.01 mg/m³) at 30°C
>Solubility: soluble in most organic solvents; soluble in water (approximately 400 mg/L)

14.2   Production and Use

Fenamiphos is used as a selective nematocide and insecticide to control nematodes, thrips, beetles, aphids, and root borers on terrestrial food crops and nonfood sites (297). It is applied by a variety of methods including broadcast, row, drench, and irrigation before or at planting time or to established plantings.

14.4   Toxic Effects

14.4.1.1 Acute Toxicity. Fenamiphos is a highly toxic organophosphate that has oral LD_{50}s of 2–100 mg/kg for rats, mice, rabbits, dogs, and guinea pigs (61, 292–294). Analytical preparations of fenamiphos are slightly more toxic than technical preparations (295). Intraperitoneal LD_{50}s are equivalent to oral LD_{50}s indicating high oral absorption (61, 292). Dermal LD_{50}s were 72–225 mg/kg for rats and rabbits (292, 296). Rabbits dermally treated with 0, 0.5, 2.5 or 10 mg/kg/d fenamiphos for 6 h/day for 21 consecutive days showed slight erythema of abraded skin that lasted 3–6 days, and inhibition of RBC and brain cholinesterase occurred at doses of 2.5 and 10 mg/kg/day. Cholinergic effects were not reported at any dose (297).

One- and 4-hour LC_{50}s and 110–175 mg/m^3 and 91–100 mg/m^3, respectively, were reported (298). When 4-hour exposures were repeated for 5 days, the LC_{50} was >28 mg/m^3 but <100 mg/m^3, indicating a cumulative effect (299). No-effect levels for RBC cholinesterase inhibition were 28 mg/m^3 (males) and 4 mg/m^3 (females) for RBCs. Male and female rats exposed to 0.03, 0.25, or 3.5 mg/m^3 fenamiphos aerosols 6 hours/day for 5 days/week for 3 weeks showed no overt cholinergic symptoms or changes in physical appearance, behavioral patterns, body weights, hematology, clinical chemistry, urinalysis, gross pathology, or organ weights (299). RBC and brain cholinesterase were unaffected at any level.

When rats were given single oral doses of 0.4, 1.6, or 2.4 mg/kg fenamiphos, four males and one female that were given 2.4 mg/kg died within 30 minutes of dosing (300). Cholinergic effects occurred among rats given 1.6 or 2.4 mg/kg, were maximum about 25 minutes after dosing, and were relatively quickly reversed. All treatment-related effects had completely resolved by 7 days after dosing. RBC cholinesterase activities were inhibited among rats given 1.6 or 2.4 mg/kg, and brain cholinesterase was not affected in any group.

There was no evidence of delayed neurotoxicity in atropinized hens that were treated with single doses of 12.5 mg/kg fenamiphos, in atropinized hens treated with 25 mg/kg twice at an interval of 3 weeks, or, in unprotected hens treated with up to 50.0 mg/kg and observed for 21 days (301).

Fenamiphos was only slightly irritating (caused mild erythema) when applied to skin of rabbits and was mildly irritating to the eye of rabbits (297). Fenamiphos did not cause either contact dermatitis or "tuberculin type allergy" in a guinea pig sensitization study (302).

14.4.1.2 Chronic and Subchronic Toxicity. Repeated intraperitoneal injections of 1 mg/kg/day fenamiphos for 60 days caused no mortality in rats, although brain and tissue cholinesterase activity was inhibited (198). Administration of 2 mg/kg/day fenamiphos for 60 days resulted in 40% and 100% mortality by 30 and 60 days, respectively, as well as significant brain and tissue cholinesterase inhibition.

No cholinergic signs or decreases in brain or RBC cholinesterase activity occurred in rats given diets that contained fenamiphos at 0, 0.36, 0.60, or 1.0 ppm for 14 weeks (approximately 0.02, 0.03, or 0.05 mg/kg/day) (297). When rats were given fenamiphos via diets that contained 0, 4, 8, 16, or 32 ppm (about 0.2, 0.4, 0.8, and 1.6 mg/kg/day) for 90 days, cholinergic signs occurred in rats given only the 32-ppm diet (297). RBC cholinesterase was inhibited in rats given 8 ppm or more. When rats were fed diets that contained 0, 1, 10, or 50 ppm fenamiphos (equivalent to about 0, 0.1, 0.7, 3.3–4.0 mg/kg/day) for 13–14 weeks, treatment-related cholinergic signs (muscle fasciculations) occurred only in high-dose females during weeks 1–3; no cholinergic signs occurred in males (303). There were no adverse ophthalmic findings. Decreases in RBC cholinesterase activity occurred in the 10- and 50-ppm group in males and in all treated groups in females, were evident at 4 weeks, and, remained at about the same level for 13 weeks. A no-observed-effect level for RBC cholinesterase inhibition extrapolated from the data was

0.4 ppm for females (approximately 0.032 mg/kg/day). No effect on brain cholinesterase was observed.

When dogs were given diets that contained 18 ppm or more for 3 months, overt cholinergic toxicity occurred (303a). No adverse effects were observed at dietary levels of 16 ppm or less (equal to about 0.45 mg/kg/day). RBC cholinesterase was inhibited in dogs fed diets that contained 2 (females only), 5 (females only), 6, 10, or 18 ppm. No adverse effects of any kind and no inhibition of RBC cholinesterase occurred in dogs given diets that contained 0 or 0.5 ppm fenamiphos for six months (equivalent to 0 and 0.01–0.02 mg/kg/day) or in dogs given diets that contained up to 1.7 ppm fenamiphos (equivalent to 0.04 mg/kg/day) for 100 days (304, 305).

14.4.1.3 Pharmacokinetics, Metabolism, and Mechanisms.

About 93–95% of an oral dose of fenamiphos was absorbed by rats and excreted by 12–15 hours after treatment (309). Forty-eight hours later, residues in tissues were greatest in liver and kidney; less in fat, gastrointestinal tract, and heart; and minimal in brain and muscle. Similar results were obtained in a whole body autoradiographic study of rats treated orally with fenamiphos which showed that oral absorption was virtually complete, distribution volume was low, and excretion was rapid and nearly complete by 8 hours after dosing (310). Pretreatment of rats for 14 days before treatment with fenamiphos did not alter absorption, distribution, or elimination patterns. Distribution following oral doses was rapid in brain, and plasma and concentrations in brain exceeded concentrations in plasma by about 1.3-fold to twofold (311). Concentrations in brain and plasma were highest 0.5 hours after dosing and declined; α and β half-lives in the brain were 2 and 100 hours, respectively, and α and β half-lives in plasma were 17.5 and 212 hours, respectively. Oral absorption of fenamiphos sulfoxide was equally rapid in pigs and reached maximum blood concentrations 2 hours after a 0.9-mg/kg dose (312). Fifty-seven percent of the administered dose was excreted within 5 hours and 90% was recovered after 48 hours. Tissue residues were minimal; only liver and kidney contained residues >0.1 ppm.

In vitro dermal absorption of fenamiphos by human and rat skin accounted for 0.42–9.95% of the applied dose and was 2.07 ± 0.33 µg/cm^2/h (human) and 3.15 ± 0.40 µg/cm^2/h (rat) (313). Dermal absorption of a granular formulation was considerably less (0.01–0.03 µg/cm^2/h) and dermal absorption of a liquid formulation was considerably more (13.0 ± 1.87 µg/cm^2/h (human) and 49.0 ± 2.69 µg/cm^2/h (rat) during 24 hours.

The major metabolic pathway for fenamiphos is oxidation and formation of the sulfoxide and sulfone analogs. Loss of the isopropyl and probably the isopropyl amine moieties are also likely. Subsequent hydrolysis, conjugation, and excretion in urine gives nonorganosoluble compounds whose molecular weights are 400–800. Rats treated intravenously or orally with fenamiphos excreted fenamiphos phenols in different stages of oxidation at the sulfur atom and their respective sulfuric acid conjugates (276).

In pigs given fenamiphos sulfoxide orally, excreted metabolites were primarily conjugated phenols of the sulfoxide and sulfone. The metabolic pathway was hydrolysis and/or oxidation followed by conjugation.

14.4.1.4 Reproductive and Developmental.

No treatment-related endocrine effects, reproductive effects, or clinical signs occurred in adults or pups given diets that contained

0, 2.5, 10, and 40 ppm fenamiphos (about 0, 0.2, 0.6–0.7, and 2.8–3.2 mg/kg/day) for two generations. Among rats fed the 40-ppm diet, F1 pups experienced decreased body weight gain during lactation, F0 and F1 females had lower body weights during lactation, terminal body weights significantly decreased in adult rats, and absolute and relative ovary weights significantly decreased. RBC cholinesterase was significantly inhibited in adult rats fed the 10-ppm diet and in 4 day old pups fed the 10-ppm diet. Brain cholinesterase was significantly inhibited in adult rats fed the 40-ppm diet but was not inhibited in 4- or 21-day-old pups fed any dietary concentration. No adverse reproductive outcomes occurred in rats given diets that contained 0, 3, 10, or 30 ppm fenamiphos (equivalent to about 0, 0.15, 1.0, and 1.5 mg/kg/day) (276). Reduced body weight gain occurred only in the F2b generation males of the 30-ppm group.

No adverse developmental outcomes occurred among offspring of rats given fenamiphos at dose levels of 0, 0.3, 1.0, and 3.0 mg/kg by gavage on gestation days 6 through 15 (276). Dams in the 3-mg/kg group exhibited cholinergic signs of toxicity within 30 minutes after dosing, and two dams in this group died. No treatment-related maternal effect were seen at lower doses.

Fenamiphos was not fetotoxic or embryotoxic at any dose in rabbits treated orally on gestation day 6–18 with 0.1, 0.3 or 1.0 mg/kg/day fenamiphos (242). However, chain fusion of sternebra was found in five fetuses in the 1.0-mg/kg/day group and was higher than controls. The authors concluded that this may have been a treatment-related abnormality. When rabbits were given 0, 0.1, 0.5, and 2.5 mg/kg/day fenamiphos by gavage on gestation days 6–18, neither visible nor measurable treatment-induced effects were observed in dams given 0.1 or 0.5 mg/kg day (276). However, 2.5 mg/kg/day produced frank maternal toxicity, evidenced by four treatment-induced deaths, decreased body weight gains, and decreased food consumption. There were no treatment-related effects in mean numbers of corpora lutea; in numbers of live or dead fetuses, litter size, or sex ratio, or in the numbers of live or resorbed fetuses. However, preimplantation loss was elevated in the 2.5-mg/kg/day group, and, in addition, the mean live pup weight was slightly reduced. Except for one malformation observed in the high-dose group, no other fetal visceral anomalies were observed.

14.4.1.5 Carcinogenesis. There was no evidence of carcinogenicity or adverse effects of any kind in dogs given a diet that contained 0.5, 1, 2, 5, and 10 ppm fenamiphos for 2 years (about 0.01, 0.025, 0.05, 0.125, or 0.250 mg/kg/day), although RBC cholinesterase activities were inhibited at 2 ppm (312a). When dogs were fed diets that contained 1, 3, or 12 ppm fenamiphos (about 0, 0.030, 0.08, and 0.3 mg/kg/day), no adverse effects of any kind occurred, except for mild anemia (decreased RBC counts, hemoglobin, hematocrit) among dogs given the 12-ppm diet (304). RBC cholinesterase activity was inhibited in dogs given 3- or 12-ppm diets, and brain cholinesterase activity was inhibited in dogs given 12-ppm diets.

There was no evidence of carcinogenic effect in rats maintained on a diet that contained 3, 10, or 30 ppm fenamiphos for two years (about 0.2, 0.67, and 2.0 mg/kg/day) (306). In the 30-ppm group, mild symptoms that reflected cholinergic toxicity were observed temporarily, and mortality in female rats increased slightly. The 3-and 10-ppm diets did not cause evident adverse effects of any kind. There was no evidence of carcinogenicity or ophthalmologic effects in rats fed diets that contained 0, 2, 10, or 50 ppm fenamiphos for 2 years (equivalent to about 0, 0.1, 0.5–0.6, or 2.4–3.4 mg/kg/day) (307). RBC cholinesterase was significantly inhibited in the 10- and 50-ppm groups. Brain cholinesterase was marginally inhibited in the 50-ppm group. Females that consumed the 50-ppm diet exhibited an increase of rough coats and alopecia "associated with stress due to severe cholinesterase inhibition" (*sic*). No other compound-related effect occurred.

There was no evidence of carcinogenicity or other adverse effect in mice given diets that contained 0, 2, 10, or 50 ppm fenamiphos for 20 months (equivalent to doses of about 0, 0.2, 1.0, and 5.0 mg/kg/day), except that mice that received 50 ppm had reduced body and organ weights, and absolute brain weights decreased at 2 ppm or more (147).

14.4.1.6 Genetic and Related Cellular Effects Studies. Fenamiphos was negative in Ames tests using *S. typhimurium* with and without metabolic activation, and negative in a preincubation reverse mutation test with a tryptophan mutant strain of *E. coli* with and without mammalian metabolic activation (276). In a forward mutation assay using Chinese hamster cells (CHO), fenamiphos induced an increase in the mutant frequency in an initial test conducted without metabolic activation. However, the response was not dose-dependent and could not be reproduced in two subsequent repeat assays. Therefore, the authors concluded that fenamiphos is not mutagenic in the CHO/HGRPT forward mutation assay. Fenamiphos did not induce an increase of sister chromatid exchange in Chinese hamster V-79 cells (308).

In a micronucleus test, there was no evidence of any mutagenic effect of fenamiphos. Nor was there any deleterious influence on erythrocyte formation as measured by the ratio of polychromatic to normochromatic erythrocytes. Fenamiphos did not have a mutagenic effect in a dominant lethal test using mice (276).

Increased chromosomal aberration rate was observed in cultures of human lymphocytes, but only in the cytotoxic range. Therefore, this evidence of mutagenicity is considered equivocal (276).

14.4.2 Human Experience

No published reports of human poisonings as a result of fenamiphos exposure were identified. However, the EPA noted that "fenamiphos has been implicated in a handler poisoning incident which resulted in hospitalizing the worker" (276).

Potential inhalation and dermal exposures to fenamiphos were estimated over 2 to 4-hour periods for six mixer-loaders or applicators involved in treating agricultural soil (1 lb a.i./acre). Inhalation exposures (measured using personal air pumps attached to the collar) were less than the detectable level of 0.001 mg/h; dermal exposures (measured using cloth patches and by collecting hand wash water) were substantially higher and ranged from about 93 to 667 μg/h (314).

14.5 Standards, Regulations, or Guidelines of Exposure

The ACGIH TLV for fenamiphos is 0.1 mg/m^3 with a skin notation (154). There is no OSHA PEL-TWA for fenamiphos. The NIOSH REL-TWA is 0.1 mg/m^3 with a skin notation. Most

ORGANOPHOSPHORUS COMPOUNDS

other countries also have an Occupational Exposure Limit of 0.1 mg/m³ with a skin notation (Australia, Belgium, Denmark, France, The Netherlands, and Switzerland).

15.0 Fensulfothion

15.0.1 CAS Number: [115-90-2]

15.0.2 Synonyms: (*O,O*-Diethyl-*O*-4(methylsulfinyl)phenyl)-phosphorothioate Dasanit; Bay 25141; S767; phosphorothioic acid *O,O*-diethyl *O*-[methylsulfinyl)phenyl] ester; Dansanit; Fonsulfothion; Terracur P; Daconit; Agricur; Chemagro 25141; diethyl *O*-(4-(methylsulfinyl)phenyl) phosphorothioate; DMSP; phenol, p-(methylsulfinyl)-, *O*-ester with *O,O*-diethyl phosphorothioate

15.0.3 Trade Names: Dasanit®; Terracur R®

15.0.4 Molecular Weight: 308.35

15.0.5 Molecular Formula: $C_{11}H_{17}O_4PS_2$

15.0.6 Molecular Structure:

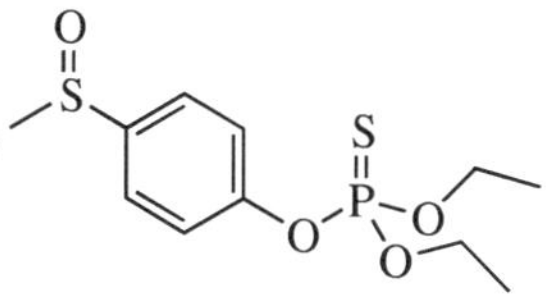

15.1 Chemical and Physical Properties

Fensulfothion is a yellow, oily liquid. Fensulfothion is oxidized readily to the sulfone and apparently isomerizes readily to the *S*-ethyl isomer

Specific gravity: 1.202 at 20°C
Boiling point: 138–141°C at 0.01 mmHg
Solubility: soluble in most organic solvents except aliphatics; slightly soluble in water (approximately 160 mg/100 mL)

15.2 Production and Use

Fensulfothion was introduced in 1957 as a systemic and contact insecticide and nematicide for use against free living and root knot nematodes. Until 1990 when its manufacture and use was discontinued, it was available as an emulsifiable concentrate, a wettable powder, a dust, and as granules (1, 155).

15.4 Toxic Effects

15.4.1.1 Acute Toxicity. Fensulfothion is an organophosphate compound that has high oral toxicity and oral $LD_{50}s$ are 1.8–10.2 mg/kg for rats (64a, 315). Intraperitoneal $LD_{50}s$ are 1.5–5.5 mg/kg, indicating that fensulfothion is well absorbed from the gastrointestinal tract. Dermal $LD_{50}s$ are 3.5–30.0 mg/kg (64a, 315). A 1-hour LC_{50} of 113 mg/m³ and a 4-hour LC_{50} of 29.5 mg/m³ were reported for rats (123). Following either oral or intraperitoneal doses, the onset of poisoning is rapid, and symptoms appear within 15 minutes, although death may be delayed. When lethal doses were given, death usually occurred within 2 hours, but after sublethal doses, the symptoms often persisted for 3 or 4 days (315).

Fensulfothion may have potentiating characteristics. An ordinarily harmless dose (550 mg/kg) of the anesthetic tricaine caused loss of righting ability in all rats and 20% mortality when it was given 1 hour after administering 2.5 mg/kg fensulfothion, a dose that inhibited carboxylesterase(s) but had no overt toxic effect (316a).

Fensulfothion did not produce neurotoxicity when given orally or intraperitoneally at doses up to 50 mg/kg to chickens protected by atropine and 2-PAM (1).

15.4.1.2 Chronic and Subchronic Toxicity. When rats were given 0.25, 0.5, or 0.75 mg/kg/day fensulfothion via intraperitoneal injection for up to 60 days, all rats given 0.75 mg/kg died within 5 days, 1 of 5 rats given 0.5 mg/kg died by day 60, and no rats given 0.25 mg/kg died within 60 days (315).

15.4.1.3 Pharmacokinetics, Metabolism, and Mechanisms. Fensulfothion is well absorbed orally as reflected by the similarity of oral and intraperitoneal LD_{50}s.

Based on findings for other phosphorothioates, it can be concluded that fensulfothion undergoes desulfuration to the corresponding oxygen analog to exert an anticholinesterase action (315). Indeed, the intraperitoneal LD_{50} of the oxygen analog in female rats was 1.2 mg/kg, equivalent to the 1.5 mg/kg observed for the parent compound. The thioether linkage of fensulfothion may also undergo oxidation to the sulfone derivative. The intraperitoneal LD_{50} of the oxygen analog sulfone was 0.9 mg/kg in female rats.

15.4.1.4 Reproductive and Developmental. Some female mice given a diet that contained 5 ppm fensulfothion died before mating (316). There was no effect on reproduction, gestation, or the lactation index of the survivors, except a slight reduction in the lactation index of third-generation pups. A level of 1 ppm had no effect. A slight nonsignificant increase of minor skeletal abnormalities occurred in rabbits that received fensulfothion during pregnancy at 0.10 mg/kg/day; no effect occurred on 0.05 mg/kg/day (316).

15.4.1.5 Carcinogenesis. When rats were provided diets that contained 1, 5, or 20 ppm fensulfothion, mortality increased in male rats given the 5- and 20-ppm diets, and body weight gain was depressed in both sexes given the 20-ppm diet (316). Plasma, RBC, and brain cholinesterase activities in females were "detectably" (*sic*) inhibited in rats given the 1-ppm diet (about 0.053 mg/kg/day).

When dogs were provided diets that contained 1, 2, or 5 ppm fensulfothion for 2 years, reduced food consumption, severe weight loss, and signs of cholinergic poisoning were initially evident in dogs given the 5-ppm diet, although food consumption increased and lost body weight was regained after the second month (316). Slight, temporary cholinergic effects and a slight reduction of cholinesterase (*sic*) occurred in dogs provided the 2-ppm diet, but not the 1-ppm diet (316).

15.4.1.6 Genetic and Related Cellular Effects Studies.
Fensulfothion did not increase the incidence of sister chromatid exchanges in V79 cells, nor did it induce cell cycle delay (308).

15.4.2 Human Experience

15.4.2.2 Clinical Cases.
Fensulfothion caused death by the following morning of a 34-year-old farmer who had applied fensulfothion to potato plants during the day. Autopsy indicated that death was due to pulmonary edema and blood cholinesterase activity which was far below normal (1). Death also occurred to a 5-year-old child who resided in a home 6 days after about 12,000 mg of fensulfothion was applied to about 10 m^2 of surface in two rooms (1). Other members of the family suffered nausea, vomiting, disorientation, diarrhea, and abdominal pain. Combined dermal and oral exposure led to severe poisoning characterized first by vomiting and weakness and by coma the next morning for a 7-year-old girl who had been playing with an empty fensulfothion container (1).

Fensulfothion poisoning from its accidental ingestion was recently reported in a family of four (317). Fensulfothion was misidentified as pepper and applied to the family's fish dinner. Shortly after the meal family members experienced nausea, vomiting, abdominal pain, and weakness. Two members had bradycardia, hypotension, and seizures. They were treated with atropine, artificially ventilated, and eventually recovered.

15.5 Standards, Regulations, or Guidelines of Exposure

Fensulfothion is not registered for use in the U.S. (145). The ACGIH TLV for fensulfothion is 0.1 mg/m^3 (154). There is no OSHA PEL-TWA. The NIOSH REL-TWA for fensulfothion is 0.1 mg/m^3. Most other countries also have an OEL of 0.1 mg/m^3 with a skin notation for fensulfothion (Australia, Belgium, France, The Netherlands, and Switzerland).

16.0 Fenthion

16.0.1 CAS Number: *[55-38-9]*

16.0.2 Synonyms:
O,O-dimethyl *O*-4-(methylmercapto)-3-methylthio)-*O*-ester with *O,O*-dimethyl phosphorothioate, phosphorothioic acid *O,O*-dimethyl *O*-(3-methyl-4(methylthio)phenyl)ester; Baycid; Baytex; Entex; Lebayeid; Queletox; Spotten; Talodex; Tiguvon; Lebaycid; *O,O*-dimethyl-*O*-[4-(methylthio)-*m*-tolyl] phosphorothioate; *O,O*-dimethyl *O*-(4-methylmercapto-3-methylphenyl) thionophosphate; *O,O*-dimethyl *O*-(3-methyl-4-methylthiophenyl) thiophosphate; *O,O*-dimethyl O-(4-methylthio-3-methylphenyl) thiophosphate; phosphotothioic acid *O,O*-dimethyl *O*[3-methyl-4-(methylthio)phenyl]ester; phosphorothioic acid *O,O*-dimethyl *O*-(4-methylthio)-*m*-tolyl ester; b 29493; bay 29493; bayer 9007; bayer 29493; bayer s-1752; *m*-cresol, 4-(methylthio)-, *O*-ester with *O,O*-dimethyl phosphorothioate; *O,O*-dimethyl *O*-4-(methylmercapto)-3-methylphenyl phosphorothioate; *O,O*-dimethyl *O*-3-methyl-4-methylthiophenyl phosphorothioate; DMTP; MPP; OMS 2; S 1752; spotton; Mosquitocide 700; Rid-a-Bird; BX-1; BX-2; cresol, 4-(methylthio)-, *O*-ester with *O,O*-dimethyl phosphorothioate; dimethyl (3-methyl-

4-(methylthio) phenyl) phosphorothionate; dimethyl methylthiotolyl phosphorothioate; dimethyl O-((4-methylmercapto)-3-methylphenyl) thionophosphate; dimethyl O-(3-methyl-4-(methylthio)phenyl) thiophosphate; dimethyl O-(4-(methylthio)-*m*-tolyl) phosphorothioate; Mercaptofos; Thiophos; 4-methylmercapto-3-methylphenyl dimethyl thiophosphate

16.0.3 Trade Names: Baycid; Baytex® Entex® Lebaycid; Tiguvon

16.0.4 Molecular Weight: 278.34

16.0.5 Molecular Formula: $C_{10}H_{15}O_3PS_2$

16.0.6 Molecular Structure:

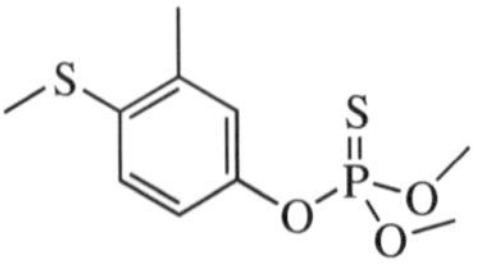

16.1 Chemical and Physical Properties

Fenthion is a yellow to tan, oily liquid

>Specific gravity: 1.250 at 20°C
>Melting point: 7.5°C
>Bioling point: 87°C at 0.01 mmHg (pure); 105°C at 0.01 mmHg (commercial grade)
>Vapor pressure: 3×10^{-5} torr at 20°C; (technical) 2.1×10^{-6} mmHg at 20°C)
>Solubility: soluble in organic solvents; nearly insoluble in water (56 mg/L)

16.1.2 Odor and Warning Properties

Fenthion has a slight garlic odor.

16.2 Production and Use

Fenthion is used primarily for livestock dermal treatments and mosquito control in residential areas. Direct dermal treatments to livestock are by spot treatment and pour-on treatment; ear tags are used for dairy and beef cattle. Mosquito control (adulticide) applications are to residential areas in Florida by ultra low volume spray (aerial and ground application) and by thermal fog (ground application) (317a).

16.4 Toxic Effects

16.4.1.1 Acute Toxicity. Fenthion is an organophosphate compound that has moderate oral toxicity and oral LD_{50}s of 215 to 615 mg/kg for rats, mice, and rabbits (64a, 317a). Deaths occurred 1–4 days postdosing (315, 318). Dermal LD_{50}s are 330–963 mg/kg for rats and rabbits (45, 64, 318, 319). Deaths occurred 2–5 days postdosing. The signs of poisoning following oral or dermal exposure develop over a period of hours but then persist for several days (318, 320).

One-hour exposure to 243 mg/m^3 fenthion caused no mortality in mice, but 20 and 40% mortality in female and male rats, respectively. One-hour LC$_{50}$s for rats range from, >1125 to 2400 mg/m^3, and, 4-hour LC$_{50}$s range from 454–2400 mg/m^3 (123, 321). When 4-hour exposures were repeated for 5 days, the LC$_{50}$ was reduced to approximately 212 mg/m^3 for males and to between 55 and 212 mg/m^3 for females (321). Acute exposures of 209 mg/m^3 resulted in deaths in 2/20 female rats and ataxia and tremors in both sexes (322, 323).

Lethal doses of fenthion, when given in fractions on successive days, are lower than they are when given as a single dose, indicating cumulative toxicity. There was 75% mortality in female rats given 50 mg/kg via gavage for 5 days (about one tenth the LD$_{50}$) and 100% mortality in female rats given 100 mg/kg for 5 days (about one sixth the LD$_{50}$) (318). The dermal LD$_{50}$ for a dose applied on 5 consecutive days was 73 mg/kg/day, whereas the single-dose dermal LD$_{50}$ was 500 mg/kg. Daily intraperitoneal doses of 10 mg/kg for 60 days caused no mortality, but doses of 20 mg/kg caused 80% mortality by 30 days, doses of 40 or 50 mg/kg caused 100% mortality by 10 days, and, a dose of 100 mg/kg caused 100% mortality by 5 days (315).

When rats were given single oral doses of 0, 20, 75, or 150 mg/kg fenthion, cholinergic signs occurred in all groups that peaked at 1.5 hours and persisted for at least 24 hours (324). When rats were given single oral dose of 0, 1, 50, or 125 mg/kg (males) and 0, 1, 75, and 225 mg/kg (females) fenthion, clinical signs of acute cholinergic toxicity occurred in mid- and high-dose rats of both sexes (325). RBC and brain cholinesterase activities were inhibited in the mid- and high-dose males and in the low-, mid- and high-dose females. A no-effect level for RBC cholinesterase of 0.7 mg/kg was derived from these data.

Dogs given a single oral dose of 220 mg/kg fenthion exhibited cholinergic signs, that disappeared by 5 days after dosing (326). RBC cholinesterase was inhibited 20 minutes after exposure, increased to pre-exposure levels by 3–12 hours after exposure, but then decreased again below control levels from 24 hours to 30 days after dosing.

One of the relatively unique toxic effects associated with fenthion is acute ocular toxicity (327). Four days after single intramuscular doses of 0.005, 0.05, 0.5, or 5.0 mg/kg fenthion, electroretinograms (ERG) in rats were supernormal (*sic*) 4 days after a dose of 25 mg/kg, the ERG was normal; and 4 days after a 100-mg/kg dose, the ERG was subnormal (327–329). Retinal acetylcholinesterase was unaffected 4 days after 0.005- or 0.05-mg/kg doses but decreased after doses of 0.5–100 mg/kg (327–329). Further study indicated that supernormal ERG changes associated with 5 mg/kg peaked at 10 days and returned to normal by two months, supernormal ERG changes associated with 25 mg/kg peaked at four days then became subnormal and recovered by two months, and ERGs associated with 50 mg/kg remained subnormal for at least 66 days (327). Single acute intraperitoneal doses of 100 mg/kg fenthion were associated with retinotoxicity and characterized by inhibition of retinal cholinesterase and temporary down-regulation of muscarinic receptors in the retina (330).

Fenthion is not considered as an eye or dermal irritant and does not cause dermal sensitization (317a).

16.4.1.2 Chronic and Subchronic Toxicity. Rats fed a daily diet that contained 300 ppm fenthion for approximately 30 days showed symptoms of organophosphate intoxication (331). When rats were given diets that contained 5, 10, 20, or 250 ppm fenthion (equivalent to daily doses of about 3, 5–6, 11–12, and 100–138 mg/kg/day) for 4 weeks, symptoms of cholinergic poisoning were mild and transient and occurred only in the 250-ppm group (318). Brain cholinesterase activities in rats from all groups were significantly inhibited. Among rats fed 0.25, 0.5, 2.5, or 5.0 mg/kg/day fenthion for 12 weeks, cholinergic signs occurred in those fed 2.5 or 5 mg/kg (331a). RBC cholinesterase activity was significantly inhibited in rats fed 0.5 mg/kg or more, and tissue cholinesterase (heart, liver) was significantly inhibited in female rats fed 0.25 mg/kg/day or more and male rats fed 0.50 mg/kg/day or more. when rats were fed diets that contained 0, 2, 25, or 125 ppm fenthion for 13 weeks (equivalent to 0, 0.13–0.17, 1.6–2.2, and 8.5–12.6 mg/kg/day, clinical cholinergic signs and a dose-related decrease in RBC and brain cholinesterase activities occurred in rats given the 25- or 125-ppm diet (317a). There were no treatment-related opthalmologic findings. When mice were fed diets that contained 0, 50, and 100 ppm fenthion for approximately (*sic*) 5 weeks, no cholinergic signs were observed in any group, but marked RBC and brain cholinesterase inhibition occurred in mice at all treatment levels (333).

When rats were exposed to 0, 1, 3, or 16 mg/m^3 fenthion aerosol for 6 h/day, 5 days/week for 3 weeks, females in the two higher group exhibited behavioral disturbances (*sic*) (334). Male rats tolerated these exposures without clinical symptoms. RBC and brain cholinesterase activity were inhibited in both sexes at 3 mg/m^3.

When rabbits were dermally treated with 5, 50, or 100 mg/kg fenthion for 6 h/day, 5 days/week for 3 weeks, no cholinergic effects occurred (317a). Inhibition of RBC cholinesterase in males and inhibition of brain cholinesterase in females occurred only at the highest dose. Dermal treatment of rabbits with 150 mg/kg according to the same paradigm caused cholinergic signs and significant RBC and brain cholinesterase inhibition (317a). When shaved (one-half abraded) rabbits were dermally treated with 5 or 25 mg/kg fenthion applied as a dilute solution in Cremophor for 21 days, RBC and brain acetylcholinesterase were inhibited at 5 mg/kg/day (317a).

16.4.1.3 Pharmacokinetics, Metabolism, and Mechanisms. Fenthion is quickly absorbed in the digestive tract, and lung, and skin and is hydrolyzed either unchanged or after enzymatic oxidation. Elimination occurs via urine and feces within 3 days. In lactating cows, fenthion was readily distributed to and eliminated in milk (350).

Peak concentrations of fenthion in milk occurred 18 or 8 hours after dermal or intramuscular exposure, respectively, and by 14–20 days after treatment accounted for 1.1% and 2.2% of the dermal or intramuscular dose, respectively. Fenthion was rapidly degraded and eliminated primarily in the urine. Following topical administration of fenthion of lactating cows, 45–55% was excreted in urine, 2–2.5% was eliminated in feces, and 1.5–2% was eliminated in milk during a period of 4 weeks (350). EPA estimated that dermal absorption would be about 20% of an applied dose (317a) based on a comparison between cholinesterase inhibition no-effect levels in rabbits treated dermally or orally which were 1 mg/kg/day and 5 mg/kg/day, respectively.

The delayed and prolonged effect from a single dose of fenthion suggests that a large portion of a dose is stored and then slowly released to be metabolized (318). Fenthion (and/or fenthion metabolites) is lipid-soluble, and there is evidence that it readily accumulates in the body. In a postmortem analysis following ingestion of fenthion, greater concentrations of fenthion were detected in human organs and fat than in blood (351).

It has been proposed that the sulfoxide and sulfone of the thioether of *fenthion* are produced before P=S to P=O oxidation takes place (352, 353). Both of these compounds given orally to rats are more toxic than fenthion itself, and their effects occur more rapidly than after fenthion. However, neither compound actively inhibits acetylcholinesterase, so presumably these compounds are oxygenated to forms that readily inhibit acetylcholinesterase (318).

16.4.1.4 Reproductive and Developmental. Decreased epididymal weight, decreased fertility, increased maternal weight gain during premating, decreased weight gain during gestation, decreased pup weight gain during lactation, and inhibition of brain acetylcholinesterase occurred in rats given a diet containing 100 ppm fenthion (about 5 mg/kg/day) for two generations (317a). Cytoplasmic vacuolation of the epithelial ductal cells of the epididymis and inhibition of RBC acetylcholinesterase occurred in parents and offspring fed 14 ppm (about 0.7 mg/kg/day). Diets that contained 1 or 2 ppm fenthion had no adverse reproductive effect. When mice were given water that contained 60 ppm fenthion (delivering doses of between 9.5–10.5 mg/kg) for five generations, there was no consistent effect on mating success, although treated mice exhibited longer periods to produce first litters in the first three generations and pup survival and growth decreased in the second, third, and fourth generations (254).

A slightly higher rate of resorptions occurred among rats given 18 mg/kg/day fenthion by gavage on gestation days 6–16 (317a). Adverse developmental effects did not occur in rats given 1 or 4.2 mg/kg/day. Cholinergic signs and decreases in body weight gain also occurred in the 18-mg/kg/day dosed pregnant rats. RBC and brain acetylcholinesterase were inhibited at 1 mg/kg/day and higher. Fetal brain acetylcholinesterase was also inhibited in the high-dose group at day 20. In rabbits given 0, 1, 2.75, or 7.5 mg/kg/day fenthion by gavage on gestation days 6 through 18, a slight increase in resorptions occurred in the 2.75- and 7.5-mg/kg/group, and increases in unossified metacarpals occurred in the 7.5-mg/kg/day group. Dams exhibited "soft stools" at 2.75 mg/kg/day, a weight gain decrease at 7.5 mg/kg/day, and inhibition of brain and RBC acetylcholinesterase at both 2.75 and 7.5 mg/kg/day (317a).

16.4.1.5 Carcinogenesis. There was no cholinergic toxicity, lens opacification, clinical chemistry effects, or hematotoxicity among rhesus monkeys that received a daily oral dose of 0.02, 0.07, or 0.2 mg/kg via corn oil gavage for 2 years (317a). However, RBC acetylcholinesterase had a threshold for inhibition at 0.07 mg/kg/day (frequent inhibition at this level up to 39% for the first 3 months of the study). More consistent inhibition was noted at 0.20 mg/kg/day.

When dogs were fed diets containing 0, 2, 5, or 50 ppm fenthion (equivalent to about 0.06, 0.3, and 1.2 mg/kg/day) for 1 year, no signs of cholinergic toxicity occurred at any dose, although the 50-ppm dose caused marked inhibition of RBC and brain cholinesterase

(335). No carcinogenicity or other toxicity occurred among dogs given diets containing 0, 3, or 10 ppm fenthion for 104 weeks or among dogs given diets that contained 30 ppm from week 1 to week 64, 60 ppm from week 65 to week 67, and 60 ppm from week 68 to week 104 (317a). However, RBC cholinesterase was inhibited at 10 ppm and higher in males and at 30 ppm and higher in females; brain cholinesterase was inhibited at 30 and 60 ppm. No cholinergic signs, tumors, or other toxicity occurred in dogs given diets that contained 0, 2, 10, or 50 ppm fenthion for 1 year (336). RBC cholinesterase activity was inhibited at 10 or 50 ppm, and brain cholinesterase was inhibited at 50 ppm.

When rats were fed diets that contained 0, 2, 3, 5, 25, or 100 ppm for 1 year, the 100-ppm diet caused decreases in body weight gain, increased mortality and inhibition of RBC, brain and submaxillary gland cholinesterase; the 25-ppm diet increased mortality in female rats and inhibition of RBC, brain, and submaxillary gland cholinesterase; the 5-ppm diet inhibited RBC cholinesterase; and the 3-ppm diet had no effect (335). No gross or microscopic lesions were observed in any group hemosiderosis in the spleen of rats fed 100 ppm. No carcinogenicity, cholinergic toxicity, hematoxicity, or alterations in clinical chemistry occurred in rats given diets that contained 3, 15, or 75 ppm fenthion (about 0.2, 1.0, or 5.0 mg/kg/day) for 24 months (586). However, the 75-ppm diet decreased body weight gain in males and slightly increased mortality in both sexes, and the 15-and 75-ppm diets significantly inhibited RBC cholinesterase in both sexes. There was no evidence of carcinogenicity in rats fed diets that contained 5, 20, or 100 ppm fenthion (about 0, 0.02–0.03, 0.8–1.3, and 5.2–7.3 mg/kg/day) for 2 years (336). However, clinical cholinergic signs, retinal degeneration, and posterior subcapsular cataract formation were observed in the 100-ppm groups; and electroretinograms were flat or suppressed in females given the 20- or 100-ppm diet. Significant RBC and brain acetylcholinesterase inhibition occurred at 5 ppm.

Subnormal ERGs occurred by 3 months and disappeared completely by 1 year among rats injected subcutaneously with 50 mg/kg fenthion once every 4 days for 1 year. Significant histopathology was evident in the retina that included disappearance of the retinal pigmentary epithelial layer, the outer nodes, inner nodes and outer granular layer, and photoreceptor cells (327, 328, 329, 337). ERGs were also extinguished and retinal degeneration was extensive in rats treated subcutaneously with 50 mg/kg twice a week for 1 year (327).

There was no evidence of carcinogenicity in rats or in female mice given fenthion in the diet at 10 or 20 ppm for 103 weeks and then observed for 0–2 additional weeks (338). Doses were 0.49 and 0.98 mg/kg/day for rats and 1.3 and 2.6 mg/kg/day for mice. Some cholinergic signs were noted at both doses in rats and mice. No evidence of carcinogenicity, cholinergic toxicity, hematotoxicity, or other toxicity was seen in mice given diets that contained 0, 0.09, 0.9, 4.6, or 25 ppm fenthion (equivalent to 0, 0.03, 0.4–0.5, 1.95–2.25, and 9.42–10.23 mg/kg/day) for 102 weeks (317a). RBC and brain cholinesterase inhibition was significant at a dietary level of 25 ppm.

16.4.1.6 Genetic and Related Cellular Effects Studies. *In vitro* tests of fenthion (339–342) and *in vivo* tests in mice (343, 344) showed no mutagenic effect of fenthion. Fenthion was negative in the dominant lethal mutation assay in male mice (345, 346) and negative or weakly clastogenic at acutely toxic doses in the mouse micronucleus test (347). Among rats treated with two oral doses of 54 mg/kg fenthion (one-fourth LD$_{50}$) 21 hours apart, there was a fivefold increase in hepatic and brain lipid peroxidation and a 3.5-fold increase in hepatic single-strand DNA breaks (348). Increased numbers of single chromatid gaps and breaks were detected in human lymphocytes incubated in variable concentrations of 98% technical grade fenthion (349).

16.4.1.7 Other: Neurological, Pulmonary, Skin Sensitization. Fenthion caused neurotoxicity that lasted from 3 to 10 days when given subcutaneously at a dose of 25 mg/kg to atropinized chickens (64). The lowest lethal dose was 40 mg/kg, indicating that neurotoxicity occurred at a sublethal dose. Atropinized hens treated with 40 mg/kg (orally) or 200 mg/kg (dermally) two times 21 days apart and observed for a total of 42 days showed no behavioral or histopathological signs of delayed type neuropathy (317a). Additionally, no evidence of pathological changes in the structure of brain or peripheral nerve indicative of delayed type neuropathy were observed in hens treated orally by gavage with 0, 084, 1.7, or 3.2 mg/kg/day for 90 days (317a).

16.4.2 Human Experience

16.4.2.2 Clinical Cases. A 26-month-old male experienced abdominal pain and vomiting for 8–18 hours after dermal (and possibly oral) exposure of a flea killer that contained fenthion. About 42 hours after exposure, he experienced respiratory arrest accompanied by miosis, hyperactive bowel sounds, pulmonary edema, and diminished muscle tone and reflexes. Therapy with atropine, 2-PAM, and ventilation saved his life, and he recovered about 32 days after exposure (1). A 71-year-old farmer who was estimated to have ingested about 83 mg/kg fenthion suffered nausea, was hospitalized unconscious after 13 hours, and died 4 days later. A 40-year old individual who drank as estimated 25 mg/kg fenthion was unconscious for about 4 days but survived (1). In another attempted suicide, a 44-year-old man was thought to have ingested 310 mg/kg fenthion. He showed only a few mild signs of poisoning when hospitalized 3 hours after ingestion. However, he experienced two relapses in spite of adequate drug therapy and required tracheotomy during recovery (1).

Fenthion blood levels and plasma cholinesterase activity were followed in a 41-year-old male who ingested an unknown amount of fenthion and died after 7 days (354). Blood fenthion concentration was 0.27 mg/L on admission (20 hours after ingestion) but rapidly increased to 0.78 mg/L coincident with worsening cholinergic symptoms. Even after 5 days, fenthion blood concentration still varied within the 0.2–0.3 mg/L level. In another case of fenthion poisoning, a 43-year-old man ingested about 30 mL of Lebaycid, corresponding to about 18 g fenthion (355). Signs of cholinergic toxicity were seen 31 hours after ingestion at which time RBC acetylcholinesterase were totally inhibited. The relatively long half-life of fenthion was illustrated in another case report of a woman who experienced severe poisoning after ingesting fenthion in a suicide attempt. A fat biopsy obtained 22 days after ingestion, after recovery from an acute cholinergic phase of poisoning but before experiencing a delayed intermediate type syndrome, indicated fat residues of about 0.15 ppm fenthion. By day 31 when all symptoms had been resolved, fat residues had decreased to zero.

Neurological symptoms that ranged from occasional tingling and numbness of the hands and feet to multiple shooting pain, back pain, numbness, and generalized muscle weakness were reported by five employees in a veterinary hospital who routinely used topical applications of a 20% fenthion solution on dogs and took no precautions to avoid dermal contract (356, 357). When the use of fenthion was discontinued, the symptoms stopped.

The characteristics of an intermediate syndrome (respiratory insufficiency, weakness of muscles, innervated by cranial nerves and weakness of proximal limb muscles) were described in four cases of fenthion poisoning (two of the people died) that occurred 2–6 days following resolution of an acute cholinergic crises and which lasted 5–18 days (358). Unusual transient dystonic movements also occurred in two of these cases. Symptoms that comprised an "intermediate syndrome" (i.e., weakness of external ocular, facial, neck, proximal limb, and respiratory muscles) and extrapyramidal signs (dystonia, rest tremor, cog-wheel rigidity, and choreoathetosis), occurred from 4 to 40 days after acute poisoning with fenthion, and resolved after 1 to 4 weeks in survivors were described in six cases of people who ingested 15–60 mL of fenthion (359).

Several publications report an association between an increased incidence of adverse visual effects (myopia and a visual disease syndrome termed "Saku" disease) and agricultural use of organophosphates, including fenthion, in Japan. The recently reviewed studies vary in quality and usefulness, but nevertheless suggest that the association is probable (327). However, only one study is available that suggests a specific link between fenthion exposure and ocular toxicity. In this study, neurological function, visual acuity, refraction, color vision, and the condition of the fundus oculi were assessed in 79 individuals who worked 5–6 h/day spraying an aqueous suspension of fenthion and were compared to equivalent observations of 100 control subjects (327). Macular changes (hypopigmentation, irregularity of background pigmentation, and dull foveal relex) were evident in 15 (19%) workers and in 3 (3%) control subjects. Other visual symptoms reported in workers who had macular change included visual impairment, reduced visual acuity, abnormal color vision, and constriction of visual fields. Pathological myopia (associated with "Saku") disease was not observed.

Symptoms reported among thirty-one workers who sprayed aqueous suspensions of fenthion (100 mg in 100 L water) by a hand-operated sprayer for 5–6 hours/day, 6 days/week included headache (56%) giddiness (44%), eye irritation (20%), anorexia and paresthesia (11%), but no signs specific for cholinergic toxicity (360). On neurological examination, two subjects had loss of ankle reflex, and one had coarse tremors. Additionally, subtle subclinical effects on psychometric tests and event related potentials were observed. However, no estimate of exposure was provided.

Fenthion administered to human volunteers at dose levels of 0.02 or 0.07 mg/kg daily for up to 4 weeks produced no physical signs or symptoms; no alterations in clinical chemistry, hematology or urinalysis; and no inhibition of RBC cholinesterase activity (361).

16.5 Standards, Regulations, or Guidelines of Exposure

Fenthion has been reregistered for use by the EPA. The ACGIH TLV for fenthion is 0.2 mg/m^3 with a skin notation (154). There is no OSHA PEL-TWA or NIOSH REL-TWA for fenthion. Other countries have Occupational Exposure Limits of 0.2 mg/m^3 for fenthion (Australia, Belgium, Germany, and Japan); others have OELs of 0.1 mg/m^3 (Denmark, The Netherlands, and Switzerland).

17.0 Fonofos

17.0.1 CAS Number: [944-22-9]

17.0.2 Synonyms: (*O*-ethyl-*S*-phenyl ethylphosphonodithioate; *O*-ethyl *S*-phenyl ethyldithiophosphonate; Ethyl *S*-phenylethylphosphonothiolthionate; Diphonate; Fonophos; *O*-ethyl *S*-phenylethylphosphonothiolothionate; ethylphosphonodithioic acid *O*-ethyl *S*-phenyl ester; Dyfonate II; N-2790; *O*-ethyl *S*-phenylethylphosphonodithioate; Stauffer N-2790; ethyl *s*-phenyl ethylphosphonodithioate; Dyphonate

17.0.3 Trade Names: Dyfonate®; Difonate; Dyphonate

17.0.4 Molecular Weight: 246.32

17.0.5 Molecular Formula: C$_{10}$H$_{15}$OPS$_2$

17.0.6 Molecular Structure:

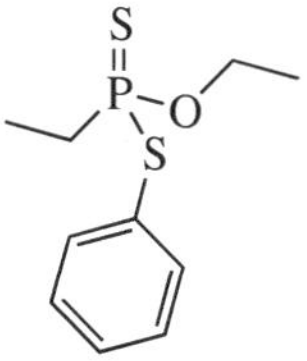

17.1 Chemical and Physical Properties

Fonofos is a light yellow liquid.

    Specific gravity: 1.154 at 20°C
    Boiling point: 130°C at 0.1 mmHg
    Flash point: >94°C, closed cup
    Vapor pressure: 0.00021 torr at 25°C
    Solubility: very slightly soluble in water (13 mg/L at 20°C); miscible with organic solvents such as kerosene, xylene, and isobutyl ketone

17.1.2 Odor and Warning Properties

Pungent, mercaptan-like odor.

17.2 Production and Use

Fonofos was introduced in 1967 for use as a soil insecticide to control corn borers, rootworms, cutworms, symphylans (garden centipedes), wireworms, and other soil and foliar pests. It is formulated as an emulsifiable concentrate and as granules. Registration of fonofos in the U.S. has been voluntarily cancelled.

17.4 Toxic Effects

17.4.1.1 Acute Toxicity. Fonofos has high oral toxicity, and oral LD$_{50}$s are 3–18.5 mg/kg for rats (64a). Cholinergic signs and death occur rapidly after lethal exposure (364). The oral LD$_{50}$ for a racemic mixture of fonofos was 14 mg/kg, which was less than the oral LD$_{50}$ of 32 mg/kg for the (S)$_p$ isomer but greater than the oral LD$_{50}$ of 9.5 mg/kg for the (R)$_p$ isomer (362). The acute i.p. LD$_{50}$ for the fonofos racemic mixture was 4.8 mg/kg, suggesting an overall detoxification role of first-pass hepatic metabolism. The dermal LD$_{50}$ for rats is 147 mg/kg and for guinea pigs is 278 mg/kg (363). Application of 0.5 ml undiluted fonofos to the skin of rabbits caused on dermal irritation, but all animals died within 24 hours (154). The 4-hour LC$_{50}$ of fonofos for rats is 900 mg/m^3, and the 1-hour LC$_{50}$ is 460 mg/m^3 (363).

Technical fonofos (0.1 mL) instilled into the eye of albino rabbits caused death during the first 24 hours after administration of the chemical, but local eye irritation was negligible (363).

17.4.1.2 Chronic and Subchronic Toxicity. Dietary feeding of fonofos of groups of dogs for 14 weeks indicated a no-observed effect level of 8 ppm (approximately 0.2 mg/kg) (120a).

17.4.1.3 Pharmacokinetic, Metabolism, and Mechanisms. Fonofos is well absorbed orally. After a single oral dose of fonofos, 98% was excreted in the urine (91%) and feces (7.4%) of rats within 96 hours. Prior exposure to fonofos did not change in excretion pattern. Tissue residues were very small and had virtually disappeared by day 16 (365). Similar results were obtained in white mice given the enantiomer of fonofos orally (365a). Fifty percent of the most toxic isomer and 95% of the less toxic isomer where eliminated within 96 hours.

Fonofos is first oxidized by microsomal enzymes to the oxon and also, by a different reaction, to O-ethyl-ethylphosphonothioic acid (ETP) and thiophenol. The oxon, in turn, is hydrolyzed to O-ethyl-ethylphosphoric acid (EOP) and thiophenol (366, 367). The oxon is not found *in vivo* due to its rapid hydrolysis (368). The other metabolites were much less toxic than the parent compound.

17.4.1.4 Reproductive and Developmental. No adverse effects were noted on overall reproductive performance at either level among the parental animals or on the numbers, well-being, or integrity of the offspring among rats fed 10 or 31.6 ppm fonofos for three generations (95).

EPA noted a fetotoxic no-observed-effect level of 1.58 mg/kg/day in rats and a fetotoxic no-observed-effect level and lowest observed effect level of 2 and 6 mg/kg/day, respectively, in mice (95, 364).

17.4.1.5 Carcinogenesis. There was no evidence of carcinogenicity when fonofos was given in the diet to rats for 105 weeks (95). This study produced a no-observed-effect level of 10 ppm (0.5 mg/kg/day) for brain acetylcholinesterase inhibition and a lowest observed effect level of 1.58 mg/kg/day based on RBC cholinesterase inhibition (95).

There was no evidence of carcinogenicity in dogs that were fed 0, 0.2, 1.5, and 12 mg/kg/day fonofos via their diet for 2 years (95). No compound-related effects were observed at 0.2 mg/kg/day; moderate (*sic*) inhibition of RBC cholinesterase, increased liver weight, tremors, lacrimation, and salivation occurred at 1.5 mg/kg/day; and, these symptoms plus microscopic lesions of the small intestines and liver occurred at 12 mg/kg/day.

17.4.2 Human Experience

Four members of a family were poisoned by pancakes mistakenly made with fonofos instead of flour; one family member died (368a). Soon after eating the pancakes, one family member developed nausea, vomiting, salivation, and sweating and was taken to a hospital where she suffered cardiorespiratory arrest. She was resuscitated and transferred to a medical center where she was artificially ventilated, exhibited muscle fasciculations, low blood pressure, pinpoint pupils, and profuse salivary and bronchial secretions. Treatment continued and she was eventually released 2 months later. Gallo and Lawryk (1) noted that, although not reported in the original paper, three other members of the family were poisoned by the pancakes, and one of them died. A fifth family member who may have mixed the batter, but did not eat pancakes, remained well.

17.5 Standards, Regulations, or Guidelines of Exposure

The registration of fonofos has been voluntarily cancelled in the United States and tolerances are being revoked. The ACGIH TLV for fonofos is 0.1 mg/m^3 with a skin notation (154). The NIOSH REL-TWA for fonophos is 0.1 mg/m^3 with a skin notation. Most other countries also have Occupational Exposure Limits of 0.1 mg/m^3 for fonophos (Australia, Belgium, France, and Switzerland).

18.0 Malathion

18.0.1 CAS Number: *[121-75-5]*

18.0.2 Synonyms:
(*O,O*-Dimethyl dithiophosphate of diethyl mercaptosuccinate; *O,O*-dimethyl-*S*-(1,2-dicarbethoxyethyl)-phosphorodithioate; diethyl [(dimethoxyphosphinothioyl)thio]butanedioate; Maldison; *O,O*-dimethyl phosphorodithioate ester of diethyl mercaptosuccinate; [(Dimethoxyphosphinothioyl)thio]butanedioic acid diethyl ester; mercaptosuccinic acid diethyl ester *S*-ester with *O,O*-dimethyl phosphorothioate; insecticide no. 4049; phosphothion; Cythion; dicarboethoxyethyl *O,O*-dimethyl phosphorodithioate; *O,O*-dimethyl *S*-(1,2-dicarbethoxyethyl) dithiophosphate; *O,O*-dimethyl *S*-(1,2-dicarbethoxyethyl)phosphorodithioate; diethyl mercaptosuccinate, *O,O*-dimethyl phosphorodithioate; 1,2-di(ethoxycarbonyl)ethyl *O,O*-dimethyl phosphorodithioate; chemathion; emmatos; karbofos, kop-thion; malagran; malamar; MLT; sadofos; *S*-(1,2-bis(carbethoxy)ethyl) *O,O*-dimethyl dithiophosphate; *S*-1,2-bis(ethoxycarbonyl)ethyl *O,O*-dimethyl dithiophosphate; calmathion; carbetox; carbethoxy malathion; carbetovur; celthion; cinexan; compound 4049; detmol ma; *S*-(1,2-di(ethoxycarbonyl)ethyl) dimethylphosphorothiolothionate; diethyl (dimethoxyphosphinothioylthio)succinate; diethyl mercaptosuccinate, *O,O*-dimethyl dithiophosphate, *S*-ester; diethyl mercaptosuccinate,

O,O-dimethyl thiophosphate; diethyl mercaptosuccinate *S*-ester with *O,O*-dimethylphosphorodithioate; diethyl mercapatosuccinic acid *O,O*-dimethyl phosphorodithioate; *O,O*-dimethyl-*S*(1,2-bis(ethoxycarbonyl)ethyl)dithiophosphate; *O,O*-dimethyl-*S*(1,2-dicarbethoxyethyl) thiothionophosphate; *O,O*-dimethyl *S*-1,2-di(ethoxycarbonyl)ethyl phosphorodithioate; *O,O*-dimethyldithiophosphate diethyl mercaptosuccinate; phosphorodithioic acid, *O,O*-dimethyl ester, *S*-ester with diethyl mercaptosuccinate; Malaspray; dicarbethoxyethyl-*O,O*-dimethyldithiophosphate; diethyl mercaptosuccinic acid, *S*-ester of *O,O*-dimethyl phosphorodithioate; dimethyl dithiophosphate of diethyl mercaptosuccinate; dimethyl phosphorodithioate of diethyl mercaptosuccinate; Ethiolacar; Etiol; Cleensheen; Lice Rid

18.0.3 Trade Names: Carbophos; Extermathion; Forthion; Fosfothion; Fyfanon; Malacide;Malatox; Maldison; Mercaptothion

18.0.4 Molecular Weight: 330.36

18.0.5 Molecular Formula: $C_{10}H_{19}O_6PS_2$

18.0.6 Molecular Structure:

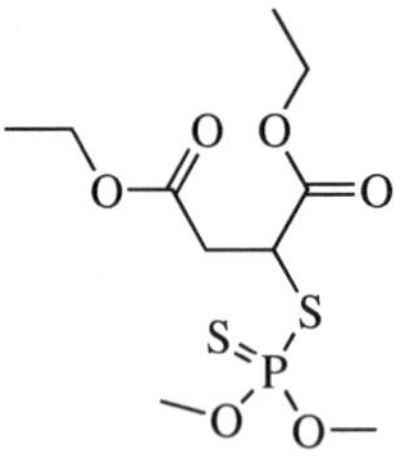

18.1 Chemical and Physical Properties

Malathion is a noncombustible, yellow to deep brown liquid. Malathion is rapidly hydrolyzed at pH > 7 or > 5 and is stable in aqueous buffered pH 5.6 solutions. It can corrode iron, steel, tinplate, lead, and copper. It is a solid below 37°C.

- Specific gravity: 1.23 at 25°C
- Melting point: 2.85–3.7°C
- Density: 1.23 at 25°C
- Boiling Point: 156-157°C at 0.7 torr
- Solubility: slightly soluble in water (145 ppm); completely soluble in alcohols, esters, ketones, ethers, aromatic solvents, and hexane, limited solubility in petroleum oils

18.1.2 Odor and Warning Properties

A mild skunk-like odor.

18.2 Production and Use

Malathion is a broad-spectrum insecticide and one of the earliest organophosphate insecticides developed. It is used to control sucking and chewing insects on fruits, vegetables, and ornamental plants. It is also used to control mosquitos, flies, household insects, animal parasites, and head and body lice.

18.4 Toxic Effects

18.4.1.1 Acute Toxicity. Malathion is an organophosphate compound that has low acute toxicity and rat oral LD_{50}s are generally in the 1,000–12,500 mg/kg range, depending on the formulation and gender tested (64, 268). An intraperitoneal LD_{50} of 750 mg/kg was reported, showing that route of exposure markedly impacts toxicity (369). This was also illustrated in a study where oral and intraperitoneal LD_{50}s in mice were 1025 and 420 mg/kg, respectively (370). Following lethal oral doses, maximal symptoms are often delayed for several hours, but death generally occurred within 2 days after poisoning (369). Technical grade malathion is more toxic than the pure product. Oral LD_{50}s for rats of 65% technical grade malathion, 90% technical grade malathion and 99% undiluted malathion were 369, 1156, and 5843 mg/kg, respectively (370). Commercial preparations of malathion vary in their toxicity due to the presence of impurities which bind to and inhibit acetylcholinesterase and also the carboxylesterase that detoxifies malathion (371–373).

A precise dermal LD_{50} for malathion has not been identified. The dermal LD_{50}s for rats for a 57% emulsifiable concentrate was greater than 4444 mg/kg (64a, 268). In rabbits, single dermal doses of up to 4 ml/kg 90% technical malathion or the 25% wettable powder caused no overt signs of toxicity, except for temporary irritation at the site of application (370). However, mortality occurred after four daily applications of 0.5 or 1 mL/kg/day and after two daily applications of 2 mL/kg. In each case, symptomology was characteristic of acute organophosphate poisoning (370).

An acute inhalation LC_{50} for malathion is not available, although a 4-hour LC_{50} of > 5200 mg/m^3 has been reported (155). However, an intravenous LD_{50} of 50 mg/kg was reported in rats (373a) which is at least 20 times smaller than reported oral LD_{50}s. In dogs, an intravenous dose of 100 mg/kg resulted in immediate and profuse salivation and tremors (370). In a 5-hour exposure of mice of 7 mg/L 95% technical grade malathion (7000 mg/m^3), there were no signs of cholinergic toxicity and no deaths (203).

Overt toxicity has rarely been identified following single, sublethal exposures to malathion, although they cause RBC and brain cholinesterase inhibition. RBC cholinesterase activity maximally decreased in rats 45 minutes after an intraperitoneal dose of 300 mg/kg (370). Repeated daily intraperitoneal doses of 300 mg/kg produced a cumulative inhibitory action on cholinesterase activity of brain, submaxillary gland, and serum in rats so that after 5 days, activities were about 30, 50, and 25% of control, respectively (369). Rats could not tolerate longer periods of treatment. Repeated administration of 200 mg/kg also progressively decreased cholinesterase activity of the brain and submaxillary gland.

An important characteristic of malathion acute toxicity is its potentiation by other organophosphates. When malathion and EPN were administered to rats separately, LD_{50}s were 1400 and 65 mg/kg, respectively. However, when malathion and EPN were administered simultaneously at a ratio of about 25:1, oral LD_{50}s were reduced to 167 and 7 mg/kg, respectively (184). The onset of symptoms for individual compounds was slow, and death usually occurred several hours after administration. However, when given together at or near the LD$_{50}$, symptoms developed much more rapidly, and death usually occurred within 1 hour. In dogs, oral doses of 2000 or 4000 mg/kg malathion alone were not fatal; however, when EPN was given simultaneously (2 or 5 mg/kg) with malathion, doses of 50, 100, and 200 mg/kg were lethal (184). Potentiation of malathion toxicity is due to the inhibition of carboxyesterase by EPN (and other organophosphates), an enzyme important in detoxifying of malathion (as well as the more toxic metabolic product malathion, malaoxon) (165, 374).

Evidence that malathion causes a paralytic type of neurotoxicity was observed in hens given 100 mg/kg or more malathion subcutaneously and observed for up to 30 days (190). Neurotoxicity, reflected by the occurrence of leg weakness that lasted for 4–14 days, occurred in atropinized chickens given single, subcutaneous doses of 100 mg/kg malathion (64a). Atropinized rats were given 600, 1000, or 2000 mg/kg malathion (88% pure) via oral gavage and were observed for signs of delayed neuropathy at 14–21 days; only the 2000-mg/kg dosed rats showed signs of gait alterations indicative of delayed neuropathy (375).

18.4.1.2 Chronic and Subchronic Toxicity. Mortality was 20, 60 and 100%, respectively, among rats given 100, 200 or 300 mg/kg/day malathion intraperitoneally for 60 days (369). Rats fed lentil diets that contained 0.95 or 6.51 ppm malathion (equivalent to about 0.06 and 0.44 mg/kg/day) exhibited no signs of cholinergic toxicity (376). However, both exposure levels were associated with increased blood urea nitrogen and increased white blood cells, and the 0.44-mg/kg/day group had decreased serum cholinesterase activity. Brain and RBC cholinesterase activity were unaffected. Similarly, mice fed soybean seeds contaminated with 7 ppm malathion of 75 days exhibited no signs of cholinergic toxicity and no effect on RBC cholinesterase activity (376).

There is evidence that malathion is a sensitzer. In a field study, 3% of workers involved in spraying malathion for mosquito control and 5% of poultry ranchers who had used malathion for at least one season showed positive reactions when malathion (95%) was applied under adhesive tape to the skin of the upper arm and allowed to remain in place for 2 days (406). In another study of ten subjects who had reported skin reactions to malathion bait, none had a reaction to patch testing of malathion. Only one exhibited a positive reaction to the bait and another had irritant reactions to both bait and malathion (407).

18.4.1.3 Pharmacokinetics, Metabolism, and Mechanisms. Approximately 90% of ingested malathion is relatively slowly absorbed and excreted in urine (1, 106, 389). Six hours after an oral dose of malathion, 75% remained in the stomach, 8% was in the small intestine, and 7% was in saliva. Thus a very small amount had been absorbed. Similar results were obtained in rats fed a lentil diet contaminated with malathion (376). After 48 hours, about 35% of the dose was excreted in urine; 45% in feces, 1.5% in exhaled air, and tissues contained about 9%. In rats given malathion orally, it is distributed to blood, adipose tissue, muscle, liver, and brain and then eliminated from these tissues at half-lives of 1.4, 2.4, 3.7, 19.4, and 17.6 days, respectively (101, 388).

Absorption of malathion via inhalation is expected to be high and elimination via urine extremely rapid and nearly complete based on pharmacokinetic studies following intravenous dosing. Among male volunteers, approximately 90% of an intravenous dose of malathion was excreted in urine within 5 days and the an elimination half-life was 3 hours (66). Thirty minutes following intravenous administration of malathion to rats most had been distributed to tissues; liver, small intestine, lung, urinary tract, and kidney accumulated extremely high levels (389). Distribution pattern at 1 and 2 hours were similar.

In mice, 25% of a 1-mg/kg dermal dose of malathion was absorbed within 60 minutes, and 67% of the dose was absorbed by 8 hours (213). Distribution of malathion was equal in liver and blood; slightly greater in urine, feces, and expired air; and greatest in the rest of the carcass at 60 minutes. By 8 hours, 30% of the dose had been excreted, 30% remained in the carcass, and about 2.5% was distributed to lungs, kidney, bladder, stomach, intestine, liver, and blood. Whole body autoradiography of rats treated dermally with a single dose of malathion indicated that after 8 hours most of the dose was equally distributed between the application site (28%), remaining skin (29%), and the small intestine and urinary bladder (23%) (389).

Among human volunteers, about 8% of an applied malathion dermal dose of 4 µg/cm^2 was absorbed by 120 hours after application to the ventral forearm (66). *In vitro* absorption by human skin was about 9% from an aqueous ethanol solution and about 0.6 to 4% from cotton sheets to which malathion had been added (389a).

Malathion is either oxidized in the liver to malaoxon by microsomal cytochrome P450 enzymes or to monoacids by a microsomal carboxyesterase. Malaoxon is the toxic metabolite of malathion that binds and inhibits acetylcholinesterase and leads to the typical cholinergic sequelae associated with organophosphate poisoning. The other products of malathion and malaoxon metabolism are detoxification products. Malaoxon is also subject to hydrolysis and carboxyesterase (105, 390, 391). There is evidence that the linkage at P–S is enzymatically broken by another cytosolic esterase as well (*A*-esterase) and forms *O,O*-dimethyl phosphorothioate (390). There is some evidence the monoacids can then be S-methylated, and that the C–S bond of either malaoxon or malathion can be further hydrolyzed (392, 393).

The rate of malathion desulfuration is not as critical as the rate of hydrolysis by carboxyesterase or hydrolysis of the C–S bond, both of which are detoxication reactions, in determining the toxicity of malathion (393). Female rats that are more sensitive to malathion toxicity have relatively lower levels of liver carboxylesterase activity (394). Further, pretreatment of mice with the microsomal inducer phenobarbital failed to decrease the mouse intraperitoneal LD$_{50}$ (i.e., failed to increase the toxicity) of malathion, whereas it did decrease the i.p. LD$_{50}$ of dimethoate which is not subjected to detoxification reactions as extensive as those for malathion (391). Additionally, oral LD$_{50}$s for rats and three strains of mice were significantly correlated with their carboxyesterase titer in liver and plasma (371). Carboxyesterase activity (also termed hydrolase B) occurs in mouse, rat, and human liver microsomal preparations and it has been shown, is markedly inhibited by a common impurity of malathion, isomalathion (372). This inhibition is associated with the potentiation of malathion toxicity (395a). Finally, it is believed that malathion and/or malaoxon inactivate carboxyesterase, so that additional dose of malathion are less effectively detoxified. This contributes to the phenomenon whereby pretreatment with malathion increases the toxicity of subsequent doses (374).

No polymorphism of carboxylesterase was detected in 12 human livers, but the range of individual activity toward malathion was about 10-fold (395). In humans, carboxyesterase is expressed only, in the liver, whereas in rodents, carboxyesterase is expressed in the serum and liver. This could provide a basis for a greater sensitivity of humans to malathion compared to rodents, although that has not been adequately explored.

18.4.1.4 Reproductive and Developmental. Viability and growth decreased in fetuses of rats fed diets that contained about 4000 ppm technical grade (95%) malathion (240 mg/kg) for 5 months (377). Pregnant rats on protein-deficient or protein-adequate diets given 500 mg/kg/day malathion orally on gestation days 6, 10, and 14 showed decreased maternal weight gain, a decreased number of implantations and live fetuses, and decreased brain acetylcholinesterase activity (392a). Protein deficiency enhanced an effect of malathion on fetal crown to rump and tail length, fetal body weight, and retardation of skeletal ossification. No detectable increases in the number of resorptions, fetal size, and external or visceral anomalies occurred in rabbits given malathion (70%) at 100 mg/kg by gavage on gestation days 7 through 12 (378).

Pregnant rats given 600 or 900 mg/kg malathion via intraperitoneal injection on gestation day 11 showed signs of maternal toxicity but did not produce malformed or low birth weight pups (379). Malathion caused a dose-dependent inhibition in brain acetylcholinesterase activity in dams and pups among rats given 138, 276, and 827 mg/kg/day malathion via intraperitoneal injection on gestation days 6 through 13 (380).

18.4.1.5 Carcinogenesis. No cholinergic toxicity or inhibition of whole blood cholinesterase activity occurred in rats fed diets that contained 100 or 200 ppm (2.5 or 6.25 mg/kg/day) for 8 weeks (194). However, when the diet contained 25 ppm (0.625 mg/kg/day) EPN, as well as 500 ppm malathion, whole blood cholinesterase was significantly inhibited. No cholinergic signs occurred in dogs fed diets that contained 25, 100, or 250 ppm malathion for 12 weeks, but RBC cholinesterase was significantly inhibited in dogs fed 250 ppm (194). In rats fed 100, 1000, or 10,000 ppm 65% technical malathion, 90% technical malathion, or 99%+ malathion in the diet (equivalent to about 6, 60, or 600 mg/kg/day) for 2 years, cholinergic toxicity was not described, but among rats given the more toxic 65% and 90% technical products, RBC and brain cholinesterase activities were inhibited in the 1000- and 10,000-ppm groups and were unaffected in the 100-ppm group (370).

There was no evidence of carcinogenicity in rats given diets that contained 4700 or 8150 ppm (about 270 mg/kg and 466 mg/kg) for 80 weeks and observed for an additional 33 weeks, in rats given diets that contained 2000 or 4000 ppm malathion (about 115 mg/kg/day and 230 mg/kg/day) for 103 weeks, or in rats given diets that contained 500 or 1000 ppm malaoxon for 103 week (482). There was no evidence of carcinogenicity in mice given diets that contained 8,000 or 16,000 malathion (about 800 and 1600 mg/kg/day) for 80 weeks and observed for an additional 14 or 15 weeks (482). During the second year, clinical signs including alopecia, rough and discolored coats, poor food consumption, hyperexcitability, and abdominal distension occurred with increasing frequency in dosed animals. A few animals appeared hyporeactive, and some had hunched appearances. During weeks 71 to 79, five high-dose females exhibited generalized body tremors.

18.4.1.6 Genetic and Related Cellular Effect Studies.
Mammalian *in vivo* and *in vitro* studies of technical or commercial grade malathion and its metabolite malaoxon show a pattern of induced chromosomal damage, as measured by increased chromosomal aberrations, sister chromatid exchanges, and micronuclei (381, 382), as well as increased mutations (383, 384). Purified (>99%) malathion gave weak or negative results in cytogenetic assays. Technical malathion was generally negative in mammalian gene mutation assay, but malaoxon was positive. Studies of human lymphocytes indicated that *in vitro* incubation with malathion increased the frequency of DNA mutations (383, 384). Malathion were positive in a modified SOS microplate assay in which the induction of β-galactosidase in *E-coli* PQ37 was used as a qualitative measure of genotoxic activity (385).

Dermal exposure caused cytogenetic damage in test animals at doses near those that produce positive results by intraperitioneal injection (381). Workers involved in a Mediterranean fruit fly aerial spraying eradication program in California in the early 1990's exhibited no increase in micronuclei formation or mutation frequency assessed by the glycophorin A (GPA) assay (386, 387). However, a significant increase in micronuclei occurred in whole blood cultures and in human lymphocytes at dose levels that also caused cytotoxicity and strong inhibition of proliferation (386, 387).

18.4.2 Human Experience

18.4.2.2 Clinical Cases.
There have been many reports of organophosphate poisoning from the intentional or accidental ingestion of malathion (1). There are a few reports of poisoning from dermal exposure of children that occurred when their hair was washed with a 50% solution of malathion to eliminate lice. Poisoning following inhalation exposures have been reported but were probably confounded by simultaneous ingestion (1). Cholinergic symptoms appear rapidly from minutes to 3 hours following ingestion, but may be delayed by 12–14 hours after dermal exposure.

Estimates of lethal doses of malathion obtained from case study reports range from 68 to 3855 mg/kg (1). Actual estimates of ingestion obtained from reports of poisonings where the quantity was approximately known indicated that a life-threatening dose is 500–1000 mg/kg (371). Doses associated with serious but sublethal toxicity have ranged from approximately <1 mg/kg (in a two-year-old boy) to 357 mg/kg (396).

Large-scale occupational poisoning by malathion occurred among sprayers, mixers, and supervisors involved in a mosquito control program that used a water-wettable powder formulation of malathion (397). An estimated 2810 cases of poisoning including five deaths were recorded during the peak month of the epidemic. Symptoms and clinical histories were consistent with organophosphate intoxication and the most severe illness were associated with the formulation that contained high levels of isomalathion and other contaminants. Exposures were primarily dermal and estimates ranged from 1 to 200 μg/cm^2. Respiratory exposure was estimated to considered less important — peak air concentrations measured during spraying were about 1.5 mg/m^3. RBC cholinesterase decreased at the end of the workday in workers, who used the two formulations that contained 2–3% isomalathion, but not in workers who used the uncontaminated formulation.

An intermediate-type syndrome characterized by weakness of proximal limb muscles, cranial nerve palsies, and respiratory depression has sometimes followed initial acute symptoms associated with malathion poisoning (1). Typical of a case that showed such delayed symptoms is that of a woman who experienced a typical cholinergic crisis about 1 hour after ingesting 300 mL malathion. She was successively treated with atropine and pralidoxime iodide but suffered a "relapse" characterized by diaphoresis, miosis, mental confusion, and respiratory dysfunction 47 hours later (398). Measurements of plasma and RBC cholinesterase activities in this patient suggested that they were 26–38% and 28–40% of normal, respectively for the 15 days following malathion ingestion. Initial plasma concentrations were quite high in this patient, and an elimination half-life of 7.6 hours was determined suggesting to the authors that because of its lipophilicity, malathion may have been retained in adipose tissues and then was released more slowly into circulation, causing the delayed effects.

No signs of cholinergic toxicity have been reported as a result of the aerial spraying of malathion to control mosquitos or the Mediterranean fruit fly (399, 400, 405). However, chronic dose rates calculated from spraying to eradicate the Mediterranean fruit fly in California were quite low — about 0.001–0.246 mg/kg/day via dermal contact, about 0.01 to 0.1 µg/kg/day via inhalation, and about 0.030–0.080 mg/kg/day via ingestion of backyard vegetables (401). Average and upper bound estimates of environmental levels of malathion during aerial spraying derived from mass deposition rates and air monitoring in affected areas were 0.09 and 0.2 µg/m^3 in air immediately after spraying, 22 and 52 mg/m^2 on outdoor surfaces, 3.8 and 9.6 µg/g in plants, and 1.5 and 3.5 µg/g in soil. Estimated average and upper bound estimates of malaoxon levels were 0.04 to 0.110 µg/m^3 in air, 0.15 and 0.46 mg/m^2 on outdoor surfaces, 0.03 and 0.08 µg/g in plants, and, 0.01 and 0.03 µg/g in soil (402).

To test the safety of malathion in controlling lice, the bodies and clothing of thirty-nine men were dusted five times a week for 8–16 weeks with talcum powder that contained 0, 1%, 5%, or 10% malathion. Complaints about odor and skin irritation were roughly proportional to dosage. No change in blood cholinesterase activity was found, except occasionally with 10% powder (403).

Exposure to malathion via dermal contact and inhalation was estimated in six entomologists who were exposed to windborne aerosols of malathion drifting downwind from generators as they moved along pastures where they were working (404). Measures of respiratory exposure were made using midget impingers at the breathing zones of the entomologists, by measuring malathion in absorbent filters in cartridge type respirators, and by measuring atmospheric concentrations. Dermal exposures were estimated from malathion contained in cotton gloves and absorbent cellulose surface patches placed in various locations on the skin. Estimates of total dermal exposures during a two week period ranged from 0.5–1.3 mg/kg (0.05–0.13 mg/kg/day assuming two 5-day weeks), and estimates of respiratory exposures ranged from 0.2–0.3 mg/kg (0.02–0.03 mg/kg/day assuming two 5-day weeks). Average air concentrations of malathion were 0.5 to 4 mg/m^3 although they ranged as high as 56 mg/m^3. None of the men exposed during the study exhibited any cholinergic clinical signs or inhibition of plasma or RBC cholinesterase activity.

18.4.2.3.4 Reproductive and Developmental. No association was found between malathion exposure and spontaneous abortion, intrauterine growth retardation, stillbirth, and most categories of congenital anomalies in a cohort of 7450 pregnancies potentially exposed to malathion applied aerially to control the Mediterranean fruit fly (405).

Five men given gelatin capsules delivering 8, 16, or 24 mg/malathion each day (equivalent to roughly 0.11, 0.23, or 0.34 mg/kg/day) for 32, 47 or 56 days, respectively, experienced no cholinergic toxicity (294). RBC cholinesterase activity was not affected in either the 8- or 16-mg/day group, but it was inhibited by exposure to 24 mg/day.

No adverse symptoms or inhibition of RBC cholinesterase occurred among volunteers exposed to dust containing 1, 5, or 10% malathion five times/week for 8 weeks (403). The estimated maximum dosage received was about 224 mg/day or 3.2 mg/kg/day.

Groups of four men received two 1-hour exposures to aerosols that contained variable amounts of malathion in unventilated rooms on each of 42 consecutive days (408). Based on the decreased weight of the aerosol containers used to spray the exposure rooms, malathion air concentrations were estimated at 0.15 g, 0.60 g, or 2.4 g of malathion per 1000 ft^3 — which is equivalent to about 5, 21, and 85 mg/m^3 malathion. No cholinergic toxicity occurred during the study, nor was RBC cholinesterase activity reliably depressed in any of the exposed subjects.

18.5 Standards, Regulations, or Guidelines of Exposure

Malathion is undergoing reregistration by the EPA (154). The ACGIH TLV for malathion is 10 mg/m^3 with a skin notation. The OSHA PEL-TWA is 15 mg/m^3 total dust with a skin notation. The NIOSH REL-TWA is 10 mg/m^3 with a skin notation. Other countries have Occupational Exposure Limits of 15 mg/m^3 (Austria, Germany, The Philippines, Thailand, and Turkey), 10 mg/m^3 (Egypt, Australia, Belgium, Finland, France, Japan, The Netherlands, and Switzerland), or 5 mg/m^3 (Denmark). In Poland the TWA is 1 mg/m^3, STEL 10 mg/m^3 (1998), and in Russia the STEL is 0.5 mg/m^3 (Jan. 93).

19.0 Methyl Parathion

19.0.1 CAS Number: [298-00-0]

19.0.2 Synonyms: Axophos; *O,O*-Dimethyl *O*-(*p*-nitrophenyl) phosphorothioate; Parathion-methyl; phosphorothioic acid *O,O*-dimethyl *O*-(4-nitrophenyl) ester; *O,O*-dimethyl *O-p*-nitrophenyl thiophosphate; dimethyl parathion; Metaphos; E 601; Penncap-M; Metafos; dimethyl 4-nitrophenyl phosphorothionate; dimethyl *p*-nitrophenyl monothiophosphate; *O,O*-dimethyl O-(*p*-nitrophenyl) thionophosphate; dimethyl *p*-nitrophenyl thionophosphate; *p*-nitrophenyldimethylthionophosphate; Dalif; nitrox 80; nitrox; wofatox; bay e-601; folidol-80; Metaphor; parathion methyl homolog; dimethyl *O-p*-nitrophenyl thiophosphate

19.0.3 Trade Names: Bladan M®; Metron®; Nitrox®; Dalf®; Diithion 63; Ketokio 52; Foidol-M®; Metacide®; Metron®; Seis-Tres 6-3; Metaspray 5E; Paraspray 6-3

19.0.4 Molecular Weight: 263.23

19.0.5 Molecular Formula: $C_8H_{10}NO_5PS$

19.0.6 Molecular Structure:

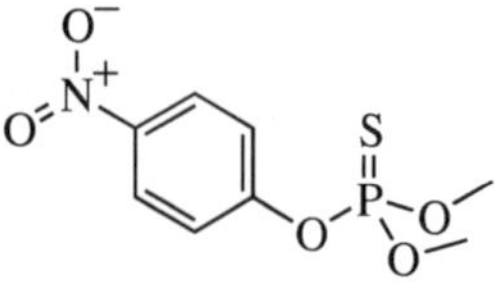

19.1 Chemical and Physical Properties

Pure methyl parathion is a white crystalline solid. The technical grade product contains about 80% methyl parathion and is a brown liquid. Methyl parathion is hydrolyzed by alkaline materials; may react with strong oxidizers; decomposes rapidly above 100°C (creating an explosion hazard) and may release fumes of dimethyl sulfide, sulfur dioxide, carbon monoxide, carbon dioxide, phosphorus pentoxide, and nitrogen oxides

> Specific gravity: 1.36 at 20°C
> Melting point: 37–38°C
> Boiling point: 143°C at 1.0 mmHg
> Vapor pressure: 0.5 torr at 20°C
> Solubility: soluble in water (0, 005 g/100 mL) ethanol, chloroform, aromatic and aliphatic solvents

19.1.2 Odor and Warning Properties

Rotten egg or garlic ordor (technical product); odor threshold 0.012 ppm.

19.2 Production and Use

Methyl parathion is a broad-spectrum insecticide and acaracide used to control boll weevils and many biting or sucking insect pests on cotton, field vegetables, rice, fruit trees, alfalfa, soy beans, forest trees, and aquatic food crops. It is also used for mosquito control. Methyl parathion is generally applied to the leaves or aerial portion of the crop by either aircraft or ground spray equipment (53). Methyl parathion is available as an emulsifiable concentrate or microencapsules (155). Methyl parathion may be formulated in combination with other pesticides (53).

19.3.5 Biomonitoring/Biomarkers

Immediately after exposure, the concentration of methyl parathion in serum is the most specific biomarker for exposure (409). Urinary concentrations of two metabolites of methyl parathion, *p*-nitrophenol and dimethyl phosphate, may also indicate exposure to methyl parathion. However, they are not unique to methyl parathion. *p*-Nitrophenol is also a breakdown product of parathion, and alkylphosphates are metabolic products of a number of organophosphates (410, 411).

19.4. Toxic Effects

19.4.1.1 Acute Toxicity. Methyl parathion is highly toxic and has oral LD_{50}s of 8–24 mg/kg for rats (64, 268, 252, 412). Dermal LD_{50}s are 6–67 mg/kg (268). Dermal LD_{10}, LD_{50}, and LD_{90}s of 506, 566, and 632 mg/kg for rats indicate a very steep dose–lethality curve (413). One-hour and 4-hour LC_{50}s of 200–287 mg/m^3 and 120 mg/m^3, respectively, were reported for rats. A 4-hour LC_{50} of $<$ 163 mg/m^3 and a 4-hour LC_{50} of 135 mg/m^3 were reported for rats for the 80% technical formulation (243). On a milligram per kilogram basis, methyl parathion via intravenous injection was up to eight times more potent than via oral gavage indicating detoxification via first-pass hepatic metabolism (414, 415). Generally, lethal doses cause death quickly within one hour.

An intraperitoneal LD_{50} of 4 mg/kg was reported for adult rats (248, 416, 417), but is $<$ 1 mg/kg and $<$ 4 mg/kg in neonatal and 12 to 13-day-old rats, respectively, showing enhanced susceptibility of young animals of methyl parathion toxicity. Subcutaneous maximum tolerated doses (MTD, the highest dose that causes no mortality) of 8 and 18 mg/kg were reported for neonate and adult mice, respectively (204).

Among rats given single oral doses of 0, 0.025, 7.5, 10 (males only), or 15 mg/kg (females only) methyl parathion, cholinergic toxicity occurred at doses of 7.5, 10, or 15 mg/kg (243). Neuropathology (focal demyelination) was also evident at these levels.

Methyl parathion caused neurotoxicity, indicated by leg weakness, that lasted from 3–28 days after dosing when given subcutaneously to atropinized chickens at a dose of 64 mg/kg, but not at a dose of 32 mg/kg (64). This shows that methyl parathion may cause a delayed neuropathy but only at doses substantially above a lethal dose and is consistent with a study submitted to EPA in which hens were given an oral dose of 215 mg/kg (243). No signs of ataxia typical of delayed neurotoxic effects were seen in treated hens, nor were neural degenerative changes observed histologically.

Immunotoxicity was observed in mice that were treated with a single oral dose of 9 mg/kg or with 0.9 or 0.4 mg/kg/day, 5 days/week for 4 weeks with the industrial product used to produce Wofatox 50 (60% methyl parathion) and that were immunized with sheep red blood cells (418). Significant increases in the number of splenic plaque forming cells (PFC) occurred among mice treated acutely with 9 mg/kg or repeatedly with 0.9 mg/kg, but not among mice treated repeatedly with 0.4 mg/kg. Immunotoxic effects were also observed in rabbits given diets that contained doses of 0.6 mg/kg/day, but not 0.2 mg/kg/day (419), and in rats given 1.25 mg/kg/day (420).

19.4.1.2 Chronic and Subchronic Toxicity. Repeated oral exposure of rabbits to methyl parathion (1.78 mg/kg/day for 28 days) was associated only with decreased body weight gain and slightly decreased plasma acetylcholinesterase activity by day 21 (421).

Cholinergic signs were not noted in rats given 0, 0.5, 1, 1.5, 2, or 2.5 mg/kg/day methyl parathion via their diet for 7 weeks followed by a 1-week observation period (421a). One female died in each of the 0.5-, 1.5- or 2-mg/kg groups, and two females died in the

2.5 mg/kg group; no deaths occurred in males. When methyl parathion was given to rats in the diet at 0, 2.5, 25, or 75 ppm (about 0, 0.12–0.16, 1.24–1.55, and 4.46–5.15 mg/kg/day) for 90 days, fourteen females and one male fed 75 ppm died by 4 weeks (243). Subjects fed 75 ppm also exhibited a variety of hematologic and clinical chemistry effects. Rats fed the 25- or 75-ppm diet exhibited inhibited RBC and/or brain cholinesterase activity. When parathion was given to rats in the diet at 0.5, 5, or 50 ppm (about 0.029–0.037, 0.295–0.365, or 3.02–3.96 mg/kg/day for 13 weeks, cholinergic toxicity and RBC and brain cholinesterase inhibition occurred (243). Among rats fed the 5-ppm diet, RBC cholinesterase inhibition occurred. No treatment related neuropathy was observed at any dose.

When mice were given 0, 2.6, 5.2, 7.8, 10.4, 13, 16.2, 32.5, or 65 mg/kg/day methyl parathion via their diet for 7 weeks followed by a 1-week observation period (421a), the only clinical signs noted were rough hair coat and arched back, but all males in the 32.5- and 65-mg/kg/day group died as did all females in the 65-mg/kg/day group. When methyl parathion was given to mice in the diet at 0, 10, 30, or 60 ppm (about 0, 2.1–2.5, 6.5–8.6, and 13.5–16.3 mg/kg/day) for 90 days, body weight decreased, absolute brain weight increased, and testes weight decreased in males given 60 ppm (243). In females fed the 60-ppm diet, brain and ovary weight decreased; in females fed the 30-ppm diet, only ovary weight decreased; no other adverse effects were recorded.

There was no evidence of cholinergic toxicity in rabbits given dermal doses of 0, 1, 5, 10, or 100 mg/kg/day methyl parathion for 21 days, although RBC cholinesterase was inhibited in rabbits given 10 or 100 mg/kg/day (243). Slight erythema and edema occurred in females at all doses. Rats exposed to methyl parathion via dermal application of 2 mg/kg/day for 30 days experienced cholinergic toxicity and 40% lethality. Treatment was associated with severe liver injury, severe renal tubular injury, and Purkinje cell necrosis, as well as a decrease in brain and RBC cholinesterase activity.

No cholinergic signs or treatment-related effects on clinical chemistry, hematologic, or urinalytic parameters occurred in dogs exposed to 0, 0.30, 1.0, or 300 mg/kg/day methyl parathion via the diet for 13 weeks (243). Dogs given 3 mg/kg/day exhibited decreased brain and RBC cholinesterase activity, and, dogs given 1 mg/kg/day exhibited decreased RBC cholinesterase activity. No adverse effects occurred in dogs given 0.3 mg/kg/day. When dogs were treated with methyl parathion in the diet at 0, 0.03, 0.3, or 3.0 mg/kg/day for 13 weeks, two high-dose males exhibited emaciation, dehydration, and thin appearance (242). RBC and brain cholinesterase activity were inhibited in high-dose males and females.

Dogs fed diets that provided 0.03, 0.1, or 0.3 mg/kg/day methyl parathion for 1 year exhibited no adverse effects at any dose other than decreased RBC cholinesterase activity at all doses (243). Cholinergic toxicity, thymic lymphoid depletion, and decreased RBC and brain cholinesterase activity occurred in dogs given 4.0 mg/kg/day via the diet for 1 year. Only decreased RBC and brain cholinesterase activity occurred in dogs given 3.5 mg/kg/day, and only decreased RBC cholinesterase activity occurred in dogs given 1.0 mg/kg/day (243). No effects occurred in dogs given 0.3 mg/kg/day, and no treatment-related effects on survival, food efficiency, hematologic and urinalytic findings, ophthalmoscopic findings, intraocular pressures, or electroretinograms occurred in any dogs at any dose (243).

When methyl parathion was given to rats via diets containing 0, 0.5, 2.5, 12.5, and 50 ppm (about 0, 0.02–0.03, 0.11–2.5, 0.53–0.70, and 2.21–3.09 mg/kg/day) for one year cholinergic signs, neuropathology (in peripheral nerve preparations), and decreased RBC and brain cholinesterase activity occurred in rats given 50 ppm (243). Among rats fed the 12.5 ppm diet, RBC and brain cholinesterase was significantly inhibited. No ocular effects were observed at any dietary level.

19.4.1.3 Pharmacokinetics, Metabolism, and Mechanisms. Absorption, distribution, and elimination of methyl parathion via either oral or inhalation exposure is rapid and extensive. Eighty-five percent of an oral dose of methyl parathion was eliminated by mice in urine by 72 hours (425). Similarly, oral methyl parathion was rapidly absorbed by rats and guinea pigs. Concentrations in blood and brain were maximal in about 1–3 hours, and it was nearly completely eliminated in urine (mostly as dimethyl phosphoric and dimethyl phosphorothioic acid) by 7 days (415). Absorption following intravenous or oral dosing of dogs with methyl parathion was 63–78% and 80–95%, respectively, and urinary elimination was rapid and extensive. Bioavailability after oral exposure was much lower (6–30%) than after intravenous exposure due to extensive metabolism and a high extraction ratio in the liver, as well as extensive serum protein binding (426, 427). Maximum serum concentrations after oral exposure occurred 2–9 hours after dosing. Following intravenous injection of dogs with methyl parathion, serum concentrations initially declined rapidly and then more slowly, and the apparent volume of distribution exceeded total body water, indicating distribution to peripheral tissues (427). The mean terminal serum half-life was 7.2 hours, although one dog had a half-life of 16.4 hours. Similar results were obtained in rats and guinea pigs treated intravenously with methyl parathion — absorption, distribution, and elimination were rapid and complete (428).

Following oral exposure, methyl parathion is distributed to blood, liver, adipose tissue, muscle, and brain (429). Distribution coefficients were highest in adipose tissue 8 days after exposure (0.99), in liver 20 days after exposure (0.17), and in brain 16 days after exposure (0.35), but they were always < 1.0, indicating no long-term accumulation of methyl parathion in tissues. Concentrations of methyl parathion are maximum 12 days after exposure in blood and liver and 8 days after exposure in adipose tissue and brain. Half-lives of elimination were 15 days for blood, 13 days for adipose tissue, 15 days for liver, and 15 days for brain (429).

Methyl parathion is oxidatively desulfurated to methyl paraoxon or dearylated to dimethyl thiophosphorothioic acid and *p*-nitrophenol via cytochrome p450 enzymes (20, 416, 417). Alternatively, methyl parathion is hydrolyzed to form *O*-methyl-*O*-*p*-nitrophenyl thiophosphate by esterase (termed A-esterase). Methyl paraoxon can also be dearylated to dimethyl phosphoric acid and *p*-nitrophenol or hydrolyzed to *O*-methyl-*O*-*p*-nitrophenyl phosphate. All of these oxidative and hydrolytic products are excreted in urine (425), except methyl paraoxon that is rarely detected in tissues. These metabolic conversions occur primarily in the liver but also occur in the lung and brain (13, 21, 434–436).

Methyl paraoxon binds to and irreversibly inhibits acetylcholinesterase. Noncatalytic, stoichiometric binding to other esterase (aliesterases) in liver and plasma also occurs, however, and may represent a significant detoxification mechanism because it may reduce the amount of methyl paraoxon that leaves the liver and/or blood to enter target tissues

(14, 21, 435). Binding to hepatic and plasma esterase may not be as significant a detoxification mechanism for methyl parathion as it is for parathion and chlorpyrifos, however, because the affinity of methyl paraoxon is considerably greater for brain acetylcholinesterase than for hepatic esterase. The reverse is true for parathion and chlorpyrifos (14). Thus, even though methyl paraoxon has a lower affinity for acetylcholinesterase than paraoxon, the relatively weaker protection afforded by the aliesterases could allow lethal levels of the hepatically generated methyl paraoxon to reach the nervous system (14).

Differences in the relative rats of toxification and detoxification reactions of methyl parathion contribute to differences in its toxicity compared to other organophosphates, as well as to differences in toxicity between sexes and among different ages (20). For example, female rats, who are slightly less sensitive to the toxicity of methyl parathion than males, metabolized less methyl parathion to methyl paraoxon than male rats in *in situ* perfused liver (21). Age-related differences in the oral LD$_{50}$ also correlated with rates of detoxification pathways of methyl paraoxon rather than with rates of direct metabolism of methyl parathion (417). Differences in the relative rates of toxification and detoxification also contribute to interspecies differences in toxicity. However, the primary factor that accounts for interspecies differences is species-specific differences in the affinity of acetylcholinesterase for methyl paraoxon (20, 437, 438).

19.4.1.4 Reproductive and Developmental. When rats were fed diets that contained 0.5, 5.0, and 25 ppm methyl parathion (0.04, 0.4, and 2 mg/kg/day) for two generations, no treatment-related histopathological effects occurred, nor were there any effects on reproductive parameters, except for maternal weight gain at 25 ppm which decreased significantly in both generations during lactation (243). Reproductive effects, however, occurred in male mice given 0, 9.4, 18.8 and 75.0 mg/kg by gavage (422). Treated animals showed a dose-related increase in the percentage of abnormal sperm. Because there was no appreciable change in testis to body weight ratio and because greater damage was induced when cells were treated as spermatocytes, the authors concluded that methyl parathion acts as a germ cell mutagen. Reproductive effects were also reported in rats treated intraperitoneally with 2.5, 3.5, 4.0, and 5.0 mg/kg methyl parathion (423). Treatment with 4.0 or 5.0 mg/kg caused a significant decrease in ovarian weight and the number of healthy follicles but no change in atretic follicles. The number and duration of estrous cycles was significantly affected at 3.5 mg/kg and higher.

When pregnant rats were given 0.3, 1.0 or 3.0 mg/kg/day methyl parathion via gavage on gestation days 6 through 15, developmental toxicity (increased postimplantation loss and embryonic resorptions, decreased fetal body weight, and delayed ossification) occurred at 3.0 mg/kg/day, but this dose was also maternally toxic (243). Pregnant rats given 1.5 mg/kg/day methyl parathion via gavage on gestation days 6 through 15 exhibited cholinergic toxicity, decreased body weight gain, and an increased number of fetal resorptions (243). When rabbits were given 0.3, 1.0, 3.0, or 5.0 mg/kg/day methyl parathion on gestation days 6 to 18, no maternal or developmental toxicity occurred, other than RBC cholinesterase inhibition at all doses 243.

Pregnant rats given 5, 10, or 15 mg/kg and pregnant mice given 20 or 60 mg/kg methyl parathion via intraperitoneal injection on gestation days 12 and 10, respectively, showed signs of toxicity at all doses (592). Some of the severely affected animals that were treated with the highest dose died. There were no fetotoxic effects in rats. The mice, however, showed more lethality and external malformations (cleft palate) at the highest dose (60 mg/kg). Among pregnant rats given 4.0 mg/kg methyl parathion on gestation days 9 or 15 or 6.0 mg/kg on gestation day 9 via intraperitioneal injection, marked RBC cholinesterase inhibition and some deaths occurred within 30 minutes after injection of 6 mg/kg (424). Embryos of both treated groups had diminished cerebral cortical cholinesterase activity, but there were no effects on fetal mortality or fetal weight and no gross anomalies or developmental defects occurred at either dose.

19.4.1.5 Carcinogenesis. Methyl parathion was not carcinogenic in rats or mice given 0, 1, or 2 mg/kg/day (rats) or 0, 4.5–8, or 9.7–16.2 mg/kg/day (mice) methyl parathion via their diet for 2 years (421a). Methyl parathion was not carcinogenic in mice given diets containing 0, 1, 7, or 50 ppm methyl parathion (0, 0.2–0.3, 1.6–2.1, and 9.2–13.7 mg/kg/day) for 2 years, although RBC and brain cholinesterase were decreased in mice given 7 or 50 ppm (243). Nor was methyl parathion carcinogenic when diets that contained 0, 0.5, 2, 5, 10, or 50 ppm methyl parathion were given to rats for 2 years (equivalent to 0.02–0.03, 0.09–0.14, 0.21–0.26, 0.46–0.71, and 2.6-5.0 mg/kg/day) (243). Cholinergic symptoms, hematologic effects, bilateral retinal degeneration, posterior subcapsular cataract, decreased body weight, increased food consumption, and some deaths occurred in rats given 50 ppm. Brain and RBC cholinesterase activity were inhibited in males given the 10- or 50-ppm diets and in females given the 50-ppm diet; RBC cholinesterase activity was inhibited in females given the 10-ppm diet. At 5 ppm, cholinergic toxicity was apparent in one female, hematologic effects were apparent in males only, and RBC cholinesterase activity decreased slightly. Neurological changes (in particular sciatic nerve degeneration) were pronounced in rats that received 50 ppm, and lesions in 5-ppm treated rats were considered significant.

19.4.1.6 Genetic and Related Cellular Effects Studies. Assays for mutagenicity of methyl parathion using prokaryotic systems have been both positive and negative. Methyl parathion was both positive and negative in *S. typhimurium* with or without metabolic activation, negative in *S. cerevisiae* for reverse mutation and gene conversion, negative for reverse mutation but positive for 5-methyl tryptophan resistance in *E. coli*, positive for mitotic recombination in *S. cerevisiae* (53, 243); positive for DNA damage in *S. typhimurium* and *E. coli*, and negative for unscheduled DNA synthesis in cultured human lung fibroblasts (53, 408a).

Increases in sister chromatid exchange were observed in Chinese hamster ovary cells with metabolic activation but not in Chinese hamster V79 cells, cultured human lymphoid cells, or lymphoma cells without metabolic activation (53).

Methyl parathion was negative in a mouse dominant lethal assay in which mice were fed diets that contained methyl parathion for 7 weeks, and a dominant lethal effect was not observed in mice given methyl parathion in drinking water for 7 weeks (53).

Increased percentages of lymphocytes that had chromosomal breaks were not found among workers from a pesticide factory that manufactured methyl parathion, but chromosomal aberrations were detected in lymphocytes of individuals who were acutely intoxicated by methyl parathion via oral ingestion, inhalation, or dermal contract (53).

19.4.2 Human Experience

19.4.2.2 Clinical Cases. Numerous cases of death and serious toxicity have been reported following acute oral exposure to methyl parathion or following a combination of acute dermal and inhalation exposure (439). Minimal lethal doses were in the range of about 5–10 mg/kg (53).

The degree of RBC acetylcholinesterase inhibition was reported in several cases of combined parathion and methyl parathion poisoning (440). Levels of RBC acetylcholinesterase activity on the day of maximum depression, which ranged from day 1 to day 9 after poisoning, ranged from about 1 to 31% of normal values. This is consistent with RBC cholinesterase activity determined in a pesticide applicator who complained of headache, light headedness, increasing malaise, insomnia, anorexia, and decreased sexual drive, that was 13% of normal (441).

Seven children displayed signs of organophosphate poisoning (lethargy, increased salivation increased respiratory secretions, pinpoint pupils, and respiratory arrest) following their apparent exposure to methyl parathion as a result of its inappropriate use as an insecticide indoors. Significant residues of methyl parathion were found in water used for drinking, orange drink mix, and indoor air (441a). Two of the children died. Upon admission to the hospital, plasma and RBC cholinesterase activities ranged from 20–70% and from 0–26% of normal, respectively.

The occurrence of symptoms consistent with an "intermediate syndrome" began 1 to 3 days after recovery from an acute cholinergic crisis precipitated by ingestion of parathion and methyl parathion (442–444). Symptoms compatible with a diagnosis of an "intermediate syndrome" were respiratory paresis, external ophthalmoparesis, ptosis, proximal limb and neck flexor muscle weakness, and absent or depressed tendon reflexes. Plasma and RBC cholinesterase activities were $<5\%$ and $<15\%$ of control values, respectively, for as long as this syndrome persisted, and the authors noted that this degree of severe cholinesterase inhibition was not observed in individuals who recovered from a cholinergic crisis without subsequently experiencing this type of syndrome.

Dermal exposure to an estimated 0.06 or 0.11 mg/kg methyl parathion on arms and hands did not result in cholinergic effects in two entomologists who entered a cotton field two hours after it was sprayed with methyl parathion, although RBC cholinesterase activities were markedly depressed in both individuals (432).

Methyl parathion air concentrations ranged from 1–30 $\mu g/m^3$ and concentrations from surface wipe samples ranged from 50–980 $\mu g/100$ cm^2 in 64 residences that had been sprayed with methyl parathion within the previous 31 days. Although absorption of methyl parathion was evident by the presence of *p*-nitrophenol in the urine of 142 individuals who lived in the affected residences, no adverse clinical effects were reported.

No cholinergic toxicity of RBC or plasma cholinesterase inhibition occurred after two male volunteers ingested 2 mg/day (0.03 mg/kg) methyl parathion for 5 days or 4 mg/day (0.06 mg/kg) for 5 days 1 to 8 weeks later. Nor were clinical signs of cholinergic toxicity apparent among other volunteers who participated in a metabolic study of methyl parathion and were given a single dose of approximately 0.06 mg/kg methyl parathion (410).

Neurological signs did not occur among men who ingested daily doses of methyl parathion that ranged from 1 to 19 mg during a 30-day period. Nor were there any uniform changes in plasma or RBC cholinesterase levels reported at any of these doses (260). By increasing the concentrations of methyl parathion administered to the same experimental population and using the same protocol, a dose that inhibited RBC cholinesterase activity by 55% in one of five subjects was 24 mg/day (76). Assuming that this individual weighed 70 kg, this would be equivalent to about 0.3 mg/kg/day. This dose and higher levels of 28 and 30 mg/day (equivalent to about 0.4 mg/kg/day) did not result in any overt clinical signs of cholinergic toxicity (76, 78).

In a study where methyl parathion (and several organophosphate pesticides) were tested for the frequency of irritant or allergic contact dermatitis in 652 subjects, one subject (described as a farmer who had occupational hand dermatitis) exhibited allergic contract dermatitis (445).

19.4.2.2.3 Pharmacokinetics, Metabolism, and Mechanisms. Rapid absorption, metabolism, and elimination of methyl parathion occurred in human volunteers given single oral doses of methyl parathion for 5 consecutive days (410). Metabolites detected in urine 24 hours after dosing included *p*-nitrophenol and dimethyl phosphate. Excretion of the primary metabolite, *p*-nitrophenol, was 60% complete in 4 hours, 86% complete in 8 hours, and 100% complete by 24 hours. Urinary excretion of dimethylphosphate followed a similar pattern and was essentially complete by 24 hours after ingestion.

Absorption of methyl parathion following combination dermal and inhalation exposures was demonstrated in a series of field studies of cotton field workers who reentered a field sprayed with methyl parathion. Urinary *p*-nitrophenol and serum methyl parathion were detected up to 12 hours after exposure, confirming that absorption of methyl parathion had occurred (409, 430, 431). Based on methyl parathion residues measured on skin, dermal exposure were probably much less than 0.1 mg/kg, and based on measured air concentrations, (< 0.2 µg/m^3) inhalation exposures probably resulted in absorbed doses much less than 0.02 mg/kg (409, 430, 431). Dermal absorption of 0.06 to 0.11 mg/kg methyl parathion adsorbed on forearms and hands of entomologists (combined with likely inhalation exposure) who worked in a cotton field 2 hours after it was treated with an ultra low volume spray was demonstrated by a concurrent, marked (40% decrease) inhibition of RBC cholinesterase activity (432).

About 5% of an applied dose of a commercial preparation of methyl parathion (*sic*) penetrated human cadaver skin placed in an *in vitro* diffusion cell during a period of 24 hours, whereas only about 1% of an applied dose of methyl parathion in acetone penetrated during 24 hours (433).

19.5 Standards, Regulations or Guidelines of Exposure

Methyl parathion is undergoing reregistration by the EPA for use as a restricted pesticide (154). The ACGIH TLV for methyl paration is 0.2 mg/m^3 with a skin notation. There is no

OSHA PEL-TWA for methyl parathion. The NIOSH REL-TWA is 0.2 mg/m^3 with a skin notation. Most other countries have Occupational Exposure Limits for methyl parathion of 0.2 mg/m^3 (Australia, Belgium, Denmark Finland, France, The Netherlands, Switzerland, and the United Kingdom) or 0.1 mg/m^3 (Hungary, Poland, and Russia).

20.0 Mevinphos

20.0.1 CAS Number: *[7786-34-7]*

20.0.2 Synonyms: Carboxymethoxy-1-methylvinyl dimethyl phosphate; *O,O*-dimethyl 1-carbomethoxy-1-propen-2-yl phosphate; 3-hydroxycrotonic acid methyl ester dimethyl phosphate; *mevinphos*; Duraphos; apavinphos; 2-carbomethoxy-1-1methylvinyl dimethyl phosphate; CMDP; *O,O*-dimethyl *O*-(2-carbomethoxy-1-methylvinyl) phosphate; dimethyl 1-carbomethoxy-1-propen-2-yl phosphate; dimethyl 2-methoxycarbonyl-1-methylvinyl phosphate; dimethyl methoxycarbonylpropenyl phosphate; dimethyl phosphate of methyl 3-hydroxy-cis-crotonate; fosdrin; gesfid; gestid; dimethyl phosphate 3-hydroxycrotonoic acid methyl ester; meniphos; menite; 2-methoxycarbonyl-1-methylvinyl dimethyl phosphate; *cis*-2-methoxycarbonyl-1-methylvinyl dimethylphosphate; *cis*-phosdrin; phosphoric acid, dimethyl ester, ester with methyl 3-hydroxycrotonate; phosphoric acid, (1-methoxycarbonylpropen-2-yl) dimethyl ester; methyl 3-((dimethoxyphosphinyl)oxy)-2-butenoate; methyl 3-hydroxycrotonoate dimethyl phosphate; (*E*)-3-hydroxycrotonic acid, methyl ester, dimethylphosphate; Mevinphos trans+cis isomers; butenoic acid, 3-((dimethoxyphosphinyl)oxy)-, methyl ester; 1-carbomethoxy-1-propen-2-yl dimethyl phosphate; crotonic acid, 3-((dimethoxyphosphinyl)oxy)-, methyl ester; dimethyl *O*-(1-carbomethoxy-1-propen-2-yl) phosphate; methoxycarbonyl-1-methylvinyl dimethyl phosphate; 1-methoxycarbonyl-1-propen-2-yl dimethyl phosphate; Methyl 3-hydroxy-alpha-crotonatedimethyl phosphate

20.0.3 Trade Names: Menite®; Mevinox®; OS-2046®; Phosdrin®; Phosfene®

20.0.4 Molecular Weight: 224.16

20.0.5 Molecular Formula: C$_7$H$_{13}$O$_6$P

20.0.6 Molecular Structure:

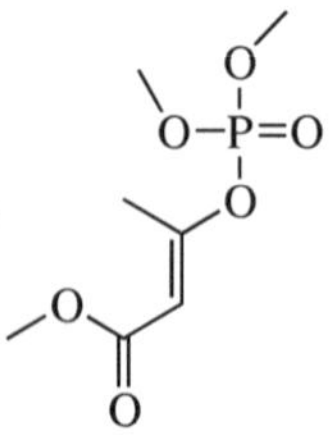

20.1 Chemical and Physical Properties

Mevinphos is a pale yellow to orange liquid. The commercial product is a mixture of cis and trans isomers; the cis isomer is also referred to as the alpha or E isomer, and the trans isomer is also referred to as the beta or Z isomer. The cis isomer (which is ten times more toxic than the trans isomer) comprises about 60% of the commercial material (445a).

Mevinphos is corrosive to black iron, drum steel, stainless steel, and brass. Mevinphos is hydrolyzed in water at pH 11 and has a half-life of 1.4 hours; as the pH is reduced, the rate of hydrolysis is also reduced, so the half-life in water at pH 6 is 120 days.

    Specific gravity: 1.25 at 20°C
Melting point: 21°C (cis); 6.9°C (trans)
Boiling point: 325°C at 760 torr
Vapor pressure: 2.9×10^{-3} torr at 21°C (trans)
Flash point: 79.44°C, open cup
Solubility: miscible with water, acetone, and benzene; slightly soluble in aliphatic hydrocarbons

20.1.2 Odor and Warning Properties

Weak Odor.

20.2 Production and Use

Mevinphos was initially registered as a pesticide in the United States in 1957 and was classified as a Restricted Use Pesticide in 1978 (297). It was formulated as a ready-to-use liquid and concentrate and applied to foliage using aerial, boom spray, and airblast equipment to control aphids, mites, thrips, and lepidopterous larvae on a wide variety of crops. Largely as a result of a high incidence of human poisonings, mevinphos is no longer registered for use in the United States (297).

20.4 Toxic Effects

20.4.1.1 Acute Toxicity. Mevinphos is an organophosphate that has extremely high oral toxicity and has LD_{50}s of 3.4–6.8 mg/kg for rats and mice (64a). The dermal LD_{50} is 4.2–4.7 mg/kg for rats, and 33.8 mg/kg for rabbits (64a, 446). In a comparative study, the oral LD_{50} for female rats was 6.0 mg/kg compared to the intraperitoneal LD_{50} of 1.5 mg/kg (446). Intraperitoneal, oral, intravenous, and subcutaneous LD_{50}s were 2.5, 12.3, 0.6, and 1.2 mg/kg, respectively, indicating that mevinphos is less toxic when administered via "hepatic" routes (intraperitoneal and oral) than via "peripheral" routes (intravenous and subcutaneous) (122). This suggests that the process of oral absorption and/or first-pass metabolism by the liver slightly decreases the acute toxicity of mevinphos. One-hour LC_{50} values are 9.8–92 mg/m^3 (155, 448).

Poisoning following oral or intraperitoneal exposure is rapid. Moderate cholinergic signs occur within 5 to 10 minutes. Peak toxicity occurs within 15 to 30 minutes, lasts about 10 minutes, and most deaths occur in 45 minutes. Symptoms develop faster following intraperitoneal exposure. Signs of distress occur in 1 minute, and death usually occurred in 10 minutes or less (446).

Mevinphos does not cause delayed neuropathy in the hen (448).

20.4.1.2 Chronic and Subchronic Toxicity.
To assess the potential for cumulative toxicity, rats were given intraperitoneal injections of 10, 25, or 50% of the intraperitoneal LD$_{50}$ (1.5 mg/kg) for 5 days/week for 17 to 22 days (446). Cumulative effects were not evident because the only sign observed was slight trembling of the head in rats that received 50% of the LD$_{50}$ immediately after injection. No deaths occurred in any group.

When rats were fed diets that contained 0, 6.3, 12.5, 50, or 100 ppm mevinphos (about 0, 0.5, 1.0, 4.0, or 8.0 mg/kg/day) for 60 days all rats fed 100 ppm died within 3 weeks (446). Rats in all dosed groups showed slight tremors. In another study, rats were fed diets that contained 0, 0.3, 2, 5, 25, 50, 100, or 200 ppm (about 0, 0.02, 0.16, 0.4, 2.0, 4.0, 8.0, or 16 mg/kg/day) for 13 to 18 weeks or a diet that contained 150 ppm for 5 weeks followed by a diet that contained 300 ppm for 2 weeks and a diet that contained 400 ppm for 6 weeks (447). Cholinergic signs were "minimal" (*sic*) at 25 ppm and increased progressively with increasing dietary level; signs were not reported in rats given the 5-ppm diet. Additionally, nonspecific, diffuse, toxic degeneration of the liver and renal tubular epithelium and degeneration of the epithelial cells that line ducts and acini of the exocrine glands occurred. RBC cholinesterase activity progressively decreased during the experimental period in rats fed diets that contained 2 ppm mevinphos or more. Brain cholinesterase was inhibited in rats fed diets of 25 ppm mevinphos or more.

When dogs were fed diets that contained 0, 0.3, 1.0, 2.5, 5.0, 75, or 200 ppm mevinphos (about 0, 0.0075, 0.025, 0.0625, 0.125, 1.875, or 5 mg/kg/day) for 14 weeks, cholinergic signs occurred in dogs given the 75- and 200-ppm diets (447). Dogs fed the 2.5- or 5-ppm diet had inhibited RBC cholinesterase activity. The 1-ppm diet had no effect on RBC cholinesterase activity. Brain cholinesterase activity was inhibited in dogs given 7.5 or 200 ppm.

20.4.1.3 Pharmacokinetics, Metabolism, and Mechanisms.
The similarity of LD$_{50}$ values and the rapid occurrence of cholinergic signs regardless of route of exposure indicates that mevinphos is well absorbed via all relevant routes of exposure. Indeed, approximately 94% of an orally administered dose to rats was absorbed, and dermal absorption of mevinphos in the rats was reportedly approximately 16.8% of an applied dose (448).

In the first 24 hours, 18% of an orally administered dose of mevinphos was excreted in the urine, at 76% was excreted in expired air. Mevinphos was distributed to all human organs within 45 minutes of ingestion of a lethal dose.

Mevinphos is rapidly hydrolyzed *in vivo* in humans to dimethyl phosphate which subsequently appears in the urine (449–451). In two cases of occupational mevinphos poisoning, peak urinary dimethylphosphate concentrations occurred within 12 hours of exposure, and elimination in urine was essentially complete within 50 hours of initial exposure (449). Initial urine concentrations of dimethyl phosphate in mevinphos-poisoned individuals wee 4.7 and 4.0 mg/L. Total excretion of dimethyl phosphate, it was estimated, was equivalent to 7.7 or 9.2 mg mevinphos. These findings are similar to that observed in two mevinphos poisonings where urine concentration of dimethyl phosphate at hospital admission was about 5 mg/L (411).

20.4.1.4 Reproductive and Developmental.
Mevinphos did not cause reproductive toxicity in rats fed a diet that contained mevinphos for two generations (448). Other

studies indicated that no developmental toxicity occurred in rats or rabbits when they were administered mevinphos during gestation (448). No other details were provided.

20.4.1.5 Carcinogenesis. No evidence of carcinogenicity was reported in chronic mice and rat studies (No other details were provided) (448).

A no-observed-effect level for inhibition of RBC cholinesterase in a chronic dog study was 0.025 mg/kg/day (448). A 1-year no-observed-effect level for inhibition of brain cholinesterase activities in rats was 0.025 mg/kg/day (lowest observed effect level = 0.35 mg/kg/day); and, a 2-year no-observed-effect level for clinical signs in rats was also 0.025 mg/kg/day (448). No other details were provided.

20.4.1.6 Genetic and Related Cellular Effects Studies. Mevinphos was mutagenic with or without metabolic activation in *S. typhimurium* and in an assay with Chinese hamster ovary cell (448). Mevinphos with or without metabolic activation also caused chromosomal aberrations in Chinese hamster cells *in vitro* but did not increase unscheduled DNA synthesis (448).

20.4.2 Human Experience

20.4.4.2 Clinical Cases. Initial symptoms of organophosphate poisoning (nausea, vomiting) occurred between 1 to 2 hours in a worker exposed while spraying greenhouse carnations and a worker exposed while cleaning out a spray bottle. Initial symptoms in both cases were followed by blurred vision, fasciculations, weakness, and miosis. Hospitalization occurred about 10 or 3 hours after exposure, respectively, at which time RBC and plasma cholinesterase activities were about 10% of normal laboratory values (449). The estimated intake of mevinphos based on quantification of dimethyl phosphate in urine was 7.7 and 9.2 mg which would be about 0.1 mg/kg, assuming a 70-kg body weight.

Widespread poisoning, apparently due mostly to dermal contact, occurred among nineteen farm workers who worked in a cauliflower field that had been sprayed 4 hours earlier with a mixture of mevinphos and phosphamidon (451a). Within minutes of beginning work, workers developed blurred vision and headache, followed by nausea, weakness, vomiting, abdominal pain, and dizziness; two workers collapsed, one into unconsciousness. Sequential determinations made during a 4-week period of recovery indicated the plasma and RBC cholinesterase activities had initially been inhibited by an average of 66% and 32%, respectively. Plasma and RBC cholinesterase activities in poisoned individuals returned to normal levels by about 57 and 66 days after poisoning, respectively (118). In another case of occupational poisoning, twenty-nine workers who were packing and cutting lettuce in a field 2 hours after it had been sprayed with mevinphos experienced eye irritation, headache, visual disturbances, dizziness nausea, vomiting, weakness, chest pain or shortness of breath, skin irritation, pruritis, eyelid fasciculation, arm fasciculation, excessive sweating, and diarrhea (118). Sequential determinations made during a 2-week period of recovery indicated that plasma and RBC cholinesterase activities had initially been inhibited by an average of 16% and 6% respectively.

Another case of widespread occupational mevinphos poisoning was that of twenty-six men who worked in nineteen different apple orchards where mevinphos was used to control apple aphids (454). Twenty-three of the workers who were exposed during mixing/loading or application of mevinphos experienced nausea, vomiting, dizziness, visual disturbances, muscle weakness, abdominal pain, headache, sweating, and salivation. Seven workers were hospitalized; four required intensive care. RBC cholinesterase activity was depressed to at least 25% below the lower limit of normal in affected workers.

Among cases of accidental poisoning by dermal contact, six children were poisoned after wearing mevinphos-contaminated trousers that had been sold as "damaged goods" (452); a 17-year-old boy experienced delayed mevinphos poisoning, which appeared 2 days after initial contact, as a result of continued wearing of contaminated clothing (453); and an orchardist experienced typical signs of anticholinesterase poisoning 10 hours after mevinphos was applied as a patch test at a dose estimated at 14 mg/kg (312). In other cases, a farmer was poisoned after his boots were accidentally filled with mevinphos solution, and four formulators experienced symptoms of cholinesterase inhibition up to 48 hours after their last contact with mevinphos (450, 451).

Severe poisoning characterized by unconsciousness, pinpoint pupils, profuse sweating, and salivation and accompanied by acute pancreatis was reported for a 37-year-old woman who intentionally ingested 200 mL mevinphos (454a).

The predominant role of dermal absorption of mevinphos in occupational settings was illustrated in a study of greenhouse workers in Finland (454b). Inhalation exposure was measured during 2 days after application of mevinphos to greenhouse plants by measuring mevinphos in green house air. Workers' dermal exposure was measured with patch and hand-wash samples. Greenhouse air concentrations in breathing zones ranged from about 0.04 µg/m^3 (detection limit) to 11.4 µg/m^3 after spraying or use of automatic foggers that resulted in maximum inhalation doses up to about 0.002 mg/kg/day. Intake from dermal exposure would have been considerably more: up to 298 µg/hr mevinphos was determined by measuring mevinphos in hand-washing water from one worker and up to 20 µg/hr mevinphos was determined using patch samples. An additional study of these workers showed that decreases in plasma and RBC cholinesterase activity correlated well with estimated dermal exposure to mevinphos reflected in hand-wash water concentration (455).

Five male volunteers were given various doses of mevinphos daily for 30 days (456). A dose of 1.0 mg/day (about 0.01 mg/kg/day) had no effect during the test period, but RBC acetylcholinesterase activity decreased by as much as 17% in two subjects during the 30-day posttest period. Doses of 1.5, 2.0, or 2.5 mg/day inhibited RBC cholinesterase in all subjects. None of these doses affected plasma cholinesterase activity. No cholinergic signs were directly related to treatment, although loose stools were occasionally reported.

Mean reductions in RBC cholinesterase and plasma cholinesterase activities of 19% and 13%, respectively, occurred in eight volunteers who ingested 0.025 mg/kg/day mevinphos for 28 days (456a, 472). No signs or symptoms of organophosphate intoxication were detectable, and no changes in standard clinical chemistry parameters, other than the reductions in cholinesterase activities, were observed at this dose.

ORGANOPHOSPHORUS COMPOUNDS 885

20.5 Standards, Regulations, or Guidelines of Exposure

The ACGIH TLV for mevinphos is 0.09 mg/m^3 with a skin notation (154). The OSHA PEL-TWA is 0.1 mg/m^3 and the NIOSH REL-TWA is 0.1 mg/m^3.

21.0 Monocrotophos

21.0.1 CAS Number: [6923-22-4]

21.0.2 Synonyms:
(*E*)-dimethyl 2-methylcarbamoyl-1-methylvinyl phosphate; (*E*)-*O,O*-dimethyl-*O*-(1-methyl-3-oxo-1-propenyl) phosphate, phosphoric acid, dimethyl (*E*)-1-methyl-3-(methylamino)-3-oxo-1-propenyl ester; Biloborn; Nuvacron; Phosphoric acid (E)-dimethyl 1-methyl-3-(methylamino)-3-oxo-1-propenyl ester; dimethyl 1- methyl-3-(methylamino)-3-oxo-1-propenyl phosphate; phosphoric acid, dimethyl 1-methyl-3-(methylamino)-3-oxo-1-propenyl ester, (*E*); Corophos; Monocil; Parryfos; dimethoxyphosphinyloxy-*N*-methyl-*cis*-crotonamide; dimethyl phosphate ester of 3-hydroxy-*N*-methyl-*cis*-crotonamide; hydroxy-*N*-methyl-*cis*-crotonamide dimethyl phosphate; phosphoric acid, dimethyl ester, ester with 3-hydroxy-*N*-methylcrotonamide

21.0.3 Trade Names: Azodrin®; Monocron®; Nuvacron®

21.0.4 Molecular Weight: 223.16

21.0.5 Molecular Formula: C$_7$H$_{14}$NO$_5$P

21.0.6 Molecular Structure:

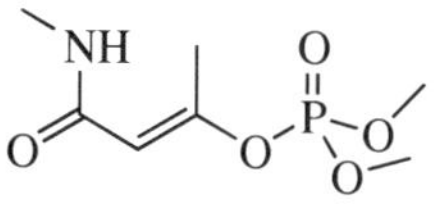

21.1 Chemical and Physical Properties

Monocrotophos is a reddish brown solid.

    Melting point: 54–55°C; 25–30°C
    Boiling point: 125°C at 0.005 mmHg
    Vapor pressure: 7×10^{-5} mmHg at 20°C
    Solubility: soluble in water, acetone, and alcohol; very slightly soluble in kerosene and diesel fuel

21.1.2 Odor and Warning Properties

Possesses a mild ester odor.

21.2 Production and Use

Monocrotophos was introduced in 1965 and used as a systemic and contact organophosphorus insecticide and acaricide to control a variety of sucking, chewing, and boring insects and spider mites on cotton, sugarcane, peanuts, ornamentals, and tobacco (155).

All registrations for monocrotophos have been cancelled in the United States but it is available in other countries as a soluble concentrate or an ultralow volume spray.

21.4 Toxic Effects

21.4.1.1 Acute Toxicity. Monocrotophos is an organophosphate that has high oral toxicity and oral $LD_{50}s$ of 17–20 mg/kg for rats (164a). Intraperitoneal, oral, intravenous, and subcutaneous $LD_{50}s$ for monocrotophos were 8.9, 14.3, 9.2, and 8.7 mg/kg, respectively, for mice, indicating that it is equally potent whether exposure is via an "hepatic" (intraperitoneal and oral) or "peripheral" (intravenous and subcutaneous) route (122). Regardless of route of exposure, lethality occurs rapidly within 5 minutes. Dermal $LD_{50}s$ are 112–129 mg/kg for rats and 270–354 mg/kg for rabbits (64a, 457). A 4-hour LC_{50} of 100 mg/m^3 and a 1-hour LC_{50} of 163–176 mg/m^3 have been reported (458).

Among mice given single oral doses of 1, 2, or 4 mg/kg monocrotophos, reduced locomotor activity occurred at 1 mg/kg and deficits in performance on a rotating rod occurred at 4 mg/kg (459). Among rats given single oral doses of 0, 2, 4, or 6 mg/kg monocrotophos, hypothermia occurred at all levels (459). Cholinergic signs occurred in rats given 6–9 mg/kg/day for 1, 3, 7, 11, or 16 days, but only until 4 to 9 days; signs were absent after 10–11 days. Repeated exposure of rats to 4.5 or 6 mg/kg/day for 16 days also caused overt cholinergic toxicity that diminished by the seventh day of exposure and illustrated accomodation to lowered cholinesterase activity. All rats given three doses of 0.45, 0.85, 1.75, or 3.5 mg/kg/day monocrotophos on alternate days exhibited mild cholinergic signs (461).

Rabbits treated dermally with 20 or 40 mg/kg/day monocrotophos for 6 h/day, 5 days/week for 3 weeks exhibited no clinical signs of toxicity (458). Monocrotophos was not irritating to eyes or skin and was not sensitizing under conditions of current standard tests in rabbits and guinea pigs (458).

Two single doses of 6.7 mg/kg monocrotophos 3 weeks apart were fatal in nine of fourteen atropinized hens (458). The hens that survived two doses did not develop signs of delayed neurotoxicity during an observation period of 3 weeks after the second dose.

Decreased body weight and reduced brain, RBC, and plasma cholinesterase activities occurred in rats given diets that contained 8 ppm monocrotophos (about 0.6 mg/kg/day) for 8 to 13 weeks (458). Body weight gain decreased in rats given diets that contained 45 or 135 ppm and blood and brain cholinesterase were inhibited in rats given diets that contained 1.5 ppm for 12 weeks (458). Mild tremors occurred in dogs given diets that contained 135 ppm, for less than 9 weeks and blood and brain cholinesterase were inhibited in dogs given 15 ppm for 13 weeks (458). In rats treated with 0.3, 0.6, or 1.2 mg/kg/day monocrotophos by gavage for 90 days, brain and whole blood cholinesterase were inhibited at all levels (458).

21.4.1.3 Pharmacokinetics, Metabolism, and Mechanisms. Monocrotophos is well absorbed orally and rapidly excreted. Sixty-three to 85% of an oral dose of monocrotophos was excreted in the urine of rats within 12–48 hours; the majority appeared within the first 6 hours (466). By 96 hours, 82%, 3%, and 6% of the dose was eliminated in urine, feces and expired air, respectively (466). An additional 5% of the dose appeared in feces within

48 hours. Most of the excreted dose in urine was present as water-soluble hydrolytic products (e.g., dimethyl phosphate), and the remainder was the parent compound or the N-hydroxymethyl amide analog of monocrotophos (160).

Human volunteers given a single intravenous injection of monocrotophos rapidly excreted 67 ± 5% of the dose in the urine within 4 to 8 hours. The elimination half-life was about 20 hours (66). When monocrotophos was applied dermally (4 μg/cm^2) to the forearm, 14± of the dose was found in the urine within 120 hours. When covered with a vaporproof film for 72 hours, 33% of the applied dose was absorbed.

The metabolism of monocrotophos involves three different reactions: N-demethylation, O-demethylation, and cleavage of the vinyl phosphate bond. Cleavage of the vinyl phosphate bond is the primary pathway resulting ultimately in excretion of resultant degradation products and conjugates in urine (466). Metabolism is not necessarily complete because significant amounts of unmetabolized monocrotophos are also excreted in urine and feces.

21.4.1.4 Reproductive and Developmental. Rats given diets that contained 10 ppm monocrotophos for two generations had lower body weights, decreased male mating index, lengthened gestation periods, fewer litters, reduced viability and lactation indices, and decreased pup weights in both generations (458). Those given diets that contained 3 ppm had lower pup weights in the second generation, and those given 0 or 0.1 ppm exhibited no adverse reproductive or developmental effects.

When rats were given 0.3, 0.6 or 1.2 mg/kg/day monocrotophos by gavage for two weeks before mating and throughout gestation and lactation, average birth weight and crown–rump length decreased in the 1.2-mg/kg/day group, but this dose was also maternally toxic. No fetotoxicity occurred among rats given 0.1, 0.3, 1, or 2 mg/kg/day by gavage on gestation days 6 through 15 in the absence of maternal toxicity which occurred at 1 mg/kg/day (458). No fetotoxicity occurred among rabbits given 0.1, 1, 3, or 6 mg/kg/day by gavage on gestation days 6 through 18 in the absence of maternal toxicity which occurred at 3 mg/kg/day (458).

21.4.1.5 Carcinogenesis. No carcinogenicity or other adverse effects occurred in mice given diets that contained 1, 2, 5, or 100 ppm monocrotophos (about 0.15, 0.30, 0.75 or 1.5 mg/kg/day) for 2 years, other than inhibition of RBC and brain cholinesterase at all levels (458). No carcinogenicity or other adverse effects occurred in rats given diets that contained 0.01, 0.03, 0.1, 1, 10, or 100 ppm (equivalent to 0.001, 0.002, 0.005, 0.05, 0.5, or 5 mg/kg/day) other than reduced weight gain at 10 ppm and higher, and RBC and brain cholinesterase inhibition at 1 ppm and higher (458).

No carcinogenicity or other adverse effects occurred in dogs given diets that contained 0.16, 1.6, or 16 ppm monocrotophos (about 0.004, 0.04, or 0.4 mg/kg/day) for 2 years, other than inhibition of RBC cholinesterase at 16 ppm and brain cholinesterase activity at 100 ppm (458).

21.4.1.6 Genetic and Related Cellular Effects Studies. Monocrotophos is consistently mutagenic in *in vitro* test systems (458). It was positive in the Ames assay (458a). A dose- and incubation-time-dependent increase in the frequency of chromosomal aberrations occurred in human lymphocytes incubated with monocrotophos. The incidence of sister chromatid exchanges also increased in a dose- and time-dependent manner.

In vivo studies also indicate that monocrotophos is mutagenic. Significant increases is chromatid breaks occurred in bone marrow cells of rats given monocrotophos (462). Chromosomal aberrations in bone marrow cells increased in rats treated orally with monocrotophos (463). Monocrotophos caused somatic mutations and also induced sex-linked recessive lethal mutations among *D. melanogaster* fed contaminated diets (464). In chicks, monocrotophos increased the occurrence of micronuclei in erythrocytes of bone marrow and peripheral blood when administered once daily for 30 days (465).

21.4.2 Human Experience

21.4.2.2 Clinical Cases. Death followed ingestion of an estimated dose of 1200 mg monocrotophos by a woman (about 20 mg/kg assuming 60 kg body weight). The highest concentrations of monocrotophos were found in brain and lung, followed by blood, kidney, and liver (467). Accidental ingestion has also resulted in coma, but patients have recovered without sequelae after prolonged hospitalization (468). A 19-year-old man splashed about 570 mL monocrotophos on his bare chest and arms. He initially experienced no symptoms but within 28 hours he vomited and experienced muscular weakness, chest pain, and blurred vision. Later, he experienced blackouts and intervening lucid intervals. By 38 hours, he could not stand and experienced dry retching, sweating, and confusion. He was hospitalized, treated, and discharged from the hospital 10 days later (469).

Monocrotophos is one of a group of organophosphate compounds associated with the "intermediate syndrome" that is characterized by sudden respiratory distress, proximal limb muscle weakness, and motor cranial nerve palsies that occur 1 to 4 days after acute poisoning. In one case, an individual was poisoned when a bottle containing monocrotophos was thrown at his head and broke, creating a laceration and spilling liquid over his head and face (470). After 6 to 7 hours, he experienced nausea, vomiting, abdominal pain, miosis, fasciculations, and excessive sweating. He was successfully treated with atropine and pralidoxime, but on the fourth day he developed respiratory distress with cyanosis and bradycardia along with weakness of limbs. He was artificially ventilated and eventually recovered. In another case, a male who had ingested monocrotophos and been treated with atropine and pralidoxime developed respiratory distress after 70–80 hours and required ventilatory support (471). He had bilateral facial paresis and weakness of the neck muscles and proximal limb muscles. Fasciculations were seen in the limbs. Deep tendon reflexes were absent in the upper limbs but were normal in the lower limbs. He gradually improved and was discharged after 16 days. In another report, an "intermediate syndrome" was described in an individual who ingested monocrotophos in a suicide attempt and in a worker who was poisoned while spraying monocrotophos (358). In both cases, a well-defined cholinergic phase was followed after 24–96 hours by respiratory difficulty and cranial nerve weakness which lasted 16 to 18 days.

Exposures to monocrotophos experienced by farmers during the spraying of cotton fields were roughly estimated at between 0.007 and 0.02 mg/kg/day based on concentrations of dimethyl phosphate excreted in urine during a 24-hour period after exposure. No overt symptomology was recorded, and RBC cholinesterase activity was unaffected (472a).

Volunteers who had worked for 5 hours in cotton fields treated with monocrotophos either 48 or 72 hours earlier showed no signs of organophosphate poisoning and no RBC cholinesterase inhibition (409).

When groups of volunteers were given oral monocrotophos at 0.0036 and 0.0057 mg/kg/day for 1 month, plasma cholinesterase was reduced, but RBC cholinesterase activity was not changed (472). Signs or symptoms of cholinergic toxicity that could be ascribed to monocrotophos were not reported in the volunteers.

21.5 Standards, Regulations, or Guidelines of Exposure

All registrations of monocrotophos in the United States have been cancelled, and tolerances are being revoked. The ACGIH TLV for monocrotophos is 0.25 mg/m^3 with a skin notation (154). There is no OSHA PEL-TWA for monocrotophos. The NIOSH REL-TWA is 0.25 mg/m^3.

22.0 Naled

22.0.1 CAS Number: [300-76-5]

22.0.2 Synonyms: Dimethyl-1,2-dibromo-2,2-dichloroethyl phosphate; Ortho 4355; Bromochlorphos, dimethyl-1,2-dibromo-2,2-dichlorethyl; Dibrom; phosphoric acid 1,2′2-dibromo-2,2-dichloroethyl dimethyl ester; Alvora; Bromex; Dibromfos; 1,2-dibromo-2,2-dichloroethyl dimethyl phosphate; dimethyl *O*-(1,2-dibromo-2,2-dichloroethyl) phosphate; Hibrom

22.0.3 Trade Names: Bromchlophos®; Bromix®; Dibrom®

22.0.4 Molecular Weight: 380.79

22.0.5 Molecular Formula: C$_4$H$_7$Br$_2$Cl$_2$O$_4$P

22.0.6 Molecular Structure:

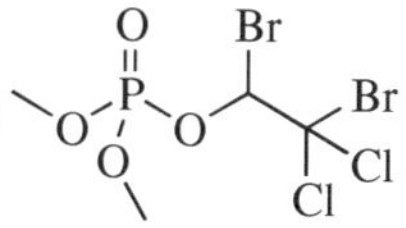

22.1 Chemical and Physical Properties

Pure naled is a white solid; the technical form is 60% pure, and it is usually obtained as a liquid. Naled is completely hydrolyzed (90% to 100%) within 48 hours at room temperature in the presence of water. It is degraded by sunlight and should be stored in lightproof containers.

Specific gravity: 1.96 at 25°C (technical)
Melting point: 26°C (pure)
Boiling point: 110°C at 0.5 torr (technical)

Vapor pressure: 0.002 torr at 20°C; 2×10^{-4} mmHg at 20°C

Solubility: practically insoluble in water; slightly soluble in aliphatic solvents; very soluble in aromatic and oxygenated solvents

22.1.2 Odor and Warning Properties

Has a slightly pungent odor.

22.2 Production and Use

Naled is a liquid insecticide and acaricide used to control mosquitoes and insects on many field crops, on plants in greenhouses, and in mushroom cultivation. It is formulated as a soluble concentrate/liquid (85% a.i.), an emulsifiable concentrate (36–58% a.i.), liquid ready to use (1–35% a.i.), ultra low volume (1–35% a.i.), dust (4% a.i.), and an impregnated collar (7–15% a.i.) Dichlorvos is frequently an impurity of technical naled.

22.4 Toxic Effects

22.4.1.1 Acute Toxicity. Naled is an organophosphate that is moderately toxic and has oral LD_{50}s of 92 to 375 mg/kg for mice and rats (62, 64a, 203). Dermal LD_{50}s of naled are 800 mg/kg for rats and, 360–390 mg/kg for rabbits (64a). A 4-hour LC_{50} of 190–200 mg/m^3 was reported for rats (471a).

When rats were given a single oral dose of 25, 100, or 400 mg/kg naled, the 400-mg/kg dose produced mortality and transient decreases in body weight gain (471a). Rats of both sexes given 100 or 400 mg/kg and females given 25 mg/kg showed marked cholinergic effects. No treatment-related neurological effects were observed 7 or 14 days after treatment at any dose level.

When rats were exposed to 3.4, 7.2, or 12.1 mg/m^3 naled for 6 hours/day, 5 days/week for 3 weeks, a dose-dependent inhibition of brain, RBC, and plasma cholinesterase occurred at all concentrations (473).

Hens treated with atropine sulfate and pralidoxime and then given two acutely toxic doses of naled 21 days apart showed clinical signs of neurotoxicity but no locomotor ataxia characteristic of delayed neurotoxicity (471a). However, axonal degeneration in the spinal cord increased in naled-treated hens.

Naled caused severe eye and dermal irritation in rabbits and was weakly positive in a skin sensitization test in guinea pigs (471a).

22.4.1.2 Chronic and Subchronic Toxicity. When rats were exposed to 0.2, 1, or 6 mg/m^3 naled for 6 hours/day, 5 days/week, exposure to 6 mg/m^3 resulted in cholinergic signs of toxicity (471a). Brain cholinesterase was inhibited at 6 mg/m^3, and RBC cholinesterase was inhibited at 1 and 6 mg/m^3.

When rats were fed diets delivering 0, 0.25, 1, 10, 100 mg/kg/day naled for 28 days, cholinergic effects occurred at 10 mg/kg/day, but not at lower doses (471a). When rats were exposed to naled by dermal application of 1, 20, or 80 mg/kg/day naled for 6 hours/day, 5 days/week for 28 days, the two highest doses were irritating to the skin (471a).

Exposure to 20 and 80 mg/kg/day also produced systemic toxicity. Body weight gain by males was depressed, RBC and brain acetylcholinesterases were inhibited at 20 and 80 mg/kg/day, liver and adrenal weights of 80-mg/kg treated females increased, and mild renal effects were observed.

When 0.4, 2.0, or 10.0 mg/kg/day naled was given to rats by gavage for 90 days, cholinergic effects occurred in three of ten high-dose females (471a). No other clinical effects occurred in either sex at any other dose level.

23.4.1.3 Pharmacokinetics, Metabolism, and Mechanisms.
No data quantifying absorption after oral, inhalation, or dermal exposures to naled are available, but absorption via all routes of exposure is expected to be nearly equal based on the relatively low oral and dermal LD_{50}s and inhalation LC_{50}s reported.

O,O-Dimethyl-2,2-dichlorovinyl phosphate (dichlorvos) is a metabolite of naled as are other hydrolytic products such as methyl phosphates (mono- and di-), *O*-methyl 2,2-dichlorovinyl phosphate (desmethyl dichlorvos), and inorganic phosphate (471a). Three metabolites were identified were in an *in vitro* study using the rat liver homogenates, dichlorvos, dichloroacetaldehyde, and bromodichloroacetaldehyde (471a).

Metabolism of naled to dimethyl phosphate was demonstrated indirectly in individuals who were outdoors during aerial spraying with a naled and temephos mixture for mosquito control (1). Urinary dimethyl phosphate concentrations increased from a maximum of 0.06 ppm to a maximum of 0.50 ppm within 3 hours after spraying.

22.4.1.4 Reproductive and Developmental.
When rats were given 2, 6, or 18 mg/kg/day naled by gavage for two generations, body weight gain was depressed at 18 mg/kg/day for F_0 males and at all dose levels for F_1 males (471a). Reproductive indexes were unaffected in both generations. Survival of pups was reduced at 18 mg/kg/day in the F_1 and F_{2b} generations, and a consistent pup weight consistently decreased during lactation in both generations.

When naled was given to pregnant rats at doses of 2, 10, or 40 mg/kg/day by gavage on gestation days 6 through 19, 40 mg/kg/day was maternally toxic produced cholinergic signs, and reduced weight gain. No developmental toxicity occurred at any dose. There was a marginal effect on resorptions at 40 mg/kg/day, but, because this dose was also maternally toxic it was not considered significant (471a). When pregnant rats were treated orally with 25 to 100 mg/kg naled on gestation days 6 through 15, no adverse teratogenic effect was evident (471b). No maternal or developmental toxicity occurred in rabbits given 0.2, 2, or 8 mg/kg/day naled by gavage on gestation days 7 through 19, although mild cholinergic signs occurred in adults at 2 mg/kg/day.

22.4.1.5 Carcinogenesis.
When dogs were given 0.2, 2.0, or 20 mg/kg/day naled by gavage for one year, cholinergic signs, increases in mineralization of spinal cord, and mild testicular degeneration in males occurred at 2 and 20 mg/kg/day (471a). RBC and brain cholinesterase activities were depressed at these same dose levels. Anemia also occurred at 2 and 20 mg/kg/day, and RBC count, hemoglobin, and hematocrit were reduced. At

20 mg/kg/day, liver and kidney weights increased but were unaccompanied by histopathological changes.

There was no evidence of carcinogenicity among rats given 0.2, 2, or 10 mg/kg/day naled by gavage for 2 years, although there was a reduction in brain and RBC cholinesterase activity at 2 or 10 mg/kg/day (471a).

Cholinesterase activity of rats treated at 0.2 mg/kg/day was unaffected. Cholinergic signs occurred on isolated occasions after dosing in four females given 10 mg/kg/day; no other adverse effects were observed at any dose.

There was no evidence of carcinogenicity when mice were administered 3, 15, or 75 mg/kg/day naled by gavage for 89 weeks (471a). The high dose was reduced to 50 mg/kg/day after 26 weeks due to cholinergic signs and high mortality. The only other treatment-related finding was a slight reduction in weight gain by males that showed a dose-related trend at the middle- and high-dose levels. Cholinesterase activity was not determined.

22.4.1.6 Genetic and Related Cellular Effects Studies. Naled was positive for gene mutation in the *S. typhimurium* reverse mutation assay but did not induce DNA damage in a *rec*-type repair test with *Proteus mirabilis* strains PG713 (*rec-*, *hcr-*) and PG273 (wild-type) (471a). The mutagenicity of naled is due to the direct alkylating ability of the parental molecule and to mutagenic metabolites (e.g., dichlorvos) generated by enzymatic splitting of the side chain. Glutathione-dependent enzymes in the S9 mix eliminate the mutagenic activity of naled completely (474).

Naled exhibited no potential to induce mutations in an *in vivo* gene mutation study (mouse spot test) that used mice given 3, 20, or 150 mg/kg/day naled by gavage on gestation days 8 to 12 (471a).

Naled produced no nuclear anomalies in a mouse bone marrow micronucleus assay in mice and had no clastogenic effect in an *in vivo* cytogenetic study in which rats were given naled by gavage (471a).

22.4.2 Human Experience

Acute symptoms following accidental or intentional poisoning by naled include typical cholinergic signs that disappear after 2 days (475).

Dermal exposures to naled caused residual papular dermatitis on the arm, glazing on the skin of the cheek, mild irritation of the neck skin, and a maculopapular eruption of the abdomen that caused a contact sensitization type dermatitis (476). Dermatitis was also caused by picking flowers sprayed with naled. In another case, contact dermatitis was reported in an aerial applicator who had used naled (477).

22.5 Standards, Regulations, or Guidelines of Exposure

The 90% technical product of naled is currently registered with EPA and is undergoing reregistration. The ACGIH TLV for naled is 3 mg/m^3 with a skin notation (154). The OSHA PEL-TWA and NIOSH REL-TWA is 3 mg/m^3 with a skin notation.

ORGANOPHOSPHORUS COMPOUNDS

23.0 Parathion

23.0.1 CAS Number: *[56-38-2]*

23.0.2 Synonyms:
O,O-Diethyl O-p-nitrophenyl phosphorothioate; DNTP; Ethyl parathion; ethyl parathion; parthion; Foliclal; Fostox; Rhodiatox; diethyl p-nitrophenyl monothiophosphate; SNP; E 605; ac 3422; Etilon; phoskil; deoxynucleoside 5′-triphosphate; Parathion-E; Aqua 9-Parathion; phosphorothioic acid O,O-diethyl-O-(4-nitrophenyl) ester; diethyl p-nitrophenyl thiophosphate; O,O-diethyl-O-(p-nitrophenyl) thionophosphate; diethylparathion; p-nitrophenol O-ester with O,O-diethylphosphorothioate; AAT; AATP; acc 3422; American Cyanamid 3422; Aralo; B 404; bay e-605; bayer e-605; bladan f; compound 3422; corothion; corthione; danthion; ecatox; fosfive; fosova; fostern; genithion; kolphos; kypthion; lirothion; murfos; nitrostygmine; niuif-100; nourithion; oleofos 20; oleoparathion; Orthosphos; panthion; Paramar; paramar 50; parathene; Parawet; pestox plus; pethion; phosphemol; phosphenol; phosphostigmine; RB; stathion; strathion; sulfos; T-47; thiophos 3422; TOX 47; vapophos; diethyl *para*-nitrophenol thiophosphate; diethyl 4-nitrophenyl phosphorothionate; diethyl p-nitrophenyl thionophosphate; drexel parathion 8E; E 605 F; e 605 forte; ekatin wf & wf ulv; ekatox; ethlon; folidol e605; folidol e & e 605; folidol oil; fosfermo; fosfex; gearphos; lethalaire g-54; oleoparaphene; OMS 19; PAC; Pacol; Paradust; rhodiasol; rhodiatrox; selephos; sixty-three special e.c.; soprathion; super rodiatox; vitrex; penncap e; thiomex; tiofos; Viran; Durathion; Thionspray No. 84; diethyl O-p-nitrophenyl phosphorothioate; Fosferno 50; Niran; O,O-diethyl-O-p-nitrophenylthiophosphate

23.0.3 Trade Names: Paraphos®; Alkron®, Alleron®, Aphamite®, Etilon®, Folidol®, Bladan® (Registered); Fosferno®, Niram®, Parapos®, Rhodiatos®

23.0.4 Molecular Weight: 291.27

23.0.5 Molecular Formula: $C_{10}H_{14}NO_5PS$

23.1 Chemical and Physical Properties

Parathion is a pale yellow liquid. Parathion is stable in acids but hydrolyzes readily in alkaline solutions. It slowly decomposes in air to paraoxon. At temperature above 120°C, parathion decomposes and may develop enough pressure to cause containers to explode. Thermal decomposition may release toxic gases such as diethyl sulfide, sulfur dioxide, carbon monoxide, carbon dioxide, phosphorus pentoxide, and nitrogen oxides.

Specific gravity: 1.26 at 25°C
Melting point: 6°C
Boiling point: 375°C at 760 mmHg
Vapor pressure: 3.78×10^{-5} torr at 20°C
Solubility: very slightly soluble in water (20 ppm); completely soluble in esters, alcohols, ketones, ethers, aromatic hydrocarbons, and animal and vegetable oils; insoluble in petroleum ether, kerosene, and spray oils

23.1.2 Odor and Warning Properties

Faint odor of garlic at temperatures above 6°C.

23.2 Production and Use

Parathion is a broad-spectrum pesticide and acaracide with a wide range of applications on many crops against many insect species. In 1992 the U.S. Environmental Protection Agency (EPA) cancelled all uses of parathion on fruit, nut, and vegetable crops. The only uses continued are those on alfalfa, barley, corn, cotton, sorghum, soybeans, sunflowers, and wheat (477a).

23.4 Toxic Effects

23.4.1.1 Acute Toxicity. Parathion is a highly toxic organophosphate compound that has oral LD_{50}s of 1–30 mg/kg for rats, mice, and guinea pigs (62, 64, 268). Dermal LD_{50}s for rats are 7–21 mg/kg (64, 268). Approximate lethal dermal doses to the rabbit ranged from 150 to 2800 mg/kg, depending on the formulation applied (252). A 4-hour LC_{50} of 32 mg/m^3 was reported for rats (123). Two-hour exposures to 3–4 mg/m^3 of a spray of commercial parathion were lethal to the female rat (252). Death occurs rapidly at lethal oral or inhalation exposures. In dermal exposures, death is often delayed, and in rabbits an "intermediate-type syndrome" occurred characterized by paralysis that affect the muscles of the neck and the extensor muscles of the forelegs ("foot drop") (252).

Intraperitoneal LD_{50}s were 3–7 mg/kg for adult rats and mice, but 1 mg/kg for neonates (Benke and Murphy 1975); and subcutaneous maximum tolerated doses (MTD, the highest dose that causes no mortality) of 2 and 18 mg/kg were reported for neonate and adult mice, respectively (204), suggesting that immature animals are more sensitive to parathion than mature animals.

When rats were given 0, 2, 4, or 7 mg/kg parathion by gavage, tremors, ataxia, and 10% mortality occurred in the 7-mg/kg group, whereas only decreased tail-pinch responsivity occurred in the 2-mg/kg group (324). When rats were given single oral doses of 0, 0.025, 0.5, 2.5, or 10.0 mg/kg/day, mortality, cholinergic toxicity, and decreased RBC and brain cholinesterase activity occurred in males given 10 mg/kg/day (477a). Males given 2.5 mg/kg had only decreased RBC cholinesterase activity. One female given 2.5 mg/kg died, and another in this group exhibited cholinergic effects and inhibition of RBC and brain cholinesterase. The time for peak effect was 4 hours and partial to full recovery occurred by 14 days.

Single-dose oral exposure to 6 mg/kg parathion caused deficits in passive avoidance learning in mice and depression in brain and RBC acetylcholinesterase activity after 0.5 h; subcutaneous exposures to 1, 2, or 4 mg/kg/day for 6 days had no effect on this type of behavior, despite a decrease in brain and RBC acetylcholinesterase activity (479). Response rates for schedule controlled behavior decreased in rats given 0.75 or 1.0 mg/kg paraoxon (the active metabolite of parathion) intraperitoneally for 3 days, but not in rats given 0.05 mg/kg. All levels of exposure, however, caused marked inhibition of brain cholinesterase (39). Similarly, among rats given 1.5, 2.5, and 4.5 mg/kg/day every other day, no sign of cholinergic toxicity occurred on days 1–5, 7–13, and 15–21, respectively, despite inhibition of brain cholinesterase (40). Rats given single oral doses of 2, 3.5, or 5 mg/kg parathion showed altered conditioned taste aversion behavior even though brain cholinesterase activities were inhibited only at 3.5 and 5 mg/kg (479a). Monkeys given a single oral dose of 2.0 mg/kg showed cholinergic toxicity, whereas monkeys given 1.5, 1.0, or 0.5 mg/kg did not (480). Performance of a visual discrimination task, however, was disrupted at 1.0 mg/kg and was associated with inhibition of blood acetylcholinesterase activity (480).

23.4.1.2 Chronic and Subchronic Toxicity. A calculated 90-dose LD_{50} was 3.1–3.5 mg/kg/day for rats given feed that contained parathion for 90 days and then observed until death (196). Using the 90-dose LD_{50} and the single-dose LD_{50} of 3–4 mg/kg, a "chronicity factor" (single-dose LD_{50}/90-dose LD_{50}) of about 1 was calculated, indicating that parathion is not expected to exhibit a cumulative toxic effect.

A diet that contained 125 ppm parathion (15.4 mg/kg/day) given to rats for 15 weeks caused severe illness, growth suppression and death, as well as severe RBC and brain cholinesterase inhibition (481). Exposure to a diet that contained 25 ppm (2.4 mg/kg/day) had no toxic effect, but RBC and brain cholinesterase activities were depressed (81). Cholinergic toxicity did not occur among rats fed diets that contained 0.05, 0.5, or 5.0 ppm parathion (about 0.005, 0.05, or 0.5 mg/kg/day) for 84 days; however, RBC cholinesterase was inhibited in the 0.05- and 0.5-mg/kg/day groups. Brain cholinesterase activity was unaffected at any exposure level. In a subchronic neurotoxicity study, rats given feed that delivered 0.05, 1.25, 2.5, or 5.0 mg/kg/day showed a RBC cholinesterase inhibition at the lowest dose tested (477a).

When rats and mice were given diets containing 5, 10, 20, 40, 80, 160, 320, 640 (mice only), or 1280 (mice only) ppm parathion for 6 weeks and then observed for two additional weeks, diets of 80 ppm or more were associated with decreased body weight and increased mortality in rats, and diets of 320 ppm or more were associated with decreased body weight and increased mortality in mice (479b). Diets that contained 5 to 40 ppm had no effect on rats, and diets that contained 5 to 160 ppm had no effect on mice.

When dogs were given 0.02, 0.08, or 8 mg/kg/day parathion via gelatin capsule, 7 days/wk for 6 months, no cholinergic signs or ocular toxicity occurred at any dose (483). However, RBC and retinal cholinesterase activities were intermittently depressed in dogs given 8 mg/kg/d.

23.4.1.3 Pharmacokinetics, Metabolism, and Mechanisms. Absorption of parathion is rapid and complete following oral exposure. Based on the plasma concentration–time curve in rabbits given 3 mg/kg parathion orally, α and β elimination rate constants are about 5 ± 2 and $0.7 \pm 0.1\ hr^{-1}$, respectively, and the β elimination half-life is about 1 ± 0.3 h (494). Urinary excretion was rapid, accounted for 46% and 85% of the administered dose 3 and 6 hours after exposure, respectively, was directly correlated with levels of parathion in plasma (494). Maximum cholinergic effects following oral exposure were slightly delayed compared to i.v. administration, and occurred 30–90 minutes after oral dosing. Similar results were observed in dogs given 10 mg/kg and in mice given 1 mg/kg parathion orally (106, 426).

Rapid absorption and elimination of parathion following oral exposure was confirmed in human volunteers given single oral doses of 1–2 mg parathion for 5 consecutive days (410). Metabolites detected in urine 24 hours after dosing included *p*-nitrophenol, diethylphosphate, and diethyl thiophosphate. Excretion of the primary metabolite, *p*-nitrophenol, was 60% complete in 4 hours, 86% complete in 8 hours, and was directly correlated with exposure. Urinary excretion of diethyl phosphate was more prolonged and reached maximum rates 4 to 8 hours after ingestion.

Absorption, distribution, and elimination are also rapid and complete following inhalation, as evidenced by results of studies where parathion was given intravenously. This is a route of exposure whose distribution and excretion characteristics are similar to those of inhalation because it allows parathion to bypass the extensive "first-pass" metabolic effect of the liver. In humans, 46% of an intravenous dose was excreted in urine by 120 hours, and an elimination half-life of 8 hours was determined (66). Intravenous administration of parathion to dogs demonstrated high serum protein binding, a very high liver extraction ratio (82–97%), rapidly decreasing serum levels, rapid excretion of 85–92% of the dose in urine by 14 hours, and a plasma half-life of 8.5 to 11.2 hours (427, 493). In atropinized rats, parathion was rapidly distributed to tissues (brain, fat, and liver) and eliminated from blood following intravenous administration; the elimination half-life was 3.4 hours (493). Absorption, distribution, and elimination were also rapid in rabbits given intravenous doses of parathion (494). The plasma concentration–time curve followed a three-compartment kinetic model, two very rapid distribution phases followed by a final slower disposition phase that had a β elimination half-life of 5 ± 3 h. Maximal cholinergic effects, including one death, occurred 10–20 minutes after injection. In 8-week-old pigs, parathion had distributed to plasma, kidney, liver, lung, brain, heart and muscle, and about 82% of the dose had been eliminated in urine by 3 hours after dosing, (495). Urinary excretion of metabolites was much lower in newborn (1–2 days old) and neonatal (1-week-old) pigs administered the same dose, and parathion tended to accumulate in newborn and neonatal tissues to a much greater extent than in the 8-week-old pigs, providing a possible partial basis for the apparent sensitivity of developing mammals to parathion (204, 417).

Dermal absorption of parathion is extensive and has been the major cause of occupational disease. In one study, humans absorbed about 20–30% of a dermally applied emulsion of parathion (approximately 2.5 mg) after 300 minutes (496), from 0.1 to 2.8% of parathion dermally applied using an absorbent pad in another (497), and 10% of an applied dose of 4 µg/cm^2 in acetone in another (66). In another study, 5 grams of 2% parathion dust was placed on the right hand and forearm of a volunteer for 2 hours, after which the hand and arm were thoroughly washed, and urinary *p*-nitrophenol (the major parathion metabolite) was monitored for 40 hours (498). Dermal absorption was indicated by increases in urinary *p*-nitrophenol that peaked after 5–6 hours and decreased to very low levels within 40 hours. Marked differences in dermal absorption depending on the anatomic site of exposure were observed among six men who were dosed with 4 µg/cm^2 ^{14}C parathion in thirteen different locations and requested not to wash the site of application for 24 hours (499). After 5 days, the following percentages of applied dose had been absorbed from each anatomic site tested: forearm: 9%; palm: 12%; ball of the foot: 14%; abdomen: 18%; back of the hand: 21%; jaw angle: 34%; postauricular area: 34%;

forehead: 36%; axilla: 64%; and scrotum: 102%. Maximum rates of absorption occurred 1–2 days after exposure but were still significant 5 days after exposure.

Location-specific absorption rates were observed in pigs in an experiment that showed that absorption rates ranged from 30–50% occluded skin to 8–25% for nonoccluded skin, depending on the site of application (500). Times of maximum excretion were 8–13 h for occluded skin and 12–17 h for nonoccluded skin. Rats treated dermally with 1.7–2.0 mg/kg (44–48 µg/cm^2) parathion reached steady state and absorbed 1.4% of the applied dose within 1 hour and 57–59% of the applied dose during a period of 168 hours (501). The skin absorption rate was 0.33 and 0.49 µg/h/cm^2, and the permeability constant was 7.5×10^{-3} and 1.0×10^{-2} cm/h for males and females, respectively. Absorbed parathion was rapidly distributed to heart, liver, and kidneys. Elimination half-times were 39.5 and 28.6 h for males and females, respectively. Mice absorbed nearly 100% of a 1-mg/kg dose of parathion from 1-cm^2 area on their shaved backs by 2 days after treatment (213). Eight hours after treatment nearly 50% of the absorbed dose was excreted in urine, and the rest was distributed primarily to intestine, liver, blood, stomach, kidney, lungs, fat, ear, spleen, and bladder.

Once absorbed, parathion is widely distributed regardless of the route of exposure. Distribution coefficients are highest in the liver (4.1–20.8) and adipose tissue (1.3–2.9) but also exceed one in the brain (1.0–1.4) and muscle (1.5–1.9), indicating retention of parathion by these tissues (101). Maximum concentrations in all tissues is reached 10–20 days after dosing. Extensive distribution of parathion to the liver was demonstrated in mice following intraperitoneal, subcutaneous, oral, or dermal exposure (106, 213, 436, 502) and in pigs following i.v. exposure (495). This is consistent with its rapid hepatic metabolism and high affinity for hepatic esterase (13, 220). Placental transfer has been demonstrated *in vitro* using term perfused human placentas (503).

Parathion is converted to paraoxon by cytochrome P450 enzymes (primarily CYP3A4, but possibly also CYP1A2 and CYP2B6 (48, 504). Binding to other cytochromes also occurs (e.g., CYP3A2, CYP2C11) and is accompanied by their inactivation (48, 504). Alternatively, parathion is dearylated to form diethyl phosphorothioic acid and *p*-nitrophenol in a reaction catalyzed by microsomal enzymes or hydrolyzed by paraoxonase (also termed A-esterase) to form *O*-ethyl-*O*-*p*-nitrophenyl thiophosphate (13). Paraoxon can also be dearylated to diethyl phosphoric acid or hydrolyzed to *O*-ethyl-*O*-*p*-nitrophenyl phosphate. *p*-Nitrophenol, the primary dearylation metabolic product formed from parathion, is eliminated in the urine and quantifying it provides an index of parathion exposure (154). These metabolic conversions occur primarily in liver, but also in the lung and brain (435, 436, 504, 505), and a significant "first-pass" metabolic effect in skin has been demonstrated (506).

Paraoxon binds to and irreversibly inhibits acetylcholinesterase. Binding to other hepatic and plasma esterases also occurs, however, and represents a significant detoxication mechanism because it prevents much of the hepatically generated paraoxon from entering the general circulation and target tissues (13, 14). Comparative differences in the rates of hepatic esterase binding, as well as rates of dearylation, have been implicated as contributors to the greater sensitivity of female rats to parathion toxicity compared to male rats (15, 19, 221), to the greater sensitivity of some tissues (e.g., brain) to parathion toxicity compared to other tissues (17), and, to the greater sensitivity of young animals to parathion toxicity compared to adults (18). Differences in the relative rates of parathion desulfuration, dearylation, hydrolysis, and esterase binding probably also contributes to interspecies differences in sensitivity to parathion toxicity; however, the primary factor that accounts for interspecies differences appears to be species-specific differences in the affinity of acetylcholinesterase for paraoxon (218, 437, 438, 507).

Interindividual differences in paraoxonase activity may contribute to interindividual differences in human sensitivity to parathion because this enzyme is polymorphic in the human population. Its expression is determined by two codominant alleles that represent high and low activity (218, 222, 508). Individuals whose paraoxonase activity is low might be expected to detoxify relatively less of an absorbed dose of parathion and therefore experience greater toxicity than individuals whose paraoxonase activity is high, although that has not been demonstrated. The low-activity phenotype is apparently more common than the high-activity phenotype. In a Southeast Asian population, 4–18% of individuals expressed the high-activity form (508). In another population, 6% were homozygous for high activity, 42% were homozygous for low activity, and 53% were heterozygous (222). Differences in baseline levels of plasma cholinesterase activity may also contribute to interindividual differences in sensitivity. Plasma cholinesterase activity in pregnant women was about 76% of the activity in nonpregnant women, and the plasma cholinesterase activity of renal patients was about 77% of normal (509).

23.4.1.4 Reproductive and Developmental. No adverse reproductive outcomes occurred in rats given feed that contained 1, 10, or 20 ppm parathion (0.05, 0.5, or 1.0 mg/kg/day) for two generations (477a). The only developmental effect observed was reduced body weight and body weight gain in pups fed the 20-ppm diet. However, excess mortality occurred among pups of second-generation rats when given diets that contained 10 ppm parathion once per (486). To further explore these findings, six additional second-generation exposed females were mated with two second-generation exposed males, and six second-generation exposed females were mated with a single second-generation control male. Only four of the twelve females produced litters. Survival of pups was 27% in litters from females mated to the exposed males, and 38% in litters mated to the control male. Thus parathion interfered with both reproductive success and developmental viability.

When rats were given 0, 0.25, 1.0, or 1.5 mg/kg/day parathion on gestation days 6 through 19, mortality occurred and body weight gain decreased in 1.5-mg/kg doses dams (477a). No fetotoxic effects were observed at any dose. When rabbits were given 0, 1, 4, or 16 mg/kg/day parathion via gavage on gestation days 7 through 19, mortality occurred and body weight gain decreased in 16-mg/kg dosed dams (477a). A decrease in litter size also occurred at the 16-mg/kg dose. Conception and litter size were not affected in wild rabbits given two oral 8-mg/kg doses of parathion 30 days apart, although brain acetylcholinesterase activity was reduced, and dosed animals had lower perirenal and kidney fat weights (487).

Intraperitoneal injection of pregnant mice with 4, 8, 10, 11, or 12 mg/kg parathion on gestation days 12, 13, and 14 caused increased resorptions and a reduction in fetal weight at doses of 8 mg/kg and higher. The highest dose was associated with 90% fetal mortality. At 4 mg/kg, fetal body weight was reduced, although the incidence of resorptions was normal (488). Intraperitoneal injection of 10 mg/kg parathion on gestation days 8, 9, and 10 had no impact on fetal weight. Intraperitoneal injection of pregnant rats with 3 or 3.5 mg/kg parathion on gestation day 11 produced increases in resorptions, decreased fetuses/litter, and reductions in fetal and placental weight (123). However, this treatment was also associated with maternal toxicity. Subcutaneous injection of pregnant rats for 4 days during the first, second, or third trimester of gestation resulted in inhibition of brain acetylcholinesterase in dams, but not in pups. However, all pups of parathion-treated rats displayed delays in development of the righting reflex (489).

Altered electrocardiographic patterns (decreased heart rates) occurred in 24-day-old progeny of spontaneously hypertensive strain rats given oral doses of 0.01, 0.1, or 1.0 mg/kg/day parathion from day 2 of pregnancy through day 15 of lactation (490). Dose-related decreases in acetylcholinesterase activity and muscarinic agonist binding occurred in the cerebral cortexes of 21- and 28-day-old rat pups treated subcutaneously with either 1.3 or 1.9 mg/kg/day parathion on postnatal days 5–20 (490a). Slight alterations also occurred in the development of memory; however, no deficits in the development of most reflex behaviors occurred. Acetylcholinesterase activity and muscarinic agonist binding also decreased in the hippocampi of 12-day-old rat pups treated subcutaneously with 0.882 mg/kg/day parathion on postnatal days 5–20 (491). Additionally, cellular disruption and necrosis occurred in the hippocampi of 21-day-old treated pups. When neonatal rats were given subcutaneous injections of 0.5, 1.0, 1.5, or 2.0 mg/kg parathion on postnatal days 8–20, doses of 1.0 mg/kg/day or more caused decreases in body weight gain, mild tremor, brain acetylcholinesterase inhibition decreases in muscarinic receptor density, and mortality (492).

23.4.1.5 Carcinogenesis. Cholinergic toxicity was not observed in dogs given feed that delivered parathion doses of 0, 0.01, 0.03, or 0.10 mg/kg/day for 12 months (477a). RBC cholinesterase activity was intermittently depressed at all doses; brain cholinesterase activity was decreased only in dogs given 0.03 mg/kg/day. When dogs were orally dosed by capsule with parathion at 0, 0.0024, 0.079, or 0.7937 mg/kg/day for 6 months, only dogs given 0.7937 mg/kg/day had decreased RBC and brain cholinesterase activity.

There was no evidence of carcinogenicity in rats given feed that contained 10, 25, or 50 ppm parathion for 64–88 weeks or in rats given 50 (about 3 mg/kg/day) or 100 ppm (about 6 mg/kg/day) parathion for 104 weeks (370, 484). Rats fed the 100-ppm diet occasionally showed peripheral tremors and irritability during the first few weeks, but were normal later. At 50 ppm no adverse effects occurred. However, when rats were given food that contained 10, 20, 50, 75 or 100 ppm parathion once a day for 1 year, those given food containing 75 or 100 ppm experienced serious poisoning and lethality, so that exposures were discontinued (486). Excess mortality occurred at 50 ppm and higher by 1 year. No signs of poisoning occurred in either the 20-ppm or 10-ppm groups.

No evidence of carcinogenicity occurred when mice were fed diets that contained 80 or 160 ppm parathion (equivalent to about 12 and 23 mg/kg/day) for 62–80 weeks and were observed for an additional 9–28 weeks (479b). Cholinergic toxicity occurred at both exposure levels. However, when rats were fed diets that averaged 23–32 and 45–63 ppm parathion (delivering doses of 1.3 or 2.6 mg/kg/day) for 13–67 weeks and observed for 9–28 additional weeks, there was an increased incidence of adrenal cortical adenomas and carcinomas in the high-dose group. Cholinergic toxicity occurred at both doses. Based on these results, the NCI concluded that the evidence for carcinogenicity was equivocal in rats (479b). Noting that the exposure duration of these experiments was less than lifetime, that adrenal cortical adenomas sometimes spontaneously arise in aged rats, and that most tumors were adenomas rather than carcinomas, the International Agency for Research on Cancer (IARC) concluded that these data provided inadequate evidence to evaluate the carcinogenicity of parathion in animals (485). No evidence of carcinogenicity occurred in another study when rats were maintained on diets that contained 0, 0.5, 5.0, and 50.0 ppm parathion for 100 (males) and 120 (females) weeks (477a).

23.4.1.6 Genetic and Related Cellular Effects Studies. Parathion was negative in the *rec*-assay (differential killing assay utilizing H17 *rec*+ and M45 *rec* − strains of *Bacillus subtilis*) and the *E. coli Pol*-assay without metabolic activation (485). In a large number of tests, it did not induce gene mutations in *E. coli, S. typhimurium, Serratia marcescens, S. cerevisiae,* or *Schizosaccharomyces pombe* with or without metabolic activation (485).

p-Nitrophenol, a metabolite of parathion, was not mutagenic to *S. marcescens* or *S. typhimurium* G46 in a mouse host-mediated assay following i.p. administration of 75 mg/kg (485). *p*-Nitrophenol, but not parathion, induced mitotic gene conversion in *S. cerevisiae.* Paraoxon reportedly in two abstracts was weakly mutagenic to *S. typhimurium* TA98 and TA1538 without metabolic activation (485) and induced lethal mutations in mice treated with 0.3 mg/kg (485).

No excess of sex-linked recessive lethal mutations was induced in *D. melanogaster* by parathion. Negative results have also been reported for the induction of unscheduled DNA synthesis by parathion in WI38 human fibroblasts, with or without uninduced mouse liver microsomal fractions (485). No dominant lethal mutation was induced in mice fed parathion for 7 weeks at 62.5, 125, or 250 mg/kg of diet or following a single i.p. injection (485).

Radiolabeled metabolites of ^{14}C parathion bound to DNA *in vitro* in the presence of a rat liver microsomal metabolizing system (485).

23.4.2 Human Experience

23.4.2.2 Clinical Cases. Parathion is a potent cholinergic poison in humans. Upon oral exposure, initial symptoms of poisoning usually occur within 1–2 h. Massive oral exposures can cause death within 5 minutes (33). Estimates of lethal parathion exposures in adults range from 120–900 mg (about 2–13 mg/kg assuming a 70-kg body weight), but are much lower in children and range from 0.1–1.3 mg/kg (1). The heightened sensitivity of children to parathion is illustrated by instances in which parathion-contaminated food was eaten by people of different ages, but death occurred mainly or exclusively among children (1, 510). Research has suggested that the sensitivity of young animals to parathion is due, at least in part, to their relatively lower levels of detoxifying enzyme activity compared to adults (18).

RBC cholinesterase activities among fifty-one mild-to-moderately poisoned individuals averaged 58–64% of normal levels, although some individuals showed essentially no cholinesterase inhibition (18). RBC cholinesterase activities among seven severe-to-fatal poisoning cases averaged 14–29% of normal levels, and all affected individuals exhibited marked depression. Clinical symptoms in all moderately and severely poisoned individuals were thoroughly documented and were characterized especially by gastrointestinal signs (abdominal pain, anorexia, diarrhea, nausea, and vomiting), fatigue, malaise, miosis or visual disturbances, headache, diarrhea, and respiratory difficulty. Giddiness, ataxia, drowsiness, paresthesia and loss of consciousness also occurred.

Occurrence of symptoms consistent with an "intermediate syndrome" began 2 days after recovery from an acute cholinergic crisis precipitated by ingestion of parathion. Symptoms included acute respiratory paresis, severe nystagmus, weakness in proximal limb muscles, and depressed tendon reflexes. Symptoms occurred in the absence of muscarinic signs, lasted for approximately 3 weeks, and were not influenced by atropine. Five instances of "intermediate syndrome" were also reported in individuals who ingested or inhaled a combination of methyl parathion and parathion (442, 443). In two reports, an intermediate syndrome was reportedly followed by axonal polyneuropathy in individuals who were severely poisoned by ingesting parathion (511).

Parathion levels in air during various operations in a parathion manufacturing plant ranged from 0.2–0.8 mg/m^3 during 6-month period. During this same period, repeated measurements were made of RBC cholinesterase activity in thirteen workers. Although there were no preexposure baseline levels of cholinesterase activity against which to compare these levels, activities markedly increased 5 months after parathion manufacture ceased, suggesting that parathion exposure had markedly depressed them in the first place (512).

Several studies simply reported depressions in cholinesterase activity associated with different work environments. Field workers or pilots involved in spraying parathion on agricultural fields have had cholinesterase activities that ranged from 60–70% of baseline and RBC activities from 30–50% of baseline (509, 513, 514).

In a poorly described report, oral intake of 0.07 mg/kg produced no clinical signs of toxicity; 0.1 mg/kg produced uneasiness, warmth, tightness of the abdomen, frequent urination, and a 12% depression in whole blood cholinesterase activity; and 0.4 mg/kg resulted in increased peristalsis, tightness of the chest, and 47% cholinesterase activity inhibition in whole blood (515).

Ten male volunteers given 0.003, 0.010, 0.025, and 0.050 mg/kg/day parathion in capsules sequentially for 3 days at each dose exhibited no signs or symptoms of toxicity at any dose. RBC cholinesterase levels were monitored before, during, and after exposure and showed no effect (516). In another study, five men given 3.0, 4.5, 6.0, or 7.5 mg/day (0.04, 0.06, 0.08, or 0.10 mg/kg/day) for up to 30 days exhibited no clinical signs. RBC cholinesterase activity decreased in three individuals to 63, 78, and 86% of their preexposure levels. RBC cholinesterase activity in all individuals recovered to preexposure levels by 37 days after exposure stopped (260). No inhibition occurred in lower doses. Women given 7.2 mg/day (0.1 mg/kg/day assuming a 60-kg body weight) orally for 5 days/week for 6 weeks exhibited no clinical signs, a 16% decrease in RBC cholinesterase activity, and a 33% decrease in whole blood cholinesterase activity (517), and men given 1 or 2 mg/kg orally for five days (0.01 or 0.03 mg/kg/day assuming a 70-kg body weight) exhibited no clinical signs and no change in RBC cholinesterase activity (410).

RBC and plasma cholinesterase activity was measured and correlated with urinary p-nitrophenol excretion for 112 hours following dermal exposure in an individual covered with 2% parathion dust and placed in a rubberized suit for $7-7\frac{1}{2}$ hours (518). No signs or symptoms of toxicity were observed. Maximum depression of plasma and RBC cholinesterase activity occurred 12–24 hours from the start of exposure but did not exceed 56 or 16% depression, respectively (518). Based on measurements of p-nitrophenol excretion, the authors estimated that 18.2 mg parathion had been absorbed over a period of 103 hours (514), which is equivalent to about 0.06 mg/kg/day, assuming a 70-kg body weight.

To assess parathion toxicity and cholinesterase inhibition following inhalation exposures, a single volunteer was exposed to vapors generated from technical parathion (*sic*) spread over a 36 in² area and heated to 105–115°F for 30 minutes/day for 4 days (515). No signs of toxicity were observed by day 4 of exposure, although RBC cholinesterase activity was 71% of preexposure baseline activity. Based on measurements of 24-hour urinary p-nitrophenol excretion, the authors estimated that 2.5 mg parathion had been absorbed each day during the 4-day period, an amount equivalent to about 0.04 mg/kg/day, assuming a 70-kg body weight. This is consistent with the occupational study noted before in which RBC cholinesterase activities rebounded 5 months after cessation of exposure to about 0.03 mg/kg/day (512).

23.5 Standards, Regulations, or Guidelines of Exposure

The ACGIH TLV for parathion is 0.1 mg/m³. The OSHA PEL-TWA is 0.1 mg/m³ with a skin notation. The NIOSH REL-TWA is 0.05 mg/m³ with a skin notation. Most other nations have Occupational Exposure Limits of 0.1 mg/m³ (Australia; Belgium, Denmark, Finland, France, Germany, Japan, The Netherlands, The Philippines, Switzerland, Thailand 0.11 mg/m², Turkey, and the United Kingdom) or 0.05 mg/m³ (Hungary and Russia).

24.0 Phorate

24.0.1 CAS Numbers: [298-02-2]

24.0.2 Synonyms: (*O,O*-Diethyl (*S*-ethylmercaptomethyl) dithiophosphate; *O,O*-diethyl (*S*-ethylthiomethyl) phosphorodithioate; phosphorodithioic acid *O,O*-diethyl *S*-[(ethylthio)methyl] ester; *O,O*-diethyl *S*-(ethylthio)methyl phosphorodithioate; *O,O*-diethyl *S*-ethylmercaptomethyl dithiophosphate; American Cyanamid 3911; EI 3911; CL 35,024; diethyl *S*-((ethylthio)methyl) phosphorodithioate

24.0.3 Trade Names: Granatox®; Rampart®; Thimet®; Timet®

24.0.4 Molecular Weight: 260.40

24.0.5 Molecular Formula: $C_7H_{17}O_2PS_3$

24.0.6 Molecular Structure:

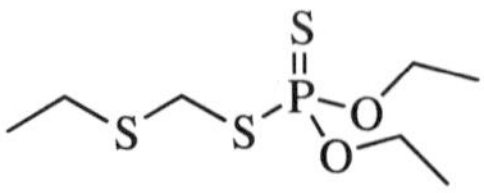

24.1 Chemical and Physical Properties

Technical phorate is a colorless to light yellow clear liquid. Phorate is stable at room temperature; it is hydrolyzed in the presence of moisture and by alkalis.

Specific gravity: 1.17 at 25°C
Melting point: −43.7°C
Boiling point: 125° to 127°C at 2 mmHg
Vapor pressure: 0.00084 mmHG at 20°C; 6.380×10^{-4} mmHg at 25°C
Saturated vapor pressure: 11.8 mg/m^3 (1.1 ppm) at 20°C
Solubility: insoluble in water, miscible with carbon tetrachloride, dibutyl phthalate, 2-methoxyethanol, dioxane, xylene, and vegetable oils

24.1.2 Odor and Warning Properties

Skunk-like odor.

24.2 Production and Use

Phorate is a systemic and contact insecticide and acaracide registered for use on corn, sugar beets, cotton, brassicas, and coffee. It is available as a 2 to 95% emulsifiable concentrate and in granular form at 6.5% to 20% active ingredient although it is currently only used in granular formulation (1). Phorate is not registered for greenhouse or indoor uses. Phorate can be applied by aircraft and ground equipment; only one application per season is allowed for most uses, although two applications are allowed for irrigated cotton, sorghum, peanuts, and sugar beets.

24.4 Toxic Effects

24.4.1.1 Acute Toxicity.
Phorate is an organophosphate that has extremely high oral, dermal and inhalation toxicity and oral LD$_{50}$s of 1–4 mg/kg, dermal LD$_{50}$s of 2–9 mg/kg, and a 1-hour LC$_{50}$ of 11–60 mg/m^3 (64a, 518a).

There are no data on primary eye or dermal irritation or on the primary dermal sensitization properties of phorate. However, the high acute toxicity of phorate would make the conduct of these types of experiments problematic (518a).

No signs of delayed neurotoxicity occurred in hens treated with two LD$_{50}$ doses (14.2 mg/kg) of phorate 21 days apart (518a).

24.4.1.2 Chronic and Subchronic Toxicity.
When rats were given diets that contained 0, 0.22, 0.66, 2, 6, 12, or 18 ppm phorate (about 0, 0.01, 0.03, 0.1, 0.3, 0.6, or 0.9 mg/kg/day) for 90 days, mortality and reduced body weight gain occurred at 12 and 18 ppm. RBC cholinesterase activity inhibition occurred at 2 ppm and higher (518a). When dogs were given capsules that delivered 0, 0.01, 0.05, 0.25, 1.25, or 2.5 mg/kg/day for 6 days/week for 13–15 weeks, dogs given 1.25 or 2.5 mg/kg/day showed typical signs of organophosphate poisoning and subsequently died (518a). RBC cholinesterase was inhibited in dogs given 0.25 mg/kg/day or more.

24.4.1.3 Pharmacokinetics, Metabolism, and Mechanisms. Phorate is well absorbed both orally and dermally, as evidenced by the low oral and dermal $LD_{50}s$. About 77% of an oral dose was excreted in the urine of rats within 24 hours, and 12% was excreted in the feces (518a). Rats given oral phorate at 2 mg/kg or six daily oral doses at 1 mg/kg/day eliminated up to 35% of the dose in urine and up to 6% in feces in 6 days. Rats treated at the rate of 1 mg/kg/day for 6 days excreted only 12% in the urine and 6% in the feces within 7 days (319). The major hydrolytic products that appeared in urine were *O,O*-diethylphosphoric acid (17%), *O,O*-diethyl phosphorothioic acid (80%), and *O,O*-diethyl phosphorodithioic acid (3%).

Rats metabolize phorate to the corresponding sulfoxide and sulfine and produce the phosphorothiolic and phosphoric acids that appear in urine (518a).

Metabolites of phorate were quantified in daily urine specimens obtained from employees of a pesticide formulating plant (184). The predominant alkyl phosphates found in urine were diethyl phosphate, diethyl phosphorothiolate, and diethyl thiophosphate.

24.4.1.4 Reproductive and Developmental. Pup survival and pup body weight decreased among rats given diets that contained 4 or 6 ppm phorate (equal to about 0.4 and 0.6–0.7 mg/kg/day) for three generations (518a). No adverse developmental or reproductive effects occurred when rats were fed diets that contained 1 or 2 ppm phorate (equal to about 0.1 or 0.2 mg/kg/day) for three generations. Slight reductions in lactation and viability indexes occurred in the second generation of mice fed diets that contained 3 ppm phorate (equal to about 0.45 mg/kg/day) for three generations (518a). No reproductive effects occurred in mice fed 0.6 or 1.5 ppm (equal to about 0.09 or 0.23 mg/kg/day).

Neither fetotoxicity nor adverse developmental effects independent of maternal toxicity occurred in progeny of rats given 0.1, 0.125, 0.2, 0.25, 0.3, 0.4, or 0.5 mg/kg/day phorate via gavage on gestation days 6 to 15 (518a). Decreased fetal weights and increased incidence of skeletal variation occurred at 0.4 mg/kg/day, and enlargement of the fetal heart occurred at 0.5 mg/kg/day, but both of these doses were also associated with maternal cholinergic signs and mortality. Similarly, a marginal increase in resorptions occurred in rats exposed to 1.94 mg/m^3 phorate for 1 hour/day on gestation days 7 through 14, but this exposure also caused maternal phorate poisoning (518a). Exposures to 0.15 or 0.40 mg/m^3 phorate for 1 hour/day on gestation days 7 through 14 had no maternal or fetal effect.

Neither fetotoxicity nor adverse developmental effects occurred in rabbits given 0.15, 0.5, 0.9, or 1.2 mg/kg/day phorate on gestation days 6 to 18 (518a).

24.4.1.5 Carcinogenesis. When dogs were given phorate via capsules at doses of 0.005, 0.01, 0.05, or 0.25 mg/kg/day for 1 year, slight body tremors, marginal inhibition of body weight gain, and RBC and brain cholinesterase inhibition occurred in males given 0.25 mg/kg/day (518a).

No evidence of carcinogenicity occurred in rats given diets that contained 0, 1, 3, or 6 ppm phorate (equal to about 0, 0.05, 0.15, or 0.3 mg/kg/day) for 2 years (518a). RBC and brain cholinesterase inhibition occurred at exposures of 3 and 6 ppm. No evidence of carcinogenicity or other adverse effects occurred in mice given diets that contained 0, 1, 3, or 6 ppm phorate (equal to about 0, 0.15, 0.45, and 0.9 mg/kg/day) for 78 weeks, other than a slight decrease in body weight gain in females that were fed 6 ppm (518a).

24.4.1.6 Genetic and Related Cellular Effects Studies. Phorate was negative for mutagenicity in *S. typhimurium*, in *E. coli*, and at the HGPRT locus in cultured Chinese hamster ovary (CHO) cells with or without metabolic activation (45). Phorate was negative in a mitotic recombination assay with *S. cerevisiae* D3 with and without metabolic activation (45). Preferential toxicity assays in DNA repair proficient and deficient strains of *E. coli* and *B. subtillis* were negative (518a).

Phorate was negative for chromosomal aberrations in a dominant lethal test in mice and did not cause chromosomal aberrations in mammalian (rat) bone marrow cells (518a).

An unscheduled DNA synthesis (UDS) assay in human fibroblasts did not show a mutagenic respone (518a).

24.4.2 Human Experience

Poisoning as a result of handling phorate-treated cotton seed led to coma, pinpoint pupils, blood-tinged, frothy sputum, and occasional convulsions in a worker (Hayes, 1963). The next day, RBC cholinesterase activities were 49% of normal. Treatment continued until day 15, at which time RBC cholinesterase activity was 24% of normal. In another study, 60% of a group of 40 workers engaged in formulating 10% phorate granules experienced cholinergic symptoms (514). Exposure levels were not reported.

In a pesticide formulating plant where phorate concentrations ranged from 0.07 to 14.6 mg/m^3, two workers experienced cholinergic toxicity (520). After appropriate treatment, recovery was prompt and uncomplicated.

24.5 Standards, Regulations, or Guidelines of Exposure

Phorate is undergoing reregistration by the EPA. The ACGIH TLV for phorate is 0.05 mg/m^3 (154). The NIOSH REL-TWA is 0.05 mg/m^3 with a skin notation and a 15-minute STEL of 0.2 mg/m^3. Most other countries have Occupational Exposure Limits of 0.05 mg/m^3 (Australia, Belgium, Denmark, France, The Netherlands, Switzerland, and the United States).

25.0 Ronnel

25.0.1 CAS Number: [299-84-3]

25.0.2 Synonyms: *O,O*-Dimethyl *O*-(2,4,5-trichlorophenyl)phosphorothioate: dimethyl trichlorophenyl thiophosphate; Trichlorometaphos; Blitex; Dermafos; dimethyl(2,4,5-trichlorophenyl)phosphorothionate; dimethyl *O*-(2,4,5-trichlorophenyl)thiophosphate; Gesektin K; Moorman's medicated RID-EZY; Nankor; phenol, 2,4,5-trichloro-, *O*-ester with *O,O*-dimethyl phosphorothioate; Remelt; *O,O*-dimethyl *O*-(2,4,5-trichlorophenyl) phosphorothioate; Rovan; Trichlorometafos; Fenclofos

25.0.3 Trade Names: Fenchlorophos®; Ectoral®; Etrolene®; Korlan®; Nankor®; Trolene®; Viozene®

25.0.4 Molecular Weight: 321.57

25.0.5 Molecular Formula: $C_8H_8Cl_3O_3PS$

25.0.6 Molecular Structure:

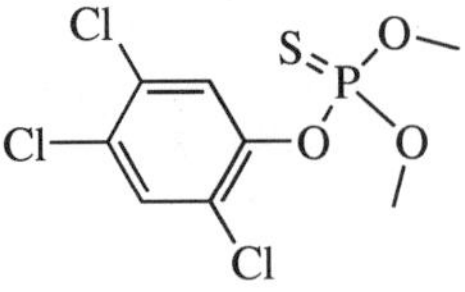

25.1 Chemical and Physical Properties

Ronnel is a white, noncombustible powder

>Specific gravity: 1.48 at 25°C
>Melting point: 41°C
>Boiling point: decomposes; 97° at 0.013 mbar
>Vapor pressure: 0.0008 torr at 25°C
>Solubility: practically insoluble in water; very soluble in acetone, carbon tetrachloride, ether, methylene chloride, toluene, and kerosene

25.2 Production and Use

Ronnel was introduced in 1954, and until 1991 when all registered uses of ronnel in the United States were cancelled by the EPA. It was used as an oral or contact insecticide to control insects that affect cattle. It has also been used as a systemic antiparasitic in humans (1).

25.4 Toxic Effects

25.4.1.1 Acute Toxicity. Ronnel is an organophosphate that has low oral toxicity and oral LD_{50}s of 1250–2630 mg/kg for rats and mice, > 500 mg/kg for dogs (larger doses caused emesis), 640 mg/kg for rabbits, and 3140 mg/kg for male guinea pigs (41, 64a). In rats given single oral doses of 250, 500, or 1000 mg/kg ronnel by gavage, RBC acetylcholinesterase was unaffected 24 hours after dosing (41).

A specific dermal LD_{50} could not be determined but was larger than 5000 mg/kg for rats (64a). The rabbit dermal LD_{50} was 1600 to 2000 mg/kg (445a). When ronnel was dermally applied to rabbits under an impervious cuff for 24 hours, one of twelve died at a dose of 1000 mg/kg, six of twelve died at a dose of 2000 mg/kg, and eight of eight died at a dose of 4000 mg/kg (41).

No cholinergic effects occurred in rats given 25, 50, 100, or 200 mg/kg ronnel via intraperitoneal injection, although brain and RBC cholinesterase were inhibited at the 200-mg/kg dose (165). Plasma carboxyesterases were markedly inhibited even at the 25-mg/kg dose. Similar effects occurred in cows given 100 mg/kg via gelatin capsule (521).

ORGANOPHOSPHORUS COMPOUNDS

When rats were given diets that contained 5, 10, or 30 ppm ronnel for 30 days or, 10, 30, 100, 300, or 500 ppm ronnel for 7 days, adverse cholinergic effects were not reported, and only the rats fed 300- or 500-ppm diets (equivalent to about 0.5, 1.0, 3.0, 10.0, 30.0, or 50.0 mg/kg/day) for 7 days exhibited significantly inhibited brain and RBC cholinesterase (165). Other esterases, however, were significantly inhibited in all rats except those fed 5-ppm diets. Overall, plasma and liver carboxyesterases were about ten times more sensitive to ronnel inhibition than brain or RBC cholinesterase. Moreover, 30-ppm diets had no effect on brain or RBC cholinesterase activity regardless of whether exposures were for 7 or 30 days, suggesting that ronnel lacks a cumulative effect. However, though the 30-ppm diet for 7 days did not inhibit brain cholinesterase, it increased the degree of cholinesterase inhibition produced by a challenge dose of malathion. The percentage inhibition of brain cholinesterase produced by 200 mg/kg malathion was three times as great as expected on the basis of additive effects in ronnel-fed rats.

A small (unspecified) amount of ronnel powder placed in the eye of a rabbit caused slight discomfort and a transient conjunctival irritation that disappeared in 48 hours (41). "A very slight hyperemia of the skin" was produced in rabbits whose skin was treated ten times during 14 days with ronnel under a gauze bandage (519).

25.4.1.2 Chronic and Subchronic Toxicity.
Albino rats that received 328 and 164 mg/kg ronnel developed cholinergic signs of poisoning within 2 weeks. Some of the rats died. No signs were observed in rats that received 16.4 and 8.2 mg/kg (522).

Among steers given diets that contained 0, 44, 88, or 176 ppm ronnel (equivalent to about 0, 1, 2, or 5 mg/kg/day), growth was promoted and thyroid function was altered (increased plasma T4) only in the 176-ppm group — no other effects were reported (523).

Repeated dermal exposure to ronnel apparently cured generalized demodicosis in eighteen of twenty dogs treated for 5–20 weeks, although it also caused lethal organophosphate poisoning in one of twenty (524).

25.4.1.3 Pharmacokinetics, Metabolism, and Mechanisms.
In rats given three acute oral doses of about 18.6 or 186.5 mg/kg/day ronnel 24 hours apart, approximately 14 and 8% of the dose was eliminated in urine as the alkyl metabolites dimethyl phosphoric acid and dimethyl phosphorothioic acid, and 41 and 47% of the dose was eliminated as 2,4,5-trichlorophenol within 48 hours of the last dose. On the third day of dosing, ronnel was distributed to body fat (370 ppm) but by the eighth day the concentration dropped significantly (0.40 ppm) (280). Maximum concentrations in fat occur after 12 hours, and residues at 7 days are equivalent and persist longest in subcutaneous and mesenteric fat, followed by spleen, kidney, and liver.

25.4.1.4 Reproductive and Developmental.
Ronnel treatment did not alter the number of total implants, live fetuses, dead fetuses, resorptions, or fetal weight in pregnant rats given 0, 400, 600, or 800 mg/kg ronnel via gavage on gestation days 6 through 15 (525). However, increases in the incidence of extra ribs were observed in fetuses from dams given 600 or 800 mg/kg.

When pregnant rabbits were given oral doses of 0, 12.5, 25, and 50 mg/kg ronnel on gestation days 6 through 18, an increase in malformed fetuses occurred in the 12.5- and

50.0-mg/kg group, an increase in fetuses that had malformations in the cardiovascular system occurred in the 50-mg/kg treated group, and fetuses that had cerebellar hypoplasia increased in the 25.5- and 50.0-mg/kg group (526).

25.4.1.5 Carcinogenesis. A single female dog given 25 mg/kg ronnel for 11 months reportedly experienced no adverse effects, although RBC and brain cholinesterase activity were below normal (41). Dogs given 0.3, 1, 3, or 10 mg/kg ronnel via their diet for 1 year exhibited no adverse effects, although RBC cholinesterase activity was inhibited in dogs given 10 mg/kg.

No evidence of carcinogenicity, cholinergic toxicity, or adverse effects were seen in dogs given diets that contained 15, 45, or 150 ppm ronnel (equivalent to daily doses of 1, 3, or 10 mg/kg/day) for 2 years (587). Neither brain nor RBC cholinesterase activities were inhibited.

There was no evidence of carcinogenicity in rats fed diets that delivered doses of 0.5, 1.5, 5, 15, or 50 mg/kg/day ronnel for 2 years (41). Among rats given 50 mg/kg/day, there was some evidence of slight granular degeneration or cloudy swelling of the parenchymal cells of the liver and cloudy swelling and vacuolation of renal tubular epithelium of the kidney. RBC and brain acetylcholinesterase were inhibited at 15 or 50 mg/kg/day.

25.4.2 Human Experience

There was no indication of sensitization among thirty men and twenty women given three patch applications of ronnel per week for 3 weeks and then challenged 2 weeks after the last application (41).

Five of twenty-one patients treated orally for creeping eruptions (larva migrans) with ronnel at 10 mg/kg/day for 5 or 10 days, reported nausea, weakness, blurred vision, and/or serpiginous ulcers (527). After ronnel was discontinued, the side effects spontaneously disappeared.

Nausea, headaches, and irritations of the throat and facial skin were occasionally reported by veterinarians who treated group infestations in cattle with pour-on applications of ronnel or other organic phosphorus pesticides (famphur, coumaphos, fenthion, trichlofon) in poorly ventilated areas. Neither plasma nor RBC cholinesterase activities were reduced in these cases, although correlations between specific exposures and cholinesterase measurements appeared imprecise. In support of this, alkyl phosphate metabolites were only occasionally found in the urine at the same times at which blood was drawn for cholinesterase determinations (528).

25.5 Standards, Regulations, or Guidelines of Exposure

Ronnel is no longer registered by the EPA for use (EPA, 1998). The ACGIH TLV for ronnel is 10 mg/m^3 (154). The OSHA PEL-TWA is 15 mg/m^3. The NIOSH REL-TWA is 10 mg/m^3. Most other countries have Occupational Exposure Limits for ronnel of 10 mg/m^3 (Australia, Belgium, Denmark: 5 mg/m^3 (Jan 93), Russia: STEL 0.3 mg/m^3 (Jan 93), France, The Netherlands, The Philippines, Switzerland, and the United Kingdom).

26.0 Sulfotepp

26.0.1 CAS Number: [3689-24-5]

26.0.2 Synonyms: Tetraethyl dithionopyrophosphate; tetraethyl dithiopyrophosphate; tetraethyl dithiopyrophosphate; tetraethyl pyrophosphorodithionate; thiodiphosphoric acid tetraethyl ester; Bladafum; Dithione; thiopyrophosphoric acid, tetraethyl ester; ASP-47; bay-g-393; bayer-e 393; bis-*O,O*-diethylphosphorothionic anhydride; bladafun; dithiodiphosphoric acid, tetraethyl ester; dithiofos; dithiophos; di(thiophosphoric) acid, tetraethyl ester; dithiotep; E393; ethyl thiopyrophosphate; lethalaire g-57; pirofos; plant dithio aerosol; plantfume 103 smoke generator; pyrophosphorodithioic acid, tetraethyl ester; pyrophosphorodithioic acid, *O,O,O,O*-tetraethyl ester; sulfatep; Tedtp; *O,O,O,O*-tetraethyl dithiopyrophosphate; tetraethyl thiodiphosphate; Dithio Insecticidal Smoke; dithiopyrophosphoric acid, tetraethyl ester; thiopyrophosphoric acid ([HO)$_2$P(S)]$_2$O), tetraethyl ester

26.0.3 Trade Names: Dithion®; Dithiophos®; Sulfotepp®; TEDP; Thiotepp®

26.0.4 Molecular Weight: 322.30

26.0.5 Molecular Formula: C$_8$H$_{20}$P$_2$S$_2$O$_5$

26.0.6 Molecular Structure:

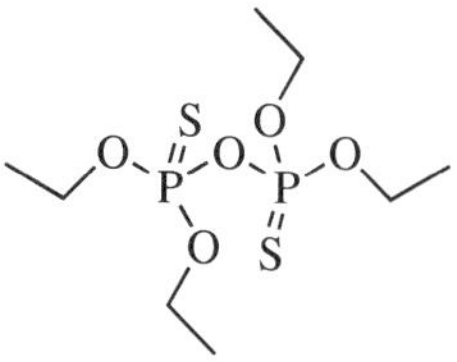

26.1 Chemical and Physical Properties

Sulfotepp is a pale yellow, noncombustible liquid.

- Specific gravity: 1.196 at 25°C
- Boiling point: 136–139°C
- Vapor pressure: 0.00017 torr at 20°C
- Solubility: slightly soluble in water (25 mg/L); soluble in alcohol and most organic solvents

26.1.2 Odor and Warning Properties

Possesses a garlic odor.

26.2 Production and Use

Sulfotepp is registered for use in greenhouses only as a fumigant formulation to control aphids, spider mites, whiteflies, and thrips (529a). It is formulated as impregnated material in smoke generators (canisters) containing 14 to 15% active ingredient. Smoke generators are placed in greenhouses and then ignited using inserted sparklers to generate a thick white smoke for fumigation.

26.4 Toxic Effects

26.4.1.1 Acute Toxicity. Sulfotepp is extremely toxic following oral, dermal or inhalation exposures. Oral LD_{50}s are 5–13.8 mg/kg for rats (529, 530). Oral LD_{50} values are 21.5–29.4 mg/kg for mice; 25 mg/kg for rabbits, 3 mg/kg for cats, and 5 mg/kg for dogs (530). In rats, maximum acetylcholinesterase inhibition occurred 24 hours after an oral dose, was greater in RBCs than in plasma, and returned to normal after 7 days (530). Intraperitoneal LD_{50}s are about one-half oral LD_{50}s for mice and rats, suggesting delayed oral absorption and/or an overall detoxifying role for the liver. Rat dermal LD_{50}s of 65 and 262 mg/kg have been reported (530). A rabbit dermal LD_{50} of 20 mg/kg has also been reported. One-hour and 4-hour LC_{50}s for rats were 160–330 mg/m^3 and 38–59 mg/m^3, respectively. One-hour and 4-hour LC_{50}s for mice were 155 mg/m^3 and 40 mg/m^3, respectively. Typical symptoms of organophosphate poisoning were observed, and animals died in 24 hours (530). Following any route of exposure, symptoms occur within 1 hour and death occurs within 24 hours; surviving animals completely recover in 1–4 days.

Sulfotepp applied to the skin of rabbits at 0.4 gm for a period of 24 hours produced no noticeable effects on the skin (530). When sulfotepp was instilled into a rabbit eyes, the conjunctiva reacted slightly but returned to normal within 24 hours (530).

Nonatropinized and atropinized hens that received single oral doses of sulfotepp that ranged from 10 to 50 mg/kg body weight exhibited no signs of ataxia or paralysis of extremities during a 4-week observation period (530).

26.4.1.2 Chronic and Subchronic Toxicity. When rats were exposed to 0, 0.89, 1.94, and 2.83 mg/m^3 sulfotepp aerosols for 6 hours/day, 5 days/week for 12 weeks, no changes in appearance, behavior, or body weight gain occurred at any dose. RBC acetylcholinesterase was not inhibited at any dose; however, plasma cholinesterase activity was inhibited in rats exposed to 2.83 mg/m^3. Absolute and relative lung weights of females exposed to 2.83 mg/m^3 were significantly higher than those of controls due to edema. No effects were observed at sulfotepp concentrations less than 2.83 mg/m^3 (530).

When rats were fed diets that contained sulfotepp at concentrations of 0, 5, 10, 20, and 50 ppm for 3 months, plasma cholinesterase activity was significantly reduced in female rats at 20 and 50 ppm and in male rats at 50 ppm. RBC acetylcholinesterase activity was significantly reduced in both sexes at 20 and 50 ppm (530). The highest dietary concentration tested for 12 weeks that was reportedly without symptoms in rats was 60 ppm; 180 ppm produced both illness and histological change (529).

When dogs were given diets that contained 0, 0.5, 3, 15, or 75 ppm sulfotepp (equivalent to about 0, 0.014, 0.11–0.12, 0.55–0.57, or 2.75–3.07 mg/kg/day) for 13 weeks, food consumption and body weight gain decreased at 75 ppm (529a). Plasma cholinesterase activity was inhibited at 3 ppm or higher, whereas RBC cholinesterase was inhibited at 15 ppm sulfotepp or higher. Occasional diarrhea and vomiting occurred in dogs at 15 ppm and were common in dogs at 75 ppm. RBC and plasma cholinesterase activity was inhibited at 75 ppm, and plasma cholinesterase was inhibited at 3 ppm and more. Brain cholinesterase was unaffected at any dose.

26.4.1.3 Pharmacokinetics, Metabolism, and Mechanisms. There are no available data on the absorption, distribution, elimination, or metabolism of sulfotepp. However, based on what is known about other organophosphates, it is likely to be well absorbed via all routes of exposures, rapidly oxidatively desulfurated to oxon derivatives, and/or hydrolyzed to dimethyl thiophosphoric acid or diethyl phosphoric acids.

26.4.1.4 Reproductive and Developmental. No embryo toxic or teratogenic effects occurred in rats given 0.1, 0.3, or 1.0 mg/kg/day sulfotepp by gavage on gestation days 9 to 15 (593). No embryo toxic or teratogenic effects occurred in rabbits given 0.1, 1.0, or 3.0 mg/kg/day on gestation days 6 to 18 (533).

26.4.1.5 Carcinogenesis. No signs of carcinogenicity or other adverse effects occurred at any dose level in mice or rats given diets that contained 0, 2, 10, or 50 ppm (equivalent to 0, 0.29, 1.43, or 7.14 mg/kg/day (mice); 0.13, 0.67 or 3.33 mg/kg/day (rats)) sulfotepp for 2 years (531). There were no signs of toxicity other than inhibition of plasma and RBC cholinesterase at 50 ppm.

26.4.1.6 Genetic and Related Cellular Effects Studies. Sulfotepp was mutagenic in *S. typhimurium* strain TA1535 with metabolic activation (78). However, sulfotepp was not mutagenic to *S. typhimurium* strains TA100, TA98, TA1537, and TA1535 with or without metabolic activation (534, 535). A micronucleus test in rats was negative (536). A dominant lethal assay, in which sulfotepp was administered orally to mice, was also negative (537).

26.4.2 Human Experience

Two men suffered from acute poisoning when spraying a mixture of sulfotepp and tetraethylpyrophosphate which had decomposed from diazinon. Symptoms were similar to those observed from organophosphorus insecticide poisonings (nausea, vomiting, burning eyes and blurred vision, difficulty breathing, headache, muscle twitching in arms and legs, and weakness). Blood cholinesterase activity showed a marked reduction but recovered 20 days after the poisoning incident in one individual and 28 days in the other individual (538).

In a review of available epidemiological information on sulfotepp EPA described an instance where a greenhouse worker experienced symptoms of organophosphate poisoning (headache, nausea, diarrhea, vomiting, cough, dizziness, sweating, fatigue, abdominal pain, anxiety, muscle aches, chest tightness, drowsiness, restlessness, shortness of breath, and excessive salivation) when applying the compound.

Skin lesions have been reported during the spraying of sulfotepp.

26.5 Standards, Regulations, or Guidelines of Exposure

Sulfotepp is undergoing reregistration by the EPA. The ACGIH TLV for sulfotepp is 0.2 mg/m^3 with a skin notation. The OSHA PEL-TWA for sulfotepp is 0.2 mg/m^3 with a skin notation; the NIOSH REL-TWA for sulfotepp is 0.2 mg/m^3 with a skin notation.

27.0 Sulprofos

27.0.1 CAS Numbers: *[35400-43-2]*

27.0.2 Synonyms: *O*-Ethyl *O*-(4-(methylthio)phenyl)-*S*-propyl phosphorodithioate; BAY NTN 9306; Phosphorodithioic acid *O*-ethyl *O*-[4-(methylthio)phenyl] *S*-propyl ester; merdafos; *O*-ethyl *O*-(4-(methylthio)phenyl) *S*-propyl phosphorodithioate; *O*-ethyl *O*-[4-(methylthio)phenyl]phosphorodithioic acid *S*-propyl ester; Bolstar 6; *O*-ethyl *O*-(4-methylthiophenyl) *S*-propyl dithiophosphate; Heliothion; ethyl *O*-(4-(methylthio)phenyl) *S*-propyl phosphorodithioate; Morpafos

27.0.3 Trade Names: Bolstar®; Helothion®

27.0.4 Molecular Weight: 322.43

27.0.5 Molecular Formula: $C_{12}H_{19}O_2PS_3$

27.0.6 Molecular Structure:

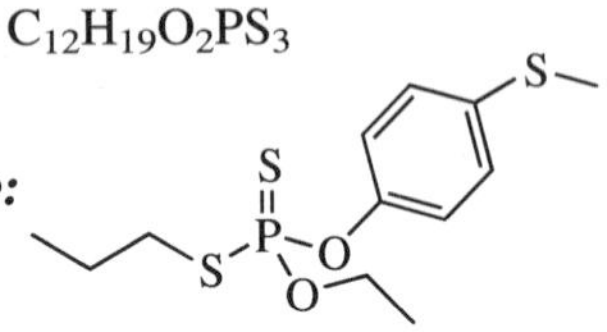

27.1 Chemical and Physical Properties

Sulprofos is a tan colored liquid that hydrolyzes in basic conditions and is stable in acid or neutral conditions.

Specific gravity: 1.20 at 20°C
Melting point: $< -50°C$
Boiling point: 125°C at 0.0075 torr (pure active ingredient) (155°C at 0.1 mmHg)
Vapor pressure: <7.88 torr at 20°C (6.3×10^{-7} mmHg at 20°C)
Solubility: soluble in organic solvents; slightly soluble in water (310 µg/L at 20°C)

27.1.2 Odor and Warning Properties

Possesses a sulfide or phosphorus-like odor.

27.2 Production and Use

Sulprofos is a selective organophosphate used to control foliar lepidopterous, dipterous, and hemipterous insects Registration of sulprofos in the United States was voluntarily cancelled by the registrant in 1997.

27.4 Toxic Effects

27.4.1.1 Acute Toxicity. Sulprofos is a moderately toxic organophosphate via oral, dermal or inhalation exposure. Oral LD_{50}s are 65–275 mg/kg and 107–304 mg/kg for female and male rats, respectively (539). Intraperitoneal LD_{50}s for male and female rats are 305 and 224 mg/kg, respectively, indicating that sulprofos is well absorbed orally (539). Oral LD_{50}s are 1831 and 1617 mg/kg in male and female mice, respectively (539).

Dermal LD$_{50}$s are 5491 and 1064 mg/kg in male and female rats, and dermal LD$_{50}$s are 820 and 994 mg/kg in male and female rabbits, respectively (539). A 4-hour LC$_{50}$ >4130 mg/m^3 and a 1-hour LC$_{50}$ >3840 mg/m^3 were reported for rats; a 4-hour LC$_{50}$ of >490 mg/m^3 was reported for mice and hamsters (539). Five 4-hour exposures of rats to 37, 94, or 259 mg/m^3 sulprofos caused cholinergic signs and inhibition of RBC and plasma cholinesterase activities.

In rats given single doses of 0, 29, 71, or 206 mg/kg sulprofos, cholinergic signs occurred in females at all dose levels and in males at the two higher doses (539). RBC cholinesterase activity was inhibited at all doses. No lesions were seen at necropsy or during histopathological examination of skeletal muscles, peripheral nerves, and tissues from the central nervous system. Following single oral exposures, the no-observed-effect-levels for plasma and brain cholinesterase inhibition were 7.5 and 30 mg/kg, respectively, in male rats (a no-observed-effect level for RBC cholinesterase inhibition was not identified). The lowest observed effect levels for plasma, RBC, and brain cholinesterase inhibition were 15, 30, and 90 mg/kg, respectively. In female rats, no-observed-effect levels for plasma, RBC, and brain cholinesterase inhibition were 4.5, 8.5, and 17 mg/kg, respectively. The lowest observed effect levels for plasma, RBC, and brain cholinesterase inhibition were 8.5, 17, and 50 mg/kg, respectively (539). In female dogs, no-observed-effect levels for plasma and RBC inhibition were 1 and 10 mg/kg, respectively. A lowest observed effect level for plasma inhibition was 5 mg/kg, but a lowest observed effect level for RBC cholinesterase inhibition was not identified. Inhibition was most severe within the first 24 hours after the acute oral dose, and brain cholinesterase was less affected than plasma and RBC cholinesterase (539).

A combination of sulprofos and malathion caused a fivefold potentiation of acute oral toxicity in rats, based on oral LD$_{50}$s (539). Similarly, a simultaneous dosing of sulprofos and azinphos-ethyl yielded a threefold potentiation of toxicity (539).

In rats exposed to 6, 14, or 74 mg/m^3 sulprofos aerosol for 6 hours/day, 5 days/week for 3 weeks, cholinergic signs ("general health impairment", trembling) occurred only at 74 mg/m^3 (539). Exposures of 6 or 14 mg/m^3 reportedly had no detrimental effects on behavior, physical appearance and growth rate, hematology, clinical chemistry, urinalysis, macroscopic pathology, or histopathology. Plasma, RBC, and brain cholinesterase activities were inhibited in males exposed to 74 mg/m^3 and in females exposed to 14 or 74 mg/m^3.

When rats were given 0.1, 1.0, and 10 mg/kg/day sulprofos by gavage for four weeks, the only parameters affected were dose- and time-dependent depressions of plasma, RBC, and brain cholinesterase enzymes (539). Plasma cholinesterase activity was inhibited in the 1.0-mg/kg/day group, and RBC and brain cholinesterase activity were inhibited in the 10-mg/kg/day group. No other treatment-related changes were seen in weight gain, blood chemistry urinalysis, gross pathology, or histopathology.

Sulprofos was not an eye or dermal irritant when tested in rabbits and did not cause dermal sensitization in guinea pigs (539).

Hens given antidote doses of pyridine-2-aldoxime methiodide and atropine sulfate followed by two doses of 65 mg/kg sulprofos (the LD$_{50}$ for the hen) 3 weeks apart exhibited no ataxia or paralysis, nor any changes in the histopathological examinations of brain, spinal cord, or sciatic nerves. There was no evidence of delayed neurotoxicity (539).

27.4.1.2 Chronic and Subchronic Toxicity. When groups of twenty male and female Sprague–Dawley rats were fed diets that contained 10, 30, 100, or 300 ppm (0.54–0.65, 3.0–3.5, and 15–17 mg/kg/day) for 90 days, reduced body weight gain occurred in females given the 300-ppm diet (539). Plasma cholinesterase activity was depressed at 30 ppm and above in both sexes. RBC cholinesterase activity was depressed at 100 ppm in males and 30 ppm in females. Brain cholinesterase activity was depressed in both sexes at 100 ppm and higher. No other effects were noted in blood parameters, urine parameters, or pathology examinations.

In rats fed diets that contained 0, 9, 47, or 226 ppm sulprofos (equivalent to 0, 0.54–0.65, 3.0–3.5, or 15–17 mg/kg/day) for 90 days, cholinergic signs occurred only in rats given the 226-ppm diet (539). Brain cholinesterase activity was inhibited at 226 ppm, RBC cholinesterase was inhibited at 47 and 226 ppm, and plasma cholinesterase was inhibited at 9 ppm (females only). No treatment-related effects were seen during gross pathology or histopathological examination of skeletal muscle, peripheral nerves, eyes, and tissues from the central nervous system.

In dogs given diets that contained 0, 10, 20, or 200 ppm (equivalent to 0, 0.26–0.32, 0.52–0.64, 5.2–6.4 mg/kg/day) for 90 days, diarrhea and regurgitation occurred at 200 ppm (539). Reduced body weights, and decreased plasma, RBC, and the brain cholinesterase activity also occurred at 200 ppm, as well as thickening of the small intestinal wall; decreased liver, kidney, brain and lung weights (males only); increased brain and liver weights, relative to body weight; increased relative lung weights (females only); and increased relative heart weights (males only). Plasma and RBC cholinesterase were also inhibited at 20 ppm. No treatment-related effects were seen in ophthalmic examinations.

No changes occurred in blood or urine parameters, behavior, clinical signs, food consumption, body weights, organ weights, mortality, or in necropsied tissues in mice given diets that contained 0, 2.5, 5, 100, 200, or 400 ppm (equivalent to 0, 0.38–43, 0.76–0.86, 15.2–17.2, 30.4–34.4, or 60.8–68.8 mg/kg/day) for 10 months (539). At 5 ppm and higher, plasma and RBC cholinesterase were inhibited and, at 200 ppm and higher.

27.4.1.3 Pharmacokinetics, Metabolism, and Mechanisms. Sulprofos is rapidly and nearly completely absorbed following oral exposures. In rats, more than 98% of an oral dose of sulprofos was eliminated within 48–72 hours after dosing (539). Rats that received one 11-mg/kg dose or ten 10-mg/kg doses of sulprofos during 15 days also excreted >96% of the dose (or last dose) within 24 hours. The primary route of excretion was through the urine, and less than 11% was excreted in the feces. Conjugated and nonconjugated phenol accounted for 11% of the administered dose, conjugated phenol sulfide accounted for 33–36% of the administered dose, and conjugated phenol sulfone accounted for 41–45% of the administered dose (539). Tissues and organs retained $\leq 2\%$ of the dose; the highest residues were in the fat, ovaries, skin, and liver. Less than 1% of the dose was expired as CO_2 or volatile organic compounds from any exposure. A previous study had demonstrated that female Sprague–Dawley rats dosed at 10 mg/kg excreted >92% within 24 hours (540).

Pigs rapidly excreted an oral dose of sulprofos. More than 95% of the dose was excreted in urine by 24 hours. The major urinary excretion metabolites were the conjugated phenol, free and conjugated phenol sulfoxide, and free and conjugated phenol sulfone. Maximum blood levels occurred 4 hours posttreatment, at which time tissue levels were < 0.01 ppm, except for 0.27- and 0.53- ppm levels in liver and kidney, respectively. Traces of sulprofos and its sulfoxide were found in the omental fat (539). All tissue residues were < 0.05 ppm of sulprofos equivalents at 48 hours posttreatment (539).

Sulprofos is involved in a series of oxidation reactions that may result in replacing the thiono sulfur with oxygen and/or the addition of first one and then two oxygen atoms to the thioether sulfur. The phosphorous-O-phenyl ester of sulprofos and any of these materials may be hydrolyzed to the free phenols, which are then conjugated and eliminated, usually in the urine. The oxidation of the thioether sulfur to its sulfoxide is catalyzed by microsomal flavin-contain monooxygenase (352, 541). The major sulprofos metabolites are phenol, phenol sulfoxide, phenol sulfone, and their conjugates. Sulprofos sulfoxide, sulprofos sulfone, O-analog and O-analog sulfone were also identified in pigs, goats and cows but not in the rat or hen (539, 540).

27.4.1.4 Reproductive and Developmental. No adverse effects on reproduction or development occurred in three generations of rats given diets that contained 0, 30, 60, or 120 ppm sulprofos (539). Third-generation parents of both sexes given the 120-ppm diet had decreased plasma and RBC cholinesterase activity, and third-generation female parents had decreased brain cholinesterase activity. Mean body weights of all treated F1a offspring were lower at weaning, but this was not evident in subsequent litters or additional generations.

No treatment-related reproductive or teratogenic effects were seen in rats given 0, 3, 10, or 30 mg/kg/day sulprofos by gavage on gestation days 6 through 15 (539). At 30 mg/kg/day, signs of maternal toxicity included clinical signs, reduced food consumption, and lower body weights. Fetal body weights were also reduced at the 30-mg/kg/day level. No maternal toxicity, embryo toxicity, or teratogenic effects occurred in rats given 0, 1, 3, or 10 mg/kg/day sulprofos on gestation days 6 through 15 (539). No developmental effects occurred in rabbits given 0, 3, 10, or 30 mg/kg/day sulprofos on gestation days 6 through 18 (539). The 30-mg/kg/day dose was maternally toxic and caused mortality, decreased weight gain, diarrhea, drowsiness, occasional salivation, and "proneness" (*sic*).

27.4.1.5 Carcinogenesis. No evidence of carcinogenicity or overt cholinergic toxicity occurred in rats given diets that contained 6, 60, or 250 ppm sulprofos (0.28–0.27, 2.46–2.74, or 10.25–11.43) for 2 years (539). Plasma, RBC, and brain cholinesterase were inhibited at 250 ppm; only plasma and RBC cholinesterase were inhibited at 60 ppm. There were no other differences in behavior, physical condition, mortality, other blood chemistry parameters, hematology, urine parameters, gross pathology, or histological findings for any treatment level other than decreased absolute and relative male kidney gonad weights at 250 ppm.

There was no evidence of carcinogenicity or overt cholinergic toxicity in dogs given diets that contained 10, 100, or 150 ppm sulprofos (0.23–0.24, 2.99–3.82, and 4.49–5.73 mg/kg/day) for 2 years (539). Plasma, RBC, and brain cholinesterase activities were inhibited at 100 ppm.

There was no evidence of carcinogenicity or other adverse effects in mice given diets that contained 2.5, 25, 200, and 400 ppm sulprofos (0.28–0.32, 2.8–3.2, 22.4–24.8, and 44.849.6 mg/kg/day) for 22 months (539). No changes were observed in food consumption, body weights, clinical signs, or mortality. Plasma and RBC cholinesterase activities decreased at dietary levels of 25 ppm, and brain cholinesterase activity decreased at 400 ppm (females only).

27.4.1.6 Genetic and Related Cellular Effects Studies. Sulprofos is neither mutagenic nor genotoxic. Sulprofos was negative in the reversion assay using *S. typhimurium* with or without metabolic activation, in the CHO/HGPRT mutation assay, in the Pol DNA repair test using *E. coli*, and did not increase the incidence of sister chromatid exchange in Chinese hamster ovary cells (308, 539). Sulprofos was negative in the micronucleus test and in a dominant lethal test using mice (539).

27.5 Standards, Regulations, or Guidelines of Exposure

Registrations for sulprofos have been cancelled by the EPA, and all tolerances are being revoked (*www.epa.gov/opsrd1/op/status.htm* accessed 10/99). The ACGIH TLV for sulprofos is 1 mg/m^3 (154). There is no OSHA PEL-TWA. NIOSH REL-TWA is 1 mg/m^3. Most other countries have Occupational Exposure Limits of 1 mg/m^3 for sulprofos (Australia, Belgium, Denmark: 1 mg/m^3, France, The Netherlands, and Switzerland).

28.0 Temephos

28.0.1 CAS Number: [3383-96-8]

28.0.2 Synonyms: O,O,O′,O′-Tetramethyl-O,O′thiodi-*p*-phenylene phosphorothioate; Temephosn; Abate; O,O′-(thiodi-4,1-phenylene)phosphorothioic acid O,O,O′,O′-tetramethyl ester; O,O′-(thiodi-4,1-phenylene)bis(O,O′-dimethylphosphorothioate); phosphorothioic acid, O,O′-(thiodi-4,1-phenylene) O,O,O′,O′-tetramethyl ester; Tempephos; O,O,O′,O′-tetramethyl O,O′-(thiodi-4,1-phenylene)phosphorothioate; temefos; Abaphos; Tetrafenphos; tetramethyl O,O′-thiodi-*p*-phenylene phosphorothioate

28.0.3 Trade Names: Abate®; Abathion®; Biothion®; Difenthos®; Difos®; Nephis 1G®; Swebate®

28.0.4 Molecular Weight: 466.46

28.0.5 Molecular Formula: $C_{16}H_{20}O_6P_2S_3$

28.0.6 Molecular Structure:

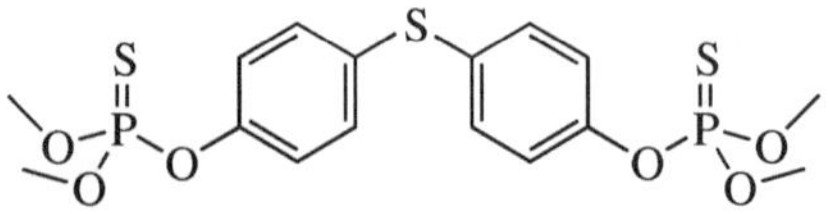

28.1 Chemical and Physical Properties

Pure temephos is a white crystalline solid; technical grade temephos (90–95% pure) is a brown, viscous liquid.

> Melting point: 30°C
> Solubility: insoluble in hexane or water; soluble in toluene, ether, dichloroethane, carbon tetrachloride, and acetonitrile

Reactivity: undergoes hydrolysis at high or low pH
Specific gravity (technical): 1.3
Vapor pressure (technical): 7.17×10^{-8} mmHg at 25°C

28.2 Production and Use

Temephos is a nonsystemic larvicide used to control mosquitoes, chironomid midges, blackflies, biting midges, moths, and sandflies. It is used on crops to control cutworms, thrips, and lygus bugs and on humans as a 2% powder to control body lice. Temephos is formulated as emulsifiable concentrates (10, 43, or 50%), wettable powders (50%) or granules (1). It can be applied by fixed-wing aircraft, helicopter, handheld sprayers, power backpack blowers, and by spoon.

28.4 Toxic Effects

28.4.1.1 Acute Toxicity. Temephos has low acute oral, dermal, and inhalation toxicity. Oral LD_{50}s are 1226–13,000 mg/kg for rats and mice although an LD_{50} value of 444 mg/kg was reported to EPA (42, 64a, 539a). Rats showed typical signs of organophosphate poisoning (*sic*) when given single oral doses of temephos as low as 500–750 mg/kg. Death of rats usually occurred after 6–7 days and after 3 days in mice. In rats given a single oral dose of 8600 mg/kg, brain acetylcholinesterase inhibition was maximum on day 3 and showed partial recovery by day 7 after dosing (542).

The dermal LD_{50} for rats was >4000 mg/kg (larger doses could not be practically applied) (42). Systemic toxicity was apparent after a much lower dermal exposure of 1200 mg/kg (1). The dermal LD_{50} in rabbits is 970–1850 mg/kg (539a). A 4-hour LC_{50} of >1300 mg/m^3 for rats was reported to EPA (539a). Air concentrations of 40 and 12.9 mg/m^3 temephos were reported as single and repeated exposures that led to cholinesterase inhibition in rats (1).

Rabbits given 100 mg/kg/day temephos for 5 days experienced organophosphate poisoning (*sic*) as well as significant RBC acetylcholinesterase inhibition (Gaines et al., 1967). Rabbits given 10 mg/kg/day for 35 days experienced significant RBC acetylcholinesterase inhibition but no signs of organophosphate poisoning. Guinea pigs were apparently resistant to temephos and showed no signs of organophosphate poisoning when given 100 mg/kg for 5 days.

When rats were given single oral doses of temephos and malathion equivalent to one-eighth the LD_{50}, a fourfold potentiation of acute toxicity occurred (42). Further, when groups of rats were given drinking water that contained 5 or 30 ppm temephos for 7 days followed by single acute intraperitoneal doses of 400 mg/kg malathion, the degree of malathion-induced brain and tissue (submaxillary gland) cholinesterase inhibition was significantly enhanced (373).

Temephos is only very slightly irritating to the eye and skin and is not a skin sensitizer (539a).

Hens given single subcutaneous doses of 1000 mg/kg temephos developed immediate leg weakness that lasted for about 26 days and died; hens given 500 mg/kg also developed immediate leg weakness that lasted for about 15 days but survived (42). All hens given temephos in their diet at concentrations of 1000 or 2000 ppm died after 30–43 days. When hens were given a diet of 500 ppm (about 15.5 mg/kg/day) no deaths occurred and leg weakness developed, but only after 30 days; when given a diet of 250 ppm, no leg weakness occurred (42).

28.4.1.2 Chronic and Subchronic Toxicity. Rats given 100 mg/kg temephos by gavage for 44 days showed "typical symptoms of organophosphate poisoning" (*sic*) after three doses and inhibition of RBC acetylcholinesterase (42). Gradual recovery from symptoms occurred while dosing progressed even though the degree of RBC acetylcholinesterase activity continued to decrease. Overt cholinergic toxicity was not evident in rats given 10 mg/kg/day, even though they also showed inhibition of RBC acetylcholinesterase. Neither overt toxicity nor RBC cholinesterase inhibition occurred among rats given 1 mg/kg/day.

All ten rats fed a diet that contained 2000 ppm temephos (150 mg/kg/day) developed organophosphate poisoning and RBC cholinesterase inhibition, and eight died between days 5 and 10 (42). No rats fed diets that contained 2, 20, or 200 ppm temephos exhibited organophosphate poisoning, although rats given the 200 ppm diet exhibited a progressive inhibition of RBC acetylcholinesterase. RBC inhibition was not evident in rats given the 20- or 2-ppm diets.

Dogs given drinking water that contained 10 or 50 ppm temephos (0.6–0.8 or 3–4 mg/kg/day) for 129 days did not exhibit overt signs of cholinergic toxicity. However, RBC acetylcholinesterase activity in dogs given 3–4 mg/kg/day was significantly inhibited.

When rats were given diets that contained 2, 6, 18, or 350 ppm temephos (0.1, 0.3, 0.9, or 17.5 mg/kg/day) for 92 days, no adverse effects of any kind (clinical signs, opthalmology, food consumption, clinical chemistry and hematology, gross or microscopic changes) occurred in any group except decreased body weight gain in females and decreased liver to body weight ratio in males fed the 350-ppm diet (539a). RBC acetylcholinesterase was significantly inhibited in rats fed either the 18- or 350-ppm diets. Brain acetylcholinesterase activity was not significantly inhibited at any dose level. When rats were fed diets that contained 0, 6, 18, or 54 ppm temephos for 90 days, RBC acetylcholinesterase activity decreased in the 18- and 54-ppm group (539a).

28.4.1.3 Pharmacokinetics, Metabolism, and Mechanisms. Oral absorption of temephos is limited. Thirty-six to 65% of a single oral dose given to rats in sesame oil was recovered unchanged in feces (543). There are no studies available on inhalation, dermal, or intravenous absorption.

Peak plasma concentrations of temephos occurred 5 to 8 hours after intubation of rats and guinea pigs with temephos (543). All but traces of absorbed temephos were eliminated in urine and feces in rats. The parent compound and small quantities of its sulfoxide metabolite were found primarily in adipose and gastrointestinal tissues. In feces and fat, temephos was mostly unchanged. Temephos appeared in rat urine as at least 13 urinary metabolites, including primarily 4,4'-thiodiphenol, 4,4'-sulfinyl diphenol, and 4,4'-sulfonyl dipheno ester conjugates. The elimination half-life was estimated at about 10 hours (543).

Although temephos clearly inhibits acetylcholinesterase, it is a much more effective inhibitor of liver carboxylesterases. When rats were given single oral doses of temephos, the dose giving 50% inhibition for three liver carboxylesterases was three to eighteen times smaller than that for RBC cholinesterase and eight to forty-five times smaller than that for brain cholinesterase (373). Carboxylesterase activity was still 30–80% of control by 10 days after dosing, suggesting its relatively slow recovery. This relationship was also apparent when rats were repeatedly exposed to temephos in drinking water. When rats were given drinking water that contained 0, 1, 3, or 5 ppm temephos for 8 weeks, RBC, brain and tissue (submaxillary gland) cholinesterase activity were unaffected, whereas liver carboxylesterases were significantly inhibited by week 2, even among 1-ppm treated rats.

28.4.1.4 Reproductive and Developmental. When rats were given diets that contained 500 ppm temephos (25 mg/kg/day) throughout mating, gestation, parturition, and lactation, there was no effect on the number of litters produced, litter size, the viability of young, or the incidence of congenital deficits (42, 539a). However, the RBC acetylcholinesterase activity of dams and healthy 21-day-old offspring decreased. No adverse reproductive or developmental effects occurred among rats fed diets that contained 0, 25, and 125 ppm temephos (equivalent to 0, 1.25, and 6.25 mg/kg/day) for three generations (539a).

No maternal or developmental toxicity occurred in rabbits given 0, 3, 10, or 30 mg/kg/day or 12.5, 35, or 50 mg/kg/day temephos orally during days 6 through 18 of gestation, although rabbits given 50 mg/kg/day exhibited decreased body weights (539a).

28.4.1.5 Carcinogenesis. EPA noted a chronic dog study (period of exposure not specified) in which significant inhibition of plasma and RBC cholinesterase occurred at dietary exposures equivalent to 12.5 mg/kg/day but not at 0.46 mg/kg/day (539a).

No evidence of carcinogenicity or other treatment-related effects occurred in rats fed a diet that contained 0, 10, 100, or 300 ppm temephos (equivalent to about 0.5, 5.0, or 15 mg/kg/day) for 2 years (539a).

28.4.1.6 Genetic and Related Cellular Effects Studies. Weakly mutagenic effects of temephos were noted in one bacterial strain (EXTOXNET). Additional tests on rabbits and on other strains of bacteria have shown that the compound is nonmutagenic.

28.4.2 Human Experience

Adverse effects were not reported among residents in the West Indies or among applicators when indoor and outdoor walls of residence and potable water were treated with temephos to control mosquitoes (1). Topical treatment with 34 to 57 g/person (about 486 to 814 mg/kg) of a 2% formulation of temephos in pyrax powder for lice control was considered both safe and effective (1).

Twenty-eight human volunteers were given daily doses of temephos in milk at an initial rate of 2 mg/day (equivalent to about 0.03 mg/kg/day) which was doubled every 3 or 4 days for a period of 4 weeks, reached the highest daily dose of 256 mg/man (about 3.7 mg/kg/day) for 5 days (544). Higher doses were impractical to administer because of an obnoxious taste. Separate volunteers were given a constant dose of 64 mg/man (about

0.9 mg/kg/day) temephos for 4 weeks. At no time were clinical symptoms reported or observed at any of the doses administered, nor was there any effect on plasma or RBC cholinesterase. The concentration of temephos in the men's urine was proportional to the dose, and temephos could still be detected at a greatly reduced concentration 3 weeks after dosing stopped.

28.5 Standards, Regulations, or Guidelines of Exposure

Temephos is undergoing reregistration by the EPA (78). The ACGIH TLV for temephos is 10 mg/m^3 (154). The OSHA PEL-TWA for temephos is 15 mg/m^3 total dust and 5 mg/m^3, respirable fraction. The NIOSH REL-TWA is 10 mg/m^3 (total dust) 5 mg/m^3, respirable. Other countries have Occupational Exposure Limits of 10 mg/m^3 (Australia, Belgium, France, The Netherlands, and Switzerland). 0.5 mg/m^3, Russia.

29.0 TEPP

29.0.1 CAS Number: [107-49-3]

29.0.2 Synonyms: Diphosphoric acid tetraethyl ester; ethyl pyrophosphate; tetraphosphoric acid tetraethyl ester; tetraethyl pyrophosphate; Nifost; Vapotone; Tetron; Killax; Moropal; Tetraethyl ester diphosphonic acid; O,O,O',O'-tetraethyl pyrophosphate; diphosphoric acid tetraethyl ester; bis-O,O-diethylphosphoric anhydride; pyrophosphoric acid, tetraethyl ester; fosvex; hexamite; kilmite 40; lethalaire g-52; lirohex; mortopal; Nifos; tetraethyl diphosphate; tetrastigmine; tetron-100

29.0.3 Trade Names: Bladan®; Fosnex®; Grisol®; HETP®; Killex®; Kilmite®; Nifos T®; Pyfos®; Tetraspa®

29.0.4 Molecular Weight: 290.20

29.0.5 Molecular Formula: $C_8H_{20}O_7P_2$

29.0.6 Molecular Structure:

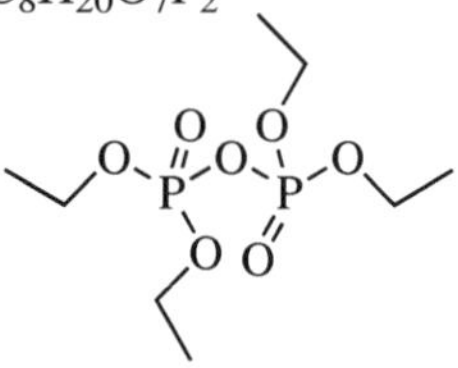

29.1 Chemical and Physical Properties

TEPP is a clear to amber hydroscopic liquid that decomposes at 170–213°C and evolves ethylene. The commercial product is 40% TEPP.

> Specific gravity: 1.185 at 20°C
> Boiling point: 124°C at 1 torr
> Vapor pressure: 1.5×10^{-4} torr at 20°C (saturates in air at 0.20 ppm)
> Solubility: miscible with water and most organic solvents except petroleum oils; hydrolyzes in water to form mono-, di- and triethyl orthophosphates; the resulting solutions are corrosive to metals

29.2 Production and Use

TEPP was introduced in 1943 as a nonsystemic aphicide and acaricide used to control aphids, spiders, mites, mealy bugs, leafhoppers, and thrips. It is commercially available as a 0.66–1.2% dust, an emulsifiable concentrate (10–40%), and as a solution in methyl chloride to be used as an aerosol (1, 155).

29.4 Toxic Effects

29.4.1.1 Acute Toxicity. TEPP is an organophosphate compound that has extremely high oral toxicity and oral $LD_{50}s$ of 1–2 mg/kg (38, 64a). A dermal LD_{50} for the male rat was 2.4 mg/kg (Gaines 1969). A 1-hour LC_{50} of 23.5 mg/m^3 and a 4-hour LC_{50} of 6.75 mg/m^3 were reported for rats (123). An intraperitoneal LD_{50} of 0.65 mg/kg was reported for rats, indicating that it is well absorbed orally; and an intravenous LD_{50} of 0.3 mg/kg was reported for rats (545). Similar results were obtained in mice — oral, dermal, intraperitoneal, and intravenous $LD_{50}s$ in mice were 3, 8, 0.83, and 0.20 mg/kg, respectively (545). In all cases, the mechanism of death was peripheral inhibition of cholinesterase in the muscles, leading to respiratory paralysis, anoxia, and terminal convulsion.

29.4.2 Human Experience

29.4.2.2 Clinical Cases. Many deaths have resulted from accidental and intentional contact with TEPP. In Japan, by 1959, TEPP had caused eight poisonings during spraying, eighteen by other accidents, and 101 by suicide attempts, ninety-nine of which were successful (1). In one case, one mouthful (*sic*) of TEPP was reported to have led to complete collapse in less than 5 minutes and death shortly afterward (1).

A pilot who spilled TEPP concentrate directly on his leg while adding it to the spray tank of his plane experienced serious poisoning (blurred vision, weakness, lightheadedness, followed by vomiting, unconsciousness, cyanosis, and frothing of foamy material from the nose and mouth) within one hour (1). He was successfully treated at a hospital and released after 50 hours. Another pilot who was covered with a spray solution containing TEPP when the hopper in his plane ruptured, collapsed almost immediately and died within 2 days despite aggressive treatment with atropine and pralidoxime (546). In another case, a 6-year-old boy died after aspilling TEPP concentrate over the front of his pants from the groin to the knees (1).

Severe poisoning (characterized by excessive salivation, breathlessness, coughing, staggering, frequent urination, and defecation) occurred among cows when a pasture was sprayed with a 1% dust of TEPP during a thermal inversion. Two cows collapsed, convulsed, and died. The owners of the cows also experienced coughing and shortness of breath (547).

In studies to evaluate the effectiveness of TEPP in treating glaucoma, a 0.01–0.0125% solution produced definite miosis in 20–30 minutes that usually lasted 24 hours (548, 549). A 0.05% solution produced maximal miosis in 20–40 min, and some pupillary constriction remained for 2 days. In addition to miosis, ciliary spasm developed and lasted 12–18 h, but there was no significant change in ocular tension. A 0.1–0.2% solution produced maximal miosis in 7–20 minutes, and it lasted more than 2–3 weeks. The higher concentration also produced twitching of the eyelids, moderate spasms of accommodation for near vision, and aching of the eye and supraorbital area. There was no change in intraocular pressure. In several glaucoma patients, 0.05–0.1% TEPP effectively lowered tension, but in two cases it was actually raised (548–550).

A maintenance dose of 8–12 mg/day TEPP (about 0.1–0.2 mg/kg/day) given in two or three divided doses by mouth was effective in treating myasthenia gravis in atropinized patients (557). Intramuscular or intravascular administration of 1 mg TEPP or more (about 0.0143 mg/kg) resulted in rapid depression of plasma cholinesterase and RBC acetylcholinesterase. Approximately four times as large a dose was required to produce a similar effect when the compound was administered orally. (From a graph presented in the study, an intramuscular dose of about 0.5 mg (about 0.007 mg/kg) reduced plasma cholinesterase and RBC acetylcholinesterase to about 20% and 75% of control, respectively.) Thus, a no-effect level for RBC acetylcholinesterase inhibition following intramuscular administration is probably < 0.007 mg/kg. Maximum depression of plasma cholinesterase occurred within 1 hour of dosing, and the maximum depression of RBC acetylcholinesterase occurred within 2 hours of dosing, regardless of the route of administration. Overt cholinergic symptoms (anorexia, vomiting, swearing, salivation, giddiness, uneasiness, headache, abdominal cramps, diarrhea, etc.) occurred after a single parenteral dose of 5 mg, after 3.6 mg for 2 days, or after 2.4 mg for 3 days. Similar results were obtained following oral dosing of 7.2 mg every 3 hours, three to five times. When symptoms appeared, they usually began suddenly about 30 minutes after the last dose of TEPP. In this same study, an average dose of 41 mg TEPP administered during a period of 5 hours or more was effective in reaching maximal or near maximal strength in eleven patients who had moderately severe and severe myasthenia gravis (550a). The average daily amount for maintaining strength was 16 mg orally in two or three divided doses. The difference that distinguished the dose of TEPP that produced a maximal increase in strength with a minimum of side effects, the dose that produced very little effect, and the dose that produced prohibitive side effects (including increased weakness) was very narrow (2–4 mg higher or lower than the optimum).

In another study of myasthenia gravis, patient's daily doses between 13 and 17 mg/day in two or three divided doses were effective (552). These authors also noted that the difference between the dose required to produce a maximal response and the dose that produced toxicity was remarkably small and ranged from 0.5 to 3 mg.

29.5 Standards, Regulations, or Guidelines of Exposure

The ACGIH TLV for TEPP is 0.05 mg/m^3 with a skin notation (154). The OSHA PEL-TWA for TEPP is 0.05 mg/m^3 with a skin notation. The NIOSH REL-TWA is 0.05 mg/m^3 with a skin notation.

30.0 Terbufos

30.0.1 CAS Number: [13071-79-9]

30.0.2 Synonyms: (S-(((1,1-Dimethylethyl)thio)methyl)-O,O-diethyl phosphorodithioate; AC 92100; phosphorodithioic acid S-[(tert-butylthio)methyl] O,O-diethyl ester;

Contraven; *S-tert*-butylthiomethyl *O,O*-diethyl phosphorodithioate; *O,O*-diethyl *S*-(((1,1-dimethylethyl)thio)methyl) phosphorodithioic acid; phosphorodithioic acid, *O,O*-diethyl *S*-[1,1-dimethylethyl)thio]methyl ester; ST 100

30.0.3 Trade Names: Counter®; Counter 15G®; ENT 27920

30.0.4 Molecular Weight: 288.42

30.0.5 Molecular Formula: $C_9H_{21}O_2PS_3$

30.0.6 Molecular Structure:

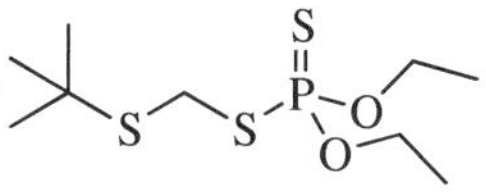

30.1 Chemical and Physical Properties

Terbufos is a clear, slightly brown liquid that hydrolyzes under alkaline conditions

Specific gravity: 1.105 at 24°C

Melting point: −29.2°C

Boiling point: 69°C at 0.1 mmHg

Vapor pressure: 3.2×10^{-4} mmHg at 25°C

Solubility: Soluble in acetone, aromatic hydrocarbons, chlorinated hydrocarbons, alcohols

30.2 Production and Use

Terbufos is an organophosphate insecticide/nematicide that is applied as a granular formulation (15 and 20% active ingredient) by soil incorporation, during planting, or postemergence to terrestrial food and feed crops. Crops treated are corn, grain, sorghum, and sugar beets. Aerial/broadcast treatment is not registered (552a).

30.4 Toxic Effects

30.4.1.1 Acute Toxicity. Terbufos is an organophosphate insecticide that has extremely high acute oral toxicity and oral LD_{50}s of 1.3–9.2 mg/kg for rats, mice, and dogs (552a). "Severe" signs of toxicity (tremors, shallow breathing, motionless crouching, and loss of righting reflex) occurred among deer mice given single oral doses of 1.69 or 2.48 mg/kg terbufos by gavage; mortality was 33% among mice given 1.69 mg/kg and 63% among mice given 2.48 mg/kg. Deaths occurred between 0.4 and 70 hours after dosing, and 90% of mice died between 2 and 9 hours after treatment.

The dermal LD_{50} for terbufos in rabbits is 0.8 to 1.1 mg/kg (552a). Dermal LD_{50}s of 123 mg/kg and 71 mg/kg were reported for rats for a formulated product in which terbufos is mixed with clay, and dermal LD_{50}s of 566 mg/kg and 238 mg/kg were reported for a formulated product in which terbufos is mixed with polymer granules (552a).

Rats inadvertently supplied with bedding contaminated with about 30 ppm terbufos developed cholinergic signs (muscle fasciculations, severe depression, exophthalmus, and ptyalism) that progressed rapidly to death (553). Exposures were most likely primarily dermal but could also have been via ingestion and inhalation. When terbufos was inadvertently mixed into cattle feed (so that doses were about 7.5 mg/kg), all heifers exposed died (554). Cows that received about one-tenth this dose (0.75 mg/kg) developed signs typical of ogonophosphate poisoning (554). The cows most "severely affected" (sic) had whole blood cholinesterase activity that averaged 0.9% of controls, and, cows that were "moderately affected" had whole blood cholinesterase activity that averaged 20% of controls. Inhibition continued for at least 30 days following ingestion of contaminated feed.

Mortality and cholinesterase activity depression (plasma, RBC, or brain not specified) occurred in rats exposed to 0.0394 mg/m^3 (552a). No other details were provided. In another study, rats were exposed to 0, 0.01, 0.02, 0.05, or 0.1 mg/m^3 terbufos for 8 h/day, 5 days/week for two weeks. Significant reductions in RBC cholinesterase activities occurred at 0.05 mg/m^3 (552a). No adverse effects occurred when rats were exposed to about 0.01, 0.025, 0.05, or 0.10 mg/m^3 terbufos for 8 h/day 5 days/week for 3 weeks, although RBC and brain cholinesterase activity decreased at the highest dose (552a).

No signs of cholinergic toxicity or depression of RBC or brain cholinesterase activity occurred at any dose in dogs given 0, 0.00125, 0.005, or 0.015 mg/kg/day terbufos by oral capsule for 28 days (552a). Decreases in RBC and brain cholinesterase activity occurred in rats given dermal doses of 5, 10, or 25 mg/kg/day (equivalent to 1, 2.5, or 6.75 mg/kg/day active ingredient) for 6 h/day, 5 days/week for 4 weeks (552a). A dermal dose of 2 mg/kg/day had no effect.

In primary eye and primary dermal irritation studies in rabbits, all animals died within 24 hours after dosing with 0.5 ml or less of terbufos (552a). No dermal sensitization study has been performed due to the acute lethality of terbufos.

Terbufos was not neurotoxic when administered to hens in a single oral dose of 40 mg/kg in an acute delayed neurotoxicity study (552a). Nor did acute delayed neuropathy occur in hens treated orally or dermally with terbufos (128).

30.4.1.2 Chronic and Subchronic Toxicity. Mesenteric and mandibular lymph node hyperplasia occurred in rats given 0.05 mg/kg/day terbufos, and liver weight and liver extramedullary hematopoiesis increased in rats given 0.025 or 0.05 mg/kg/day terbufos via their diet for 13 weeks (552a).

30.4.1.3 Pharmacokinetics, Metabolism, and Mechanisms. A metabolism study of rats indicated that 83% of a single administration of 0.8 mg/kg terbufos was excreted in the urine in the form of metabolites and 3.5% in the feces during 168 hours. No terbufos accumulated in tissues (552a).

30.4.1.4 Reproductive and Developmental. No adverse reproductive effects occurred in rats given 0.0125 or 0.05 mg/kg/day terbufos for three generations (552a). When rats were given diets that contained 0, 0.5, 1, or 2.5 ppm terbufos for three generations, those fed 2.5 ppm (0.2 mg/kg/day) showed decreased pregnancy rate and male fertility and decreased body weight gain and lower pup weights during lactation (552a). No maternal or developmental effects occurred among rats fed 1 ppm (about 0.08–0.09 mg/kg/day).

Early fetal resorptions, the number of litters that had two or more resorptions, and postimplantation losses increased among rats given 0.2 mg/kg/day terbufos dose by gavage on gestation days 6 through 15 (552a). No adverse developmental effects occurred in rats given 0.05 or 0.1 mg/kg/day. A slight reduction in fetal body weight and an increase in resorptions occurred among rabbits given 0.5 mg/kg/day by gavage on gestation days 7 through 19 (552a). No adverse developmental effects occurred in rabbits given 0.05, 0.1, or 0.25 mg/kg/day.

30.4.1.5 Carcinogenesis. No signs of cholinergic toxicity occurred at any level in dogs given 0, 0.015, 0.06, 0.09, or 0.12 mg/kg/day terbufos via capsule for 1 year (552a). However, initial higher doses of 0.24 and 0.48 mg/kg/day had been reduced after the first 6–8 weeks due to cholinergic-related behavioral signs, reduced food consumption and weight gain, depressed hematology parameters, and gross changes of congestion, edema and necrosis in the gastrointestinal tract. RBC and brain cholinesterase activities were inhibited in dogs given 0.09 and 0.12 mg/kg/day.

No adverse cholinergic effects occurred in rats fed diets that contained 0.125, 0.5, or 1.0 ppm terbufos for 1 year (0.007–0.009, 0.028–0.036, and 0.055–0.071 mg/kg) (552a). Brain cholinesterase activity was reduced among rats given the 1-ppm diet. There was no evidence of neoplastic activity among rats given diets that delivered doses of 0, 0.0125, 0.05, or 0.1 mg/kg/day (which was raised to 0.2 mg/kg/day at week 6, to 0.4 mg/kg/day at week 12, and reduced back to 0.2 mg/kg/day at week 16) for 2 years (552a). Mortality and exopthalmia increased in rats given 0.05 mg/kg/day. RBC cholinesterase inhibition occurred at all doses, and brain cholinesterase inhibition occurred at the 0.05-mg/kg/day dose and higher.

When dietary doses of 0, 0.45, 0.9, or 1.8 mg/kg/day were administered to CD-1 mice for 18 months, mortality and reduction in weight gain increased at the highest dose (552a). There was no evidence of neoplastic activity.

30.4.1.6 Genetic and Related Cellular Effects Studies. Terbufos was positive in a dominant lethal study using rats (552a). However, it was not mutagenic in a variety of other studies tested to cytotoxic levels. Terbufos was negative in the Ames reversion assay with *S. typhimurium* and *E. coli* strains and in the CHO/HGPRT assay (552a). Terbufos did not cause structural chromosomal aberrations in Chinese hamster ovary cells in culture or in an *in vitro* cytogenetics assay in rats, was negative in the rat hepatocyte primary culture/DNA repair test, and did not alter DNA repair in *S. typhimurium* and *E. coli* strains (552a).

30.4.2 Human Experience

Among eleven farmers who applied a formulated terbufos product while planting corn, dermal exposure (assessed using gauze patches) was estimated at an average of 72 µg/h, and respiratory exposure (assessed using personal monitoring pumps) was estimated at an average of 11 µg/h (range 2.8–27.4 µg/h). Exposure duration averaged 7.4 hours. No alkyl phosphates were detected in urine, and no significant depression of RBC cholinesterase activity occurred. Assuming an 8 hour day, 100% absorption, and a

10-m³/day inhalation rate, a respiratory dose of 11 μg/hr would be achieved by exposure to roughly 0.009 mg/m³ terbufos. Thus, this study suggests a human no-observed-effect level ≥0.009 mg/m³ (551).

30.5 Standards, Regulations, or Guidelines of Exposure

Terbufos is undergoing re-registration by the EPA. There is no ACGIH TLV, OSHA PEL-TWA, or NIOSH REL-TWA for terbufos.

31.0 Trichlorfon

31.0.1 CAS Number: [52-68-6]

31.0.2 Synonyms: O,O-Dimethyl 1 (2,2,2-trichloro-1-hydroxyethyl)-phosphonate; DEP; Chlorofos; Metrifonate; DETF; Chlorophos; (2,2,2-trichloro-1-hydroxyethyl)-phosphonic acid dimethyl ester; O,O-dimethyl 1-hydroxy-2,2,2-trichloroethylphosphonate; Bayer L 13/59; Vermicide Bayer 2349; TCF; dimethoxy-2,2,2-trichloro-1-hydroxyethylphosphine oxide; O,O-dimethyl-(1-hydroxy-2,2,2-trichloro)ethyl phosphate; dimethyl 1-hydroxy-2,2,2-trichloroethyl phosphonate; O,O-dimethyl 1-oxy-2,2,2-trichloroethyl phosphonate; dimethyl trichlorohydroxyethyl phosphonate; 1-hydroxy-2,2,2-trichloroethylphosphonic acid dimethyl ester; methyl chlorphos; trichlorophon; trinex; aerol 1; agroforotox; anthon; bay 15922; bayer 15922; bilarcil; bovinox; britten; briton; cekufon; chlorak; chloroftalm; chlorophthalm; chloroxyphos; ciclosom; combot equine; Danex; depthon; dipterax; dipterex 50; diptevur; ditrifon; dylox-metasystox r; Dyrex; Dyvon; equino-acid; equino-aid; flibol e; fliegenteller; forotox; foschlor; foschlor 25; foschlor r; foschlor r-50; hypodermacid; leivasom; loisol; masoten; mazoten; metifonate; metriphonate; neguvon a; phoschlor; phoschlor r50; polfoschlor; ricifon; ritsifon; satox 20wsc; soldep; sotipox; trichlorphon fn; tugon fly bait; Tugon; volfartol; votexit; WEC 50; wotexit; dimethyl (2,2,2-trichloro-1-hydroxyethyl)phosphonate; Briten; dimethyl (2,2,2-trichlorohydroxyethyl) phosphonate; Foschlorine; Metrifonatum; OMS-0800

31.0.3 Trade Names: Dylox®; Dipterex®; Proxol®; Neguvon®

31.0.4 Molecular Weight: 257.44

31.0.5 Molecular Formula: $C_4H_8Cl_3O_4P$

31.0.6 Molecular Structure:

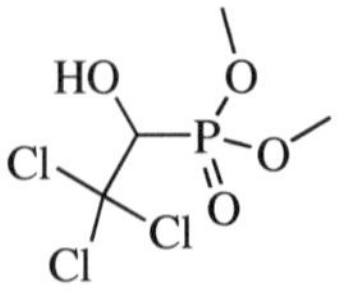

31.1 Chemical and Physical Properties

Pure trichlorfon is a white crystalline solid

    Specific gravity: 1.73
    Melting point: 83–84°C

Boiling point: 100°C

Vapor pressure: 7.8×10^{-6} mmHg at 20°C

Solubility: readily soluble in dichloromethane, 2-propanol; soluble in toluene; nearly insoluble in *n*-hexane; slightly soluble in water (1–5 g/100 ml at 21°C)

31.2 Production and Use

Trichlorfon is used to control a wide variety of lepidopteran larvae, white grubs, mole crickets, sod webworms, leaf miners, stink bugs, ants, and other nuisance pests on outdoor turf and ornamentals; to indoor control flies, ants and roaches; for mound treatment for harvester ants; and to control cattle grubs and cattle lice. It is available as a soluble powder, granules, or bait. For turf, ornamental, or nursery use, it is applied via mechanical or hand-held sprayers, spreaders (for granular formulations), or irrigation systems; for indoor or outdoor perimeter use, it is applied as a soluble powder in water through hand-held sprayers or is applied as the solid; and for livestock, it is poured on from a cup or dipper (554a).

Trichlorfon (termed metrifonate when used therapeutically) has also been widely used as an antihelminthic to treat schistosomiasis (558) and has been extensively investigated as a treatment for Alzheimer's disease (556, 557).

31.4 Toxic Effects

31.4.1.1 Acute Toxicity. Trichlorfon is a moderately toxic organophosphate compound that has oral LD_{50}s of 136–630 mg/kg for rats and 727–866 mg/kg for mice (62, 64a). The dermal LD_{50} for trichlorfon was more than 2000–2800 mg/kg (64a, 81). Thus trichlorfon is considerably less potent on a body weight basis via dermal contact than via ingestion. Four-hour LC_{50}s of 533–1300 mg/m^3 have been reported for rats (123). Death occurs unusually quickly—within 5–15 minutes after dosing (81, 558a). If animals survive, recovery is equally quick and occurs within a few hours.

The single acute oral dose that causes 50% inhibition (ED_{50}) of brain, RBC, and plasma cholinesterase activities in 3-month-old rats was determined and compared to ED_{50}s for dichlorvos, the major *in vivo* nonenzymatic breakdown product of trichlorfon. The oral ED_{50} values for trichlorfon were 90 mg/kg for brain and 80 mg/kg for RBC and plasma cholinesterase, whereas the oral ED_{50} values for dichlorvos were 8 mg/kg for brain and 6 mg/kg for RBC and plasma cholinesterase (559). Thus, trichlorfon is one-tenth to one-fifteenth as potent a cholinesterase inhibitor as dichlorvos.

When rats were given single oral doses of 10, 30, and 100 mg/kg trichlorfon, the 100-mg/kg dose of trichlorfon caused salivation, tremor, diarrhea, ptosis, flat body, decreased body temperature, and decreased pentylenetetrazole seizure threshold, whereas the 30- or 10-mg/kg doses did not (560).

Trichlorfon is unusual among organophosphate pesticides in that several animal studies have demonstrated a beneficial effect of acute low doses of trichlorfon on certain behaviors in rats. For example, in old rats, improvements in spatial reference memory function in the water maze and in passive avoidance learning occurred after acute oral treatment with 10–

30 mg/kg trichlorfon (561, 562). Performance in another behavioral task (Morris water escape) was optimally improved among young adult rats given 10–30 mg/kg trichlorfon or 0.03 mg/kg dichlorvos, the breakdown product of trichlorfon (563), and among rats given 3–30 mg/kg trichlorfon (564). Higher acute doses (80 mg/kg) were required to significantly increase the concentration of cortical acetylcholine and to improve "object recognition" in aged rats — a 30 mg/kg dose had no effect on these measures even though it caused significant acetylcholinesterase inhibition (565). Acute oral treatment with trichlorfon (5–15 mg/kg) has also ameliorated deficits in certain behaviors (water maze performance; passive avoidance tasks) caused by scopolamine or lesions of the basal forebrain (566).

Trichlorfon was moderately irritating to the eye and was a moderate contact allergen in the skin (554a).

31.4.1.2 Chronic and Subchronic Toxicity.
Trichlorfon does not have a cumulative effect. Repeated intraperitoneal doses of 50 mg/kg/day trichlorfon (about one-fourth of the i.p. LD_{50}) for a period of 60 days had no effect on mortality. But repeated doses of 100 mg/kg/day (about one-half LD_{50}) produced 40% mortality by 60 days, and repeated doses of 150 mg/kg produced 100% mortality by 60 days (558a).

Cholinergic toxicity was not observed among rats fed diets that contained 1, 5, 25, or 125 ppm (equivalent to doses of 0.088, 0.39, 2.4, or 11.3 mg/kg/day) for 1 to 13 weeks (81). When rats were given diets that contained 100, 500, or 2500 ppm trichlorfon (about 6–7, 31–35, and 165–189 mg/kg/day) for 13 weeks, cholinergic toxicity and inhibition of RBC and brain cholinesterase activities occurred in male rats given 2500 ppm, and RBC and brain cholinesterase inhibition occurred in females given 500 or 2500 ppm (63).

No signs of cholinergic or other toxicity were seen in dogs fed 42 mg/kg/day trichlorfon for 6 days/week for 3 months (1) or in dogs given diets that contained 50, 200, or 500 ppm trichlorphon for 12 weeks (84). The dietary level of 500 ppm (about 10.5 mg/kg/day) inhibited RBC cholinesterase activity (84).

No cholinergic toxicity occurred when trichlorfon was administered dermally to rabbits for 15 days (4 days/week for 3 weeks) at doses of 0, 100, 300, or 1000 mg/kg/day, although RBC cholinesterase activity was inhibited at 300 mg/kg but not at 100 mg/kg/day (554a).

There is no evidence that trichlorfon causes delayed neuropathy. Hens that received acute subcutaneous doses of 100 or 300 mg/kg trichlorfon showed no visible sign of acute neurotoxicity. Brain, spinal cord, and plasma cholinesterase activities were significantly inhibited, but no significant reductions of neurotoxic esterase activity in these tissues were observed. Subcutaneous doses of 100 mg/kg trichlorfon every 3 days for a total of six doses caused little or no sign of overt neurotoxicity. No inhibition of spinal cord or brain NTE occurred. When trichlorfon was given to hens at doses of 0, 3, 9, or 18 mg/kg/day for 90 days, there were no overt indications of response characteristic of delayed neurotoxicity, although, histologically, a slight effect on nervous tissue, characterized as axonal degeneration was present in hens that received 18 mg/kg/day (554a).

31.4.1.3 Pharmacokinetics, Metabolism, and Mechanisms. Trichlorfon is rapidly and completely absorbed via oral exposure and is extremely rapidly metabolized and excreted. Once absorbed following a 133 mg/kg oral dose, trichlorfon was not detected, but dichlorvos (a nonenzymatic breakdown product of trichlorfon) was distributed primarily in blood, adipose tissue, muscle, and liver. Peak concentrations occurred 1,7,7, and 7 days after exposure, respectively. Dichlorvos was undetected at 20 days. Calculated half-lives in blood, adipose tissue, muscle, and liver were 7, 11, 10, and 12 days, respectively (388).

Trichlorfon and dichlorvos levels were followed in plasma and RBCs of seven individuals given single oral doses of 7.5–10 mg/kg trichlorfon repeated after 2 weeks for treating schistosomiasis (574). The relationship of dichlorvos to trichlorfon in plasma and RBCs was about 1%. A biphasic curve for the elimination of trichlorfon in Plasma developed; the first phase had a half-life of 0.4 to 0.6 hours, and the second phase had a half-life of about 3 hours. Clearance of trichlorfon was primarily due to formation of dichlorvos.

Regardless of whether rats were given single or repeated oral doses or single intravenous doses of trichlorfon, 80–90% of the dose was excreted within 24 hours. The major route of excretion was via the urine, followed by feces and expired air. About 1 to 2% of the dose was found in the tissues after 96 hours.

Several groups of investigators examined the pharmacokinetics of trichlorfon in humans in conjunction with assessing its efficacy as a treatment for schistosomiasis and Alzheimer's disease. These studies confirmed that trichlorfon (and its decomposition product dichlorvos) is very rapidly absorbed and cleared in humans. After acute oral treatment of healthy male volunteers with a 2, 5, 7.5, or 12 mg/kg dose of trichlorfon (metrifonate), the maximum blood concentration of trichlorfon was obtained between 12 minutes and 2 hours and the half-life in blood was about 2 hours (575, 576). The concentrations of dichlorvos, the nonenzymatic breakdown product of trichlorfon, closely followed those of trichlorphon at a constant ratio of about 1 to 100. The concentrations of trichlorfon were detectable up to 8 hours, but those of dichlorvos had fallen below the level of detection by then. Both plasma and RBC cholinesterases were readily inhibited and were still low after 24 hours — none of the volunteers complained of side effects (and the half-life of the breakdown product, dichlorvos, was about 3.8 hours).

Pharmacokinetic analysis of both trichlorfon and dichlorvos in blood on the first and sixth day of a 21-day treatment regimen indicated that, regardless of maintenance dose (0.25 to 1.0 mg/kg/day), the half-lives of elimination were the same (about 2 or 3 hours for metrifonate and dichlorvos, respectively). This confirms that there was little or no accumulation of either of these compounds from long-term administration (556). Blood concentrations of dichlorvos were approximately 2% of the parent (trichlorfon) compound concentrations.

The predominant metabolic pathway that involves cleavage of the P–C phosphonate bond that generates trichloroethanol and dimethyl phosphate which are then excreted in urine (576a). Quantitatively minor pathways of metabolism include demethylation (to dimethyl trichlorfon) and (nonenzymatic) dehydrochlorination to dichlorvos which is rapidly metabolized to dichloroethanol and dimethyl phosphate which are then excreted in urine.

Trichlorfon itself does not act significantly as a cholinesterase inhibitor, but induces cholinesterase inhibition through its hydrolytic degradation product, dichlorvos (559).

31.4.1.4 Reproductive and Developmental. When rats were fed diets that contained 0, 300, 1000, or 3000 ppm for three generations, 3000 ppm (about 150 mg/kg/day) caused a marked decrease in the pregnancy rate and early deaths in pups, 1000 ppm (about 50 mg/kg/day) caused reduced numbers of pups per litter and weight of individual pups, and 300 ppm (about 15 mg/kg/day) had no detectable effect on reproduction. There was no indication of teratogenesis, even at dosages that were highly toxic (1). When rats were fed diets that contained 0, 150, 500, or 1750 ppm (about 0, 15, 50, and 175 mg/kg/day) for two generations, pulmonary and renal lesions in the F1 generation and adverse reproductive outcomes (dilated renal pelvises in F1 pups and decreased F1 pup weight on days 7 and 21) occurred at 1750 ppm (554a).

Decreases in brain and RBC cholinesterase activities and an increased incidence of abortion occurred among rabbits given 35 or 110 mg/kg/day trichlorfon by gavage on gestation days 6 through 18 (554a). Increased numbers of resorptions, decreased fetal body weights (males), and delayed ossification occurred among rabbits given 110 mg/kg/day. Decreased cholinesterase activity (*sic*) and reduced ossification of skulls, vertebrae, and sternebrae in fetuses occurred among rats fed diets that contained 2500 ppm trichlorfon (equivalent to 45 mg/kg/day) on gestation days 6 through 15 of gestation (554a). Dietary exposures of 500 or 1125 ppm (about 102 or 227 mg/kg/day) had no adverse effect.

Trichlorfon was teratogenic when given to pregnant CD rats by gavage at 480 mg/kg/day on gestation days 6 through 15, but not when given only on gestation days 8 or 10 alone. Signs of cholinesterase inhibition occurred in dams after each dose. Teratogenic responses (increased fetal death and stunted and malformed fetuses) also occurred in hamsters after administration of 300 or 400 mg/kg/day on gestation days 7 through 11, but not after 200 mg/kg/day. Embryotoxicity, but not teratogenicity, occurred after administration of 400 mg/kg/day only on day 8 of gestation. In mice, there was a significant increase in the incidence of cleft palate following treatment with 600 mg/kg/day on gestation days 10 through 14 and on days 12 through 14 (567).

Congenital ataxia and tremor occurred in piglets of sows treated with trichlorfon between days 45 and 63 of pregnancy (1). Usually all of the piglets in any given litter were affected, and many died. Autopsy revealed marked cerebellar and spinal hypoplasia. Retrospective study showed that the smallest dose capable of causing this effect was 56 mg/kg. In another study, cerebellar hypoplasia occurred in piglets from sows treated with 60 mg/kg on day 55 or on days 55 and 70 of gestation (568). Similar cerebellar lesions were observed in guinea pigs treated with trichlorfon during prenatal development. Dose–response studies showed that 100 mg/kg trichlorfon given to guinea pigs on 3 consecutive days during days 40–50 of gestation resulted in offspring that had brain hypoplasia, ataxia, and tremors (569). When trichlorfon (125 mg/kg/day) or dichlorvos (15 mg/kg/day) was administered to guinea pigs between day 42 and 46 of gestation, offspring exhibited severe reductions in brain weight that was most pronounced in the cerebellum, medulla oblongata, thalamus/hypothalamus, and quadrigemina (570).

ORGANOPHOSPHORUS COMPOUNDS

31.4.1.5 Carcinogenesis. When trichlorfon was administered to monkeys via Tang orange drink at doses of 0, 0.2, 1.0, or 5.0 mg/kg/day for 6 days/week for 10 years, RBC and brain cholinesterase activities decreased in those given 0.2 mg/kg/d. Monkeys given 5.0 mg/kg/day exhibited decreased body weight, anemia, and transitory signs of cholinesterase inhibition (pupillary constriction, muscle fasciculation, and diarrhea; females only) (554a).

There were no effects on mortality or body weights nor was there any cholinergic toxicity in dogs given diets that contained 0, 50, 250, 500, or 1000 ppm (equivalent to 0, 1.25, 6.25, 12.5, or 25 mg/kg/day, respectively) for 1 year (554a). There was mild to moderate enlargement of the spleen in dogs fed the 1000-ppm diet, as well as congestion of the spleen and lymphoid atrophy and foci of inflammatory liver cells. Dogs given the 500- or 1000-ppm diet exhibited decreases in RBC cholinesterase activity. When dogs were fed diets that contained 0, 50, 200, 800, or 3200 ppm for 4 years, dogs fed the 50-ppm diet were unaffected; dogs fed the 200-ppm diet had depressed RBC cholinesterase activity; dogs fed the 800- or 3200-ppm diet had depressed RBC cholinesterase activity, reduced food intake, retarded body weight gain, and increased mortality; and, dogs fed the 3200-ppm diet also exhibited tremors, cramps, and salivation (588).

When rats were fed diets that contained 0, 50, 100, 200, 250, 400, 500, or 1000 ppm for 17–24 months, no treatment-related effects occurred in those fed 50 to 250 ppm (588). Serum cholinesterase activity was depressed in rats fed 500 ppm trichlorfon, and serum, RBC, and tissue (submaxillary gland) cholinesterase activities were depressed in rats fed 1000 ppm trichlorfon. Histopathological results that suggested the occurrence of mammary tumors, the absence of primary follicles and primitive ova, and necrotizing arteritis in rats fed 400, 500, and 1000 ppm; and tubular and roblastomas, focal aspermatogenesis, decreased growth rats, and decreased survival in rats fed 1000-ppm were equivocal (345). However, in another study, when rats were fed diets that contained 0, 50, 250, 500, or 1000 ppm trichlorfon for 24 months, no treatment-related effects other, than whole blood cholinesterase depression at 1000 ppm occurred (588). There was no increase in incidence of either benign or malignant tumors, including mammary tumors. Nor was there any indication of cystic atrophy, tubular hyperplasia in the ovaries, or aspermatogenesis.

When rats were fed diets that contained 0, 100, 300, 1514 (males), or 1750 ppm (females) trichlorfon (equivalent to about 0, 4.4–5.8, 13.3–17.4, and 75–94 mg/kg/d) for 24 months, the females given 1750 ppm exhibited rough hair coats, granular kidneys, foci in lungs, decreased body weight gain, and anemia, but no evidence of oncogenicity (554a). The males given 1514 ppm exhibited paleness and hunched backs, thickened enlarged duodenum, thickened, granular nonglandular stomachs, decreased body weight gain, anemia, and an increase in the incidence of benign pheochromocytomas and mononuclear cell leukemia. The EPA Office of Pesticide Programs (OPP) Carcinogenicity Peer Review Committee, however, concluded that these lesions were not compound-related.

When rats were fed diets that contained 2500 ppm trichlorfon (equivalent to 129–159 mg/kg/day), they exhibited signs of cholinergic toxicity, decreased body weight and body weight gain, decreased RBC parameters (hematocrit, hemoglobin, RBC count and MCV), hypercholesterolemia, increased serum hepatic enzymatic (SAP, AST, ALT, GGT) activity, and decreased brain and RBC cholinesterase activity. Nonneoplastic lesions included duodenal hyperplasia, gastritis, pulmonary hyperplasia and inflammation, nasolacrimal inflammation, hepatocellular hyperplasia and vacuolation, chronic nephropathy, and dermal lesions. There was an increase in the incidence of alveolar/bronchiolar adenomas in males, renal tubular adenomas in males, and alveolar/bronchiolar carcinomas in females that were not statistically significant, but the EPA noted that they were outside the historical control range for these types of tumors. There was no compound-related increase in the incidence of either benign pheochromocytomas or mononuclear cell leukemia (554a).

Cholinergic signs and inhibition of brain and RBC cholinesterase activities occurred in all mice given diets that contained 0, 300, 900, or 2700 ppm trichlorfon (equivalent to 0, 45, 135, and 405 mg/kg/day) for 24 months (554a). Mortality in female mice was significantly related to dose. There was an increase in lung tumors in low- and mid-dose females, but not in high-dose females. Therefore, the EPA OPP Carcinogenicity Peer Review Committee concluded that the tumors were not dose-related and trichlorfon was not carcinogenic in this study.

31.4.1.6 Genetic and Related Cellular Effects Studies. Trichlorfon is sometimes mutagenic in *in vitro* systems (554a). In gene mutation assays with *S. typhimurium*, trichlorfon was weakly mutagenic at toxic concentrations with or without activation (554a). Trichlorfon produced base-pair substitution mutations in *S. typhimurium*, although mutagenic activity decreased in the presence of S9 (594). Trichlorfon was not mutagenic with or without activation in one gene mutation assay with *S. cerevisiae* (554a).

In other bacterial or prokaryotic assays, trichlorfon was positive for DNA damage and repair in *S. typhimurium* but was negative in relative toxicity assays with *E. coli* and *B. subtilis* strains (554a). In a DNA damage and repair study conducted with *S. Cerevisiae*, trichlorfon was positive for mitotic recombination with and without S9 activation (554a). In a recombinant DNA study, trichlorfon did not inhibit the growth of *Bacillus subtilis* (554a).

The genotoxity of trichlorfon in mammalian cells has also been equivocal. Trichlorfon induced significant increases in mutation frequencies with or without activation in an *in vitro* cytogenetic study in mammalian cells (type not specified) (554a). Trichlorfon was inactive in inducing unscheduled DNA synthesis in rat hepatocytes up to levels of severe cytotoxicity (554a). At cytotoxic levels of 1000 µg/mL, trichlorfon was associated with a marginal but significant increase in sister chromatid exchange in Chinese hamster ovary cells (1). Trichlorfon was clastogenic in human lymphocytes in the absence of S9 activation (554a) and also reportedly induced sister chromatid exchange in human lymphocytes. In other studies, trichlorfon reportedly slightly increased the frequency of chromosomal abnormalities in cultured human lymphocytes (571).

Under physiological conditions, trichlorfon alkylates DNA (through its conversion to dichlorvos which is more easily demethylated), and an increased rate of chromosomal aberrations in bone marrow cells of Syrian hamsters and mice treated with trichlorfon has been observed. In mice that received a single dose of trichlorfon, liver DNA adducts increased maximally 6 hours after dosing, were substantially reduced 24 hours after dosing, and were undetectable by 48 hours after dosing (584a). Increases in sister chromatid exchanges occurred in the bone marrow of mice treated 24 hours earlier with 120 mg/kg, but not with 30 or 50 mg/kg trichlorfon (584). Some researchers have obtained positive effects in dominant lethal mutation tests with mice (595), but some have not (571). No chromosomal aberrations occurred in the bone marrow or spermatogonia of mice treated with acute doses of 100 mg/kg trichlorfon, and dominant lethal mutations did not significantly increase in mice given 5 ppm trichlorfon for 5 days/week for 7 weeks (571).

Significant and persistent increases (lasting at least 180 days after exposure) in aneuploidy in lymphocytes of individuals who attempted suicide by using trichlorfon (Ditriphon-50) were reported (572). Trichlorfon also produced aneuploidy in genetically engineered human lymphoblastoid cell lines (573).

31.4.2 Human Experience

31.4.2.2 Clinical Cases. Trichlorfon has been associated with numerous cases of accidental or incidential poisonings (1, 572). Gallo and Lawryk (1) reviewed 379 cases of trichlorfon poisoning and found that in some cases (3%), acute poisonings were accompanied by "mental disturbance" (described as loss of memory and problem-solving ability, delirium, depression and anxiety, psychomotor stimulation, hallucinations, and paranoid delusions); and in other cases (21%), poisoning was accompanied by a delayed type polyneuropathy. It is not clear, however, whether these symptoms were due to trichlorfon itself or to some contaminant in the ingested material. Estimates of dosages associated with poisonings suggested that ingestion of 40 mg/kg causes light to moderate poisoning, ingestion of 80–700 mg/kg causes severe poisoning, and ingestion of 30,000–90,000 mg/kg causes death (1).

Trichlorfon-contaminated fish was implicated as the cause of a cluster of congenital abnormalities that occurred in a Hungarian village in 1989–1990 (584a). Eleven of fifteen live births, were affected by congenital abnormalities (four with Down's syndrome) and six were twins. Examination of this group along with two negative control groups (mothers and their children born in the same village in 1987–1988, and mothers and children born in 1989–1990 in a nearby village) and a positive control group (mothers and their children who had congenital abnormalities from the same village) indicated that all mothers of affected children had eaten "contaminated" fish during pregnancy, whereas only about one-third of mothers from all control groups combined had. Other potential causes for the cluster, such as known teratogens, familial inherited disorders, and consanguinity were excluded as contributing causes. The "contaminated" fish were obtained from a pond which had been heavily treated with trichlorfon to eradicate parasites. The content of trichlorfon in fish was 0.15 to 0.26 mg/kg and, it was estimated was as high as 100 mg/kg.

In a poorly described study, air concentrations of trichlorfon (dipterex) and blood cholinesterase activity were monitored among employees who worked in a trichlorfon (dipterex) packing facility in China (577). Estimates of mean 8-hour time-weighted average air concentrations (aerosol or dust not specified) ranged from about 0.2 to 0.6 mg/m^3 estimated dermal exposure (mainly to face, neck, and hands a 44 cm^2 area) ranged from about 0.4 to 7.2 µg/cm^2, and blood cholinesterase activity decreased to about 23% of preexposure baseline levels after about 2-month's work. The occurrence of overt cholinergic effects was evidently not examined. Unfortunately, specific types of exposures were not thoroughly described, and it was not possible to distinguish between the relative contribution of inhalation and dermal exposures to an internal dose.

Trichlorfon has been widely used as an antihelminthic (1, 578). The formulations of trichlorfon that have been used for treating people infested by worms have frequently produced mild, rapidly reversible side effects (560). The dose ordinarily employed is 5 to 15 mg/kg given orally three times at intervals of 2 weeks. (555, 579). When used as an antihelminthic, a total dose as high as 37.7 mg/kg produced very mild poisoning, and dose of 10 mg/kg/day had no untoward effects. However, mild poisoning was reported among other persons when they received daily doses as low as 5 mg/kg/day for 12 days. A single dose at the rate of 10 mg/kg was tolerated (1).

Trichlorfon has been proposed as a treatment for Alzheimer's disease based on its ability to elevate brain levels of acetylcholine because of its relatively long lasting inhibition of acetylcholinesterase. Clinical trials in Alzheimer's disease patients suggest that maintaining RBC cholinesterase activity at levels that are 40–60% of predose baseline through daily dosing with trichlorfon can significantly improve "cognitive ability" (557, 580). Effective therapeutic doses were a 2 mg/kg/day loading dose for 2 weeks followed by a maintenance dose of 0.65 mg/kg/day. This regimen yielded significant RBC cholinesterase inhibition and "mild and transient" cholinergic adverse effects that were primarily gastrointestinal. Lower doses—a 0.5 mg/kg/day loading dose for 2 weeks followed by a 0.2 mg/kg/day maintenance dose—significantly inhibited RBC cholinesterase activity but did not affect other outcome measures.

In a 21-day study, Alzheimer's disease patients received loading doses of 1.5, 2.5, 4.0, or 4.0 mg/kg/day for 6 days followed by maintenance doses of 0.25, 0.40, 0.65, or 1.0 mg/kg/day, respectively, for 15 days (556). After 21 days of treatment, RBC acetylcholinesterase inhibition occurred at all doses. In another study designed to evaluate the safety and tolerability of relatively high loading doses followed by lower maintenance doses and to determine the maximum tolerated dose of trichlorfon, groups of probable Alzheimer's disease patients were administered either 2.5 mg/kg/day for 14 days followed by 4.0 mg/kg/day for 3 days, then 2.0 mg/kg/day for 14 days or 2.5 mg/kg/day for 14 days followed by 1.5 mg/kg/day for 35 days (581). RBC acetylcholinesterase inhibition occurred in all groups. Moderate to severe cholinergic effects (muscle cramps, abdominal discomfort, headache, muscle weakness, generalized moderate to severe muscle cramps, weakness, inability to resume daily activities, and coordination difficulties) occurred in six of eight patients given the higher doses (4.0 mg/kg/day for 3 days and 2.0 mg/kg/day for 14 days); mild to moderate cholinergic effects (gastrointestinal disturbances, muscle cramps, and lightheadedness/dizziness) occurred among patients given the lower maintenance dose (1.5 mg/kg/day).

In some cases, weekly rather than daily dosing has been used. When doses were 2.5, 5, 7.5 and 15 mg/kg/week, no side effects occurred at 2.5 mg/kg, but did occur in patients who received 15 mg/kg (582). Nausea, vomiting and diarrhea were most commonly reported. Sixteen patients had electroencephalographic (EEC) abnormalities. When a weekly dose of about 5 mg/kg was given for 3 or 4 weeks followed by a weekly dose of 2.1 mg/kg for up to 6 months, adverse effects were reported as "mild and transient" (583).

ORGANOPHOSPHORUS COMPOUNDS

When patients were given 2 mg/kg/day trichlorfon for 5 days and 0.95 mg/kg on day 6, followed by 2.9 mg/kg weekly, adverse effects were uncommon (583).

31.5 Standards, Regulations, or Guidelines of Exposure

Trichlorphon is undergoing reregistration by the EPA. There is no ACGIH TLV, OSHA PEL-TWA, or NIOSH REL-TWA for trichlorphon.

SUMMARY

Tables 95.1–95.3 summarize toxicity data for compounds listed in text.

BIBLIOGRAPHY

1. M. A. Gallo and N. J. Lawryk, Organic phosphorus pesticides. In W. J. Hayes and E. R. Laws Jr., Eds., *Handbook of Pesticide Toxicology, Classes of Pesticides Ed*, Acadamic Press Inc., New York, pp. 1040–1049 (1991).
2. H. W. Chambers, organophosphorus compounds, an overview. In J. E. Chambers and P. E. Levi, Eds., *Organophosphates. Chemistry, Fate and Effects*, Academic Press, Inc., San Diego CA, 1992, pp. 3–18.
3. Environmental Protection Agency (EPA), *Recognition and Management of Pesticide Poisonings.* EPA 735-R-98-0.03, 1999.
4. T. C. Marrs, Organophosphate poisoning, *Pharmac. Ther.* **58**, 51–63 (1993).
5. International Programme on Chemical Safety (IPCS), Organophosphorus insecticides, a general introduction. *Environmental Health Criteria* **63**, 177 (1986).
6. D. J. Ecobichon, Pesticides. In C. D. Klaassen, Ed., Casarett and Doull's Toxicology. *The Basic Science of Poisons*, 5th Ed., McGraw-Hill, New York, 1996, pp. 643–689.
7. D. J. Ecobichon, Ed., *Occupational Hazards of Pesticide Exposure. Sampling, Monitoring, Measuring*, Taylor Francis, Philadelphia PA, 1999.
8. World Health Organization (WHO), *Biological Monitoring of Chemical Exposure in the Workplace. Guidelines*, Selected pesticides, 1996, Chapt. 5.
9. J. E. Storm, K. K. Rozman, and J. Doull, Occupational Exposure Limits (OELs) for 30 organophosphate pesticides based on inhibition of red blood cell (RBC) acetylcholinesterase, Toxicology **150**, 1–29 (2000).
9a. EPA, Office of Pesticide Programs, Status summary of the organophosphate review process, http://www.epa.gov/pesticides/op/status.htm (accessed October 2000).
10. B. E. Mileson, J. E. Chambers, W. L. Chen, et al., Common mechanism of toxicity, a case study of organophosphorus pesticides. *Tox. Sci.* **41**, 8–20 (1998).
11. Environmental Protection Agency (EPA), *Science Policy on the Use of Data on Cholinesterase Inhibition for Risk Assessment. Office of Pesticide Programs*, FAX-On-Demand (202-401-0527). Item 6022, 1998.
12. B. Ballantyne and T. C. Marrs, Overview of the biological and clinical aspects of organophosphates and carbamates. In B. Ballantyne and T. C. Marrs, Eds., *Clinical and Experimental Toxicology of Organophosphates and Carbamates*, Butterworth-Heinemann Ltd, Oxford, 1992, pp. 3–14.

12a. D. M. Maxwell and D. E. Lenz, Structure-activity relationship and anticholinesterase activity. In B. Ballantyne and T. C. Marrs, Eds.,*Clinical and Experimental Toxicology of Organophosphates and Carbamates*, Butterworth-Heinemann, Oxford, 1992, pp. 47–58.

13. H. Chambers, B. Brown, and J. E. Chambers, Noncatalytic detoxication of six organophosphorous compounds by rat liver homogenates. *Pest. Biochem. Physiol.* **36**, 308–315 (1990).

14. J. E. Chambers and R. L. Carr, Inhibition patterns of brain acetylcholinesterase and hepatic and plasma aliesterases following exposures to three phosphorothionate insecticides and their oxons in rats. *Fund. Appl. Toxicol.* **21**, 111–119 (1993).

15. J. E. Chambers et al., Role of detoxication pathways in acute toxicity levels of phosphorothionate insecticides in the rat. *Life Sci.* **54**, 1357–1364 (1994).

16. T. Ma and J. E. Chambers, A kinetic analysis of hepatic microsomal activation of parathion and chlorpyrifos in control and phenobarbital-treated rats. *J. Biochem. Toxicol.* **10**, 63–68 (1995).

17. A. L. Pond, H. W. Chambers, and J. E. Chambers, Organophosphate detoxication potential of various rat tissues via a-esterase and aliesterase activities. *Toxicol. Letters* **78**, 245–252 (1995).

18. T. T. Atterberry, W. T. Burnett, and J. E. Chambers, Age-related differences in parathion and chlorpyrifos toxicity in male rats, target and nontarget esterase sensitivity and cytochrome P450-mediated metabolism. *Toxicol. Appl. Pharmacol.* **147**, 411–418 (1997).

19. M. T. Chambers, T. Ma, and J. E. Chambers, Kinetic parameters of desulfuration and dearylation of parathion and chlorpyrifos by rat liver microsomes. *Food Chem. Toxicol.* **32**, 763–767 (1994).

20. R. A. Neal and K. P. DuBois, Studies on the mechanism of detoxification of cholinergic phosphorothioates. *J. Pharmacol. Exp. Ther.* **148**, 185–192 (1965).

21. H. X. Zhang and L. G. Sultatos, Biotransformation of the organophosphorus insecticides parathion and methyl parathion in male and female rat livers perfused *in situ*. *Drug Metab. Disp.* **19**, 473–477 (1991).

22. K. Steenland, B. Jenkins, R. G. Ames, et al., Chronic neurological sequelae to organophosphate poisoning. *Am. J. Public Health* **84**, 731–736 (1994).

23. E. Savage, T. Keefe, L. Mounce, et al., Chronic neurological sequelae of acute organophosphate pesticide poisoning. *Arch. Environ. Health* **43**, 38–45 (1988).

24. L. Rosenstock, M. Keifer, and W. Daniell et al., Chronic central nervous system effects of acute organophosphate pesticide intoxication. *Lancet* **338**, 223–227 (1991).

25. S. Gershon and F. H. Shaw, Psychiatric sequelae of chronic exposure to organophosphorus insecticides. *Lancet* **1**, 1371–1374 (1961).

26. D. R. Metcalf and J. H. Holmes, EEG, psychological, and neurological alterations in humans with organophosphorus exposure. *Ann NY Acad. Sci.* **160**, 357–365 (1969).

27. A. Hirshberg, and Y. Lerman, Clinical problems in organophosphate insecticide poisoning The use of a computerized information system. *Fund. Appl. Toxicol.* **4**, S2009–2214 (1984).

27a. J. H. Holmes and M. D. Gao, Observations on acute and multiple exposure to anticholinesterase agents. *Trans. Am. Clin. Climatol Assoc.* **68**, 86–103 (1957).

28. C. S. Miller and H. C. Mitzer, Chemical sensitivity attributed to pesticide exposure versus remodeling. *Arch. Environ. Health* **50**, 119–129 (1995).

29. L. Ritter, Report of a panel on the relationship between public exposure to pesticides and cancer. *Cancer* **80**, 2019–2033 (1997).

30. B. N. Ames and L. S. Gold, Environmental pollution, pesticides, and the prevention of cancer: Misconceptions. *FASEB Journal* **11**, 1041–1052 (1997).

31. H. R. Wolfe et al., Exposure of fertilizer mixing plant workers to disulfoton. *Bull. Environ. Contam. Toxicol.* **20**, 79–86 (1978).
32. W. F. Durham and H. R. Wolfe, Measurment of the exposure of workers to pesticides. *WHO Bull.* **26**, 75–91 (1962).
33. W. F. Durham and W. J. Hayes, Organic phosphorus poisoning and its therapy. *Arch. Environ. Health* **5**, 27–53 (1962).
34. W. F. Durham, H. R. Wolfe, and J. W. Elliott, Absorption and excretion of parathion by spraymen. *Arch. Environ. Health* **24**, 381–387 (1972).
35. J. Griffith and R. C. Duncan, Exposure of agricultural workers to anticholinesterases. In B. Ballantyne and T. C. Marrs, Eds., *Clinical and Experimental Toxicology of Organophosphates and Carbamates*, Butterworth-Heinemann Ltd, Oxford, 1992, pp. 339–345.
36. C. M. Riley and C. J. Wiesner, On-target and off-target deposition. In D. Ecobichon, Ed., *Occupational Hazards of Pesticide Exposure. Sampling, Monitoring, Measuring*, Taylor & Francis, Philadelphia, PA, 1999, pp. 9–50.
37. E. Thorpe, A. B. Wilson, and K. M. Dix, Teratological studies with dichlorvos vapor in rabbits and rats. *Arch. Toxicol.* **30**, 29–38 (1972).
38. J. P. Frawley, E. C. Hagan, and O. G. Fitzhigh, A comparative pharmacological and toxicological study of organic phosphate-anticholinesterase compounds. *J. Pharmacol. Exper. Ther.* **105**, 156–165 (1952).
39. R. L. Carr and J. E. Chambers, Acute effects of the organophosphate paraoxon on schedule-controlled behavior and esterase activity in rats: Dose-response relataionships. *Pharmacol. Biochem. Behav.* **40**, 929–936 (1991).
40. D. A. Jett et al., Differential regulation of muscarinic receptor subtypes in rat brain regions by repeated injections of parathion. *Toxicol. Letters* **73**, 33–41 (1994).
41. D. D. McCollister, F. Oyen, and V. Rowe, Toxicological studies of O,O-Dimethyl-O-(2,4,5-trichlorophenyl) phosphorothioate (ronnel) in laboratory animals. *J. Agric. Food Chem.* **7**, 689–693 (1959).
42. T. B. Gaines, R. Kimbrough, and E. R. Laws, Toxicology of abate in laboratory in animals. *Arch. Environ. Health* **14**, 283–288 (1967).
43. A. C. Nostrandt, S. Padilla, and V. Moser, The relationship of oral chlorpyrifos effects on behavior, cholinesterase inhibition, and muscarinic receptor density in rats. *Pharmacol. Biochem. Behav.* **58**, 15–23 (1997).
43a. J. D. Atterberry et al., *Arch. Environ. Health* **3**, 112–121 (1961).
44. S. Padilla, V. Z. Wilson, and P. J. Bushnell, Studies on the correlation between blood cholinesterase inhibition and 'target tissue' inhibition in pesticide-treated rats. *Toxicol.* **92**, 11–25 (1994).
45. S. Padilla, *Toxicology* **122**, 215–220 (1995).
46. M. M. Winthrobe and G. R. Lee, Hematologic alterations. In M. M. Winthrobe, et al. Eds., *Harrison's Principles of Internal Medicine*, McGraw Hill, New York, 1974, pp. 28–38.
47. G. T. Sidell, Soman and Sarin: Clinical manifestations and treatment of accidental poisoning by organophosphates. *Clin. Toxicol.* **7**, 1–17 (1974).
48. A. M. Butler and M. Murray, Biotransformation of parathion in human liver, participation of CYP3A4 and its inactivation during microsomal parathion oxidation. *J. Pharmacol. Exp. Ther.* **280**, 966–973 (1997).

49. L. G. Sultatos, L. D. Minor, and S. D. Murphy, Metabolic activation of phosphorothionate pesticides: Role of the liver. *J. Pharmacol. Exp. Ther.* **232**, 624–628 (1985).

50. L. G. Sultatos, Metabolic activition of the organophosphorus insecticides chlorpyrifos and fenitrothion by perfused rat liver. *Toxicology* **68**, 1–9 (1991).

51. B. Nutley and J. Cocker, Biological monitoring of workers occupationally exposed to organophosphorus pesticides, *Pest. Sci.* **38**, 315–322 (1993).

52. S. J. Reid and R. R. Watts, A method for the determination of dialkyl phosphate residues in urine. *J. Anal. Toxicol.* **5**, 126–132 (1981).

53. Agency for Toxic Substances and Disease Registry (ATSDR), *Toxicological Profile for Methyl Parathion*, U.S. Department of Health and Human Services, 1992.

54. Agency for Toxic Substances and Disease Registry (ATSDR), *Toxicological Profile for Disulfoton*, U.S. Department of Health and Human Services, 1995.

55. Agency for Toxic Substances and Disease Registry (ATSDR), *Toxicological Profile for Diazinon*, U.S. Department of Health and Human Services, 1996.

56. Agency for Toxic Substances and Disease Registry (ATSDR), *Toxicological Profile for Chlorpyriphos*, U.S. Department of Health and Human Services, 1997.

57. Agency for Toxic Substances and Disease Registry (ATSDR), *Toxicological Profile for Dichlorvos*, U.S. Department of Health and Human Services, 1977.

57a. C. A. Franklin et al., *J. Toxicol. Environ. Health* **7**, 715–731 (1981).

58. G. Carrier and R. C. Brunet, A toxicokinetic model to asses the risk of azinphosmethyl exposure in humans through measures of urinary elimination of allyl phosphates. *Toxicol. Sci.* **47**, 23–32 (1999).

59. C. Aprea et al., Biological monitoring of exposure to organophosphorus insecticides by assay of urinary alkylphosphates: Influence of protective measures during manual operations with treated plants. *Arch. Occup. Environ. Health* **66**, 333–338 (1994).

60. S. A. McCurdy et al., Assessment of azinphosmethyl exposure in California peach harvest workers. *Arch. Environ. Health* **49**, 289–296 (1994).

61. Toxicological studies with the compound BAY 68138, unpublished report. Bayer AG, Wuppertal, Germany (1967).

62. T. J. Haley et al., Estimation of the LD1 and extrapolation for five organothiophosphate pesticides. *Arch. Toxicol.* **34**, 102–109 (1975).

62a. D. H. Snow and A. D. J. Watson, *Pathol. Aust. Vet. J.* **49**, 113–119 (1973).

63. I. P. Sheets et al., Subchronic neurotoxicity screening studies with six organophosphate insecticides: An assessment of behavior and morphology relative to cholinesterase inhibition. *Fund. Appl. Toxicol.* **35**, 101–119 (1997).

64. A. N. Worden et al., Toxicity of guthion for the rat and dog. *Toxicol. Appl. Pharmacol.* **24**, 405–412 (1973).

64a. T. B. Gaines, Acute toxicity of pesticides. *Toxicol. Appl. Pharmacol.* **14**, 515–534 (1969).

65. National Cancer Institute (NCI), *Bioassay of Azinphosmethyl for Possible Carcinogenicity*. Technical Report Series No. 24. NCI Carcinogenesis, National Cancer Institute, Bethesda, MD, 1978.

66. R. J. Feldman and H. I. Maibach, Percutaneous penetration of some pesticides and herbicides in man. *Toxicol. Appl. Pharmacol.* **28**, 126–132 (1974).

67. S. D. Murphy and K. P. DuBois, Enzymatic conversion of the divethoxy ester of benzotrizine dithiophosphoric acid to an anticholinesterase agent. *J. Pharmacol. Exp. Ther.* **119**, 572–583 (1957).

68. Environmental Protection Agency (EPA), *Dichlorvos (DDVP), Risk Assessment Issues for the FIFRA Science Advisory Panel*, Background document for FIFRA SAP July 8, 1998.

69. C. A. Franklin, Estimation of dermal exposure to pesticides and its use in risk assessment. *Can. J. Physiol. Pharmacol.* **62**, 1037–1039 (1984).

70. J. F. Kraus et al., Physiological response to organophosphate residues in field workers. *Arch. Environ. Contamin.* **5**, 471–485 (1977).

71. V. Drevenkar et al., *Arch. Environ. Contam. Toxical.* **20**, 417–422 (1991).

72. G. E. Quinby, K. C. Walker, and W. F. Durham, Public health hazards involved in the use of organic phosphorus insecticides in cotton culture in the Delta area of Mississippi. *J. Econ. Entomol.* **6**, 831–838 (1958).

73. Z. Jegier, Exposure to guthion during spraying and formulation. *Arch. Environ. Health* **8**, 565–569 (1964).

74. D. M. Richards et al., A controlled field trial of physiological responses to organophosphate residues in farm workers. *J. Environ. Pathol. Toxicol.* **2**, 493–512 (1978).

75. J. A. Rider et al., Toxicity of parathion, systox, octamethyl pyrophosphoramide, and methyl parathion in man. *Toxicol. Appl. Pharmacol.* **14**, 603–611 (1969).

76. J. A. Rider, J. I. Swader, and E. J. Puletti, Methyl parathion and guthion anticholinesterase effects in human subjects. *Fed. Proc.* **29**, 349 (1970).

76a. J. A. Rider et al., *Fed. Proc.* **30**, 443 (1971).

77. J. A. Rider, J. I. Swader, and E. J. Puletti, Anticholinesterase toxicity studies with methyl parathion, guthion and phospdrin in human subjects. *Fed. Proc.* **30**, 443 (1971).

79. D. B. Davies and B. J. Holub, Comparative subacute toxicity of dietary diazinon in the male and female rat. *Toxicol. Appl. Pharmacol.* **54**, 359–367 (1980).

80. D. B. Davies and B. J. Holub, Toxicological envaluation of dietary diazinon in the rats. *Arch. Environ. Contam. Toxicol.* **9**, 637–650 (1980).

81. E. F. Edson and D. N. Noakes, The comparative toxicity of six organophosphorus insecticides in the rat. *Toxicol. Appl. Pharmacol.* **2**, 523–539 (1960).

82. R. B. Bruce, J. W. Howard, and J. R. Elsea, Toxicity of O,O-diethyl O-(2-isopropyl-6-methyl-4-pyrimidyl) phosphorothioate (diazinon). *Agri. Food Chem.* **3**, 1017–1021 (1955).

83. A. R. Singh, G. C. Mccormick, and A. T. Arthur, 90-Day oral toxicity study in rats, unpublished study dated August 4, 1990 from Ciba-Geigy Corporation, EPA Guidelines no. 82-1. Research Department, Pharmaceuticals Divison. Summit, N. J. *EPA-40815003*, 1988.

84. M. W. Williams, H. N. Fuyat, and O. G. Fitzhugh, The subacute toxicity of four organic phosphates to dogs. *Toxicol. Appl. Pharmacol.* **1**, 1–7 (1959).

85. F. L. Earl et al., Diazinon toxicity-comparative studies in dogs and miniature swine. *Toxicol. Appl. Pharmacol.* **18**, 285–295 (1971).

86. F. R. Kirchner, One-two year oral toxicity study in rats, unpublished reports no. EPA-41942002 Basel, Switzerland, 1991.

87. National Cancer Institute (NCI), *Bioassay of Diazinon for Possible Carcinogenicity*, Carcinogenisis Technical Report Series No. 137, 1979.

88. G. Woodard, M. Woodard, and H. T. I. Cronin, Safety evaluation of the pesticide diazinon by a two-year feeding trial in Rhesus monkeys. *Fed. Proc.* **27**, 597 (1968).

89. V. A. Green, Effects of pesticides on rat and chick embryo. *Trace Subst. Environ. Health-III*; 183–209 (1970).

90. T. B. Barnes, 90-day oral toxicity study in dogs, unpublished Report No. EPA-40815004 Basel, Switzerland, 1988.

91. S. B. Harris and J. F. Holson, *A Teratology Study in New Zealand White Rabbits*, Science Applications, Inc. Unpublished Report No. 801205 Basel, Switzerland, 1981.
92. J. F. Robens, Teratological studies of carbaryl, diazinon, norea, disulfram and thiran in small laboratory animals. *Toxicology and Applied Pharmacology* **15**, 152–163 (1969).
93. P. K. Dobbins, Organic phosphate insecticides as teratogens in the rat. *J. Florida M. Assoc.* **54**, 452–456 (1967).
94. A. M. Hoberman et al., Transplacental inhibition of esterases in fetal brain following exposure to the organophosphate diazinon. *Teratology* **19**, 30A–31A (1979).
95. Environmental Protection Agency (EPA). Integrated Risk Information System. www.epa.gov/ngispgm3/iris. (accessed 10/99).
96. D. B. McGregor, A. Brown, P. Cattanach et al., Responses of the L5178Y tk+/tk- mouse lymphoma cell forward mutation assay, III. 72 coded chemicals. *Environ. Mol Mutagen* **12**, 85–154 (1988).
97. T. C. Marshall, H. W. Dorough, and H. E. Swim, Screening of pesticides for mutagenic potential using *Salmonella typhimurium* mutants. *J. Agric. Food Chem.* **24**, 560–563 (1976).
98. W. Mucke, K. O. Alt, and H. O. Esser, Degradation of ^{14}C-labeled diazinon in the rat. *J. Agri. Food Chem.* **18**, 208–212 (1970).
99. F. Iverson, D. L. Grant, and J. Lacroix, Diazinon metabolism in the dog. *Bull. Environ. Contam. Toxicol.* **13**, 611–618 (1975).
100. T. Nakatsugawa, N. M. Tolman, and P. A. Dahm, Oxidative degradation of diazinon by rat liver microsomes. *Biol. Pharmacol.* **18**, 685–688 (1969).
101. R. Garcia-Repetto, D. Martinez, and M. Repetto, Coefficient of distribution of some organophosphorous pesticides in rat tissue. *Vet. Human Toxicol.* **37**, 226–229 (1995).
102. H. X. Wu, C. Evreux-Gros, J. Descotes, Diazinon toxicokinetics, tissue distribution and anticholinesterase activity in the rat. *Biomed. Environ. Sci.* **9**, 358–369 (1996).
103. R. C. Wester et al., Percutaneous absorption of diazinon in humans. *Food Chem. Toxicol.* **31**, 569–572 (1993).
104. R. S. H. Yang, E. Hodgson, and W. C. Dauterman, Metabolism *in vitro* of diazinon and diazoxon in rat liver. *J. Agri. Food Chem.* **19**, 10–13 (1971).
105. M. J. J. Ronis and T. M. Badger, Toxic interactions between fungicides that inhibit ergosterol biosynthesis and phosphorothioate insecticides in the male rat and bobwhite quail (*Colinus virginianus*). *Toxicology and Applied Pharmacology* **130**, 221–228 (1995).
106. S. M. Ahdaya, R. J. Monroe, and F. E. Guthrie, Absorption and distribution of intubated insecticides in fasted mice. *Pest. Biochem. Physiol.* **16**, 38–46 (1981).
107. L. S. Bichile et al., Acute reversible cerebellar signs after diazinon poisoning (letter). *J. Assoc. Physicians India* **31**, 745–746 (1983).
108. V. N. Kabrawala, R. M. Shah, and G. G. Oza, Diazinon poisoning (a study of 25 cases). *Indian Pract.* **18**, 716–717 (1965).
109. M. R. Limaye, Acute organophosphorus compound poisoning — A study of 76 necropsies. *J. Indian Med. Assoc.* **47**, 492–498 (1965).
110. A. Poklis, F. W. Kutz, and J. F. Sperling, A fatal diazinon poisoning. *Foresic Sci. Int.* **15**, 135–140 (1980).
111. A. E. DePalma, D. S. Kwalick, and N. Zukerberg, Pesticide poisoning in children. *J. Am. Med. Assoc.* **211**, 1979–1981 (1970).
112. A. J. Dagli, J. S. Moss, and W. A. Shaikh, Acute pancreatitis as a complication of diazinon poisoning. A case report. *J. Assoc Physicians India* **29**, 794–795 (1981).

113. H. W. Klemmer, E. R. Reichert, and W. L. Yauger Jr., Five cases of intentional ingestion of 25 percent diazinon with treatment of recovery. *Clin. Toxicol.* **12**, 435–444 (1978).

114. E. R. Reichert et al., Diazinon poisoning in eight members of related households. *Clin. Toxicol.* **11**, 5–11 (1977).

115. I. Hirschy et al., Diazinon poisoning in Hawaii. *Mortality and Morbidity Weekly Report* **19**, 130–131 (1970).

115a. L. Wecker, R. E. Mrak, and W. D. Dettborn, *J. Environ. Path. Toxicol. Oncol.* **6**, 171–175 (1985).

116. T. English, E. F. Ellis, and J. Ackerman, Organic phosphate poisoning-Cleveland, OH. *Morbidity and Mortality Weekly Report.* **19**, 397–404 (1970).

117. E. Stalberg et al., Effect of occupational exposure to organophosphorus insecticides on neuromuscular function. *Scan. J. Work Environ. Hlth* **4**, 255–261 (1978).

118. M. J. Coye et al., Clinical confirmation of organophosphate poisoning by serial cholinesterase analyses. *Arch. Int. Med.* **147**, 438–442 (1987).

118a. E. D. Richter et al., *Arch. Environ. Health* **47**, 135–138 (1992).

119. H. S. Lee, Acute pancreatitis and organophosphate poisoning—a case report and review. *Singapore Med. J.* **30**, 599–601 (1989).

120. N. Maizlish et al., A behavioral evaluation of pest control workers with short-term, low-level exposure to the organophosphate diazinon. *Am. J. Indus. Med.* **12**, 153–172 (1987).

120a. EPA, Organophosphorus pesticides: Documents for Dichlorvos (http//www.epa.gov.// pesticides/op/ddvp.htm) (accessed 10/99).

121. W. F. Durham et al., Studies on the toxicity of *O,O*-dimethyl-2-2-dichlorovinyl phosphate (DDVP). *AMA Arch. Indus. Health* **15**, 340–349 (1957).

122. J. Natoff, Influence of the route of exposure on the acute toxicity of cholinesterase inhibitors. *Eur. J. Toxicol.* **3**, 363–367 (1970).

123. G. Kimmerle and D. Lorke, Toxicology of insecticidal organophosphates. *Pflanzenschutz-Nachrichten* **21**, 111–142 (1968).

124. G. Schmidt et al., Effects of dichlorvos (DDVP) inhalation on the activity of acetylcholineterase in the bronchial tissue of rats. *Arch. Toxicol.* **42**, 191–198 (1979).

125. B. J. Dean and E. Thorpe, Studies with dichlorvos vapor in dominant lethal mutation tests on mice. *Arch. Toxicol.* **30**, 51–59 (1972).

126. B. A. Schwetz, H. D. Ioset, and B. K. Leong, Teratogenic potential of dichlorvos given by inhalation and gavage to mice and rabbits. *Teratology* **20**, 383–387 (1979).

127. R. F. Witter et al., Studies on the safety of DDVP for the disinection of commerical aircraft. *Bull. WHO* **24**, 635–642 (1961).

128. B. M. Francis, R. L. Metcalf, and L. G. Hansen, Toxicity of organophosphorus esters of laying hens after oral and dermal administration. *J. Environ. Sci. Health* **20**, 73–95 (1985).

128a. S. Ishikawa and M. Miyata, Development of myopia following chronic organophosphate pesticide intoxication, an epidemiological and experimental study. In W. H. Merigen and B. Weiss, *Neurotoxicity of the Visual System*, Raven Press, New York, 1980, pp. 233–254.

129. A. Ueda, K. Aoyama, and F. Manda, Delayed-type allergenicity of triforine (saprol(R)). *Contact Dermatitis* **31**, 140–145 (1994).

130. National Cancer Institute (NCI), *Bioassay of Dichlorvos for Possible Carcinogenicity (CAS No. 62-73-7)*. Carcinogenesis Technical Report Series No. 10, 1977.

131. National Toxicology Program (NTP), *Toxicology and Carcinogenesis Studies of Dichlorvos in F344/N Rats and B6C3F$_1$ Mice* (gavage studies), NTP Technical Report TR324, NIH Publication NO. 89-2398, 1989.

132. E. H. Timmons et al., Dichlorvos effects on estrous cycle onset in the rat. *Lab. Animal Sci.* **25**, 45–47 (1975).
133. J. A. Collins, M. A. Schooley, and V. K. Singh, The effect of dietary dichlorvos on swine reproduction and viability of their offspring. *Toxicol. Applied Pharmacol.* **19**, 377 (1971).
134. J. H. Mennear, Dichlorvos, a regulatory conundrum. *Reg. Toxicol. Pharmacol.* **27**, 265–272 (1998).
134a. R. N. Schop, M. H. Nardy, and M. T. Goldberg, *Fund. Appl. Toxicol.* **15**, 666–675 (1990).
135. A. Tungul, A. M. Bonin, S. He, R. S. Baker, Micronuclei induction by dichlorvos in the mouse skin. *Mutagenesis* **6**, 405–408 (1991).
136. D. Blair et al., Dichlorvos. A 2-year inhalation carcinogenesis study in rats. *Arch. Toxicol.* **35**, 281–294 (1976).
137. D. H. Hutson and E. C. Hoadley, The metabolism of [^{14}C-methyl]dichlorvos in the rat and the mouse. *Xenobiotica* **2**, 107–116 (1972).
138. D. H. Hutson and E. C. Hoadley, The comparative metabolism of [^{14}C-vinyl] dichlorvos in animals and man. *Arch. Toxicol.* **30**, 9–18 (1972).
139. D. H. Hutson, E. C. Hoadley, and B. A. Peckering, The metabolic fate of [vinyl-^{14}C] dichlorvos in the rat after oral and inhalation exposure. *Xenobiotica* **1**, 593–611 (1971).
140. A. S. Wright, D. H. Hudson, and M. F. Wooder, The chemical and mechanical reactivity of dichlorvos. *Arch. Toxcol.* **42**, 1–18 (1979).
141. J. E. Loeffler, J. C. Potter, and S. L. Scordelis, Long term exposure of swine to 14C Dichlorvos atmosphere. *J. Agri. Food Chem.* **24**, 367–371 (1976).
142. R. Traverso, A. Moretto, and M. Lotti, Human serum "A" esterases hydrolysis of *O,O*-dimethyl-2,2-dichloro vinyl phosphate. *Biochem Pharmacol.* **38**, 671–676 (1989).
143. E. Reiner, V. Simeon, and M. Skrinjaric-Spolijar, Hydrolysis of *O,O*-dimethyl-2,2-dichlorovinyl phosphate (DDVP) by esterases in parasitic helminths, and in vertebrate plasma and erythrocytes. *Comp Biochem. Physiol.* **66**, 149–152 (1980).
144. J. R. Playfer et al., Genetic polymorphism and interethnic variability of plasma paroxonase activity. *J. Med. Gen.* **13**, 337–342 (1976).
145. D. Blair, E. C. Hoadley, and D. H. Hutson, The distribution of dichlorvos in the tissues of mammals after its inhalation or intravenous administration. *Toxicol. Appl. Pharmacol.* **31**, 243–253 (1975).
146. Dzwonkowska and Hubner, *Arch. Toxicol.* **58**, 152–156 (1986).
147. Technical fenamiphos (NEMACUR) oncogenicity study in mice, unpublished report, Bayer, Stilwell, KS, 1982.
148. C. G. T. Mathias, Persistent contact dermatitis from the insecticide dichlorvos. *Contact Dermatitis* **9**, 217–218 (1983).
149. R. E. Gold et al., Dermal and respiratory exposure to applicators and occupants of residences treated with dichlorvos (DDVP). *J. Econom. Ent.* **77**, 430–436 (1984).
149a. J. S. Leary et al., *Arch. Environ. Health* **29**, 308–314 (1974).
149b. M. Menz, H. Luetkemeir, and K. Sachesse, *Arch. Environ. Health* **28**, 72–76 (1974).
150. E. C. Vigliani, Exposure of newborn babies to VAPONA insecticide, Tenth annual meeting. *Toxicol. Appl. Pharmacol.* **19**, 184 (1971).
151. W. A. Rasmussen et al., Toxicological studies of DDVP for disinection of aircraft. *Aerosp. Med.* **34**, 593–600 (1963).
152. J. A. Rider, H. C. Moeller, and E. J. Puletti, Continuing studies on anticholinesterase effect of methyl parathion, initial studies with guthion, and determination of incipient toxicity level of dichlorvos in humans. *Fed. Proc. Fed. Am. Soc. Exp. Biol.* **26**, 427 (1967).

152a. A. C. Bager et al., *Toxicol. Appl. Pharmacol.* **41**, 389–394 (1977).

153. M. B. Slomka and C. H. Hine, Clinical pharmacology of dichlorvos. *Acta Pharm. et Toxicol.* **49**, 105–108 (1981).

154. American Conference of Governmental Industrial Hygienists (ACGIH). *Documentation of Threshold Limit Values and Biological Exposure Indices*, 6th ed., Cincinnati, OH, 1991.

155. R. T. Meister, *Farm Chemical Handbook'95*. Meister Publishing Company, 1995.

156. I. C. Maxwell and P. M. Le Quesne, Neuromuscular effects of chronic administration of two organophosphorus insecticides to rats. *NeuroTox* **3**, 1–10 (1982).

157. J. S. Bus and J. E. Gibson, Bidrin, perinatal toxicity and effect on the development of brain acetylcholinesterase and choline acetyltransferase in mice. *Fd. Cosmet. Toxicol.* **12**, 313–322 (1974).

158. A. Nishio and E. M. Uyeki, Induction of sister chromatid exchange in Chinese hamster ovary cells by organophosphate insecticides and their oxygen analogs. *J. Toxicol. Environ. Health* **8**, 939–946 (1981).

159. D. L. Bull and D. A. Lindquist, Metabolism of 3-hydroxy-*N*-methyl-*cis*-crotonamide dimethyl phosphate (azodrin) by insects and rats. *J. Agr. Food Chem.* **14**, 105–109 (1966).

160. R. E. Menzer and J. E. Casida, Nature of toxic metabolites formed in mammals, insecticides, and plants from 3-(dimethoxyphosphinylosy)-*N,N*-dimethyl-*cis*-crotonamide and its *N*-erthyl analog. *J. Agr. Food Chem* **13**, 102–112 (1965).

161. E. M. Lores, D. E. Bradway, and R. F. Moseman, Organophosphorus pesticide poisonings in humans, determination of residues and metabolites in tissues and urine. *Arch. Environ. Health* **33**, 270–276 (1978).

162. R. Perron and B. B. Johnson, Insecticide poisoning. *New Engl. J. Med.* **281**, 274–275 (1969).

163. R. A. Warriner, A. S. Nies, and W. J. Hayes, Severe organophosphate poisoning complicated by alcohol and turpentine ingestion. *Arch. Environ. Health* **32**, 203–205 (1977).

164. J. P. Frawley et al., Toxicologic investigations on Delnav. *Toxicol. Appl. Pharmacol.* **5**, 605–624 (1963).

165. S. D. Murphy and K. L. Cheever, Effect of feeding insecticides, inhibition of carboxyesterase and cholinesterase activities in rats. *Arch. Environ. Health* **17**, 749–758 (1968).

166. B. W. Arthur and J. E. Casida, Biological activity and metabolism of Hercules AC-528 components in rats and cockaroaches. *J. Econ. Ent.* **52**, 20–27 (1959).

167. G. L. Kennedy, J. P. Frawley, and J. C. Calandra, Multigeneration reproductive effects of three pesticides in rats. *Toxicol. Appl. Pharmacol.* **25**, 589–596 (1973).

168. W. H. Harned and J. E. Casida, Dioxiathion metabolites, photoproducts, and oxidative degradation products. *J. Ag. Food Chem.* **4**, 689–699 (1976).

169. C. R. Angle and J. Wermers, Accidental organophosphate poisoning: Even flea-dip needs safety packaging. *New Eng. J. Med.* **290**, 1031–1032 (1974).

170. C. R. Angle and J. Wermers, Human poisoning with flea-dip concentrate. *J. Onc. Vet. Med. Assoc.* **165**, 174–175 (1974).

171. Summary of unpublished studies, Bayer Corporation, Stilwell, Kansas, 1991.

172. T. J. Bombinski and K. P. DuBois, Toxicity and mechanism of action of di-syston. *A.M.A. Arch. Indus, Health* **17**, 192–199 (1958).

173. B. W. Schwab and S. D. Murphy, Induction of anticholinesterase tolerance in rats with doses of disulfoton that produce no cholinergic signs. *J. Toxicol. Environ. Health* **8**, 199–204 (1981).

174. B. W. Schwab, L. G. Costa, and S. D. Murphy, Muscarinic receptor alternations as a mechanism of anticholinesterase tolerance. *Toxicol. Appl. Pharmacol.* **71**, 14–23 (1993).

175. J. Llorens et al., Characterization of disulfoton-induced behavioral and neurochemical effects following repeated exposure. *Fund. Appl. Toxicol.* **20**, 163–169 (1993).
176. Thio-demeton/oral toxicity to mice/dietary administration for three months. Huntington Research Centre, England, unpublished report 1972.
177. W. B. Stavinoha, L. C. Ryan, and P. W. Smith, Biochemical effects of an organophosphorus cholinesterase inhibitor on the rat brain. *Ann NY Acad. Sciences* **160**, 378–382 (1969).
178. EPA Organophosphorus pesticides documents for disulfoton (http://www.epa.gov./pesticides/op/disulfoton.htm) (accessed 10/99)
179. H. Suzuki and S. Ishikawa, Ultrastructure of the ciliary muscle treated by organophosphate pesticide in beagle dogs. *Br. J. Opthal.* **58**, 931–940 (1974).
180. S. S. O. Uga and K. Mukuno, Histopathological study of canine optic nerve retina treated by organophosphate pesticide. *Invest Ophthalmol Vis. Sci.* **16**, 877–881 (1977).
181. Bayer, unpublished report. Technical grade disulfoton, a chronic toxicity feeding study in the beagle dog. Bayer Corporation, Stilwell, Kansas, 1997.
182. L. C. Ryan, B. R. Endecott, and G. D. Hanneman, Effects of an organophosphorus pesticide on reproduction in the rat. Department of Transportation, Federal Aviation Administration, Office of Aviation Medicine, AD 709327, 1970.
183. J. T. Stevens, R. E. Stitzel, and J. J. McPhillips, Effects of anticholinesterase insecticides on hepatic microsomal metabolism. *J. Pharmacol. Exp. Ther.* **181**, 576–583 (1972).
184. C. D. Brokopp, J. L. Wyatt, and J. Gabica, Dialkyl phosphates in urine samples from pesticide formulators exposed to disulfoton and phorate. *Bull. Environ. Contam. Toxicol.* **26**, 524–529 (1981).
185. Y. H. Atallah, W. P. Cahill, and D. M. Whitacre, Exposure of pesticide applicators and support personnel to *O*-ethyl-(4-nitrophenyl)phenylphosphonothioate (EPN). *Arch Environ. Contam. Toxicol.* **11**, 219–225 (1982).
186. H. Hattori, U. Suzuki, and T. Yasuoka, Identification of quantitation of disulfoton in urine and blood of a cadaver by gas chromatography/mass spectrometry. *Nippon Noigaku Zasshi* **36**, 411–413 (1982).
186a. K. Futagami et al., *J. Toxicol.-Clin. Toxicol.* **33**, 151–155 (1995).
187. E. P. Savage, J. J. R. Bagby, and L. Mounce, Pesticide poisoning in rural Colorado. *Rocky Mt. Med. J.* **68**, 29–33 (1971).
188. M. Yashiki, T. Kojima, and M. Ohtani, Determination of disulfoton and its metabolites in the body fluids of a di-syston intoxication case. *Forensic Sci. Int.* **48**, 145–154 (1990).
189. J. Brodeur and K. P. DuBois, Comparison of acute toxicity of anticholinesterase insecticides to weanling and adult male rats. *Proc. Soc. Exp. Biol. Med* **114**, 509 (1963).
190. W. F. Durham, T. B. Gaines, and W. J. Hayes, Paralytic and related effects of certain organic phosphorus compounds. *A.M.A. Arch. Indus. Health* **15**, 326–330 (1956).
191. B. M. Francis, Effects of structure, atropine, and dosing regimen on the delayed neurotoxicity of the insecticide EPN. *Bull. Environ. Contam. Toxicol.* **38**, 283–288 (1987).
192. A. H. El-Sabae et al., Neurotoxicity of organophosphorus insecticides leptophos and EPN. *J. Environ. Sci. Health B* **12**, 269–288 (1977).
193. J. P. Frawley and N. N. Fuyat, Effect on low dietary levels of parathion and systox on blood cholinesterase of dogs. *Pest. Toxicol.* **5**, 346–348 (1957).
194. J. P. Frawley et al., Marked potentiation in mammalian toxicity from simultaneous administration of two anticholinesterase compounds. *J. Pharmacol. Exp. Ther.* **121**, 96–106 (1957).
194a. S. D. Murphy and K. P. DuBois, *Proc. Soc. Exp. Biol. Med.* **96**, 813–818 (1957).

195. S. D. Murphy and K. P. DuBios, The influence of various factors on the enzymatic conversion of organic thiophosphates to anticholinesterase agents. *J. Pharmacol. Exp. Ther.* **12**, 194–202 (1958).

196. W. J. Hayes, Jr., The 90-Dose LD$_{50}$ and a chronicity factor as measures of toxicity. *Toxicol. Appl. Pharmacol.* **11**, 327–335 (1967).

197. K. P. DuBios and F. K. Kinoshita, Influence of induction of hepatic microsomal enzymes by phenobarbitail on toxicity of organic phosphate insecticides. *Proc. Soc. Exp. Biol. Med.* **129** (3), 699–702 (1968).

198. The subacute parenteral toxicity of Bay 68138 to rat, unpublished report, Bayer, Stilwell, KS, 1968.

199. R. J. Nolan et al., Chlorpyrifos: Pharmacokinetics in human volunteers. *Toxicol. Appl. Pharmacol.* **73**, 8–15 (1984).

200. D. E. Bradway, T. M. Shafik, and E. M. Lores, Comparison of cholinesterase activity, residue levels and urinary metabolite excretion of rats exposed to organophosphorus pesticides. *J. Agric. Food Chem.* **25**, 1353–1358 (1977).

200a. J. E. Bakke, V. J. Feil, and C. E. Price, Rat urinary metabolites from *O,O*-diethyl-*O*-(3,5,6-trichloro-2-pyridyl) phosphorothioate. *J. Environ. Sci. Health B* **3**, 225–230 (1976).

201. G. N. Smith, B. S. Watson, and F. S. Fischer, Investigations on Dursban insecticide. Metabolism of [^{36}Cl] *O,O*-diethyl *O*-3,5,6-trichloro-2-pyridyl phosphorothioate in rats. *J. Agri. Food Chem.* **15**, 132–138 (1967).

202. S. B. McCollister et al., Studies of the acute and long-term oral toxicity of chlorpyrifos (*O,O*-diethyl-*O*-(3,5,6-trichloro-2-phridyl) phosphorothioate). *Food Cosmet. Toxicology* **12**, 45–61 (1974).

203. P. E. Berteau and D. A. Wallace, A comparison of oral and inhalation toxicities of four insecticides to mice and rats. *Bull. Environ. Contam. Toxicol.* **19**, 113–120 (1978).

203a. V. C. Moser and S. Padilla, Age-and gender-related differences in the time course of behavioral and biochemical effects produced by oral chlorpyrifos in rats. *Toxicol. Applied Pharmacol.* **149**, 107–119 (1998).

204. C. N. Pope et al., Comparision in *in vitro* cholinesterase inhibition in neonatal and adult rats by three organophosphorothioate insecticides. *Toxicology* **68**, 51–61 (1991).

205. C. N. Pope et al., Long-term neurochemical and behavioral effects induced by acute chlorpyrifos treatment. *Pharmacol. Biochem. Behav.* **42**, 251–256 (1992).

206. T. K. Chakraborti, J. D. Farrar, and C. N. Pope, Comparative neurochemical and neurobehavioral effects of repeated chlorpyrifos exposures in young and adult rats. *Pharmacol. Biochém. Behav.* **46**, 219–224 (1993).

207. P. J. Bushnell, C. N. Pope, and S. Pandilla, Behavioral and neurochemical effects of acute chlorpyrifos in rats: Tolerance to prolonged inhibition of cholinesterase. *J. Pharmacol. Exp. Ther.* **266**, 1007–1017 (1993).

208. J. Cohn and R. C. Macphail, Chlorpyrifos produces selective learning deficits in rats working under a schedule of repeated acquisition and performance. *J. Pharmacol. Exp. Ther.* **283**, 312–320 (1997).

209. J. L. Mattson et al., Single-dose and 13-week repeated-dose neurotoxicity screening studies of chlorpyrifos insecticide. *Food and Chemical Toxicology* **34**, 393–405 (1996).

210. R. A. Corley et al., Chlorpyrifos, a 13-week nose-only vapor inhalation study in Fischer 344 rats. *Fund. Appl. Toxicol.* **13**, 616–618 (1989).

211. W. J. Breslin, Evaluation of the developmental and reproductive toxicity of chlorpyriphos in the rat. *Fund. Applied Toxicol.* **29**, 119–130 (1996).

212. A. L. Hayes, R. A. Wise, and F. W. Weir, Assessment of occupational exposure to organophosphates in pest control operators. *Am. Indus. Hyg. Assoc. J.* **41**, 568–575 (1980).

213. P. V. Shah, R. J. Monroe, and F. E. Guthrie, Comparative rates of dermal penetration of insecticides in mice. *Toxicol. Appl. Pharmacol.* **59**, 414–423 (1981).

214. S. A. Soliman et al., Six-month daily treatment of sheep with neurotoxic organophosphorus compounds *Toxicity. Appl. Pharmacol.* **69**, 417–431 (1983).

215. L. G. Sultatos and S. D. Murphy, Hepatic microsomal detoxification of the organophosphates paraoxon and chlorpyrifos oxon in the mouse. *Drug Metab. Disp.* **11**, 232–238 (1983).

216. L. G. Sultatos and S. D. Murphy, Kinetic analyses of the microsomal biotransformation of the phosphorothioate insecticides chlorpyrifos and parathion. *Fund. Appl. Toxicol.* **3**, 16–21 (1983).

217. L. C. Costa et al., Serum paraoxonase and its influence on paraoxon and chlorpyrifos-oxon toxicity in rats. *Toxicol. Appl. Pharmacol.* **103**, 66–76 (1990).

218. W.-F. Li, L. G. Costa and C. E. Furlong, Serum paraoxonase status, a major factor in determining resistance to organophosphates. *J. Toxicol. Environ. Hlth.* **40**, 337–346 (1993).

219. J. E. Chambers and H. W. Chambers, Oxidative desulfuration of chlorpyrifos, chlorpyrifos-methyl, and leptophos by rat brain and liver. *Biochem. Toxicol.* **4**, 201–203 (1989).

220. T. B. Griffin, F. Coulstan, and D. D. McCollister, Studies of the relative toxicities of chlorpyrifos and chlorpyrifosmethyl in man. *Toxicol. Applied Pharmacol.* **37**, 105 (1976).

221. T. Ma and J. E. Chambers, Kinetic parameters of desulfuration and dearylation of parathion and chlorpyrifos by rat liver microsomes. *Food Chem. Toxicol.* **32**, 763–767 (1994).

222. D. N. Nevin et al., Paraoxonase genotypes, lipoprotein lipase activity and HDL. *Arterosler. Thromb. Vasc. Biol.* **16**, 1243–1249 (1996).

223. C. E. Furlong et al., Role of genetic polymorphism of human plasma paraoxonase/arylesterase in hydrolysis of the insecticide metabolites chlorpyrifos oxon and paraoxon. *Amr. J. Hum. Genet.* **43**, 230–238 (1988).

224. C. E. Furlong et al., Spectrophotometric assays for the enzymatic hydrolysis of the active metabolites of chlorpyrifos and parathion by plasma paraoxonase/arylesterase. *Anal. Biochem.* **180**, 242–247 (1989).

225. R. J. Richardson, Assessment of the neurotoxic potential of chlorpyrifos relative to other organophosphorus compounds, a critical review of the literature. *J. Toxicol. Environmental Health* **44**, 135–165 (1995).

226. F. Jitsunari et al., Determination of 3,5,6-trichloro-2-pyridinol levels in the urine of termite control workers using chlorpyrifos. *Acta Med. Okayama* **43**, 299–306 (1989).

227. M. Lotti et al., Inhibition of lymphocytic neuropathy target esterase predicts the development of organophosphate-induced delayed polyneuropathy. *Arch. Toxicol.* **59**, 176–179 (1986).

228. G. M. Tush and M. I. Anstead, Pralidoxime continuous infusion in the treatment of organophosphate poisoning. *Annals Pharmaco.* **31**, 441–444 (1997).

229. J. Joubert et al., Acute organophosphate poisoning presenting with choreo-athetosis. *Clin. Toxicol.* **22**, 187–191 (1984).

230. B. S. Selden and S. C. Curry, Prolonged succinylcholine-induced paralysis in organophosphate insecticide poisoning. *Annals Emergency Medicine* **16**, 215–217 (1987).

231. L. A. Aiuto, S. G. Pavlakis, and R. A. Boxer, Life-threatening organophosphate-induced delayed polyneuropathy in a child after accidental chlorpyrifos ingestion. *J. Ped.* **122**, 658–660 (1993).

232. H. J. deSilva, P. S. Sanmuganathan, and N. Senanayake, Isolated bilateral recurrent laryngeal nerve paralysis, a delayed complication of organophosphorus poisoning. *Hum. Exp. Toxicol.* **13**, 171–173 (1994).

233. J. G. Kaplan et al., Sensory neuropathy associated with dursban (chlorpyrifos) exposure. *Neurology* **43**, 2193–2196 (1993).

234. J. D. Sherman, Organophosphate pesticides neurological and respiratory toxicity. *Toxicol. Indus. Health* **11**, 33–39 (1995).

235. J. D. Sherman, Chlorpyrifos (Dursban) - associated birth defects, report of four cases. *Arch. Environmental Health* **51**, 5–8 (1996).

236. J. D. Sherman, Dursban revisited: Birth defects, U.S. Environmental Protection Agency, and Centers for Disease Control. *Arch. Environmental Health* **52**, 332–333 (1997).

237. F. E. Brenner et al., R. R. Morbidity among employees engaged in the manufacture or formulation of chlorpyrifos. *Br. J. Indus. Med.* **46**, 133–137 (1989).

238. C. J. Burns et al., Update of the morbidity experience of employees potentially exposed to chlorpyrifos. *Occup. Environ. Med.* **55**, 65–70 (1998).

239. R. G. Ames et al., Health symptoms and occupational exposure to flea control products among California pet handlers. *J. Am. Indus. Hyg. Assoc.* **50**, 466–472 (1989).

240. R. Silvestri, J. A. Himes, and G. T. Edds, Repeated oral administration of coumaphos in sheep, interactions of coumaphos with bishydroxycoumarin, trichlorfon, and phenobarbital sodium. *Am. J. Vet. Res.* **36**(3), 289–292, (1975).

241. EPA *Reregistration Eligibility Deceseon Coumaphos*, List 0018, 1993.

242. EPA, Human Health Risk Assessment Azinophos-methyl 1999, (http://www.epa.gov/pesticides/op/azm.htm) (accessed 10/2000)

243. EPA, Organophosphorous Pesticides, Methyl Parathion (http://www.epa.gov/pesticides/op/methyl-parathion.htm) (accessed 10/99).

244. EPA, Organophosphorous Pesticides, Coumaphos (http://www.epa.gov/pesticides/op/coumaphos.htm) (accessed 10/99).

245. R. A. Bellows et al., Effects of organic phosphate systemic insecticides on bovine embryonic survival and development. *Am. J. Vet. Res.* **35**(8), 1133–1140 (1975).

246. J. K. Malik et al., Biotransformation and disposition of the coumaphos metabolite 3-chloro-4-methyl-(4-^{14}C)-7-hydroxycoumarian in rats. *Arch. Toxicol.* **48**, 51–59 (1981).

247. T. Fang et al., Coumaphos intoxications mimic food poisoning. *Clin Toxicol.* **33**, 699–703 (1995).

248. K. P. DuBois and J. M. Coon, Toxicology of organic phosphorus-containing insecticides to mammals. *A. M. A. Arch. Indus. Hyg. Occup. Med.* **6**, 9–13 (1952).

249. C. B. Shaffer and B. West, The acute and subacute toxicity of technical *O,O*-Diethyl *S*-2-diethylaminoethyl phosphorothioate hydrogen oxalate (Tetram[1]). *Toxicol. Appl. Pharmacol.* **2**, 1–13 (1960).

250. W. B. Deichmann and R. Rakoczy, Toxicity and mechanism of action of systox. *AMA Arch. Indus. Health* **11**, 324–331 (1955).

251. K. P. DuBios, S. D. Murphy, and D. R. Thrush, Toxicity and mechanism of action of some metabolites of systox. *AMA Arch. Ind. Health* **13**, 606–612 (1956).

252. W. B. Deichmann, W. Pugliese, and J. Cassidy, Effects of dimethyl and diethyl paranitrophenyl thiophosphate on experimental animals. *Indus. Hyg. Occup. Med.* **5**, 44–51 (1952).

253. J. M. Barnes and F. A. Denz, The reaction of rats to diets containing octamethyl pyrophosphoramide (schradan) and *O,O*-diethyl-*S*-ethylmercaptoethanol thiophosphate ("Systox"). *Brit. J. Indus. Med.* **11**, 11–19 (1954).

254. C. H. Budreau and R. P. Singh, Teratogenicity and embryotoxicity of demeton and fenthion in CF#1 mouse embryos. *Toxicol. Applied Pharmacol.* **24**, 324–332 (1973).

255. K. S. Khera, Q. N. La Ham, and H. C. Grice, Toxic effects induced by inoculation of EPN and systox into duck eggs. *Toxicol. Appl. Pharmacol.* **7**, 488 (1965).

256. K. S. Khera et al., Toxic effects in ducklings hatched from embryos inoculated with EPN or systox. *Food Cosmet. Toxicol.* **3**, 581–586 (1965).

257. C. M. Menzie, *Metabolism of Pesticides*. U.S. Dept. Interior, Bureau of Sport Fisheries and Wildlife, Publication 127. Washington, DC, U.S. Government Printing Office, 1969.

258. R. B. March et al., Metabolism of systox in the white mouse and American cockroach. *J. Encon. Ent.* **48**, 355–363 (1955).

259. Y. S. Kagan, Y. I. Kundiev, and M. A. Trotsenko, Work hygiene in the use of systemic organophosphorus insecticides, second communication. *Gig. Sanit.* **23**, 2–32. (1958).

260. J. A. Rider et al., Toxicity of parathion, systox, octamethyl pyrophosphoramide, and methyl parathion in man. *Toxicol. Appl. Pharmacol.* **14**, 603–611 (1969).

261. D. F. Heath and M. Vandekar, Some spontaneous reactions of O,O-Dimethyl S-ethylthioethyl phosphorothiolate and related compounds in water and on storage, and their effects on the toxicological properties of the compounds. *Biochem. J.* **67**, 187–201 (1957).

262. International Programme on Chemical Safety (IPSC), *Environmental Health Criteria 197 Demeton-S-methyl*. World Health Organization, Geneva, 1997.

263. M. Vandekar, The toxic properties of demeton-methyl ("metasystox") and some related compounds. *Br. J. Ind. Med.* **15**, 158–167 (1958).

264. M. B. Abou-Donia, Toxicokinetics and metabolism of delayed neurotoxic organophosphorus esters. *Neuro Toxicol.* **4** (1), 113–130 (1983).

265. M. R. Hegazy, Poisoning by meta-isosystox in sprayment and in accidentally exposed patients. *Br. J. Ind. Med.* **22**, 230–235 (1965).

266. I. H. Redhead, Poisoning on the farm, report of a case of organophosphorus poisoning. *The Lancet* **1**, 686–688 (1968).

267. R. D. Jones, Organophosphate poisoning at a chemical packaging company. *Br. J. Ind Med.* **39**, 377–381 (1982).

268. T. B. Gaines, The acute toxicity of pesticides to rats. *Toxicol. Appl. Pharmacol.* **2**, 88–99 (1960).

269. Z. Vasilic et al., The metabolities of organophosphorus pesticides in urine as an indicator of occupational exposure. *Toxicol. Environ. Chem.* **14**, 11–127 (1987).

270. J. E. Chambers and H. W. Chambers, Time course of inhibition of acetylcholinesterase and aliesterases following parathion and paraoxon exposures in rats. *Toxicol. Appl. Pharmacol.* **103**, 420–429 (1990).

271. E. M. Boyd et al., *Chem. Toxicol.* **2**, 295–302 (1969).

272. C. S. Skinoner and W. W. Kilgore, *J. Toxicol. Environ. Health* **9**, 491–497 (1982)

273. NCI, *Bioassays* National Technical Information Service, Springfield, VA, 1979.

274. M. B. Abou-Donia et al., *Toxicol. Appl. Pharmacol.* **68**, 54–65 (1983).

275. S. Homma, *Rinsho Ganka* **27**, 1163–1164 (1972).

276. V. Drevenkar et al., *Chem. Biol. Interactions* **87**, 315–322 (1993).

277. H. C. Hodge et al., Studies of the toxicity and of the enzyme kinetics of ethyl p-nitrophenyl thionobenzene phosphonate (EPN). *J. Pharm. Exp. Ther.* **112**, 29–39 (1954).

278. Y. Hosaka and A. Yamaura, Electroencephalograms in a case of acute fetal intoxication due to an organophosphorus pesticide. *Rinsho Noha (Clin Electroencephalogr.)* **18**(10), 655–656, (1976).

279. J. M. Charles and J. Farmer, Disposition of an oral dose of O-ethyl-O-p-nitrophenyl phenylphosphonothioate (EPN) in the rat. *Toxicol. Appl. Pharmacol.* **48**, A38 (1979).

280. R. L. Chrzanowski and A. G. Jelinek, Metabolism of phenyl carbon-14-labeled O-ethyl O-(4-nitrophenyl) phenylphosphonothioate in the rats and in hens at toxic and subtoxic dose levels. *J. Agric. Food Chem.* **29**, 580–587 (1981).

281. M. B. Abou-Donia, B. L. Reichert, and M. A. Ashry, The absorption, distribution, excretion, and metabolism of a single oral dose of O-ethyl O-4-nitrophenyl phenylphosphonothioate. *Toxicol. Appl. Pharmacol.* **70** (1), 18–28 (1983).

282. S. Sugiyama et al., NAD-coupled enzymatic oxidation of O-ethyl O-p-nitrophenyl phenyphosphonothioate (EPN) to its oxygen analog with liver microsomes of rats. *Jpn. J. Pharmacol.* **37**, 242–252 (1995).

283. M. Hitchcock and S. D. Murphy, Enzymatic reduction of O,O-(4-nitrophenyl) phosphorothioate, O,O-Diethyl O-(4-nitrophenyl) phosphate, and O-ethyl O-(4-nitrophenyl) benzene thiophosphate by tissues from mammals, birds, and fishes. *Biochem. Pharmacol.* **16**, 1801–1811 (1967).

284. C. S. Forsyth and J. E. Chambers, Activation and degradation of the phosphorothionate insecticides parathion and EPN by rat brain. *Biochem. Pharmacol.* **38**, 1597–1608 (1989).

284a. BAY 68 138 toxicological studies, unpublished report. Bayer AG, Wuppertal, Germany, 1971.

285. K. P. DuBois, F. K. Kinoshita, and J. P. Frawley, Quantitative measurement of inhibition of aliesterases, acylamidase, and cholinesterase by EPN and Delnav. *Toxicol. Appl. Pharmacol.* **12**, 273–284 (1968).

286. C. S. Petty, Organic phosphate insecticide poisoning. *Am. J. Med.* **24**, 467–470 (1958).

287. H. Shiozaki et al., Two cases of autopsies victims of acute intoxication by organophosphorus pesticides, EPN and dichlorvos. *Iryo (Med. Treatment)* **31**(2), 161–163 (1977).

288. J. A. Rider et al., A study of the anticholinesterase properties of EPN and malathion in human volunteers. *Clin. Res.* **7**, 81–82 (1959).

289. Z. Ni et al., Induction of micronucleus by organophosphorus pesticides both in *in vivo* and *in vitro*. *Hua Hsi I Ko Ta Hsueh Hsueh Pao* **24**, 82–86 (1993).

290. S. P. Bhunya and G. B. Jena, Evaluation of genotoxicity of a technical grade organophosphate insecticide, tafethion (ethion), in chicks. *In-Vivo* **8**, 1987–1879 (1994).

291. R. D. Mosha, N. Gyrd-Hansen, and P. Nielsen, Fat of ethion in goats after intravenous, oral and dermal administration. *Pharmacol. Toxicol.* **67**, 246–251 (1990).

292. The acute toxicity and anticholinesterase action of Bayer 68138. unpublished report, Bayer, Stilwell, KS, 1967.

293. Determination of the acute toxicity (LD50) in rats unpublished report, Bayer AG, Wuppertal, Germany, 1975.

294. H. C. Moeller and J. A. Rider, Plasma and red blood cell cholinesterase activity as indications of the threshold of incipient toxicity of ethyl-p-nitrophenyl thionobenzenephosphonate (EPN) and malathion in human beings. *Toxicol. Appl. Pharmacol.* **4**, 123–130 (1962).

295. Comparative oral toxicity in rats of several impurities and a technical compound of NEMACUR with analytical grade NEMACUR, unpublished report, Bayer, Stilwell, KS, 1972.

296. Ref. 178, (insert ethion for (10/99) disulfaton).

297. Ref. 178, (insert 7 examples (10/99) for disulfaton).

298. Acute inhalation toxicity study with nemacur active ingredient on rats, unpublished report. Bayer AG, Wuppertal, Germany, 1972.

299. NEMACUR active ingredient (SRA 3886) subacute inhalational toxicity study on rats, unpublished report. Bayer AG, Wuppertal, Germany, 1979.

300. SRA 3886 (Common Name, Fenmiphos) Acute oral neurotoxicity screening study in Wistar rats. Unpublished report, Bayer AG, Wuppertal, Germany, 1995.

301. SRA 3886 Technical (common name, fenamiphos) delayed neurotoxicity studies on hens following acute oral administration. unpublished report, Bayer AG, Wuppertal, Germany, 1987.

302. Fenamiphos. Dermal sensitization study in the guinea pigs, unpublished report, Bayer, Stilwell, KS, 1983.
303. SRA 3886 (Common Name, Fenaminphos) Subchronic neurotoxicity screening study in Wistar rats (thirteen week administration in the diet). unpublished report, Bayer AG, Wuppertal, Germany, 1996.
304. Chronic feeding toxicity study of technical grade fenamiphos (NEMACUR) with dogs, unpublished report, Bayer, Stilwell, KS, 1991.
305. Ninety-day cholinesterase study on dogs with fenamiphos in diet, unpublished reports, Bayer, Stilwell, KS, 1983.
306. BAY 68 138 Chronic toxicological studies on dogs, unpublished study, Bayer AG, Wuppertal, Germany, 1972.
307. Combined chronic toxicity/oncogencity study of technical fenamiphos (NEMACUR) with rats, unpublished report, Bayer, Stilwell, KS, 1986.
308. H. H. Chen, S. R. Sirianni, and C. C. Huang, Sister chromatid exchanges in Chinese hamster ovary cells treated with seventeen organophosphorus compounds in the presence of a metabolic activation system. *Environ. Mut.* **4**, 621–624 (1982).
309. The metabolic fate of ethyl-r-(methylthio)-m-tolyl isopropylphosphoramidate (BAY 68138), ethyl 4-(methyl-sulfinyl)-m-tolyl isopropylphosphoramidate (BAY 68138 sulfoxide), and ethyl 4-(methylsulfonyl)-m-tolyl-isopropylphosphoramidate (Bay 68138 sulfone) by white rats, unpublished report. Bayer, Stilwell, KS, 1969.
310. (Phenyl-1-14C) NEMACUR, whole body autoradiographic distribution of the radioactivity in the rat, unpublished report, Bayer, Stilwell, KS, 1988.
311. A. K. Salama, M. A. Radwan, and F. I. El-Shahawi, Pharmacokinetic profile and anticholinesterase properties of phenamiphos in male rats. *J. Environ. Sci. Health B* **27**, 307–323 (1992).
312. A. Bell, R. Bames, and G. R. Simpson, Cases of absorption and poisoning by the pesticide "Phosdrin." *Med. J. Aust.* **1**, 178–180 (1968).
312a. Bayer 68138, Chronic toxicological studies on rats, Wuppertal, Germany, 1972, unpublished.
313. J. J. M. Van de Sandt and J. P Groten, *In vitro* percutaneous absorption study with nemacur using human and rat skin, TNO Nutrition and Food Research Institute, 1996.
314. J. B. Knaak, K. C. Jacobs, and G. M. Wang, Estimating the hazard to humans applying nemacular 3EC with rat dermal-dose ChE response data. *Bull. Environ. Contam. Toxicol.* **37**, 159–163 (1986).
315. K. P. DuBois and F Kinoshita, Acute toxicity and anticholinestease action of O,O-Dimethyl O-[4-(methylthio)-m-tolyl] phosphorothioate (DMTP, Baytex) and related compounds. *Toxicol. Appl. Pharmacol.* **6**, 86–95 (1964).
316. Food and Agriculture Organization/World Health Organization (FAO/WHO), *1972 Evalutions of Some Pesticide Residues in Food.* WHO Pesticide Residues Series No. 2, WHO, Geneva, 1973.
316a. S. D. Cohen, Mechanisms of toxicological interactions involving organophosphate insecticides. *Fund. Appl. Toxicol.* **4**, 315–324 (1984).
317. C. Greenaway and P. Orr, Selected topics, toxicology — a foodborne outbreak causing a cholinergic syndrome. *J. Emerg. Med.* **14**, 339–344 (1996).
317a. EPA, Organophosphorous Pesticides, Fenthion (http://www.epa.gov/pesticides/op/fenthion.htm) (accessed 10/99).

318. J. I. Francis and J. M. Barnes, Studies on the mammalian toxicity of fenthion. *Bull. WHO* **29**, 205–212 (1963).

319. J. S. Bowman and J. E. Casida, Further studies on the metabolism of thimet by plants, insects, and mammals. *J. Econ. Entomol.* **51**, 838–843 (1958).

320. Studies on the toxicity and anticholinesterase action of Bayer 29493. unpublished report, Bayer, Stilwell, KS, 1959.

321. Acute inhalation toxicity studies, unpublished report Bayer AG, Wuppertal, Germany, 1978.

322. Acute inhalation toxicity study with BAYTEX technical in rats, unpublished report, Bayer, Stillwell, KS.

323. Acute one hour inhalation toxicity study with technical grade BAYTEX in rats, unpublished report, Bayer, Stilwell, KS, 1987.

324. V. C. Moser, Comparisons of the acute effects of a cholinesterase inhibitors using a neurobehavioral screening battery in rats. *Neurotoxicol. Teratol.* **17**, 617–625 (1995).

325. E1752 (Common Name, Fenthion) Acute oral neurotoxicity screening study in Wistar rats. unpublished report, Bayer AG, Wuppertal, Germany, 1997.

326. K. Sakaguchi et al., Effects of fenthion, isoxathion, dichlorvos and propaphos on the serum cholinesterase isoenzyme patterns of dogs. *Vet. Hum. Toxicol.* **39**, 1–5 (1997).

327. B. Dementi, Ocular effects of organophosphates A historical perspective of saku disease. *J. Appl. Toxicol.* **14**, 119–129 (1994).

328. H. Imai, (Toxicity of organophosphorus pesticides (fenthion) on the retina, electroretinographic and biochemical study.) *J. Jpn. Opthalmol. Soc.* **78**, 1–10 (1974) (Japanese).

329. H. Imai, (Studies on the ocular toxicity of organophosphorus pesticides. Part I. electroretinographic and biochemical study on rats after a single administration of fenthion). *J. Jpn. Opthalmol. Soc.* **78**, 163–172 (1974) (Japanese).

330. P. Tandon et al., Fenthion produces a persistent decrease in muscarinic receptor function in the adult rat retina. *Toxicol. Appl. Pharmacol.* **125**, 271–280 (1994).

331. M. Kawai et al., The effects of organophosphorus compounds on the eyes of experimental animals. *Boei Eisei* **23**, 1–10 (1976).

331a. K. Shimamoto and K. Hattori, *Acta. Med. Univ. Kioto* **40**, 163–171 (1969).

332. *EPA, Status of Pesticides on Registration, Registration and Special Review* EPA 738-R-98-002, 1998.

333. 1752 Technical (common name, fenthion) range finding study to determine the maximum tolerated dose MTD of B6C3F$_1$ mice, unpublished report, Bayer AG, Wuppertal, FRG, 1989.

334. Fenthion (S 1752, the active ingredient of LEBAYCID and BAYTEX) subacute inhalation study on rats, unpublished report, Bayer AG, Wuppertal, FRG, 1979.

335. Chronic oral toxicity of Bayer 29493 to male and female rats, unpublished report. Bayer, Stilwell, KS, 1963.

336. Combined chronic toxicity/oncogenicity study of technical grade fenthion (BAYTEX) with rats, unpublished report, Bayer, Stillwell, KS, 1990.

337. M. Miyata, H. Imai, and S. Ishikawa, Experimental retinal pigmentary degeneration by organophosphorus pesticides in rats. *Excerpta Medica. Int. Congr. Ser.* **450**(1), 901–902 (1979).

338. National Cancer Institute (NCI), *Bioassay of Fenthion for Possible Carcinogenicity*, National Technical Information Service, Springfield, VA, 1979.

339. P. J. Hanna and K. F. Dyer, Mutagenicity of organophosphorus compounds in bacteria and *drosophila. Mutat. Res.* **28**, 405–420 (1975).

340. F. Oesch, unpublished report, University of Maine, 1977.
341. V. F. Simmon, A. D. Mitchell, and T. A. Jorgenson, Evaluation of selected pesticides and chemical mutagens. *EPA-600/1-77-028, NTIS Pub. No. PB-268647*, National Technical Information Service, 1977.
342. V. F. Simmons, D. C. Poole, and E. S. Riccio et al., In vitro mutagenicity and genotoxicity assays of 38 pesticides. *Environ. Mutagen.* **1**, 142–143 (1979).
343. H. Imai, Toxicity of organophosphorus pesticide (fenthion) on the retina electroretinographic and biochemical study. *Acta. Soc. Ophthalmol. (Japan)* **79**, 1067–1076 (1975) (Japanese).
344. E 1752 micronucleus test on the mouse, unpublished report, Bayer AG, Wuppertal, Germany, 1990.
345. S 1752 dominant lethal study on male mice to test for mutagenic effects, unpublished report. Bayer AG, Wuppertal, Germany, 1978.
346. E 1752 dominant lethal test on the male mouse, unpublished report, Bayer AG, Wuppertal, Germany, 1997.
347. E 1752 *Salmonella* microsome test, unpublished report. Bayer AG, Wuppertal, Germany, 1990a.
348. D. Bagchi et al., *In vitro* and *in vivo* generation of reactive oxygen species, DNA damage and lactate dehydrogenase leakage by selected pesticides. *Toxicol.* **104**, 129–140 (1995).
349. M. V. U. Rani and M. S. Rao, *In vitro* effect of fenthion on human lymphocytes. *Bull. Environ. Contam. Toxicol.* **47**, 316–320 (1991).
350. C. O. Knowles and B. W. Arthur, Metabolism of and residues associated with dermal and intramuscular application of radiolabeled fenthion to diary cows. *J. Econ. Entomol.* **59**, 1346–1352 (1966).
351. A. M. Tsatsakis et al., Experiences with acute organophosphate poisonings in Crete. *Vet. Human Toxicol.* **38**, 101–107 (1996).
352. R. E. Tynes and E. Hodgson, Magnitude of involvement of the mammalian flavin-containing monooxygenase in the microsomal oxidation of pesticides. *J. Agr. Food Chem.* **33**, 471–479 (1985).
353. K. Venkatesth, P. E. Levi, and E. Hodgson, The flavin-containing monooxygenase of mouse kidney. *Biochem. Pharmacol.* **42**, 1411–1420 (1991).
354. M. R. Brunetto et al., Observation on a human intentional poisoning case by the organophosphorus insecticide fenthion. *Invest. Clin.* **33**, 89–94 (1992).
355. P. Mahieu et al., Severe and prolonged poisoning by fenthion, significance of the determination of the anticholinesterase capacity of plasma. *J. Toxicol. Clinical Toxicol.* **19**, 425–432 (1982).
356. P. D. Lichty and R. W. Hartle, Health hazard evaluation report no. HETA-83-373-1501. In *Hazard Evaluations and Technical Assistance Branch*, NIOSH, US Dept of Health and Human Services, Gainesville, GA, 1984.
357. R. L. Metcalf et al., Neurologic findings among workers exposed to fenthion in a veterinary hospital-Georgia. *Morbidity and Mortality Weekly Report* **34**, 402–403 (1985).
358. N. Senanayake and L. Karalliedde, Neurotoxic effects of organophosphorus insecticides. *N. Engl. J. Med.* **316**, 761–763 (1987).
359. N. Senanayake and P. S. Sanmuganathan, Extrapyramidal manifestations complicating organophosphorus insecticide poisoning. *Human Experimental Toxicology* **14**, 600–604 (1995).

360. U. K. Misra, M. Prasad, and C. M. Pandey, A study of cognitive functions and event related potentials following organophosphate exposure. *Electromyogr. Clin. Neurophysiol* **34**, 197–203 (1994).

361. Safety evaluation of fenthion in human volunteers, unpublished report. Bayer, Stilwell, KS, 1979.

362. P. W. Lee, R. Allahyari, and T. R. Fukuto, Studies on the chiral isomers of fonofos and fonofos oxon — 1. Toxicity and antiesterase activities. *Pesticide Biochemistry and Physology* **8**, 146–157 (1978).

363. R. J. Weir and L. W. Hazelton, Organic phosphates. In *Patty's Industrial Hygeine and toxicology*. Eds., G. D. Clayton and F. E. Clayton. Wiley & Sons, New York, 1982, pp. 4820–4822.

364. Environmental Protection Agency (EPA), *Fonofos. Office of Drinking Water Health Advisories Drinking Water Health Advisory, 443-547*, 1989.

365. L. J. Hoffman, I. M. Ford, and J. J. Menn, Dyfonate metabolism studies - I. Absorption, distribution, and excretion of O-ethyl S-phenyl ethylophosphonodithioate in rats. *Pest. Biochem. Physiol.* **1**, 349–355 (1971).

365a. P. W. Lee, R. Allahyari, and T. R. Fukuto, Studies on the cliral isomers of fonofos and fonofos oxon - III. *In vivo* metabolism. *Pesticide Biochemistry and Physiology* **9**, 23–32 (1978).

366. J. B. McBain, I. Yamamot, and J. E. Casida, Mechanism of activation and deactivation of dyfonate (O-ethyl-S-phenyl ethylphosphonodithioate) by rat liver microsomes. *Life Sci.* **10**, 947–954 (1971).

367. J. B. McBain, I. Yamamot, and J. E. Casida, Oxygenated intermediate in peracid and microsomal oxidations of the organophosphonothioate insectide dyfonate. *Life Sci.* **10**, 1311–1319 (1971).

368. J. B. McBain, L. J. Hoffman, and J. J. Menn, Dyonate metabolism studies — 11. Metabolic pathway of O-ethyl S-phenyl ethylphosphonodithiate in rats. *Pest. Biochem. Physiol.* **1**, 356–365 (1971).

368a. T. D. Dressel et al., *Ann. Surg.* **189**, 199–204 (1979).

369. K. P. DuBois et al., Studies on the toxicity and mechanism of action of some new insecticidal thionophosphates. *Indus. Hyg. Occup. Med.* **8**, 350–358 (1953).

370. L. W. Hazleton and E. G. Holland, Toxicity of malathion summary of mammalian investigations. *A.M.A Arch. Indus. Hyg. Occup. Med.* **8**, 399–405 (1953).

371. R. E. Talcott, N. M. Mallipudi, and T. R. Fukuto, Malathion carboxylesterase titler and its relationship to malathion toxicity. *Toxicol. Applied Pharmacol.* **50**, 501–504 (1977).

372. R. E. Talcott et al., Inactivation of esterases by impurities isolated from technical malathion. *Toxicol. Appl. Pharmacol.* **49**, 107–112 (1979).

373. S. D. Murphy and K. L. Cheever, Carboxylesterase and cholinesterase inhibition in rats Abate and interaction with malathion. *Arch. Environ. Health* **24**, 107–117 (1972).

373a. E. C. Hagan, *Fed. Proc.* **12**, 327 (1953).

374. S. D. Murphy, Malathion inhibition of esterases as a determinant of malathion toxicity. *J. Pharmacol. Exp. Ther.* **156**, 352–365 (1967).

375. M. Enrich, B. S. Jortner, and S. Padilla, Comparison of the relative inhibition of acetylcholinesterase and neuropathy target esterase in rats and hens given cholinesterase inhibitors. *Fund. Appl. Toxicol.* **24**, 94–101 (1995).

376. M. T. Akay et al., Bioavailability and toxicological potential of lentil-bound residues of malathion in rats. *J. Environ. Sci. Hlth B* **27**(4), 325–340 (1992).

377. K. Kalow and A. Martin, Second-generation toxicity of malathion in rats. *Nature* **192**, 464–465 (1961).

378. M. G. A. Machin and W. G. McBride, Teratological study of malathion in the rabbit. *J. Toxicol. Environ. Health* **26**, 249–253 (1989).

379. R. D. Kimbrough and T. B. Gaines, Effect of organic phosphorous compounds and alkylating agents in the rat fetus. *Arch. Environ. Health* **16**, 805–808 (1968).

380. M. D. Matthews, and K. S. Devi, Effect of malathion, estradiol-17-beta and progesterone on ascorbic acid metabolism in prenatal rats and their pups. *Vet. Human Toxicol.* **35**, 6–10 (1993).

381. P. Flessel, P. J. Quintana, and K. Hopper, Genetic toxicity of malathion, a review. *Environ. Mol. Mutagen.* **22**, 7–17 (1993).

382. S. M. Amer, M. A. Fahmy, and S. M. Donya, Cytogenetic effect of some insecticides in mouse spleen. *J. Appl. Toxicol.* **16**, 1–3 (1996).

383. J. M. Pluth et al., Molecular bases of hprt mutations in malathion-treated human tlymphocytes. *Mutat. Res.* **397**, 137–148 (1998).

384. J. M. Pluth et al., Increased frequency of specific genomic deletion resulting from *in vitro* malathion exposure. *Cancer Res.* **56**, 2393–2399 (1996).

385. J. A. Venkat et al., Relative genotoxic activities of pesticides evaluated by a modified SOS microplate assay. *Environ. Mol. Mutagen.* **25**, 67–76 (1995).

386. N. Titenko-Holland et al., Genotoxicity of malathion in human lymphocytes assessed using the micronucleus assay *in vitro* and *in vivo*, a study of malathion-exposed workers. *Mutat. Res.* **388**, 85–95 (1997).

387. G. C. Windham et al., Genetic monitoring of malathion-exposed agricultural workers. *Am. J. Indus. Med.* **33**, 164–174 (1998).

388. R. Garcia-Repetto, D. Martinez, and M. Repetto, Malathion and dichlorvos toxicokinetics after the oral administration of malathion and trichlorfon. *Vet. Human Toxicol.* **37**, 306–309 (1995).

389. M. A. Saleh et al., Determination of the distribution of malathion in rats following various routes of administration by whole-body electronic autoradiography. *Toxicol. Indus. Health* **13**, 751–758 (1997).

389a. R. C. Wester, P. Quan, and H. I. Maibach, *Food Chem. Toxicol.* **34**, 731–734 (1996).

390. V. M. Bhagwat and B. V. Ramachandran, Malathion A and B esterases of mouse liver. II. Effect of EPN *in vitro* and *in vivo*. *Biochem. Pharmacol.* **24**, 1727–1729 (1975).

391. R. E. Menzer and N. H. Best, Effect of phenobarbital on the toxicity of several organophosphorus insecticides. *Toxicol. Applied Pharmacol.* **13**, 37–42 (1968).

392. M. Mahajna, G. B. Quistad, and J. E. Casida, S-Methylation of O,O-dialkly phosphorodithioic acids, O,O,S- trimethyl phosphorodithioate and phosphorothiolate as metabolites of dimethoate in mice. *Chem. Res. Toxicol.* **9**, 1202–1206 (1996).

392a. S. Prabhakaran, F. Shameem, and K. S. Devi, Influence of protein deficiency on hexachlorocyclohexane and malathion toxicity in pregnant rats. *Vet. Hum. Toxicol.* **35**, 429–433 (1933).

393. F. C. G. Hoskin and J. E. Walker, Malathion as a model for the enzymatic hydrolysis of the neurotoxic agent, VX. *Bull. Environ. Contam. Toxicol.* **59**, 9–13 (1997).

394. J. Brodeur and K. P. DuBois, Ali-esterase activity and sex difference in malathion toxicity. *Federation Proceedings for the Federation of the American Society of Experimental Biology* **23**, 200 (1964).

395. M. Hosokawa et al., Interindividual variation in carboxylesterase levels in human liver microsomes. *Drug Metab. Disp.* **23**, 1022–1027 (1995).

395a. N. Umetsu et al., Effect of impurities on the mammalian toxicity of technical malathion and acephate. *J. Toxicol. Environ. Health* **7**, 481 (1977).

396. T. Namba, M. Greenfield, and D. Grob, Malathion poisoning - a fatal case with cardiac manifestations. *Arch. Environ. Health* **21**, 533–541 (1970).

397. J. Baker et al., Epidemic malathion poisoning in Pakistan malaria workers. *The Lancet* **1**, 31–34 (1978).

398. K. Futagami et al., Relapse and elevation of blood urea nitrogen in acute fenitrothion and malathion poisoning. *Int. J. Clin. Pharmacol. Ther.* **34**, 453–456 (1996).

399. A. L. Gardner and R. E. Iverson, The effect of aerially applied malathion of an urban population. *Arch. Environ. Health* **16**, 823–826 (1968).

400. E. Kahn et al., Assessment of acute health effects from the medfly eradication project in Santa Clara County, California. *Arch. Environ. Health* **47**, 279–284 (1992).

401. M. A. Marty et al., Assessment of exposure to malathion and malaoxon due to aerial application over urban areas of southern California. *J. Exposure Anal. Environ. Epi.* **4**, 65–81 (1994).

402. M. A. Bradman et al., Malathion and malaoxon environmental levels used for exposure assessment and risk characterization of aerial applications to residential areas of southern California, 1989–1990. *J. Exp. Anal. Environ. Epidem.* **4**, 49–63 (1994).

403. W. J. Hayes et al., Safety of malathion dusting powder for louse control. *Bull WHO* **22**, 503–514 (1960).

404. D. Culver, P. Caplan, and G. S. Batchelor, Studies of human exposure during aerosol application of malathion and chlorthion. *A.M.A. Arch. Indus. Health* **13**, 37–50 (1956).

405. D. C. Thomas et al., Reproductive outcomes in relation to malathion spraying in the San Francisco Bay Area, 1981–1982. *Epidemiology* **3**, 32–39 (1992).

406. T. H. Milby and W. L. Epstein, Allergic contract sensitivity to malathion. *Arch. Environ. Health* **9**, 434–437 (1964).

407. H. M. Schanker et al. Immediate and delayed type hypersensitivity to malathion. *Annals of Allergy* **69**, 526–528 (1992).

408. H. H. Golz, Controlled human exposures to malathion aerosols. *A.M.A. Arch. Indus. Heath* **19**, 52–59 (1959).

409. G. W. Ware et al., Establishment of reentry intervals for organophosphate-treated cotton fields based on human data: III. 12 to 72 hours post-treatment exposure to monocrotophos, ethyl- and methyl parathion. *Arch. Environ. Contam. Toxicol.* **3**, 289–306 (1975).

410. D. P. Morgan et al., Urinary excretion of paranitrophenol and alkyl phosphates following ingestion of methyl or ethyl parathion by human subjects. *Arch. Environ. Contam. Toxicol.* **6**, 159–173 (1977).

411. J. E. Davies and J. C. Peterson, Surveillance of occupational, accidental, and incidental exposure to organophosphate pesticides using urine alkyl phosphate and phenolic metabolite measurements. *Ann. N. Y. Aca. of Sc.* **837**, 257–268 (1997).

412. T. J. Haley et al., Estimation of the LD_1 and extrapolation of the $LD_{0.1}$ for five organothiophosphate pesticides. *Eur. J. Toxicol.* **8**, 229–235 (1975).

413. D. Oritz et al., Acute toxicological effects in rats treatment with a mixture of commerically formulated products containing methyl parathion and permethrin. *Ecotoxicol. Environ. Safety* **32**, 154–158 (1995).

414. J. Miyamoto et al., Studies on the mode of action of organophosphorus compounds. Part II. Inhibition of mammalian cholinesterase *in vivo* following the administration of sumithion and methylparathion. *Agri. Bio. Chem.* **27**, 669–676 (1963).
415. J. Miyamoto et al., Studies on the mode of action of organophosphorus compounds. Part I. Metabolic fate of P^{32} labeled sumithion and methylparathion in guinea pig and white rat. *Agri. Bio. Chem.* **27**, 381–389 (1963).
416. G. M. Benke et al., Comparative toxicity, anticholinesterase action and metabolism of methyl parathion and parathion in sunfish and mice. *Toxicol. Appl. Pharmacol.* **28**, 97–109 (1974).
417. G. M. Benke and S. D. Murphy, The influence of age on the toxicity and metabolism of methyl parathion and parathion in male and female rats. *Toxicol. Appl. Pharmacol.* **31**, 254–269 (1975).
418. L. Institöris et al., Immunotoxic effects of MPT-IP containing 60% methylparathion in mice. *Hum. Exp. Toxicol.* **11**, 11–16 (1992).
419. J. D. Street and R. P. Sharma, Alteration of induced cellular and humoral immune responses by pesticides and chemicals of environmental corcern, quantitative studies of immunosuppression by DDT, aroclor 1254, carbaryl, carbofuran, and methyl parathion. *Toxicol. Appl. Pharmacol.* **32**, 587–602 (1975).
420. A. I. Shtenberg and R. M. Dzhunusova, (Suppression of the immunobiological reactivity of the animal organism under the effect of certain phospho-organic pesticides) (in Russian) *Biull. Eksp. Biol. Med.* **65**, 86–88 (1968).
421. J. J. Ceron, C. G. Panizo, and A. Montes, Toxicological effects in rabbits induced by endosulfan, lindane, and methylparathion representing agricultural byproducts contamination. *Bull. Environ. Contam. Toxicol.* **54**, 235–265 (1995).
421a. NCI, Bioassay of methyl parathion for pesuber carcinogenicity TR-157, 1979.
422. G. Mathew, K. K. Vijalaxmi, and M. A. Rahiman, Methyl parathion-induced sperm shape abnormalities in mouse. *Mut. Res.* **280**, 169–173 (1992).
423. P. Dhondup and B. B. Kaliwal, Inhibition of ovarian compensatory hypertrophy by the administration of methyl parathion in hemnicastrated albino rats. *Repro. Toxicol.* **11**, 77–84 (1994).
424. S. A. Fish, Organophosphorous cholinesterase inhibitors and fetal development. *Am. J. of Ob. Gyn.* **96**, 1148–1153 (1966).
425. R. M. Hollingworth, R. L. Metcalf, and T. R. Fukuto, The selectivity of sumithion compared with methyl parathion metabolism in the white mouse. *J. Agric. Food Chem.* **15**, 242–249 (1967).
426. R. A. Braeckman et al., Toxicokinetics of methyl parathion and parathion in the dog after intravenous and oral administration. *Arch. Toxicol.* **54**, 71–82 (1983).
427. R. A. Braeckman et al., Kinetic analysis of the fate of methyl parathion in the dog. *Arch. Toxicol.* **43**, 263–271 (1980).
428. J. Miyamoto, Studies on the mode of action of organophosphorus compounds. Part III. Activation and degradation of sumithion and methylparathion in mammals *in vivo*. *Agri. Bio. Chem.* **28**, 411–421 (1964).
429. R. Garcia-Repetto, D. Martínez, and M. Repetto, Biodisposition study of the organophosphorus pesticide, methyl- parathion. *Bull. Environ. Contam. Toxicol.* **59**, 901–908 (1997).

430. G. W. Ware et al., Establishment of reentry intervals for organophosphate-treated cotton fields based on human data-II. Azodrin, ethyl and methyl parathion. *Arch. Env. Contam. Toxicol.* **2**, 117–129 (1974).

431. G. W. Ware et al., Establishment of reentry intervals for organophosphate-treated cotton fields based on human data: I. Ethyl- and methyl parathion. *Arch. Env. Contam. Toxicol.* **1**, 48–59 (1973).

432. S. J. Nemec, P. L. Adkisson, and H. W. Dorough, Methyl parathion adsorbed on the skin and blood cholinesterase levels of persons checking cotton treated with ultra-low-volume sprays. *J. Econ. Ent.* **61**, 1740–1742 (1968).

433. P. Sartorelli et al., *In vitro* percutaneous penetration of methyl parathion from a commerical formulation through the human skin. *Occup. Environ. Med.* **54**, 524–525 (1997).

434. J. S. de Lima et al., Methyl parathion activation by a partially purified rat brain fraction. *Toxicol. Lett.* **87**, 53–60 (1996).

435. J. E. Chambers, H. W. Chambers, and J. E. Snawder, Target site bioactivation of the neurotoxic organophosphorus insecticide parathion in partially hepatectomized rats. *Life Science* **48**, 1023–1029 (1991).

436. T. M. Soranno and L. G. Sultatos, Biotransformation of the insecticide parathion by mouse brain. *Toxicol. Letters* **60**, 27–37 (1992).

437. C. Wang and S. D. Murphy, Kinetic analysis of species difference in acetylcholinesterase sensitivity to organophosphate insecticides. *Toxicol. Appl. Pharmacol.* **66**, 409–419 (1982).

438. J. R. Kemp and K. B. Wallace, Molecular determinants of the species-selective inhibition of brain acetylcholinesterase. *Toxicol. Appl. Pharmacol.* **104**, 246–258 (1990).

439. I. G. Fazekas, Über die makroskopischen und mikroskopischen Veränderungen bei der Wofatox-Vergiftung (methyl-parathion) (macroscopic and microscopic changes in wofatox (parathion-methyl) intoxication). *Z. Rechtsmedizin* **68**, 189–194 (1971).

440. J. L. Willems et al., Cholinesterase reactivation in organophosphorus poisoned patients depends on the plasma concentrations of the oxime pralidoxime methylsulphate and of the organophosphate. *Arch. Toxicol.* **67**, 79–84 (1993).

441. R. McConnell et al., Monitoring organophosphate insecticide-exposed workers for cholinesterase depression. *Am. Coll. Occup. Environ. Med.* **34**, 34–37 (1992).

441a. A. Dean et al., *MMWR* **33**, 592–594 (1984).

442. J. De Bleecker et al., Intermediate syndrome due to prolonged parathion poisoning. *Acta Neural. Scand.* **86**, 421–424 (1992).

443. J. De Bleecker et al., Prolonged toxicity with intermediate syndrome after combined parathion and methyl parathion poisoning. *Clin. Toxicol.* **30**, 333–345 (1992).

444. J. L. De Bleecker, Transient opsoclonus in organophosphate poisoning. *Acta Neurol. Scand.* **86**, 529–531 (1992).

445. P. Lisi, S. Caraffini, and D. Assalve, Irritation and sensitization potential of pesticides. *Contract Dermatitis* **17**, 212–218 (1987).

445a. R. E. Gosselin, R. P. Smith, and H. C. Hodge, Section II, Ingredients index. In, *Clinical Toxicology of Commerical Products.* Williams & Wilkins, Baltimore, 1984, pp. 292.

446. J. Kodama, M. S. Morse, and H. H. Anderson, Comparative toxicity of two vinyl-substituted phosphates. *AMA Arch. Ind. Hyg. Occup. Med.* **9**, 45–61 (1954).

447. F. P. Cleveland and J. F. Treon, Response of experimental animals to phosdrin insecticide in their daily diets. *Agri. Food Chem.* **9**, 484–488 (1961).

448. R. C. Cochran et al., Risks from occupational and dietary exposure to mevinphos. *Rev. Environ. Contam. Toxicol.* **146**, 1–24 (1996).

449. J. H. Holmes, J. H. G. Starr, and R. C. Hanisch, Short-term toxicity of mevinphos in man. *Arch. Environ. Health* **29**, 84–89 (1974).

450. E. R. Reichert, H. W. Klemmer, and T. J. Haley, A note on dermal poisoning from mevinphos and parathion. *Clin. Toxicol.* **12**, 33–35 (1978).

451. T. J. Haley, E. R. Reichert, and H. W. Klemmer, Acute human poisoning with parathion and mevinphos in man. *Clin. Toxicol.* **12**, 33–35 (1978).

451a. M. D. Whorton and D. L. Obrinsky, *J. Toxicol. Environ. Health*, **11**, 347–354 (1983).

452. M. C. Warren, J. P. Conrad, and J. J. Bocian, Clothing-borne epidemic. *J. Am. Med. Assoc.* **184**, 266–268 (1963).

453. J. Brachfeld and M. R. Zavon, Organic phosphate (phosdrin) intoxication report of a case and the results of treatment with 2-PAM. *Arch. Environ. Health* **11**, 859–862 (1965).

454. C. Sagerser et al., Occupational pesticide poisoning in apple orchards in Washington, 1993. *Mortality and Morbidity Weekly Report* **42**, 993–995 (1994).

454a. C. Hsiaso et al., *Clin. Toxicol.* **34**, 343–347(1996).

454b. J. Kangas et al., *Am. Ind. Hyg. Assoc. J.* **54**, 150–157(1993).

455. A. Jauhiainen et al., Biological monitoring of workers exposed to mevinphos in greenhouses. *Bull. Environ. Contam. Toxicol.* **49**, 37–43 (1992).

456. J. A. Rider, E. J. Puletti, and J. I. Swader, The minimal oral toxicity level for mevinphos in man. *Toxicol. Appl. Pharmcol.* **32**, 97–100 (1975).

456a. M. M. Verberk and H. J. A. Salle, *Toxicol. Appl. Pharm.* **42**, 351–358 (1977).

457. A. Janardhan, B. A. Rao, and P. Sisodia, Species variation in acute toxicity of monocrotophos and methyl benzimidazole carbamate. *Indian J. Pharmacol.* **18**, 102–103 (1986).

458. T. Skripsky and R. Loosli, Toxicology of monocrotophos. *Rev. Environ. Contam. Toxicol* **139**, 13–39 (1994).

458a. M. Moriya et al., *Mutat. Res.* **16**, 185–216 (1983).

459. S. N. Mandhane and C. T. Chopde, Neurobehavioral effects of acute monocrotophos administration in rats and mice. *Indian J. Pharmacol.* **27**, 245–249 (1995).

460. K. V. Swamy, R. Ravikumar, and P. M. Mohan, Assessment of behavioral tolerance to monocrotophos toxicity in albino rats. *Indian J. Pharmacol.* **25**, 24–29 (1993).

461. W. D. Ratnasooriya, L. D. C. Peiris, and Y. N. A. Jayatunga, Analgesic and sedative action of monocrotophos following oral administration in rats. *Med. Sci. Res.* **23**, 401–402 (1995).

462. N. Adhikari and I. S. Grover, Genotoxic effects of some systemic pesticides: In vivo chromosomal aberrations in bone marrow cells in rats. *Environ. Mol. Mut.* **12**, 235–242 (1988).

463. P. Venna and P. B. Murthy, Effect of starvation on organophosphorus pesticide induced genotoxicity in rats. *Int. J. Food Sciences Nutrition* **45**, 71–77 (1994).

464. N. K. Tripathy and K. K. Patnaik, Studies on the genotoxicity of monocrotophos in somatic and germ-like cells of *drosophila*. *Mutat. Res.* **278**, 23–29 (1992).

465. G. B. Jena and S. P. Bhunya, Thirty day genotoxicity study of an organophosphate insecticide, monocrotophos, in a chick *in vivo* test system. *In-Vivo* **6**, 527–530 (1992).

466. W. Mucke, Metbolism of monocrotophos in animals. *Rev. Environ Contam. Toxicol.* **139**, 59–65 (1994).

467. H. P. Gelbke and H. J. Schlicht, Fatal poisoning with a plant protective containing monocrotophos, dodine and dinocap. *Toxicol. Eur. Res.* **1**, 181–184 (1978).
468. J. Przezdziak and W. Wisneiwska, A case of acute organophosphate poisoning. *Wiad. Lek.* **28**, 1093–1095 (1975).
469. R. E. Simson, G. R. Simpson, and D. J. Penney, Poisoning with monocrotophos, an organophosphorous pesticide. *Med. J. Aust.* **2**, 1013–1016 (1969).
470. J. B. Peiris, R. Fernando, and K. D. Abrew, Respiratory failure from severe organophosphate toxicity due to absorption through the skin. *Forensic Sci. Int.* **36**, 251–253 (1988).
471. A. Mani, M. S. Thomas, and A. P. Abraham, Type II paralysis or intermediate syndrome following organophosphorous poisoning. *J. Assoc. Phys. India* **40**, 542–544 (1992).
471a. EPA, Organophosphorous Pesticides, Naled (http://www.epa.gov/pesticides/op/naled.htm) (accessed 10/99).
471b. K. Khera et al., *J. Environ. Sci. Health* B **14**, 563–577 (1979).
472. M. M. Verberk, Incipient cholinesterase inhibition in volunteers ingesting monocrotophos or mevinphos for one month. *Toxicol. Appl. Pharmacol.* **42**, 345–350 (1977).
472a. R. Kummer and N. J. Van Sittert, Field studies on health effects from the application of two organophosphorus insecticide formulations by hand-held ULV to cotton. *Toxicol. Letters* **33**, 7–24 (1986).
473. J. R. Rittenhouse, *Three-Week Aerosol Inhalation Toxicity Study of Chevron Naled Technical in Rats — Preliminary Data Release.* Chevron Environmental Health Center, Richmond, CA, 1985.
474. R. Braun et al., Activity of organophosphorus insecticides in bacterial tests for mutagenicity and DNA repair B direct alkylation versus metabolic activation and breakdown. *Chem.-Biol. Interactions* **43**, 361–370 (1983).
475. M. Huelse and P. Federspil, Gleichgewichtsstoerungen nach Insektizidvergiftung (Alkyl Phosphate). (Disturbances of equilibrium due to poisoning by organophosphorus insecticides). *HNO* **23**, 185–189 (1975).
476. W. F. Edmundson and J. E. Davies, Occupational dermatitis from naled. *Arch. Environ. Health* **15**, 89–91 (1967).
477. D. L. Mick, T. D. Gartin, and K. R. Long, A case report: Occupational exposure to the insecticide naled. *J. Iowa Med. Soc.* **60**, 395–396 (1970).
477a. EPA, Organophosphorous Pesticides, Ethyl Parathion (http://www.epa.gov/pesticides/op/ethyl-parathion.htm) (accessed 10/99).
478. W. B. Deichmann, P. Brown, and C. Downing, Unusual protective action of a new emulsifier for the handling of organic phosphates. *Science* **116**, 221 (1952).
479. L. Reiter, G. Talens, and D. Woolley, Acute and subacute parathion treatment, effects on cholinesterase activities and learning in mice. *Toxicol. Appl. Pharmacol.* **25**, 582–588 (1973).
479a. J. Roney et al., *Pharmacol. Biochem. Behav.* **24**, 737–742 (1986).
479b. NCI, Bioassay for parathion for possible carcinogenicity, 1979.
480. L. W. Reiter, G. M. Talens, and D. E. Woolley, Parathion administration in the monkey, time course of inhibition and recovery of blood cholinesterases and visual discrimination performance. *Toxicol. Appl, Pharmacol.* **33**, 1–13 (1975).
481. E. F. Edson and D. N. Noakes, The comparative toxicity of six organophosphorus insecticides in the rat. *Toxicol. Appl. Pharmacol.* **2**, 523–539 (1960).

482. National Cancer Institute (NCI), *Bioassay of Malathion for Possible Carcinogenicity*, Technical Report Series No. 192. DHEW (NIH) Pub. No. 79-1748, National Cancer Institute, Bethesda, MD, 1979.
483. J. E. Atkinson et al., Assessment of ocular toxicity in dogs during 6 months' exposure to a potent organophosphate. *J. Appl. Toxicol.* **14**, 145–152 (1994).
484. L. W. Hazelton and E. G. Holland, Pharamacology and toxicology of parathion. *Adv. Chem Ser.* **1**, 31–38 (1950).
485. International Agency for Research on Cancer (IARC), *IARC Monographs on the Evaluation of the Carcinogenic Risk of Chemicals to Humans. Miscellaneous Pesticides* **30**, 153–181 (1983).
486. J. M. Barnes and F. A. Denz, The chronic toxicity of *p*-nitrophenyl diethyl thiophosphate (E605) a long-term feeding experiment with rats. *J. Hyg.* **49**, 430–441 (1951).
487. W. E. Montz, Jr., R. L. Kirkpatrick, and P. F. Scanlon, Parathion effects on reproductive characteristics and vital organ weights of female cottontail rabbits (*Sylvilagus floridanus*). *Bull. Environ. Contam. Toxicol.* **33**, 484–490 (1984).
488. R. D. Harbison, Comparative toxicity of some selected pesticides in neonatal and adult rats. *Toxicol. Appl. Pharmacol.* **32**, 443–446 (1974).
489. G. Talens and G. Woolley, Effects of parathion administration during gestation in the rat on development of the young. *Proc. West. Pharmacology Soci.* **16**, 141–145 (1973).
490. R. Deskin et al., Parathion toxicity in perinatal rats born to spontaneously hypertensive dams. *J. Environ. Path. Toxicol.* **2**, 291–300 (1978).
490a. C. R. Stamper et al., *Neurotoxicol. Teratol.* **10**, 261–266 (1988).
491. B. Veronesi and C. Pope, The neurotoxicity of parathion-induced acetylcholinesterase inhibition in neonatal rats. *Neuro Toxicol.* **11**, 465–482 (1990).
492. C. Dvergsten and R. B. Meeker, Muscarinic cholinergic receptor regulation and acetylcholinesterase inhibition in response to insecticide exposure during development. *Int. J. Develop. Neurosci.* **12**, 63–75 (1994).
493. D. A. Eigenberg, T. L. Pazdernik, and J. Doull, Hemoperfusion and pharmacokinetic studies and parathion and paraoxon in the rat and dog. *Drug Metab. Disp.* **11**, 336–370 (1983).
494. M. Peña-Egido et al., Urinary excretion kinetics of *p*-nitrophenol following oral administration of parathion in the rabbit. *Arch. Toxicol.* **62**, 351–354 (1988).
495. P. Nielsen et al., Disposition of parathion in neonatal and young rats. *Pharmacol. Toxicol.* **68**, 233–237 (1991).
496. T. Fredriksson, Percutaneous absorption of parathion and paraoxon. *Arch. Environ. Health* **3**, 37–70 (1961).
497. W. F. Durham and H. R. Wolfe, An additional note regarding measurement of the exposure of workers to pesticides. *Bull. WHO* **29**, 279–281 (1963).
498. A. J. Funckes, S. Miller, and W. J. Hyes, Jr., Initial field studies in Upper Volta with dichlorvos residual fumigant as a malaria eradication technique. *Bull. WHO* **29**, 243–246 (1963).
499. H. I. Maiback et al., Regional variation in percutaneous penetration in man. *Arch. Environ. Health* **23**, 208–211 (1971).
500. G. L. Oiao, S. K. Chang, and J. E. Riviere, Effects of anatomical site and occlusion on the percutaneous absorption and residue pattern of 2,6-(ring-14C) parathion *in vivo* in pigs. *Toxicol. Appl. Pharmacol.* **122**, 131–138 (1993).

501. J. B. Knaak et al., Percutaneous absorption and dermal dose-cholinesterase response studies with parathion and carbaryl in the rat. *Toxicol. Applied Pharmacol.* **76**, 252–263 (1984).

502. T. Fredriksson and J. K. Bigelow, Tissue distributin of P^{32}-labeled parathion. *Arch. Environ. Health* **2**, 63–67 (1961).

503. O. Benjaminov et al., Parathion transfer and acetylcholinesterase activity in an *in-vitro* perfused term human placenta. *Vet. Hum. Toxicol.* **34**, 10–12 (1992).

504. W. J. Hayes, Jr., Organic phosphorus pesticides, *Pesticides Studied in Man*. Williams and Wilkins, Baltimore, MD, 1982, pp. 358–359.

505. F. Lessire, Relationship between parathion and paraoxon toxicokinetics, lung metabolic activity, and cholinesterase inhibition in guinea pig and rabbit lungs. *Toxicol. Appl. Pharmacology* **138**, 201–210 (1996).

506. G. L. Qiao and J. E. Rivier, Significant effects of application site and occulsion on the pharmacokinetics of cutaneous penetration and biotransformation of parathion *in vivo* in swine. *J. Pharm. Sci.* **84**, 425–432 (1995).

507. J. A. Johnson and K. B. Wallace, Species-related differences in the inhibition of brain acetylcholinesterase by paraoxon and malaoxon. *Toxicol. Appl. Pharmacol.* **88**, 234–241 (1987).

508. A. C. Roy et al., Serum paraoxonase polymorphism in three populations of Southeast Asia. *Hum. Hered.* **41**, 265–269 (1991).

509. P. Sanz et al., Red blood cell and total blood acetylcholinesterase and plasma pseudocholinesterase in humans, observed variances. *Clin. Toxicol.* **29**, 81–90 (1991).

510. K. Kanagaratnam, W. H. Boon, and T. K. Hoh, Parathion poisoning from contaminated barley. *Lancet* **1**, 538–542 (1960).

511. R. Besser, L. Gutmann, and L. S. Weilemann, Polyneuropathy following parathion poisoning. *J. Neurol., Neurosurg. Psych.* **56**, 1135–1136 (1993).

512. H. V. Brown and A. F. Bush, Parathion inhibition of cholinesterase. *Indus. Hyg. Occup. Med.* **1**, 633–636 (1950).

513. A. M. Osorio et al., Investigation of a fatality among parathion applicators in California. *Am. J. Indus. Med.* **20**, 533–546 (1991).

514. S. K. Kashyap, Health surveillance and biological monitoring of pesticide formulators in India. *Toxicol. Letters* **33**, 107–114 (1986).

515. W. V. Hartwell, J. R. G. Hayes, and A. J. Funckes, Respiratory exposure of volunteers to parathion. *Arch. Environ. Health* **8**, 820–825 (1964).

516. J. A. Rider et al., The effect of parathion on human red blood cell and plasma cholinesterase. *A.M.A. Arch. Indus. Health* **18**, 441–445 (1958).

517. E. F. Edson, Summaries of toxicological data. No-effect levels of three organophosphates in the rats, pig, man. *Food Cos. Toxicol.* **2**, 311–316 (1964).

518. G. R. Hayes, Jr., A. J. Funckes, and W. V. Hartwell, Dermal exposure of human volunteers to parathion. *Arch. Environ. Health* **8**, 829–833 (1964).

518a. EPA, Organophosphorous Pesticides, Phorate (http://www.epa.gov/pesticides/op/phorate.htm) (accessed 10/99).

519. D. D. McCollister, F. Oyen, and V. Rowe, Toxicological studies of *O,O*-Dimethyl-*O*-(2,4,5-trichlorophenyl) phosphorothioate (ronnel) in laboratory animals. *J. Agric. Food Chem.* **7**, 689–693 (1959).

520. R. J. Young, Phorate intoxication at an insecticide formulation plant. *Am. Ind. Hyg. Assoc.* **40**, 1013–1016 (1979).

521. F. W. Plapp and J. E. Casida, Bovine metabolism of organophosphorous insecticides, metabolic fate of O,O-dimethyl O-(2,4,5-trichlorophenyl) phosphorothioate in rats and a cow. *J. Agric. Food Chem.* **6**, 662–667 (1958).

522. I. N. Gladenko and L. I. Stuk, Toxicity of trolene. *Veterinaria (Moscow)* **48**, 95–96 (1972).

523. T. S. Rumsey, H. Tao, and J. Bitman, Effects of ronnel on growth, endocrine function and blood measurements in steers and rats. *J. Anim. Sci.* **53**, 217–225 (1981).

524. B. B. Baker et al., Evaluation of topical application of ronnel solution for generalized demodicosis in dogs. *J. Ame. Vet. Med. Assoc.* **168**, 1105–1107 (1976).

525. K. S. Khera, C. Whalen and G. Angers, Teratogenicity study on pyrethrum and rotenone (natural origin) and ronnel in pregnant rats. *J. Toxicol. Environ. Health* **10**, 111–119 (1982).

526. I. Nafstad et al., Teratogenic effects of the organophosphorus compound fenchlorphos in rabbits. *Acta Veterinaria Scand.* **24**, 295–304 (1983).

527. J. E. Balthrop, Ronnel in creeping eruption. *J. Florida Med. Assoc.* **53**, 820–821 (1966).

528. V. B. Beat and D. P. Morgan, Evaluation of hazards involved in treating cattle with pour-on organophosphate insecticides. *J. Am Vet. Med. Assoc.* **170**, 812–813 (1977).

529. A. J. Lehman, Chemicals in foods, a report to the association of the food and drug officials on current developments. Part II, Pesticides, section III. Subacute and chronic toxicity. *Assoc. Food Drug Off. USQ Bull.* **16**, 47–53 (1952).

529a. Ref. 178 (insert Sulfotep for disulfaton).

530. G. Kimmerle and O. R. Klimmer, Acute and subchronic toxicity of sulfotep. *Arch. Toxicol.* **33**, 1–16 (1974).

531. Sulfotep — chronic toxicity study in mice, unpublished report No. 10954 Bayer AG, Wuppertal, FRG, 1982.

532. Dominant lethal test on male mouse to evaluate S276 for mutagenic potential, unpublished report, Bayer AG, Wuppertal Germany, 1980.

533. Sulfotep — teratogenicity study in rabbits, unpublished report no. 12906 bayer AG, Wuppertal, FRG, 1984.

534. Sulfotep — Ames test. unpublished report no. 17982, Bayer AG, Wuppertal, FRG, 1989.

535. Sulfotep — Ames Test, unpublished report, Institute of Pharmacology, University of Mainz, 1977.

536. Sulfotep — Micronucleus test, unpublished report No. 7917 Bayer AG, Wuppertal, FRG, 1978.

537. Sulfotep — Dominant lethal test, unpublished report no. 8286 Bayer AG, Wuppertal, FRG, 1979.

538. S. A. Soliman et al., Two acute human poisoning cases resulting from exposure to diazinon transformation products in Egypt. *Arch. Environ. Health* **37**, 207–212 (1982).

539. R. D. Jones, *Sulprofos, Toxicological Assessment*, Bayer Agriculture Division, 1994.

539a. EPA Organophosphorous Pesticides, Temephos (http://www.epa.gov/pesticides/op/temephos.htm) (accessed 10/99).

540. D. L. Bull and G. W. Ivie, Metabolism of O-ethyl O-14(methylthio) phenyl S-propyl phosphorodithioate (BAY NTN 9306) by white rats. *J. Agr. Food Chem.* **24**, 143–146 (1976).

541. N. P. Hajjar and E. Hodgson, Sulfoxidation of thioether-containing pesticides by the flavin-adenine dinucleotide-dependent monoxygenase of pig liver microsomes. *Biochem. Pharmacol.* **31**, 745–752 (1982).

542. J. V. Rao, A. N. Swamy, and S. Yamin, Rat brain acetylcholinesterase response to monocrotophos and abate. *Bull. Environ. Contam. Toxicol.* **48**, 850–856 (1992).

543. R. C. Blinn, Metabolic fate of abate insecticide in the rat. *J. Agri. Food Chem.* **17**, 118–122 (1969).

544. E. R. Laws et al., Toxicology of abate in volunteers. *Arch. Environ. Health* **14**, 289–291 (1967).

545. *Registry to Toxic Effects of Chemical Substances*, RTECS, 1998.

546. P. W. Smith, W. B. Stavinoha, and L. C. Ryan, Cholinesterase inhibition in relation to fitness to fly. *Aerosp. Med.* **39**, 754–758 (1968).

547. G. E. Quinby and G. M. Doornink, Tetraethyl pyrophosphate poisoning following airplane dusting. *J. Am. Med. Assoc.* **191**, 95–100 (1965).

548. W. M. Grant, Miotic and antiglaucomatous activity of tetraethyl pyrophosphate in human eyes. *Arch. Ophthalmol.* **39**, 576–586 (1948).

549. W. G. Marr and D. Grob, Some ocular effects of a new anticholinesterase agent, tetraethyl pyrophosphate (TEPP) and its use in the treatment of chronic glaucoma. *Am. J. Ophthalmol.* **33**, 904–908 (1950).

550. W. M. Upholt et al., Visual effects accompany TEPP-induced miosis. *A.M.A. Arch. Opthamology* **56**, 123–134 (1956).

550a. D. Grob and M. Harvey, *Bull. Johns Hopkins Hosp.* **84**, 532–567 (1949).

551. J. M. Devine, G. B. Kinosluta, and R. P. Peterson, Farm workers exposure to terbufos [phosphorodithioic acid, S-(tert-butylthio) methyl O,O-diethyl ester] during planting operations of corn. *Arch. Environ. Contam. Toxicol.* **15**, 113–119 (1986).

552. C. T. Stone and J. C. Rider, Treatment of myasthenia gravis. *J. Am. Med. Assoc.* **141**, 107–111 (1949).

552a. EPA, Organophosphorus Pesticides, Terbufos (http://www/epa.gov/pesticides/op/terbufos.htm) (accessed 10/99).

553. S. V. Gibson et al., Organophosphate toxicity in rats associated with contaminated bedding. *Laboratory Animal Sciences* **37**, 789–791 (1987).

554. H. J. Boermans et al., Effect of terfubos poisoning on the blood cholinesterase and hematological values in a dairy herd. *Can. Vet. J.* **26**, 350–353 (1985).

554a. EPA, Organophosphorous Pesticides, Trichlorofon (http://www.epa.gov/pesticides/op/trichlorofon.htm) (accessed 10/99).

555. L. C. Pettigrew et al., Pharmacokinetics, pharmacodynamics, and safety and metrifonate in patients with Alzheimer's disease. *J. Clin. Pharmacol.* **38**, 236–245 (1998).

556. A. S. V. Burgen, C. A. Keele, and D McAlpine, Tetraethyl pyrophosphate in myasthenia gravis. *Lancet* **254**, 519–521 (1948).

557. J. L. Cummings et al., Metrifonate treatment of the cognitive deficits of Alzheimer's disease. *Neurology* **50**, 1214–1221 (1998).

558. J. M. Jewsbury, Metrifonate in schistosomiasis — Therapy and prophylaxis. *Acta Pharmacol. Toxicol.* **49**(Suppl. V), 123–130 (1981).

558a. K. P. DuBois and G. J. Cotter, Studies on the toxicity and mechanism of action of dipterex. *AMA Arch. Indus. Health* **11**, 53–60 (1955).

559. V. Hinz, S. Grewig, and B. H. Schmidt, Metrifonate and dichlorvos: Effects of a single oral administration on cholinesterase activity in rat brain and blood. *Neurochem. Res.* **21**, 339–345 (1996).

560. B. Holmstedt et al., Metrifonate summary of toxicological and pharmacological information available. *Arch. Toxicol.* **41**, 3–29 (1978).

561. P. Riekkinen, Jr., et al., Metrifonate improves spatial navigation and avoidance behavior in scopolamine-treated, medial septum-lesioned and aged rats. *Eur. J. Pharmacol.* **309**, 121–130 (1996).

562. M. Riekkinen, B. H. Schmidt, and P. Riekkinen, Jr., Subchronic treatment increases the duration of the cognitive enchancement induced by metrifonate. *Eur. J. Pharmacol.* **338**, 105–110 (1997).

563. F. J. Van der Staay, V. C. Hinz, and B. H. Schmidt, Effects of metrifonate on escape and aviodance learning in young and aged rats. *Behav. Pharmacol.* **7**, 56–64 (1996).

564. M. W. Jann, Preclinical pharmacology of metrifonate. *Pharmacotherapy* **18**, 55–67 (1998).

565. C. Scali et al., Effect of subchronic administration of metrifonate on extracellular acetylcholine levels in aged F344 rats. *Biol. Psychiatry* **42**, 121S (1997).

566. A. Itoh et al., Effects of metrifonate on memory impairment and cholinergic dysfunction in rats. *Eur. J. Pharmacol.* **322**, 11–19 (1997).

567. R. E. Staples and E. H. Goulding, Dipterex teratogenicity in the rat, hamster, and mouse when given by gavage. *Environ. Health Perspect.* **30**, 105–113 (1979).

568. A. M. Pope et al., Trichlorfon-induced congenital cerebeller hypoplasia in neonatal pigs. *J. Vet. Med. Assoc.* **189**, 781–783 (1986).

569. G. N. Berge, I. Nofstad, and F. Fonnum, Prenatal effects of trichlorfon on the guinea pig brain. *Arch. Toxicol.* **59**, 30–55 (1986).

570. A. Mehl et al., The effect of trichlorfon and other organophosphates on prenatal brain development in the guinea pig. *Neurochem. Res.* **19**, 569–574 (1994).

571. M. Moutschen-Dahmen and N. Degrave, Metrifonate and dichlorvos, cytogenetic investigations. *Acta Pharmacol. Toxicol.* **49**(Suppl. V), 29–39 (1981).

572. A. Czeizel, Phenotypic and cytogenetic studies in self-poisoned patients. *Mutat. Res.* **313**, 175–180 (1994).

573. A. T. Doherty et al., A study of the aneugenic activity of trichlorphon detected by contromere-specific probes in human lymphoblastoid cell lines. *Mutat. Res.* **372**, 221–231 (1996).

574. I. Nordgren et al., Levels of metrifonate and dichlorvos in plasma and erythrocytes during treatment of schistosomiasis with Bilarcil. *Acta Pharmacol. Toxicol.* **49**(Suppl. V), 79–86 (1981).

575. Y. A. Abdi and T. Villén, Pharmacokinetics of metrifonate and its rearrangement product dichlorvos in whole blood. *Pharmacol. Toxicol.* **68**, 137–139 (1991).

576. Y. Aden-Abdi et al., Metrifonate in healthy volunteers: Interrelationship between pharmacokinetic properties, cholinesterase inhibition and side-effects. *Bull. WHO* **68**, 731–736 (1990).

576a. W. Dedek, *Acta Pharm. Toxicol.* **49**, (Suppl. V), 40–50 (1981).

577. X. Hu et al., Toxicity of dipterex, a field study. *Br. J. Ind. Med.* **43**, 414–419 (1986).

578. W. N. Snellen, Therapeutic properties of metrifonate. *Acta Pharmacol. Toxicol.* **49**(Suppl. V), 114–117 (1981).

579. A. G. Gilman et al., *Goodman and Gilman's The Pharmacological Basis of Therapeutics*, 8th ed. MacMillan Publishing Company, New York, 1990.
580. J. C. Morris et al., Metrifonate benefits congnitive, behavioral, and global function in patients with Alzheimer's disease. *Neurology* **50**, 1222–1230 (1998).
581. N. R. Culter et al., Safety and tolerability of metrifonate in patients with Alzheimer's disease, results of a maximum tolerated dose study. *Life Sci.* **16**, 1433–1441 (1998).
582. D. A. Drachman and P. Leber, Treatment of Alzheimer's disease searching for a breakthrough, settling for less. *N. Engl. J. Med.* **336**, 1245 (1997).
583. R. E. Becker, J. A. Colliver, S. J. Markwell et al., Double-blind, placebo-controlled study of metrifonate, and acetylcholinesterase inhibitor, for Alzheimers disease. *Alzheimer Dis. Assoc. Disorders* **10**, 124–131 (1996).
584. S. R. Mardical et al., *Mut. Res.* **300**, 135–140 (1991).
584a. A. F. Badawi et al., *Cancer Lett.* **75**, 167–173. (1993).
584b. A. E. Czeizel et al., *Lancet* **341**, 539–542 (1993).
585. National Cancer Institute (NCI), *Bioassay of Dioxathion for Possible Carcinogenicity (CAS No. 78-34-2)*, Carcinogenesis Technical Report Series 125, 1978.
586. Bay 29 493 Chronic Toxicity Study on Rats (Two-year Feeding Experiment) unpublished report Bayer AG, Wuppertal, Germany, 1977.
587. A. N. Worden, P. R. B. Noel, and L. E. Mawdsley-Thomas, Effect of ronnel after chronic feeding to dogs. *Toxicol. Appl. Pharmacol.* **23**, 109 (1972).
588. Mechemer, Chronic toxicity of metrifonate. *Acta Pharmacol. Toxicol.* **49**(Suppl. V), 15–28 (1981).
589. M. M. Deacon et al., Embryotoxicity and fetotoxicity of orally administered chlorpyrifos in mice. *Toxicol. Appl. Pharm.* **54**, 31–40 (1980).
590. J. M. Spyker and D. I. Avery, Neurobehavioral effects of prenatal exposure to the organophosphate diazinon in mice. *J. Toxicol. Environ. Health* **3**, 989–1002 (1977).
591. K. D. Courtney et al., Teratogenic evaluation of the pesticides baygon, carbofuran, dimethoate and EPN. *J. Environ. Sci. Health B* **20**, 373–406 (1985).
592. T. Tanimura, T. Katsuya, and H. Nishimura, Embryotoxicity of acute exposure to methyl parathion in rats and mice. *Arch. Environmental Health* **15**, 609–613 (1967).
593. Sulfotep teratogenicity study in rats, unpublished report no. 9171 Bayer AG, Wuppertal, FRG, 1980.
594. C. Barrueco et al., Mutagenesis, **6**(1) 71–76 (1991).
595. G. W. Fischer et al., *Chem. Biol. Interact* **19**(2) Nov 205–213 (1977).
596. E. Zeiger et al., *Environ Mutagen* 1988, 11 Suppl 12:158.

Subject Index

Abate. *See* Temephos
ABS. *See* Acrylonitrile-butadiene-styrene
Acetal resins – polyoxymethylene, 597–600
   chemical and physical properties, 575t–576t, 597–599, 597t, 598t
   production and use, 599, 639t
   toxic effects, 599–600
Acetate cotton. *See* Cellulose acetate
Acetose. *See* Cellulose acetate
Acetylated cellulose. *See* Cellulose triacetate
Acetyl cellulose. *See* Cellulose acetate
ACM. *See* Acrylic elastomer
Acrilafil. *See* Styrene-acrylonitrile
Acrylic, production and use, 639t
Acrylic acid, hydroxypropyl ester. *See* Propylene glycol monoacrylate
Acrylic acid polymer. *See* Polyacrylic acid
Acrylic elastomer, 449–450
   chemical and physical properties, 427t, 449
   exposure standards, 450
   production and use, 449–450
Acrylics polymers, 496–497
Acrylonitrile-butadiene rubber, 425, 442–443
   chemical and physical properties, 427t, 442
   exposure standards, 443–444
   production and use, 442–443
   toxic effects, 443
      experimental studies of, 443
      human experience, 443
Acrylonitrile-butadiene-styrene, 504–505
   chemical and physical properties, 488t, 504
   production and use, 504–505, 577t, 639t
   toxic effects, 505

Acrylonitrile-styrene co-polymer. *See* Styrene-acrylonitrile
Adulsin. *See* Methyl cellulose
Alkyd resins, 590–592
   additives, 592
   production and use, 577t, 590–591, 639t
   toxic effects, 591–592
Allyl mercaptan, 719–720
   chemical and physical properties, 683t, 719
   genetic and cellular effects, 720, 733
   pharmacokinetics, metabolism, and mechanisms, 719–720
   production and use, 719
   toxic effects, 719–720
      experimental studies of, 719–720
Allyl *n*-octyl sulfoxide, 744
   chemical and physical properties, 684t
   toxic effects, 744
Allyl polymers, 592–593
Almolose. *See* Carboxymethyl cellulose
Alvyl. *See* Polyvinyl alcohol
Aminoplastics, 644
Amyl mercaptan. *See* n-Pentyl mercaptan
tert-Amyl mercaptan, 704–705
   acute toxicity, 705
   chemical and physical properties, 682t, 704
   production and use, 705
   toxic effects, experimental studies of, 705
Amyl thiolalcohol. *See* n-Pentyl mercaptan
Aniline/phenol/formaldehyde, 638
Aramid, 542–543, 561–563
   production and use, 578t
Aromatic copolyester, production and use, 640*t*

967

SUBJECT INDEX

Aromatic polyamides, 561–563
  chemical and physical properties, 545t–546t, 561–562
  production and use, 562, 640t
  toxic effects
    experimental studies of, 550t–551t, 562–563
    human experience, 563
Axophos. See Methyl parathion
Azinphos-methyl, 784–790
  acute toxicity, 770t, 786, 789
  biomonitoring/biomarkers, 789
  carcinogenesis, 774t, 788–789
  chemical and physical properties, 785
  chronic and subchronic toxicity, 786–787, 789–790
  exposure assessment, 785
  exposure standards, 790
  pharmacokinetics, metabolism, and mechanisms, 787
  production and use, 785
  reproductive and developmental effects, 776t, 787–788
  toxic effects, 786–790
    human experience, 789–790

Bagolax. See Methyl cellulose
BAY 18436. See Demeton-S-methyl
BAY 25141. See Fensulfothion
Baycid. See Fenthion
BAY NTN 9306. See Sulprofos
BAY SRA 3886. See Fenamiphos
Baytex. See Fenthion
Bear skunk. See n-Butyl mercaptan
Benzenemethanethiol. See Benzyl mercaptan
Benzenethiol. See Phenyl mercaptan
Benzosulfonazole. See Benzothiazole
Benzothiazole, 749–750
  acute toxicity, 750
  chemical and physical properties, 684t, 749
  genetic and cellular effects, 750
  production and use, 749
  toxic effects, 750
    experimental studies of, 750
Benzothiazole derivatives, 749–758
2(3H)-Benzothiazolethione. See 2-Mercaptobenzothiazole
Benzyldisulfide. See Dibenzyl disulfide
Benzyl mercaptan, 721–722
  acute toxicity, 722
  chemical and physical properties, 683t, 721–722
  chronic and subchronic toxicity, 722
  genetic and cellular effects, 722
  production and use, 722
  toxic effects, 722
    experimental studies of, 722
Bexan. See Styrene-acrylonitrile
BGAP. See Butylene glycol adipic acid polyester
Bidrin. See Dicrotophos
Bisphenol A/epichlorohydrin/formaldehyde, 638
Blitex. See Ronnel
Bolstar. See Sulprofos
BPG 400. See Polypropylene glycol butyl ethers
BPG 800. See Polypropylene glycol butyl ethers
BR. See Polybutadiene
Bromochlorphos. See Naled
Butadiene rubber, chemical and physical properties, 429f
Butane, 1,1'-sulfonylbis-. See n-Dibutyl sulfone
Butanediol, 41–47
  chemical and physical properties, 41, 42t
  production and use, 41–42
  toxic effects, 42
1,2-Butanediol, 42–43
  acute toxicity, 42–43
  genetic and cellular effects, 43
  pharmacokinetics, metabolism, and mechanisms, 43
1,3-Butanediol, 43–45
  acute toxicity, 43–44
  carcinogenesis, 45
  chronic and subchronic toxicity, 44
  human experience, 45
  pharmacokinetics, metabolism, and mechanisms, 44–45
  reproductive and developmental effects, 45
1,4-Butanediol
  acute toxicity, 46
  chemical and physical properties, 46
  chronic and subchronic toxicity, 46
  pharmacokinetics, metabolism, and mechanisms, 46
  reproductive and developmental effects, 46
2,3-Butanediol, 47
  acute toxicity, 47
1,3-Butanediol diacrylate, 350–351
  acute toxicity, 351
  chemical and physical properties, 351
  toxic effects, 351
    experimental studies of, 351
1-Butanethiol. See n-Butyl mercaptan
2-Butanethiol. See sec-Butyl mercaptan
N-tert-Butyl-2-benzothiazolesulfenamide, 754–755
  acute toxicity, 754
  chemical and physical properties, 684t, 754
  genetic and cellular effects, 755
  production and use, 754

SUBJECT INDEX

reproductive and developmental effects, 755
toxic effects, 754–755
  experimental studies of, 754–755
  human experience, 755
Butylene glycol, ethers of, 331–336
1,2-Butylene glycol. *See* 1,2-Butanediol
1,3-Butylene glycol. *See* 1,3-Butanediol
1,4-Butylene glycol. *See* 1,4-Butanediol
Butylene glycol adipic acid polyester, 351–352
  chronic and subchronic toxicity, 352
  pharmacokinetics, metabolism, and mechanisms, 352
  production and use, 351
  reproductive and developmental effects, 352
  toxic effects, 352
    experimental studies of, 352
Butylene glycol mono-*n*-butyl ether, 334–335
  acute toxicity, 335
  chemical and physical properties, 274t
  exposure standards, 335
  toxic effects, 335
    experimental studies of, 335
Butylene glycol monoethyl ether, 333–334
  acute toxicity, 334
  chemical and physical properties, 274t
  exposure standards, 334
  toxic effects, 333–334
    experimental studies of, 333–334
Butylene glycol monomethyl ether, 331–333
  acute toxicity, 332–333
  chemical and physical properties, 274t
  exposure standards, 333
  toxic effects, 332–333
    experimental studies of, 332–333
Butyl glycol ether. *See* Ethylene glycol mono-*n*-butyl ether
n-Butyl mercaptan, 696–698
  acute toxicity, 697
  chemical and physical properties, 682t, 696
  chronic and subchronic toxicity, 697
  exposure assessment, 696
  exposure standards, 686t, 698
  genetic and cellular effects, 698
  pharmacokinetics, metabolism, and mechanisms, 698
  production and use, 696
  reproductive and developmental effects, 698
  toxic effects, 697–698
    experimental studies of, 697–698
    human experience, 698
sec-Butyl mercaptan, 699–700
  acute toxicity, 700
  chemical and physical properties, 682t, 699–700

exposure assessment, 700
production and use, 700
toxic effects, 700
  experimental studies of, 700
tert-Butyl mercaptan, 700–702
  acute toxicity, 701
  chemical and physical properties, 682t, 700
  chronic and subchronic toxicity, 701
  exposure assessment, 701
  genetic and cellular effects, 702
  production and use, 701
  reproductive and developmental effects, 702
  toxic effects, 701–702
    experimental studies of, 701–702
Butyl mercaptans, 682t, 696–702
p-tert-Butylphenol-formaldehyde, 638
Butyl rubber, 447–448
  chemical and physical properties, 427t, 429f, 447–448
  exposure standards, 448
  production and use, 448, 577t, 639t
n-Butyl sulfide. *See* Dibutyl sulfide
n-Butyl sulfone. *See* *n*-Dibutyl sulfone

Carbowax. *See* Polyethylene glycol methyl ethers
Carboxymethoxy-1-methylvinyl dimethyl phosphate. *See* Mevinphos
Carboxymethyl cellulose, 541–542
  chemical and physical properties, 541
  environmental impact, 542
  production and use, 541
  sodium salt (See Sodium carboxymethyl cellulose)
  toxic effects, 541
Casein, production and use, 639t
Celacol m. *See* Methyl cellulose
Cellophane. *See* Cellulose
Cellosize. *See* Hydroxyethyl cellulose
Celluloid, 534
Cellulose, 521–530
  acute toxicity, 527–528, 529t
  carcinogenesis, 528
  chemical and physical properties, 522, 522t–524t
  exposure assessment, 527
  exposure standards, 530
  neurological, pulmonary, skin sensitization effects, 528–530
  production and use, 522–527, 578t, 639t
  toxic effects, 527–530
    experimental studies of, 527–530
    human experience, 530
Cellulose acetate, 530–531
  acute toxicity, 529t

SUBJECT INDEX

Cellulose acetate (*Continued*)
  chemical and physical properties, 522t–526t, 530–531
  production and use, 531, 578t, 639t
  toxic effects, 531
  human experience, 531
Cellulose acetate butyrate, 533
  chemical and physical properties, 523t, 533
  production and use, 533, 639t
Cellulose carboxymethyl ether. *See* Carboxymethyl cellulose
Cellulose CM. *See* Carboxymethyl cellulose
Cellulose ether. *See* Ethyl cellulose
Cellulose methylate. *See* Methyl cellulose
Cellulose nitrate, 533–535
  acute toxicity, 529t
  chemical and physical properties, 522t, 525t–526t, 533–534
  environmental impact, 535
  exposure assessment, 534
  production and use, 534, 639t
  toxic effects, 529t, 534–535
    experimental studies of, 534
    human experience, 534–535
Cellulose propionate, production and use, 639t
Cellulose sodium glycolate. *See* Sodium carboxymethyl cellulose
Cellulose triacetate, 531–532
  chemical and physical properties, 523t, 525t–526t, 531–532
  production and use, 532
  toxic effects, 529t, 532
    experimental studies of, 532
Cellulosics, 521–542
  other polysaccharides, polyamides, and polyimides, 521–568
Cetyl mercaptan. *See* n-Hexadecyl mercaptan
Chlorinated polyether, production and use, 639t
Chlorinated polyethylene, 450–451
  chemical and physical properties, 427t, 450
  neurological, pulmonary, skin sensitization effects, 451
  production and use, 450–451
  toxic effects, 451
Chlorofos. *See* Trichlorfon
Chloromethyloxirane rubber. *See* Epichlorohydrin rubber
Chloropropyl n-octyl sulfoxide, 743–744
  chemical and physical properties, 684t, 744
  production and use, 744
  toxic effects, 744
Chlorosulfonated polyethylene, 446–447
  carcinogenesis, 447

chemical and physical properties, 427t, 446
exposure standards, 447
production and use, 446, 639t
toxic effects, 433t, 446–447
  human experience, 447
Chlorpyrifos, 790–798
  acute toxicity, 770t, 792–793
  biomonitoring/biomarkers, 796
  carcinogenesis, 774t, 795–796
  chemical and physical properties, 791
  chronic and subchronic toxicity, 793, 797–798
  clinical cases, 796–798
  exposure assessment, 792
  exposure standards, 798
  genetic and cellular effects, 796
  pharmacokinetics, metabolism, and mechanisms, 794–795
  production and use, 791
  reproductive and developmental effects, 776t, 795
  toxic effects, 792–798
    human experience, 796–798
CMC. *See* Carboxymethyl cellulose
CMC sodium salt. *See* Sodium carboxymethyl cellulose
Collodion. *See* Cellulose nitrate
Coumaphos, 798–802
  acute toxicity, 770t, 799–800
  carcinogenesis, 774t, 801
  chemical and physical properties, 799
  chronic and subchronic toxicity, 800
  clinical cases, 802
  exposure standards, 802
  genetic and cellular effects, 801–802
  pharmacokinetics, metabolism, and mechanisms, 800–801
  production and use, 799
  reproductive and developmental effects, 776t, 801
  toxic effects, 799–802
    human experience, 802
CPE. *See* Chlorinated polyethylene
CR-39, 592–593
CSM. *See* Chlorosulfonated polyethylene
Cyclohexanethiol. *See* Cyclohexyl mercaptan
N-Cyclohexyl-2-benzothiazolesulfenamide, 755–757
  acute toxicity, 755–756
  chemical and physical properties, 684t, 755
  genetic and cellular effects, 756–757
  production and use, 755
  reproductive and developmental effects, 756
  toxic effects, 755–757
    human experience, 757
Cyclohexyl mercaptan, 720–721
  acute toxicity, 720–721

SUBJECT INDEX

chemical and physical properties, 683t, 720
exposure standards, 721
production and use, 720
toxic effects, 720–721
    experimental studies of, 720–721
DDVP. *See* Dichlorvos
1-Decanethiol. *See* n-Decyl mercaptan
n-Decyl mercaptan, 711
  chemical and physical properties, 682t, 711
  exposure standards, 711
  production and use, 711
  toxic effects, experimental studies of, 711
DEG. *See* Diethylene glycol
DEGDA. *See* Diethylene glycol diacrylate
DEGDN. *See* Diethylene glycol dinitrate
Demethicone. *See* Dimethyl silicone
Demeton, 802–806
  acute toxicity, 770t, 803
  carcinogenesis, 774t, 805
  chemical and physical properties, 802–803
  chronic and subchronic toxicity, 804
  clinical cases, 805
  exposure standards, 806
  genetic and cellular effects, 805
  pharmacokinetics, metabolism, and mechanisms, 804
  production and use, 803
  reproductive and developmental effects, 776t, 804–805
  toxic effects, 803–805
    human experience, 805
Demeton-*S*-methyl, 806–811
  acute toxicity, 770t, 807
  biomonitoring/biomarkers, 809
  carcinogenesis, 774t, 808–809
  chemical and physical properties, 806
  chronic and subchronic toxicity, 807
  clinical cases, 809–811
  exposure standards, 811
  genetic and cellular effects, 809
  pharmacokinetics, metabolism, and mechanisms, 808
  production and use, 807
  reproductive and developmental effects, 776t, 808
  toxic effects, 807–811
    human experience, 809–811
Demetox. *See* Demeton-*S*-methyl
Demox. *See* Demeton
DEP. *See* Trichlorfon
DETF. *See* Trichlorfon
DGBE. *See* Diethylene glycol mono-*n*-butyl ether

DGdiME. *See* Diethylene glycol dimethyl ether
DGEE. *See* Diethylene glycol monoethyl ether
DGEEA. *See* Diethylene glycol monoethyl ether acetate
DGHE. *See* Diethylene glycol mono-*n*-hexyl ether
DGIBE. *See* Diethylene glycol mono-isobutyl ether
DGME. *See* Diethylene glycol monomethyl ether
DGMEA. *See* Diethylene glycol monomethyl ether acetate
DGPE. *See* Diethylene glycol mono-*n*-propyl ether
Diallyl disulfide
  genetic and cellular effects, 733
  pharmacokinetics, metabolism, and mechanisms, 720
Diallyl sulfide, pharmacokinetics, metabolism, and mechanisms, 720
Diazinon, 811–817
  acute toxicity, 770t, 812–813
  carcinogenesis, 774t, 815
  chemical and physical properties, 812
  chronic and subchronic toxicity, 813
  clinical cases, 816–817
  exposure standards, 817
  genetic and cellular effects, 815–816
  pharmacokinetics, metabolism, and mechanisms, 813–814
  production and use, 812
  reproductive and developmental effects, 776t, 814–815
  toxic effects, 812–817
    human experience, 816–817
Diazitol. *See* Diazinon
Dibenzyl disulfide, 738–739
  chemical and physical properties, 683t
  toxic effects, 739
Dibutyl sulfide, 735
  chemical and physical properties, 683t, 735
  production and use, 735
  toxic effects, 735
n-Dibutyl sulfone, 740–741
  acute toxicity, 741
  chemical and physical properties, 683t, 741
  genetic and cellular effects, 741
  toxic effects, 741
Dichlorvos, 817–824
  acute toxicity, 818–819
  biomonitoring/biomarkers, 822
  carcinogenesis, 774t, 821–822
  chemical and physical properties, 818
  chronic and subchronic toxicity, 819–820
  clinical cases, 822–823
  exposure standards, 823–824
  genetic and cellular effects, 822

Dichlorvos (*Continued*)
  pharmacokinetics, metabolism, and mechanisms, 820
  production and use, 818
  reproductive and developmental effects, 777t, 820–821
  toxic effects, 818–823
    human experience, 822–823
Dicrotophos, 824–827
  acute toxicity, 770t, 825
  carcinogenesis, 774t, 826
  chemical and physical properties, 824–825
  chronic and subchronic toxicity, 825
  exposure standards, 826–827
  genetic and cellular effects, 826
  pharmacokinetics, metabolism, and mechanisms, 825
  production and use, 825
  reproductive and developmental effects, 777t, 826
  toxic effects, 825–826
    human experience, 826
N,N-Dicyclohexylbenzothiazolesulfenamide, 757–758
  acute toxicity, 757–758
  carcinogenesis, 758
  chemical and physical properties, 684t
  production and use, 757
  toxic effects, 757–758
    experimental studies of, 757–758
Diene elastomers, 431–455
1,2-Diethoxyethane. *See* Ethylene glycol diethyl ether
Diethyl cellosolve. *See* Ethylene glycol diethyl ether
O,O-Diethyl O-(3-chloro-4-methyl-2-oxo-2H-1-benzopyran-7-yl) phosphorothioate. *See* Coumaphos
O,O-Diethyl O-2-diethyl O-2-isopropyl-4-methyl-6-pyrimidinyl thiophosphate. *See* Diazinon
Diethyl [(dimethoxyphosphinothioyl)thio]butanedioate. *See* Malathion
Diethylene glycol
  acute toxicity, 14
  carcinogenesis, 16
  chemical and physical properties, 2f, 12–13, 13f
  chronic and subchronic toxicity, 14–15
  epidemiology studies, 16
  exposure assessment, 13
  exposure standards, 16–17
  genetic and cellular effects, 16
  pharmacokinetics, metabolism, and mechanisms, 15–16
  production and use, 13
  toxic effects, 13–16
    experimental studies of, 14–16
    human experience, 16
Diethylene glycol diacrylate, 343–344
  acute toxicity, 343
  chemical and physical properties, 343
  genetic and cellular effects, 344
  neurological, pulmonary, skin sensitization effects, 344
  toxic effects, 343–344
    experimental studies of, 343–344
    human experience, 344
Diethylene glycol dimethyl ether, 210–213
  acute toxicity, 211
  chemical and physical properties, 210
  chronic and subchronic toxicity, 211
  exposure assessment, 210
  exposure standards, 213
  genetic and cellular effects, 213
  pharmacokinetics, metabolism, and mechanisms, 211–212
  production and use, 210
  reproductive and developmental effects, 212–213
  toxic effects, 210–213
    experimental studies of, 211–213
Diethylene glycol dinitrate, 373–376
  acute toxicity, 375
  chemical and physical properties, 366t
  chronic and subchronic toxicity, 375–376
  exposure assessment, 374
  exposure standards, 376
  neurological, pulmonary, skin sensitization effects, 376
  pharmacokinetics, metabolism, and mechanisms, 376
  toxic effects, 374–376
    experimental studies of, 375–376
Diethylene glycol 1,4-dioxane ether, chemical and physical properties, 179t
Diethylene glycol divinyl ether, 214
  toxic effects, 214
Diethylene glycol ethyl vinyl ether, 214–215
  chemical and physical properties, 179t, 214
  exposure standards, 215
  production and use, 215
  toxic effects, 215
Diethylene glycol mono-*n*-butyl ether, 201–207
  acute toxicity, 203–204, 203t
  chemical and physical properties, 179t, 202
  chronic and subchronic toxicity, 204
  clinical cases, 206
  environmental impact, 206–207
  exposure assessment, 74, 202

SUBJECT INDEX

exposure standards, 206
genetic and cellular effects, 205–206
neurological, pulmonary, skin sensitization effects, 206
odor and warning properties, 202
pharmacokinetics, metabolism, and mechanisms, 78t, 204–205
production and use, 76t, 202
reproductive and developmental effects, 205
toxic effects, 202–206
  experimental studies of, 203–206
  human experience, 206
Diethylene glycol mono-*n*-butyl ether acetate, 241–243
  acute toxicity, 241–242, 242t
  chemical and physical properties, 229t
  chronic and subchronic toxicity, 242
  exposure assessment, 241
  pharmacokinetics, metabolism, and mechanisms, 242
  production and use, 76t, 241
  toxic effects, 241–243
    experimental studies of, 241–242
    human experience, 243
Diethylene glycol monoethyl ether, 193–199
  acute toxicity, 194–195, 195t
  carcinogenesis, 198
  chemical and physical properties, 179t
  chronic and subchronic toxicity, 196–197
  exposure assessment, 74, 193
  exposure standards, 199
  genetic and cellular effects, 199
  neurological, pulmonary, skin sensitization effects, 199
  pharmacokinetics, metabolism, and mechanisms, 78t, 197–198, 199
  production and use, 76t, 193
  reproductive and developmental effects, 191t, 193t, 198
  toxic effects, 194–199
    experimental studies of, 194–199
    human experience, 199
Diethylene glycol monoethyl ether acetate, 239–240
  acute toxicity, 240
  chemical and physical properties, 229t
  pharmacokinetics, metabolism, and mechanisms, 240
  production and use, 76t, 240
  toxic effects, 240
    experimental studies of, 240
Diethylene glycol mono-*n*-hexyl ether, 208–210
  acute toxicity, 209
  chemical and physical properties, 179t

exposure assessment, 208
genetic and cellular effects, 209–210
production and use, 76t
toxic effects, 208–209
  experimental studies of, 209–210
Diethylene glycol mono-isobutyl ether, 207–208
  chemical and physical properties, 179t
  exposure assessment, 207
  production and use, 207
  toxic effects, 207–208
Diethylene glycol monomethyl ether, 187–192
  acute toxicity, 188
  chemical and physical properties, 179t
  chronic and subchronic toxicity, 188–189
  exposure assessment, 74, 187
  exposure standards, 192
  genetic and cellular effects, 192
  pharmacokinetics, metabolism, and mechanisms, 78t, 189–190, 192
  production and use, 75t, 187
  reproductive and developmental effects, 81, 190–192, 191t, 193t
  toxic effects, 187–192
    experimental studies of, 188–192
    human experience, 192
Diethylene glycol monomethyl ether acetate, 238–239
  acute toxicity, 239
  chemical and physical properties, 229t
  pharmacokinetics, metabolism, and mechanisms, 239
  toxic effects, 239
    experimental studies of, 239
Diethylene glycol monomethylpentyl ether, 215–216
  production and use, 215
  toxic effects, 216
Diethylene glycol mono-*n*-propyl ether, 199–201
  acute toxicity, 200–201
  chemical and physical properties, 179t
  exposure assessment, 200
  exposure standards, 201
  neurological, pulmonary, skin sensitization effects, 201
  production and use, 76t
  toxic effects, 200–201
    experimental studies of, 200–201
Diethylene glycol phenyl ether, exposure assessment, 74
O,O-Diethyl-*S*-ethylmercaptoethyl dithiophosphate. *See* Disulfoton
O,O-Diethyl (*S*-ethylmercaptomethyl) dithiophosphate. *See* Phorate

SUBJECT INDEX

O,O-Diethyl *O* (and *S*)-2-(ethylthio)ethyl phosphorothioate mixture. *See* Demeton
O,O-Diethyl *O-p*-nitrophenyl phosphorothioate. *See* Parathion
Diethyl sulfide, 733–734
 chemical and physical properties, 683t, 733
 exposure assessment, 734
 production and use, 734
 toxic effects, 734
Diethyl sulfoxide, 743
 chemical and physical properties, 684t
 toxic effects, 743
O,O-Diethyl *O*-(3,5,6-trichloro-2-pyridinyl) phosphorothioate. *See* Chlorpyrifos
Diglyme. *See* Diethylene glycol dimethyl ether
2,3-Dihydrothiirene. *See* Ethylene sulfide
1,2-Dihydroxybutane. *See* 1,2-Butanediol
1,3-Dihydroxybutane. *See* 1,3-Butanediol
1,4-Dihydroxybutane. *See* 1,4-Butanediol
1,2-Dihydroxyethane. *See* Ethylene glycol
1,3-Dihydroxypropane. *See* 1,3-Propanediol
N,N-Diisopropyl-2-benzothiazolesulfenamide, 753–754
 acute toxicity, 753
 chemical and physical properties, 684t, 753
 genetic and cellular effects, 753
 production and use, 753
 toxic effects, 753–754
  experimental studies of, 753
  human experience, 754
Dimethoxyethane. *See* Ethylene glycol dimethyl ether
O,O-Dimethyl-1-carbomethoxy-1-propen-2-yl phosphate. *See* Mevinphos
Dimethyl-1,2-dibromo-2,2-dichloroethyl phosphate. *See* Naled
O,O-Dimethyl *O*-2,2-dichlorvinyl dimethyl phosphate. *See* Dichlorvos
O,O-Dimethyl-*O*-(3-dimethylamino-1-methyl-3-oxo-1-propenyl) phosphate. *See* Dicrotophos
Dimethyl disulfide, 731–733
 acute toxicity, 732
 chemical and physical properties, 683t, 732
 chronic and subchronic toxicity, 732–733
 genetic and cellular effects, 733
 production and use, 732
 reproductive and developmental effects, 733
 toxic effects, 732–733
  experimental studies of, 732–733
N-(1,1'-Dimethylethyl)-2-benzothiazolesulfenamide. *See* *N-tert*-Butyl-2-benzothiazolesulfenamide
(*E*)-Dimethyl 2-methylcarbamoyl-1-methylvinyl phosphate. *See* Monocrotophos

O,O-Dimethyl *O*-(*p*-nitrophenyl) phosphorothioate. *See* Methyl parathion
Dimethylol urea. *See* Urea-formaldehyde
Dimethylpolysiloxane hydrolyzate. *See* Silicone
2,2-Dimethyl-1,3-propanediol diacrylate, 352–353
 acute toxicity, 353
 carcinogenesis, 354
 chemical and physical properties, 337t
 genetic and cellular effects, 354
 neurological, pulmonary, skin sensitization effects, 354
 toxic effects, 353–354
  experimental studies of, 353–354
1,1-Dimethyl-1-propanethiol. *See* *tert*-Amyl mercaptan
Dimethyl silicone, 658–664
 acute toxicity, 661
 carcinogenesis, 662–663
 chemical and physical properties, 659
 chronic and subchronic toxicity, 661
 environmental impact, 664
 epidemiology studies, 663–664
 exposure assessment, 660
 exposure standards, 664
 genetic and cellular effects, 662
 pharmacokinetics, metabolism, and mechanisms, 661–662
 production and use, 659
 reproductive and developmental effects, 662
 toxic effects, 660–664
  experimental studies of, 660–662
Dimethyl sulfide, 730–731
 chemical and physical properties, 683t, 731
 exposure assessment, 731
 production and use, 731
 toxic effects, 731
Dimethyl sulfone, 739–740
 acute toxicity, 740
 chemical and physical properties, 683t, 740
 genetic and cellular effects, 740
 production and use, 740
 toxic effects, 740
  experimental studies of, 740
O,O-Dimethyl 1 (2,2,2-trichloro-1-hydroxyethyl)-phosphonate. *See* Trichlorfon
O,O-Dimethyl *O*-(2,4,5-trichlorophenyl)phosphorothioate. *See* Ronnel
Diolice. *See* Coumaphos
Dioxane, 177–187
 acute toxicity, 180–181, 180t, 185
 carcinogenesis, 182–183, 186
 chemical and physical properties, 178, 179t
 chronic and subchronic toxicity, 181

SUBJECT INDEX

epidemiology studies, 186
exposure assessment, 178–180
exposure standards, 108t–109t, 187
genetic and cellular effects, 183–184, 186
neurological, pulmonary, skin sensitization effects, 184–185
odor and warning properties, 178
pharmacokinetics, metabolism, and mechanisms, 181–182, 185–186
production and use, 178
reproductive and developmental effects, 182
toxic effects, 180–186
  experimental studies of, 180–185
  human experience, 185–186
2,3-*p*-Dioxanedithion *S,S*-bis-(*O,O*-diethyl phosphorodithioate). *See* Dioxathion
Dioxathion, 826–830
  acute toxicity, 770t, 828–829
  carcinogenesis, 774t, 829–830
  chemical and physical properties, 827
  chronic and subchronic toxicity, 829
  clinical cases, 830
  exposure standards, 830
  genetic and cellular effects, 830
  pharmacokinetics, metabolism, and mechanisms, 829
  production and use, 827–828
  reproductive and developmental effects, 777t, 829
  toxic effects, 828–830
    human experience, 830
Dipentenedimercaptan. *See* D-Limonene dimercaptan
Diphenyl sulfone, 742–743
  chemical and physical properties, 683t, 742
  production and use, 743
  toxic effects, 743
Diphonate. *See* Fonofos
Diphosphoric acid tetraethyl ester. *See* TEPP
Dipofene. *See* Diazinon
Di-*n*-propyl disulfide, 734–735
  acute toxicity, 734–735
  chemical and physical properties, 683t, 734
  genetic and cellular effects, 735
  toxic effects, 734–735
Dipropylene glycol, 35–36
  acute toxicity, 36
  chemical and physical properties, 35
  chronic and subchronic toxicity, 36
  exposure assessment, 35
  production and use, 35
  toxic effects, 35–36
    experimental studies of, 36
    human experience, 36

Dipropylene glycol allyl ethers, 326–327
  acute toxicity, 327
  chemical and physical properties, 327
Dipropylene glycol butyl ether acetate, production and use, 278t
Dipropylene glycol ethyl ether acetate, production and use, 278t
Dipropylene glycol isobutyl ethers, 327–329
  acute toxicity, 328–329
  chemical and physical properties, 328
  genetic and cellular effects, 329
  odor and warning properties, 328
  toxic effects, 328–329
    experimental studies of, 328–329
Dipropylene glycol methyl ether acetate, production and use, 278t
Dipropylene glycol mono-*n*-butyl ether, 314–317
  acute toxicity, 315
  chemical and physical properties, 274t, 314
  chronic and subchronic toxicity, 315–316
  exposure standards, 317
  genetic and cellular effects, 316
  neurological, pulmonary, skin sensitization effects, 316
  production and use, 278t
  reproductive and developmental effects, 316
  toxic effects, 315–317
    experimental studies of, 315–316
    human experience, 317
Dipropylene glycol mono-*tertiary*-butyl ether, 317–321
  acute toxicity, 318–319
  chemical and physical properties, 317
  chronic and subchronic toxicity, 319–320
  exposure assessment, 317
  exposure standards, 321
  genetic and cellular effects, 321
  neurological, pulmonary, skin sensitization effects, 321
  production and use, 278t, 317
  toxic effects, 318–321
    experimental studies of, 318–321
Dipropylene glycol monoethyl ether, 311–314
  acute toxicity, 312–313
  chemical and physical properties, 274t
  chronic and subchronic toxicity, 313
  exposure standards, 314
  genetic and cellular effects, 314
  neurological, pulmonary, skin sensitization effects, 314
  production and use, 278t
  toxic effects, 312–314
    experimental studies of, 312–314

Dipropylene glycol monomethyl ether, 307–311
    acute toxicity, 309
    chemical and physical properties, 274t
    chronic and subchronic toxicity, 309–310
    exposure standards, 311, 312t
    genetic and cellular effects, 311
    neurological, pulmonary, skin sensitization effects, 311
    odor and warning properties, 308
    pharmacokinetics, metabolism, and mechanisms, 310
    production and use, 278t, 308
    reproductive and developmental effects, 310
    toxic effects, 308–311
        experimental studies of, 309–311
        human experience, 311
Dipropylene glycol monomethyl ether acetate, 360–362
    acute toxicity, 361–362
    chemical and physical properties, 338t
    exposure assessment, 361
    exposure standards, 362
    genetic and cellular effects, 362
    neurological, pulmonary, skin sensitization effects, 362
    pharmacokinetics, metabolism, and mechanisms, 362
    production and use, 276t
    reproductive and developmental effects, 362
    toxic effects, 361–362
        experimental studies of, 361–362
Dipropylene glycol propyl ether, production and use, 278t
Dipropylene glycol propyl ether acetate, production and use, 278t
Disulfoton, 830–836
    acute toxicity, 770t, 831–832
    biomonitoring/biomarkers, 835
    carcinogenesis, 774t, 834–835
    chemical and physical properties, 831
    chronic and subchronic toxicity, 832–833
    clinical cases, 836
    exposure standards, 836
    genetic and cellular effects, 835
    pharmacokinetics, metabolism, and mechanisms, 833
    production and use, 831
    reproductive and developmental effects, 777t, 834
    toxic effects, 831–836
        human experience, 836
2,3-Dithiabutane. *See* Dimethyl disulfide
Dithioglycol. *See* 1,2-Ethanedithiol
Dithione. *See* Sulfotepp

Dithiosystox. *See* Disulfoton
Divinyl sulfone, 741–742
    chemical and physical properties, 683t, 742
    genetic and cellular effects, 742
    toxic effects, 742
DMDS. *See* Dimethyl disulfide
DMS. *See* Dimethyl sulfide
DNTP. *See* Parathion
1-Dodecanethiol. *See* n-Dodecyl mercaptan
t-Dodecanethiol. *See* tert-Dodecyl mercaptan
n-Dodecyl mercaptan, 711–714
    acute toxicity, 712
    chemical and physical properties, 683t, 712
    chronic and subchronic toxicity, 713–714
    exposure standards, 714
    genetic and cellular effects, 714
    production and use, 712
    reproductive and developmental effects, 714
    toxic effects, 712–714
        experimental studies of, 712–714
        human experience, 714
tert-Dodecyl mercaptan, 714–717
    acute toxicity, 715–716
    chemical and physical properties, 683t, 715
    chronic and subchronic toxicity, 716
    environmental impact, 716–717
    exposure assessment, 715
    genetic and cellular effects, 716
    production and use, 715
    reproductive and developmental effects, 716
    toxic effects, 715–716
        experimental studies of, 715–716
DPGBE. *See* Dipropylene glycol mono-n-butyl ether
DPGEE. *See* Dipropylene glycol monoethyl ether
DPGME. *See* Dipropylene glycol monomethyl ether
DPTB. *See* Dipropylene glycol mono*tertiary*-butyl ether
Duraphos. *See* Mevinphos
Dursban. *See* Chlorpyrifos
DYME. *See* Diethylene glycol dimethyl ether

ECTFE. *See* Ethylene chlorotrifluoroethylene copolymer
EGBE. *See* Ethylene glycol mono-n-butyl ether
EGBEA. *See* Ethylene glycol mono-n-butyl ether acetate
EGdiEE. *See* Ethylene glycol diethyl ether
EGdiME. *See* Ethylene glycol dimethyl ether
EGDN. *See* Ethylene glycol dinitrate
EGEEA. *See* Ethylene glycol monoethyl ether acetate
EGEHE. *See* Ethylene glycol mono-2-ethylhexyl ether
EGHE. *See* Ethylene glycol mono-n-hexyl ether

SUBJECT INDEX

EGIBE. See Ethylene glycol monoisobutyl ether
EGiPE. See Ethylene glycol monoisopropyl ether
EGME. See Ethylene glycol monomethyl ether
EGMEA. See Ethylene glycol monomethyl ether acetate
EGPE. See Ethylene glycol mono-n-propyl ether
EGPhE. See Ethylene glycol monophenyl ether
EGTBE. See Ethylene glycol mono-tert-butyl ether
EGVE. See Ethylene glycol monovinyl ether
EHD. See 2-Ethyl-1,3-hexanediol
Elastomeric polyester, 587
Elastomers, 425–455
    analysis and specifications, 429–430
    chemical and physical properties, 427t–428t, 429f
    exposure standards, 431
    production and use, 577t
    toxic effects, 430–431
Elvanol. See Polyvinyl alcohol
EPDM. See Ethylene-propylene
Epichlorohydrin rubber, 451–452
    chemical and physical properties, 427t, 429f, 451–452
    exposure standards, 452
    production and use, 452
    toxic effects, 452
1,2-Epithiopropane. See Propylene sulfide
EPM. See Ethylene-propylene
EPN, 836–842
    acute toxicity, 770t, 837–838
    carcinogenesis, 774t, 841
    chemical and physical properties, 837
    chronic and subchronic toxicity, 839
    clinical cases, 841–842
    exposure standards, 842
    neurological, pulmonary, skin sensitization effects, 841
    pharmacokinetics, metabolism, and mechanisms, 839–840
    production and use, 837
    reproductive and developmental effects, 777t, 840
    toxic effects, 837–842
        human experience, 841–842
Epoxy, production and use, 577t, 639t
ETFE. See Ethylene tetrafluoroethylene copolymer
1,2-Ethanediol. See Ethylene glycol
1,2-Ethanediol monoacetate. See Ethylene glycol monoacetate
1,2-Ethanedithiol, 724–726
    acute toxicity, 725–726
    chemical and physical properties, 683t, 725
    environmental impact, 726
    exposure assessment, 725
    genetic and cellular effects, 726

production and use, 725
toxic effects, 725–726
    experimental studies of, 725–726
Ethanesulfonyl chloride, 747–748
    acute toxicity, 748
    chemical and physical properties, 684t, 748
    toxic effects, experimental studies of, 748
Ethanethiol. See Ethyl mercaptan
Ethanethiol, 2-(ethylthio)-S-ester with O,O-dimethyl phosphorothioate. See Demeton-S-methyl
Ethanox. See Ethion
Ethion, 842–845
    acute toxicity, 770t, 843–845
    carcinogenesis, 774t, 845
    chemical and physical properties, 843
    chronic and subchronic toxicity, 844
    exposure standards, 845
    genetic and cellular effects, 845
    pharmacokinetics, metabolism, and mechanisms, 844
    production and use, 843
    reproductive and developmental effects, 777t, 844
    toxic effects, 843–845
        human experience, 845
Ethohexadiol. See 2-Ethyl-1,3-hexanediol
Ethoxyacetic acid, pharmacokinetics, metabolism, and mechanisms, 79t
2-Ethoxyethanol. See Ethylene glycol monoethyl ether
1-(2-Ethoxyethoxy)-2-vinyl-oxyethane. See Diethylene glycol ethyl vinyl ether
Ethyl acrylate polymer, 515
    chemical and physical properties, 488t, 515
2-Ethylbutoxypropanol, mixed isomers. See Propylene glycol monoethylbutyl ether, mixed isomers
Ethyl cellulose, 535–536
    acute toxicity, 529t
    chemical and physical properties, 525t–526t, 535
    exposure assessment, 535
    production and use, 535, 639t
    toxic effects, 529t, 536
Ethylcyclohexyldithiol, 726
    acute toxicity, 726
    chemical and physical properties, 683t, 726
    toxic effects, experimental studies of, 726
Ethylene/acrylic, chemical and physical properties, 429f
Ethylene chlorotrifluoroethylene copolymer, 475–476
    chemical and physical properties, 428t, 475
    exposure assessment, 476

SUBJECT INDEX

Ethylene chlorotrifluoroethylene
  copolymer (*Continued*)
  production and use, 475
  toxic effects, human experience, 476
Ethylene dimercaptan. *See* 1,2-Ethanedithiol
Ethylene episulfide. *See* Ethylene sulfide
Ethylene glycol, 1–12
  acute toxicity, 4, 4f, 10–11
  chemical and physical properties, 2, 2f
  chronic and subchronic toxicity, 5–6, 11
  clinical cases, 10–12
  environmental impact, 12
  exposure assessment, 3
  exposure standards, 12
  odor and warning properties, 2
  pharmacokinetics, metabolism, and mechanisms, 6–9, 7f, 11–12
  production and use, 2–3
  reproductive and developmental effects, 9–10
  toxic effects, 4–12
    experimental studies of, 4–10
    human experience, 10–12
Ethylene glycol diacetate, 340–341
  acute toxicity, 341
  chemical and physical properties, 337t, 340
  chronic and subchronic toxicity, 341
  exposure standards, 341
  reproductive and developmental effects, 341
  toxic effects, 340–341
    experimental studies of, 341
Ethylene glycol diacrylate, 341–342
  acute toxicity, 342
  chemical and physical properties, 342
  exposure standards, 342
  genetic and cellular effects, 342
  neurological, pulmonary, skin sensitization effects, 342
  toxic effects, 342
    experimental studies of, 342
Ethylene glycol diethyl ether, 175–177
  acute toxicity, 176–177
  chemical and physical properties, 86t
  exposure assessment, 176
  exposure standards, 177
  odor and warning properties, 175
  production and use, 175
  reproductive and developmental effects, 177
  toxic effects, 176–177
    experimental studies of, 176–177
Ethylene glycol dimethyl ether, 172–175
  acute toxicity, 173
  chemical and physical properties, 86t
  exposure assessment, 173

  exposure standards, 175
  genetic and cellular effects, 174
  neurological, pulmonary, skin sensitization effects, 174–175
  odor and warning properties, 173
  reproductive and developmental effects, 173–174
  toxic effects, 173–175
    experimental studies of, 173–175
    human experience, 175
Ethylene glycol dinitrate, 365–373
  acute toxicity, 367, 372–373
  chemical and physical properties, 366, 366t
  chronic and subchronic toxicity, 367–368
  clinical cases, 372–373
  exposure assessment, 366
  exposure standards, 373, 374t
  genetic and cellular effects, 371
  pharmacokinetics, metabolism, and mechanisms, 368–371, 369f
  production and use, 366
  toxic effects, 366–373
    experimental studies of, 367–371
    human experience, 371–373
Ethylene glycol monoacetate, 338–340
  acute toxicity, 339
  chemical and physical properties, 337t
  chronic and subchronic toxicity, 340
  exposure standards, 340
  reproductive and developmental effects, 340
  toxic effects, 338–340
    experimental studies of, 339–340
Ethylene glycol mono-*n*-butyl ether, 136–157
  acute toxicity, 139–140, 139t, 152–155
  carcinogenesis, 82, 150–151
  chemical and physical properties, 86t
  chronic and subchronic toxicity, 141–144
  clinical cases, 152–156
  environmental impact, 157
  epidemiology studies, 156–157
  exposure assessment, 74, 137–138
  exposure standards, 108t–109t, 157
  genetic and cellular effects, 151–152
  neurological, pulmonary, skin sensitization effects, 152, 156
  odor and warning properties, 136
  PBPK model, 83, 85f, 146–147, 155–156
  pharmacokinetics, metabolism, and mechanisms, 78t, 79t, 96t, 129t, 144–149, 145t, 146t, 155–156
  production and use, 75t, 137
  reproductive and developmental effects, 149–150, 156–157
  toxic effects, 81–82, 138–157
    experimental studies of, 138–152

SUBJECT INDEX

human experience, 152–157
Ethylene glycol mono-*tert*-butyl ether, 159–160
  acute toxicity, 159
  exposure assessment, 159
  exposure standards, 160
  toxic effects, 159
    experimental studies of, 159
    human experience, 159
Ethylene glycol mono-*n*-butyl ether acetate, 236–238
  acute toxicity, 237–238
  chemical and physical properties, 229t, 237
  chronic and subchronic toxicity, 238
  exposure assessment, 237
  exposure standards, 108t–109t, 238
  pharmacokinetics, metabolism, and mechanisms, 238
  production and use, 75t, 237
  toxic effects, 237–238
    experimental studies of, 237–238
    human experience, 238
Ethylene glycol monoethyl ether, 110–124
  acute toxicity, 112–114, 122
  carcinogenesis, 120
  chemical and physical properties, 86t
  chronic and subchronic toxicity, 114–115, 122
  clinical cases, 121–124
  exposure assessment, 74, 111–112
  exposure standards, 108t–109t, 124
  genetic and cellular effects, 82–83, 120–121
  odor and warning properties, 111
  PBPK model, 83, 117
  pharmacokinetics, metabolism, and mechanisms, 78t, 79t, 92f, 96t, 115–118, 122–123
  production and use, 75t, 111
  reproductive and developmental effects, 81, 118–120, 123–124
  toxic effects, 81, 112–124
    experimental studies of, 112–121
    human experience, 121–124
Ethylene glycol monoethyl ether acetate, 232–236
  acute toxicity, 233–234
  chemical and physical properties, 229t
  chronic and subchronic toxicity, 234
  clinical cases, 236
  epidemiology studies, 236
  exposure assessment, 232–233
  exposure standards, 108t–109t, 236
  pharmacokinetics, metabolism, and mechanisms, 234, 236
  production and use, 75t, 232
  reproductive and developmental effects, 235
  toxic effects, 233–236
    experimental studies of, 233–235

human experience, 235–236
Ethylene glycol mono-2-ethylhexyl ether, 165–167
  acute toxicity, 166–167
  chemical and physical properties, 86t
  exposure assessment, 166
  exposure standards, 167
  production and use, 165
  toxic effects, 166–167
    experimental studies of, 166–167
Ethylene glycol mono-2,4-hexadiene ether, 164–165
  acute toxicity, 165
  chemical and physical properties, 164
  exposure assessment, 164
  exposure standards, 165
  production and use, 164
  toxic effects, 165
    experimental studies of, 165
Ethylene glycol mono-*n*-hexyl ether, 161–164
  acute toxicity, 162–163
  chemical and physical properties, 86t
  chronic and subchronic toxicity, 163
  exposure assessment, 161
  exposure standards, 164
  genetic and cellular effects, 163–164
  pharmacokinetics, metabolism, and mechanisms, 96t, 163
  production and use, 75t, 161
  reproductive and developmental effects, 163
  toxic effects, 161–164
    experimental studies of, 162–164
    human experience, 164
Ethylene glycol monoisobutyl ether, 157–159
  acute toxicity, 158
  chemical and physical properties, 86t
  exposure assessment, 157–158
  exposure standards, 158–159
  toxic effects
    experimental studies of, 158
    human experience, 158
Ethylene glycol monoisopropyl ether, 132–136
  acute toxicity, 132–134, 133t
  chemical and physical properties, 86t, 132
  chronic and subchronic toxicity, 134
  exposure standards, 108t–109t, 136
  genetic and cellular effects, 136
  odor and warning properties, 132
  pharmacokinetics, metabolism, and mechanisms, 134–135, 135f
  production and use, 132
  reproductive and developmental effects, 135
  toxic effects, 132–136
    experimental studies of, 132–136
    human experience, 136

Ethylene glycol monomethyl ether, 84–85
  acute toxicity, 87–89
  carcinogenesis, 101
  chemical and physical properties, 86t
  chronic and subchronic toxicity, 89–91
  clinical cases, 103–106
  epidemiology studies, 106–107
  exposure assessment, 74, 87
  exposure standards, 107, 108t–109t
  genetic and cellular effects, 101–102
  immunologic effects, 82
  neurological, pulmonary, skin sensitization effects, 102–103
  odor and warning properties, 85
  PBPK model, 83, 84f, 92–96, 104–106
  pharmacokinetics, metabolism, and mechanisms, 78t, 79t, 91–98, 93t, 94t, 96t, 104–106
  production and use, 75t, 87
  reproductive and developmental effects, 81, 98–101
  toxic effects, 81, 87–107
    experimental studies of, 87–103
    human experience, 103–107
Ethylene glycol monomethyl ether acetate, 228–232
  acute toxicity, 230
  chemical and physical properties, 229t
  clinical cases, 231
  epidemiology studies, 231
  exposure assessment, 228
  exposure standards, 232
  genetic and cellular effects, 231
  neurological, pulmonary, skin sensitization effects, 231
  pharmacokinetics, metabolism, and mechanisms, 230–231
  reproductive and developmental effects, 231
  toxic effects, 228–231
    experimental studies of, 230–231
    human experience, 231
Ethylene glycol mono-2-methylpentyl ether, 160–161
  chemical and physical properties, 86t, 160
  exposure assessment, 160
  exposure standards, 161
  toxic effects, 160
Ethylene glycol monomethylphenyl ether, 171–172
  exposure assessment, 172
  production and use, 172
  toxic effects, 172
Ethylene glycol monophenyl ether, 168–171
  acute toxicity, 169–170
  chemical and physical properties, 86t
  chronic and subchronic toxicity, 170

exposure assessment, 74, 169
exposure standards, 171
genetic and cellular effects, 171
neurological, pulmonary, skin sensitization effects, 171
odor and warning properties, 168
pharmacokinetics, metabolism, and mechanisms, 96t, 170
production and use, 75t, 169
reproductive and developmental effects, 170–171
toxic effects, 169–171
  experimental studies of, 169–171
  human experience, 171
Ethylene glycol mono-*n*-propyl ether, 124–132
  acute toxicity, 126–127
  chemical and physical properties, 86t, 125
  chronic and subchronic toxicity, 127–128
  exposure assessment, 125
  exposure standards, 132
  genetic and cellular effects, 130–131
  neurological, pulmonary, skin sensitization effects, 131
  pharmacokinetics, metabolism, and mechanisms, 128–129, 129t
  production and use, 75t, 125
  reproductive and developmental effects, 130
  toxic effects, 125–131
    experimental studies of, 125–131
    human experience, 131
Ethylene glycol mono-2,6,8-trimethyl-4-nonyl ether, 167–168
  chemical and physical properties, 86t, 167
  exposure assessment, 167
  exposure standards, 168
  production and use, 167
  toxic effects, 168
Ethylene glycol monovinyl ether, 107–110
  acute toxicity, 110
  exposure standards, 110
  pharmacokinetics, metabolism, and mechanisms, 110
  toxic effects, 110
Ethylene glycols, nitrate esters of, 365–383
Ethylene-propylene, 440–442
  chemical and physical properties, 427t, 429f, 441
  production and use, 441, 577t
  toxic effects, 441–442
    human experience, 442
Ethylene propylene rubber, production and use, 639*t*
Ethylene sulfide, 735–736
  acute toxicity, 736
  carcinogenesis, 736
  chemical and physical properties, 683t, 736

SUBJECT INDEX

production and use, 736
toxic effects, 736
   experimental studies of, 736
Ethylene tetrafluoroethylene copolymer, 473
   chemical and physical properties, 428t, 473
   production and use, 473
Ethylene-vinyl acetate, production and use, 640t
Ethyl glyme. See Ethylene glycol diethyl ether
2-Ethyl-1,3-hexanediol, 53–55
   acute toxicity, 54
   chemical and physical properties, 53
   chronic and subchronic toxicity, 54
   exposure assessment, 53
   exposure standards, 55
   pharmacokinetics, metabolism, and mechanisms, 55
   production and use, 53
   toxic effects, 53–55
      experimental studies of, 54
      human experience, 54–55
Ethyl hexylene glycol. See 2-Ethyl-1,3-hexanediol
Ethyl mercaptan, 690–692
   acute toxicity, 691–692
   chemical and physical properties, 682t, 690
   chronic and subchronic toxicity, 692
   exposure assessment, 691
   exposure standards, 686t, 692
   genetic and cellular effects, 692
   pharmacokinetics, metabolism, and mechanisms, 692
   production and use, 691
   toxic effects, 691–692
      exposure standards, 691–692
O-Ethyl O-(4-(methylthio)phenyl)-S-propyl phosphorodithioate. See Sulprofos
O-Ethyl O-p-nitrophenyl phenylphosphonothioate. See EPN
Ethyl parathion. See Parathion
O-Ethyl-S-phenyl ethylphosphonodithioate. See Fonofos
Ethyl pyrophosphate. See TEPP
Ethyl sulfide. See Diethyl sulfide
Ethyl thioether. See Diethyl sulfide

Fatty acid polymers, 561–563
Fenamiphos, 845–851
   acute toxicity, 770t, 846–847
   carcinogenesis, 774t, 849–850
   chemical and physical properties, 846
   chronic and subchronic toxicity, 847–848
   exposure standards, 850–851
   genetic and cellular effects, 850

   pharmacokinetics, metabolism, and mechanisms, 848
   production and use, 846
   reproductive and developmental effects, 777t, 848–849
   toxic effects, 846–850
      human experience, 850
Fenchlorophos. See Ronnel
Fensulfothion, 851–853
   acute toxicity, 851–852
   carcinogenesis, 774t, 852
   chemical and physical properties, 851
   chronic and subchronic toxicity, 852
   clinical cases, 853
   exposure standards, 853
   genetic and cellular effects, 853
   pharmacokinetics, metabolism, and mechanisms, 852
   production and use, 851
   reproductive and developmental effects, 777t, 852
   toxic effects, 851–853
      human experience, 853
Fenthion, 853–861
   acute toxicity, 770t, 854–855
   carcinogenesis, 774t, 857–858
   chemical and physical properties, 854
   chronic and subchronic toxicity, 856
   clinical cases, 859–860
   exposure standards, 860–861
   genetic and cellular effects, 858–859
   neurological, pulmonary, skin sensitization effects, 859
   pharmacokinetics, metabolism, and mechanisms, 856–857
   production and use, 854
   reproductive and developmental effects, 778t, 857
   toxic effects, 854–860
      human experience, 859
FEP. See Fluorinated ethylene propylene copolymer
FKKM. See Perfluoro rubber
FKM. See Vinylidene fluoride-hexafluoropropylene
Fluorinated ethylene propylene, production and use, 639t
Fluorinated ethylene propylene copolymer, 470–472
   chemical and physical properties, 428t, 471–472
   neurological, pulmonary, skin sensitization effects, 472
   production and use, 472
   toxic effects, 472
      human experience, 472
Fluoroelastomers, production and use, 639t
Fluoropolymers, production and use, 639t

Fluorosilicone, chemical and physical properties, 429f
FMVE. See Perfluoroalkoxy copolymer
Fonofos, 861–862
  acute toxicity, 770t, 862
  carcinogenesis, 774t , 862–863
  chemical and physical properties, 861
  chronic and subchronic toxicity, 862
  exposure standards, 863
  pharmacokinetics, metabolism, and mechanisms, 862
  production and use, 861
  reproductive and developmental effects, 778t, 862
  toxic effects, 862–863
    human experience, 863
Formaldehyde
  phenol polymer (See Phenol-formaldehyde resin)
  urea polymer urea (See Urea-formaldehyde)
Fostacryl. See Styrene-acrylonitrile
FPM. See Vinylidene fluoride-hexafluoropropylene
Furan polymers/furan resins, 650–651
  production and use, 577t, 650–651
  toxic effects, 651
Furfuryl alcohol resins. See Furan polymers/furan resins

Glycol dinitrate. See Ethylene glycol dinitrate
Glycol esters, diesters, and ether esters, 336–365
  chemical and physical properties, 336, 337t, 338t
  exposure assessment, 336
  production and use, 336
  toxic effects, 336
Glycol ethers
  background, 73, 74t
  of butylene, 331–336
  carcinogenesis, 82–83
  environmental impact, 83
  exposure assessment, 74–77
  genetic and cellular effects, 82–83
  hematologic effects, 81–82
  immunologic effects, 82
  pharmacokinetic models, 83
  pharmacokinetics, metabolism, and mechanisms, 77–80, 78t, 79t, 80f
  production and use, 73–74, 75t–76t
  of propylene, 271–331
    background, 271, 272t
    chemical and physical properties, 273t, 274t
    exposure assessment, 275
    pharmacokinetics, metabolism, and mechanisms, 280–281
    production and use, 271–275, 276t–277t, 278t
    toxic effects, 275–280, 279t

  reproductive and developmental effects, 81
  toxic effects, 81–83
Glycol monoacetate. See Ethylene glycol monoacetate
Glycols, 1–64
  chemical and physical properties, 2f
Glyme. See Ethylene glycol dimethyl ether
Glyme-1. See Ethylene glycol diethyl ether
Guncotton. See Cellulose nitrate
Guthion. See Azinphos-methyl

Heptanethiol. See Heptyl mercaptan
Heptyl mercaptan, 706–707
  chemical and physical properties, 682t, 706
  exposure assessment, 706
  exposure standards, 707
  production and use, 706
  toxic effects, experimental studies of, 707
Hexadecanethiol. See n-Hexadecyl mercaptan
n-Hexadecyl mercaptan, 717–718
  chemical and physical properties, 683t, 717
  exposure assessment, 718
  exposure standards, 718
  production and use, 717
  toxic effects, experimental studies of, 718
Hexanedioic acid, polymer with butanediol. See Butylene glycol adipic acid polyester
1-Hexanethiol. See Hexyl mercaptan
Hexylene glycol. See 2-Methyl-2,4-pentanediol
Hexyl mercaptan, 705–706
  acute toxicity, 706
  chemical and physical properties, 682t, 705
  exposure assessment, 706
  exposure standards, 706
  production and use, 705
  toxic effects, experimental studies of, 706
HIPS. See Polystyrene
Hydroquinone, bis(2-hydroxyethyl) ether. See 2,2'-[1,4-Phenylenebis(oxy)]bisethanol
Hydroxyethyl cellulose, 539–540
  chemical and physical properties, 525t–526t, 539
  production and use, 539
  toxic effects, 540
2-Hydroxy-1-propyl acrylate. See Propylene glycol monoacrylate
Hydroxypropyl cellulose, 540–541
  chemical and physical properties, 540
  production and use, 540
  toxic effects, 540
Hydroxypropyl methyl cellulose, 538–539
  chemical and physical properties, 538
  production and use, 538
  toxic effects, 539

SUBJECT INDEX

Hypromellose. *See* Hydroxypropyl methyl cellulose

Ionomers, 422
   production and use, 640t
Isoamyl mercaptan, 703–704
   chemical and physical properties, 682t, 704
   toxic effects, 704
      experimental studies of, 704
Isobutoxyethanol. *See* Ethylene glycol monoisobutyl ether
Isobutyl mercaptan, 698–699
   chemical and physical properties, 682t, 699
   exposure assessment, 699
   toxic effects, 699
      experimental studies of, 699
N-Isopropyl-2-benzothiazolesulfenamide, 752–753
   acute toxicity, 752
   chemical and physical properties, 684t
   genetic and cellular effects, 752
   production and use, 752
   toxic effects, 752–753
      experimental studies of, 752
      human experience, 752–753
Isopropyl mercaptan, 694–695
   acute toxicity, 695
   chemical and physical properties, 682t, 695
   production and use, 695
   toxic effects, 695
      experimental studies of, 695
      human experience, 695
Ixan. *See* Polyvinylidene chloride

Kaurit S. *See* Urea-formaldehyde

Latex. *See* Dimethyl silicone
Lauryl mercaptan. *See* n-Dodecyl mercaptan
D-Limonene dimercaptan, 726–727
   acute toxicity, 727
   chemical and physical properties, 683t, 727
   exposure assessment, 727
   genetic and cellular effects, 727
   production and use, 727
   toxic effects, experimental studies of, 727
Linear terephthalate polyesters, 581–585
Liquid crystal polymers, production and use, 640t

Malathion, 863–871
   acute toxicity, 770t, 865–866
   carcinogenesis, 774t, 868
   chemical and physical properties, 864
   chronic and subchronic toxicity, 866
   clinical cases, 869–871

   exposure standards, 871
   genetic and cellular effects, 869
   pharmacokinetics, metabolism, and mechanisms, 866–868
   production and use, 864–865
   reproductive and developmental effects, 778t, 868, 871
   toxic effects, 865–871
      human experience, 869–871
Maldison. *See* Malathion
MBT. *See* 2-Mercaptobenzothiazole
Melamine-formaldehyde resins, 648–650
   acute toxicity, 649
   chronic and subchronic toxicity, 649
   exposure assessment, 648–649
   production and use, 577t, 639t, 648
   reproductive and developmental effects, 649
   toxic effects, 649–650
      experimental studies of, 649
      human experience, 649–650
Meldane. *See* Coumaphos
Mercaptans, 685–730
Mercaptobenzene. *See* Phenyl mercaptan
2-Mercaptobenzothiazole, 750–752
   acute toxicity, 750–751
   carcinogenesis, 751
   chemical and physical properties, 684t
   genetic and cellular effects, 751
   neurological, pulmonary, skin sensitization effects, 751
   production and use, 750
   reproductive and developmental effects, 751
   toxic effects, 750–752
      experimental studies of, 750–751
      human experience, 751–752
1-Mercaptobutane. *See* n-Butyl mercaptan
3(or 4)-Mercaptocyclohexaneethanethiol. *See* Vinyl cyclohexene-derived dimercaptan
3-Mercapto-beta-4-dimethyl-cyclohexaneethanethiol. *See* D-Limonene dimercaptan
1-Mercaptododecane. *See* n-Dodecyl mercaptan
Mercaptoethane. *See* Ethyl mercaptan
Mercaptomethane. *See* Methyl mercaptan
1-Mercaptopropane. *See* Propyl mercaptan
2-Mercapto propane. *See* Isopropyl mercaptan
Mesyl chloride. *See* Methanesulfonyl chloride
Methacrylic acid methyl ester. *See* Methyl methacrylate
Methallyl *n*-octyl sulfoxide, 744–745
   chemical and physical properties, 684t
   toxic effects, 745
p-Methane-2,9-dithiol. *See* D-Limonene dimercaptan

SUBJECT INDEX

Methanesulfonic acid, 745–746
  acute toxicity, 745–746
  chemical and physical properties, 684t, 745
  environmental impact, 746
  exposure assessment, 745
  genetic and cellular effects, 746
  production and use, 745
  toxic effects, 745–746
    experimental studies of, 745–746
Methanesulfonyl chloride, 746–747
  chemical and physical properties, 684t, 747
  environmental impact, 747
  production and use, 747
  toxic effects, 747
Methanethiol. *See* Methyl mercaptan
Methoxyacetic acid, pharmacokinetics, metabolism, and mechanisms, 79*t*
4-Methoxybutanol. *See* Butylene glycol monomethyl ether
2-Methoxy ethanol. *See* Ethylene glycol monomethyl ether
2-Methoxyethanol acetate. *See* Ethylene glycol monomethyl ether acetate
Methoxypolyethylene glycols. *See* Polyethylene glycol methyl ethers
1-Methoxypropan-2-ol. *See* Propylene glycol monomethyl ether
2-Methyl-2-butanethiol. *See* tert-Amyl mercaptan
3-Methyl-1-butanethiol. *See* Isoamyl mercaptan
Methyl cellulose, 536–537
  chemical and physical properties, 537
  environmental impact, 538
  exposure assessment, 537
  production and use, 537
  toxic effects, 537–538
Methyl disulfide. *See* Dimethyl disulfide
(1-Methyl-1,2-ethanediyl) bisoxybispropanol. *See* Tripropylene glycol
N-(1-Methylethyl)-2-benzothiazolesulfenamide. *See* *N*-Isopropyl-2-benzothiazolesulfenamide
Methyl ethylene glycol. *See* Propylene glycol
Methyl glycol. *See* Ethylene glycol monomethyl ether
Methyl mercaptan, 687–690
  acute toxicity, 688–689, 689
  chemical and physical properties, 682t, 687
  chronic and subchronic toxicity, 689–690
  clinical cases, 689–690
  exposure assessment, 688
  exposure standards, 686t, 690
  pharmacokinetics, metabolism, and mechanisms, 689
  production and use, 688
  toxic effects, 688–690
    experimental studies of, 688–689
    human experience, 689–690
Methyl methacrylate, 514–515
  chemical and physical properties, 488t, 514–515
  exposure standards, 515
  production and use, 514
  toxic effects, 514–515
Methyl parathion, 871–880
  acute toxicity, 770t, 873
  biomonitoring/biomarkers, 872–873
  carcinogenesis, 774t, 877
  chemical and physical properties, 872
  chronic and subchronic toxicity, 873–875
  clinical cases, 878–879
  exposure standards, 879–880
  genetic and cellular effects, 877–878
  pharmacokinetics, metabolism, and mechanisms, 875–876, 879
  production and use, 872
  reproductive and developmental effects, 778t, 876–877
  toxic effects, 873–879
    human experience, 878–879
2-Methyl-2,4-pentanediol, 48–51
  acute toxicity, 49–50
  chemical and physical properties, 49
  chronic and subchronic toxicity, 50
  exposure assessment, 49
  exposure standards, 51
  pharmacokinetics, metabolism, and mechanisms, 50
  production and use, 49
  toxic effects, 49–50
    experimental studies of, 49–50
    human experience, 50
2-Methyl-2,4-pentanediol diacetate, 354
  acute toxicity, 354
  chemical and physical properties, 337t
  toxic effects, 354
2-Methyl-1-propanethiol. *See* Isobutyl mercaptan
2-Methyl-2-propanethiol. *See* tert-Butyl mercaptan
2-Methyl-2-propene-1,1-diol diacetate, 352–353
  acute toxicity, 352–353
  chemical and physical properties, 337t
  toxic effects, 352–353
    experimental studies of, 352–353
Methyl sulfone. *See* Dimethyl sulfone
Methylsulfonic acid. *See* Methanesulfonic acid
2-Methylthiirane. *See* Propylene sulfide
Methylthiomethane. *See* Dimethyl sulfide
Metrifonate. *See* Trichlorfon
Mevinphos, 880–885
  acute toxicity, 770t, 881

SUBJECT INDEX

carcinogenesis, 774t, 883
chemical and physical properties, 880–881
chronic and subchronic toxicity, 882
clinical cases, 883–884
exposure standards, 885
genetic and cellular effects, 883
pharmacokinetics, metabolism, and mechanisms, 882
production and use, 881
reproductive and developmental effects, 779t, 882–883
toxic effects, 881–884
human experience, 883–884
Mixed polyglycols, 57–64
MMA. See Methyl methacrylate
MME. See Methyl methacrylate
Monocrotophos, 885–889
acute toxicity, 770t, 886
carcinogenesis, 774t, 887
chemical and physical properties, 885
clinical cases, 888–889
exposure standards, 889
genetic and cellular effects, 887–888
pharmacokinetics, metabolism, and mechanisms, 886–887
production and use, 885–886
reproductive and developmental effects, 779t, 887
toxic effects, 886–889
human experience, 888–889
Monoglyme. See Ethylene glycol dimethyl ether
Monopropylene glycol allyl ethers, 326–327
acute toxicity, 327
chemical and physical properties, 327
Monopropylene glycol isobutyl ethers, 327–329
acute toxicity, 328–329
chemical and physical properties, 328
genetic and cellular effects, 329
odor and warning properties, 328
toxic effects, 328–329
experimental studies of, 328–329
MPEGS. See Polyethylene glycol methyl ethers
MSA. See Methanesulfonic acid
MsCl. See Methanesulfonyl chloride

Naled, 889–893
acute toxicity, 770t, 890
carcinogenesis, 775t, 891–892
chemical and physical properties, 889–890
chronic and subchronic toxicity, 890–891
exposure standards, 892
genetic and cellular effects, 892
pharmacokinetics, metabolism, and mechanisms, 891

production and use, 890
reproductive and developmental effects, 779t, 891
toxic effects, 890–892
human experience, 892
Natrosol. See Hydroxyethyl cellulose
Navadel. See Dioxathion
NBR. See Acrylonitrile-butadiene rubber
Nemacur. See Fenamiphos
Neoprene, chemical and physical properties, 429f
Nifos. See TEPP
Nitrate esters of ethylene and propylene glycols, 365–383
Nitrile barries resins, production and use, 640t
Nitrile rubber
chemical and physical properties, 429f
production and use, 577t
Nitrocellulose. See Cellulose nitrate
n-Nonanethiol. See n-Nonyl mercaptan
tert-Nonanethiol. See tert-Nonyl mercaptan
n-Nonyl mercaptan, 709–710
chemical and physical properties, 682t
exposure standards, 710
toxic effects, experimental studies of, 709
tert-Nonyl mercaptan, 710
chemical and physical properties, 682t, 710
toxic effects, experimental studies of, 710
NPGDA. See 2,2-Dimethyl-1,3-propanediol diacrylate
Nuvacron. See Monocrotophos
Nylon, 542–543, 552–555
chemical and physical properties, 548t, 552, 553t
environmental impact, 555–556
production and use, 553, 577t, 578t, 639t
toxic effects, 554–555
experimental studies of, 548t–549t, 554
human experience, 555
Nylon 6, 555–556
chemical and physical properties, 544t–545t, 555
production and use, 555–556
toxic effects, 548t–549t, 552t, 556, 559f
Nylon 11, 556–557
chemical and physical properties, 544t–545t, 556
production and use, 556
toxic effects, 549t
Nylon 12, 557
chemical and physical properties, 544t–545t, 557
production and use, 557
toxic effects, 548t–549t

SUBJECT INDEX

Nylon 66, 557–558
 chemical and physical properties, 544t–545t, 558
 toxic effects
  experimental studies of, 548t–549t, 558, 559f, 559t
  human experience, 558
Nylon 610, 558–560
 chemical and physical properties, 548t–549t, 559
 production and use, 559
 toxic effects, 559f, 560
Nylon 612, 560–561
 chemical and physical properties, 544t–545t, 560
 production and use, 560
 toxic effects, 560

1-Octadecanethiol. *See* n-Octadecyl mercaptan
n-Octadecyl mercaptan, 718–719
 chemical and physical properties, 683t, 718
 exposure assessment, 718
 exposure standards, 719
 production and use, 718
 toxic effects, experimental studies of, 719
n-Octyl mercaptan, 707–708
 acute toxicity, 707–708
 chemical and physical properties, 682t, 707
 exposure assessment, 707
 exposure standards, 708
 production and use, 707
 toxic effects, experimental studies of, 707–708
tert-Octyl mercaptan, 708–709
 chemical and physical properties, 682t, 708
 production and use, 708
 toxic effects, experimental studies of, 708–709
Olefin, 431–455
 diene elastomers, and vinyl halides, 425–476
  analysis and specifications, 429–430
  chemical and physical properties, 427t–428t, 429f
  exposure standards, 431
  introduction, 425–431
  toxic effects, 430–431
Olefin resins, 413–422
 chemical and physical properties, 414t
Onion oil. *See* Propyl allyl disulfide
Organic polymers, 638–657
Organic sulfur compounds. *See* Sulfur compounds, organic
Organophosphorus compounds, introduction, 767–784
 acute toxicity, 768–772, 770t
 analytical methods, 784
 biomonitoring/biomarkers, 782–784
 carcinogenesis, 773, 774t–775t
 exposure assessment, 775–784
 intermediate syndrome, 772
 organophosphate-induced delayed neuropathy, 773
 production and use, 768
 reproductive and developmental effects, 773, 776t–781t
 treatment, 771–772
Ortho 4355. *See* Naled
2,2′-Oxybis-ethanol. *See* Diethylene glycol
Oxybis propanol. *See* Dipropylene glycol
2,2′-Oxydiethanol. *See* Diethylene glycol

Parathion, 893–902
 acute toxicity, 770t, 894–895
 carcinogenesis, 775t, 899–900
 chemical and physical properties, 893–894
 chronic and subchronic toxicity, 895
 clinical cases, 900–902
 exposure standards, 902
 genetic and cellular effects, 900
 pharmacokinetics, metabolism, and mechanisms, 895–898
 production and use, 894
 reproductive and developmental effects, 779t, 898–899
 toxic effects, 894–902
  human experience, 900–902
PB. *See* Polybutadiene; Polybutylene
PBD. *See* Polybutadiene
PBI. *See* Polybenzimidazole
PBT. *See* Polybutylene terephthalate
PCM. *See* Perchloromethyl mercaptan
PCTFE. *See* Polychlorotrifluoroethylene
PDMS. *See* Polydimethylsiloxanes
PDS. *See* Di-n-propyl disulfide
PE. *See* Polyethylene
PEG. *See* Polyethylene glycols; Polyoxyethylene
PEI. *See* Polyetherimides
Pentamethylene glycol. *See* 1,5-Pentanediol
1,5-Pentanediol, 48
 chemical and physical properties, 48
 production and use, 48
 toxic effects, 48
1-Pentanethiol. *See* n-Pentyl mercaptan
1,5-Pentylene glycol. *See* 1,5-Pentanediol
n-Pentyl mercaptan, 702–703
 chemical and physical properties, 682t, 703
 exposure assessment, 703
 production and use, 703
 toxic effects, experimental studies of, 703
Perchloromethyl mercaptan, 728–730
 acute toxicity, 729
 chemical and physical properties, 683t, 728–729

SUBJECT INDEX

chronic and subchronic toxicity, 730
exposure assessment, 729
exposure standards, 686t, 730
production and use, 729
toxic effects, 729–730
    experimental studies of, 729–730
    human experience, 730
Perfluoroalkoxy copolymer, 472
    chemical and physical properties, 428t, 472
    production and use, 472
Perfluoro rubber, 448–449
    chemical and physical properties, 427t, 448
    production and use, 449–450
Perlon. *See* Nylon 6
PFA. *See* Perfluoroalkoxy copolymer
P-F-R-2. *See* Phenol-formaldehyde resin
PGBE. *See* Propylene glycol mono-*n*-butyl ether
PGBEE. *See* Propylene glycol butoxyethyl ether
PGDN. *See* Propylene glycol dinitrate
PGEE. *See* Propylene glycol monoethyl ether
PGME. *See* Propylene glycol monomethyl ether
PGPE. *See* Propylene glycol*n*-monopropyl ether
PGPhE. *See* Propylene glycol phenyl ether
PGTBE. *See* Propylene glycol mono-*tertiary*-butyl ether
Phenol-formaldehyde resin, 638–644
    acute toxicity, 641
    chronic and subchronic toxicity, 642–643
    clinical cases, 642–644
    exposure assessment, 641
    neurological, pulmonary, skin sensitization effects, 643–644
    production and use, 638–641, 639t
    reproductive and developmental effects, 641–642
    toxic effects, 641–644
        experimental studies of, 641–642
        human experience, 642–644
Phenolics, 638, 639*t*–640*t*
Phenoxy, production and use, 639*t*
2,2′-[1,4-Phenylenebis(oxy)]bisethanol, 227–228
    acute toxicity, 227–228
    chemical and physical properties, 227
    exposure standards, 228
    toxic effects, 227–228
        experimental studies of, 227–228
Phenyl glycol. *See* Styrene glycol
Phenyl mercaptan, 722–724
    acute toxicity, 723–724
    chemical and physical properties, 683t, 723
    chronic and subchronic toxicity, 724
    exposure assessment, 723
    exposure standards, 686t, 724
    genetic and cellular effects, 724

pharmacokinetics, metabolism, and mechanisms, 724
production and use, 723
toxic effects, 723–724
    experimental studies of, 723–724
    human experience, 724
Phenylmethyl mercaptan. *See* Benzyl mercaptan
Phenylphosphonothoic acid, *O*-ethyl *O*- o-nitrophenyl ester. *See* EPN
Phenyl sulfone. *See* Diphenyl sulfone
Phorate, 902–905
    acute toxicity, 770t, 903
    carcinogenesis, 775t, 904–905
    chemical and physical properties, 903
    chronic and subchronic toxicity, 903
    exposure standards, 905
    genetic and cellular effects, 905
    pharmacokinetics, metabolism, and mechanisms, 904
    production and use, 903
    reproductive and developmental effects, 779t, 904
    toxic effects, 903–905
        human experience, 905
Phosphorodithioic acid, *O,O*-dimethyl-*S*-((4-oxo-1,2,3-benzotirazin-3 (4H)-yl)methyl) ester. *See* Azinphos-methyl
PIR. *See* Polyisocyanurate foam
Plexiglas. *See* Polymethyl methacrylate
Pluracol V-10, TP-410, and TP-740 polyols, 63–64
    chemical and physical properties, 63, 63t
    exposure standards, 64
    production and use, 63
    toxic effects, 63, 63t
Pluronic polyols, 60–62
    acute toxicity, 61, 61t
    carcinogenesis, 62
    chemical and physical properties, 60, 60t
    chronic and subchronic toxicity, 61, 62t
    exposure standards, 62
    production and use, 60
    reproductive and developmental effects, 61–62
    toxic effects, 60–62
PMM. *See* Perchloromethyl mercaptan
Polyacrylamide, 508–511
    chemical and physical properties, 488t, 509
    environmental impact, 511
    exposure standards, 511
    production and use, 509–510
    toxic effects, 510–511
        experimental studies of, 433*t*, 510
        human experience, 510–511
Polyacrylate, chemical and physical properties, 429*f*

Polyacrylic acid, 511–512
  chemical and physical properties, 488t, 512
  production and use, 512
Polyacrylonitrile, 500–502
  carcinogenesis, 501, 502
  chemical and physical properties, 488t, 500
  production and use, 500–501, 639t
  toxic effects, 433t, 501–502
    experimental studies of, 501
    human experience, 502
Polyalkylene glycols. See Ucon fluids
Polyamide. See Nylon
Polyamide 6. See Nylon 6
Polyamide 11. See Nylon 11
Polyamide 12. See Nylon 12
Polyamide 66. See Nylon 66
Polyamide 610. See Nylon 610
Polyamide 612. See Nylon 612
Polyamides, 521–568, 542–568
  analysis and specifications, 543
  aromatic, 561–563
  chemical and physical properties, 542–543, 544t–546t, 547f
  exposure standards, 543
  fire and toxicity of smoke, 547–552
  production and use, 543
  toxic effects, 543, 548t–551t, 552t
Polybenzimidazole, 656–657
  production and use, 657
  toxic effects, 657
Polybutadiene, 438–440
  carcinogenesis, 440
  chemical and physical properties, 427t, 439
  chronic and subchronic toxicity, 440
  exposure assessment, 440
  exposure standards, 440
  production and use, 439, 577t
  toxic effects, 440
    experimental studies of, 440
Polybutylene, 420–421
  chemical and physical properties, 414t, 421
  production and use, 421, 640t
  toxic effects, 421
Polybutylene ether glycol, 587
Polybutylene glycols, 47
  chemical and physical properties, 47
  production and use, 47
  toxic effects, 47
Polybutylene terephthalate, 585–587
  chemical and physical properties, 586
  elastomeric polyester, 587
  toxic effects, 586–587
Polycarbonate of bisphenol A
  chemical and physical properties, 574t–575t
  production and use, 639t
Polycarbonate resins, 593–597
  chemical and physical properties, 593
  exposure standards, 597
  plastic processing fumes, 596–597
  production and use, 593–595
  toxic effects, 595–597
Polychloroprene, production and use, 577t, 639t
Polychlorotrifluoroethylene, 474–475
  chemical and physical properties, 428t, 474–475
  exposure assessment, 475
  exposure standards, 475
  production and use, 475
  toxic effects, 475
Polydimethylsiloxane, 658
  acute toxicity, 661
  carcinogenesis, 662–663
  chronic and subchronic toxicity, 661
  environmental impact, 664
  epidemiology studies, 663–664
  exposure assessment, 660
  exposure standards, 664
  genetic and cellular effects, 662
  pharmacokinetics, metabolism, and mechanisms, 661–662
  production and use, 659
  reproductive and developmental effects, 662
  toxic effects, 660–664
    experimental studies of, 660–662
Polydimethylsiloxane rubber, production and use, 659–660
Polyester fibers, 581–585
  carcinogenesis, 585
  production and use, 578t, 581–582
  properties and applications, 574t–575t, 582–583
  structure and processing of film, 582
  toxic effects, 583–585
Polyester resins, unsaturated, 587–590
  chemical and physical properties, 587–588
  genetic and cellular effects, 590
  production and use, 577t, 588, 639t
  toxic effects, 588–590
Polyesters, 573–607
  chemical and physical properties, 573, 574t–576t
  exposure standards, 580–581
  fire and toxicity of smoke, 581
  human responses, 579–580
  linear terephthalate, 581–585
  polyethers, and related polymers, 573–607
    chemical and physical properties, 573, 574t–576t
    exposure standards, 580–581

SUBJECT INDEX

fire and toxicity of smoke, 581
  human responses, 579–580
  production and use, 573, 577t–578t
  specifications and test methods, 573
  toxic effects, 578–579
production and use, 573, 577t–578t
specifications and test methods, 573
toxic effects, 578–579
Polyether, 597–600
  chemical and physical properties, 575t–576t
  chlorinated, production and use, 639t
Polyetheretherketone, production and use, 640t
Polyetherimides, 563–568
  chemical and physical properties, 563–564
  exposure standards, 567–568
  production and use, 564–566, 640t
  toxic effects
    experimental studies of, 566–567
    human experience, 567
Polyethersulfone, production and use, 640t
Polyethersulfone 200P, chemical and physical properties, 576t
Polyethylene, 413–417
  carcinogenesis, 417
  chemical and physical properties, 414t, 415–416
  production and use, 416, 577t, 639t
  toxic effects, 416–417
    experimental studies of, 416–417, 433t
    human experience, 417
Polyethylene chlorosulfonate, chemical and physical properties, 429f
Polyethylene glycol methyl ethers, 226–227
  acute toxicity, 226–227
  pharmacokinetics, metabolism, and mechanisms, 227
  toxic effects, 226–227
    experimental studies of, 226–227
Polyethylene glycols, 21–26
  acute toxicity, 23, 23t
  chemical and physical properties, 21, 22t
  chronic and subchronic toxicity, 24, 25t
  exposure assessment, 22
  pharmacokinetics, metabolism, and mechanisms, 24–26
  production and use, 21–22
  toxic effects, 22–26
    experimental studies of, 23–26
    human experience, 26
Polyethylene terephthalate, 581–585
  carcinogenesis, 585
  production and use, 581–582, 639t
  properties and applications, 574t–575t, 582–583
  structure and processing of film, 582

toxic effects, 583–585
Polyglycol 11 and 15 series, 57
  chemical and physical properties, 57, 57t
  exposure standards, 57
  production and use, 57
  toxic effects, 57, 58t
Poly(2-hydroxethyl methacrylate) hydrogel, 512–513
  chemical and physical properties, 513
  production and use, 513
Polyimides, 563–568
  chemical and physical properties, 563–564
  exposure standards, 567–568
  production and use, 564–566, 640t
  toxic effects
    experimental studies of, 566–567
    human experience, 567
Polyisobutylene, 421
Polyisocyanurate foam, 624
Polyisoprene, 431–434
  chemical and physical properties, 427t, 432
  clinical cases, 434
  production and use, 432, 577t, 639t
  toxic effects, 432–434
  human experience, 434
Polymer industry, 398–399, 577t–578t
Polymers
  organic, 638–657
  sulfur, 606–607
  synthetic, introduction, 397–413
    characteristics, 399
    classification, 399–401
    dose-response relationship, 402
    environmental impact, 412–413
    estrogenic effect, 403
    exposure assessment, 410–412
    fire and, 403–404
    polymer industry, 398–399
    smoke and, 404–405
    smoke toxicity, 406–407
      significance of, 408–410
      tests for, 407–408
    toxic effects, 401–402
    toxic fire hazard, 410
Polymethyl methacrylate, 505–508
  carcinogenesis, 507–508
  chemical and physical properties, 488t, 506
  clinical cases, 508
  epidemiology studies, 508
  isotactic [25188-98-1], 506
  production and use, 506–507
  syndiotactic [25188-97-0], 506
  toxic effects, 507–508

SUBJECT INDEX

Polymethyl methacrylate (*Continued*)
   experimental studies of, 507–508
   human experience, 508
Polymethylpentene, 421
   production and use, 640t
Poly(4-methylpentene-1), production and use, 640*t*
Polyoxyethylene, 600–602
   acute toxicity, 601
   chemical and physical properties, 575t–576t, 601
   repeated dose toxicity, 601–602
   reproductive and developmental effects, 602
   toxic effects, 601–602
      human experience, 602
Polyoxymethylene. *See* Acetal resins – polyoxymethylene
Polyphenylene oxide-based resin, 602–606
   chemical and physical properties, 575t–576t, 603
   production and use, 603–604, 640t
   toxic effects, 604–606
Polyphenylene sulfide, 607
   production and use, 640t
Poly(*p*-phenylene sulfide), 606–607
   chemical and physical properties, 576t, 607
   toxic effects, 607
Poly(propanediols). *See* Polypropylene glycols
Polypropylene, 417–420
   carcinogenesis, 419–420
   chemical and physical properties, 414t, 418–419
   neurological, pulmonary, skin sensitization effects, 420
   odor and warning properties, 419
   production and use, 419, 577t, 639t
   toxic effects, 419–420, 433t
      human experience, 420
Polypropylene glycol butyl ethers, 329–331
   acute toxicity, 330–331
   chemical and physical properties, 330, 330t
   chronic and subchronic toxicity, 331
   exposure assessment, 330
   production and use, 330
   toxic effects, 330–331
      experimental studies of, 330–331
Polypropylene glycols, 38–41
   acute toxicity, 39–40
   chemical and physical properties, 38, 38f
   chronic and subchronic toxicity, 40
   exposure assessment, 39
   pharmacokinetics, metabolism, and mechanisms, 41
   production and use, 39
   toxic effects, 39
      experimental studies of, 39–40

      human experience, 41
Polysaccharides, 521–568
Polystyrene, 496–500
   carcinogenesis, 499
   chemical and physical properties, 488t, 498
   environmental impact, 500
   exposure standards, 499
   production and use, 498–499, 577t, 639t
   toxic effects, 499
      experimental studies of, 499
      human experience, 499
Polysulfone of bisphenol A, 606–607
   chemical and physical properties, 576t, 607
   production and use, 640t
   toxic effects, 607
Poly-Tears. *See* Hydroxypropyl methyl cellulose
Polytetrafluoroethylene, 467–470
   carcinogenesis, 469
   chemical and physical properties, 428t , 468
   exposure assessment, 468–469
   exposure standards, 470, 471f
   pharmacokinetics, metabolism, and mechanisms, 469
   production and use, 468
   toxic effects, 469–470
      experimental studies of, 433*t,* 469
      human experience, 469–470
Polytetramethylene terephthalate. *See* Polybutylene terephthalate
Polyurethane, 619–638
   acute toxicity, 628–631, 630t
   carcinogenesis, 635
   chronic and subchronic toxicity, 631–633, 633–635
   clinical cases, 633–636
   epidemiology studies, 636–637
   exposure assessment, 627–628
   exposure standards, 637–638
   genetic and cellular effects, 633
   production and use, 577t, 619–627, 639t
   toxic effects, 628–638
      experimental studies of, 628–633
      human experience, 633
Polyurethane coatings and adhesives, 625–626
   clinical cases, 635–636
Polyurethane elastomers, 454–455, 625, 626*f*
   chemical and physical properties, 429f, 454
   exposure standards, 455
   production and use, 454–455, 577t
   toxic effects, 455
Polyurethane elastomers *[9009-54-5],* 577*t*
Polyurethane fibers, 626, 626*f*

SUBJECT INDEX

clinical cases, 636
Polyurethane foams, 623–625
  acute toxicity, 628–631, 630t
  carcinogenesis, 635
  chemical and physical properties, 623–625, 624t
  chronic and subchronic toxicity, 631–633, 633–635
  epidemiology studies, 636–637
  exposure standards, 637–638
  genetic and cellular effects, 633
  toxic effects, human experience, 633–637
Polyurethanes, miscellaneous organic polymers, and silicones, 619–664
  overview, 619
Polyvinyl acetate, 489–492
  alcohol, and derivate polymers, 487–496
    chemical and physical properties, 488t
    exposure standards, 489
    production and use, 487–489
    toxic effects, 489
  carcinogenesis, 491
  chemical and physical properties, 488t, 490
  environmental impact, 492
  exposure standards, 491–492
  production and use, 490–491, 577t, 639t
  toxic effects, 433t, 491
Polyvinyl acetate-butyl acrylate, 488t, 492
Polyvinyl acetate ethylene, 489–492
  carcinogenesis, 491
  chemical and physical properties, 488t, 490
  environmental impact, 492
  exposure standards, 491–492
  production and use, 490–491
  toxic effects, 491
Polyvinyl acetate 1-ethylhexyl acetate, 489–492
  carcinogenesis, 491
  chemical and physical properties, 490
  environmental impact, 492
  exposure standards, 491–492
  production and use, 490–491
  toxic effects, 491
Polyvinyl acetate-2-ethylhexyl acrylate, 488t, 492
Polyvinyl acetates, 487–496
Polyvinyl alcohol, 487, 492–496
  carcinogenesis, 495
  chemical and physical properties, 488t, 493–494
  chronic and subchronic toxicity, 495
  environmental impact, 496
  exposure standards, 496
  production and use, 494
  toxic effects, 433t, 494–495
    experimental studies of, 495

human experience, 495
Polyvinyl chloride, 455–463
  carcinogenesis, 460, 462
  chemical and physical properties, 427t, 456
  chronic and subchronic toxicity, 459
  clinical cases, 462
  environmental impact, 463
  epidemiology studies, 462–463
  exposure assessment, 458
  exposure standards, 463
  genetic and cellular effects, 462
  neurological, pulmonary, skin sensitization effects, 460–461, 462
  pharmacokinetics, metabolism, and mechanisms, 459–460
  production and use, 456–458, 577t, 639t
  reproductive and developmental effects, 463
  toxic effects, 433t, 458–465
    human experience, 462–463
Polyvinyl fluoride, 428t, 466–467
  chemical and physical properties, 428t, 466–467
  production and use, 467
  toxic effects, 467
Polyvinylidene chloride, 463–465
  chemical and physical properties, 427t, 464
  exposure standards, 465
  production and use, 464–465, 639t
  toxic effects, 433t, 465
Polyvinylidene fluoride, 473–474
  chemical and physical properties, 428t, 473–474
  exposure assessment, 474
  production and use, 474
  toxic effects, 474
  human experience, 474
Polyvinylpyrrolidone, 651–656
  acute toxicity, 653
  carcinogenesis, 654–655
  chemical and physical properties, 652
  chronic and subchronic toxicity, 653–654
  clinical cases, 655–656
  exposure assessment, 652
  genetic and cellular effects, 655
  pharmacokinetics, metabolism, and mechanisms, 654, 655–656
  production and use, 652
  reproductive and developmental effects, 654
  toxic effects, 653–656
    experimental studies of, 653–655
    human experience, 655–656
Polyviol. *See* Polyvinyl alcohol
PP. *See* Polypropylene
Primary acetate. *See* Cellulose triacetate

1,2-Propanediol. *See* Propylene glycol
1,3-Propanediol, 33–35
  acute toxicity, 34
  chemical and physical properties, 33, 33t
  chronic and subchronic toxicity, 34
  exposure assessment, 34
  pharmacokinetics, metabolism, and mechanisms, 34
  production and use, 34
  reproductive and developmental effects, 34
  toxic effects, 34
    experimental studies of, 34
Propanesulfonyl chloride, 748–749
  acute toxicity, 749
  chemical and physical properties, 684t, 748
  toxic effects, experimental studies of, 749
1-Propanethiol. *See* Propyl mercaptan
2-Propanethiol. *See* Isopropyl mercaptan
2-Propene-1-thiol. *See* Allyl mercaptan
2-Propenoic acid, oxydi-2,1-ethanediyl. *See* Diethylene glycol diacrylate
2-Propenoic acid, 2,2-dimethyl-1,3-propanediyl ester. *See* 2,2-Dimethyl-1,3-propanediol diacrylate
2-Propenoic acid, 1,2-ethanediyl ester. *See* Ethylene glycol diacrylate
2-Propenoic acid, 1-methyl-1,3-propanediyl ester. *See* 1,3-Butanediol diacrylate
2-Propenyl propyl disulfide. *See* Propyl allyl disulfide
1-Propoxy-2-propanol. *See* Propylene glycol*n*-monopropyl ether
Propyl allyl disulfide, 737–738
  chemical and physical properties, 683t, 737
  exposure assessment, 738
  exposure standards, 686t, 738
  production and use, 737–738
  toxic effects, 738
    human experience, 738
Propyl disulfide. *See* Di-*n*-propyl disulfide
Propylene glycol, 26–33
  acute toxicity, 27, 33
  carcinogenesis, 32
  chemical and physical properties, 2f, 26–27
  chronic and subchronic toxicity, 28–31
  clinical cases, 33
  ethers of, 271–331
    background, 271, 272t
    chemical and physical properties, 273t, 274t
    exposure assessment, 275
    pharmacokinetics, metabolism, and mechanisms, 280–281
    production and use, 271–275, 276t–277t, 278t
    toxic effects, 275–280, 279t
  exposure assessment, 27

  genetic and cellular effects, 32
  pharmacokinetics, metabolism, and mechanisms, 31, 31f, 33
  production and use, 27
  reproductive and developmental effects, 32
  toxic effects, 27–33
    experimental studies of, 27–32
    human experience, 32–33
Propylene glycol butoxyethyl ether, 305–306
  chemical and physical properties, 273t
  production and use, 306
  toxic effects, 306
Propylene glycol dinitrate, 378–383
  acute toxicity, 379–381, 380t
  chemical and physical properties, 366t
  chronic and subchronic toxicity, 381–382
  exposure assessment, 379
  exposure standards, 382–383, 383t
  pharmacokinetics, metabolism, and mechanisms, 382
  toxic effects, 379–383
    experimental studies of, 379–382
    human experience, 382
Propylene glycol monoacrylate, 346–349
  acute toxicity, 347–348
  chemical and physical properties, 347
  exposure standards, 349, 350t
  genetic and cellular effects, 349
  neurological, pulmonary, skin sensitization effects, 349
  odor and warning properties, 347
  toxic effects, 347–349
    experimental studies of, 347–349
    human experience, 349
Propylene glycol mono-*n*-butyl ether, 297–300
  acute toxicity, 297–298
  chemical and physical properties, 273t
  chronic and subchronic toxicity, 298–299
  exposure standards, 300
  genetic and cellular effects, 299
  neurological, pulmonary, skin sensitization effects, 300
  production and use, 276t
  reproductive and developmental effects, 299
  toxic effects, 297–300
    experimental studies of, 297–300
Propylene glycol mono-*tertiary*-butyl ether, 300–304
  acute toxicity, 301–302
  chemical and physical properties, 273t
  chronic and subchronic toxicity, 302–303
  exposure assessment, 300
  exposure standards, 304
  genetic and cellular effects, 303–304

SUBJECT INDEX

neurological, pulmonary, skin sensitization effects, 304
pharmacokinetics, metabolism, and mechanisms, 303
production and use, 277t, 300
reproductive and developmental effects, 303
toxic effects, 301–304
  experimental studies of, 301–304
Propylene glycol monobutyl ether acetate, production and use, 276t
Propylene glycol mono-*tert*-butyl ether acetate, production and use, 277t
Propylene glycol monoethylbutyl ether, mixed isomers, 304–305
  chemical and physical properties, 305
  toxic effects, 305
Propylene glycol monoethyl ether, 290–293
  acute toxicity, 291–292
  chemical and physical properties, 273t
  chronic and subchronic toxicity, 292
  genetic and cellular effects, 293
  production and use, 276t, 291
  reproductive and developmental effects, 292–293
  toxic effects, 291–293
    experimental studies of, 291–293
Propylene glycol monoethyl ether acetate, 363–365
  acute toxicity, 364
  chronic and subchronic toxicity, 364
  exposure assessment, 363
  genetic and cellular effects, 365
  neurological, pulmonary, skin sensitization effects, 365
  pharmacokinetics, metabolism, and mechanisms, 364–365
  production and use, 276t
  toxic effects, 364–365
    experimental studies of, 364–365
Propylene glycol monoisopropyl ether, 296
  chemical and physical properties, 273t
  toxic effects, 296
Propylene glycol monomethyl ether, 281–290
  acute toxicity, 279t, 280, 282–284
  carcinogenesis, 288
  chemical and physical properties, 273t
  chronic and subchronic toxicity, 284–285
  clinical cases, 289
  exposure assessment, 282
  exposure standards, 289–290, 290t
  genetic and cellular effects, 288
  odor and warning properties, 281
  pharmacokinetics, metabolism, and mechanisms, 280–281, 285–287, 286t, 289
  production and use, 276t, 281

reproductive and developmental effects, 287–288
toxic effects, 275–281, 282–289
  experimental studies of, 282–288
  human experience, 288–289
Propylene glycol monomethyl ether acetate, 360–362
  acute toxicity, 279t, 280, 361–362
  chemical and physical properties, 338t
  exposure assessment, 361
  exposure standards, 362
  genetic and cellular effects, 362
  neurological, pulmonary, skin sensitization effects, 362
  pharmacokinetics, metabolism, and mechanisms, 362
  production and use, 276t
  reproductive and developmental effects, 362
  toxic effects, 361–362
    experimental studies of, 361–362
Propylene glycol monophenyl ether acetate, production and use, 277t
Propylene glycol-*n*-monopropyl ether, 293–296
  acute toxicity, 294–295
  chemical and physical properties, 273t
  chronic and subchronic toxicity, 295
  production and use, 276t
  toxic effects, 294–295
    experimental studies of, 294–295
Propylene glycol monopropyl ether acetate, production and use, 276t
Propylene glycol phenyl ether, 306–307
  acute toxicity, 307
  chemical and physical properties, 273t, 306
  chronic and subchronic toxicity, 307
  exposure standards, 307
  genetic and cellular effects, 307
  production and use, 277t
  toxic effects, 307
    experimental studies of, 307
Propylene glycols, nitrate esters of, 365–383
Propylene sulfide, 736–737
  chemical and physical properties, 683t, 737
  genetic and cellular effects, 737
  production and use, 737
  toxic effects, 737
Propyl glycol. *See* Ethylene glycol mono-*n*-propyl ether
Propyl mercaptan, 692–694
  acute toxicity, 693–694
  chemical and physical properties, 682t, 693
  exposure assessment, 693
  exposure standards, 694
  production and use, 693
  toxic effects, 693–694

Propyl mercaptan (*Continued*)
    experimental studies of, 693–694
    human experience, 694
PS. *See* Polystyrene
PTFE. *See* Polytetrafluoroethylene
PUR. *See* Polyurethane foams
PVA. *See* Polyvinyl acetate; Polyvinyl alcohol
PVAL. *See* Polyvinyl alcohol
PVC. *See* Polyvinyl chloride
PVDC. *See* Polyvinylidene chloride
PVDF. *See* Polyvinylidene fluoride
PVP. *See* Polyvinylpyrrolidone
Pyroxiline. *See* Cellulose nitrate

Rampart. *See* Phorate
Rayon. *See* Cellulose
Repellent 1207. *See* Chloropropyl *n*-octyl sulfoxide
Ronnel, 905–908
    acute toxicity, 770t, 906–907
    carcinogenesis, 775t, 908
    chemical and physical properties, 906
    chronic and subchronic toxicity, 907
    exposure standards, 908
    pharmacokinetics, metabolism, and mechanisms, 907
    production and use, 906
    reproductive and developmental effects, 779t, 907–908
    toxic effects, 906–908
    human experience, 908
RSH. *See* Mercaptans
Rubber. *See* Polyisoprene

SAN. *See* Styrene-acrylonitrile
Saran. *See* Polyvinylidene chloride
SBR. *See* Styrene butadiene rubber
Silicone, 452–453, 657–658
    chemical and physical properties, 427t, 429f, 452–453, 658
    environmental impact, 453–454
    exposure standards, 453
    production and use, 453, 639t, 658
    toxic effects, 453
        experimental studies of, 453
        human experience, 453
Silicone elastomers, production and use, 659–660
Silicone fluids and resins, production and use, 659
Silicone rubber. *See* Dimethyl silicone
Silicones, 657–664
Sodium carboxymethyl cellulose, 536
    chemical and physical properties, 524t, 525t–526t, 536
    production and use, 536

toxic effects, 536
Stearyl mercaptan. *See n*-Octadecyl mercaptan
Styrene-acrylonitrile, 502–504
    chemical and physical properties, 488t, 503
    production and use, 503, 577t, 639t
Styrene-butadiene, production and use, 577t, 639t
Styrene butadiene rubber, 425, 434–438
    biomonitoring/biomarkers, 436
    chemical and physical properties, 427t, 429f, 435
    epidemiology studies, 436–438
    exposure standards, 438
    production and use, 435–436
    toxic effects, 436–438
        human experience, 436–438
Styrene glycol, 55–56
    acute toxicity, 56
    chemical and physical properties, 55
    exposure assessment, 55
    genetic and cellular effects, 56
    pharmacokinetics, metabolism, and mechanisms, 56, 56f
    production and use, 55
    toxic effects, 56
        experimental studies of, 56
Sulfides, 730–739
1,1′-Sulfinylbisethane. *See* Diethyl sulfoxide
Sulfole 90. *See tert*-Nonyl mercaptan
Sulfole 120. *See tert*-Dodecyl mercaptan
Sulfonates, 739–749
Sulfones, 739–743
Sulfonium chlorides, 739–749
1,1′-Sulfonylbisbenzene. *See* Diphenyl sulfone
Sulfonylbismethane. *See* Dimethyl sulfone
Sulfotepp, 909–911
    acute toxicity, 770t, 910
    carcinogenesis, 775t, 911
    chemical and physical properties, 909
    chronic and subchronic toxicity, 910
    exposure standards, 911
    genetic and cellular effects, 911
    pharmacokinetics, metabolism, and mechanisms, 911
    production and use, 909
    reproductive and developmental effects, 780t, 911
    toxic effects, 910–911
        human experience, 911
Sulfoxides, 739–749
Sulfur compounds, organic, 681–758
    chemical and physical properties, 681–685, 682t–684t
    exposure standards, 685, 686t
Sulfur polymers, 606–607
Sulprofos, 912–916

SUBJECT INDEX

acute toxicity, 770t, 912–913
carcinogenesis, 775t, 915–916
chemical and physical properties, 912
chronic and subchronic toxicity, 914
exposure standards, 916
genetic and cellular effects, 916
pharmacokinetics, metabolism, and mechanisms, 914–915
production and use, 912
reproductive and developmental effects, 780t, 915
toxic effects, 912–916
TDAC. *See* Triethylene glycol diacetate
TEGDN. *See* Triethylene glycol dinitrate
Temephos, 916–920
acute toxicity, 770t, 917–918
carcinogenesis, 775t, 919
chemical and physical properties, 916–917
chronic and subchronic toxicity, 918
exposure standards, 920
genetic and cellular effects, 919
pharmacokinetics, metabolism, and mechanisms, 918–919
production and use, 917
reproductive and developmental effects, 780t, 919
toxic effects, 917–920
human experience, 919–920
TEPP, 920–922
chemical and physical properties, 920
clinical cases, 921–922
exposure standards, 922
production and use, 921
toxic effects, 921–922
Terbufos, 922–926
acute toxicity, 770t, 923–924
carcinogenesis, 775t, 925
chemical and physical properties, 923
chronic and subchronic toxicity, 924
exposure standards, 926
genetic and cellular effects, 925
pharmacokinetics, metabolism, and mechanisms, 924
production and use, 923
reproductive and developmental effects, 780t, 924–925
toxic effects, 923–926
human experience, 925–926
Tetradecyl mercaptan, 717
chemical and physical properties, 683t, 717
production and use, 717
toxic effects, 717
Tetraethyl dithionopyrophosphate. *See* Sulfotepp
Tetraethylene glycol, 19–20

chemical and physical properties, 20
exposure assessment, 20
production and use, 20
toxic effects, 20
Tetraethylene glycol diethyl ether, 225–226
acute toxicity, 226
chemical and physical properties, 226
exposure standards, 226
toxic effects, 226
experimental studies of, 226
Tetraethylene glycol monophenyl ether, 225
chemical and physical properties, 225
toxic effects, 225
Tetraethylene glycol monovinylethyl ether, 224–225
acute toxicity, 224
chemical and physical properties, 224
exposure standards, 225
toxic effects, 224
experimental studies of, 224
O,O,O',O'-Tetraethyl-S,S'-methylene diphosphorodithioate. *See* Ethion
Tetraglycol. *See* Tetraethylene glycol
O,O,O',O'-Tetramethyl-O,O'thiodi-p-phenylene phosphorothioate. *See* Temephos
Tetraphosphoric acid tetraethyl ester. *See* TEPP
Texanol. *See* 2,2,4-Trimethyl-1,3-pentanediol monoisobutyrate
TGBE. *See* Triethylene glycol mono-n-butyl ether
TGEE. *See* Triethylene glycol monoethyl ether
TGME. *See* Triethylene glycol monoethyl ether; Triethylene glycol monomethyl ether
Thermoplastic elastomers, 425–426
production of, U.S., 577t
Thermoplastic polyester, production and use, 640t
Thermosetting elastomers, 425–426
production of, U.S., 577t
1-Thia-3-azaindene. *See* Benzothiazole
Thiacyclopropane. *See* Ethylene sulfide
2-Thiapropane. *See* Dimethyl sulfide
Thiirane. *See* Ethylene sulfide
methyl (See Propylene sulfide)
Thiobenzyl alcohol. *See* Benzyl mercaptan
1,1'-Thiobisbutane. *See* Dibutyl sulfide
1,1'-Thiobisethane. *See* Diethyl sulfide
Thiobismethane. *See* Dimethyl sulfide
Thiocarbonyl tetrachloride. *See* Perchloromethyl mercaptan
Thioethers. *See* Sulfides
Thiophenol. *See* Phenyl mercaptan
TMPD glycol. *See* 2,2,4-Trimethyl-1,3-pentanediol
alpha-Toluenethiol. *See* Benzyl mercaptan
TPE. *See* Thermoplastic elastomers
TPGBE. *See* Tripropylene glycol mono-n-butyl ether

SUBJECT INDEX

Trichlorfon, 926–935
  acute toxicity, 770t, 927–928
  carcinogenesis, 775t, 931–932
  chemical and physical properties, 926–927
  chronic and subchronic toxicity, 928
  clinical cases, 933–935
  exposure standards, 935
  genetic and cellular effects, 932–933
  pharmacokinetics, metabolism, and mechanisms, 929–930
  production and use, 927
  reproductive and developmental effects, 780t, 930
  toxic effects, 927–935
    human experience, 933–935
Trichlorometaphos. See Ronnel
Trichloromethyl sulfur chloride. See Perchloromethyl mercaptan
Triethylene glycol, 17–19
  acute toxicity, 18
  chemical and physical properties, 2f, 17
  chronic and subchronic toxicity, 18–19
  exposure assessment, 18
  pharmacokinetics, metabolism, and mechanisms, 19
  production and use, 17–18
  reproductive and developmental effects, 19
  toxic effects, 18–19
    experimental studies of, 18–19
    human experience, 19
Triethylene glycol diacetate, 344–345
  acute toxicity, 345
  chemical and physical properties, 337t
  reproductive and developmental effects, 345
  toxic effects, 345
    experimental studies of, 345
Triethylene glycol dinitrate, 376–378
  acute toxicity, 377–378, 377t
  chemical and physical properties, 366t
  exposure assessment, 377
  exposure standards, 378
  neurological, pulmonary, skin sensitization effects, 378
  toxic effects, 377–378
    experimental studies of, 377–378
Triethylene glycol divalerate, 345–346
  acute toxicity, 346
  chemical and physical properties, 346
  toxic effects, 346
    experimental studies of, 346
Triethylene glycol mono-n-butyl ether, 222–224
  acute toxicity, 223
  chemical and physical properties, 179t

  pharmacokinetics, metabolism, and mechanisms, 78t
  production and use, 76t, 223
  reproductive and developmental effects, 223–224
  toxic effects, 223–224
    experimental studies of, 223–224
    human experience, 224
Triethylene glycol monoethyl ether, 220–222
  acute toxicity, 221
  chemical and physical properties, 179t
  chronic and subchronic toxicity, 221
  exposure standards, 222
  pharmacokinetics, metabolism, and mechanisms, 78t, 221, 222
  production and use, 76t, 220
  reproductive and developmental effects, 195t, 222
  toxic effects, 221–222
    experimental studies of, 221–222
    human experience, 222
Triethylene glycol monomethyl ether, 216–220
  acute toxicity, 217
  chemical and physical properties, 179t
  chronic and subchronic toxicity, 217
  exposure assessment, 74
  genetic and cellular effects, 219
  neurological, pulmonary, skin sensitization effects, 219–220
  pharmacokinetics, metabolism, and mechanisms, 78t, 217–218, 220
  production and use, 76t, 216
  reproductive and developmental effects, 193t, 195t, 218–219
  toxic effects, 216–220
    experimental studies of, 217–220
    human experience, 220
Triethylene glycol phenyl ether, exposure assessment, 74
Triethylene glycol propyl ether, production and use, 76t
Triglycol diacetate. See Triethylene glycol diacetate
Triisobutyl mercaptan, 702
  acute toxicity, 702
  chemical and physical properties, 702
  toxic effects, 702
    experimental studies of, 702
Trimethylene glycol. See 1,3-Propanediol
2,2,4-Trimethyl-1,3-pentanediol, 51–53
  acute toxicity, 51–52
  chemical and physical properties, 51
  chronic and subchronic toxicity, 52
  exposure assessment, 51
  pharmacokinetics, metabolism, and mechanisms, 52

production and use, 51
reproductive and developmental effects, 52
toxic effects, 51–52
  experimental studies of, 51–52
  human experience, 52
2,2,4-Trimethyl-1,3-pentanediol diisobutyrate, 358–360
  acute toxicity, 358–359
  chemical and physical properties, 337t
  chronic and subchronic toxicity, 359–360
  exposure assessment, 358
  exposure standards, 360
  neurological, pulmonary, skin sensitization effects, 359, 360
  pharmacokinetics, metabolism, and mechanisms, 359
  production and use, 358
  toxic effects, 358–360
    experimental studies of, 358–359
    human experience, 359–360
2,2,4-Trimethyl-1,3-pentanediol monoisobutyrate, 354–358
  acute toxicity, 356, 357
  chemical and physical properties, 337t
  clinical cases, 357
  exposure assessment, 355–356
  exposure standards, 358
  genetic and cellular effects, 357
  neurological, pulmonary, skin sensitization effects, 357
  production and use, 355
  reproductive and developmental effects, 356–357
  toxic effects, 356–357
    experimental studies of, 356–357
    human experience, 357
Tripropylene glycol, 36–38
  acute toxicity, 37
  chemical and physical properties, 37
  production and use, 37
  toxic effects, 37
    experimental studies of, 37
Tripropylene glycol allyl ethers, 326–327
  acute toxicity, 327
  chemical and physical properties, 327
Tripropylene glycol allyl ethers [1331-17-5], 326–327
Tripropylene glycol ethyl
  propyl ether, production and use, 278t
  propyl ether acetate, production and use, 278t
Tripropylene glycol isobutyl ethers, 327–329
  acute toxicity, 328–329
  chemical and physical properties, 328
  genetic and cellular effects, 329

odor and warning properties, 328
toxic effects, 328–329
  experimental studies of, 328–329
Tripropylene glycol methyl ether acetate, production and use, 278t
Tripropylene glycol mono-n-butyl ether, 325–326
  acute toxicity, 325–326
  chemical and physical properties, 274t
  exposure standards, 326
  genetic and cellular effects, 326
  production and use, 278t
  toxic effects, 325–326
    experimental studies of, 325–326
Tripropylene glycol monoethyl ether, 324
  chemical and physical properties, 274t, 324
  exposure standards, 324
  toxic effects, 324
Tripropylene glycol monomethyl ether, 321–324
  acute toxicity, 322–323
  chemical and physical properties, 274t
  chronic and subchronic toxicity, 323
  exposure standards, 324
  genetic and cellular effects, 323–324
  pharmacokinetics, metabolism, and mechanisms, 323
  production and use, 278t
  reproductive and developmental effects, 323
  toxic effects, 322–324
    experimental studies of, 322–324
Tripropylene glycol monomethyl ether acetate, 360–362
  acute toxicity, 361–362
  chemical and physical properties, 338t
  exposure assessment, 361
  exposure standards, 362
  genetic and cellular effects, 362
  neurological, pulmonary, skin sensitization effects, 362
  pharmacokinetics, metabolism, and mechanisms, 362
  production and use, 276t
  reproductive and developmental effects, 362
  toxic effects, 361–362
    experimental studies of, 361–362
TSE. See Thermosetting elastomers
TXIB. See 2,2,4-Trimethyl-1,3-pentanediol diisobutyrate

Ucon fluids, 58–59
  acute toxicity, 58, 59t
  carcinogenesis, 59
  chemical and physical properties, 58
  chronic and subchronic toxicity, 58–59

Ucon fluids (*Continued*)
  pharmacokinetics, metabolism, and mechanisms, 59
  toxic effects, 58–59
Urea-formaldehyde, 644–648
  acute toxicity, 645–646
  chronic and subchronic toxicity, 646–647
  exposure standards, 648
  neurological, pulmonary, skin sensitization effects, 647
  production and use, 639t, 644–645
  toxic effects, 645–647
    experimental studies of, 645–646
    human experience, 646–647
Urethanes. *See* Polyurethane

Valeric acid, triethylene glycol diester. *See* Triethylene glycol divalerate
Viclan. *See* Polyvinylidene chloride
Vinol. *See* Polyvinyl alcohol
Vinyl cyclohexene-derived dimercaptan, 727–728
  acute toxicity, 727
  chemical and physical properties, 683t, 727
  genetic and cellular effects, 727
  production and use, 727
  toxic effects, experimental studies of, 727
Vinyl halides, 455–476
Vinylidene chloride-acrylonitrile, 428*t*, 466
Vinylidene chloride-methylacrylate, 466
  chemical and physical properties, 427t, 466
Vinylidene chloride-vinyl chloride, 465
  chemical and physical properties, 427t , 465
  production and use, 465
Vinylidene fluoride-hexafluoropropylene, 444–446
  acute toxicity, 445–446
  chemical and physical properties, 427t, 444
  exposure assessment, 445
  exposure standards, 446
  neurological, pulmonary, skin sensitization effects, 445–446
  production and use, 444–445
  toxic effects, 445–446
    experimental studies of, 445–446
Vinyl sulfone. *See* Divinyl sulfone

Chemical Index

Abate. *See* Temephos
ABS. *See* Acrylonitrile-butadiene-styrene
Acetal resins – polyoxymethylene *[9002-81-7]*, 575t–576t , 597–600, 597t, 598t, 639t
Acetate cotton. *See* Cellulose acetate
Acetose. *See* Cellulose acetate
Acetylated cellulose. *See* Cellulose triacetate
Acetyl cellulose. *See* Cellulose acetate
ACM. *See* Acrylic elastomer
Acrilafil. *See* Styrene-acrylonitrile
Acrylic acid, hydroxypropyl ester. *See* Propylene glycol monoacrylate
Acrylic acid polymer. *See* Polyacrylic acid
Acrylic elastomer *[9003-32-1]*, 427t, 449–450
Acrylonitrile-butadiene rubber *[9003-18-3]*, 427t, 442–443
Acrylonitrile-butadiene-styrene *[9003-56-9]*, 488t, 504–505, 577t, 639t
Acrylonitrile-styrene co-polymer. *See* Styrene-acrylonitrile
Adulsin. *See* Methyl cellulose
Allyl mercaptan *[870-23-5]*, 683t, 719–720
Allyl *n*-octyl sulfoxide *[3868-44-8]*, 684t, 744
Almolose. *See* Carboxymethyl cellulose
Alvyl. *See* Polyvinyl alcohol
Amyl mercaptan. *See* *n*-Pentyl mercaptan
tert-Amyl mercaptan *[1679-09-0]*, 682t, 704–705
Amyl thiolalcohol. *See* *n*-Pentyl mercaptan
Aniline/phenol/formaldehyde *[24937-74-4]*, 638
Axophos. *See* Methyl parathion
Azinphos-methyl *[86-50-0]*, 770t , 774t , 776t, 784–790

Bagolax. *See* Methyl cellulose
BAY 18436. *See* Demeton-S-methyl
BAY 25141. *See* Fensulfothion
Baycid. *See* Fenthion
BAY NTN 9306. *See* Sulprofos
BAY SRA 3886. *See* Fenamiphos
Baytex. *See* Fenthion
Bear skunk. *See* *n*-Butyl mercaptan
Benzenemethanethiol. *See* Benzyl mercaptan
Benzenethiol. *See* Phenyl mercaptan
Benzosulfonazole. *See* Benzothiazole
Benzothiazole *[95-16-9]*, 684t, 749–750
2(3*H*)-Benzothiazolethione. *See* 2-Mercaptobenzothiazole
Benzyldisulfide. *See* Dibenzyl disulfide
Benzyl mercaptan *[100-53-8]*, 683t, 721–722
Bexan. *See* Styrene-acrylonitrile
BGAP. *See* Butylene glycol adipic acid polyester
Bidrin. *See* Dicrotophos
Bisphenol A/epichlorohydrin/formaldehyde *[28906-96-9]*, 638
Blitex. *See* Ronnel
Bolstar. *See* Sulprofos
BPG 400. *See* Polypropylene glycol butyl ethers
BPG 800. *See* Polypropylene glycol butyl ethers
BR. *See* Polybutadiene
Bromobutyl rubber *[68441-14-5]*, 447
Bromochlorphos. *See* Naled
Butane, 1,1'-sulfonylbis-. *See* *n*-Dibutyl sulfone
1,2-Butanediol *[584-03-2]*, 42–43
1,3-Butanediol *[107-88-0]*, 43–45
1,4-Butanediol *[110-63-4]*, 46

999

CHEMICAL INDEX

2,3-Butanediol *[513-85-9]*, 47
Butanediol *[25265-75-2]*, 41–47, 42*t*
1,3-Butanediol diacrylate *[19485-03-1]*, 350–351
1-Butanethiol. *See n*-Butyl mercaptan
2-Butanethiol. *See sec*-Butyl mercaptan
N-tert-Butyl-2-benzothiazolesulfenamide *[95-31-8]*, 684*t*, 754–755
1,2-Butylene glycol. *See* 1,2-Butanediol
1,3-Butylene glycol. *See* 1,3-Butanediol
1,4-Butylene glycol. *See* 1,4-Butanediol
Butylene glycol adipic acid polyester *[9080-04-0]*, 351–352
Butylene glycol monoethyl ether *[111-73-9]*, 274*t*, 333–334
Butylene glycol monomethyl ether *[111-32-0]*, 274*t*, 331–333
Butyl glycol ether. *See* Ethylene glycol mono-*n*-butyl ether
n-Butyl mercaptan *[109-79-5]*, 682*t*, 686*t*, 696–698
sec-Butyl mercaptan *[513-53-1]*, 682*t*, 699–700
tert-Butyl mercaptan *[75-66-1]*, 682*t*, 700–702
p-tert-Butylphenol-formaldehyde *[25085-50-1]*, 638
Butyl rubber *[9010-85-9]*, 427*t*, 447–448, 639*t*
n-Butyl sulfide. *See* Dibutyl sulfide
n-Butyl sulfone. *See n*-Dibutyl sulfone

Carbowax. *See* Polyethylene glycol methyl ethers
Carboxymethoxy-1-methylvinyl dimethyl phosphate. *See* Mevinphos
Carboxymethyl cellulose *[9000-11-7]*, 541–542
Carboxymethyl cellulose, sodium salt. *See* Sodium carboxymethyl cellulose
Celacol m. *See* Methyl cellulose
Cellophane. *See* Cellulose
Cellosize. *See* Hydroxyethyl cellulose
Celluloid *[8050-88-2]*, 534
Cellulose *[9004-34-6]*, 521–530, 522*t*–524*t*, 578*t*, 639*t*
Cellulose acetate *[9004-35-7]*, 522*t*–526*t*, 529*t*, 530–531, 578*t*, 639*t*
Cellulose acetate butyrate *[9004-36-8]*, 523*t*, 533, 639*t*
Cellulose carboxymethyl ether. *See* Carboxymethyl cellulose
Cellulose CM. *See* Carboxymethyl cellulose
Cellulose ether. *See* Ethyl cellulose
Cellulose methylate. *See* Methyl cellulose
Cellulose nitrate *[9004-70-0]*, 522*t*, 525*t*–526*t*, 529*t*, 533–535, 639*t*
Cellulose sodium glycolate. *See* Sodium carboxymethyl cellulose
Cellulose triacetate *[9012-09-3]*, 523*t*, 525*t*–526*t*, 529*t*, 531–532

Cetyl mercaptan. *See n*-Hexadecyl mercaptan
Chlorinated polyethylene *[9002-86-2]*, 427*t*, 450–451
Chlorobutyl rubber *[68081-82-3]*, 447
Chlorofos. *See* Trichlorfon
Chloromethyloxirane rubber. *See* Epichlorohydrin rubber
Chloropropyl *n*-octyl sulfoxide *[3569-57-1]*, 684*t*, 743–744
Chlorosulfonated polyethylene *[68037-39-8]*, 427*t*, 433*t*, 446–447, 639*t*
Chlorpyrifos *[2921-88-2]*, 770*t*, 774*t*, 776*t*, 790–798
CMC. *See* Carboxymethyl cellulose
CMC sodium salt. *See* Sodium carboxymethyl cellulose
Collodion. *See* Cellulose nitrate
Coumaphos *[56-72-4]*, 770*t*, 774*t*, 776*t*, 798–802
CPE. *See* Chlorinated polyethylene
CR-39 *[142-22-3]*, 592–593
CSM. *See* Chlorosulfonated polyethylene
Cyclohexanethiol. *See* Cyclohexyl mercaptan
N-Cyclohexyl-2-benzothiazolesulfenamide *[95-33-0]*, 684*t*, 755–757
Cyclohexyl mercaptan *[1569-69-3]*, 683*t*, 720–721

DDVP. *See* Dichlorvos
1-Decanethiol. *See n*-Decyl mercaptan
n-Decyl mercaptan *[143-10-2]*, 682*t*, 711
DEG. *See* Diethylene glycol
DEGDA. *See* Diethylene glycol diacrylate
DEGDN. *See* Diethylene glycol dinitrate
Demethicone. *See* Dimethyl silicone
Demeton *[8065-48-3]*, 770*t*, 774*t*, 776*t*, 802–806
Demeton-*S*-methyl *[919-86-8]*, 770*t*, 774*t*, 776*t*, 806–811
Demetox. *See* Demeton-*S*-methyl
Demox. *See* Demeton
DEP. *See* Trichlorfon
DETF. *See* Trichlorfon
DGBE. *See* Diethylene glycol mono-*n*-butyl ether
DGdiME. *See* Diethylene glycol dimethyl ether
DGEE. *See* Diethylene glycol monoethyl ether
DGEEA. *See* Diethylene glycol monoethyl ether acetate
DGHE. *See* Diethylene glycol mono-*n*-hexyl ether
DGIBE. *See* Diethylene glycol mono-isobutyl ether
DGME. *See* Diethylene glycol monomethyl ether
DGMEA. *See* Diethylene glycol monomethyl ether acetate
DGPE. *See* Diethylene glycol mono-*n*-propyl ether
Diallyl disulfide *[2179-57-9]*, 720, 733
Diallyl sulfide *[592-88-1]*, 720

Diazinon *[333-41-5]*, 770*t*, 774*t*, 776*t*, 811–817
Diazitol. *See* Diazinon
Dibenzyl disulfide *[150-60-7]*, 683*t*, 738–739
Dibutyl sulfide *[544-40-1]*, 683*t*, 735
n-Dibutyl sulfone *[589-04-9]*, 683*t*, 740–741
Dichlorvos *[62-73-7]*, 770*t*, 774*t*, 777*t*, 817–824
Dicrotophos *[141-66-2]*, 770*t*, 774*t*, 777*t*, 824–827
N,N-Dicyclohexylbenzothiazolesulfenamide *[4979-32-2]*, 684*t*, 757–758
1,2-Diethoxyethane. *See* Ethylene glycol diethyl ether
Diethyl cellosolve. *See* Ethylene glycol diethyl ether
O,O-Diethyl *O*-(3-chloro-4-methyl-2-oxo-2*H*-1-benzopyran-7-yl) phosphorothioate. *See* Coumaphos
O,O-Diethyl *O*-2-diethyl *O*-2-isopropyl-4-methyl-6-pyrimidinyl thiophosphate. *See* Diazinon
Diethyl [(dimethoxyphosphinothioyl)thio]butanedioate. *See* Malathion
Diethylene glycol *[110-46-6]*, 2*f*, 12–17
Diethylene glycol diacrylate *[4074-88-8]*, 343–344
Diethylene glycol dimethyl ether *[111-96-6]*, 210–213
Diethylene glycol dinitrate *[693-21-0]*, 366*t*, 373–376
Diethylene glycol divinyl ether *[764-99-8]*, 214
Diethylene glycol ethyl vinyl ether *[10143-53-0]*, 214–215
Diethylene glycol mono-*n*-butyl ether *[112-34-5]*, 76*t*, 78*t*, 201–207
Diethylene glycol mono-*n*-butyl ether acetate *[124-17-4]*, 76*t*, 229*t*, 241–243, 242*t*
Diethylene glycol monoethyl ether *[111-90-0]*, 74, 76*t*, 78*t*, 179*t*, 191*t*, 193–199, 193*t*, 195*t*
Diethylene glycol monoethyl ether acetate *[112-15-2]*, 76*t*, 229*t*, 239–240
Diethylene glycol mono-*n*-hexyl ether *[112-59-4]*, 76*t*, 179*t*, 208–210
Diethylene glycol mono-isobutyl ether *[18912-80-6]*, 179*t*, 207–208
Diethylene glycol monomethyl ether *[111-77-3]*, 75*t*, 179*t*, 187–192, 193*t*
Diethylene glycol monomethyl ether acetate *[629-38-9]*, 229*t*, 238–239
Diethylene glycol monomethylpentyl ether *[10143-56-3]*, 215–216
Diethylene glycol mono-*n*-propyl ether *[6881-94-3]*, 76*t*, 179*t*, 199–201
O,O-Diethyl-*S*-ethylmercaptoethyl dithiophosphate. *See* Disulfoton

O,O-Diethyl *(S*-ethylmercaptomethyl) dithiophosphate. *See* Phorate
O,O-Diethyl *O* (and *S*)-2-(ethylthio)ethyl phosphorothioate mixture. *See* Demeton
O,O-Diethyl *O-p*-nitrophenyl phosphorothioate. *See* Parathion
Diethyl sulfide *[352-93-2]*, 683*t*, 733–734
Diethyl sulfoxide *[70-29-1]*, 684*t*, 743
O,O-Diethyl *O*-(3,5,6-trichloro-2-pyridinyl) phosphorothioate. *See* Chlorpyrifos
Diglyme. *See* Diethylene glycol dimethyl ether
2,3-Dihydrothiirene. *See* Ethylene sulfide
1,2-Dihydroxybutane. *See* 1,2-Butanediol
1,3-Dihydroxybutane. *See* 1,3-Butanediol
1,4-Dihydroxybutane. *See* 1,4-Butanediol
1,2-Dihydroxyethane. *See* Ethylene glycol
1,3-Dihydroxypropane. *See* 1,3-Propanediol
N,N-Diisopropyl-2-benzothiazolesulfenamide *[95-29-4]*, 684*t*, 753–754
Dimethoxyethane. *See* Ethylene glycol dimethyl ether
O,O-Dimethyl-1-carbomethoxy-1-propen-2-yl phosphate. *See* Mevinphos
Dimethyl-1,2-dibromo-2,2-dichloroethyl phosphate. *See* Naled
O,O-Dimethyl *O*-2,2-dichlorvinyl dimethyl phosphate. *See* Dichlorvos
O,O-Dimethyl-*O*-(3-dimethylamino-1-methyl-3-oxo-1-propenyl) phosphate. *See* Dicrotophos
Dimethyl disulfide *[624-92-0]*, 683*t*, 731–733
N-(1,1′-Dimethylethyl)-2-benzothiazolesulfenamide. *See* *N-tert*-Butyl-2-benzothiazolesulfenamide
(E)-Dimethyl 2-methylcarbamoyl-1-methylvinyl phosphate. *See* Monocrotophos
O,O-Dimethyl *O-(p*-nitrophenyl) phosphorothioate. *See* Methyl parathion
Dimethylol urea. *See* Urea-formaldehyde
Dimethylpolysiloxane hydrolyzate. *See* Silicone
2,2-Dimethyl-1,3-propanediol diacrylate *[2223-82-7]*, 337*t*, 352–353
1,1-Dimethyl-1-propanethiol. *See* *tert*-Amyl mercaptan
Dimethyl silicone *[9016-00-6]*, 658–664
Dimethyl sulfide *[75-18-3]*, 683*t*, 730–731
Dimethyl sulfone *[67-71-0]*, 683*t*, 739–740
O,O-Dimethyl 1 (2,2,2-trichloro-1-hydroxyethyl)-phosphonate. *See* Trichlorfon
O,O-Dimethyl *O*-(2,4,5-trichlorophenyl)phosphorothioate. *See* Ronnel
Diolice. *See* Coumaphos

1002 CHEMICAL INDEX

Dioxane *[123-91-1]*, 108t–109t, 177–187, 179t
2,3-*p*-Dioxanedithion *S,S*-bis-(*O,O*-diethyl phosphorodithioate). *See* Dioxathion
Dioxathion *[78-34-2]*, 770t, 774t, 777t, 826–830
Dipentenedimercaptan. *See* D-Limonene dimercaptan
Diphenyl sulfone *[127-63-9]*, 683t, 742–743
Diphonate. *See* Fonofos
Diphosphoric acid tetraethyl ester. *See* TEPP
Dipofene. *See* Diazinon
Di-*n*-propyl disulfide *[629-19-6]*, 683t, 734–735
Dipropylene glycol *[25265-71-8]*, 35–36
Dipropylene glycol allyl ethers *[1331-17-5]*, 326–327
Dipropylene glycol mono-*n*-butyl ether *[29911-28-2]*, 274t, 278t, 314–317
Dipropylene glycol mono-*tertiary*-butyl ether *[132739-31-2]*, 278t, 317–321
Dipropylene glycol monoethyl ether *[15764-24-6; 300025-38-8]*, 274t, 278t, 311–314
Dipropylene glycol monomethyl ether *[34590-94-8]*, 274t, 278t, 307–311, 312t
Dipropylene glycol monomethyl ether acetate *[88917-22-0]*, 276t, 338t, 360–362
Disulfoton *[298-04-4]*, 770t, 774t, 777t, 830–836
2,3-Dithiabutane. *See* Dimethyl disulfide
Dithioglycol. *See* 1,2-Ethanedithiol
Dithione. *See* Sulfotepp
Dithiosystox. *See* Disulfoton
Divinyl sulfone *[77-77-0]*, 683t, 741–742
DMDS. *See* Dimethyl disulfide
DMS. *See* Dimethyl sulfide
DNTP. *See* Parathion
1-Dodecanethiol. *See* *n*-Dodecyl mercaptan
t-Dodecanethiol. *See* *tert*-Dodecyl mercaptan
n-Dodecyl mercaptan *[112-55-0]*, 683t, 711–714
tert-Dodecyl mercaptan *[25103-58-6]*, 683t, 714–717
DPGBE. *See* Dipropylene glycol mono-*n*-butyl ether
DPGEE. *See* Dipropylene glycol monoethyl ether
DPGME. *See* Dipropylene glycol monomethyl ether
DPTB. *See* Dipropylene glycol mono*tertiary*-butyl ether
Duraphos. *See* Mevinphos
Dursban. *See* Chlorpyrifos
DYME. *See* Diethylene glycol dimethyl ether

ECTFE. *See* Ethylene chlorotrifluoroethylene copolymer
EGBE. *See* Ethylene glycol mono-*n*-butyl ether
EGBEA. *See* Ethylene glycol mono-*n*-butyl ether acetate
EGdiEE. *See* Ethylene glycol diethyl ether
EGdiME. *See* Ethylene glycol dimethyl ether

EGDN. *See* Ethylene glycol dinitrate
EGEEA. *See* Ethylene glycol monoethyl ether acetate
EGEHE. *See* Ethylene glycol mono-2-ethylhexyl ether
EGHE. *See* Ethylene glycol mono-*n*-hexyl ether
EGIBE. *See* Ethylene glycol monoisobutyl ether
EGiPE. *See* Ethylene glycol monoisopropyl ether
EGME. *See* Ethylene glycol monomethyl ether
EGMEA. *See* Ethylene glycol monomethyl ether acetate
EGPE. *See* Ethylene glycol mono-*n*-propyl ether
EGPhE. *See* Ethylene glycol monophenyl ether
EGTBE. *See* Ethylene glycol mono-*tert*-butyl ether
EGVE. *See* Ethylene glycol monovinyl ether
EHD. *See* 2-Ethyl-1,3-hexanediol
Elvanol. *See* Polyvinyl alcohol
EPDM. *See* Ethylene-propylene
Epichlorohydrin rubber *[106-89-8]*, 427t, 429f, 451–452
1,2-Epithiopropane. *See* Propylene sulfide
EPM. *See* Ethylene-propylene
EPN *[2104-64-5]*, 770t, 774t, 777t, 836–842
ETFE. *See* Ethylene tetrafluoroethylene copolymer
1,2-Ethanediol. *See* Ethylene glycol
1,2-Ethanediol monoacetate. *See* Ethylene glycol monoacetate
1,2-Ethanedithiol *[540-63-6]*, 683t, 724–726
Ethanesulfonyl chloride *[594-44-5]*, 684t, 747–748
Ethanethiol. *See* Ethyl mercaptan
Ethanethiol, 2-(ethylthio)-*S*-ester with *O,O*-dimethyl phosphorothioate. *See* Demeton-*S*-methyl
Ethanox. *See* Ethion
Ethion *[563-12-2]*, 770t, 774t, 777t, 842–845
Ethohexadiol. *See* 2-Ethyl-1,3-hexanediol
2-Ethoxyethanol. *See* Ethylene glycol monoethyl ether
1-(2-Ethoxyethoxy)-2-vinyl-oxyethane. *See* Diethylene glycol ethyl vinyl ether
Ethyl acrylate polymer *[9003-32-1]*, 488t, 515
2-Ethylbutoxypropanol, mixed isomers. *See* Propylene glycol monoethylbutyl ether, mixed isomers
Ethyl cellulose *[9004-57-3]*, 525t–526t, 529t, 535–536, 639t
Ethylcyclohexyldithiol *[28679-10-9]*, 683t, 726
Ethylene chlorotrifluoroethylene copolymer *[25101-45-5]*, 428t, 475–476
Ethylene dimercaptan. *See* 1,2-Ethanedithiol
Ethylene episulfide. *See* Ethylene sulfide
Ethylene glycol *[107-21-1]*, 1–12, 2f
Ethylene glycol diacetate *[111-55-7]*, 337t, 340–341
Ethylene glycol diacrylate *[2274-11-5]*, 341–342

Ethylene glycol diethyl ether *[629-14-1]*, 86*t*, 175–177
Ethylene glycol dimethyl ether *[110-71-4]*, 86*t*, 172–175
Ethylene glycol dinitrate *[628-96-6]*, 365–373, 366*t*
Ethylene glycol monoacetate *[542-59-6]*, 337*t*, 338–340
Ethylene glycol mono-*n*-butyl ether *[111-76-2]*, 108*t*–109*t*, 136–157
Ethylene glycol mono-*tert*-butyl ether *[7580-85-0]*, 159–160
Ethylene glycol mono-*n*-butyl ether acetate *[112-07-2]*, 236–238
Ethylene glycol monoethyl ether *[110-80-5]*, 75*t*, 78*t*, 79*t*, 92*f*, 96*t*, 110–124
Ethylene glycol monoethyl ether acetate *[111-15-9]*, 75*t*, 108*t*–109*t*, 229*t*, 232–236
Ethylene glycol mono-2-ethylhexyl ether *[1559-35-9]*, 86*t*, 165–167
Ethylene glycol mono-2,4-hexadiene ether *[27310-21-0]*, 164–165
Ethylene glycol mono-*n*-hexyl ether *[112-25-4]*, 86*t*, 96*t*, 161–164
Ethylene glycol monoisobutyl ether *[4439-24-1]*, 86*t*, 157–159
Ethylene glycol monoisopropyl ether *[109-59-1]*, 132–136, 135*f*
Ethylene glycol monomethyl ether *[109-86-4]*, 84–107, 108*t*–109*t*
Ethylene glycol monomethyl ether acetate *[110-49-6; 32718-56-2]*, 228–232, 229*t*
Ethylene glycol mono-2-methylpentyl ether *[29290-45-7]*, 86*t*, 160–161
Ethylene glycol monophenyl ether *[122-99-6]*, 86*t*, 168–171
Ethylene glycol mono-*n*-propyl ether *[2807-30-9]*, 86*t*, 124–132
Ethylene glycol mono-2,6,8-trimethyl-4-nonyl ether *[10137-98-1]*, 86*t*, 167–168
Ethylene glycol monovinyl ether *[764-48-7]*, 107–108
Ethylene-propylene *[9010-79-1]* (copolymer), 427*t*, 429*f*, 440–442, 577*t*
Ethylene-propylene *[25038-37-3]* (terpolymer), 440
Ethylene sulfide *[420-12-2]*, 683*t*, 735–736
Ethylene tetrafluoroethylene copolymer *[54302-05-5]*, 428*t*, 473
Ethyl glyme. *See* Ethylene glycol diethyl ether
2-Ethyl-1,3-hexanediol *[94-96-2]*, 53–55
Ethyl hexylene glycol. *See* 2-Ethyl-1,3-hexanediol
Ethyl mercaptan *[75-08-1]*, 682*t*, 686*t*, 690–692
O-Ethyl *O*-(4-(methylthio)phenyl)-*S*-propyl phosphorodithioate. *See* Sulprofos

O-Ethyl *O*-*p*-nitrophenyl phenylphosphonothioate. *See* EPN
Ethyl parathion. *See* Parathion
O-Ethyl-*S*-phenyl ethylphosphonodithioate. *See* Fonofos
Ethyl pyrophosphate. *See* TEPP
Ethyl sulfide. *See* Diethyl sulfide
Ethyl thioether. *See* Diethyl sulfide

Fenamiphos *[22224-92-6]*, 770*t* , 774*t* , 777*t*, 845–851
Fenchlorophos. *See* Ronnel
Fensulfothion *[115-90-2]*, 774*t* , 777*t*, 851–853
Fenthion *[55-38-9]*, 770*t*, 774*t*, 778*t*, 853–861
FEP. *See* Fluorinated ethylene propylene copolymer
FKKM. *See* Perfluoro rubber
FKM. *See* Vinylidene fluoride-hexafluoropropylene
Fluorinated ethylene propylene copolymer *[25067-11-2]*, 428*t*, 470–472
FMVE. *See* Perfluoroalkoxy copolymer
Fonofos *[944-22-9]*, 770*t*, 774*t*, 778*t*, 861–862
Formaldehyde, phenol polymer. *See* Phenol-formaldehyde resin
Formaldehyde, urea polymer urea. *See* Urea-formaldehyde
Fostacryl. *See* Styrene-acrylonitrile
FPM. *See* Vinylidene fluoride-hexafluoropropylene
Furan polymers/furan resins *[25212-86-6]*, 650–651
Furfuryl alcohol resins. *See* Furan polymers/furan resins

Glycol dinitrate. *See* Ethylene glycol dinitrate
Glycol monoacetate. *See* Ethylene glycol monoacetate
Glyme. *See* Ethylene glycol dimethyl ether
Glyme-1. *See* Ethylene glycol diethyl ether
Guncotton. *See* Cellulose nitrate
Guthion. *See* Azinphos-methyl

Heptanethiol. *See* Heptyl mercaptan
Heptyl mercaptan *[1639-09-4]*, 682*t*, 706–707
Hexadecanethiol. *See* *n*-Hexadecyl mercaptan
n-Hexadecyl mercaptan *[2917-26-2]*, 683*t*, 717–718
Hexanedioic acid, polymer with butanediol. *See* Butylene glycol adipic acid polyester
1-Hexanethiol. *See* Hexyl mercaptan
Hexylene glycol. *See* 2-Methyl-2,4-pentanediol
Hexyl mercaptan *[111-31-9]*, 682*t*, 705–706
HIPS. *See* Polystyrene
Hydroquinone, bis(2-hydroxyethyl) ether. *See* 2,2′-[1,4-Phenylenebis(oxy)]bisethanol
Hydroxyethyl cellulose *[9004-62-0]*, 525*t*–526*t* , 539–540

2-Hydroxy-1-propyl acrylate. *See* Propylene glycol monoacrylate
Hydroxypropyl cellulose *[9004-64-2]*, 540–541
Hydroxypropyl methyl cellulose *[9004-65-3]*, 538–539
Hypromellose. *See* Hydroxypropyl methyl cellulose

Isoamyl mercaptan *[541-31-1]*, 682*t*, 703–704
Isobutoxylethanol. *See* Ethylene glycol monoisobutyl ether
Isobutyl mercaptan *[513-44-0]*, 682*t*, 698–699
N-Isopropyl-2-benzothiazolesulfenamide *[10220-34-5]*, 684*t*, 752–753
Isopropyl mercaptan *[75-33-2]*, 682*t*, 694–695
Ixan. *See* Polyvinylidene chloride

Kaurit S. *See* Urea-formaldehyde

Latex. *See* Dimethyl silicone
Lauryl mercaptan. *See* n-Dodecyl mercaptan
D-Limonene dimercaptan *[4802-20-4]*, 683*t*, 726–727

Malathion *[121-75-5]*, 770*t*, 774*t*, 778*t*, 863–871
Maldison. *See* Malathion
MBT. *See* 2-Mercaptobenzothiazole
Melamine-formaldehyde resins *[9003-08-1]*, 639*t*, 648–650
Meldane. *See* Coumaphos
Mercaptobenzene. *See* Phenyl mercaptan
2-Mercaptobenzothiazole *[149-30-4]*, 684*t*, 750–752
1-Mercaptobutane. *See* n-Butyl mercaptan
3(or 4)-Mercaptocyclohexaneethanethiol. *See* Vinyl cyclohexene-derived dimercaptan
3-Mercapto-beta-4-dimethyl-cyclohexaneethanethiol. *See* D-Limonene dimercaptan
1-Mercaptododecane. *See* n-Dodecyl mercaptan
Mercaptoethane. *See* Ethyl mercaptan
Mercaptomethane. *See* Methyl mercaptan
1-Mercaptopropane. *See* Propyl mercaptan
2-Mercapto propane. *See* Isopropyl mercaptan
Mesyl chloride. *See* Methanesulfonyl chloride
Methacrylic acid methyl ester. *See* Methyl methacrylate
Methallyl *n*-octyl sulfoxide *[4886-36-6]*, 684*t*, 744–745
p-Methane-2,9-dithiol. *See* D-Limonene dimercaptan
Methanesulfonic acid *[75-75-2]*, 684*t*, 745–746
Methanesulfonyl chloride *[124-63-0]*, 684*t*, 746–747
Methanethiol. *See* Methyl mercaptan
4-Methoxybutanol. *See* Butylene glycol monomethyl ether

2-Methoxy ethanol. *See* Ethylene glycol monomethyl ether
2-Methoxyethanol acetate. *See* Ethylene glycol monomethyl ether acetate
Methoxypolyethylene glycols. *See* Polyethylene glycol methyl ethers
1-Methoxypropan-2-ol. *See* Propylene glycol monomethyl ether
2-Methyl-2-butanethiol. *See* *tert*-Amyl mercaptan
3-Methyl-1-butanethiol. *See* Isoamyl mercaptan
Methyl cellulose *[9004-67-5]*, 536–537
Methyl disulfide. *See* Dimethyl disulfide
(1-Methyl-1,2-ethanediyl) bisoxybispropanol. *See* Tripropylene glycol
N-(1-Methylethyl)-2-benzothiazolesulfenamide. *See* *N*-Isopropyl-2-benzothiazolesulfenamide
Methyl ethylene glycol. *See* Propylene glycol
Methyl glycol. *See* Ethylene glycol monomethyl ether
Methyl mercaptan *[74-93-1]*, 682*t*, 686*t*, 687–690
Methyl methacrylate *[80-62-6]*, 488*t*, 514–515
Methyl parathion *[298-00-0]*, 770*t*, 774*t*, 778*t*, 871–880
2-Methyl-2,4-pentanediol *[107-41-5]*, 48–51
2-Methyl-2,4-pentanediol diacetate *[1637-24-7]*, 337*t*, 354
2-Methyl-1-propanethiol. *See* Isobutyl mercaptan
2-Methyl-2-propanethiol. *See* *tert*-Butyl mercaptan
2-Methyl-2-propene-1,1-diol diacetate *[10476-95-6]*, 337*t*, 352–353
Methyl sulfone. *See* Dimethyl sulfone
Methylsulfonic acid. *See* Methanesulfonic acid
2-Methylthiirane. *See* Propylene sulfide
Methylthiomethane. *See* Dimethyl sulfide
Metrifonate. *See* Trichlorfon
Mevinphos *[7786-34-7]*, 770*t*, 774*t*, 779*t*, 880–885
MMA. *See* Methyl methacrylate
MME. *See* Methyl methacrylate
Monocrotophos *[6923-22-4]*, 770*t*, 774*t*, 779*t*, 885–889
Monoglyme. *See* Ethylene glycol dimethyl ether
Monopropylene glycol allyl ethers *[1331-17-5]*, 326–327
MPEGS. *See* Polyethylene glycol methyl ethers
MSA. *See* Methanesulfonic acid
MsCl. *See* Methanesulfonyl chloride

Naled *[300-76-5]*, 770*t*, 775*t*, 779*t*, 889–893
Natrosol. *See* Hydroxyethyl cellulose
Navadel. *See* Dioxathion
NBR. *See* Acrylonitrile-butadiene rubber
Nemacur. *See* Fenamiphos
Nifos. *See* TEPP
Nitrocellulose. *See* Cellulose nitrate

n-Nonanethiol. *See n*-Nonyl mercaptan
tert-Nonanethiol. *See tert*-Nonyl mercaptan
n-Nonyl mercaptan *[1455-21-6]*, 682*t*, 709–710
tert-Nonyl mercaptan *[25360-10-5]*, 682*t*, 710
NPGDA. *See* 2,2-Dimethyl-1,3-propanediol diacrylate
Nuvacron. *See* Monocrotophos
Nylon 6 *[25038-54-4]*, 544*t*–545*t*, 548*t*–549*t*, 555–556
Nylon 12 *[24937-16-4]*, 544*t*–545*t*, 549*t*, 557
Nylon 66 *[32131-17-2]*, 557–558, 559*f*, 559*t*
Nylon 610 *[9008-66-6]*, 548*t*–549*t*, 558–560
Nylon *[63428-83-1]*, 542–543, 548*t*, 552–555, 553*f*, 577*t*, 578*t*, 639*t*

1-Octadecanethiol. *See n*-Octadecyl mercaptan
n-Octadecyl mercaptan *[2885-00-9]*, 683*t*, 718–719
n-Octyl mercaptan *[111-88-6]*, 682*t*, 707–708
tert-Octyl mercaptan *[141-59-3]*, 682*t*, 708–709
Onion oil. *See* Propyl allyl disulfide
Organic sulfur compounds. *See* Sulfur compounds, organic
Ortho 4355. *See* Naled
2,2'-Oxybis-ethanol. *See* Diethylene glycol
Oxybis propanol. *See* Dipropylene glycol
2,2'-Oxydiethanol. *See* Diethylene glycol

Parathion *[56-38-2]*, 770*t*, 775*t*, 779*t*, 893–902
PB. *See* Polybutadiene; Polybutylene
PBD. *See* Polybutadiene
PBI. *See* Polybenzimidazole
PBT. *See* Polybutylene terephthalate
PCM. *See* Perchloromethyl mercaptan
PCTFE. *See* Polychlorotrifluoroethylene
PDMS. *See* Polydimethylsiloxanes
PDS. *See* Di-*n*-propyl disulfide
PE. *See* Polyethylene
PEG. *See* Polyethylene glycols; Polyoxyethylene
PEI. *See* Polyetherimides
Pentamethylene glycol. *See* 1,5-Pentanediol
1,5-Pentanediol *[111-29-5]*, 48
1-Pentanethiol. *See n*-Pentyl mercaptan
1,5-Pentylene glycol. *See* 1,5-Pentanediol
n-Pentyl mercaptan *[110-66-7]*, 682*t*, 702–703
Perchloromethyl mercaptan *[594-42-3]*, 683*t*, 686*t*, 728–730
Perfluoroalkoxy copolymer *[26655-00-5]*, 428*t*, 472
Perlon. *See* Nylon 6
PFA. *See* Perfluoroalkoxy copolymer
P-F-R-2. *See* Phenol-formaldehyde resin
PGBE. *See* Propylene glycol mono-*n*-butyl ether
PGBEE. *See* Propylene glycol butoxyethyl ether

1005

PGDN. *See* Propylene glycol dinitrate
PGEE. *See* Propylene glycol monoethyl ether
PGME. *See* Propylene glycol monomethyl ether
PGPE. *See* Propylene glycol *n*-monopropyl ether
PGPhE. *See* Propylene glycol phenyl ether
PGTBE. *See* Propylene glycol mono-*tertiary*-butyl ether
Phenol-formaldehyde resin *[9003-35-4]*, 638–644, 639*t*
2,2'-[1,4-Phenylenebis(oxy)]bisethanol *[104-38-1]*, 227–228
Phenyl glycol. *See* Styrene glycol
Phenyl mercaptan *[108-98-5]*, 683*t*, 686*t*, 722–724
Phenylmethyl mercaptan. *See* Benzyl mercaptan
Phenylphosphonothoic acid, *O*-ethyl *O*-o-nitrophenyl ester. *See* EPN
Phenyl sulfone. *See* Diphenyl sulfone
Phorate *[298-02-2]*, 770*t*, 775*t*, 779*t*, 902–905
Phosphorodithioic acid, *O,O*-dimethyl-*S*-((4-oxo-1,2,3-benzotirazin-3 (4H)-yl)methyl) ester. *See* Azinphos-methyl
PIR. *See* Polyisocyanurate foam
Plexiglas. *See* Polymethyl methacrylate
PMM. *See* Perchloromethyl mercaptan
Polyacrylamide *[9003-05-8]*, 433*t*, 488*t*, 508–511
Polyacrylic acid *[9003-01-4]*, 488*t*, 511–512
Polyacrylonitrile *[25014-41-9]*, 433*t*, 488*t*, 500–502, 639*t*
Polyalkylene glycols. *See* Ucon fluids
Polyamide. *See* Nylon
Polyamide 6. *See* Nylon 6
Polyamide 11. *See* Nylon 11
Polyamide 12. *See* Nylon 12
Polyamide 66. *See* Nylon 66
Polyamide 610. *See* Nylon 610
Polyamide 612. *See* Nylon 612
Polybenzimidazole *[26985-65-9]*, 656–657
Polybutadiene *[9003-17-2]*, 427*t*, 438–440, 577*t*
Polybutylene *[9003-29-6]*, 414*t*, 420–421, 640*t*
Polybutylene, isotactic *[25036-29-7]*, 420
Polybutylene glycols *[25190-06-1]*, 47
Polybutylene terephthalate *[24968-12-5; 26062-94-2; 30965-26-5]*, 574*t*–575*t*, 585–587
Polychlorotrifluoroethylene *[9002-83-9]*, 428*t*, 474–475
Polydimethylsiloxanes *[9016-00-6; 63148-62-9; 63394-02-5]*, 658
Polyester fibers *[25038-59-9]*, 574*t*–575*t*, 581–585
Polyethylene *[9002-88-4]*, 413–417, 414*t*, 433*t*, 577*t*, 639*t*
Polyethylene glycol methyl ethers *[9004-74-4]*, 226–227
Polyethylene glycols *[25322-68-3]*, 21–26, 22*t*, 25*t*

1006   CHEMICAL INDEX

Polyethylene terephthalate *[25038-59-9]*, 574*t*–575*t*, 581–585, 639*t*
Poly(2-hydroxethyl methacrylate) hydrogel *[25249-16-5]*, 512–513
Polyisoprene *[9003-31-0]*, 427*t*, 431–434, 639*t*
Polymethyl methacrylate *[9011-14-7]*, 488*t*, 505–508
Polyoxymethylene. *See* Acetal resins – polyoxymethylene
Polyphenylene oxide-based resin *[9041-80-9; 25134-01-4]*, 575*t*–576*t*, 602–606, 640*t*
Poly(*p*-phenylene sulfide) *[53027-72-8; 9016-75-5]*, 606–607
Polyphenylene sulfide *[9016-75-5]*, 607, 640*t*
Poly(propanediols). *See* Polypropylene glycols
Polypropylene *[9003-07-0]*, 414*t*, 417–420, 433*t*, 577*t*, 639*t*
Polypropylene, isotactic *[25085-53-4]*, 417, 418
Polypropylene glycol butyl ethers *[9003-13-8]*, 329–331
Polypropylene glycols *[25322-69-4]*, 38–41, 38*f*
Polystyrene *[9000-53-6]*, 488*t*, 496–500, 577*t*, 639*t*
Polysulfone of bisphenol A *[25135-51-7]*, 576*t*, 606–607
Poly-Tears. *See* Hydroxypropyl methyl cellulose
Polytetrafluoroethylene *[9002-84-0]*, 428*t*, 467–470, 471*f*
Polytetramethylene terephthalate. *See* Polybutylene terephthalate
Polyurethane *[9009-54-5]*, 619–638, 639*t*
Polyurethane elastomers *[9009-54-5]*, 427*t* , 429*f*, 454–455
Polyvinyl acetate *[9003-20-7]*, 433*t*, 488*t* , 489–492, 577*t*, 639*t*
Polyvinyl acetate-butyl acrylate *[25067-01-0]*, 488*t*, 492
Polyvinyl acetate ethylene *[24937-78-8]*, 488*t*, 489–492
Polyvinyl acetate 1-ethylhexyl acetate *[25067-02-7]*, 489–492
Polyvinyl acetate-2-ethylhexyl acrylate *[25067-02-1]*, 488*t*, 492
Polyvinyl alcohol *[9002-89-5]*, 433*t*, 488*t* , 492–496
Polyvinyl chloride *[9002-86-2]*, 427*t* , 433*t*, 455–463, 577*t*, 639*t*
Polyvinyl fluoride *[24981-14-4]*, 428*t*, 466–467
Polyvinylidene chloride *[9002-85-1]*, 427*t* , 433*t*, 463–465, 639*t*
Polyvinylidene fluoride *[24937-79-9]*, 428*t*, 473–474
Polyvinylpyrrolidone *[9003-39-8]*, 651
Polyviol. *See* Polyvinyl alcohol
PP. *See* Polypropylene
Primary acetate. *See* Cellulose triacetate

1,2-Propanediol. *See* Propylene glycol
1,3-Propanediol *[504-63-2]*, 33–35, 33*t*
Propanesulfonyl chloride *[10147-36-1]*, 684*t* , 748–749
1-Propanethiol. *See* Propyl mercaptan
2-Propanethiol. *See* Isopropyl mercaptan
2-Propene-1-thiol. *See* Allyl mercaptan
2-Propenoic acid, 2,2-dimethyl-1,3-propanediyl ester. *See* 2,2-Dimethyl-1,3-propanediol diacrylate
2-Propenoic acid, 1,2-ethanediyl ester. *See* Ethylene glycol diacrylate
2-Propenoic acid, 1-methyl-1,3-propanediyl ester. *See* 1,3-Butanediol diacrylate
2-Propenoic acid, oxydi-2,1-ethanediyl. *See* Diethylene glycol diacrylate
2-Propenyl propyl disulfide. *See* Propyl allyl disulfide
1-Propoxy-2-propanol. *See* Propylene glycol*n*-monopropyl ether
Propyl allyl disulfide *[2179-59-1]*, 683*t* , 686*t*, 737–738
Propyl disulfide. *See* Di-*n*-propyl disulfide
Propylene glycol *[57-55-6]*, 2*f*, 26–33, 31*f*
Propylene glycol butoxyethyl ether *[124-16-3]*, 273*t*, 305–306
Propylene glycol dinitrate *[6423-43-4]*, 366*t*, 378–383
Propylene glycol monoacrylate *[999-61-1* (1-acrylate)*; 2918-23-2* (2-acrylate)*]*, 346–349, 350*t*
Propylene glycol mono-*n*-butyl ether *[5131-66-8; 29387-86-8]*, 276*t*, 297–300
Propylene glycol mono-*tert*-butyl ether *[57018-52-7]*, 273*t*, 277*t*, 300–304
Propylene glycol monoethylbutyl ether, mixed isomers *[26447-42-7]*, 304–305
Propylene glycol monoethyl ether acetate *[54839-24-6* (predominant α iosmer)*; 57350-24-0* (2-ethoxy-1-propyl-acetate or β isomer)*; 98516-30-4* (mixed isomers)*]*, 276*t*, 338*t*, 363–365
Propylene glycol monoethyl ether *[1569-02-4* (α-isomer)*; 52125-53-8* (mixture of α and β isomers)*]*, 276*t*, 290–293
Propylene glycol monoisopropyl ether *[3944-36-3; 29387-84-6; 110-48-5]*, 273*t*, 296
Propylene glycol monomethyl ether *[107-98-2; 1320-67-8]*, 273*t*, 276*t*, 279*t*, 281–290, 286*t*, 290*t*
Propylene glycol monomethyl ether acetate *[108-65-6]*, 279*t*, 338*t*, 360–362
Propylene glycol-*n*-monopropyl ether *[1569-01-3* (α-isomer)*; 30136-13-1* (mixture)*]*, 273*t*, 276*t*, 293–296

Propylene glycol phenyl ether *[770-35-4]*, 273t, 277t, 306–307
Propylene sulfide *[1072-43-1]*, 683t, 736–737
Propyl glycol. *See* Ethylene glycol mono-*n*-propyl ether
Propyl mercaptan *[107-03-9]*, 682t, 692–694
PS. *See* Polystyrene
PTFE. *See* Polytetrafluoroethylene
PUR. *See* Polyurethane foams
PVA. *See* Polyvinyl acetate; Polyvinyl alcohol
PVAL. *See* Polyvinyl alcohol
PVC. *See* Polyvinyl chloride
PVDC. *See* Polyvinylidene chloride
PVDF. *See* Polyvinylidene fluoride
PVP. *See* Polyvinylpyrrolidone
Pyroxiline. *See* Cellulose nitrate

Rampart. *See* Phorate
Rayon. *See* Cellulose
Repellent 1207. *See* Chloropropyl *n*-octyl sulfoxide
Ronnel *[299-84-3]*, 770t, 775t, 779t, 905–908
RSH. *See* Mercaptans
Rubber. *See* Polyisoprene

SAN. *See* Styrene-acrylonitrile
Saran. *See* Polyvinylidene chloride
SBR. *See* Styrene butadiene rubber
Silicone *[63148-62-9]*, 427t, 429f, 452–453, 639t, 657–658
Silicone rubber. *See* Dimethyl silicone
Sodium carboxymethyl cellulose *[9004-32-4]*, 524t, 525t–526t, 536
Stearyl mercaptan. *See* *n*-Octadecyl mercaptan
Styrene-acrylonitrile *[9003-54-7]*, 488t, 502–504, 577t, 639t
Styrene butadiene rubber *[9003-55-8]*, 427t, 434–438
Styrene glycol *[93-56-1]*, 55–56, 56f
1,1'-Sulfinylbisethane. *See* Diethyl sulfoxide
Sulfole 90. *See* *tert*-Nonyl mercaptan
Sulfole 120. *See* *tert*-Dodecyl mercaptan
1,1'-Sulfonylbisbenzene. *See* Diphenyl sulfone
Sulfonylbismethane. *See* Dimethyl sulfone
Sulfotepp *[3689-24-5]*, 770t, 775t, 780t, 909–911
Sulprofos *[35400-43-2]*, 770t, 775t, 780t, 912–916

TDAC. *See* Triethylene glycol diacetate
TEGDN. *See* Triethylene glycol dinitrate
Temephos *[3383-96-8]*, 770t, 775t, 780t, 916–920
TEPP *[107-49-3]*, 770t, 775t, 780t, 920–922
Terbufos *[13071-79-9]*, 770t, 775t, 780t, 922–926
Tetradecyl mercaptan *[2079-95-0]*, 683t, 717
Tetraethyl dithionopyrophosphate. *See* Sulfotepp

Tetraethylene glycol *[112-60-7]*, 19–20
Tetraethylene glycol diethyl ether *[4353-28-0]*, 225–226
Tetraethylene glycol monophenyl ether *[36366-93-5]*, 225
O,O,O',O'-Tetraethyl-S,S'-methylene diphosphorodithioate. *See* Ethion
Tetraglycol. *See* Tetraethylene glycol
O,O,O',O'-Tetramethyl-O,O'thiodi-*p*-phenylene phosphorothioate. *See* Temephos
Tetraphosphoric acid tetraethyl ester. *See* TEPP
Texanol. *See* 2,2,4-Trimethyl-1,3-pentanediol monoisobutyrate
TGBE. *See* Triethylene glycol mono-*n*-butyl ether
TGEE. *See* Triethylene glycol monoethyl ether
TGME. *See* Triethylene glycol monoethyl ether; Triethylene glycol monomethyl ether
1-Thia-3-azaindene. *See* Benzothiazole
Thiacyclopropane. *See* Ethylene sulfide
2-Thiapropane. *See* Dimethyl sulfide
Thiirane. *See* Ethylene sulfide
Thiirane, methyl. *See* Propylene sulfide
Thiobenzyl alcohol. *See* Benzyl mercaptan
1,1'-Thiobisbutane. *See* Dibutyl sulfide
1,1'-Thiobisethane. *See* Diethyl sulfide
Thiobismethane. *See* Dimethyl sulfide
Thiocarbonyl tetrachloride. *See* Perchloromethyl mercaptan
Thiophenol. *See* Phenyl mercaptan
TMPD glycol. *See* 2,2,4-Trimethyl-1,3-pentanediol
alpha-Toluenethiol. *See* Benzyl mercaptan
TPE. *See* Thermoplastic elastomers
TPGBE. *See* Tripropylene glycol mono-*n*-butyl ether
Trichlorfon *[52-68-6]*, 770t, 775t, 780t, 926–935
Trichlorometaphos. *See* Ronnel
Trichloromethyl sulfur chloride. *See* Perchloromethyl mercaptan
Triethylene glycol *[112-27-6]*, 2f, 17–19
Triethylene glycol diacetate *[111-21-7]*, 337t, 344–345
Triethylene glycol dinitrate *[111-22-8]*, 366t, 376–378
Triethylene glycol mono-*n*-butyl ether *[143-22-6]*, 222–224
Triethylene glycol monoethyl ether *[112-50-5]*, 76t, 78t, 179t, 195t, 220–222
Triethylene glycol monomethyl ether *[112-35-6]*, 76t, 78t, 179t, 193t, 195t, 216–220
Triglycol diacetate. *See* Triethylene glycol diacetate
Triisobutyl mercaptan *[25103-58-6]*, 702
Trimethylene glycol. *See* 1,3-Propanediol
2,2,4-Trimethyl-1,3-pentanediol *[144-19-4]*, 51–53

CHEMICAL INDEX

2,2,4-Trimethyl-1,3-pentanediol diisobutyrate *[6846-50-0]*, 337t, 358–360
2,2,4-Trimethyl-1,3-pentanediol monoisobutyrate *[77-68-9; 77-68-9; 25265-77-4]*, 337t, 354–358
Tripropylene glycol *[1638-16-0]*, 36–38
Tripropylene glycol mono-*n*-butyl ether *[57499-93-1; 55934-93-5]*, 274t, 278t, 325–326
Tripropylene glycol monoethyl ether *[20178-34-1; 75899-69-3]*, 274t, 324
Tripropylene glycol monomethyl ether *[20324-33-8; 25498-49-1]*, 274t, 278t, 321–324
TSE. *See* Thermosetting elastomers
TXIB. *See* 2,2,4-Trimethyl-1,3-pentanediol diisobutyrate

Urea-formaldehyde *[9011-05-6]*, 639t, 644–648

Urethanes. *See* Polyurethane

Valeric acid, triethylene glycol diester. *See* Triethylene glycol divalerate
Viclan. *See* Polyvinylidene chloride
Vinol. *See* Polyvinyl alcohol
Vinyl cyclohexene-derived dimercaptan *[37241-32-0]*, 683t, 727–728
Vinylidene chloride-acrylonitrile *[9010-76-8]*, 428t, 466
Vinylidene chloride-methylacrylate *[25038-72-6]*, 427t, 466
Vinylidene chloride-vinyl chloride *[9011-06-7]*, 427t, 465
Vinylidene fluoride-hexafluoropropylene *[9011-17-0]*, 427t, 444–446
Vinyl sulfone. *See* Divinyl sulfone

Ref
RA
1229
.P38

SOUTH UNIVERSITY
709 MALL BLVD.
SAVANNAH, GA 31406